The Pacific Northwest

Plant Locator

2000-2001

Oregon, Washington & Idaho Sources of the Plants You've Been Looking For

Compiled by Susan Hill and Susan Narizny

BLACK-EYED SUSANS PRESS
www.blackeyedsusanspress.com

DEDICATED TO ALL THOSE GARDENERS LIKE OURSELVES FOR WHOM PLANT LUST IS A CHRONIC AFFLICTION

COVER PHOTO: PHORMIUM 'Sundowner' (a New Zealand flax cultivar)
Picture taken at Gossler's Nursery, Springfield, OR
by Doug Hill

COVER DESIGN: Cornerstone Graphics, Portland, OR

Second Edition

Printed in the United States of America

FOR INFORMATION ABOUT THIS BOOK, OR TO ORDER COPIES, PLEASE COMMUNICATE WITH THE PUBLISHER AT:
BLACK-EYED SUSANS PRESS
PMB 227
6327-C SW Capitol Highway
Portland, Oregon 97201-1937
Fax: (503) 245-4638 Email: bespress@aol.com
Website: www.blackeyedsusanspress.com

RAVE REVIEWS
for *THE PACIFIC NORTHWEST PLANT LOCATOR 1999-2000*

"The postman brought the copy of this excellent publication to the door today. It is very impressive indeed."
- **John Stockdale**, publisher of the CD-rom edition of Great Britain's *RHS Plant Finder*.

"This book seems sure to become a well-thumbed, dirt-smudged classic."
- **Anne Nelson**, *Southwest Community Connection*, Portland OR

"This reference will surely become a valued tool for plant lovers throughout the region and beyond...If you are a nursery, large or small, do the horticultural community a favor and submit your plant list."
- **Stephanie Feeney**, author of *The Northwest Gardener's Resource Guide.*

"...this directory should prove indispensable not only for finding specific trees, shrubs or flowers, but as a guide to all the specialty nurseries we are blessed with in the Northwest."
- **Valerie Easton**, library manager for the Elisabeth C. Miller Library, Center for Urban Horticulture in Seattle, writing in the *Seattle Times*

"We love your effort and want to carry it in our specialty nursery."
- **Karen Souza**, Fremont Gardens

"What a huge undertaking! And an extraordinarily useful one. I've spent some time going through it and am very impressed with the range of cultivars represented. It is amazing how much more is available now than there was 10 years ago – but…the problem has been connecting gardeners with the sources for the plant(s) they want to buy."
- **Marie McKinsey**, webmother, www.nwgardening.com

"The book is also helpful in sorting out the myriad nomenclature mysteries of hort-dom. I had been bewildered when one of my favorite species of roses, *Rosa rubrifolia,* disappeared from the rose catalogs. Looking it up in the *Plant Locator*, I found it has been reassigned the name of *Rosa glauca.* Not only do I now know its proper name, I have 15 sources for the plant, should I wish to give it as a gift, or should my own cherished specimen ever go belly up and need replacing."
- **Anne Nelson**, *Southwest Community Connection*, Portland OR

"This year's Book of Lists Award goes to The Pacific Northwest Plant Locator."
- **Sharon Wootton**, *The Herald*, Everett, WA

"In a totally unscientific and arbitrary test of the book, I looked up several plants that caught my eye in Scotland this summer, and the *Plant Locator* had them listed. If the book proves to be popular with gardeners, the authors plan to update the book annually. Here's hoping it becomes a best seller!"
- **Lynn Lustberg**, *Garden Showcase*

"Here is an excellent resource book for finding those special plants to team up with our hostas and shade plants, or any other plant your garden desires."
- ***Northwest Hosta & Shade Gardening Society Newsletter***

"My heartfelt and oh so grateful congratulations on your book. Since picking it up on Saturday I've referred to it THREE times for Oregonian articles and each time found what I needed. I'm amazed."
- **Ketzel Levine**, National Public Radio's Doyenne of Dirt

"It really gets used."
- **Vern Nelson**, garden writer, columnist, and lecturer, Portland OR

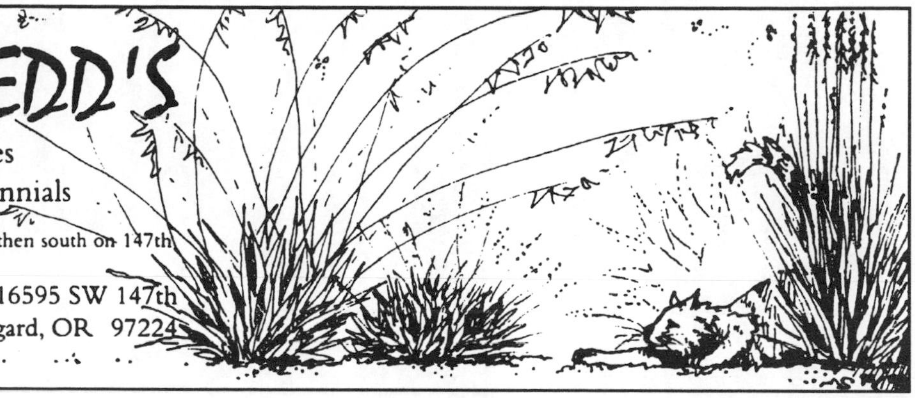

TABLE OF CONTENTS

FOREWARD
by
Lucy Hardiman

The black-eyed Susans have done it again! The second edition of *The Pacific Northwest Plant Locator* is testimony to the collective vision of Susan Hill and Susan Narizny, Oregon plant devotees. Publication of the first edition, the 1999-2000 *Plant Locator,* was risky business at best. The two Susans, taking a cue from Britain's plant reference bible, *The RHS Plant Finder,* set out to provide Northwest gardeners with a a practical, sensible, user-friendly guide to sourcing plants in our region. The overwhelming success and acceptance of that first edition have paved the way for the compilation and production of this second edition.

The avowed faithful rejoiced when the first edition rolled off the presses, and it was astounding to watch the rate at which the book was purchased by plant aficionados. Garden writers were among the first to spread the word about the value of *The Plant Locator* in assisting gardeners, landscape contractors and designers, and nursery people in locating plants, many of which are available only from small, specialty growers. *The Pacific Northwest Plant Locator* allows interested and curious hort-heads to quickly and efficiently find nurseries that grow and sell the plants for which they yearn. It saves countless hours of frustration trying to remember where one might have seen or read about that special plant, without which one's garden would not be complete. Gardeners are notorious for their acquisitive and obsessive tendencies. If the plant exists, we have an inexplicable urge and longing to find it, regardless of the cost.

It was through our common passion for plants and our involvement in the Hardy Plant Society of Oregon that I first met the two Susans. Susan Narizny owned her own nursery and grew interesting and out-of-the-ordinary perennials with a special focus on hardy fuchsias. In a previous life, Susan Hill and her husband owned and operated a garden design business in California. Their happy union brought diverse perspectives into the research and development of the *Plant Locator*. The result is a compendium that lists plants from Abelia to Ziziphus that make the plant nerd swoon.

The Pacific Northwest Plant Locator occupies a place of honor on my bookshelf, although it is very rarely found in its proper place. More often than not, it's in a pile on my desk, lying open next to the computer, or beside the bed, where it provides scintillating nighttime reading. It has become one of the most well-used resources in my very large library, the resource of choice whether I am looking for plants for a client or verifying spelling or correct nomenclature for an article about gardens or gardening.

I can only applaud the two Susans for recognizing the need for this kind of resource, taking years out of their lives to compile the information, and for having the courage to go out on a limb and self publish a book for plant lovers. Congratulations.

Lucy Hardiman
Perennial Partners
April 2000

INTRODUCTION AND ACKNOWLEDGEMENTS

Almost every day, someone asks one of us, "Do you know where I can get such and such a plant?" Until now, our answers have been quite limited by our memories of the catalogues we've thumbed excitedly through, of the many nurseries we've visited, or even of responses to the questions we've asked in the gardens of fellow enthusiasts we've visited over the years. And we have asked ourselves, "Why isn't there a plant sourcebook for the Pacific Northwest so that gardeners here can find local sources for plants they want?" We finally came to the conclusion that, to answer these questions, we would have to compile such a resource ourselves. Thus *The Pacific Northwest Plant Locator* was born.

Pacific Northwest nurseries are increasingly being recognized throughout the greater gardening community for the quality and variety of their plants, and many are nationally and internationally known and respected. We have also found, during the hours we've spent compiling this book, that virtually every kind of specialty nursery exists here as well. And, for those of us fortunate enough to live here, in a gardening Eden, there is a plethora of walk-in nurseries which carry a wide variety of both the usual and the rare. This book attempts to find as many as possible and to list the many different perennial plants available in our region. We have not sought to include annual flowers or vegetables, or seed sources; the huge volume of these, their ephemeral nature, the ease with which they are found at practically any retailer, and the confusion of their names make cataloguing them a gigantic and unwieldy task. However, when nurseries have included a few annuals in their lists, we have added them to the book. Also, we do list perennial vegetables such as potato tubers and edible members of the onion family.

Although we have kept the focus of the book in Oregon, Washington, and Idaho, we envision that gardeners all over the West, perhaps in other parts of the US, and other parts of the world even, may find this book useful to help them in their search for intriguing plants. We also envision that nursery owners will use this book to find sources of plants they would like to grow, to see which plants they might introduce into circulation that are not yet generally available, and to sort out nomenclature and spelling which have been puzzling to them for the plants they carry.

We intend to issue new editions of this book regularly to help gardeners keep abreast of plants being offered, and to encourage nursery owners who have not yet listed their plants to participate in future ones. Those interested in being included may contact us at one of the addresses on the back of the title page.

We would like to acknowledge and thank all the nursery owners who have contributed their plant lists, information, and advertisements; without their cooperation, this book could never have been possible. They have given us more, however, than their cooperation; many of them have shared with us their enthusiasm for our project and encouraged us in our endeavor to make the *Plant Locator* a reality. We are especially grateful for the accolades and advice given us by all those who have communicated with us about the first edition; you will find examples of their comments on the back cover and the first page. We are indebted to Lucy Hardiman for her enthusiastic encouragement from the moment we mentioned the project to her, her efforts to create recognition for the book throughout the horticultural community, and for her foreword to this second editon. We are also indebted to: Paul Black for assisting us with correct Iris nomenclature; Monrovia Nursery and Terra Nova Nursery for correct listings of their introductions; Dr. Simon Thornton-Wood and Mike Grant of the Royal Horticultural Society, Wisley, UK, for helping us sort out numerous nomenclatural questions; Gail Austin for current Hemerocallis names; John Bosshardt for legal advice; and Stephanie and Larry Feeney of the *Northwest Gardeners' Resource Directory* and Diana Petty for publishing advice. Our heartfelt thanks go to Doug Hill for creating the art for our black-eyed Susan logo and for the cover photo for this edition, to John Narizny for sharing his computer and his design ideas, to Kevin Narizny and Laurel Narizny for resolving computer and formatting crises, and to Jennifer Kuhn, Doug Hill, and Laurel Narizny for proofreading assistance. To our families, who have had to put up with a surfeit of botanical Latin and months of neglect and abandonment, we extend our love and gratitude for their support of us during the book-birthing process.

HOW TO USE THIS BOOK

The *Plant Locator* is divided into five sections. The first, and largest, contains the **Plant Listing** in alphabetical order by genus, species, subspecies or variety, and cultivar (cultivated variety) name, followed by codes for the nurseries that carry the plants. This list is in order by the Latin names for the plants because it is essential that the plants be identified by accepted botanical nomenclature to ensure, as much as is possible, that nurseries providing the same plant are listed together. For those who are new to, or uncomfortable with, Latin names, we have added an extensive **Common Name Index** to this edition (section five). These common names are keyed to the plants found in the **Plant Listing** (section one). We also suggest that a general reference book, like the *Sunset Western Garden Book*, or *Wyman's Gardening Encyclopedia*, available in any library, is an excellent tool to help you begin to identify which common and Latin names go together. And, reputable local nurseries can help you identify which plant names apply to the plants you want.

Space considerations have prompted us to use codes for the nursery names which follow the plant listings. We have tried to make the codes logical, so that once you are acquainted with a nursery's code, you will easily recognize it when you see it again. To this end, we have made most of the codes the first three letters of the nursery's name; when more than one nursery has the same first three letters, we have used additional beginning letters from other parts of the name. We have also used the small letters "o", "w", and "i" before the three-letter nursery code to indicate whether the nursery is located in Oregon, Washington, or Idaho so that you may more easily find nurseries close to you.

The second section, **Nursery/Code Index**, lists **all** participating nurseries alphabetically by nursery name, followed by their codes. You will notice that some nurseries do not have code names; these nurseries did not send us a list of plants they have available but did wish to be included in the directory. Since they have no listings in the directory of plants, they have no need for codes. Details for these nurseries are to be found in section four, **Additional Nurseries**.

Section three, **Nurseries With Codes,** lists, alphabetically by code, the nurseries which provided us with catalogs or plant lists. All the nursery descriptions, both here and in section four, give information about location, whether sales are by mail or walk-in, phone numbers, email addresses, and websites, whether catalogs are available, hours, acceptance of credit cards, and what stock the nursery carries or specializes in. The information in these entries is what was provided by nursery owners rather than written by us.

USING THE BOOK IS EASY. Look up the Latin name of the plant you wish to buy, for instance **PSEUDOTSUGA** ***menziesii***, the Douglas fir (you might first go to the common name index to look under Fir, Douglas for the Latin name). Under the genus **PSEUDOTSUGA** you will find a list of all the species carried by participating nurseries, and under ***menziesii*** a list of the cultivars available. Across from the species or cultivar you want is a list of codes indicating which nurseries sell it. Turn to section three, **Nurseries With Codes**, and look up the code to find out where to purchase the plant. If you want to find out which plants a specific nursery carries, first go to the **Nursery/Code Index**, look up the code for the nursery name, and then turn to the plant listings you are interested in to see if that nursery sells them.

We wish to make clear that if you can't find a plant in the **Plant Listing,** either you may not have the correct name, or it may not be available at any of the nurseries participating in this book. In cases where plants may commonly be found in the trade under alternate names, we have attempted to add "see" references from the incorrect names to the correct ones when we know them. Also, plants may be regionally available at nurseries not listing their plants with us, in which case we have no way of locating them for you. Nurseries list their plants here free of charge; if you have a favorite nursery that doesn't appear in the book, please encourage the owners to participate in future editions of the *Plant Locator.*

ABBREVIATIONS, TYPE STYLES, AND SYMBOLS

Plants are listed in the directory as follows:

GENUS NAME (bold, all caps); ***species***, or specific epithet (bold, small letters, italics); ***subspecies*** or ***variety*** (bold, small letters, italics); 'Cultivar Name' (non-bold, initial letters capitalized, single quotes)

e.g., **AQUILEGIA *vulgaris*** var. ***stellata*** 'Nora Barlow'

Additional elements in the listings are:

+	(before a name) means that the plant is a graft hybrid genus
aff.	affinis (allied to)
ch	name has changed from last edition's listing
co	confused name; same plant is offered under two or more distinct names
do	double-flowered plant
ex.	from
f.	forma (a botanical form)
fruit	species grown for edible fruit
g.	grex
hort.	plant is of garden origin, rather than a naturally occurring species or the result of a breeding program
last listed 99/00	plant was listed in the previous edition of the *Plant Locator* but is no longer offered by a participating nursery
nut	species grown for edible nut
qu	questionable name
®	a registered trademark name (appears after the cultivar name)
sp.	species
sp. aff.	a plant similar to the named species
ssp.	subspecies
™	a trademarked name (appears after the cultivar name)
ub	unverified breeder name
un	unverified name (we were unable to find this name in any reference)
va	variegated plant
var.	varietas (a botanical variety)
x	a cross between different genera, species, or cultivars (a hybrid)

Some cultivar names are not in quotes. These are generally trade names, names applied to plants by people in the nursery trade to give them sales appeal. They are not the same names given to the plants by the individuals who bred or introduced them. These latter names, "breeders' names," are considered, by current conventions, to be the correct names for the plants, since they were given when the plant was created or introduced into commerce. We have changed our treatment of breeder names in this edition. Formerly, we either listed nurseries under the breeder name (e.g. **WEIGELA** 'Olympiade') with a see reference from the trade name (**WEIGELA** Rubidor or Briant Rubidor) or, in the case of Roses, under the trade name (e.g. **ROSA** Belle Story). Now, all trade/breeder names are listed together, with no quotes around the trade name and quotes around the breeder name (hence **ROSA** Belle Story / 'Auselle' or **WEIGELA** Briant Rubidor / 'Olympiade'), followed by codes for the nurseries which carry them. Readers can now see the actual name of the plant as it was introduced, as well as the name under which it has become known in nurseries. This is particularly helpful in identifying plants which have been given multiple trade names, and verifying that plants under two different names are actually the same. For example, the rose Fourth of July is also known as Crazy For You (its European trade name). Because both of these names are tied to the breeder name, 'Wekroalt', we know that these are the same plant. Thus we list **ROSA** Fourth of July / 'Wekroalt' and we list a "see **R**. Fourth of July" reference after **ROSA** Crazy For You.

Another common example of the breeder name/trade name controversy occurs with breeder names which have been translated from another language into English. Possibly the most commonly encountered plant for which this occurs is **SEDUM** Autumn Joy, which was introduced by a German breeder as 'Herbstfreude' (Herbst=Autumn, freude=Joy). In order to acknowledge the breeder's right to the plant name, convention now demands that the original name be listed as the correct one, so Autumn Joy appears without quotes with a "see" reference to the correct cultivar name, 'Herbstfreude.' This happens in many instances for German, and, increasingly, Japanese introductions.

Other names not in quotes are: descriptions or names of colors, which are printed in small letters (e.g., blue form, large form); sex of plant for those which are dioecious (having male and female reproductive organs on separate plants – kiwis and ginkgos are examples of this); specific groups of plants that are not actually cultivar names (e.g., **ALSTROEMERIA** Dr.

Salter's hybrids); group names, where there are a number of individuals with similar attributes which in the past have been listed under the same cultivar name, but are often seed grown (e.g., purple-leaved cotinus, formerly **COTINUS *coggygria*** 'Atropurpurea,' now listed as **COTINUS *coggygria*** Rubrifolius Group); and places or people from which plants were procured (e.g., ex. Tel Aviv, and ex. T.C. Smale).

When nurseries have indicated that they are selling seed grown stock, we have followed the species or cultivar listing with the words "seed grown" (e.g., **DAHLIA *merckii*** 'Hadspen Star' seed grown). This indicates that the nursery is not offering plants that are divisions or cuttings of parent stock, but rather are growing new plants from seed of the parent. Seed grown stock can vary, as opposed to cuttings or divisions, which are vegetatively propagated and which produce clones of the parent. Therefore, botanically speaking, these plants are not the same cultivar, although they will usually greatly resemble it. This difference has been quite evident, for instance, in plants of **HEUCHERA *micrantha*** var. ***diversifolia*** 'Palace Purple.' This cultivar has been extensively propagated by seed in the trade, and many of us have bought plants listed as 'Palace Purple' which have differed noticeably, so that we didn't get quite what we thought we would. This can be either disappointing or exciting, if we find new attractive variations that are ours alone. In any case, the buyer deserves to know as much as possible how true to the expected form his plant will be; asking nursery people how their plants are propagated will increase the buyer's knowledge of his purchase and help avoid the "Caveat Emptor" situations that can occur.

When a plant has been listed only as "sp." rather than a species name, this indicates that a nursery has been unable to identify which species it is. Generally speaking, this does not indicate that the listing refers to a grouping of more than one species, but rather to a single, unidentified one. "Sp. aff." followed by a species name indicates that the species has not been positively identified but is similar to the species listed.

You will also see groups of letters and numbers after some species (e.g., **CAMELLIA *japonica*** HC 970230). These are collection codes, designations applied to species by those who first collected seed from them in the wild and who have subsequently grown a stock of the plant. These plants are not yet named, and are designated by the collection code as a plant that is somehow different from the commonly seen members of the same species. They may also become legitimately named varieties, subspecies or cultivars of the parent species. Please refer to individual nursery catalogs for the meaning of these codes, including the names of the collectors.

A NOTE ABOUT ALPHABETIZATION: for purposes of listing plants in alphabetical order, we have ignored the abbreviated designations such as sp., var., x, and f. and have interfiled names based on their first letters; thus **ACER *x leucoderme*** comes between **ACER *laxiflorum*** and **ACER *lobelii*.**. Collection code designations are also interfiled with other names by the letters which come before the numbers. Plain numbers (e.g. 74-46) are filed at the end of the name list for the species. Apostrophes are also ignored, so that **IRIS** Santana comes between Santa and Santa's Helper. Hyphens, however, have not been ignored but have been treated as spaces, so that **ACER** *palmatum* 'Ao-shidare' comes before 'Aoba-jo.' The designation "hort.," meaning that a plant has originated in a garden environment, is also ignored in alphabetizing. In cases of group names, the group name comes before cultivars with similar names (e.g. **CEDRUS *atlantica*** Glauca Group comes before **CEDRUS *atlantica*** 'Glauca Fastigiata').

NOMENCLATURE NOTES

When we talked with friends and nursery owners about our plans to compile this book, their first question was often "What are you going to do about nomenclature?" This is a topic about which much could be written, and indeed has been written in the *RHS Plant Finder*, which we would suggest that the reader use as a supplementary reference. The following is an explanation of how we have handled nomenclature in our book.

In our innocence, we thought we could just look up all the plant names we have encountered in nursery catalogs in various references like the *Plant Finder*. While this is in the main what we have done, it is not that simple. We have found many, many variations of names for what we believe are the same plants (one reason to compile a reference such as this is to help sort out this confusion), many plants too new to be listed in references or completely different from what is being sold elsewhere in the country or the world, and many cross-references, as botanists continue to change their assessments of what the proper names of certain plants are.

We have attempted to list all plants as their names have been given to us, although we have standardized, following the references below, minor and obvious variations. We have used the following sources (and the catalogues of plant introducers) to check the correctness of names given to us (see **Bibliography** for further information):

The RHS Plant Finder
The New RHS Dictionary Index of Garden Plants
Hortus Third
The AHS A-Z Encyclopedia of Garden Plants
Sunset Western Garden Book
American Daffodil Society Abridged Classified List of Daffodil Names
American Dahlia Society Composite Listing of Dahlias and *Classification and Handbook of Dahlias*
American Iris Society Registry
Eureka (checklist of Hemerocallis)
The Color Encyclopedia of Ornamental Grasses
Conifers: The Illustrated Encyclopedia
The Genus Hosta
Rhododendron Hybrids
Guidebook to Available Rhododendrons
Maples For Gardens
Peonies
Roses
A Book of Salvias
The Random House Book of Herbs

In general, with few exceptions, we have decided to follow international nomenclatural standards as set forth in the *RHS Plant Finder*. These exceptions deal mainly with British and American spelling differences, such as leaving out the letter u in words like as "favourite" (we have changed the spelling of the genus **BUDDLEJA** in this edition to reflect international standards rather than American spelling). We recognize that there will be correct cultivar names with words such as "favourite" and we are gradually finding and correcting them. For some genera, such as **HOSTA** and **SAXIFRAGA**, we have begun to add parenthetical references to species groups listed with some cultivar names and will continue to add more of these species references in future editions. Other exceptions will be mentioned in the notes on individual genera.

When a plant is too new to appear in a reference source, or when we have been unable to verify a listed name, we have entered its name, as it has been given to us by the nursery, with no cross references, although, again, obvious minor variations between nursery listings have been standardized. All names which we have been unable to verify using our source materials have been designated with "un" to the left of the name; therefore, if a name has no code next to it, we have verified it as correct (unless it is merely descriptive, such as **ERIOGONUM *umbellatum*** Bear Tooth Pass, or dwarf form, in which case we have printed it as given to us by the nursery). This should be of assistance to nursery owners in checking names of their plant stocks. When "qu" appears to the left of the name, we have reason to believe that this name may be incorrect; an example would be mixing Latin and English words in a name, such as 'Double Variegata', which is strictly against international nomenclatural convention. When "co" appears, such as next to **ROSA** 'Archiduc Joseph,' there is evidence that other plants are being sold under this name.

Some plants have "see" references saying "last listed 99/00". This means that the plant was offered by a nursery in the first edition of the *Plant Locator* (1999-2000) but is no longer carried by any of the current nurseries. We have left these names in only when we could verify their correctness; unverifiable names of plants no longer available have been dropped.

NOTES FOR INDIVIDUAL GENERA AND SPECIES

In the following notes, the species in question has become incorrectly agglomerated with other distinct plants, and we cannot tell from the nurseries' listings which plant they have. Proper names indicate the individual who first associated that species name with a particular plant. Usually, the first person to do so is considered to have named "the true species"; the others have misidentified their particular plants and this misidentification has stuck. These notes indicate the proper separation of plants into their correct designations. "Hort." indicates that the plant is of garden origin, not the true species, but the true species exists and is a valid name, although we have not included it here. Most of the information found here has been gleaned from *The RHS Plant Finder 1999-2000,* which has the botanical resources of the Royal Horticultural Society behind it.

ACAENA *adscendens*: There are four "see" references, as follows: Vahl, see **A. *magellanica*** ssp. ***laevigata***; Margery Fish, see **A. *affinis***; hort. see either **A. *magellanica*** ssp. ***magellanica,*** or **A. *saccaticupula*** 'Blue Haze'.

ACER *lobelii*: Bunge, see **A. *turkestanicum.***

ACER *palmatum* 'Aka-shime-no-uchi'**:** according to *Maples For Gardens*, this cultivar is rare, and "many plants believed to be [this one] are actually **A. *p.*** 'Atrolineare', with which it may be confused."

ACER *palmatum* 'Sessilifolium'**:** There are two forms of this cultivar: dwarf see **A**. ***p***. 'Hagoromo'; and tall, see **A**. ***p***. 'Koshimino'.

ALLIUM *aflatuenense*: hort., see **A. *hollandicum*** or **A. *stipitatum*.**

ANDROSACE *mucronifolia*: hort., see **A. *sempervivoides*.**

ANEMONE *x fulgens*: many nurseries that list this plant actually offer cultivars of **A. *coronaria.***

ANEMONE *japonica*: see either **A. *x hybrida*** or **A. *hupehensis*.**

ANISODONTEA *x hypomadara*: hort., see **A. *capensis*.**

ARALIA *chinensis*: hort., see **A. *elata*.**

ARTEMISIA *canescens*: hort., see **A. *alba*** 'Canescens'; Willd., see **A. *armeniaca*.**

ARTEMISIA *schmidtiana*: References disagree as to whether 'Silver Mound' is a cultivar or common name. The appearance and habit of the species incline us to believe that Silvermound or Silver Mound is a common name, not a cultivar.

ASTER *dumosus*: The *RHS Plant Finder* states, "Many of the asters listed under **A**. ***novi-belgii*** contain varying amounts of **A. *dumosus*** blood in their parentage. It is not possible to allocate these to one species or the other and they are therefore listed under **A. *novi-belgii.***"

ASTILBE 'Purple Cats' and 'Purpurkerze'**:** We believe that the cultivar 'Purple Cats' may be a translation of a misinterpretation ('Purpurkätze') of 'Purpurkerze', which means "purple candle" and would refer to the shape of the astilbe flower.

AZOLLA *caroliniana*: non Willd., see **A. *mexicana***; Willd., see **A. *filiculoides*.**

BERBERIS *gagnepainii*: hort., see **B. *gagnepainii*** var. ***lanceifolia*.**

CALTHA *polypetala*: hort., see **C. *palustris*** var. ***palustris*.**

CAREX *glauca*: Scopoli, see **C. *flacca*** ssp. ***flacca*.**

CAREX *nigra*: According to *The Color Encyclopedia of Ornamental Grasses*, "most plants sold by this name in the U.S. are correctly **C. *flacca*.** The individual flowers of **C. *flacca*** have three stigmas; flowers of **C. *nigra*** have two stigmas."

CAREX *morrowii*: hort., see **C. *oshimensis*** or **C. *hachijoensis*.**

CHIONODOXA *luciliae*: hort., see **C. *forbesii*.**

NOTES FOR INDIVIDUAL GENERA AND SPECIES

CISTUS *crispus*: hort., see **C. *x pulverulentus*.**

CISTUS *maculatus*: according to *Hortus III,* “a name applied to blotched variants of those spp. [species] in which petals are sometimes blotched, sometimes unblotched.” Not listed as a valid name in other sources.

CLEMATIS ‘Blue Boy’**:** Diversifolia Group, see **C. *x eriostemon*** ‘Blue Boy’; Lanuginosa/Patens Group, see **C.** ‘Elsa Spaeth’.

CLEMATIS *chrysocoma*: According to the *RHS Plant Finder,* “the true C. ***chrysocoma*** is a non-climbing erect plant with dense yellow down on the young growth, still uncommon in cultivation”; hort., see **C. *montana*** var. ***sericea*.**

CLEMATIS *grata*: hort., see **C. *jouiniana*.**

CLEMATIS Masquerade**:** Viticella Group, see **C.** ‘Maskarad’.

CLEMATIS *orientalis*: hort., see **C. *tibetana*** ssp. ***vernayi*.**

CLEMATIS *paniculata*: Thunberg, see **C. *terniflora***; Gmelin, see **C. *indivisa*.**

CLEMATIS *virginiana*: Hooker, see **C. *ligusticifolia***; hort., see **C. *vitalba*.**

CODONOPSIS *tangshen*: misapplied, see **C. *rotundifolia*** var. ***angustifolia*.**

CORTADERIA *richardii*: hort., see **C. *fulvida*.**

COTONEASTER *microphyllus*: hort., see **C. *purpurascens*.**

COTONEASTER *microphyllus* var. ***cochleatus*:** (Franch.) Rehd. & Wils, see **C. *cochleatus***; misapplied, see **C. *cashmiriensis*.**

COTONEASTER *microphyllus* var. ***thymifolius*:** hort., see **C. *linearifolius***; (Lindl.) Koehne, see **C. *integrifolius*.**

CRATAEGUS *crus-galli*: hort., see **C. *persimilis*** ‘Prunifolia’.

CROCOSMIA *aurea*: hort., see **C. *x crocosmiiflora*** ‘George Davison’ Davison.

CROCOSMIA ‘Citronella’**:** misapplied, see **C.** 'Honey Angels'; J. E. Fitt, see **C. *x crocosmiiflora*** 'Citronella'. However, according to the *Plant Finder,* “the true plant of this name has a dark eye... The plant usually offered may be more correctly **C.** ‘Gerbe d’Or’.”

CROCOSMIA *x crocosmiiflora* ‘George Davison’**:** hort, see **C. *x crocosmiiflora*** 'Golden Glory' or **C. *x c*.** 'Sulphurea'.

CROCOSMIA x ***crocosmiiflora*** ‘Queen Alexandra’**:** according to the *Plant Finder,* “as many as 20 different clones are grown under this name. It is not clear which is correct.”

DIASCIA *flanaganii*: see either **D. *stachyoides*** or **D. *vigilis*.**

DICENTRA *eximia*: hort., see **D. *formosa*.**

DIGITALIS *lamarckii*: hort., see **D. *lanata*.**

ECHINOPS *ritro*: hort., see **E. *bannaticus*.**

EPILOBIUM *californicum*: Haussknecht, see **ZAUSCHNERIA *californica*** ssp. ***angustifolia***; hort., see **ZAUSCHNERIA *californica***

ERIGERON *pyrenaicus*: Rouy, see **ASTER *pyrenaeus***; hort., see **E. *alpinus*.**

EUPHORBIA *longifolia*: D Don, see **E. *donii***; hort., see **E. *cornigera***; Lamarck, see **E. *mellifera*.**

NOTES FOR INDIVIDUAL GENERA AND SPECIES

EUPHORBIA *wallichii*: Kohli, see **E. *cornigera***; misapplied, see **E. *donii***.

FUCHSIA lycoides: hort., see **F.** 'Lycoides'.

FUCHSIA 'Pumila': According to the *Plant Finder*, "plants under this name might be **F. *magellanica*** var. ***pumila.***"

FUCHSIA *rosea*: Ruiz & Pav., see **F. *lycioides*** Andrews; hort., see **F.** 'Globosa'.

FUCHSIA 'Tiny Tim'**:** According to *The New A-Z on Fuchsias,* there are two distinct cultivars of this name, Walker, 1966, and Gagnon, 1967; the former has flowers colored purple and red, the latter has medium blue flowers turning first lavender and then fuchsia-colored.

GAULTHERIA 'Miz Thang'**:** The nursery that lists this plant indicates that they have given it this name, not knowing what its correct name is.

GERANIUM *platypetalum*: misapplied, see **G. *x magnificum***; Franchet, see **G. *sinense***.

HEBE *buxifolia* 'Nana': There are two distinct cultivars with this name, one identified by (Benth.) Ckn. & Allan.

HEDERA *canariensis*: There are two forms of this, one Willd., one hort.

HELLEBORUS *atrorubens*: hort., see **H. *orientalis*** Lamarck ssp. ***abchasicus*** Early Purple Group.

HELLEBORUS *orientalis*: Jim and Audrey Metcalfe, Helleborus breeders, have explained that although the true species, ***orientalis,*** does exist, virtually all the **H. *orientalis*** sold are hybrids and therefore fall correctly under **H. *hybridus***. The RHS has only very recently decided to adopt **H. *hybridus*** for these hybrids; this is a change from our first edition.

HOSTA 'Blue Angel'**:** misapplied, see **H. *sieboldiana*** var. ***elegans.***

HYDRANGEA: under species ***macrophylla*** we have, where we could find the information, added the designations H (hortensia or mophead) or L (lacecap) after the cultivar names, which used to be classified as **HYDRANGEA *macrophylla*** var. ***macrophylla*** and **H. *m.*** var. ***normalis***, respectively. Please note that other species and cultivars may have mophead and lacecap type flowers (e.g., **H.** 'Preziosa', which is a hortensia and **H. *serrata*** 'Grayswood', which is a lacecap).

IRIS: We have increased the number of groups under which we list iris cultivar names over those in the *Plant Finder*. For instance, we have added names under Spuria hybrids rather than interfiling them with the bearded iris cultivars; even though there may be some confusion between **I. *orientalis*** and **I. *spuria*** hybrid plants, we feel that most people consider these to be Spurias. Also, we have rejected the group "Californian hybrids," not an American designation as the adjective used here is always "California." We have listed these plants under Pacific Coast Hybrids, which is how they are known in our region, and maintained a list of cultivars under **I. *douglasiana*** as well. We have also grouped Cal-Sibe and Sino-Siberian cultivars under those designations. We are considering adding the notations for types of bearded irises (e.g. TB for tall-bearded), as is done in the *RHS Plant Finder*, in future editions.

KERRIA *japonica*: double, see **K. *japonica*** 'Pleniflora'; single, see **K. *japonica*** 'Simplex'.

KNIPHOFIA *galpinii:* hort., see **K. *triangularis*** ssp. ***triangularis***.

KOELERIA *cristata*: see either **K. *macrantha*** or **K. *pyramidata***.

LAGERSTROEMIA: cultivars named for Native American tribes are hybrids between spp. ***indica*** and ***fauriei***, bred for mildew resistance.

LAVANDULA *x intermedia* 'Arabian Night'**:** According to the *Plant Finder*, "plants under this name might be **L. *x intermedia*** 'Impress Purple'."

LAVANDULA *spica*: According to the *Plant Finder*, "this name is classed as a name to be rejected (*nomen rejiciendum*) by the *International Code of Botanical Nomenclature*." See **L. *angustifolia***, **L. *latifolia***, or **L. *x intermedia***.

NOTES FOR INDIVIDUAL GENERA AND SPECIES

LAVANDULA 'Twickel Purple': According to the *Plant Finder*, "two cultivars are sold under this name, one a form of **L. x *intermedia***, the other of **L. *angustifolia***."

LAVANDULA vera: DC., see **L. *angustifolia***; hort., see **L. *x intermedia*** Dutch Group.

LIRIOPE *exiliflora* Silvery Sunproof: misapplied, see **L. *spicata*** 'Gin-ryu' or **L. *muscari*** 'Variegata'.

MISCANTHUS *sinensis* 'Yaku Jima': According to *The Color Encyclopedia of Ornamental Grasses*, "not a clonal cultivar, but a name used for several very similar, compact, narrow-leaved plants, usually less than 5 ft. tall. Diminutive forms of **M. *sinensis*** are common on the Japanese island of Yaku Jima (or Yakushima). Staff of the U.S. National Arboretum, Washington, D.C., first brought 'Yaku Jima' seedlings to the United States that were variable but similar in being small-sized and narrow-leaved. The name 'Yaku Jima' has been used by nurseries as a catch-all for these seedlings and their progeny."

NEMESIA *fruticans*: name uncertain according to *Sunset Western Garden Book*; not found in other references.

NOTHOFAGUS *procera*: Oerst., see **N. *nervosa***; misapplied, see **N. *x alpina*** (*RHS Plant Finder 2000-2001)*

NYMPHAEA 'Hollandia': misapplied, see **N.** 'Darwin'.

NYMPHAEA 'Rembrandt': misapplied, see **N.** 'Météor'.

OENOTHERA *odorata*: Hook. & Arn., see **O. *biennis***; hort., see **O. *stricta*** or **O. *glaziouana***.

OLEARIA *x mollis*: hort., see **O. *ilicifolia***.

OLEARIA *x scilloniensis*: hort., see **O. *stellulata*** DC.

ORIGANUM *heracleoticum*: L., see **O. *vulgare*** ssp. ***hirtum***; hort., see **O. *applei***.

PAEONIA: We have added the designations H (herbaceous) and T (tree) after cultivar names where we had information as to which type they were.

PELTANDRA virginica: Schott, see **P. *undulata***.

PENSTEMON 'Sour Grapes': hort., see **P.** 'Stapleford Gem'.

PICEA *pungens* 'Foxtail': We believe this may be the same as **P.** 'Iseli Foxtail', but cannot verify it. The Porterhowse catalogue states that **P. *p***. 'Foxtail' "was found at Iseli Nursery in Oregon and named by Don Howse around 1976."

PIERIS *japonica* 'Variegata': hort., see **P. *japonica*** 'White Rim'.

PLEIONE *yunnanensis*: hort., see **P. *bulbocodioides*** 'Yunnan'.

POLEMONIUM *foliosissimum*: hort., see **P. *archibaldiae***.

POLYGONATUM *multiflorum*: hort., see **P. *x hybridum***.

PRIMULA *x juliana*: previously listed as **P.** Pruhonicensis Hybrids according to the *RHS Index of Garden Plants*; however, the *Plant Finder* currently lists these as stand-alone cultivars.

PRIMULA 'Wanda': Nurseries listing this cultivar may be carrying Wanda Supreme Series seedlings under the name 'Wanda'. 'Wanda' is a cultivar with small dark purplish-red flowers and purplish-green leaves; the seed series has similar leaves and flower size, but the flowers come in a range of colors.

PSEUDOSASA *amabilis*: hort., see **Arundinaria *tecta***.

QUERCUS *prinus*: Engelm., see **Q. *montana***.

RHODODENDRON: We have interfiled azalea cultivar names within the rhododendron list in this edition, a change from

the first edition where we grouped azaleas separately. We now have added either EA (evergreen azalea), DA, (deciduous azalea), or A (undetermined type) after azalea names, both cultivars and species. Where we had the information, we have also added Ad, designating azaleodendrons, a rhododendron/azalea cross.

RHODODENDRON 'China Doll'**:** According to *Greer's Guidebook to Available Rhododendrons,* the cultivar of this name is correctly a yellow-flowered hybrid from Australia. A white-flowered 'China Doll' found in the U.S. trade is now renamed 'Senator Henry M. Jackson'.

RHODODENDRON *dryophyllum*: hort., see **R. *phaeochrysum*** var. ***levistratum***; Balfour & Forrest, see **R. *phaeochyrsum*** var. ***phaeochrysum.***

RHODODENDRON *litangense*: According to *Greer's Guidebook to Available Rhododendrons*, this species "is now considered synonymous with ***impeditum.***" The author makes a case, however, for a ***litangense*** which "is different from the plant that is grown under the name of ***impeditum*** in the U.S."

RHODODENDRON *ponticum*: This species is correctly a rhododendron; however, in the past there has been a species ***ponticum*** which was an azalea as well. The azalea is now classified as **R. *luteum*.**

RHUS *glabra* 'Laciniata'**:** hort., see **R. *x pulvinata*** Autumn Lace Group.

ROSA: Many rose cultivar names are listed with no quotes, as explained previously under **Abbreviations, Type Styles, and Symbols**. As stated before, these are trade names (followed by breeder names in quotes). They may be followed by the designation ® or ™, indicating that the name is trademarked. In future editions, we will consider also adding additional parenthetical species notes, as in other genera, and the type of rose designations (e.g., "HT" for hybrid tea) to the cultivar names.

ROSA 'Archiduc Joseph'**:** misapplied, see **R.** 'General Schablikine' according to the *Plant Finder*. According to Peter Beales, "in the USA the rose...grown under the name 'Archiduc Joseph' in the UK is known and sold as 'Mons. Tillier'."

ROSA 'Baroness Rothschild'**:** HP (hybrid perpetual), see **R.** 'Baronne Adolph de Rothschild'; hybrid tea, see **R.** Baronne Edmond de Rothschild / 'Meigriso'.

ROSA Eden Rose®**:** from Meilland Roses, is not to be confused with Eden/Eden Rose 88/Eden Climber, all of which are one and the same rose, also from Meilland Roses ('Meiviolin'). There exists also 'Climbing Eden Rose'.

ROSA Intrigue**:** Two roses of this name exist, dark red Intrigue / 'Korlech' (also known as Lavaglut and Lavaglow) from Kordes, and reddish-purple Intrigue / 'Jacum' from Jackson & Perkins.

ROSA Playtime**:** Two roses of this name exist, Playtime / 'Korsaku', and Playtime™ / 'Morplati' (we are unable to verify colors of either rose).

SALIX *caprea* var. ***pendula*:** female, see **S. *caprea*** 'Weeping Sally'; male, see **S. *caprea*** 'Kilmarnock' (m).

SALIX *japonica*: hort., see **S. *babylonica*** 'Lavalleei'.

SALVIA *officinalis* 'Aurea'**:** According to the *Plant Finder*, "**S. *officinalis*** var. ***aurea*** is a rare variant of the common sage with leaves entirely of gold. It is represented in cultivation by the cultivar 'Kew Gold'. The plant usually offered as **S. *officinalis*** 'Aurea' is the gold variegated sage **S. *officinalis*** 'Icterina'."

SALVIA *sclarea* var. ***turkestanica*:** According to the *Plant Finder*, "plants in gardens under this name are not **S. *sclarea*** var. ***turkistaniana*** of Mottet."

SALVIA *viscosa*: Sesse & Moc., see **S. *riparia*.**

SALVINIA *rotundifolia*: Willd., see **S.** *auriculata*; hort., see **S. *minima*.**

SAXIFRAGA: The nomenclature for this genus has recently changed from the typical listing of cultivars under species names to cultivar names followed by their species names in parentheses.

NOTES FOR INDIVIDUAL GENERA AND SPECIES

SEDUM *nevii*: According to the *Plant Finder*, "the true species is not in cultivation. Plants under this name are usually either **S. *glaucophyllum*** or occasionally **S. *beyrichianum*.**"

SEMPERVIVUM *x hausemannii*: According to *Hortus III*, this is actually **S. *x fauconnettii***, "but the name has also been used for **S. *x barbulatum*.**"

SEMPERVIVUM *villosum*: Lindl. non Ait., see **AEONIUM *spathulatum***; Ait. non Lindl., see **AICHRYSON *villosum.***

SENECIO *greyi*: hort., see **BRACHYGLOTTIS** (Dunedin Group) 'Sunshine'; Hooker, see **BRACHYGLOTTIS *greyi.***

SORBUS *discolor*: hort., see **S. *commixta*.**

SORBUS *hybrida*: hort., see **S. *x thuringiaca*.**

SORBUS *pohuashanensis*: hort., see **S. *x kewensis*.**

SORBUS *scopulina*: hort., see **S. *aucuparia*** 'Fastigiata'; Hough non Greene, see **S. *decora*.**

SORBUS *sitchensis*: Piper pro parte., see **S. *scopulina*.**

SPIRAEA *x bumalda*: Cultivars formerly under this name are now listed under **S. *japonica***, and 'Bumalda' is now a cultivar under **S. *japonica*** also.

STREPTOCARPUS *x hybridus*: Cultivars formerly listed under this name are now stand-alone cultivars under **STREPTOCARPUS.**

TEUCRIUM *chamaedrys*: hort., see T. x ***lucidrys*.**

THYMUS *x citriodorus*: misapplied, see **T. *x citriodorus*** 'Aureus'.

THYMUS 'Silver Posie': According to the *Plant Finder*, "the cultivar name 'Silver Posie' is applied to several different plants, not all of them **T. *vulgaris*.**"

TRICYRTIS Hototogisu: According to the *Plant Finder*, "this is the common name applied generally to all Japanese tricyrtis and specifically to **T. *hirta*.**"

VIOLA *labradorica*: hort., see **V. *riviniana*** Purpurea Group.

DISCLAIMER

We wish to note that we cannot be held responsible for errors of omission or commission. We have striven to be as accurate as possible in listing plant names and the nurseries that carry them, as well as the nursery information provided by owners, and to this end have proofread the manuscript several times with great care. However, in a volume of this size and detail, a few errors are bound to occur, and we apologize for them. We actively solicit both corrections and comments about plant nomenclature from interested readers, since it is our goal to produce the most accurate and readable listing possible. Please see the back of the title page for information about communicating with us.

Additionally, we can make no claim that the plants actually offered for sale by any listed nursery are indeed the plants they say they are, nor can we assure readers that nurseries will have stock of plants that they list. We urge readers to check with nurseries about specific plant availability before they visit, and to inquire of the nursery as to the specific features of the plant if there is any reason to be suspicious that plants sold under this name may indeed be something else (an example of this is **PENSTEMON** 'Sour Grapes,' which may or may not be the plant originally named 'Sour Grapes'; it is often confused in the trade with **P.** 'Stapleford Gem').

BIBLIOGRAPHY

AHS A-Z Encyclopedia of Garden Plants, The
Christopher Brickell and Judith D. Zuk, editors, New York, Dorling Kindersley, 1997.

Andersen Horticultural Library's Source List of Plants and Seeds
Edited by Richard T. Issacson, University of Minnesota, 1996.

Book of Salvias, A
Betsy Clebsch, Portland, Timber Press, 1997.

Color Encyclopedia of Ornamental Grasses, The
Rick Darke, Portland, Timber Press, 1999.

Conifers, The Illustrated Encyclopedia, Volumes 1 – 2
D. M. van Gelderen and J. R. P. van Hoey Smith, Portland, Timber Press, 1996.

Eureka! 1999 National Daylily Locator
Ken and Kay Gregory, Granite Falls, NC.

Genus Host, The
W. George Schmid, Portland, Timber Press, 1997.

Greer's Guidebook to Available Rhododendrons Species and Hybrids
Harold E. Greer, Eugene, Offshoot Publications, 1996.

Hortus Third
Revised and expanded by the staff of the Liberty Hyde Bailey Hortorium, Cornell University, New York, Macmillan, 1976.

Maples for Gardens
C. J. van Gelderen and D. M. van Gelderen, Portland, Timber Press, 1999.

Peonies
Allan Rogers, Portland, Timber Press, 1995.

Random House Book of Herbs, The
Roger Phillips and Nicky Foy, New York, Random House, 1990.

Rhododendron Hybrids, Second Edition
Homer E. Salley and Harold F. Greer, Portland, Timber Press, 1992.

RHS Plant Finder (1999-2000), *The*
Tony Lord, editor, compiled by the Royal Horticultural Society, London, Dorling Kindersley, 1999.

Roses
Peter Beales, New York, Henry Holt and Company, 1992.

Sunset Western Garden Book
Edited by the editors of Sunset Books and Sunset Magazine, Menlo Park, Sunset Publishing, 1995.

PLANT LISTING

	ABELIA	
	chinensis	oFor
va	Confetti / 'Conti'	oFor oGos wPir
	'Edward Goucher'	oFor oAls oGar oSho oBlo oGre oTPm oWnC oCir wCoN oSle wCoS
	floribunda	wHer
	x grandiflora	oAls oGar oSho oBlo wCCr oTPm oThr oTal oWhS wAva oSle wCoS
va	***x grandiflora*** 'Francis Mason'	oFor oJoy oGar oSho wCCr oWnC oCir
	x grandiflora 'Prostrata'	oGar oSho
	x grandiflora 'Sherwoodii'	oFor oGar oGoo oBlo oWnC
	x grandiflora 'Sunrise'	oGos wCol oSho oGre
	x grandiflora 'Variegata'	see **A. *x grandiflora*** 'Francis Mason'
	ABELIOPHYLLUM	
	distichum	oFor wHer wCul oGos oGar oSho oBov oGre wKin oUps
	distichum Roseum Group	wHer
	ABELMOSCHUS	
	manihot	oHug
	ABIES	
	alba	oFor wCSG wRav
	alba 'Compacta'	oPor
	alba 'Green Spiral'	oPor
	alba 'Pendula'	oPor
	alba 'Pyramidalis'	oPor oRiv
	alba 'Schwarzwald'	oPor
	amabilis	oFor wBlu oRiv wRav
	amabilis Hoyt witches' broom	oPor
	amabilis 'Spreading Star'	wHer oPor
	x arnoldiana	oPor
	balsamea	oFor wKin oBRG wRav
un	***balsamea*** var. ***balsamea*** 'Eugene Gold'	oPor
	balsamea 'Nana'	wHer oPor oGos oAls oGar oSis oRiv oSho oGre oWnC wSta wAva wCoS
un	***balsamea*** 'Old Ridge'	oPor
	balsamea 'Piccolo'	oPor
	balsamea 'Prostrata'	oGre
un	***balsamea*** 'Quinten Spreader'	oPor
	balsamea 'Verkade's Prostrate'	oPor
un	***balsamea*** 'Wolcott Pond'	oPor
	bornmuelleriana	see **A. *nordmanniana*** ssp. ***equi-trojani***
	borisii-regis	oFor oPor
	bracteata	oFor oPor oRiv oGre wCCr
un	***bracteata*** 'Corbin'	oPor
	cephalonica	oPor wCCr
	cephalonica 'Meyer's Dwarf'	oPor
	chensiensis	last listed 99/00
	chensiensis x ***balsamea***	oRiv
	cilicica	oFor oPor oRiv
	cilicica witches' broom Hunnewell	oPor
	concolor	oFor oGar oRiv wCCr wKin wPla oSle wCoS
	concolor 'Archer's Dwarf'	oPor
	concolor 'Argentea'	oFor oPor oGos wCol oGar oRiv oGre oWnC wSta wCoN
un	***concolor*** 'Big Shot'	oPor
un	***concolor*** 'Birthday Broom'	oPor
un	***concolor*** 'Blue Cloak'	oPor
	concolor 'Candicans'	see **A. *concolor*** 'Argentea'
	concolor 'Compacta'	oPor oRiv oWnC
	concolor 'Conica'	oPor
un	***concolor*** 'FWP Kinky'	oPor
	concolor 'Gable's Weeping'	oFor oPor
un	***concolor*** 'Glenmore'	oPor
un	***concolor*** 'La Veta'	oPor
	concolor 'Masonic Broom'	last listed 99/00
	concolor 'Piggelmee'	oPor
	concolor 'Pyramidalis'	oPor
un	***concolor*** 'Sherwood Blue'	oPor
un	***concolor*** 'Sidekick'	oPor
	concolor Violacea Group	oPor
	concolor 'Wattezii'	oPor
	concolor 'Wattezii Prostrate'	last listed 99/00
	concolor 'Wintergold'	oPor
un	***concolor*** 'Woodstock'	oPor
	delavayi	oPor
	durangensis var. ***coahuilensis***	oPor
	equi-trojani	see **A. *nordmanniana*** ssp. ***equi-trojani***
	fabri EDHCH 97118	wHer
	fargesii	oFor oPor
	firma	oFor oPor
un	***firma*** 'Halgren'	oPor
un	***firma*** 'Sol'	oPor
	forrestii var. ***georgei***	oPor
	forrestii var. ***smithii***	wCol
	fraseri	oFor wCSG oGar oRiv oSho wRav
	fraseri 'Kline's Nest'	oPor
	fraseri 'Prostrata'	wHer
	fraseri 'Vashti'	oPor
	gamblei	oPor
	georgei	see **A. *forrestii*** var. ***georgei***
	georgei var. ***smithii***	see **A. *forrestii*** var. ***smithii***
	'Gold Traum'	see **A. *koreana*** 'Golden Dream'
	grandis	oFor wBlu wWoB wNot wShR wCSG oBos oRiv oSho wKin oWnC wPla oAld wFFl wRav oSle
	grandis 'Aurea'	oPor
	grandis 'Compacta'	oPor
	grandis 'Johnson'	oPor
	guatemalensis	oPor
	holophylla	oFor oPor
	homolepis	oPor wRav
	homolepis 'Prostrata'	oPor
	homolepis 'Tomomi'	oPor
	homolepis 'Variegata'	oPor
	kawakamii	oPor
	koreana	oFor oPor oEdg oRiv wTGN wCoN wRav
	koreana 'Aurea'	see **A. *koreana*** 'Flava'
	koreana 'Blaue Zwo'	oPor
un	***koreana*** 'Blauer Eskimo'	oPor
	koreana 'Blauer Pfiff'	oPor
	koreana 'Cis'	oPor
un	***koreana*** 'Doni Tajusho'	oPor
	koreana 'Flava'	wHer oPor wCol oRiv
un	***koreana*** 'Freudenburg'	oPor
	koreana 'Gait'	oPor
un	***koreana*** 'Gelbbunt'	oPor
un	***koreana*** 'Glauca'	oPor oWnC
	koreana 'Golden Dream'	oPor oGos oGar
	koreana Goldener Traum	see **A. *koreana*** 'Golden Dream'
	koreana 'Green Carpet'	oPor
	koreana HC 970295	wHer
	koreana 'Horstmann'	oPor
qu	***koreana*** 'Horstmann's Silberlocke'	oFor wHer oGar oRiv oGre oWnC wWhG
	koreana 'Inverleith'	oPor
un	***koreana*** 'Lippetal'	oPor
un	***koreana*** 'Nanaimo'	oPor wClo
	koreana 'Oberon'	oPor
	koreana 'Piccolo'	oPor
	koreana 'Pinocchio'	oPor
	koreana 'Prostrate Beauty'	oPor
un	***koreana*** 'Silber Reif'	oPor
un	***koreana*** 'Silber Schweig'	oRiv
	koreana 'Silberkugel'	oPor
	koreana 'Silberlocke'	oPor
	koreana 'Silbermayers'	oPor oRiv oGre
	koreana 'Silberperl'	oPor
	koreana 'Silberzwerg'	oPor
	koreana 'Silver Show'	oPor
	koreana 'Starker's Dwarf'	oPor oRiv
un	***koreana*** 'Taiga'	oPor
	lasiocarpa	oFor wWoB oAls oGar oBos oRiv oSho wKin wPla oAld wAva wCoN wRav oSle wCoS
	lasiocarpa var. ***arizonica***	oFor wShR wAva wRav
	lasiocarpa 'Arizonica Compacta'	oFor
un	***lasiocarpa*** 'Arizonica Compacta Glauca'	oPor oRiv
	lasiocarpa 'Compacta'	oPor
	lasiocarpa 'Duflon'	oPor oGos
un	***lasiocarpa*** 'Glauca Compacta'	oGre
	lasiocarpa 'Green Globe'	oPor oGos oGre
un	***lasiocarpa*** 'Lopalpun'	oPor
	lasiocarpa 'Mulligan's Dwarf'	oPor
un	***lasiocarpa*** 'Utah'	oPor
	magnifica	oFor
	magnifica var. ***shastensis***	oFor
	mariesii	oPor
	marocana	see **A. *pinsapo*** var. ***marocana***
	nephrolepis	oFor oPor wRav
	nobilis	see **A. *procera***
	nordmanniana	oFor oGar wCCr wCoN wRav
	nordmanniana 'Barabits' Compact'	oPor
un	***nordmanniana*** 'Berlin'	oPor
un	***nordmanniana*** 'Brandt'	oPor
	nordmanniana ssp. ***equi-trojani***	oFor oPor wRav
	nordmanniana ssp. ***equi-trojani*** 'Barney'	oPor
	nordmanniana 'Golden Spreader'	oPor oRiv
un	***nordmanniana*** 'Munsterland'	oPor
un	***nordmanniana*** 'Prostrata'	oPor
	nordmanniana 'Tortifolia'	oPor

un	***nordmanniana*** 'Trautman	oPor
	nordmanniana witches' broom Hunnewell	oPor
	numidica	oPor
	numidica 'Pendula'	oPor
	pardei	oPor
	pindrow	oFor oPor wCol wCCr wRav
	pinsapo	oFor oPor oRiv wCCr wTGN wCoN wRav
	pinsapo 'Aurea'	oPor wCol oRiv
	pinsapo 'Glauca'	oFor oPor oRiv wSta
	pinsapo 'Hamondii'	oPor
	pinsapo 'Horstmann'	oPor oGos oRiv
	pinsapo 'Kelleriis'	oPor
	pinsapo var. ***marocana***	oPor oRiv wCCr
	procera	oFor wBlu oPor wWoB wNot wShR oBos oRiv wKin oWnC wFFl wRav wCoS
	procera 'Aurea'	oRiv
	procera 'Blaue Hexe'	oPor
un	***procera*** 'Frijsenborg'	oEdg oRiv
	procera Glauca Group	oFor oPor oGar oRiv
un	***procera*** 'Glauca Pendula'	wSta
	procera 'Glauca Prostrata'	oPor wClo
	procera 'Jeddeloh'	oPor
	procera 'La Graciosa'	oPor oRiv
un	***procera*** 'Pendula'	wSta
	procera 'Sherwoodii'	oGre oPor
un	***procera*** 'Stanley's Select'	oPor oGre
qu	***pungens*** var. ***glauca***	wShR
	recurvata	oFor oPor
	recurvata var. ***ernestii***	oPor
	religiosa	oFor oPor
	sachalinensis	last listed 99/00
	sachalinensis var. ***mayriana***	oPor
	sibirica	oFor oPor
	spectabilis	oPor
	veitchii	oFor oPor
un	***veitchii*** 'Glauca'	oPor
	veitchii 'Hedergott'	oPor
un	***veitchii*** var. ***veitchii***	wClo
un	***veitchii*** weeping form	oPor
	vejarii	oPor
	ABROMEITIELLA	
	brevifolia	oRar
	chlorantha	see **A. *brevifolia***
	ABRUS	
	precatorius	oOEx
	ABUTILON	
un	'Afterglow'	oCis
un	'Apollo'	oCis
	'Apricot'	oAls
un	Bella F. mixed	oGar
va	'Cannington Sonia'	wHer
un	'Cathedral Bells'	oCis
un	'David'	oCis
un	'Dennis'	oCis
un	'Evening Glow'	oNat
un	'Freckles'	oCis
un	'Halo'	oCis wAva
un	'Hot Sex Wax'	oCis
un	'Huntington Pink'	wHer oNat wAva
un	*x **hybridum*** 'Red Dwarf'	oTrP
un	*x **hybridum*** 'Voodoo'	wHer
	'Kentish Belle'	oNat
un	'Linda Vista Peach'	oNat
	'Louis Marignac'	wHer
un	'Louis Sasson'	oCis
	megapotamicum	oTrP wHer oNat oMis wAva
un	'Melon Delight'	wHer
un	***menziesii***	oTrP
un	'Mobile Pink'	oNat oCis
	'Moonchimes'	wHer oNat oCis wAva
	'Nabob'	oNat oCis
un	'New One'	wHer wAva
un	'New Orleans'	oCis
	pictum	wAva
qu	***pictum aureomaculatum***	oNat
va	***pictum*** 'Thompsonii'	oUps
un	'Porosa While'	oCis
	'Roseum'	oGar
va	'Savitzii'	wWoS
un	'Seashell'	oCis
	'Silver Belle'	wHer
un	'Snowfall'	wHer
va	'Souvenir de Bonn'	oTrP wHer oCis wAva
	striatum	see **A. *pictum***
un	'Sunset'	oCis
un	'Tangelo'	wAva
	'Tangerine'	oNat oCis
	'Tangerine Belle'	see **A.** 'Tangerine'
un	'Thomas Hobbs'	wHer oCis
un	'Tom Trillium'	oCis
	trailing yellow	wAva
un	'Vesuvius'	wHer oNat oCis
	'Vesuvius Red'	oGar
un	'Victor Reiter'	wHer oCis
un	'Victor's Folly'	oCis
un	'Victory'	oCis
	vitifolium	wHer wCCr
	vitifolium 'Veronica Tennant'	oGos
	'Wisley Red'	wHer
	ACACIA	
	baileyana	last listed 99/00
	boormanii	oFor wCCr
	buxifolia	last listed 99/00
	caffra	oTrP
	dealbata	oFor
	dealbata subalpina	oTrP
	farnesiana	oFor
	frigescens	last listed 99/00
	greggii	oFor
	howittii	see **A. *verniciflua***
un	***hubbardiana***	oTrP
un	***lasiocalyx***	oTrP
	longifolia	oFor
	melanoxylon	oFor
	mucronata	oFor
	nigricans	last listed 99/00
un	***pinguifolia***	oTrP
	pravissima	oFor oTrP wCCr
	redolens	oFor
	retinodes	wCCr
	riceana	oFor
	rotundifolia	last listed 99/00
	rubida	wCCr
un	***siculiformis***	oTrP
	sophorae	last listed 99/00
	stenophylla	wCCr
	verniciflua	last listed 99/00
un	***wrightii***	oFor
	ACAENA	
	adscendens (see Nomenclature Notes)	wWoS
	caesiiglauca	wCCr
	glauca	see **A. *caesiiglauca***
	inermis	oTrP
	microphylla	oTrP oJoy
	microphylla Copper Carpet	see **A. *microphylla*** 'Kupferteppich'
	microphylla 'Kupferteppich'	wWoS oHed
	novae-zelandiae	oTrP oJoy wTGN
	ovalifolia HCM 98004	wHer
	saccaticupula	oTrP
	saccaticupula 'Blue Haze'	oJoy oHed wCSG wRob
	sp. HCM 98189	wHer
	ACANTHOLIMON	
	androsaceum	see **A. *ulicinum***
	glumaceum	oSis
	ulicinum	oSis
	venustum	oSis
	ACANTHOPANAX	see **ELEUTHEROCOCCUS**
	ACANTHUS	
	balcanicus	see **A. *hungaricus***
	caroli-alexandri	oJoy oAls
	hirsutus f. ***roseus***	last listed 99/00
	hungaricus	wHer oJoy wHig oAls wCSG oGar oGre wPir wBWP
	longifolius	see **A. *hungaricus***
	mollis	oFor wCul wGAc oNat oJoy oAls wCSG oAmb oMis oGar oGoo oSho oSha oBlo oGre wPir wCCr wHom wMag wTGN wBox wBWP wCoN wEde oUps wCoS
un	***mollis*** 'Aureum'	oJoy
	mollis 'Hollard's Gold	wHer wEde
	mollis 'Oak Leaf'	oGar
	spinosus	wCul oJoy oEdg wHig wCSG oGar oNWe wTGN
	spinosus Spinosissimus Group	wHer oNWe oGre
	ACCA	
	sellowiana	oFor oGar oSho oWhi oGre wDav wCCr oWnC wSte
	sellowiana 'Coolidge'	oOEx
	sellowiana 'Improved Coolidge'	see **A. *sellowiana*** 'Coolidge'

	sellowiana 'Nazemetz'	oOEx
un	sellowiana 'Trask'	oOEx
un	sellowiana 'Unique'	wGra
	ACER	
	acuminatum	oFor wRav
un	aidzuense	oFor oRiv
	albopurpurascens	last listed 99/00
	argutum	oFor wHer oRiv wRav
	barbinerve	oFor wFWB oRiv oGre
	buergerianum	oFor wFWB oAmb oGar oRiv oCis oWhi oGre wCCr oWnC wSta wCoN wRav
va	buergerianum 'Goshiki-kaede'	wFWB
	buergerianum 'Mino-yatsubusa'	wFWB wWhG
	buergerianum 'Miyasama-yatsubusa	wFWB
un	buergerianum 'Naruto-kaede'	wFWB oGre
	buergerianum var. ningpoense	oFor oWhi
	buergerianum var. ningpoense DJCH 98001	wHer
	buergerianum 'Tancho'	last listed 99/00
	buergerianum 'Wako-nishiki'	wFWB
	campbellii	oFor
	campbellii ssp. flabellatum	wCol
	campbellii ssp. flabellatum DJHC 800	wHer
	campbellii ssp. flabellatum var. yunnanense	last listed 99/00
	campbellii ssp. sinense	see A. sinense
	campestre	oFor wFWB oRiv oSho oGre wCoN wRav oSle
va	campestre 'Carnival'	oFor oGar
	campestre 'Compactum'	see A. campestre 'Nanum'
	campestre 'Nanum'	oFor wFWB oGos oRiv oWhi
	campestre 'Postelense'	oFor wHer oRiv
va	campestre 'Pulverulentum'	oFor wHer oRiv
	campestre 'Queen Elizabeth'	oRiv
	campestre 'Royal Ruby'	oFor oWhi
	capillipes	oFor oAls oDan oRiv oSho oGre wCCr wCoN wRav oSle
	capillipes HC 970726	wHer
	cappadocicum	oFor oRiv oWhi wCCr wRav oSle
	cappadocicum 'Aureum'	last listed 99/00
	cappadocicum ssp. lobelii	wCCr
	carpinifolium	oFor wCol oAmb oRiv oGre wCCr
	carpinifolium HC 970719	wHer
	caudatifolium	oFor oWhi wCCr
	caudatifolium B&SWJ 3114	wHer
	caudatifolium 'Variegatum'	wHer
	caudatum	oRiv
	caudatum	last listed 99/00
	caudatum ssp. ukurunduense	oFor wHer wFWB oGre
	cinerascens	oWhi
	cinnamomifolium	see A. coriaceifolium
	circinatum	oFor wFWB wThG wWoB wBur wClo wNot oDar oAls wShR oGar oBos oRiv oSho oWhi oBlo wRai oGre wWat wCCr oWnC oTri wTGN wSta wPla oAld oCir wAva wCoN wFFI wRav oTin oSle wSnq wCoS
un	circinatum 'Glen-del'	wFWB
	circinatum 'Little Gem'	wHer wFWB wCol oDan oRiv oGre
	circinatum 'Monroe'	wHer wFWB oGos wCol oDan oRiv oGre oTin
un	circinatum 'Pacific Fire'	wFWB
	circinatum 'Sunglow'	wFWB
	cissifolium	oFor oGar oGre wCCr
	cissifolium HC 970723	wHer oRiv
	coriaceifolium	oFor wHer oRiv oCis
	crataegifolium	wRav
	crataegifolium HC 970656	wHer
un	crataegifolium 'Me uri ko fuba'	oFor oGar oWhi oGre
va	crataegifolium 'Veitchii'	wFWB oRiv
	davidii	oFor wFWB oEdg oDan oCis oGre wCCr oWnC wAva wCoN wRav
	davidii DJHC 179	wHer
	davidii EDHCH 97263	wHer
un	davidii 'Hansha Suru'	wFWB
un	davidii 'Hughes Variegate'	see A. davidii 'Hansha Suru'
un	davidii 'Scarlet Forest'	oWhi
	davidii 'Serpentine'	oRiv oCis
	davidii x tegmentosum	wHer
	distylum	oRiv
	divergens	oFor
	elegantulum	oFor wHer wCol oRiv wRav
	erianthum	oRiv
	fabri	oFor wHer oRiv oCis
	forrestii	oFor oRiv oWhi oSle
	forrestii DJHC 798	wHer
	x freemanii 'Armstrong'	see A. rubrum 'Armstrong'
	x freemanii Autumn Blaze®/ 'Jeffersred'	oFor oGar oBlo wKin oWnC
	x freemanii 'Autumn Flame'	see A. rubrum 'Autumn Flame'
	ginnala	see A. tataricum ssp. ginnala
	glabrum	oFor wFWB wThG wWoB wShR oRiv wWat wRav oSle
	glabrum ssp. douglasii	oFor wPla
	grandidentatum	see A. saccharum ssp. grandidentatum
	griseum	wHer wFWB oGos wSwC wClo oMac oDar oAls oAmb oGar oDan oRiv oWhi oBlo oBov oGre oWnC wTGN wSta wAva wCoN wRav wCoS
un	griseum 'Gingerbread'	oFor
	grosseri	wCCr wSta wRav
	grosseri var. hersii	oFor
	heldreichii	wCol
	henryi	oFor wHer wFWB oRiv oGre wAva wCoN wRav
	japonicum	oFor wCoN wRav
	japonicum 'Aconitifolium'	oFor wHer wFWB oGos wClo oAmb oSho oWhi oGre oWnC wSta wCoN
un	japonicum f. aconitifolium 'Maiku jaku'	wWhG
	japonicum 'Attaryi'	oWhi oGre
	japonicum 'Green Cascade'	wFWB wClo oAmb oGar oGre wWhG oTin
un	japonicum 'Henny Bigleaf'	oGar
un	japonicum 'Lovett'	oGar
	japonicum 'Ogurayama'	see A. shirasawanum 'Ogurayama'
un	japonicum 'Oregon Fern'	wFWB
	japonicum 'O-taki'	oGar wWhG
	japonicum 'Vitifolium'	oFor oGos oAmb oWhi oGre oWnC wWhG
	kawakamii	see A. caudatifolium
	laxiflorum	wCCr
	x leucoderme	see A. saccharum ssp. leucoderme
	lobelii (see Nomenclature Notes)	last listed 99/00
	longipes	oFor wHer
	macrophyllum	oFor wWoB wBur wNot wShR oGar oBos oRiv wRai wWat oAld wFFI oSle wCoS
	macrophyllum 'Kimballiae'	oFor
	macrophyllum 'Seattle Sentinel'	oFor
	mandschuricum	oFor wHer wFWB oEdg oAmb oDan oRiv oGre
	mandschuricum HC 970382	wHer
	maximowiczianum	oFor
	maximowiczii	oFor oRiv
	metcalfii	oRiv
	micranthum	oRiv wRav
	miyabei	oRiv oGre
	mono	oFor wFWB oEdg oRiv wRav
	mono f. amibiguum	oAmb
	mono HC 970162	wHer
va	mono 'Hoshiyadori'	wFWB
	monspessulanum	oFor wHer wFWB oAmb oRiv oGre wCCr wRav
	morifolium HC 970023	wHer
	negundo	wFWB oBlo wCoN oSle wCoS
va	negundo 'Flamingo'	oFor oGar oRiv oGre oWnC oThr wCoN
	negundo 'Kelly's Gold'	oFor oGos oRiv oGre
	negundo 'Sensation'	oFor oRiv oWhi
	negundo 'Variegatum'	wCSG wCoN
	negundo var. violaceum	wHer oRiv oGre
	nigrum	see A. saccharum ssp. nigrum
	nikoense	see A. maximowiczianum
	ningpoense	see A. buergerianum ssp. ningpoense
	nipponicum	wHer
un	Norwegian Sunset®	wKin
	oblongum	oFor oRiv
	oliverianum	oFor oRiv wCCr wRav
	oliverianum ssp. formosanum B&SWJ 3037	wHer
	opalus ssp. obtusatum	oFor
	orientale	see A. sempervirens
un	Pacific Sunset™	oFor oGar oSho oWnC
	palmatum	oFor wFWB oAls oAmb oGar oRiv oGre wCCr wTGN wSta oThr wAva wRav oSle wCoS
	palmatum 'Aka Shigitatsusawa'	oFor wFol wFWB oAmb oRiv oWhi oGre oWnC wWhG oTin
	palmatum 'Aka Shime-no-uchi' (see Nomenclature Notes)	wFWB
	palmatum 'Akaji-nishiki'	see A. truncatum 'Akaji nishiki'
	palmatum 'Akita-yatsubusa	wFol
	palmatum 'Ao-kanzashi	wFWB oWhi oGre oTin
	palmatum 'Ao-shidare'	wFWB

	palmatum 'Aoba-jo'	wFol wFWB wClo wWhG
un	***palmatum*** 'Aocha-nishiki'	oGre
	palmatum 'Aoshime-no-uchi'	see **A. *palmatum*** 'Shinobugaoka'
	palmatum 'Aoyagi'	oFor wFWB oGar oWhi oGre oTin
	palmatum 'Arakawa'	wFWB wClo oAmb oGre oTin
	palmatum 'Aratama'	last listed 99/00
va	***palmatum*** 'Asahi-zuru'	wFWB wClo oTin
	palmatum 'Atrolineare'	oFor wFWB oWhi wWhG
	palmatum f. ***atropurpureum***	oFor wFWB wSwC oEdg oWhi oGre wTGN wAva
	palmatum 'Atropurpureum'	wFWB oAls oGar oRiv wTGN wRav wCoS
	palmatum 'Aureum'	last listed 99/00
un	***palmatum*** 'Autumn Moon'	see **A. *shirasawanum*** 'Autumn Moon'
	palmatum 'Autumn Red'	oTin
	palmatum 'Azuma-murasaki'	oGre oWnC
un	***palmatum*** 'Barrie Bergman'	wFWB
	palmatum 'Beni-fushigi'	wCol oGar oWhi oGre oWnC oTin
	palmatum 'Beni-hime'	wFol oGre wWhG
un	***palmatum*** 'Beni-hoshi'	oFor wFol wFWB oGre
	palmatum 'Beni-kagami'	oGre
	palmatum 'Beni-kawa'	oFor wFWB wClo oAmb oSis wWhG
	palmatum 'Beni-komachi'	oFor wFWB oRiv oGre wWhG oTin
	palmatum 'Beni-komachi' sport	wFWB
un	***palmatum*** 'Beni-komo-no-su'	wFol oGre
	palmatum 'Beni-maiko'	wFWB oRiv oGre wWhG wWel
	palmatum 'Beni-otake'	wFol oGar oGre wWhG oTin
va	***palmatum*** 'Beni-schichihenge'	oFor wFol wFWB wClo oGar oRiv oGre oWnC wWhG oTin wWel
	palmatum 'Beni-shidare'	oWnC
un	***palmatum*** 'Beni-shi-en'	oGre wWhG
	palmatum 'Beni-shigitatsu-sawa'	see **A. *palmatum*** 'Aka Shigitasusawa'
va	***palmatum*** 'Beni-tsukasa'	oFor wFWB wWhG oTin wWel
un	***palmatum*** 'Beni-ubi-gohan'	oGre
un	***palmatum*** 'Beni-yatsubusa'	wFWB
	palmatum 'Berry Dwarf'	wFWB
	palmatum 'Bloodgood'	oFor wFWB oAls oAmb oGar oRiv oWhi oBlo oGre oTPm oWnC wTGN wSta oThr wWhG wCoS wWel
	palmatum 'Bonfire'	see **A. *truncatum*** 'Akaji-nishiki'
un	***palmatum*** 'Bonnie Bergman'	wFWB
	palmatum 'Boskoop Glory'	wFWB oGar oWhi oGre oWnC
	palmatum 'Brandt's Dwarf'	wFol oGre
	palmatum 'Brocade'	wFWB
un	***palmatum*** 'Burgundy Flame'	oTin
	palmatum 'Burgundy Lace'	wFWB oAls oGar oRiv oGre oWnC wSta oTin
va	***palmatum*** 'Butterfly'	wFWB wCol oAls oAmb oGar oDan oBlo oGre oWnC wSta oThr wWhG oTin wCoS wWel
un	***palmatum*** 'Calico'	wCol oTin
un	***palmatum*** 'Chantilly Lace'	wFWB oGre
un	***palmatum*** 'Chiba'	oRiv
va	***palmatum*** 'Chirimen-nishiki'	wFWB
	palmatum 'Chishio'	see **A. *palmatum*** 'Shishio'
	palmatum 'Chishio Improved'	see **A. *palmatum*** 'Shishio Improved'
	palmatum 'Chitoseyama'	wFWB oGar wWel
	palmatum 'Coonara Pygmy'	oFor wFol wClo oGar oGre oWnC wWhG oTin wWel
	palmatum 'Corallinum'	oFor wFol wFWB oRiv oGre wWhG oTin wWel
un	***palmatum*** 'Crimson Prince'	oGre
	palmatum 'Crimson Queen'	see **A. *palmatum*** var. ***dissectum*** 'Crimson Queen'
	palmatum 'Crippsii'	oGre
un	***palmatum*** 'Curtis Strapleaf'	oWhi
	palmatum 'Deshojo'	oGre
	palmatum var. ***dissectum***	oRiv oBlo wSta oCir
un	***palmatum*** var. ***dissectum*** 'Baby Lace'	oGre wWhG wWel
	palmatum var. ***dissectum*** 'Baldsmith'	wFWB wClo oGre
	palmatum var. ***dissectum*** 'Beni-shidare'	see **A. *palmatum*** 'Beni-shidare'
un	***palmatum*** var. ***dissectum*** 'Corbin'	wWel
	palmatum var. ***dissectum*** 'Crimson Queen'	oFor wFWB oAls oAmb oGar oGre oTPm oWnC wWhG oTin wCoS wWel
	palmatum var. ***dissectum*** Dissectum Atropurpureum Group	wFWB oWnC wTGN wAva
	palmatum var. ***dissectum*** 'Dissectum Flavescens'	wFWB oGar oGre wWel
	palmatum var. ***dissectum*** 'Dissectum Nigrum'	wFWB oAls oAmb oGre oWnC wSta wWhG wCoS wWel
	palmatum var. ***dissectum*** 'Dissectum Palmatifidum'	oGre oWnC
	palmatum var. ***dissectum*** 'Dissectum Rubrifolium'	wFWB wBox
	palmatum var. ***dissectum*** 'Dissectum Variegatum'	last listed 99/00
	palmatum var. ***dissectum*** Dissectum Viride Group	oFor wFWB wClo oAls oAmb oGar oDan oSho oGre oWnC wSta oThr wWhG wCoS wWel
un	***palmatum*** var. ***dissectum*** 'Ever Dark'	oGar oGue
	palmatum var. ***dissectum*** 'Ever Red'	see **A. *palmatum*** var. ***dissectum*** 'Dissectum Nigrum'
un	***palmatum*** var. ***dissectum*** 'Filiferum Purpureum'	oGar
va	***palmatum*** var. ***dissectum*** 'Filigree'	see **A. *palmatum*** 'Filigree'
qu	***palmatum*** var. ***dissectum*** 'Filigree Lace'	oWnC
	palmatum var. ***dissectum*** 'Germaine's Gyration'	oWhi oGre oWnC
va	***palmatum*** var. ***dissectum*** 'Goshiki-shidare'	see **A. *palmatum*** 'Goshiki-shidare'
un	***palmatum*** var. ***dissectum*** 'Green Mist'	wWhG wWel
	palmatum var. ***dissectum*** 'Inaba-shidare'	oFor wFWB oAls oAmb oGar oRiv oSho oWhi oGre oTPm oWnC oGue wWhG oTin wWel
	palmatum var. ***dissectum*** 'Lionheart'	wWhG
	palmatum var. ***dissectum*** 'Orangeola'	oFor wFWB oRiv oWhi oGre oWnC wWhG wWel
	palmatum var. ***dissectum*** 'Ornatum'	oWnC
	palmatum var. ***dissectum*** 'Red Dragon'	see **A. *palmatum*** 'Red Dragon'
	palmatum var. ***dissectum*** 'Sunset'	wFWB
	palmatum var. ***dissectum*** 'Tamukeyama'	see **A. *palmatum*** 'Tamukeyama'
un	***palmatum*** var. ***dissectum*** 'Toyama-nishiki'	wFWB wCol oWnC wWhG oTin wWel
	palmatum var. ***dissectum*** 'Viridis'	see **A. *palmatum*** var. ***dissectum*** Dissectum Viride Group
	palmatum var. ***dissectum*** 'Waterfall'	see **A. *palmatum*** 'Waterfall'
un	***palmatum*** 'Ebony'	oWnC
	palmatum 'Elegans'	wFWB wClo oAmb wWel
un	***palmatum*** 'Elizabeth'	wWel
un	***palmatum*** 'Emerald Lace'	wFol wFWB
	palmatum Emperor I ™ / 'Wolff'	oRiv wWel
	palmatum 'Ever Autumn'	wWel
	palmatum 'Ever Red'	see **A. *palmatum*** var. ***dissectum*** 'Dissectum Nigrum'
un	***palmatum*** 'Fall's Fire'	wFWB
un	***palmatum*** 'Fascination'	oGre
	palmatum 'Filiferum Purpurea'	oWhi
va	***palmatum*** 'Filigree'	wFWB oAmb oGar oGre wWhG oTin wWel
	palmatum 'Fireglow'	wFWB wClo oAmb oDan oBlo oGre oWnC wSta oThr wWhG wWel
	palmatum 'Fjellheim'	wFWB
un	***palmatum*** 'Flavescens'	see **A. *palmatum*** var. ***dissectum*** 'Dissectum Flavescens'
	palmatum 'Flavescens Dissectum'	see **A. *palmatum*** var. ***dissectum*** 'Dissectum Flavescens'
un	***palmatum*** 'Forever Autumn'	wWhG
	palmatum 'Garnet'	wFWB oGos wClo oAmb oGar oRiv oGre oTPm oWnC wSta wWhG oTin wCoS wWel
	palmatum 'Garyu'	wFWB
	palmatum 'Goddard's Prostrate'	wFWB
va	***palmatum*** 'Goshiki-kotohime'	oFor wFWB wClo wCol oSis oRiv oGre oWnC wWhG oTin
va	***palmatum*** 'Goshiki-shidare'	oFor wFWB oRiv oGre
un	***palmatum*** 'Green Cascade'	oWnC
un	***palmatum*** 'Green Filigree'	wWel
un	***palmatum*** 'Green Hornet'	wFWB oGre
un	***palmatum*** 'Green Star'	wFWB
	palmatum 'Green Trompenburg'	last listed 99/00
qu	***palmatum*** 'Ground Cover'	oTin
	palmatum 'Hagoromo'	oTin
	palmatum 'Hanami-nishiki'	wFWB wCol oGre oTin wWel
un	***palmatum*** 'Hanazono-nishiki'	wFWB oTin
va	***palmatum*** 'Harusame'	wWhG wWel
va	***palmatum*** 'Hazeroino'	wFWB
	palmatum HC 970617	wHer
	palmatum 'Hessei'	wWel
va	***palmatum*** 'Higasayama'	oFor wFWB oAmb oGre oTin
	palmatum 'Hogyoku'	oFor wFWB oAmb oGar oRiv oGre wWhG wWel
	palmatum 'Hohman's Variegated'	wFol
un	***palmatum*** 'Hoshi-kuzu'	oTin
	palmatum 'Hupp's Dwarf'	wFol oGre
	palmatum 'Ibo-nishiki'	oRiv wWel
	palmatum 'Ichigyoji'	oFor wFWB oGre oWnC wWhG oTin wWel
	palmatum 'Iijima-sunago'	oAmb oGar oWhi oGre wWhG oTin wWel

	palmatum 'Inaba-shidare'	see **A. *palmatum*** var. ***dissectum*** 'Inaba-shidare'
	palmatum 'Inazuma'	oWhi oGre oWnC wWhG
un	***palmatum*** 'Issai-nishiki'	wFWB oRiv
un	***palmatum*** 'Itame-shidare'	wFWB oTin
	palmatum 'Itami-nishiki'	wFol
un	***palmatum*** 'Japanese Sunrise'	wFWB oGre wWhG
un	***palmatum*** 'Jerry Schwartz'	wFWB
	palmatum 'Jiro-shidare'	wFWB oGar oWhi oGre oWnC wWhG
un	***palmatum*** 'Johnnies Pink'	wWel
un	***palmatum*** 'Julian'	see **A. *palmatum*** 'Pendulum Julian'
va	***palmatum*** 'Kagero'	oFor oGre
va	***palmatum*** 'Kagiri-nishiki'	wFol wFWB oWhi oGre oWnC
	palmatum 'Kamagata'	oFor wFol wFWB wClo oGar oGre oWnC wWhG oTin wWel
va	***palmatum*** 'Karasugawa'	wFol wFWB wCol oRiv oWnC oTin
	palmatum 'Kasagiyama'	oFor wFol wFWB oAmb oSis oWhi wWhG oTin
	palmatum 'Kasen-nishiki'	wFWB
	palmatum 'Kashima'	oFor wFWB oRiv oWnC oTin
	palmatum 'Katsura'	oFor wFWB oAmb oGar oRiv oGre oWnC wWhG oTin
	palmatum 'Ki-hachijo'	wClo oWhi wWhG
un	***palmatum*** 'Killarney'	wFWB
	palmatum 'Kinran'	oWnC wWel
	palmatum 'Kinshi'	oGre oWnC
	palmatum 'Kiyohime'	oFor wFWB oRiv
un	***palmatum*** 'Kiyohime-yatsubusa'	oGar oGre oTin
	palmatum 'Kogane-nishiki'	wFWB oGre oTin
un	***palmatum*** 'Kogane-sakae'	wFWB
	palmatum 'Komache-hime'	wFWB
	palmatum 'Koreanum'	wFWB oGre
	palmatum 'Koshibori-nishiki'	wFWB oWnC wWhG oTin
	palmatum 'Koshimino'	oGre oTin
	palmatum 'Koto-ito-komachi'	oFor
	palmatum 'Koto-maru'	wFol wFWB
	palmatum 'Koto-no-ito'	wFol wFWB oAmb oRiv oWhi oGre wWhG oTin wWel
	palmatum 'Kotohime'	wFWB wClo
un	***palmatum*** 'Kotohime-yatsubusa'	oTin
un	***palmatum*** 'Kurabeyama'	wFWB wWhG wWel
	palmatum 'Kurui-jishi'	oFor wFWB oRiv oGre wWhG oTin
	palmatum 'Kurui-jishiki'	last listed 99/00
	palmatum 'Linearilobum'	wWhG
un	***palmatum*** 'Linearilobum Atrolineare'	wFol oGar
	palmatum 'Lozita'	wFWB wWhG
	palmatum 'Lutescens'	last listed 99/00
	palmatum 'Maiko'	last listed 99/00
	palmatum 'Mai-mori'	oGre
	palmatum 'Mama'	wFWB oGre oTin wWel
un	***palmatum*** 'Marakumo'	wFWB
	palmatum 'Margaret Bee'	wFWB
	palmatum 'Masamurasaki'	last listed 99/00
va	***palmatum*** 'Masukagami'	wFol wWel
	palmatum 'Matsu-ga-e'	wFol wFWB oRiv
	palmatum 'Matsukaze'	wFWB oTin
	palmatum ssp. ***matsumurae***	wFWB wWhG wWel
	palmatum 'Matsuyoi'	wFWB
	palmatum 'Mikawa-yatsubusa'	wFol wFWB oGos wClo oGre oWnC wWhG oTin wWel
	palmatum 'Mioun'	wFWB
	palmatum 'Mirte'	oGre
	palmatum 'Mizu-kuguri'	oTin
	palmatum 'Mizuho-beni'	wFWB
	palmatum 'Monzukushi'	wFWB
	palmatum 'Moonfire'	oFor wFWB oAmb oGar oThr wWel
un	***palmatum*** 'Morning Star'	wFWB
un	***palmatum*** 'Mossman's Clone'	oTin
un	***palmatum*** 'Mount Lehman'	wClo
un	***palmatum*** 'Murasaki-hime'	wFol oTin
	palmatum 'Murasaki-kiyohime'	oFor wFWB wClo oAmb oGar oRiv oWnC wWhG wWel
	palmatum 'Murashino'	wFWB
	palmatum 'Mure-hibari'	wWhG wWel
	palmatum 'Murogawa'	last listed 99/00
un	***palmatum*** 'Musa-murasaki'	oGre
un	***palmatum*** 'Musashino'	oGre
un	***palmatum*** 'Myoi'	oGar oWhi wWhG
	palmatum 'Nanase-gawa'	wFWB
	palmatum 'Nishiki-gawa'	wFWB oGre oTin
	palmatum 'Nishiki-momiji'	oAmb wWel
un	***palmatum*** 'Nishiki-sho'	wFWB oTin
	palmatum 'Nomura'	oTin
un	***palmatum*** 'Noshi-beni'	oThr
un	***palmatum*** 'Novum'	wFWB oGre
	palmatum 'Nuresagi'	oAmb oGar wWhG wWel
	palmatum 'O-jishi'	wFol wClo
	palmatum 'O-kagami'	oFor wFWB oAmb oGar oRiv oWhi wWhG
	palmatum 'O-nishiki'	wFWB oWnC wWhG oTin wWel
un	***palmatum*** 'Octopus'	wFWB
	palmatum 'Ogon-sarasa'	wFWB oWhi oGre wWhG
	palmatum 'Ogurayama'	see **A. *shirasawanum*** 'Ogurayama'
	palmatum 'Okukuji-nishiki'	wFol
	palmatum 'Okushimo'	oFor wFWB wClo oAls oAmb oRiv oWhi oGre oWnC oTin wWel
	palmatum 'Omato'	wFWB wWhG oTin wWel
	palmatum 'Omurayama'	wFol wFWB wClo oSis oRiv oGre wWhG
	palmatum 'Orange Dream'	wFWB wWel
	palmatum 'Orangeola'	see **A. *palmatum*** var. ***dissectum*** 'Orangeola'
un	***palmatum*** 'Oregon Ever Red'	wFWB
	palmatum 'Oregon Sunset'	wFWB oGar oWhi oTin
va	***palmatum*** 'Orido-nishiki'	oFor wHer wFWB wClo oAls oAmb oGar oRiv oWhi oGre oWnC oThr wWhG oTin wWel
	palmatum 'Ornatum'	wFWB
	palmatum 'Osakazuki'	oFor wFol wFWB oAmb oGar oRiv oWhi oGre oWnC oTin wWel
	palmatum 'Oshio beni'	wWhG
	palmatum 'Oshu-beni'	wFWB oAls oAmb oGar oWnC
	palmatum 'Oshu-shidare'	wFWB wWhG
	palmatum 'Oto-hime'	wFol wFWB oTin
	palmatum 'Otome-zakura'	wFWB oAmb oGar oWhi oTin
va	***palmatum*** 'Peaches and Cream'	wFol wFWB wCol wWhG wWel
un	***palmatum*** 'Pendulum Angustilobum'	wFWB
	palmatum 'Pendulum Julian'	wFWB oGre oTin wWel
	palmatum 'Pine Bark'	see **A. *palmatum*** 'Nishiki-gawa'
	palmatum 'Pixie'	wFWB oWhi oGre wWhG oTin wWel
un	***palmatum*** 'Raraflora'	oGre
un	***palmatum*** 'Red Bamboo'	wFWB
	palmatum 'Red Dragon'	oFor wFol wFWB wClo oAls oAmb oGar oRiv oSho oGre oTPm oWnC oThr wWhG
un	***palmatum*** 'Red Emperor'	oFor wFol oRiv oGre
un	***palmatum*** 'Red Feather'	wFWB
	palmatum 'Red Filigree Lace'	wFol wFWB oGos oGar oGre wWhG oTin wWel
	palmatum red laceleaf contorted	oAls
	palmatum 'Red Pygmy'	oFor wFol wFWB wClo oAmb oGar oWhi oGre oWnC wWhG oTin wWel
un	***palmatum*** 'Red Ribbonleaf'	wSta
	palmatum 'Red Select'	oGre oThr oTin
	palmatum 'Red Spider'	oGre
un	***palmatum*** 'Red Wood'	wFWB wClo oRiv wWhG
qu	***palmatum*** 'Rosa Variegatum'	wFWB
	palmatum 'Roseomarginatum'	see **A. *palmatum*** 'Kagiri-nishiki'
	palmatum 'Rough Bark'	see **A. *palmatum*** 'Arakawa'
qu	***palmatum*** 'Rubrifolium'	wFWB oWnC
	palmatum 'Ryuzu'	wFWB oTin
	palmatum 'Sa-otome'	wFWB
va	***palmatum*** 'Sagara-nishiki'	oFor wFWB oGre wWhG oTin
	palmatum 'Samidare'	last listed 99/00
	palmatum 'Sango-kaku'	oFor wFWB oDar oAls oAmb oGar oRiv oSho oBlo oGre oWnC wTGN wSta oThr wWhG oTin wCoS wWel
	palmatum 'Saoshika'	wFWB
	palmatum 'Sazanami'	wFWB wClo oWnC oTin wWel
	palmatum 'Scolopendriifolium'	wFWB oAmb wWhG wWel
un	***palmatum*** 'Scolopendriifolium Rubrum'	oFor wFWB oGre oWnC
	palmatum 'Seigen'	wFWB
un	***palmatum*** 'Seigen Rubrum'	wFWB
	palmatum 'Seiryu'	oFor wHer oAls oAmb oGar oDan oRiv oBlo oGre oWnC wSta oThr wWhG oTin wCoS wWel
	palmatum 'Sekimori'	oFor wFWB oGre oWnC wWel
	palmatum 'Sekka-yatsubusa'	wClo oGre oTin wWel
	palmatum 'Sessilifolium' (see Nomenclature Notes)	wFWB
	palmatum 'Shaina'	oFor wFol wFWB oGos wClo wCol oAmb oGar oSis oRiv oGre oTPm oWnC wWhG wWel
	palmatum 'Sharp's Pygmy'	oFor wFol wWhG oTin
	palmatum 'Sherwood Flame'	oFor wFWB oAmb oGar oGre wWhG oTin wWel
	palmatum 'Shigarami'	wFWB wWhG wWel
va	***palmatum*** 'Shigitatsu-sawa'	wFWB wWhG oTin wWel
	palmatum 'Shigure-bato'	wFWB oAmb wWhG oTin
	palmatum 'Shikageori-nishiki'	wFWB
	palmatum 'Shime-no-uchi'	last listed 99/00
	palmatum 'Shindeshojo'	wHer wFWB oAls oAmb oRiv oGre oWnC wBox oTin wWel
	palmatum 'Shinobugaoka'	wFWB oWhi oGre wWhG oTin

	palmatum 'Shinonome'	oTin
un	***palmatum*** 'Shiranami'	wFWB
un	***palmatum*** 'Shishi-yatsubusa'	wFWB
	palmatum 'Shishigashira'	oFor wFWB wClo oAmb oRiv oWhi oBlo oGre oWnC wTGN wSta wWhG oTin wWel
	palmatum 'Shishio'	wFWB oGre wSta
	palmatum 'Shishio Improved'	oFor wFol oAls oAmb oGre oTin wWel
un	***palmatum*** 'Shishiohime'	wFol wFWB wCol
	palmatum 'Shojo'	oGre
	palmatum 'Shojo-nomura'	oFor wFWB oGar oWhi oGre wWhG wWel
	palmatum 'Shojo-shidare'	oFor wFWB
	palmatum 'Skeeters'	wFWB oGre oTin
un	***palmatum*** 'Spring Delight'	oClo oWnC wWhG wWel
	palmatum 'Suminagashi'	oFor wFWB oAmb oGar oWhi oGre oWnC wWhG oTin wWel
un	***palmatum*** 'Superbum'	oGre oWnC
	palmatum 'Taimin-nishiki'	wFWB oGre
	palmatum 'Takinogawa'	last listed 99/00
un	***palmatum*** 'Tak's Fine Leaf'	oTin
	palmatum 'Tamahime'	oFor wFol wFWB wCol wWhG oTin
	palmatum 'Tamukeyama'	oFor wFWB oAls oGar oSis oSho oGre oWnC oThr wWhG wWel
	palmatum 'Tana'	wFWB
	palmatum 'Tennyo-no-hoshi'	wFWB oAmb oGre oTin
	palmatum 'The Bishop'	wFWB
	palmatum threadleaf	oWnC
un	***palmatum*** 'Tim's Dwarf'	oTin
	palmatum tiny leaf	wFWB oGre oTin
	palmatum 'Tiny Tim'	wFWB oRiv
	palmatum 'Trompenburg'	oFor wFWB oAls oAmb oGar oWhi oGre wWhG wWel
	palmatum 'Tsuchigumo'	wFWB oAmb oGre oTin
	palmatum 'Tsukomo'	last listed 99/00
	palmatum 'Tsukubane'	wFWB
	palmatum 'Tsukushigata'	oFor wFol wFWB wWhG
	palmatum 'Tsuma-beni'	wWhG oTin
	palmatum 'Tsuma-gaki'	wFWB oAmb oWnC wWhG oTin
	palmatum 'Tsuri-nishiki'	wFWB wWel
	palmatum 'Ueno-yama'	wFWB oGre oTin
va	***palmatum*** 'Ukigumo'	wFWB oGos wClo wCol oAmb oGar oGre oWnC wBox wWhG oTin wWel
	palmatum 'Ukon'	wFWB wClo oRiv wWhG wWel
	palmatum 'Umegae'	wFWB
	palmatum 'Utsu-semi'	wFWB wWhG
un	***palmatum*** 'V. Corbin'	wFWB wWhG
un	***palmatum*** 'Variegatum'	wFWB
va	***palmatum*** 'Versicolor'	wFWB wWhG oTin
	palmatum 'Villa Taranto'	wFol wFWB wClo oAmb oGre wWhG wWel
un	***palmatum*** 'Vitifolium'	wFWB
	palmatum 'Volubile'	oGre oTin wWel
	palmatum 'Wabito'	wFol wFWB oAmb
	palmatum 'Washi-no-o'	last listed 99/00
	palmatum 'Waterfall'	wFWB oGar oRiv oGre oWnC oTin wWel
	palmatum 'Watnong'	wFWB
un	***palmatum*** 'Whitney'	wWhG
un	***palmatum*** 'Willow Leaf'	wFWB
	palmatum 'Wou-nishiki'	see **A. *palmatum*** 'O-nishiki'
	palmatum 'Yasemin'	wFWB
un	***palmatum*** 'Yashio'	wFWB
un	***palmatum*** 'Yellow Bird'	wCol
	palmatum 'Yezo-nishiki'	oFor wFWB oGar oWhi
	palmatum 'Yuba-e'	wFWB
	palmatum 'Yugure'	oGre
	palmatum 'Yurihime'	wFWB oTin
	paxii	oRiv oCis wCCr
	pectinatum	oFor oRiv
	pectinatum ssp. ***forrestii***	see **A. *forrestii***
	pectinatum ssp. ***laxiflorum***	see **A. *laxiflorum***
	pensylvanicum	oFor wHer wFWB oRiv oGre wAva wRav
	pensylvanicum 'Erythrocladum'	oFor oRiv
	pentaphyllum	oGos oRiv oCis wWhG
	pictum	see **A. *mono***
	pilosum var. ***stenolobum***	oFor oEdg oRiv
	platanoides	wFWB oBlo wKin oWnC wAva oSle
	platanoides 'Columnare'	wKin oWnC
	platanoides 'Crimson King'	oFor oDar oAls oGar oSho oBlo oGre wKin oWnC wTGN wSta oSle wCoS
	platanoides 'Crimson Sentry'	oFor oGar oSho oBlo oWnC wCoS
	platanoides 'Deborah'	oFor oWnC
va	***platanoides*** 'Drummondii'	oFor oGar oBlo oGre wTGN wCoS
	platanoides 'Emerald Lustre'	oGar
	platanoides 'Emerald Queen'	oAls oBlo wKin oWnC wCoS
	platanoides 'Globosum'	oGar wKin wCoS
	platanoides 'Olmsted'	oGre
	platanoides 'Oregon Pride'	wKin
	platanoides Princeton Gold / 'Prigo'	oFor oAls oGar
	platanoides 'Royal Red'	oFor oBlo oWnC wTGN wCoS
	platanoides 'Silver Variegated'	wWel
	platanoides 'Variegatum'	wKin
	pseudoplatanus	wFWB wAva oSle
	pseudoplatanus 'Atropurpureum'	oFor oGar oSho
	pseudoplatanus 'Brilliantissimum'	wFWB oGos oGre wWel
va	***pseudoplatanus*** 'Leopoldii'	oFor oGre
un	***pseudoplatanus*** 'Puget Pink'	wHer
	pseudoplatanus 'Spaethii'	see **A. *pseudoplatanus*** 'Atropurpureum'
	pseudoplatanus f. ***variegatum***	oFor
	pseudosieboldianum	oFor wHer wFWB wBCr wClo oGar oRiv oGre wAva wRav wWhG
un	***pseudosieboldianum*** var. ***tatsiense*** HC 970209	wHer
	pulverulentum	see **A. *campestre*** 'Pulverulentum'
	pycnanthum	last listed 99/00
	robustum	oFor
	rubrum	wFWB wClo wShR oRiv wKin wAva
	rubrum 'Armstrong'	oFor oGar oDan oRiv oWnC wCoS
	rubrum 'Autumn Flame'	oFor oAls oGar oDan oRiv oBlo oWnC wCoS
	rubrum 'Autumn Radiance'	oGar
	rubrum 'Bowhall'	oGar oWnC wWel
	rubrum 'Embers'	wCoS
un	***rubrum*** 'Fairview Flame'	oAls
	rubrum 'Indian Summer'	see **A. *rubrum*** 'Morgan'
un	***rubrum*** 'Landsburg'	wKin
	rubrum 'Morgan'	oGar
	rubrum October Glory®	oAls oGar oRiv oSho oWnC wCoS
un	***rubrum*** Pacific Sunset™	oAls
	rubrum Red Sunset™/ 'Franksred'	oFor oAls oGar oSho oBlo oGre wKin oWnC wSta wCoS wWel
	rubrum 'Scarlet Sentinal'	oGar
	rufinerve	oFor wFWB oGar oRiv
	rufinerve 'Albolimbatum'	see **A. *rufinerve*** 'Hatsuyuki'
va	***rufinerve*** 'Hatsuyuki'	wHer wFWB
un	***rufinerve*** 'Winter Gold'	wFWB
	saccharinum	wFWB oRiv wKin oWnC wCoN wCoS
	saccharinum 'Silver Queen'	oGar
	saccharum	oFor wFWB wBCr wBur oEdg wShR wKin oWnC wTGN wAva wCoN oSle wWel
	saccharum 'Bonfire'	last listed 99/00
	saccharum 'Commemoration'	last listed 99/00
	saccharum ssp. ***grandidentatum***	oFor wHer wFWB oEdg oRiv oWhi wCCr
	saccharum 'Green Mountain'	oFor oGar oBlo wCoS
	saccharum 'Laciniatum'	oFor
	saccharum 'Legacy'	oGar wRai wCoS
	saccharum ssp. ***leucoderme***	oGre
	saccharum ssp. ***nigrum***	oFor
	saccharum ssp. ***nigrum*** 'Greencolumn'	oFor
	saccharum 'Sweet Shadow'	oFor oGar oDan oWhi
	'Scanlon'	oGar
	semenovii	see **A. *tataricum*** ssp. ***semenowii***
	sempervirens	oSis oRiv wCCr
	shirasawanum	oFor wCol oGar oDan oWhi oTPm
	shirasawanum 'Aureum'	oFor oGos wClo oGar oSis oGre oWnC wCoN wWhG oTin wWel
	shirasawanum 'Autumn Moon'	wFWB oGos oGre oWnC wWel
	shirasawanum 'Microphyllum'	wFWB
	shirasawanum 'Ogurayama'	wFWB oGos oWnC wWel
	shirasawanum 'Palmatifolium'	wFWB
	sieboldianum	oFor wFWB oGar wRav
	sieboldianum 'Sode-no-uchi'	oFor oGre wWhG oTin wWel
	sikkimense ssp. ***metcalfii***	see **A. *metcalfii***
	sinense	oFor wHer wCCr
	spicatum	oFor
	stachyophyllum	see **A. *tetramerum***
	sterculiaceum	oFor
	sterculiaceum ssp. ***franchetii***	oFor wCCr
	sterculiaceum ssp. ***franchetii*** EDHCH 97161	wHer
un	***takesimense***	oFor
	taronense	wFWB
	tataricum	oFor wFWB oRiv
	tataricum ssp. ***ginnala***	oFor wFWB oRiv wKin wCoN oSle
un	***tataricum*** ssp. ***ginnala*** 'Compacta'	oFor oRiv
un	***tataricum*** ssp. ***ginnala*** 'Emerald Elf'	oFor
	tataricum ssp. ***ginnala*** 'Flame'	wFWB wBCr oGar oGre wKin oWnC
	tataricum ssp. ***ginnala*** Red Rhapsody®/ 'Mondy'	oSho
	tataricum ssp. ***semenowii***	oFor

	Name	Sources
	tegmentosum	oFor wFWB oEdg oRiv oGre wAva wRav
	tegmentosum HC 970081	wHer
un	tegmentosum 'White Tigress'	oGos
	tenuifolium	wFWB
	tetramerum DJHC 297	wHer
	tonkinense	wHer
	trautvetteri	oFor
	triflorum	oFor wHer wFWB oGos oAmb oDan oCis oGre wRav
	truncatum	oFor wFWB wBCr oGar oRiv oGre wCCr wRav wWel
	truncatum 'Akaji-nishiki'	wFWB oAmb oGre wWhG oTin wWel
	tschonoskii	oFor wFWB oRiv
	tschonoskii ssp. koreanum	oFor wHer oEdg oGre
	villosum	see A. sterculiaceum
	wilsonii	last listed 99/00
	ACERIPHYLLUM	see **MUKDENIA**
	ACHILLEA	
	ageratifolia	oSis wFai
	ageratum	oGoo
	ageratum 'W. B. Childs'	oHed
	Anthea / 'Anblo'	oFor
	'Apfelbluete'	oFor oJoy oWnC oSec
	Appleblossom	see A. 'Apfelbluete'
	clavennae	oAls wFai wSnq
	'Coronation Gold'	oFor oSis wFai
	'Credo'	oSho oGre
	Debutante hybrids	wTho
	decolorans	see A. ageratum
	'Fanal'	oGoo
	'Feuerland'	oFor wWoS oAmb oGar oGoo wFai oWnC wEde
	filipendulina	oGoo oSha iGSc
	filipendulina 'Gold Plate'	oGar oUps
	Fireland	see A. 'Feuerland'
	'Great Expectations'	see A. 'Hoffnung'
	'Hoffnung'	oFor oGoo oSec
	'Huteri'	oHed
	x jaborneggii	wFGN
	x kellereri	oHed oSis
qu	'Kelly'	oSho
	'Lachsschoenheit'	oFor wCul oJoy oAls oGar oGoo oWnC
	x lewisii 'King Edward'	oSis
	'Martina'	oFor
	'Maynard's Gold'	see A. tomentosa 'Aurea'
	millefolium	wNot oAls wShR wCSG oBos iGSc oCrm oBar wPla oAld wAva wFFl wWld oSle
	millefolium apricot	oSec
un	millefolium 'Borealis'	wWoS oAls
	millefolium 'Cerise Queen'	wCul wTho oGar iGSc wMag iArc oUps
un	millefolium 'Cherry Queen'	oFor
un	millefolium 'Christel'	oFor
	millefolium 'Colorado'	oGar oSev oGre oUps
	millefolium 'Fire King'	wHer
un	millefolium 'Heidi'	oFor wCul oGar wFai oEga
un	millefolium 'Kirschkoenigin'	oSis
	millefolium 'Lavender Beauty'	see A. millefolium 'Lilac Beauty'
un	millefolium 'Lavender Deb'	wCul
un	millefolium 'Lavender Lady'	oGar
	millefolium 'Lilac Beauty'	oSec
un	millefolium 'Oertal's Rose'	oWnC
un	millefolium 'Orange Queen'	oFor wSnq
	millefolium 'Paprika'	oFor wCul oAls oSha oGre oWnC wMag oCir wSnq
un	millefolium 'Proa'	oGoo
	millefolium 'Red Beauty'	oFor wCul oJoy oAls oSho oEga
	millefolium f. rosea	wFGN oGoo
un	millefolium 'Sawa Sawa'	oFor wWoS
un	millefolium 'Snow Taller'	wCul
un	millefolium 'Topas'	oUps
	millefolium 'White Beauty'	oFor wMag
un	'Moonbeam'	wAva
	'Moonlight'	oNWe
	'Moonshine'	oFor wCul oAls wFGN oGar oGoo oSho wFai oWnC oSec oSle
	'Moonwalker'	oFor oSev oWnC
	ptarmica	wShR oGoo oSec
	ptarmica 'Angel's Breath'	oWnC
	ptarmica 'Ballerina'	oFor oEga
	ptarmica 'Major'	oCir
qu	ptarmica 'Nana Ballerina'	oFor
do	ptarmica The Pearl Group	oFor wCul wFGN wCSG oSha wFai oWnC wTGN oCir
do	ptarmica The Pearl Group 'Boule de Neige' (clonal)	wHer

	Name	Sources
un	'Rodney's Choice'	oGoo
	Salmon Beauty	see A. 'Lachsschoenheit'
	'Schwellenburg'	oFor wFai wSnq
un	sibirica var. camtschatica 'Love Parade'	oSev
un	sibirica 'Kamtschaticum'	oFor oUps
	Summer Pastels Group	oFor wTho oAls oSis wFai iGSc wHom oWnC wMag oCir wAva oUps
	Summer Pastels Group 'Summer Pink'	oSis
	Summer Pastels Group 'Summer Red'	oSis
	'Summerwine'	wWoS
	'Taygetea'	wCul oSec
	'Terracotta'	oFor wWoS oDar oJoy oGar oGoo oNWe oSho oGre wFai wAva wSnq
	tomentosa	oJoy oGoo oBlo wFai iGSc
	tomentosa 'Aurea'	oFor oSis oUps
	umbellata	oGoo
	'Walther Funcke'	w WoS wSnq
	'Wesersandstein'	oFor wCul oAls oSho oWnC wMag
	Weser River Sandstone	see A. 'Wesersandstein'
	'Wilczekii'	oSis
	ACHLYS	
	triphylla	wHer wThG oNat oRus oBos oAld wFFl
	ACHNATHERUM	
	brachytricha	see **CALAMAGROSTIS** brachytricha
	ACHYRANTHES	
	bidentata	last listed 99/00
	ACIDANTHERA	see **GLADIOLUS**
	ACINOS	
	alpinus	oGoo iGSc oBar
	thymoides	last listed 99/00
	ACONITUM	
	alboviolaceum	wHer
qu	'Album'	wFai
	bartlettii	last listed 99/00
	'Bressingham Spire'	wCul
	x cammarum 'Bicolor'	wCol oGar wNay wFai oWnC wMag wAva oUps
	carmichaelii	oAls wCSG oGar oNWe oSho oGre wFai wBox wSte
	carmichaelii 'Arendsii'	oFor wCul oNat wHig oHed wMag oUps
	carmichaelii 'Barker's Variety'	see A. carmichaelii Wilsonii Group 'Barker's Variety'
	carmichaelii Wilsonii Group	oJil
	carmichaelii Wilsonii Group 'Barker's Variety'	oHed
	carmichaelii Wilsonii Group 'Spaetlese'	wTGN
un	chiisanensis HC 970400	wHer
	episcopale	wHer
	fischeri	see A. carmichaelii
	hemsleyanum	wHer
	henryi	wHig
	'Ivorine'	oFor wHer wCol wHig oGar wNay wRob oGre wMag
	japonicum	last listed 99/00
	lamarckii	see A. lycoctonum ssp. neapolitanum
un	'Late Crop'	oJoy
un	sp. aff. longicassidatum HC 970142	wHer
	lycoctonum	wTGN
	lycoctonum ssp. lycoctonum	oJoy
	lycoctonum ssp. neapolitanum	oJil wCol wCSG wNay oGre
	lycoctonum ssp. vulparia	wCul oNWe
	napellus	wCul wCSG oGar wTGN wAva oEga
	napellus albus	see A. napellus ssp. vulgare 'Albidum'
	napellus carneum	see A. napellus ssp. vulgare 'Carneum'
	napellus 'Rubellum'	oGar wFai
	napellus ssp. vulgare 'Albidum'	oJoy wNay wMag wTGN
	napellus ssp. vulgare 'Carneum'	wHig wMag
	'Newry Blue'	oJoy oRus oGar wFai
	paniculatum 'Roseum'	oAls wCSG oSho
	sp.	oRus
	sp. DJHC 98351	wHer
	sp. DJHC 98410	wHer
	sp. DR9625	wCol
	'Spark's Variety'	oFor wCul wCol oGar oNWe oSho wMag
	'Stainless Steel'	oHed wNay wRob
un	sp. aff. staintonii HWJCM 226	wHer
	'Tall Blue'	oJoy
	variegatum	last listed 99/00
	volubile	see A. hemsleyanum
	vulparia	see A. lycoctonum ssp. vulparia
un	zabellum	wHer
	ACORUS	
	calamus	oFor oTrP wSoo oGoo oSsd wFai iGSc oWnC wBox oUps

	calamus 'Variegatus'	wHer wCli wCul wSoo wWal wGAc oHed wCSG oHug oGar oGoo oSsd oSho oOut oWnC oTri oUps
	gramineus	oTrP oAls oHed wShR oHug oGar oDan wFai wCoN
	gramineus 'Licorice'	oTrP wWoS oDar oAls oGoo wFai oCrm
va	***gramineus*** 'Masamune'	oBRG
un	***gramineus*** 'Minimus Aureus'	see **A. *gramineus*** 'Pusillus Aureus'
va	***gramineus*** 'Oborozuki'	wWoS wWin
va	***gramineus*** 'Ogon'	oFor wHer wCli wCul oJoy oAls oAmb oHug oGar oSsd oSis oGre wBox oSec oUps wSte
	gramineus 'Pusillus'	oFor oTrP wCli oJoy oSis wSta
un	***gramineus*** 'Pusillus Aureus'	oFor wCli oJoy oSis oNWe wWin wBox
un	***gramineus*** 'Tanimanoyuki'	oFor wHer
	gramineus 'Variegatus'	oFor wHer wCri oJoy oHug oGar oGoo oSsd oGre wWin wCCr wHom wTGN wBox oBRG wEde oSle
va	***gramineus*** 'Yodo-no-yuki	wHer
	ACROCARPUS	
	fraxinifolius	oTrP
	ACTEA	
	asiatica HC 970495	wHer
	pachypoda	wHer oRus oNWe oGre
	rubra	oHan oJoy oRus oBos wFFl wSte
	rubra Russell Graham form	oRus
	spicata	wHer
	ACTINIDIA	
	arguta 'Ananasnaja' female	wBCr wBur wClo oGar oWhi wRai oOEx wPug
	arguta 'Anna'	see **A. *arguta*** 'Ananasnaja'
un	***arguta*** 'C-1' sex unknown	wPug
un	***arguta*** 'C-2' male	wPug
un	***arguta*** 'C-3' sex unknown	wPug
un	***arguta*** 'C-4' sex unknown	wPug
un	***arguta*** 'C-5' male	wPug
un	***arguta*** 'Chico' female	wPug
un	***arguta*** 'Cordifolia' female	wPug
un	***arguta*** 'Cordifolia 1563-51'	wPug
un	***arguta*** 'Cornell' male	wPug
un	***arguta*** x ***deliciosa*** 'H-1' female?	wPug
un	***arguta*** x ***deliciosa*** 'H-2' sex unknown	wPug
un	***arguta*** x ***deliciosa*** 'H-3' male	wPug
un	***arguta*** x ***deliciosa*** 'H-4' sex unknown	wPug
un	***arguta*** x ***deliciosa*** 'H-5' sex unknown	wPug
un	***arguta*** x ***deliciosa*** 'H-6' female?	wPug
un	***arguta*** x ***deliciosa*** 'H-7' male	wPug
un	***arguta*** x ***deliciosa*** 'H-8' sex unknown	wPug
un	***arguta*** x ***deliciosa*** 'H-9' sex unknown	wPug
un	***arguta*** x ***deliciosa*** 'H-10' sex unknown	wPug
un	***arguta*** x ***deliciosa*** 'H-12' sex unknown	wPug
un	***arguta*** DJH 317	wHer
	arguta female	wBCr oAls oGar oGre
un	***arguta*** 'Geneva'	oGre
un	***arguta*** 'Geneva 2' female	wPug
un	***arguta*** 'Hardy Red'	oGre
	arguta 'Issai' self pollinating	wBur wClo oAls oGar oWhi oOEx oGre wTGN wPug
un	***arguta*** 'Jumbo'	wClo wRai oOEx oGre
un	***arguta*** 'Ken's Red' female	wBur wRai oOEx wPug
	arguta male	wBCr wBur wClo oAls oGar oWhi wRai oOEx oGre oWnC wPug
un	***arguta purpurea*** female	wPug
un	***arguta purpurea*** male	wPug
un	***arguta*** '74-49'	oGre oWnC
un	***arguta*** '119-40' self pollinating	wPug
	callosa male	wPug
	chinensis	see **A. *deliciosa***
un	***chrysantha*** J611 male	wPug
un	***chrysantha*** J612 female	wPug
	coriacea	wHer
un	***deliciosa*** 'Abbott' female	wPug
un	***deliciosa*** 'Allison'	see **A. *deliciosa*** 'Abbott'
	deliciosa 'Blake' self pollinating	wPug
un	***deliciosa*** 'California' male	wClo wPug
un	***deliciosa*** 'Canton' female, red	wPug
un	***deliciosa*** 'Elmwood' female	oGre wPug
un	***deliciosa*** 'First Emperor' female, yellow	wPug
un	***deliciosa*** 'Fuzzy' male	wRai
un	***deliciosa*** 'Gracie' female	wPug
	deliciosa 'Hayward' female	wBur oAls oGar oWnC
qu	***deliciosa*** 'Hayward' male	oAls oWnC
	deliciosa male	oGar oGre
un	***deliciosa*** 'Mandarin' female, orange	wPug
un	***deliciosa*** 'Matsu' male	wPug

	deliciosa 'Matua' male	wBur wPug
un	***deliciosa*** 'Monty' female	wPug
un	***deliciosa*** 'Saanich 12' female	wPug
	deliciosa 'Saanichton' female	wClo oWhi wRai oOEx wTGN
un	***deliciosa*** 'Saanichton 12'	wBur
	deliciosa 'Tomuri' male	wPug wCoS
un	***deliciosa*** UNC female, yellow	wPug
un	***deliciosa*** UNC male	wPug
	deliciosa variegated	wPug
un	***deliciosa*** 'Vincent'	wCoS
un	***eriantha*** UBC-1 female	wPug
un	***eriantha*** L2F2 male	wPug
un	***eriantha*** L2C1 female	wPug
un	***hemsleyana*** female	wPug
un	***hemsleyana*** 'UBC-1' female	wPug
	kolomikta	oFor wHer wCul oGos wBur wClo oHed wCSG oDan oOEx oUps
	kolomikta 'Arctic Beauty'	see **A. *kolomikta***
un	***kolomikta*** 'Aromatnaya' female	wPug
un	***kolomikta*** 'Broadmoor' male	wPug
	kolomikta female	oGar oGre
un	***kolomikta*** 'Gil' male	wPug
	sp. aff. ***kolomikta*** HC 970144	wHer
	kolomikta 'Krupnopladnaya' female	wPug
	kolomikta male	oFor wClo oGar wRai oGre oWnC
un	***kolomikta*** 'Northwoods' male	wPug
un	***kolomikta*** 'Sentabraskaya' female	wPug
un	***kolomikta*** September Sun tm	wRai oWnC
un	***kolomikta*** '142-38' female	wPug
un	***latifolia*** M5A3 male	wPug
un	***latifolia*** M6A0 female	wPug
un	***macrosperma***	oGre
un	***macrosperma*** Arnold Arboretum, male	wPug
	melanandra male	wPug
	melanandra sex unknown	wPug
un	***melanandra*** '1064-79' female	wPug
	pilosula	wHer
	polygama	oGre
	polygama female	wHer
un	***polygama*** HC 970436	wHer
	polygama male	wHer
un	***polygama*** 'UW-1' male	wPug
un	***polygama*** 'UW-2' female	wPug
	purpurea	wClo
un	***purpurea*** 'Hardy Red'	oWnC
un	***purpurea*** 'Ken's Red'	see **A. *arguta*** 'Ken's Red'
	purpurea male	oWnC
	rubricaulis var. ***coriacea***	see **A. *coriacea***
	rufa B&SWJ 3525	wHer
	sp. DJHC 505	wHer
	sp. EDHCH 97198	wHer
	sp. EDHCH 97258	wHer
	ACTINOMERIS	
	squarrosa	see **VERBESINA *alternifolia***
	ACTINOSTROBUS	
	pyramdalis	oFor
	ADANSONIA	
	digitata	oRar
un	***fony***	oRar
un	***grandidieri***	oRar
un	***za***	oRar
	ADENIA	
	digitata	oRar
un	***glauca***	oRar
un	***keramanthus***	oRar
un	***perriei***	oRar
	spinosa	oRar
un	***venenata***	oRar
	ADENIUM	
	obesum	oRar
	obesum ssp. ***oleifolium***	oRar
	obesum ssp. ***somalense***	oRar
	ADENOCARPUS	
	decorticans	oFor wCCr
	ADENOPHORA	
un	'Amethyst Chimes'	oJoy
	bulleyana	last listed 99/00
	confusa	oBlo wHom
un	***kurilensis***	wCol
	lamarckii HC 9770317	wHer
	liliifolia	oFor wTho wCSG
	polyantha	last listed 99/00
	stricta	oJoy
	stricta ssp. ***sessilifolia***	wHer
	uehatae	last listed 99/00

un	**ADESMIA**	
	sp. HCM 98137-B	wHer
	ADIANTUM	
	aleuticum	wFol wNot wFan oRus wShR oSis oGre oBRG wCoN wFFl wSte wSnq
	aleuticum dwarf ecotype	see **A. *pedatum*** var. ***subpumilum*** f. ***minimum***
	capillus-veneris	oFor oRus wTGN
	capillus-veneris 'Fimbriatum'	oSis
	capillus-veneris hardy Michigan form	oSis
	'Fragrans'	see **A. *raddianum*** 'Fragrantissimum'
	hispidulum	oFor wTGN
	pedatum	oFor wThG wWoB oAls oRus oGar oDan oBos oNWe wNay wFai wPir wHom oTri wTGN oAld oCir
	pedatum ssp. ***aleuticum***	see **A. *aleuticum***
	pedatum var. ***subpumilum*** f. ***minimum***	wFan oRus oGar
	raddianum 'Fragrantissimum'	oUps
	raddianum 'Gracillimum'	oUps
	raddianum 'Pacific Maid'	oUps
	venustum	wThG oRus oNWe wRob oBov oGre wPir
	ADINA	
	rubella	oFor oGre
	ADLUMIA	
	fungosa	oNWe
	ADONIS	
	aestivalis	iGSc
	amurensis	wNay
	vernalis	last listed 99/00
	ADROMISCHUS	
	bolusii	oSqu
	cristatus	oSqu
	maculatus	last listed 99/00
	rotundifolius	oSqu
un	***sinus*** 'Alexandria'	oSqu
	tricolor	oSqu
	AECHMEA	
	fasciata	oGar
	AEGOPODIUM	
	podagraria	oSho iArc oCir
	podagraria 'Variegatum'	oFor wTho oAls wCSG oGoo oGre wHom oCrm oBar wTGN oSec wAva oSle
	AEONIUM	
	arboreum 'Atropurpureum'	oSqu
	ciliatum	oSqu
	haworthii	oSqu
	haworthii 'Variegatum'	last listed 99/00
un	'Silver Edge'	oSqu
	tabuliforme	last listed 99/00
	'Zwartkop'	last listed 99/00
	AESCHYNANTHUS	
	'Mona'	oGar
	pulcher	oGar
	radicans	oGar
	AESCULUS	
	californica	oFor oEdg wShR wCCr
	californica Oregon collection	oCis
	x carnea	oEdg oGre wTGN oSle wCoS
	x carnea 'Briotii'	oFor oAls oGar oSho oWhi oGre oWnC wWel
un	***x carnea*** 'Fort McNair'	oFor oGar oWhi oGre
	x carnea 'O'Neill Red'	oFor oGar oGre oWnC
	flava	oFor oGar
	georgiana	see **A. *sylvatica***
	glabra	oFor wCCr
	glabra 'October Red'	oFor oGar
	hippocastanum	wShR oGar oGre oSle
do	***hippocastanum*** 'Baumannii'	oFor
	indica	wCCr
	octandra	see **A. *flava***
	parviflora	oFor oWhi wWel
un	***parviflora*** 'Rogers'	oFor
	parviflora f. ***serotina***	oFor oGre
	pavia	oFor oEdg oGar oGre wCCr wCoS wWel
	sylvatica	oFor
	turbinata	oFor
	x woerlitzensis	oFor
	AETHIONEMA	
	caespitosum	oSis
	coridifolium	oGar oSqu
	grandiflorum	oSis
	iberideum	oSis
	oppositifolium	oSis
	'Warley Rose'	oFor oSis
	AGAPANTHUS	
	africanus	wCSG oGar oSho wCCr oUps wCoS wWel
	africanus 'Albus'	oGar oWnC
qu	***albus***	oMis oSho wWel
qu	'Bell'	oAls
	'Blue Baby'	last listed 99/00
	'Blue Globe'	wMag
	'Blue Triumphator'	oEga
	'Bressingham Blue'	wSwC
	'Bressingham White'	oGre
	campanulatus var. ***albidus***	wHer
	campanulatus 'Cobalt Blue'	oGos oGar oGre wFai
	'Donau'	oWnC
qu	'Gayle's Lavender'	wMag
	'Gayle's Lilac'	oJoy oGre
un	'Getty White'	wMag
	Headbourne hybrids	wHer oNat oRus oMis oGar oGre wCCr wSta wCoN oUps wSte wWel
	Headbourne hybrids, blue	oDar oRus
	Headbourne hybrids, white	oDar oRus
	inapertus ssp ***hollandii***	last listed 99/00
	'Joyful Blue'	oJoy
	'Kingston Blue'	wHer
	'Lilliput'	oWnC wWel
	'Loch Hope' seed grown	last listed 99/00
	Midknight Blue®/ 'Monmid'	oAls oHed oRus oWnC wWel
	'Midnight'	oDar
qu	Mooreanus hybrids	oRus
un	Mosswood hybrids	wHer
	orientalis	see **A. *pracox*** ssp. ***orientalis***
un	Pearl hybrids	oGos oGar
	'Peter Pan'	oFor oNat oAls oGar oSho oGre oWnC wTGN oUps
	'Pinocchio'	wWoS oSho
	praecox ssp. ***orientalis***	oAls
	'Queen Ann'	oGar oWnC
	'Rancho White'	oGre
	'Snowdrop'	oJoy
	sp. variegated	oHed
do	'Storm Cloud'	oRus
	'Streamline'	wHer
va	'Tinkerbell'	wHer oGos wCol oHed wWel
un	'Verwood Dorset' seed grown	wHer
	'White Superior'	wMag
	AGAPETES	
	'Ludgvan Cross'	oBov
	serpens	oRar
	AGARISTA	
	populifolia	oFor wHer oCis wCCr
	AGASTACHE	
	aurantiaca	last listed 99/00
	aurantiaca 'Apricot Sunrise'	oNat oGoo oSis wFai wPir wHom wMag wSnq
	aurantiaca hybrids	wHer
	barberi	oGoo
	barberi 'Tutti-frutti'	oFor oNat oGoo oWnC oSec oSle
	'Blue Fortune'	oFor oAls wPir oWnC wMag wEde wSnq
	cana	oFor oGoo wFai wCCr oWnC oSle
	cana 'Heather Queen'	wWoS oUps
un	***coccinea***	oGoo
	'Firebird'	oGoo oSis wEde wSnq
	foeniculum	oNat oJoy oAls wFGN oGar oGoo iGSc oCrm oBar oSec wNTP wWld oUps
	foeniculum 'Alba'	oGoo
	foeniculum 'Fragrant Delight'	wHom oUps
	mexicana	oGoo wPir
	mexicana hybrids	wHer oGoo
	mexicana purple	wCCr
un	***mexicana*** 'Toronjil Morado' seed grown	wHer wWoS
	nepetoides	iGSc
	'Pink Panther'	oJoy oAls oWnC wMag oCir
	rugosa	oSis oSec
	rupestris	oFor w WoS oAls oAmb oSis wPir wSnq
un	***rupestris*** 'Sunset'	oFor
	'Tangerine Dreams'	wWoS
	urticifolia	wWld
	AGATHOSMA	
	ovata	last listed 99/00
	AGAVE	
	americana	oSto oUps
	americana 'Variegata'	oUps
	angustifolia 'Marginata'	oEdg
	colimana	oRar
un	***coucui*** BLM 1397	oRar
un	***deserti*** var. ***simplex***	oRar

un *eduardii* oRar
un *kerchovei* BLM 1284 oRar
potatorum var. *verschaffeltii* oSqu
schidigera BLM 0680 oRar
un *shawii* var. *goldmanii* oRar
utahensis oSqu
utahensis var. *utahensis* oCis
xylonacantha BLM 0605 oRar

AGERATINA see **EUPATORIUM**

AGLAONEMA
crispum 'Marie' oGar
'Silver Queen' oGar

AGONIS
flexuosa last listed 99/00

AGRIMONIA
eupatoria oGoo iGSc

AGROPYRON
magellanicum see **ELYMUS** *magellanicus*

AGROSTEMMA
blood red iArc

AILANTHUS
altissima wCoN

AJANIA
pacifica oHed wCCr wHom wTGN wSte
pacifica 'Silver 'n' Gold' see **A.** *pacifica*
un *pacifica* 'Yellow Splash' oFor

AJUGA
un 'Chocolate Chip' wHer oHed oNWe wRob oGre
un *lobata* wHer
qu 'Min Crispa Red' oUps
pyramidalis 'Metallica Crispa' oJoy oHed oAmb oNWe oSha oUps
qu *pyramidalis* 'Metallica Crispa Purpurea' oSis
reptans wSta
reptans 'Alba' oWnC
reptans bronze oSho iGSc wSta wAva wCoS
reptans 'Bronze Beauty' oUps oSle
reptans burgundy oSqu
va *reptans* 'Burgundy Glow' wWoS oDar wCSG oTwi oSqu wFai oWnC wWhG oUps oSle
reptans 'Burgundy Lace' see **A.** *reptans* 'Burgundy Glow'
reptans 'Catlin's Giant' wHer oJoy wHig oAls oOut wRob oSqu oGre wFai wCCr oWnC oSec oSle
reptans 'Compacta' oSqu
reptans 'Crispa' wFai
reptans 'Jungle Beauty' oWnC
un *reptans* 'Mahogany' wWhG
va *reptans* 'Multicolor' wAva
un *reptans* 'Pink Delight' wHer oSqu
reptans 'Pink Elf' wAva wWhG
reptans 'Pink Surprise' oAls oSha oGre oSle
reptans 'Purple Brocade' oSis oSho oWnC wWhG
un *reptans* 'Silver Beauty' wWoS oNat oUps
un *reptans* 'Silver Queen' oJoy wRob oSle

AKEBIA
longeracemosa BSWJ 3606 wHer
x pentaphylla oGar oGre
quinata oFor wCul wCri wClo wCol wCSG oGar oDan oSho oOut oBlo oOEx wPir oCrm wTGN wSta wCoN oUps oSle wSnq wCoS wWel
un *quinata* 'Alba' oGre
quinata HC 970420 wHer
un *quinata* Purple Bouquet™ wRai
un *quinata* 'Rosea' oGar oGre
un *quinata* 'Shirobana' oGar oSho wRai oWnC wWel
un *quinata* 'Silver Bells' wTGN wBox
quinata white wCoN
trifoliata oFor oGre wSnq
un *trifoliata* Deep Purple™ wRai
trifoliata DJH 433 wHer
trifoliata EDHCH 97248 wHer

ALANGIUM
chinense oFor oRiv oCis oWnC wBox
platanifolium oFor oDan
platanifolium var. *macrophyllum* wHer
sp. EDHCH 97256 wHer

ALBIZIA
julibrissin oFor oTrP oDar oAls oGar oRiv oSho oBlo oGre iGSc wBox wWhG wCoS
julibrissin var. *alba* oTrP wCCr
julibrissin 'E. H. Wilson' oFor
un *julibrissin* 'Fan Silk' oGre oWnC
un *julibrissin* 'Red Silk' wCoN
julibrissin f. *rosea* oTrP oGar wCCr oWnC
qu 'Red Fan' oBlo
sp. oOEx

ALBUCA
un *africana* oRar
un *setosa* wGra
shawii wHer

ALCEA
ficifolia oNat
pallida wFGN
do Peaches 'n' Dreams tm oCir
rosea wCul wCri wFai iArc
rosea apricot wAva
do *rosea* Chater's Double Group oWnC oCir
do *rosea* Chater's Double Group 'Chater's Double Apricot' oWnC
do *rosea* Chater's Double Group 'Chater's Double Magenta' oWnC
do *rosea* Chater's Double Group 'Chater's Double Maroon' wMag
do *rosea* Chater's Double Group 'Chater's Double Purple' oAls oWnC
do *rosea* Chater's Double Group 'Chater's Double Red' oAls oWnC wMag
do *rosea* Chater's Double Group 'Chater's Double Rose' oAls
do *rosea* Chater's Double Group 'Chater's Double Salmon' oAls oWnC
do *rosea* Chater's Double Group 'Chater's Double White' oWnC wMag
do *rosea* Chater's Double Group 'Chater's Double Yellow' oAls oWnC wMag
un *rosea* 'Country Romance' oCir
un *rosea* 'Double Appleblossom' oUps
rosea double apricot wFGN oUps
rosea double pink oCir
un *rosea* 'Fordhook Giant' oCir
rosea 'Indian Spring' wMag
un *rosea* Indian Summer mixed wFGN
rosea 'Nigra' wCul oNat wHom wAva oSle
rosea Powder Puffs mixed wFGN
un *rosea* 'Simplex' oNat
rosea single pink oCir
un *rosea* 'The Watchman' oFor wFGN oUps
rugosa oMis
un *rugosa-stellulata* oJil
un *simplex* wMag oSle

ALCHEMILLA
alpina wCul oRus wCSG oGoo oSis oOut oGre wFai oWnC wTGN wBox oSle
conjuncta wHig wPir
ellenbeckii wCul oJoy oAls wFGN oGoo wCoN
epipsila wWoS
erythropoda oFor wWoS wCul oRus oAmb oMis wNay oGre wFai wBWP
faeroensis wHig
faeroensis var. *pumila* oFor
glaucescens wHer wCul
mollis oFor wCul wCri wSwC oNat wHig oAls oHed oRus wFGN wCSG oMis oGar oSho oSha oTwi wNay oGre wPir wHom oWnC wMag oBar wTGN wSta wBWP oCir wAva wCoN oEga oUps
mollis 'Auslese' oFor wWoS oJoy oSis
qu *mollis* 'Improved' oAls oGar
mollis 'Robusta' wBox
un *mollis* 'Thriller' oFor oGar oBlo
un *pectinata* oFor wHer
un *pyranaica* wCul
saxatilis oHed oRus
vulgaris see **A.** *xanthochlora*
xanthochlora oAls oGoo iGSc oCrm oBar wNTP

ALETRIS
farinosa oCrm

ALEURITES
moluccana last listed 99/00

ALISMA
plantago-aquatica wWat oWnC oTri
plantago-aquatica var. *parvifolium* oSsd
subcordatum see **A.** *plantago-aquatica* var. *parvifolium*

ALLAMANDA
cathartica wWel

ALLIUM
acuminatum wHer oHan
aflatunense (see Nomenclature Notes) wHer oWoo
albidum see **A.** *denudatum*
albopilosum see **A.** *cristophii*

	ampeloprasum	oAls wFGN wIri
un	amphibolum	oFor
	atropurpureum	wCul wIri
	azureum	see A. caeruleum
	bulgaricum	see NECTAROSCORDUM siculum ssp. bulgaricum
	caeruleum	oDan wIri
	carinatum ssp. pulchellum	wIri
	carinatum ssp. pulchellum f. album	wHer
	cepa	oAls oGoo
un	cepa Aggregatum Group 'Brittany Red'	wIri
un	cepa Aggregatum Group 'Dutch Yellow'	wIri
	cepa Aggregatum Group gray shallot	wIri
un	cepa Aggregatum Group 'Holland Red'	wIri
	cepa Aggregatum Group yellow multipliers	wIri
un	cepa 'Candy'	wIri
un	cepa 'First Edition'	wIri
	cepa Proliferum Group	wWoS wFai iGSc wHom
	cepa Proliferum Group Catawissa onion	wIri
	cepa Proliferum Group Egyptian walking onion	wIri
	cepa red	wCoS
un	cepa 'Red Burgermaster'	wIri
un	cepa 'Red Wethersfield'	wIri
un	cepa 'Sweet Red'	wIri
un	cepa 'Texas Supersweet'	wIri
un	cepa 'Walla Walla'	wIri wCoS
	cepa white	wCoS
un	cepa 'White Ebenezer'	wIri
un	cepa 'White Granex'	wIri
	cepa 'White Sweet Spanish'	wIri
	cepa yellow	wCoS
un	cepa 'Yellow Granex'	wIri
un	cepa 'Yellow Rock'	wIri
un	cepa 'Yellow Sweet Spanish'	wIri
	cernuum	oHan wCul oJoy oRus wShR oGoo oBos wFai wWat wCCr oWnC wWld
	cernuum album	wHer
	cernuum dark form	last listed 99/00
	cernuum large form	wCol
	cowanii	see A. neapolitanum Cowanii Group
	cristophii	wCul oNat oRus oDan oNWe wFai oWoo oSec wIri
	cyaneum	oJoy wCol wHig
	cyaneum blue form	oSis
	cyaneum miniature blue form	oSis
	cyaneum sky blue form	wHer oSec
	cyathophorum var. farreri	oSis
	cyrillii	wCCr
un	denudatum ssp. caucasicum	oSis
	elatum	see A. macleanii
	flavum	wHer wCul oGoo wCCr
	geyeri	oFor
	giganteum	oWoo wIri
	'Gladiator'	wHer
	'Globemaster'	oNWe wIri
	hollandicum	wIri
	hollandicum 'Purple Sensation'	wHer wCul oRus oGar oDan oNWe wFai
ch	'Ivory Queen'	last listed 99/00
	karataviense	oDan oSis oGre wIri
	macleanii	oRus
	mairei	wCol
	'Mars'	wIri
	moly	oRus oWoo wAva wIri
	multibulbosum	see A. nigrum
	neapolitanum	oWoo wAva wIri
	neapolitanum Cowanii Group	oRus
	neapolitanum 'Grandiflorum'	oNWe
	nigrum	wCul oRus wIri
	nutans	oFor wCCr
	oreophilum	oDan oWoo wIri
	ostrowskianum	see A. oreophilum
	paniculatum	last listed 99/00
	porrum	oAls wFGN
	x proliferum	see A. cepa Proliferum Group
	pulchellum	see A. carinatum ssp. pulchellum
	'Rien Poortvliet'	wIri
	rosenbachianum	wIri
	rosenbachianum 'Album'	wIri
	roseum var. bulbiferum	wRoo oRus
	roseum 'Grandiflorum'	see A. roseum var. bulbiferum
	sativum	oAls
un	sativum 'Asian Tempest'	wIri
un	sativum 'Brown Tempest'	wIri
un	sativum 'Burgundy'	wIri
un	sativum 'Chesnok Red'	wIri
un	sativum 'Early Italian Purple'	wIri
un	sativum 'German Porcelain'	wIri
un	sativum 'German Red'	wIri
un	sativum 'Inchelium Red'	wIri
un	sativum 'Italian Late'	wIri
un	sativum 'Korean Red'	wIri
un	sativum 'Musik'	wIri
un	sativum 'Nootka Rose'	wIri
un	sativum 'Rosewood'	wIri
un	sativum 'Silver Rose'	wIri
un	sativum 'Spanish Roja'	wIri
	saxatile	oSec
	schoenoprasum	wTho oTDM oAls wFGN wCSG oGoo iGSc wHom oWnC oBar wBox oSec oCir oUps oSle
	schoenoprasum fine-leaved	wNTP
	schoenoprasum 'Forescate'	wWoS
	schoenoprasum 'Grolau'	iGSc
un	schoenoprasum 'Purly'	wNTP
	schubertii	oDan oNWe wIri
	senescens	wHig oGoo wCCr
	senescens ssp. montanum	wCol
	senescens ssp. montanum var. glaucum	oFor wWoS oAls iGSc oWnC oSec
	siculum	see NECTAROSCORDUM siculum
	sikkimense	oSqu
	sphaerocephalon	wHer wCul oNat wIri
	stipitatum	oRus
	tanguticum	wHig
qu	taquetti	wCSG
	thunbergii	wCol
	thunbergii 'Ozawa'	oSis
	triquetrum	wIri
	tuberosum	wCul wTho oTDM oAls wFGN oGoo oOut wFai iGSc wHom oWnC oCrm oBar wNTP oCir oUps oSle
	tuberosum mauve	iGSc
	turkestanicum	wHig
	unifolium	oRus wIri
	wallichianum	see A. wallichii
	wallichii EDHCH 97058	wHer
	wallichii HWJCM 232	wHer
	ALLOCASUARINA	
	littoralis	last listed 99/00
	verticillata	last listed 99/00
	ALLUADIA	
un	ascendens	oRar
	dumosa	oRar
	humbertii	oRar
	montagnacii	oRar
	procera	oRar
	ALLUADIOPSIS	
	fiherensis	oRar
	ALNUS	
	cordata	oFor oRiv wCCr
	cremastogyne	oRiv
un	crispa ssp. sinuata	see A. viridis ssp. sinuata
	firma	last listed 99/00
	firma var. multinervis	see A. pendula
	glutinosa	oRiv
	glutinosa 'Imperialis'	oFor oGos oHed oAmb oGar oRiv
	hirsuta	oFor oRiv
	incana	oSle
	incana 'Aurea'	oFor
	incana ssp. incana	oRiv wCCr wPla
	incana 'Laciniata'	oFor oRiv
	japonica	wCCr
	jorullensis	oFor
	maritima	last listed 99/00
	maximowiczii	oFor oRiv
	nepalensis	oRiv wCCr
	nitida	oRiv
	oregana	see A. rubra
	pendula	oRiv
	rhombifolia	oRiv oSle
	rubra	oFor wWoB wBur wNot wShR oGar oBos oRiv wWat wCCr wPla oAld wFFl oSle
	rugosa	oRiv
	serrulata	see A. rugosa
	sinuata	see A. viridis ssp. sinuata
	sitchensis	see A. viridis ssp. sinuata
	x spaethii	oFor oRiv
	tenuifolia	see A. incana ssp. incana
	viridis	oRiv
	viridis ssp. sinuata	oFor wWoB wShR oRiv wWat wPla wFFl

	ALOCASIA	
	x amazonica	oGar
	sp. black stemmed	oOEx
	ALOE	
	aristata	oSqu
un	***aristata*** var. ***montanum***	oRar
	barbadensis	see **A. *vera***
	boylei	oRar
	branddraaiensis	oRar
	brevifolia	oSqu
	candelabrum	oRar
	castanea	oRar
	chabaudii DR 2124	oRar
un	***christiani***	oRar
	cooperi	oRar
	cryptopoda	oRar
un	***davyana*** var. ***subolifera***	oRar
	descoingsii x ***haworthioides***	oSqu
un	***dominella***	oRar
	ferox	oRar
	greatheadii	oRar
	haworthioides	last listed 99/00
un	***inyangensis***	oRar
	jucunda	last listed 99/00
un	'Lizard Lips'	oSqu
	lutescens	oRar
	marlothii	oRar
	ortholopha	oRar
	parvibracteata	oRar
	peglerae	oRar
	saponaria	oRar
	striata	oRar
	thraskii	oRar
un	***transvallensis***	oRar
	variegata	last listed 99/00
	vera	oTDM oAls wFGN oGar oGoo iGSc wHom oCrm oBar oUps
	zebrina	oRar
	ALOINOPSIS	
	spathulata	oDan
	ALOPECURUS	
	pratensis 'Aureovariegatus'	oDar
	pratensis 'Aureus'	oFor oJoy oAls oSho wWin oWnC wBox oSec oUps
	ALOYSIA	
	triphylla	wWoS oTDM oAls wFGN oGoo wFai iGSc wHom oWnC oCrm oBar wNTP oUps oSle
	ALPINIA	
	japonica	oCis
	ALSEUOSMIA	
un	***vaccinacea***	wHer
	ALSTROEMERIA	
un	'Apricot Blush'	wHom
	aurantiaca	see **A. *aurea***
	aurea	oFor wCSG oSec
	aurea deep orange	oRus
	aurea gold	oRus
	aurea HCM 98117	wHer
un	'Coast Alpine'	oRus
	dark shades	oJoy
	deep purple	wHom
	diluta	last listed 99/00
	Doctor Salter's hybrids	last listed 99/00
un	***hookeri*** ssp. ***hookeri***	wHer
	hybrid, purple-lilac	oRus
	hybrids	oAls wFGN wAva
un	'Inca Koya'	oRus
	ligtu hybrids	oRus oGar wTGN
un	Pacific Sunset hybrids	oRus oGar wEde
	paupercula	last listed 99/00
qu	***peruviana***	oSec
	presliana ssp. ***australis***	last listed 99/00
	psittacina	oGoo oSis oNWe oGre oSec
un	***psittacina*** 'Variegata'	oFor wHer oNWe oSec wSte
	pulchella	see **A. *psittacina***
	pulchra	oRus
un	Sunset Series	oBlo
	'Sweet Laura'	wWoS oRus oGar oBlo
	'Tricolor'	see **A. *pulchra***
	violacea	see **A. *paupercula***
	ALTERNANTHERA	
	lehmannii	oOEx
un	***rubra***	oHug wTGN
	ALTHAEA	
	cannabina	oFor
	officinalis	oAls wFGN oGoo wFai wCCr iGSc wHom oCrm oBar oSec oSle
	officinalis 'Romney Marsh'	wHer
	rugosa	see **ALCEA *rugosa***
	ALYOGYNE	
	hakeifolia	last listed 99/00
	huegelii	last listed 99/00
	huegelii 'Santa Cruz'	oGar
	ALYSSOIDES	
un	'Lost April'	iArc
	utriculata	oSis
	ALYSSUM	
un	***caespitosum*** Jurasek List	wMtT
	idaeum	oSis
	moellendorfianum	oSis
	montanum	oUps
	montanum 'Berggold'	iArc
	montanum Mountain Gold	see **A. *montanum*** 'Berggold'
	propinquum	oSis
	purpureum	wMtT
	saxatile	see **AURINIA *saxatilis***
	spinosum	wEde
	spinosum 'Roseum'	oSis
	tortuosum	oSis
	AMARANATHUS	
un	***cruentus*** x ***powellii***	iGSc
	X AMARCRINUM	
	memoria-corsii	wTGN
	AMARYLLIS	
	belladonna	oDan wTGN
	AMELANCHIER	
	alnifolia	wBCr wThG wWoB wBur wShR oBos oRiv wWat wCCr wKin oTri wBox wPla oAld wAva oSle
un	***alnifolia*** var. ***humptulipensis***	wFFl
un	***alnifolia*** 'Northline'	oOEx
	alnifolia var. ***pumila***	oSis
ch	***alnifolia*** 'Regent'	last listed 99/00
	alnifolia 'Smokey'	wRai oOEx
	alnifolia 'Smokey' seed grown	wBur
un	***alnifolia*** 'Success'	wRai
un	***alnifolia*** 'Thiessen'	oOEx
un	***asplenifolia***	oRiv
	canadensis	wCul oRiv wAva wSnq
	x grandiflora	wBCr oOEx
	x grandiflora 'Autumn Brilliance'	wRai wWel
un	***x grandiflora*** 'Forest Prince'	oFor
	x grandiflora 'Robin Hill'	oFor
ch	***x grandiflora*** 'Strata'	last listed 99/00
	laevis	oFor oRiv
	laevis 'Cumulus'	oFor
	lamarckii	wWel
	stolonifera	oFor
	AMMOBIUM	
	alatum	oGoo
	AMOMUM	
un	***subulatum***	oOEx
	AMORPHA	
	canescens	oFor oUps
	fruticosa	oFor
	nana	last listed 99/00
	AMORPHOPHALLUS	
un	'Balang'	oOEx
	konjac	see **A. *rivierei***
un	'Naxi'	oOEx
	rivierei DJHC 970580	wHer
	sp. ex Amazon	oOEx
un	'Szechuanese'	oOEx
	AMPELOPSIS	
	aconitifolia	oFor wHer oGar oDan oGre
	arborea	oFor
	brevipedunculata	see **A. *glandulosa*** var. ***brevipedunculata***
	cordata	oFor
	delavayana	oFor wCCr
	glandulosa var. ***brevipedunculata***	oTrP wCul oHed wCSG oGar oGre wCCr wCoN oUps wWel
va	***glandulosa*** var. ***brevipedunculata*** 'Elegans'	oFor oTrP wCul wClo wCSG oGar wRob wRai oGre oWnC wTGN wAva oEga wWel
	glandulosa var. ***brevipedunculata*** HC 970582	wHer
	humulifolia	oFor oGar oGre
	japonica	oFor oGre
	megalophylla	oFor wHer

	Plant	Sources
	AMSONIA	
	ciliata	oDan
	hubrichtii	wHer oNat oJoy oRus oMis oNWe oGre
	illustris	oJoy oRus
	jonesii	oDan
	orientalis	oAls
	tabernaemontana	oHed oMis oSis wFai wBox
	tabernaemontana var. *salicifolia*	oJoy wNay
	WFF Selection	oSec
	ANACAMPSEROS	
	alstonii	oRar
un	*australiana*	oRar
	telephiastrum	oSqu
	ANACYCLUS	
	depressus	see A. *pyrethrum* var. *depressus*
un	*maroccanus* 'Pointe'	oJoy
	pyrethrum var. *depressus*	oWnC
	pyrethrum var. *depressus* 'Silberkissen'	oAls
	ANAGALLIS	
	monellii	last listed 99/00
	monellii 'Pacific Blue'	wCSG
	tenella 'Studland'	oHed
	ANAPHALIS	
	alpicola	oSis
	margaritacea	oHan wNot wShR oGoo wWat wPla oAld wFFl wWld
	margaritacea 'Neuschnee'	oSis
	margaritacea New Snow	see A. *margaritacea* 'Neuschnee'
un	*sinica* 'Moor's Silver'	wHer
	triplinervis 'Sommerschnee'	oHed
	triplinervis Summer Snow	see A. *triplinervis* 'Sommerschnee'
	ANCHUSA	
	azurea	oNWe wCoN
	azurea 'Dropmore'	oFor
	azurea 'Loddon Royalist'	oAls
	officinalis	oGoo iGSc
	ANDRACHNE	
	colchica	oFor wHer
	ANDROMEDA	
	glaucophylla	oFor
	polifolia	oAls oGar oBlo oSqu oTPm oWnC wSta
	polifolia 'Blue Ice'	oFor wWoS oGos oAls oGar oGre wTGN oGue
un	*polifolia* 'Chuo Red'	oSis
	polifolia 'Kirigamine'	oFor oSis
	polifolia 'Macrophylla'	oSis
	polifolia 'Nana'	oFor oGar wSta
	polifolia 'Nana Alba'	oSis
	polifolia pink	oAls oGue
qu	*variegata*	oRiv
	ANDROPOGON	
	gerardii	oFor wShR oBRG
	gerardii 'Pawnee'	wTGN
un	**ANDROPYRON**	
	magellanicum	oOut
	magellanicum 'Blue Tango'	oOut
	ANDROSACE	
un	'Callisto'	wMtT
	carnea	last listed 99/00
	carnea ssp. *brigantiaca*	wMtT
un	*carnea* 'Lavenderglow'	wMtT
	carnea ssp. *rosea*	wMtT
	chamaejasme	last listed 99/00
	cylindrica x *hirtella*	last listed 99/00
	foliosa	wCol
	halleri	see A. *carnea* ssp. *rosea*
	hedraeantha	last listed 99/00
	x heeri	wMtT
	hirtella	wMtT
	jacquemontii	see A. *villosa* var. *jacquemontii*
un	'Jupiter'	wMtT
	laevigata var. *ciliolata*	wMtT
	laevigata 'Gothenburg'	wMtT
	laevigata var. *laevigata*	wMtT
	laevigata var. *laevigata* 'Packwood'	wMtT
	lanuginosa	oHed oSis oSqu
un	'Millstream'	oSis wMtT
	montana	wMtT
	mucronifolia (see Nomenclature Notes)	wMtT
	muscoidea	last listed 99/00
	nivalis	last listed 99/00
	x pedemontana	last listed 99/00
	pyrenaica	wMtT
un	*pyrenaica* 'Rosea'	wMtT
un	'Rhapsody'	wMtT
	rigida DJHC 372	wHer
	rotundifolia	oNWe
	sarmentosa 'Chumbyi'	oSis oSqu
	sarmentosa 'Sherriff's'	oSis
	sempervivoides	wCol oSis
	sericea Jurasek list	last listed 99/00
	spinulifera DJHC 336	wHer
	strigillosa	wMtT
	vandellii	wMtT
un	'Venus'	wMtT
	villosa	last listed 99/00
	villosa var. *jacquemontii*	oSis
	ANDRYALA	
	agardhii	last listed 99/00
	ANEMARRHENA	
	asphodeloides	wCol
un	**ANEMATHELE**	
	lessoniana	wHer wWin
	ANEMONE	
	altaica	oFor oRus
	biarmiensis	see A. *narcissiflora* ssp. *biarmiensis*
	blanda	wCul wThG wRoo oNWe wAva
	blanda blue	wSnq
	blanda var. *rosea*	wSnq
	blanda 'White Splendour'	wSnq
	canadensis	oFor oHan wCul oRus
do	*coronaria* Saint Brigid Group	wRoo
	cylindrica	oHan
	deltoidea	oBos
	demissa DJHC 340	wHer
	drummondii	oRus wMtT
	flaccida	last listed 99/00
	x fulgens (see Nomenclature Notes)	wRoo
	hupehensis	oFor oTwi oWnC oUps
	hupehensis 'Alba'	see A. *x hybrida* 'Honorine Jobert'
	hupehensis 'Hadspen Abundance'	wWoS oGar wRob
	hupehensis var. *japonica*	oRus oGar
	hupehensis var. *japonica* 'Bressingham Glow'	wMag wSnq
	hupehensis var. *japonica* 'Pamina'	wWoS wGAc oNat wCSG oGar oGre wMag
	hupehensis var. *japonica* pink	oBov
	hupehensis var. *japonica* 'Prinz Heinrich'	oFor wHer oNat oAls wCSG oAmb oGar wRob oBlo wMag
ch	*hupehensis* 'September Charm'	see A. *x hybrida* 'September Charm'
	x hybrida	oBov wPir wCoN
	x hybrida 'Alba'	see A. *x hybrida* 'Honorine Jobert'
	x hybrida 'Alice'	oFor wWoS oNat oAls wRob wMag
	x hybrida double pink	oSle
	x hybrida 'Honorine Jobert'	oFor wHer wWoS wCul oNat oAls wCSG oAmb oMis oGar wRob oGre wFai wHom wMag wEde wSte wSnq
	x hybrida 'Koenigin Charlotte'	oFor wHer wWoS oEdg oGar oBlo wFai wMag wTGN wSte
	x hybrida 'Kriemhilde'	oFor oJoy
	x hybrida 'Loreley'	oGar
	x hybrida 'Margarete'	wCul oJoy oGar wMag
	x hybrida 'Max Vogel'	last listed 99/00
	x hybrida mixed	wCSG
	x hybrida 'Monterosa'	last listed 99/00
	x hybrida pink	oGar
	x hybrida Prince Henry	see A. *hupehensis* var. *japonica* 'Prinz Heinrich'
	x hybrida Queen Charlotte	see A. *x hybrida* 'Koenigin Charlotte'
	x hybrida 'Richard Ahrens'	wWoS oJoy
	x hybrida 'Rosenschale'	last listed 99/00
ch	*x hybrida* 'September Charm'	oFor wHer wWoS oDar oEdg oMis oGar oBlo oWnC wMag
	x hybrida 'Whirlwind'	oFor wWoS wCul oNat oDar oAls oAmb oGar wRob oBlo oGre oWnC wMag wBox wSnq
	x hybrida white	oGar oSle
ch	*japonica* (see Nomenclature Notes)	
	x lesseri	wMag
	leveillei	wHer
	x lipsiensis	oSis wMtT
	multifida	last listed 99/00
	multifida var. *globosa*	oHan
	multifida HCM 98101	wHer
	multifida 'Rubra'	last listed 99/00
	narcissiflora	last listed 99/00
	narcissiflora ssp. *biarmiensis* DBG list	wMtT
	nemorosa	wCul wCol wCSG oSis oNWe
	nemorosa 'Alba'	last listed 99/00
do	*nemorosa* 'Alba Plena'	wHer oHed oSis
	nemorosa 'Allenii'	hoSis oNWe oGre wSte

qu	nemorosa 'Allionii'	wCol
	nemorosa blue	oGre
	nemorosa 'Blue Bonnet'	wCri
do	nemorosa 'Blue Eyes'	last listed 99/00
	nemorosa 'Bracteata'	wCol oSis oNWe oGre
do	nemorosa 'Bracteata Pleniflora'	oHed
	nemorosa 'Buckland'	wHer
	nemorosa 'Green Fingers'	wCri
	nemorosa 'Hilda'	wCri
	nemorosa lavender form	oHed
	nemorosa 'Leeds' Variety'	wHer
un	nemorosa 'Lismore Blue'	wHer
	nemorosa 'Lychette'	oSis oNWe oGre
	nemorosa pink	wHer
	nemorosa 'Robinsoniana'	wHer oRus oSis oNWe oGre wFai wSte
	nemorosa 'Rosea'	oRus
do	nemorosa 'Vestal'	wHer wCol oNWe
	nemorosa 'Virescens'	oRus
	nemorosa 'Viridiflora'	wHer
	nemorosa white	oNWe
	nemorosa white aging to pink	wCol
	nemorosa 'Wilks' White'	wCri
	obtusiloba	oNWe
	occidentalis	oFor
	oregana	see A. quinquefolia var. oregana
	parviflora	wMtT
	patens	see PULSATILLA patens
	polyanthes	wCol
	pulsatilla	see PULSATILLA vulgaris
	quinquefolia var. oregana	oBos
	ranunculoides	last listed 99/00
	ranunculoides 'Flore Pleno'	see A. ranunculoides 'Pleniflora'
	ranunculoides large flowered form	last listed 99/00
do	ranunculoides 'Plenaflora'	oSis
un	ranunculoides ssp. ranunculoides	oSis
	rivularis	oFor oGar oNWe oGre wFai wBox
	rivularis DJHC 98059	wHer
	robustissima	see A. tomentosa 'Robustissima'
	rupicola DJHC 98179	wHer
	x seemannii	see A. x lipsiensis
	sylvestris	oFor wCol oRus oAmb oGar oGre oWnC wMag wEde oUps oSle
	sp. aff. tomentosa DJHC 596	wHer
	tomentosa 'Robustissima'	oFor wCul oJoy oAls oGar wMag oInd
	trifolia	wHer wCol
do	trifolia 'Semiplena'	wHer
	virginiana	oFor
	vitifolia	last listed 99/00
	vitifolia B&SWJ 245	last listed 99/00
	vitifolia 'Robustissima'	see A. tomentosa 'Robustissima'
	ANEMONELLA	
	thalictroides	wHer oHan wCol oRus
	thalictroides 'Cameo'	wCol oNWe
do	thalictroides 'Oscar Schoaf'	last listed 99/00
ch	thalictroides 'Schoaf's Double Pink'	see A. thalictroides 'Oscar Schoaf'
	ANEMOPSIS	
	californica	oCrm
	ANETHUM	
	graveolens	wTho oTDM oAls wCSG wFai iGSc wHom oWnC oUps
un	graveolens 'Bouquet Dwarf'	wFGN
	graveolens 'Dukat'	wNTP
	graveolens 'Fern Leaved'	oAls wFGN wHom wNTP
	ANGELICA	
	archangelica	wCul oAls wFGN oGoo wFai iGSc oWnC oCrm oBar oUps
	arguta	wWld
	atropurpurea	oGoo oSec
	gigas	oFor wCul wCol wFGN oNWe wFai oCrm oBar oUps
	gigas HC 970404	wHer
	pachycarpa	oNWe oGre
	polymorpha sinensis	oAls oBar wTGN
	pubescens	oOEx
qu	sinensis	oCrm
un	stricta var. atropurpurea	oJil
un	stricta 'Purpurea'	oNWe oGre
	'Vicar's Mead' seed grown	wHer
	ANGOPHORA	
un	melanoxylon	wCCr
	ANIGOZANTHOS	
	flavidus	oTrP oGoo oCir
	ANISODONTEA	
un	biflora	oTrP
	capensis	oTrP oGar

ch	capensis 'Tara's Pink'	oHed oGar wTGN
	x hypomadara (see Nomenclature Notes)	last listed 99/00
	scabrosa	oHed
	ANNONA	
	cherimola	oOEx
	ANOMATHECA	
	laxa	oFor oTrP wWoS oSis
	laxa var. alba	oNWe
	laxa blue flowered	oSis
	laxa red-eyed white form	oSis
	ANREDERA	
	baselloides	oRar
un	ramosa BLM 1254	oRar
	sp. BLM 0475	oRar
	ANTENNARIA	
	dioica	oFor iGSc
qu	dioica 'Minima Rubra'	oSis
	dioica 'Nyewoods Variety'	oHed oSis
ch	dioica var. rosea	see A. rosea
	dioica 'Rubra'	oAls
	neglecta var. gaspensis	oSis
	rosea	oJoy wShR oSis oGre
	sp.	oBos
	ANTHEMIS	
ch	biebersteinii	see A. marschalliana ssp. biebersteinii
	carpatica 'Karpatenschnee'	oSis
	cupaniana	last listed 99/00
	marschalliana ssp. biebersteinii	oFor oSis
	nobilis	see CHAMAEMELUM nobile
	sancti-johannis	oSis
	tinctoria	wCSG oGoo wFai iGSc
	tinctoria 'E. C. Buxton'	oFor
	tinctoria 'Kelwayi'	oAls oWnC
	tinctoria 'Sauce Hollandaise'	wHig oHed oAmb oGar oSis oOut oGre wBox
	tinctoria 'Wargrave Variety'	wHer oInd
	ANTHERICUM	
	liliago	oJoy wHig oMis
	ramosum	oJoy wHig
	ANTHOXANTHUM	
	odoratum	wFGN oGoo iGSc oCrm
	ANTHRISCUS	
	cerefolium	oAls wFGN iGSc oCrm
	cerefolium Brussels Winter	see A. cerefolium 'D'Hiver de Bruxelles'
	cerefolium 'D'Hiver de Bruxelles'	wNTP
	sylvestris 'Ravenswing'	oHed oNWe oGre
	ANTHYLLIS	
	montana	oFor
	montana 'Rubra'	oFor
	vulneraria	oGoo
	vulneraria ssp. alpestris	oSis
	vulneraria var. coccinea	oNWe
un	vulneraria ssp. pulchella	oSis
un	vulneraria 'Rubra'	oSis
	ANTIGONON	
	leptopus	oTrP
	ANTIRRHINUM	
	braun-blanquetii	oMis oInd
	hispanicum 'Avalanche'	oCis
	hispanicum ssp. hispanicum roseum	oHed
va	majus 'Taff's White'	wWoS oCis
	molle	oSis
qu	repens dark pink/white eye	wTGN
qu	repens purple	wTGN
qu	repens white/pink throat	wTGN
un	'Royal Bride'	oCir
	sempervirens	oSis
	APIOS	
	americana	oFor wGAc oGoo oOEx
	APOCYNUM	
	cannabinum	iGSc
un	**APOLLONIAS**	
	barbujana	wCCr
	APONOGETON	
	distachyos	wGAc oHug oGar oSsd oWnC oTri
	APOROCACTUS	
	conzattii	last listed 99/00
	flagriformis	oSqu
	APOROHELIOCEREUS	
un	'Hazel'	oSqu
un	**APOROPHYLLUM**	
	'Beautie'	oSqu
	'Evita'	oSqu
	'Karen'	oSqu
	'Moonlight'	oSqu

	'Oakleigh Conquest'	oSqu
	'Royale'	oSqu
	'Temple Glow'	oSqu
	'Vivide'	oSqu
	APTENIA	
	cordifolia	oTrP
	AQUILEGIA	
un	'All Gold'	oOut
	alpina	wTho oJoy oSha oBlo oSqu oGre
	'Alpine Blue'	oBlo
	anemoniflora	oFor wTGN
	atrata	wCul oJoy
un	'Back Porch'	oMis
	bertolonii	oHan
	Biedermeier Group	oJoy oOut wHom wWhG oSle
	'Blue Jay' (Songbird Series)	oBlo oWnC
	'Blue Star ' (Star Series)	wMag
	blue/white selection	oHed
un	'Bluebird' (Songbird Series)	oSho oWnC wTGN wWhG
	caerulea	see **A. *coerulea***
	canadensis	oFor oRus oGoo oBlo wBWP
	canadensis 'Corbett'	oSis
un	***candida***	oTrP
	'Cardinal' (Songbird Series)	oWnC wWhG
	chaplinei	oSis
	chrysantha	oFor oHan oAmb oGoo
	chrysantha var. ***hinckleyana***	last listed 99/00
un	***chrysophylla***	wWld
	clematiflora	see **A. *vulgaris*** var. ***stellata***
	coerulea	oFor oHan oGoo wCCr wPla
qu	'Colorado' (Swan Series) violet	wWhG
qu	'Colorado' (Swan Series) white	wWhG
	'Crimson Star'	oFor oWnC wMag
	desertorum	last listed 99/00
	discolor	oSis oNWe
un	'Double Pantomime'	wTho wHom
	'Dove' (Songbird Series)	oMis oSis oWnC wTGN
	Dragonfly hybrids	wTho oSis oBlo wWhG
	eximia	wHer
un	'Fame White'	oWnC wTGN
	flabellata f. ***alba***	oHed
	flabellata 'Blue Angel'	oEdg
	flabellata Cameo Series	oGar oSho wWhG
	flabellata 'Cameo Blue and White' (Cameo Series)	oAls
	flabellata 'Cameo Blush' (Cameo Series)	oAls
	flabellata 'Cameo Pink and White' (Cameo Series)	oAls
un	***flabellata*** 'Cameo Rose and White' (Cameo Series)	oAls
	flabellata 'Ministar'	oAmb oGar oGre oUps
	flabellata 'Nana'	see **A. *flabellata*** var. ***pumila***
	flabellata 'Nana Alba'	see **A. *flabellata*** var. ***pumila*** f. ***alba***
	flabellata var. ***pumila***	oFor wCul oGar oSho
	flabellata var. ***pumila*** f. ***alba***	oFor oWnC
	flabellata var. ***pumila*** f. ***kurilensis***	wMtT
	flabellata purple	oJoy
	formosa	oHan wWoB wNot oRus wShR oGar oGoo oBos wFai wWat wMag oTri oAld wFFl wWld oSle wSte
	formosa var. ***truncata***	oSec
	fragrans	oHed oRus oGar wFai
	fragrans mixed seedlings	oRus
	'Goldfinch' (Songbird Series)	oJoy oSho
un	***grahamii***	wMtT
	'Hensol Harebell'	wCul oCis
	hinckleyana	see **A. *chrysantha*** var. ***hinckleyana***
	hybrid, variegated	oGar
	'Irish Elegance'	oMis oDan wHom
	jonesii	wMtT
	karelinii	oMis
	'Lavender and White' (Songbird Series)	oWnC
un	'Leprechaun Gold'	oFor
	longissima	oHed wCCr
	McKana Group	wTho wMag oCir
	'Mellow Yellow'	wCul oMis oGar wCCr
	'Mellow Yellow' seed grown	wHer
	Music Series	oWnC iArc oUps wSnq
	'Old Bucket'	oMis
	olympica	last listed 99/00
	'Pink Bonnet'	last listed 99/00
	pubescens	oNWe
	'Red Hobbit'	last listed 99/00
un	Remembrance ™	wWhG
	'Robin' (Songbird Series)	wTGN wWhG
	rockii	oTrP wCol oNWe
	rockii DJHC 98388	wHer
	'Roman Bronze'	oGar
	saximontana	oHan oSis oNWe
	sp. aff. ***saximontana***	oRus
	saximontana blue	oHed
	scopulorum	oSis wMtT
	shockleyi	oHan oSis
	skinneri	wHer oHed
	Songbird Series	wWhG
	sp. HCM 98084	wHer
	'Sunburst Ruby' seed grown	last listed 99/00
un	'Tequila Sunrise'	iArc
	triternata	oHan oNWe
	viridiflora	wFGN wSte
un	***viridiflora*** 'Chocolate Soldier'	oFor oSis
	vulgaris	oAls oTwi
	vulgaris 'Adelaide Addison'	wFai
un	***vulgaris*** 'Beatrix Farrand' seed grown	wHer
	vulgaris 'Clematiflora'	see **A. *vulgaris*** var. ***stellata***
do	***vulgaris*** 'Double Pleat'	oJoy oDan
	vulgaris 'Double Pleat' blue	oSha
do	***vulgaris*** var. ***flore-pleno***	wCul
	vulgaris 'Grandmother's Garden'	oMis oGre
	vulgaris 'Heidi'	oAls
un	***vulgaris*** 'Lime Frost'	oFor wWoS oGar oCir wSnq
	vulgaris 'Macedonica' seed grown	last listed 99/00
	vulgaris 'Magpie'	see **A. *vulgaris*** 'William Guiness'
un	***vulgaris*** 'Pink Petticoat'	oSec
	vulgaris Pom Pom Series	oMis
	vulgaris 'Silver Edge' seed grown	wHer
	vulgaris var. ***stellata***	wCul oSis oNWe oCir
do	***vulgaris*** var. ***stellata*** Barlow mixed	oCir wWhG
do	***vulgaris*** var. ***stellata*** 'Black Barlow'	oFor oMis oGar oSha oWnC iArc
do	***vulgaris*** var. ***stellata*** 'Blue Barlow'	oFor oAls oMis oWnC
do	***vulgaris*** var. ***stellata*** 'Nora Barlow'	oFor voAls wCSG oGre wFai oWnC wMag
	vulgaris var. ***stellata*** 'Ruby Port'	oFor wSwC oAls oHed oMis oWnC
va	***vulgaris*** 'Variegata'	see **A. *vulgaris*** Vervaeneana Group
va	***vulgaris*** Vervaeneana Group	wHer wCul oJoy oNWe oOut wCCr wHom
va	***vulgaris*** Vervaeneana Group pink and red	wHer
va	***vulgaris*** Vervaeneana Group white	wHer
va	***vulgaris*** Vervaeneana Group 'Woodside'	see **A.** Vervaeneana Group
	vulgaris 'William Guiness'	oSec
	'White Star' (Star Series)	oSha
	Woodside Variegated mix	see **A. *vulgaris*** Vervaeneana Group
	ARABIS	
	alpina ssp. ***caucasica***	oAls oWnC
qu	***alpina*** ssp. ***caucasica*** ssp. ***brevifolia***	oSis
do	***alpina*** ssp. ***caucasica*** 'Flore Pleno'	oSis
un	***alpina*** ssp. ***caucasica*** 'Red Sensation'	oWnC
	alpina ssp. ***caucasica*** ***rosea***	wWhG
	alpina ssp. ***caucasica*** 'Schneehaube'	wTho oAls wWhG oSle
	alpina ssp. ***caucasica*** Snowcap	see **A. *alpina*** ssp. ***caucasica*** 'Schneehaube'
	alpina ssp. ***caucasica*** 'Variegata'	oSis oOut oWnC wTGN wWhG oUps oSle
	androsacea	oJoy oSis
	x arendsii 'Compinkie'	oWnC wTGN oUps
	aubrietoides	oJoy wMtT
	blepharophylla 'Fruelingszauber'	wTho oAls oSis wAva wWhG oUps
	blepharophylla Spring Charm	see **A. *blepharophylla*** 'Fruehlingszauber'
	bryoides olympica	oSis
	caucasica	see **A. *alpina*** ssp. ***caucasica***
	ferdinandi-coburgi 'Old Gold'	oFor oSis wFai
	ferdinandi-coburgi 'Variegata'	see **A. *procurrens*** 'Variegata'
	procurrens	last listed 99/00
un	***procurrens*** 'Glacier'	oSqu
	procurrens 'Variegata'	oFor oSis oSec wAva oUps
	x sturii	oFor oJoy oSqu wSta
	ARACHNIODES	
	simplicior	oFor oRus oBRG
un	***simplicior*** 'Variegata'	wSnq
	ARALIA	
	cachemirica	wHer
	californica	wHer wCol oGar oCrm wBWP
	chinensis (see Nomenclature Notes)	oFor wCoN
	chinensis DJHC 655	wHer
	continentalis HC 970077	wHer
	cordata HC 970214	wHer
	elata	oFor oGos oEdg oGar wCoN
	elata 'Aureovariegata'	wHer oGos oDan wWel
	elata 'Silver Umbrella'	wHer wWel
	elata 'Variegata'	wHer oGos oDan wWel
	racemosa	oGoo oUps
	sp. EDHCH 97206	wHer
	spinosa	wCCr

	Plant	Sources
	ARAUCARIA	
	araucana	oFor oTrP wBur oDar oAls oGar oRiv oSho oWhi oBlo wRai oOEx oGre wCCr wCoN wCoS wWel
	bidwillii	oTrP
	heterophylla	oGar
	ARAUJIA	
	sericifera	oRar
	ARBUTUS	
	arizonica	wCCr
	Marina'	oFor wHer oGos oGar oRiv oGre oWnC wCoN wSte wWel
	menziesii	oFor wWoB wBur wNot oDar oAls oGar oDan oBos oRiv oGre wWat oWnC wCoN wRav oSle
	unedo	oFor oAls oRiv wRai oOEx wCCr wTGN wCoN wSte
	unedo 'Compacta'	oFor oAls oGar oRiv oSho oGre wPir oWnC wSta wWel
	unedo f. ***rubra*** from seed	wCCr
	ARCTERICA	see **PIERIS**
	ARCTIUM	
	lappa	oGoo oCrm oBar
	ARCTOSTAPHYLOS	
	canescens	oFor oGar
	columbiana	oFor wClo oBos wWat
	densiflora 'Howard McMinn'	oFor oSho
	glauca	oCis
	hookeri	wShR wCCr
	manzanita	wCCr
	x media 'Wood's Red'	oFor wMtT
	nevadensis	oFor oTri
	nevadensis x ***canescens***	oFor oGar wCCr
un	***nevadensis*** 'Chipeta'	oSis
	nummularia	last listed 99/00
	nummularia hybrid	oFor
	Oregon State hybrid	oFor wCCr
	'Pacific Mist'	oFor
	pajaroensis 'Warren Roberts'	last listed 99/00
	patula	oFor oGar
	stanfordiana ssp. ***bakeri***	last listed 99/00
	uva-ursi	wThG wWoB wNot oAls wShR oGoo oBos oBlo wWat iGSc oCrm oTri wPla oCir
	uva-ursi x ***andersonii***	wCCr
un	***uva-ursi*** 'Big Bear'	oFor
un	***uva-ursi*** 'Emerald Carpet'	oFor oGar oWnC wWel
un	***uva-ursi*** 'Green Star'	wSta
	uva-ursi 'Massachusetts'	oFor oAls oGar oRiv oSho oBlo wRai oGre wCCr oWnC wSta oSle wCoS wWel
un	***uva-ursi*** 'Mendocino'	wCCr
un	***uva-ursi microphylla***	oFor
un	***uva-ursi*** 'Mitsch's Selected Form'	oSis
	uva-ursi 'Point Reyes'	oFor wCul oBlo oWnC wTGN
	uva-ursi 'Radiant'	oFor oWnC
	uva-ursi 'Thymifolia'	oFor oSis
	uva-ursi 'Vancouver Jade'	oFor oAls oGar oSis oSho oGre wSta wSte
	uva-ursi 'Vulcan's Peak'	oSis
	uva-ursi 'Wood's Compact'	oSis wSta
un	***uva-ursi*** 13	wCCr
	viscida	oFor
	ARCTOTIS	
	fastuosa 'Zulu Prince'	oGar
	x hybrida 'Apricot'	wHer
	x hybrida 'China Rose'	wHer
	x hybrida 'Flame'	last listed 99/00
	x hybrida Harlequin hybrids white	oSis
	x hybrida 'Mahogany'	wHer
	x hybrida 'Pink'	wHer
	x hybrida 'Red Devil'	wHer
	x hybrida 'Wine'	oSis
	venusta	oUps
	ARDISIA	
	japonica	wHer oAls oGar oCis oSho oGre wTGN wWel
un	***japonica*** 'Chirimen'	wHer oGar oSho
un	***mamillata***	oTrP
	ARENARIA	
	balearica	oTrP oSis oBov oGre wWhG
	'Blue Cascade'	wMtT
un	***cretica*** var. ***stygia***	wMtT
	montana	oFor wCul oAls oSis oGre iArc wWhG
un	***norvegica*** ssp. ***norvegica***	oSis
	obtusiloba	see **MINUARTIA *obtusiloba***
	rubella	see **MINUARTIA *rubella***
	sp. Wallowa Mountains	oSis wMtT

	Plant	Sources
	tetraquetra	oSis
	tetraquetra ssp. ***amabilis***	oSis wMtT
	tetraquetra var. ***granatensis***	see **A. *tetraquetra*** ssp. ***amabilis***
	ARGEMONE	
	sp.	wSte
	ARGYRANTHEMUM	
	'Bridesmaid'	wHer
	'Champagne'	last listed 99/00
	'Comtesse de Chambord'	wHer
	frutescens mixed	oGar
	'Fuji Sundance'	last listed 99/00
	'Golden Treasure'	wHer
	'Mike's Pink'	last listed 99/00
do	'Pink Australian'	wHer
	'Rising Sun'	wHer
un	'Silver Lady'	oGar
un	'Sugar Buttons'	oGar
	'Summer Melody'	oGar
un	'Starlight'	wHer
	'Weymouth Pink'	wHer
	'Weymouth Surprise'	last listed 99/00
	ARGYROCYTISUS	see **CYTISUS**
	ARIOCARPUS	
	agavoides	oRar
	fissuratus	oRar
un	***fissuratus*** var. ***hintonii*** SB 1568	oRar
	fissuratus var. ***lloydii***	oRar
	kotschoubeyanus	oRar
	kotschoubeyanus 'Albiflorus'	oRar
qu	***kotschoubeyanus*** var. ***macdowellii***	oRar
	retusus	oRar
	retusus var. ***elongatus***	see **A. *retusus***
	retusus var. ***furfuraceus***	see **A. *retusus***
	retusus San Rafael form	oRar
	scapharostrus	oRar
	trigonus	oRar
	ARISAEMA	
	amurense AEG 57	oNWe
	amurense ssp. ***robustum***	wCri
	amurense ssp. ***robustum*** DJH 061	wHer
	angustatum var. ***peninsulae*** HC 970182	last listed 99/00
	angustatum var. ***peninsulae*** f. ***variegatum*** HC 970197	wHer
	candidissimum	oNWe
	concinnum	oNWe
	sp. aff. ***concinnum*** DJHC 579	wHer
	concinnum DJHC 727	wHer
	consanguineum	wHer oJil wCol oNWe oOEx wRho oGre
	consanguineum AEG 89	oNWe
	consanguineum DJHC 009	last listed 99/00
	consanguineum DJHC 161 B	wHer
	consanguineum EDHCH 97039	wHer
	consanguineum Himalayan form	last listed 99/00
	sp. aff. ***consanguineum*** PJ 082	oNWe
	costatum	last listed 99/00
	dracontium	oTrP oRus oDan
	erubescens	wCol
	exappendiculatum brown form	last listed 99/00
	fargesii	last listed 99/00
	flavum	oRed oNWe oGre
	flavum abbreviatum	see **A. *flavum***
	flavum CHAD 1292	last listed 99/00
	flavum Giant form AEG 81	oNWe
	flavum glossy leaf form	last listed 99/00
	flavum 66 84 193, AEG 72	oNWe
	formosanum BSWJ 3703	wHer
	franchetianum	wHer
	franchetianum DJHC 447	wHer
	sp. aff. ***griffithii*** HWJCM 147	last listed 99/00
	heterophyllum	last listed 99/00
	heterophyllum DJH 154	wHer
	intermedium HWJCM 444	wHer
	sp. aff. ***intermedium*** HWJCM 080	last listed 99/00
	jacquemontii	wHer wCri
	japonicum	see **A. *serratum***
un	***kelung-insularis*** BS&WJ 3704	wHer
	kiushianum	wHer
un	***negishii***	wHer
	nepenthoides	wHer
	ochraceum	see **A. *nepenthoides***
	ringens	oGar oNWe
	ringens B&SWJ 1522	wHer
	sp. aff. ***robustum***	oNWe
	robustum HC 970183	wHer
	saxatile	last listed 99/00

	sazensoo	oNWe
	serratum	last listed 99/00
	serratum HC 970630	wHer
	serratum silver leaf pattern	wHer
	sikokianum	oRed oNWe oGre
	sikokianum GBG 107	oNWe
	sikokianum HC 970572	wHer
	sikokianum RSBG	last listed 99/00
un	***sikokianum*** x ***takedae***	oNWe
	sp. DJHC 560	wHer
	sp. HC 970499	wHer
	sp ex Tibet	oNWe
	speciosum	last listed 99/00
	speciosum var. ***sikkimense***	oRed
	taiwanense	wHer wWoS oNWe
	taiwanense B&SWJ	last listed 99/00
	taiwanense silver leaf	wHer
	tortuosum	wHer oNWe
	tortuosum hardy, giant form	oNWe
	tortuosum HWJCM 561	last listed 99/00
	tosaense	wHer
	triphyllum	oFor oTrP wThG oNat oEdg wCol oRus oAmb oMis oRed oDan oAld
	sp. aff. ***verrucosum*** HWJCM 161	last listed 99/00
	yamatense var. ***sugimotoi*** HC 970744	wHer
	yunnanense DJHC 011	wHer
	sp. aff. ***yunnanense*** PJ164	oNWe
	ARISARUM	
	proboscideum	oTrP wHer wWoS wThG wCol oHed oDan oSis oNWe
	ARISTOLOCHIA	
	californica	wHer oRiv
	clematitis	oFor
	durior	see **A. *macrophylla***
	elegans	see **A. *littoralis***
	littoralis	oFor oTrP oUps
	longa ssp. ***paucinervis***	wHer
	macrophylla	oFor oGar
	manshuriensis HC 970157	last listed 99/00
	ARISTOTELIA	
	chilensis	wHer
	serrata	oFor
	ARMERIA	
	alliacea	wCCr oWnC wCoN
	'Bee's Ruby'	last listed 99/00
	caespitosa	see **A. *juniperifolia***
	girardii	wCoN
	juniperifolia	wTho oDar oJoy oSis wFai wWhG
	juniperifolia 'Alba'	wCSG oSis
	juniperifolia 'Bevan's Variety'	oJoy
	maritima	wTho oJoy wShR oGar oSho oGre wSta oCir wAva wCoN wWld oUps
	maritima 'Alba'	oFor oWnC oCir oSle
	maritima 'Bloodstone'	oMis oGar oSho oWnC wCoS
	maritima 'Cotton Tail'	wWoS oGar oSho
	maritima Duesseldorf Pride	see **A. *maritima*** 'Duesseldorfer Stolz'
	maritima 'Duesseldorfer Stolz'	oDar oJoy oGar oDan wWhG
	maritima 'Pink Lusitanica'	wBWP
un	***maritima*** 'Rubrifolia'	wWoS oHed oGar oSis oNWe
	maritima 'Snowball'	wCCr
	maritima 'Spendens'	oGar oSho oWnC oUps
un	***maritima*** 'Victor Reiter'	oFor oDan oSqu oCir
	maritima 'Vindictive'	oWnC
va	'Nifty Thrifty'	oHed
	plantaginea	see **A. *alliacea***
	pseudarmeria	oMis
un	***purpurascens***	oMis
	ARMORACIA	
	rusticana	oAls wCSG oGar oGoo oSho wFai iGSc
	rusticana 'Variegata'	oFor oEdg wCol oGar wRob oSec
	ARNICA	
	chamissonis	oGoo iGSc oCrm
	cordifolia	wPla
	montana	oCrm oBar oUps
	sp.	wWld
	ARONIA	
	arbutifolia	wCul
	arbutifolia 'Brilliantissima'	oFor oGar oRiv oGre wWel
	melanocarpa	oFor wBCr wBur oRiv oSho oOEx oGre wKin
	melanocarpa 'Autumn Magic'	wCul wBur oDan wTGN
	melanocarpa var. ***elata***	oGar oGre
un	***melanocarpa*** 'Nero'	wRai wBox
	melanocarpa 'Viking'	wClo wRai wWel
	x prunifolia	wCCr

	ARRACACIA	
	xanthorrhiza	oOEx
	ARRHENATHERUM	
	elatius ssp. ***bulbosum***	wSwC wWhG wCoS
	elatius ssp. ***bulbosum*** 'Variegatum'	oFor wHer wWoS wCli wCul oAls oGoo oOut wWin oWnC oTri wTGN oBRG oCir
	ARTEMISIA	
	abrotanum	oAls wCSG oGoo wFai iGSc oBar wNTP
un	***abrotanum*** 'Limoneum'	oFor oAls
	abrotanum silver leaved	wFGN oGoo
un	***abrotanum*** 'Tangerine'	wFai iGSc
	absinthium	wFGN oGoo wFai iGSc oCir wCoN oUps
	absinthium 'Huntingdon'	oFor wHer wFGN oSec
	absinthium 'Lambrook Silver'	oSis wFai
	afra	oGoo
	alba	oBar
	albula 'Silver King'	see **A. *ludoviciana*** ssp ***mexicana*** var. ***albula***
un	***alclockii***	wMtT
	annua	wFGN iGSc oCrm wNTP oUps
	arborescens	oGoo
	assoana	see **A. *caucasica***
	campestris ssp. ***borealis***	oSqu
	campestris ssp. ***borealis wormskioldii***	see **A. *campestris*** ssp. ***borealis***
	camphorata	see **A. *alba***
	cana	see **SERIPHIDIUM *canum***
	canescens (see Nomenclature Notes)	wCul
	caucasica	oSis wFai
	caucasica caucasica JJH909461	wMtT
un	'Doreen'	wWoS
	douglasiana	wWld
	dracunculus	wWoS wCul oTDM oJoy wFGN wCSG wRai wFai iGSc wHom oWnC wNTP wCoN oUps
	dracunculus dracunculoides	oGoo
	dracunculus var. ***sativa***	oAls oGoo oBar
un	***dracunculus*** 'Texas'	wHom
	frigida	oGoo wFai wPla wCoN
	glacialis	wMtT
	indica	oGoo
un	***indica*** 'Variegata'	wCol
	lactiflora	oGoo wFai
	lactiflora Guizhou Group	oFor wHer wCul oNat oDar oJoy wCol wHig oAls oHed oRus oMis oGar oGoo oOut oWnC oSec oCir wAva oUps oSle wSte
	ludoviciana	oGoo wPla wCoN
	ludoviciana ssp ***mexicana*** var. ***albula***	oFor oDar oAls wFGN oGar oGoo wFai wHom oWnC oBar oUps
	ludoviciana 'Silver Frost'	wCul oSis
	ludoviciana 'Silver King'	see **A. *ludoviciana*** ssp ***mexicana*** var ***albula***
	ludoviciana 'Silver Queen'	oFor wFGN oGoo oGre wFai wHom oBar
	ludoviciana 'Valerie Finnis'	oFor wHer wWoS wCul oNat oAls wFGN oGoo oSis oOut wFai wHom oWnC oSec
	nova	see **SERIPHIDIUM *novum***
	nutans	see **SERIPHIDIUM *nutans***
	pedemontana	see **A. *caucasica***
	pontica	wFGN oSis wFai wCoN
	'Powis Castle'	oFor wHer wWoS wCul wSwC oNat oAls wFGN oGar oGoo oSho oGre wFai wCCr wHom oWnC oBar wTGN oSec oInd wBWP oWhS wCoN wEde oUps oSle wSnq
	pycnocephala 'David's Choice'	oWnC wSnq
	schmidtiana	oJoy oSqu
	schmidtiana 'Nana'	wWhG
	schmidtiana 'Silver Mound' (see Nomenclature Notes)	oFor oAls oGar oBlo oGre wHom oWnC wTGN oSec oCir oEga oUps oSle
	'Silver Brocade'	see **A. *stelleriana*** 'Boughton Silver'
	'Silver Dust'	see **SENECIO *cineraria*** 'Silver Dust'
	stelleriana	wCSG wCoN
	stelleriana 'Boughton Silver'	wTho oAls oGoo oSho oSqu wFai wHom oWnC wTGN iArc oCir oUps
	stelleriana 'Silver Brocade'	see **A. *stelleriana*** 'Boughton Silver'
	tridentata	see **SERIPHIDIUM *tridentatum***
	versicolor	oSis
un	***versicolor*** 'Seafoam'	oFor wWoS wWhG
	vulgaris	oGoo wFai iGSc oCrm oBar oSec
va	***vulgaris*** 'Cragg-Barber Eye'	wWoS
	ARTHROPODIUM	
	candidum maculatum	oNWe
un	***candidum purpureum*** 'Nanum'	oTrP oJoy
	cirratum	last listed 99/00
	milleflorum	wHer

	Name	Sources
	ARUM	
	concinnatum	wHer
	concinnatum ssp. *albispathum*	wHer
	concinnatum ex T. C. Smale	last listed 99/00
	creticum	oDan
	cyrenaicum	last listed 99/00
	dioscoridis	last listed 99/00
	hygrophilum ex Tel Aviv	last listed 99/00
	italicum	wHer oJoy wCol wCSG oGre wSta wAva wCoN
	italicum 'Green Marble'	last listed 99/00
	italicum hybrid	oNWe
	italicum ssp. *italicum* x *maculatum*	wHer
	italicum ssp. *italicum* 'Marmoratum'	oTrP oGos oNWe oGre
ch	*italicum* ssp. *italicum* 'Tiny'	last listed 99/00
	italicum 'Pictum'	see **A.** ***italicum*** ssp. ***italicum*** 'Marmoratum'
	italicum 'Spotted Jack'	oNWe
	palaestinum	last listed 99/00
	purpureospathum ex PB 51	last listed 99/00
	ARUNCUS	
	aethusifolius	oFor wCul wGAc oJoy wCol oRus wCSG oMis oGar oCis oTwi wNay wRob oSqu wFai oUps
	aethusifolius HC 970263	wHer
	dioicus	oFor oNat wNot oJil oAls oRus wShR wCSG oGar oBos oSho oSha wNay oBlo oGre wFai wWat wCCr oWnC wMag wFFl oUps wSte wSnq
	dioicus DJHC 598	wHer
	dioicus dwarf	oRus
un	*dioicus* 'Francie'	oGar
	dioicus var. *kamtschaticus*	last listed 99/00
	dioicus var. *kamtschaticus* HC 970496	wHer
	dioicus 'Kneiffii'	oFor oAmb wNay
	dioicus sylvestris	see **A.** ***dioicus***
	dioicus 'Zweiweltenkind'	last listed 99/00
	sylvestris	see **A.** ***dioicus***
	ARUNDINARIA	
	amabilis	see **PSEUDOSASA** ***amabilis***
	anceps	see **YUSHANIA** ***anceps***
	atropurpurea	see **SASAELLA** ***masumuneana***
	auricoma	see **PLEIOBLASTUS** ***auricomus***
	falcata	see **DREPANOSTACHYUM** ***falcatum***
	falconeri	see **HIMALAYACALAMUS** ***falconeri***
	fortunei	see **PLEIOBLASTUS** ***variegatus***
	funghomii	see **SCHIZOSTACHYUM** ***funghomii***
	gigantea	oNor wBea oTBG
un	*gigantea* 'Mary Sims'	wBea
	gigantea tecta	see **A.** ***tecta***
	graminea	see **PLEIOBLASTUS** ***gramineus***
	maling	see **YUSHANIA** ***maling***
	pygmaea	see **PLEIOBLASTUS** ***pygmaeus***
	tecta	oTra oNor wBea
	variegata	see **PLEIOBLASTUS** ***variegatus***
	viridistriata	see **PLEIOBLASTUS** ***auricomus***
	ARUNDO	
	donax	wCli wGAc oGar oOut oGre oTri wTGN wCoN
	donax 'Albo Striata'	see **A.** ***donax*** var. ***versicolor***
	donax 'Variegata'	see **A.** ***donax*** var. ***versicolor***
	donax var. *versicolor*	wHer wWoS wCli oGos oAls oSsd oCis oOut wWin oWnC wBox wCoN wSte
	ASARINA	
un	'Amethyst Pink'	oUps
	barclayana	see **MAURANDYA** ***barclayana***
	erubescens	see **LOPHOSPERMUM** ***erubescens***
	procumbens yellow	oGar
	ASARUM	
	arifolium	oFor oTrP wHer wThG oRus oGar oGre
	blumei	wHer
	campaniforme	wHer
	canadense	oFor wHer oRus oGar oGoo oGre wFai iGSc oCir
	caudatum	oFor oHan wThG wWoB oNat wNot oJoy wCol oRus wShR oSis oBos oSho wFai wWat oCrm oTri wPla oAld wCoN wFFl oSle wSte
un	*caudatum* f. *album*	wHer wCol oRus oTri
un	*caudatum* var. *cardiophyllum*	oGos
qu	*caudigerrelum*	oCis
	caudigerum	oCis
un	*caudigerum* var. *cardiophyllum*	wHer
	caulescens	wHer oSis
	chinense	wHer
	delavayi	oCis
	europaeum	oFor oGos oRus oSis oNWe wRob coHon oWnC wMag iArc
	hartwegii	oSis
	lemmonii	oSis
	magnificum	wHer
	maximum	wHer oCis
un	*muramatsui*	wHer
un	*naniflorum* 'Eco Décor'	wHer
	nipponicum	wHer
un	*proforonotum*	oCis
	shuttleworthii	oRus oAmb
un	'Silver Falls'	wAva
	splendens	oFor oGos oRus oCis
	takaoi	wHer wCol
	ASCLEPIAS	
	'Cinderella'	oFor oJoy oAls oGar oSho wAva oUps
	curassavica	oGoo
un	*curassavica* 'Red Butterfly'	iArc oUps
un	*curassavica* 'Silky Gold'	oGoo oUps
	eriocarpa	oGoo
	fascicularis	wNot
	incarnata	oFor wCul oJoy oAls oGoo oSha wFai oSle
	incarnata 'Ice Ballet'	oFor oGar oGoo wFai oUps
	incarnata 'Soulmate'	oFor oGar oUps
	speciosa	wNot oGoo
	syriaca	oFor oJoy oAls oGoo oBlo
	tuberosa	oFor wThG oRus oMis oGar oGoo oSho iGSc oBar wAva oUps wSnq
	tuberosa Gay Butterflies Group	oAls oEga
	tuberosa 'Hello Yellow'	oUps
	verticillata	oGoo
	ASIMINA	
un	*angustifolia*	oRiv
un	*incarna*	oRiv
un	*obovata*	oRiv
un	*parviflora*	oRiv
	triloba	oFor wBur wClo oEdg oRiv wRai oOEx wTGN
un	*triloba* 'Campbell #1'	wBur
un	*triloba* 'Davis'	wBur oOEx
un	*triloba* 'Ford Amend'	wRai
un	*triloba* 'Mary Foos Johnson'	oOEx
un	*triloba* 'Mitchell'	last listed 99/00
un	*triloba* 'Overleese'	wBur wRai oOEx
un	*triloba* 'Pennsylvania Golden'	wBur wClo oGre
un	*triloba* 'Prolific'	wBur wClo oGar wBox
un	*triloba* 'Rebecca's Gold	wBur wRai oOEx
un	*triloba* 'Sunflower'	wBur wClo oGar wRai oOEx oGre wBox
un	*triloba* 'Sweet Alice'	oOEx
un	*triloba* 'Taylor'	wBur
un	*triloba* 'Taytoo'	oOEx
un	*triloba* 'Wells'	wBur oGar wRai oOEx oGre
	ASPARAGUS	
	africanus	oTrP
	densiflorus 'Myersii'	oTrP
	densiflorus Sprengeri Group	oTrP oGar oWnC
	densiflorus 'Sprengeri Compacta'	oGar
	officinalis	wBCr wClo wCoS
un	*officinalis* 'EC 157' male	oGar
un	*officinalis* 'Jersey Knight'	oAls oSho wRai
	officinalis 'Mary Washington'	oAls
un	*officinalis* 'Sweet Purple'	oSho
	plumosus	see **A.** ***setaceus***
	setaceus	oGar
	ASPERULA	
	aristata	oNWe wMtT
	gussonei	oSis wMtT
	odorata	see **GALIUM** ***odoratum***
	sintenisii	wMtT
	suberosa	oSis
	ASPHODELINE	
	liburnica	last listed 99/00
	lutea	wSwC oJoy oAls oDan oNWe oSho wFai wCCr
	ASPHODELUS	
	albus	oJil
	ASPIDISTRA	
	elatior	oTrP oGar wPir wCoS
	sp.	oCis
	ASPLENIUM	
	adulterinum	wFol
	ebenoides	oFor oJoy oGar oSqu oGre oBRG
	forisiense	wFol
un	*kobayashii*	wFol
	nidus	oGar

un	***pekinense***	wFol
	pinnatifidum	wFol
	platyneuron	oTrP
	scolopendrium	oFor oJoy wFan oRus oGar oSis oSqu oGre oBRG wSnq
un	***scolopendrium*** 'Augustifolia'	oRus
	scolopendrium Cristatum Group	oRus oBRG
	scolopendrium 'Kaye's Lacerated'	oBRG
	scolopendrium Laceratum Group	wFan oRus wTGN
	trichomanes Incisum Group	wFol
	ASTARTEA	
	fascicularis	last listed 99/00
	ASTELIA	
	chathamica	wHer
	ASTER	
	albescens	wHer
	albescens EDHCH 97237	wHer
	x alpellus 'Triumph'	oFor
	alpinus	oSis oSho
	alpinus Dark Beauty	see **A. *alpinus*** 'Dunkle Schoene'
	alpinus 'Dunkle Schoene'	oSis wMag
	alpinus 'Goliath'	wTho oSis oCir
	alpinus 'Happy End'	oSis
	alpinus Trimix	iArc
	amellus 'King George'	wWoS oHed wMag
	amellus 'Veilchenkoenigin'	oHed
	amellus Violet Queen	see **A. *amellus*** 'Veilchenkoenigin'
	asperulus	wHer
un	'Blue Butterfly'	oJoy
	carolinianus	oFor wHer
un	'Cassie'	wAva
	chilensis	oFor oAld
	'Chorister'	oFor
	'Climax'	wCul
	'Coombe Fishacre'	wHer wWoS oHed oGar wFai
	cordifolius	oFor
	cordifolius 'Ideal'	wHer
	cordifolius 'Silver Spray'	wFai
	'Dark Pink Star'	oGar
	divaricatus	oFor oHan wCul wCSG oGoo wMag wAva
	divaricatus Raiche form	wHer
	dumosus (see Nomenclature Notes)	
	dwarf selection # 1	wCol
	ericoides	last listed 99/00
	ericoides 'Blue Star'	oFor
	ericoides 'Erlkoenig'	wHer oJoy
	ericoides Esther	last listed 99/00
	ericoides 'Monte Cassino'	see **A. *pringlei*** 'Monte Cassino'
	ericoides 'Pink Cloud'	wMag
	ericoides f. ***prostratus*** 'Snow Flurry'	oFor oHed
	ericoides 'White Heather'	wHer
un	***fendleri*** 'My Antonia'	oFor
	foliaceus	wShR
	x frikartii	wHom
	x frikartii 'Flora's Delight'	oFor wMag wSnq
	x frikartii 'Jungfrau'	wHer
	x frikartii 'Moench'	oFor wHer wSwC oNat oJoy oAls oHed wCSG oAmb oMis oGar wRob oBlo wFai oWnC oEga oUps wSnq
	x frikartii Wonder of Staffa	see **A. *x frikartii*** 'Wunder von Staffa'
	x frikartii 'Wunder von Staeffa'	oFor
un	'Gladys'	wAva
un	***hallii***	oHan oRus
	'Kylie'	wHer
	laevis	oFor oGoo
	laevis 'Calliope'	wHer oGar oCis oNWe oGre
	lateriflorus 'Horizontalis'	oFor oUps
	lateriflorus 'Lady in Black'	oFor wWoS oNat oAmb oGar oNWe oGre wSnq
	lateriflorus 'Lovely'	oFor oWnC
	lateriflorus 'Prince'	oFor wHer wWoS wSwC oAls oHed oRus oDan oSis oNWe wRob oGre wSte
un	'Lilac Blue Admiral'	oJoy
	'Little Carlow'	wHer wWoS wCul oHed oGar oNWe wFai
	'Little Dorrit'	wHer wAva
	macrophyllus 'Albus'	wHer
un	'Melba'	wAva
	modestus	wNot oAls oBos oAld
qu	'Morro-Partridge'	oSec
	nemoralis	oHug
	novae-angliae	oGoo oDan wMag wWld oUps
	novae-angliae 'Alma Poetschke'	see **A.** 'Andenken an Alma Poetschke'
	novae-angliae 'Andenken an Alma Poetschke'	oFor wCul oAls wCSG oGar oSis oGre wFai wMag wTGN oUps
un	***novae-angliae*** 'Cape Cod'	oWnC oCir
	novae-angliae 'Harrington's Pink'	oAls
	novae-angliae 'Hella Lacy'	oFor wHer oAls wCSG oGar oGre
	novae-angliae 'Honeysong Pink'	oFor oNat oJoy oAls oWnC wMag
	novae-angliae 'Mrs. S. T. Wright'	wHer
	novae-angliae 'Purple Dome'	oFor oJoy oAls oGar oOut oWnC
	novae-angliae Red Star	see **A. *novae-angliae*** 'Roter Stern'
	novae-angliae 'Roter Stern'	oAls
	novae-angliae 'Rudelsburg'	wWoS
	novae-angliae September Ruby	see **A. *novae-angliae*** 'Septemberrubin'
	novae-angliae 'Septemberrubin'	oFor oNat oJoy oAls oRus wCSG wMag wTGN wAva oUps
	novae-angliae 'Treasure'	wHer wCul
	novi-belgii	oSec
	novi-belgii 'Alert'	oFor wCul oRus oDan oOut oGre
	novi-belgii 'Alice Haslam'	oFor oSis wTGN wSnq
un	***novi-belgii*** 'Ariel'	wCSG
un	***novi-belgii*** 'Astee Karmijn'	oGar
un	***novi-belgii*** 'Astee Milka'	oGar
un	***novi-belgii*** 'Astee Roset'	oGar
un	***novi-belgii*** 'Augenweide'	wEde
	novi-belgii 'Blue Lagoon'	oGre
un	***novi-belgii*** 'Bonny Blue'	oFor
un	***novi-belgii*** 'Celeste'	oGar
	novi-belgii 'Coombe Radiance'	wWoS
	novi-belgii 'Coombe Ronald'	wHer
	novi-belgii 'Coombe Violet'	wCul
	novi-belgii 'Countess of Dudley'	wWoS
	novi-belgii 'Daniela'	wWoS oWnC
	novi-belgii 'Diana'	oRus
	novi-belgii 'Elsie Dale'	wHer
	novi-belgii 'Jean'	last listed 99/00
	novi-belgii 'Jenny'	oFor wCSG wMag
	novi-belgii 'Judith'	oGar
	novi-belgii 'Lady in Blue'	wWoS oJoy oAls oWnC wMag
un	***novi-belgii*** 'Lambada'	oGar
un	***novi-belgii*** 'Loke Viking'	oGar
	novi-belgii 'Marie Ballard'	oNat
un	***novi-belgii*** 'Monarch'	wMag
	novi-belgii 'Mount Everest'	wMag
un	***novi-belgii*** 'Nesthakchen'	oFor oWnC wMag oSle
	novi-belgii 'Newton's Pink'	oWnC
	novi-belgii 'Niobe'	oSis oGre
un	***novi-belgii*** 'Odin Viking'	oGar
	novi-belgii 'Patricia Ballard'	oJoy wCSG oGar
un	***novi-belgii*** 'Pink Bouquet'	oFor oAls wCSG oSis
	novi-belgii 'Professor Anton Kippenberg'	oFor wCul wSwC oJoy oAls oRus oMis oGar oGoo oDan oSis oWnC wSnq
un	***novi-belgii*** 'Purple Viking'	oGar
	novi-belgii 'Rosenwichtel'	wAva
	novi-belgii 'Royal Opal'	oAls
	novi-belgii 'Royal Ruby'	oRus
	novi-belgii 'Schoene von Dietlikon'	wCSG oGar
	novi-belgii 'Starlight'	oGre
	novi-belgii 'Sterling Silver'	wCul
un	***novi-belgii*** 'Sun Star'	oGar
un	***novi-belgii*** 'Tiny Tot'	oSis
un	***novi-belgii*** 'Violet Carpet'	wCul oSis
un	***novi-belgii*** 'White Fairy'	oGoo
	novi-belgii 'White Ladies'	oJoy
	novi-belgii 'White Swan'	oRus wCSG oGar oWnC
	novi-belgii 'Winston S. Churchill'	oFor oGar
un	***novi-belgii*** 'Wood's Light Blue'	oAls oSis
un	***novi-belgii*** 'Wood's Pink'	oAls oWnC
un	***novi-belgii*** 'Wood's Purple'	wWoS
un	***novi-belgii*** 'Zwergenhimmel'	oGre
un	***oblongifolius*** 'October Skies'	oFor
	'Ochtendgloren'	wHer
un	'Our Latest One'	wHer
	patens	oGoo
	'Pearl Star'	last listed 99/00
	pilosus var. ***demotus***	wHer
	'Pink Star'	last listed 99/00
un	'Porzallan'	wHer
	pringlei 'Monte Cassino'	oFor wWoS oOut oGre wFai oVBI
un	***pringlei*** 'Pink Cloud'	wWoS
	puniceus	oGoo
un	'Purple Monarch'	oJoy
qu	***radulinus***	oBos
	'Ringdove'	wCul
un	'Rose Serenade'	oFor
	savatieri	oFor wTCS
un	***savatieri*** 'Variegata'	wSnq
	sedifolius 'Snow Flurry'	see **A. *ericoides*** f. ***prostratus*** 'Snow Flurry'
	sibiricus	wMtT

	sp. DR97 111	wCol
	spectabilis	oGoo wSnq
	subspicatus	wCCr oAld wWld
un	'Sun Rose'	wCSG wMag
un	'Sungal'	oGar
	tataricus 'Jindai'	wHer
	thomsonii 'Nanus'	last listed 99/00
	tongolensis	oGoo
	tongolensis 'Napsbury'	oWnC
	tongolensis 'Wartburg Star'	see **A. *tongolensis*** 'Wartburgstern'
qu	***tongolensis*** 'Wartberg Star' pink	wHom
	tongolensis 'Wartburgstern'	oFor wTho oAls oSis wTGN oUps
	'White Climax'	wHer wCul
	ASTERANTHERA	
	ovata	oBov
	ovata HCM 98212	wHer
	ASTERISCUS	
	'Gold Coin'	see **A. *maritimus***
	maritimus	oGar wTGN
	ASTILBE	
	'America'	oSho wRob
	'Aphrodite'	wCul oGar oSho wNay wRob
	x arendsii	oSho
	x arendsii 'Amethyst'	oDar oAls wCSG oGar oTwi wNay wRob oGre
	x arendsii 'Anita Pfeifer'	oAls oSho
un	***x arendsii*** 'Bella'	oMis
	x arendsii 'Bergkristall'	wRob
	x arendsii 'Brautschleier'	oFor wCul oAls oGar oSho wRob oGre wFai oWnC wMag
	x arendsii 'Bressingham Beauty'	wWoS oGar wRob oWnC wMag wAva
	x arendsii Bridal Veil	see **A. *x arendsii*** 'Brautschleier'
	x arendsii 'Bumalda'	wNay wRob wFai
	x arendsii 'Catherine Deneuve'	see **A. *x arendsii*** 'Federsee'
	x arendsii 'Cattleya'	oFor oDar wHig oGar oTwi wNay wRob wSnq
un	***x arendsii*** 'Cotton Candy'	wNay wRob
un	***x arendsii*** 'Darwin's Favorite'	wNay
	x arendsii 'Diamant'	oDar oAls wNay
	x arendsii Diamond	see **A. *x arendsii*** 'Diamant'
	x arendsii Elizabeth Bloom / 'Eliblo'	wNay wRob
	x arendsii 'Ellie van Veen"	oGar wNay wSnq
	x arendsii 'Erica'	oGar oSho wNay oBlo oGre wSnq
	x arendsii 'Fanal'	oFor wCul oGar oSho oTwi wNay wRob oBlo wFai wMag wAva wSnq wCoS
	x arendsii 'Federsee'	oGar oTwi wRob oBRG oCir
	x arendsii 'Feuer'	oFor wNay
	x arendsii Fire	see **A. *x arendsii*** 'Feuer'
un	***x arendsii*** 'Flamingo'	wNay
	x arendsii 'Gertrud Brix'	oGar
ch	***x arendsii*** 'Gladstone'	see **A.** 'W. E. Gladstone'
	x arendsii Glow	see **A. *x arendsii*** 'Glut'
	x arendsii 'Glut'	oFor oGar oSho wRob oGre oWnC
	x arendsii 'Granat'	wNay wRob wMag
	x arendsii 'Grete Puengel'	wNay
	x arendsii Hyacinth	see **A. *x arendsii*** 'Hyazinth'
	x arendsii 'Hyazinth'	oGar oWnC wMag oHou
	x arendsii 'Irrlicht'	wRob
	x arendsii 'Kvele'	wNay
	x arendsii mixed	wCSG wMag
	x arendsii 'Obergaertner Juergens'	wWoS wNay wRob
	x arendsii pink	oGar
ch	***x arendsii*** 'Queen of Holland'	see **A.** 'Queen of Holland'
qu	***x arendsii*** 'Red Cattleya'	oDar wMag
	x arendsii 'Rotlicht'	wRob
	x arendsii Sister Theresa	see **A. *x arendsii*** 'Zuster Theresa'
	x arendsii 'Snowdrift'	wNay wRob oWnC oHou wSnq
ch	***x arendsii*** 'Spartan'	see **A. *x arendsii*** 'Rotlicht'
	x arendsii 'Spinell'	oGar wRob oBlo
	x arendsii 'Venus'	wRob
ch	***x arendsii*** 'Washington'	see **A.** 'Washington'
	x arendsii 'Weisse Gloria'	oGar
	x arendsii white	oMis
	x arendsii 'Zuster Theresa'	wNay
	'Atrorosea'	wNay
	'Avalanche'	wNay wRob wMag
	'Betsy Cuperus'	wNay wRob
	biternata	wNay wRob
	'Bonn'	wRob
	'Bremen'	oGar oCir
	'Bronce Elegans'	wCul oGar oNWe oSho wNay wRob
	Bronze Elegance	see **A.** 'Bronce Elegans'
	chinensis	wShR
	chinensis var. ***davidii*** HC 970083	wHer
un	***chinensis*** var. ***davidii*** 'King Albert'	oTwi wRob
	chinensis DJHC 519	wHer
	chinensis 'Finale'	wCul oSho wNay wRob oHou wSnq
	chinensis 'Intermezzo'	oTwi wNay wRob
	chinensis peach	oSho
	chinensis pink dwarf	wHom
	chinensis var. ***pumila***	oFor wCul wHig oRus oGar oSis oSho oTwi oOut wNay wRob oBov oGre oWnC wMag wTGN oSle wSnq
	chinensis var. ***pumila*** 'Serenade'	wNay wRob
qu	***chinensis*** 'Purple Cats' (see Nomenclature Notes)	oNat oWnC oHou
	chinensis 'Purpurkerze'	wNay oGre
	chinensis var. ***taquetii***	oGar wCCr wHom
	chinensis var. ***taquetii*** 'Bunter Zauber'	oTwi
	chinensis var. ***taquetii*** 'Purpurlanze'	wHig wRob
	chinensis var. ***taquetii*** 'Superba'	oFor wCul oDar oGar oSho oOut wNay wFai wMag wAva oSle wSnq
	chinensis 'Veronica Klose'	oAls oGar wNay
	chinensis 'Visions'	oFor wHig oGar wRob oBRG wSnq
un	***congesta***	wNay
	x crispa 'Lilliput'	oNWe wNay wRob
	x crispa 'Perkeo'	wHig oGar wNay wRob wMag wSnq
	'Darwin's Dream'	wSnq
	davidii	see **A. *chinensis*** var. ***davidii***
	'Deutschland'	oFor wCul wHig wCSG oGar oTwi wNay oBlo oGre oWnC
	'Dunkellachs'	wNay wRob wSnq
	'Elisabeth'	oGar wNay wRob wSnq
	'Emden'	wRob
	'Etna'	oFor wCul wNay wRob wFai wSnq
	'Europa'	oFor oGar wNay wRob oGre oWnC
	glaberrima var. ***saxatilis***	oNWe oTwi wNay oBov
	'Gladstone'	see **A.** 'W. E. Gladstone'
	grandis	last listed 99/00
	grandis DJHC 98411	wHer
	'Hennie Graafland'	oFor wWoS wCul oAls oGar wNay wRob oBlo wMag oHou oUps wSnq
qu	'Incerry'	oFor
	'Inshriach Pink'	wWoS wHig oGar oSis wNay wRob
	'Jo Ophorst'	wWoS wNay
	'Koblenz'	oNWe wNay
	koreana HC970052	wHer
	'Kriemhilde'	wNay
un	'Louie'	wHig
	'Maggie Daley'	oGar wNay wRob
	'Mainz'	wNay wRob
qu	'Moerheim's Glory'	wNay wRob
	'Montgomery'	oGar oSho oTwi wRob oGre wSnq
	Ostrich Plume	see **A.** 'Straussenfeder'
	'Professor van der Wielen'	wCul oNat wNay wRob wTGN
	pumila	see **A. *chinensis*** var. ***pumila***
	'Purple Candle'	see **A. *chinensis*** 'Purpurkerze'
	'Purple Lance'	see **A. *chinensis*** var. ***taquetii*** 'Purpurlanze'
ch	'Queen of Holland'	wNay wRob oWnC
	'Red Sentinel'	oAls oMis oGar oTwi oOut wNay wRob wSnq
	'Rheinland'	oFor oNat oGar wRob oBlo oWnC oHou wSnq
	x rosea 'Peach Blossom'	oFor wCul wHig oAls oGar wNay wRob oBlo wMag wAva oUps wSnq
	simplicifolia	wCol
	simplicifolia 'Bronze Elegance'	see **A.** 'Bronce Elegans'
un	***simplicifolia*** 'Bronze Queen'	oSis
	simplicifolia 'Carnea'	wNay
	simplicifolia 'Darwin's Snow Sprite'	wNay
	simplicifolia 'Jacqueline'	oSho wNay
	simplicifolia 'Praecox'	wRob
	simplicifolia 'Praecox Alba'	oTwi wRob
	sp. DJHC 586	wHer
	'Sprite'	oFor wCul wCol wHig oGar oSis oNWe wNay wRob wTGN wSnq
	'Straussenfeder'	oFor wCul wGAc oGar oTwi oOut wNay wRob wMag wTGN
	taquetii	see **A. *chinensis*** var. ***taquetii***
	'Vesuvius'	oFor oOut wNay
ch	'W. E. Gladstone'	oGar wNay oWnC
ch	'Washington'	wNay wRob oGre
	'Willie Buchanan'	wHig oSho wNay wRob oSqu wMag
	ASTILBOIDES	
	tabularis	oRus oGar oNWe wNay oGre oBRG
	ASTRAGALUS	
	angustifolius	oSis
	crassicarpus	oFor oUps
	ASTRANTIA	
	carniolica	wHig

	carniolica major	see **A. *major***
	carniolica var. ***rubra***	see **A. *major rubra***
	hybrids	oRus
	major	wCul oJoy wCol wHig oHed wCSG oMis oGar oNWe oBlo wFai wCoN oUps
	major alba	wWoS oJoy oGar oDan oSis oOut wNay wFai wBox
	major 'Claret'	wEde
	major 'Hadspen Blood'	wWoS oGos wCol oHed oDan oNWe wRob
	major ssp. ***involucrata*** 'Shaggy'	oNat oNWe oGre
	major ssp. ***involucrata*** 'Shaggy' seed grown	wHer
	major 'Lars'	wWoS oAls oGar wNay wSnq
	major mixed	wRob
	major pink/red	oNWe
un	***major*** 'Pink Star'	wHig
	major 'Primadonna'	wWoS oNWe wNay wFai
	major red selection	oNWe
	major Rose Symphony	see **A. *major*** 'Rosensinfonie'
	major rosea	wNay
	major 'Rosensinfonie'	oGar oSis
	major rubra	oFor wWoS oJoy wHig oAls oOut oSev wAva
	major 'Ruby Cloud'	oJoy
	major 'Ruby Wedding'	oGar oWnC
	major 'Sunningdale Variegated'	wWoS oHed oNWe wNay wSnq
	major white	last listed 99/00
	major white to pink	oNWe oGre
	maxima	wCul wCol wHig
	maxima 'Alba'	oWnC
un	***maxima*** 'Rosea'	wWoS
	'Rainbow'	oAls oGar
	'Rubra'	see **A. *major rubra***
un	**ASTROLEPIS**	
	sp. (sinuata/windhamii)	wFan
	ASTROPHYTUM	
	asterias	oRar
	ASYNEUMA	
un	***linifolium*** ssp. ***linifolium***	wMtT
	ATHROTAXIS	
	cupressoides	oFor wHer oDan oRiv
	laxifolia	oFor wHer wWel
	ATHYRIUM	
un	'Branford's Beauty'	wFan
	cyclosorum	see **A. *filix-femina*** var. ***sitchense***
	filix femina	oFor wWoB wNot oJil oRus wShR oGar oBos oSho oSqu oGre oTrP wTGN oBRG oAld wCoN wSte wSnq
	filix-femina congestum cristatum	oRus
	filix-femina Cristatum Group	wSnq
	filix-femina Fancy Fronds strain	wFan oSis
	filix-femina 'Frizelliae'	wFol wFan
	filix-femina 'Minutissimum'	oSis
	filix-femina var. ***sitchense***	wFan wFFl
un	***filix-femina*** 'Vernoniae Cristatum'	oFor oRus oSis oSqu oGre oBRG
	goeringianum 'Pictum'	see **A. *niponicum*** var. ***pictum***
	niponicum	wFol wFan
	niponicum var. ***pictum***	oFor wFol wSwC oJoy wCol wFan oAls oRus oAmb oGar oSis oNWe oOut wNay oBlo oSqu oGre wHom oWnC oBRG wCoN wSnq
un	***niponicum*** var. ***pictum*** 'Ursula's Red'	oGar wRob
un	***niponicum*** var. ***pictum*** 'Wildwood Twist'	oFor
	otophorum	oFor wFol wFan oRus oGar oTPm oBRG wSnq
	pycnocarpon	see **DIPLAZIUM *pycnocarpon***
	thelypteroides	see **DEPARIA *acrostichoides***
	vidalii	wFan
	ATRACTYLODES	
	macrocephala	oCrm
un	**ATRICHUM**	
	selwynii	wFFl
	ATRIPLEX	
	canescens	oFor wPla
	AUBRIETA	
	'Aureovariegata'	wWoS oHed
	Bengal hybrids	oSis
	'Blue Cascade'	wWhG
	'Cascade Purple'	see **A.** 'Purple Cascade'
	'Cascade Red'	see **A.** 'Red Cascade'
	deltoidea Variegata Group	oFor oSis wWhG
un	***deltoidea*** 'Variegata Blue'	oGre
	'Doctor Mules'	oSis
	'Gloriosa'	oSis
	gracilis	oSis
	'Gurgedyke'	oFor oSis
	'Hendersonii'	wTho
un	'Lamb's Brilliant	oFor oCir
	'Leichtlinii'	oFor wTho oSis
	pinardii	wMtT
	'Purple Cascade'	oAls oWnC wTGN wAva wWhG
un	'Purple Gem'	oSis
un	'Purple Heart'	oFor wTGN oUps
	'Red Cascade'	oAlsoGar wTGN wAva wWhG oUps
un	'Rokey's Purple'	oFor wTGN oUps
	'Royal Blue' (Royal Series)	wFai
	'Royal Velvet'	oSis
	'Whitewell Gem'	wTho oAls
	AUCUBA	
	himalaica	wHer
va	***japonica*** 'Crotonifolia' female	last listed 99/00
un	***japonica*** 'Emily Rose'	wWel
	japonica 'Fructu Albo'	wCCr
va	***japonica*** 'Gold Dust' female	oAls oBlo oWnC
un	***japonica*** 'Gold Spot'	oBlo
	japonica 'Goldstrike' male	oGar oSho oGre oWnC wTGN wWel
	japonica male	wCCr
	japonica 'Marina'	oCis
	japonica 'Mr. Goldstrike'	see **A. *japonica*** 'Goldstrike'
	japonica 'Nana Rotundifolia' female	oJoy
va	***japonica*** 'Picturata' male	oAls oGar oSho oGre oWnC wTGN wAva wCoS wWel
	japonica 'Rozannie' female/male	wHer oGos oGre
	japonica sawtoothed	see **A. *japonica*** 'Serratifolia'
	japonica 'Serratifolia'	oAls oGar oSho oWnC wWel
un	***japonica*** 'Spiker'	oGar
va	***japonica*** 'Variegata' female	oGar wWel
	serrata	see **A. *japonica*** 'Serratifolia'
	AURINIA	
	saxatilis	oGar wTGN oUps
	saxatilis 'Compacta'	oAls
	saxatilis 'Dudley Nevill Variegated'	oSis
	saxatilis Gold Ball	see **A. *saxatilis*** 'Goldkugel'
	saxatilis 'Gold Dust'	wTho oEga wWhG
	saxatilis 'Goldkugel'	oSho oWnC oSle
un	***saxatilis*** 'Mountain Gold'	oAls
	saxatilis 'Sulphurea'	wWhG
	saxatilis 'Sunny Border Apricot'	oFor
qu	***saxatilis*** 'Sunnybrook Apricot'	oSec
	saxatilis 'Tom Thumb'	oSis
	AUSTROCEDRUS	
	chilensis	oFor
	AVENA	
un	***sativa*** 'Cayouse'	oAls
	AVERRHOA	
	carambola	last listed 99/00
	AZARA	
	dentata	oFor wHer oRiv oCis
	integrifolia	oFor
	lanceolata	oFor oRiv oSho oSec
	lanceolata HCM 98163	wHer
	microphylla	oFor wHer oGar oRiv oCis oBov oGre oSec wSte wWel
	microphylla 'Variegata'	wHer oCis oSec wSte
	petiolaris	wCCr
	petiolaris HCM 98066	wHer
	serrata	oFor oRiv oCis oSho wCCr
	AZOLLA	
	caroliniana (see Nomenclature Notes)	oTrP
	sp.	wGAc oHug oSsd
	AZORELLA	
	trifurcata 'Nana'	oFor wMtT
	AZORINA	
	vidalii	oUps
	BABIANA	
	pulchra	oFor
	stricta	oFor
	BACCHARIS	
	halimifolia	oFor
	halimifolia 'Twin Peaks'	oFor
	magellanica	wHer
	patagonica	oFor
	pilularis	oBos
un	***pilularis*** 'Al's Blue'	oCis
	pilularis var. ***consanguinea***	oFor wCCr
	sp. HCM 98031	wHer
	sp. HCM 98099	wHer
	BACOPA	
	amplexicaulis	see **B. *caroliniana***
	caroliniana	oHug oGar

	'Deep Rose'	wTGN
qu	'Giant Snowflake'	wTGN
un	*lenageria*	wGAc oGar
	monnieri	oCrm
	BALDELLIA	
	ranunculoides f. *repens*	oHug wTGN
	BALLOTA	
	acetabulosa	oHed oSis
	'All Hallows Green'	wWoS
	nigra	oNat oGar
va	*nigra* 'Archer's Variegated'	oFor wWoS oGoo oSec
	nigra 'Variegata'	see **B. *nigra*** 'Archer's Variegated'
	pseudodictamnus	wCSG
un	*pseudodictamnus* 'Nana'	oSis
	BALSAMORHIZA	
	deltoidea	oFor wShR wWld
	sagittata	wPla
	BAMBUSA	
qu	*dissimulator*	oTra
un	*maligensis*	oTra wBea
	multiplex 'Alphonse Karr'	oTra wCli wBea oOEx oTBG
	multiplex 'Fernleaf'	oTra wBea wBox
	multiplex 'Golden Goddess'	oTra oGar oGre wBox
un	*multiplex* 'Green Alphonse Karr'	oTra
	multiplex var. *riviereorum*	oNor wBea wBox oTBG
	multiplex 'Silverstripe'	oTra oNor wBea
	multiplex 'Stripe Stem Fernleaf'	oTra
	multiplex 'Wang Tsai'	see **B. *multiplex*** 'Fernleaf'
un	*mutabilis*	wBea
	oldhamii	oTra wBea
un	*pachinensis*	oTra
un	*sinospinosa*	oTra
	textilis	oTra
	tuldoides	oTra
	ventricosa	oTra wCli wBea oTBG
un	*ventricosa* 'Kimmei'	oTra
	vulgaris 'Vittata'	oTra wCli
	BANISTERIOPSIS	
	caapi	oOEx
	BANKSIA	
	littoralis	last listed 99/00
	marginata	oTrP
	occidentalis	last listed 99/00
	saxicola	oFor
	serrata	oTrP
	spinulosa	last listed 99/00
	BAPTISIA	
	australis	oFor wHer wCul wSwC oNat oAls oHed oRus wFGN wCSG oMis oGoo oNWe oSha wFai wMag oBar wTGN oSec iArc oWhS oCir wCoN oUps wSte
un	*australis minor*	oUps
	australis white	oMis
	bracteata	oFor
	lactea	wHer oSec
	leucantha	see **B. *lactea***
	leucophaea	see **B. *bracteata***
	'Purple Smoke'	wHer oHed oNWe
	tinctoria	oFor
un	*viridis*	wCul
	BASHANIA	
	fargesii	oTra wCli wBlu wBea wBmG
	BAUHINIA	
	monandra	oTrP
	natalensis	oTrP
	petersiana	oTrP
un	**BAUMEA**	
	rubiginosa	wHer oHug oWnC
	rubiginosa 'Variegata'	wGAc oGar oSsd wTGN
	BEAUCARNEA	see **NOLINA**
	BEGONIA	
	boliviensis	oHed
	boliviensis B 83.0252	oRar
	cordifolia	last listed 99/00
un	sp. aff. *filipes* BLM 0522	oRar
	fuchsioides	oTrP
	grandis	see **B. *grandis*** ssp. ***evansiana***
	grandis ssp. *evansiana*	oFor wWoS oNat oRus oRed oDan oSis oSho wRob oBov oGre wFai oSec wSnq
	grandis ssp. *evansiana* var. *alba*	oFor wHer wWoS oRus oSis wRob wSnq
	sp. aff. *grandis* ssp. *evansiana* EDHCH 97322	wHer
	grandis ssp. *evansiana* HC 970628	wHer
un	sp. aff. *grisea*	oRar
	x hiemalis	oGar
	Non Stop Series	wTho oGar
	partita	oRar
	peltata	oRar
	'Pin-up'	wTho
qu	*rieger*	oGar
	sinensis	wHer oSec
	sp. DJHC 580	wHer
	sutherlandii	wHer oSis oSec wSte
	BELAMCANDA	
	chinensis	oEdg oGoo oGre iGSc oCrm oBar wTGN wCoN oSle
	chinensis HC 970594	wHer
	BELLIS	
	perennis	wCoN
	perennis 'Dresden China'	last listed 99/00
	perennis 'Pink Buttons'	wWhG
	perennis Pomponette Series	wTho oAls oWnC oCir
un	*perennis* 'Red Button'	wWhG
	perennis 'Rob Roy'	wWoS oWnC
	perennis 'Super Enorma'	oWnC
un	*perennis* 'White Button'	wWhG
	BELLIUM	
	bellidioides	oSis
un	*cahruliescens*	wCul
	minutum	last listed 99/00
	BERBERIDOPSIS	
	corallina	oGos
	BERBERIS	
	aggregata	oFor
	amurensis	oFor wHer
	aquifolium	see **MAHONIA *aquifolium***
	aristata	oFor oRiv
	'Arthur Menzies'	see **MAHONIA *x media*** 'Arthur Menzies'
un	*barandana*	wHer
	buxifolia 'Nana'	see **B. *buxifolia*** 'Pygmaea'
	buxifolia 'Pygmaea'	oFor wCul oAls oGar oRiv oBlo
	calliantha	wHer oGos
	x carminea 'Pirate King'	oFor
	chinensis	oFor oRiv
un	*cristata*	oOEx
	darwinii	oFor wHer oDar oEdg oRiv oBlo oGre
	dasystachya	oFor
	dictyophylla	oFor wHer oRiv
	dictyota	see **MAHONIA *dictyota***
	dielsiana	oFor oRiv
	empetrifolia	wHer
	x frikartii 'Amstelveen'	oFor
	gagnepainii (see Nomenclature Notes)	wHer
	gagnepainii 'Chenault'	see **B. *x hybridogagnepainii*** 'Chenaultii'
	gilgiana	oFor oRiv
	x gladwynensis 'William Penn'	oFor oGar oSho oBlo oWnC wTGN
	gyalaica	oFor oRiv
	hookeri	oFor oRiv
	x hybridogagnepainii 'Chenaultii'	oFor
	x interposita 'Wallich's Purple'	oFor
	jamesiana	oRiv
	julianae	oFor oAmb oGar oRiv oTPm
un	*x kewensis*	oCis
	koreana	oFor oRiv
	lempergiana	last listed 99/00
	linearifolia	wHer
	linearifolia 'Orange King'	wHer
	x lologensis 'Apricot Queen'	oWnC
	lycium	oFor
	x media 'Parkjuweel'	oFor
	x media 'Red Jewel'	oFor
	x mentorensis	oFor oWnC
	mitifolia	oFor
	morrisonenis	oFor oRiv
	nervosa	see **MAHONIA *nervosa***
	x ottawensis f. *purpurea*	wCCr
va	*x ottawensis* 'Silver Miles'	oFor
	prattii	oFor oRiv oSho wCCr oSec
	pruinosa	wHer wCol
	regeliana	oFor
	repens	see **MAHONIA *repens***
	replicata	oGos oCis
	replicata hort.	wHer
	shensiana	last listed 99/00
	sherriffii	wHer oGos
	sieboldii	oFor wHer
	sp.	wCul
	x stenophylla 'Claret Cascade'	oFor
	x stenophylla 'Corallina Compacta'	wCul oBRG
	x stenophylla 'Irwinii'	oFor wHer oGos oBlo oGre oWnC wSta

	Name	Sources
	x stenophylla 'Nana'	oSho
qu	***x stenophylla*** 'Nana Compacta'	oFor oDar
un	'Tara'	oFor
	temolaica	last listed 99/00
	thunbergii	oGar
	thunbergii f. ***atropurpurea***	oFor wCul wBCr oDar oGar oRiv wFai wKin oWnC wSta oGue wWel
	thunbergii 'Atropurpurea Nana'	oFor wCul oDar oAls oGar oRiv oSho oBlo oGre wKin oTPm oWnC wTGN wSta oBRG wAva oGue oSle wWel
	thunbergii 'Aurea'	oFor wHer wCul oDar oAls oHed oAmb oGar oNWe oGre wFai oTPm oGue oUps wWel
un	***thunbergii*** 'Aurea Nana'	wHer
	thunbergii 'Bagatelle'	oFor oGos oGre wWel
un	***thunbergii*** 'Concord'	oFor
	thunbergii 'Crimson Pygmy'	see **B. *thunbergii*** 'Atropurpurea Nana'
un	***thunbergii*** 'Crimson Velvet'	oFor wCul oGar
un	***thunbergii*** 'Gentry'	oGar
	thunbergii 'Gold Ring'	see **B. *thunbergii*** 'Golden Ring'
	thunbergii Golden Nugget™/ 'Monlers'	oAls oGar oWnC
	thunbergii 'Golden Ring'	oFor wHer wCul oDar wCSG oAmb oDan wAva oGue wWel
	thunbergii 'Helmond Pillar'	oFor wHer oGos oGar oDan oCis oGre wPir oBRG wWel
un	***thunbergii*** 'Marshall Upright'	oFor
va	***thunbergii*** 'Rose Glow'	oFor wCul oDar oAls oGar oRiv oSho oBlo oSqu oGre wKin oTPm oWnC wTGN wAva oGue wCoS wWel
un	***thunbergii*** 'Royal Burgundy'	wTGN
	thunbergii 'Sparkle'	oFor oAls oGar wTGN
	thunbergii Sunsation™/ 'Monry'	oAls oGar oWnC wWel
un	***thunbergii*** 'Thornless'	oFor
un	***tischleri***	oFor
	triacanthophora	see **B. *wisleyensis***
	vernae	oFor
	verruculosa	oFor wCul oGar oGre oTPm oWnC
un	***virescens***	oFor
un	***vulgaris*** 'Royal Cloak'	wHer oGos oGar oGre
un	***vulgaris*** 'Velvet Cloak'	oSec
	sp. aff. ***wallichiana***	wHer
	wilsoniae	oFor oCis
	wilsoniae 'Ace'	wHer
	wisleyensis	oFor oTPm
	BERCHEMIA	
	racemosa	oFor
	scandens	oFor
	BERGENIA	
	'Abendglut'	wHig wTGN wCoS
	'Baby Doll'	wHig oGar oGre wBox oBRG
	'Beethoven'	oNWe oWnC
un	Blackthorn hybrids	wHer
qu	'Bressingham'	oSto
	'Bressingham Ruby'	wWoS oGar oSev wFai wBox oBRG
	'Bressingham Salmon'	last listed 99/00
	'Bressingham White'	wWoS oGar
	'Britten'	wHer wWoS oRus
	ciliata	oSto wHig oGar oNWe oOEx wFai wBox
un	***ciliata*** var. ***ciliata***	oHed
	cordifolia	oFor oGar oBlo wFai wMag wSta oInd iArc wAva oSle
	cordifolia light pink	oSto
	cordifolia 'Redstart'	oGar oWnC wBox oInd oUps
	cordifolia Winter Glow	see **B. *cordifolia*** 'Winterglut'
un	***cordifolia*** 'Winterglut'	oFor w WoS oJoy oSev oGre
	crassifolia	wTGN oCir oUps
	'Eroica'	wWoS
	'Evening Glow'	see **B.** 'Abendglut'
	'Glockenturm'	oAls
un	***latifolia***	wMag
	ligulata	see **B. *ciliata***
	'Morgenroete'	oWnC wBox oBRG
	new hybrids	oUps
un	***omeiensis***	wHer oRus
	'Perfect'	oSto wHig
	purpurascens	wWoS wCCr oWnC wSnq
un	***purpurascens*** 'Cramond'	wHig oSis
	purpurascens DJHC 98206	wHer
	'Red Beauty'	oJoy
	Red Bloom	see **B.** 'Rotblum'
	'Rotblum'	oFor w WoS oAls oGar oSho wFai
	'Silberlicht'	oFor oSto wHig oGar oGre oBRG
	Silver Light	see **B.** 'Silberlicht'
	stracheyi	oHed
	stracheyi Alba Group	wHig
	'Sunningdale'	last listed 99/00
	BERKHEYA	
	multijuga	last listed 99/00
	BERLANDIERA	
	lyrata	oFor oMis oGoo wSte
	BESSEYA	
	alpina	last listed 99/00
	wyomingensis	wMtT
	BETULA	
	alba	see **B. *pendula***
	albosinensis	oDar wCol oGar oRiv oGre
	albosinensis var. ***septentrionalis***	oFor
	alleghaniensis	oFor oRiv wCCr
	apoiensis	oFor
	costata	oRiv wCCr
un	'Crimson Frost'	oFor oGre
	davurica	oFor oDar oEdg oGre wCCr
	davurica HC 970136	wHer
	ermanii	oFor oGar wCCr
	fontinalis	see **B. *occidentalis***
	glandulifera	see **B. *pumila***
	glandulosa	oRiv
	grossa	oFor oRiv
	jacquemontii	see **B. *utilis*** var. ***jacquemontii***
	lenta	oFor oRiv
	lenta ssp. ***uber***	oFor
	luminifera	oFor oDar oEdg oRiv
	mandshurica var. ***japonica***	oFor wBCr oRiv wKin
	maximowicziana	last listed 99/00
	nana	oFor oSis
	nigra	oFor wBCr oGar oRiv
un	***nigra*** 'Fox Valley'	oFor oSho
	nigra 'Heritage'	oFor oDar oEdg oAmb oDan oGre oWnC wTGN wSte
	nigra 'Little King'	oAmb
	occidentalis	oFor oBos wCCr wPla
	papyrifera	oFor wWoB wShR oGar oRiv wWat wKin oWnC wTGN wPla wFFl
	pendula	wBCr oGar oBlo wKin wTGN oSle
	pendula 'Atropurpurea'	see **B. *pendula*** 'Purpurea'
un	***pendula*** 'Burgundy Wine'	oHed oGar
	pendula clump	oAls
	pendula 'Dalecarlica'	see **B. *pendula*** 'Laciniata'
	pendula 'Fastigiata'	wSta
	pendula 'Laciniata'	oFor oAls wCSG oWhi wKin oWnC wCoS
	pendula Purple Rain®/ 'Monle'	oGar oWnC wCoS
	pendula 'Purpurea'	oFor oRiv oBlo oGre oSle wCoS
	pendula 'Youngii'	oFor oAls oGar oSho oBlo wKin oWnC wTGN wSta wCoS
	platyphylla	oEdg
	platyphylla var. ***japonica***	see **B. *mandshurica*** var. ***japonica***
	platyphylla 'Whitespire'	oFor
	populifolia	oFor wKin
	pubescens	oFor
	pumila	oFor oRiv oSle
	saposhnikovii	oFor
	schmidtii	oRiv
	tianschanica	oFor wCCr
	'Trost's Dwarf'	oFor wFWB oGar oSis oRiv oSho
un	***turkestanica***	oFor oGre
	uber	see **B. *lenta*** ssp. ***uber***
	utilis	oFor
	utilis HWJCM 260	wHer
	utilis var. ***jacquemontii***	oFor oDar oEdg oAls oGar oDan oSho oWhi wRai oGre wCCr wKin oWnC wTGN wBox wSta oSle wCoS wWel
	utilis var. ***jacquemontii*** 'Grayswood Ghost'	oGos
	utilis var. ***jacquemontii*** 'Jermyns'	oGos
un	***utilis*** 'Yunnan'	oGos
	BIARUM	
	carduchorum	last listed 99/00
	tenuifolium	last listed 99/00
	BIDENS	
	heterophylla	wHer
un	'Peters Gold Carpet'	wTGN
	pilosa	oCrm
	BIGNONIA	
	capreolata	oFor oSho
un	***capreolata*** 'Dragon Lady'	oGar
un	***capreolata*** 'Tangerine Beauty'	oAls oGar
	BILLARDIERA	
	longiflora	wHer oHed oGre
	longiflora fructu-albo	wHer

	BILLBERGIA	
	nutans	wCSG
	BISCHOFIA	
un	*policarpa*	wHer oGar oSho wBox
	BLECHNUM	
un	*australe*	wFan
	chilense	wHer wSte
	cordatum	see **B. *chilense***
	nudum	last listed 99/00
	sp. aff. *occidentale*	wFan
	penna-marina	oFor wCul wThG wCol oRus oNWe oBov wBox
	penna-marina ssp. *alpinum*	wFan
	penna-marina 'Cristatum'	wRob
un	*penna-marina* ssp. *penna-marina*	wFan
	spicant	oFor wFol wThG wWoB wNot oJoy wFan oAls oRus wShR oGar oDan oBos oSho oGre wPir wWat oTri wTGN oBRG oAld wCoN wFFl wRav oSle wSte wSnq
un	*spicant* 'Redwoods Giant'	wFan
	BLEPHILIA	
	ciliata	oGoo
	BLETILLA	
	ochracea	last listed 99/00
	striata	oFor oTrP oGar oDan oWnC oSec wSta wCoN
	striata alba	see **B. *striata*** var. ***japonica*** f. ***gebina***
un	*striata* 'Innocence'	oDan
	striata var. *japonica* f. *gebina*	oGar oRed oSis oGre
	striata var. *japonica* pink	oGre
un	*striata rosea*	oRed oSis
	striata 'Variegata'	oFor
	'Yokohama'	oRed
	BOEHMERIA	
	nivea	wFGN oCrm
	nivea HC 970601	wHer
	platanifolia HC 970438	wHer
	BOLAX	
	glebaria	see **AZORELLA *trifurcata***
	gummifera	wSta
qu	*gummifera* 'Nana'	oSis
	BOLBOSCHOENUS	
	maritimus	wWat
	BOLTONIA	
	asteroides var. *latisquama* 'Nana'	oFor oAls oSis
	asteroides 'Pink Beauty'	oFor oJoy wCSG oSis oWnC wMag oInd oUps
	asteroides 'Snowbank'	oJoy oAls wCSG oGar oSis wRob oGre oWnC wTGN oInd
	BOMAREA	
	sp. T75-9 Mexico	oCis
	BOMBAX	
	ellipticum	see **PSEUDOBOMBAX *ellipticum***
un	*palmeri*	oRar
	BORAGO	
	laxiflora	see **B. *pygmaea***
	officinales	oAls wFai iGSc wHom oWnC wNTP oSle
	pygmaea	oSec
	BORINDA	
un	*fungosa*	oTBG
	BORONIA	
	megastigma	last listed 99/00
	megastigma 'Lutea'	oFor
	BOUGAINVILLEA	
	'Barbara Karst'	oGar oWnC
	x buttiana 'Enid Lancaster'	oWnC
	x buttiana 'Rosenka'	oGar oWnC
	'California Gold'	see **B. *x buttiana*** 'Enid Lancaster'
	'Elizabeth Angus'	oGar
	glabra	oGar
	'James Walker'	oGar wWel
	Oo-La-La®/ 'Monka'	oWnC wWel
	Purple Queen™/ 'Moneth'	oGar oWnC wWel
va	'Raspberry Ice'	oGar wCoS
	'San Diego Red'	oGar
	'Southern Rose'	oGar
	spectabilis 'Lavender Queen'	oGar
	BOUTELOUA	
un	*curtipendula* 'Trailway'	oFor
	gracilis	oFor wShR oOut oSec
	BOWIEA	
	volubilis	oRar
	BOYKINIA	
	aconitifolia	wHer wHig
	elata	see **B. *occidentalis***
	major	oFor oHan wHig oRus
	occidentalis	oRus oBos oAld
	BRACHYCHITON	
	acerifolius	oTrP
un	*acuminata*	oRar
	australis	last listed 99/00
	gregorii	oTrP
	populneus	oTrP oRar
	rupestris	oTrP
	BRACHYGLOTTIS	
	compacta	wHer
	Dunedin Group	oCis
	(Dunedin Group) 'Sunshine'	wCCr wMag
	'Leonard Cockayne'	wCCr
	monroi	oCis
	BRACHYSCOME	
	iberidifolia	oCir
	iberidifolia Splendour Series 'Blue Splendour'	oUps
un	'New Amethyst'	wTGN
	BRACHYSTACHYUM	
un	*densiflorum villosum*	oNor
	BRACHYSTEGIA	
un	*glaucescens*	oTrP
	BRACHYSTELMA	
un	*australe*	oRar
	barberae	oRar
	foetidum	oRar
un	*meyerianum*	oRar
un	*nanum*	oRar
un	*thunbergii*	oRar
	BRAHEA	
	armata	oFor wDav
	dulcis	last listed 99/00
	edulis	wDav
	nitida	last listed 99/00
	pimo	last listed 99/00
un	**BRETSCHNEIDERA**	
	sinensis	oFor oGar oCis oSho oWnC wBox
	BREYNIA	
	disticha	oGar
un	**BRIPTURUM**	
	symnosum	oNWe
	BRIZA	
	media	oFor wWoS wCli oAls oGar oGoo oOut wRob wWin oWnC wTGN oSec oCir wSte
	BRODIAEA	
	congesta	see **DICHELOSTEMMA *congestum***
	coronaria	oBos
	elegans HHS 93-106	last listed 99/00
	hyacinthina	see **TRITELEIA *hyacinthina***
	lactea	see **TRITELEIA *hyacinthina***
	laxa	see **TRITELEIA *laxa***
	BROMUS	
va	*inermis* 'Skinner's Gold'	oFor oHed oDan
	BROUSSONETIA	
	papyrifera	oFor oCis oGre
	BRUGMANSIA	
	aurea	oGar oOEx
un	'Belle Blanche'	wHom
	x candida 'Double White'	oGar wSte
	x candida 'Ecuador Pink'	oNat
	x candida pink	wMag
	x candida white	oGar
	double white	oNat wMag wAva wGra
	double yellow	wMag
	x insignis light pink form	wCol
	x insignis white	wHer
un	'Jamaican Yellow'	wHer
	orange	wAva
	pink	wAva
	sanguinea	wHer oCrm wGra
	sanguinea Huanto	oOEx
	suaveolens	oGar
un	*suaveolens* 'Belle Blanche'	oGar
un	*suaveolens* 'Peaches & Cream'	oNat oGar
un	'Sunray'	oNat
	variegated white	wGra
	versicolor	oTrP
	versicolor 'Charles Grimaldi'	wSte
	white	wMag wGra
	yellow	wMag
	BRUNFELSIA	
	australis	oOEx
	jamaicensis	oOEx

	lactea	oOEx
	BRUNNERA	
	macrophylla	oFor wWoS wCul oNat wHig oAls oRus oAmb oGar oSho wNay wRob wFai wMag oEga
	macrophylla Aluminium Spot / 'Langtrees'	oFor wHer oGos oJil oAls oRuswNay wRob oGre wMag wSte
	macrophylla 'Betty Bowring'	last listed 99/00
va	***macrophylla*** 'Dawson's White'	wGAc oGar oNWe wNay wRob wFai wBox wSte
va	***macrophylla*** 'Hadspen Cream'	oNat
	macrophylla 'Langtrees'	see **B. *macrophylla*** Aluminium Spot/ 'Langtrees'
	macrophylla 'Variegata'	see **B. *macrophylla*** 'Dawson's White'
	BRYONIA	
	dioica	iGSc oCrm
	BUCHLOE	
	dactyloides	wBox
ch	**BUDDLEIA**	see **BUDDLEJA**
	BUDDLEJA	
	alternifolia	oGar oGoo oCis
	alternifolia 'Argentea'	wHer wCul oAmb oDan wFai oUps
	asiatica	oFor oInd
	colvilei	last listed 99/00
	colvilei 'Kewensis'	oHed oCis
	crispa	wPir
	davidii	oGar wFai wKin oCrm oInd oGue oUps
	davidii 'African Queen'	oFor oJoy
	davidii var. ***alba***	wCCr
	davidii 'Black Knight'	oFor wHer wCul wSwC oNat oDar oAls wCSG oGar oGoo oDan oRiv oSho oSha oGre oWnC wMag wBox oInd oWhS oCir wAva wWhG oUps oSle wWel
un	***davidii*** 'Bonnie'	oFor wWoS
	davidii 'Border Beauty'	oFor wWoS oJoy
un	***davidii*** 'Burgundy'	oFor wWoS oJoy
un	***davidii*** 'Butterfly Ball'	oHed
	davidii 'Charming'	oFor oAls oGar oGoo oGre
un	***davidii*** 'Cornwall Blue'	wWoS
	davidii 'Dartmoor'	oFor wHer wWoS wCul wSwC oNat oHed oGar oGoo oWnC wMag oInd oCir
un	***davidii*** 'DeNiro'	wCSG
	davidii 'Dubonnet'	oFor wWoS oNat oJoy oAls wWel
	davidii 'Empire Blue'	oFor wWoS wCul oAls
	davidii 'Fascinating'	wHer
	davidii 'Glasnevin'	see **B. *davidii*** 'Glasnevin Blue'
	davidii 'Glasnevin Blue'	oHed
	davidii 'Gonglepod'	oHed
va	***davidii*** 'Harlequin'	oFor wHer wWoS wCul oGos wSwC oJoy oEdg oAls wCSG oGar oGoo oNWe oSho wRob oGre oWnC wMag oSec wWhG oUps wSte wWel
	davidii 'Harlequin Alba'	see **B. *davidii*** 'White Harlequin'
	davidii 'Ile de France'	oFor wHer oJoy
qu	***davidii*** 'Indigo'	wCul oRiv
qu	***davidii*** 'Indigo Pink'	oJoy
	davidii 'Lochinch'	see **B.** 'Lochinch'
un	***davidii*** 'Miss Ellen'	oFor wWoS oInd
	davidii 'Nanho Blue'	oDar oAls oGar oGoo oDan wTGN oCir wAva wWhG
	davidii 'Nanho Purple'	wSwC oAls oGar wRai oWnC wMag wTGN oCir
un	***davidii*** 'Nanho White'	wAva
qu	***davidii*** var. ***nanhoensis*** 'Indigo'	wHer
	davidii var. ***nanhoensis*** Petite Indigo™/ 'Mongo'	oFor oJoy oAls oGar oSho
	davidii var. ***nanhoensis*** Petite Plum™/ 'Monum'	oFor oAls oGar oGre oSec wWhG
un	***davidii*** var. ***nanhoensis*** 'Petite Purple'	oNat oAls wCSG oSis oUps
	davidii var. ***nanhoensis*** Petite Snow™/ 'Monite'	oGar oSho
un	***davidii*** 'Niche's Choice'	oGar
un	***davidii*** 'Northlake'	wCCr
	davidii 'Opera'	oDar
	davidii 'Orchid Beauty'	wWoS
	davidii 'Pink Charming'	see **B. *davidii*** 'Charming'
un	***davidii*** 'Pink Perfection'	oNat wHom oWnC oCir
un	***davidii*** 'Potter's Purple'	oFor wWoS oEdg
	davidii 'Princeton Purple'	oJoy
	davidii 'Purple Prince'	oFor wHer wCul oAls oGar oGoo oSha
un	***davidii*** 'Red Plume'	oFor
	davidii 'Royal Red'	oFor wHer wCul oJoy oAls wCSG oGar oOut oGre oWnC wMag oCir wAva wWhG oUps oSle wWel
qu	***davidii*** 'Royal Violet'	oSec
un	***davidii*** 'Rudy's Rainbow'	oGar
	davidii 'Snow Bank'	oFor oInd
un	***davidii*** 'Trewithen Blue'	wWoS
	davidii variegated	wFai
un	***davidii*** 'Violet Message'	wWoS
	davidii 'White Bouquet'	oFor wCul oNat oAls oGar oGoo oGre
va	***davidii*** 'White Harlequin'	wHer oJil
	davidii 'White Profusion'	wWoS oDar oJoy oAls wCSG oGar wMag wTGN oCir wWhG wWel
un	'Ellen's Blue'	wHer
	fallowiana var. ***alba***	oHed oGoo wCCr
	fallowiana var. ***alba*** ACE 2481	oHed
	forrestii	wCCr
	globosa	oFor oDar wCSG oGar oGoo oRiv
un	***hemsleyana***	wHer
	x intermedia	oGoo oSec
	japonica	oFor wCul oHed wCSG oAmb oGar wTGN
	limitanea	wCCr
	lindleyana	oFor wPir
	'Lochinch'	oFor wHer wWoS wCul wCri wSwC oDar oJoy oHed oGar oGoo wRai oSec wBWP oWhS oUps wSnq wSnq wWel
	macrostachya	wCCr
	myriantha	oHed
	nappii	wHer oHed
	nivea	oFor wCul oHed
un	***nivea*** ssp. ***nivea***	wHer
	nivea var. ***yunnanensis*** DJHC 482	wHer
	paniculata	oHed
	x pikei	oSec
	x pikei 'Hever'	oAls oDan
	'Pink Delight'	oFor wHer wCul oDar oAls wCSG oGar oOut wMag wTGN wAva wWhG oUps oSle wWel
	salviifolia	oHed wAva
	sp. DJHC 631A	wHer
un	'Vashon Skies'	wCCr
	venenifera	wCCr
	'West Hill'	wCCr
	x weyeriana 'Golden Glow'	oJoy wCCr oUps
un	***x weyeriana*** 'Honeycomb'	oFor wWoS oEdg oGre oInd
	x weyeriana 'Sungold'	oFor wHer wWoS wCul oNat oDar oAls wCSG oGar oGoo wFai wHom oWnC wMag oSec oInd oWhS oCir wAva wWhG oUps
un	'White Ball'	wHer
	BUGLOSSOIDES	
	purpurocaerulea	wHer
	BULBOCODIUM	
	vernum	wHer
	BUNIUM	
	bulbocastanum	oOEx
	BUPLEURUM	
un	***chinensis***	oOEx
	fruticosum	oFor wHer wPir
	rotundifolium	oCrm
	BURSERA	
un	***aloexylon*** BLM 0781	oRar
un	***arida*** BLM 1318	oRar
un	***ariensis***	oRar
un	***asplenifolia*** BLM 1289	oRar
un	***biflora***	oRar
un	***biflora*** BLM 0793	oRar
un	***bipinnata***	oRar
un	***copalifera***	oRar
un	***cuneata***	oRar
un	***discolor*** BLM 1338	oRar
un	***excelsa*** var. ***acutidens*** BLM 0174	oRar
un	***fagaroides***	oRar
un	***galeottiana*** BLM 1272	oRar
un	***glabrifolia*** BLM 1306	oRar
un	***graveolens*** BLM 1391	oRar
un	***hindsiana***	oRar
un	***jorulensis*** BLM 1329	oRar
un	***kerberii*** BLM 0674	oRar
un	***lancifolia*** BLM 1069	oRar
	microphylla	oRar
un	***odorata***	oRar
un	***palmeri*** BLM 0651	oRar
un	***schlectendalii***	oRar
	simaruba	oRar
un	***submoniloformis***	oRar
un	***suntui***	oRar
un	***vejar-vasquesii*** BLM 1340	oRar

	Name	Sources
un	***xochepalensis*** BLM 1339	oRar
	BUTIA	
un	***arenicola***	oOEx
	bonnetii	oOEx
	capitata	oTrP oSho oOEx wDav
	BUTOMUS	
	umbellatus	oFor wGAc oHug oGar oSsd oTri
	BUXUS	
	aurea	see **B. *sempervirens***
un	'Cynthia'	wBox
	'Green Mountain'	oGar oSho oWnC wBox wWel
	'Green Velvet'	oFor oGar wWel
un	***microphylla*** 'Asiatic'	oGar
un	***microphylla*** 'Aureomarginata'	oBRG
	microphylla 'Compacta'	oGos oSis oBov wSta oBRG
	microphylla 'Curly Locks'	oFor
	microphylla 'Faulkner'	oWnC oBRG
	microphylla var. ***insularis***	see **B. *sinica*** var. ***insularis***
	microphylla var. ***japonica***	wCCr oWnC oBRG wCoN
	microphylla var. ***japonica*** 'Green Beauty'	oGar oGre oWnC wBox wSta oBRG wCoN wCoS wWel
	microphylla var. ***japonica*** 'Morris Dwarf'	oSqu wBox
	microphylla var. ***japonica*** 'Morris Midget'	oFor oSis wBox oBRG
	microphylla var. ***japonica*** 'National'	wHer wBox
un	***microphylla*** var. ***japonica*** 'Wintergreen'	oFor oGar oSho wBox
	microphylla 'John Baldwin'	oGos wBox
	microphylla 'Kingsville'	see **B. *microphylla*** 'Compacta'
	microphylla var. ***koreana***	see **B. *sinica*** var. ***insularis***
	microphylla var. ***sinica***	see **B. *sinica***
	sempervirens	oAls oGar oRiv oGre wCCr oTPm oWnC wBox wSta wCoN oSle wWel
	sempervirens 'Argenteovariegata'	oBov
va	***sempervirens*** 'Aurea Pendula'	last listed 99/00
	sempervirens 'Aureovariegata'	oGar oGre wPir
	sempervirens dwarf variegated form	oSis
va	***sempervirens*** 'Elegantissima'	wBox
un	***sempervirens*** 'Golden Swirl'	oCis
	sempervirens 'Graham Blandy'	oFor wHer oGos oAls oGar oSis oSho oGre oWnC wBox oBRG wWel
	sempervirens 'Kingsville Dwarf'	see **B. *microphylla*** 'Compacta'
	sempervirens 'Latifolia Maculata'	wHer
un	***sempervirens*** 'Nana Variegata'	wHer
	sempervirens 'Prostrata'	wSta
	sempervirens 'Rosmarinifolia'	wHer wCol wCCr
	sempervirens 'Rotundifolia'	wSta
un	***sempervirens*** 'Schmidt'	wWel
	sempervirens 'Suffruticosa'	oAls oGar oGoo oSho oSqu oGre wPir iGSc oTPm oWnC wBox wSta wNTP wCoN wWel
	sempervirens 'Vardar Valley'	oFor oTPm wBox
	sempervirens 'Variegata'	oFor wSwC oRiv iGSc oTPm oWnC oCir
	sempervirens 'Waterfall'	wHer
un	***sempervirens*** 'Watnong'	oBRG
un	***sempervirens*** 'Wonford Paige'	oBRG
	sinica	oBov
	sinica fastigiate form	oSis
	sinica var. ***insularis***	oFor oGar oRiv oGre wBox oBRG wCoN
	sinica var. ***insularis*** 'Justin Brouwers'	wHer
	sinica var. ***insularis*** 'Tide Hill'	oSis wWel
	sinica var. ***insularis*** 'Winter Gem'	oAls oRiv oGre wKin oWnC wCoN wWel
	wallichiana	wHer oSec
	CABOMBA	
	caroliniana	oGar
	CAESALPINIA	
	gilliesii	wCCr
	CALAMAGROSTIS	
	x acutiflora 'Karl Foerster'	oFor wHer wWoS wCli oDar oJoy oGar oDan oSho wRob oGre wFai wWin wHom oWnC oTri wTGN oBRG oUps oSle wSnq
va	***x acutiflora*** 'Overdam'	oFor wHer wWoS wCli oJoy oAls oGar oDan oOut oGre wFai wWin oWnC wBox oUps wSte wSnq
	x acutiflora 'Stricta'	wHer wHig wRob
un	***x acutiflora*** 'Tricolor'	oSec
	arundinacea	oCis wBWP
	brachytricha	oFor wCli oNat oDar oGar wWin wSta oBRG wSnq
	CALAMINTHA	
	ascendens	see **C. *sylvatica*** ssp. ***ascendens***
	cretica	wFGN oGoo
	grandiflora	oFor oJoy wFGN oGoo wFai oBar
	grandiflora 'Variegata'	oNat oAls oGoo oSis oWnC oSec
	nepeta	wFai wCCr wBWP
	nepeta ssp. ***glandulosa*** 'White Cloud'	oHed oSis
	nepeta ssp. ***nepeta***	oFor

	Name	Sources
un	***nepeta*** ssp. ***nepeta*** 'Alba'	oFor
	nepetoides	see **C. *nepeta*** ssp. ***nepeta***
	officinalis	iGSc
	sylvatica ssp. ***ascendens***	iGSc
	CALANTHE	
	brevicornu	last listed 99/00
	herbacea	oRed
	mannii	oRed
	puberula	last listed 99/00
	reflexa	oRed
	tricarinata	oRed
	CALATHEA	
	amabalis	see **CTENANTHE *amabalis***
	roseopicta	oGar
	CALCEOLARIA	
	alba	wHer
	biflora	oSis
	crenatiflora	last listed 99/00
	ericoides	wHer
	falklandica	oHed
	helianthemoides	last listed 99/00
	Herbeohybrida Group 'Bright Bikinis'	oGar
	'John Innes'	oFor oGar oSis
un	'Kentish Beauty'	oCis
	'Kentish Hero'	wHer oHed oMis
	perfoliata	wHer
	pinifolia	wMtT
	sp. HCM 98082	wHer
	tenella	wHer
	CALEA	
	zacatechichi	iGSc
	CALENDULA	
	officinalis	wFai iGSc
	officinalis Bon Bon mixed	oUps
un	'Resina'	wNTP
	CALIBANUS	
	hookeri	oRar
	CALLA	
	aethiopica	see **ZANTEDESCHIA *aethiopica***
	palustris	wGAc oSsd
	CALLICARPA	
	americana	wCSG oGar
	americana var. ***lactea***	oFor
	bodinieri	oTrP wCCr
	bodinieri var. ***giraldii***	wHer wSta
	bodinieri var. ***giraldii*** 'Profusion'	oFor wHer wCul oGos wSwC oDar oAls oAmb oGar oDan oSho oGre wFai wTGN wBox wCoN oUps wSte wCoS wWel
	dichotoma	oGar oGre oBRG
un	***dichotoma albifructus***	oFor
un	***dichotoma*** 'Issai'	oFor oGre oUps wWel
un	***dichotoma*** 'Winterthur'	wWoS
un	***formosana***	wCCr
un	'Honeysong'	wCol
	japonica	oFor oRiv
	japonica 'Leucocarpa'	oFor oRiv oGre wCoN
	mollis	oFor
	mollis HC 970589	wHer
	rubella	oTrP
	CALLIRHOE	
	digitata	oRus
	involucrata	oFor oJoy oSis oSec oUps
	CALLISIA	
	fragrans	oTrP
	CALLISTEMON	
	acuminatus	oFor oRiv
	citrinus	oFor oTrP oSho wCCr oCir
	citrinus 'Splendens'	oGre
	linearis	oFor
un	***montanus***	oFor
	pallidus	oFor
	salignus	oFor oGre
	salignus var. ***australis***	wCCr
	sieberi	oCis
	sieberi pink	wCCr
	speciosus	last listed 99/00
	subulatus	oHed oCis wCCr
	viminalis	oTrP
	viminalis 'Little John'	oGar
	viridiflorus	oFor oCis oSho wCCr
	CALLITRIS	
	columellaris	oTrP
	drummondii	oTrP
	oblonga	oFor
	rhomboidea	oFor

	CALLUNA	
	vulgaris	iGSc
un	***vulgaris*** 'Aberdeen'	wHea
	vulgaris 'Alba Aurea'	wHea
	vulgaris 'Alba Carlton'	wHea
	vulgaris 'Alba Jae'	wHea oGar oTPm oWnC
do	***vulgaris*** 'Alba Plena'	oTPm
un	***vulgaris*** 'Alice Knight'	wHea
	vulgaris 'Alison Yates'	wHea
	vulgaris 'Allegretto'	wHea
	vulgaris 'Allegro'	wHea oGar oBlo oGre wTGN
	vulgaris 'Alportii'	oWnC
	vulgaris 'Alys Sutcliffe'	wHea
	vulgaris 'Amilto'	wHea
	vulgaris 'Amy'	wHea
	vulgaris 'Anneke'	wHea
do	***vulgaris*** 'Annemarie'	wHea oGre
	vulgaris 'Anthony Davis'	last listed 99/00
do	***vulgaris*** 'Applecross'	wHea
	vulgaris 'Aurea'	oTPm wSta
	vulgaris 'Aureifolia'	wSta
	vulgaris 'Autumn Glow'	wHea
	vulgaris 'Barbara Fleur'	wHea
	vulgaris 'Barja'	wHea
	vulgaris 'Barnett Anley'	wHea
	vulgaris 'Battle of Arnhem'	wHea
un	***vulgaris*** 'Bayport'	wHea
	vulgaris 'Beoley Crimson'	wHea oGar oWnC
	vulgaris 'Beoley Gold'	wHea oGar wCCr oWnC
	vulgaris 'Blazeaway'	wSta
	vulgaris 'Bonfire Brilliance'	wHea
	vulgaris 'Boreray'	wHea
	vulgaris 'Boskoop'	wHea
	vulgaris 'Bradford'	last listed 99/00
	vulgaris 'Branchy Anne'	oWnC
	vulgaris 'Bray Head'	wHea
	vulgaris 'Brightness'	wHea oWnC
do	***vulgaris*** 'Brita Elisabeth'	wHea
un	***vulgaris*** 'Bronze Beauty'	wHea
un	***vulgaris*** 'California'	oSis
	vulgaris 'Carmen'	last listed 99/00
	vulgaris 'Carole Chapman'	last listed 99/00
	vulgaris 'Catherine Anne'	wHea
	vulgaris 'Celtic Gold'	wHea
un	***vulgaris*** 'Chase White'	wHea
	vulgaris 'Citronella'	wHea
	vulgaris 'Clare Carpet'	wHea
	vulgaris 'Coby'	wHea
	vulgaris 'Con Brio'	wHea
	vulgaris 'Copper Glow'	wHea
	vulgaris 'Corbett Red'	oFor wHea oAls oGar oSis oGre oTPm oWnC
un	***vulgaris*** 'Corbett's White'	wHea
	vulgaris 'Cottswood Gold'	wHea
do	***vulgaris*** 'County Wicklow'	wHea oAls oGar oSho oTPm oWnC
do	***vulgaris*** 'Cramond'	wHea
	vulgaris 'Cream Steving'	wHea
	vulgaris 'Crimson Glory'	last listed 99/00
	vulgaris 'Crowborough Beacon'	wHea
	vulgaris 'Cuprea'	wHea oWnC oBRG
	vulgaris 'Dainty Bess'	wHea oSqu oWnC
qu	***vulgaris*** 'Dainty Bess Minor'	oSis
do	***vulgaris*** 'Dark Beauty'	wHea oGre wTGN
do	***vulgaris*** 'Dark Star'	wHea
	vulgaris 'Darkness'	wCCr
	vulgaris 'Dart's Amethyst'	wHea
	vulgaris 'Dart's Flamboyant'	wHea
	vulgaris 'Dart's Parakeet'	wHea
	vulgaris 'Dart's Squirrel'	wHea
	vulgaris 'David Hutton'	wHea
do	***vulgaris*** 'Devon'	wHea
	vulgaris 'Drum-ra'	wHea
	vulgaris 'Dunnet Lime'	wHea
	vulgaris 'Dunnydeer'	wHea
	vulgaris 'Easter-bonfire'	wHea
	vulgaris 'Edith Godbolt'	wHea
un	***vulgaris*** 'Ellie Barbour'	wHea
do	***vulgaris*** 'Else Frye'	oFor
do	***vulgaris*** 'Elsie Purnell'	wHea oAls oGar oWnC
	vulgaris 'Fairy'	wHea
	vulgaris 'Falling Star'	wHea
	vulgaris 'Feuerwerk'	wHea
	vulgaris 'Finale'	wHea
	vulgaris 'Findling'	wHea
	vulgaris 'Firefly'	wHea oGre wTGN
	vulgaris 'Firestar'	wHea
	vulgaris 'Flamingo'	wHea wTGN
	vulgaris 'Fortyniner Gold'	wHea
	vulgaris 'Foxhollow Wanderer'	oWnC
	vulgaris 'Foxii Nana'	oSis oTPm wSta
	vulgaris 'French Grey'	wHea
	vulgaris 'Glendoick Silver'	wHea
	vulgaris 'Glenfiddich'	wHea
	vulgaris 'Glenlivet'	wHea
	vulgaris 'Glenmorangie'	wHea
do	***vulgaris*** 'Gold Hamilton'	wHea
	vulgaris 'Gold Haze'	wHea
	vulgaris 'Gold Knight'	wHea
	vulgaris 'Gold Kup'	wHea
	vulgaris 'Golden Carpet'	wHea
	vulgaris 'Golden Feather'	wHea oWnC wSta
	vulgaris 'Goldsworth Crimson'	oWnC
	vulgaris 'Goldsworth Crimson Variegated'	wHea
un	***vulgaris*** 'Green Cardinal'	wHea oGar
	vulgaris 'Grey Carpet'	wHea
	vulgaris 'Grijsje'	wHea
	vulgaris 'Grizzly'	wHea
	vulgaris 'Guinea Gold'	wHea
do	***vulgaris*** 'H. E. Beale'	wHea oGar oGre oTPm
	vulgaris 'Hamlet Green'	wHea
	vulgaris 'Hammondii Aureifolia'	wHea
	vulgaris 'Hammondii Rubrifolia'	wHea
	vulgaris 'Harlekin'	wHea
	vulgaris 'Hibernica'	wHea
	vulgaris 'Hiemalis'	last listed 99/00
	vulgaris 'Highland Rose'	wHea
	vulgaris 'Hilda Turberfield'	wHea
	vulgaris 'Hillbrook Orange'	wHea
	vulgaris 'Holstein'	wHea
	vulgaris 'Hoyerhagen'	wHea
	vulgaris 'Hugh Nicholson'	wHea
	vulgaris 'Humpty Dumpty'	oSis
	vulgaris 'Indian Thick Rug'	wHea
do	***vulgaris*** 'Isobel Hughes'	wHea
do	***vulgaris*** 'J. H. Hamilton'	wHea oSqu oTPm oWnC
	vulgaris 'Jan'	wHea
	vulgaris 'Jan Dekker'	wHea
	vulgaris 'Japanese White'	wHea
do	***vulgaris*** 'Jimmy Dyce'	wHea
	vulgaris 'John F. Letts'	oSis wSta
	vulgaris 'Johnson's Variety'	last listed 99/00
	vulgaris 'Joy Vanstone'	wHea
do	***vulgaris*** 'Juno'	last listed 99/00
	vulgaris 'Kerstin'	wCul wHea oGre
do	***vulgaris*** 'Kinlochruel'	wHea oWnC
	vulgaris 'Kirby White'	wHea
	vulgaris 'Lambstails'	wHea
	vulgaris 'Lime Glade'	wHea
	vulgaris 'Long White'	last listed 99/00
	vulgaris 'Lyndon Proudley'	wHea
un	***vulgaris*** 'MacDonald of Glencoe'	wHea
	vulgaris 'Mair's Variety'	last listed 99/00
	vulgaris 'Marion Blum'	wHea
	vulgaris 'Marleen'	wHea
	vulgaris 'Martha Hermann'	last listed 99/00
un	***vulgaris*** 'Mayfair'	wHea
	vulgaris 'Mazurka'	wHea
do	***vulgaris*** 'Mick Jamieson'	wHea
	vulgaris 'Minima'	oSis
	vulgaris 'Minty'	wHea
	vulgaris 'Mousehole'	last listed 99/00
un	***vulgaris*** 'Mrs. E. D. Maxwell'	oWnC
	vulgaris 'Mrs. Pat'	oTPm
	vulgaris 'Mrs. Ronald Gray'	oFor wHea oAls oGar oSqu wSta
	vulgaris 'Mullion'	last listed 99/00
	vulgaris 'Multicolor'	wHea wSta
do	***vulgaris*** 'My Dream'	wHea
	vulgaris 'Naturpark'	wHea
	vulgaris 'Nordlicht'	wHea
	vulgaris 'Olive Turner'	wHea
un	***vulgaris*** 'Orange Beauty'	wHea
	vulgaris 'Orange Queen'	wHea oWnC
	vulgaris 'Pat's Gold'	wHea
	vulgaris 'Penhale'	wHea
	vulgaris 'Perestrojka'	wHea
do	***vulgaris*** 'Peter Sparkes'	oFor wHea oGar oTPm oWnC oBRG
un	***vulgaris*** 'Pretty Pink'	oGar
	vulgaris 'Prostrate Orange'	wHea
	vulgaris 'Pygmaea'	wHea oSis wSta
	vulgaris 'Pyramidalis'	last listed 99/00

do	***vulgaris*** 'Radnor'	wHea oGar wCCr oWnC
	vulgaris 'Red Carpet'	wHea
do	***vulgaris*** 'Red Favorit'	wHea
	vulgaris 'Red Fred'	wHea
	vulgaris 'Red Haze'	last listed 99/00
	vulgaris 'Red Pimpernel'	wHea
do	***vulgaris*** 'Red Star'	wHea
	vulgaris 'Red Wings'	wHea
	vulgaris 'Redbud'	wHea
	vulgaris 'Reini'	wHea
	vulgaris 'Robber Knight'	wHea
	vulgaris 'Robert Chapman'	oFor wCul wHea oAls oGar oDan oWnC wSta oBRG wWel
	vulgaris 'Roland Haagen'	wHea
	vulgaris 'Romina'	wHea
	vulgaris 'Roodkapje'	wHea
	vulgaris 'Rosalind'	wHea
	vulgaris 'Ross Hutton'	wHea
	vulgaris 'Roter Oktober'	wHea oGre wTGN
	vulgaris 'Ruby Slinger'	wHea
do	***vulgaris*** 'Ruth Sparkes'	wSta
	vulgaris 'Saint Nick'	oBRG
	vulgaris 'Sally Anne Proudley'	oWnC
do	***vulgaris*** 'Schurig's Sensation'	last listed 99/00
	vulgaris 'Serlei'	wCul wHea
	vulgaris 'Serlei Purpurea'	wHea
	vulgaris 'Sesam'	wHea
	vulgaris 'Silver King'	last listed 99/00
	vulgaris 'Silver Knight'	wHea oAls oGar oSis oSho oBlo wCCr
	vulgaris 'Silver Queen'	oFor wHea oSis oTPm
	vulgaris 'Silver Spire'	wHea
	vulgaris 'Sir John Charrington'	wHea oSis oTPm
	vulgaris 'Sister Anne'	wHea oAls oGar oSqu oWnC
	vulgaris 'Soay'	last listed 99/00
	vulgaris 'Spicata'	wHea
	vulgaris 'Spitfire'	wHea
	vulgaris 'Spring Cream'	oFor wHea oGar oBlo oWnC
	vulgaris 'Spring Glow'	wHea
	vulgaris 'Spring Torch'	oFor wHea oAls oGar oGre wCCr oTPm oWnC wTGN wSta
do	***vulgaris*** 'Strawberry Delight'	wHea
	vulgaris 'Sunrise'	wHea
	vulgaris 'Sunset'	wHea oWnC
	vulgaris 'Tenuis'	wHea
do	***vulgaris*** 'Tib'	oFor wHea oAls oGar oTPm oWnC wTGN
	vulgaris 'Tom Thumb'	wHea oSis
	vulgaris 'Torogay'	wHea
	vulgaris 'Tricolorifolia'	wHea
	vulgaris 'Velvet Fascination'	wHea
	vulgaris 'Visser's Fancy'	wHea
do	***vulgaris*** 'White Coral'	wHea
	vulgaris 'White Lawn'	wHea oSis oSqu oWnC wSta
	vulgaris 'White Queen'	wHea
	vulgaris 'Wickwar Flame'	wHea oGar oWnC wTGN
	vulgaris 'Wingates Gem'	wHea
	vulgaris 'Winter Chocolate'	oFor oSqu oWnC
	vulgaris 'Winter Red'	oGre
	vulgaris 'Yvette's Gold'	last listed 99/00
	vulgaris 'Yvette's Silver'	last listed 99/00
	CALOCEDRUS	
	decurrens	oFor oPor wWoB wNot oDar oAls wShR oGar oRiv oSho oWhi oGre wWat wCCr oWnC wAva wCoN wRav oSle wSte wWel
va	***decurrens*** 'Aureovariegata'	oPor oRiv oWnC
	decurrens 'Berrima Gold'	last listed 99/00
	decurrens 'Depressa'	last listed 99/00
	macrolepis	wHer
	CALOCEPHALUS	
	brownii	see **LEUCOPHYTA *brownii***
	CALOCHORTUS	
	luteus 'Golden Orb'	wGAc wWld
	CALONYCTION	see **IPOMOEA**
	CALOTHAMNUS	
	pinifolius	oTrP wGra
	CALTHA	
	asarifolia	oSle
	biflora	wCol
	leptosepala	wCol
	palustris	oFor wSoo wGAc oHug oGar oSsd oGre oWnC wTGN wCoN
	palustris var. ***alba***	wGAc oHug
do	***palustris*** 'Flore Pleno'	oAls oHug oGar oSsd oNWe
un	***palustris*** var. ***himalensis*** f. ***alba***	wHer
	palustris 'Monstruosa'	oRus
do	***palustris*** 'Multiplex'	oFor wGAc
	palustris var. ***palustris***	wHer wCol
do	***palustris*** var. ***palustris*** 'Plena'	wGAc oSis oWnC
	polypetala (see Nomenclature Notes)	oHed
	polypetala hort.	see **C. *palustris*** var. ***palustris***
	CALYCANTHUS	
qu	***carolinus***	oRiv
	chinensis	see **SINOCALYCANTHUS *chinensis***
	fertilis	see **C. *floridus*** var. ***glaucus***
	floridus	oFor oEdg wShR oGar oRiv oSho oWhi wRai oOEx oGre oWnC wTGN wCoN oSle wWel
	floridus var. ***glaucus***	oFor
	occidentalis	oFor wShR oRiv oCis oGre wSte
un	**CALYLOPHUS**	
	drummondii	oSis
	sp. Dallas County, Texas	oSis
	CAMASSIA	
	cusickii	oRus oGar oWoo
	esculenta	see **C. *quamash***
	leichtlinii hort.	see **C. *leichtlinii*** ssp. ***suksdorfii***
	leichtlinii 'Alba'	see **C. *leichtlinii*** ssp. ***leichtlinii***
	leichtlinii blue	see **C. *leichtlinii*** ssp. ***suksdorfii***
	leichtlinii Blue Danube	see **C. *leichtlinii*** ssp. ***suksdorfii*** 'Blauwe Donau'
	leichtlinii ssp. ***leichtlinii***	oRus wTGN
do	***leichtlinii*** 'Semiplena'	wHer
	leichtlinii ssp. ***suksdorfii***	oFor wRoo oRus wShR oSsd oBos oNWe oSho wWat oAld wWld
	leichtlinii ssp. ***suksdorfii*** 'Alba' seed grown	wHer
	leichtlinii ssp. ***suksdorfii*** 'Blauwe Donau'	oDan
	quamash	oFor wHer oHan wThG oRus oSsd oBos oOEx oWoo wWat oTri wTGN wPla wWld
	quamash 'Blue Melody'	last listed 99/00
	quamash 'Orion'	oDan
	CAMELLIA	
	'Barbara Clark'	wHer
	chrysantha	see **C. *nitidissima*** var. ***nitidissima***
	'Coral Delight'	oGre oWnC
	'El Dorado'	oGre
	'Freedom Bell'	oGar
	grijsii	oCis oGre
	hiemalis	oGre wCoN
	hiemalis 'Chansonette'	oFor oAls oGre
	hiemalis 'Showa-no-sakae'	oGar wWhG
un	Ice Angels tm Series	wWhG
	japonica 'Adolphe Audusson'	oBlo
	japonica 'Akashigata'	oFor
un	***japonica*** 'Allie Blue'	oGre
	japonica 'Berenice Boddy'	oGos
	japonica 'Betty Sheffield Supreme'	oGre
un	***japonica*** 'Black Prince'	oGar
	japonica 'Blood of China'	oGre wWhG
	japonica 'Bob Hope'	oGar oWnC wWhG
	japonica 'Bokuhan'	oAls oGre
un	***japonica*** 'Brilliant'	oAls
	japonica 'Brushfield's Yellow'	oGre wWhG
	japonica 'C. M. Hovey'	oGre oWnC wWhG
	japonica 'C. M. Wilson'	oGar
un	***japonica*** 'Captain John Sutter'	oAlsoGar
	japonica 'Carter's Sunburst'	oGar wWhG
	japonica 'Chandleri Elegans'	see **C. *japonica*** 'Elegans'
	japonica Chinese cultivars	oCis
	japonica 'Colonel Firey'	see **C. *japonica*** 'C. M. Hovey'
	japonica 'Coquettii'	oGar oGre wWhG
un	***japonica*** 'Daikagura Variegata'	oGar
un	***japonica*** 'Dainty California'	oAls oGar oSho wWhG
un	***japonica*** 'Dantel's Supreme'	oGar
	japonica 'Debutante'	oGar oBlo oWnC wWhG
	japonica 'Doctor Burnside'	oGar
un	***japonica*** 'Eleanor McGown'	oGar oSho wWhG
	japonica 'Elegans'	oGar oGre
	japonica 'Elegans Splendor'	wWhG
	japonica 'Elegans Variegata'	oGar wWhG
un	***japonica*** 'Elizabeth Labfy'	oBlo
	japonica 'Emperor of Russia'	oGre
un	***japonica*** 'Finlandia'	oGar oGre
	japonica 'Finlandia Variegated'	oGre wWhG
	japonica 'Fred Sander'	oGar oGre
	japonica 'Glen 40'	see **C. *japonica*** 'Coquettii'
	japonica 'Governor Mouton'	oFor
	japonica 'Grand Sultan'	oGre
	japonica 'Grandiflora Rosea'	oGre
	japonica 'Hagoromo'	oGar oGre wWhG
	japonica 'Hawaii'	oBlo

	japonica Herme	see **C. *japonica*** 'Hikarugenji'
	japonica 'Hikarugenji'	oAls oGar oWnC wWhG
	japonica 'In the Pink'	wWhG
	japonica 'Jordan's Pride'	see **C. *japonica*** 'Hikarugenji'
	japonica 'Joseph Pfingstl'	oAls
un	***japonica*** 'Junior Miss'	oGre
	japonica 'Kick-off'	oAls
un	***japonica*** 'King Size'	oAls oGre wWhG
	japonica 'Kramer's Supreme'	oAls oGar oGre oWnC wWhG
	japonica 'Kumasaka'	oAls oGre oWnC wWhG
	japonica Lady Clare	see **C. *japonica*** 'Akashigata'
	japonica 'Lady Vansittart'	oGre
	japonica 'Magic City'	oGre wWhG
	japonica 'Magnoliiflora'	see **C. *japonica*** 'Hagoromo'
un	***japonica*** 'Maiden of Promise'	oGre
un	***japonica*** 'Margie'	wTGN
	japonica 'Mathotiana Supreme'	oGar wWhG
un	***japonica*** 'Moshio'	oGre
	japonica 'Mrs. Charles Cobb'	oAls oGar oWnC wWhG
	japonica 'Mrs. Tingley'	oGar
	japonica 'Nina Avery'	oGre wWhG
	japonica 'Nuccio's Gem'	oGar oSho oGre wWhG
	japonica 'Nuccio's Jewel'	wWhG
	japonica 'Nuccio's Pearl'	oAls oGar wWhG
	japonica 'Otome'	oAls oGar
un	***japonica*** 'Pearl Maxwell'	oGar oWnC
un	***japonica*** 'Pink Parade'	oAlsoGar wWhG
	japonica 'Pink Perfection'	see **C. *japonica*** 'Otome'
	japonica 'Pope Pius IX'	see **C. *japonica*** 'Prince Eugene Napoleon'
	japonica 'Prince Eugene Napoleon'	oGar oBlo
	japonica 'Purity'	see **C. *japonica*** 'Shiragiku'
	japonica 'R. L. Wheeler'	wWhG
un	***japonica*** 'Rainy Sun'	oGre
	japonica 'Rosea Superba'	oGre
un	***japonica*** 'Scented Treasure'	oGar oSho oGre
	japonica 'Scentsation'	oGar oWnC
un	***japonica*** 'Sen Sekai'	wCol
	japonica 'Shiragiku'	oAls wWhG
	japonica 'Shiro Chan'	oGar wWhG
	japonica 'Silver Waves'	oGar wWhG
un	***japonica*** 'Sir Victor Davies'	oGre
un	***japonica*** 'Spring's Promise'	oGre
un	***japonica*** 'Sunset Oaks'	oAls oSho oGre
un	***japonica*** 'Surprise Bouquet'	oAls
	japonica 'Tinsie'	see **C. *japonica*** 'Bokuhan'
	japonica 'Ville de Nantes'	oSho oGre wWhG
un	***japonica*** 'White Mermaid'	wHer oGos oGar oDan oWnC wBox
	'Nicky Crisp'	wHer
	nitidissima var. ***nitidissima***	oCis
un	***nitidissima*** var. ***nitidissima*** 'Phaeopubisperma'	oCis
	oleifera	oFor wHer oGos oCis oGre oWnC
	'Polar Ice'	wClo oGre
	reticulata	wCoN
un	***saluenensis*** 'Chimes'	wCCr
un	***saluenensis*** 'Rose Bowl'	wCCr
	sasanqua	wCSG wCoN
	sasanqua 'Apple Blossom'	oAls oGar wSta wWhG
un	***sasanqua*** 'Aurumi Gawa'	wSta
un	***sasanqua*** 'Beni Tsubake'	wCCr
	sasanqua 'Bonanza'	oGar wWhG
un	***sasanqua*** 'Briar Rose'	wCCr
	sasanqua 'Cleopatra'	oGos oGar oBlo oGre wSta
un	***sasanqua*** 'Hana-jiman'	oGar oSho oGre wSta
	sasanqua 'Jean May'	oGar oGre oWnC wWhG
	sasanqua 'Kanjiro'	oAls oGar oBlo oWnC wWhG
	sasanqua 'Mine-no-yuki'	oFor oAls oGar oSho wCCr wWhG
	sasanqua 'Narumigata'	wSta
	sasanqua 'Rainbow'	oGre
	sasanqua 'Setsugekka'	oFor oGar oGre oWnC wSta wWhG
un	***sasanqua*** 'Shishifukujin'	wCCr
	sasanqua 'Shishigashira	oFor oGar oSho oGre oWnC wWhG
un	***sasanqua*** 'Showa-no-sake'	oBlo wSta
un	***sasanqua*** 'Tago-no-tsuki'	wCCr
un	***sasanqua*** 'Texas Star'	wCCr
	sasanqua 'White Doves'	see **C. *sasanqua*** 'Mine-no-yuki'
	sasanqua 'Yuletide'	see **C. *vernalis*** 'Yuletide'
	'Scentuous'	last listed 99/00
un	'Sierra Spring'	wWhG
	sinensis	wHer wBur oCis oSho oOEx oGre wCCr iGSc
un	***sinensis*** 'Blushing Maiden'	wHer oGos oGar oRiv oCis wRai oGre oWnC oCrm
un	***sinensis*** 'Rosea'	oCis
un	***sinensis*** Teabreeze™	wHer oGar oCis wRai oGre oCrm
un	***sinensis*** 'Variegata'	wHer
	'Snow Flurry'	oGre
	'Swan Lake'	oGar oWnC
	'Tom Knudsen'	oAls oGar oWnC wWhG
	transarisanensis	oCis
	transnokoensis	oCis
	vernalis 'Ginryu'	wCCr
	vernalis 'Hiryu'	oGos
	vernalis 'Yuletide'	oAls oGar oRiv oSho oBlo oWnC wSta wWhG wSte
	x williamsii 'Anticipation'	wHer oGos oGar
	x williamsii 'Ballet Queen'	wHer oGre
	x williamsii 'Brigadoon'	oGar wWel
	x williamsii 'Daintiness'	wHer
	x williamsii 'Debbie'	wHer wWel
	x williamsii 'Donation'	wHer oGos wClo oGre wWel
	x williamsii 'E. G. Waterhouse'	last listed 99/00
	x williamsii 'Elsie Jury'	wHer
	x williamsii 'J. C. Williams'	oGos oGre wCCr wWhG
	x williamsii 'Jury's Yellow'	wHer oGos wClo oGar oGre wWel
	x williamsii 'Mary Christian'	wCCr
	x williamsii 'November Pink'	wCCr
	x williamsii 'Rendezvous'	wHer
	x williamsii 'Ruby Wedding'	oGre
	x williamsii 'Taylor's Perfection'	oGar wWel
	x williamsii 'Water Lily'	last listed 99/00
un	'Winter's Charm'	oGre
un	'Winter's Dream'	oGre
un	'Winter's Hope'	oGre
	'Winter's Interlude'	oFor oGre
un	'Winter's Rose'	oGre
un	'Winter's Star'	oFor oGar oGre
un	'Winter's Waterlily'	oGre
	CAMPANULA	
	alliariifolia	wCul oJoy oRus oMis
	alliariifolia 'Ivory Bells'	see **C. *alliariifolia***
	alpestris	wMtT
va	'Balchiniana'	oHed
	barbata	oNWe
	bellidifolia	last listed 99/00
	betulifolia	oSis
	'Birch Hybrid'	oFor oJoy oAls oSis oSho wRob wHom oWnC wTGN wWhG oSle
	bluemelii Jurasek List	wMtT
	'Bumblebee'	wMtT
	'Burghaltii'	last listed 99/00
	caespitosa	see **C. *cespitosa***
	cespitosa	wMtT
	calycanthema	see **C. *medium*** 'Calycanthema'
	carpatica	oFor wTho oNWe oBlo
	carpatica f. ***alba*** 'Weisse Clips'	oFor oAls oGar oSis oWnC wMag wTGN oEga wWhG
	carpatica 'Blaue Clips'	oSis wMag oCir oEga wWhG
	carpatica blue	iArc
	carpatica Blue Clips	see **C. *carpatica*** 'Blaue Clips'
un	***carpatica*** 'Blue Uniform'	oWnC
	carpatica 'Chewton Joy'	wWoS
un	***carpatica*** 'Deep Blue Clips'	oAls
	carpatica 'Graham Giant'	wMtT
un	***carpatica*** 'Light Blue Clips'	oEga
	carpatica var. ***turbinata***	wMtT
	carpatica var. ***turbinata*** 'Karl Foerster'	wWoS
	carpatica var. ***turbinata*** 'Wheatley Violet'	wWoS
	carpatica White Clips	see **C. *carpatica*** f. ***alba*** 'Weisse Clips'
un	***carpatica*** 'White Uniform'	oWnC
	cashmeriana	oSis
	chamissonis	wCol oSis oNWe
	chamissonis 'Oyobeni'	oSis
	choruhensis	oHed
	choruhensis Jurasek list	wMtT
	cochleariifolia	oTrP oAls oSis oSec oSle
	cochleariifolia var. ***alba***	oSis
	cochleariifolia var. ***alba*** 'Bavaria White'	wCul oWnC
	cochleariifolia 'Bavaria Blue'	wCul oWnC
	cochleariifolia 'Elizabeth'	see **C. *cochleariifolia*** 'Elizabeth Oliver'
do	***cochleariifolia*** 'Elizabeth Oliver'	oFor wWoS oJoy wCol wMtT oOut wRob
	cochleariifolia var. ***pallida*** 'Miranda'	oFor wWoS oSis
do	***cochleariifolia*** 'R.B. Loder'	wMtT
	collina	oJoy oSis
un	***dzaaku***	wMtT
	'Elizabeth'	wWoS wHig oAls oHed
	'Faichem Lilac'	oNWe
	fenestrellata	oSis
	finitima	see **C. *betulifolia***
	garganica	wEde wWhG

un	garganica 'Birchwood Gold'	oHed
	garganica 'Dickson's Gold'	wHer wWoS wCol oAls oHed oGar oSis wRob
	glomerata	wTGN wWhG
	glomerata var. acaulis	oFor oMis oSis
	glomerata var. alba	wCul oJoy oSha
	glomerata var. alba 'Schneekrone'	wWoS oWnC wTGN oSle
	glomerata blue	iArc
	glomerata 'Caroline'	wWoS wHig wRob
	glomerata Crown of Snow	see C. glomerata var. alba 'Schneekrone'
	glomerata var. dahurica	oJoy oMis
	glomerata 'Joan Elliott'	oFor wWoS wCul oAls oGar oWnC wTGN oSle
	glomerata 'Superba'	oFor wTho oDan oSha oWnC wAva oEga
	grossekii	oFor
	x haylodgensis	see C. x haylodgensis 'Plena'
do	x haylodgensis 'Plena'	oSis wMtT
do	x haylodgensis 'Warley White'	oSis
	hercegovina 'Nana'	wMtT
	herminii	last listed 99/00
	'Hilltop Snow'	wMtT
un	'Hot Lips'	wWoS oNWe
	incurva	oSis oNWe wEde
	isophylla	wSta
	isophylla 'Alba'	last listed 99/00
	isophylla 'Mayi'	oSis
	kemulariae	oFor wWoS oSis
	'Kent Belle'	wWoS oHed oGar wTGN wSte
un	kinokawamae	oJoy
	lactiflora	oFor oJil oMis
	lactiflora 'Blue Cross'	wWoS
un	lactiflora 'Bright Eyes'	wWoS
	lactiflora 'Loddon Anna'	wWoS oNat oHed wRob
	lactiflora 'Pouffe'	wWoS
	lactiflora 'Prichard's Variety'	wHer oNWe wMag
	lactiflora 'White Pouffe'	wRob wMag
	latifolia	wCul
	latifolia 'Alba'	oNWe
	latifolia var. macrantha	wMag
	latiloba 'Hidcote Amethyst'	wHig oWnC
	'Lynchmere'	wWoS oSis wMtT
un	'Mai Blyth'	wMtT
	medium	wTho oMis oBlo
	medium blue	oWnC
	medium 'Calycanthema'	oUps
	medium double mix	wMag
	medium lilac	oWnC
	medium pink	oAls
	medium rose	iArc
	medium single mix	wMag
	medium white	wWhG
	muralis	see C. portenschlagiana
	nitida	see C. persicifolia var. planiflora
	patula ssp. abietina	last listed 99/00
	patula ssp. abietina JCA 260.210	wMtT
	persicifolia	oFor wCul wCri wTho wHig oAls wCSG oNWe wCCr wSta oCir wAva oEga
	persicifolia alba	oFor oAls oAmb oSha wMag oCir
do	persicifolia 'Alba Coronata'	wHer
do	persicifolia 'Blue Bloomers'	oNWe
do	persicifolia 'Boule de Neige'	wWoS
	persicifolia 'Chettle Charm'	wWoS oDar wHig oAls oHed oGar oOut wTGN
	persicifolia double blue	oJoy wHig oHed
	persicifolia double white	oFor
	persicifolia 'Gawen'	see C. persicifolia 'Hampstead White'
	persicifolia 'George Chiswell'	see C. persicifolia 'Chettle Charm'
	persicifolia 'Grandiflora'	last listed 99/00
	persicifolia 'Grandiflora Alba'	oUps
do	persicifolia 'Hampstead White'	wWoS wHig oHed wPir
do	persicifolia 'Moerheimii'	oJoy oGar
	persicifolia var. planiflora	oSis wMtT
	persicifolia var. planiflora f. alba	oSis wMtT
do	persicifolia 'Pride of Exmouth'	wHer
	persicifolia 'Telham Beauty'	oFor oJoy oAmb oGar oWnC wTGN
un	persicifolia 'Victoria'	wWoS
do	persicifolia 'Wortham Belle'	wWoS oHed
	pilosa	see C. chamissonis
	piperi 'Marmot Pass'	last listed 99/00
	piperi 'Townsend Ridge'	last listed 99/00
	planiflora	see C. persicifolia var. planiflora
	portenschlagiana	oFor oAls oRus oSho oSqu oWnC oSec wEde oUps oSle
	portenschlagiana 'Resholdt's Variety'	wHer wWoS oHed wMtT

	poscharskyana	wSwC oNat oAls oRus wCSG oMis oGar oSqu wHom wSta iArc wAva wWhG
	poscharskyana 'Blauranke'	wWoS
	poscharskyana 'Blue Gown'	wWoS wCol oSis
	poscharskyana 'E. H. Frost'	wHer wWoS
un	poscharskyana 'Multiplicity'	wWoS
	poscharskyana 'Stella'	wWoS oSis
	prenanthoides	oMis
	primulifolia	wRob wCCr
	x pseudoraineri	wMtT
	pulla	last listed 99/00
	x pulloides	oSis wMtT
	punctata	oFor
	punctata 'Alba'	see C. punctata f. albiflora
	punctata f. albiflora	oHed oRus
	punctata f. albiflora 'Nana Alba'	wWoS wRob
un	punctata 'Alina's Double'	wHer
un	punctata 'Cherry Bells'	oFor wHer wWoS oNat oJoy wHig oAls oAmb oOut wRob oGre wTGN wWhG oSle wSte
	punctata var. hondoensis	last listed 99/00
	punctata 'Rosea'	oSis
	punctata f. rubriflora	oRus oMis oGar oSho oSec
un	punctata 'Wedding Bells'	wWoS wHig wHom
	pusilla	see C. cochleariifolia
	pyramidalis	oAls wAva
	raineri	last listed 99/00
	rotundifolia	oFor oHan wTho oJoy oRus wShR oMis oGar oBlo iArc wWld oSle
	rotundifolia var. alba	oRus
	rotundifolia ssp. arctica 'Mount Jotunheimen'	see C. rotundifolia 'Jotunheimen'
	rotundifolia 'Jotunheimen'	wMtT
un	rotundifolia 'Ned's White'	wMtT
	rotundifolia 'Olympica'	oAls wHom
	rubriflora	see C. punctata f. rubriflora
un	'Samantha'	wWoS wCol oHed oNWe
	saxifraga	last listed 99/00
	scabrella	wMtT
	scouleri	oBos
un	'Sojourner'	wMtT
un	suanetica	wMtT
	takesimana	oFor wWoS oNat oAls wRob wMag wBWP wWhG oUps oSle
	takesimana alba	oAls
	takesimana alba DJH 197	wHer
un	takesimana 'Beautiful Trust'	wHer
	thyrsoides	last listed 99/00
	tommasiniana	last listed 99/00
	trachelium	wEde
	trachelium var. alba	wEde
do	trachelium 'Alba Flore Pleno'	oHed
do	trachelium 'Bernice'	oFor wHer wWoS wHig oHed oDan
	tridentata	oNWe
	troegerae Jurasek List	wMtT
	'Tymonsii'	oSis
	'Van-Houttei'	wHer
	CAMPANULA X SYMPHYANDRA	
	C. punctata x S. ossetica	oHed
	CAMPSIS	
	flava	see C. radicans f. flava
	grandiflora	oFor
un	grandiflora 'Morning Calm'	oGar
un	'Indian Summer'	oGos
	radicans	oFor wCul oAls wCSG oGar oGoo oSho oWhi wKin wAva wCoN oUps
	radicans 'Flamenco'	oFor wWoS oAmb oSho oUps oSle
	radicans f. flava	oFor oAls oGar oSho oGre oWnC wWel
un	radicans 'Judy'	oGre
un	radicans 'Minnesota Red'	oAls oGar
	radicans 'Yellow Trumpet'	see C. radicans f. flava
	x tagliabuana	wCoN wCoS
un	x tagliabuana 'Guilfoylei'	oGre
	x tagliabuana 'Madame Galen'	oGar oSho oSle wWel
	CAMPTOTHECA	
	acuminata	oGre oWnC
	CAMPYLOTROPIS	
	macrocarpa	wCCr
	CANNA	
un	achira	oFor
un	achira 'Esmeralda's'	oOEx
un	achira 'Maximo's Purple'	oOEx
un	'Ambassador'	wCSG
un	'Angel Pink'	oGar
va	'Bengal Tiger'	see C. 'Striata'

	'Black Knight'	oHug oGar oDan
	'Chinese Coral'	wCSG
	'City of Portland'	oGar oDan
	'Cleopatra'	oHug oGar
va	'Durban'	oGar wFai wBox
	dwarf yellow	wHer
	flaccida	oHug
un	'Futurity Pink'	wCSG
	x generalis 'Aureostriatus'	see **C.** 'Striata'
	x generalis 'Pretoria'	see **C.** 'Striata'
	glauca	wHer
un	'Grande'	wHer
un	'Harvest Yellow'	oGre
	indica 'Purpurea'	last listed 99/00
	indica west Mexico	oCis
	'Intrigue'	wHer oGar oCis oGre wFai
	'King City Gold'	oGar
un	'La Buffa'	oCis
un	Longwood Hybrids	wGAc oSsd
	'Lucifer'	oNat oJoy oGar oDan
un	'Minerva'	oGar oOut oGre wFai wBox wSte
	musifolia	wHer
un	'North Star'	wCSG oGar
	orange water	oHug
	'Panache'	wHer oGar oOut oGre wFai wSte
un	'Peach Blush'	wCSG
	'Pfitzer's Chinese Coral'	see **C.** 'Chinese Coral'
	'Picasso'	oNat oGar
un	'Pink Futurity'	oNWe
un	'Pink President'	wCSG
	'Pink Sunburst'	wGAc oGre
	pink water	oHug
	'President'	oJoy oGar oGre wWel
	'Pretoria'	see **C.** 'Striata'
	'Ra'	oGar
	'Red King Humbert'	see **C.** 'Roi Humbert'
un	'Red Velvet'	oGar oGre wBox
	red water	oHug
	'Richard Wallace'	oGar
	'Roi Humbert'	oGar
	'Salmon'	oGar
un	'Shining Pink'	oGar
un	'Shining Sun'	oGar
va	'Striata'	wHer wGAc oNat oJoy wCol oHug oGar oSsd oDan oNWe oOut oGre wFai wHom wWel
va	'Striped Beauty'	oHug oGar
va	'Stuttgart'	oGos wCol oGar
	'The President'	see **C.** 'President'
	Tropicanna / 'Phasion'	oAls oGar oNWe oSho oGre wWel
	'Wyoming'	oNat oJoy oGar oDan oNWe oGre wWel
	'Yellow Humbert'	oGar

CAPPARIS

	spinosa	last listed 99/00
	spinosa var. ***inermis***	iGSc

CAPSICUM

	annuum Conoides Group	oGar
	frutescens	oCrm
	pubescens	oOEx

CARAGANA

	arborescens	oFor wBCr oRiv wKin oWnC wPla wCoN oSle
	arborescens 'Pendula'	oAls
	arborescens 'Walker'	oFor wKin
	frutex	oFor
	frutex 'Globosa'	oFor
	microphylla	last listed 99/00

CARALLUMA

	baldratii	oRar
	dummeri	see **PACHYCYMBIUM *dummeri***
	hexagona	oRar
	shadhbana	see **C. *hexagona***

CARDAMINE

un	***angulata***	oBos
	diphylla	wHer oRus
	enneaphyllos	wHer
	heptaphylla	wHer
	laciniata	wHer
	macrophylla	wHer
	x maxima	wHer
	pentaphyllos	wHer
	pratensis	wHer
do	***pratensis*** 'Edith'	wHer
do	***pratensis*** 'Flore Pleno'	oJil oNWe
do	***pratensis*** 'Salzach'	wHer wEde
do	***pratensis*** 'William'	last listed 99/00
	quinquefolia	wHer
	raphanifolia	wHer
	trifolia	wHer oJil wCol
	waldsteinii	wHer

CARDIANDRA

	alternifolia :HC 970549	wHer

CARDIOCRINUM

	cathayanum	wHer
un	***cordatum*** var. ***cordatum***	wHer
	cordatum var. ***glehnii***	last listed 99/00
	giganteum	oMac oRus oDan
	giganteum var. ***giganteum***	wHer
	giganteum var. ***yunnanense***	oRus
	giganteum var. ***yunnanense*** DJHC 797	wHer

CAREX

	acuta	wBox
	albula	see **C. *comans***
un	***aperta***	wWat
	berggrenii	wHig oOut
	brunnea	oTrP wCli
	brunnea 'Variegata'	wCli
	buchananii	wHer wCli oDar oJoy oAmb oMis oGar oSho oOut wWin wPir wHom oWnC wTGN wBox oSec oCir oUps wSnq
un	***buchananii*** 'Red Fox'	oFor
	buchananii 'Viridis'	wSwC oOut wWin
	caryophyllea 'The Beatles'	oFor wHer oOut
	comans	oTrP wHer wCri oGar oDan wWin
	comans bronze	wCli oNat oDar oJoy oOut wNay wEde
un	***comans*** 'Bronze Curls'	oOut
	comans 'Frosted Curls'	wWoS wCli wCul wCri oDar wCol oHed oAmb oGar oSho wWin wHom oWnC wTGN wBox oBRG wEde wSnq
	comans 'Frosty Curls'	see **C. *comans*** 'Frosted Curls'
	conica 'Hime-kan-suge'	see **C. *conica*** 'Snowline'
	conica 'Marginata'	see **C. *conica*** 'Snowline'
va	***conica*** 'Snowline'	wCli oSis oOut wWin oWnC
	conica 'Variegata'	see **C. *conica*** 'Snowline'
ch	***dolichostachya*** Gold Fountains	see **C. *dolichostachya*** 'Kaga-nishiki'
va	***dolichostachya*** 'Kaga-nishiki'	wHer wWoS oJoy oHed oGar oCis oOut wNay wRob wWin oWnC wBox wSte wSnq
	elata	wFai
va	***elata*** 'Aurea'	wHer wCli wGAc oNat oJoy oAls oHed oAmb oMis oGar oNWe oOut wNay wRob oGre wFai wWin oUps oSle wSnq
	elata 'Bowles' Golden'	see **C. *elata*** 'Aurea'
	elata 'Knightshayes'	wWoS oJoy oGar oNWe oOut oSec
un	***filiosa***	oTrP oHed
	firma 'Variegata'	wMtT
	flacca	wCCr wSte
	flacca ssp. ***flacca***	oGre
	flagellifera	oFor oTrP wHer wCli oJoy oGar oDan oNWe oSho wCCr wHom oWnC wBox wSnq
	flava	wWin wBox
ch	***glauca*** (see Nomenclature Notes)	wCli wSoo oDar oHug wWin wBox oSec wEde
	grayi	oFor oTrP wCli oSho wWin wSte
un	***hendersonii***	wNot
ch	'Ice Dance'	oFor wHer wWoS wCli oGos wSwC oAls oGar oOut wNay wWin oWnC oUps oSle wSte wSnq
	'Kan-suge'	see **C. *morrowii***
un	***lyngbyei***	wWat
un	***macrocephala***	wShR wWat
	morrowii (see Nomenclature Notes)	oTrP wHom
un	***morrowii*** 'Aureomarginata'	oBRG
	morrowii 'Aureovariegata'	see **C. *morrowii*** 'Goldband'
	morrowii var. ***expallida***	see **C. *morrowii*** 'Variegata'
	morrowii 'Goldband'	wWoS wCli wSwC oNat oRus wCSG oGar oDan oGre wFai wWin oTPmwTGN wBox oSec oBRG wWhG oSle wCoS
ch	***morrowii*** 'Ice Dance'	see **C.** 'Ice Dance'
	morrowii 'Variegata'	oDar oJoy oHed oMis oGar oOut wFai wWin wBox oSec wSta oCir
	muskingumensis	oFor wCli oDar oSis oCis wWin wBox wSnq
un	***muskingumensis*** 'Ice Fountains'	wWoS wCol wWin
	muskingumensis 'Oehme'	wHer wWoS oJoy wCol oHed wEde wSnq
	muskingumensis 'Wachtposten'	oOut wSnq
	nigra (see Nomenclature Notes)	oFor wGAc oDar oGar oSsd oSqu oWnC wBox
	nigra 'Variegata'	oJoy oSis oSho oOut oSec wSnq
	nudata	wWin

	obnupta	wSoo wWoB wNot wShR oHug oSsd oBos wWat oTri oAld
	ornithopoda 'Variegata'	oFor wHer oSis
va	oshimensis 'Evergold'	wCul oDar oOut wNay oBlo wWin oTri wBox oBRG oUps wSte wSnq
	pansa	wShR
	pendula	oFor oDar oGar wWin wSte
	petriei	oFor oOut oSec
	phyllocephala	last listed 99/00
va	phyllocephala 'Sparkler'	wHer oOut
	plantaginea	wCol oGar wRob
	pseudocyperus	wWin wSnq
	riparia	last listed 99/00
	riparia 'Variegata'	wHer wWin oWnC oSec
un	rostrata	wWat
	secta	wSnq
	secta var. tenuiculmis	wSnq
un	siderosticha 'Island Brocade'	wWoS oGos wCol oDan wNay wRob wWin
un	siderosticha 'Spring Snow'	wNay wWin
	siderosticha 'Variegata'	oFor wCli wSwC oJoy wCol oAls oHed oGar oSis oCis oNWe oSho oOut wNay wRob oGre wWin wSte
	sp. T95-10 Mexico	oCis
	sp. Uganda	wHer
un	stipata	oSsd wWat oTri
un	tenuisecta 'Bronzina'	wWin
	testacea	oTrP wHer wCli wSwC oJoy oAls oOut oGre wFai wWin wPir wHom
	testacea 'Old Gold'	oAls oGar wEde
	texensis	wBox
	trifida	wCCr
	umbrosa	oDar
un	velebit humilis	wBox
	CARICA	
	x heilbornii	wRai oOEx
	pentagona	see C. x heilbornii
	pubescens	oOEx
	CARLINA	
	acaulis	oGoo
	acaulis ssp. simplex	oJoy
	vulgaris	oGoo
	CARMICHAELIA	
	arborea	oFor
	CARPENTERIA	
	californica	oGre wCoN wSte
	californica 'Elizabeth'	oFor oGos oGar oSho
	CARPINUS	
	betulus	oFor oAls oWnC
	betulus 'Fastigiata'	oGar oRiv oWnC wBox wSta wWel
	betulus 'Frans Fontaine'	oGar oGre
un	betulus 'Globosa'	oWnC wWel
un	betulus 'Heterophylla'	oFor
	betulus 'Pendula'	oFor
	betulus 'Purpurea'	oFor oWhi
	betulus 'Pyramidalis'	see C. betulus 'Fastigiata'
	betulus 'Quercifolia'	oWhi
	caroliniana	oFor
	cordata	oFor oRiv oGre
	cordata DJH 353	wHer
	coreana	oFor oAmb oDan oRiv oGre
	coreana DJH 284	last listed 99/00
	henryana	oFor oRiv
	japonica	oFor wClo oDar oEdg oAls oAmb oGar oDan oRiv oGre wAva wWel
	sp. aff. kawakamii B&SWJ 3404	wHer
	koreana	see C. coreana
	laxiflora	last listed 99/00
	laxiflora DJH 376	wHer
	mollicoma	oFor
	omeiensis	wHer
	orientalis	oFor oRiv
	polyneura	oFor
	tschonoskii	last listed 99/00
	turczaninowii	oFor oAmb oDan oRiv oGre wWel
	viminea	oFor
	CARPODETUS	
	serratus	oFor
	CARTHAMUS	
	tinctorius	oUps
	CARUM	
	carvi	oAls iGSc oWnC oUps
	copticum	iGSc
	CARYA	
	aquatica	oFor
	glabra	last listed 99/00
	illinoinensis	oFor wBCr wBur
un	illinoinensis 'Carlson #3'	wBur
un	illinoinensis 'Cornfield'	wBur
un	illinoinensis 'Kanza'	wBur
un	illinoinensis 'Lucas'	wBur
un	illinoinensis 'Pawnee'	wBur
un	illinoinensis 'Snaps'	wBur
	laciniosa	wBCr wBur oOEx
un	laciniosa 'Campbell'	wBur
un	laciniosa 'Fayette'	wBur
	laciniosa 'Henry'	wBur
	myristiciformis	last listed 99/00
	ovata	oFor wBCr wBur oOEx
	tomentosa	oFor
	CARYOPTERIS	
	x clandonensis	wCCr oSec wCoN
	x clandonensis 'Arthur Simmonds'	oFor w WoS
un	x clandonensis 'Black Knight'	wCSG
	x clandonensis 'Blue Mist'	wCul oJoy wCSG oGar oGre oWnC wEde
	x clandonensis 'Dark Knight'	see C. x clandonensis 'Dark Night'
	x clandonensis 'Dark Night'	oFor wCul oDar oAls oHed oMis oGar oSis oGre wFai wPir oWnC wTGN oUps wWel
	x clandonensis 'Heavenly Blue'	oFor oSho oOut wMag
	x clandonensis 'Kew Blue'	wCSG
	x clandonensis 'Longwood Blue'	wWoS wCul oAls oGar oGoo oGre wMag oEga oUps
	x clandonensis 'Worcester Gold'	oFor wHer wWoS oNat oDar oJoy oEdg oAls oMis oGar oGoo oDan oSis oNWe wRob oGre wFai wPir oWnC wBox oSec wBWP wEde oUps
	divaricata	oFor
	incana	last listed 99/00
un	incana 'Alba'	wHer
	incana pink form	oNWe
	odorata	last listed 99/00
	CASIMIROA	
un	edulis 'Edgehill'	oOEx
un	edulis 'Lemon Gold'	oOEx
un	edulis 'Leroys'	oOEx
un	edulis 'McDill'	oOEx
	edulis 'Suebelle'	oOEx
un	edulis 'Sunrise'	oOEx
	CASSIA	
	corymbosa	see SENNA corymbosa
	hebecarpa	see SENNA hebecarpa
	lindheimerana	see SENNA lindheimerana
	tomentosa	see SENNA multiglandulosa
	CASSINIA	
un	amoensis	oCis
	fulvida	see C. leptophylla ssp. fulvida
	leptophylla	wCCr
	leptophylla ssp. fulvida	oHed oCis
	leptophylla ssp. vauvilliersii	wCri
	vauvilliersii	see C. leptophylla ssp. vauvilliersii
	sp. x sp.	wHer
	CASSIOPE	
	'Askival Stormbird'	wMtT
	'Badenoch'	last listed 99/00
	'Edinburgh'	wMtT
	lycopodioides	wMtT
	lycopodioides 'Beatrice Lilley'	wMtT
	lycopodioides var. globularis	last listed 99/00
	mertensiana	wMtT
	mertensiana pink	wMtT
	'Muirhead'	wMtT
	stelleriana	wMtT
	CASTANEA	
un	'Bear Creek Chinese' seed grown	wBCr
un	'Bisalta # 3'	wBur
un	'Bisalta # 3' seed grown	wBur
un	'Campbell # 1' seed grown	wBCr
un	'Colossal'	wBur oWhi wRai
un	'Colossal' seed grown	wBCr wBur oWhi wRai
	dentata	wBCr wBur wRai oOEx
un	'Douglas Hybrid' seed grown	wBCr
un	'Gellatly Dark European' seed grown	wBCr
	'Layeroka'	wBCr wBur
	'Layeroka' seed grown	wBCr wBur
un	'Manoka' seed grown	wBCr
un	'Maraval'	wBur
	'Marigoule'	wBur
un	'Marron du Var'	wBur
	mollissima	wBur oWhi wRai oOEx
un	'Myoka'	oWhi
un	'Nevada'	wBur

un	'Precoce Migoule'	wBur
	sativa	oFor wBur
un	*sativa* 'Schrader'	wRai
un	'Skioka'	wBCr wBur wRai
un	'Skioka' seed grown	wBCr wBur
un	'Skookum'	wBur
un	'Skookum' seed grown	wBCr wBur
un	'Sleeping Giant' seed grown	wBCr
	CASTANOPSIS	
	chrysolepis	see **CHRYSOLEPIS** *chrysophylla*
	chrysophylla	see **CHRYSOLEPIS** *chrysophylla*
	cuspidata	oCis oSho
	sp. DJHC 98287	wHer
	CASTILLEJA	
un	*parviflora*	oBos
	sp.	wThG wPla
	CASUARINA	
	cristata	last listed 99/00
	equisetifolia	oTrP
	littoralis	see **ALLOCASUARINA** *littoralis*
	stricta	see **ALLOCASUARINA** *verticillata*
	CATALPA	
	bignonioides	oFor oEdg wShR oGre
	bignonioides 'Aurea'	oFor oGos oGar oDan oGre wWel
	bignonioides 'Nana'	oFor
	bungei	oFor oGar
	x erubescens 'Purpurea'	oGos
	ovata	oFor
	speciosa	oFor oGar wKin wTGN
va	*speciosa* 'Pulverulenta'	last listed 99/00
	CATANANCHE	
	bicolor	see *C. caerulea* 'Bicolor'
	caerulea	wTho oJoy oAls oHed wCSG oMis oGoo oGre wHom iArc wAva oEga oSle
	caerulea 'Alba'	oFor
	caerulea 'Bicolor'	oMis
	CATHA	
	edulis	oOEx
	CATTLEYA	
	hybrids	oGar
	CAULOPHYLLUM	
	thalictroides	iGSc oCrm
	CAYRATIA	
un	sp. aff. *thomsonii* DJHC 501	wHer
	CEANOTHUS	
	americanus	oFor
	'Blue Cushion'	wCCr
	'Blue Jeans'	wCCr
	'Concha'	oFor wHer oGar wPir
	cordulatus	oFor
	cuneatus	oFor
	cuneatus var. *rigidus*	last listed 99/00
	'Dark Star'	oFor oHed wWel
	x delileanus 'Gloire de Versailles'	wHer wWoS wCul wCCr
	x delileanus 'Henri Desfosse'	oHed
	'Frosty Blue'	oFor
un	'Gerda Isenberg'	wCCr
	gloriosus	oFor oGar oOut oBlo oGre wMag wTGN wSte
	gloriosus 'Anchor Bay'	oHed
	gloriosus 'Emily Brown'	wCCr
	gloriosus 'Point Reyes'	see *C. gloriosus*
	sp. aff. *griseus*	oHed
	griseus var. *horizontalis*	wShR
	griseus var. *horizontalis* 'Yankee Point'	oHed oUps
	hearstiorum	last listed 99/00
	impressus	oFor oBlo
un	*impressus* 'Vandenberg'	oJoy oHed
	integerrimus	oFor
	'Italian Skies'	oFor
	'Joyce Coulter'	wWel
	'Julia Phelps'	oFor wCul oHed
	lemmonii	oFor
	x lobbianus	oGar
un	'Louis Edmonds'	oFor
	x mendocinensis	wCCr
	'Mountain Haze'	wCul
	ovatus	last listed 99/00
	x pallidus 'Marie Simon'	oFor wCul oHed wCSG
	prostratus	wHer
	'Ray Hartman'	wWoS oGar wWel
ch	*rigidus*	see *C. cuneatus* var. *rigidus*
	sanguineus	oFor wWat wPla
	'Snow Flurries'	oFor
	thyrsiflorus	wShR oAld wWld
	thyrsiflorus var. *repens*	wCCr
	thyrsiflorus 'Skylark'	oFor
un	*thyrsiflorus* 'Snowball'	oCis
	x veitchianus	wHer oBlo wCCr
	velutinus	wThG wWoB wWat wPla
un	'Victoria'	oFor wHer wClo oJoy oAls oHed wShR oGar oRiv oSho oOut wCCr oWnC wTGN wCoS wWel
	CEDRELA	
	sinensis	see **TOONA** *sinensis*
	CEDRONELLA	
	canariensis	wWoS oAls oGoo iGSc wHom oCrm
	CEDRUS	
	atlantica	oEdg oRiv oWnC wTGN wSta wAva wCoN wRav wCoS
un	*atlantica* 'Argentea Fastigiata'	oGre
	atlantica 'Aurea'	wCol oWnC wSta
un	*atlantica* 'Aurea Robusta'	oPor wCol
	atlantica 'Fastigiata'	oPor oRiv wSta
	atlantica Glauca Group	oPor oDar oAls oGar oDan oRiv oBlo oGre oTPm oWnC wSta wAva oGue
	atlantica 'Glauca Fastigiata'	oPor
	atlantica 'Glauca Pendula'	oFor oPor oDar wCol oGar oDan oRiv oBlo oGre wSta wAva wWel
un	*atlantica* 'Horstmann'	oPor
un	*atlantica* 'Liliput'	oPor
un	*atlantica* 'Morocco'	wSta
	atlantica 'Pendula'	oAls oTPm oWnC wAva
un	*atlantica* 'Pendula' serpentine	oWnC
un	*atlantica* 'Sahara Frost'	wCol
un	*atlantica* 'Sahara Ice'	wCol
	atlantica 'Silberspitz'	oPor
	atlantica 'Uwe'	oPor
	brevifolia	oFor oPor oRiv oWnC wSta
	brevifolia 'Epstein'	oPor
un	*brevifolia* 'Treveron'	oPor wClo wCol oGre
	deodara	oFor oDar oEdg oAls wShR oGar oRiv oSho oWhi oBlo oTPm oWnC wTGN wSta oCir wAva wCoN wRav wCoS
va	*deodara* 'Albospica'	oFor oRiv oGre
	deodara 'Aurea'	oDar oAls oDan oRiv oSho oGre wSta
	deodara 'Aurea Pendula'	oFor oWnC wSta
	deodara 'Blue Snake'	oPor
un	*deodara* 'Bush's Electra'	oPor wCol
	deodara 'Cream Puff'	oFor oPor wClo oSis oGre wWel
un	*deodara* 'Deep Cove'	oPor wClo oWnC
	deodara 'Descanso Dwarf'	oPor oSis wWel
un	*deodara* 'Devinely Blue'	oPor wClo oJoy wCol oGre oWnC wSta wWel
	deodara 'Divinely Blue'	see *C. deodara* 'Devinely Blue'
	deodara 'Eisregen'	oPor
un	*deodara* 'Eiswinter'	oPor
	deodara 'Emerald Spreader'	oPor wCol
	deodara 'Feelin' Blue'	oPor wClo
	deodara 'Glauca'	oRiv
un	*deodara* 'Gold Cascade'	oPor
	deodara 'Gold Cone'	oPor oGar oSis oWnC
	deodara 'Gold Mound'	oPor
	deodara 'Golden Horizon'	oPor
un	*deodara* 'Harvest Gold'	oPor
un	*deodara* 'Hollandia'	oPor oSis
	deodara 'Karl Fuchs'	oPor
	deodara 'Kashmir'	oFor oRiv wWel
	deodara 'Klondyke'	oPor wClo
	deodara 'Lime Glow'	oPor
un	*deodara* 'Limelight'	oPor
	deodara 'Pendula'	oPor wClo oGar wSta
	deodara 'Polar Winter'	oPor
	deodara 'Prostrata'	oFor
	deodara 'Prostrate Beauty'	oPor
	deodara 'Pygmy'	oPor oGos oGre
un	*deodara* 'Raywood's Contorted'	oPor
	deodara 'Repandens'	oPor
	deodara 'Shalimar'	oFor oPor oTPm
	deodara 'Silver Mist'	oFor oPor wClo wCol
un	*deodara* 'Snow Sprite'	oPor wWel
	deodara twisted	wWel
	deodara 'Verticillata Glauca'	oPor wSta
un	*deodara* 'Vink's Golden'	oPor
	deodara 'Well's Golden'	oPor oGre
un	*deodara* 'White Imp'	oFor oPor oWnC
	libani	oFor wCCr oWnC wCoN oGue wCoS
ch	*libani* ssp. *atlantica*	see *C. atlantica*
ch	*libani* ssp. *brevifolia*	see *C. brevifolia*
	libani ssp. *libani*	wSta

un	***libani*** ssp. ***libani*** 'Beacon Hill'	oPor wClo wCol
un	***libani*** ssp. ***libani*** 'Blue Angel'	oPor
	libani ssp. ***libani*** 'Comte de Dijon'	oPor
un	***libani*** ssp. ***libani*** 'Eugene'	oPor
	libani ssp. ***libani*** 'Glauca Pendula'	oPor wTGN wSta
	libani ssp. ***libani*** 'Golden Dwarf'	last listed 99/00
	libani ssp. ***libani*** 'Green Knight'	oPor oGre
	libani ssp. ***libani*** 'Green Prince'	oPor wClo
	libani ssp. ***libani*** Nana Group	oFor oPor oGre wSta
	libani ssp. ***libani*** 'Pendula'	oDar oGar wSta
	libani ssp. ***libani*** 'Sargentii'	oPor oGre
un	***libani*** ssp. ***libani*** 'Stenacoma'	oPor
	CEIBA	
un	***acuminata*** BLM 0172	oRar
	aesculifolia BLM 0198	oRar
	pentandra	oRar
	CELASTRUS	
	orbiculatus	oFor oGar wFai
	orbiculatus 'Diana' female	oGre
	orbiculatus female	oFor
	orbiculatus 'Hercules' male	oGre
un	***rosthornianus***	oFor
	scandens	oFor wShR oGar oWnC wAva oSle
	scandens female	oGar oGre
	scandens male	oGre
	CELMISIA	
	alpina	wHer
	gracilenta	oCis
	CELSIOVERBASCUM	see **VERBASCUM**
	CELTIS	
	africana	oTrP
	australis	oOEx wCCr
	caucasica	oFor wCCr
	glabrata	oFor
	julianae	oFor
	laevigata	oFor
	occidentalis	oFor wBCr oGar oWnC
	reticulata	oFor wCCr
	sinensis	oCis
un	***sinensis*** 'Pendula'	wHer
	sp.	oOEx
un	***tenuifolia***	oFor
un	***yunnanensis***	wCCr
	CENTAUREA	
	dealbata	oNat oSho wHom wMag wAva
qu	***dealbata*** 'Persian Cornflower'	wBox
	dealbata 'Rosea'	oWhS
	dealbata 'Steenbergii'	oFor
	depressa	wHer
	hypoleuca 'John Coutts'	wHig
	macrocephala	oGoo oDan wFai oSle
	montana	oFor oTrP oAls wShR wCSG oSha oBlo wHom oWnC wAva oSle
	montana alba	last listed 99/00
	nigra ssp. ***rivularis***	oFor wHer oRus oSle
	pulcherrima	last listed 99/00
	simplicicaulis	oHed
	CENTAURIUM	
	erythraea	oCrm
	CENTELLA	
	asiatica	oAls oGoo oOEx oCrm oBar
	CENTRADENIA	
un	'Spanish Shawl'	wTGN
	CENTRANTHUS	
	ruber	oFor oNat wFGN wCSG oGoo oGre wHom wMag oCir oEga
	ruber albiflorus	see **C. *ruber*** 'Albus'
	ruber 'Albus'	oFor oEdg oRus oGoo wFai oWnC
	ruber atrococcineus	oRus
un	***ruber atrococcineus*** 'Coral'	oRus
	ruber var. ***coccineus***	oAls oGar oWnC wBWP oSle
	CEPHALANTHUS	
un	***gigantea***	wCCr
	occidentalis	oFor wGAc oHug oGoo oDan oRiv wCCr
	CEPHALARIA	
	alpina	wHig oSqu
	gigantea	oFor wCul oMis wBWP
	leucantha	oSec wBWP
	radiata	last listed 99/00
	sp. aff. ***scabra***	oSis
	CEPHALOCEREUS	
	palmeri	see **PILOSOCEREUS *leucocephalus***
	CEPHALOPHYLLUM	
	alstonii	oSqu
	CEPHALOTAXUS	
	fortunei	oFor wSte
	harringtonia var. ***drupacea***	oFor wHer oWnC
	harringtonia 'Duke Gardens'	oFor wHer oPor
	harringtonia 'Fastigiata'	oFor oPor oSis wWel
	harringtonia var. ***harringtonia*** HC 970574	wHer
	harringtonia var. ***koreana***	oFor
	harringtonia var. ***koreana*** HC 970357	wHer
	harringtonia var. ***nana*** HC 970390	wHer
	harringtonia 'Prostrata'	wHer
	harringtonia var. ***sinensis***	see **C. *sinensis***
	oliveri	oCis
	sinensis	wHer oPor oGre wSte wWel
	CEPHALOTUS	
	follicularis	wOud
	CERASTIUM	
	alpinum var. ***lanatum***	oJoy
	bierbersteinii	see **C. *tomentosum***
	tomentosum	wTho oAls wSta oCir wCoN oEga oSle wCoS
	tomentosum var. ***columnae***	oSis
	tomentosum 'Silberteppich'	last listed 99/00
un	**CERATOIDES**	
	lanata	wPla
	CERATONIA	
	siliqua	oFor
	CERATOPHYLLUM	
	demersum	wGAc oHug oGar oWnC
	CERATOSTIGMA	
	griffithii	oFor oAmb oWnC oUps
	plumbaginoides	oFor wCul oNat oAls wCSG oGar oSis oSho oBlo oGre wPir oWnC wTGN wAva oEga oUps
	willmottianum	oGar wPir oUps
	CERCIDIPHYLLUM	
	japonicum	oFor wWoB wSwC wClo oEdg oAls oGar oRiv oBlo oGre wCCr oWnC wTGN wSta wAva wCoN wSte wCoS wWel
	japonicum 'Heronswood Globe'	wHer oGos oAmb
	japonicum f. ***pendulum***	oGos oAmboGar oRiv oGre oWnC wTGN wCoN
un	***japonicum*** 'Strawberry'	wWel
	magnificum	wCoN
	magnificum 'Pendulum'	oGos oGar oGre
	CERCIS	
	canadensis	oFor wBCr wSwC oAls wShR oGar oRiv oSho oWhi oGre wKin wTGN wSta wCoN oSle wCoS
	canadensis 'Alba'	oFor oDan oBlo wWel
un	***canadensis*** 'Appalachian Red'	oGre wWel
	canadensis 'Flame'	oFor oGre
	canadensis 'Forest Pansy'	oFor wClo oDar oAls oDan oRiv oSho oBlo oGre wPir oWnC wTGN wSta wWhG wSte wCoS wWel
un	***canadensis*** ssp. ***mexicanus***	oGre
	canadensis var. ***occidentalis***	oFor oDan oRiv oGre wCCr wCoN
un	***canadensis*** var. ***occidentalis alba***	oFor oGar
	canadensis 'Pendula'	oFor
	canadensis 'Royal White'	oGre
	canadensis 'Rubye Atkinson'	oFor oGar oGre
	canadensis 'Silver Cloud'	oFor oGar oWnC
	canadensis var. ***texensis*** 'Oklahoma'	see **C. *reniformis*** 'Oklahoma'
	chinensis	oFor oGar oGre wCCr oWnC wTGN oSec wCoN wRav
	chinensis 'Avondale'	oFor oGos oGre oWnC wWel
un	***gigantea***	oFor
	griffithii	oFor wCCr
	occidentalis	see **C. *canadensis*** var. ***occidentalis***
	racemosa	oGre
	reniformis 'Oklahoma'	oFor wClo oBlo oGre oWnC
	reniformis 'Texas White'	oFor
	siliquastrum	oFor oGre wCCr
un	***yunnanensis***	oTrP oCis
	CERCOCARPUS	
	betuloides	see **C. *montanus*** var. ***glaber***
	ledifolius	oFor wShR oBos wPla
	lediformis	see **C. *ledifolius***
	montanus	oFor oBos oTri wPla
	montanus var. ***glaber***	oFor
	CEREUS	
	peruvianus	see **C. *uruguayanus***
	uruguayanus	oOEx
	CERINTHE	
	'Kiwi Blue'	oGar oUps

	major	wSte
	major 'Purpurascens'	wSwC oAls wCSG oNWe wSte
	CEROPEGIA	
	caffrorum	see **C. *linearis***
	cimiciodora	oRar
	haygarthii	oSqu
	linearis	oSqu
un	***linearis*** var. ***debilis***	oRar
	linearis ssp. ***woodii***	oRar
	radicans ssp. ***smithii***	oSqu
	stapeliiformis	oRar
	CESTRUM	
un	***diurnum*** x ***nocturnum***	oCis
	'Newellii'	wHer oSho
	nocturnum	oTrP iGSc
	parqui HCM 98073	wHer
un	***smithii***	oCis
	CHAENOMELES	
	cathayensis	oFor
un	'Cherry Red'	oAls
un	'Iwai Nishiki'	oGre
	japonica	last listed 99/00
	japonica var. ***alpina***	oFor
qu	***japonica*** var. ***alpina*** 'Pygmaeus'	oSis
un	***japonica*** 'Chojuraku' bonsai	oGre
un	***japonica*** 'Clark's White'	oGar
un	***japonica*** 'Hime' bonsai	oGre
	japonica 'Minerva'	oWnC
	japonica 'Orange Delight'	oFor oGre oWnC
	japonica var. ***pygmaea***	oFor
	japonica 'Rubra'	oGre
	japonica 'Scarff's Red'	oGre
un	***japonica*** 'Super Red'	oGar oWnC
un	'Pink Beauty'	oFor
un	'Red Chariot'	oFor
	speciosa	wCSG
un	***speciosa*** 'Chojubai'	oGre
	speciosa 'Contorta'	oFor wHer oGos wCol oHed oAmb oGar wRai
do	***speciosa*** 'Falconnet Charlet'	oFor
	speciosa 'Nivalis'	oFor
	speciosa 'Red Chief'	oFor oGar
	speciosa 'Spitfire'	oFor oSho oGre
	speciosa 'Toyo Nishiki'	oFor oDar oGar oSho wRai oOEx oGre oTPm wTGN wWel
do	***x superba*** 'Cameo'	oFor wCul oGos oAls oGar oSho wRai oOEx oGre oTPm oBRG wAva wCoS
	x superba 'Hollandia'	wCul oGar
	x superba 'Jet Trail'	oFor wCul oGar oOEx oTPm oWnC
	x superba 'Pink Lady'	oFor oGar
	x superba 'Texas Scarlet'	oFor oAls oGar oSho oBlo oOEx oGre wKin oTPm oWnC
un	Victory™	wRai
	CHAEROPHYLLUM	
	hirsutum 'Roseum'	wHer oNWe oGre wBox
	CHAMAEBATIARIA	
	millefolium	oFor
	CHAMAECEREUS	
	sylvestrii	see **ECHINOPSIS *chamaecereus***
	CHAMAECYPARIS	
	funebris	see **CUPRESSUS *funebris***
	lawsoniana	oFor oDar wShR oBos wWat wCCr wCoN wRav oSle
	lawsoniana 'Alumii'	oAls wShR wCoS
	lawsoniana 'Aurea'	oPor
	lawsoniana 'Barabits' Globe'	oWnC
un	***lawsoniana*** 'Barry's Silver'	oWnC
	lawsoniana 'Blue Surprise'	oGre oWnC
	lawsoniana 'Columnaris'	wCCr
	lawsoniana 'Ellwoodii'	oFor oDar oAls oBlo
	lawsoniana 'Ellwood's Nymph'	last listed 99/00
	lawsoniana Ellwood's Pillar / 'Flolar'	oWnC
	lawsoniana 'Ellwood's Pygmy'	oWnC
va	***lawsoniana*** 'Ellwood's White'	oWnC
	lawsoniana 'Erecta Aurea'	wCCr
	lawsoniana 'Erecta Viridis'	wCCr
	lawsoniana 'Forsteckensis'	last listed 99/00
	lawsoniana 'Glauca'	wCCr
	lawsoniana 'Gold Flake'	oPor
	lawsoniana 'Golden King'	oGar
	lawsoniana 'Golden Showers'	oFor oBlo
	lawsoniana 'Intertexta'	last listed 99/00
	lawsoniana 'Kilmacurragh'	wCCr
	lawsoniana 'Knowefieldensis'	last listed 99/00
	lawsoniana 'Konijn's Silver'	last listed 99/00
	lawsoniana 'Krameri'	oBRG
	lawsoniana 'Lemon Queen'	last listed 99/00
	lawsoniana 'Lutea Nana'	oAls oGar
	lawsoniana 'Lycopodioides'	wCCr
	lawsoniana 'Miki'	oWnC
	lawsoniana 'Mimima Aurea'	oFor oSis oWnC
	lawsoniana 'Nana'	oJoy
	lawsoniana 'Nestoides'	oGar
	lawsoniana 'Nidiformis'	oWnC
	lawsoniana 'Oregon Blue'	oFor oPor oRiv
	lawsoniana 'Pelt's Blue'	oBRG
	lawsoniana 'Pembury Blue'	last listed 99/00
	lawsoniana 'Pygmaea Argentea'	oFor
	lawsoniana 'Rijnhof'	oPor
va	***lawsoniana*** 'Silver Queen'	oFor
va	***lawsoniana*** 'Silver Threads'	oGre
va	***lawsoniana*** 'Silver Tip'	oSis
un	***lawsoniana*** 'Snow Queen'	oGre
	lawsoniana 'Stewartii'	oPor
	lawsoniana 'Sunkist'	oBRG wWel
	lawsoniana 'Van Pelt'	see **C. *lawsoniana*** 'Pelt's Blue'
	nootkatensis	oFor oBlo wCCr oBRG wCoN wRav wCoS
	nootkatensis 'Aurea'	oPor oGar
va	***nootkatensis*** 'Aureovariegata'	oPor
	nootkatensis 'Compacta'	oPor
un	***nootkatensis*** 'Compacta Glauca'	oSis
	nootkatensis 'Glauca'	oWnC
un	***nootkatensis*** 'Glauca Compacta'	oFor
un	***nootkatensis*** 'Glauca Nana'	oPor
un	***nootkatensis*** 'Glauca Pendula'	oBRG
	nootkatensis 'Green Arrow'	oFor wHer oWnC
	nootkatensis 'Pendula'	oFor oPor wClo oAls oAmb oGar oGre oWnC wSta wAva wCoS
un	***nootkatensis*** 'Torulosa'	wSta
va	***nootkatensis*** 'Variegata'	oFor wClo oJoy oGar
	obtusa	wWoB oAls wCCr wCoN
un	***obtusa*** 'Arneson's Dwarf'	oPor
	obtusa 'Aurea'	oAls oGar oSho wSta oBRG wCoS
un	***obtusa*** 'Aurea Nana'	wHer wCoS
	obtusa 'Bess'	oPor
un	***obtusa*** 'Blue Feathers'	oFor oPor wClo oAls oGar oSis oSho oWnC oBRG
	obtusa var ***breviramea***	wHer
un	***obtusa*** 'Bright Gold'	oTPm
	obtusa 'Caespitosa'	last listed 99/00
	obtusa 'Chabo-yadori'	oFor wHer oPor wClo oAls oGar oSis oGre oWnC wWel
	obtusa 'Chilworth'	last listed 99/00
un	***obtusa*** 'Compact Pyramid'	oPor
	obtusa 'Compacta'	wClo oGar oTPm wSta wAva
	obtusa 'Confucius'	oFor wHer oPor oGar oSis oTPm oWnC wAva
	obtusa 'Contorta'	wHer oPor
	obtusa 'Coralliformis'	oFor oPor oAls oAmb oGar oBlo wSta oGue oSle
	obtusa 'Crippsii'	oPor wClo oTPm wSta
	obtusa 'Cyanoviridis'	oAls oGue
	obtusa 'Dainty Doll'	oFor wHer oPor oGre wWel
	obtusa 'Densa'	see **C. *obtusa*** 'Nana Densa'
un	***obtusa*** 'Diane Verkade'	oPor
	obtusa 'Drath'	oPor
	obtusa 'Elf'	last listed 99/00
un	***obtusa*** 'Elmwood'	wClo
	obtusa 'Ericoides'	wHer
	obtusa 'Erika'	oPor wTGN
	obtusa 'Fernspray Gold'	oFor wHer oPor oGar oRiv oTPm oWnC wWel
	obtusa 'Filicoides'	oPor oJoy oGar oRiv wKin oWnC oBRG wWel
	obtusa 'Flabelliformis'	oPor oBRG
un	***obtusa*** 'Gold Fern'	oWnC oBRG
	obtusa 'Golden Drop'	wHer oPor oGar wWel
	obtusa 'Golden Fairy'	oSis
va	***obtusa*** 'Golden Filament'	last listed 99/00
	obtusa 'Golden Nymph'	last listed 99/00
	obtusa 'Golden Sprite'	oPor oSis oGre
	obtusa 'Gracilis'	wClo oJoy oGar oGre wPir oTPm oWnC wAva wWel
	obtusa 'Gracilis Aurea'	wHer oPor
un	***obtusa*** 'Gracilis Compacta'	oAls
	obtusa 'Graciosa'	see **C. *obtusa*** 'Loenik'
un	***obtusa*** 'Hahari'	wHer
	obtusa 'Hage'	oPor oGar oGre
un	***obtusa*** 'JR'	oPor
	obtusa 'Juniperoides'	wHer oPor oGos oSis oBRG

	obtusa 'Kamarachiba'	oPor
	obtusa 'Kanaamihiba'	wHer oSis oGre oBRG
	obtusa 'Kojolcohiba'	last listed 99/00
	obtusa 'Kosteri'	oFor wHer oPor oJoy oAls oGar oSis oBlo oGre oTPm oWnC wSta oBRG wAva wWel
un	***obtusa*** 'Kosteri' fast form	oGar
un	***obtusa*** 'Lemon Twist'	wHer
	obtusa 'Leprechaun'	oPor oGar
	obtusa 'Little Markey'	last listed 99/00
	obtusa 'Loenik'	oPor oGar
	obtusa 'Lougheed'	oGre
	obtusa 'Lutea Nova'	oPor
	obtusa 'Lycopodioides'	oRiv wSta
	obtusa 'Lycopodioides Aurea'	last listed 99/00
un	***obtusa*** 'Lynn's Golden Ceramic'	oPor
va	***obtusa*** 'Mariesii'	oFor wHer oPor wClo oAls oGar oBRG
	obtusa 'Mastin'	oSis
un	***obtusa*** 'Meroke'	oGre oBRG
un	***obtusa*** 'Meroke Twin'	oPor
	obtusa 'Minima'	oTPm
	obtusa 'Nana'	oPor oJoy oAls oGar oSis oRiv oBlo wSta wWhG
	obtusa 'Nana Aurea'	oPor oJoy
qu	***obtusa*** 'Nana Aurea Lutea'	wWhG
qu	***obtusa*** 'Nana Black Beauty'	oJoy
	obtusa 'Nana Densa'	oBRG
	obtusa 'Nana Gracilis'	oFor wHer oPor wClo oJoy oAls oGar oSho oGre oTPm oWnC oBRG wAva wWel
qu	***obtusa*** 'Nana Gracilis Compacta'	oJoy
	obtusa 'Nana Lutea'	oFor wHer oPor wClo oAls oGar oSis oRiv oGre oTPm oWnC wSta wWel
qu	***obtusa*** 'Nana Nana'	wSta
	obtusa 'Nana Snowkist'	see **C. *obtusa*** 'Snowkist'
un	***obtusa*** 'Oregon Crested'	oPor oSis
un	***obtusa*** 'Pagoda Green'	wWel
	obtusa 'Paul's Select'	oPor
	obtusa 'Prostrata'	last listed 99/00
	obtusa 'Pygmaea'	oBRG
	obtusa 'Pygmaea Aurescens'	wHer oPor oJoy oAls oGar oGre
qu	***obtusa*** 'Pygmaea Chimo-hiba'	oPor
	obtusa 'Reis Dwarf'	oFor oPor oAls oGar oSis oGre oBRG
	obtusa 'Repens'	oPor oGre oBRG
	obtusa 'Rigid Dwarf'	oPor
	obtusa 'Saffron Spray'	oPor
un	***obtusa*** 'Shogun's Gold'	wCol
va	***obtusa*** 'Snowflake'	oPor oGre
va	***obtusa*** 'Snowkist'	oFor oPor wClo
	obtusa 'Spiralis'	oFor oPor oAls oWnC
	obtusa 'Split Rock'	oFor oPor oGar oSis oGre wSta oBRG
un	***obtusa*** 'Starkist'	oBRG
	obtusa 'Stoneham'	oPor oSis
	obtusa 'Tempelhof'	oPor oGar oGre oWnC oBRG
	obtusa 'Tetragona Aurea'	oFor wHer oPor wClo wSta oBRG
va	***obtusa*** 'Tonia'	oPor
	obtusa 'Topsy'	oPor
	obtusa 'Torulosa'	see **C. *obtusa*** 'Coralliformis'
un	***obtusa*** 'Torulosa Nana'	oAls oGar
	obtusa 'Verdon'	oPor oJoy oAls oGar oSis oTPm oBRG wWel
un	***obtusa*** 'Wells'	oRiv
un	***obtusa*** 'Willamette Elegance'	oPor wTGN
	obtusa 'Wykoff'	oPor
	pisifera	oRiv wCoN
un	***pisifera*** 'Arnhem'	oPor
	pisifera 'Aurea'	wWhG
un	***pisifera*** 'Aurea Compacta'	oAls
	pisifera 'Baby Blue'	oPor oGre oWnC
	pisifera 'Boulevard'	oFor oPor oJoy oAls oBlo oTPm oWnC oBRG oTal
	pisifera 'Compacta'	wHer
	pisifera 'Compacta Variegata'	wHer oPor
un	***pisifera*** 'Creamball'	oFor oPor wClo oGar oGre oBRG wWel
	pisifera 'Curly Tops'	oPor oGre
	pisifera 'Cyanoviridis'	oGar oGre wSta wAva wWel
	pisifera 'Devon Cream'	oGar
	pisifera 'Filifera'	oGar wSta wCoS wWel
	pisifera 'Filifera Aurea'	oAls oSis oRiv oBlo oGre wAva wWel
un	***pisifera*** 'Filifera Aurea Nana'	oAls oGar oSis oTPm wSta oBRG wWel
va	***pisifera*** 'Filifera Aureovariegata'	oFor oPor oSis oWnC
	pisifera 'Filifera Gold Spangle'	see **C. *pisifera*** 'Gold Spangle'
	pisifera 'Filifera Golden Mop'	see **C. *pisifera*** 'Golden Mop'
un	***pisifera*** 'Filifera Mops'	see **C. *pisifera*** 'Golden Mop'
	pisifera 'Filifera Nana'	oSis
un	***pisifera*** 'Filifera Nana Aurea'	oWnC
	pisifera 'Filifera Sungold'	see **C. *pisifera*** 'Sungold'
	pisifera 'Gold Dust'	see **C. *pisifera*** 'Plumosa Aurea'
un	***pisifera*** 'Gold Pin Cushion'	oWnC
	pisifera 'Gold Spangle'	oPor oGre oTPm
	pisifera 'Golden Mop'	oFor oPor oAls oGar oGre oTPm wWel
un	***pisifera*** 'Golden Pincushion'	oPor oSqu wWel
	pisifera 'Golden Plumosa'	see **C. *pisifera*** 'Plumosa Aurea'
	pisifera 'Hime-sawara'	oSis
qu	***pisifera*** 'Juniperoides Aurea'	oAls oGar oBRG
un	***pisifera*** 'Lemon Thread'	oFor wHer oPor oSis oBRG wWel
un	***pisifera*** 'Lime Tart'	oPor
un	***pisifera*** 'Lutescens'	oPor oJoy
un	***pisifera*** 'Mikko'	oBRG oPor
	pisifera 'Minima'	wClo oAls
	pisifera 'Minima Aurea'	wTGN
qu	***pisifera*** 'Minima Compacta Aurea'	oJoy
	pisifera 'Nana'	oAls oGar
	pisifera 'Nana Aureovariegata'	wHer
	pisifera 'Plumosa'	oBlo
va	***pisifera*** 'Plumosa Albopicta'	oPor
	pisifera 'Plumosa Aurea'	oPor oAls oGar oRiv
	pisifera 'Plumosa Aurea Compacta'	oGar
	pisifera 'Plumosa Aurea Nana'	wHer
	pisifera 'Plumosa Compressa'	oFor wClo oSis oGre oTPm oBRG
un	***pisifera*** 'Plumosa Compressa Aurea'	oPor
	pisifera 'Plumosa Flavescens'	last listed 99/00
qu	***pisifera*** 'Plumosa Juniperoides'	oPor wClo
qu	***pisifera*** 'Plumosa Nana Compacta'	oWnC
un	***pisifera*** 'Plumosa Nana Variegata'	oAls oGar
	pisifera 'Pygmaea'	oPor wSta
va	***pisifera*** 'Silver Lode'	oPor oAls oGar
va	***pisifera*** 'Snow'	oFor wClo oGre oWnC oBRG
un	***pisifera*** 'Snow Reversion'	oJoy
	pisifera 'Spaan's Cannonball'	oPor
	pisifera 'Squarrosa'	oRiv wCCr
	pisifera 'Squarrosa Aurea'	oGre
	pisifera 'Squarrosa Dumosa'	wAva
	pisifera 'Squarrosa Intermedia'	oPor oTPm wAva
un	***pisifera*** 'Squarrosa Intermedia Nana'	oSis
	pisifera 'Squarrosa Lombarts'	oPor
	pisifera 'Squarrosa Sulphurea'	oFor oBRG
	pisifera 'Sungold'	oPor oAls oGar oGre wWel
	pisifera 'Tama-himuro'	oPor oWnC
un	***pisifera*** 'Tatsunami Hiba'	oPor
	pisifera 'True Blue'	oPor oGre
	pisifera 'Tsukomo'	oPor oGar oSis oTPm wSta
	pisifera 'White Pygmy'	oPor
	thyoides	oFor oAls oBlo wCCr wCoN
	thyoides 'Andelyensis'	oPor oRiv
un	***thyoides*** 'Andelyensis Conica'	oGar oSis oGre
	thyoides 'Aurea'	oFor oGre
un	***thyoides*** 'Blue Sport'	oPor oGre
	thyoides 'Conica'	oPor
	thyoides 'Ericoides'	oPor oAls oGar oSis oSho oGre
	thyoides 'Glauca'	oFor
un	***thyoides*** 'Glauca Pendula'	oFor oGre
	thyoides 'Heatherbun'	oPor wClo oJoy oGar oSqu oGre oTPm oBRG oGue
	thyoides 'Little Jamie'	oPor oSis oGre
	thyoides 'Meth Dwarf'	wHer oPor wClo oGre oBRG
	thyoides Red Star	see **C. *thyoides*** 'Rubicon'
	thyoides 'Rubicon'	oPor oSis oWnC oBRG
	thyoides 'Top Point'	oPor oGre oWnC oBRG
va	***thyoides*** 'Variegata'	oFor
	CHAMAECYTISUS	
	glaber	oFor
	purpureus	oGre
	purpureus 'Atropurpureus'	oFor oGre
	ratisbonensis	oFor
	CHAMAEDAPHNE	
	calyculata	oFor
	calyculata 'Nana'	oGos oSis
	CHAMAEDOREA	
	cataractarum	oGar
	elegans 'Bella'	oGar
	CHAMAELIRIUM	
	luteum	wCol
	CHAMAEMELUM	
	nobile	oAls wFGN oGoo wFai iGSc wHom oWnC oCrm oBar oSec wNTP oCir oUps
do	***nobile*** 'Flore Pleno'	wCul oJoy oHed wFGN oGoo oWnC oSle
	nobile 'Treneague'	wFGN
	CHAMAEROPS	
	humilis	oSho wDav

	CHASMANTHE	
	aethiopica	oTrP
	bicolor	oTrP
	floribunda	oTrP
	CHASMANTHIUM	
	latifolium	oFor wWoS wCli oNat oDar oJoy wHig oAls oRus oAmb oGar oGoo oDan oOut oGre wWin wPir wHom wTGN wBox oBRG wWhG wSnq
	CHASMATOPHYLLUM	
	masculinum	oFor oSqu
	CHEILANTHES	
	argentea	oRus oGar oGre oBRG
	lanosa	wFan oRus oSis
	siliquosa	oGar
	tomentosa	wFol oGos oRus oGre oBRG
	CHEIRANTHUS	see **ERYSIMUM**
	CHELIDONIUM	
	majus	oFor oGoo wFai iGSc oCrm
	majus 'Laciniatum Flore Pleno' seed grown	wHer
	CHELONE	
	glabra	oFor oRus oGoo wNay
	lyonii	oFor oRus oSec
un	***lyonii*** 'Hot Lips'	oFor wHer wWoS oNat oUps
	obliqua	oFor oJoy oAls wCSG oGar oSho wNay oGre oWnC wMag oUps
	obliqua var. ***alba***	see **C. *glabra***
	CHELONOPSIS	
un	***yagiharana***	wHer
	CHENOPODIUM	
	ambrosioides	oAls wFai iGSc wHom oBar
	bonus-henricus	oGoo wFai iGSc
	CHIASTOPHYLLUM	
un	***hirsutum*** 'Roseum'	oOut wFai
	oppositifolium	wShR wRob
va	***oppositifolium*** 'Jim's Pride'	wHer wWoS wMtT wRob
	CHILOPSIS	
	linearis	oFor
un	***linearis*** 'Burgundy'	oFor oGar
	CHIMAPHILA	
	maculata	wThG
	menziesii	wFFl
	umbellata	wThG oBos oTri wFFl
	CHIMONANTHUS	
	nitens	oFor
	praecox	oFor wCol oAls oDan oRiv oCis wCoN wSnq wWel
	praecox 'Concolor'	oFor wHer oGos oGar oCis oGre oWnC wSte
	praecox var. ***luteus***	oGos wWel
	zhejiangensis	oFor
	CHIMONOBAMBUSA	
	marmorea	oTra wCli wBea wBox wBam
	marmorea 'Variegata'	oTra wCli wBea oOEx wBox
	quadrangularis	oTra wCli wBea
	quadrangularis 'Svow'	oTra wCli wBea
	tumidissinoda	oTra wCli wBea
	CHIONANTHUS	
	retusus	oFor oEdg oAls oAmb oGar oDan oWhi oGre oWnC wCoN
	retusus var. ***serrulatus***	oFor
	virginicus	oFor wHer wCul oGos wSwC oAls oGar oDan oRiv oWhi oBlo oGre wCCr wCoN oUps wWel
	CHIONOCHLOA	
	conspicua	wHer
	CHIONODOXA	
	forbesii 'Pink Giant'	oWoo wAva
	luciliae (see Nomenclature Notes)	wRoo oWoo
	luciliae 'Alba'	oWoo
	X CHITALPA	
	tashkentensis	oBlo oSec
	tashkentensis 'Morning Cloud'	oJoy oSho oWnC
	tashkentensis 'Pink Dawn'	oFor oSho oGre
	CHLIDANTHUS	
	fragrans	oGar
	CHLORANTHUS	
	serratus	last listed 99/00
	CHLOROGALUM	
	pomeridianum	oFor wCol
	CHLOROPHYTUM	
	comosum	oGar
	comosum 'Variegatum'	oGar
va	***comosum*** 'Vittatum'	oGar
un	**CHOERIOSPONDIA**	
	sapondin	oOEx
	CHOISYA	
	arizonica	oCis
	'Aztec Pearl'	wHer oNat oHed oAmb oCis oSho wWel
	dumosa	oHed
	ternata	oFor oAls oGar oRiv oSho oBlo oGre oWnC oCir wCoN wCoS wWel
	ternata Sundance / 'Lich'	oGos oAmb oGar oCis oNWe oGre oWnC wWel
	CHONDROPETALUM	
	tectorum	last listed 99/00
	CHORISIA	
	speciosa	oRar
	CHRYSALIDOCARPUS	
	lutescens	see **DYPSIS *lutescens***
	CHRYSANTHEMUM	
	'Anastasia'	last listed 99/00
	'Apricot'	last listed 99/00
	balsamita	see **TANACETUM *balsamita***
	cinerariifolium	see **TANACETUM *cinerariifolium***
	'Clara Curtis'	oFor wCul oAls oWnC wTGN
	'Emperor of China'	last listed 99/00
	grandiflorum	oGar
	'Hillside Sheffield'	oFor wSwC oGar oWnC oSle
	'Innocence'	last listed 99/00
	leucanthemum	see **LEUCANTHEMUM *vulgare***
	'Mary Stoker'	oFor oAls oWnC
	maximum	see **LEUCANTHEMUM *x superbum***
	'Mei-kyo'	wHer
	nipponicum	see **NIPPONANTHEMUM *nipponicum***
	pacificum	see **AJANIA *pacifica***
	parthenium	see **TANACETUM *parthenium***
	'Pink Ice'	oFor oCis
	'Sheffield Pink'	see **C.** 'Hillside Sheffield'
	x superbum	see **LEUCANTHEMUM *x superbum***
	weyrichii	oSis
un	***weyrichii*** 'White Bomb'	oSis
	yezoense	oFor
	CHRYSOGONUM	
	virginianum	oFor wCol oHed oRus
	CHRYSOLEPIS	
	chrysophylla	oBos wWat
	CHRYSOPSIS	
	mariana	oFor
	CHRYSOSPLENIUM	
	davidianum	wHer oNWe
	CHRYSOTHAMNUS	
	nauseosus	oFor wPla
	viscidiflorus	oFor wPla
	CHUSQUEA	
un	***circinata***	oTra
	coronalis	oTra oTBG
	culeou	oTra
	culeou 'Breviglumis'	see **C. *culeou*** 'Tenuis'
	culeou seedlings	oTra
	culeou 'Tenuis'	wBea wBox
un	***cumingii***	oTra
un	***foliosa***	oTra
un	***galleottiana***	oTra
	liebmannii	oTra
un	***mimosa***	oTra
	nigricans	oTra wBea wBox
un	***pittieri***	oTra wBea
un	***subtilis***	oTra
un	***tomentosa***	oTra
	valdiviensis	oTra
	CICHORIUM	
	intybus	oCrm
	CIMICIFUGA	
ch	***acerina***	see **C. *japonica***
	americana	wCul oRus
un	'Black Hills Beauty'	wMag
un	'Black Negligee'	wWoS oNWe
	dahurica	wNay
	dahurica HC 970065	wHer
	elata	oRus
	japonica	oFor wRob
	laciniata	wHer
un	***mairei*** DJHC 113	wHer
	racemosa	oFor oHan wThG oHed oRus wNay wRob wFai iGSc wMag oUps
	racemosa atropurpurea	see **C. *simplex*** var. ***simplex*** Atropurpurea Group
	racemosa var. ***cordifolia***	see **C. *rubifolia***

	ramosa	see C. *simplex* var. *simplex* Prichard's Giant'
	rubifolia	wHer oRus wNay wRob
	simplex	wCSG
	simplex var. *matsumurae*	wCul
	simplex var. *matsumurae* 'Elstead'	wHer
	simplex var. *matsumurae* 'Frau Herms'	wHer
	simplex var. *matsumurae* 'White Pearl'	oAls oRus oGar wNay wRob oGre wFai wTGN wBox
	simplex var. *simplex* Atropurpurea Group	oFor wCul wGAc wCSG oGar oDan wNay wRob wFai wTGN oEga
	simplex var. *simplex* 'Brunette'	oFor wHer wWoS oGos oEdg oHed oRus oAmb oGar oNWe wNay wRob wMag oBRG oHou wCoN oUps
	simplex var. *simplex* 'Brunette' seed grown	oNWe
un	*simplex* var. *simplex* 'Hillside Black Beauty'	oAls wRob oGre
	simplex var. *simplex* 'Prichard's Giant'	oFor wNay oCrm
	simplex var. *simplex* 'Prichard's Giant' seed grown	wHer
	CINNAMOMUM	
	burmanii	oOEx
	camphora	oCis oOEx wCCr iGSc
un	*chekianjensis*	wHer oCis
	japonicum	oFor oGre oWnC
un	*prorectum*	oCis
	CIRCAEA	
	alpina	wFFl
	CIRSIUM	
	japonicum 'Rose Beauty'	oSec
	occidentale	oSec
	CISSUS	
	antarctica	oGar
	discolor	oGar
	incisa	see C. *trifoliata*
	rhombifolia	oGar
	rhombifolia 'Ellen Danica'	oGar
	sicyoides	oRar
	sp.	oOEx
	striata	oFor
	tetrastigma	see TETRASTIGMA *voinierianum*
	trifoliata BLM 0258	oRar
	tuberosa BLM 0544	oRar
	CISTUS	
	x aguilarii	oJoy
	'Anne Palmer'	oJoy
	'Barnsley Pink'	see C. 'Grayswood Pink'
ch	'Blanche'	see C. *ladanifer* 'Blanche'
	'Chelsea Bonnet'	wHer
	clusii	last listed 99/00
	x corbariensis	see C. *x hybridus*
	creticus	wFGN wPir wCCr oUps
	creticus ssp. *creticus*	wSwC wSnq
	creticus ssp. *incanus*	oFor wCSG oGar
	x crispatus 'Warley Rose'	wHer oDar oHed wCCr
	crispus (see Nomenclature notes)	wHer
un	*crispus* 'Juliet'	wCSG
	x dansereaui	wHer
	x dansereaui 'Decumbens'	oFor oSis
	'Doris Hibberson'	wCul
	'Elma'	wHer oJoy oGre wPir oWnC wTGN wSte
	'Grayswood Pink'	wHer wSwC
	heterophyllus	wCCr
ch	*hirsutus*	see C. *inflatus*
	x hybridus	oFor wCul oAls wShR wCSG oGar oDan oRiv oGre wPir oWnC wTGN wSta wAva wCoS wWel
	incanus	see C. *creticus* ssp. *incanus*
	incanus ssp. *creticus*	see C. *creticus* ssp. *creticus*
ch	*inflatus*	wCSG
	ladanifer	oDar oJoy wFGN oGar oGoo oDan oSis oGre wPir wCCr oCrm oUps wWel
ch	*ladanifer* 'Blanche'	oHed
qu	*ladanifer* 'Maculatus'	oGar
	ladanifer Palhinhae Group	see C. *ladanifer* var. *sulcatus*
	ladanifer var. *sulcatus*	oFor
	laurifolius	wCSG wPir
	x lusitanicus	see C. *x dansereaui*
	maculatus (see Nomenclature notes)	oSho wCoS
	monspeliensis	wCCr
	palhinhae	see C. *ladanifer* var. *sulcatus*
	'Peggy Sammons'	wWoS oHed
	populifolius	oJoy wCCr
	populifolius ssp. *major*	wCCr
qu	*prostratus*	oUps
	x pulverulentus 'Sunset'	oFor wHer wWoS oJoy oHed wFGN oGar oSis oCis wPir wCCr wSta
	x purpureus	oFor wHer wCul oAls wFGN wCSG oGar oRiv oSho wTGN wCoS wWel
	x purpureus 'Alan Fradd'	wHer
	x purpureus 'Betty Taudevin'	see C. *x purpureus*
	x purpureus 'Brilliancy'	oJoy oAls oSho wWel
	salviifolius	oJoy oSis wPir
	salviifolius 'Prostratus'	oGar oSho oWnC
	'Silver Pink'	wWoS oJoy oHed oGar wCCr
	sintenisii	wHer
	x skanbergii	oFor wCul oDar oJoy oHed wFGN oGar oBlo wCCr oWnC oCir oUps oSle wWel
un	'Snow White'	wHer wSnq
	'Victor Reiter'	oSis
	X CITROFORTUNELLA	
un	Citrangequat 'Mr. John's Longevity'	oOEx
un	Citrangequat Sinton	oOEx
un	Citrangequat 'Thomasville'	oOEx
	floridana 'Eustis'	oOEx
	microcarpa	wRai
un	*microcarpa* 'Keraji Mandarin'	oOEx
	microcarpa 'Tiger'	wBox
	X CITRONCIRUS	
un	Citrangadin, Glen	oOEx
un	Citremon	oFor
un	Citrumelo	oOEx
	webberi	oOEx
un	*webberi* Morton	oOEx
un	*webberi* 'Snow Sweet'	oOEx
	CITRUS	
	aurantiifolia	wCoS
	aurantiifolia 'Bearss'	see C. *latifolia* 'Bearss'
un	*aurantiifolia* Kaffir Lime	wBox
	hystrix	oOEx
un	Ichandarin, Shangjuan	oOEx
un	Ichandarin, Sudachi	oOEx
un	Ichandarin, Yuzu	oOEx
	ichangensis	oOEx
un	sp. aff. *kulu*	oOEx
	latifolia 'Bearss'	oGar wRai oWnC wCoS
un	*limon* 'Dwarf Lisbon'	oWnC
	limon 'Eureka'	oWnC
	limon 'Lisbon'	oGar wCoS
	medica var. *digitata*	oGar oOEx
ch	*medica* var. *sarcodactylis*	see C. *medica* var. *digitata*
un	*x meyeri* 'Improved Meyer'	oAls oGar wRai oWnC wCoS wWel
	x meyeri 'Meyer'	oOEx
un	Nansho Daidai	oOEx
un	*oboboidea*	oOEx
un	Papeda, Khasi	oOEx
un	*x paradisi* 'Dwarf Star Ruby'	oWnC
	x paradisi 'Oroblanco'	oGar oOEx
	x paradisi 'Star Ruby'	oGar oOEx
	reticulata 'Dancy'	oGar oOEx wCoS
	reticulata 'Kinnow'	oOEx
un	*reticulata* Mandarin Group 'Changsha'	oOEx
	reticulata Mandarin Group 'Clementine'	wRai
un	*reticulata* Mandarin Group 'Snow Picked Mini Mandarin'	oOEx
ch	*reticulata* Satsuma Group	see C. *unshiu*
un	*sinensis* 'Campbell'	oWnC wCoS
un	*sinensis* 'Dwarf Washington'	oWnC
	sinensis 'Midknight'	oGar
	sinensis 'Moro Blood'	oGar wRai oOEx
	sinensis 'Robertson'	oOEx
	sinensis 'Sanguinelli'	oOEx
	sinensis 'Trovita'	oOEx
	sinensis 'Washington'	oGar wRai wCoS
	sp.	oCis
	unshiu 'Owari'	oOEx
	CLADRASTIS	
	kentukea	oFor oEdg oGar oDan oSho oGre oWnC wCoN wWel
	kentukea 'Rosea'	oFor oWhi oGre
	lutea	see C. *kentukea*
un	**CLAOPODIUM**	
	crispifolium	wFFl
	CLARKIA	
	amoena	oBos
	CLAYTONIA	
	lanceolata	oBos
	megarhiza var. *nivalis*	wMtT
	parvifolia	see NAIOCRENE *parvifolia*
	perfoliata	oBos

	sibirica	oAls oBos wFFl
	virginica	wThG
	CLEMATIS	
	'Abundance'	oJoy oExu wTGN
	Alabast™/ 'Poulala'	oJoy oExu
	'Alba Luxurians'	oJoy oExu oGar
do	'Albina Plena'	oJoy
	'Aljonushka'	wHer oJoy oExu
	'Allanah'	oJoy oExu oWnC
	alpina	oFor wHer wCul wPir oWnC
	alpina blue	oGar
	alpina 'Columbine'	oExu
	alpina 'Constance'	oJoy oExu oHed oGar
	alpina 'Frances Rivis'	wHer oJoy
	alpina 'Frankie'	wCul
	alpina 'Jacqueline du Pre'	wHer oExu oHed
	alpina 'Odorata'	oExu
	alpina 'Pamela Jackman'	oFor wHer oJoy oExu wTGN
	alpina 'Pink Flamingo'	oExu
	alpina 'Ruby'	wHer oJoy oGre wTGN
	alpina ssp. ***sibirica***	last listed 99/00
	alpina 'Willy'	wHer wCul oJoy oGar oDan oGre wTGN
	'Anita'	oExu
	Anna Louise™/ 'Evithree'	oJoy oGre
	apiifolia HC 970168	wHer
	'Arabella'	oJoy oExu
	Arctic Queen ™/ 'Evitwo'	oJoy oGar oGre wWel
	aristata	oFor
	armandii	oFor wCul oGar oSho oBlo oGre wFai wPir oWnC wTGN wSta wCoN wWhG wSte wCoS wWel
	armandii 'Apple Blossom'	last listed 99/00
	armandii 'Snowdrift'	oJoy oAls oGar
	x aromatica	oJoy
	'Asao'	oJoy oExu oGar oGre
	'Ascotiensis'	oJoy oGar oGre
	'Barbara Dibley'	last listed 99/00
	'Barbara Jackman'	oJoy oExu
	'Bees' Jubilee'	oJoy oGar oGre oHou wWhG
	'Bella'	oJoy
	'Belle of Woking'	oJoy oExu oGar
	'Betty Corning'	wWoS oJoy oExu
	'Bill MacKenzie'	oHed
	'Blekitny Aniol'	oJoy
	'Blue Belle'	wWoS wCul
	'Blue Bird'	wCul oGre wTGN
	'Blue Boy' (see Nomenclature Notes)	
	'Blue Dancer'	oExu oHed
un	'Blue Jay'	oSho
un	Blue Light™	wWhG
	Blue Moon™ / 'Evirin'	wWoS
	brachyura HC 970101	wHer
	brevicaudata	oFor
	'Brunette'	oExu
	sp. aff. ***buchananiana*** HWJCM 507	wHer
	'Burford Variety'	last listed 99/00
	campaniflora	oFor wHer wCri
qu	'Candida'	oJoy oWnC wWel
	'Capitaine Thuilleaux'	see **C.** 'Souvenir du Capitaine Thuillleaux'
	'Cardinal Wyszynski'	see **C.** 'Kardynal Wyszynski'
	'Carnaby'	oJoy oGar wTGN wWhG oUps
	x cartmanii 'Joe'	oSis
	chiisanensis	wHer wCol
	chiisanensis HC 970372	wHer
	chiisanensis 'Lemon Bells'	oExu oNWe
	chrysocoma (see Nomenclature Notes)	oFor wCSG oGar
un	***chrysocoma*** var. ***glabrescens***	oExu
	chrysocoma var. ***sericea***	see ***C. montana*** var. ***sericea***
	cirrhosa	last listed 99/00
	cirrhosa var. ***balearica***	oExu
	cirrhosa 'Freckles'	oJoy oExu
	columbiana	wHer oSis
	'Comtesse de Bouchaud'	oJoy oExu oGar oSho oGre oWnC oHou wWhG
	connata	last listed 99/00
	connata DJHC 98265	wHer
	'Countess of Lovelace'	oJoy wWhG
	crispa	wHer wCol
	'Daniel Deronda'	oJoy oExu oGar oGre oHou
	'Dawn'	oGar
	'Doctor Ruppel'	oFor oJoy oGar oGre oWnC wTGN oHou wWhG
	'Duchess of Albany'	wCul oJoy oExu oNWe oGre
	'Duchess of Edinburgh'	oFor wHer wCul oJoy oExu oGar oBlo oGre wTGN wWhG oUps
	x durandii	wHer wWoS wCul oJoy oExu
	'Edith'	oJoy oGre
	'Edomurasaki'	wHer oJoy
	'Elsa Spaeth'	oJoy oExu wCSG oGar oGre oHou wWel
	'Emilia Plater'	oExu
	x eriostemon	oJoy
	x eriostemon 'Blue Boy'	wHer oExu
	x eriostemon 'Hendersonii'	oNWe
	'Ernest Markham'	oJoy oGar oGre oWnC oHou wWhG
	'Etoile de Malicorne'	last listed 99/00
	'Etoile Rose'	oJoy oExu
	'Etoile Violette'	oFor wHer wWoS wCul oJoy oExu oHed oGar oSho oBlo oGre wTGN wWhG wWel
	'Fair Rosamond'	oJoy
	fargesii	see ***C. potaninii***
	x fargesioides	see **C.** 'Paul Farges'
	fasciculiflora	wHer oCis
	'Fireworks'	oJoy oGar oGre oHou
	flammula	oFor oExu oGar oDan
	'Floralia'	oGar
qu	***florida*** 'Alba Plena'	oWnC
do	***florida*** 'Flore Pleno'	oJoy
	florida 'Sieboldii'	wHer
	fremontii	wHer wCol
un	***fruticosa*** 'Mongolian Gold'	wCol
	'Fuji-musume'	last listed 99/00
	'General Sikorski'	oJoy oExu oGar oGre wWhG
	gentianoides	wHer
	'Gillian Blades'	oFor oJoy oExu oGar wWhG
	'Gipsy Queen'	oJoy oExu oGar oBlo oGre oHou wWhG
	glauca	see ***C. intricata***
	glycinoides	oFor
	'Golden Harvest'	last listed 99/00
	Golden Tiara®/ 'Kugotia'	oGar oSho
	grata (see Nomenclature Notes)	oFor
	'Gravetye Beauty'	wHer oJoy oExu oNWe
	'Guernsey Cream'	oJoy oExu wCSG oGre
	'Guiding Star'	oGar
	'Gypsy Queen'	see **C.** 'Gipsy Queen'
	'H. F. Young'	oJoy oExu oGar oGre
	'Hagley Hybrid'	wCul oJoy oExu oGar oGre oWnC wWhG
	'Haku-okan'	wHer oJoy oExu oGar oGre wWhG
	'Helios'	oJoy
	'Helsingborg'	wHer oJoy oExu oHed oGar
	'Henryi'	oJoy wShR oGar oGre wTGN oHou wWhG
	heracleifolia	oGar wSte
	heracleifolia var. ***davidiana***	wCul oExu oGre wTGN
	heracleifolia var. ***davidiana*** HC 970373	wHer
	hirsutissima	wHer
	hookeriana	oFor
	'Huldine'	oExu oGre
	'Hybrida Sieboldii'	oJoy oGar oGre oWnC oHou oWhS wWhG wWhG
	indivisa	wWhG
	integrifolia	wCul wCol oExu oGar oSis wFai oWnC oSec
	integrifolia 'Alba'	oExu oNWe
	integrifolia var. ***albiflora***	wHer oSis
un	***integrifolia*** 'Caerulea'	wHer oExu oMis oGar
	integrifolia 'Durandii'	see ***C. x durandii***
un	***integrifolia*** 'Hakurei'	oJoy oMis
un	***integrifolia*** 'Hanajima'	oJoy
	integrifolia 'Olgae'	last listed 99/00
	integrifolia 'Rosea'	oJoy oExu
	intricata	last listed 99/00
	'Jackmanii'	oFor wCul oJoy oExu wCSG oMis oGar oSho oBlo oGre oWnC wTGN oHou oWhS wWhG wCoS
	'Jackmanii Alba'	wCul
	'Jackmanii Superba'	oExu oGar oUps
un	sp. aff. ***japonica*** complex DJHC 594	wHer
	'John Warren'	oJoy oGar oGre oHou wWhG
	Josephine™ / 'Evijohill'	wWhG
	x jouiniana	oFor
	x jouiniana 'Mrs. Robert Brydon'	wHer wCol oExu
	x jouiniana 'Praecox'	wHer wClo oExu oGar
	'Kakio'	oJoy oExu oGar oGre wTGN oHou
	'Kardynal Wyszynski'	oExu oBlo wTGN
	'Kasugayama'	last listed 99/00
	'Kathleen Dunford'	oJoy oExu oGar oGre
	'Kathleen Wheeler'	last listed 99/00
	'Ken Donson'	oWnC
	'Kermesina'	oFor wCul oJoy oExu oGar oNWe oGre
	'Kiri Te Kanawa'	oJoy

	koreana DJH 340	wHer
	'Kugotia'	see C. Golden Tiara®/ 'Kugotia'
	'Lady Betty Balfour'	wHer oJoy oExu oGar
	'Lanuginosa Candida'	oFor wCul oGar oGre oWhS
	'Lasurstern'	oJoy oExu oGar oBlo
	'Lawsoniana'	oExu oGar
	Liberation™ / 'Evifive'	oFor oGar oGre
qu	ligufolia	wWat
	ligusticifolia	oFor oHan wNot wShR wPla
	'Lincoln Star'	oJoy oWnC
	'Little Joe'	oCis
	'Little Nell'	wWoS wCul wRob
	'Lord Nevill'	oJoy
	'Louise Rowe'	wHer wWel
	'Lunar Lass'	oCis
	'Lunar Lass' seed grown	oNWe
	'Luther Burbank'	oJoy
	macropetala	oJoy oExu oGar oBlo
	macropetala blue	oWnC
	macropetala 'Blue Bird'	wHer
	macropetala 'Jan Lindmark'	wHer oJoy oGre oWhS
	macropetala 'Lagoon'	oGre
	macropetala hort. 'Maidwell Hall'	wHer wCul oGre wTGN
	macropetala 'Markham's Pink'	wCul wTGN
	macropetala 'White Lady'	wCul
	'Madame Julia Correvon'	wWoS oJoy oExu oGar oSho oWnC wWel
	'Madame le Coultre'	see C. 'Marie Boisselot'
	'Margaret Hunt'	wWhG
	'Margot Koster'	wWoS oJoy oExu
	'Marie Boisselot'	oJoy oExu oGar oGre wWhG
	marmoraria	last listed 99/00
	marmoraria hybrid	oNWe
	'Masquerade' (see Nomenclature Notes)	oJoy oExu oGar oGre
	'Minuet'	oJoy oExu oNWe oBlo
	'Miss Bateman'	oJoy oBlo oHou
	montana	wHer oExu
	montana alba	see C. montana
	montana 'Fragrant Spring'	oExu wWhG
	montana f. grandiflora	oFor wHer oWnC wTGN wWhG
un	montana 'Jenny'	wHer
	montana 'Peveril'	oExu
do	montana 'Pleniflora'	oExu
	montana var. rubens	wCul wCol oExu wCSG oGar oBlo oGre oWnC wTGN wCoS wWel
do	montana var. rubens 'Broughton Star'	wHer oJoy
	montana var. rubens 'Continuity'	oExu
	montana var. rubens 'Elizabeth'	oExu oGar
	montana var. rubens 'Freda'	wHer oJoy oExu
do	montana var. rubens 'Marjorie'	wHer wCul oJoy oExu oGar oGre
	montana var. rubens 'Mayleen'	oJoy oExu
	montana var. rubens 'Odorata'	wHer oGar
	montana var. rubens 'Pink Perfection'	oJoy oExu oGar oGre wTGN wWhG
	montana var. rubens 'Superba'	oJoy wWhG
	montana var. rubens 'Tetrarose'	oFor wHer wCul oJoy oGre wPir wTGN
	montana var. sericea	wHer wCul oExu oGar oGre
	montana var. wilsonii	wHer oJoy oExu oGar oGre wWel
	'Mrs. Cholmondeley'	oJoy oGre oWnC
	'Mrs. George Jackman'	oJoy
	'Mrs. N. Thompson'	oJoy oExu oGar oGre oHou wWhG
	'Mrs. P. B. Truax'	oJoy oExu oGar
un	'Mrs. P. T. James'	wHer
	'Mrs. T. Lundell'	oJoy
	'Multi Blue'	oJoy oGar oGre wTGN oHou
	'Nadezhda	oJoy
	'Nelly Moser'	wCul oJoy oExu wCSG oGar oGre oWnC wTGN oHou wWhG oUps wCoS wWel
	'Nicolai Rubtsov'	oGre oWnC
	'Niobe'	oJoy oExu oGar oGre oWnC wTGN oHou oUps wWel
	occidentalis var. dissecta	wHer
	orientalis (see Nomenclature Notes)	oFor oGar
	orientalis 'Burford'	see C. 'Burford Variety'
	orientalis var. orientalis	last listed 99/00
	'Pagoda'	wHer
	'Pamjat Serdtsa'	wHer oExu
	paniculata (see Nomenclature Notes)	wCul oBlo wTGN
	'Paul Farges'	wHer
	'Perle d'Azur'	wHer oJoy oExu oGar oNWe
	'Perrin's Pride'	wTGN
	Petit Faucon™ / 'Evisix'	oJoy oExu oGar
	'Peveril Pearl'	oJoy
	'Pink Champagne'	see C. 'Kakio'
	'Pink Fantasy'	oBlo wTGN
	pitcheri	wHer
	'Polish Spirit'	oFor wHer wWoS oJoy oExu oGar oNWe oGre oWnC wWhG
	potaninii	oGar oDan
	potaninii var. potaninii	wHer wRob
	potaninii var. souliei	see C. potaninii var. potaninii
	'Prince Charles'	oJoy oExu
	'Princess Diana'	oJoy
	'Proteus'	oJoy oGar wTGN
	'Pruinina'	oJoy
un	pseudopoganandra	wHig
	purpurea plena elegans	see C. viticella 'Purpurea Plena Elegans'
	'Ramona'	see C. 'Hybrida Sieboldii'
	ranunculoides	wCol
	sp. aff. ranunculoides DJHC 494	wHer
	recta	oFor oNWe
un	recta 'Lime Close'	wHer
	recta 'Purpurea'	wCul oHed oGar oNWe oOut wWel
	rehderiana	wHer oExu oSis wSte
	'Rhapsody'	wWel
	'Richard Pennell'	oJoy oExu oGre
	'Roguchi'	oJoy wCol oNWe wRob
	'Roko-Kolla'	oJoy
	'Romantika'	oJoy oExu
	'Rosy O'Grady'	wHer oGre
	'Rouge Cardinal'	oFor wWoS oJoy oExu oGar oGre oHou oWhS wWhG
	'Royal Velours'	oFor wHer oJoy oExu
	Royal Velvet™/ 'Evifour'	oJoy oExu oGre
	'Royalty'	oJoy wCSG oGar oWnC
	'Ruby Glow'	last listed 99/00
un	'Sano no Murasaki'	oJoy
	'Sealand Gem'	last listed 99/00
	'Serenata'	wHer oExu
	serratifolia	oFor
	'Sho-un'	wHer
	'Silver Moon'	wHer
un	'Sinee Plamia'	oJoy
	'Sir Trevor Lawrence'	oJoy
	'Snow Queen'	oGar
	'Sodertalje'	oExu
	souliei var. fargesii	see C. potaninii var. potaninii
	'Souvenir du Capitaine Thuilleaux'	oGre
	sp. B&SWJ 3494	wHer
	sp. DJHC 299	wHer
	sp. DJHC 312	wHer
	sp. DJHC 368	wHer
	sp. DJHC 98022	wHer
	sp. DJHC 98128	wHer
	sp. EDHCH 97092	wHer
	sp. Honshu, Itsukushigaharu Hgts.	oNWe
	sp. Kyushu, Mt. Aso, Mt. Kujo	oNWe
	sp. N Sichuan, Jiou Zhaigou, Riz Hai Gully	oNWe
	sp. N Sichuan, SE of Songpan	oNWe
	spooneri	see C. montana var. sericea
	stans	wHer oMis wRob
	'Star of India'	oJoy oExu oGar oBlo oWnC oHou
	'Sunset'	oJoy oGar oGre oWnC
	'Sylvia Denny'	oJoy
	tangutica	oFor wCul wShR wRob oGre oWnC wTGN wBWP wSte
	tangutica 'Gravetye Variety'	wHer oExu oGar wWhG
	tangutica 'Lambton Park'	oJoy
un	tangutica 'My Angel'	oExu
	tangutica 'Radar Love'	oMis
un	tenuifolia	oHed
	terniflora	oFor wHer oJoy oExu oGar oGre wWel
	terniflora HCM 98053	wHer
	'The President'	wCul oJoy oExu oGar oGre oWnC wMag wTGN wWhG wWel
un	thibetica var. akebioides DJHC 98380	wHer
	tibetana	oFor
	tibetana ssp. vernayi 'Orange Peel'	oSis
	sp. aff. tongluensis HWJCM 76	wHer
	x triternata 'Rubromarginata'	oJoy oExu
	'Twilight'	oJoy
	x vedrariensis 'Rosea'	wHer
	'Venosa Violacea'	wHer oJoy oExu oGar oGre oWnC wWhG
	'Veronica's Choice'	oJoy oWnC
	'Victoria'	last listed 99/00
	'Ville de Lyon'	wHer wCul oJoy oExu wCSG oGar oGre oHou oWhS wWhG
	Vino™/ 'Poulvo'	oFor
	'Viola'	oJoy
	viorna	wHer oNWe
	viorna hybrids	wCol

	virginiana (see Nomenclature Notes)	oGar
	vitalba	oGar oSho oSle
	viticella	oFor oExu oGar wCCr wWhG
do	***viticella*** 'Purpurea Plena Elegans'	oFor wHer wCul oJoy oExu oGar wMag wTGN
	'Viticella Rubra'	last listed 99/00
	'Voluceau'	oJoy oExu
	'Vyvyan Pennell'	wCul oJoy
	'Wada's Primrose'	oGre
	'Warsaw Nike'	see **C.** 'Warszawska Nike'
	'Warszawska Nike'	oJoy oExu oBlo wWel
	'White Swan'	wHer
un	'Will Barron'	oExu
	'Will Goodwin'	oJoy oExu oGar oGre wTGN wWhG
	'William Kennett'	oGar
	'Yukikomachi'	oJoy
	CLEMATOCLETHRA	
	sp. EDHCH 97133	wHer
	CLEOME	
	hassleriana	wSte
	CLERODENDRUM	
	bungei	oFor wHer oHed oCis oGre wMag wTGN wCoN
	myricoides 'Ugandense'	oTrP wGra
	thomsoniae	oTrP oGar oUps
	trichotomum	oFor wCul oGos oGar oDan oRiv oSho oWhi oGre wFai wPir wCCr wSta oWhS wCoN wSnq wWel
	trichotomum var. ***fargesii***	oGre wSte
	trichotomum HC 970231	wHer
	trichotomum HC 970591	wHer
	ugandense	see **C. *myricoides*** 'Ugandense'
	CLETHRA	
	acuminata	oFor wHer oEdg oAmb oRiv oGre wAva
	alnifolia	oFor oGar oDan oSho oBlo wAva wCoN wSte
	alnifolia 'Alba'	oGre
un	***alnifolia*** 'Fern Valley Pink'	oFor
	alnifolia 'Hummingbird'	oFor wHer wWoS oGos oAmb oGar oRiv oGre wFai wPir wWel
	alnifolia 'Paniculata'	oFor wCul oAmb oGar oSho oGre wWel
	alnifolia 'Pink Spire'	oFor wHer oDar oJoy oGar oRiv oSho oGre wFai
	alnifolia 'Rosea'	oFor
	alnifolia 'Ruby Spice'	oFor wHer wWoS oGos oEdg oGar oSho oGre wFai oUps wCoS wWel
un	***alnifolia*** 'September Beauty'	wWoS
un	'Anne Bidwell'	wHer
	barbinervis	oFor oGos oEdg oRiv oSho oWhi oGre wCCr wSte wWel
	barbinervis HC 970718	wHer
	fargesii	oFor oGos
	monostachya	oGos
	paniculata	see **C. *alnifolia*** 'Paniculata'
	tomentosa	oFor
	CLEYERA	
	japonica	oHed oCis
	japonica HC 970611	wHer
	CLIANTHUS	
	formosus	wGra
	puniceus	wGra
	puniceus 'Albus'	oFor oTrP wGra
	puniceus 'Pink Flamingo'	see **C. *puniceus*** 'Roseus'
	puniceus 'Red Cardinal'	see **C. *puniceus***
	puniceus 'Roseus'	wGra
	puniceus 'White Heron'	see **C. *puniceus*** 'Albus'
	CLIFTONIA	
	monophylla	oRiv
	CLINOPODIUM	
	vulgare	oGoo iGSc
	CLINTONIA	
	andrewsiana	wHer
	uniflora	wThG wFFI
	CLIVIA	
	miniata Belgian hybrids	oGar
	COBAEA	
	scandens	oGar oUps wSte
	scandens f. ***alba***	oGar
	COCCINIA	
	indica	see **C. *grandis***
	grandis	oRar
	rehmannii	oRar
	COCCULUS	
	carolinus	oFor
	laurifolius	oFor
	orbiculatus	oFor
	orbiculatus HC 970001	wHer
	trilobus	see **C. *orbiculatus***
	CODIAEUM	
	variegatum var. ***pictum*** 'Norma'	oGar
	CODONOPSIS	
	bhutanica	wHer
	cardiophylla	wHer
	clematidea	oFor oDan wEde oUps
	sp. aff. ***forrestii*** DJHC 98305	wHer
	lanceolata	wHer
	sp. aff. ***lanceolata*** HC 970661	wHer
	ovata	oJoy
	pilosula	wHer oCrm oBar
	tangshen (see Nomenclature Notes)	oOEx wFai
	tangshen EDHCH 97334	wHer
	tubulosa	last listed 99/00
	tubulosa DJHC 537	wHer
	ussuriensis	see **C. *lanceolata***
	COFFEA	
	arabica	oGar iGSc
	COIX	
	lacryma-jobi	wFai
	COLCHICUM	
	'The Giant'	oSec
	COLEONEMA	
	pulchrum	last listed 99/00
	COLEUS	see **PLECTRANTHUS, SOLENOSTEMON**
	COLLETIA	
	cruciata	see **C. *paradoxa***
	paradoxa	wCCr
	COLLINSONIA	
	canadensis	oCrm
	COLLOMIA	
	grandiflora	oBos
	COLOBANTHUS	
	quitensis	wMtT
	COLOCASIA	
	antiquorum	see **C. *esculenta***
	'Black Princess'	see **C.** 'Black Magic'
	esculenta	oTrP wGAc oHug oSsd
un	***esculenta*** 'Black Jade'	oGar
	esculenta 'Black Magic'	wHer wWoS wGAc oHug oGar oSsd
	esculenta 'Fontanesii'	oHug oGar oSsd
	esculenta 'Illustris'	wGAc oHug
un	***rubra***	oHug
	COLQUHOUNIA	
	coccinea var. ***mollis***	see **C. *coccinea*** var. ***vestita***
	coccinea var. ***vestita***	oFor
	COLUTEA	
	arborescens	oFor
	COMARUM	see **POTENTILLA**
	COMMELINA	
	dianthifolia	last listed 99/00
	tuberosa Coelestis Group	last listed 99/00
	COMMIPHORA	
un	***pyracanthoides***	oRar
	COMPTONIA	
	peregrina	wHer wPir
	CONRADINA	
	verticillata	last listed 99/00
	CONSOLIDA	
	ajacis blue	oNWe
un	***ajacis*** 'White Cloud'	wFGN
	ambigua	see **C. *ajacis***
	'Blue Cloud'	wFGN
	CONVALLARIA	
	majalis	oFor wHig oRus oGar wRob wFai wHom wMag oCrm wTGN wAva wCoN oUps wSte
	majalis 'Albostriata'	wHig
do	***majalis*** 'Flore Pleno'	oSis wRob
	majalis var. ***rosea***	wHer wHig oRus oSho
	majalis 'Variegata'	oHed oNWe wHom
un	***sarmentosa***	wHer
	CONVOLVULUS	
	cneorum	oFor wHer wCri oJoy wCSG oAmb oGar oSis oCis oSho oUps wSte
	cneorum JJH collection	oCis
	mauritanicus	see **C. *sabatius***
	sabatius	oDan oSis wWel
	COPROSMA	
	x kirkii 'Kirkii Variegata' female	wHer
va	***repens*** 'Pink Splendour' male	last listed 99/00

	robusta	wCCr
	COPTIS	
	laciniata	oBov oTri
	CORALLOCARPUS	
un	bainesii	oRar
	CORDYLINE	
	australis	wCCr
un	australis 'Atropurpurea'	oTrP
un	australis 'Red Sensation'	wHer
	banksii	oFor
	baueri	oGar wWel
	indivisa	oAls oGar oSev wTGN wWel
	indivisa 'Rubra'	wTGN
	COREOPSIS	
un	alpina 'Alba'	oSis
	auriculata	last listed 99/00
	auriculata 'Nana'	oFor oGar oDan wTGN
un	auriculata 'Zamfir'	oSis
	Baby Sun	see C. 'Sonnenkind'
	'Goldfink'	oAls
qu	grandiflora 'Brown Eyes'	iArc
	grandiflora 'Domino'	oEga
	grandiflora 'Early Sunrise'	oFor oAls oGar oBlo oGre wHom oWnC oSec iArc oCir oEga wWhG
	grandiflora Flying Saucers / 'Walcoreop'	oAls oAmb oWnC wWhG
	grandiflora 'Mayfield Giant'	last listed 99/00
	lanceolata	oGoo
	lanceolata 'Sterntaler'	oSis
	mutica	wCCr
	rosea	oFor oJoy oAls oMis oGoo oSho oSha wHom oWnC wMag wTGN oSec iArc oEga wWhG oUps
un	rosea 'Alba'	oFor oSec
	rosea 'American Dream'	oFor
	'Sonnenkind'	wTho oGar oOut wHom oWnC oCir wAva wWhG oSle
	'Sunburst'	wTho
	'Sunray'	oAls oGar oSho oSha oGre oWnC
un	'Tequila Sunrise'	oAls oNWe oSho wWhG wSnq
	tripteris	oFor oGoo
	verticillata	wShR oMis
	verticillata 'Golden Gain'	last listed 99/00
	verticillata 'Golden Shower'	see C. verticillata 'Grandiflora'
	verticillata 'Grandiflora'	oFor oGar
	verticillata 'Moonbeam'	oFor wCul wSwC oNat oDar oJoy oAls wCSG oAmb oMis oGar oSho oOut wRob oGre wHom oWnC wMag wTGN oSec oCir wAva oEga wWhG oUps oSle wSnq
	verticillata 'Zagreb'	oFor wCul oJoy oAls oGar oSho oGre oWnC oSec wAva oEga wWhG oUps oSle
	CORIANDRUM	
	sativum	wTho oAls wFGN wFai iGSc wHom oWnC
	sativum 'Cilantro'	wFGN wHom oCir
	sativum 'Santo'	wFGN
	CORIARIA	
	myrtifolia	oFor
	ruscifolia HCM 98054	wHer
	ruscifolia HCM 980178	wHer
	terminalis var. xanthocarpa	oNWe
	CORNUS	
	alba 'Argenteomarginata	see C. alba 'Elegantissima'
	alba 'Argenteovariegata'	see C. alba 'Variegata'
	alba 'Aurea'	oNWe wBox wWel
	alba 'Bailhalo'	see C. alba Ivory Halo / 'Bailhalo'
un	alba 'Bud's Yellow'	oFor w WoS wWel
va	alba 'Elegantissima'	wHer wCul oGar oDan oNWe oBlo wRai oGre wPir wCCr oWnC wTGN oCir wAva wSte wWel oFor wWoS wCul oGar oOut wKin
va	alba 'Gouchaultii'	oFor wHer oDar oGar oDan wWel
	alba Ivory Halo / 'Bailhalo'	w WoS wCul oGar wWel
	alba 'Kesselringii'	oFor
	alba 'Red Gnome'	oFor
	alba 'Siberian Pearls'	oFor oGos
	alba 'Sibirica'	oFor wCul wBCr wCSG oWnC wMag
	alba 'Sibirica Variegata'	oFor
va	alba 'Spaethii'	oFor wHer oGos oRiv oGre wWel
	alba 'Variegata'	oSho
	alternifolia	oFor oRiv oGre wKin wAva wCoN oSle
va	alternifolia 'Argentea'	oFor oGos oGar oSho oGre wTGN
	alternifolia 'Variegata'	see C. alternifolia 'Argentea'
	amomum	oFor oRiv wCCr wBox
	baileyi	see C. stolonifera 'Baileyi'
	bretschneideri	oFor
	canadensis	oFor wHer wThG wWoB oAls oRus oGar oSis wRai wWat wMag oTri wTGN wPla oAld wCoN oSle wSnq wWel
	capitata	oFor oRiv oCis wCCr oWnC wCoN
un	capitata 'Mountain Moon'	oCis oGre oWnC wBox
	chinensis	last listed 99/00
	chinensis DJHC 819	wHer
	controversa	oFor oEdg wCol oGar oRiv oGre oWnC wCoN oSle
	controversa EDHCH 97173	wHer
	controversa HC 970086	wHer
un	controversa 'Janinie'	wWel
un	controversa 'June Snow'	wWel
	controversa 'Pagoda'	oGos
	controversa 'Variegata'	oFor wHer oGos wClo oEdg oGar oRiv oGre wWel
	'Eddie's White Wonder'	oFor oGar oBlo oWnC wSta wCoN wWel
	florida	oFor wShR oGar oWhi oWnC wSta oThr wAva wCoN oGue
	florida budded	oThr
	florida 'Cherokee Brave'	oGar oWnC wWel
	florida 'Cherokee Chief'	oGar oBlo oGre oWnC oThr wCoS wWel
	florida 'Cherokee Daybreak'	oGar oWnC
	florida 'Cherokee Princess'	oGar oGre oWnC wTGN wWel
un	florida 'Cherokee Sunset'	oFor oGar wWel
	florida 'Cloud Nine'	oFor oGar oGre oWnC wWel
	florida 'First Lady'	last listed 99/00
	florida 'Golden Nugget'	oFor oGre
	florida 'Hohman's Gold'	wSta oThr
	florida 'Pendula'	oFor wSta oThr
va	florida 'Pink Flame'	last listed 99/00
	florida 'Pygmaea'	oFor
va	florida 'Rainbow'	oWnC
	florida reddish bud	oThr
	florida 'Royal Red'	oFor
	florida f. rubra	oFor oGar oSho oGre oWnC wSta oThr wCoS
	florida variegated	oThr
va	florida 'Welchii'	oFor
	foemina	see C. stricta
	glabrata	wCCr
	hemsleyi	wCCr
	hessei	oGre
	kousa	wThG wWoB wBur wClo oGar oCis oWhi oBlo oGre wAva wCoN oGue wRav oSle wCoS
un	kousa var. angustata	wHer oCis oGre
un	kousa 'Autumn Rose'	oFor oGre
	kousa 'Beni-fuji'	oGos
un	kousa Big Apple™ (fruit)	wBur wRai oWnC
	kousa var. chinensis	oFor wBCr wSwC oDar oAmb oGar oRiv oOEx oBov oGre oWnC wSta oThr wAva wWhG wWel
	kousa var. chinensis 'China Girl'	wSta
	kousa var. chinensis dwarf pink	oGar
un	kousa var. chinensis 'Honros'	oGre
	kousa var. chinensis 'Milky Way'	oFor wCul oGos wBur oGar oWnC wSta wAva
	kousa 'Elizabeth Lustgarten'	oFor oGre wWel
un	kousa 'Fireworks'	wCol
un	kousa 'Gold Cup'	oGar
va	kousa 'Gold Star'	oGos wCol oGre
	kousa HC 970266	wHer
un	kousa 'Heart Throb'	oFor wCol wWel
un	kousa 'Highland'	oGar
un	kousa 'Lemon Ripple'	wWhG
qu	kousa 'Lustgarten'	wCol
	kousa 'Lustgartgen Weeping'	oGar oRiv oGre
un	kousa 'Moonbeam'	oFor oGre
	kousa 'National'	oFor oGar oGre oWnC
	kousa 'Radiant Rose'	oFor oGar oDan
	kousa 'Rosabella'	see C. kousa 'Satomi'
un	kousa 'Ruby Slippers'	wWhG
	kousa 'Satomi'	oFor oGos wBur wClo oGar oRiv oCis oSho oBlo oGre oWnC wTGN wSta wAva wWhG wCoS wWel
va	kousa 'Snowboy'	oFor oGos wCol
	kousa 'Snowflake'	oGos oGar oGre wWhG
un	kousa 'Steeple'	wCol oGar
	kousa 'Summer Majesty'	oGar
	kousa 'Summer Stars'	oGar wWel
va	kousa 'Sunsplash'	wWel
va	kousa 'Temple Jewel'	oGos wWhG
	kousa 'Triple Crown'	wWhG
	kousa 'Weaver's Weeping'	oFor wHer oGar oRiv wWhG

un	kousa 'Wolf Eyes'	wCol oGre
	macrophylla	last listed 99/00
	macrophylla HC 970227	wHer
	mas	oFor wCul wBCr wThG wBur wClo oEdg wCSG oGar oDan oRiv oWhi oBlo oOEx oWnC wSta wWhG wWel
va	mas 'Aurea'	oFor wHer wWel
un	mas 'Elegant' (fruit)	oGre wBox
	mas 'Golden Glory'	oFor oRiv oGre wWel
un	mas 'Jellico' (fruit)	wRai
un	mas 'Oleg'	wSte
un	mas Pioneer™ (fruit)	wRai oGre wBox
un	mas Red Star™ (fruit)	oGre wWel
	mas yellow fruited (fruit)	wRai
un	mas 'Yevgeny'	wSte
	nuttallii	oFor wWoB wBur wShR oBos oRiv oGre wWat wPla oAld wRav
	nuttallii 'Colrigo Giant'	oFor oGar oSho wCoN
	nuttallii 'Gold Spot'	oFor oWhi oGre wSta
	occidentalis	wHer oRiv wFFl
	officinalis	oFor oEdg oDan oRiv wCCr oWnC oBRG
un	omeiense Summer Passion™	oGar oDan oCis oGre oWnC wBox
	paniculata	see C. racemosa
	paucinervis	oFor
	pumila	oFor
	racemosa	oFor oGar
	x rutgersiensis	wCoN
	x rutgersiensis 'Aurora'	oFor wClo oGre oWnC wWel
	x rutgersiensis Celestial™ / 'Rutdan'	oFor oGre oWnC
	x rutgersiensis Constellation™ / 'Rutcan'	oFor oGre oWnC
	x rutgersiensis Galaxy™	see C. x rutgersiensis Celestial™ / 'Rutdan'
	x rutgersiensis 'Ruth Ellen'	oFor
un	x rutgersiensis 'Stardust'	oGre
	x rutgersiensis 'Stellar Pink'	oFor wClo oWnC wSta wWel
	sanguinea	oFor wRai wCoN
	sanguinea 'Compressa'	see C. hessei
	sanguinea 'Midwinter Fire'	wHer oGos oGar oRiv wRob oGre wAva
	sanguinea 'Winter Beauty'	oFor oRiv
	sanguinea 'Winter Flame'	see C. sanguinea 'Winter Beauty'
	sericea	see C. stolonifera
	sessilis	oFor
	sp. DJHC 231	wHer
	stolonifera	oFor oHan wBlu wBCr wThG wWoB wBur wClo oNat oAls wShR oGar oDan oBos oGre wWat wCCr wKin oTri wSta wPla oAld wCoN oSle wSte wSnq
un	stolonifera 'Allemans Compact'	oGos
	stolonifera 'Baileyi'	wCul oGar oGre oWnC
un	stolonifera 'Cardinal'	wTGN wWel
	stolonifera 'Flaviramea'	oFor wHer wCul wWoB oGar oDan oRiv oOut oBlo oGre wCCr wKin oWnC wSta oCir wAva wWel
	stolonifera 'Hedgerows Gold'	oGos wCol oHed oNWe
	stolonifera 'Isanti'	oFor oGar oRiv oWnC wWel
	stolonifera 'Kelsey Dwarf'	see C. stolonifera 'Kelseyi'
	stolonifera 'Kelseyi'	oFor oAls oGar oRiv oBlo wCCr oWnC wTGN wSta oBRG oCir oSle wWel
	stolonifera ssp. occidentalis	see C. occidentalis
	stolonifera 'Silver and Gold'	oFor oGos oHed oRiv oGre
	stolonifera 'Sunshine'	oGos wCCr
qu	stolonifera 'Variegata'	oFor oAls
	stricta	oRiv
	x unalaschkensis	wShR wFFl
	walteri	oFor oRiv
	COROKIA	
	cotoneaster	oDar oAls oHed oAmb oRiv oCis oNWe oSho wPir oTPm wTGN oBRG wCoN
un	cotoneaster 'Anton's Dwarf'	oCis
	cotoneaster 'Little Prince'	oCis
	macrocarpa	oFor oRiv
	x virgata	oRiv oBRG
	x virgata 'Bronze King'	wHer
	x virgata 'Red Wonder'	oCis
	x virgata 'Sunsplash'	oHed oCis
	CORONILLA	
	emerus	see HIPPOCREPIS emerus
	valentina ssp. glauca 'Variegata'	oFor
	CORREA	
	alba	oTrP wHer
	backhouseana	oBov
un	'Dawn in Santa Cruz'	oFor
	'Dusky Bells'	oHed oCis
	'Ivory Bells'	oHed
	'Marian's Marvel'	oHed oCis
	pulchella	wHer oJoy oBov
	pulchella 'Alba'	wCri
	pulchella 'Mission Bells'	oFor
	pulchella 'Orange Flame'	oCis
un	'Sister Dawn'	oCis
	CORTADERIA	
	richardii (see Nomenclature Notes)	wHer wCCr wCoS
	selloana	oFor oDar oAls oGar oGoo oSho oBlo oGre wHom oTPm oWnC wTGN wSta oCir wWhG
va	selloana 'Aureolineata'	oDar oGre wPir
	selloana 'Carnea'	oWnC
un	selloana 'Elegans'	oGre
	selloana 'Gold Band'	see C. selloana 'Aureolineata'
	selloana Ivory Feathers® / 'Pumila'	oAls oGar oDan oGre oWnC wTGN
	selloana pink	oFor wHom wCoN
	selloana 'Pumila'	see C. selloana Ivory Feathers® / 'Pumila'
	selloana 'Rosea'	oTri
qu	selloana rubra	oGre
	selloana Sun Stripe® / 'Monvin'	oFor oAls oGar
	selloana white	wCoN
	CORTUSA	
	matthioli ssp. pekinensis	oMis
	CORYDALIS	
un	'AMF'	see C. flexuosa Award of Merit Form
un	'Blackberry Wine'	oFor wCol
	bulbosa ssp. densiflora	see C. solida ssp. incisa
	buschii	wMtT
	cheilanthifolia	oFor wWoS wCul oHed oRus wRob oSec wCoN
un	edulis	oTrP oNWe
	elata	oFor wHer wWoS wSwC wCol oRus oAmb oDan oNWe wMtT oOut oGre wAva
un	flexuosa Award of Merit Form	wHer wWoS oGar oNWe wMtT oOut oGre wBox
	flexuosa 'Blue Panda'	oFor wHer wWoS wCol oHed oRus oGar oDan oSis wMtT oBlo oGre wFai wHom wMag wTGN wWhG
	flexuosa 'China Blue'	oFor wHer wWoS wSwC wGAc oNat oJoy wCol oAls oHed oMis oGar oDan oNWe oSho wRob oGre wFai wHom wMag wTGN wAva oEga oSle wSte
un	flexuosa 'Golden Panda'	wCol
	flexuosa 'Nightshade'	wHer
	flexuosa 'Pere David'	wHer wWoS oJil oJoy oAls oHed oRus wFGN oGar oOut oGre wTGN wBox wCoN wSte
	flexuosa 'Purple Leaf'	oFor wHer wWoS oNat wCol oHed oRus oGar oDan oCis oOut wRob oGre wFai oHon wMag wAva
un	leucanthema DJHC 752	wHer
	linstowiana	wHer wMtT
	lutea	oFor wCul oNat wCol oAls oRus oMis oGar oBlo oGre wFai wAva wCoN wWld wWhG
un	macrantha	oRus
	ochroleuca	oRus oMis oGar oSis oNWe wMag
	ophiocarpa	oFor wTGN
un	'Royal Purple'	wWhG
	saxicola	oRus oNWe
	scouleri	wHer oBos oTri oAld
	smithiana	oNWe oGre
	solida	wHer oSis
	solida from Penza	oNWe wMtT
	solida 'Harkov'	wMtT
	solida ssp. incisa	oSis
	solida f. transsylvanica Sunset Strain	wMtT
	sp. ex Dufu Temple	wCol oRus
	thalictrifolia	see C. saxicola
	tomentella	last listed 99/00
	CORYLOPSIS	
	glabrescens	oFor oWhi wWel
	glabrescens var. gotoana	oRiv
un	glaucophylla	oGos
	gotoana	see C. glabrescens var. gotoana
	multiflora	oCis
	pauciflora	oFor wHer oGos wClo oGar oDan oRiv oGre wPir wSta oBRG wCoN wWel
	platypetala	see C. sinensis var. calvescens
	sinensis	oFor oGos oGar oRiv oGre wCCr
	sp. aff. sinensis	wRho
	sinensis var. calvescens	oFor oGar oRiv oGre wWel
	sinensis var. calvescens f. veitchiana	oFor oAls oGar oDan oRiv oGre wPir wSnq
	sinensis var. sinensis	oGos oGre wWel
	sinensis var. sinensis DJHC 98320	wHer

	sinensis var. sinensis 'Spring Purple'	last listed 99/00
	sinensis var. willmottiae	see C. sinensis var. sinensis
	spicata	oFor oGos wClo oAls wCSG oGar oDan oRiv oSho oBov oGre wFai wPir wCCr wSta oBRG wCoN wWel
	spicata red leaf	wWel
	veitchiana	see C. sinensis var. calvescens f. veitchiana
	willmottiae	see C. sinensis var. sinensis
	'Winterthur'	last listed 99/00
	CORYLUS	
	americana	oFor wBCr wBur
	americana x avellana 'Rutter'	wBCr
	americana x avellana 'Skinner'	wBCr
	avellana (nut)	wBlu wBCr wKin wCoS
	avellana 'Atropurpurea'	see C. avellana 'Fuscorubra'
	avellana 'Contorta'	oFor wHer wBur wRai oOEx wTGN wSta wAva wCoN wWhG wWel
	avellana 'Fuscorubra' (nut)	wCoN
	colurna	oFor wBCr wBur oOEx oGre wKin
	x colurnoides	wRai
	x colurnoides Chinese	wBCr
un	x colurnoides 'Gellatly Turkish'	wBCr
	x colurnoides 'Red Leaf'	wBCr
	cornuta	wShR wWat oAld
un	cornuta x avellana	wRai
un	cornuta x avellana 'Big Red'	wBCr
un	cornuta x avellana 'Gellatly'	wBCr wBur
un	cornuta x avellana 'Peace River Cross'	wBCr
	cornuta var. californica	oFor wWoB wNot wCCr oTri wFFl
	maxima (nut)	oFor wBur wCSG
	maxima 'Barcelona'	wBCr wBur oWhi wRai wVan
	maxima 'Butler'	wBCr wClo oWhi wVan
	maxima 'Du Chilly'	wVan
	maxima 'Ennis'	wClo oWhi
un	maxima 'Fortin'	oOEx
	maxima Halle Giant	see C. maxima 'Halle'sche Riesennuss'
	maxima 'Halle'sche Riesennuss"	wBCr wBur wRai
	maxima 'Lewis'	wBur
	maxima 'Purpurea'	wCoN
	maxima 'Red Filbert'	oFor wBur oOEx
	maxima 'Rote Zeller'	see C. maxima 'Red Filbert'
	maxima 'Tonne de Giffon'	wBur
	COSMOS	
	atrosanguineus	wWoS oJoy oEdg oHed wFGN wCSG oMis wHom wMag wTGN oCir wAva wSnq
	sulphureus Bright Lights mixed	oUps
	COSTUS	
	cuspidatus	last listed 99/00
	igneus	see C. cuspidatus
	COTINUS	
	coggygria	oFor oAls oGar oDan oNWe oWhi wCCr wCoN wWhG oSle
	coggygria var. atropurpureum	see C. coggygria Purpureus Group
	coggygria 'Nordine'	wCul oGre
	coggygria 'Pink Champagne'	oFor oGar oGre oWnC wWhG wWel
	coggygria Purpureus Group	oFor oDar oGar oDan oWhi wPir wTGN wAva wCoS
	coggygria 'Royal Purple'	oFor wCul oGos wSwC oDar wCSG oGar oRiv oGre wKin oWnC wTGN oSec oThr wSte wWel
	coggygria 'Velvet Cloak'	oFor wWoS oGar oGre wWel
	'Grace'	oFor wHer oGos oAmb oGre wWel
	obovatus	oAmb oRiv oThr wCoN wSte
	COTONEASTER	
	acutifolius	oFor oRiv wKin wPla
	adpressus 'Little Gem'	oFor oDar oDan oSis oSqu oTPm wSta oCir
	adpressus 'Tom Thumb'	see C. adpressus 'Little Gem'
	affinis	wCCr
	apiculatus	oFor oAls oTPm
	atropurpureus 'Variegatus'	wCul oAmb oGar oBRG
	bullatus	oFor
	bullatus DJHC 98487	wHer
	bullatus f. floribundus	see C. bullatus
	bullatus var. macrophyllus	see C. rehderi
	buxifolius	wCCr
	Canadian Creeper™ / 'Moner'	last listed 99/00
	congestus	oSho oTPm
	cooperi	oRiv
	dammeri	oAls oGar oRiv oBlo oTPm oWnC wTGN wSta wCoN wCoS wWel
qu	dammeri 'Cooperi'	oGos oBRG
	dammeri 'Lowfast'	oAls oBlo oTPm
	dammeri 'Mooncreeper'	oFor oGar
qu	dammeri 'Moonwalker'	wSta
	dammeri 'Streibs Findling'	see C. procumbens
	dielsianus	oFor
	divaricatus	oFor
	sp. aff. divaricatus DJHC 644	wHer
	franchetii	oFor oGar oBlo wCCr oCir
	franchetii DJHC 98362	wHer
	frigidus	oFor
	glabratus	wCCr
qu	glauca 'Nana'	oBlo
	glaucophyllus	oFor wCul oSho
	harrovianus	wCCr
	hebephyllus	oFor
	henryanus	wCCr
	horizontalis	oTrP oGar oRiv wCoN
un	horizontalis 'Cheney'	oTPm oSec
un	horizontalis var. horizontalis	oBlo
	horizontalis perpusillus	see C. perpusillus
	horizontalis 'Variegatus'	see C. atropurpureus 'Variegatus'
	'Hybridus Pendulus'	see C. salicifolius 'Pendulus'
	hymalaicus	oDan
	integerrimus	wKin
	lacteus	oFor wCul oDar oAls oGar oSho oBlo wCCr oTPm wWel
qu	lacteus 'Red Clusterberry'	oAls
	lindleyi	oFor
	linearifolius	oGre
	microphyllus (see Nomenclature Notes)	oGar oBlo oGre oTPm wSta
	microphyllus var. cochleatus (see Nomenclature Notes)	oFor
un	microphyllus 'Cooperi'	oDar wCSG oGar oTPm wSta
un	microphyllus 'Emerald Spray'	oFor wTGN
un	microphyllus 'Himalaya'	oDar
	microphyllus HWJCM 112	wHer
	microphyllus 'Streibs Findling'	see C. procumbens
un	microphyllus 'Timblina'	wSta
	microphyllus var. thymifolius (see Nomenclature Notes)	oFor wCul oGar oDan wWel
	moupinensis EDHCH 97245	wHer
	multiflorus	oFor
	nitens	oFor
	parneyi	see C. lacteus
	perpusillus	oGar oRiv oTPm oWnC
un	perpusillus 'Variegata'	wWel
	procumbens	oFor wHer oNWe oSqu wWel
	racemiflorus var. songaricus	oFor
	radicans 'Eichholz'	oFor
	rehderi	oFor
	sp. aff. roseus HWJCM 571	wHer
	salicifolius	oRiv
	salicifolius 'Gnom'	wHer wSta
	salicifolius 'Pendulus'	wHer wWel
	salicifolius 'Repens'	oFor oGar oSho oGre oWnC
	salicifolius 'Rothschildianus'	oFor wHer oSho
	salicifolius 'Scarlet Leader'	oBlo
	simonsii	oFor wCCr
	sp. DJHC 029	wHer
	sp. EDHCH 97293	wHer
	splendens	oFor wCCr
	x suecicus 'Coral Beauty'	oFor oRiv oBlo oGre oTPm oWnC wBox oCir
	turbinatus	oFor
	x watereri	wCCr
	zabelii	oFor
	COTULA	
	coronopifolia	wGAc oSsd
un	dinariaca	oTrP
un	filifolia	oTrP
	lineariloba	oHed
	minor	see LEPTINELLA minor
	perpusilla	see LEPTINELLA pusilla
	pyrethrifolia	see LEPTINELLA pyrethrifolia
	squalida	see LEPTINELLA squalida
	sp.	wCol
	COTYLEDON	
	ladysmithiensis	oSqu
	orbiculata	oNWe
	orbiculata DBG list	last listed 99/00
	papillaris	oSqu
un	undulata superba	oSqu
	COWANIA	
qu	mexicana var. stansburiana	oFor
	stanburyana	wPla
	CRAIBIODENDRON	
	yunnanense	wHer

	CRAMBE	
	cordifolia	wHer wWoS wCul oAls wCSG oAmb oGar oNWe wMag wBWP oWhS
	maritima	wHer wCul oAls wCSG oGar oNWe oOEx
	CRASSULA	
	anomala	oSqu
	arborescens	oSqu
	argentea	see **C. *ovata***
	barklyi	last listed 99/00
un	'Christina'	oSqu
un	***conjuncta***	oSqu
un	'Coralita'	oSqu
un	'Crosby's Red'	oSqu
	dubia	oSqu
un	'Emerald'	oSqu
	falcata	oSqu
un	'Gollum'	oSqu
un	'High Voltage'	oSqu
	lycopodioides	see **C. *muscosa***
	x marchandii	oSqu
un	'Moon Glow'	oSqu
	muscosa	oSqu
	muscosa 'Variegata'	oSqu
	ovata	oGar
	perforata 'Variegata'	oSqu
	remota	oSqu
un	***rubescens*** JJH 94-01695	oCis
	setulosa var. ***curta***	oSis
un	'Springtime'	oSqu
	streyi	last listed 99/00
un	'Sunset'	oSqu
	teres	see **C. *barklyi***
un	'Tom Thumb'	oSqu
un	***tricolor***	oSqu
un	***volkensis***	oSqu
un	'Waves'	oSqu
	+ CRATAEGOMESPILUS	
	dardarii 'Asnieresii'	see **C.** 'Jules d'Asnieres'
	'Jules d'Asnieres'	oFor
	CRATAEGUS	
	aestivalis	see **C. *opaca***
	ambigua	oFor
	arnoldiana	oFor
	'Autumn Glory'	oGar
	azarolus	wRai
	brachyacantha	oFor
	calpodendron	wCCr
	chlorosarca	last listed 99/00
	chungtienensis	wCol
	columbiana	oFor
	crus-galli (see Nomenclature Notes)	last listed 99/00
	crus-galli 'Inermis'	oFor oSle
un	Crusader®/ 'Cruzam'	wKin
	cuneata	last listed 99/00
	douglasii	oFor wWoB wBur wNot wShR oBos wWat wCCr oTri wPla oAld oSle
	erythropoda	oFor
	laevigata	last listed 99/00
	laevigata 'Crimson Cloud'	oGar wKin
do	***laevigata*** 'Paul's Scarlet'	oFor oAls oGar oSho oBlo oWnC wTGN wCoS
	x lavalleei	wWel
	x lavalleei 'Carrierei'	oAls
	mollis	oFor
	monogyna	oCrm
	monogyna 'Compacta'	oGos wCol
	monogyna 'Flexuosa'	oGar
	opaca	oFor
un	***opaca*** 'Big Red'	oOEx
un	***opaca*** 'Elite'	oOEx
un	***opaca*** 'Golden Farris'	oOEx
un	***opaca*** 'Goldie'	oOEx
un	***opaca*** 'Goliath'	oOEx
un	***opaca*** 'Reliable'	oOEx
un	***opaca*** 'Super Spur'	oOEx
un	***opaca*** 'T.O. Warren Super Berry'	oOEx
un	***opaca*** 'Warren's Opaca'	oOEx
un	***opaca*** 'Yellow'	oOEx
	oxyacantha	see **C. *laevigata***
	pedicellata	wCSG
	phaenopyrum	oFor wBCr oGar oDan oWnC wRav wWel
un	***phaenopyrum*** 'Lustre'	oFor oGar
	pinnatifida	oFor oOEx
	pinnatifida var. ***major***	oFor
	sp. DJHC 319	wHer

	sp. dwarf form	oRiv
	viridis 'Winter King'	oSle
	CRAWFURDIA	
	speciosa HWJCM 135	last listed 99/00
	CRINODENDRON	
	hookerianum	oGos oCis
	hookerianum HCM 98183	wHer
	patagua	oFor oCis
	X CRINODONNA	see **X AMARCRINUM *memoria-corsii***
	CRINUM	
	americanum	wGAc oHug oSsd
un	***angustifolium*** 'Tahitian Gold'	oTrP
	asiaticum	oTrP
	asiaticum DJHC 970606	wHer
	CRITHMUM	
	maritimum	oGoo
	CROCOSMIA	
	aurea (see Nomenclature Notes)	oTrP oCis
	Bressingham Beacon / 'Blos'	wWoS
un	'Challa'	wWoS
	'Citronella' (see Nomenclature Notes)	oJoy oAls oNWe wTGN oBRG
	'Coleton Fishacre'	see **C. *x crocosmiiflora*** 'Gerbe d'Or'
	x crocosmiiflora	oFor oGar
	x crocosmiiflora 'Babylon'	oGar oBRG
	x crocosmiiflora 'Carmin Brilliant'	wHer oGar oGre oBRG
	x crocosmiiflora 'Constance'	oFor wWoS oGar oBRG
	x crocosmiiflora 'Eastern Promise'	wHer
	x crocosmiiflora 'Emily McKenzie'	wWoS oNat oDar oJoy oAls wCSG oMis oDan oNWe oGre wHom wTGN oBRG wSnq
	x crocosmiiflora 'George Davison' (see Nomeclature Notes)	wWoS oSho
	x crocosmiiflora 'Gerbe d'Or'	oGre
	x crocosmiiflora 'James Coey'	oGar oGre wFai
un	***x crocosmiiflora*** 'Kathleen'	oJoy
	x crocosmiiflora 'Lady Hamilton'	wHer wWoS
	x crocosmiiflora 'Meteore'	oSis
	x crocosmiiflora 'Norwich Canary'	oFor oTrP wGAc oJoy oGre wFai wHom wBon
	x crocosmiiflora 'Queen Alexandra' (see Nomenclature Notes)	wWoS
	x crocosmiiflora 'Solfaterre'	oTrP wHer wWoS wSwC oDar oHed oDan oNWe oOut oGre wFai wCoN
	x crocosmiiflora 'Star of the East'	last listed 99/00
	x crocosmiiflora 'Venus'	oFor wWoS oJoy oGar
	'Emberglow'	wHer wWoS oGos oGar oSho oGre wFai oBRG oEga wSnq
	Golden Fleece	see **C. *x crocosmiiflora*** 'Gerbe d'Or'
	'Jenny Bloom'	wSwC oJoy oGar oBRG
	'Jupiter'	oFor wWoS wCul wCSG oGar wFai wHom oEga wSte
	'Lady Wilson'	see **C. *x crocosmiiflora*** 'Norwich Canary'
	'Lucifer'	oFor wWoS wCul wCri wSwC oNat oDar wCol oAls oRus wCSG oMis oGar oSis oNWe oSho oSha oGre wFai wCCr wHom wMag wTGN wSta oInd oBRG oCir wCoN oEga oUps oSle wSnq
un	'Malahide'	wHer
	'Mars'	wHer oGos
	masoniorum	oFor oTrP wSwC oJoy wCSG wHom wSta
un	***masoniorum*** 'Seven Seas'	oOut
un	'Olympic Sunrise'	wHer
	'Plaisir'	last listed 99/00
	pottsii	oTrP
qu	'Severn Seas'	oSle
	'Severn Sunrise'	wHer
	CROCUS	
	x luteus 'Golden Yellow'	oWoo
	'Mammoth Yellow'	see **C. *luteus*** 'Golden Yellow'
	sativus	wWoS
	speciosus	wFai
	vernus 'Flower Record'	wRoo
	vernus 'Jeanne d'Arc'	wRoo oWoo
	vernus 'Joan of Arc'	see **C. *vernus*** 'Jeanne d'Arc'
	vernus 'Pickwick'	wRoo oWoo
	vernus 'Remembrance'	wRoo oWoo
	yellow	wRoo
	CRYPTANTHUS	
	acaulis	oGar
	CRYPTOGRAMMA	
	crispa	oBos
	CRYPTOMERIA	
	fortunei	see **C. *japonica*** var. ***sinensis***
	japonica	oFor oAls wCCr
	japonica Araucarioides Group	wWel

un	***japonica*** f. ***atropurpurea***	oGar
	japonica 'Bandai-sugi'	oPor oGre
	japonica 'Birodo-sugi'	oPor
	japonica 'Black Dragon'	oFor wHer oPor wClo wCol oAls oGar oSis oGre
un	***japonica*** 'Bloomers Witch's Broom'	oPor
	japonica 'Compacta'	wHer
	japonica 'Compressa'	wHer oGar oSis oGre oBRG
un	***japonica*** 'Congesta'	oPor oGre
	japonica 'Cristata'	oFor oPor wClo wCol oSis
	japonica 'Dacrydioides'	wHer oPor
	japonica Elegans Group	oFor wHer wCri oPor wClo oJoy wShR oGar oRiv oSho oOut oBlo oGre oWnC wTGN wSta wAva wCoN wWhG wWel
	japonica 'Elegans Aurea'	wHer oPor wClo oRiv oNWe oGre oWnC wWhG wWel
	japonica 'Elegans Compacta'	oPor wClo oAls oGar oWhi
	japonica 'Elegans Nana'	wHer
	japonica 'Globosa'	wHer
	japonica 'Globosa Nana'	oPor
un	***japonica*** 'Gyoku Ryu'	wHer
un	***japonica*** 'Hino-sugi'	oPor
	japonica 'Jindai-sugi'	oFor wHer oPor wCol oGre
	japonica Kelly's fastigiate	oPor
	japonica 'Kilmacurragh'	oPor
un	***japonica*** 'Kitayama'	wWel
va	***japonica*** 'Knaptonensis'	oPor wClo oSis oBRG
	japonica 'Koshyi'	oPor oWnC
	japonica 'Little Diamond'	last listed 99/00
	japonica 'Littleworth Dwarf'	see **C. *japonica*** 'Littleworth Gnom'
	japonica 'Littleworth Gnom'	last listed 99/00
	japonica 'Lobbii Nana'	see **C. *japonica*** 'Nana'
un	***japonica*** 'Mankichi'	oPor
	japonica 'Nana'	oFor wHer oAls oGar oDan oGre wAva wCoN
	japonica 'Nana Albispica'	oPor
	japonica 'Osaka-tama'	oPor
	japonica 'Pygmaea'	oPor wCoN
un	***japonica*** 'Ryoku Gyoku'	oPor
	japonica 'Sekkan'	wHer
	japonica 'Sekkan-sugi'	oFor oPor wClo wCol oAls oGar oGre oWnC wWel
	japonica var. ***sinensis***	oFor
	japonica 'Spiralis'	oFor wHer oPor wCol oAls oGar oSis oSho oBov oGre
	japonica 'Spiraliter Falcata'	wHer oPor wCol oSis
un	***japonica*** 'Taisho Tama'	wCol
	japonica 'Tansu'	oFor wHer oPor wClo oAls oGar oRiv oBlo oGre oWnC wSta oBRG
	japonica 'Tenzan'	oPor oGos oGre
	japonica 'Vilmoriniana'	oFor oPor oAls oGar oDan oSho oBlo oBRG wCoN
un	***japonica*** 'Yellow Twig'	wHer
	japonica 'Yokohama'	last listed 99/00
	japonica 'Yoshino'	oFor wClo oTPm
	CRYPTOSTEGIA	
	grandiflora	oTrP
	CRYPTOTAENIA	
	japonica	oGoo oNWe
	japonica f. ***atropurpurea***	oFor wHer wGAc oJoy wHig oRus oDan oSho oOut wRob wFai wBox
	CTENANTHE	
	amabalis	oGar
	lubbersiana	last listed 99/00
	CUCUBALUS	
	baccifer EDHCH 97280	last listed 99/00
	CUCURBITA	
un	***ficifolia*** var. ***grande***	oOEx
	CUMINUM	
	cyminum	last listed 99/00
	CUNILA	
	origanoides	oUps
	CUNNINGHAMIA	
	konishii	oFor wHer oRiv
	lanceolata	oFor oTrP wHer wShR oGar oRiv oCis oSho oWhi oGre wCoN wWel
	lanceolata dwarf	oCis oGre
	lanceolata 'Glauca'	oFor oPor oRiv oCis oOut oGre oWnC wTGN wSta wCoN
	unicaniculata	see **C. *lanceolata***
	CUPHEA	
	aequipetala	last listed 99/00
	cyanaea	oHed
	cyanaea hirtella	see **C. *hirtella***
	cyanaea pink form	wHer

un	***glutinosa***	wCCr
	hirtella	oHed
	hyssopifolia	oTrP oGoo iGSc
	ignea	oTrP
	ignea form	wHer
un	'Lavender Face'	wTGN
un	'Lavender Lace'	oGar
un	***melvilla***	oTrP
	X CUPRESSOCYPARIS	
	leylandii	oFor wClo oAls oBlo oGre wSta wCoS wWel
	leylandii 'Castlewellan'	oFor oAls oGar oSho oBlo oGre wCCr oWnC wSta wCoN wWel
un	***leylandii*** 'Contorta'	oFor
	leylandii Emerald Isle®/ 'Moncal'	oAls oGar oSho oWnC wCoN wWel
	leylandii 'Golconda'	wHer
	leylandii 'Gold Rider'	oFor oPor
va	***leylandii*** 'Harlequin'	last listed 99/00
	leylandii 'Naylor's Blue'	oFor oAls oGar oSho oGre wCCr wCoN wWel
	leylandii 'Picturesque'	oPor
	leylandii 'Robinson's Gold'	wHer
va	***leylandii*** 'Silver Dust'	oFor oPor
	ovensii	wCCr
	CUPRESSUS	
ch	***abramsiana***	see **C. *goveniana*** var. ***abramsiana***
un	***andeleyensis***	oRiv
	arizonica	wCoN
	arizonica var. ***arizonica*** 'Arctic'	last listed 99/00
	arizonica var. ***glabra***	wCSG oSho
	arizonica var. ***glabra*** 'Blue Ice'	oFor wHer oJoy oDan oSqu oGre oTPm wSta
	arizonica var. ***glabra*** 'Blue Pyramid'	wHer oGre wWel
un	***arizonica*** var. ***glabra*** 'Carolina Sapphire'	wHer
	arizonica var. ***nevadensis***	oFor
un	***assamica***	oFor
	atlantica	see **C. *sempervirens*** var. ***atlantica***
un	***austrotibeticus*** LS & E	wHer
	bakeri	oFor wHer wCCr
	cashmeriana	oFor wHer oRiv
	chengiana	oFor
un	***chengiana*** var. ***kansuensis***	oFor
	duclouxiana	wCol oRiv oCis
	forbesii	wCCr
	funebris	wCCr
	glabra	see **C. *arizonica*** var. ***glabra***
	goveniana	oFor wCCr
ch	***goveniana*** var. ***abramsiana***	wCCr
	goveniana var. ***pygmaea***	oFor
un	***jiangeensis***	wHer
	lusitanica	oFor
	lusitanica var. ***benthamii***	oPor
	macnabiana	oFor
	macrocarpa	oFor wShR oGar
	macrocarpa 'Coneybearii Aurea'	last listed 99/00
	macrocarpa 'Donard Gold'	oRiv
	macrocarpa 'Fine Gold'	oRiv
	macrocarpa 'Greenstead Magnificent'	oRiv wWel
	macrocarpa 'Lutea'	last listed 99/00
	sargentii	oFor wCCr
	sempervirens	oWhi wCoN wCoS
	sempervirens var. ***atlantica***	oPor
	sempervirens var. ***dupreziana***	wHer
	sempervirens 'Glauca'	oGar oSho oWnC wCoS wWel
	sempervirens var. ***sempervirens***	see **C. *sempervirens*** Stricta Group
	sempervirens Stricta Group	oFor
	sempervirens 'Swane's Golden'	oGar oSho oGre wWel
	sempervirens 'Worthiana'	wHer
un	***tongmaiensis***	oFor
	torulosa	oFor wHer
	CURCULIGO	
	orchioides	oTrP
	CURCUMA	
un	***australasica*** 'Aussie Plume'	oDan
	domestica	see **C. *longa***
	longa	oOEx iGSc oCrm
un	***petiolata*** 'Emperor'	oDan
	CUSSONIA	
	spicata	oRar
	CYANANTHUS	
	lobatus large flowered form	wHer
	microphyllus	wMtT
	CYANOTIS	
	kewensis	oSqu
	somaliensis	last listed 99/00

	Plant	Sources
	CYATHEA	
	cooperi	wFan
	CYATHODES	
	colensoi	last listed 99/00
un	**CYATHULA**	
	officinalis	oCrm
	CYCAS	
	revoluta	oGar oCis
	CYCLAMEN	
	africanum	oHan
	balearicum	oHan
	cilicium	oHan oRus oSis
	coum	oFor oHan wThG oRus oGar oSis oSho oHon
	coum 'Album'	oHan oSis
	coum ssp. ***coum***	wMtT
un	***coum*** 'Rubrum'	oRus
	cyprium	oHan
	europaeum	see **C. *purpurascens***
	graecum	oHan
	hederifolium	oFor oHan wCri wThG oNat oJil oJoy wCol oRus wCSG wPle oGar oSis oNWe oGre oHon oCir wAva wWhG
	hederifolium 'Album'	oFor oHan oRus oGar oSis
	hederifolium var. ***hederifolium*** f. ***albiflorum***	oHed
	hederifolium rose pink	oRus
	intaminatum	oHan
	libanoticum	oHan
	mirabile	oHan
	neopolitanum	see **C. *hederifolium***
	persicum	oHan oGar
	purpurascens	oHan oRus
	repandum	oHan
	CYCLANTHERA	
	brachystachya	oOEx
	explodens	see **C. *brachystachya***
	pedata	oOEx
un	***spinosa***	oOEx
	CYCLOBALANOPSIS	see **QUERCUS**
	CYCLOCARYA	
	paliurus	oSho oGre
	CYDONIA	
	oblonga	oFor wBur oWnC
un	***oblonga*** 'Aromatnaya'	wClo wRai
un	***oblonga*** 'Cooke's Jumbo'	oOEx
un	***oblonga*** 'Havran'	wRai
un	***oblonga*** 'Kaunching'	wBur
un	***oblonga*** 'Orange'	wRai oOEx
un	***oblonga*** 'Pineapple'	oAls oOEx
un	***oblonga*** 'Pineapple' dwarf	oGar
un	***oblonga*** 'Perfume'	oOEx
un	***oblonga*** 'Rich's'	oFor
un	***oblonga*** 'Smyrna'	wClo oOEx
un	***oblonga*** 'Van Damen'	wRai
	CYMBALARIA	
	aequitriloba	oSis oWnC wCoN
	hepaticifolia	wHer
	muralis	oFor oTrP wTho wThG oDar oAls wFai oCir wCoN
un	***muralis*** 'Globosa'	oGoo
	muralis 'Nana Alba'	oSis
	CYMBOPOGON	
	citratus	wWoS oAls wFGN oGoo wRai wHom oBar wSte
un	***flexuosus***	oGoo iGSc oCrm oBar wNTP oUps
	nardus	wFai iGSc wHom oCir
	CYNANCHUM	
un	***decaisnianum***	oRar
un	***forrestii***	wCol
	macrolobum Descoings 28263	oRar
	marnieranum	oRar
	sp.	oRar
	CYNARA	
	cardunculus	oAls oGar oGoo oGre wFai iGSc wHom oSec oUps
	cardunculus 'Cardy'	wGAc oOut
	cardunculus 'Florist Cardy'	oAls
	cardunculus Scolymus Group 'Green Globe'	wCSG oGar
	scolymus	see **C. *cardunculus*** Scolymus Group
	CYNOGLOSSUM	
	grande	oBos
	nervosum	oAls
	wallichii	last listed 99/00
	CYPELLA	
	coelestis	oTrP wHer
	plumbea	see **C. *coelestis***
	CYPERUS	
	albostriatus	oTrP
	albostriatus 'Variegatus'	oTrP oHug oUps
	alternifolius	see **C. *involucratus***
	diffusus	see **C. *albostriatus***
un	***giganteus***	wGAc oHug oWnC
	haspan	see **C. *papyrus*** 'Nanus'
	involucratus	oTrP wGAc oHug oGar oSsd oWnC oUps
un	***involucratus*** 'Giganteus'	oGar
	involucratus 'Gracilis'	oTrP wGAc oHug oSsd oWnC
un	***involucratus*** 'Haspen Viviparous'	oTrP oHug oGar oSsd
un	***involucratus variegatus***	oHug oUps
	longus	wSoo wGAc oHug oSsd oWnC
	papyrus	oTrP wGAc oHug oSsd
	papyrus 'Nanus'	wGAc oHug
un	***strictus***	oUps
un	***strigosus***	oSsd
	CYPHOSTEMMA	
	juttae	oRar
un	***sandersonii***	oRar
	CYPRIPEDIUM	
	calceolus	wThG
	flavum	oRed
un	***formosanum***	oRed
un	***franchetii***	oRed
	henryi	oRed
	japonicum	oRed
	macranthum	oRed
	margaritaceum	oRed
un	***segawae***	oRed
	tibeticum	oRed
	CYRILLA	
	racemiflora	oFor oRiv oGre
	CYRTANTHUS	
	elatus	wGra
	hybrid red	wGra
	mackenii var. ***cooperi***	wGra
	obliquus	wGra
	speciosus	see **C. *elatus***
un	**CYRTOCARPA**	
	procera	oRar
	CYRTOMIUM	
	caryotideum	wFan oGar wNay oSqu oGre oBRG
	falcatum	oFor oAls wSte
	falcatum 'Rochfordianum'	wFol wFan oRus oSqu wTGN wSnq
	fortunei	oFor wFan oSqu oBRG
un	***fortunei*** var. ***intermedium***	wFol wCol wFan
un	***lonchitoides***	wFol
	macrophyllum	oRus
	macrophyllum mini	oRus
	CYSTOPTERIS	
	bulbifera	oRus
	fragilis	oRus
	CYTISUS	
	battandieri	oFor oGos oDar oEdg oSho oGre wSte wWel
	'Burkwoodii'	oFor wHer oGar oTPm oWnC
	decumbens	oFor
	'Dorothy Walpole'	oWnC
	'Hollandia'	oFor oAls oBlo
	'La Coquette'	oFor
	'Lena'	oAls
	'Moonlight'	oFor oAls oGar oBlo oWnC
	nigricans	last listed 99/00
	x praecox	see **C. *x praecox*** 'Warminster'
	x praecox 'Allgold'	oFor
	x praecox 'Warminster'	oTPm
	purpureus	see **CHAMAECYTISUS *purpureus***
	Red Favorite	see C. 'Roter Favorit'
	red hybrid	oUps
	'Roter Favorit'	oFor
	scoparius ssp. ***maritimus***	last listed 99/00
	scoparius var. ***prostratus***	see **C. *scoparius*** ssp. ***maritimus***
	spachianus	see **GENISTA *x spachiana***
	'Zeelandia'	wHer
	DABOECIA	
	cantabrica	oGar wCoN
	cantabrica f. ***alba***	oFor wHea oAls oGar wTGN
	cantabrica 'Arielle'	wHea
	cantabrica 'Atropurpurea'	oFor oGar
	cantabrica 'Atropurpurea' seedling	wHea
	cantabrica 'Bicolor'	wHea

	cantabrica 'Cinderella'	wHea
	cantabrica 'Covadonga'	wHea
	cantabrica 'Creeping White'	last listed 99/00
	cantabrica 'Globosa Pink'	wHea
	cantabrica 'Harlequin'	wHea
	cantabrica 'Lilacina'	wHea
	cantabrica 'Pink'	oGar
	cantabrica 'Polifolia'	wHea
	cantabrica 'Praegerae'	wHea
va	***cantabrica*** 'Rainbow'	wHea
	cantabrica ssp. ***scotica*** 'William Buchanan'	wHea oAls oGar
	cantabrica ssp. ***scotica*** 'William Buchanan Gold'	oFor
un	***cantabrica*** 'Silver Bells'	wHea
	cantabrica 'Waley's Red'	wHea
	cantabrica 'Wijnie'	wHea
	x scotica	see **D. *cantabrica*** ssp. ***scotica***
	DACRYCARPUS	
	dacrydioides	oFor wHer
	DACRYDIUM	
	bidwillii	see **HALOCARPUS *bidwillii***
	cupressinum	oFor
	franklinii	see **LAGAROSTROBOS *franklinii***
	DACTYLORHIZA	
	fuchsii	oRed
	DAHLIA	
	'A La Mode'	wCon oFre oSwa
	'Abridge Natalie'	wCon
	'Alabaster'	oFre
	'Alena Rose'	wCon
	'Ali Oop'	oSwa
un	'Alice Underdahl'	wCon
un	'Alise'	wCon
	'Aljo'	wCon
	'All Triumph'	wCon oFre oSwa
	'Alloway Candy'	wCon
	'Almand Joy'	wCon wSea oFre oSwa
	'Almand Supreme'	wSea
	'Alpen Cardinal'	oFre
	'Alpen Cherub'	wCon
	'Alpen Jean'	wCon
	'Alpen Magic'	wSea
	'Alpen Snowflake'	wSea
	'Alpowa'	wCon
	'Alva's Lilac'	wSea oFre
	'Amanda Jarvis'	oFre oSwa
	'Amber Queen'	oSwa
	'Amorangi Joy'	wCon
	'Amy K'	wCon
	'Anatol'	last listed 99/00
	'Andrea'	oSwa
	'Andrew David'	wCon
un	'Andy D'	wCon
	'Angel Face'	oGar
	'Angel's Dust'	oSwa
	'Anglian Water'	wCon
un	'Anna Mari'	wCon
	'Apple Blossom'	oSwa
	'Apricot'	wSea
	'Apricot Parfait'	wSea
	'April Dawn'	wCon wSea oSwa
	'Arabian Night'	wSea oFre oSwa
	'Asahi Chohji'	wCon
	'Aslan'	wCon
	'Athalie'	wSea
un	'Aurora'	wCon
	'Awaikoe'	oFre oSwa
	'B. J. Tillotson'	wSea
	'B-Man'	oSwa
	'Baarn Bounty'	oFre oSwa
	'Baby Red'	oSwa
un	'Baby Yellow'	oSwa
un	'Baha Sun'	wSea
	'Ballerina'	oSwa
	'Bambino'	oSwa
	'Barbara Schell'	oFre
	'Barbarry Ball'	wCon wSea
	'Barbarry Banker'	wCon wSea
	'Barbarry Dominion'	last listed 99/00
	'Barbarry Esquire'	wCon
	'Barbarry Gem'	wSea
	'Barbarry Indicator'	wCon
	'Barbarry Pinky'	last listed 99/00
	'Bashful'	oSwa
	'Be-a-Sport'	wSea oSwa
	'Be Bop'	oSwa
	'Becca'	wSea
	'Bedford Blush'	oSwa
	'Bednall Beauty'	wHer oNWe
	'Bellefleur'	wCon
	'Berwick Wood'	wCon
	'Beryl K'	wCon
	'Betty Anne'	oSwa
	'Betty Bowen'	wSea
	'Beulah Ruth'	wCon
	'Bingo'	wCon wSea
un	'Binky'	oFre oSwa
	'Bishop of Llandaff'	wHer wWoS wCon wSea oSwa oHed oGar oNWe oGre
	'Bitsa'	wSea
	'Bitsy'	oSwa
	'Black Narcissus'	wSea oFre
	'Black Satin'	oSwa
	'Blackie'	oFre
un	'Blackie Not-So-Black'	oFre
	'Blended Beauty'	wCon
	'Bliss'	wCon wSea oSwa
	'Bloom's Betty'	wCon
	'Bloom's Graham'	wCon wSea
	'Bloom's Irene'	wCon wSea
	'Bloom's Kehala'	wCon
	'Bloom's Wildwood'	last listed 99/00
	'Bloom's XL'	wCon
un	'Blue Star'	oFre
	'Blue Streak'	wCon
	'Bo-de-o'	wSea
	'Bold Accent'	oSwa
	'Bonaventure'	wSea
un	'Bonne Easter'	wSea
	'Bonne Esperance'	wSea oFre oSwa
un	'Boone'	wCon
un	'Border Choice'	oSwa
	'Brandon James'	oFre oSwa
	'Brian R'	wCon
	'Brian Ray'	oSwa
	'Bright Star'	wSea oFre oSwa
	'Britt's Choice'	wCon
	'Brooke Nicole'	oSwa
	'Brookside Cheri'	oFre oSwa
	'Brookside J. Cooley'	wSea
	'Brookside Snowball'	wSea oFre oSwa
	'Brushstrokes	wCon oSwa
	'Bubblegum'	oSwa
	'Buffie G'	oSwa
	'Burma Gem'	oFre oSwa
	'Cabo Bella'	wSea
	'Cabo Perla'	wSea
	'Caboose'	oSwa
	'Camano Ariel'	wCon
	'Camano Messenger'	wCon
	'Camano Thunder'	wCon
	'Cameo'	oSwa
un	'Cameo Lavender'	oFre
	'Cameo Peach'	oSwa
	'Canby Centennial'	oSwa
	'Candy Cane'	wCon oSwa
un	'Candy Red'	oFre
	'Cap'n Herschel'	wSea
	'Caproz Jerry Garcia'	wSea
	'Caproz My Maria'	wSea
	'Cara Tina'	wCon
un	'Carl Chilson'	wCon
	'Catherine Remus'	oFre
	'Center Court'	oSwa
	'Cha Cha'	oFre oSwa
	'Chapron'	wCon
	'Charlie Too'	last listed 99/00
	'Chee'	oFre
	'Cheerio'	wCon
un	'Cherry Drop'	oFre oSwa
un	'Cherry Orange'	oFre
	'Cherry Wine'	wSea
	'Cherubino'	oFre oSwa
	'Chessy'	last listed 99/00
	'Cheyenne'	wCon oFre oSwa
	'Cheyenne Chieftain'	wCon
	'Chickadee'	oSwa
	'Chilson's Pride'	wCon wSea oFre oSwa
	'Chimacum Topaz'	wCon

	'China Doll'	oSwa
un	'Chou'	oSwa
	'Christmas Carol'	oFre
	'Christmas Star'	oFre
	'Clara Marie'	oFre
	'Classic A 1'	wCon
	'Clint's Climax'	wCon oFre
	'Clyde's Choice'	oSwa
	'Co Co'	wCon
	'Color Magic'	wSea
	'Comet'	wCon
	'Connecticut Dancer'	wSea
	'Copper Glow'	wSea
	'Coral Baby'	oSwa
	'Coral Frills'	oFre oSwa
	'Coral Gypsy'	oSwa
	'Coralie'	wSea
	'Cornel'	last listed 99/00
	'Crazy Legs'	oFre oSwa
	'Creve Coeur'	wCon wSea
	'Crichton Cherry'	wCon
	'Crichton Honey'	wSea oFre oSwa
	'Crossfield Ebony'	wCon oFre oSwa
	'Croydon Ace'	oFre
	'Crystal Ann'	wCon
	'Cuddles'	oSwa
	'Curly Flame'	wCon
	'Curly Que'	oFre oSwa
	'Cynthia Louise'	oSwa
	'Czar Willo'	wCon oFre
	'Daisy'	wSea
	'Daleko Adonis'	wCon oSwa
	'Dana'	wCon
	'Dana Iris'	wSea
	'Daniel Edward'	oFre oSwa
	'Danjo Doc'	wCon
	'Danum Cupid'	wCon
	'Dare Devil'	oSwa
	'Dark Magic'	wCon wSea oSwa
	'Dark Prince'	wCon
	'Dauntless'	wCon wSea
	'Davenport Honey'	wCon
	'Deidra K'	wSea
	'Deliah'	wCon
un	'Denishka'	wSea
un	'Denver Nugget'	wSea
	'Derek Sean'	wCon oSwa
	'Desiree'	wCon
	'Devon Joy'	oFre
un	'Dire Devil'	oSwa
	'Dizzy'	oSwa
	'Donny C'	oFre oSwa
	'Don's Delight'	wCon
un	'Dori'	wSea
	'Double Trouble'	oFre oSwa
	'Dr. Les'	oFre oSwa
	'Drummer Boy'	wSea
	'Duet'	wCon wSea oFre oSwa
	'Duke of Earl'	wCon
	'Dustin Williams'	oSwa
	'Dutch Baby'	wCon wSea
	'Edinburgh'	last listed 99/00
	'Edith Mueller'	oSwa
un	'Edna'	wSea
	'Edna C'	oSwa
	'Eldon Wilson'	wCon
	'Eleanor'	oSwa
	'Elizabeth Hammett'	wCon wSea
	'Elizabeth Marie'	wCon
un	'Ella's Cream Puff'	oFre
	'Ellen Henry'	wCon
	'Ellen Huston'	wHer oFre oGar
	'Elma Elizabeth'	wCon wSea
	'Elmbrook Rebel'	wCon
	'Elsie Huston'	wSea
	'Elvira'	wSea
	'Emma's Coronet'	wCon
	'Emory Paul'	wCon oSwa
	'Enrique'	wSea
	'Envy'	oSwa
	'Erin Ann'	wSea
	'Excentric'	oSwa
	'Fairy Queen'	oFre
	'Fairy Tale'	oSwa
	'Falcon's Future'	wSea
	'Fantasia'	wCon
	'Fascination'	wCon oSwa
un	'Fascination II'	wCon
	'Fatima'	wCon oFre oSwa
	'Fidalgo Beauty'	wCon
	'Fidalgo Blacky'	wCon
	'Fidalgo Butterball'	wCon
	'Fidalgo Clown'	wCon
	'Fidalgo Julie'	wSea
	'Fidalgo Lisa'	wCon
	'Fidalgo Magic'	wSea
	'Fidalgo Nugget'	wCon
	'Fidalgo Snowman'	wCon
	'Fidalgo Supreme'	wCon
	'Fidalgo White Mafolie'	wCon
	'Figurine'	wSea
	'Fiona Stewart'	wSea
un	'Fire Glow'	oFre
	'Fire Magic'	oFre oSwa
	'First Kiss'	oSwa
	'First Love'	last listed 99/00
	'Flying Saucer'	oFre
	'Folklorico'	wSea
	'Fool's Gold'	oSwa
	'Forncett Furnace'	wHer
	'Foxy Lady'	oFre oSwa
	'Frank Holmes'	oFre oSwa
	'Frank Hornsey'	wSea
	'Frau Louise Mayer'	oSwa
	'Free Spirit'	oSwa
	'French Doll'	oSwa
	'Frigoulet'	wCon
un	'Funfair'	wCon
	'Funny Face'	oFre oSwa
	'Fuzzy Wuzzy'	oSwa
	'Gargantuan'	wCon
	'Gary'	oFre
	'Gay Princess'	wCon oSwa
	'Geerlings Yellow'	wCon
	'Gerrie Hoek'	oSwa
un	'Gerrie Scott'	wCon
	'Gertrude Martinck'	oSwa
un	'Giant Choh'	wSea
	'Giggles'	oSwa
un	'Gil's Pastel'	wCon
	'Ginger Willo'	wCon oFre
	'Gingeroo'	oSwa
	'Gitt's Attention'	oSwa
	'Gitt's Perfection'	oSwa
un	'Glasscock'	wCon
	'Glen Echo'	wCon
	'Glenbank Joy'	wSea
	'Glenbank Twinkle'	oSwa
	'Glenplace'	wCon oFre
	'Glorie van Heemstede'	wSea
	'Gold Ball'	oFre oSwa
un	'Gold Brandon'	oFre
un	'Gold Cactus'	oFre
	'Golden Egg'	oSwa
	'Golden Heart'	wCon
	'Golden Years'	wCon
un	'Goldwing'	wSea
	'Gonzo Grape'	oSwa
	'Goodwill 90'	wCon
	'Grace Rushton'	wCon
	'Grand Prix'	last listed 99/00
	'Grand Willo'	wCon
un	'Great White'	wHer oGar
	'Gregory Stephen'	oSwa
	'Grenidor Pastelle'	wSea
	'Gretchen Ann'	wCon
	'Gypsy Kiss'	wSea
	'Hakuyoh'	oSwa
	'Hallmark'	wSea
	'Hamari Accord'	wCon
un	'Hamilton Candy'	wCon
	'Hamilton Lillian'	wCon wSea
	'Happy Face'	oSwa
	'Harriet Collins'	wCon
un	'Harvest Moon'	wCon
	'Hayley Jayne'	wSea
	'He Man'	wSea
	'Heather Marie'	oFre oSwa
	'Hee Haugh'	oFre oSwa
	'Helen Richmond'	oSwa

DAHLIA

	'Herbert Smith'	oSwa
un	'Heronista'	wHer
	'Hi Lite'	oFre oSwa
	'Hiliary Dawn'	wCon
	'Hillcrest Amour'	wSea
	'Hillcrest Delight'	wCon
	'Hillcrest Fiesta'	wCon
	'Hillcrest Hillton'	wCon
	'Hissy Fitz'	oSwa
	'Honey Dew'	oSwa
	'Honka'	wCon oFre
	'Hot Stuff'	oSwa
un	'Hot Tin Roof'	oFre
	'Houtaikoh'	oSwa
	'Hulin's Carnival'	wSea oFre oSwa
	'Idaho Red'	wCon
	'Ike'	wSea
	'Imperial Wine'	oSwa
	imperialis 'Alba'	last listed 99/00
	'Indian Sunset'	oSwa
	'Inflammation'	oSwa
	'Inland Dynasty'	wSea
un	'Inland National'	oFre
	'Innocence'	wCon wSea oSwa
	'Intensity'	wCon
	'Iola'	wCon
	'Irene Florence'	wCon
	'Irene's Pride'	wCon
	'Island Blaze'	wCon
	'Island Dawn'	wSea
	'Islander'	last listed 99/00
	'Ivory Palaces'	wCon
	'J J Hazel'	wCon
	'Jabberbox'	oSwa
	'Jack O' Lantern'	oFre oSwa
	'Jackie S'	wCon
	'Jamboree'	oSwa
un	'James Albin'	oSwa
un	'Janet Jean'	wCon
un	'Janna'	wSea
	'Japanese Bishop'	wSea oFre oSwa
	'Japanese Waterlily'	wCon
	'Jazzy'	oSwa
	'Jean Enersen'	wSea
	'Jenna'	oSwa
	'Jennie'	wCon
	'Jennifer Lynn'	oSwa
un	'Jennifer 2'	oFre
	'Jerita'	wCon
	'Jersey's Beauty'	wSea
	'Jescot Julie'	last listed 99/00
	'Jessie G'	wCon wSea
	'Jil'	oSwa
	'Jitterbug'	oFre oSwa
un	'Jo'	wCon
un	'Jo Anna Pettit'	oSwa
un	'Jocelyn G'	wCon
	'Joe K'	wSea
	'Johann'	wCon
	'John Bramlett'	wSea
	'Jorja'	wCon
	'Joshua P'	wSea
	'Joy Ride'	oSwa
un	'Joyce'	oFre
	'Joyce Marie'	wCon
	'Juanita'	wSea oFre oSwa
un	'Julia Pendly'	wCon
	'Just Peachy'	wCon wSea
un	'Just So'	oSwa
	'Justin'	wCon
	'Kalalock Sunset'	last listed 99/00
	'Kasasagi'	wCon oSwa
un	'Kasuga'	oFre oSwa
	'Katherine Temple'	wCon
un	'Kathy Parker'	wSea
	'Keewatin Pioneer'	wCon
	'Keith H'	wCon wSea
un	'Keiwera Gold'	wSea
	'Kellie Ann'	wCon
un	'Kelly'	wSea
	'Kenora Canada'	wSea
	'Kenora Challenger'	wCon
	'Kenora Christmas'	wSea
	'Kenora Fireball'	wSea
	'Kenora Lisa'	wCon wSea
	'Kenora Ontario'	wSea
	'Kenora Sunburst'	last listed 99/00
	'Kenora Valentine'	wSea
	'Kenora Wildfire'	wSea
	'Ken's Flame'	wCon
	'Ken's Rarity'	wCon
	'Keri Blue'	wCon
un	'Keri Dancer'	wCon
un	'Keri Fruit Salad'	wCon
	'Keri Quill'	wCon
	'Keri Smokey'	wCon
	'Kevin N'	last listed 99/00
	'Kidd's Climax'	wCon oSwa
	'Kingston'	wCon
un	'Kingston Queen'	wHer
	'Koinonia'	wCon
	'Koko Puff'	oSwa
	'Koppertone'	oSwa
	'Kracker Jac'	oSwa
	'La Paloma'	wSea
	'Lady Esther'	oFre
	'Lady in Red'	wSea
	'Lady Lael'	wCon
	'Lady Linda'	wSea oFre oSwa
	'L'Ancresse'	oSwa
un	'L'Andresse'	wCon
	'Last Dance'	oFre oSwa
	'Lauren Michelle'	wCon wSea oSwa
	'Lavender Athalie'	wSea
	'Lavender Chiffon'	oFre oSwa
	'Lavender Freestyle'	wSea
un	'Lavender Kerkrade'	wSea
	'Lavender Line'	wCon
un	'Lavender Mom'	oFre
	'Lavender Ruffles'	oSwa
	'Lawrence Welk'	oSwa
	'Lemon Candy'	oSwa
	'Lemon Cane'	wCon
	'Lemon Kiss'	oSwa
	'Lemon Meringue'	wCon
	'Lemon Shiffon'	oSwa
	'Lemon Swirl'	wCon
	'Lemon Tart'	oSwa
	'Lil Buddy'	wCon
	'Lilac Mist'	oSwa
un	'Lillian Stewart'	wCon
	'Lindy'	oSwa
	'Lisa'	wCon oFre
	'Little J'	wCon
	'Little Lamb'	oFre oSwa
	'Little Laura'	wCon
	'Little Magic'	oFre oSwa
	'Little Matthew'	last listed 99/00
	'Little Scottie'	wCon
	'Little Showoff'	wCon
	'Little Snowdrop'	oFre
	'Lois V'	last listed 99/00
	'Lollipop'	oFre oSwa
	'Lorilli Dawn'	wCon
	'Love Potion'	oSwa
	'Ludwig Helfert'	oSwa
	'Lula Pattie'	wCon wSea oSwa
	'Lune de Cap'	wCon
	'Lupin Bernie'	oFre
	'Lupin Candy'	oFre
	'Lutt Witchen'	oFre
	'Lydia Suckow'	wCon
un	'Maarn'	oSwa
	'Madame de Rosa'	wSea
	'Madelaine Ann'	wCon
	'Magic Moment'	wSea oSwa
	'Magically Dun'	oSwa
	'Magnificat'	wSea
	'Maisie Mooney'	wCon
	'Majestic Athalie'	wSea
	'Maki'	oSwa
un	'Mantequilla'	wSea
	'Mardy Gras'	wCon oSwa
	'Margaret Duross'	oSwa
	'Margaret Ellen'	oSwa
	'Marie Schnugg'	wCon oFre
un	'Marika'	wCon
	'Mark Lockwood'	wCon
	'Marlene Joy'	wCon wSea oFre oSwa
	'Marmalade'	oSwa

	'Marshmellow Sky'	wCon
	'Martin's Yellow'	last listed 99/00
un	'Marvelous Mel'	wCon
un	'Marvelous Sal'	wCon
	'Mary'	wCon
	'Mary Evelyn'	oFre oSwa
	'Mary Hammett'	wSea
	'Mary J'	wCon
	'Mary Jennie'	wCon oSwa
	'Mary Jo'	wSea oSwa
	'Mary Lee McNall'	oSwa
	'Mary Morris'	wCon
	'Mary Munns'	oSwa
	'Master David'	wCon
	'Master Michael'	oFre
	'Match'	oFre
un	'Matchless'	wCon
	'Matchmaker'	wCon wSea oFre oSwa
	'Matilda Huston'	oFre oSwa
	'Mauve Climax'	wCon
	'Melinda Jane'	wCon
	'Melissa M'	oSwa
	merckii	wHer oHed
	merckii 'Edith Eddleman'	last listed 99/00
	'Michael C'	wSea
	'Michelle K'	wCon
un	'Michelle Mignot'	oSwa
	'Mickey'	oSwa
	'Midnight Dancer'	oSwa
	'Midnight Moon'	oSwa
un	'Midnight Sun'	oFre
un	'Mies'	oSwa
un	'Miranda'	oFre
	'Miss Muffit'	wSea oFre
	'Miss Rose Fletcher'	wCon oSwa
	'Mom's Special'	wCon oFre
	'Mona Lisa'	wCon
	'Monkstown Diane'	last listed 99/00
	'Moonfire'	wHer
	'Moonglow'	wCon
	'Moonstruck'	oSwa
	'Moray Susan'	wCon
	'Morgenster'	wCon oFre
	'Morning Star'	wCon
	'Mr. Brett'	wCon
	'Mr. Larry'	oSwa
	'Mrs. Santa Claus'	wSea
	'Mrs. T'	wSea
un	'Muriel'	wCon
	'Murray Petite'	wCon
	'My Joy'	oFre oSwa
	'Mystique'	oSwa
	'Nagel's Solidite'	wSea
	'Narnia'	wCon
	'Needles'	oSwa
	'Nenekazi'	wCon
	'Neon Splendor'	oSwa
	'Nepos'	wCon
	'Nettie'	wCon oFre oSwa
	'New Greatness'	wSea
	'Nick's Pick'	oSwa
	'Nicky K'	oSwa
	'Nicola Higgo'	wSea
	'Night Editor'	wCon
	'Nijinski'	wSea oFre oSwa
	'Nita'	oFre
	'Noreen'	wCon wSea
	'Normandy Splendor'	wSea
	'Northern Lights'	wSea
	'Northland Primrose'	wCon
	'Nutley Sunrise'	wSea oFre oSwa
	'Obsession'	oSwa
un	'Ollie White'	wCon
	'Omnibus'	wCon
	'Optic Illusion'	oFre oSwa
	'Orange Cushion'	oFre
	'Orange Jewel'	wCon
	'Orange Julius'	oFre oSwa
un	'Orange Mode'	oFre
un	'Orange Santa'	oFre
	'Orchid Lace'	oFre oSwa
	'Oregon Reign'	oSwa
	'Orkney'	wCon
	'Otto's Thrill'	oFre
	'Papageno'	oSwa

	'Pape's Pink'	wSea
	'Parakeet'	oSwa
	'Pari Taha Sunrise'	oFre oSwa
	'Park Princess'	wSea oFre oSwa
	'Paroa Gillian'	wCon
	'Party Girl'	wCon
	'Passion'	wCon
	'Pat Fearey'	wCon
	'Patches'	oFre oSwa
	'Pat'n'Dee'	wCon
	'Patty'	oVBI
	'Patty Cake'	oSwa
	'Paul Smith'	wCon wSea
	'Pazazz'	oSwa
	'Peaches'	oFre oSwa
un	'Peachy'	oFre
	'Pee Gee'	wCon
	'Peek-a-boo'	oSwa
	'Pennsgift'	last listed 99/00
	'Pensford Marion'	wCon
	'Peppermint'	oFre
	'Peppermint Candy'	oFre
	'Pineapple Lollipop'	wCon wSea
	'Pinelands Pal'	wCon
	'Pinelands Pam'	wCon
	'Pinelands Princess	wCon
un	'Pink Amanda'	oFre
un	'Pink Ball'	oSwa
un	'Pink Blush'	oFre
	'Pink Carnival'	oFre
	'Pink Gingham'	oSwa
	'Pink Kerkrade'	wSea
	'Pink Mona'	wCon
	'Pink Orchid'	oFre
	'Pink Parfait'	oFre oSwa
	'Pink Pastelle'	wCon
	'Pink Satin'	oSwa
	'Pink Shirley Alliance'	wCon wSea oFre
	'Piperoo'	oFre
	'Pipsqueak'	oSwa
	'Playboy'	wSea oSwa
	'Plum Pretty'	oSwa
	'Pocrates'	wCon
	'Pooh'	oSwa
	'Prime Minister'	oFre
	'Prime Time'	oSwa
	'Prince Valiant'	wCon
un	'Princess Debbie'	oFre
	'Prinses Beatrix'	oFre
	'Puget Sparkle'	wCon
un	'Purple Debbie'	oFre
	'Purple Gem'	oFre oSwa
	'Purple Haze'	oSwa
	'Purple Imp'	oFre oSwa
	'Purple Royalty'	wSea
	'Purple Splash'	wSea
	'Purple Taiheijo'	oFre oSwa
	'Randi Dawn'	oSwa
	'Rare Beauty'	wCon
	'Raz-Ma-Taz'	oFre oSwa
	'Rebecca Lynn'	oFre oSwa
	'Red Garnet'	oSwa
un	'Red Pygmy'	oSwa
un	'Red Ringo'	oFre
	'Red Velvet'	wSea oFre
un	'Red Warrior'	wSea oFre
	'Redd Devil'	oSwa
	'Reddy'	oFre
	'Rheinderen'	wSea
	'Richard Rodgers'	wCon
	'Ringo'	wCon oFre oSwa
	'Rip City'	wSea oFre oSwa
	'Ripples'	oSwa
	'River Road'	wCon wSea
	'Robert Too'	wCon
	'Rock a Bye'	oSwa
	'Romance'	oFre oSwa
un	'Roodkapje'	oSwa
	'Rosalie K'	oFre
	'Rose Toscano'	wSea
	'Rosemary Webb'	wCon wSea oFre
un	'Rosita'	wCon oSwa
un	'Rosy Morn'	wCon
	'Rosy Wings'	oSwa
	'Rothesay Reveller'	wSea

	'Rothesay Superb'	oSwa
	'Roundabout'	oSwa
	'Rowdy'	oSwa
	'Ruby Red'	wCon
	'Ruddy'	wCon
	'Ruskin Diane'	wSea
	'Ruskin Gypsy'	oSwa
	'Ruth Elaine'	wCon
	'Ruthie G'	wSea
	'Ryan C'	oSwa
	'Rynfou'	oSwa
	'Salmon Beauty'	wCon
	'Sam Herst'	oSwa
	'San Luis Rey'	oSwa
	'Santa Claus'	wCon
un	'Santa Rosalia'	wSea
un	'Sarah Ann'	wCon
	'Sassy'	wSea oFre
un	'Scarborough Ace'	wSea
un	'Scarborough Brilliant'	wCon
	'Scarborough Centenary'	wSea
	'Scura'	wSea
	'Sea-Amarillo'	wSea
	'Sea-Baile'	wSea
un	'Sea Breeze'	oFre
	'Sea-Canela'	wSea
	'Sea-Cascabel'	wSea
	'Sea-Chiquita'	wSea oFre
	'Sea-Clown'	wSea
	'Sea-Electra'	wSea
	'Sea-Encantadorita'	wSea
	'Sea-Fuego'	wSea
	'Sea-Gesche'	wSea
	'Sea-Helado'	wSea
	'Sea-Miss'	wSea
	'Sea-Nugget'	wSea
	'Sea-Oro'	wCon wSea
	'Sea-Padre Jim'	wSea
	'Sea-Precious'	wSea
	'Sea-Rosebud'	wSea
	'Sea-Sparrow'	wSea
	'Sea-Sprite'	wSea
	'Seduction'	oSwa
	'Senior Ball'	wCon
	'September Morn'	oSwa
	'Serenade'	oSwa
	'Shadow Cat'	wCon
un	'Shannon'	oVBI
	'Sharon Ann'	wSea
	'Sheabird'	wCon
	'Sheba'	wCon
	'Sheila Mooney'	wCon
	'Sherwood's Peach'	oFre oSwa
	'Shirley Alliance'	oFre
	'Show 'N' Tell'	wCon wSea oFre oSwa
	'Show Off'	oSwa
	'Siemen Doorenbos'	oFre oSwa
	'Silhouette'	oSwa
	'Silver Slipper'	wSea
	'Silverado'	oSwa
	'Simplicity'	oSwa
	'Sir Rom'	wCon
un	'Sisa'	oSwa
	'Sister Fulwood'	oSwa
	'Skyrocket'	wSea
	'Skywalker'	oFre oSwa
	'Small World'	last listed 99/00
	'Smokey Gal'	oFre
	'Sneezy'	oSwa
	'Snickerdoodle'	oSwa
	'Snoho Beauty'	wCon
	'Snoho Betty'	wCon wSea
	'Snow Country'	oSwa
	'Snow White'	wCon
	'Snowbound'	oSwa
	'Snowflake'	oSwa
	'So Dainty'	wCon
	'Sonja Heinie'	wCon
un	'Sonny Boy'	wCon
	'Sparkler'	oSwa
	'Spartacus'	wCon wSea oSwa
	'Spellbreaker'	oSwa
	'Splitfire'	oSwa
	'Sporty Cheerio'	wCon
un	'Sporty Tyler'	wCon
	'Staci Erin'	last listed 99/00
	'Stacy Rachelle'	oFre oSwa
	'Star Lite'	oFre oSwa
	'Starbrite'	wCon
	'Starchild'	oFre oSwa
un	'Stella'	oFre oSwa
	'Stella J'	wCon
	'Stellyvonne'	wCon
	'Sterling Silver'	wCon wSea oFre oSwa
	'Stoneleigh Cherry'	wCon
	'Stoneleigh Joyce'	wCon
	'Sugar Cane'	wCon oSwa
	'Summer's End'	wCon oSwa
	'Sun Burst'	oFre
	'Sundown'	oSwa
	'Sunny'	oSwa
	'Sunstruck'	oFre oSwa
	'Surprise'	oSwa
	'Susan Willo'	wCon oFre
	'Swan Lake'	oFre
	'Swan's Desert Storm'	wSea oSwa
	'Swan's Discovery'	oSwa
	'Swan's Glory'	oSwa
	'Swan's Gold Medal'	oFre oSwa
	'Swan's Olympic Flame'	oSwa
	'Swan's Sunset'	oSwa
	'Sweet Dreams'	oFre oSwa
	'Sweet Sixteen'	wCon
	'Tahiti Sunrise'	wCon wSea oSwa
	'Tally ho'	oNWe
	'Tamara Marston'	wCon
	'Tanjoh'	oSwa
	'Tartan'	wSea
	'Tasogare'	wSea oSwa
un	'Teamarie Butterscotch'	wCon
	'Ted's Choice'	wSea oSwa
	'Tempest'	oSwa
	'Tennessee Rose'	wSea
	tenuicaulis	wHer
	'Tequilla Sunrise'	oSwa
	'Texas Wild Thang'	wSea
	'Thelma Clements'	wSea
	'Thomas A. Edison'	oSwa
	'Thriller'	oSwa
	'Tickled Pink'	oSwa
	'Tiffany Lynn'	oFre
	'Tijuana'	wSea
	'Tiny Dancer'	oSwa
un	'Tiny May'	wCon
	'Todd H'	wCon
	'Tom Yano'	last listed 99/00
	'Tomo'	wCon
	'Tootsie'	wCon
	'Top Honor'	wCon
	'Topmix Violetta'	wSea oFre
	'Touch of Class'	oSwa
	'Touche'	oSwa
	'Trengrove Tauranga'	wSea
	'Trevor'	wCon
	'Tropic Sun'	oSwa
	'Tui Orange'	wSea
	'Tweedie Pie'	oFre oSwa
	'Twinkle Toes'	wSea
un	'Twinks'	oSwa
	'Uchuu'	oFre
	'Union Gap'	wCon
un	'Ursla Senzone'	wCon
un	'Utrecht'	oFre
	'Utmost'	wCon
un	'Velda Inez'	wCon
un	'Vera Seyfang'	oSwa
	'Vera's Elma'	wCon
	'Vernon Rose'	wCon wSea
	'Vicky Jackson'	wCon
un	'Violetta'	wSea
	'Vivian B'	wCon
	'VJ'	wCon
	'Voodoo'	oSwa
	'Walter Hardisty'	wSea oFre oSwa
	'Wanda's Capella'	wSea oFre
	'Watercolors'	oSwa
un	'Weigen'	oSwa
	'Wendy's Place'	wCon
	'Western Gal'	oFre
	'Western Goldtip'	oFre

un	'What's It'	oFre
	'Wheels'	oSwa
	'White Alva's'	wSea
	'White Ballet'	wSea
un	'White Carnival'	oFre
	'White Fawn'	oSwa
	'White Linda'	wSea
	'White Lilliput'	oFre
	'White Nettie'	oFre oSwa
	'White Rustig'	wCon
	'Who Dun It'	oSwa
	'Wicky Woo'	wCon oFre oSwa
	'Wildcat'	oSwa
	'Wildman'	wCon oSwa
	'Wildwood Glory'	wCon
	'Wildwood Marie'	wCon wSea
	'Wildwood Swirls'	wSea
	'Willo Borealis'	wCon
	'Willo Violet'	oFre
	'Wine & Roses'	wCon
	'Winnie'	wSea oFre
un	'Winter Ice'	oFre
	'Worton Anne'	wSea
	'Worton Bluestreak'	wSea oSwa
	'Wowie'	oSwa
	'Yara Falls'	wCon
	'Yellow Bird'	wCon oFre oSwa
un	'Yellow C'	oFre
	'Yellow Climax'	oSwa
	'Yellow Hammer'	wHer
un	'Yoro Kobi'	oSwa
	'Yvonne'	wCon wSea oVBI
un	'Zakuro'	wSea
	'Zorro'	wSea oFre oSwa
	DAISWA	see **PARIS**
	DAMNACANTHUS	
	indicus	wHer
	DANAE	
	racemosa	wHer
	DAPHNE	
	alpina	oNWe wMtT
	arbuscula	oSis
	bholua	wHer oGos
	bholua Darjeeling form	oCis
	x burkwoodii	wCoN
un	***x burkwoodii*** 'Briggs Moonlight'	oGos
va	***x burkwoodii*** 'Carol Mackie'	oFor wWoS oGos wClo oAls wCSG oGar oSis oGre oWnC wCoN
	x burkwoodii 'Lavenirei'	oSis
un	***x burkwoodii*** 'Silveredge'	oGre
	x burkwoodii 'Somerset'	oFor wWoS oGos wClo oAls oGar oSho oGre oWnC wCoN wWel
	caucasica	oFor oSis wWel
	cneorum	oGar wCoN
	cneorum 'Eximia'	oSis
un	***cneorum*** 'Leila Haines'	oSis
	cneorum 'Major'	last listed 99/00
	cneorum var. ***pygmaea*** 'Alba'	oSis
	cneorum 'Rose Glow'	oFor oAls oSis oSho oGre oWnC wTGN
	cneorum 'Ruby Glow'	wSwC oGar wCoS
	cneorum 'Variegata'	wHer oSho wMtT oGre
	collina	see **D. *sericea*** Collina Group
	genkwa	wHer wCol oSis oRiv wCoN
	genkwa large flowered form	oSis
	giraldii	wCol
	jasminea Delphi form	oSis
	kosaninii	wMtT
	laureola	oNWe oWhS
	laureola ssp. ***philippi***	last listed 99/00
un	'Lawrence Crocker'	oFor wHer wCol oSis
	longilobata	last listed 99/00
	x manteniana	oSis
	mezereum	oFor wHer oDar wCSG oGar oSis oGre wAva wCoN wWel
	mezereum f. ***alba***	wHer
un	***mezereum*** var. ***alpina***	wMtT
	x napolitana	last listed 99/00
	odora	oNat oSho oGre wCCr oTPm oThr wCoN oUps wCoS wWel
	odora f. ***alba***	wCSG oWnC
	odora 'Aureomarginata'	oFor oGos oAls wCSG oGar oGre wPir oWnC wSte wWel
	odora var. ***leucantha***	see **D. *odora*** f. ***alba***
	odora 'Marginata'	see **D. *odora*** 'Aureomarginata'
	pontica	last listed 99/00
	retusa	see **D. *tangutica*** Retusa Group
	'Rossetii'	oSis
	sericea Collina Group	oSis
	tangutica	wHer wCol oNWe
	tangutica Retusa Group	oSis oNWe wMtT
	x thauma	oSis
	DAPHNIPHYLLUM	
	himalense var. ***macropodum***	oFor oGos oSho oGre wCCr oWnC wBox
	himalense var. ***macropodum*** HC 970612	wHer
	humile	last listed 99/00
	DARLINGTONIA	
	californica	wOud oGar
	DARMERA	
	peltata	wHer oHan wCul oMac wCol oAls oRus oAmb oGar wRob wNay wCoN
	peltata dwarf form	wHer
	DASYLIRION	
	longissimum	wCCr
	longissimum UBCG 94-0840	oCis
	sp.	oFor
	wheeleri	wHer oCis
	DASYPHYLLUM	
	diacanthoides	wHer
	DATISCA	
	cannabina	wSte
	DATURA	
	inoxia	oNWe oCrm wSte
un	***inoxia*** 'Alba'	oNat
	metel	oCrm
	meteloides	see **D. *inoxia***
	stramonium	iGSc
	white	wGra
	DAUCUS	
	carota	iGSc
	DAVALLIA	
	solida var. ***fejeensis***	oGar
	DAVIDIA	
	involucrata	oFor wHer oGos oEdg wCol oGar oDan oRiv oCis oGre oWnC wBox wCoN wSte wWel
un	***involucrata*** 'Sonoma'	wWel
	DEBREGEASIA	
un	***edulis***	oCis
	DECAISNEA	
	fargesii	oFor oGos oEdg oRiv oCis oSho oOEx oGre oWnC wBox wCoN wSte wWel
	fargesii DJHC 809	wHer
	DECUMARIA	
	barbara	oFor wHer oGos
	sinensis	oCis
	DEINANTHE	
	bifida	last listed 99/00
	bifida HC 970678	wHer
	DELONIX	
	regia	oRar
	DELOSPERMA	
	aberdeenense	oSis
un	***aberdeenense*** 'Abbey Snow'	oSis
	ashtonii	oFor oSis oCis
un	***basuticum*** DBG/NARGS 206	wMtT
	brunnthaleri	oSev
	brunnthaleri pink	oFor
	brunnthaleri yellow	oFor
	congestum	oJoy oSis
un	***congestum*** 'Gold Nugget'	oFor
	congestum PK 205	oCis
	congestum PK 206	oCis
	cooperi	oFor oSto oAls oAmb oGar oDan oSis oSev wFai oWnC wTGN oEga oUps oSle
	cooperi dwarf	oCis
	daveyi	oSis
	floribundum ex. Springfontein	oCis
un	***floribundum*** 'Starburst'	oFor oAls oSis
un	sp. aff. ***laevisae*** JJH 94-01655	oCis
un	***lavesii***	oSis
	lineare	wRob
	nubigenum	oFor oJoy oSto oGar oSis oSqu oWnC oUps oSle
un	***nubigenum*** 'Lesotho'	oSis
un	***obtusum***	oSis
un	***sphalmanthoides***	wMtT
	sutherlandii	last listed 99/00
	sp. aff. ***sutherlandii***	oJoy
	sutherlandii JJH 94-01638	oCis
	sp. DBG/NARGS206	wMtT

	sp. pink	oNWe
	sp. upright orange	oSto
	sp. upright pink	oSto
	sp. upright yellow	oSto
	sp. white	wMtT
	DELPHINIUM	
	'Alice Artindale'	wHer
	Astolat Group	wFGN wMag wTGN wAva wWhG
	x bellamosum	oWnC wTGN
	Belladonna Group	oUps
	Belladonna Group 'Capri'	last listed 99/00
	Belladonna Group 'Casa Blanca'	oWnC
	Belladonna Group 'Voelkerfrieden'	last listed 99/00
un	'Beverly Hills'	iArc
	Black Knight Group	oMis oWnC wMag wTGN iArc wAva wWhG
	Blue Bird Group	oGoo wMag wWhG
	'Blue Jay'	oWnC wMag
	'Blue Mirror'	oGar oSho oBlo oWnC
qu	'Butterfly'	wTGN
	cardinale	wWld
	cashmerianum	wHer
	ceratophorum CLD 893	wHer
	'Cherry Blossom'	oAls wFGN
	Connecticut Yankees Group	wWoS wHom
	'Dreaming Spires'	last listed 99/00
	dwarf blue	wFGN
	elatum	last listed 99/00
un	***elatum*** 'Blushing Bride'	wWoS oNat
	elatum 'Harlekijn'	oNat
un	***elatum*** 'Innocence'	wWoS oNat
	elatum mixed hybrids	oNat
qu	***elatum*** New Millenium hybrids	oNat
un	***elatum*** 'Royal Aspirations'	wWoS oNat
	English hybrids	oJoy wMag
	English hybrids blue	oNWe
	English hybrids pink	oNWe
	Galahad Group	oMis oGoo oWnC wMag wWhG
	grandiflorum 'Blue Butterfly'	wSwC oGar wHom oWnC wWhG
un	***grandiflorum*** 'Blue Elf'	oGar
qu	***grandiflorum*** 'Butterfly Compactum'	oUps
	grandiflorum PM 104	wHer
	Guinevere Group	oGoo oWnC wMag wTGN
	King Arthur Group	wTGN wWhG
	Magic Fountains Series	wTho oAls oGar oNWe wFai wHom oWnC iArc oCir wAva oEga wWhG oUps
un	'Magic Fountains Dark Blue' (Magic Fountains Series)	oAls oGar oWnC
un	'Magic Fountains Lavender' (Magic Fountains Series)	oAls oWnC
un	'Magic Fountains Lilac Pink' (Magic Fountains Series)	oAls oGar oWnC
un	'Magic Fountains Pure White' (Magic Fountains Series)	oAls oWnC
	'Magic Fountains Sky Blue' (Magic Fountains Series)	oAls oGar oWnC
un	'Magic Fountains White' (Magic Fountains Series)	oGar
	menziesii	oBos
	nudicaule	wWhG
un	***nudicaule*** 'Laurin'	wWhG
	nuttallianum	oBos
	Pacific Giants	see **D.** Pacific hybrids
	Pacific hybrids	wTho wCSG oGar wFai oEga oUps
	'Red Rocket'	wWoS
	sp. aff. ***requienii***	wCol
un	'Royal Blue'	oWhS
	x ruysii 'Pink Sensation'	wMag
un	'Sky Blue'	oSho
	sp. DJHC 496	wHer
	Summer Skies Group	wMag
un	'Summerfield Marjory x 'Apollo'	oNWe
	tatsienense	oJoy
	tricorne	wThG
	trolliifolium	oBos oAld
	trolliifolium blue	oNWe
	DENDRANTHEMA	**SEE** ***Chrysanthemum***
	DENDROCALAMUS	
un	***asper***	oTra
	latiflorus	see **BAMBUSA *oldhamii***
un	***membranaceous***	oTra
	strictus	oTrP oTra
	DENDROMECON	
	rigida	wCCı

	DENDROPANAX	
un	***morbifera*** HC 970313	wHer
	DENNSTAEDTIA	
	punctilobula	oRus wSte
	DENTARIA	
	diphylla	see **CARDAMINE *diphylla***
	DEPARIA	
	acrostichoides	oTri
	DESCHAMPSIA	
	cespitosa	oFor wCli wShR wCSG oSho wWat oWnC wPla oAld wWld wSnq
	cespitosa Bronze Veil	see **D.** 'Bronzeschleier'
	cespitosa 'Bronzeschleier'	wWoS wWin oTri oBRG wSnq
	cespitosa 'Fairy's Joke'	see **D. *cespitosa*** var. ***vivipara***
	cespitosa Golden Veil	see **D. *cespitosa*** 'Goldschleier'
	cespitosa 'Goldgehaenge'	oFor oJoy oGar oDan oSis oGre
	cespitosa 'Goldschleier'	oHed wBox
	cespitosa 'Northern Lights'	oFor wWoS wCli oGos oNat oJoy wCol wHig oAls oHed oGar oDan oSis wRob wWin wTGN wEde oUps wSte wSnq
	cespitosa 'Schottland'	oFor wSnq
	cespitosa 'Tautraeger'	last listed 99/00
	cespitosa var. ***vivipara***	oFor wCli oDar oUps
	flexuosa 'Aurea'	see **D. *flexuosa*** 'Tatra Gold'
	flexuosa 'Tatra Gold'	wWoS wCli wCol oGar oSis wWin wCCr wSnq
	DESFONTAINIA	
	spinosa	oGos oCis oBov
	spinosa HCM 98118	wHer
	DESMANTHUS	
	illinoensis	oFor
	DESMODIUM	
	canadense	oFor
	elegans	oFor oHed
	tiliifolium	see **D. *elegans***
	DEUTZIA	
	calycosa DJHC 98368	wHer
	corymbosa	oFor wHer wCul
	crenata	oFor wCoN
	crenata var. ***nakaiana***	oSis
	crenata var. ***nakaiana*** 'Nikko'	oFor wHer oNat oHed oAmb oGar oRiv oGre wFai oWnC wTGN oCir wEde oUps wWel
	x elegantissima	wCoN
	x elegantissima 'Rosealind'	oFor
	gracilis	oAls oGar oRiv wKin wCoN wWel
	gracilis 'Marmorata'	oFor
	gracilis 'Monzia'	see **D.** Pink-a-Boo®/ 'Monzia'
	gracilis 'Variegata'	see **D. *gracilis*** 'Marmorata'
	x hybrida	wCoN
	x hybrida 'Contraste'	last listed 99/00
	x hybrida 'Magicien'	wHer wWoS oGos oHed oAmb oNWe oGre wCCr oUps
	x hybrida 'Mont Rose'	oFor oGos
	x kalmiiflora	oFor wCul
	x lemoinei 'Compacta'	oFor
	x magnifica	oFor oGar
	sp. aff. ***parviflora*** DJH 361	wHer
	Pink-a-Boo®/ 'Monzia'	oAls oGar oGre oWnC wCoS
un	'Pink Minor'	oGos oGre
	'Pink Pompon'	see **D.** 'Rosea Plena'
	pulchra	wCCr
	pulchra B&SWJ 3435	wHer
	x rosea	wHer wCoN
	x rosea 'Carminea'	oFor
do	'Rosea Plena'	oFor oGre
	scabra	wCoN
	scabra 'Codsall Pink'	oFor oGar oGre oWnC
	sp. aff. ***scabra*** HC 970679	wHer
do	***scabra*** 'Pride of Rochester'	oFor wHer wCSG
	scabra 'Variegata'	oFor oGar oGre wWel
	setchuenensis	wCCr
	setchuenensis var. ***corymbiflora***	wHer
	sp. DJHC 095	wHer
un	'Summer Snow'	wRob
	DIANELLA	
	caerulea	oTrP
	revoluta	oTrP
	tasmanica	wHer
	tasmanica 'Variegata'	oHed oSho
	DIANTHUS	
	Allwoodii Alpinus Group	wTho oSha wFai
	alpinus	oSis
	alpinus 'Albus'	oSis
	alpinus 'Joan's Blood'	wWhG

	'Amarinth'	last listed 99/00
	amurensis 'Siberian Blue'	last listed 99/00
	'Aqua'	wTGN
un	***arenarius*** 'Snow Flurries'	wWoS
un	'Aunt Rose'	wWoS
	barbatus	wTho oCir
un	***barbatus*** 'Darkest of All'	oNWe
un	***barbatus*** 'Deepest Maroon'	wCul
un	***barbatus*** 'Double Dwarf'	oGar
un	***barbatus*** 'Homeland'	oEga
	barbatus Indian Carpet mixed	oAls
un	***barbatus*** Midget mixed	oUps
un	***barbatus*** 'Newport Pink'	oEga
un	***barbatus*** Single Midget mixed	oEga
	barbatus 'Sooty'	wHer
	barbatus 'Wee Willie'	oUps
	barbatus white	oEga
	'Bath's Pink'	oFor oAls oGar oOut oWnC oCir
	'Bat's Double Red'	wWoS wCul wFGN oGoo oWnC wTGN
	'Beatrix'	wWoS
un	'Black King'	oAls
	'Blue Hills'	oHed wFGN
un	'Bridal Veil'	oGoo
	'Brympton Red'	last listed 99/00
	caesius	see **D. *gratianopolitanus***
	'Camilla'	last listed 99/00
un	'Candy Dish'	wWoS
	caryophyllus	oAls wFGN oBlo wFai oBar oCir
	caryophyllus Chabaud Series	oCir
	caryophyllus Grenadin Series	oAls
un	***caryophyllus*** 'Grenadin Pink' (Grenadin Series)	wMag
un	***caryophyllus*** 'Grenadin Red (Grenadin Series)	wMag
	'Charles Musgrave'	see **D.** 'Musgrave's Pink'
	chinensis	last listed 99/00
	chinensis HC 970024	wHer
un	'Chumbley Farrer'	wCul
un	'Crimson Treasure'	oJoy oSis
	'Dad's Favourite'	wWoS oHed
	'Danielle Marie'	last listed 99/00
	deltoides	wCul wShR oSha wWhG
	deltoides 'Albus'	wCul
un	***deltoides*** 'Arctic Fire'	wTho oSis oSha wWhG oUps
	deltoides 'Brilliant'	wTho oSha oWnC wMag
	deltoides 'Erectus'	oSis
	deltoides Flashing Light	see **D. *deltoides*** 'Leuchtfunk'
	deltoides 'Leuchtfunk'	wTho oOut oBlo wMag wBWP wWhG
	deltoides 'Nelli'	oSis
un	***deltoides*** 'Rose Feather'	oSha
	deltoides white	oSha
	deltoides 'Zing Rose'	iArc oUps
un	'Desmond'	wFGN
	'Doris'	last listed 99/00
	'Dottie'	oGoo
qu	Double Gaiety mixed	oCir
	erinaceus	wMtT
	'Essex Witch'	wFGN
	'Fenbow Nutmeg Clove'	last listed 99/00
	'Firewitch'	oGar oSqu oEga oUps
un	'French Kiss'	wWoS
	freynii	oSis
un	'Frost Fire'	oAls
	glacialis	last listed 99/00
	'Gloriosa'	oGoo
	'Gold Dust'	wHom
	gratianopolitanus	wTho oAls oGoo oGre wFai
un	***gratianopolitanus*** 'Bourbon'	oJoy oWnC
un	***gratianopolitanus*** 'Frosty Fire'	wFGN oWnC
	gratianopolitanus 'Karlik'	oSis
un	***gratianopolitanus*** 'Little Boy Blue'	wWoS wFGN oWnC
un	***gratianopolitanus*** 'Mountain Mist'	wWoS
	gratianopolitanus 'Petite'	oSqu
	gratianopolitanus 'Rosenfeder'	last listed 99/00
	gratianopolitanus 'Tiny Rubies'	oFor wTho oAls wFGN oGar oGoo oSis oSho oSqu oGre oWnC wTGN oCir wAva wWhG
	haematocalyx Jurasek List	wMtT
	'Her Majesty'	last listed 99/00
	'Hope'	last listed 99/00
	'Horatio'	wWhG
	'Houndspool Ruby'	wWoS
	Ideal Series	oUps
	'Inchmery'	wWoS wCul oGoo
	'Inshriach Dazzler'	wMtT

	'Itsaul White'	oFor
un	'King of the Blacks'	wCul
	knappii	oFor oDan
un	***knappii*** 'Yellow Beauty'	iArc
	'La Bourboule'	wFGN oSis oSqu
	'La Bourboule Albus'	oSis
	'Laced Hero'	wWoS wCul
	'Laced Romeo'	wFGN oGoo
un	'Lady Glenda'	wMag
	'Lady Granville'	wWoS wCul oGoo
	'Little Bobby'	wWoS oGar wWhG
	'Little Jock'	oJoy wFGN oGar oGoo oSqu oCir
	'London Delight'	wWoS oGoo
	'London Lovely'	last listed 99/00
	'Margaret Curtis'	oJoy oAls
un	'Minimounds'	wMtT
un	'Mom's Cinnamon Pink'	oJoy
	monspessulanus	wCCr
	'Mrs. Sinkins'	oGoo
	'Musgrave's Pink'	wWoS oHed
	myrtinervius	oSis
	nardiformis	oSis
	noeanus	see **D. *petraeus*** ssp. ***noeanus***
	'Oakington'	oJoy
	'Old Clove Red'	last listed 99/00
un	'Old Spice'	wWoS
	'Painted Beauty'	wWoS
	'Paisley Gem'	last listed 99/00
un	Parfait hybrids	oUps
	pavonius	oSis
	'Peppermint Patty'	last listed 99/00
	petraeus ssp. ***noeanus***	oSis
	'Pheasant's Eye'	wWoS
	'Pike's Pink'	oJoy oSis oSqu oWnC oCir
	'Pink Mrs. Sinkins'	last listed 99/00
	plumarius	wCCr
un	***plumarius*** 'Christine'	wWoS
	pontederae	last listed 99/00
	'Prairie Pink'	wWoS wFGN
un	'Price's White'	wWoS
	'Queen of Sheba'	oGoo
	'Rachel'	wWoS wMtT
	'Red Velvet'	last listed 99/00
	'Rose de Mai'	oJoy oGoo wMag
	simulans	oJoy oSis
un	'Sonata'	oJoy
	'Sops-in-wine'	wWoS wCul oJoy wFGN
ch	'Spottii'	see **D.** 'Spotty'
	'Spotty'	oAls wFGN oGar oGoo oSis wHom oWnC oEga wWhG
	'Spring Beauty'	wTho oBlo wMag
un	'Starfire'	wSta
un	'Sternkissen'	oAls
	superbus	oSqu
	'Sweetheart Abbey'	last listed 99/00
un	***sylvestris*** ssp. ***longicaulis***	oSis
	'Ursula Le Grove'	wWoS
	'Velvet and Lace'	wHom
	'Waithman Beauty'	wWoS

DIASCIA

	'Andrew'	oJoy
qu	'Apple Blossom'	oJoy
	barberae 'Blackthorn Apricot'	oJoy oMis oGre wMag wTGN oUps
	barberae 'Ruby Field'	oFor oJoy oAls oSis oGre wFai wMag oCir oUps
	Coral Belle / 'Hecbel'	wWoS oJoy oHed oInd
qu	'Elliot's Variety'	oJoy
	'Emma'	wHer wWoS oJoy oInd
	fetcaniensis	wHer oJoy oAls oMis oGar oGre oInd
	flanaganii (see Nomenclature Notes)	wWoS
	'Hector Harrison'	see **D.** 'Salmon Supreme'
	'Hector's Hardy'	oJoy
	integerrima	wHer oJoy oSec
	integerrima 'Blush'	wHer oAls oInd
	'Jacqueline's Joy'	wHer wWoS oHed oCir
	'Kate'	wHer
	'Lady Valerie'	wHer wWoS oAls
un	'Langthorn's Lavender'	wHer wWoS oWnC oUps
	'Lilac Belle'	oJoy
	'Lilac Mist'	wHer
	'Little Charmer'	oHed
un	'Longhorn Purple'	oHed oMis
	'Louise'	wHer oInd
	'Lucy'	oJoy oHed
	'Paula'	wHer

	purpurea	wHer
	'Red Start'	wWoS oHed
	rigescens	oHed wCSG
un	'Rose Queen'	oInd
	'Ruby's Pink'	last listed 99/00
	'Rupert Lambert'	oHed
	'Salmon Supreme'	oFor wWoS oJoy oWnC
	sp.	oJoy
	'Strawberry Sundae'	wTGN
	'Twinkle'	wHer oJoy oAls oHed
	vigilis	oJoy oHed oNWe
	vigilis 'Jack Elliott'	wHer oAls oHed oInd
	'Wendy'	wHer
	DICENTRA	
	'Adrian Bloom'	oGar oSho wRob wTGN wAva wSnq
	'Bacchanal'	oGar wRob wMag wWhG wSnq
	'Boothman's Variety'	see **D.** 'Stuart Boothman'
	canadensis	wCol oSis
	'Coldham'	wHer
	cucullaria	oRus oSis oAld
	eximia (see Nomenclature Notes)	oFor oNat oGar oBlo oWnC wCoN oUps
	eximia 'Alba'	see **D. *eximia*** 'Snowdrift'
	eximia E.C. seedling	oRus
	eximia 'Snowdrift'	oFor wCul oAls oRus oGar oGre wWhG
	eximia 'Zestful'	wCul oAls oGar oGre
	formosa	oFor oHan wCul wThG wWoB wNot oRus wShR oMis oGar oBos oOut wWat wHom oTri oBRG oAld wAva wCoN wFFl wWld wWhG oUps oSle wSte
	formosa alba	wHer
	formosa 'Aurora'	oGar
	formosa 'Sweetheart'	oMis oSis wRob
	'Langtrees'	wWoS wHig oHed oGar
	'Luxuriant'	oFor oAls oGar oSho oBlo oGre wMag wTGN oSec wCoN wSnq
	macrantha	wHer oNWe oGre
un	'Ruby Marr'	oRus
	scandens	wHig oMis oGar oCis oNWe wSte
	'Silver Smith'	oSec
	Snowflakes / 'Fusd'	oGar
	spectabilis	oFor wWoS wCul wSwC oNat oDar oAls oGar oNWe oSho oSha oBlo oGre wFai wHom oWnC wTGN oSec oCir wAva wCoN wWhG oSle wSnq wCoS
	spectabilis 'Alba'	wWoS wCul oNat oAls oGar oNWe oBlo oGre wHom oWnC wTGN
	spectabilis 'Goldheart'	wHer oNat wFai
	'Stuart Boothman'	oFor wWoS wCul wCol oGar oSho oOut wRob oGre
	DICHELOSTEMMA	
	congestum	wWld
	DICHORISANDRA	
	thyrsiflora	oTrP
	DICHROA	
	febrifuga	wHer oHed oCis
	febrifuga HWJCM 515	wHer
	DICHROMENA	
	colorata	see **RHYNCHOSPORA *colorata***
	DICKSONIA	
	antarctica	wFan oGar wDav wWel
	fibrosa	last listed 99/00
un	**DICRANUM**	
	sp.	wFFl
	DICTAMNUS	
	albus	oAls wCoN oUps
	albus var. *caucasicus* 'Albiflorus'	oGar
	albus var. *purpureus*	oJoy wTGN
	albus var. *purpureus* pink	oFor
	fraxinella	see **D. *albus*** var. ***purpureus***
	DIDIEREA	
	madagascarensis	oRar
	trollii	oRar
	DIERAMA	
	argyreum	wHer
	cooperi	wHer oHed
	dracomontanum	oTrP wHer wCul oJil oJoy oDan oNWe
	dwarf hybrid	oDan
un	*galpinii*	wHer
	igneum	last listed 99/00
	latifolium	wHer oNWe
	pauciflorum	wHer wCol oNWe
	pendulum	oTrP wHer oHed oBRG
	pulcherrimum	oFor oTrP wHer wCul wCri oGos oNat oDar oJoy wCol oAls oHed oRus oAmb oGar oDan oCis oNWe oOut oGre wFai wPir wCCr wBox wCoN wGra wSte
	pulcherrimum var. *album*	last listed 99/00
	pulcherrimum var. *album* seed grown	wHer
un	*pulcherrimum* Carrabine Strain	wHer
un	*pulcherrimum* Darkest Purple Strain	wHer
	pulcherrimum Donard hybrids	see **D. *pulcherrimum*** Slieve Donard hybrids
un	*pulcherrimum* Flaring Tips Strain	wHer
	pulcherrimum Slieve Donard hybrids	oMis
	reynoldsii	last listed 99/00
	robustum	wHer oDan oNWe
	sp.	oMac
	DIERVILLA	
	lonicera	last listed 99/00
un	*lonicera* 'Copper'	oFor
un	*rivularis* 'Summer Stars'	oFor
	sessilifolia	oFor
	sessilifolia 'Butterfly'	oGos oGre
	DIETES	
	bicolor	oTrP wCCr
	iridioides	oTrP oGar
	vegeta	see **D. *iridioides***
	DIGITALIS	
	alba	see **D. *purpurea*** f. ***albiflora***
	ambigua	see **D. *grandiflora***
un	'Apricot Beauty'	wTho
	ferruginea	wCul oJoy oRus
	ferruginea 'Gigantea'	wBWP
	grandiflora	oFor wHer oHan wCul oNat oRus oAmb oWnC wMag
	grandiflora 'Carillon'	oGar oSis
	'John Innes Tetra'	oJoy oNWe oSec
	laevigata	oRus
	lamarckii (see Nomenclature Notes)	oSec
	lanata	oHan oRus wBWP
	lutea	oHan wCul wHig oRus wPir wBWP
	x mertonensis	oFor wCul wTho oAls oHed oCis oSho wFai oWnC wMag wTGN oSec iArc oEga oUps
	obscura	oRus oSis oNWe
	parviflora	wHer wCul oJoy oRus oSec
	purpurea	wFGN wShR oBos iGSc wHom wMag oAld
	purpurea f. *albiflora*	wTho oGar wHom oWnC wMag
	purpurea apricot	see **D. *purpurea*** 'Sutton's Apricot'
	purpurea Excelsior Group	oGar oBlo oWnC wBWP oCir oEga oSle
	purpurea Foxy Group	oAls oGar oBlo
	purpurea Gloxinioides Group 'The Shirley'	oAls
	purpurea ssp. *heywoodii*	wHig oHed oNWe oWnC oSec wBWP
	purpurea 'Pam's Choice'	oNWe oGre
	purpurea 'Primrose Carousel'	oHed oCir
	purpurea 'Sutton's Apricot'	wCul oAls oAmb oGar oWnC wMag wBox oCir
	sibirica	wCCr
un	*thapsi* 'Spanish Peaks'	oFor oSis
	viridiflora	wCCr
	DIMOCARPUS	
	longan	oOEx
	DIONAEA	
	muscipula	wOud oGar
un	*muscipula* 'Akai Ryu'	wOud
	muscipula 'Red Dragon'	see **D. *muscipula*** 'Akai Ryu'
	DIONYSIA	
	aretioides	last listed 99/00
	aretioides 'Gravetye'	last listed 99/00
	DIOON	
	edule	oOEx
	spinulosum	oRar oOEx
	DIOSCOREA	
	batatas	oFor wRai oOEx iGSc oCrm
	bulbifera	oRar
	cotinifolia	oRar
	deltoidea HWJCM 023	wHer
un	*dumetorum*	oTrP oRar
	elephantipes	oRar
un	*elephantipes* x *paniculata*	oRar
un	*globosa*	oRar
un	*godawari*	oFor
	hastifolia	oRar
un	*hirtiflora*	oRar
un	*japonica*	oFor oOEx
un	*mexicana*	oRar
un	*paniculata*	oRar
un	*penangensis*	oOEx
un	*pentaphylla* Vikhai Getta	oOEx
un	*pentaphylla* Yao-shan-yao	oFor oOEx

	quinqueloba HC 970060	wHer
un	***remotiflora*** BLM 0650	oRar
un	***sansibarensis***	oRar
	sp. (Amazonian Air Potato)	oOEx
	sp. BLM 0376	oRar
	sp. BLM 0418	oRar
	sp. BLM 0722	oRar
	sp. (Godawari Mt. Yam)	oOEx
	sp. (Traitung's Yam)	oOEx
un	***sylvatica***	oTrP oRar
	villosa	oOEx iGSc oCrm
	DIOSPYROS	
	austroafricana	oTrP
	duclouxii	oOEx
	kaki 'Chocolate'	last listed 99/00
	kaki 'Fuyu'	oAls oSho oWhi wRai oOEx wWel
un	***kaki*** 'Giant Fuyu'	oGar oGre
un	***kaki*** 'Gionbo'	oOEx
un	***kaki*** 'Great Wall'	oOEx
	kaki 'Hachiya'	oGar wRai oOEx
un	***kaki*** 'Hana Fuyu'	oOEx
un	***kaki*** 'Honan Red'	oOEx
un	***kaki*** 'Izu'	wBur wClo wRai oOEx
un	***kaki*** 'Jiro'	wBur
un	***kaki*** 'Nishimuri Wase'	oOEx
un	***kaki*** 'Saijo'	wBur oOEx
un	***kaki*** 'Tae-bon-si'	oOEx
	lotus	oRiv oOEx
	rhombifolia	oFor
	virginiana	oFor wBCr oRiv
un	***virginiana*** 'Early Golden'	wBur wRai oOEx oGre oWnC
un	***virginiana*** 'Garretson'	wBur oOEx oWnC
	virginiana male	wBur wRai oWnC
un	***virginiana*** 'Meader'	wBur wClo wRai oOEx
	DIPELTA	
	sp. aff. ***floribunda*** DJHC 789	last listed 99/00
	yunnanensis	wCol
	sp. aff. ***yunnanensis*** DJHC 104	wHer
	DIPHYLLEIA	
	cymosa	wHer
	grayi	last listed 99/00
	DIPLACUS	see **MIMULUS**
	DIPLADENIA	see **MANDEVILLA**
	DIPLARRHENA	
	latifolia	last listed 99/00
	DIPLAZIUM	
	pycnocarpon	oRus
	DIPOGON	
	lignosus	oTrP
	DIPSACUS	
un	***japonicus***	oOEx
	DIPTERONIA	
	sinensis	oFor oRiv oCis wCCr
	DIRCA	
	palustris	oRiv
	DISANTHUS	
	cercidifolius	oFor wHer oGos wClo oAmb oGar oDan oRiv oSho oGre wCoN wWhG wSte wWel
	DISCARIA	
	serratifolia HCM 98098	wHer
	DISELMA	
	archeri	wHer
	DISPOROPSIS	
	arisanensis	last listed 99/00
un	***fusca-picta***	wHer oCis
	pernyi	oFor wSwC wCol oRus oGar oCis wRob
un	***pernyi*** 'Bill Baker Form'	oJoy
	pernyi DJHC 735	wHer
	DISPORUM	
un	***bodineri***	wHer
un	***cantoniense*** 'Aureovariegata'	wHer
	cantoniense var. ***cantoniense*** B&SWJ 5290	wHer
	sp.aff. ***cantoniense*** DJHC 724	wHer
	sp.aff. ***cantoniense*** HWJCM 069	wHer
	flavens	wCol oSis oNWe
	hookeri	oRus oBos oAld wFFl wSte
	hookeri var. ***oreganum***	wHer wNot wCol
	lanuginosum	oFor wHer
	maculatum	wThG
	nantauense B&SWJ	wHer
un	***palla variegata***	oRus
	sessile	oRus
	sessile 'Variegatum'	wCol oRus oDan oNWe oSho
	sessile 'Variegatum' broad leaf form	wHer
	sessile 'Variegatum' narrow leaf form	wHer
	smilacinum	last listed 99/00
	smilacinum 'Aureovariegatum'	wHer
	smilacinum HC 970422	wHer
	smithii	wHer oHan wCol oRus oGar oSis oBos oTri oAld wFFl wSte
	uniflorum HC 970012	wHer
	viridescens HC 970422	last listed 99/00
	DISTYLIUM	
	myricoides	wHer oGos oCis oSho oGre wPir oWnC wBox
un	***myricoides*** 'Lucky Charm'	wWel
	racemosum	oFor oGos oGar oCis oSho oGre wPir oWnC wBox wWel
un	***racemosum*** 'Akebono'	wHer
un	***racemosum*** 'Guppy'	wHer
	DOCYNIA	
	delavayi	oOEx
	indica	last listed 99/00
un	***indica*** King Haw	oOEx
un	***indica*** Queen Haw	oOEx
	DODECATHEON	
	alpinum	last listed 99/00
	clevelandii ssp. ***insulare***	last listed 99/00
	clevelandii ssp. ***patulum***	oSis
	dentatum	wCol wPle oSis
	hendersonii	wThG wCol oSis oNWe oTri
	jeffreyi	wHer wCol oRus wWld
	jeffreyi 'Rotlicht'	oRus oOut wBox
	meadia	oFor wHer wThG oJoy wCol oRus oSis oAld
	meadia f. ***album***	oFor wHer oRus oGar oSis
	meadia 'Goliath'	oDan
	meadia red hybrids	oSis wTGN
un	***meadia*** 'Roseum'	oGar
	poeticum	oBos
	pulchellum	wCol oRus oBos
	redolens	last listed 99/00
	sp.	wShR
	sp. Illinois Valley	oSis
	DODONAEA	
un	***filiformis***	oFor
	viscosa 'Purpurea'	wCCr
	DOMBEYA	
	burgessiae	oTrP
	calantha	see **D. *burgessiae***
	goetzenii	see **D. *torrida***
	torrida	oTrP
	DOODIA	
	caudata	wFan
	media	oGos oGre oBRG
	DORONICUM	
	caucasicum	see **D. *orientale***
	columnae	wCSG oWnC
	'Little Leo'	wWoS oWnC oEga
	orientale 'Finesse'	oFor oEga
	orientale 'Magnificum'	oFor oAls oGar oGre oWnC oUps
	sp.	oRus
	DORSTENIA	
un	***carnulosa***	oRar
	crispa	see **D. *foetida***
	foetida	oRar
un	***naroc***	oRar
	sp. aff. ***psilurus*** MES 422	oRar
	sp. B & L 877	oRar
	sp. LAV 23457	oRar
	DORYANTHES	
	excelsa	oTrP
	DORYCNIUM	see **LOTUS**
	DOUGLASIA	
un	***idahoensis***	wMtT
	laevigata	see **ANDROSACE *laevigata***
	montana	see **ANDROSACE *montana***
	nivalis	see **ANDROSACE *nivalis***
	DOVYALIS	
	caffra	oTrP
un	**DOYEREA**	
	emetocathartica BLM 0175	oRar
	DRABA	
	aizoides	oSis
	cappadocica	last listed 99/00
un	***condensatum***	oSis
	dedeana	wMtT oSqu
	densifolia NNS93-224	last listed 99/00
	hispanica	wMtT

	incana Stylaris Group	last listed 99/00
	ossetica var. racemosa	wMtT
	rigida	oSis
	rigida var. bryoides	oSis
	rosularis	oSis
	sibirica	oSis
	stylaris	see D. incana Stylaris Group
	DRACAENA	
	cincta 'Tricolor'	oGar
	deremensis	see D. fragrans Deremensis Group
va	fragrans Deremensis Group 'Lemon Lime'	oGar
va	fragrans Deremensis Group 'Warneckei'	oGar
	fragrans 'Janet Craig'	oGar
un	fragrans 'Janet Craig Compact'	oGar
	indivisa	see CORDYLINE indivisa
va	marginata	oGar
va	marginata 'Colorama'	oGar
un	'Spike'	oGar
	DRACOCEPHALUM	
	argunense	oHed
	botryoides	oSis
	bullatum	wCol
un	sp. aff. discolor JJH9209138	wMtT
	forrestii	oSis
	grandiflorum	oFor
	sp. aff. paulsenii JJH9308132	wMtT
	ruyschianum	oSis
	wendelboi	oSis
	DRACUNCULUS	
	canariensis	last listed 99/00
	vulgaris	wHer oJoy oNWe wAva
	DREGEA	
	sinensis	wHer
	DREPANOSTACHYUM	
	falcatum	last listed 99/00
	glomeratum	see HIMALAYACALAMUS falconeri var. glomerata
	hookerianum	see HIMALAYACALAMUS hookerianus
	khasianum	oTra wBea
	DRIMIOPSIS	
	kirkii	oRar
	maculata	oTrP oRar
	DRIMYS	
	lanceolata	oFor wHer oCis wSte
	winteri	oFor wHer oCis oSho wSte
	winteri var. chilensis	oFor
	DROSANTHEMUM	
	floribundum	oSqu
	DROSERA	
	adelae	oGar
	aliciae	wOud
	binata ssp. dichotoma giant	wOud
	binata ssp. dichotoma small red form	wOud
un	binata ssp. dichotoma T form	wOud
	binata 'Extrema'	wOud
	binata 'Multifida'	wOud
	binata T form	wOud
	x californica 'California Sunset'	wOud
	capensis	oGar
	capensis alba	wOud
	capensis narrow-leaved	wOud
	capensis wide-leaved	wOud
	capillaris	wOud
	filiformis var. filiformis	wOud
un	filiformis var. filiformis 'Florida Giant'	wOud
	filiformis x intermedia	wOud
	filiformis var. tracyi	wOud
	intermedia	last listed 99/00
	'Marston Dragon'	wOud
	rotundifolia	last listed 99/00
	rotundifolia compact form	wOud
	spathulata	oGar
	DRYAS	
	octopetala	wCul wCol oSis oBos
	octopetala 'Minor'	oSis
	x suendermannii	wMtT
	DRYOPTERIS	
qu	sp. aff. abbreviata	oRus
	affinis ssp. cambrensis	last listed 99/00
	affinis 'Congesta Cristata'	wFan
	affinis Crispa Group	oRus oGre oBRG
	affinis 'Crispa Barnes'	oFor wFan
	affinis 'Crispa Congesta'	see D. affinis 'Crispa Gracilis'
	affinis 'Crispa Gracilis'	wCol wFan oRus
	affinis 'Cristata'	wFan oRus oGar oBRG
	affinis 'Cristata Angustata'	wFan
	affinis 'Cristata The King'	see D. affinis 'Cristata'
	affinis 'Revolvens'	wFol wFan wNay oBRG
	affinis 'Stableri'	see D. x complexa 'Stablerae'
	affinis 'Stableri Crisped'	see D. x complexa 'Stablerae Crisped'
	arguta	oRus
	atrata	see D. cycadina
	austriaca	see D. dilatata
un	bissetiana	wFol wFan
	blanfordii	wFan
	carthusiana	wFan oRus wNay
	celsa	oFor wFan oRus
	championii	last listed 99/00
	x complexa	wFol oSqu
un	x complexa 'Robust'	oGar oBRG
	x complexa 'Stablerae'	oFor wFan oRus
qu	x complexa 'Stablerae Crisped'	wFan wRob
	corleyi	last listed 99/00
	crassirhizoma	wFan
	crispifolia	last listed 99/00
	cristata	wFan
	cycadina	oFor wFan oRus oGar oSqu oTPm oUps
	dickinsii	last listed 99/00
un	dickinsii 'Incisa'	wFol
	dilatata	oGar oBos oTri oAld
un	dilatata 'Compacta'	oBRG
	dilatata 'Crispa Whiteside'	wFan oRus
	dilatata 'Grandiceps'	oRus
	dilatata 'Jimmy Dyce'	wFan oRus oGar oSis oGre oBRG
	dilatata 'Lepidota Cristata'	wFan oRus oGar oSis oSqu oGre oBRG
un	dilatata 'Recurvata'	oFor oRus oGar oSis oSqu oGre oBRG
un	dilatata 'Recurved Form'	wFan
	erythrosora	oFor wCul wFol oJoy wFan oRus oGar oSis oNWe oOut oSqu oGre wPir wHom oTPm oWnC wTGN oBRG wSte wSnq
	erythrosora 'Prolifera'	wFan oRus oBRG
un	erythrosora 'Prolifera Whirley-top'	wCol wFan
	expansa	wNot wFan oRus oBRG wFFl
	filix-mas	wHom
	filix-mas 'Barnesii'	wFan oRus oGar oGre
	filix-mas 'Crispa Congesta'	see D. affinis 'Crispa Gracilis'
	filix-mas 'Crispa Cristata'	oRus
	filix-mas 'Crispatissima'	oRus oGar oGre oBRG
	filix-mas 'Cristata Jackson'	wCol
	filix-mas 'Cristata Martindale'	wFan
	filix-mas 'Furcans'	oRus
	filix-mas Grandiceps Group	wFan oRus oGar oBRG
un	filix-mas 'Lepidota Polydactyla'	oFor
	filix-mas 'Linearis Congesta'	last listed 99/00
	filix-mas 'Linearis Polydactyla'	wFan oRus oGar wNay oSqu oTPm oBRG wSnq
un	filix-mas 'Undulata Robusta'	oFor oRus oGar wSnq
	fuscipes	last listed 99/00
	goldieana	wCol wFan oRus oNWe
	hondoensis	wCol wFan
un	koidzumiana	wFol
un	kuratae	wFan
	lacera	last listed 99/00
	lepidopoda	wFol oRus
	ludoviciana	last listed 99/00
	marginalis	oFor oRus oGar wNay oGre oTri
un	namegatae	wFol
un	niponensis	wFol
	noveboracensis	see THELYPTERIS noveboracensis
un	pacifica	wFol
un	polylepis	wCol wFan
un	pseudo-filix-mas	wFan oRus
	purpurella	oRus
un	pycnopteroides	wFol
	x remota	oFor wFol wFan oRus oGre
	sabae	last listed 99/00
un	saxifraga	wFol
	scottii	wFan oRus oBRG
	sieboldii	wFan oRus oBRG
	spinulosa	see D. carthusiana
	stewartii	oFor wFan oRus oGar wNay
un	sublacera	wFan
	thelypteris	see THELYPTERIS palustris
	tokyoensis	wFan
	uniformis	oRus
	uniformis 'Cristata'	wFan wRob
	varia	last listed 99/00
	wallichiana	oFor wFan oRus oGar oSqu oGre oBRG wSte wSnq
un	wallichiana 'Molten Lava'	wFan

	DUCHESNEA	
	indica	oRus oWnC
va	***indica*** 'Harlequin'	oFor
	DUDLEYA	
	attenuata ssp. ***orcuttii***	oSqu
	cymosa ssp. ***pumila*** NNS95-228	wMtT
un	'White Sprite'	oSqu
un	**DULICHIUM**	
	arundinaceum	wSoo wGAc oHug oGar oSsd oWnC
	DUMASIA	
	truncata HC 970717	last listed 99/00
	DURANTA	
un	'Golden Edge'	oFor
	DUVALIA	
	modesta	oRar
	DYMONDIA	
	margaretae	oTrP
	DYPSIS	
	decaryi	oGar
	lutescens	oGar
	ECCREMOCARPUS	
	scaber	oHed oGoo wSte
	scaber f. ***carmineus***	oHed
un	***scaber*** 'Tresco'	oHed
qu	Tresco mixed	oUps
	ECHEVERIA	
	albicans	oRar
	sp. aff. ***atropurpurea*** BLM 0834	oRar
un	'Barbillion'	oSqu
	bella	oSqu
	'Black Prince'	last listed 99/00
un	'Cass's Hybrid'	oSqu
	coccinea BLM 0707	oRar
un	'Culibra'	oSqu
un	'Deren-oliver'	oSqu
	derenbergii	oSqu
un	'Dondo'	oSqu
	'Doris Taylor'	last listed 99/00
un	'Firelight'	oSqu
un	'Giant Mexican Firecracker'	oSqu
	gibiflora BLM 0545	oRar
	globulosa	see **E. *secunda***
un	'Gusto'	oSqu
un	'Hyalina'	oSqu
un	'Icicle'	oSqu
	lindsayana x ***pulidonis***	oSqu
un	'Lola'	oSqu
un	'Mauna Loa'	oSqu
	minima	last listed 99/00
un	***multicaulis***	oSqu
	nodulosa	oSqu
	'Paul Bunyon'	last listed 99/00
	pulidonis	oSqu
	pulvinata	oSqu
un	'Ramillette'	oSqu
un	***runyonii*** var. ***runyonii***	oSqu
un	***runyonii*** 'Topsy Turvy'	oSqu
	secunda	oSqu
	secunda var. ***glauca***	oSqu
un	'Set-oliver'	oSqu
un	'Set-sprucii'	oSqu
	setosa	oSqu
un	***subsessilis***	oSqu
un	'Sunburst'	oSqu
un	'Tippy'	oSqu
	turgida	oSqu
un	'Violet Queen'	oSqu
un	***viridissima***	oHed oRar
	'Warfield Wonder'	last listed 99/00
un	'Zipper'	oSqu
	ECHIDNOPSIS	
	repens	oRar
	ECHINACEA	
	angustifolia	oFor oGoo wFai iGSc oBar oUps
	pallida	oFor oRus wFai wWld oUps oSle
	paradoxa	oFor oMis oGoo wFai iGSc oCrm oUps
	purpurea	oFor oEdg oRus wFGN wCSG oGar oGoo oNWe oSho oOut oBlo oGre iGSc wHom oCrm oBar oSec wNTP wWld oUps
un	***purpurea*** 'Alba'	wCul wTho wFGN oGar
	purpurea 'Bravado'	oFor oNat oGar wHom oEga wWhG oUps
	purpurea 'Bright Star'	see **E. *purpurea*** 'Leuchtstern'
un	***purpurea*** 'Cygnet White'	wWoS
	purpurea 'Kim's Knee-High'	wWhG
	purpurea 'Leuchtstern'	oFor oWnC oUps oSle
	purpurea 'Magnus'	oFor wCul wSwC oDar oAls wCSG oMis oGar oDan oSho oOut wRob wFai oWnC wMag wTGN iArc oCir oEga wWhG oUps wSnq wCoS
un	***purpurea*** 'Ovation'	wTho
	purpurea 'White Swan'	oFor oAls wCSG oAmb oMis oGar oGoo oDan oSho oOut oBlo oGre wFai iGSc wHom oWnC wMag iArc oCir oEga wWhG oUps oSle wSnq
	tennesseensis	oGoo iGSc oCrm oUps
	ECHINOCEREUS	
	fendleri	oRar
	leucanthus	oRar
	pectinatus var. ***rigidissimus***	see **E. *rigidissimus***
	rigidissimus	oOEx
	schmollii	oRar
	ECHINODORUS	
	cordifolius	oHug wTGN
un	***cordifolius*** 'Marble Queen'	wGAc oHug oGar
	grandiflorus	wGAc oHug oGar
	muricatus	see **E. *grandiflorus***
	ECHINOPS	
	bannaticus	oRus
	bannaticus 'Blue Globe'	wWoS wHom
	bannaticus 'Blue Glow'	see **E. *bannaticus*** 'Blue Globe'
	bannaticus 'Taplow Blue'	oFor oAls
	exaltatus	wCul oGoo
	ritro (see Nomenclature Notes)	oAls oRus wShR wCSG oGar oSha wFai
	ritro ssp. ruthenicus	oSec
	sp.	oRus
	sphaerocephalus	oGoo
	sphaerocephalus 'Arctic Glow'	oFor wCul oAls oGar wSnq
	ECHINOPSIS	
	chamaecereus	oTrP oSqu
	ECHIUM	
	candicans	wCCr
	fastuosum	see **E. *candicans***
	italicum	wBWP
	pininana	wHer wBWP
	russicum	wHer
	russicum JA 432 300	oNWe
	EDGEWORTHIA	
	chrysantha	oFor wHer oGos wClo oDan oCis oGre oWnC wWel
un	***chrysantha*** 'Gold Rush'	oGar oGre
	chrysantha 'Rubra'	oCis
	gardneri	wHer oCis
	papyrifera	see **E. *chrysantha***
un	***racemosa***	oSho
	EDRAIANTHUS	
	graminifolius	oSis
	pumilio	oSis wMtT
	EGERIA	
	densa	oHug oGar
	EHRETIA	
	dicksonii	oFor
	EICHHORNIA	
	crassipes	wSoo wGAc oHug oGar oSsd oWnC
	ELAEAGNUS	
	angustifolia	oFor wCul wBCr oEdg oGre oWnC wBox wPla wCoN oSle
	angustifolia 'Red King'	oFor
	commutata	oFor wCoN
	x ebbingei	oFor oGar oSho wCoN
	x ebbingei 'Gilt Edge'	oGar oDan oSho oGre wWel
	multiflora	wCul wBur wClo oWhi oOEx
	multiflora 'Raintree Select'	wRai
un	***multiflora*** Sweet Scarlet™	wRai oGre wSte
	parvifolia	oWhi
	pungens	wPir wCCr
	pungens 'Aurea'	wCCr
un	***pungens*** 'Clemson Variegated'	wHer
	pungens 'Fruitlandii'	oGar oGre oWnC wWel
va	***pungens*** 'Maculata'	oFor wHer wCul oGos wPir wAva
	'Quicksilver'	wHer wWoS oHed
	umbellata	wBCr oOEx wKin wPla
	umbellata 'Cardinal'	last listed 99/00
	ELAEOCARPUS	
un	***decipiens***	oFor oGar oGre oWnC
	ELEOCHARIS	
	dulcis	wGAc oHug oSsd oWnC
un	***dulcis*** 'Variegatus'	wCli
un	***montevidensis***	wGAc oWnC
un	***ovata***	oSsd wWat
	palustris	oSsd wWat oTri oAld

un	***radicans***	oAls
	tuberosa	see **E. *dulcis***
	ELETTARIA	
	cardamomum	iGSc
	ELEUTHEROCOCCUS	
	gracilistylus	oOEx
	henryi	wHer
un	***senticosus*** var. ***coreanus*** HC 970349	wHer
	sieboldianus	oFor oCis
	sieboldianus 'Aureomarginatus'	oFor
	sieboldianus 'Variegatus'	oGos oGar oNWe oGre wPir oUps
	sp. DJHC 604	wHer
	sp. DJHC 642	wHer
	ELLISIOPHYLLUM	
qu	***bipinnatum***	oTrP
	pinnatum	wWoS
	pinnatum B&SWJ 197	wHer
	ELODEA	
	canadensis	wGAc oWnC
	densa	see **EGERIA *densa***
	ELSHOLTZIA	
	stauntonii	oFor
	stauntonii 'Alba'	last listed 99/00
	ELYMUS	
	arenarius	see **LEYMUS *arenarius***
	condensatus	see **LEYMUS *condensatus***
	glaucus	see **E. *hispidus***
	hispidus	wPla oAld oCir wSnq wCoS
	magellanicus	wHer wWoS wSwC oNat oCis wWin wSnq
	EMBLICA	see **PHYLLANTHUS *emblica***
	EMBOTHRIUM	
	coccineum	oFor oAmb oRiv wCCr
	coccineum HCM 98181	wHer
	coccineum HCM 98193	wHer
	coccineum HCM 98218	wHer
	coccineum HCM 98220	wHer
	coccineum Lanceolatum Group	last listed 99/00
	EMMENOPTERYS	
	henryi	oFor wHer oGos oCis
	EMPETRUM	
	eamesii ssp. ***hermaphroditum***	oSis
	nigrum	wHer
un	**ENADENIUM**	
	gossweilerii B81.876	oRar
	ENDYMION	see **HYACINTHOIDES**
	ENKIANTHUS	
	campanulatus	oEdg oDan oRiv oSho oGre wCCr oWnC wSta wCoN wWel
	campanulatus 'Bovees # 12'	oBov
	campanulatus var. ***campanulatus*** f. ***albiflorus***	oBov
un	***campanulatus*** 'Hollandia Red'	oAmb oGar oDan wWel
	campanulatus 'Red Bells'	oFor wHer oGos wWel
	campanulatus red form	oBov
qu	***campanulatus*** 'Rubens'	oGre
	campanulatus var. ***sikokianus***	oGos oGre
	cernuus var. ***matsudae*** f. ***rubens***	wHer oDan oRiv oSho oBRG wSte
	deflexus	wHer wCCr
	perulatus	oRiv oBov oGre wCCr wWel
	ENSETE	
	ventricosum	oGar
	ventricosum 'Maurelii'	oGar wWel
	ENTADA	
	sp.	oOEx
	ENTANDROPHRAGMA	
	caudatum	oRar
	ENTELEA	
	arborescens	last listed 99/00
	EOMECON	
	chionantha	wCol oRus oAmb oBov oGre
	EPHEDRA	
	americana var. ***andina***	oFor
	major	last listed 99/00
	minima	oCis oOEx oUps
	monosperma JJH 930860	last listed 99/00
	nebrodensis	see **E. *major***
	nevadensis	oFor oGoo oOEx iGSc
	sinica JJH 9308149	last listed 99/00
	viridis	oFor iGSc
	EPIGAEA	
	repens	wMtT oBov
	EPILOBIUM	
	angustifolium	oFor wNot oRus oGoo oBos wWat iGSc oAld
	angustifolium var. ***album***	wHer

	californicum (see Nomenclature Notes)	see **ZAUSCHNERIA *californica***
	canum	see **ZAUSCHNERIA *californica*** ssp. ***cana***
	dodonaei	wHer oSec
	EPIMEDIUM	
	acuminatum DJHC 841	wHer
	acuminatum L575	wHer
	alpinum	wHer wCri wCol oRus
un	***alpinum*** 'Shrimp Girl'	oRus oOut wRob oGre
	Asiatic hybrids	last listed 99/00
	'Beni-kujaku'	wHer wNay
	'Black Sea'	wCol
un	***brachyrrhizum***	wCol wNay
un	'Butterscotch'	wNay
	x cantabrigiense	wHig oRus oSis wRob wNay oGre wBox
un	***chlorandrum*** DJHC 705	wHer
	davidii	wHig oNWe
	davidii x ***acuminatum***	wHer
	davidii EMR 4125	wNay
	diphyllum	wHer wNay
un	***diphyllum*** 'Roseum'	wCol wNay
	dolichostemon	wHer
	'Enchantress'	wHer wCol
un	***epsteinii*** cc. 940255	wHer
	franchetii	wHer
	grandiflorum	oNWe wRob wNay wPir wCoN wEde
	grandiflorum 'Album'	oRus
	grandiflorum 'Crimson Beauty'	wHer oRus oNWe
	grandiflorum f. ***flavescens***	last listed 99/00
	grandiflorum higoense	see **E. *higoense***
un	***grandiflorum*** 'Irene'	wMtT
	grandiflorum 'Koji'	wHer
	grandiflorum ssp. ***koreanum***	wHer
	grandiflorum 'Lilafee'	wHer wWoS oGos wCol oHed oRus oGar oNWe oOut wRob wNay wFai wBox
	grandiflorum 'Nanum'	last listed 99/00
	grandiflorum 'Rose Queen'	oFor wWoS oRus oNWe oSho wRob wNay oGre
	grandiflorum 'Saxton Purple'	wHer
un	***grandiflorum*** 'Silver Queen'	wRob
un	***grandiflorum*** 'Sirius'	wHig oNWe
un	***grandiflorum*** 'Snow Queen'	oNWe
un	***grandiflorum*** 'Tama-no-Genpei'	wHer
	grandiflorum f. ***violaceum***	oSis
	grandiflorum 'White Queen'	wHer
un	***grandiflorum*** 'Yubae'	wHer
	higoense	wHig oHed oNWe
	'Kaguyahime'	last listed 99/00
	leptorrhizum	wHer wCol wHig wNay
	'Little Shrimp'	oGos wHig
un	***myrianthum***	wCol wNay
	ogisui OG 91.001	last listed 99/00
	x omeiense 'Stormcloud' OG 82.002	wHer
un	'Orion'	wCol
	pauciflorum OG 92.123	wHer
	x perralchicum	oSis
	x perralchicum 'Frohnleiten'	wHer wWoS wCol oRus oNWe wRob wNay oGre wBox
	x perralchicum 'Wisley'	wHer wCSG
	perralderianum	wCul oGre
	perralderianum 'Weihenstephan'	wHer
	pinnatum ssp. ***colchicum***	wWoS wCul wCol oRus oAmb oSis wRob wNay oGre wFai wEde wSte
	pubescens	last listed 99/00
	pubescens OC 91.003	wHer
	pubigerum	oFor oRus wRob wNay oGre wBox
	rhizomatosum	wHer
	x rubrum	oFor wHer wCul oGos wCol wHig oAls oHed oRus oGar oSis oNWe oSho oOut wRob wNay oBov wFai oWnC wMag wTGN wBox oUps
	sagittatum	wHer
	sempervirens	wHer wCol
	setosum	oHed
	stellulatum 'Wudang Star' L 1193	wHer
	sulphureum	see **E. *x versicolor*** 'Sulphureum'
	x versicolor 'Neosulphureum'	wHer oHed
	x versicolor 'Sulphureum'	oFor wWoS wCul oGos wCol wHig oAls oRus oMis oGar oSis oNWe oOut wRob wNay oGre wFai wAva wEde wSte
	x versicolor 'Versicolor'	wHer
	x warleyense	oFor wHer oSis oNWe wMtT oOut wNay
	x warleyense Orange Queen	see **E. *x warleyense*** 'Orangekoenigin'
	x warleyense 'Orangekoenigin'	wHer wNay
un	***x youngianum*** 'John Gallagher'	wHer wMtT
un	***x youngianum*** 'Kozakura'	oSis

	Plant	Sources
	x youngianum 'Merlin'	wHer wCol
	x youngianum 'Niveum'	oFor wCul oGos oMis oGar oSis oNWe oOut wRob wNay oGre
	x youngianum 'Roseum'	oGos wCol oAls oRus oGar wRob wNay wTGN oEga
un	***x youngianum*** 'Violicum'	oGos
	x youngianum 'Yenomato'	wCol oSis oNWe wRob
	EPIPACTUS	
	gigantea	oFor oTrP oSis
	gigantea 'Serpentine Night'	oGos wCol
	EPIPREMNUM	
	aureum	oGar
	EPITHELANTHA	
	micromeris var. ***pachyrhiza*** SB 325	oRar
	pachyrhiza	see **E. *micromeris*** var. ***pachyrhiza***
	EQUISETUM	
un	'Caribou River'	oTrP
un	***diffusum***	oWnC
	hyemale	oFor wSoo wGAc wNot oSsd iGSc oWnC oTri
	hyemale var. ***affine***	oWnC
	scirpoides	oFor oTrP oDar oHug oSsd
	scirpoides contorta	oTrP wFan
	ERAGROSTIS	
	curvula	wSnq
	spectabilis	wSnq
	trichodes	oFor oGoo
	trichodes 'Bend'	oOut oWnC
	ERANTHIS	
	hyemalis	oWoo
	EREMOCITRUS	
	glauca	oOEx
	EREMOCITRUS X CITRUS	
	E. ***glauca*** x C. ***meyeri***	oOEx
	EREMURUS	
	bungei	see **E. *stenophyllus*** ssp. ***stenophyllus***
	himalaicus	oRus
	x isabellinus 'Cleopatra'	oRus
	x isabellinus 'Pinokkio'	oRus
	x isabellinus Ruiter hybrids	oRus oGar
	x isabellinus Shelford hybrids	oAls oRus
	pink	wMag
	robustus	oRus
	salmon	wMag
	stenophyllus ssp. ***stenophyllus***	oFor oRus oGar oSho
	white	wMag
	ERIANTHUS	see **SACCHARUM**
	ERICA	
	arborea	oSho wCCr
	arborea var. ***alpina***	wHea
	arborea 'Estrella Gold'	wCul wHea
	arborea 'Spring Smile'	wHea
	australis 'Holehird'	wHea
	australis 'Riverslea'	wHea
	canaliculata	oHed
	carnea	wHea
	carnea 'Accent'	wHea
	carnea 'Adrienne Duncan'	wHea
	carnea 'Alan Coates'	wHea oWnC
	carnea 'Altadena'	wHea
	carnea 'Ann Sparkes'	wHea oSis
	carnea 'Atrorubra'	wHea
	carnea 'Aurea'	oBlo oWnC
	carnea 'Barry Sellers'	wHea
	carnea 'Bell's Extra Special'	wHea
	carnea 'Beoley Pink'	wHea
un	***carnea*** 'Bright Jewel'	wHea
	carnea 'Challenger'	wHea
	carnea 'Christine Fletcher'	wHea
	carnea 'Clare Wilkinson'	wHea oSqu
	carnea 'December Red'	oFor wHea oSho
	carnea 'Dommesmoen'	last listed 99/00
	carnea 'Early Red'	wHea
	carnea 'Eileen Porter'	wHea
	carnea 'Golden Starlet'	wHea
	carnea 'Gracilis'	last listed 99/00
	carnea 'Heathwood'	wHea
	carnea 'Hilletje'	wHea
	carnea 'Jennifer Anne'	wHea
	carnea 'King George'	oFor oAls oGar oSho oTPm oWnC wSta
	carnea 'Lesley Sparkes'	last listed 99/00
	carnea 'Lohse's Rubin'	wHea
	carnea 'Loughrigg'	wHea oGar oSis oWnC wSta
	carnea 'March Seedling'	wHea oGar oWnC
	carnea 'Myretoun Ruby'	oFor wHea oTPm oWnC wTGN

	Plant	Sources
	carnea 'Pink Cloud'	wHea
	carnea 'Pink Spangles'	oFor wHea oGar oWnC wSta
	carnea 'Pirbright Rose'	wHea oAls
	carnea 'Porter's Red'	wHea oAls oGar oSis
	carnea 'Prince of Wales'	wHea
	carnea 'Queen Mary'	wHea
	carnea 'R.B. Cooke'	wHea
	carnea 'Red Rover'	last listed 99/00
	carnea 'Rosalie'	wHea
	carnea 'Rotes Juwel'	wHea
	carnea 'Rubinteppich'	wHea
	carnea 'Ruby Glow'	wHea oGar oTPm
	carnea 'Schneekuppe'	wHea
	carnea 'Sherwood Creeping'	wHea
un	***carnea*** 'Sherwood Early Red'	wHea wSta
	carnea 'Sherwoodii'	see **E. *carnea*** 'Sherwood Creeping'
	carnea 'Snow Queen'	wSta
	carnea 'Spring Cottage Crimson'	wHea
	carnea 'Springwood Pink'	oFor wHea wSwC oGar oSho oWnC wSta oBRG
	carnea 'Springwood White'	wHea wSwC oGar oSho oWnC wSta
	carnea 'Startler'	oGar
	carnea 'Sunshine Rambler'	wHea
	carnea 'Thomas Kingscote'	wHea oSqu
	carnea 'Treasure Trove'	wHea
	carnea 'Vivellii'	oFor wHea oAls oGar oSho oTPm wSta
	carnea 'Walter Reisert'	wHea
	carnea 'Wanda'	wHea
	carnea 'Wentwood Red'	wHea
	carnea 'Westwood Yellow'	wHea
	carnea 'Winter Sports'	wHea
	ciliaris 'Corfe Castle'	wHea
	ciliaris 'David McClintock'	wHea
	ciliaris 'Globosa'	wHea
	ciliaris 'Maweana'	wHea oSis
	ciliaris 'Mrs. C.H. Gill'	wHea
	ciliaris 'Ram'	wHea
	cinerea f. ***alba***	wHea
	cinerea 'Alba Minor'	oSis
	cinerea 'Ashdown Forest'	wHea
	cinerea 'Atropurpurea'	wHea oAls oGar
	cinerea 'Atrorubens'	wHea oWnC
	cinerea 'Atrosanguinea'	wHea
	cinerea 'Baylay's Variety'	wHea
	cinerea 'C. D. Eason'	wHea oAls oGar oSho oBlo oTPm oWnC
	cinerea 'C. G. Best'	wHea
	cinerea 'Cevennes'	wHea
	cinerea 'Cindy'	wHea
	cinerea 'Coccinea'	oSis
	cinerea 'Contrast'	wHea
	cinerea 'Eden Valley'	wHea
	cinerea 'Fiddler's Gold'	wHea
	cinerea 'Frances'	wHea
	cinerea 'G. Osmond'	last listed 99/00
	cinerea 'Golden Drop'	wHea oSis oWnC
	cinerea 'Golden Hue'	wHea wSta
	cinerea 'Golden Sport'	wHea
	cinerea 'Gurnsey Plum'	last listed 99/00
	cinerea 'Hardwick's Rose'	last listed 99/00
	cinerea 'Iberian Beauty'	last listed 99/00
	cinerea 'Knap Hill Pink'	last listed 99/00
	cinerea 'Lilacina'	wHea
	cinerea 'Lime Soda'	wHea
	cinerea 'Miss Waters'	wHea
	cinerea 'Mrs. Ford'	wHea
	cinerea 'P. S. Patrick'	wHea
	cinerea 'Pallas'	oGar
	cinerea 'Pentreath'	wHea
	cinerea 'Pink Ice'	wHea
	cinerea 'Purple Beauty'	wHea oWnC
	cinerea 'Rosabella'	wHea
	cinerea 'Rosea'	wHea
	cinerea 'Rozanne Waterer'	wHea
	cinerea 'Splendens'	wHea
	cinerea 'Velvet Night'	wHea oGar oSis oTPm
un	***cinerea*** 'Violacea'	wHea
	cinerea 'Vivienne Patricia'	wHea oGar
	x darleyensis 'Ada S. Collings'	wHea
	x darleyensis 'Alba'	see **E. *x darleyensis*** 'Silberschmelze'
	x darleyensis 'Archie Graham'	wHea
	x darleyensis 'Arthur Johnson'	wHea oGar
un	***x darleyensis*** 'Cross Puzzle'	wHea
	x darleyensis 'Darley Dale'	wHea oGar oSho
	x darleyensis 'Dunreggan'	wHea

	*x **darleyensis*** 'Furzey'	wHea wSwC oGar oSis oSho oBlo oGre oTPm oWnC wCoS
	*x **darleyensis*** 'George Rendall'	wHea oGar oSis
	*x **darleyensis*** 'Ghost Hills'	oFor wHea oGar oTPm oWnC
	*x **darleyensis*** 'J. W. Porter'	wHea oWnC
	*x **darleyensis*** 'Jack H. Brummage'	wHea oWnC
	*x **darleyensis*** 'Jenny Porter'	wHea oWnC
	*x **darleyensis*** 'Kramer's Rote'	wHea oGar oWnC wTGN
	*x **darleyensis*** 'Margaret Porter'	wHea oGar oSis oGre
	*x **darleyensis*** 'Mary Helen'	wHea
	*x **darleyensis*** 'Silberschmelze'	wHea oGar
	*x **darleyensis*** 'Spring Surprise'	wHea
	*x **darleyensis*** 'W. G. Pine'	wHea
	*x **darleyensis*** 'White Glow'	wHea
	*x **darleyensis*** 'White Perfection'	wHea
	erigena 'Brian Proudley'	wHea
	erigena 'Brightness'	last listed 99/00
	erigena 'Ewan Jones'	wHea
	erigena 'Golden Lady'	oFor oSis
	erigena 'Hibernica'	wHea
	erigena 'Irish Dusk'	oFor wHea oGar oWnC
	erigena 'Maxima'	wHea
un	***erigena*** 'Mediterranean Pink'	oAls oGar oSho oBlo oTPm oWnC oGue wCoS
un	***erigena*** 'Mediterranean White'	oAls oSho oBlo oTPm oWnC oGue wCoS
	erigena 'Superba'	wHea oGar
	erigena 'W.T. Rackliff'	wHea
un	*x **griffithsii*** 'Elegant Spike'	wHea
	*x **griffithsii*** 'Heaven Scent'	wHea
	*x **griffithsii*** 'Valerie Griffiths'	wHea
	lusitanica	wHea
	lusitanica 'George Hunt'	wHea
	mackayana ssp. ***andevalensis***	last listed 99/00
	mackayana 'Doctor Ronald Gray'	wHea
do	***mackayana*** 'Maura'	last listed 99/00
	mackayana 'Shining Light'	wHea
	*x **praegeri***	see **E.** ***x stuartii***
	scoparia	wCCr
	scoparia ssp. ***scoparia*** 'Minima'	wHea
un	***sicula*** var. ***libanotica***	oSis
	spiculifolia	oFor wHea oSis
	spiculifolia x ***bergiana***	wHea
	*x **stuartii*** 'Irish Lemon'	oFor wHea oGar oSis oSqu oWnC
	*x **stuartii*** 'Irish Orange'	wHea
	*x **stuartii*** 'Stuartii'	wHea
	terminalis	wHea wCCr
	terminalis 'Thelma Woolner'	wHea
	tetralix 'Alba'	wHea oSqu oWnC
	tetralix 'Alba Mollis'	wHea oSis oWnC
	tetralix 'Alba Praecox'	wTGN
	tetralix 'Con Underwood'	wSta
	tetralix 'Daphne Underwood'	last listed 99/00
	tetralix 'Darleyensis'	oSis
	tetralix 'George Frazer'	wHea oAls oGar
	tetralix 'Hookstone Pink'	wHea
	tetralix 'Ken Underwood'	last listed 99/00
	tetralix 'Mollis'	last listed 99/00
	tetralix 'Pink Glow'	wHea oSis
	tetralix 'Pink Pepper'	wHea
	tetralix 'Pink Star'	oSis
	tetralix 'Rosea'	wHea
	tetralix 'Swedish Yellow'	wHea oSis
	tetralix 'Tina'	last listed 99/00
	vagans 'Birch Glow'	oSis
	vagans 'Cornish Cream'	wHea
	vagans 'Diana Hornibrook'	last listed 99/00
	vagans 'Golden Triumph'	last listed 99/00
	vagans 'J. C. Fletcher'	wHea
	vagans 'Lyonesse'	wHea
	vagans 'Miss Waterer'	wHea
	vagans 'Mrs. D.F. Maxwell'	wHea oAls oGar oTPm oWnC wSta
	vagans 'Pyrenees Pink'	wHea
	vagans 'Saint Keverne'	wCCr oWnC
	vagans 'Yellow John'	wHea
	*x **veitchii*** 'Exeter'	wHea
	*x **veitchii*** 'Pink Joy'	wHea
	*x **watsonii*** 'Cherry Turpin'	wHea
	*x **watsonii*** 'Dawn'	oFor wHea oAls oGar oSho oSqu oTPm oBRG
	*x **watsonii*** 'Dorothy Metheny'	wHea
	*x **watsonii*** 'H. Maxwell'	wHea
	*x **watsonii*** 'Mary'	wHea
	*x **watsonii*** 'Pink Pacific'	wHea
un	*x **watsonii*** 'Pink Pearl'	wHea
	*x **watsonii*** 'Truro'	wHea wSta

	*x **williamsii*** 'Cow-y-Jack'	wHea
	*x **williamsii*** 'David Coombe'	wHea
	*x **williamsii*** 'Gwavas	wHea
	*x **williamsii*** 'Ken Wilson'	wHea
	*x **williamsii*** 'P.D. Williams	wHea oWnC
	ERIGERON	
	acer var. ***debilis*** NNS96-86	wMtT
	aurantiacus	oFor
	aureus 'Canary Bird'	wMtT
	Azure Fairy	see **E.** 'Azurfee'
	'Azurfee'	oFor wCSG oWnC wTGN
	basalticus	wMtT
	'Blue Beauty'	oFor
un	'Chameleon'	wMtT
	chrysopsidis 'Grand Ridge'	last listed 99/00
	compositus	oSis oBos
	compositus JCS9811	wMtT
	compositus pink forms	wMtT
	Darkest of All	see **E.** 'Dunkelste Aller'
	'Dunkelste Aller'	oNat oWnC wTGN oUps
	elegantulus	wMtT
	'Foersters Liebling'	last listed 99/00
	glaucus	oAls oBos oWnC wSnq
un	***glaucus*** 'Olga'	oSis
	'Goat Rocks'	wMtT
	karvinskianus	wWoS wTho oDar oHed oGar wPir oSec wBWP iArc oCir oEga oUps
ch	***karvinskianus*** 'Profusion'	see **E.** ***karvinskianus***
	leiomerus	wMtT
	montanensis	see **E.** ***ochroleucus***
	ochroleucus	wMtT
	ochroleucus var. ***scribneri*** NNS97-103	wMtT
	peregrinus	oBos wWld
	Pink Jewel	see **E.** 'Rosa Juwel'
	'Prosperity'	last listed 99/00
	pyrenaicus (see Nomenclature Notes)	wMtT
	'Rosa Juwel'	oFor oWnC oUps
	scopulinus	oSis
	speciosus	wCSG
un	***umbelliferum*** 'Meadow Muffin'	wHer
	uncialis var. ***conjugans*** JCA1.323.720	wMtT
	ERINACEA	
	anthyllis	oSis
	pungens	see **E.** ***anthyllis***
	ERINUS	
	alpinus	oFor oAls oSis oNWe
	alpinus var. ***albus***	oSis
	alpinus 'Carmineus'	last listed 99/00
	alpinus 'Picos de Europa'	oSis
	ERIOBOTRYA	
	deflexa	oOEx
un	***dubia***	oOEx
	japonica (fruit)	oFor oTrP wBur wCCr
un	***japonica*** 'Advance'	oOEx
un	***japonica*** 'Big Jim'	oOEx
	japonica 'Champagne'	oOEx
	japonica 'Gold Nugget'	last listed 99/00
	japonica 'MacBeth'	oOEx
	ERIOGONUM	
	elatum	oFor
	grande var. ***rubescens***	see **E.** ***latifolium*** var. ***rubescens***
	kennedyi ssp. ***alpigenum*** NNS93-276	wMtT
	latifolium var. ***rubescens***	oHed
	ochrocephalum NNS95-246	wMtT
	ovalifolium	wMtT
	siskiyouense	wMtT
	umbellatum	oFor wShR wPla
un	***umbellatum*** 'Alturas Red'	oSis
	umbellatum Bear Tooth Pass	oSis
	umbellatum dwarf form	oSis
	umbellatum var. ***minus*** NNS95-249	wMtT
un	***umbellatum*** 'Silver Lake'	oSis
un	***umbellatum*** 'Siskiyou Gold'	oSis
un	***umbellatum*** 'Whiskey Peak'	oSis
	ERIOPHORUM	
	angustifolium	wGAc oHug oGar oSsd oWnC oSec
	spissum	see **E.** ***vaginatum***
	vaginatum	see **SCIRPUS** ***fauriei*** var. ***vaginatus***
	ERIOPHYLLUM	
	lanatum	oFor oJoy oBos oSho wRob oAld wWld
	lanatum 'Bella'	wBWP
	lanatum 'Siskiyou'	oSis
	ERODIUM	
	carvifolium	oJoy
	castellanum JC 941	oNWe

	Name	Sources
	chamaedryoides	see **E. *reichardii***
	cheilanthifolium	oSis
	chrysanthum	oJoy oHed oSis
	chrysanthum pink form	oHed
	glandulosum	oTrP
	glandulosum 'Roseum'	oHed oSis
	x kolbianum 'Natasha'	oHed
	macradenum	see **E. *glandulosum***
	manescaui	wCri wHig oDan
un	'Marion'	oHed
	pelargoniiflorum	wHer
	petraeum ssp. ***crispum***	see **E. *cheilanthifolium***
	petraeum 'Roseum'	see **E. *glandulosum*** 'Roseum'
	reichardii	oAls wCoN
un	***reichardii*** 'Roseum'	oFor oTrP oSis wSta
	rupestre	oHed
	supracanum	see **E. *rupestre***
	x variabile	oUps
	x variabile 'Bishop's Form'	oHed wAva
do	***x variabile*** 'Flore Pleno'	wTGN
	ERUCA	
	vesicaria ssp. ***sativa***	oAls wFai
	ERYNGIUM	
	agavifolium	wCul oEdg wHig oSis oCis oSho oSle
	agavifolium HCM 98048	wHer
	alpinum	wHer wGAc oSis
	alpinum 'Blue Star'	wWoS oWnC
	alpinum 'Superbum'	last listed 99/00
	amethystinum	wCul wCSG oGar wPir oSec oUps
	billardieri	last listed 99/00
	bourgatii	wHig oGoo oGre wPir
	caeruleum	oJoy
	dichotomum	wCul
	foetidum	oAls iGSc
	giganteum	oFor wHer wCul wCri oJoy oGoo oNWe wBox
	giganteum 'Silver Ghost' seed grown	last listed 99/00
	maritimum	wFai
	'Miss Willmott's Ghost'	see **E. *giganteum***
	x oliverianum	wBox
	pandanifolium	oJoy oCis
	pandanifolium var. ***lasseauxii***	wHer
	planum	oJoy wHig oMis oGoo oCis oBlo
	planum 'Blaukappe'	wWoS wSnq
	planum Blue Cap	see **E. *planum*** 'Blaukappe'
	planum 'Flueela'	wHer
	proteiflorum	oMis oCis
	sp. T72-2 Mexico	oCis
	x tripartitum	wCul wGAc oJoy wHig oCis oSho wPir wBox
	umbelliferum	wHer oCis
	variifolium	wGAc oJoy wHig oMis oGar oNWe wBox
	yuccifolium	oFor wCul oJoy oSho iGSc oCrm oUps oSle
	x zabelii	wHer
	ERYSIMUM	
	alpinum	see **E. *hieraciifolium***
	'Aunt May'	wHer wWoS
	'Bowles' Mauve'	oFor wWoS oAls wCSG oMis oGar oSis wCCr wHom oWnC wMag wTGN oCir wAva
	cheiri	wFai
	cheiri 'Golden Bedder' (Bedder Series)	oBlo
	'Constant Cheer'	wHer oHed
	'Glowing Embers'	wWoS
	helveticum	oSis
	hieraciifolium	oFor
	'John Codrington'	wHer wWoS oAls
	'Julian Orchard'	wHer wWoS oHed
	kotschyanum	wTho oAls oWnC
	linifolium	oUps
	linifolium 'Variegatum'	wWoS oNat oHed oSis wHom oWnC wTGN oUps
	nivale	oHan
un	'Orange Beauty'	oGar
	'Orange Flame'	oJoy oAls oSis wMag
	'Perry's Pumpkin'	wHer
	'Plant World Gold'	wHer
un	'Prince of Purple'	oWnC
	pulchellum	wMtT
un	'Sunlight'	oJoy oAls
	'Wenlock Beauty'	oFor wWoS wCul oNat oJoy oSis oNWe oWnC
	ERYTHRINA	
	americana BLM 0723	oRar
	caffra	last listed 99/00
	coralloides	see **E. *americana***
	flabelliformis	oRar
un	***herbacea*** var. ***herbacea***	oRar
	lysistemon	oRar
	poeppigiana	oRar
	sp. BLM0688	oRar
	sp. BLM 0963	oRar
	sp. BLM 1220	oRar
	zeyheri	oRar
	ERYTHRONIUM	
	americanum	oTrP wHer wThG
	californicum	wThG
	californicum 'White Beauty'	oNWe
	citrinum	oSis
	'Citronella'	oSis
	dens-canis	wThG
	dens-canis 'Rose Queen'	last listed 99/00
	dens-canis 'Snowflake'	last listed 99/00
	hendersonii	oSis
	'Kondo'	last listed 99/00
	oregonum	oRus oBos oNWe oTri oAld
	'Pagoda'	wHer oRus oSis
	revolutum	wThG oRus oNWe oSho oAld
	revolutum 'Rose Beauty'	oSis
	tuolumnense	wThG oNat oRus oSis
	ESCALLONIA	
	'Alice'	wCCr
	'Apple Blossom'	oFor oGar oRiv oSho oBlo oGre oWnC wTGN oWhS wAva wWel
qu	***biflora*** var. ***compacta*** 'Alba'	oCis
	x exoniensis	wHer
	x exoniensis 'Frades'	oFor oAls oGar oSho wCCr wTGN wCoS wWel
	x exoniensis 'Garland Dwarf'	oGar
	x exoniensis hybrid pink	oAls
un	***x exoniensis*** 'Pink Princess'	oBlo oGre oBRG oThr
	fradesii	see **E. *x exoniensis*** 'Frades'
	'Iveyi'	wHer wSte
	laevis	wCCr
	laevis 'Gold Brian'	oFor oGre
	'Newport Dwarf'	oGar oSho wCoS wWel
	'Pride of Donard'	oFor oGar oSho wWel
	'Red Dream'	wHer
	'Red Elf'	oGar oRiv wWel
	revoluta	oFor
	rosea	wHer
	rubra	oFor oSis wCCr oWnC wSta oBRG
	rubra dwarf form	oBlo
	rubra HCM 98067	wHer
	rubra 'Ingramii'	wCCr
un	***rubra*** 'Nana'	wSta
	'Slieve Donard'	wHer wCCr
	sp. aff. ***virgata*** HCM 98049	wHer
	ESCHSCHOLZIA	
	californica pink form	wCul
	EUCALYPTUS	
	aggregata	oFor oGar
	alpina	last listed 99/00
	archeri	oFor oTrP wHer oRiv oCis wRai
	bridgesiana	iGSc
	brookeriana	last listed 99/00
un	***calicola***	oTrP
	camphora	oFor
	cinerea	oFor oTrP oGar oGre iGSc oWnC wCoN oUps
	cinerea 'Pendula'	oGoo
	citriodora	oTrP oGoo iGSc wHom oUps
	coccifera	oFor wHer wSte
	cordata	oFor
	crenulata	oTrP
	dalrympleana	oFor oDar oGar wRai oOEx oGre wBox wSte
	delegatensis	wCCr
	dives	oTrP
	glaucescens	oTrP wCCr wSte
	globulus	iGSc
	globulus bicostata	last listed 99/00
	gregsoniana	oFor wSte
	gunnii	oFor wHer wCul wBur oDar oGar oRiv oGre wFai wPir wBox wCoN wSte wCoS
	gunnii divaricata	oTrP wCCr wSte
	johnstonii	oFor
	kitsoniana	last listed 99/00
	kruseana	last listed 99/00
	kybeanensis	oTrP

	leucoxylon	oTrP
	marginata dwarf	last listed 99/00
	melliodora	oGoo
	moorei	last listed 99/00
	moorei nana	oFor
un	***morrisbyi***	oFor
	neglecta	oFor oTrP oCis
	nicholii	oFor
	niphophila	see **E. *pauciflora*** ssp. ***niphophila***
	nitens	wCCr
	nova-anglica	oFor
	orbifolia	last listed 99/00
un	***oreades***	oFor
	ovata	oFor
	parvifolia	oFor oTrP wCCr wSte
	parvula	see **E. *parvifolia***
	pauciflora	last listed 99/00
	pauciflora ssp. ***debeuzevillei***	oFor oTrP wSte
	pauciflora ssp. ***niphophila***	oFor oTrP wCul wBur oDar oCis oGre wPir wCCr iGSc wWel
un	***pauciflora*** ssp. ***niphophila*** 'Mt. Hotham'	oTrP
	perriniana	oFor oGre wCCr wSte
	platypus	last listed 99/00
un	***polyandra*** DJH 422	wHer
	polyanthemos	oRiv wDav wHom
	pulverulenta	oFor wSte
	radiata	iGSc
	regnans	wCCr
un	***rigidula***	oTrP
	rubida	oTrP wSte
un	***scoparia***	oFor
	stellulata	oFor oGre wSte
	subcrenulata	oFor
qu	***uncinata***	oTrP
	urnigera	oFor oSho
	viminalis	oFor oTrP wSte
	EUCOMIS	
	autumnalis	oTrP wHer oDan oCis oRar oVBI
	bicolor	oTrP oGar oRar
	comosa pink	oGar
un	***comosa*** 'Sparkling Burgundy'	wHer wCol
	montana	oTrP
	pole-evansii	wHer
un	'Rosa'	wTGN
	EUCOMMIA	
	ulmoides	oFor oRiv oWhi oOEx oWnC oSle
	EUCRYPHIA	
	cordifolia	last listed 99/00
	glutinosa	oFor oRiv oBov wCCr wSta
	glutinosa 'Flora Plena'	see **E. *glutinosa*** Plena Group
	glutinosa 'Nana'	last listed 99/00
do	***glutinosa*** Plena Group	last listed 99/00
	x intermedia	oGos
	x intermedia 'Rostrevor'	oFor wHer wCCr
	lucida	oFor oGos oRiv oBov
	lucida 'Pink Cloud'	oGos
	x nymansensis	wHer oRiv oSho wCoN wWel
	x nymansensis 'Mount Usher'	oFor oGar
	x nymansensis 'Nymansay'	oFor oGos
	EUGENIA	
	uniflora	oOEx
	EUMORPHIA	
	sericea	oCis
	EUONYMUS	
	alatus	oAls wShR wCSG oRiv oBlo wKin oWnC wSta oCir
	alatus var. ***apterus***	wHer
un	***alatus*** Chicago Fire™ / 'Timber Creek'	oAls wTGN
	alatus 'Compactus'	oFor oDar oGar oDan oRiv oSho oBlo oGre wKin oWnC wTGN wSta oGue wWhG oSle wCoS wWel
	alatus 'Monstrosus'	oFor wHer oGar
un	***alatus*** 'Rudy Hagg'	oFor
un	***alatus*** var. ***striatus*** DJH 172	wHer
	americanus	oFor oSho
	'Boxhill Banshee'	wBox
	bungeanus	wHer
un	***bungeanus*** 'Pink Lady'	oFor oGre
	europaeus	oFor oRiv wCCr oWnC wWel
	europaeus 'Aldenhamensis'	oFor wHer
un	***europaeus*** 'Red Ace'	oGos wCSG
un	***europaeus*** 'Red Cap'	oFor wWel
	europaeus 'Red Cascade'	wCul oGre
	fortunei	wCSG
un	***fortunei*** 'Aureovariegatus'	oSho
	fortunei Blondy / 'Interbolwi'	oFor
va	***fortunei*** 'Canadale Gold'	last listed 99/00
	fortunei 'Coloratus'	oRiv oGre oUps
va	***fortunei*** 'Emerald Gaiety'	oFor oAls wCSG oGar oRiv oSho wKin oTPm oWnC wTGN wBox wSta oCir oUps wWel
va	***fortunei*** 'Emerald 'n' Gold'	wCul oAls oGar oRiv oNWe oSho oGre wCCr wKin oTPm oWnC oThr wWel
	fortunei 'Gold Spot'	see **E. *fortunei*** 'Sunspot'
va	***fortunei*** Golden Prince	oGar oWnC
va	***fortunei*** 'Harlequin'	oFor oAls oGar oRiv oGre wTGN oUps wWel
	fortunei 'Ivory Jade'	oGar oSho oCir wWel
	fortunei 'Kewensis'	oFor wCul oDar oJoy oRiv oSho
un	***fortunei*** 'Moonshadow'	oGar
	fortunei var. ***radicans***	oRiv
un	***fortunei*** 'Silver Edge'	oTPm
va	***fortunei*** 'Silver Queen'	oAls
va	***fortunei*** 'Sunspot'	oFor oBlo oGre wCCr oWnC
	fortunei 'Variegatus'	oGar
	hamiltonianus ssp. ***sieboldianus***	oFor oGre
	japonicus	oGar
	japonicus 'Albomarginatus'	oAls oTPm
	japonicus 'Aureomarginatus'	oGar oRiv oTPm oWnC
	japonicus 'Aureovariegatus'	see **E. *japonicus*** 'Ovatus Aureus'
va	***japonicus*** 'Aureus'	wCoS
un	***japonicus*** 'Bekomasaki'	wHer oAls oGar oSis oSho oSec oBRG wWel
un	***japonicus*** 'Butterscotch'	oAls
	japonicus 'Chollipo'	wWel
	japonicus 'Gold Heart'	oSho
	japonicus 'Grandifolius'	oGar
un	***japonicus*** Green Spire	see **E. *japonicus*** 'Bekomasaki'
un	***japonicus*** 'Hines Gold'	oWnC
	japonicus 'Microphyllus'	oAls oGar oSho oBlo oWnC wSta oCir
	japonicus 'Microphyllus Albovariegatus'	oGar oSho oGre oTPm oCir oSle
un	***japonicus*** 'Microphyllus Butterscotch'	oWnC
	japonicus 'Microphyllus Variegatus'	see **E. *japonicus*** 'Microphyllus Albovariegatus'
va	***japonicus*** 'Ovatus Aureus'	oAls wCSG oGar
	japonicus 'Silver King'	oGar oGre oWnC oBRG
	japonicus Silver Princess™ / 'Moness'	oAls oGar oSho
	japonicus 'Variegatus'	see **E. *japonicus*** 'Microphyllus Albovariegatus'
	kiautschovicus 'Manhattan'	oFor oGar oBlo wWel
	latifolius	oFor wHer
un	'Lemon Queen'	wBox
	nanus var. ***turkestanicus***	oFor wHer
	obovatus	oFor
	occidentalis	wNot oRiv wCCr
	phellomanus	oFor
	planipes	oGos
	'Rokojo'	oFor oDar oSis oRiv wTGN wWel
	sachalinensis	see **E. *planipes***
	sp. DJHC 462	wHer
un	***vagans***	wHer
	EUPATORIUM	
un	***aromaticum*** 'Jocius Variegated'	wCol oOut
un	'Bartered Bride'	oFor oRus wSnq
	cannabinum	oFor oGoo
	cannabinum 'Album'	wHer
do	***cannabinum*** 'Flore Pleno'	oFor wHer wCul oGar wNay oGre wSnq
	capillifolium 'Elegant Feather'	wHer wCol oOut oSec
	chinense DJHC 474	wHer
	coelestinum	oGar oGoo oOut wSnq
	fistulosum	oFor wCul wCol oGoo
un	***fistulosum*** 'Album'	wCol
	fistulosum 'Atropurpureum'	wBWP
	fistulosum selection	wWoS
	hyssopifolium	oFor wSnq
un	***leucolepsis***	oGoo
un	'Little Red'	oGar oOut oGre wFai
	perfoliatum	wFGN oGoo iGSc oCrm oBar
	purpureum	oAls oRus wFGN oGoo wFai iGSc oCrm oBar iArc oUps
	purpureum ssp. ***maculatum***	oAls wRob oUps
	purpureum ssp. ***maculatum*** 'Atropurpureum'	wWoS oRus oMis oGar oNWe oOut wNay oCir oSle
	purpureum ssp. ***maculatum*** 'Gateway'	oFor wHer wWoS wCol oRus oAmb wSnq
	purpureum selection	oUps
un	'Red Giant'	oGar oDan oOut oGre
	rugosum	oFor oMis

	rugosum 'Chocolate'	oFor wHer wWoS wCul oNat oDar wCol wHig oHed wCSG oAmb oDan oCis oNWe oSho oOut wNay oGre wFai wTGN wBox oUps wSnq
	variable 'Variegatum'	wHer
	EUPHORBIA	
un	***albipollinifera*** ES 1056 A12	oRar
	alluaudii	oRar
	amygdaloides	wCSG oDan wMag
	amygdaloides 'Purpurea'	oFor wWoS wCul oGos oJoy oAls oHed oRus oGar oDan oSis wRob wNay oGre wPir oSec wAva wSte wSnq
	amygdaloides var. ***robbiae***	oFor wHer wWoS wCul oGos wSwC wHig oAls oHed oRus wFGN wCSG oAmb oMis oGar oSho wRob oBov oGre wPir wCCr oSle wSte wSnq
	amygdaloides 'Rubra'	see **E. *amygdaloides*** 'Purpurea'
	balsamifera ssp. ***balsamifera***	oRar
	barnhartii	oRar
un	Beth Chatto hybrids	wHer
un	***californica*** var. ***californica***	oRar
un	***californica*** var. ***hindisiana*** BLM 0873	oRar
	candelabrum	oRar
	cap-saintemeniensis	oRar
	characias	wCul
va	***characias*** ssp. ***characias*** 'Burrow Silver'	oHed oCis
	characias ssp. ***characias*** 'Humpty Dumpty'	wCul wFGN oGar
	characias dwarf form	oSec
	characias 'Goldbrook' seed grown	wHer
	characias 'Portuguese Velvet'	wWoS wSwC oHed oNWe wSnq
	characias 'Portuguese Velvet' seed grown	wHer
	characias ssp. ***wulfenii***	oFor oGos wSwC oNat oDar oJoy oHed wCSG oMis oGar wFai wPir wTGN wBox oCir oEga oSle
un	***characias*** ssp. ***wulfenii*** 'Canyon Gold'	wWoS
	characias ssp. ***wulfenii*** 'Jimmy Platt'	oAls oDan
	characias ssp. ***wulfenii*** 'John Tomlinson'	last listed 99/00
	characias ssp. ***wulfenii*** 'John Tomlinson' seed grown	wHer
	characias ssp. ***wulfenii*** 'Lambrook Gold'	wSnq
un	'Cherokee'	oJoy oGar oNWe
un	***collectoides*** BLM 0689	oRar
	'Compton Ash'	see **E.** 'Copton Ash'
un	***confinalis*** var. ***confinalis***	oRar
un	***confinalis*** var. ***rhodesica***	oRar
un	'Copton Ash'	wHer wHig oHed oMis oNWe
	corallioides	wPir
	cornigera	wWoS oHed
	corollata	oMis
	cyparissias	oFor wCul oRus wCSG oGar oGre wFai
	cyparissias 'Clarice Howard'	wWoS oSho oGre wPir wBox
	cyparissias 'Fens Ruby'	wHer wCul wCol wHig oAls oHed oSqu wSnq
	cyparissias 'Orange Man'	wHer wWoS oJoy oSec wSnq
un	***decaryi*** var. ***decaryi***	oRar
un	***decaryi*** var. ***spirostacha***	oRar oSqu
un	***didieroides***	oRar
	donii	last listed 99/00
	donii Dixter form	wHer
	dulcis 'Chameleon'	oFor wHer wWoS wCul wSwC oNat oJil oDar oJoy wCol wHig oAls oHed oRus wFGN wCSG oMis oGar oSis oCis oNWe oSho oOut wNay oGre wFai oWnC wTGN oSec wAva wEde oEga oUps oSle wSte wSnq
un	***dulcis*** 'Mocha Gekko'	oJoy
	enormus	oRar
	epithymoides	see **E. *polychroma***
	Excaliber / 'Froeup'	oHed
	fimbriata	oRar
	flanaganii	last listed 99/00
	francoisii	oRar
	fruticosa blue form	last listed 99/00
	'Garblesham Enchanter'	wHer
	globosa	oRar
	'Golden Foam'	see **E. *stricta***
	gorgonis	oRar
	grandicornis	oRar
	greenwayi	oRar
	griffithii	oBlo
	griffithii 'Dixter'	wHer wWoS oJoy oHed oRus oGar oGre
	griffithii 'Fern Cottage'	wHer wWoS
un	***griffithii*** 'Fire Charm'	oAls
	griffithii 'Fireglow'	oFor wWoS oNat oDar oJoy oGar oDan oNWe oSho oTwi oGre wFai oSec wAva oUps wSnq
un	***griseola*** var. ***mashonica***	oRar
	groenewaldii	oRar
un	***guiengola*** BLM 0825	oRar
	halipedicola	oRar
	hedyotoides	oRar
un	***heterophylla*** 'Yokoi's White'	wWoS
un	***horrida*** var. ***nova***	oSqu
	ingens	oGar
	'Jade Dragon'	wWoS oGos wSwC wCSG oAmb wNay wSte wSnq
un	***lagunillarum*** BLM 1404	oRar
	lathyris	wCul iGSc wHom
	longifolia (see Nomenclature Notes)	oNWe
	x martinii	oFor wHer wCul oGos oJoy oAls oRus wFGN oMis oGar oDan oOut wNay wPir oWnC wMag wTGN wBox oCir oSle wSte wSnq wWel
un	***x martinii*** 'Red Martin'	wWoS oGos oDar wHig oHed oRus oNWe oGre wFai
un	***mcvaughii*** BLM 0673	oRar
	mellifera	wWoS wCCr
	milii	oGar
	monteiroi	oRar
	myrsinites	oFor wHer wCul oDar oSto oHed wCSG oMis oGar oDan oSis oUps wSnq
	neriifolia	oRar
un	***oaxacana***	oRar
	oblongata	last listed 99/00
	palustris	oFor oHed oNWe wSnq
	pfersdorfii	see **E. *submammillaris***
un	***platyclada*** var. ***hardyi***	oRar
un	***platyclada*** var. ***platyclada***	oRar
un	***platyclada*** var. ***platyclada*** B70.060	oRar
	polychroma	wHer wCul oAls oHed wCSG oMis oGar oDan oNWe oSho oOut oTwi oBlo oGre wFai wMag wTGN wAva oEga
	polychroma 'Candy'	wWoS oAls
	polychroma 'Major'	wHer
	polychroma 'Midas'	wHer oWnC
	polychroma 'Purpurea'	see **E. *polychroma*** 'Candy'
	x pseudovirgata	oHed
	pteroneura BLM 0626 & BLM 0826	oRar
	pulcherrima red	oGar
qu	***purpurea***	oAmb oGar
	robbiae	see **E. *amygdaloides*** var. ***robbiae***
	samburensis	oRar
	schillingii	wHer oHed oGar oNWe wBox
un	***schlectendalii*** BLM 0390	oRar
	schoenlandii	oRar
	seguieriana ssp. ***niciciana***	wWoS oHed oSis
	sikkimensis	oFor oJoy oHed oGar oGre wSte
	sp. HCM 98184	wHer
	sp. Madgascar	oRar
	spinosa	oSis
	squarrosa	oRar
	stellaespina	oRar
	stellata	oRar
	stricta	wWoS
	stricta 'Golden Foam'	see **E. *stricta***
	stygiana	wHer
	submammillaris	oSqu
	susannae	oRar
un	***tanquahuete*** BLM 0666	oRar
un	***tortirama***	oRar
	transvaalensis	oRar
	viguieri var. ***ankarafantsiensis***	oRar
	villosa	oJoy wBox
	virosa	oRar
	wallichii (see Nomenclature Notes)	wCul oNWe
	wulfenii	see **E. *characias*** ssp. ***wulfenii***
un	'Wundulate'	oSqu
	xantii BLM 0872	oRar
	EUPHORIA	
	longan	see **DIMOCARPUS *longan***
	EUPTELEA	
	pleiosperma	oFor oGre wPir wBox
	polyandra	oFor wCCr
un	**EURHYNCHIUM**	
	oreganum	wFFI
	praelongum	wFFI
	EURYA	
	emarginata	last listed 99/00

	emarginata HC 970595	wHer
	japonica	wCCr
	japonica 'Winter Wine'	oCis
	EURYOPS	
	pectinatus 'Munchkin'	oGar oSho
	pectinatus 'Viridis'	oAls oGar oSho wTGN wCoS wWel
	EUSCAPHIS	
	japonica	oCis
	EVODIA	
	danielii	see **TETRADIUM *daniellii***
	hupehensis	see **TETRADIUM *daniellii*** Hupehense Group
	EVOLVULUS	
	pilosus 'Blue Daze'	last listed 99/00
	EXACUM	
	affine	oGar
	EXBUCKLANDIA	
	populnea DJHC 549	last listed 99/00
	EXOCHORDA	
	giraldii	oFor oTrP
	giraldii var. ***wilsonii***	oFor
	korolkowii	wCCr
	x macranatha	oFor
	x macranatha 'The Bride'	oFor wHer wCul oAls oHed oAmb oGar oGre wCoN wWel
	FABIANA	
	imbricata	oCis
	imbricata alba	oHed oCis
	imbricata f. ***violacea***	oFor wHer oHed
	FAGUS	
	crenata	oAmb oRiv
	grandifolia	last listed 99/00
	orientalis	oFor
	sylvatica	oFor wBlu wBur wShR oAmb oGar oRiv oGre wSta wAva oSle wWel
	sylvatica 'Ansorgei'	oFor oRiv
	sylvatica 'Atropunicea'	see **F. *sylvatica*** Atropurpurea Group
	sylvatica Atropurpurea Group	oFor wBCr wBur oEdg oGar oRiv wRai oGre wKin wTGN wAva oSle
	sylvatica Atropurpurea Group 'Swat Magret'	oFor
	sylvatica 'Atropurpurea Pendula'	see **F. *sylvatica*** 'Purpurea Pendula'
	sylvatica 'Black Swan'	oFor oGos
	sylvatica 'Cockleshell'	oFor
	sylvatica 'Cristata'	last listed 99/00
	sylvatica 'Dawyck'	oRiv oBlo oGre oWnC wCoN
	sylvatica 'Dawyck Gold'	oFor wBox
	sylvatica 'Dawyck Purple'	oFor oDan
	sylvatica 'Fastigiata'	see **F. *sylvatica*** 'Dawyck'
	sylvatica var. ***heterophylla*** 'Aspleniifolia'	last listed 99/00
	sylvatica var. ***heterophylla*** f. ***laciniata***	oRiv wSta wCoN
	sylvatica 'Interrupta'	oFor
	sylvatica 'Luteovariegata'	oFor
	sylvatica 'Pendula'	oFor wClo oAls oRiv wKin oWnC wSta wCoN
	sylvatica 'Purple Fountain'	oFor oAls oBlo oWnC wSta
	sylvatica purpurea	see **F. *sylvatica*** Atropurpurea Group
	sylvatica 'Purpurea Pendula'	oFor wClo oGar oDan oWhi wSta oThr wCoN wWel
va	***sylvatica*** 'Purpurea Tricolor'	oGar
	sylvatica 'Quercifolia'	oFor oWhi
	sylvatica 'Quercina'	oFor
	sylvatica 'Red Obelisk'	oFor wClo oGar oDan oWhi oGre wWel
	sylvatica 'Riversii'	oFor oRiv oBlo oThr
	sylvatica 'Rohan Gold'	last listed 99/00
	sylvatica 'Rohanii'	oFor oRiv wSta
	sylvatica 'Roseomarginata'	see **S. *sylvatica*** 'Purpurea Tricolor'
	sylvatica 'Tortuosa'	oFor
	sylvatica 'Tortuosa Purpurea'	oFor
	sylvatica 'Tricolor'	oGos wClo oDar oAls oRiv oBlo oGre oWnC wTGN wSta oThr wCoN
	FALLOPIA	
	baldschuanica	oFor wCul oAls wCSG oGar oGre iGSc oWnC wTGN oWhS wCoS
	japonica var. ***compacta*** 'Variegata'	oFor wWoS oGar wHom
	japonica 'Crimson Beauty'	wHer
un	***japonica*** 'Devon's Cream'	wWoS
un	***japonica*** 'Spectabilis'	oGre
	japonica 'Variegata'	oFor oNWe oOut wRob wFai
	multiflora	oFor wFGN oGoo oOEx iGSc oCrm oBar
	FALLUGIA	
	paradoxa	oFor wShR oSho wCCr wPla
	FARFUGIUM	
va	***japonicum*** 'Aureomaculatum'	oFor wHer oGos oAmb oDan oCis wNay
	japonicum 'Crispatum'	oFor wHer wWoS oEdg oCis oNWe oOut wNay oGre wTGN
	japonicum 'Cristatum'	see **F. *japonicum*** 'Crispatum'
	FARGESIA	
un	***angustissima***	oTra wBea oTBG
	denudata	wCli oTBG
	dracocephala	oTra wCli oNor oGos wBea wBmG oTBG wBam
un	***dracocephala variegata***	oTBG
un	***fortis***	wBea
	fungosa	oTra wCli oNor wBea
	murieliae	oTra wCli oNor oGos wBea wRai wBmG oTBG
	murieliae seedlings	oTra
	nitida	oFor oTra wHer wSus wCli oNor wBlu oGos wBea oGar oSho wRai oOEx oGre wBox wBmG oTBG wBam
	nitida 'Anceps'	oTBG
	nitida 'Eisenach'	oTra oNor wBmG oTBG
	nitida 'Ems River'	oTBG
un	***nitida*** 'McClure'	oTBG
	nitida 'Nymphenburg'	wCli oNor wRai oTBG
	robusta	wCli wBea wBmG
	utilis	oTra wCli oNor wBea wBmG
	FASCICULARIA	
	bicolor	last listed 99/00
un	***bicolor*** ssp. ***bicolor***	wHer
	X FATSHEDERA	
	lizei	oFor oGar oSho oBlo wSta wCoN
un	***lizei*** 'Argenteovariegata'	oCis
va	***lizei*** 'Aurea'	wHer
un	***lizei*** 'Aureomaculata'	oSho
	lizei 'Aureopicta'	see **F. *lizei*** 'Aurea'
un	***lizei*** 'Aureovariegata'	oCis
un	***lizei*** 'Curly'	oCis
qu	***lizei*** 'Media-Picta'	oFor oGar
	FATSIA	
un	***corymbosa***	oCis
	japonica	oAls oGar oSho oBlo oGre oWnC wTGN wSta wCoN wCoS wWel
	japonica 'Variegata'	oCis
	FAUCARIA	
	lupina	last listed 99/00
	FEIJOA	see **ACCA**
	FELICIA	
	amelloides	oGar wHom oCir
	amelloides variegated	wHom
	FENESTRARIA	
	rhopalophylla	last listed 99/00
	FERRARIA	
	crispa	wCSG
	undulata	see **F. *crispa***
	FERULA	
	communis	wCol oDan wBWP
	communis 'Gigantea'	see **F. *communis***
	FESTUCA	
	amethystina	oRus wCCr oWnC oUps
un	***amethystina*** 'Rainbow'	oOut
	amethystina 'Superba'	oGar oOut oGre wWin oSec wSnq
	cinerea	last listed 99/00
	gautieri 'Pic Carlit'	wCli
	glauca	oAls wCSG oGar wCCr wHom wSta wPla oCir wCoN oEga oUps
	glauca 'Blaufink'	wWin
	glauca 'Blauglut'	oFor oGar oWnC wBox oSle wSnq
	glauca 'Blausilber'	oWnC wBox
un	***glauca*** 'Boulder Blue'	wWin
	glauca 'Daeumling'	wWin
	glauca 'Elijah Blue'	oFor wCli wCul wSwC oDar oJoy oAls oRus oGar oSis oSho oOut oBlo oGre wFai wWin wHom oWnC wTGN wBox oBRG wEde wWhG oUps wSnq wCoS
	glauca 'Golden Toupee'	wHer wWoS oAls oOut oSqu wWin wSnq
	glauca 'Meerblau'	wWin oWnC oBRG
	glauca 'Minor'	wMtT
un	***glauca*** 'Sarah's Blues'	oTPm
	glauca Sea Urchin	see **F. *glauca*** 'Seeigel'
	glauca 'Seeigel'	oRus oSqu wSnq
	glauca Tom Thumb	see **F. *glauca*** 'Daeumling'
	idahoensis	wShR wWat wPla
	mairei	oFor
	ovina	oSho
	ovina glauca	see **F. *glauca***
	ovina 'Tetra Gold'	wWoS
	rubra	wCSG

	tenuifolia	wCli
	valesiaca var. ***glaucantha***	last listed 99/00
	FIBIGIA	
	clypeata	oSis
	FICUS	
	arnottiana	oRar
	benghalensis	oRar
un	***benjamina*** 'Jacqueline'	oGar
	benjamina 'Variegata'	oGar
	binnendijkii 'Alii'	oGar
	carica (fruit)	wCoN
	carica 'Alma'	oOEx
un	***carica*** 'Armenian'	oOEx
un	***carica*** 'Ballard's VO5'	oOEx
un	***carica*** 'Black Jack'	oOEx oWnC
	carica 'Black Mission'	oSho
	carica 'Brown Turkey'	oEar wBur oAls oGar oSho oBlo wRai oOEx oWnC wCoS
	carica 'Celeste'	oOEx
	carica 'Conandria'	oOEx
	carica 'Deanna'	oOEx
	carica 'Desert King'	oEar wBur wClo oGar wRai wTGN
un	***carica*** 'Early Violet'	oOEx
qu	***carica*** 'Everbearing'	oOEx
un	***carica*** 'Excel'	oOEx
un	***carica*** 'Flanders'	oOEx
un	***carica*** 'Galbun'	oOEx
un	***carica*** 'Gillette'	oOEx
un	***carica*** 'Hardy Chicago'	oOEx
	carica 'Italian Everbearing'	oSho
	carica 'Italian Honey'	see **F. *carica*** 'Lattarula'
un	***carica*** 'Jenkins Estate'	oEar
	carica 'Kadota'	oOEx oWnC
un	***carica*** 'King'	oAls oOEx
	carica 'Lattarula'	oEar wBur oAls oGar wRai
un	***carica*** 'Magnolia'	oEar oOEx
un	***carica*** 'Mary Lane'	oOEx
	carica 'Mission'	oOEx oWnC
un	***carica*** 'Negronne'	oEar wBur wClo oGar oOEx
un	***carica*** 'Nero'	oOEx
	carica 'Neveralla'	oEar wRai oOEx
un	***carica*** 'Oregon Prolific'	oAls oGar
	carica 'Panachee'	wRai oOEx
un	***carica*** 'Pasquale'	oOEx
	carica 'Peter's Honey'	oEar wRai oWnC
un	***carica*** 'Petite Negri'	oGar oOEx oGre
un	***carica*** x ***pumila***	oCis
un	***carica*** 'Royal Vineyard'	oOEx
	carica 'Rutara'	see **F. *carica*** 'Peter's Honey'
un	***carica*** 'Saint Anthony'	oOEx
	carica 'Tena'	oOEx
un	***carica*** 'Tennessee Mountain'	oOEx
un	***carica*** 'Vern's Brown Turkey'	wWel
un	***carica*** 'Verte'	oOEx
un	***carica*** 'Violette de Bordeaux'	see **F. *carica*** 'Negronne'
	carica 'White Marseilles'	oOEx
	cordata	oTrP oRar
un	***glumosa***	oRar
	goldmanii BLM 0110	oRar
	infectoria	see **F. *virens***
	ingens	oRar
	lyrata	oGar
	microcarpa BLM 0404	oRar
qu	'Natasha'	oGar
	palmeri BLM 0214	oRar
	petiolaris	oRar
	pumila	oTrP oGar
	pumila 'Minima'	oTrP
	pumila 'Quercifolia'	oTrP
	pumila 'Variegata'	oGar
	religiosa	last listed 99/00
	sarmentosa var. ***nipponica***	oSis
	virens	oRar
qu	'Zigzag'	oGar
	FILIPENDULA	
	camtschatica	oHan wNay
	camtschatica HC 970498	wHer
un	'Carmine'	wCol
ch	***digitata*** 'Nana'	see **F. *multijuga***
un	***glaberrima***	wCol
	hexapetala	see **F. *vulgaris***
	hexapetala 'Flore Pleno'	see **F. *vulgaris*** 'Multiplex'
	'Kahome'	oFor wWoS oTwi
	kamtchatica	see **F. *camtschatica'***
qu	'Kokome'	wCol wRob wNay
	multijuga	oFor wCul oSho oUps
	palmata 'Elegans'	see **F. *purpurea*** 'Elegans'
ch	***palmata*** 'Elegantissima'	see **F. *purpurea*** 'Elegans'
	palmata 'Nana'	see **F. *multijuga'***
	purpurea	oFor wCul oWnC
	purpurea f. ***albiflora***	oFor
	purpurea 'Elegans'	wCul wHig oGar oSho oOut wRob wNay wMag wTGN wBox oUps
	purpurea 'Nana'	oNWe oGre
	purpurea 'Plena'	wNay
	rubra	oFor oTwi
	rubra 'Magnifica'	see **F. *rubra*** 'Venusta'
	rubra 'Venusta'	oFor wCul oNat oJoy wCol oAls oGar oDan oSho wRob wNay oGre wFai wMag wTGN oEga oUps
	rubra 'Venusta Magnifica'	see **F. *rubra*** 'Venusta'
	ulmaria	oGoo wFai iGSc oCrm oBar
	ulmaria 'Aurea'	wHer wWoS oJil wCol oAls oHed oMis oDan oNWe oOut wRob wNay oSqu oGre wFai wBox oSec
	ulmaria 'Aureovariegata'	see **F. *ulmaria*** 'Variegata'
do	***ulmaria*** 'Flore Pleno'	oFor wCul wNay
un	***ulmaria*** 'Nana'	oHan
	ulmaria 'Plena'	see **F. *ulmaria*** 'Flore Pleno'
	ulmaria 'Variegata'	oFor wWoS wCul wGAc oJil wHig oHed oGar oNWe oOut wNay oGre wFai wBox
	vulgaris	oFor oHan wCul oRus oGar oTwi wFai iGSc wCoN
	vulgaris 'Flore Pleno'	see **F. *vulgaris*** 'Multiplex'
do	***vulgaris*** 'Multiplex'	oFor wHig wNay oSqu
	FIRMIANA	
	platanifolia	see **F. *simplex***
	simplex	oFor oTrP oGar oRiv oCis wCCr oWnC wBox wCoN
	FITTONIA	
	albivenis Verschaffeltii Group	oGar
	FITZROYA	
	cupressoides	oFor wHer oAls oGar oRiv oBRG
	FOCKEA	
	angustifolia	oRar
	edulis	oRar
	multiflora	oRar
	FOENICULUM	
	vulgare	wCSG oGoo wFai iGSc wHom oUps
	vulgare var. ***azoricum***	oAls oWnC
	vulgare var. ***dulce***	oAls wFGN wFai
	vulgare nigrum	see **F. *vulgare*** 'Purpureum'
	vulgare 'Purpureum'	wCul oAls wFGN oGoo wFai iGSc wHom oCrm oBar oSec wEdc
	vulgare 'Smokey'	oGar
	FOKIENIA	
	hodginsii	oPor oRiv oCis
	FONTANESIA	
un	***phyllyreoides*** ssp ***fortunei*** 'Titan'	oFor
	FORESTIERA	
	acuminata	oFor
ch	***neomexicana***	see **F. *pubescens***
	pubescens	oFor
	FORSYTHIA	
	'Arnold Dwarf'	oFor oGre oWnC
	'Beatrix Farrand'	oFor oGar oSho oBlo
va	'Fiesta'	oFor wHer wCol oGar oGre wTGN
	'Gold Tide'	see **F.** Maree d'Or / 'Courtasol'
va	'Golden Times'	wHer
	x intermedia	wCoS
un	***x intermedia*** 'Goldleaf'	oGre wWel
	x intermedia 'Karl Sax'	oFor oBlo wCCr oWnC
	x intermedia 'Lynwood'	oFor oJoy oAls oGar oBlo oGre oTPm oWnC wSta oSle wWel
	x intermedia 'Lynwood Gold'	see **F. *x intermedia*** 'Lynwood'
	x intermedia 'Meadowlark'	oFor oGre wKin
	x intermedia 'Minigold'	oFor oDar
	x intermedia 'Spring Glory'	oJoy oGar oWnC
un	***x intermedia*** 'Sunrise'	oTPm wWel
	x intermedia 'Variegata'	oFor
	Maree d'Or / 'Courtasol'	oFor oGar oGre wWel
un	'New Hampshire Gold'	oFor
	'Northern Gold'	oFor
	'Northern Sun'	oFor
	ovata	wCCr
	ovata 'Ottawa'	oDar
	suspensa	oFor oGar oCrm
	suspensa dwarf weeping	oCis
	suspensa 'Nymans'	wCCr
	'Tremonia'	oFor oDar

	viridissima	oGar oCis
	viridissima 'Bronxensis'	wCul wCol oGar oRiv oGre wKin wWhG
un	viridissima 'Klein'	wWoS
	viridissima var. koreana	wCCr
	viridissima 'Robusta'	wCCr
	FORTUNEARIA	
	sinensis	wHer wCCr
	FORTUNELLA	
	hindsii (fruit)	last listed 99/00
	margarita (fruit)	oGar
	margarita 'Nagami' (fruit)	oAls wRai oWnC
	obovata (fruit)	oOEx
	FOTHERGILLA	
un	'Beaver Creek'	oGre
	gardenii	oFor wCul wSwC oAls oAmb oGar oRiv oBlo oGre wTGN
	gardenii 'Blue Mist'	wHer oGos oEdg oGar oDan oRiv oGre wFai wAva wWel
un	gardenii 'Eastern Form'	oGos
un	gardenii 'Jane Platt'	wHer oGos oGre
	major	wHer oGos wCoN
	major Monticola Group	oFor oBov oGre
	monticola	see F. major Monticola Group
	'Mount Airy'	oFor wHer wCul oGos oDar oEdg oRiv oWhi oGre wTGN oBRG wAva wCoN wSte wSnq wWel
	FOUQUIERIA	
	burragei	oRar
	columnaris	oRar
	diguetii BLM 0210	oRar
	fasciculata B74.0427	oRar
	formosa	oRar
un	macdougalii BLM 0107	oRar
	splendens	oRar
	FRAGARIA	
	x ananassa (fruit) 'Benton'	oAls oGar oSho wRai
	x ananassa (fruit) 'Fern'	oAls oGar
	x ananassa (fruit) 'Hecker'	oSho
	x ananassa (fruit) 'Hood'	wClo oAls oGar oSho
un	x ananassa (fruit) 'Peach Sized'	wCSG
un	x ananassa (fruit) 'Puget Reliance'	wClo
un	x ananassa (fruit) 'Puget Summer'	wRai
	x ananassa (fruit) 'Quinault'	oGar wTGN
	x ananassa (fruit) 'Rainier'	wBur oAls oGar
un	x ananassa (fruit) 'Seascape'	oAls oGar wCoS
	x ananassa (fruit) 'Selva'	wClo oGar
	x ananassa (fruit) 'Sequoia'	oSho
	x ananassa (fruit) 'Shuksan'	wClo oAls wRai
un	x ananassa (fruit) 'Sumas'	oGar
	x ananassa (fruit) 'Totem'	oAls
un	x ananassa (fruit) 'Tribute'	oAls oGar oSho
	x ananassa (fruit) 'Tristar'	wBur wClo oAls oGar oSho wRai wCoS
	californica	oFor
un	'Capron'	wRai
	chiloensis	oFor wCul wWoB oRus wShR wCSG oBos wRai wWat oTri oSle
	chiloensis 'Chaval'	last listed 99/00
	chiloensis 'Variegata'	oSho
un	'Frau Meize Schindler'	wHer
un	'Lila Diamond'	oGre
	'Lipstick'	oFor wWoS oDar oAls oSis wRai oGre oWnC wCoN oSle wCoS
un	'Mt. Omei'	oFor
un	nilgerrensis EDHCH 97059	wHer
un	nipponicum	oFor
un	'Perfumata de Tortona'	wRai
	Pink Panda / 'Frel'	oFor wCul wCri wFGN wCSG oGre wSta wAva wCoS
un	'Pink Shades'	wTGN
	'Red Ruby'	wTGN
	vesca (fruit)	oFor wFGN oGoo oBos wFai wWat wFFl
un	vesca 'Albicarpa'	see F. vesca 'Fructu Albo'
	vesca 'Alexandra'	oWnC
	vesca 'Alpine Yellow'	wRai
un	vesca crinita	wNot
	vesca 'Fructu Albo'	oSis
	vesca 'Pineapple Crush'	oHan
	vesca 'Ruegen'	wRai
un	vesca 'Ruegen Improved'	oHan oGre
	vesca 'Variegata'	wHer wWoS oDar oRus wTGN oInd oUps
	virginiana	wWoB wShR wWat wPla wFFl
un	virginiana var. platypetala	wNot oBos
	FRANCOA	
	appendiculata	oFor
	appendiculata HCM 98044	wHer
	ramosa	oHed oMis wAva
	sonchifolia	oFor oJoy oHed oRus oMis
	sonchifolia purple form	oHed
	FRANKLINIA	
	alatamaha	oFor oGos oDan oGre oWnC wCoN wWel
	FRAXINUS	
	americana	oFor
	americana 'Autumn Applause'	oGar
	americana 'Autumn Purple'	oFor oDan oSho wKin oWnC
	angustifolia	wCCr oWnC
	angustifolia 'Flame'	see F. angustifolia 'Raywood'
	angustifolia 'Raywood'	oFor oAls oGar oSho oBlo oGre wWel
	anomala	last listed 99/00
	bungeana	oFor
	chinensis	oFor wCCr
	cuspidata	oFor
	dipetala	oFor
	excelsior 'Aurea' (see Nomenclature Notes)	oFor oWhi
un	excelsior 'Golden Desert'	oGar oSho oWnC
	excelsior 'Hessei'	oFor
	excelsior 'Jaspidea' (see Nomenclature Notes)	oFor
	excelsior 'Pendula'	oFor
	latifolia	oFor wWoB wBur wNot wShR oGar oBos oRiv wWat wCCr oAld oSle
	longicuspis var. sieboldiana	see F. sieboldiana
	nigra 'Fallgold'	wKin
	oregona	see F. latifolia
	ornus	oFor oGre wCCr
	oxycarpa	see F. angustifolia
	paxiana	wCCr
	pennsylvanica	wBCr oEdg wCoN
un	pennsylvanica 'Cimmaron'	oFor oAls oGar oBlo
un	pennsylvanica 'Leprechaun'	oSho
	pennsylvanica 'Marshall's Seedless'	oGar oBlo oWnC wCoS
	pennsylvanica 'Patmore'	oWnC
	pennsylvanica 'Summit'	oWnC
	pennsylvanica 'Variegata'	oGre
	quadrangulata	oFor oRiv oGre
	sieboldiana	oFor
	sieboldiana HC 970172	wHer
	FREMONTODENDRON	
	'California Glory'	oFor
	californicum	oRiv
	'Pacific Sunset'	oRiv
	FRITILLARIA	
	affinis var. tristulus	oSis oBos
	camschatcensis	oOEx
	crassifolia ssp. kurdica	last listed 99/00
	imperialis 'Lutea'	oWoo
	imperialis 'Rubra'	oWoo
	lanceolata	see F. affinis var. tristulus
	meleagris	oNWe oGre oWoo
	meleagris 'Alba'	oNWe
	michailovskyi	oWoo
	persica	oWoo
	recurva	oSis
	FUCHSIA	
	'Abbe Farges'	oDel wDBF
	'Abigail'	oDel
	'Admiration'	last listed 99/00
	'Alice Doran'	oDel
	'Alice Hoffman'	wDBF
	'Alice Kling'	oDel
	'Alison Ewart'	oDel
	'Alison Patricia'	oDel
	'Alison Sweetman'	wHer
	alpestris	oDel oJoy
un	'Alsace'	oDel
	'Amapola'	oDel wDBF
	ampliata	oDel
	'Amy Lye'	oDel
un	'Amy Marie'	wDBF
	'Andenken an Heinrich Henkel'	oDel
un	'Angel Eyes'	wDBF
	'Angel's Dream'	oDel
	'Angel's Flight'	wDBF
	'Anita'	oDel
	'Applause'	oDel
	arborescens	oDel
	'Archie Owen'	wDBF
un	'Arctic Night'	oDel
	'Army Nurse'	oDel oGar wSnq
un	'Astoria'	oDel

	'Atomic Glow'	wDBF
	'Aurea'	see F. ***magellanica*** var. ***gracilis*** 'Aurea'
	'Australia Fair'	oDel
un	'Autumn Orange'	oDel
	'Autumnale'	oDel wDBF
	ayavacensis	oDel
	'Aztec'	oDel
	'Baby Ann'	wSnq
	'Baby Blue Eyes'	wDBF
	'Baby Bright'	oDel
	'Baby Lilac'	oDel
	x bacillaris 'Reflexa'	oDel wDBF
un	'Bagdad'	oJoy oSis oRiv wDBF
	'Balkonkoenigin'	oDel
un	'Ballerina Blau'	oDel
	'Barbara'	oDel
	'Bashful'	oDel
	'Beacon'	oDel wSnq
	'Beacon Rosa'	oDel oAls oGar
	'Beatrice Burtoft'	oDel
un	'Beauty-n-Red'	wDBF
	'Beauty of Clyffe Hall'	oJoy
	'Beauty of Trowbridge'	oDel
	'Beauty Queen'	wDBF
un	'Becky Pike'	wDBF
	'Bee Keesey'	oDel
	'Bell Buoy'	wDBF
	'Bella Rozella'	oDel wTGN
	'Bellbottoms'	oDel
	'Belle of Salem'	wDBF
	'Ben Jammin'	oDel
	'Berba's Coronation'	oDel
	'Bergnimf'	oDel
	'Bernadette'	oDel
un	'Bernisserstein'	oDel
	'Bertha Gadsby'	oDel
un	'Bert's Arendnistje'	oDel
	'Beth Robley'	last listed 99/00
	'Bette Sibley'	oDel
	'Betty Jean'	wDBF
un	'Betty Wass'	oDel
	'Beverley'	oDel
	'Bicentennial'	oDel wDBF
un	'Big Mama'	wHer wSwC
un	'Bill Kennedy'	oDel
	'Billy Green'	oDel
	'Black Prince'	wDBF oCir wSte
	'Black Princess'	oDel
	'Bland's New Striped'	wSnq
	'Blood Donor'	oDel
	'Blue Boy'	oDel wDBF
	'Blue Bush'	oDel
	'Blue Eyes'	wDBF oCir
un	'Blue Louie'	oDel
	'Blue Mirage'	oDel
	'Blue Satin'	oDel wDBF
	'Blue Tit'	oDel
	'Blue Veil'	last listed 99/00
	'Blue Waves'	oDel
	'Bluette'	oDel
	'Blush of Dawn'	wDBF
un	'Bobby's Redwing'	oDel
	'Bobolink'	oDel
	boliviana	see F. ***sanctae-rosae***
	boliviana (Carriere) var. ***alba***	oDel
	'Bon Accorde'	oFor oDel oGar oWnC wDBF
	'Bonanza'	wHom
	'Bonnie Doan'	oDel
	'Border Queen'	oDel wSnq
	'Bouquet'	oDel
	'Bow Bells'	oDel
	brevilobis	oDel
	'Brian C. Morrison'	oDel
	'Brian Stannard'	oDel
	'Brighton Belle'	oDel
	'Briony Caunt'	oDel
	'British Jubilee'	oDel
	'Brookwood Belle'	oDel
	'Brutus'	oDel wSnq
un	'Bud Cole'	oJoy wDBF
	'Buenos Aires'	oDel
	'Bugle Boy'	oDel
	'Buttercup'	wDBF
	'Caesar'	oDel
	'Caledonia'	oJoy wDBF
	'California'	oDel wDBF
	'Cambridge Louie'	oDel
	'Cameron Ryle'	oDel
	campos-portoi	oDel
	'Cara Mia'	oDel
	'Cardinal'	oDel wDBF wSnq
	'Carmel Blue'	oDel
un	'Carmel Gray'	oDel
	'Caroline'	last listed 99/00
un	'Carrie Lou'	wDBF
	'Cascade'	oDel wDBF oCir
un	'Ceil Peller'	oDel
	'Celebration'	oDel
	'Celia Smedley'	oDel oJoy wDBF
un	'Century 21'	wDBF
	'Chang'	oDel
	'Chantry Park'	oDel
	'Charming'	oDel
	'Checkerboard'	oDel oHed wDBF
un	'Checkered Lady'	oDel
	'Cheers'	oDel wDBF
un	'Cherry Pop'	oDel
un	'Chicken House'	oDel
	'Chillerton Beauty'	oFor wHer oDel oJoy oAls oGar oGre wHom wSnq
	'Chiquita Maria'	wDBF
un	'Chloe'	wDBF
	'Christine Bamford'	oDel
	'Christmas Elf'	oJoy wDBF
	cinerea	oDel
un	'Cinnamon'	oDel wDBF
	'Circus Spangles'	oDel oSev
	'Citation'	oDel
	'City of Leicester'	oDel
	'City of Portland'	oDel wDBF
	'Claire de Lune'	oDel
	'Cliantha'	oDel
	'Cloth of Gold'	oDel wSnq
	'Cloverdale Pearl'	oDel oJoy
	'Coachman'	oDel
	coccinea	oDel
	x colensoi	oDel
un	'Columbia'	oDel
	'Connie'	oDel
	'Conspicua'	oDel
	'Constance'	oDel
	'Contramine'	oDel
	'Coombe Park'	oDel
	'Coquet Bell'	oDel
un	'Coral Shells'	oDel
	'Coralle'	oDel
	'Corallina'	oDel oGre wDBF
	cordifolia (Bentham)	oDel
	'Corsair'	oDel
	'Cotta Bella'	oDel
	'Cotta Fairy'	oDel
	'Cotta Vino'	oDel
	'Cotton Candy'	oDel
	'Countess of Aberdeen'	oDel
	'Court Jester'	oDel
	crassistipula	oDel
	'Crinkley Bottom'	oDel
	'Crosby Soroptimist'	oDel
un	'Cunning'	oDel oGar
	'Curtain Call'	oDel oCir
	'Cymon'	oDel
un	'Cyndy Pike'	oDel
	'Cyndy Robyn'	oDel
	'Dancing Flame'	oDel
	'Daniel Lambert'	oDel
	'Danielle'	wDBF
	'Dark Eyes'	oDel wDBF oCir
	'Dark Secret'	oDel
	'David'	wWoS oDel oJoy oGar oGre wDBF wSnq
un	'De Ell'	oDel
	'Debby'	oDel
	'Deben'	oDel
	'Dee Dee'	oDel
	'Deep Purple'	oDel wTGN
	'Delta's Bride'	oDel
	'Delta's Dream'	oDel
	'Delta's Groom'	oDel
	'Delta's Night'	oDel
	'Delta's Parade'	oDel
un	'Delta's Robijn'	oDel

	'Delta's Wonder'	oDel
	denticulata	oDel
	'Desert Sunset'	oJoy
	'Deutsche Perle'	oDel
	'Devonshire Dumpling'	wHer oDel
	'Diana Wright'	oDel
	'Dick Swinbank'	oDel
	'Dimples'	oJoy wDBF
	'Dirk van Delen'	oDel
	'Display'	oDel oJoy wDBF wSnq
un	'Doctor David Chan'	oDel
	'Doctor Foster'	oDel
	'Doctor Olson'	oDel wDBF
	'Dollar Princess'	oDel oGar oGre wHom wDBF
un	'Dolly'	wDBF
un	'Dolly Roach'	wDBF
	'Dominyana'	oDel
	'Don Peralta'	oDel
	'Doreen Redfern'	oDel
un	'Doreen Stroud'	oDel
	'Doris Joan'	oDel
un	'Doris Tate'	oDel
	'Dorothy'	oDel
	'Dorothy M. Goldsmith'	oDel
	'Dorothy Shields'	oDel
un	'Dottie'	oDel
un	'Double Otto'	oDel oJoy oGar wRai wDBF wSte wSnq wCoS
	'Drama Girl'	oDel
	'Drame'	oDel
	'Duchess of Albany'	oDel
	'Dulcie Elizabeth'	oDel
un	'Dusky Blue'	oDel wDBF
	'Dusky Rose'	oDel wDBF
	'Dutch Mill'	last listed 99/00
	'Easterling'	oDel
	'Edith'	oDel
	'Edith Jack'	oDel
	'Edna May'	oDel
	'Edwin J. Goulding'	oDel
un	'Eileen Marie'	wDBF
	'El Camino'	wDBF
	'El Cid'	oDel
	'Eleanor Leytham'	oDel
	'Eleanor Rawlins'	wHer
	'Elfriede Ott'	oDel wDBF
	'Elsie Downey'	oDel
	'Emberglow'	wDBF
	'Emma Louise'	oDel
	'Empress of Prussia'	oDel
	'Enchanted'	wDBF
	encliandra ssp. ***tetradactyla***	oDel
	'Enfant Prodigue'	oDel
	'Enstone'	see **F.** ***magellanica*** var. ***molinae*** 'Enstone'
	erecta	see **F.** 'Bon Accorde'
	'Erecta Novelty'	see **F.** 'Bon Accorde'
	'Eric's Hardy'	oDel
	'Ernest Rankin'	oDel
	'Ernestine'	oDel
	'Ernie Bromley'	oDel
	'Estelle Marie'	oDel
	'Eureka Red'	oDel oSev wDBF
	'Eusebia'	oDel
un	'Exoniensis'	oDel oSis
	'Fabian Franck'	oDel
un	'Fairy Dancer'	oDel
	'Fairytales'	oDel
	'Falklands'	oDel
	'Falling Stars'	oDel wDBF
	'Fanfare'	oDel
	'Feather Duster'	oDel wDBF
	'Fergie'	oDel
	'Fey'	oDel
	'Fiery Spider'	oDel
	'Fifi'	oDel wDBF
	'Fiona'	wDBF
	'Fire Opal'	wDBF
	'Firecracker'	wHom wDBF
	'First Love'	oDel wDBF
	'First Success'	oDel wDBF
	'Flash'	oDel oSis wSnq
	'Flat Jack o'Lancashire'	oDel
un	'Flex'	oDel
	'Flirtation Waltz'	oDel
un	'Florabelle'	wDBF
	'Fluorescent'	oDel
	'Fly-by-night'	oDel
	'Flying Saucer'	oDel
un	'Fontaine'	wDBF
	'Foolke'	oDel
	'Fort Bragg'	wDBF
	'Foxgrove Wood'	last listed 99/00
	'Frank Sanford'	oDel wDBF
	'Frank Saunders'	oDel
	'Frank Unsworth'	wDBF
un	'Frost-n-Fire'	oDel
	'Fuchsia Fan'	oDel
	'Fuchsiade '88'	oDel
	'Fuchsiarama '91'	oDel
	fulgens	oDel
	fulgens rubra grandiflora	see **F.** 'Rubra Grandiflora'
	'Galadriel'	oDel wDBF
	'Garden News'	oDel
	'Garden Week'	oDel
	'Gartenmeister'	see **F.** Gartenmeister Bonstedt'
	'Gartenmeister Bonstedt'	oDel oJoy oGar wHom wDBF oCir
	'Gary Wayne'	oDel
	'Gay Fandango'	oDel
	'Gazebo'	oDel wDBF
	gehrigeri	oDel
	'General Charles de Gaulle'	oDel
	'Genii'	wHer oHed oNWe
	'George Barr'	oDel
un	'George Roach'	wDBF
	'Geraldine'	oDel
	'Gerharda's Aubergine'	oDel
	'Gingham Girl'	wDBF
	'Girls Brigade'	oDel
	'Gladiator'	oDel
	glaziouana	oDel
	'Glenby'	wDBF
	'Globosa'	oDel oJoy wDBF
	Glowing Lilac	oDel wDBF
	'Golden Anniversary'	oDel
	'Golden Arrow'	oDel
	'Golden Gate'	oDel
	'Golden Girl'	oDel
	'Golden Marinka'	oSev
	'Golden Swingtime'	oDel
	'Gordon's China Rose'	oDel wDBF
	'Gottingen'	oDel
	'Gracilis'	see **F.** ***magellanica*** var. ***gracilis***
	'Grand Prix'	oDel
	'Grandma Sinton'	oDel
	'Grayrigg'	last listed 99/00
	'Greenpeace'	oDel
	'Greta'	oDel
	'Groene Kan's Glorie'	oDel wDBF
	'Grumpy'	oDel
	'Guinevere'	oDel
	'Gypsy Girl'	oDel
	'Gypsy Prince'	wDBF
	'Halsall Belle'	oDel
	'Hanna'	oDel
	'Happy Fellow'	oDel
	'Harbour Bridge'	oDel
	hartwegii	oDel
	'Hatschbachii'	oDel wSnq
	'Haute Cuisine'	oDel
	'Hawaiian Night'	oDel
	'Hawkshead'	wHer wCul oDel oAls oHed oCis oGre wSnq
	'Hayward'	oDel
	'Heidi Weiss'	wDBF
	'Herbe de Jacques'	oDel
	'Hermiena'	oDel wDBF
	'Heron'	oInd
	'Hidcote Beauty'	oDel wDBF
	'Hollydale'	oDel oGar wDBF
	'Howlett's Hardy'	oDel
	'Hula Girl'	oDel wDBF
	'Humboldt Holiday'	oDel
	'Huntsman'	oDel
	'Ian Leedham'	oDel
	'Ice Maiden'	oDel
un	'Ida's Delight'	oDel
	'Imperial Crown'	wDBF
	'Impudence'	oDel
	'Ina Jo Marker'	oDel wDBF
	'Indian Maid'	oDel wDBF

	'Inferno'	wDBF
	'Insulinde'	oDel
un	'Irish Cup'	oDel
	'Isis'	oFor wWoS oAls oGar oSis oRiv oCis wDBF oBRG wSte wSnq
un	'Island Sunset'	oGar oSho wDBF wAva
	'Isle of Purbeck'	oDel
un	'Isle of Wight'	oDel
	'Jack Shahan'	oDel wDBF oCir
	'James Lye'	oDel
	'James Travis'	oDel
	'Jane Elizabeth'	oDel
un	'Janessa'	wDBF
	'Janice Revell'	oDel
	'Jean Pidcock'	oDel
un	'Jean Temple'	oDel wDBF
un	'Jenessa'	oDel
	'Jenny Sorensen'	oDel
	'Jessimac'	oDel
	'Jim Muncaster'	oDel
	'Jingle Bells'	oDel oJoy wDBF wSnq
	'Joan Cooper'	oDel
	'Joan Leach'	oSec
	'Joan Smith'	oDel
	'John Maynard Scales'	oDel
	'Joy Bielby'	oDel
	'Joy Patmore'	oDel
	'Joyce Maynard'	oDel
	'Julie Horton'	oDel wDBF
un	'Jump for Joy'	oDel
	'June Bride'	oDel oJoy oGar oGre wDBF oCir
	'Jupiter'	oDel wDBF
	'Jupiter 70'	oJoy
	'Kathleen Smith'	last listed 99/00
	'Katinka'	oDel
un	'Kay Radford'	oDel
	'Kay Riley'	oDel
	'Kegworth Delight'	oDel
un	'Kelley Denise'	oDel
	'Kelly Jo'	oDel
	'Ken Goldsmith'	oDel
	'Keystone'	oDel
	'King of Siam'	oDel
un	'Kitty O'Day'	oDel
	'Kolding Perle'	oDel
	'La Campanella'	oDel
un	'La Costa'	wDBF
	'Lady Boothby'	oDel
	'Lady Isobel Barnett'	oDel
	'Lady Thumb'	oDel wDBF wSnq
	'Laurie'	oDel
un	'Lavalou'	oDel
un	'Lavender Beauty'	wDBF
	'Lavender Cascade'	oDel
	'Lechlade Gorgon'	oDel
	'Lechlade Magician'	wHer oDel
	'Lechlade Violet'	oDel
	'Leicestershire Silver'	oDel
	'Lena'	oDel oJoy wDBF
	'Lena Dalton'	oJoy
un	'Leo Goetelen'	oDel
	'Leonora'	wDBF
	'Leverkusen'	oDel wDBF
	'Liebriez'	wHer oDel
	'Lindisfarne'	oDel
	'Lisi'	oDel
	'Little Beauty'	oDel oJoy wDBF wSnq
un	'Little Darling'	oJoy wDBF
	'Little Fellow'	oDel
un	'Little Giant'	oGar oSis oBlo wSta
	'Little Jewel'	oDel wDBF
	'Little Ronnie'	oDel
un	'Little Snow Queen'	oDel
	'Little Witch'	oDel
	'Loeky'	oDel
	'Logan Garden'	see **F.** ***magellancia*** 'Logan Woods'
un	'Lohn der Liebe'	oDel
	'Lolita'	oDel
	'Lord Byron'	oDel oGar wDBF
	'Lord Lonsdale'	oDel
	'Lorna Hercherson'	oDel
	'Lothario'	oDel
	'Lottie Hobby'	oDel oJoy wFGN wDBF wSnq
	'Louise Emershaw'	oDel wDBF
un	'Louise Rooney'	wDBF

	'Love in Bloom'	oDel
	'Love Knot'	oDel
	'Lucky 13'	wDBF
	lycioides (see Nomenclature Notes)	oDel
	'Lye's Unique'	oDel
	'Lynn Ellen'	wDBF
	'Lynne Marshall'	oDel
	'Machu Picchu'	oDel wDBF
	macrostema	see **F.** ***magellanica*** var. ***macrostema***
	'Madame Cornelissen'	oGar wDBF wSnq
	'Maddy'	oDel
	magdalenae	last listed 99/00
	magellanica	oFor wCri oDel oNat oJoy oAls wCSG oGoo oBlo wPir wCCr wSta wAva wCoN oUps
	magellanica 'Alba'	see **F.** ***magellanica*** var. ***molinae***
un	***magellanica*** 'Alpina'	wWoS
	magellanica var. ***gracilis***	oGar wDBF
	magellanica var. ***gracilis*** 'Aurea'	oFor wWoS oDel oJoy oHed oGar oSis wMag wBox wDBF wAva wSnq
va	***magellanica*** var. ***gracilis*** 'Tricolor'	oJoy oHed oGoo oGre wDBF
	magellanica var. ***gracilis*** 'Variegata'	oFor oDel oHed oGar wHom wDBF wSnq
	magellanica HCM 98152	wHer
	magellanica 'Logan Woods'	oDel
	magellanica var. ***macrostema***	wSwC oJoy oGar oInd
qu	***magellanica*** 'Maiden's Blush'	oFor oJoy wDBF
	magellanica var. ***molinae***	wWoS wCul wCri wSwC oDel oHed oGar wCCr wHom wAva wSte wSnq
	magellanica var. ***molinae*** 'Enstone'	oDel
	magellanica var. ***molinae*** 'Golden Sharpitor'	wHer wWoS oSec wSnq
va	***magellanica*** var. ***molinae*** 'Sharpitor'	wHer wWoS wSnq
	magellanica var. ***molinae*** 'Sharpitor Aurea'	see **F.** ***magellanica*** var. ***molinae*** 'Golden Sharpitor'
	magellanica var. ***pumila*** (see Nomenclature Notes)	oJoy oGar
	magellanica 'Thompsonii'	wSnq
va	***magellanica*** 'Versicolor'	wHer oSis wSte
	'Major Heaphy'	oDel
	'Mama Bleuss'	wDBF
	'Mancunian'	oDel
	'Mantilla'	oDel wDBF
	'Marcus Graham'	last listed 99/00
	'Margaret'	oDel
	'Margaret Brown'	oDel
	'Margaret Pilkington'	oDel
	'Margaret Tebbit'	oDel
	'Margery Blake'	oDel
un	'Margie Griffith'	oDel
	'Maria Merrills'	oDel
un	'Marie Eileen'	oDel
	'Marietta'	oDel
	'Marin Glow'	oDel
	'Marinka'	oDel wDBF
	'Martha Brown'	last listed 99/00
	'Martha Franck'	oDel
	'Mary'	oDel oGar wDBF
	'Mary Ellen'	wDBF
	'Mary Ellen Guffey'	last listed 99/00
	'Mary Fairclo'	last listed 99/00
	'Mary Joan'	oDel
	'Mary Miloni'	oDel
	'Mary Poppins'	oDel
un	'Mary's Beauty'	oDel
	'Masquerade'	wDBF
	'Maytime'	wDBF
	'Meadowlark'	wDBF
	'Melanie'	oDel
	'Mephisto'	oDel oJoy wDBF
	'Michael Kurtz'	oDel
	microphylla	oDel oHed oCis oBlo
	microphylla ssp. ***hidalgensis***	oDel
un	***microphylla*** 'Variegata'	oHed
	'Miep Aalhuizen'	oDel
	'Millrace'	oDel
	'Miniature Jewels'	wDBF
	'Minnesota'	oDel wDBF
	minutifolia	oTrP
	'Miss Aubrey'	wDBF
	'Miss California'	oDel wDBF
	'Miss Debbie'	wDBF
	'Miss San Diego'	oDel
	'Mission Bells'	oDel oGar wDBF
	molinae alba	see **F.** ***magellanica*** var. ***molinae***

	'Monsieur Thibaut'	oDel
	'Monterey'	oDel
	'Montezuma'	oDel
	'Moon Glow'	oDel wDBF
	'Morgenrood'	oDel
	'Morning Light'	wDBF
	'Mr. A. Huggett'	oDel
	'Mrs. John D. Fredericks'	oDel wDBF
	'Mrs. Lovell Swisher'	oDel
	'Mrs. Marshall'	oDel
	'Mrs. Popple'	oDel oJoy oAls wDBF
un	'Mrs. Shirley Lorance'	wDBF
	'Mrs. Victor Reiter'	wDBF
	'Mrs. W. P. Wood'	oDel wSnq
	'Mrs. W. Rundle'	wDBF
	'Multa'	oDel
	'My Oh My'	oDel
	'Nancy'	wCul wSnq
	'Nancy Lou'	oDel
	'Navy Blue'	oDel
	'Nellie Nuttall'	last listed 99/00
	'Neopolitan'	oDel
	'Nettala'	oDel wDBF
	'Nicis Findling'	oDel
	'Nicola Jane'	oDel
	nigricans	last listed 99/00
	'Niula'	oDel
	'Northway'	oDel
	'Novella'	oDel wDBF
	'Nunthorpe Gem'	oDel
	'Ocean Mist'	oDel
un	'Old Fashioned'	oJoy
un	'Old Glory'	oDel
un	'Ole'	oDel
	'Olympic Sunset'	oDel
un	'Orange Dream'	oDel
	'Orange Drops'	wDBF
	'Orange Queen'	wDBF
un	'Orange Spider'	wDBF
	'Oregon'	oDel wDBF
	'Orient Express'	oDel
	Oriental Flame	oDel
	'Oriental Lace'	oDel oAls wSnq
	'Ortenburger Festival'	oDel
	'Other Fellow'	oDel wDBF
	'Otto'	oDel oJoy
	'Our Darling'	oDel
	'Pacquesa'	oDel
un	'Pamela'	oDel
	paniculata	oDel
	'Panylla Prince'	oDel
	'Papoose'	oDel oGre wTGN wDBF wSnq
	'Party Frock'	oDel
	parviflora (Lindley)	oDel
	'Patio Princess'	oDel
	'Patricia Ann'	oDel
un	'Pat's Dream'	wSwC oJoy oGar oGre oWnC wDBF wAva
un	'Patty Lou'	oDel
	'Paula Jane'	oDel
un	'Pauline McFarland'	oDel
	'Peachy'	oDel
	'Pee Wee Rose'	oDel
	'Peppermint Candy'	oDel wDBF
	'Peppermint Stick'	oDel oJoy wDBF
un	'Perky'	oDel
	'Perry Park'	oDel
	'Personality'	oDel
	'Peter Pan'	wCul oJoy oGar oGre wTGN wDBF
un	'Petra'	oDel
	'Phyllis'	oDel
	'Pinch Me'	oDel wDBF
	'Pink Cloud'	oDel
	'Pink Delight'	wDBF
	'Pink Fairy'	oDel
	'Pink Galore'	oDel
	'Pink Jade'	oDel oGar
	'Pink Marshmallow'	oDel wHom wDBF oCir
	'Pink Panther'	oDel
un	'Pink Parade'	wDBF
	'Pink Pearl'	wHer
	'Pink Rain'	oDel
	'Pink Snow'	oDel
	'Pink Temptation'	oDel
	'Pinto de Blue'	oDel
un	'Plum Glory'	oDel
un	'Plum Perfect'	oDel
	'Plum Pudding'	oDel
	'Plumb-bob'	oDel
	'Postiljon'	oDel
	'Powder Puff'	oDel
	'President'	oDel wSnq
	'President George Bartlett'	oDel
	'President Leo Boullemier'	wHer
	'President Moir'	oDel
	'President Roosevelt'	wDBF
un	'Pride of Eugene'	oDel
	'Prince of Orange'	oDel
	procumbens	oTrP wHer wWoS oDel oJoy oHed wDBF wSnq
un	***procumbens*** 'Variegata'	wHer oHed oCis
	'Prosperity'	oDel
	'Pumila' (see Nomenclature Notes)	oFor oDel oRiv wDBF wSnq
	'Purbeck Mist'	oDel
	'Purple Heart'	last listed 99/00
	'Purple Rain'	oDel
	'Purple Sage'	oDel wDBF
un	'Purpur Klokje'	oDel
	'Pussy Cat'	oDel
	'Put's Folly'	oDel
	'Quasar'	oDel wDBF
un	'Quasar Lady'	oDel
	'Queen Elizabeth'	oDel
un	'Queen Esther'	oDel wSnq
un	'Queen of Naples'	oDel wDBF
	'R. A. F.'	wDBF
	'Rachel'	wDBF
	'Raggedy Ann'	wDBF
	'Rainbow'	last listed 99/00
	'Rambling Rose'	wDBF
	'Randy'	oDel
	'Raspberry'	oDel wDBF
un	'Raspberry Punch'	oDel wDBF
un	'Raspberry Twist'	oDel
	'Ratatouille'	oDel
	'Reading Show'	oDel
	'Red Devil'	oDel wDBF
	'Red Imp'	oDel
un	'Red Rain'	oDel
	'Red Shadows'	oDel wDBF
	'Red Spider'	oDel wDBF
	'Reflexa'	see **F. *x bacillaris*** 'Reflexa'
	regia	wHer wDBF
	regia var. ***alpestris***	see **F. *alpestris***
	regia ssp. ***regia***	oDel
	regia ssp. ***reitzii***	oDel wSnq
	regia ssp. ***serrae***	oDel
	'Remembrance'	oDel
	'Remus'	oDel
	'Riant'	oDel
	'Riccartonii'	oFor wSwC oDel oGar wMag wTGN wDBF oUps wSnq
un	'Rita Sklar'	oDel
un	'Robert Sharpe'	oDel
	'Rocket Fire'	oDel
	'Romance'	wDBF
	'Ronald L. Lockerbie'	oDel
	'Roos Breytenbach'	oDel
	'Rosalie Rooney'	wDBF
	'Rose Churchill'	oDel
	'Rose Fantasia'	oDel
	'Rose Lace'	oDel
	'Rose of Castile'	oDel oGre
	'Rose of Monterey'	oDel
	rosea (see Nomenclature Notes)	wSnq
un	'Rosetta'	wSnq
	Rosy Ruffles	oDel wDBF
	'Rough Silk'	oDel
	'Roy Walker'	oDel
	'Royal Air Force'	see **F.** 'R. A. F.'
	'Royal Mosaic'	oDel wTGN
	'Royal Pink'	wDBF
	'Royal Robe'	oDel wDBF
	'Royal Velvet'	oDel
	'Rubra Grandiflora'	oDel
	'Ruby Wedding'	last listed 99/00
	'Ruffles'	wDBF
	'Rufus'	oDel wDBF wSnq
	'Ruthie'	wDBF
un	'Salmon Butterfly'	oDel
un	'Samantha'	oDel

	'Sampson's Delight'	oDel
	'San Leandro'	oDel wDBF
	'San Mateo'	wDBF
un	'San Pasqual'	oDel
	sanctae-rosae	oDel
	'Santa Clara'	oDel wDBF
un	'Santa Claus'	oDel oJoy oGar oSis oBlo wTGN wSta wDBF wSte wSnq
	'Santa Cruz'	wDBF
	'Scarlet Ribbons'	oDel
	'Schiller'	oDel
	'Schnabel'	oDel
un	'Schneckeri'	oDel
un	'Scintillation'	wDBF
	'Sealand Prince'	wSnq
un	'Seattle Blue'	oDel
un	'Seaview Sunset'	oDel
	'Sebastopol'	oDel
un	'Senorita'	oDel wSnq
un	'September Morgan'	oDel
	'Seventh Heaven'	oDel wDBF
	'Shady Lady'	oDel
	'Sharon'	oDel
un	'Sharpton's'	oGar
	'Sheila Crooks'	oDel
	'Shelley Lyn'	wDBF
	'Shellford'	oDel wDBF
un	'Shirley Lorance'	oDel
un	'Sid Drapkin'	oDel
	'Silver Pink'	oGar
	'Silver Queen'	oDel
	'Silverdale'	oDel
	'Simon J. Rowell'	oDel
	'Sister Ann Haley'	oDel
	'Sister Ginny'	oDel
	'Sleepy'	oDel
	'Sleigh Bells'	oDel
	'Smokey Mountain'	oDel wDBF
	'Snow Burner'	oDel
	'Snowcap'	oDel
	'Snowfire'	oDel wDBF
	'Snowy Summit'	oDel wDBF
	'So Big'	wDBF
un	'Society Belle'	oDel
	Software	oDel wDBF
	Son of Thumb	wHer
	'Sophisticated Lady'	wDBF
	'South Coast'	oDel
	'South Gate'	oDel wDBF
	'Space Shuttle'	oDel
	'Speciosa'	oDel
	splendens	oDel wDBF
	'Squadron Leader'	oDel
un	'Stanford'	oDel
	'Stanley Cash'	wDBF
un	'Star of Blue'	oDel
	'Star of Pink'	oDel
un	'Starburst'	oDel wDBF
	'Stardust'	wDBF
	'Starry Trails'	wDBF
un	'Stephen Adam'	oDel
un	'Stewart Martin'	oDel
	'Strawberry Fizz'	oDel
	'String of Pearls'	oDel
	'Suikerbossie'	oDel wSnq
	'Sunny Skies'	oDel
va	'Sunray'	oDel wDBF
	'Sunset'	oDel
	'Sunshine'	wSwC oDel oGar wSnq
	'Super British'	oDel
	'Supersport'	wDBF
	'Surprise'	oDel oJoy wDBF wSnq
qu	'Swan Lee'	wDBF
	'Swanley Gem'	oDel
	'Swanley Yellow'	oDel
	'Sweet Sixteen'	oDel
	'Swingtime'	oDel wDBF oCir
	'Swiss Miss'	oDel
	'Tangerine'	oDel wDBF
un	'Tanila Ann'	oDel
	'Tarra Valley'	oDel
un	'Tasman Sea'	oDel
un	'Ted Paskesen'	oDel
un	'Ted Sweetman'	oDel
	'Tempo Doelo'	oDel
	'Temptation'	oDel
	'Tennessee Waltz'	wDBF
	'Texas Longhorn'	oDel
	'Texas Star'	oDel
	'Thalia'	oDel
	'Think Pink'	oDel wDBF
	'Thomasina'	oDel wDBF
	'Thunderbird'	oDel
	thymifolia	wDBF
	thymifolia ssp. ***thymifolia***	oDel
un	'Timothy Brian'	oDel
	tincta	see **F. *vargarsiana***
	'Ting-a-ling'	wDBF
	'Tinker Bell'	oDel wDBF
	'Tiny Rose'	wDBF
	'Tiny Tim' (see Nomenclature Notes)	oGre
	'Tom Thumb'	oFor oTrP oDel oJoy oGar oSis oWnC wDBF wSnq
va	'Tom West'	oDel wDBF
	'Tom Woods'	wDBF
	'Tony Porter'	oDel
	'Torch'	oDel
	'Torvill and Dean'	oDel
	'Trase'	oDel
	'Traudchen Bonstedt'	oDel
	'Treasure'	oDel
	'Tricolor'	see **F. *magellanica*** var. ***gracilis*** 'Tricolor'
	'Trisha'	oDel
	'Troubadour'	oDel
	'Trudy'	oDel
	'Trumpeter'	oDel
un	'Tukwila'	wSta
	'Tumbling Waters'	oDel
	'Twirling Square Dancer'	oDel
	'Two Tiers'	oDel
	'Ullswater'	oDel
	'Uncle Charley'	oDel
	'Uncle Jules'	oDel
	'Uncle Mike'	oDel
	'Unique'	wDBF
	'Vanessa Jackson'	oDel
	vargarsiana	oDel
	'Variegated Lottie Hobbie'	last listed 99/00
	'Venus Victrix'	oDel
	venusta	last listed 99/00
un	'Vesta Walters'	oDel
	'Vielliebchen'	oDel oGre wDBF wSnq
	'Vincent van Gogh'	oDel wDBF
	'Viola'	oDel oGar
	'Vivienne Thompson'	oDel
	'Voltaire'	oDel
	'Voodoo'	oDel wDBF
	vulcanica (Andre)	see **F. *ampliata***
un	'Waconda Queen'	oDel wDBF
un	'Waconda Star'	wDBF
un	'Walkers Painted Desert'	oDel
	'Walz Bella'	oDel
	'Walz Blauwkous'	oDel
	'Walz Bruintje'	oDel
	'Walz Gigolo'	oDel
	'Walz Harp'	oDel
	'Walz Jubelteen'	oDel
	'Walz Waardin'	oDel
	'Washington Centennial'	oDel wDBF
	'Welsh Dragon'	oDel
	'Wendy Leedham'	oDel
	'Wendy's Beauty'	oDel
un	'White Churchill'	wDBF
qu	'White Encliandra'	wDBF
un	'White Eyes'	wDBF
	'White Spider'	wDBF
	'Whiteknights Amethyst'	oDel
	'Whiteknights Pearl'	wHer oDel oAls oMis wSnq
	'Whiteknights Ruby'	oDel
un	'Will Gibbs'	oDel
un	'Will Rogers'	oDel
	'Willie Tamerus'	oDel
un	'Winchester Cathedral'	oDel
	'Winston Churchill'	oDel oGar wDBF
	'Wood Violet'	oDel
	'Yolanda Franck'	oDel
	'Yonder Blue'	oDel
	'Ziegfield Girl'	oDel wDBF

GAILLARDIA

	aristata	see **G. *x grandiflora***

	'Bremen'	oWnC wTGN
	'Burgunder'	wTho oAls oHed oAmb wHom oWnC iArc
	'Dazzler'	oGar oWnC oCir
	Goblin	see **G.** 'Kobold'
	'Goldkobold'	last listed 99/00
	x grandiflora	wThG wNot oAls wShR oBos wPla wWld
	x grandiflora 'Burgundy'	see **G.** 'Burgunder'
	'Kobold'	oFor wTho oAls oMis oGar oBlo wFai wHom oWnC iArc oEga oUps wCoS
	'Mandarin'	last listed 99/00
	Monarch Group	oGar
qu	'The Sun'	oAls
	GALANTHUS	
	elwesii	wCoN
	ikariae Latifolius Group	wHer
	nivalis	wRoo oWoo wCoN
do	***nivalis*** 'Flore Pleno'	wHer oBov oWoo
	nivalis 'Viridapicis'	wHer
	'S. Arnott'	wHer
	GALAX	
	aphylla	see **G. *urceolata***
	urceolata	oBov wCCr wCoN
	GALEGA	
	x hartlandii 'Alba'	last listed 99/00
	x hartlandii 'Lady Wilson'	oJoy oGar oDan oNWe oGre wFai wBox
	GALIUM	
	mollugo	oGoo
	odoratum	oFor wWoS oNat oDar oAls wFGN oGar oGoo wFai wHom oCrm oBar wTGN wSta wNTP oCir wCoN wWhG oUps oSle wSnq
	verum	wFGN oGoo wFai iGSc
	GALTONIA	
	candicans	oFor wHer wCol oDan oNWe
	princeps	oFor wHer
	viridiflora	wHer oEdg oNWe
	GALVEZIA	
	speciosa	oGoo
	GARCINIA	
un	***acuminata***	oOEx
	GARDENIA	
	augusta	oGar oOEx
	augusta 'Aimee'	oGar oWnC
	augusta 'August Beauty'	oWnC wCoS
un	***augusta*** 'Chuck Hayes'	oGar
	augusta First Love	see **G. *augusta*** 'Aimee'
	augusta 'Mystery'	oWnC wWel
	augusta 'Radicans'	oWnC
	augusta 'Radicans Variegata'	oWnC
	augusta 'Veitchiana'	oAls oGar iGSc oWnC wWel
un	'Chuck Hays'	oCis
	fortunei	oCis
	jasminoides	see **G. *augusta***
un	'Kleim's Hardy'	oCis wCoN wCoS wWel
	GARRYA	
	buxifolia	oFor
	elliptica	oFor oGar oBos oRiv oSho wTGN wCoN
	elliptica 'Evie' (male)	oFor wHer oRiv wWel
	elliptica 'James Roof' (male)	oFor oAmb oRiv oGre
	fremontii	oFor oRiv oAld wCoN
	x issaquahensis	oRiv wCCr
un	***x issaquahensis*** 'Carl English'	oFor
	GASTERIA	
	armstrongii	last listed 99/00
	bicolor var. ***liliputana***	oSqu
	disticha	last listed 99/00
un	***gracilis*** variegated	oSqu
	liliputana	see **G. *bicolor*** var. ***liliputana***
ch	***nigricans***	see **G. *disticha***
un	***poellnitziana***	oSqu
	verrucosa	last listed 99/00
	X GAULNETTYA	see **GAULTHERIA**
	GAULTHERIA	
	antipoda 'Adpressa'	oBov
	crassa	last listed 99/00
	cuneata	wHer
	eriophylla	wHer
	forrestii DJHC 477	wHer
	fragrantissima	wThG oBos
	fragrantissima HWJCM 149 (previously listed as HWJCM 159)	wHer
	griffithiana	wHer
	hispida	wHer
	hookeri	last listed 99/00
	insana	wCCr
	insana HCM 98199	wHer
	itoana B&SWJ 3406	wHer
	leucocarpa	see **G. *pumila*** var. ***leucocarpa***
	littoralis	oFor
	macrostigma	wHer
	miqueliana	oFor wHer oBlo oGre wSta oBRG wWel
	'Miz Thang' (see Nomenclature Notes)	wHer
	mucronata	wHer wSta wCoN
	mucronata 'Alba' (female)	oFor wHer oNWe
	mucronata pink	wCCr
	mucronata red	oGar
	mucronata 'Rosea' (female)	oFor
	mucronata 'Thymifolia' (male)	wHer
	mucronata white	wCCr
un	***myrtilloides*** f. ***alba***	wHer
un	***myrtilloides*** f. ***rubra***	wHer
	nummularioides	wHer wCul oSis oBov wSta
	nummularioides minor	wHer oBov
	ovalifolia	see **G. *fragrantissima***
	phillyreifolia	wHer wCCr
	procumbens	oFor wWoS wCul wWoB wBur oAls oRus oGar oGoo oBlo wRai oOEx oSqu oGre wTGN oSec wSta oBRG wAva wWhG wCoS
	pumila	oFor wHer wCCr
	pumila coll. Thermas de Chillan, seed grown	wHer
	pumila var. ***leucocarpa***	wHer
	pyroloides	last listed 99/00
un	***rubra***	wWel
	schultesii	wRho
	semi-infera	wHer
	shallon	oFor wBlu wThG wWoB wBur wNot oAls wShR oGar oBos oBlo wRai oOEx oSqu oGre wWat wCCr oTri wSta oAld oCir wAva wFFl wRav wSnq wWel
un	***shallon*** 'Snoqualmie Pass'	wHer
	sinensis	wHer oBov
	tetramera	last listed 99/00
	thymifolia	last listed 99/00
	veitchiana	wHer
	x wisleyensis ruby fruited	oBov
	x wisleyensis 'Wisley Pearl'	oFor wHer oGos oNWe
	GAURA	
	lindheimeri	oFor wCul oJoy wCSG oMis oGar oDan oSho oSha wRob oBlo wFai wPir wCoN oUps
va	***lindheimeri*** 'Corrie's Gold'	oFor wHer wSwC oJoy oAls wFGN oAmb oSis wRob wFai wPir oWnC wMag wTGN oSec oInd oUps
un	***lindheimeri*** 'Dauphine'	oSis
un	***lindheimeri*** 'Franz Valley'	oFor wWoS
un	***lindheimeri*** 'Golden Speckles'	oJoy
va	***lindheimeri*** 'Jo Adela'	last listed 99/00
	lindheimeri 'Siskiyou Pink'	oFor wHer wWoS wSwC oNat oDar oJoy oAls oHed oAmb oMis oDan oSis oNWe oSho oOut wFai wPir wHom oWnC wTGN oSec oInd oCir oUps oSle
	lindheimeri 'The Bride'	wWoS wHom
	lindheimeri 'Whirling Butterflies'	oFor oNat oTDM oDar oAls oSis wHom oWnC wMag wTGN oSec oCir
	GAYLUSSACIA	
	baccata	oFor
	GAZANIA	
	'Freddie'	wHer
	krebsiana	oNWe
un	***linearis*** 'Colorado Gold'	oFor
	GELSEMIUM	
	rankinii	oGar oSho oWnC
	sempervirens	oGar wCCr oWnC wCoS
	sempervirens 'Pride of Augusta'	oFor
	GENISTA	
	aetnensis	oFor wHer wCCr
	dalmatica	see **G. *sylvestris***
	lydia	oFor oAls oGar oSis oBlo wKin oTPm oWnC
	pilosa	oAls oGar oBlo wWel
	pilosa 'Goldilocks'	oFor
	pilosa 'Vancouver Gold'	oFor oGar oSis oCis oSho oGre oTPm oWnC wSta
	pulchella	oFor oSis
	saggitalis	oFor
	x spachiana	oFor
	sylvestris	oSis
	tinctoria	oFor oGoo
do	***tinctoria*** 'Flore Pleno'	last listed 99/00

	tinctoria 'Royal Gold'	oFor wKin
	villarsii	see G. pulchella
	GENTIANA	
	acaulis	last listed 99/00
	acaulis f. alba	oWnC
	acaulis hybrids	oSis
	acaulis 'Rannoch'	last listed 99/00
	'Alex Duguid'	see G. farreri 'Duguid'
	andrewsii	oRus
	angustifolia	oRus
	asclepiadea	oRus oNWe wEde
	asclepiadea var. alba	wHer oRus oGar oGre
	asclepiadea 'Phyllis' seed grown	wHer
	asclepiadea 'Rosea'	last listed 99/00
	clusii JJH 87112	last listed 99/00
	cruciata	oFor oJoy wBox
	dahurica	oFor oAls
	decumbens	oGre
	dinarica	oFor
	farreri JCA4.418.810	wMtT
	farreri 'Duguid'	wMtT
	gracilipes	oFor oAls oMis oWnC
	grombczewskii	oFor
	lagodechiana	see G. septemfida var. lagodechiana
	lutea	last listed 99/00
	x macaulayi 'Kingfisher'	wMtT
	makinoi 'Royal Blue'	oFor oSis
	olivieri	wMtT
	paradoxa	wEde
un	paradoxa 'Blauer Herold'	oFor wWoS oMis oSis
	parryi	oDan
	septemfida	oFor oRus oSis wWld
	septemfida var. lagodechiana	wWoS oJoy oEdg oGar wMag
	septemfida var. lagodechiana 'Select'	last listed 99/00
	sino-ornata	wCol wMtT
	sino-ornata 'Edith Sarah'	wMtT
	sp. HC 970409	wHer
	straminea	last listed 99/00
	'Strathmore'	wWoS wMtT
	ternifolia	wMtT
	tibetica	oFor oRus oGoo
	tibetica DJHC 98079	wHer
	triflora var. japonica	oSev
un	triflora var. japonica 'Alba' seed grown	wHer
	triflora var. montana	oRus
un	uehihamel	wHer
	verna	wHer oSis oNWe
	GENTIANOPSIS	last listed 99/00
un	**GERADANTHUS**	
	macrorhiza	oRai
	GERANIUM	
	aconitifolium	see G. rivulare
	albanum	last listed 99/00
	anemonifolium	see G. palmatum
	'Ann Folkard'	wHer wWoS wCul oGos wSwC oNat wCol wHig oAls oHed oRus oAmb oGar oDan oSis oNWe wRob oGre wMag wTGN wBox wSte
	'Anne Thomson'	wHer oNWe oOut
	aristatum	wHig
	asphodeloides	wHer
	asphodeloides 'Prince Regent'	wHer
	'Baby Blue'	wHer wHig oAmb
	'Bertie Crug'	wEde
	'Blue Cloud'	wHer
	'Brookside'	oFor wWoS wSwC oGar oDan oSho oOut oBlo
	canariense	wHer
	x cantabrigiense	wCul oRus oSqu wCoN oSle wSte wSnq
	x cantabrigiense 'Biokovo'	oFor wHer wWoS wCul wSwC oNat wHig oAls oHed oRus oMis oGar wRob oSqu wFai oWnC wMag wTGN wBox oHou oUps oSle wSte wSnq
	x cantabrigiense 'Cambridge'	oFor wHer wHig oRus oSho wMag wTGN wSnq
un	x cantabrigiense 'Jan's'	oHou
	x cantabrigiense 'Karmina'	oFor oNat wHig oInd oHou
	x cantabrigiense 'St Ola'	wHer wHig
	'Chantilly'	wHer wHig
	cinereum	wCoN wCoS
	cinereum 'Ballerina'	oFor wWoS wCul wHig oAls oRus wCSG oGar oDan oSis oNWe oSho wMag wWhG
	cinereum 'Laurence Flatman'	oFor wWoS wHig oAls oRus oOut wFai wMag wTGN wBox oHou oSle
	cinereum 'Siskiyou Selection'	oSis

	cinereum var. subcaulescens	oFor wCul oJoy oRus oGar oSis
	cinereum var. subcaulescens 'Giuseppii'	wWoS
	cinereum var. subcaulescens 'Splendens'	oFor oAls wCSG oGar
	clarkei x collinum 'Kashmir Pink'	wWoS
	clarkei x collinum 'Kashmir Purple'	wWoS wHig oRus wAva
	clarkei x collinum 'Kashmir White'	wHer wCul wHig oRus wCSG wFai wMag
	collinum	last listed 99/00
	'Cricklewood'	wCri
	dalmaticum	oFor oRus wCSG oSis wRob wTGN wAva wWhG
	dalmaticum 'Album'	wCul wHig oRus
	'Dilys'	wHer wHig oSho
	'Diva'	last listed 99/00
	'Elizabeth Ross'	wHig
	endressii	oSho wMag wTGN wCoN wEde
	eriostemon	see G. platyanthum
	gracile	wHer
	gracile 'Blush'	wHer wHig
	grandiflorum	see G. himalayense
	grandiflorum var. alpinum	see G. himalayense 'Gravetye'
	grevilleanum	see G. lambertii
	gymnocaulon	wHer
	harveyi	wHer oJoy oHed oSis wTGN
	himalayense	wWoS wCul oNat wHig oMis oGar oSis wFai wHom wMag wTGN wCoN oUps oSle
	himalayense var. alpinum	see G. himalayense 'Gravetye'
	himalayense 'Birch Double'	see G. himalayense 'Plenum'
	himalayense 'Gravetye'	wWoS oRus wCSG oGre wFai
	himalayense 'Irish Blue'	wHer wCul wHig
do	himalayense 'Plenum'	wHer wWoS wCul wSwC oDar oJoy wHig oHed oRus wCSG oGar oSis oNWe wMag oHou wCoN wEde oUps
	ibericum	wCul oRus
	incanum	oNWe
	'Johnson's Blue'	oFor wHer wWoS wCul wSwC oNat oDar wHig oAls oHed oRus wCSG oGar oNWe oSha wRob oBlo oGre wFai wHom oWnC wMag wTGN wBox oCir wAva wCoN oUps wSnq
	'Joy'	wHer wWoS wHig
	koreanum DJH 347 seed grown	wHer
	kotschyi var. charlesii	last listed 99/00
	lambertii	last listed 99/00
	libani	oHed
	'Little Gem'	wHig
	macrorrhizum	oFor oNat oAls oRus oDan oTwi wRob oGre wFai oWnC wMag wTGN wBWP oCir wCoN oUps oSle wSte wSnq
	macrorrhizum 'Album'	oFor wHer wCul wHig oRus wSnq
	macrorrhizum 'Bevan's Variety'	oFor wCul oNat wHig oRus oAmb oWnC wMag wTGN oHou wSnq
	macrorrhizum 'Czakor'	wHer wHig
	macrorrhizum x dalmaticum	wCri
	macrorrhizum 'Ingwersen's Variety'	oFor wCri wHig oHed oRus wCSG oGar oSha oSqu wMag wBox
	macrorrhizum 'Lohfelden'	wHer wHig
	macrorrhizum 'Pindus'	wHig
un	macrorrhizum 'Purpurot'	oRus
	macrorrhizum 'Ridsko'	wHig
	macrorrhizum 'Spessart'	oFor wWoS oNWe wMag wTGN
	macrorrhizum 'Variegatum'	wHer wCol
	macrorrhizum 'Velebit'	wHig
un	macrorrhizum 'Walter Ingwersen'	see G. macrorrhizum 'Ingwersen's Variety'
	macrorrhizum 'White-Ness'	wHig
	macrostylum	last listed 99/00
	maculatum	wHer oRus oGoo iGSc
	maculatum f. albiflorum	oFor wHer wHig oRus oMis oWnC wMag wTGN wEde
	maculatum 'Chatto'	oRus oMis wTGN
	x magnificum	wHer wWoS wCul oNat wHig oHed oRus oGar oSho wRob oBlo oGre wFai oWnC wTGN wBox wEde oSle wSte
	malviflorum	wFai
	x monacense	wWoS wMag
	x monacense var. anglicum	wWoS wHig
	x monacense 'Muldoon'	wHer
	nepalense var. thunbergii	see G. thunbergii
qu	nigrum	wCul
	'Nimbus'	wHer wWoS oGar
qu	nivalis	oTwi
	nodosum	wCul wCri oJoy wHig oRus wTGN wBWP oSle
	nodosum 'Svelte Lilac'	wHer
	'Oh My God Pass' DBG 121	last listed 99/00

	oreganum	oHan
	orientalitibeticum	wWoS wCul wCri oNat wHig oHed oGre wMag wBox
	'Orkney Pink'	wWoS wHig wEde
	x oxonianum 'A. T. Johnson'	oFor wWoS wCri wHig oRus oGar oWnC wTGN
	x oxonianum 'Bressingham Delight'	wWoS oGar wTGN
	x oxonianum 'Claridge Druce'	oNat wHig oAls oRus wCSG oGar wRob oBlo oGre wFai oWnC wMag wTGN oWhS wEde oUps oSle
	x oxonianum 'Coronet'	wHer
	x oxonianum 'David McClintock'	last listed 99/00
	x oxonianum 'Hollywood'	wHer oDar oAls
	x oxonianum 'Julie Brennan'	last listed 99/00
un	***x oxonianum*** 'Katherine Adele'	wHer wWoS wHig
	x oxonianum 'Lady Moore'	wHer
	x oxonianum 'Lambrook Gillian'	wHer
	x oxonianum 'Miriam Rundle'	wHig
	x oxonianum 'Old Rose'	wHer wWoS wHig
un	***x oxonianum*** 'Pearl Boland'	oFor wHer wWoS oGar oDan oUps
	x oxonianum 'Phoebe Noble'	wHer wWoS wHig
	x oxonianum 'Rebecca Moss'	wHer wCul oJoy
	x oxonianum 'Rose Clair'	wWoS wCul oJoy
	x oxonianum 'Rosenlicht'	oFor wWoS oNat wHig oWnC wMag wTGN oHou oSle
	x oxonianum 'Sherwood'	wHer wWoS wHig
do	***x oxonianum*** 'Southcombe Double'	wHer wWoS
	x oxonianum 'Southcombe Star'	wHer wWoS wHig
un	***x oxonianum*** 'Sue Cox'	wHig
	x oxonianum f. ***thurstonianum***	wHer wWoS wHig
	x oxonianum 'Wageningen'	wWoS
	x oxonianum 'Walter's Gift'	wHer oDar wHig
	x oxonianum 'Wargrave Pink'	oFor wWoS wCul oNat oDar oJoy wHig oAls oRus wCSG oGar oDan oSho oGre wFai oWnC wMag oHou wAva wWhG oSle
	x oxonianum white form	wHer
	x oxonianum 'Winscombe'	wWoS oDar oNWe oGre
	palmatum	wHer wCri oDan
	palustre	wHig wTGN
	'Patricia'	wHer oGos wSwC oEdg wCol wHig oAls oHed oGar oDan oOut oGre wTGN wSte
	phaeum	wWoS wCul wSwC oJil oRus wCSG oBlo wFai wMag wBox
	phaeum 'Album'	oFor wCul wHig oHed wFai wTGN
	phaeum 'Calligrapher'	wWoS
un	***phaeum*** 'Darkest of All'	wHig
	phaeum 'Joan Baker'	wHer wWoS wHig
	phaeum 'Langthorn's Blue'	oNat
	phaeum 'Lily Lovell'	wHer wWoS oNat wHig oRus oNWe
	phaeum 'Little Boy'	wHer
un	***phaeum*** 'Margaret Wilson'	wHer
un	***phaeum*** 'Mrs. Withey Price'	wCol wHig
	phaeum 'Mourning Widow'	wHer wAva oSle
	phaeum var. ***phaeum***	oJoy
un	***phaeum punctatum***	wCul wCri
qu	***phaeum*** 'Purpureum'	wTGN
un	***phaeum*** 'Rainforest Selection'	wHig
	phaeum 'Samobor'	oFor wHer wWoS oJoy wHig oHed oRus oNWe oOut oGre wTGN wSte
	phaeum silver/white	oRus
va	***phaeum*** 'Taff's Jester'	wHer wWoS
	phaeum 'Variegatum'	wWoS oGos oGar oOut wRob
	'Phillippe Vapelle'	oFor wHer wWoS oGos wHig oHed oRus oAmb oGar oOut wSte
	platyanthum	wHig wCSG
	platypetalum (see Nomenclature Notes)	wHig oGar
	platypetalum (Fisch. & C. A. Mey.) 'Georgia Blue'	wHer
	pratense	oFor wCri oRus oSha oGre wTGN wBWP wAva wCoN oUps
	pratense f. ***albiflorum***	wHer wHig wBWP
	pratense f. ***albiflorum*** 'Silver Queen'	wHer wWoS
	pratense 'Caeruleum Flore Pleno'	see **G. *pratense*** 'Plenum Caeruleum'
	pratense Midnight Reiter Strain	wHer wCri oGos wSwC oJil oDan oNWe oOut
	pratense 'Mrs. Kendall Clark'	wHer wCul wCri wHig oAls wCoN
do	***pratense*** 'Plenum Caeruleum'	oFor wHer wHig wTGN
do	***pratense*** 'Plenum Violaceum'	wHer wHig oHed oNWe wTGN
	pratense roseum	wHig
	pratense 'Splish-splash'	oUps
	pratense 'Striatum'	wHer wHig
un	***pratense*** 'Victor Reiter Junior'	wWoS oDan
	pratense Victor Reiter Strain	wHer oGos oJil wHig oRus oGar oNWe
	pratense 'Wisley Blue'	wHer
	procurrens	wCri wHig wFai

	psilostemon	oFor wWoS wCul wHig oHed oRus oGar oDan wRob wFai wMag wTGN wBox wAva wSte
	psilostemon 'Bressingham Flair'	last listed 99/00
	pylzowianum	wHer wHig
	pyrenaicum 'Bill Wallis'	wHig oNWe
	reflexum	wHer
	regelii	wHer oHed
	renardii	oFor wWoS wCul oNat wHig oHed oRus oMis oGar oDan oSis wFai oWnC wTGN wBox oSle wSte wSnq
	renardii 'Tcschelda'	wHig
	renardii 'Whiteknights'	wHer
	x riversleaianum 'Mavis Simpson'	oFor wWoS wCul oJoy oAls oHed oMis oDan oNWe oSha wFai wTGN wBox oSle wSnq
	x riversleaianum 'Russell Prichard'	oFor wWoS oHed oNWe wFai wMag oHou
	rivulare	wHig
	robertianum	wFai
	robustum	last listed 99/00
	sp. aff. ***robustum***	oNWe
	rubifolium	wHer
	ruprechtii	wHig
	'Salome'	oFor wHer wWoS wEde
	sanguineum	oFor oTrP wCul oDar oRus wCSG oGar oNWe oOut wMag wTGN wBox wBWP oHou wCoN wCoS
	sanguineum 'Album'	oFor wHer wCul wHig oAls oRus wCSG oAmb oMis oGar oSis wRob wFai wMag wTGN wAva wEde oUps wSnq
	sanguineum 'Alpenglow'	wHig oSis wRob oBlo
	sanguineum 'Ankum's Pride'	last listed 99/00
un	***sanguineum*** 'Appleblossom'	wHig
	sanguineum 'Bloody Graham'	wHig
	sanguineum 'Cedric Morris'	wWoS wHig oSis
un	***sanguineum*** 'Connie's Variety'	oJoy wHig
	sanguineum 'Elsbeth'	wWoS wHig oHed oGre wBox
	sanguineum 'Farrer's Form'	oHed
	sanguineum 'Glenluce'	wHer wHig
	sanguineum 'Holden'	wHer
	sanguineum 'John Elsley'	oFor wWoS wHig oGar oOut oGre wTGN wWhG oSle
	sanguineum 'Jubilee Pink'	oJoy
	sanguineum var. ***lancastrense***	see **G. *sanguineum*** var. ***striatum***
	sanguineum 'Max Frei'	oFor wWoS oNat oJoy wHig oRus oGar oWnC wTGN oUps
	sanguineum 'Max Frei' seed grown	wEde
	sanguineum 'Nanum'	wHer wHig
un	***sanguineum*** 'New Hampshire'	see **G. *sanguineum*** 'New Hampshire Purple'
	sanguineum 'New Hampshire Purple'	oFor wHer wWoS oNat oJoy wHig oAls oRus oGar oSis oSqu oGre wFai oWnC wTGN wEde oUps oSle wSnq
	sanguineum 'Nyewood'	oJoy wHig
	sanguineum 'Purple Flame'	wHig oRus
	sanguineum 'Shepherd's Warning'	wWoS oSho
	sanguineum var. ***striatum***	oFor wWoS wCul oJoy wHig oAls oHed oRus oMis oGar oSis wRob oWnC wMag wTGN wBox wSte
	sanguineum var. ***striatum*** Splendens'	wAva
	sanguineum 'Vision'	wCSG
	sessiliflorum ssp. ***novae-zelandiae*** 'Nigricans'	oAls oHed oOut wSnq
	sessiliflorum 'Rubrum'	oGre
	sessiliflorum x ***traversii*** ex Merrist Coll.	oNWe
	sinense	wHer
	'Sirak'	wHer wWoS oNWe
	soboliferum	last listed 99/00
un	***sp.*** Pamir Mountains	wWoS
	'Spinners'	wHer wHig oHed
	'Stanhoe'	wHer wWoS wCul oGos wSwC oDar oHed oGar wSte
qu	***stapfianum***	wWoS
	'Sue Crug'	wWoS
un	'Sugar Plum'	oGar
	swatense	wHig
	sylvaticum	last listed 99/00
	sylvaticum 'Album'	wHig
	sylvaticum 'Amy Doncaster'	wWoS oHed
	sylvaticum 'Baker's Pink'	wHer
un	***sylvaticum*** 'Lilac Time'	wWoS
	sylvaticum 'Mayflower'	oFor wWoS oJoy wHig wRob wTGN
	sylvaticum 'Silva'	wHer wWoS
	sylvaticum ssp. ***sylvaticum*** var. wanneri	wHig
un	'Terra Franche'	oFor wWoS
	thunbergii	oRus

	traversii var. ***elegans***	oHed
	tuberosum	wHer oHed
	versicolor	wHer wHig oDan wCCr
	viscosissimum	oFor oHan wPla
	wallichianum	wHig
	wallichianum 'Buxton's Variety'	oJil oNWe wEde
	wallichianum Buxton's Variety Strain	wHer
	wallichianum selected pink, seed grown	wHer
	wallichianum 'Syabru' seed grown	wHer
	wlassovianum	wHer wCri wHig
	yoshinoi	last listed 99/00
	GEUM	
	aleppicum HC 970307	last listed 99/00
	'Beech House Apricot'	wWoS oHed
	'Borisii'	oFor oSec wAva oSle
	chiloense	wCoN
	'Coppertone'	wWoS oSqu
	'Dolly North'	oFor
	'Feuermeer'	oHed oWnC
	Fire Lake	see **G.** 'Feuermeer'
	'Georgenburg'	oFor oDar oJoy oRus
	'Lady Stratheden'	oFor wCul oBlo oWnC
	macrophyllum	oHan wNot oBos wFFl
	montanum	oNWe
	'Mrs. J. Bradshaw'	oFor wCul oTDM oAls oMis oBlo oWnC wMag wTGN oCir wAva wCoN oEga oUps
	rivale	oFor oHan wCul oJoy oGoo oSis
	rivale 'Album'	oFor oCis
	rivale 'Leonard's Variety'	oFor wCul oMis oCis wBWP
	'Starker's Magnificum'	oFor wCul oOut wFai
	sp. HCM 98108	wHer
	triflorum	oFor wPir wWld oUps
	triflorum var. ***ciliatum***	oBos
	triflorum var. ***ciliatum*** NNS97-139	wMtT
	urbanum	oGoo
	GEVUINA	
	avellana HCM 98055	wHer
	GIBASIS	
	geniculata	oGar
	GILLENIA	
	trifoliata	oFor oHed oNWe
	GINKGO	
	biloba	oFor oTrP wBur wClo oEdg oAls wShR oGoo oRiv oCis wRai oOEx oGre iGSc wKin oCrm wTGN wBox wSta wBWP wAva wCoN wRav oUps
	biloba 'Autumn Gold' male	oFor wBur oGar oRiv oGre oWnC wTGN wCoN wCoS wWel
un	***biloba*** 'Canopy'	wCol
un	***biloba*** 'Chase Manhattan'	oGre wCoN
un	***biloba*** 'Chichi'	oFor wBur wCol oAmb oDan oRiv wWel
	biloba 'Fairmount' male	oFor oRiv
qu	***biloba*** 'Fall Gold'	oCis
	biloba 'Fastigiata' male	oFor oWhi
	biloba female	oFor
un	***biloba*** 'Gresham'	wCoN
un	***biloba*** 'Jade Butterflies'	oGos wCol oGar oRiv oWnC wSte wWel
un	***biloba*** 'Kew'	oFor
un	***biloba*** 'Magyar'	oFor wBur oRiv oCis oGre
	biloba 'Mayfield' male	wBur oDan oRiv oGre
	biloba Pendula Group	oFor wBur oAmb oGar oRiv oGre
un	***biloba*** 'Pendula Rowe'	oRiv
	biloba 'Princeton Sentry' male	oFor oGar oRiv oCis oGre oWnC wWel
	biloba 'Salem Dandy' male	wRai
	biloba 'Salem Lady' female	wRai
	biloba 'Saratoga' male	oFor wBur wCol oAmb oDan oRiv oSho oBlo oGre wCoN
un	***biloba*** Shangri-la	wBur oGar oRiv oCis oGre
un	***biloba*** 'Spring Grove'	wCoN
	biloba 'Tubifolia'	oFor wCol oDan oRiv wWel
un	***biloba*** 'Windover Gold'	oGre
	GLADIOLUS	
un	'Advance'	wCon
	'Applause'	wCon
un	'Blue Mount'	wCon
un	'Blue Sky'	wCon
	callianthus	wHom
	callianthus 'Murieliae'	oNat
un	'Chanticleer'	wCon
un	'China Doll'	wCon
	communis	oCis
	communis ssp. ***byzantinus***	wHer oNat
	dalenii	wHer
un	'Daydream'	wCon
un	'Fiesta'	wCon
un	***x gandavensis*** 'Boone'	wHer
un	'Good News'	wCon
un	'Grand Slam'	wCon
un	'Green'	wCon
un	'Her Majesty'	wCon
	'High Style'	wCon
	Homoglad hybrids	wHer
un	'Honey Gold'	wCon
	illyricus	oFor wHer
	italicus	wHer
	'Jacksonville Gold'	last listed 99/00
un	'Lady Di'	wCon
un	'Land O'Lakes'	wCon
un	'Maestro'	wCon
un	'Matchpoint'	wCon
un	'Minnesota Rose'	wCon
un	'Monte Negro'	wCon
	'Nova Lux'	wCon
	palustris	oFor
	papilio	wHer wCul wHig wCSG
	'Plum Tart'	wCon
	'Priscilla'	wCon
un	'Rapid Red'	wCon
un	'Red Majesty'	wCon
un	'Summer Rose'	wCon
	'White Friendship'	wCon
un	'White Prosperity'	wCon
	'Wind Song'	wCon
	'Wine and Roses'	wCon
	GLAUCIDIUM	
	palmatum	wHer oEdg
	GLAUCIUM	
	corniculatum	wBWP
	flavum	wCul wBWP wCoN
	GLECHOMA	
	hederacea	wFai oWnC wTGN
	hederacea 'Variegata'	oSle
	GLEDITSIA	
	caspica	oFor wCCr
	triacanthos 'Imperial'	oWnC
	triacanthos f. ***inermis***	wBCr wKin wPla
	triacanthos f. ***inermis*** 'Aurea'	see **G. *triacanthos*** 'Sunburst'
	triacanthos 'Rubylace'	oFor
	triacanthos 'Shademaster'	oWnC wCoN
	triacanthos 'Skyline'	oGar wCoN
un	***triacanthos*** 'Suncole'	oDar
	triacanthos 'Sunburst'	oFor oAls oGar oBlo wKin wTGN wCoN
un	***vestita***	oCis oWnC
	GLOBBA	
	winitii	oTrP
	GLOBULARIA	
	cordifolia	oSis oNWe
	incanescens	oSis
un	'Indubium'	oCis
	meridionalis	oSis
	nana	see **G. *repens***
	nudicaulis	oSis
	punctata	oSis
	repens	oSis
	trichosantha	oFor
	GLORIOSA	
	superba	oUps
	superba 'Rothschildiana'	oGar oVBI
	GLUMICALYX	
	sp.	oCis
	GLYCERIA	
un	***borealis***	wWat
	maxima var. ***variegata***	wCli wSoo oGar oSsd oNWe oWnC
	GLYCYRRHIZA	
	echinata	last listed 99/00
	glabra	oAls wFGN wFai iGSc oBar oUps
	uralensis	oGoo oOEx oCrm oBar
	GLYPTOSTROBUS	
	lineatus	see **G. *pensilis***
	pensilis	oPor oRiv
un	**GONATANTHUS**	
	pumilus HWJCM 047 green	wHer
	GONIOLIMON	
	incanum 'Blue Diamond'	oJoy oWnC oUps
	tataricum	oFor oAls oGar oGoo oSho wFai oEga
	tataricum var. ***angustifolium***	oSha
	tataricum 'Woodcreek'	oWnC
	GOODYERA	
	oblongifolia	wThG oSis oBos wFFl
	pubescens	oTrP wThG

	Name	Sources
	GORDONIA	
	lasianthus	oFor oRiv
	GOSSYPIUM	
	arboreum	oTrP
	GRAPTOPETALUM	
un	'Ghosty'	oSqu
	macdougalii	last listed 99/00
un	*paraguayense bernalense*	oSqu
	X GRAPTOSEDUM	
un	'California Sunset'	oSqu
un	'Francesco Baldi'	oSqu
un	'Vera Higgins'	oSqu
	X GRAPTOVERIA	
un	'Kew'	oSqu
un	'Margaret Rose'	oSqu
	'Silver Star'	oSqu
un	'Titubans'	oSqu
	GRATIOLA	
	officinalis	oGoo
	GREENOVIA	
	aurea	oSqu
	GREVILLEA	
	'Bronze Rambler'	wGra
	'Canberra Gem'	oSho
	juniperina f. *sulphurea*	last listed 99/00
	lanigera 'Compacta'	oWnC
	lanigera 'Mt. Tamboritha'	see **G. *lanigera*** 'Compacta'
	'Noeli'	oTrP wGra
un	'Pink Pearl'	oCis
	robusta	last listed 99/00
	rosmarinifolia dwarf pink form	wGra
	victoriae	wHer oCis wCCr
	GREWIA	
	biloba	oFor
	caffra	oTrP
	GRINDELIA	
	camporum	oGoo
	integrifolia	wWld
	robusta	see **G. *camporum***
	stricta	oGoo
	GRISELINIA	
	littoralis	wCCr
	littoralis 'Variegata'	wHer wCCr wSte
un	*racemosa* HCM 98174	wHer
	GUNNERA	
	chilensis	see **G. *tinctoria***
	flavida	last listed 99/00
	magellanica	oTrP oGos oDar oJoy oGre
	magellanica HCM 98188-A	wHer
	manicata	oTrP wWoS oJoy oAls wCSG oAmb oGar oNWe wNay oGre oTri wTGN wCoN wSnq wWel
	monoica	wHer
	prorepens	wHer wWoS
qu	*reptans*	oTrP
	tinctoria	oGos wGAc oDar oRus oHug oSho wRob oGre wCCr wBox wCoN wWel
	tinctoria HCM 98011	wHer
	GYMNASTER	
	savatieri	see **ASTER *savatieri***
	GYMNOCARPIUM	
	dryopteris	wFol wNot wFan oRus oBRG wFFl
	dryopteris 'Plumosum'	oRus wRob wSte
	GYMNOCLADUS	
	chinensis	oFor
	dioica	oFor wBCr oEdg oGar wCCr
	GYPSOPHILA	
	aretioides	oSis wMtT
	aretioides 'Caucasica' JJH970723	wMtT
	briquetiana	oSis
	caucasica	see **G. *aretioides*** 'Caucasica'
	cerastioides	wCul oJoy oHed oAmb oSis wMtT wCoN oSle
qu	'Double Snow'	wMag
	'Festival Happy'	oFor
	'Festival Pink'	oFor oEga wWhG
un	'Festival White'	oEga wWhG
	libanotica	last listed 99/00
	nana	wMtT
	pacifica	wMag wWhG
	paniculata	oJoy wCoN
do	*paniculata* 'Bristol Fairy'	oAls oGar oGoo oWnC wMag oEga
do	*paniculata* 'Compacta Plena'	wHer wCul
	paniculata 'Perfekta'	wCul oGar wMag
do	*paniculata* 'Pink Fairy'	wCul oAls oGar oCir
do	*paniculata* 'Schneeflocke'	wCul oJoy oWnC
	paniculata Snowflake	see **G. *p.*** 'Schneeflocke'
	paniculata 'Viette's Dwarf'	oFor wWoS oGar
	paniculata white	oCir
	repens	oWnC
	repens 'Alba'	wTho oJoy wWhG
	repens 'Rosea'	oFor oAls oSis oWnC wWhG
	'White Festival'	oFor
	HABENARIA	
	blephariglottis	oTrP
	ciliaris	oTrP wThG
	HABERLEA	
	ferdinandi-coburgii	wCol
	rhodopensis	oBov
	HABRANTHUS	
	robustus	wHer
	tubispathus	wHer
	HACKELIA	
un	*uncinata*	wCSG
	HACQUETIA	
	epipactis	oRus oSis
	HAKEA	
un	*orthorrhyncha*	wGra
	HAKONECHLOA	
	macra	wCli oJoy wCol oAls oGar wRob oBlo wFai wWin wMag wSte
	macra 'Albostriata'	see **H. *macra*** 'Albovariegata'
	macra 'Albovariegata'	wHer wWoS oJoy wCol oGar oDan oSis oOut wNay wRob oGre wWin
	macra 'Aureola'	oFor wHer wWoS wCli wCul wCri oGos wSwC wGAc oJoy wHig oAls oHed oRus wCSG oAmb oMis oGar oDan oSis oNWe oOut wNay wRob oGre wWin wHom wTGN wBox oSec wCoN wWhG oUps wSte wSnq
un	*macra* 'Aureovariegata'	wSte
	HALESIA	
	carolina	oFor oEdg oAls oAmb oGar oRiv oGre wCoN
un	*carolina* 'Rosea'	oDan oGre
	diptera	oFor
	diptera var. *magniflora*	wHer oRiv oGre
	monticola	oFor oGar oDan oRiv oSho oWhi oGre
	monticola var. *vestita* f. *rosea*	oGre wWel
	tetraptera	see **H. *carolina***
	X HALIMIOCISTUS	
	sahucii	last listed 99/00
	wintonensis	oFor wHer wSwC oGar oSis oSho oGre wTGN oCis wSte
	wintonensis 'Merrist Wood Cream'	wHer wWoS oJil oNWe wCCr
	HALIMIUM	
	alyssoides	wCCr
	calycinum	wHer
	commutatum	see **H. *calycinum***
	halimifolium	last listed 99/00
	lasianthum f. *concolor*	wHer
	ocymoides	wHer oSis
	HALIMODENDRON	
	halodendron	last listed 99/00
	HALLERIA	
	lucida	last listed 99/00
	HALOCARPUS	
	bidwillii	wHer
	HALORAGIS	
	erecta 'Wellington Bronze'	wWoS wHig oSis wCCr oCis
	HAMAMELIS	
	'Brevipetala'	oGre
un	'February Gold'	oRiv
	x intermedia 'Allgold'	oGos oGre
	x intermedia 'Arnold Promise'	oFor wHer oGos wCol oAmb oGar oDan oRiv oSho oBlo oGre oTPm wSta wWhG wWel
	x intermedia 'Barmstedt Gold'	wWel
	x intermedia 'Carmine Red'	last listed 99/00
	x intermedia 'Diane'	oFor wHer wClo wCol oAls oAmb oGar oDan oRiv oBlo oGre wSta wCoN wWhG wSte wWel
	x intermedia 'Feuerzauber'	oRiv oGre wCoN
	x intermedia Fire Charm	see **H. *x intermedia*** 'Feuerzauber'
	x intermedia 'Fire Cracker'	oFor oGre
	x intermedia 'Hiltingbury'	oGos oGre
	x intermedia 'Jelena'	oFor wHer oGos wCol oAmb oGar oDan oRiv oWhi oBlo oGre oWnC wSta wCoN wWel
	x intermedia 'Luna'	oGos

	x intermedia 'Moonlight'	oGos oGre
	x intermedia 'Orange Beauty'	oGos oGre
	x intermedia 'Pallida'	oFor oAls oGar oRiv oGre oWnC wSte wWel
	x intermedia 'Primavera'	oFor oGos wClo oGar oRiv oGre wWel
	x intermedia 'Ruby Glow'	oFor oGos oRiv oGre
	x intermedia 'Sunburst'	oGos wClo oGre
	x intermedia 'Westerstede'	oGos oGre wSte
	x intermedia 'Winter Beauty'	oGos oGre
	japonica 'Arborea'	oGos oGre
	japonica var. ***flavopurpurascens***	oGos oGre
	japonica 'Zuccariniana'	oFor oGos oGre
un	***mexicana***	oGos
	mollis	wTGN wCoN wSte wWel
	mollis 'Coombe Wood'	oGos oGre
	mollis 'Goldcrest'	oFor
	vernalis	oFor oGar oRiv oCrm
	vernalis 'Christmas Cheer'	last listed 99/00
	vernalis 'Lombart's Weeping'	wHer oGos oGre
un	***vernalis*** 'Purpurea'	oGos
	vernalis 'Sandra'	oGos oGre
	virginiana	oFor oAls wShR oGoo oRiv oGre iGSc wSta wCoN wSte wCoS

HANABUSAYA

	asiatica HC 970090	wHer

HAPLOPAPPUS

un	***mucronatus***	oSis

HATIORA

	gaertneri	oGar
	rosea	oSqu
	salicornioides	oSqu

HAWORTHIA

	arachnoidea	oSqu
	asperula	oSqu
	x cuspidata	oSqu
	cymbiformis	oSqu
	fasciata	last listed 99/00
	limifolia	last listed 99/00
	margaritifera	last listed 99/00
	margaritifera 'Variegata'	oSqu
	mirabilis	last listed 99/00
	mundula	see **H. *mirabilis***
	parksiana	last listed 99/00
	planifolia	see **H. *cymbiformis***
	pygmaea	see **H. *asperula***
	reinwardtii	oSqu
	resendeana	see **H. *reinwardtii***
	reticulata hurlingii	oSqu
un	***retusa acuminata*** 'Variegata'	oSqu
	subattenuata	see **H. *margaritifera***
	translucens	oSqu
	truncata	last listed 99/00
	turgida var. ***pallidifolia***	last listed 99/00
un	***viscosa*** 'Variegata'	oSqu
	xiphiophylla	last listed 99/00

HEBE

	albicans	wAva
un	***albicans*** 'Alameda'	oHed
un	***albicans*** 'Oswego'	oHed
	albicans 'Sussex Carpet'	wHer oDan oSis
	'Alicia Amherst'	oFor oBlo
	'Amy'	oNat oJoy wCSG oGar oDan oUps oSle
	x andersonii 'Variegata'	wHer
	'Autumn Glory'	oFor wSwC oJoy oAls oGar wCCr wCoN
	'Autumn Joy'	oCir
	x bidwillii	oCis
un	'Blue Elf'	oGar oSho
un	'Blue Mist'	oJoy oAls oGar oCis oBRG wCoS
	bollonsii	oCis
	'Bowles's Hybrid'	wHig
	buxifolia (see Nomenclature Notes)	wSwC oGar oOut oBlo oGre wCCr oWnC wTGN oBRG wCoN oUps oSle
	buxifolia 'Nana' (see Nomenclature Notes)	oAls wWel
un	***buxifolia*** 'Repens'	wSta
	'Caledonia'	oJoy wCol oGar oSis oCir
	canterburiensis	wAva
	'Carl Teschner'	see **H.** 'Youngii'
	carnosula	wSwC oBlo oUps
	'Coed'	oCis
	'Colvos Hybrid'	wCCr
	cupressoides	wWoS oJoy wPir wSta oCis
	cupressoides 'Boughton Dome'	oFor oGar oDan oSis oCir oUps
	decumbens	oFor oSis oGre
	'Dorothy Peach'	see **H.** 'Watson's Pink'
	'Emerald Green'	oSis
	glaucophylla	wCul oJoy oAls oGar wPir oBRG oCir oSle wWel
	'Glaucophylla Variegata'	oWnC
	'Great Orme'	wHer oJoy wAva
	'Hagley Park'	oJoy
un	'Hanna'	oGar
un	'Highgrove'	wHer
	hulkeana	wCCr
un	'Lavender Charm'	oUps
	'Macewanii'	oSis oCis
	mackenii	see **H.** 'Emerald Green'
	'Margery Fish'	see **H.** 'Primley Gem'
	'Margret'	oGre wCoN
	'McKean'	oGar oOut wTGN oBRG oCir oUps
qu	***mckeanii***	oJoy oOut
	'Mrs. Winder'	wWoS oCis oUps
	'Nicola's Blush'	wCSG oGar oUps
	ochracea	wSta oBRG
	ochracea 'James Stirling'	wHer oSis oBRG wWel
	odora	wHer oJoy
	parviflora	see **H.** 'Bowles's Hybrid'
	parviflora var. ***angustifolia***	oJoy wCCr
	'Patty's Purple'	oGos wSwC oJoy oGar oSho oBlo oGre oCir wCoN oUps oSle wWel
	'Paula'	oJoy
	'Pimeba'	oSis
	pimeleoides 'Quicksilver'	oSis wPir wCCr oWnC
	pinguifolia 'Pagei'	wCul wSwC oNWe oGre
un	'Pinocchio'	oUps
un	'Powis Castle Blue'	oCis
	'Primley Gem'	oJoy oSis
	prostrata	wHer wAva
	'Purple Picture'	oJoy oAls oHed oGar oSis oCis oUps
	rakaiensis	last listed 99/00
	recurva	wHer oHed oWnC
	'Red Edge'	wSwC
un	***rigidula***	oGar oWnC oCis
	salicifolia	last listed 99/00
un	***salicifolia*** 'Violet Snow'	oCis
	'Silver Beads'	wWoS
	silver green	oCir
	speciosa	oFor
	speciosa 'Tricolor'	see **H. *speciosa*** 'Variegata'
va	***speciosa*** 'Variegata'	oFor wWoS wSwC oJoy oGar wTGN oUps oSle
un	***sutherlandii***	oCis
	topiaria	oUps
	traversii	oGar wCCr oCis
un	'Veronica Lake'	oJoy
	'Watson's Pink'	wAva
un	'Whipcord'	oCis
	'White Gem'	wHer wCul
	'Wiri Charm'	oUps
	'Youngii'	oFor wWoS oJoy wShR oCis oCir

HECHTIA

	podantha BLM 0638	oRar

HEDERA

	canariensis (see Nomenclature Notes)	wHer oSho wPir oWnC wTGN
va	***canariensis*** hort. 'Gloire de Marengo'	wHer oSho
	canariensis hort. 'Variegata'	see **H. *canariensis*** hort. 'Gloire de Marengo'
	colchica	oSqu
un	***colchica*** 'Batami'	wHer
	colchica 'Dentata'	oSqu
	colchica 'Dentata Variegata'	last listed 99/00
	colchica 'My Heart'	see **H. *colchica***
va	***colchica*** 'Sulphur Heart'	oFor wWel
	helix	oAls oGar oBlo wMag wSta wCoN
va	***helix*** 'Adam'	oSqu
	helix 'Amberwaves'	oSqu
	helix 'Angularis'	oSqu
	helix 'Angularis Aurea'	oSqu
un	***helix*** 'Anita'	oSqu
un	***helix*** 'Ann Gamie'	wCCr
	helix 'Appaloosa'	oSqu
	helix 'Arapahoe'	oSqu
	helix 'Asterisk'	oSqu
	helix 'Baby Face'	oSqu
un	***helix*** 'Baby Gold Dust'	oSqu
	helix var. ***baltica***	oSho oBlo oSqu oWnC wCoN
un	***helix*** 'Bettina'	oSqu
	helix 'Big Deal'	oSqu
un	***helix*** 'Blarney'	oSqu
	helix 'Boskoop'	oSqu
	helix 'Brigette'	see **H. *helix*** 'California'
un	***helix*** 'Brimstone'	oSqu

HEDERA

va	***helix*** 'Bruder Ingobert'	oSqu
	helix 'Buttercup'	oSis oSqu
	helix 'Butterflies'	oSho oWnC
va	***helix*** 'Caecilia'	oSqu
	helix 'Calico'	see **H. *helix*** 'Schaefer Three'
	helix 'California'	oSqu wSta
	helix 'California Fan'	oHed oSqu
va	***helix*** 'California Gold'	oGoo oSqu
	helix 'Carolina Crinkle	oSqu
un	***helix*** 'Cathedral'	wSta
un	***helix*** 'Chalice'	oSqu
	helix 'Chicago'	oSqu
	helix 'Christian'	see **H. *helix*** 'Direktor Badke'
	helix 'Chrysanna'	oSqu
	helix 'Cockle Shell'	oSqu
	helix 'Congesta'	oAls oSqu
	helix 'Conglomerata'	oSqu
un	***helix*** 'Connecticut Yankee'	wCCr
	helix 'Crinolette'	oSqu
	helix 'Curley-Q'	see **H. *helix*** 'Dragon Claw'
	helix 'Curleylocks'	see **H. *helix*** 'Manda's Crested'
va	***helix*** 'Curvaceous'	oSqu
	helix 'Cuspidata Minor'	see **H. *hibernica*** 'Cuspidata Minor'
un	***helix*** 'Dark Knight'	oSqu
	helix 'Dicke von Stauss'	oSqu
	helix 'Direktor Badke'	oSqu
un	***helix*** 'Dolly'	oSqu
	helix 'Donerailensis'	last listed 99/00
	helix 'Dragon Claw'	oSqu wSta
	helix 'Duckfoot'	oAls oSqu oWnC wMag wTGN wSta
un	***helix*** 'Duckfoot' variegated	oHed
	helix dwarf gold	wSta
	helix dwarf green	wSta
un	***helix*** 'Eclipse'	oSqu
va	***helix*** 'Elfenbein'	oSqu
	helix 'Emerald Gem'	see **H. *helix*** 'Angularis'
	helix 'Emerald Globe'	oSqu
	helix 'Emerald Jewel'	see **H. *helix*** 'Pittsburgh'
	helix 'Erecta'	oSis oSqu
	helix 'Erin'	see **H. *helix*** 'Pin Oak'
va	***helix*** 'Eva'	oSqu
un	***helix*** 'Excalibur'	oSqu
	helix 'Fallen Angel'	oSqu
	helix 'Fan'	wHer oSqu wCCr oCis
va	***helix*** 'Fantasia'	oSqu wSta
	helix 'Filigran'	wHer oSqu
	helix 'Flamenco'	oSqu
	helix 'Fluffy Ruffles'	oSqu
	helix 'Garland'	oSqu
	helix 'Gavotte'	oSqu
va	***helix*** 'Gertrud Stauss'	oAls
va	***helix*** 'Glacier'	oAls oSis noSqu oWnC wSta
	helix 'Gold Dust'	oGoo oSqu
va	***helix*** 'Goldchild'	oAls oGoo oSho noSqu oWnC
va	***helix*** 'Goldcraft'	wCCr
un	***helix*** 'Golden Carpet'	oSqu
un	***helix*** 'Golden Fleece'	oSqu
va	***helix*** 'Golden Ingot'	oGoo oSqu
un	***helix*** 'Golden Nugget'	oSqu
va	***helix*** 'Golden Snow'	oSqu
	helix 'Goldheart'	see **H. *helix*** 'Oro di Bogliasco'
	helix 'Green Feather'	oSqu
	helix 'Green Finger'	see **H. *helix*** 'Tres Coupe'
un	***helix*** 'Green Globe'	oSqu
	helix 'Green Ripple'	oSqu
	helix 'Green Spear'	see **H. *helix*** 'Spear Point'
qu	***helix*** 'Hahns'	oGar wMag wCoN wCoS
va	***helix*** 'Harald'	oSqu
	helix 'Harry Wood'	see **H. *helix*** 'Modern Times'
un	***helix*** 'Heartleaf'	oGoo
un	***helix*** 'Hedgehog'	wWoS wCol oHed oOut
	helix 'Helena'	oSqu
	helix 'Helvetica'	oSqu
un	***helix*** 'Henriette'	oSqu
	helix 'Ingrid'	see **H. *helix*** 'Harald'
	helix 'Innuendo'	oSqu
un	***helix*** 'Irish Gold'	oSqu
	helix 'Irish Lace'	oSqu
	helix 'Itsy Bitsy'	see **H. *helix*** 'Pin Oak'
	helix 'Ivalace'	oGoo oSqu
	helix 'Jasper'	oSqu
va	***helix*** 'Jubilee'	oSqu
un	***helix*** 'Kaleidescope'	oSqu
	helix 'Knuelch'	oSqu
un	***helix*** 'Kobold'	oSqu
va	***helix*** 'Kolibri'	oSqu
	helix 'Kurios'	oSqu
	helix 'La Plata'	oSqu
qu	***helix*** 'Lace'	wCoS
un	***helix*** 'Lady Frances'	oSqu
	helix 'Lalla Rookh'	oSqu
un	***helix*** 'Latina'	oSqu
va	***helix*** 'Lemon Swirl'	oSqu
	helix 'Leo Swicegood'	oSqu
un	***helix*** 'Lilliput'	oSqu
	helix 'Limey'	oSqu
va	***helix*** 'Little Diamond'	oSis oSqu
	helix 'Little Witch'	oSqu
	helix 'Malvern'	oSqu
	helix 'Manda Fringette'	oSqu
	helix 'Manda's Crested'	oSqu wCCr
un	***helix*** 'Marginata of Hibbard'	oSqu
	helix 'Marie-Luise'	oSqu
un	***helix*** 'Marilyn'	oSqu
un	***helix*** 'Mariposa'	oSqu
	helix 'Melanie'	wHer
	helix 'Merion Beauty'	oSqu
	helix 'Midget'	oSqu
va	***helix*** 'Mini Ester'	oSqu
	helix 'Minima'	see **H. *helix*** 'Donerailensis'
va	***helix*** 'Minor Marmorata'	oSqu
va	***helix*** 'Misty'	oSqu
	helix 'Modern Times'	oSqu
un	***helix*** 'Moon Beam'	oSqu
va	***helix*** 'Mrs. Pollock'	oSqu
	helix 'Needlepoint'	oAls oSho oSqu wBox wSta wCoN oSle
	helix 'Neptune'	oSqu
un	***helix*** 'Nice Guy'	oSqu
	helix 'Olive Rose'	oSqu
va	***helix*** 'Oro di Bogliasco'	wWoS wCol oAls oHed oAmb oGoo oSis oSqu oGre wCCr oCis oBRG oSle
va	***helix*** 'Paper Doll'	oSho oSqu
va	***helix*** 'Parasol'	oSqu
	helix 'Pedata'	oGar oSqu
	helix 'Perkeo'	oSqu
	helix 'Pin Oak'	oGoo oSqu
	helix 'Pirouette'	oSqu
	helix 'Pittsburgh'	oSqu
	helix 'Pixie'	oSqu wCoS
	helix f. ***poetarum***	oAls oSho oBRG
	helix 'Poetica'	see **H. *helix*** f. ***poetarum***
un	***helix*** 'Prima Donna'	oSqu
va	***helix*** 'Rauschgold'	oSqu
va	***helix*** 'Reef Shell'	oSqu
	helix 'Ritterkreuz'	oSqu
va	***helix*** 'Romanze'	oSqu
	helix 'Sagittifolia Variegata'	oSqu
va	***helix*** 'Schaefer Three'	oGoo oSqu
	helix 'Shamrock'	oGoo oSqu wSta
	helix 'Shannon'	oSqu
un	***helix*** 'Silbermove'	oSqu
va	***helix*** 'Silver King'	oSqu
un	***helix*** 'Silver Lining'	oSqu
	helix 'Sinclair Silverleaf'	oSqu
	helix 'Small Deal'	last listed 99/00
	helix 'Spear Point'	last listed 99/00
	helix 'Spetchley'	oSqu
	helix 'Spinosa'	oSqu
un	***helix*** 'Star'	oSqu
	helix 'Stuttgart'	oSqu
	helix 'Sunrise'	oSqu
un	***helix*** 'Tanja'	oSqu
	helix 'Teardrop'	oAls oSqu
	helix 'Telecurl'	oGoo oSqu
un	***helix*** 'Temptation'	oSqu
	helix 'Thorndale'	oSqu
	helix 'Tiger Eyes'	oSqu
un	***helix*** 'Tomboy'	oSqu
	helix 'Tres Coupe	oSqu oWnC
	helix 'Triton'	oGoo
va	***helix*** 'Tussie Mussie'	oSqu
	helix 'Variegata'	oGar
	helix 'Walthamensis'	oSqu
un	***helix*** 'Webfoot'	oSis
un	***helix*** 'White Marble'	wSta
va	***helix*** 'Williamsiana'	oSqu
	helix 'Wilson'	oSqu wSta
	helix 'Woerneri'	last listed 99/00
va	***helix*** 'Zebra'	oSqu
	hibernica 'Cuspidata Minor'	oSqu

va	***hibernica*** 'Sulphurea'	oSqu
	nepalensis 'Marbled Dragon'	wHer
un	***nepalensis*** var. ***sinensis*** 'Boaxing Star' DJHC 711	wHer
	HEDYCHIUM	
	coccineum	oCis
	sp. aff. ***coccineum***	wHer
	coccineum 'Tara'	wHer
	coronarium	oTrP oHug oOEx
	densiflorum 'Assam Orange'	wHer
	sp. aff. ***densiflorum*** HWJCM 073	wHer
	densiflorum 'Stephen'	last listed 99/00
	flavescens	oOEx
	forrestii	wHer
	gardnerianum	oTrP oOEx oCis
	greenei	oTrP oCis
un	'Kinkaku'	wHer
un	'Peach Beauty'	wWoS
	pradhanii	wHer
	spicatum	wHer
	yunnanense	last listed 99/00
	sp. aff. ***yunnanense*** DJHC 459	wHer
	HEDYSARUM	
	coronarium	last listed 99/00
	HEIMIA	
	myrtifolia	oTrP
	salicifolia	oFor wCCr iGSc
	HELENIUM	
	autumnale	oFor oHan oGoo
un	***autumnale*** 'Flammenrud'	wCSG
	autumnale reddish brown	oGoo
	bigelovii	last listed 99/00
	'Bruno'	oGar
	'Coppelia'	oGar
	'Crimson Beauty'	wRob
	'Die Blonde'	oHed
	'Dunkelpracht'	oHed
	'Feuersiegel'	oNat oHed oWnC
	'Flammendes Kaethchen'	oHed
	hoopesii	oFor oJoy oGoo
un	'Kugelsonne'	oJoy
	'Kupferzwerg'	oHed
	'Moerheim Beauty'	oFor wCul wSwC oNat oAls oHed oMis wTGN oSle
qu	'New Hybrid'	oGar
	puberulum	wCCr
	Red and Gold	see **H.** 'Rotgold'
	'Riverton Beauty'	oAls
	'Rotgold'	oEdg oAls iArc oSle
	'The Bishop'	wSwC
	'Zimbelstern'	wCul oJoy oHed oGar oWnC wBox
	HELIANTHEMUM	
	'Annabel'	oFor oAls oGar oDan oWnC wTGN wWhG
	'Apricot'	oFor
	'Ben Heckla'	wHer
	'Ben Ledi'	oUps
	'Ben More'	oUps
	'Ben Nevis'	wWhG
un	'Boule de Feu'	oFor wHer wCul oAmb wTGN
un	'Bright Spot'	wHer wWoS oSis
un	'Burgundy Dazzler'	oSis
do	'Cerise Queen'	wHer
	'Cheviot'	oFor wWoS
un	'Dazzler'	oFor oAls oGar oWnC oEga wWhG oUps
	double apricot	wCul oGar oSis
do	'Double Orange'	oFor wMag
un	'Double Peach'	oFor
un	'Eloise'	wHer oSis
	'Fire Dragon'	wCul
	'Fireball'	see **H.** 'Mrs. C. W. Earle'
	'Flame'	oFor
	'Georgeham'	wWoS
un	'Golden Nugget'	oFor
	'Henfield Brilliant'	oFor wHer wSwC oHed oGar wCCr oWnC wWhG oUps
do	'Mrs. C.W. Earle'	oSis
	'Mrs. Moules'	wSwC oGar
	mutabile	wTho oEga
	oelandicum ssp. ***alpestre***	oSis
un	***oelandicum*** ssp. ***alpestre*** 'Serpyllifolium'	oSis
un	***oelandicum*** var. ***pencillatum***	oSis
	orange	oFor
	'Orange Surprise'	wHer oSis
un	'Peach'	oFor oAls oGar oSis wWhG
	pilosum	wCCr

	'Raspberry Ripple'	oFor oAls oGar oSis
	'Rhodanthe Carneum'	oFor wHer wCul wSwC oAls oHed oGar oDan oSho wCCr oWnC wMag wWhG oUps
un	'Rose Glory'	oFor oAls oEga
do	'Rose of Leeswood'	oSis
	'Shot Silk'	last listed 99/00
un	'Saint Mary's'	oFor wHer wCul oAmb wWhG
un	'Sunfleck'	wCul
	'The Bride'	oHed
	'Voltaire'	oSis
	white	oFor
	'Wisley Pink'	see **H.** 'Rhodanthe Carneum'
	'Wisley Primrose'	oFor wCul oHed oGar oSis oSho oWnC wWhG
	'Wisley White'	last listed 99/00
	HELIANTHUS	
	angustifolius	wCul
	decapetalus	wPir
	decapetalus 'Flore Pleno'	last listed 99/00
	'Lemon Queen'	oFor wHer wCul wSwC oAls oHed oGar oGoo oOut
	maximilianii	oFor oGar oGoo
	microcephalus	last listed 99/00
	x multiflorus 'Flore Pleno'	oFor
	tuberosus	wBCr wCSG
	tuberosus red	wBCr
	HELICHRYSUM	
	angustifolium	see **H. *italicum***
un	***arenarium*** ssp. ***aucheri***	wMtT
	argyrophyllum	last listed 99/00
un	***aucheri***	wHer
	basalticum	oCis
qu	***compactum***	wMtT
	'County Park Silver'	see **OZOTHAMNUS** 'County Park Silver'
	frigidum	last listed 99/00
	heldreichii Hythe form	oCis
	hookeri	see **OZOTHAMNUS *hookeri***
	italicum	oAls wFGN oGoo wFai iGSc oWnC oCrm oBar wNTP
	italicum ssp. ***microphyllum***	oAls wFGN oGoo oBar
	italicum 'Nanum'	see **H. *italicum*** ssp. ***microphyllum***
	italicum ssp. ***serotinum***	wAva
un	***italicum*** 'Wood's'	wFGN oBar
un	'Licorice Splash'	wTGN
	milfordiae	last listed 99/00
	'Mo's Gold'	see **H. *argyrophyllum***
	sp. aff. ***pagophilum*** JJ&JH9401713	last listed 99/00
	petiolare	oHed wHom oUps wCoS
	petiolare 'Dargan Hill Monarch'	oJoy
	petiolare 'Limelight'	wHer oNat oHed oUps
un	***petiolare*** 'Nanum'	wHer
	petiolare 'Variegatum'	wHer oNat oHed
	petiolatum	see **H. *petiolare***
un	'Petite Licorice'	wTGN
	'Schwefellicht'	wHer wHom
	selago	see **OZOTHAMNUS *selago***
	sessilioides	wMtT
qu	'Silver Brocade'	oCir
un	'Silver Light'	wHom
	silver variegated	wHom
	Sulphur Light	see **H.** 'Schwefellicht'
qu	'White Licorice'	wTGN
	HELICTOTRICHON	
	sempervirens	oFor wHer wCli wCul wSwC oNat oDar oJoy wHig oAls oHed oAmb oGar oSis oSho oOut oBlo wFai wWin wPir oTPm oTri wBox oSec wSta oBRG oCir wCoN wWhG oUps oSle wSte wSnq
	sempervirens 'Saphirsprudel'	oFor wHer wWoS oJoy wCol oOut oGre wWin wSnq
	sempervirens Sapphire	see **H. *sempervirens*** 'Saphirsprudel'
	HELIOPSIS	
	helianthoides	oFor
	helianthoides var. ***scabra*** 'Sommersonne'	oAls oGar iArc
	helianthoides Summer Sun	see **H. *helianthoides*** var. ***scabra*** 'Sommersonne'
	'Lorraine Sunshine'	wWoS oDar oGar
	HELIOTROPIUM	
	arborescens	wHom wCoS
	arborescens 'Album'	wFGN wHom
un	'Fragrant Delight'	wTho oAls wFGN oGar wHom
qu	'Irene'	oGar
	'Marine'	oGar wNTP
	'White Lady'	oHed

	HELLEBORUS	
	argutifolius	oFor wCul oGos wGAc oHon oNat oJoy wHig oAls oHed oRus wCSG oAmb oGar oDan oSis oBov oGre wFai wPir oCis wCoN wEde wWld
	argutifolius hybrid	oGre wBox
un	***argutifolius*** Janet Starnes Strain	wHer oGos oHon oGar oDan oSis oOut oGre wBox oCis
	argutifolius x ***lividus***	oHon oGre
	argutifolius Pacific Frost Strain	wHer oGar
	argutifolius variegated form	oNWe
	atropurpureus	see **H. *atropurpureus***
	atrorubens (see Nomenclature Notes)	wHer oAls oNWe
	croaticus	wHer
	cyclophyllus	wCul
	cyclophyllus coll. Macedonia	wHer
	dumetorum	last listed 99/00
	foetidus	wCul oHon oJoy wCol wHig oAls oRus wCSG oGar oDan oSis oOut oBov wPir wBox wBWP wCoN
	foetidus fragrant	oHon
	foetidus 'Green Giant' seed grown	wHer
	foetidus 'Miss Jekyll's Scented' seed grown	wHer
	foetidus 'Ruth' seed grown	wHer
un	***foetidus*** 'Sienna' seed grown	wHer
	foetidus 'Sopron'	oNWe
	foetidus 'Sopron' seed grown	wHer
un	***foetidus*** 'Tros-os-Montes' seed grown	wHer
	foetidus Wester Flisk Group	wHer oHon oAls oRus oNWe wPir oCis
	hybridus ssp. ***abchasicus*** Early Purple Group	oRus
	hybridus apricot parent	oNWe
	hybridus Ballard's Group, red	wHer
un	***hybridus*** 'Best of Blacks'	oCis
	hybridus black	oGar oGre wBox
	hybridus black parent	oGar oNWe
	hybridus black purple	oRus
	hybridus blue black	oGos
	hybridus blue black parent	oNWe
	hybridus blue black, ex. Graham Birkin	oRus
un	***hybridus*** 'Concord'	oGar wBox
	hybridus Cricklewood doubles	wCri
	hybridus dark purple	oGos oHon
	hybridus deep red parent	oNWe
un	***hybridus*** Green Group	wHer
	hybridus Hedgerows hybrids	oHed
	hybridus hybrids	oHon wCol oRus oDan oNWe oBlo wSta wCoN wEde oEga wSte
un	***hybridus*** 'Medallion'	oSis
	hybridus mixed colors	oHon oNat wHig oGar
un	***hybridus*** Phedar Strain	oHed
un	***hybridus*** Phillip Curtis Strain	wFai
un	***hybridus*** Picotee Group	wHer
	hybridus pink	oHon oJoy oGar oNWe oOut oGre wBox
	hybridus pink double parent	oNWe
	hybridus pink, ex. Graham Birkin	oRus
un	***hybridus*** Pink Group	wHer
	hybridus pink parent	oNWe
	hybridus pink spotted	oRus oNWe
	hybridus pink spotted, ex. Graham Birkin	oRus
un	***hybridus*** Pink Strain	oSis
	hybridus plum	oJoy
	hybridus purple	oHon oGar oSis oNWe oOut oGre wBox
	hybridus purple black	oNWe
	hybridus purple black parent	oNWe
un	***hybridus*** Purple Group	wHer
	hybridus purple parent	oNWe
	hybridus red	oAls oRus oGar oNWe oOut wBox wCoN
	hybridus red, ex. Graham Birkin	oRus
	hybridus 'Red Mountain'	oAls wNay wWhG
	hybridus red parent	oNWe
	hybridus red with white edge parent	oNWe
	hybridus rose	oJoy
un	***hybridus*** Royal Heritage Strain	wWoS oGar oDan oOut wNay oGre wFai wTGN wCoN wWhG
	hybridus slate	oGar wBox
un	***hybridus*** Slate Group	wHer
	hybridus slate parent	oNWe
	hybridus spotted	oAls
	hybridus white	oHon oRus oGar oNWeoOut oGre wBox
	hybridus white, spotted	oHon
un	***hybridus*** White Group	wHer
	hybridus white parent	oGar oNWe
	hybridus white spotted, ex. Graham Birkin	oRus
	hybridus white spotted parent	oNWe
	hybridus white Ushba seedling parent	oNWe
	hybridus white with red veins parent	oNWe
	hybridus yellow	oGos oRus oGar oNWe oOut oGre wBox
un	***hybridus*** Yellow Group	wHer
	hybridus yellow parent	oNWe
	istriacus	see **H. *multifidus*** ssp. ***istriacus***
	lividus	wHer oGos oHon oRus oNWe wEde
	lividus ssp. ***corsicus***	see **H. *argutifolius***
	multifidus ssp. ***bocconei***	last listed 99/00
	multifidus ssp. ***istriacus***	wHer
	niger	oFor wCul oGos oHon oAls oRus oGar oSis oNWe oSho oWnC wMag wCoN wWhG oUps
	niger Blackthorn Group	oGre
	niger ssp. ***macranthus***	oFor
	niger Sunrise Group PHE 5.2/5.3	wHer
	niger Sunset Group PHE 5.6	wHer
	niger 'White Magic'	oGos oGre wWel
	odorus	wCul oNWe
	odorus WM 9414	wHer
	orientalis (see Nomenclature Notes)	oFor wCul wCri oHon oJoy oEdg oAls oRus wCSG oAmb oGar oSho oOut oGre wPir oWnC wMag oCir wAva oUps wCoS
	purpurascens	wHer oAls oNWe oOut wNay oWnC wMag
un	***purpurascens*** 'Red Power'	oSho
	x sternii	oHon oNWe oGre
	x sternii Blackthorn Group	wHer oGos oHon oGar oOut oGre
	x sternii 'Boughton Beauty'	wCul
un	***x sternii*** Raitz Strain	wHer
	thibetanus	oNWe
	torquatus	oNWe
	torquatus red	oNWe
	torquatus white, red veins	oNWe
	viridis	wCul oHon
	viridis ssp. ***viridis***	wHer
	HELONIAS	
	bullata	oSis
	HELONIOPSIS	
	orientalis	oBov
	HELWINGIA	
	chinensis	oRiv oCis
	chinensis broad leaf form	wHer oCis
	chinensis narrow leaf form	wHer oCis
	japonica DJHC 495	wHer
	HEMEROCALLIS	
	'Addie Branch Smith'	wSno
	'Admiral's Braid'	oMid oGai
	'Affair to Remember'	oGai
	'African Diplomat'	oMid
	'African Grape'	oMid
	'Ah Youth'	oCap
	'Alan Adair'	oGai
	'Alaqua'	oMid
	'Albuquerque Memory'	oGai
	'Alec Allen'	oGai
	'All American Tiger'	oGai
	'All Fired Up'	oMid
	'Almond Puff'	oGai
	'Alpine Mist'	oMid oGai
	'Alvin Sholar'	oCap
	'Always Afternoon'	oMid oGai oBRG
	'Amber Love'	oMid
	'American Revolution'	wSno oGar
	'Amy Christine'	oGai
	'Amy Michelle'	oGai
	'Andrew Christian'	oGai
	'Angel Artistry'	oMid
	'Angel of Light'	oGai
	'Ann Blocher'	oGai
	'Ann Kelley'	wSno
	'Anna Warner'	oNat
	'Annie Welch'	wSno
	'Antique Treasure'	oGai
	'Anzac	oGar
un	'Apache Scout'	oGai
	'Apollodorus'	wSno
	'Apple'	wSno
	'Apricot'	oGar
	'Apricot Jade'	oMid
	'Apricot Surprise'	oNat
	'Arabian Magic'	oMid
	'Arctic Snow'	oCap oGai
	'Arthur Moore'	oGai
	'Artist Etching'	oGai
	'Asian Artistry'	wSno

	'Asian Pearl'	oGai
	'Atlanta Debutante'	oCap oGai
	'Atlanta Lamplighter'	oGai
	'Atlanta Orchid Mist'	oMid
	'Atlantis'	oGai
	'Attacapa'	oGai
	'Attribution'	oGai
	'August Cheer'	last listed 99/00
	'Autumn Lace'	oWnC
	'Autumn Red'	oFor
	'Aviance'	oGai
	'Aztec Chalice'	oCap oMid
	'Aztec Gold'	oFor
	'Baby Betsy'	oGar oGai oBRG
	'Baby Blues'	oMid
	'Baja'	oGar oGai
	'Bald Eagle'	oBRG
	'Ballentrea'	oGai
	'Baltimore Oriole'	wBox
	'Bamboo Ruffles'	oGai
	'Banana Cream Beauty'	oMid
	'Barbara Mitchell'	oGai
	'Barbary Corsair'	oFor wSno oAls oWnC oBRG
	'Beat the Barons'	oCap oGai
	'Beautiful Edgings'	oMid
	'Beauty to Behold'	oGai
	'Becky Lynn'	oBRG
un	'Belgian Lace'	oGai
	'Belle Amber'	wSno oGai
	'Beloved Ballerina'	oGai
	'Ben Adams'	oMid oGai
	'Ben Lee'	oGai
	'Benchmark'	oCap oGai
un	'Beppie'	wWoS
	'Bertie Ferris'	wWoS wSno oGar oTwi oGai oBRG wSnq
	'Best Kept Secret'	oGai
	'Best of Friends'	wBox oBRG
	'Better Believe It'	wTLP
	'Betty Benz'	oGai
do	'Betty Woods'	oCap
	'Beulah Stevens'	oGai
	'Beverly Center'	oGai
	'Big Apple'	oGai
	'Big Bird'	oMid wBox oBRG
	'Bimini Run'	oGai
	'Bisque'	oCap
	'Bite Size'	oGai
	'Bitsy'	wBox oBRG
	'Bittersweet Holiday'	oGai
	'Bittersweet Honey'	wWoS
	'Black Ambrosia'	oGai
	'Black Eye'	oMid
	'Black Eyed Stella'	oFor oWnC wBox
	'Black Eyed Susan'	oMid
	'Black on Black'	oGai
	'Black Plush'	oGai
	'Black Watch'	oGai
	'Blessed Again'	oCap
	'Blessing'	last listed 99/00
	'Blonde Is Beautiful'	oMid
	'Blood Spot'	oGai
	'Blue Moon Rising'	oMid
	'Blue Sheen'	oGai
	'Blushing JoAnn'	oCap
	'Bobbie Gerold'	last listed 99/00
	'Bogie and Becall'	oGai
	'Bold One'	oMid oGai oBRG
	'Bold Tiger'	oMid oGai
	'Bonanza'	oFor wSno oGar oSho oTwi oBlo wBox
	'Bone China'	last listed 99/00
	'Bonnie Corley'	oGai
	'Booger'	oMid
	'Bookmark'	oGai
	'Border Lord'	oGai
	'Borgia Queen'	oGai
	'Brand New Lover'	oGai
	'Brass Buckles'	oBRG
un	'Breath Catcher'	oGai
	'Breathless Beauty'	wSno oGar oBlo
	'Breed Apart'	oGai
	'Brent Gabriel'	oGai
	'Bridal Suite'	oGai
	'Bright Butterflies'	oMid
	'Broadway Bold Eyes'	oMid
	'Broadway Gal'	oGai
	'Broadway Image'	oMid
	'Broadway Pink Slippers'	oCap oMid
	'Brocaded Gown'	oGai oBRG
	'Brutus'	wSno
	'Bubbling Over'	oMid
	'Bunny Eyes'	oGai
	'Burgundy Bud'	oGre oBRG
	'Burling Street'	oCap
	'Burmese Sunlight'	oGai
	'Buttered Popcorn'	oCap
	'Butterfly Encounter'	oGai
	'Butterfly Kisses'	oBRG
	'Butterpat'	oBRG
	'Butterpat Twist'	oGai
	'Butterscotch Charm'	oGai
	'Butterscotch Ruffles'	oGai
	'Button Bee'	oGai
	'Byzantine Emperor'	oGai
	'Calgary Stampede'	oMid
	'California Sunshine'	oCap oMid oGai
	'Caliphs Robes'	oGai
	'Cameroons'	oMid oGai
	'Canadian Border Patrol'	oMid
	'Candy Apple'	wWoS
	'Candy Floss'	oGai
	'Candy Sweet'	oCap
	'Cantique'	oBRG
	'Cape Cod'	oBRG
	'Caprician Fiesta'	oGai
	'Caprician Honey Gold'	oCap oGai
	'Cardinal Feathers'	oGai
	'Carey Quinn'	oGar oTwi
	'Caring Friends'	last listed 99/00
un	'Carmel Glaze'	oCap
	'Carmen Marie'	wSno
	'Carolicolossal'	oGre oBRG
un	'Carolisalmon'	oGai
	'Carolyn Criswell'	oGai wBox
	'Carolyn Hendrix'	oGai
	'Carousel Princess'	oGai
	'Cartwheels'	oAls
un	'Catherine Irene'	oGai
	'Catherine Neal'	oGar oGai
	'Catherine Woodbery'	oNat oAls oGar wBox wSta wTLP
	'Cavaliers Gold'	oGai
	'Caviar'	oGai
	'Celebrity Elite'	oCap wBox
	'Chamonix'	wBox
	'Charles Johnston'	oGai
	'Charlie Pierce Memorial'	oCap oGai
	'Charm Light'	oCap
	'Chartwell'	oGai
	'Chateau Blanc'	oGai
	'Chateau DeFleur'	oGai
	'Cherished Treasure'	oGai
	'Cherries Are Ripe'	oMid
	'Cherry Chapeau'	oMid
	'Cherry Cheeks'	oGar
	'Chicago Apache'	oFor oGar wBox
	'Chicago Blackout'	wWoS
	'Chicago Brave'	wSno
	'Chicago Cattleya'	wWoS oMid
	'Chicago Gold Coast'	wBox
	'Chicago Knockout'	wSno oBRG
	'Chicago Peach'	oGar
	'Chicago Petticoats'	wBox oBRG
	'Chicago Picotee Lace'	wSno oGai
	'Chicago Royal Robe'	oGar oGai
	'Chicago Silver'	oGar
	'Chicago Star'	wBox
	'Chicago Sunrise'	wBox
	'Chicago Violet'	wSno
	'Children's Festival'	wWoS
	'Chinatown'	oGai
	'Chinese Scholar'	oMid
	'Chinese Temple Flower'	oGai
un	'Chinook Winds'	oGai
	'Chorus Line'	oGai oBRG
	'Chorus Line Kid'	oMid
	'Christmas Carol'	oAls wEde
	'Cimarron Knight'	oMid oGai
	'Cinderella's Dark Side'	oMid
	'Cinnamon Sweets'	oMid
	citrina	oCis
	'Classy Cast'	oCap

	'Clemenceaux'	oGai
	'Cleopatra'	oMid
	'Clincher'	wSno
	'Cloverdale'	oGai
	'Coburg Preview'	oGai
	'College Try'	wSno
	'Comic Strip'	last listed 99/00
	'Coming up Roses'	oGai
	'Common Sense'	wSno oGai
	'Condilla'	oMid oBRG
	'Cookie Monster'	wBox
	'Copperhead'	oGai
	'Coral Dawn'	oGai
	'Corky'	oGar oGai wSnq
	'Corryton Pink'	wBox
	'Cortez Cove'	oGai
	'Cosmic Caper'	oGai
	'Country Charmer'	last listed 99/00
	'Country Club'	oFor oBRG
	'Country Lane'	oBRG
	'Country Melody'	wBox
	'Court Magician'	oMid oGai
	'Courtly Love'	oGai
	'Courts of Europe'	oMid
	'Coventry Countess'	oMid
	'Cranberry Baby'	oGre oGai oBRG
	'Cranberry Cove'	oGai
un	'Cream Parfait'	oBRG
	'Creative Art'	last listed 99/00
	'Creative Edge'	oGai
	'Crimson Joy'	oCap
	'Crinkled Bouquet'	last listed 99/00
	'Crown Royal'	oGai
	'Crystal Cupid'	oMid oGai oBRG
	'Custard Candy'	oMid oBRG
	'Dacquiri'	oCap
	'Dallas Star'	wBox
	'Damascene'	oGai
	'Dancing Shiva'	oBRG
	'Daring Dilemma'	oMid oGai
	'Dark Star'	oCap oGai
	'Dark Symmetry'	oGai
	'Darrell'	oGai
	'Date Book'	oAls
	'Dawn Kennon'	oGai
	'Dazzle'	oGai
un	'Dead Eye Dick'	oGai
	'Dead Ringer'	oMid oGai
	'Decatur Apricot'	oCap wSno
	'Decatur Cherry Smash'	oGai
un	'Decatur Rainbow'	oGai
	'Decatur Sun'	oGai
un	'Decatur Sundae'	oGai
	'Deep Pools'	oGai
	deep red	oNat
	'Deja Vu'	wSno
un	'Delightfully Pink'	oMid
un	'Deloris Gould'	oCap
	'Dempsey Everblooming Gold'	oCap
	'Derby Bound'	oGai
	'Derrick Cane'	oMid
	'Desdemona'	wSno
	'Designer Gown'	oGai
	'Designer Jeans'	oMid oGai
	'Destiny's Child'	oMid oGai
	'Dewey Roquemore'	oGai
	'Diamond Ice'	oMid
	'Diva Assoluta'	oCap oGai
	'Divertissment'	oGai
	'Doll House'	oGai
	'Doma Knaresborough'	oMid
	'Dominic'	oNat wSno oGar oGai oBRG
	'Don Stevens'	oMid
	'Dorethe Louise'	oMid
	'Dorothy Lambert'	wBox
	'Doryt Moss'	oGai
	'Dotted Petals'	oMid
	'Double Attraction'	oGai
	'Double Cranberry Ruffles'	oGai
	'Double Daffy'	oMid
	'Double Ethel'	wSno oGai
	'Double Grapette'	oCap
	'Double Ladybug Ra'	oGai
	'Double Low'	oMid
	double orange	oSle
	'Double Overtime'	oGai
	'Double Pink Treasure'	oGai
	'Double Puff'	oGai
	'Double Red Royal'	oGai
	'Double Sunburst'	oGai
	'Double Talk'	oDar
	'Dragon Mouth'	wBox
	'Dragon's Eye'	oMid oGai
	'Dragons Treasure'	oGai
	'Dream Baby'	wBox
	'Dream Book'	oGai
	'Dream Date'	oBRG
	'Dream Legacy'	oMid
	'Dream Team'	oGai
	'Dublin Elaine'	wSno wBox
	'Duke's Daughter'	oGar
	'Dunedin'	oMid
	dwarf peach	wBCr
	dwarf yellow	oCir
	'Earl Roberts'	oGai
	'Earth Angel'	oMid
	'Ed'	last listed 99/00
	'Ed Murray'	oGar oGai oBRG
	'Edith Anne'	oCap oGai
	'Edna Spalding'	oDar
	'Eenie Fanfare'	oGar oGai
	'Eenie Weenie'	oFor oGar oGai wTGN wBox wSnq
	'El Padre'	oGai
	'El Tigre'	oMid
	'Eleanor Marcotte'	oGai
un	'Elegant Era'	oGai
	'Elfin Aristocrat'	oGai
	'Elfin Stella'	oCap
	'Elizabeth Salter'	oGai
	'Elizabeth's Magic'	oGai
	'Ellen Christine'	oGai
	'Elsie Spalding'	oGai
	'Emerald Joy'	oGai
	'Emperor Butterfly'	oGai
	'Emperor's Dragon'	oGai
	'Enchanted Butterfly'	oMid
	'Esther Peery'	oMid
	'Etched in Gold'	oGai
	'Eva'	wSno
	'Evening Enchantment'	oMid oGai
	'Evening Gown'	oFor wBox
	'Ever So Ruffled'	oMid
	'Excitable'	oGai
	'Exotic Candy'	oMid
	'Exotic Echo'	oBRG
	'Eye-Yi-Yi'	oCap
	'Eyed Radiance'	oMid
	'Ezekiel'	oGai
un	'Fairy'	wMag
	'Fairy Bonnet'	wWoS
	'Fairy Charm'	oCap oGar oBRG
	'Fairy Filigree'	oMid
	'Fairy Finery'	oGai
	'Fairy Jester'	oGai
	'Fairy Tale Pink'	oCap wSno oAls oGre oGai wBox oBRG wSnq
	'Fall Farewell'	oGai
	'Fanciful Finery'	wBox
	'Far Niente'	oGai
	'Farmer's Daughter'	oGai
	'Favorite Things'	oGai
	'Feather Down'	oGai
un	'Fields Red'	wBox
	'Fifty Springs'	oGai
	'Final Touch'	oCap oGar oGai
un	'Firelight'	oSec
	'Firepower'	oGai
	'Firestorm'	wBox
	'First Noel'	oGai
	'Flamenco Queen'	oGai
	'Flaming Poppa'	wSno
	'Flasher'	oGre oGai oWnC oBRG
	flava	see **H. *lilioasphodelus***
	'Flirty Edna'	oMid
	'Florentine Prince'	wSno
	'Florentine Silk'	oGai
	'Floyd Cove'	oMid
un	'Flyin' High'	oNat
	'Flying Saucer'	wSno
	'Fond Hope'	oBRG

	'Forever Stella'	oGai
	'Forsyth Cream Puff'	oCap
un	'Forsyth Heart's Delight'	oMid oGai
	'Forsyth Jimny Cricket'	oGai
	'Forsyth Kate Update'	oGai
	'Forsyth Lemon Drop'	oCap oGai
	'Forsyth Pearl Drops'	last listed 99/00
	'Forty Second Street'	oGai
	'Foxfire Light'	oGar
	'Fragrant Light'	oCap
	'Frances Joiner'	oMid oGai
	'Frandean'	wSno
	'Frank Gladney'	oGai
	'Frans Hals'	oFor wCul wSno oGar oTwi wMag wTLP
	'Fran's Knobby'	oGai
	'French Pavilion'	oGai
	'Frivolous Frills'	wSno
	'Frontier Days'	oGai
	'Frosty Beauty'	oGar wMag wWhG
	'Full Moon Magic'	oGai
	fulva	oTwi
	fulva 'Green Kwanso'	oBRG
	fulva 'Kwanzo Variegata'	wHig
	'Fuzz Bunny'	oCap
	'Galena Moon'	wMag
	'Garden Goddess'	oGai
un	'Garnet'	oGar
	'Gay Cravat'	wSno oGai oBRG
	'Gemini'	wSno oGai
	'Gentle Humor'	oGai
	'Gentle Rose'	oMid
	'Gentle Shepherd'	wSno oGar oBRG
	'Geometrics'	oGai
	'George Caleb Bingham'	wSno wSta
	'Georgia O'Keefe'	oGai
	'Gertrude Condon'	oBRG
	'Ginger Whip'	oBRG
	'Gingerbread Man'	oGar
	'Gladys Campbell'	oMid
	'Gleeman Song'	oGai
	'Glittering Gown'	oGai
	'Gloria Blanca'	oGai
	'Glorious Temptation'	last listed 99/00
	'Glory in Ruffles'	oBRG
	'Golden Cheer'	oGai
	'Golden Corduroy'	wSno
	'Golden Dewdrop'	oBRG
	'Golden Gift'	wMag
un	'Golden Girl'	wMag
	'Golden Prize'	oAls wTLP
	'Golden Scroll'	oCap oGai
	'Golden Voice'	oGai
	'Goolagong'	oBRG
	'Gothic Window'	oGai
	'Graceful Eye'	oGai
	'Graceland'	oMid oGai
	'Grand Palais'	oGai
un	'Grandeur Dreams'	oMid
	'Grape Eyes'	oGai
	'Grape Magic'	wSno oGai
	'Grape Velvet'	wWoS oBRG
	'Great Connections'	oGai
	'Grecian Key'	wWoS
	'Greek Goddess'	oGai
	'Green Eyes Wink'	wSno
	'Green Flutter'	oAls oGar wBox oBRG
	'Green Glitter'	oGai
	'Green Ice'	oGai
	'Guardian Angel'	wSno
	'Gypsy Pixie'	oBRG
un	'Haji Baba'	oGai
un	'Hall's Pink'	oFor wCul oAls oGar oSho oBlo oWnC wMag wWhG oSle
	'Happy Bandit'	oMid
	'Happy Returns'	oCap oGar oGre oGai oBRG wSnq
	'Harmonic Convergence'	oGai
	'Harry Barras'	oGai
—	'Hawk'	oBRG
un	'Hawthorne Rose'	wSno
	'Hazel Sawyer'	oBlo
	'Heart's Glee'	oGai
	'Heather Green'	oGai
	'Heather Pink'	last listed 99/00
	'Heaven Can Wait'	oCap
un	'Heavenly Angel'	wBox
	'Heavenly Treasure'	oCap
	'Heavens Rejoice'	wBox
	'Hello There'	oBRG
	'High Sierra'	oBRG
	'Highland Lord'	wSno oGai
	'Holiday Delight'	oGai
	'Holly Herrema'	oGar
	'Holy Moses'	oCap
	'Home Cooking'	oMid
	'Hot Bronze'	oGai
	'Houdini'	oGai oBRG
	'Humdinger'	oMid
	'Hundredth Anniversary'	oGai
	'Hunter's Torch'	oGai
	'Hyperion'	oAls oGar wFai wMag wSta wTLP oBRG wSnq
un	'Hyperion Supreme'	wBox
	'Ice Carnival'	oMid oGar oTwi oBRG
	'Ida Duke Miles'	oGai
	'Ideal Affair'	oGai
	'If'	oMid oGai
	'Imperial Guard'	oGai
un	'Imperial Lustre'	oGai
	'In a Heartbeat'	oGai
	'Incognito'	oBRG
	'Indian Love Call'	wWoS
	'Indian Sky'	oGai
	'Indian Weaving'	oMid
	'Indigo Moon'	oMid oGai
	'Indy Charmer'	wSno
	'Indy Love Song'	oGai
	'Indy Rhapsody'	oGai
	'Inner View'	oGai
	'Integrity'	oGai
	'Irish Belleek'	oGai
	'Irish Elf'	wSno oBRG
	'Irish Glory'	wSno oDar
	'Irish Ice'	oGai
	'Irish Spring'	oGai
	'Irving Shulman'	oMid
	'Isaiah'	oCap wSno oGre oBRG
	'Ivory Dawn'	oGai
	'Jacob'	oBRG
	'Jamaica Rose'	oGai
	'Jambalaya'	oGai
	'James Marsh'	oCap oGai
	'Jane Angus'	oBRG
	'Janice Brown'	oCap oMid oGai
	'Jan's Twister'	oMid
	'Jason Mark'	oMid oGai
	'Jason Salter'	oGai
	'Jean Wise'	oGai
	'Jedi Dot Pearce'	oMid oGai
	'Jedi Irish Spring'	wSno oMid
	'Jedi Lacy Lady'	oGai
	'Jen Melon'	oCap
	'Jersey Spider'	wBox
	'Jeune Tom'	oCap wSno
	'Jim Jim'	oGai
	'Jim Watson'	oGai
	'Jo Barbre'	oBRG
	'Joan Senior'	oGar oGai wBox oBRG wSnq
	'Jock Randall'	oGai
	'Joe Marinello'	oMid
	'Jovial'	wSno oBRG
	'Joylene Nichole'	oCap oMid oGai
	'Judith'	oGre oBRG
	'Julia's Choice'	wSno
un	'June Bug'	oFor
	'Just a Bit'	oGai
	'Kallista'	oGai
	'Karen My Love'	oGai
	'Karen Sue'	oMid
	'Kate Carpenter'	oCap oGai
	'Kathleen Salter'	oGai
	'Kecia'	wSno oGai
	'Kenan'	oGai
	'Kent's Favorite Two'	oMid
	'Kimmswick'	oCap
	'Kindly Light'	oBRG
	'King Alfred'	oGai
un	'Kissy Face'	wSno
	'Kitten Richardson'	last listed 99/00
	'Kwanso'	see **H. *fulva*** 'Green Kwanso'
	'Kwanso Flore Pleno'	see **H. *fulva*** 'Green Kwanso'

	'La Mer'	oGai oBRG
	'Lacquered Urn'	oGai
	'Lady Pharaoh'	oGai
	'Ladybug's Two Moons'	oGai
	'Lake Norman Spider'	oCap oMid
	'Lake Norman Sunset'	oMid
	'Lamplighter'	oGai
	'Last Accomplishment'	wSno oGai oBRG
	'Latin Lover'	oGai
	'Laugh And Sing'	oBRG
	'Lauren Leah'	last listed 99/00
	'Lavender Aristocrat'	wSno
un	'Lavender Beauty'	oSec
	'Lavender Bonnet'	oMid
	'Lavender Deal'	oBRG
	'Lavender Dew'	wSno
un	'Lavender Minutia'	oMid
	'Lavender Patina'	last listed 99/00
	'Lavender Plicata'	oMid
	'Lavender Tonic'	last listed 99/00
	'Lee Gates'	oGai
	'Leeba Orange Crush'	wSno
un	'Lemon Chiffon'	oNat
	'Lemon Flurry'	oGai
	'Lemon Lace'	oCap wBox
	'Lemon Lollypop'	oFor oCap oMid wBox
	'Leprechaun's Wealth'	oCap oGai
	'Lexington Avenue'	oGai
	'Lighter Side'	oGai
	'Lights Of Detroit'	oGai
	'Lil Ledie'	oBRG
	'Lilac Snow'	last listed 99/00
	lilioasphodelus	wCul wCri oDar wHig wCSG oGar oBRG wSte
	'Limoges Porcelain'	oGai
	'Little Blackie'	oGai
	'Little Brandy'	oGai
	'Little Brave'	wSno
	'Little Bumble Bee'	oFor wWoS
	'Little Business'	wSno oGai wBox oBRG
	'Little Cadet'	wBox
	'Little Cameo'	oGai
	'Little Christine'	wSno
	'Little Cobbler'	wSno
	'Little Dandy'	oGai
	'Little Dreamer'	oMid
	'Little Fantastic'	oNat oGai oBRG
	'Little Fat Dazzler'	oMid oBRG
	'Little Grapette'	oCap wSno oAls oMid wTLP oBRG
	'Little Heavenly Angel'	wSno
	'Little Isaac'	oCap
un	'Little Kiss'	oGai
	'Little Maggie'	oBRG
	'Little Monica'	oGai oBRG
	'Little Nadine'	last listed 99/00
	'Little Pink Fluff'	oNat oGai
	'Little Pink Slippers'	oCap wSno
	'Little Print'	last listed 99/00
	'Little Red Hen'	wSnq
un	'Little Red Hot'	oGai
	'Little Rich'	wSno
	'Little Romance'	last listed 99/00
	'Little Squiz'	oCap
	'Little Wart'	oBRG
	'Little Wayne'	oCap
	'Little Wine Cup'	oAls oGai oWnC oBRG
	'Little Zinger'	oGai oBRG
	'Lord Jeff'	oMid
	'Louise Boswell'	oMid
	'Louise Manelis'	oCap wSno wBox oBRG
	'Louise Mercer'	oGai
	'Love Me'	oGai
	'Love Those Eyes'	oMid
	'Loving Memories'	wSno
	'Lucille Watkins'	oCap
	'Lullaby Baby'	oCap wSno oGar oGre oBRG
	'Lunar Sea'	oGai
	'Lustrous Jade'	oGai
	'Luxury Lace'	wTLP
	'Lynn Hall'	oGai
	'Lyttleton'	oBRG
	'Mae Graham'	wBox oBRG
	'Maggie McDowell'	oGai
	'Magic Carpet Ride'	oGai
	'Magic Filigree'	oMid oGai
	'Magic Masquerade'	last listed 99/00
	'Magic Obsession'	oGai
	'Magic Potion'	oMid
	'Magnifique'	oGai
un	'Maja'	oSho
	'Majestic Morning'	oGai
	'Malaysian Monarch'	wSno oMid oGai
un	'Manilla Moon'	oAls
	'Marathon Dancer'	oMid
	'Marcia Fay'	oBRG
	'Marcia Flynt'	oCap
un	'Mardi Gras Mask'	oGai
un	'Margaret Perry'	oFor oAls oGar
un	'Marie Beckman'	oGai
	'Marie Hooper Memorial'	oMid
	'Mariska'	oCap wSno
	'Mary Ethel Anderson'	oMid
	'Mary Todd'	wBox
	'Mascara'	oGai
un	'Magic Masquerade'	oMid
	'Matisse'	oMid
	'Mauna Loa'	oGai
un	'May Basket'	oGai
	'May Colvin'	oGar
	'May May'	oBRG
	'May Unger'	oGai
	'Medieval Splendor'	oMid
	'Melon Balls'	wSno
	'Melon Patch'	oCap
un	'Merlot'	oSec
	'Merlot Rouge'	oMid
	'Merry Makers Serenade'	oMid
	'Merry Miss'	oMid
	'Metaphor'	wSno oGai oBRG
	'Michelle My Belle'	oGai
	'Midnight Magic'	oMid oGai
	'Midnight Oil'	oMid
	'Midwest Gem'	oMid
	'Millie Midge'	wSno
	'Ming Porcelain'	oCap oGai oBRG
	'Ming Snow'	oMid
	'Ming Toy'	oFor
	'Mini Pearl'	oCap oAls oBRG
	'Mini Stella'	oWnC
	sp. aff. ***minor*** HC 970389	wHer
	'Minstrel Boy'	wSno
	'Miracle Maid'	oGai
un	'Miss Mary Mary'	oGar
	'Mokan Butterfly'	oCap oGai
	'Monica Marie'	oCap oMid
	'Monkey'	oBRG
	'Monkey Maker'	oGai
	'Moonlight Mist'	oCap oGai
	'Moonlit Masquerade'	oMid
	'Morning Dawn'	oFor oBRG
	'Mountain Violet'	wSno
	'My Belle'	last listed 99/00
	'My Eye Elsie'	oMid
	'My Hope'	wBox
	'My Melinda'	oGai
	'Mynelle's Starfish'	oMid
	'Mysterious Veil'	wSno
	'Mystery Valley'	wWoS
	'Mystical Rainbow'	oMid
	'Nacogdoches Bing Cherry'	oGai
	'Nacogdoches Pansy'	oGai
	'Nagasaki'	wSno oGai
un	'Nancy Barnett'	oBRG
	'Nanuq'	oMid
	'Naomi Ruth'	wWoS
	'Navajo Princess'	oGai
	'Neal Berrey'	oMid
	'Nebuchadnezzar's Furnace'	oGai
	'New Series'	oMid
	'Nicholas'	wBox
	'Niece Beverly'	oGai
	'Night Beacon'	wSno oMid oGai
	'No Nonsense'	wSno
	'Nordic Night'	oGai
	'Northern Lemon Star'	oGai
	'Northfield'	wSno
	'Notorious'	oGai
	'Nottingham'	oGai
	'Numinous Moments'	wSno
	'Obsessed'	oMid

	'Obsidian'	oMid
	'Ocean Ice'	oGai
	'Ocean Rain'	last listed 99/00
	'Old Port'	oGai
	'Old Tangiers'	oGai
	'Ole Ole'	oMid
	'Olive Bailey Langdon'	oMid oGai oBRG
	'Oliver Dragon Tooth'	oGai
	'Olive's Odd One'	oMid
	'Olympic Gold'	oMid
	'On Cloud Nine'	oGai
	'Open Hearth'	oBRG
un	'O'Plenty'	oBRG
	orange	wBCr oCir
	'Orange Piecrust'	oGai
	'Orange Velvet'	oMid
	'Orchid Candy'	oMid
	'Oregon Mist'	oBRG
	ORP 16	wSno
	ORP 41	wSno
	ORP 50	wSno
	'Outrageous'	last listed 99/00
	'Paca Mere'	oMid oGai
un	'Pacific Sunset'	oCap
	'Paige's Pinata'	oMid oGai
	'Palace Pearls'	last listed 99/00
	'Pandora's Box'	oCap wSno oBRG
	'Pantaloons'	oMid
	'Papal Guard'	oGai
	'Paper Butterfly'	oMid oGai
	'Papillon'	oGai
	'Paprika Velvet'	wSno
	'Parade Of Peacocks'	last listed 99/00
	'Pardon Me'	wWoS oCap oGar oGre oGai oBRG
	'Party Queen'	oMid
	'Pasqueflower'	oMid
	'Pastel Classic'	oGai
	'Pastures of Pleasure'	oCap oGai
	'Patchwork Puzzle'	oCap oMid
	'Patio Parade'	oGar
	'Patrician Heritage'	oGai
	'Pauper Prince'	oGai
	'Peach Candy'	oMid
	'Peach Fairy'	wWoS
	'Peach Horizon'	oMid
	'Peach Whisper'	oGai
	'Peacock Maiden'	oMid
	'Pearl Chiffon'	oGai
	'Penny Earned'	oGai
	'Penny's Worth'	oGai
	'Perennial Pleasure'	wBox
	'Persian Market'	oGre wBox oBRG
	'Persian Shrine'	wBox
	'Phoenician Ruffles'	oMid oGai
	'Pimiento Pepper'	oMid
	pink	wMag
	'Pink Attraction'	wBox
	'Pink Canary'	oGai
	'Pink Damask'	wCSG oGar oWnC
	'Pink Embers'	wSno oTwi oWnC
	'Pink Lavender Appeal'	wSno
	'Pink Monday'	oCap
un	'Pink Pearl'	oSec
	'Pink Playmate'	oCap oGai
	'Pink Puff'	oSho
	'Pirate Lord'	oGai
	'Pirate's Patch'	oMid oGai
	'Pittance'	wSno
	'Pittsburgh Golden Triangle'	oGai
	'Pixie Beauty'	oBRG
	'Pixie Parasol'	oCap wSno wMag wSnq
	'Pleasant Edging'	oMid
	'Pleasant Hill Pink'	oCap
	'Plum Candy'	oGai
	'Poinciana Regal'	oGai
	'Pojo'	wSno
	'Poogie'	oGai
	'Porcelain Caress'	oGai
	'Porcelain Finery'	oMid
qu	'Prairie'	wMag
	'Prairie Bells'	oDar
	'Prairie Blue Eyes'	oFor oGai wBox wTLP oBRG
	'Prairie Moonlight'	oGar oGai
	'Prairie Sunburst'	wBox
	'Precious Love'	wSno

	'Precious One'	wBox
	'Preppy Pink'	oCap
	'Prester John'	oGai wBox
	'Pride Of Massachusetts'	oMid
	'Princess Ellen'	oMid
	'Princess Kaiulani'	oGai
un	'Princess Margaret'	oBRG
un	'Princeton Blush'	oGai
	'Princeton Pink'	oGai
	'Priscilla's Dream'	oGai
	'Ptarmigan'	oMid
	'Puddin'	see **H.** 'Brass Buckles'
	'Pumpkin Kid'	oMid
	purple	oGar
	purple bicolor	oSho
	'Purple Peppermint'	oMid
	'Purple Pinwheel'	oMid
	'Purple Pittance'	oCap
un	'Purple Plumage'	oGai
	'Purple Rain Dance'	oMid
	'Purple Storm'	oMid oGai
	'Purple Waters'	oFor oBlo wTLP
	'Quaker Bonnet'	oGai
	'Quaking Aspen'	wBox
	'Queens Delight'	oGai
	'Queensland'	oMid
	'Quinn Buck'	oCap
	'Ra Hansen'	oGai
	'Rachel My Love'	oCap wSno oGai
	'Radiant Beams'	wBox
	'Radiant Ruffles'	oGai
	'Rainbow Round'	oGai
	'Rajah'	oSho oBlo
un	'Rapscallion'	oGai
	'Raspberry Candy'	oGai
	'Raspberry Pixie'	oCap oGar wTLP oBRG
un	'Raspberry Shortcake'	oBRG
	'Rayon de Miel'	last listed 99/00
	'Razzle'	oGai
	'Real Wind'	wSno oGai
	'Reba My Love'	oGai
	'Red Grace'	oGai
	'Red Magic'	oFor oAls wWhG
	'Red Rain'	oGai
	'Red Rhapsody'	oGai
	'Red Ribbons'	oMid wBox oBRG
	'Red Rum'	wSno oGai
qu	'Red Select'	wMag
	'Red Thrill'	oGai
	'Red Tide'	oBRG
	'Red Volunteer'	oGai wBox
	'Regency Masquerade'	oGai
	'Renee'	oGai
	'Respighi'	oMid
	'Revealing Beauty'	oGai
	'Rhine Maiden'	oGai
	'Ribbon Candy'	oGai
	'Riptide'	oCap
	'Rocket City'	oGar
	'Rogue's Masquerade'	oCap
	'Roman Renaissance'	oGai
	'Romantic Fool'	oGai
un	'Rosé'	oCap
	'Rose Emily'	oCap
	'Rose Festival'	wBox
	'Rose Joy'	oMid
	'Rose Petticoat'	oGai oBRG
	'Rose Talisman'	oCap
	'Rosella Sheridan'	oBRG
	'Rosie Meyer'	oBRG
	'Rosy Fragrance'	oCap
	'Royal Braid'	oMid
	'Royal Charm'	oMid
	'Royal Corduroy'	oGai
	'Royal Exchequer'	oGai
	'Royal Fanfare'	oBRG
	'Royal Palace Prince'	wSno
	'Royal Saracen'	oGai
	'Royal Sprite'	oGai
	'Royal Thornbird'	oMid
	'Ruby Catherine'	oGai
	'Ruby Spider'	oCap oGai
	'Ruby Throat'	oGai
	'Ruffled Apricot'	oCap wSno oGai wBox oBRG
	'Ruffled Cutie'	wBox

HEMEROCALLIS

	'Ruffled Panties'	oMid
	'Ruffled Sunburst'	oMid
	'Rumble Seat Romance'	oMid
	'Russian Rhapsody'	oCap wSno oWnC
	'Sabra Salina'	oMid
un	'Sacred Song'	wBox
	'Samantha Lucretia'	wSno
	'San Juan Sunset'	oGai
	'Sanford Double Violet'	oGai
	'Sanford House'	wSno
	'Sari'	oBRG
un	'Savory Rouge'	oBRG
	'Scarlet Chalice'	oGai
	'Scarlet Orbit'	oMid oGre oBRG
	'Scarlet Tanager'	wBox wSnq
	'Scarlock'	oGai
	'Scatterbrain'	oGai
	'Schoolgirl'	oGar oWnC
	'Screech Owl'	oBRG
	'Sea Urchin'	oGai
	'Sebastian'	oCap oGai
	'Second Thoughts'	oGai
	'Seductor'	last listed 99/00
	'Seductress'	oGai oBRG
	'See Here'	oGai
	'Seminole Dream'	oGai
un	'September Gold'	oGar
	'Serena Dark Horse'	oMid
	'Serena Lady'	oGai
	'Serena Sunburst'	oGai
	'Serene Madonna'	oGai
	'Serengeti'	oGai
	'Shadow Dance'	oGai
	'Shadowed Pink'	oGai
	'Shady Lady'	wSno
	'Shaman'	wBox
	'Shari Harrison'	oBRG
	'Shark's Tooth'	oGai
	'Sherwood Chief'	last listed 99/00
	'Sherwood Gladiator'	oDan oGai oBRG
un	'Sherwood Late Show'	oGai
	'Sherwood Paradise Garden'	oCap oGai
	'Sherwood Serenity'	oCap oGai
un	'Sherwood Soldier'	oGai
	'Sherwood Sweet Honey'	oCap oGai
	'Sherwood Sweet Oliva'	oCap
	'Shinto Etching'	oMid
un	'Shortee'	oGar
	'Siamese Royalty'	oGai
	'Silent Prayer'	oGai
	'Silent Thunder'	oGai
	'Silk And Honey'	oGai
	'Silk Stockings'	oGai
	'Silken Touch'	oGai
	'Siloam Amazing Grace'	oGai
	'Siloam Art Work'	last listed 99/00
	'Siloam Arthur Kroll'	oMid
	'Siloam Baby Talk'	oBRG
	'Siloam Bertie Ferris'	oMid
	'Siloam Bo Peep'	oCap oGai oBRG
	'Siloam Bouquet'	last listed 99/00
un	'Siloam Bumblebee'	wBox
	'Siloam Button Box'	wSno oGai wBox
	'Siloam Byelo'	oCap
	'Siloam Candy Girl'	oGai
	'Siloam Charles Stinnett'	oGai
	'Siloam Cinderella'	oBRG
	'Siloam David Kirchhoff'	oFor oCap oMid oGai
	'Siloam Doodlebug'	oMid oGai
do	'Siloam Double Classic'	oFor oGai
	'Siloam Dream Baby'	oGai
	'Siloam Ethel Smith'	wSno oMid wBox
	'Siloam French Doll'	oMid oGre oBRG
	'Siloam French Lace'	oGai
	'Siloam French Marble'	wSno
	'Siloam Grace Stamile'	oMid oGai
	'Siloam Gumdrop'	oCap oGai
	'Siloam Harold Flickinger'	oCap
	'Siloam Jandee'	oMid oGai
	'Siloam Jim Cooper'	oGai
	'Siloam John Yonski'	oGai
	'Siloam June Bug'	wSno oGai wBox
	'Siloam Justine Lee'	oMid
	'Siloam Little Angel'	oGai
	'Siloam Little Girl'	oCap wSno oGai oBRG
	'Siloam Merle Kent'	oGai
	'Siloam New Toy'	oCap oGai
	'Siloam Nugget'	oMid oGai oBRG
	'Siloam Olin Frazier'	oGai
	'Siloam Pee Wee'	oGai
	'Siloam Pink Glow'	oFor
	'Siloam Plum Tree'	wSno oBRG
	'Siloam President Barnes'	oMid
	'Siloam Prissy'	oMid
qu	'Siloam Red'	oGar
	'Siloam Red Ruby'	oGai
	'Siloam Red Toy'	oGre oBRG
	'Siloam Ribbon Candy'	oMid oGai
	'Siloam Robbie Bush'	oGai
	'Siloam Royal Prince'	wWoS
	'Siloam Russell Morgan'	oCap
	'Siloam Shocker'	oMid
	'Siloam Show Girl'	wSno oGai
	'Siloam Sugar Time'	oMid
	'Siloam Tee Tiny'	wSno oMid oGai
	'Siloam Tiny Jewel'	oMid
	'Siloam Tiny Tim'	oGre oBRG
	'Siloam Tiny Toy'	oGai oBRG
	'Siloam Toy Time'	wBox oBRG
	'Siloam Ury Winniford'	oMid oGai wBox
	'Siloam Virginia Henson'	oMid oGai
	'Siloam Wendy Glawson'	oCap
	'Silver Ice'	last listed 99/00
	'Silver Sprite'	oGai
	'Sinbad Sailor'	oGai
	'Singapore Peach'	wSno
	'Singing Sixteen'	wSno
	'Sings The Blues'	wSno oGai
	'Sir Lancelot'	oGai
	'Sirocco'	wSno wBox
	'Sister Mildred'	wSno oGai
	'Sleigh Ride'	oBRG
	'Small Town Girl'	oMid
	'Small World'	oGai
	'Smarty Pants'	oCap wSno
	'Smash Hit'	oGai
	'Smuggler's Gold'	oMid
	'Snow Ballerina'	oGai
	'Snow Valley'	wBox
un	'Snow White'	wMag
	'Snowy Eyes'	oBRG
	'Snowy Owl'	oGai
	'So Excited'	oGai
	'So Lovely'	wSno oBRG
	'So Sweet'	oCap oGai
	'Soft Lavender'	last listed 99/00
un	'Soft Summer Night'	oGai
	'Sombrero Way'	oGai
	'Someone Special'	oCap
	'Sophisticated Miss'	last listed 99/00
	'Sounds Of Silence'	oCap oMid
	'Southern Love'	oCap
	'Southern Mistress'	oGai
	sp.	wBlu
	sp. from Asia Minor	oOEx
	'Spanish Glow'	oCap
	'Sparkplug'	oMid
	'Speak of Angels'	oFor oCap
	'Spell Fire'	last listed 99/00
	'Spider Breeder'	oBRG
	'Spider Miracle'	oBRG
	'Spiderman'	oBRG
	'Spray Of Pearls'	oGai
	'Staci Cox'	oGai
	'Stafford'	wCul
	'Star Of Fantasy'	oMid
	'Stella de Oro'	oFor wCul wCri wBCr wSwC oCap wSno oDar oAls oMid oMis oGar oSho oBlo wRai wFai oGai oWnC wMag wTGN wBox wSta wTLP oBRG oCir wEde wWhG oSle wSnq
	'Stoplight'	oGai
	'Storm Spell'	oMid
	'Strawberry Candy'	oMid oGai oBRG
	'Strawberry Hill'	oMid
	'Strawberry Ice'	oGai
	'Strawberry Swirl'	oMid oGai
	'Streaker'	wWhG
	'Strutter's Ball'	oCap oMid oGai
	'Study In Scarlet'	oGai
	'Sue Rothbauer'	oNat

	'Sugar Candy'	oGai
	'Sugar Cookie'	oGai
	'Sugar Rush'	oGai
	'Sultans Ruby'	oGai
	'Summer Wine'	oFor oGar oGai
	'Sun Locket'	oBRG
un	'Sun Parade'	oMid
un	'Sundae View'	oBRG
	'Sunday Gloves'	oNat wSno oMid oGai
un	'Sunray'	oBlo
	'Sunset Whisper'	oCap
	'Super Double Delight'	oGai
	'Super Prize'	oGai
	'Super Purple'	oGai
	'Superlative'	oGai
	'Suwanee Sweetheart'	oGai
	'Suzie Wong'	wBox
un	'Sweet Afternoon'	oGai
	'Sweeter Music'	oBRG
	'Swirling Water'	oCap wSno oGai
	'Swiss Mint'	oGai
	'Take a Peak'	oMid
	'Tali Queen'	oGai
	'Tallyman'	oGai
	'Tapestry of Dreams'	oMid
	'Teahouse Geisha'	oGai
	'Tender Love'	oCap wBox oBRG
	'Tender Shepherd'	wSno
un	'Tetraploid Joan Senior'	oCap
	'Tetrina's Daughter'	oFor wWoS oGar wBox
un	'Texas Red'	wBox
	'Thais'	oGai
	'Thomas Lee'	wSno
	'Thunderbird Feathers'	oMid
	'Tia'	oGai
	'Tiffany Gold'	oMid
	'Tigerling'	oMid
	'Time Lord'	oGai
	'Time Window'	oGai
	'Timeless Fire'	oMid oGai
	'Timeless Romance'	oGai
	'Tiny Tiki'	oGai
	'Tiny Toon'	oGai
	'Tiny Trumpeter'	oSec
	'Tixie'	oGai
	'Tobi Gene'	wBox
	'Todd Monroe'	oBRG
	'Toltec Sundial'	wSno oBRG
	'Too Marvelous'	oGai
	'Tootsie Rose'	last listed 99/00
	'Top Show Off'	oMid
	'Touch Of Class'	oGai
	'Touch of Magic'	oMid
un	'Towhee'	oGai
	'Toy Circus'	oBRG
	'Toyland'	oGar
	'Trade-Last'	oCap
	'Trahlyta'	oGai
	'Tree Swallow'	wSno
	'Tropical Doll'	oGai
	'Tropical Snow'	oGai
	'Trudy Harris'	oCap
	'Tuscawilla Pink Joe'	oGai
	'Tuscawilla Princess'	oMid
	'Twilight Crepe'	oGai
	'Twilight Swan'	oGai
	'Unique Purple'	wSno oGre oGai oBRG
	'Valedictorian'	oGai
	'Vanilla Fluff'	oGai
	'Vanilla Ruffles'	last listed 99/00
	'Vein of Riches'	oMid
	'Vendetta'	oGai
	'Vera Biaglow'	oCap oGai
	'Victorian Collar'	oCap
	'Vintage Bordeaux'	last listed 99/00
	'Viracocha'	wSno
	'Vohann'	oCap
	'Volcanic Explosion'	oMid
	'War March'	oGai
	'Warrior Prince'	oGai
	'Water Wheel'	oGai
	'Water Witch'	oGar
	'Waterford'	oGai
	'We Thank Thee'	oGre oBRG
	'Wedding Band'	oMid

	'Wee Wizard'	oGai
	'Well Of Souls'	oGai
	'When I Dream'	oCap
	'Whimsical'	oDar
	'Whispering Halo'	oGai
	'White Opal'	oGai
	'White Parasol'	oMid
	'White Temptation'	oCap oMid oGai
	'White Triangle'	wBCr
	'White Zone'	oMid
un	'Whitehouse Lady'	oMid
	'Whooperie'	oCap oGai
	'Wide Fantasy'	oGai
	'Wild One'	oMid
	'Will Return'	oMid
	'Windsor Castle'	oBRG
	'Wine Delight'	oGar oSho oBlo oGai
	'Wineberry Candy'	oMid
	'Winsome Cherub'	oGai
	'Wistful Moment'	oGai
	'Witch Hazel'	wSnq
	'Witch Stitchery'	oGai
	'Witch's Thimble'	oMid
	'Wood Duck'	oBRG
	'Woodside Amethyst'	last listed 99/00
	'Woodside Ruby'	oGai
	'Wynnson'	oMid oGai
	'Xia Xiang'	oMid
	'Yardmaster'	oCap
	'Yazoo Mildred Primos'	wSno
	yellow	oSle
	'Yellow Bouquet'	oMid oGai
	'Yellow Explosion'	oCap oGai wBox
	'Yellow Lollipop'	oGar oBRG
un	'Yellow Lollipop II'	wCul
	'Yellowstone'	oFor oGar wWhG
	'Yesterday Memories'	oCap
	'Yo Yo Champ'	oGai
	'You Devil'	oMid
	'Zagora'	wSno
	'Zinfandel'	oGai
un	**HEMIA**	
	salicifolia	oOEx
	HEMIPHRAGMA	
	heterophyllum HWJCM 172	last listed 99/00
	HEPATICA	
	acutiloba	wHer oRus oSis
	americana	oFor oRus oSis oBov oUps
	henryi	wHer
	x media	last listed 99/00
	nobilis	oNWc
qu	***nobilis*** var. ***americana***	wHer
	nobilis var. ***japonica***	oNWe
	nobilis var. ***japonica*** white	oNWe
un	***nobilis*** var. ***nobilis***	wHer
un	***nobilis*** var. ***nobilis*** pink	wHer
	nobilis pink	wCol
	transsilvanica	wHer
un	***villosa*** 'Purple Leaf'	oRus
	HEPTACODIUM	
	miconioides	oFor wHer oGos oJil oGar oCis wSte
un	***microphyllus***	oGre
	HERACLEUM	
	lanatum	oBos oGre oAld
	minimum 'Roseum'	oNWe
	HERMODACTYLUS	
	tuberosus	wCul
	HERNIARIA	
	glabra	oAls oAmb oMis wFai oCrm wTGN oCir oUps
	HESPERALOE	
	parviflora	oSho wCCr oWnC
	HESPERANTHA	
	falcata	last listed 99/00
	HESPERIS	
	matronalis	oFor oGar wFai oUps
	matronalis var. ***albiflora***	oFor oSec oUps
	HETEROCENTRON	
	elegans	wTGN
	HETEROMELES	see **PHOTINIA**
	HEUCHERA	
	americana	oGoo
	americana Dale's Strain	wTho oSho oSha
	'Amethyst Mist'	oAmb oGar wFai wMag
	'Autumn Haze'	wWoS

HEUCHERA

un 'Brandon Pink' oAls oGar wNay wMag
Bressingham hybrids oAls oGar wHom oWnC iArc oCir oEga wWhG
qu ***x brizoides*** 'Boule de Feu' oWnC
x brizoides 'Chatterbox' oGar
x brizoides 'June Bride' oFor
un 'Bronze Beacon' wSnq
'Can-Can' wHer wWoS wCol wCSG oDan oSis oOut wNay
'Canyon Pink' oSis
'Cappuccino' oFor wWoS oSis
'Cascade Dawn' oGar wRob
'Cathedral Windows' last listed 99/00
'Champagne Bubbles' wWoS oGre oBRG
Charles Bloom / 'Chablo' last listed 99/00
'Checkers' wWoS wNay
'Cherries Jubilee' oFor oEdg oHed oAmb oGar oSis oNWe wAva
'Chiqui' oSis
chlorantha oHan oBos
un 'Chocolate Mist' wMag
'Chocolate Ruffles' oFor wHer oJoy oHed oGar
'Chocolate Veil' oFor wCul wSte
'Coral Bouquet' oHed oDan oSis oNWe
un 'Crimson Curls' wTGN wWhG
cylindrica oHan wCul wWld
un ***cylindrica*** var. ***glabella*** 'Siskiyou' oSis
cylindrica 'Greenfinch' oGar oDan wBWP
un ***cylindrica*** 'Red Bud' oSis
'Ebony & Ivory' wHer
'Eco Magnififolia' oSis
un 'Eden's Aurora' wNay
'Eden's Joy' wNay
'Eden's Mystery' wNay
'Eden's Shine' wNay
elegans NNS95-289 last listed 99/00
'Emperor's Cloak' oMis
un 'Fackel' oCir
Firefly see **H.** 'Leuchtkaefer'
'Firesprite' oSho wAva
'Fireworks' wWoS wNay
'Garnet' oFor
glabra oHan wThG
'Green Spice' oGre
grossulariifolia wWld
grossulariifolia dwarf form oSis
un ***hirsutissima*** 'Santa Rosa' wHer wWoS wNay
'Huntsman' oAls oHed
hybrid oRus
'Jack Frost' wWoS oSis oSqu oBRG
'Lace Ruffles' oFor wCul
'Leuchtkaefer' wWoS oAls oNWe oSho oOut oWnC oUps wSnq
'Mardi Gras' wCol
micrantha oHan wNot oAls oRus wShR oBos oTri oAld wFFl wWld
micrantha var. ***diversifolia*** Bressingham Bronze / 'Absi' oGar wWhG
micrantha var. ***diversifolia*** 'Palace Purple' oFor wCul wTho oAls wCSG oGar oSho oSha oBlo wFai wHom oWnC wMag wTGN oInd oCir wAva wCoN wWhG oUps wSnq
'Mint Frost' oEdg oAls oGar oSho oOut wNay wRob oEga
un 'Mint Julep' wCol oAmb
va 'Monet' wWoS wCol oGar wNay oGre wEde wSnq
'Montrose Ruby' wWoS oRus
un 'Mrs. Barbara Toogood' oSqu
'Mount Saint Helens' oGar wFai
'Northern Fire' wHer wNay
'Oakington Jewel' wWoS oSis oWnC oBRG wWhG
'Palace Passion' wWoS oBRG
un 'Peacock Feather' wCol
'Persian Carpet' wNay wRob
'Petite Pearl Fairy' wWoS wCol
'Pewter Moon' oFor oGar wWhG
'Pewter Veil' wWoS wCul oEdg wCol oGar wFai
pilosissima wCCr
'Plum Pudding' wHer wWoS wSwC oJoy wCol oAls oGar oDan oOut oSle wSnq
pulchella oRus
'Purple Petticoats' wHer wWoS wRob
'Purple Sails' oBRG
Rain of Fire / 'Pluie de Feu' oWnC
'Raspberry Regal' oAls wFai
'Regal Robe' last listed 99/00
'Ring of Fire' oFor oNWe wNay
un ***rubescens*** var. ***alpicola*** oHan
'Ruby Ruffles' wFGN oSis
'Ruby Veil' oFor oAls wFGN oEga wSte
'Ruffles' oFor wCul wHig oSle
sanguinea oAls wCSG wSta wAva
sanguinea 'Frosty' wNay
un ***sanguinea*** 'Milky Way' oSis
sanguinea 'Sioux Falls' wTho
va ***sanguinea*** 'Snow Storm' oFor oSis oEga
sanguinea 'Spangles' wNay
sanguinea 'Splendens' oFor
qu ***sanguinea*** 'Splendens Milky Way' oGar
va ***sanguinea*** 'Splish Splash' oHed wRob wAva
un 'Silver Scrolls' wWoS
'Silver Shadows' wWoS oSev wTGN
'Silver Veil' wWoS
'Smokey Rose' wWoS wRob
un 'Snow Angel' oFor wWhG
un 'Sonata Rose' wHig
'Stormy Seas' wWoS oAls
'Strawberry Swirl' wCol oOut oGre oBRG oSle
'Veil of Passion' wWoS
'Velvet Night' oFor wWoS oJoy oEdg oGar oDan oGre wTGN wAva oEga
un 'Vivid' oAls oWnC
un 'Wendy' oHed
'Whirlwind' wCol
un 'White Marble' oSis
'White Spires' wWoS oHed oOut
'Winter Red' oSis

X HEUCHERELLA

alba 'Bridget Bloom' oFor oAls oHed oGar wRob
alba 'Rosalie' oHed oSis
'Burnished Bronze' wWoS wCol oDan
un 'Checkered White' oSis
'Cinnamon Bear' wWoS
'Cranberry Ice' oFor wWoS
'Crimson Clouds' wWoS
'Kimono' wWoS
'Pink Frost' oFor oJoy wCol oAls oGar wMag oBRG
'Silver Streak' wRob
un 'Snow White' oFor oJoy
'Viking Ship' oEdg wCol oHed oGar oDan oNWe wNay wRob

X HIBANOBAMBUSA

tranquillans wBea wBam
va ***tranquillans*** 'Shiroshima' oFor oTra wSus wCli oNor wBea oSho wBmG oTBG

HIBISCUS

arnottianus last listed 99/00
coccineus oFor oSsd oOut oWnC
cooperi see **H. *rosa-sinensis*** 'Cooperi'
hamabo oTrP
qu ***lasiolepis*** oSec
manihot see **ABELMOSCHUS *manihot***
'Morning Glory' oNat
moscheutos oMac oHug oDan oWnC wWel
moscheutos 'Anne Arundel' oAls oGar
qu ***moscheutos*** Baltimore mixed oUps
moscheutos 'Blue River II' oFor oAls
moscheutos 'Lady Baltimore' oGar wWel
moscheutos 'Lord Baltimore' oAls oGar
moscheutos Disco Belle Series oNat oAls wTGN
moscheutos Disco Belle pink (Disco Belle Series) oMis oUps
moscheutos Disco Belle red (Disco Belle Series) oMis oUps
moscheutos Disco Belle white (Disco Belle Series) oMis oGar oUps
un ***moscheutos*** 'Mallow Marvels' oFor
moscheutos ssp. ***palustris*** wGAc
moscheutos Southern Belle Group oGar oUps
un ***moscheutos*** 'The Clown' wWel
mutabilis 'Ruber' oWnC
un 'Old Yella' oGre
palustris see **H. *moscheutos*** ssp. ***palustris***
rosa-sinensis oNat
rosa-sinensis 'Brilliant' oGar
un ***rosa-sinensis*** 'Cherie' oGar
va ***rosa-sinensis*** 'Cooperi' oNat
un ***rosa-sinensis*** 'Fire Wagon' oGar
rosa-sinensis 'Helene' oFor oGre wKin

	Plant	Sources
	rosa-sinensis Itsy Bitsy® Kona Princess / 'Monria'	oGar
un	***rosa-sinensis*** 'Rosea'	oNat
un	***rosa-sinensis*** 'Santana'	oGar
un	***rosa-sinensis*** 'The Path'	oNat
un	'Sweet Caroline'	oAls oGar oGre
	syriacus	oFor wSta
	syriacus 'Aphrodite'	oAls oGar oSho oGre wKin wWel
do	***syriacus*** 'Ardens'	oGar oWhi
un	***syriacus*** 'Banner'	oWnC
	syriacus Blue Bird	see **H. *syriacus*** 'Oiseau Bleu'
	syriacus 'Blushing Bride'	oJoy oAls oGar oSho oBlo oWnC
	syriacus 'Collie Mullens'	oAls oGar oSho wPir wKin oWnC wCoS
	syriacus 'Diana'	oFor wCul oAls oGar oWhi oGre wPir wKin oWnC wMag oThr oUps wWel
	syriacus double blue	oWnC
	syriacus double pink	oWnC
	syriacus double red	oWnC
un	***syriacus*** 'Freedom'	wTGN
	syriacus 'Helene'	oAls oGar oSho wWel
	syriacus 'Lucy'	oGar oGre
va	***syriacus*** 'Meehanii'	oSho
	syriacus 'Minerva'	oFor oGar oDan oWhi oGre wKin wCoS wWel
	syriacus 'Oiseau Bleu'	oFor wCul oAls oMis oGar oSis oBlo oGre wKin oGue wWel
	syriacus 'Paeonaeflorus'	oGar
	syriacus 'Purpurea'	oBlo
	syriacus 'Red Heart'	wCul oGar oWhi oBlo wPir wKin oWnC oGue wCoS wWel
	syriacus 'Variegatus'	see **H. *syriacus*** 'Meehanii'
	syriacus 'Woodbridge'	oGar oGoo oGre wWel
	trionum	oGar wSte
un	'Turn of the Century'	oAls oGar oGre
	HIERACIUM	
	aurantiacum	see **PILOSELLA *aurantiaca***
	maculatum	oFor
	HIEROCHLOE	
	occidentalis	oGoo
	odorata	oFor oTrP wWoS oGoo iGSc oCrm oSec
	HIMALAYACALAMUS	
un	***asper***	oTra wBea
	falconeri	oTra wBea
	falconeri 'Damarapa'	oTra wBea
un	***falconeri*** var. ***glomerata***	wBea
	hookerianus	oTra oNor wBea
un	***intermedius***	oTra wBea
un	***porcatus***	oTra
	HIPPEASTRUM	
un	'Amoretta'	oGar
un	'Floris Hecker'	oGar
un	'Surprise'	oGar
	'White Dazzler'	oGar
	HIPPOCREPIS	
	emerus	oFor
	HIPPOPHAE	
	rhamnoides	oFor wBCr wBur oRiv oOEx oWnC wWel
un	***rhamnoides*** 'Frugana'	wClo oOEx
un	***rhamnoides*** 'Hergo'	oOEx
	rhamnoides 'Leikora' female	wBur wClo oGar wRai oOEx wWel
	rhamnoides male	wBur wClo wRai oOEx wWel
un	***rhamnoides*** 'Otradnaya'	wBur
un	***rhamnoides*** 'Russian Orange'	wWel
un	***rhamnoides*** 'Sprite'	oFor
un	***rhamnoides*** 'Star of Altai' female	wRai
	salicifolia	oFor
	HOHERIA	
un	'Snowflurry'	oCis
	HOLBOELLIA	
un	'China Blue'	oAls wWel
	coriacea	oFor oTrP wHer wCul oGos oAmb oGar oSho oBov oGre oCis oBRG wSte
	fargesii	oFor oCis
	fargesii DJHC 506	wHer
	fargesii DJHC 98217	wHer
	latifolia	oCis
	latifolia HWJCM 008	wHer
	sp. aff. ***latifolia*** EDHCH 97338	wHer
	sp.	oOEx
	HOLCUS	
	mollis 'Albovariegatus'	oFor oGre wWin
	mollis 'Variegatus'	see **H. *mollis*** 'Albovariegatus'
	HOLODISCUS	
	discolor	oFor wWoB wThG wNot oJil wShR oGar oBos oRiv oGre wWat wCCr oTri wPla oAld wAva wCoN wFFI wRav
	dumosus	oFor oHan
	HOMALOCLADIUM	
	platycladum	oTrP wGra
	HOMALOMENA	
un	***lindenii*** 'Emerald Gem'	oGar
un	**HOMALOTHECIUM**	
	fulgescens	wFFI
	HOMOGLOSSUM	see **GLADIOLUS**
	HOODIA	
	bainii	oRar
	gordonii	oRar
	macrantha	oRar
un	***pilifera***	oRar
	HORDEUM	
un	***brachyanthemum***	oAld
	jubatum	oGoo
	HORMINUM	
	pyrenaicum	last listed 99/00
	HOSTA	
va	'Abba Dabba Do'	oWaW oMid oGar wTow wNay oGre
	'Abby'	oMid wTow wNay oHou
	'Abiqua Ariel'	oWaW wNay
	'Abiqua Aries'	oWaW
	'Abiqua Blue Edger'	oWaW wCol wTow wNay
	'Abiqua Blue Krinkles'	oWaW wNay
	'Abiqua Blue Madonna'	wNay
	'Abiqua Delight'	oWaW wCol oMid wNay wRob
	'Abiqua Drinking Gourd'	oFor wWoS oWaW oZOT oMid oGar oNWe wNay oGre
	'Abiqua Elephant Ears'	oWaW
	'Abiqua Gold Shield'	oWaW
	'Abiqua Ground Cover'	oWaW
	'Abiqua Hallucination'	wTow
va	'Abiqua Moonbeam'	oWaW oCap oMid wTow wNay wRob
	'Abiqua Paradigm'	see **H.** 'Paradigm'
	'Abiqua Parasol'	oWaW
	'Abiqua Recluse'	wWoS oWaW oCap oMid wTow wNay wRob oWnC
	'Abiqua Trumpet'	oFor oWaW oZOT wTow wNay
	aequinoctiiantha	wTow wRob
	'Akarana'	wNay
	'Albomarginata' ***(fortunei)***	oWaW oAls oRus oMid oGar oTwi wNay oBlo wFai oHou iArc oCir oUps oSle
un	'Alcebana Kukurin'	wNay
va	'Allan P. McConnell'	oWaW oJoy oAls oRus oMid oGar oSis oTwi wNay oGre wMag wEde
	'Alpine Aire'	last listed 99/00
	'Alvatine Taylor'	oWaW wNay wRob
un	'Amber Tiara'	wNay
un	'American Dream'	oMid wTow wNay
un	'Angel Feathers'	oWaW wTow wNay
	'Anne Arett'	wNay wRob
va	'Antioch'	oWaW oDar wHig oMid oGar wNay oGre
	'Aoki' ***(fortunei)***	wNay
	'Aoki Variegated' ***(fortunei)***	wNay
	'Aphrodite'	oWaW wCol oMid oGar wNay wRob
	'Apple Green'	oWaW
	'Aqua Velva'	oWaW oJoy oMid wNay wRob
un	'Archangel'	wRob
	'Arctic Circle'	oJoy oMid wNay
un	'Aristocrat'	wNay wRob
	'Asian Pearl'	wCol
	'Aspen Gold'	oWaW oJoy oMid wTow wNay wRob
un	'August Blue'	oGre
	'August Moon'	oFor oDar oJoy oAls wCSG oMid oGar wTow oSho wNay wRob oGre oWnC wMag oBRG wAva wEde wWhG wSnq
	'Aureomaculata'	see **H. *fortunei*** var. ***albopicta***
	'Aureomarginata' ***(montana)***	oWaW oMid oGar wTow oTwi wNay wRob wMag wEde
	'Aureomarginata' ***(ventricosa)***	oWaW oMid oGar oTwi wNay wRob wMag
va	'Aurora Borealis'	wCol oMid oNWe wNay wRob oHou
un	'Austin Dickinson'	oTwi wNay wRob
	'Aztec Treasure'	oMid wNay
	'Azure Snow'	oFor oWaW oZOT oMid wTow wNay wRob
	'Baby Bunting'	oWaW oMid wNay wRob
	'Banyai's Dancing Girl'	oWaW wNay wRob
	'Beauty Substance'	wTow
	'Bengee'	wRob
	'Bennie McRae'	wNay

	'Betcher's Blue'	oMid wNay
	'Bethel Big Leaf'	wRob
	'Betsy King'	wNay
	'Bette Davis Eyes'	oWaW wNay
	'Betty'	wRob
	'Big Boy'	wNay
	'Big Daddy'	oCap oAls oMid oGar wNay oGre wMag
	'Big John'	wNay
	'Big Mama'	oMid wTow wNay wRob
	'Big Sam'	oMid
	'Bigfoot'	oJoy wTow wNay wRob
va	'Bill Brincka'	wNay wRob
un	'Birchwood Blue Beauty'	wNay wRob
	'Birchwood Elegance'	oJoy oMid wRob
	'Birchwood Gem	wNay
	'Birchwood Gold'	oWaW oJoy wHig oGar wNay
	'Birchwood Parky's Gold'	oAls oMid wNay oBlo oGre
	'Birchwood Ruffled Queen'	oMid wNay
	'Bitsy Gold'	wTow
	'Bizarre'	wNay
un	'Black Foot'	wNay
	'Black Hills'	oWaW oMid wNay wRob
	'Blonde Elf'	oWaW oMid oTwi wNay
ch	'Blue Angel' (see Nomenclature Notes)	oFor oWaW oJoy oMid oGar wTow oTwi wNay wRob oBlo oHou
	'Blue Arrow'	oMid wTow wNay
	'Blue Bayou'	wNay
	'Blue Belle'	see **H.** Tardiana Group 'Blue Belle'
	'Blue Betty Lou'	oWaW oJoy oMid wNay
	'Blue Blazes'	oMid wNay wRob
	'Blue Blush'	see **H.** Tardiana Group 'Blue Blush'
	'Blue Boy'	oFor oWaW oJoy oRus oMid
	'Blue Cadet'	oWaW oJoy wCSG oAmb oMid oGar oTwi oOut wNay oGre wMag wAva wEde oSle
	'Blue Cup'	oFor
	'Blue Danube'	see **H.** Tardiana Group 'Blue Danube'
	'Blue Diamond'	see **H.** Tardiana Group 'Blue Diamond'
	'Blue Dimples'	see **H.** Tardiana Group 'Blue Dimples'
	'Blue Edger'	wNay
un	'Blue for You'	wRob
un	'Blue Gown'	wNay
	'Blue Heart'	oAls
	'Blue Heaven'	wNay
	'Blue Ice'	wTow wNay wRob
	'Blue Jay'	wRob oWnC
	'Blue Lady'	wNay
	'Blue Mammoth'	oMid oGar wNay
	'Blue Moon'	see **H.** Tardiana Group 'Blue Moon'
un	'Blue Plisse'	wNay
	'Blue Ripples'	wNay wRob
	'Blue Rock'	oWaW wNay
	blue seedlings	wAva
	'Blue Seer'	oFor wNay wRob oHou
va	'Blue Shadows'	oWaW oMid wNay wRob
	'Blue Skies'	see **H.** Tardiana Group 'Blue Skies'
un	'Blue Splendor'	oGar wRob
	'Blue Tips	oWaW wNay
	'Blue Troll'	wNay
	'Blue Umbrellas'	oMid oGar wTow oSho wNay wRob
	'Blue Veil'	wNay
	'Blue Velvet'	wRob
	'Blue Vision'	oWaW oMid wNay wRob
	'Blue Wedgwood'	see **H.** Tardiana Group 'Blue Wedgwood'
	'Blue Whirls'	oWaW wNay wRob
un	'Bobbie Sue'	wNay
	'Bold and Brassy'	wRob
va	'Bold Edger'	wWoS oWaW oMid wNay
va	'Bold Ribbons'	oWaW oZOT oMid oGar wNay
	'Bold Ruffles'	wRob
va	'Borwick Beauty'	oMid wNay wRob
	'Bountiful'	wRob
un	'Brandywine'	wNay
un	'Brave Amherst'	wNay
un	'Bravo M'	oMid
un	'Brenda's Beauty'	oMid
	'Bressingham Blue'	oCap oTwi wNay wMag
	'Bridegroom'	wNay wRob
	'Brigham Blue'	see **H.** ***sieboldiana***
	'Bright Glow'	see **H.** Tardiana Group 'Bright Glow'
va	'Bright Lights'	wWoS oWaW oMid wNay
va	'Brim Cup'	oFor oWaW oMid wTow wNay oWnC
	'Brookwood Blue'	last listed 99/00
	'Brother Ronald'	see **H.** Tardiana Group 'Brother Ronald'
	'Buckshaw Blue'	oWaW wNay wRob
	'Buckwheat Honey'	wNay wRob
	'Bunchoko'	see **H.** 'Ginko Craig'
va	'Butter Rim'	oJoy
	'Caerula'	wMag
	'Calypso'	oMid wNay
	'Camelot'	see **H.** Tardiana Group 'Camelot'
	'Candy Hearts'	oWaW oAls wCSG oGar oTwi wNay oGre oWnC oHou
	capitata	wNay
	capitata DJH 313	wHer
	sp. aff. ***capitata*** HC 970378	wHer
un	'Carder Blue'	oJoy wNay
va	'Carnival'	oWaW oMid wNay wRob
va	'Carol'	oWaW oGar oTwi wNay
	'Carrie'	oWaW wNay
ch	'Carrie Ann'	see **H.** 'Carrie'
	'Cartwheels'	oMid
va	'Celebration'	oMid wNay
	'Chameleon'	oJoy wNay
va	'Change of Tradition'	last listed 99/00
va	'Chantilly Lace'	wNay wRob
	'Chartreuse Wedge'	oWaW wRob
	'Chartreuse Wiggles'	oMid wNay wRob
va	'Cheatin Heart'	wNay wRob
	'Cherry Berry'	oFor wWoS oWaW wTow wNay oWnC
va	'Chinese Sunrise'	oWaW oZOT wHig oMid wTow oNWe oTwi wNay oWnC
	'Chiquita'	oWaW oMid wNay
va	'Choko Nishiki'	oFor oWaW oMid oDan wTow wNay wRob
va	'Christmas Tree'	oWaW oJoy wCol oMid oGar oDan oNWe wNay wRob oBRG oHou
va	'Citation'	oWaW wRob
	'City Lights'	oMid wNay wRob
	'Clarence'	wNay
	clausa	wNay
	clausa var. ***normalis***	wRob
	clausa var. ***normalis*** HC 970341	wHer
un	'Cody'	wNay
	'Collector's Banner'	wCol
	'Collector's Choice'	oWaW wCol wNay
va	'Color Glory'	oMid oGar wNay
	'Colossal'	oJoy oMid wNay wRob
	'Columbus Circle'	oMid wRob
va	'Coquette'	oWaW oMid wRob oWnC
un	'Cotillion'	wRob
	'Counter Point'	oMid wNay
un	'Cowrie'	wNay
	'Craigs Temptation'	oWaW oTwi wNay
va	'Crepe Suzette'	oWaW oMid wNay
	'Crested Reef'	wNay
	'Crested Surf'	oWaW wNay wRob
	'Crinoline Petticoats'	wNay
	crispula	oFor oMid wNay
va	'Crown Jewel'	wNay
	'Crown Prince'	oMid wNay
un	'Crown Royalty'	wNay
va	'Crowned Imperial'	oAls oTwi wNay wRob
va	'Crusader'	oWaW oMid wTow wNay wRob oHou
	'Curls'	oMid wNay wRob
un	'Curtain Call'	oTwi
	'Cynthia'	oWaW oJoy wCol wRob
un	'Dana Nicolette'	wNay
va	'Dark Star'	oMid wTow wNay wRob
	'Dark Victory'	wNay
	'Daybreak'	oWaW wTow wNay wRob oHou
	decorata	wNay
	'Deluxe Edition'	oWaW
	'Devon Blue'	see **H.** Tardiana Group 'Devon Blue'
	'Devon Gold'	wNay
	'Devon Green'	wNay wRob
	'Devon Mist'	wNay wRob
	'Devon Tor'	wNay wRob
va	'Dew Drop'	wNay
	'Dewline'	wNay
va	'Diamond Tiara'	oWaW oCap oAls oMid wTow wNay wRob
va	'Don Stevens'	oWaW oMid wNay wRob
	'Donahue Piecrust'	oWaW wNay wRob
	'Dorothy'	wNay
	'Dorset Blue'	oWaW oMid oGar wNay wRob wBox
	'Dorset Charm'	see **H.** Tardiana Group 'Dorset Charm'
	'Dorset Flair'	see **H.** Tardiana Group 'Dorset Flair'
	'Doubloons'	oMid wNay wRob
	'Drummer Boy'	oWaW oRus
	'Duchess'	wNay
	'Duke'	wNay
va	'DuPage Delight'	oMid wRob

	'Dylan's Dilly'	oWaW wTow
	'Edge of Night'	oWaW oMid wNay wRob
	'Edina Heritage'	wNay
	'Edward Wargo'	wNay
va	'El Capitan'	wWoS oWaW oMid wNay wRob
	elata	wNay wRob
	'Elatior'	oWaW oMid wTow oTwi wNay wRob
	'Elegans'	see **H.** ***sieboldiana*** var. ***elegans***
un	'Elephant Burgers'	wNay
	'Elisabeth'	wNay
va	'Elizabeth Campbell'	wNay
	'Ellen Carder'	wNay
va	'Ellerbroek'	oCap oJoy oTwi oGre
	'Elvis Lives'	wWoS oWaW oTwi wNay wRob wEde
	'Embroidery'	wTow
	'Emerald Carpet'	wNay
un	'Emerald Edger'	wNay
un	'Emerald Necklace'	wNay
va	'Emerald Tiara'	oWaW oMid wTow wNay wRob
va	'Emily Dickinson'	oWaW oJoy oGar oSis oTwi wNay wRob
	'Eric Smith'	see **H.** Tardiana Group 'Eric Smith'
un	'Eternity'	wNay
	'Evelyn McCafferty'	wNay
	'Excitation'	oWaW wNay
va	'Fair Maiden'	oFor oMid wTow wNay oWnC
	'Fall Bouquet'	wNay
	'Fall Emerald'	wNay
	'Fan Dance'	oFor oMid wNay wRob
	'Fantasia'	wRob
	'Fantastic'	oMid
	'Fascinator'	oJoy oMid wNay wRob
	'Feather Boa'	oWaW wCol oMid wNay
un	'Filagree'	wRob
	'Finlandia' ***(kikutii)***	wNay
va	'Fire and Ice'	wTow oSho wNay wRob oWnC
un	'Flame Stitch'	oMid wTow wNay
	'Fleeta Brownell Woodroffe'	wNay
	'Fleetas Blue'	wNay
	'Floradora'	oWaW oJoy wNay
	'Flower Power'	oWaW oMid wNay
	fluctuans 'Variegated'	see **H.** 'Sagae'
	'Fond Hope'	oWaW
	'Fool's Gold'	last listed 99/00
va	'Formal Attire'	wWoS oWaW oMid wNay wRob
	fortunei	wMag
	fortunei albomarginata	see **H.** 'Albomarginata' ***(fortunei)***
	fortunei var. ***albopicta***	oAls oRus oGar oTwi wNay oBlo oHou iArc oLSG
	fortunei var. ***albopicta*** f. ***aurea***	oTwi wNay
	fortunei 'Aoki'	see **H.** 'Aoki' ***(fortunei)***
	fortunei 'Aoki Variegated'	see **H.** 'Aoki Variegated' ***(fortunei)***
	fortunei aurea	see **H.** ***fortunei*** var. ***albopicta*** f. ***aurea***
	fortunei var. ***aureomaculata***	see **H.** ***fortunei*** var. ***albopicta***
	fortunei var. ***aureomarginata***	oWaW oDar oAls oGar wNay wRob oBlo wEde oUps
	fortunei 'Francee'	see **H.** 'Francee' ***(fortunei)***
	fortunei var. ***gigantea***	see **H.** ***montana***
	fortunei var. ***hyacinthina***	oFor oDar oAls oGar wTow oTwi wNay oWnC wMag oBRG oHou iArc wEde oLSG wWhG oUps oSle
	fortunei var. ***obscura***	wNay wEde
	'Fortunei Viridis'	wNay
un	'Foundling'	wNay
un	'Fountainette'	wNay
un	'Fragrant Blue'	oWaW oZOT oJoy oMid wTow wNay wRob
va	'Fragrant Bouquet'	oWaW oZOT oAls oAmb oMid oGar wTow oOut wNay wRob oGre wMag wEde
	'Fragrant Gold'	oJoy oMid wNay wRob
va	'Francee' ***(fortunei)***	oFor oCap oDar wCol oAls oAmb oMid oGar oSis oTwi wNay wRob oGre wFai oWnC wMag wTGN oBRG oHou wAva wEde wWhG oSle wCoS
va	'Frances Williams'	oFor wSwC oCap oDar oAls oRus oMid oGar wNay oGre wMag wTGN oHou wWhG oUps oSle
un	'Frances Williams Squash Edge'	wNay
va	'Fresh'	wNay
un	'Fried Bananas'	wWoS oMid wNay wRob
	'Fried Green Tomatoes'	oMid oGar oTwi wNay
va	'Fringe Benefit'	oWaW oJoy wNay wRob oHou
va	'Frosted Jade'	oWaW oJoy oMid wTow oTwi wNay wRob
un	'Fuji'	oFor
	'Fused Veins'	wRob
va	'Gaiety'	wNay
	'Gaijin'	oMid wNay
va	'Gala'	wNay
un	'Galaxy'	oMid
un	'Garnet Prince'	wNay wRob
va	'Gay Blade'	oMid wRob
va	'Geisha'	oWaW oMid wNay wRob
	'Gene Summers'	wNay
	'Geneva Stark'	oWaW
va	'Gilt Edge'	wNay
	'Gingee'	oWaW
va	'Ginko Craig'	oWaW oAls oRus oMid oGar wNay oGre oWnC oHou
va	'Gloriosa'	oWaW wNay wRob
	'Glory'	oWaW oMid wNay wRob
	'Gold Cup'	oWaW
	'Gold Drop'	oJoy oAls oRus oMid oSis wTow wNay oWnC oUps
	'Gold Edger'	oWaW oZOT oJoy oAls oMid oGar wTow oSho oTwi wNay wRob oBlo oGre wMag oHou wAva
	'Gold Regal'	oWaW wTow oTwi wNay wRob
	'Gold Regal' sport	wRob
	'Gold Seer'	wRob
va	'Gold Standard'	oWaW oCap oDar oJoy oAls oRus oMid oGar wTow oSho oTwi oOut wNay wRob oBlo oGre oWnC wMag wAva wEde wSnq
va	'Goldbrook Glamour'	wRob
	'Goldbrook Glimmer'	see **H.** Tardiana Group 'Goldbrook Glimmer'
va	'Goldbrook Gratis'	wRob
	'Goldbrook Grayling'	wRob
	'Goldbrook Grebe'	wRob
	'Golden Anniversary'	wNay
	'Golden Bullion'	oMid wNay wRob oGre
un	'Golden Empress'	wNay
	'Golden Fascination'	oMid wNay
un	'Golden Friendship'	wNay wRob
un	'Golden Guernsey'	oWaW wCol oMid wNay wRob
	'Golden Haze'	oMid
	'Golden Medallion'	oWaW wNay wMag
un	'Golden Memories'	wNay
	'Golden Nugget'	oTwi wNay wRob
	'Golden Oriole'	oMid wNay
	'Golden Plum'	wNay
	'Golden Prayers'	oRus oGar wNay
	'Golden Scepter'	oWaW oJoy oMid wTow wNay wRob
	'Golden Sculpture'	oWaW oMid oGar wNay
	'Golden Spades'	wNay
	'Golden Spider'	oMid wNay
	'Golden Sunburst'	oFor oAls wNay oWnC
	'Golden Teacup'	wNay wRob
va	'Golden Tiara'	oFor oWaW oCap oJoy wCol wHig oAls oMid oGar oSis wTow oSho oTwi oOut wNay wRob oBlo oGre oWnC wMag wTGN oHou iArc wAva wWhG oSle wSnq
	'Golden Torch'	wNay
	'Golden Waffles'	oMid wNay wRob
	'Goliath'	oMid wNay
	'Good as Gold'	oMid wNay
	'Gosan Gold Midget'	wNay
un	'Gosan Gold Mist'	wTow wNay
	'Gosan Hildegarde'	wNay
un	'Gosan Leather Strap'	wTow wNay
	'Gosan Mina'	wNay
un	'Gosan Shining'	wTow
un	'Gosan Sunproof'	wTow
	'Gracillima Variegated'	see **H.** 'Vera Verde'
	'Granary Gold'	oMid wNay wRob
un	'Grand Forks'	wNay
un	'Grand Marmalade'	wNay
	'Grand Master'	oWaW oJoy wNay
un	'Grand Slam'	wRob
va	'Grand Tiara'	oWaW oMid oTwi wNay wRob
	'Gray Cole'	wNay
	'Great Desire'	wNay
va	'Great Expectations'	oWaW oZOT wGAc oJoy oAls oMid oGar wTow oTwi wNay wRob oGre wTGN wBox oHou wAva wEde
	'Green Acres'	oMid wNay wRob
	'Green Angel'	wNay wRob
	'Green Eyes'	oWaW wNay
	'Green Formal'	wNay
	'Green Fountain'	oWaW oTwi wNay wRob
un	'Green Gables'	wNay
va	'Green Gold'	oAls wNay wRob
	'Green Piecrust'	oWaW oMid wNay

	'Green Platter'	oMid
	'Green Sheen'	oWaW oMid wNay wRob
	'Green Smash'	last listed 99/00
un	'Green Suite'	oSle
	'Green Velveteen'	oWaW wNay
	'Green Wedge'	oAls
va	'Green with Envy'	wRob
	green with white edge	oNWe
	'Grey Piecrust'	wNay
va	'Ground Master'	oSho oTwi wNay
	'Ground Sulphur'	oWaW wTow wNay
va	'Guacamole'	wWoS oWaW oAls oMid oGar wTow oTwi wNay
un	'Guardian Angel'	oWaW wTow wNay wRob
	'Hadspen Blue'	see **H.** Tardiana Group 'Hadspen Blue'
	'Hadspen Hawk'	see **H.** Tardiana Group 'Hadspen Hawk'
	'Hadspen Heron'	see **H.** Tardiana Group 'Hadspen Heron'
va	'Haku-chu-han'	wNay wRob
	'Halcyon'	see **H.** Tardiana Group 'Halcyon'
	'Happiness'	see **H.** Tardiana Group 'Happiness'
	'Happy Hearts'	oWaW wNay wRob oBRG
	'Harmony'	see **H.** Tardiana Group 'Harmony'
	'Harvest Dandy'	wNay wRob
	'Harvest Dawn'	wRob
	'Harvest Glow'	wNay
	'Hazel'	wNay
	'Heart Ache'	oMid wRob
un	'Heart Throb'	wNay
	'Heartleaf'	oWaW
un	'Heart's Content'	wRob
va	'Heartsong'	wNay
	'Helen Doriot'	wNay
	'Heliarc'	oWaW wNay
	helonioides hort. f ***albopicta***	see **H. *rohdeifolia***
	'Hertha'	last listed 99/00
un	'Hi Ho Silver'	wNay
un	'Hidden Cove'	wNay
	'High Kicker'	wNay
	'High Noon'	oMid wNay wRob
va	'Hilda Wassman'	wNay wRob
	'Hirao No. 59'	last listed 99/00
un	'Hirao Elite'	wNay
	'Hirao Majesty'	oWaW oMid wNay wRob
	'Hirao Splendor'	oMid wNay wRob
	'Hirao Supreme'	oWaW oMid oTwi wNay wRob
un	'Hirao Zeus'	wNay
	'Holiday White'	oMid
	'Hollys Dazzler'	wNay wRob
un	'Hollys Gold'	wRob
	'Hollys Honey'	wNay
	'Hollys Shine'	wRob
	'Honey'	wNay
	'Honey Moon'	wNay
	'Honeybells'	oFor oCap oAls oRus oMid oGar oTwi oOut wNay wRob oGre wMag wSta wAva oUps oSle wSnq
va	'Honeysong'	oMid wTow wNay wRob
va	'Hoosier Harmony'	oMid wTow wNay
	'Hoosier Homecoming'	oMid
un	'Hope'	wTow
	'Housatonic'	wNay
	'Hyacinthina'	see **H. *fortunei*** var. ***hyacinthina***
	'Hydon Sunset'	wNay wRob
	hypoleuca	wNay wRob
un	'Immense'	wNay wRob
	'Inaho'	oMid wRob
un	'Indiana Knight'	wNay wRob
va	'Inniswood'	oFor oWaW wCol oMid oTwi wNay
	'Invincible'	oWaW oJoy wCol oMid wTow oTwi wNay wRob wEde
	'Iona'	wNay wRob
va	'Iron Gate Delight'	oMid
	'Iron Gate Glamor'	oJoy oMid wNay wRob
va	'Iron Gate Supreme'	oMid
un	'Island Charm'	oWaW wTow wNay wRob
	'Island Forest Gem'	wNay
un	'Iszat U Doc'	wNay wRob
	'Iwa'	see **H.** 'Iwa Soules'
	'Iwa Soules'	oWaW wRob
	'Jack of Diamonds'	oMid wNay
	'Jade Beauty'	oMid wNay
	'Jade Cascade'	oFor oWaW oJoy oDan wTow wNay wRob
un	'Jade Lancer'	oWaW wNay
	'Jade Scepter'	wNay
	'Jambeliah'	wTow

va	'Janet'	oJoy oMid oTwi wNay wRob
	japonica	oRus
qu	'Jester'	wNay oBlo
	'John Wargo'	wNay
	'Jolly Green Giant'	oMid wNay
	jonesii	last listed 99/00
un	'Joseph'	wNay
un	'Julia'	oMid wRob
	'Julie Morss'	wNay wRob
	'Jumbo'	wNay
	'June'	see **H.** Tardiana Group 'June'
	'June Beauty'	wNay
va	'Just So'	oMid wNay
	'Kabitan'	see **H. *sieboldii*** f. ***kabitan***
un	***karishima***	oRus
un	'Katherine'	oDan
	'Katherine Lewis'	oFor oWaW
	'Kathryn Lewis'	oTwi wNay
va	'Kifukurin' ***(kikutii)***	oWaW oMid
un	'Kifukurin Hyuga' ***(kikutii)***	oTwi wNay wRob
	'Kifukurin Ko Mame'	oWaW wTow oOut wNay wRob
	'Kifukurin Ubatake' ***(pulchella)***	wNay wRob
	kikutii	oMid
	kikutii var. ***caput-avis***	wNay wRob
	kikutii var. ***polyneuron***	wNay
	kikutii 'Pruinosa'	wNay
	kikutii var. ***yakusimensis***	oWaW oJoy wNay wRob
un	***kikutii*** var. ***yakusimensis*** 'Gosan'	wNay
un	***kikutii WB*** x ***longipes WB***	wNay
	'Kinbotan'	wRob
	'King James'	oJoy wRob
	'King Michael'	oJoy wRob
	'King Tut'	oWaW oMid wTow wNay wRob
	'Kingfisher'	wNay
	'Kinkaku'	oWaW wTow wNay wAva
	'Kirishima'	wRob
	'Kisuji'	wNay
un	'Kit Kat'	wRob
un	'Kiwi Black Magic'	wNay
un	'Kiwi Blue Baby'	wNay
un	'Kiwi Blue Cup'	wNay
un	'Kiwi Canoe'	wNay
un	'Kiwi Cream Edge'	wNay wRob
un	'Kiwi Emerald Isle'	wNay
un	'Kiwi Gold Rush'	wNay wRob
un	'Kiwi Hippo'	wNay
un	'Kiwi Minnie Gold'	wNay
un	'Kiwi Parasol'	wNay
un	'Kiwi Spearmint'	wNay
un	'Kiwi Sunshine'	wNay
un	'Kiwi Treasure Trove'	wNay wRob
	kiyosumiensis DJH423	wNay
	kiyosumiensis DJH423 seed grown	wHer
	'Klopping Variegated'	wNay wRob
va	'Knockout'	oWaW oMid wNay
un	'Komodo Dragon'	oTwi wNay
un	'Kong'	wNay
qu	'Korean Minor'	oWaW oMid
qu	'Koreana Variegated'	oMid
va	'Krossa Cream Edge'	wSwC oAls
	'Krossa Regal'	oFor oWaW oZOT oCap oAls oRus oMid oGar oDan wTow oTwi wNay wRob oBlo oGre wMag oBRG oHou oCir wEde wWhG oSle
va	'Lacy Belle'	oWaW wTow wRob
	'Lady-in-Waiting'	oWaW wRob
va	'Lady Isobel Barnett'	oWaW oMid wNay wRob oHou
	laevigata	wTow wNay wRob
	'Lakeport Blue'	oWaW oMid wNay wRob
	'Lakeside Accolade'	wNay wRob
un	'Lakeside April Snow'	wNay wRob
	'Lakeside Black Satin'	oTwi wNay
un	'Lakeside Blue Cherub'	oTwi wNay
un	'Lakeside Cha Cha'	wNay wRob
un	'Lakeside Delight'	wNay wRob
un	'Lakeside Kaleidoscope'	wNay
un	'Lakeside Little Gem'	wNay
un	'Lakeside Lollipop'	wNay
un	'Lakeside Looking Glass'	wNay
un	'Lakeside Neat Petite'	wNay
un	'Lakeside Ninita'	wRob
un	'Lakeside San Kao'	wNay
	'Lakeside Shadow'	wNay
va	'Lakeside Symphony'	oMid
	lancifolia	oFor oJoy wHig wNay oHou oSle

qu	***lancifolia minor***	oWaW
qu	***lancifolia*** 'Numor'	oWaW
	'Large Marge'	wCol
	'Lauman Blue'	oWaW wNay
	'Lauman Garden Blue'	see **H.** 'Lauman Blue'
un	'Leading Lady'	wNay
	'Leather Sheen'	oMid wNay wRob
	'Lee Armiger'	wRob
	'Lemon Chiffon'	wRob
	'Lemon Delight'	wTow wNay wRob
	'Lemon Lime'	oWaW oZOT oJoy wHig oAls oMid wTow wNay
	'Lemon Twist'	wNay wRob
	'Lemonade'	oMid wNay
va	'Leola Fraim'	wWoS oWaW oMid oTwi wNay wRob wEde
	'Leviathan'	wNay
	'Lights Up'	wRob
	'Lime Krinkles'	wNay
	'Lime Shag'	oMid wRob
un	'Limey Lisa'	wNay wRob
	'Little Aurora'	oWaW oJoy wHig oHed oGar oSis wNay wRob oGre
	'Little Blue'	oWaW wNay
un	'Little Cyn'	wRob
un	'Little Doll'	wRob
	'Little Jim'	wNay
	'Little Razor'	wNay
un	'Little Sunspot'	wNay
va	'Little White Lines'	wNay
va	'Little Wonder'	wWoS oWaW oMid wNay wRob
	longipes	oBov
un	***longipes aurea***	oWaW
	longipes 'Golden Dwarf'	last listed 99/00
	longipes f. ***hypoglauca***	wTow wNay
un	***longipes*** f. ***hypoglauca*** 'Loyalist'	wRob
	longipes f. ***viridipes***	see **H. *crispula***
	longissima	oWaW wRob
va	'Louisa'	wHig
un	'Love Burst'	wCol
	'Love Pat'	oFor oWaW oJoy wCol oRus oMid oGar wNay wRob wMag oHou
un	'Loyalist'	oZOT wTow wNay wRob wWhG
va	'Lucy Vitals'	oFor oWaW oMid oTwi wNay wRob
un	'Lunar Magic'	oMid
un	'Lunar Orbit'	oMid wTow wRob
un	'Lyme Regis'	wRob
	'Mackwoods 23'	oMid
	'Maekawa'	oWaW
	'Magic Carpet'	wRob
un	'Mama Mia'	wNay wRob
un	'Maraschino Cherry'	wNay
un	'Margie'	wNay
	'Margie Weissenberger'	wNay wRob
	'Marilyn'	oMid wNay
	'Marquis'	wNay
	'Maruba Iwa'	oWaW oMid wTow wNay wRob
	'Mary Jo'	oMid
va	'Mary Marie Ann'	wNay wRob
va	'Masquerade'	oWaW oJoy wCol oMid wNay
	'Mastodon'	wNay wRob
un	'Maui Buttercups'	oMid
	'Maya'	oWaW wNay
	'Mediovariegata'	see **H. *undulata*** var. ***undulata***
un	'Medusa'	wNay wRob
	'Mentor Gold'	oWaW oMid oTwi wNay oHou
	'Metallic Sheen'	oMid wNay
	'Metallica'	wNay
	'Midas Touch'	oMid wTow wNay wRob
	'Middle Ridge'	oMid wNay
va	'Midwest Magic'	oMid wNay wRob
	'Mikado'	wNay
un	'Mikawa-no-yuki'	wNay wRob
	'Mildred Seaver'	oWaW oMid wTow oTwi wNay wRob oHou
va	'Millie's Memoirs'	oMid wNay
un	'Ming Treasure'	oMid
	'Minnie Klopping'	oMid
	minor	oWnC
un	***minor*** 'Gosan'	wNay
un	***minor*** 'Nakai'	oGar
va	'Minuteman'	wWoS oWaW oAls oGar wTow oTwi wNay wRob wSnq
	'Mischief'	wNay wRob
un	'Miss Petite'	wCol
	'Misty Waters'	oMid
va	'Moerheim'	oCap oMid
un	'Mohegan'	wNay
un	'Monitor'	wNay
un	'Monopoly'	wNay
	montana	oWaW oZOT oMid oSho oTwi wNay
	montana aureomarginata	see **H.** 'Aureomarginata' ***(montana)***
un	***montana*** 'Chirifu Tochigi'	wRob
	montana f. ***macrophylla***	oWaW wNay
	'Montreal'	wNay
va	'Moon Glow'	wNay
va	'Moon River'	oWaW oMid wNay wRob
un	'Moon Waves'	wNay
un	'Moongate Flying Saucer'	wNay
un	'Moongate Little Dipper'	oTwi wNay
va	'Moonlight'	oWaW oMid oTwi wNay
un	'Moonlight Sonata'	wRob
	'Moscow Blue'	wNay
un	'Mostly Ghostly'	oFor wTow wNay
	'Mount Fuji'	oMid
va	'Mount Hope'	oGar wNay wRob
	'Mount Royal'	wNay
un	'Mountain Green'	oWaW
va	'Mountain Snow'	oWaW wCol oMid wNay oHou
	'Mountain Sunrise'	oMid wNay
	'Mr. Big'	oFor oWaW wTow wAva
	'Muriel Seaver Brown'	wNay
un	'My Friend Nancy'	wRob
	'Nagaeto'	wNay wRob
	nakaiana	oWaW oMid wNay
	'Nakaimo'	oWaW
va	'Nancy Lindsay'	wNay
un	'National Arboretum'	wNay
	'Neat and Tidy'	oWaW oMid wNay
va	'Neat Splash Rim'	wRob
	'New Wave'	oMid wNay
un	'Niagara Falls'	wNay
va	'Night before Christmas'	oWaW oMid oGar wTow oTwi wNay wRob oGre oBRG
	nigrescens	oWaW oJoy oMid wTow oTwi wNay wRob
va	'North Hills'	oJoy wNay
un	'North Pacific High'	wNay
un	'Northern Exposure'	oJoy oMid oGar wNay wRob wEde
va	Northern Halo™	oAls oMid oGar
un	'Northern Mystery'	wRob
va	Northern Star™	oMid
va	Northern Sunray™	oMid
	'Not So'	oWaW wNay
	'Obsession'	wRob
	'Ogon Koba'	wNay
un	'Ogon Tsushima'	wRob
un	'O'Harra'	wNay
	okamotoi	wTow
	'Olive Bailey Langdon'	oMid wNay
un	'Olympic Magic'	wNay
	'On Stage'	see **H.** 'Choko Nishiki'
	opipara	last listed 99/00
	'Opipara Koriyama'	oWaW wNay
	'Osprey'	see **H.** Tardiana Group 'Osprey'
	'Oxheart'	oWaW wNay oGre
	pachyscapa	wNay wRob
	'Pacific Blue Edger'	oWaW wNay wRob wAva
	'Pacific Lace'	wCol
	'Pacific Sunlight'	wNay wRob
	'Paintbrush'	wNay wRob
	'Painted Lady'	wNay
va	'Pandora's Box'	oMid wTow wNay wRob
va	'Paradigm'	oWaW wCol oMid wNay wRob oWnC
	'Paradise Joyce'	wNay
	'Paradise Power'	wTow
	'Pastures'	see **H.** 'Pastures New'
	'Pastures New'	oWaW wNay wRob
va	'Patrician'	wCol wNay wRob
va	'Patriot'	oFor oWaW oZOT wGAc oCap oAls oMid oGar wTow wNay wRob oGre oWnC wMag oBRG oHou wWhG
un	'Pauley'	wRob
	'Pauline Brac'	wNay
va	'Paul's Glory'	oFor oWaW oJoy oMid oGar wTow oTwi wNay wRob oWnC
va	'Peace'	oWaW oMid wTow oTwi wNay wRob
	'Pearl Lake'	oWaW oCap oMid oGar oTwi wNay oGre oHou wEde
	'Peedee Elfin Bells'	wNay

	'Peedee Gold Flash'	oWaW wRob
	'Pelham Blue Tump'	wNay wRob
	'Permanent Wave'	wNay wRob
	'Perry's True Blue'	oWaW wNay
	'Peter Pan'	oWaW
un	'Phantom'	oTwi wNay wRob
	'Photo Finish'	wRob
	'Piecrust Power'	wWoS
	'Piedmont Gold'	oJoy oMid oNWe wNay wRob
un	'Pilgrim'	wNay
	'Pineapple Poll'	wWoS oMid wNay
un	'Pink Panther'	wNay
	'Pioneer'	oJoy wNay
	'Pixie Power'	wNay wRob
va	'Pizzazz'	oFor oWaW oJoy oMid wNay wRob wMag oBRG oHou
	'Placemat'	wNay
	plantaginea	oFor oWaW oZOT oDar oJoy wCol wHig oAls oMid oGar wNay wRob wMag
	plantaginea var. ***grandiflora***	see **H.** ***plantaginea*** var. ***japonica***
	plantaginea var. ***japonica***	oAls wCSG wAva
va	'Platinum Tiara'	oMid oTwi wNay wRob
	'Pollyanna'	wNay
va	'Pooh Bear'	oMid wNay
	'Popo'	oWaW wNay wRob
	'Potomac Pride'	oAls
un	'Praying Hands'	wTow
	'President's Choice'	last listed 99/00
	'Princess of Karafuto'	oWaW wNay
	'Purbeck Ridge'	wNay
	'Purple and Gold'	oMid wNay
	'Purple Bouquet'	wNay
	'Purple Dwarf'	oWnC
	'Purple Lady Finger'	oWaW oTwi wNay wRob
un	'Purple Passion'	wTow
	'Purple Profusion'	wNay oHou
	pycnophylla	wTow
va	'Queen Josephine'	oWaW oMid wNay wRob
un	'Queen of Islip'	wNay
un	'Quill'	wNay
	'Quilted Hearts'	wNay
	'Quilted Skies'	wNay
	'Radiance'	last listed 99/00
va	'Radiant Edger'	oWaW oMid wNay wRob
	'Raleigh Remembrance'	oMid
va	'Rascal'	oMid wTow oTwi wNay
	'Raspberry Sorbet'	wWoS oMid wNay
un	'Recluse'	oGre
	rectifolia	wHer wNay oSle
	rectifolia 'Chionea'	wNay
	rectifolia HC 970472	wHer
va	'Regal Splendor'	oFor oWaW oJoy oMid wTow oTwi wNay wRob wEde
	'Reginald Kaye'	wNay
va	'Resonance'	oJoy wNay
va	'Reversed'	wNay
va	'Rhapsody'	oMid wNay wRob
un	'Rhapsody in Blue'	wNay
un	'Richland Blue'	wNay
	'Richland Gold'	oMid wNay
un	'Richmond'	wRob
un	'Richmond Blue'	wRob
un	'Rippled Honey'	wNay wRob
	'Rippling Waves'	oWaW wNay
	'Rising Sun'	oMid wRob
	'River Nile'	wNay
va	'Robert Frost'	wWoS oWaW oMid wTow wNay wRob
	'Robusta'	see **H.** ***sieboldiana*** var. ***elegans***
	'Rock Princess'	oSis
un	'Rocky Mountain High'	wRob
va	***rohdeifolia***	oWaW wNay
	'Rosanne'	last listed 99/00
	'Rough Waters'	wNay
	'Royal Quilt'	oJoy wNay
	'Royal Standard'	oFor oWaW oCap oAls oMid oGar oTwi oOut wNay wRob oGre oWnC wMag wSta oBRG oHou iArc oCir oLSG wWhG wSnq
	'Royalty'	oMid wNay
	'Ruffles'	oWaW
	'Rugosa'	last listed 99/00
	rupifraga	oWaW oMid wNay wRob
	rupifraga 'Urajiro'	see **H.** 'Urajiro Hachijo''
	'Ryan's Big One'	oGar oNWe
	'Sagae'	oFor oWaW oZOT oMid wTow oTwi wNay wRob wEde
	'Saint Elmo's Fire'	oFor oWaW oMid oTwi wNay
	'Saishu Jima'	oWaW oJoy
un	'Saishu Jima Closed'	wRob
un	'Saishu Yahato Sito'	wRob
	'Salute'	oWaW wTow wNay wRob
	'Samual Blue'	oWaW oMid wNay
va	'Samurai'	oJoy oGar wNay wRob
	'Satin Beauty'	oMid
	'Savannah'	oWaW wNay wRob
	'Scooter'	oWaW oMid wNay wRob
un	'Sea Angel'	wRob
un	'Sea Beacon'	wRob
	'Sea Bunny'	wNay wRob
un	'Sea Dragon'	wNay
va	'Sea Dream'	oWaW oMid wNay wRob
	'Sea Drift'	oMid wRob
	'Sea Fire'	oZOT oMid wNay
un	'Sea Gold Dust'	wNay wRob
	'Sea Gold Star'	wNay
un	'Sea Grotto'	oMid wNay
un	'Sea Hero'	wNay wRob wEde
	'Sea Lightning'	oWaW oMid
un	'Sea Lotus'	wMag
	'Sea Lotus Leaf'	oJoy oMid wNay wRob
va	'Sea Mist'	oWaW wRob
	'Sea Monster'	oMid wNay wRob
	'Sea Octopus'	wNay
	'Sea Sapphire'	oWaW oZOT oMid wNay wRob
un	'Sea Slate'	wNay
va	'Sea Thunder'	oWaW oMid oTwi wNay wRob oWnC
	'Sea Waves'	oJoy
	'Sea Wiggles'	wNay
	'Sea Yellow Sunrise'	oMid wNay
va	'Second Wind'	oWaW oMid wNay wRob
	'See Saw'	oJoy oMid wNay
	seedlings	wAva
va	'September Sun'	wWoS oWaW wCol oMid wNay wRob
un	'September Surprise'	wNay
	'Serendipity'	oWaW oRus oMid wNay
va	'Shade Fanfare'	oFor oWaW oZOT oCap oAls oMid oGar wNay wRob oBlo oWnC oHou wWhG
	'Shade Master'	oMid wNay
va	'Sharmon'	oFor wNay
	'Sherborne Profusion'	see **H.** Tardiana Group 'Sherborne Profusion'
	'Sherborne Songbird'	see **H.** Tardiana Group 'Sherborne Songbird'
	'Sherborne Swan'	see **H.** Tardiana Group 'Sherborne Swan'
	'Sherborne Swift'	last listed 99/00
	sp. aff. ***shikokiana*** HC 970566	wHer
un	'Shining Image'	wNay wRob
	'Shining Tot'	oWaW wNay
un	'Shiny Penny'	wNay wRob
va	'Shogun'	oWaW oJoy oGar oNWe oTwi wNay wRob
	'Showboat'	oMid wTow wRob
	sieboldiana	oMac oRus wCSG oGar oSho oOut oGre oSle
	sieboldiana var. ***elegans***	oFor oWaW oCap oDar oJoy oAls oMid oGar oDan wTow oTwi wNay oBlo oGre wFai wMag wSta oBRG wAva wWhG oUps
qu	***sieboldiana*** 'Gigantea'	oWaW
	sieboldiana var. ***mira***	oJoy wTow wRob
	sieboldii var. ***alba***	wNay
	sp. aff. ***sieboldii*** HC 970697	wHer
va	***sieboldii*** f. ***kabitan***	oWaW oZOT wNay wRob oHou
	sieboldii f. ***spathulata***	oMid wNay wRob
	'Silver Bowl'	wNay
un	'Silver Falls'	oRus
va	'Silver Lance'	last listed 99/00
	'Silverado'	wNay wRob
	'Silvery Slugproof'	see **H.** Tardiana Group 'Silvery Slugproof'
va	'Sitting Pretty'	wNay
	'Skookumchuck'	wNay wRob
va	'Snow Cap'	oMid wNay oBRG
va	'Snow Crust'	wNay wRob
va	'Snow Flakes'	oGar wNay oGre wMag oHou
	'Snow Mound'	last listed 99/00
va	'Snow White'	oMid
	'Snowbound'	wNay
	'Snowden'	oJoy wNay wRob
	'So Big'	wNay wRob
	'So Sweet'	oFor wWoS oWaW oJoy wCol oAls oMid oGar oTwi oOut wNay oGre oWnC wMag oBRG wEde wSnq
	'Soft Shoulders'	wTow wRob
	'Something Blue'	oWaW oMid oNWe wNay

un	'Southern Comfort'	oMid
	sp. DJHC 98421	wHer
	sp. HC 970644	wHer
	'Spacious Skies'	oFor oWaW
	'Sparkling Burgundy'	oWaW wNay wRob
	'Sparky'	wCol wRob
	'Spartan Gem'	wNay
	'Special Gift'	oWaW wNay
	'Spilt Milk'	oFor oMid wTow wNay wRob oWnC
	'Spinach Patch'	oWaW oMid wNay
un	'Spingarn's Japan'	wNay
va	'Spinners' ***(fortunei)***	wNay
un	'Splashed Leather'	wNay
	'Splish-Splash'	wNay
va	'Spritzer'	oWaW wNay wRob
	'Spun Sulphur'	wNay
	'Squash Casserole'	wTow
	'Starburst'	oMid wNay
un	'Stetson'	oTwi wNay wRob
va	'Stiletto'	oFor wWoS oWaW wCol oAls oMid oGar wNay wRob oHou
un	'Stirfry'	wNay wRob
va	'Striptease'	oFor oWaW oMid wTow wNay wRob
va	'Sugar and Cream'	oWaW oJoy oMid oGar wNay wRob oHou wSnq
	'Sugar Plum Fairy'	oJoy oAls wCSG wNay oHou
va	'Sultana'	oWaW oZOT oMid wTow oTwi wNay wRob oWnC
	'Sum and Substance'	oFor wWoS oWaW oZOT oCap oJoy oAls oMid oGar oDan wTow oNWe oSho oTwi wNay wRob oGre oWnC wMag wTGN oBRG oHou wAva wEde wWhG oUps oSle
un	'Sum It Up'	wNay wRob
	'Summer Fragrance'	oWaW oMid wNay wRob
un	'Summer Joy'	oFor wTow wNay wRob
va	'Summer Music'	oFor oWaW oMid oGar wTow oTwi wNay wRob oBRG
	'Sun Glow'	oWaW
	'Sun Power'	wWoS oWaW oZOT oJoy oMid oGar wTow wNay wRob
un	'Sunami'	wNay
va	'Sundance'	wNay wRob
	'Sunlight'	wNay
	'Sunlight Sister'	oFor oWaW oDan wNay wRob
	'Sunny Smiles'	oWaW wNay
	'Sun's Glory'	wRob
	'Super Bowl'	oMid wRob
va	'Super Nova'	oMid wNay wEde
un	'Surprised by Joy'	wNay
un	'Suzanne'	oAls wNay
un	'Sweet Bo Peep'	wNay wRob
	'Sweet Marjorie'	oMid wNay wRob
	'Sweet Standard'	oMid
un	'Sweet Sunshine'	oTwi wNay wRob
	'Sweet Susan'	wNay wRob wMag
	'Sweet Tater Pie'	oMid
	'Sweet Winifred'	wNay wRob
va	'Sweetie'	oFor oWaW wNay wRob
	'Swirling Hearts'	wRob
	'Swoosh'	wNay
qu	***takahashii*** 'Gosan'	wNay wRob
	'Tall Boy'	oJoy oMid wNay
	'Tall Twister'	wNay
va	'Tamborine'	oWaW oMid wTow wNay wRob
	Tardiana Group 'Blue Belle'	oMid wNay
	Tardiana Group 'Blue Blush'	oMid wNay wRob
	Tardiana Group 'Blue Danube'	oWaW wNay
	Tardiana Group 'Blue Diamond'	wNay wRob
	Tardiana Group 'Blue Dimples'	oWaW oMid wNay
	Tardiana Group 'Blue Moon'	oWaW oMid oTwi wNay wRob
	Tardiana Group 'Blue Skies'	wNay
	Tardiana Group 'Blue Wedgwood'	oWaW wCSG oGar wTow wRob oGre oHou wAva wEde
	Tardiana Group 'Bright Glow'	oWaW wNay
	Tardiana Group 'Brother Ronald'	oMid wNay wRob
	Tardiana Group 'Camelot'	oFor oWaW oMid wNay oGre oWnC
	Tardiana Group 'Devon Blue'	wNay wRob
	Tardiana Group 'Dorset Charm'	wNay wRob
	Tardiana Group 'Dorset Flair'	oWaW wNay wRob
	Tardiana Group 'Eric Smith'	oWaW wTow wRob
va	Tardiana Group 'Goldbrook Glimmer'	wRob
	Tardiana Group 'Hadspen Blue'	oWaW oCap oMid wNay wMag
	Tardiana Group 'Hadspen Hawk'	oWaW oMid wNay
	Tardiana Group 'Hadspen Heron'	oWaW oJoy wNay wRob

	Tardiana Group 'Halcyon'	oFor wWoS oWaW wSwC oCap oJoy wHig oAls oRus oGar wTow oNWe oTwi wNay wRob oGre oWnC oHou oCir wWhG wSnq
	Tardiana Group 'Happiness'	wNay
	Tardiana Group 'Harmony'	oMid wNay
va	Tardiana Group 'June'	oFor oWaW oZOT oJoy oAls oMid oGar wTow oTwi wNay wRob oGre
	Tardiana Group 'Osprey'	oMid oTwi wNay wRob
	Tardiana Group 'Sherborne Profusion'	wNay wRob
	Tardiana Group 'Sherborne Songbird'	wNay
	Tardiana Group 'Sherborne Swan'	wNay wRob
	Tardiana Group 'Silvery Slugproof'	wRob
	Tardiana Group 'Wagtail'	wNay wRob
	tardiflora	oWaW oRus wNay wRob
	tardiva	wRob
qu	***tardiva*** 'Gosan'	wNay
un	***tardiva striata***	wNay
un	'Tattle Tale Gray'	wNay
	'Tattoo'	wTow
	'Temple Bells'	oMid wNay wRob
	'Tenryu'	oWaW oMid wNay wRob
	'Thomas Hogg'	see **H.** ***undulata*** var. ***albomarginata***
	tibae	wNay
	'Tiddlywinks'	wRob
	'Tiny Tears'	wWoS oWaW oMid wMtT wNay wRob
	tokudama	wHig oAls wNay wMag
	tokudama f. ***aureonebulosa***	wWoS oWaW oMid oGar wNay wRob
	tokudama 'Carder'	oWaW oGar wNay oGre
va	***tokudama*** f. ***flavocircinalis***	wWoS oWaW oAls oMid oSho wNay wRob wMag oHou
	tokudama f. ***flavoplanata***	wNay
	'Tokudama Hime'	last listed 99/00
	'Tokudama Ogon Hime'	last listed 99/00
qu	***tokudama*** 'Zager'	wNay
va	'Torchlight'	oAls wNay wRob
	'Tortifrons'	wNay wRob
un	***tosana***	wNay
un	'Tossed Salad'	wNay
	'Tot Tot'	oWaW wNay wRob
	'Trail's End'	wNay
un	'Tranquility'	wNay
	'Treasure'	wNay wRob
	'True Blue'	oWaW wNay
	tsushimensis	last listed 99/00
	'Twilight'	oGar wNay wRob
va	'Twist of Lime'	oWaW wNay wRob
	'Uguis'	wNay wRob
	'Ultraviolet Light'	wNay wRob
	undulata	wHig wMag
	undulata var. ***albomarginata***	oFor oDar oGar oTwi wNay
	undulata var. ***erromena***	wNay
	undulata 'Mediovariegata'	see **H.** ***undulata*** var. ***undulata***
va	***undulata*** var. ***undulata***	oCap oAls oGar oOut oGre oHou iArc oSle wSnq
va	***undulata*** var. ***univittata***	wNay
	undulata 'Variegata'	see **H.** ***undulata*** var. ***undulata***
	'Urajiro Hachijo'	wRob
	'Uzu No Mai'	wCol wTow wNay wRob
	'Valentine Lace'	oWaW oMid wNay
un	'Valerie's Vanity'	oMid wNay
un	'Van Wade'	oWaW oTwi wNay wRob
	'Vanilla Cream'	oWaW oTwi wNay oHou
	variegated	oAls
	ventricosa	oWaW oAls oGar oTwi wNay oLSG
	ventricosa var. ***aureomaculata***	oWaW wNay
	ventricosa aureomarginata	see **H.** 'Aureomarginata' ***(ventricosa)***
	'Venucosa'	oMid wNay
	'Venus'	oWaW wRob
	venusta	wHig oRus oTwi wNay
	venusta dwarf form	oSis
	venusta HC 970249	wHer
	venusta 'Portor'	oWaW
va	'Vera Verde'	oJoy oMid wNay
	'Verkade's No. 1'	wNay
va	'Verna Jean'	oWaW wTow wNay wRob
va	'Veronica Lake'	oWaW wCol oAls oMid wTow oTwi wNay wRob
un	'Versailles Blue'	wNay
va	'Viette's Yellow Edge'	wNay
	'Wagon Wheels'	wNay
	'Wagtail'	see **H.** Tardiana Group 'Wagtail'
va	'Wahoo'	wRob
	'Walden Green'	oWaW
	'Warwick Cup'	wNay
un	'Warwick Curtsey'	wNay wRob

un	'Warwick Delight'	wNay wRob
un	'Warwick Edge'	wNay wRob
	'Warwick Essence'	wNay
	'Waterford'	wNay
qu	'Waving Ruffles'	wMag
va	'Waving Winds'	wRob
un	'Waving Wuffles'	wNay
	'Wheaton Blue'	oAls oMid oGar wNay wRob wEde
	'Whirligig'	wRob
va	'Whirlwind'	oWaW oMid wTow oTwi wNay wRob
un	'White Border'	wNay
	'White Caps'	wTow
va	'White Christmas'	oMid oGar wNay wRob
	'White Edger'	wNay
	'White Knight'	wNay
	'White On'	oFor oWaW wRob
	'White Ray'	oWaW oMid
	'White Vision'	oWaW oMid
va	'Wide Brim'	oWaW oZOT oCap oDar oJoy oAls oMid oGar oTwi wNay wRob oBlo oGre oWnC oBRG oHou wEde oLSG wWhG oSle wCoS
	'Wild River Gold'	see **H.** 'Wind River Gold'
un	'Willy Nilly'	wWoS oWaW oMid wNay wRob
	'Wind River Gold'	oWaW wCol wNay wRob
	'Winfield Blue'	wRob
	'Winning Edge'	oMid
	'Witches Brew'	wNay
un	'Wolverine'	wTow oTwi wNay wRob
	'Woodland Green'	wNay
	'Wrinkles and Crinkles'	wNay wRob
qu	'Yakushima'	wRob
	'Yakushima-mizu' ***(gracillima)***	oMid wNay wRob
un	'Yellow Bird'	wRob
	'Yellow Boa'	oWaW
va	'Yellow River'	oWaW oMid wTow oTwi wNay wRob
va	'Yellow Splash'	oRus
va	'Yellow Splash Rim'	oMid wNay wRob oWnC
un	'Yellow Submarine'	oWaW
	'Yellow Waves'	wRob
	yingeri	oWaW wCol wNay
va	'Zager White Edge'	oWaW oMid wNay
	'Zounds'	oFor oWaW oMid wTow oTwi wNay wRob oGre oHou wAva wCoS
	HOUTTUYNIA	
	cordata	oTrP wGAc oHug oGar oBlo wRai iGSc oWnC wCoN
va	***cordata*** 'Chameleon'	wWoS oDar oAls wCSG oGar oSsd oDan iGSc wHom oTri wTGN oHou oUps wCoS
do	***cordata*** 'Flore Pleno'	oFor oGar oUps
	cordata Variegata Group	oFor oHug oWnC
	HOVENIA	
	dulcis	oFor oRiv oOEx oGre wTGN
	dulcis HC 970240	wHer
	HOWEA	
	forsteriana	oGar
	HOYA	
	carnosa	oGar
	HUERNIA	
un	***beniensis***	oRar
	clavigera	oRar
	hislopii	oRar
	loeseneriana	oRar
	longituba ssp. ***cashalensis*** DMC 41588	oRar
	macrocarpa ssp. ***concinna*** Barad-11634D	oRar
	pendula ex IAS 58	oRar
	primulina	oRar
	saudi-arabica	oRar
	somalica	oRar
	sp. L 13103	oRar
	stapelioides	oRar
	verekeri PD 071	oRar
	zebrina PRA 47P	oRar
	HUMULUS	
	lupulus	wFGN oGoo wFai iGSc wHom wCoN oSle
	lupulus 'Aureus'	wCul oJoy wCSG oGar oGre wFai wPir wHom wTGN oInd wCoN wWel
un	***lupulus*** 'Bullion'	oFor
un	***lupulus*** 'Cascade'	oFor oGar oFrs wHom
un	***lupulus*** 'Centennial'	oFrs
un	***lupulus*** 'Chinook'	oFrs
	lupulus 'Fuggle'	oFrs
	lupulus 'Hallertauer'	oFrs
un	***lupulus*** 'Kent Golding'	oFrs
un	***lupulus*** 'Liberty'	wMag

un	***lupulus*** 'Magnum'	oFrs
un	***lupulus*** 'Mt. Hood'	oFrs
un	***lupulus*** 'New Zealand Hallertauer.	oFrs
un	***lupulus*** 'Northern Brewer'	oFrs
un	***lupulus*** 'Nugget'	oFor oGar oFrs
un	***lupulus*** 'Pacific Gem'	oFrs
un	***lupulus*** 'Perle'	oFrs
un	***lupulus*** 'Saaz'	oFrs
un	***lupulus*** 'Sterling'	oFrs
un	***lupulus*** 'Willamette'	oAls oGar oFrs
ch	**HUTCHINSIA**	see **PRITZELAGO**
	HYACINTHOIDES	
	hispanica	wRoo
	non-scripta	wBox
	HYACINTHUS	
	orientalis 'Amethyst'	wRoo oWoo
	orientalis 'Amsterdam	wRoo
	orientalis 'Anna Liza'	wRoo
	orientalis 'Bismarck'	wRoo
un	***orientalis*** 'Blue Ice'	oGar
	orientalis 'Blue Jacket'	wRoo oWoo
	orientalis 'Carnegie'	wRoo
	orientalis 'City of Haarlem'	wRoo oGar
	orientalis 'Delft Blue'	oGar oWoo
	orientalis 'Fondant'	oGar
	orientalis 'Gipsy Queen'	wRoo oGar
do	***orientalis*** 'Hollyhock'	oWoo
	orientalis 'Jan Bos'	oGar oWoo
	orientalis 'King of the Blues'	wRoo oWoo
	orientalis 'Lady Derby'	oWoo
	orientalis 'Peter Stuyvesant'	oGar
	orientalis 'Pink Pearl'	wRoo oGar oWoo
	orientalis 'Queen of the Pinks'	wRoo
	orientalis 'White Pearl'	wRoo oWoo
	'Woodstock'	oGar
	HYDRANGEA (SEE NOMENCLATURE NOTES)	
	anomala ssp. ***anomala*** DJHC 792	wHer
	anomala ssp. ***anomala*** HWJCM 010	wHer
	anomala ssp. ***petiolaris***	oFor wCul wClo oDar oAls oHed wCSG oGar oDan oRiv oSho oOut oGre oWnC wTGN wSta oBel oBRG oCir wWhG oUps wSnq wCoS wWel
	anomala ssp. ***petiolaris*** HC 970226	wHer
un	***anomala*** ssp. ***petiolaris*** 'Mirranda'	oGre oBRG
	arborescens	oGoo
	arborescens 'Annabelle'	oFor wWoS wCul oGos wCou wClo oDar oJoy oGar oOut oGre oWnC wMag oBel wWhG wSnq wWel
	arborescens 'Grandiflora'	wCou
	arborescens ssp. ***radiata***	last listed 99/00
	aspera	oFor oGre wTGN wSte
	aspera Kawakamii Group	wHer wWoS
	aspera 'Macrophylla'	wCul oGos wSwC oAls wSnq
	aspera 'Mauvette'	oGos
	aspera ssp. ***robusta***	oAls
	aspera 'Rocklon'	wHer oAmb
	aspera ssp. ***sargentiana***	wHer oGar oRiv oWhi oWnC wBox oBel wWel
	aspera ssp. ***strigosa***	oGar
	aspera ssp. ***strigosa*** DJHC 721	wHer
	aspera Villosa Group	wWoS oGos wClo oAls oHed oAmb oGar oRiv oNWe oOut wRob oGre oWnC oBel wSnq
	aspera Villosa Group DJHC 594 B	wHer
	heteromalla	oFor wCCr
	heteromalla Bretschneideri Group	wCol
	heteromalla DJHC 493 (last listed as DJHC 793)	wHer
	heteromalla HWJCM 180	wHer
un	***hirta***	wHer
	integrifolia	wHer
	involucrata	wHer wCou wCCr
	involucrata x ***aspera***	wHer
	involucrata dwarf form	wHer
	involucrata HC 970542	wHer
do	***involucrata*** 'Hortensis'	wHer
un	***involucrata*** 'Plena'	wHer
un	***involucrata*** 'Tama Aziasi'	wCou oBel
un	***involucrata*** 'Yokudanka'	wHer
	lobbii	wHer
	luteovenosa	wCCr
	luteovenosa HC 970561	wHer
	luteovenosa 'Variegata'	wHer
	macrophylla	wShR wCSG oSho wMag wBox wSta
un	***macrophylla*** 'Adria' (H)	wClo

	macrophylla 'All Summer Beauty' (H)	oJoy
	macrophylla 'Alpengluehen' (H)	oFor oJoy oWnC wMag oBel wSnq
	macrophylla 'Altona' (H)	oFor wCou wCSG oBel
do	***macrophylla*** 'Amethyst' (H)	oFor wCul wCou wCSG oWnC wBox wWhG wSnq wWel
	macrophylla 'Ami Pasquier' (H)	oFor wCul wCou
	macrophylla 'Ayesha' (H)	oFor wHer wCul wCou wClo oNat oJil oJoy oAls oHed wCSG oAmb oGar oDan oNWe oSho oOut wRob oWnC wMag wTGN oBel oBRG wWhG
	macrophylla 'Beaute Vendomoise' (L)	wHer wCou
	macrophylla 'Blauer Prinz' (H)	wCou
	macrophylla 'Blaumeise' (L)	oFor wCul wCou oJoy oAls oGar oWnC wBox oBel wWel
	macrophylla blue (H)	oGar
	macrophylla 'Blue Billow' (L)	oGos wCou oSis oSev oGre oBel wWhG wSnq
un	***macrophylla*** 'Blue Danube' (H)	wHer wCou wWhG
	macrophylla Blue Prince (H)	see **H.** ***macrophylla*** 'Blauer Prinz'
	macrophylla 'Blue Wave' (L)	see **H.** ***macrophylla*** 'Mariesii Perfecta'
	macrophylla 'Blushing Pink' (H)	oFor
	macrophylla 'Bodensee' (H)	wHer oGar
un	***macrophylla*** 'Bottstein' (H)	wCou oJoy
	macrophylla 'Bouquet Rose' (H)	oGre
	macrophylla Buttons 'N Bows™ / 'Monrey'	oAls oSho wWel
un	***macrophylla*** 'Coerulea' (L)	wHer oBel
do	***macrophylla*** 'Domotoi' (H)	oFor wCou oJoy oAls oGar oGre wBox oGue
un	***macrophylla*** 'Dooley'	oDar
	macrophylla 'Enziandom' (H)	oJoy oGar oWnC wBox
	macrophylla 'Europa' (H)	oFor wCou wCSG
	macrophylla Fasan ™ (L)	wWoS wCou oDar wBox
	macrophylla 'Firelight' (H)	see **H.** ***macrophylla*** 'Leuchtfeuer'
	macrophylla 'Forever Pink' (H)	oAls oGar oWnC
un	***macrophylla*** 'French Royal Blue'	wCSG
un	***macrophylla*** 'Freudenstein' (L)	wHer oJoy
	macrophylla 'Frillibet' (H)	wHer
	macrophylla 'Gartenbaudirektor Kuhnert' (H)	wCou oJoy oGar oGre oWnC wMag oUps wWel
	macrophylla 'Generale Vicomtesse de Vibraye' (H)	oFor wCul wCou wBox
	macrophylla 'Geoffrey Chadbund' (L)	see **H.** ***macrophylla*** 'Moewe'
	macrophylla 'Gertrud Glahn' (H)	wCou oJoy oGar oGre wBox
	macrophylla 'Glowing Embers' (H)	oFor wHer wClo oNat oDar oAls oGar oSho oGre oWnC wTGN wSta oCis oBRG oGue wWhG wWel
	macrophylla 'Goliath' (H)	wCou oJoy oGre oWnC wMag
un	***macrophylla*** 'Hadsbury' (L)	wHer
	macrophylla 'Hamburg' (H)	wCul wCou oGar oGre oWnC wMag wSnq
	macrophylla 'Harlequin' (H)	wHer
un	***macrophylla*** 'Heavenly Blue' (H)	oNat
	macrophylla 'Heinrich Seidel' (H)	wCul wCou oGre wMag
un	***macrophylla*** 'Hobella'	oGre wCoS wWel
	macrophylla 'Hoernli' (H)	wCul wCou
	macrophylla 'Holstein' (H)	wCou wBox
un	***macrophylla*** 'Hovaria'	wMag
do	***macrophylla*** 'Izu-no-hana' (L)	wHer
un	***macrophylla*** 'Jennifer' (H)	oJoy
un	***macrophylla*** 'Jogasaki' (L)	wHer
	macrophylla 'Kardinal' (L)	wWel
	macrophylla 'Kluis Superba' (H)	wCul wCou wCSG wSnq
	macrophylla 'La France' (H)	wCou oGre
un	***macrophylla*** 'La Marne' (H)	wCou
	macrophylla 'Lanarth White' (L)	oFor wHer wCul wCri wCou oJoy oAls oHed wCSG oAmb oGar oGre oWnC wBox oBRG wWel
	macrophylla 'Le Cygne' (H)	oFor wHer
un	***macrophylla*** 'Lemon Wave' (L)	wWoS oGos oJoy oOut oGre oWnC oBel wWhG
	macrophylla 'Leuchtfeuer' (H)	oJoy
	macrophylla 'Libelle' (L)	wHer wCou oJoy oGre wBox oBel
	macrophylla 'Lilacina' (L)	oFor wCul wCou oJoy wCSG oDan oGre wWhG wWel
	macrophylla var. ***macrophylla***	see **H.** ***macrophylla***
	macrophylla 'Maculata' (L)	oFor oAmb wKin oWnC wWhG
un	***macrophylla*** 'Madame Baardse' (H)	oGar oGre
	macrophylla 'Madame Emile Mouillere' (H)	wWoS wCul wCou oNat oJoy oHed oAmb oGar oGre oWnC oGue wWhG wSnq wWel
un	***macrophylla*** 'Madame Faustin Travouillon' (H)	wCul wCou
un	***macrophylla*** 'Mandschurica' (H)	wHer wCou
	macrophylla 'Marechal Foch' (H)	wCul wCou wCSG oGre
	macrophylla 'Mariesii' (L)	oFor wCou oNat oJoy wCSG oGar oGre oBRG oGue wWel
	macrophylla 'Mariesii Perfecta' (L)	wCul wCou oDar oJoy wCSG oMis oGar oOut oGre oBel wWhG wSnq wWel
	macrophylla 'Mariesii Variegata' (L)	wWoS oNat oJoy wCol oAls oDan oSis oWnC oBRG oUps wSnq wCoS wWel
	macrophylla 'Masja' (H)	wHer wWoS oJoy oAls oGre
	macrophylla 'Mathilda Gutges' (H)	wCou oAls wCSG oGar wMag
un	***macrophylla*** 'Merritt's Beauty'	oAls oGre oBRG wCoS wWel
un	***macrophylla*** 'Merritt's Blue' (H)	wHer oAls oGar oWnC wWel
un	***macrophylla*** 'Merritt's Pride' (H)	oGar
un	***macrophylla*** 'Merritt's Supreme' (H)	oFor oGar oOut wTGN oUps wWel
un	***macrophylla*** 'Merveille' (H)	oJoy oGar oWnC
	macrophylla 'Miss Belgium' (H)	wHer wCul oGos wCou oGre oWnC wMag oBRG
	macrophylla 'Moewe' (L)	wHer wWoS wCul wCou oGar oGre oCis
	macrophylla 'Mousmee' (H)	oFor oJoy oGre oBel
un	***macrophylla*** 'Mousseline' (H)	wCou
un	***macrophylla*** 'Nachtigall' (L)	wHer wWoS oAmb oGar
	macrophylla 'Niedersachsen' (H)	wCul wCou
	macrophylla 'Nightingale' (L)	see **H.** 'Nachtigall'
	macrophylla 'Nigra' (H)	oFor oJil oDar oAls oHed oAmb oGar wRob oGre oWnC wMag wTGN wBox wWel
	macrophylla 'Nikko Blue' (H)	wCou oDar oJoy oAls wShR wCSG oGar oGoo oSho oGre wKin oWnC wMag wTGN oCir oGue wWhG wSnq wCoS wWel
ch	***macrophylla*** var. ***normalis***	oAmb oWnC
	macrophylla var. ***normalis*** blue	oUps
	macrophylla var. ***normalis*** pink	oUps
un	***macrophylla*** 'Ogonba-Azisai' (H)	wHer
un	***macrophylla*** 'Oregon Pride'	oGre wWel
un	***macrophylla*** 'Oregon Spring'	oOut
	macrophylla 'Otaksa' (H)	oGar wMag wBox
un	***macrophylla*** 'Paris' (H)	oJoy oAls oGar oSho
	macrophylla 'Parzifal' (H)	wHer wCou
	macrophylla 'Pia'	see **H.** ***macrophylla*** Pink Elf ® / 'Pia'
qu	***macrophylla piamina*** 'Winning Edge'	oGre
un	***macrophylla*** 'Piaminia' (H)	wCou wSnq
un	***macrophylla*** 'Pieta' (H)	oSis
	macrophylla pink	oGar
un	***macrophylla*** 'Pink Beauty' (H)	oFor
	macrophylla Pink Elf ® / 'Pia' (H)	oFor oTrP wHer wWoS oGos oJoy oAls oGar oSis oSho wRob oSev oGre oWnC wMag wBox wWhG oUps wCoS wWel
un	***macrophylla*** 'Pink Lace' (H)	oFor
	macrophylla Pink 'N Pretty™ / 'Monink' (H)	oGar oGre oWnC wCoS
	macrophylla 'Preziosa'	see **H.** 'Preziosa'
va	***macrophylla*** 'Quadricolor' (L)	wHer wCul
	macrophylla Red 'N Pretty tm / 'Monred'	oAls oWnC
un	***macrophylla*** 'Red Star'	oGre oBRG wWel
	macrophylla 'Regula' (H)	oJoy oAls oGre oBRG
un	***macrophylla*** 'Revelation' (H)	oSis
un	***macrophylla*** 'Rose Supreme' (H)	oJoy
un	***macrophylla*** 'Rotdrossel' (L)	wHer
un	***macrophylla*** 'Royal Purple' (H)	oFor
un	***macrophylla*** 'Schenkenburg' (H)	last listed 99/00
	macrophylla 'Sea Foam' (L)	last listed 99/00
un	***macrophylla*** 'Sensation' (H)	oGar wWhG
	macrophylla Sister Theresa	see **H.** ***macrophylla*** 'Soeur Therese'
un	***macrophylla*** 'Skips' (H)	wHer
	macrophylla 'Soeur Therese' (H)	wHer oJoy oGar oGre oWnC wBox wWel
	macrophylla 'Souvenir du President Paul Doumer' (H)	oFor wCul wCou wBox wWhG
	macrophylla 'Taube' (L)	oFor wHer wCou oJoy oAmb oGar wWel
	macrophylla Teller Blau (L)	oGre
	macrophylla Teller Blue	see **H.** ***macrophylla*** Teller Blau
	macrophylla Teller Pink	see **H.** ***macrophylla*** Teller Rosa
	macrophylla Teller Red	see **H.** ***macrophylla*** Teller Rot
	macrophylla Teller Rosa (L)	oGre
	macrophylla Teller Rot (L)	oFor oGos oGre
	macrophylla Teller White	see **H.** ***macrophylla*** 'Libelle'
	macrophylla 'Toedi' (H)	oJoy oGar
	macrophylla 'Tokyo Delight' (L)	oFor wWoS oJoy wCCr wWhG
	macrophylla 'Tovelit'	oGue
	macrophylla 'Tricolor' (L)	wCou oMis oGar oOut oWnC oSec oBel oGue
	macrophylla 'Trophee' (H)	wCou oJoy
	macrophylla 'Variegata'	see **H.** ***macrophylla*** 'Maculata'
	macrophylla 'Variegata Wave Hill'	see **H** ***macrophylla*** 'Wave Hill'
	macrophylla 'Veitchii' (L)	oFor wHer wCou oNat oGre wSnq
un	***macrophylla*** 'Wave Hill' (L)	wHer oJoy oOut oGre

un	macrophylla 'Weidner's Blue' (L)	oAls oGar oWnC
un	macrophylla 'Weidner's Pink' (L)	oAls oGar wTGN
	macrophylla white	oAls oGar oGre wWel
	macrophylla 'White Swan' (H)	see H. macrophylla 'Le Cygne'
	macrophylla 'White Wave' (L)	wCul wCou oJoy
un	macrophylla 'Winning Edge'	oFor
	macrophylla yellow leaf	wWel
	macrophylla 'Zaunkoenig'	wCul wCou
	paniculata Angel's Blush® / 'Ruby'	oGar oWnC wWhG wCoS wWel
	paniculata B&SWJ 3802	wHer
	paniculata 'Brussels Lace'	wHer wWoS oGos oGre
	paniculata 'Burgundy Lace'	oGos oGar oBel
	paniculata 'Floribunda'	wHer
	paniculata 'Grandiflora'	oFor wWoS wCul wCou wClo oDar oAls oGar oGoo oRiv oBlo oGre wKin oWnC wMag wTGN oBel oBRG oGue oSle wWel
	paniculata 'Grandiflora' tree form	wCul
	paniculata HC 970618	wHer
	paniculata 'Kyushu'	wHer wCul wSnq wWel
	paniculata Pee Gee	see H. paniculata 'Grandiflora'
un	paniculata 'Peewee'	oFor wHer
	paniculata Pink Diamond / 'Interhydia'	oFor wWoS oGos oDar oGar wTGN wWel
	paniculata 'Ruby'	see H. paniculata Angel's Blush® / 'Ruby'
	paniculata 'Tardiva'	oFor oGos oDar oAls oGar oDan oOut oGre wMag oBel oBRG oSle wWel
	paniculata tree form	oBel
	paniculata 'Unique'	oFor wHer oAmb oGar oGre oBel wSnq wWel
	paniculata 'White Lace'	wHer
	paniculata 'White Moth'	oFor wHer oGos oGre
	petiolaris	see H. anomala ssp. petiolaris
	'Preziosa'	oFor wHer wCul wCou oNat oDar oJoy oAls oHed wCSG oAmb oGar oNWe wRob oGre oWnC wMag oBel oBRG wWhG wSnq wWel
qu	'Preziosa Pink Beauty'	oGar
	quelpartensis HC 970252	wHer
	quercifolia	oFor wCul wCou oNat oAls oMis oGar oDan oRiv oSho oBlo oGre wFai wKin oSec oBel oThr wWhG oSle wSte wSnq
un	quercifolia 'Alice'	oFor wHer oDar oEdg oWnC oUps wWel
	quercifolia 'Flore Pleno'	see H. quercifolia Snow Flake™
	quercifolia 'Harmony'	wCou
	quercifolia 'Pee Wee'	oFor wSwC wClo oEdg wCSG oGre wSnq wWel
	quercifolia 'Sike's Dwarf'	wHer oGos oDar oAmb oGre wTGN
do	quercifolia Snow Flake™	oFor wCul oGos wCou oHed wTGN wCoS wWel
	quercifolia Snow Queen / 'Flemygea'	oFor wHer oGos oDar oAmb oGar oGre wWel
	robusta	oAls wCCr
	sargentiana	see H. aspera ssp. sargentiana
	scandens ssp. liukiuensis	last listed 99/00
	seemannii	oFor wHer
	serrata	wCri oAls oBel
	serrata 'Acuminata'	see H. serrata 'Bluebird'
un	serrata 'Akishino Temari'	wHer
un	serrata 'Amagi Amacha'	wHer
	serrata 'Beni-gaku'	wHer oGos wClo oJoy oAls oAmb oGar oGre wTGN oBel wSnq wWel
	serrata 'Blue Deckle' (L)	oFor wHer oGar
un	serrata 'Blue Lace'	wCul oGre
	serrata 'Bluebird'	oFor wCul oGos wCou oJoy oAls oGar oDan oSis wRob oWnC wTGN oBel wSnq
un	serrata 'Chandelier'	wCol
	serrata chinensis	wHer
	serrata 'Grayswood'	oFor wCul wCou oJoy oGre wBox wSnq wWel
un	serrata 'Hanabi'	wHer
un	serrata 'Hinyosumi'	oGar
un	serrata 'Iyo-Shiboro'	wHer
	serrata 'Jogasaki'	last listed 99/00
	serrata 'Kiyosumi'	wHer wTGN
	serrata koreana	last listed 99/00
un	serrata 'Midoriboshi-Temari'	wHer
	serrata 'Miranda' (L)	wHer
do	serrata 'Miyama-yae-murasaki'	wHer
	serrata 'Preziosa'	see H. 'Preziosa'
	serrata 'Rosalba'	oGos
va	serrata 'Shichidanka-nishiki'	wHer oHed
un	serrata 'Shinonomei'	wHer
	serrata 'Shirofugi'	wHer
do	serrata 'Shirotae'	wHer
un	serrata 'Shishiba'	wHer
un	serrata 'Tama Azisai'	wBox
	serrata 'Tiara'	wHer oHed
	serrata 'Uzu Azisai'	wHer
un	serrata 'Yae no Amacha'	wHer
	serratifolia HCM 98056	wHer
	sikokiana HC 970689	wHer
	villosa	see H. aspera Villosa Group
	HYDRASTIS	
	canadensis	oTrP wWoS oGoo wFai iGSc oCrm oBar
	HYDROCHARIS	
	morsus-ranae	oSsd oWnC
	HYDROCLEYS	
	commersonii	see H. nymphoides
	nymphoides	wGAc oHug oSsd oWnC
	nymphoides giant	oSsd
	HYDROCOTYLE	
	asiatica	see CENTELLA asiatica
un	bonariensis	wGAc
	ranunculoides	oTri
un	sibthorpioides 'Crystal Confetti'	oHug
un	sieboldii 'Variegata'	wHer
	sp.	oHug
	umbellata	oHug oGar
	verticillata	wSoo wGAc oWnC
	HYDROPHYLLUM	
un	tenuipes	wNot oAld wFFl
un	tenuipes 'Apparition'	wCol
	HYGROPHILA	
un	polysperma 'Tropic Sunset'	oGar
un	**HYLLOSTACHYS**	
	nigra	oOEx
un	**HYLOCOMIUM**	
	splendens	wFFl
	HYMENANTHERA	see MELICYTUS
	HYMENOCALLIS	
	x festalis	oGar
	liriosome	wGAc
	narcissiflora	oVBI
	HYPERICUM	
	anagalloides	oBos
	androsaemum	oFor wCul wCSG oBov
	androsaemum 'Albury Purple'	oFor wWoS wCul wSwC oNat oJil oJoy oAls oHed wCSG oGar oNWe wRob oGre wTGN oCis oUps wWel
	androsaemum 'Glacier'	oFor wSwC oNat oJoy oDan wRob oGre wHom
va	androsaemum f. variegatum 'Mrs. Gladis Brabazon'	wHer wWoS oGar oNWe oSho
	balearicum	wCCr
	bellum pale form	wHer
	calycinum	oAls wCSG oSho oWnC wSta oCir wCoN oSle wCoS
	delphicum	wCul
qu	empetrifolium nanum	oJoy oAls oWnC
	empetrifolium ssp. oliganthum	oSis
	forrestii	oFor wCCr
un	fortuneanum 'Purple Fountain'	oFor oGos oCis
	fragile	see H. olympicum f. minus
	frondosum	oFor oNWe
	frondosum Sunburst™	oAls oGar oCis
	'Hidcote'	oFor wCul oJoy oGar oBlo oBov oGre wCoN wWel
	hookerianum	oFor
	x inodorum 'Elstead'	oFor wCul oJoy wCSG
va	x inodorum 'Summergold'	oFor wHer wWoS wCul oJoy oHed oGar oDan oSis oNWe oGre wHom wMag
	kalmianum	oFor oGar
un	kalmianum 'Ames'	wWoS
	kamtschaticum	last listed 99/00
	kiusianum var. yakusimense	last listed 99/00
	kouytchense	oFor wWoS
	lancasteri	oFor oHed
	x moserianum	oBlo oLSG
va	x moserianum 'Tricolor'	oFor wWoS oJoy oAls wCSG oGar oDan oGre oCir oUps
	oblongifolium	oFor
	olympicum	wCul oJoy
	olympicum f. minus	oSis oWnC
	olympicum f. minus 'Variegatum'	oHed
qu	patulum 'Variegatum'	oFor oJoy oGre oCis
	perforatum	oAls iGSc oWnC
un	perforatum 'Medizinal'	oCrm oBar
un	perforatum 'Topas'	oGoo oUps
	prolificum	last listed 99/00
	pseudopetiolatum var. yakusimense	see H. kiusianum var. yakusimense

	'Rowallane'	wHer wCSG wCCr
	sp. DJHC 584	wHer
	'Sungold'	see **H. *kouytchense***
	yezoense	last listed 99/00
	HYPOCYRTA	see **NEMATANTHUS**
	HYPOESTES	
va	***phyllostachya***	oGar
	HYPOXIS	
	hirsuta	oRus oSis
	HYSSOPUS	
	officinalis	oFor oAls oGoo wFai wCCr iGSc oWnC wNTP wCoN oUps
	officinalis f. ***albus***	iGSc
	officinalis ssp. ***aristatus***	wCul oGoo oBar
	officinalis dwarf blue	oCir
	officinalis dwarf pink	oCir
qu	***officinalis*** 'Nana'	wBox
	officinalis pink	oAls
un	***officinalis*** 'Pink Delight'	oFor wFGN
un	***officinalis*** 'Pink Sprite'	iGSc
	officinalis ssp. ***roseus***	wFGN oGoo
	officinalis 'Sissinghurst'	wFGN
un	***seravachanicus***	oNWe
	HYSTRIX	
	patula	oFor oWnC
	IBERIS	
	candolleana	see **I. *pruitii*** Candolleana Group
	gibraltarica	oNWe
	hybrid pink	oAls
	jucunda	see **AETHIONEMA *coridifolium***
	pruitii Candolleana Group	oSis
	saxitilis 'Pygmaea'	see **I. *sempervirens*** 'Pygmaea'
un	***sayana***	oSis
	sempervirens	wTho oGar wMag wBox wSta oCir wWhG
un	***sempervirens*** 'Alexander's White'	oOut oWnC oUps
	sempervirens 'Autumn Snow'	oSis
	sempervirens 'Little Gem'	see **I. *sempervirens*** 'Weisser Zwerg'
	sempervirens 'Purity'	oAls oGar oWnC oEga
	sempervirens 'Pygmaea'	oGre oWnC
	sempervirens 'Schneeflocke'	oGar oSho oWnC wMag wCoN wCoS
	sempervirens Snowflake	see **I. *sempervirens*** 'Schneeflocke'
	sempervirens 'Snowmantle'	oLSG
	sempervirens 'Weisser Zwerg'	oGar oBlo oSqu oGre wMag oCir wCoN oLSG wCoS wWel
un	'Snow White'	wMag iArc
	IBERVILLEA	
un	***fusiformis*** BLM 0391	oRar
un	***hypoleuca*** BLM 0808	oRar
	lindheimeri BLM 0257	oRar
un	***maxima*** BLM 0661	oRar
	sonorae BLM 0040	oRar
un	***sonorae*** var. ***peninsularis***	oRar
un	***tenuisecta*** BLM 0060	oRar
ch	**IBOZA**	see **TETRADENIA**
un	**ICHNOLEPIS**	
	tuberosa	oRar
	IDESIA	
	polycarpa	oFor oTrP oGar oRiv oSho oGre wCCr oWnC wBox
ch	**IDRIA**	see **FOUQUIERIA**
	ILEX	
	x altaclerensis 'Eldridge'	wCCr
va	***x altaclerensis*** 'Golden King' female	wCCr
	x altaclerensis 'Hendersonii'	wCCr
qu	***x altaclerensis*** 'Nobilis Picta'	wCCr
	x altaclerensis 'Wilsonii' female	last listed 99/00
	'Apollo' male	oFor wHer
	aquifolium	oWnC
	aquifolium 'Alaska' female	oFor
	aquifolium 'Angustifolia' female	oRiv oBlo wCCr wSta
va	***aquifolium*** 'Argentea Marginata' female	oAls oGar oWnC wCoS
un	***aquifolium*** 'Aureopicta'	wCCr
	aquifolium 'Bacciflava' female	oFor wCCr
un	***aquifolium*** 'Beaconwood'	wCCr
un	***aquifolium*** 'Beautyspra'	oFor
	aquifolium 'Ciliata Major' female	oFor oGar
un	***aquifolium*** 'Crispa Aurea'	oFor
	aquifolium 'Crispa' male	oRiv
	aquifolium female	oGar
va	***aquifolium*** 'Ferox Argentea' male	oFor oGar wCCr wSta
	aquifolium 'Ferox Argentea Marginata'	see **I. *aquifolium*** 'Ferox Argentea'
	aquifolium 'Ferox Aurea'	last listed 99/00
	aquifolium 'Flavescens' female	wHer
un	***aquifolium*** 'Gold Butterfly'	oFor
	aquifolium Gold Coast r / 'Monvila'	oGar
va	***aquifolium*** 'Golden Milkboy' male	wHer
	aquifolium 'Hastata' male	wHer
	aquifolium 'Little Bull'	last listed 99/00
un	***aquifolium*** 'Osgood'	oFor
un	***aquifolium*** 'Phantom Gold'	oFor
	aquifolium 'Pinto'	oFor
un	***aquifolium*** 'Pixie'	oFor
un	***aquifolium*** 'Polka Dot'	oFor
	aquifolium 'Recurva' male	wHer
	aquifolium red early	oGar
va	***aquifolium*** 'Rubricaulis Aurea' female	oRiv wCCr
	aquifolium 'San Gabriel'	oGar
un	***aquifolium*** 'Scotia'	wCCr
va	***aquifolium*** 'Silver Queen' male	oGar wWel
un	***aquifolium*** 'Teufel's Female'	oGar
un	***aquifolium*** 'Teufel's Silver Variegated'	oGar
	aquifolium 'Teufel's Zero'	oFor
un	***aquifolium*** 'Twenty Below'	oFor
	aquifolium variegated	oRiv oSho
un	***aquifolium*** 'Yule Glow'	oFor
	x aquipernyi 'San Jose' female	oFor oAls
va	***attenuata*** 'Sunny Foster' female	oFor wCCr
un	'Aurora'	wHer
un	'Bonfire'	oGre
	'Brilliant'	wCCr
	'China Boy' male	oSho
	'China Girl' female	oSho
	chinensis	see **I. *purpurea***
	ciliospinosa	wHer
	'Clusterberry' female	oFor
	coriacea	oFor
	cornuta 'Berries Jubilee'	oAls oWnC
	cornuta 'Burfordii' female	oFor oAls oGar oBlo wWel
	cornuta 'Burfordii Nana'	see **I. *cornuta*** 'Dwarf Burford'
	cornuta 'Dazzler' female	oFor oAls oBlo
	cornuta 'Dwarf Burford' female	oAls oGar oBRG
	cornuta x ***latifolia***	oFor
va	***cornuta*** 'O. Spring' female	oFor wHer
	cornuta 'Rotunda' female	oGar
	cornuta 'Willowleaf' female	oAls oGar oSho oWnC
	crenata	oRiv wSta
un	***crenata*** 'Aurea'	wSta
un	***crenata*** 'Aurora'	wSta
	crenata 'Bee Hive'	oBRG
	crenata 'Bennett's Compact' male	oWnC oBRG wWel
	crenata 'Bullata'	see **I. *crenata*** 'Convexa'
un	***crenata*** 'Cole's Hardy'	wCCr
	crenata 'Compacta'	see **I. *crenata*** 'Bennett's Compact'
un	***crenata*** 'Conners'	wCCr
	crenata 'Convexa' female	oFor wWoB oAls oGar oSho oBlo oGre oTPm oWnC wBox wSta oCir wSte wWel
	crenata 'Dwarf Pagoda' female	oGos oAmb oRiv oNWe oBRG
	crenata 'Glory' male	oGar oGre wSta
	crenata 'Golden Gem' female	last listed 99/00
	crenata 'Green Dragon'	oFor oRiv oSqu
	crenata 'Green Island' male	oAls oGar oSho oBlo oGre oTPm oWnC wSta
	crenata 'Helleri' female	oFor wHer oAls oGar oBlo oTPm oWnC wSta oBRG wWel
	crenata 'Hetzii' female	oGre oUps
un	***crenata*** 'Jersey Pinnacle'	oFor
	crenata 'Mariesii' female	oRiv oBov wSta oBRG wWel
qu	***crenata*** 'Nana Compacta'	wSta
	crenata 'Northern Beauty'	oAls wSte wWel
un	***crenata*** 'Pincushion'	oFor oSqu
va	***crenata*** 'Shiro-fukurin' female	wHer
	crenata 'Sky Pencil' female	wHer oGos wPir wCCr oCis wWel
un	***crenata*** 'Sky Sentry'	oGar oDan oRiv oGre oWnC wBox wWel
un	***crenata*** 'Vardar Valley'	oTPm
	crenata f. ***watanabeana*** female	wHer wCCr
	cyrtura	oRiv
	decidua	last listed 99/00
un	***decidua*** 'Council Fire'	oFor
un	***decidua*** 'Pocahontas'	oFor
un	***decidua*** 'Red Escort	oFor
	decidua 'Warren's Red' female	oFor oRiv oBov
	dimorphophylla	oBRG
un	***diplosperma*** DJHC 801	wHer
	dipyrena	wCCr
	'Doctor Kassab' female	oFor
	Ebony Magic™	oGar oSho
qu	'Edward Goucher'	oRiv
un	'Emily Brunner'	oFor
	fargesii	last listed 99/00
	'Foster Number 2'	oFor oBlo

	Plant	Sources
	glabra 'Compacta' female	oGar
	glabra f. ***leucocarpa*** 'Ivory Queen'	last listed 99/00
un	***glabra*** 'Shamrock'	oGre
	'Harvest Red' female	oFor
un	'Holly Hedge'	wHer
	integra	wCCr
un	***integra*** 'Green Shadow'	wHer
	integra x ***rugosa***	wCCr
	laevigata female	oFor
	latifolia	oFor wHer oRiv oGre oCis
	Little Rascal®/ 'Mondo' male	oGar wWel
	'Mary Nell' female	oFor wHer oSho
un	***x meserveae*** Berri-Magic™	oWnC
	x meserveae Blue Angel® female	wCul oGar oGre
	x meserveae 'Blue Boy' male	oFor oAls oGar oGre
	x meserveae 'Blue Girl' female	oFor oAls oGar
	x meserveae Blue Prince® male	oFor oGar
	x meserveae Blue Princess® female	oFor oGar oUps
	muchagara	wCCr
	myrtifolia	oFor
	'Nellie R. Stevens' female	oGar oBlo oGre wCCr
	opaca	last listed 99/00
	opaca 'Canary'	oFor
un	***opaca*** 'Cardinal'	oFor
	opaca 'Jersey Knight' male	last listed 99/00
un	***opaca*** 'Lake City'	oFor
un	***opaca*** 'North Wind'	wHer
	opaca 'Old Heavy Berry'	oFor
	opaca 'Rotundifolia"	wBox
	paraguariensis	oCrm
	pedunculosa	oRiv wCCr
	pedunculosa female	oFor
	pedunculosa male	oFor
	sp. aff. ***pernyi*** DJHC 802	wHer
	purpurea	wHer oDan oRiv wCCr oWnC oCis wWel
	'Rock Garden' female	oFor wHer
	rotunda	oBlo
	rotundifolia	see **I. *opaca*** 'Rotundifolia'
un	'Sentinel'	oBlo
	'September Gem' female	wHer wCCr
	serrata	oFor
	serrata HC 970693	wHer
	serrata male	oFor
un	***shennongjianensis*** NA 49246C	wHer
	'Sparkleberry' female	oFor wHer oGre
	'Tanager' female	oFor
	verticillata	oFor wCCr wTGN
	verticillata 'Afterglow' female	oFor wHer oGar oGre
	verticillata f. ***aurantiaca*** female	oFor wHer oGre
un	***verticillata*** 'Autumn Glow'	oFor oGar
un	***verticillata*** 'Berry Nice'	oFor
un	***verticillata*** 'Bonfire'	oFor
	verticillata 'Bright Horizon'	oGre
	verticillata 'Cacapon'	oFor wHer oGre
	verticillata 'Dwarf Male'	wHer
un	***verticillata*** 'Early Bright'	oGre
	verticillata 'Early Male'	oFor
	verticillata 'Jim Dandy' male	oFor wHer oGar oGre oWnC
	verticillata 'Late Male'	wHer oAmb
	verticillata 'Maryland Beauty'	oFor wHer oGre
	verticillata 'Nana' female	oFor wHer oGar
	verticillata 'Oosterwijk'	wHer
	verticillata 'Red Sprite'	see **I. *verticillata*** 'Nana'
	verticillata 'Southern Gentleman' male	oFor wHer oGre
	verticillata 'Stop Light' female	oGre
	verticillata 'Sunset' female	oFor
un	***verticillata*** 'Winter Gold'	oFor wHer oGre wWel
	verticillata 'Winter Red' female	oFor oAmb oGar oGre oUps
	vomitoria	oBlo oWnC
	vomitoria 'Pendula'	wHer
un	***vomitoria*** 'Shillings'	oRiv
	vomitoria 'Stoke's Dwarf'	oSho
	vomitoria 'Wiggin's Yellow'	wHer
un	***vomitoria*** 'Will Fleming'	wHer
un	***vomitoria*** 'Xanthocarpa'	oFor
	x wandoensis	last listed 99/00
un	'William Hawkins'	oCis
	yunnanensis female	wHer
	ILIAMNA	see **SPHAERALCEA**
	ILLICIUM	
	anisatum	oGos oBov wCCr oCis
	floridanum	oFor wHer
un	***floridanum*** 'Alba'	wWel
un	***floridanum*** 'Coosa Red'	oRiv
un	***floridanum*** 'Haley's Comet'	wHer

	Plant	Sources
	floridanum pink	oRiv
	floridanum white	oRiv
	henryi	oFor oCis
un	***lanceolatum***	wHer oCis
un	***parviflorum***	oFor oRiv
	IMPATIENS	
	balsamina	oCrm
	glandulifera	wCSG
	niamniamensis	oTrP
	noli-tangere	oBos
	omeiana	wHer oRus oCis
	IMPERATA	
	cylindrica	oTrP wCSG oCir
	cylindrica 'Red Baron'	see **I. *cylindrica*** 'Rubra'
	cylindrica 'Rubra'	oFor wWoS wCli wTho oNat oDar oJoy oAls oHed oAmb oMis oGar oSis oNWe oOut wNay wRob oBlo oGre wFai wWin wHom oTPm oWnC oTri wTGN wSta oBRG wCoN wWhG oUps wSte wSnq wCoS
	INCARVILLEA	
	arguta	oFor wCol oNWe oGre oSec oUps
	arguta DJHC 98300	wHer
	delavayi	oFor oMis oGar oBlo oGre wTGN oWhS oEga oUps
un	***delavayi alba***	oMis
	delavayi 'Snowtop'	oFor wCSG oGar oGre
	emodi	last listed 99/00
	forrestii	oNWe
	mairei	oSis
	mairei DJHC 98328	wHer
	mairei var. ***grandiflora***	oJoy oSis
	olgae	oTrP oUps
	sinensis	oUps
	sinensis 'Cheron'	oFor oSec oUps
	zhongdianensis	wCol oSis oNWe
	INDIGOFERA	
	amblyantha	oNWe
	australis	last listed 99/00
	decora f. ***alba***	oSis
	gerardiana	see **I. *heterantha***
	heterantha	oSis
	kirilowii	oFor wHer
	pendula ACE 1148	last listed 99/00
	sp. blue	oOEx
	INDOCALAMUS	
	latifolius	oTra wBea oGre wBmG oTBG
	latifolius 'Hopei'	wBea
	latifolius 'Hopei' seed grown	oTBG
	longiauritus	oFor wBam
	solidus	wBmG
	tessellatus	oTra wSus wCli oNor wBea oGar wBox wBam wBmG oTBG
un	***tessellatus*** 'Hamadae'	wCli
	INULA	
	dysenterica	see **PULICARIA *dysenterica***
	ensifolia	oFor oAls wCSG oGoo oOut oWnC wBWP
	glandulosa	see **I. *orientalis***
	helenium	oFor wCul wFGN wCSG oGoo wFai iGSc oCrm
un	***helenium*** 'Goliath'	oBar
	hookeri	wBWP
	magnifica	wHer
	orientalis	oFor
	racemosa	wBWP
	royleana	wHer
	verbascifolia	oSis
	IOCHROMA	
	cyaneum	wHer
	IPHEION	
	'Rolf Fiedler'	wHer oHed oSis
	uniflorum	oTrP oNat oNWe oSec wSta
	uniflorum 'Froyle Mill'	oHed oSis
	uniflorum 'Wisley Blue'	wWoS oHed oSis oGre
	IPOMOEA	
	aculeata	see **I. *alba***
	alba	oUps wSte
un	***batatas*** 'Beauregard'	wIri
	batatas 'Blackie'	wWoS wTGN oUps
	batatas 'Bush Porto Rico'	wIri
	batatas 'Centennial'	wIri
un	***batatas*** 'Georgia Jets'	wIri
un	***batatas*** 'Margarita'	wWoS oUps
un	***batatas*** 'Pink Frost'	wWoS

un	***batatas*** 'Tricolor'	oUps
	batatas 'Vardaman'	wIri
un	***batatas*** 'White Triumphs'	wIri
un	***bolusiana***	oRar
un	***decasperma*** BLM 0074	oRar
	leptophylla	oRar
	lobata	oUps
un	***lobata*** 'Exotic Love'	oUps
	x multifida	oUps
un	***platensis***	oRar
	pubescens BLM 0053	oRar
un	***purpurea*** 'Grampa Ott' seed grown	wSte
	quamoclit	wHom wSte
	'Scarlett O'Hara'	oUps
	sp.	oRar
	sp. BLM 1092	oRar
	sp. BLM 1128	oRar
	sp. BLM 1409	oRar
un	'Terrace Lime'	wTGN
	tricolor	oUps
	tricolor 'Flying Saucers'	oUps
	tricolor 'Heavenly Blue	oUps
	tricolor 'Heavenly Blue' seed grown	wSte
un	***tricolor*** Mount Fuji mix	oUps

IRIS (SEE NOMENCLATURE NOTES)

'A L'Orange'	oMid
'Aaron's Rod'	oMid
'Abbey Road'	oSch wAit
'About Last Night'	oMid
'About Town'	oMid
'Abridged Version'	wAit
'Abstract Art'	last listed 99/00
'Acapulco Sunset'	wAit
'Acclamation'	wWal
'Ace Royale'	oMid
'Acoma'	oSch wAit oCoo
'Act Three'	oMid
'Add It Up'	oMid wAit
'Added Value'	oMid
'Adobe Rose'	oCoo
'Adorable Diva'	last listed 99/00
'Adorable Rose'	oMid
'Advance Design'	wAit
'Affaire'	last listed 99/00
'Affluence'	oCoo
'After Eight'	wAit
'After the Ball'	oCoo
'After the Dawn'	oCoo
'After the Storm'	wAit
'Afternoon Delight'	wWal oSch oCoo
'Agatha Christie'	oMid
'Aggressively Forward'	oMid
'Aichi-no-kagayaki'	oFor
'Ajax the Less'	wAit
'Alabaster Unicorn'	oMid
'Aladdin's Flame'	wAit
'Alaskan Seas'	oSch
'Alene's Other Love'	last listed 99/00
'Alice Briscoe'	oSch
'Alice Goodman'	oSch
'Alien Mist'	oMid
'Alizes'	oSch wAit oCoo
'All Abuzz'	oMid
'All American'	wAit
'All Dressed Up'	oMid oCoo
'All Silent'	oMid
'All That Jazz'	wWal
'Allison Elizabeth'	wAit
'Almost Camelot'	last listed 99/00
'Almost Dark'	last listed 99/00
'Almost Eden'	oMid
'Almost Heaven'	oMid wAit
'Almost Paradise'	oCoo
'Alpine Castle'	wWal
'Alpine Journey'	wWal
'Alpine Lake'	oSch wAit
'Alpine Region'	oMid
'Alpine Storm'	oMid
'Alpine Summit'	wAit
'Alpine Twilight'	oMid
'Altered States'	oMid
'Altruist'	wWal oSch oCoo
'Amadora'	wAit
'Amazon Bride'	wWal
'Amber Artisan'	last listed 99/00
'Amber Snow'	last listed 99/00
'Amber Tamour'	oCoo
'Ambroisie'	oMid
'American Classic'	oSch oMid wAit
'American Sweetheart'	wWal
'America's Cup'	wWal wAit
'Amethyst Dancer'	wAit oCoo
'Amigo'	wAit
'Amour'	oCoo
'Anagram'	last listed 99/00
'And Royal'	last listed 99/00
'Andalou'	oMid
'Andiamo'	oMid
'Angel Heart'	wAit
'Angel's Call'	oMid
'Angel's Touch'	wWal
'Angels in Flight'	oMid wAit
'Anna Belle Babson'	wAit oCoo
'Anne Gaddie'	oSch
'Anne Murray'	oMid
'Answered Prayers'	oMid
'Anvil of Darkness'	oMid
'Anxious'	oCoo
'Anything Goes'	oMid oCoo
'Aphrodisiac'	wWal oSch
'Aplomb'	oSch oMid wAit
'Apollo'	wRoo oGar
'Apollo One'	wAit
'Apollodorus'	oSch
'Apollo's Touch'	oSch
'Approachable'	oMid oCoo
'Apricot Ambrosia'	oMid
'Apricot Drops'	oMid wAit
'Apricot Fizz'	oMid
'Apricot Fringe'	wWal
'Apricot Frosty'	wAit
'Apricot Party'	oMid
'Apricot Topping'	oMid
'April in Paris'	wAit
'Aquatic Alliance'	wAit
'Arc de Triomphe'	oMid
'Arctic Age'	oSch
'Arctic Bliss'	oCoo
'Arctic Express'	oSch oMid wAit
'Arctic Fox'	oMid
'Arctic Lavender'	wAit
'Arizona Citrus'	oMid
'Armada'	oCoo
'Armageddon'	wWal oSch
'Around Midnight'	wWal oMid
'Art Center'	wWal oMid
'Art Deco'	oSch oMid wAit
'Art Faire'	wWal
'Art Nouveau'	last listed 99/00
'Art Quake'	oMid wAit
'Art School Angel'	oMid wAit
'Artful'	oMid
'Artist's Whim'	oSch oMid
'Ascent of Angels'	wAit
'Ascii Art'	oMid
'Ask Alma'	oSch wAit
'Aspire'	oMid
'Astrachanica Kalmikij'	wAit
'Astrid Cayeux'	wWal
'Astro Flash'	wWal
'Atlanta Belle'	oMid
'Atomic Flame'	oMid
'Attention Getter'	wAit
attica	oHed
'Auld Rose'	wWal
'Aunt Lucy'	wAit
'Aura Light'	oMid wAit
'Aurora's Blush'	last listed 99/00
'Austrian Garnets'	wAit
'Autumn Circus'	oSch oMid wAit oCoo
'Autumn Grandeur'	wAit oCoo
'Autumn Maple'	wAit
'Autumn Tryst'	wAit
'Autumn Wings'	oCoo
'Avalon Sunset'	wWal oSch oMid wAit
'Awakening'	oMid wAit
'Awesome'	wWal
'Awesome Blossom'	oMid
'Az Ap'	oSch
'Aztec Burst'	wAit

IRIS

'Aztec King'	wWal
'Azure Icicle'	last listed 99/00
'Azzurra'	wAit
'Babbling Brook'	wWal wAit
'Baboon Bottom'	oMid wAit
'Baby Blessed'	oSch
'Baby Grand'	last listed 99/00
'Bachelor Party'	oMid
'Baci'	oMid
'Back Street Affair'	oMid wAit
'Baklava'	oMid
'Bal Masque'	oMid wAit
'Balch Springs'	oSch wAit
'Ballerina Girl'	oMid
'Ballerina Princess'	wAit
'Ballet Lesson'	oMid wAit
'Ballyhoo'	wWal
'Baltic Blue'	oMid
'Baltic Star'	oMid wAit
'Bama Baby'	last listed 99/00
'Bamba'	oMid
'Banana Frappe	oCoo
'Bangladesh'	oMid
'Bangles'	oMid wAit
'Bantam'	oSch
'Barn Dance'	oMid
'Batik'	wWal oSch oMid wAit oCoo
'Batman'	wWal
'Battle Royal'	wAit
'Baubles and Beads'	oMid wAit
'Bay Watch'	oMid
'Be Happy'	wAit
'Beachgirl'	wWal
'Beachwood Buzz'	oMid
'Beautiful Baby'	oCoo
'Beautiful Vision'	wWal oSch
'Beaux Arts'	wWal
'Bedford Lilac'	oSch wAit
'Bedtime Story'	oSch
'Bee Mused'	oMid
'Bee's Knees'	last listed 99/00
'Before the Storm'	wWal oSch oMid wAit oCoo
'Beginner's Luck'	oMid
'Beguine'	wWal oCoo
'Behind Closed Doors'	oMid wAit
'Behold Your Muse'	oMid
'Being Busy'	oMid
'Bell Tempo'	wWal
'Berry Me Not'	oCoo
'Bertwistle'	last listed 99/00
'Best Bet'	wWal wAit
'Best Man'	oCoo
'Bet Twice'	oMid
'Betty Dunn'	oMid wAit
'Betty Simon'	wWal
'Betwixt'	oSch
'Beverly Sills'	wWal oSch wAit oCoo
'Bewilderbeast'	oSch oMid wAit
'Beyond'	wWal
'Bicentennial'	wWal
'Big Boss'	oMid
'Big Dipper'	wWal
'Big Sky'	oSch
'Big Squeeze'	oMid
'Billie the Brownie'	wAit
'Billionaire'	last listed 99/00
'Birthday Greetings'	oMid wAit
'Birthday Song'	wWal
'Birthday Surprise'	oMid
'Bistro'	oMid
'Bit O' Magic'	last listed 99/00
'Bittersweet Joy'	last listed 99/00
'Black Andromeda'	last listed 99/00
'Black as Night'	wWal wAit oCoo
'Black Butte'	last listed 99/00
'Black Cherry Delight'	oSch
'Black Dragon'	wWal
'Black Falls'	oMid
'Black Flag'	wWal
'Black Hills Gold'	wWal oCoo
'Black Orpheus'	wWal
'Black Sergeant'	wWal
'Black Tic Affair'	wWal oMid wAit oCoo
'Blackbeard'	oSch wAit
'Blackcurrant'	wAit
'Blackout'	wWal oCoo
'Blast'	oSch oMid wAit
'Blazing Light'	wWal oCoo
'Blazing Sunrise'	oMid
'Blended Frills'	oMid
'Blenheim Royal'	oSch oCoo
'Blessed Again'	wAit
'Bleu de Mer'	oCoo
'Blonde Bombshell'	oMid
'Blue Again'	oMid
'Blue Aristocrat'	oSch
'Blue Baron'	wWal
'Blue Cheer'	oMid wAit
'Blue Chip Pink'	wWal
'Blue Chip Stock'	oMid
'Blue Crusader'	oSch oMid
'Blue Denim'	wWoS
un 'Blue Diamond' (Dutch)	wRoo
'Blue Eyed Blond'	oSch wAit
'Blue Field' (Dutch)	wRoo
'Blue for You'	oMid
'Blue It Up'	oCoo
'Blue Line'	wAit
'Blue Luster'	wWal
'Blue Note Blues'	oCoo
'Blue Ribbon' (Dutch)	see **I.** 'Blue Field' (Dutch)
'Blue Rill'	wAit
'Blue Skirt Waltz'	oSch
'Blue Staccato'	oCoo
'Blue Suede Shoes'	oSch oMid wAit
'Blueberry Ice'	oMid
'Bluebird in Flight'	oSch
'Blueblood Yellow'	last listed 99/00
'Bluelight Yokohama'	oMid
'Blush of Youth'	oMid
'Blushes'	oSch
'Blushing Duchess'	wAit
'Blushing Pink'	wWal
'Boisterous'	oMid
'Bold Accent'	wWal
'Bold Crystal'	wAit
'Bold Fashion'	oSch oMid
'Bold Imp'	wAit
'Bold Look'	oSch
'Bold Print'	oSch
'Bold Stroke'	oMid wAit
'Bonus Lite'	oMid
'Bonus Mama'	last listed 99/00
'Boo'	oGar
'Boogie Man'	oCoo
'Boogie Woogie'	oSch oMid wAit oCoo
'Boom Boom Bunny'	oSch wAit
'Boomerang'	last listed 99/00
'Boop Eyes'	see **I.** Sino-Siberian hybrid 'Boop Eyes'
'Boot Hill'	oMid
'Bordeaux Pearl'	oMid
'Border Bandit'	oSch
'Border Control'	oMid
'Border Music'	oMid
'Borderline'	wWal
'Born at Dawn'	oMid
'Born Beautiful'	oMid
'Born to Exceed'	oMid
'Boss Tweed'	wWal wAit
'Boudoir'	oMid wAit
'Bountiful Harvest'	wWal oSch wAit
'Boutique Fashion'	oCoo
'Boy Friend'	wWal
'Boy Next Door'	oMid
'Boysenberry Buttercup'	oMid wAit
'Braggadocio'	oSch oMid wAit
'Brambleberry'	oMid
'Brandy'	wWal
'Brash'	wAit
'Bravado'	wWal
'Brazen Beauty'	last listed 99/00
'Brazenberry'	oMid
'Brazilian Holiday'	wWal oSch oMid wAit
'Bread and Wine'	wWal
'Breakers'	wWal oSch oCoo
'Breezes'	last listed 99/00
'Bridal Gown'	wWal
'Bridal Passion'	wWal
'Bride's Halo'	wAit
'Bright Child'	oMid

	'Bright Reflection'	wWal
	'Bright Vision'	wAit
	'Brighten the Corner'	wAit
	'Brilliant Display'	oMid
	'Bristo Magic'	wWal
	'Broad Grin'	oSch
	'Broadway'	oSch
	'Broadway Baby'	oSch wAit oCoo
	'Broadway Jo'	wAit
	'Brocaded Gown'	wWal
	'Broken Dreams'	oMid wAit
	'Bronzette Star'	last listed 99/00
	'Brown Lasso'	oSch wAit
	'Bubble Bath'	oCoo
	'Bubble Dancer'	oMid wAit
	'Bubbling Along'	oMid wAit
	'Bubbly Mood'	wWal oCoo
	'Buckwheat'	last listed 99/00
	'Buddy Boy'	oMid
	'Buffer Zone'	oMid
	'Bugsy'	oMid wAit
	'Buisson de Roses'	oMid
	bulleyana	oFor oTrP wCol
	bulleyana DJHC 223	wHer
	'Bumblebee Deelite'	oSch wAit
	'Bunnicula'	oSch wAit
	'Burgundy Bubbles'	oCoo
	'Burgundy Party'	last listed 99/00
	'Burst'	oMid
	'Burst of Blue'	oMid
	'Busy Being Blue'	oCoo
	'Busy Signal'	last listed 99/00
	'Butter Pecan'	oSch oCoo
	'Butter Rings'	oMid wAit
	'Butterfingers'	oMid wAit
qu	'Butterflies'	wBox
	'By Your Leave'	oMid oCoo
	'Bye Bye Blues'	oMid wAit
	'Cabaret Royale'	wWal oCoo
	'Cabbage Patch Kid'	wWal
	'Cachet'	oMid
	'Cailet'	oMid
	'Cajun Beauty'	wWal
	'Cajun Queen'	oMid
	'Cajun Rhythm'	oSch oMid
	'Cajun Spices'	oSch wAit
	Cal-Sibe hybrids 'Golden Waves'	oGar wAit
	Cal-Sibe hybrids 'Pacific Smoothie'	wAit
	Cal-Sibe hybrids 'Pacific Starprint'	wAit
	Cal-Sibe hybrids 'Pacific Waves'	wWal
	'Calculated Grace'	oMid
	'Caldron Fire'	last listed 99/00
	'Calico Cat'	oSch wAit
	'Calico Kid'	wAit
	'Caliente'	oCoo
	'California Style'	oSch wAit
	'Call Waiting'	oMid
	'Calling Card'	oCoo
	'Calliope Magic'	oMid
	'Calm Sea'	oSch oCoo
	'Calm Stream'	oMid
	'Calypso Mood'	wWal
	'Camelot Rose'	wWal
	'Camp Fire'	wAit
	'Can Can Dancer'	oMid wAit
	'Can Do'	wAit
	'Candy Apple'	last listed 99/00
	'Candy Floss'	oSch
	'Candy Queen'	last listed 99/00
	'Cannington Delight'	last listed 99/00
	'Cannonball'	wWal
	'Can't Wait'	oMid
	'Cantina'	oMid wAit
	'Cape Ferrelo'	see **I. *douglasiana*** 'Cape Ferrelo'
	'Cape Sebastian'	see **I. *douglasiana*** 'Cape Sebastian'
	'Capital City Jazz'	oMid
	'Capricious'	oCoo
	'Capricorn Cooler'	wWal
	'Captain Indigo'	oMid
	'Captain's Joy'	wWal oMid
	'Caption'	oCoo
	'Captivating'	oMid
	'Captive Sun'	oMid wAit
	'Captured Spirit'	oSch
	'Caramba'	oSch
	'Carats'	oMid
	'Caribbean Dream'	wWal oCoo
	'Caribee'	oCoo
	'Carnival Song'	last listed 99/00
	'Carnival Time'	wWal
	'Carousel Waltz'	oMid
	'Carriage Trade'	wWal
	'Carriwitched'	oMid
	'Carved Crystal'	wWal
un	'Casablanca' (Dutch)	wRoo
	'Cascade Springs'	wAit oCoo
	'Cascading Rainbow'	oMid
	'Casino Belle'	wWal
	'Casino Queen'	wWal
	'Cast a Spell'	oMid
	'Castaway'	oMid
	'Catalyst'	wWal
	'Catch a Wave'	oMid wAit
	'Cats Reign'	oMid
	'Cayenne Pepper'	wWal
	'Cee Cee'	oMid wAit
	'Cee Jay'	oSch wAit
	'Celebration Song'	wWal oSch oMid wAit oCoo
	'Celestial Flame'	oCoo
	'Celsius'	last listed 99/00
	'Celtic Harp'	oMid
	'Centre Court'	wWal
	'Ceremonium'	oSch
	'Chalk Talk'	wWal
	'Champagne and Caviar'	oMid wAit
	'Champagne Elegance'	wWal oSch oMid oGar wAit oCoo
	'Champagne Encore'	oMid wAit
	'Champagne Frost'	oMid wAit
	'Champagne Junior'	oMid
	'Champagne Waltz'	wWal oSch oMid wAit oCoo
	'Champagne Wishes'	oMid wAit
	'Chance Encounter'	oMid
	'Change of Pace'	wWal oSch wAit
	'Change your Ways'	oCoo
	'Changing Winds'	last listed 99/00
	'Chanted'	oSch wAit
	'Chanteuse'	wWal
	'Chantilly Dancer'	oMid
	'Chapeau'	wWal
	'Chapel Bells'	wWal
	'Chardonnay'	wAit
	'Charisma'	oCoo
	'Charmed'	last listed 99/00
	'Charmed Circle'	wWal
	'Charter Oak'	wWal
	'Chartres'	oMid
	'Chartreuse Bounty'	wCul
	'Chasing Rainbows'	oMid oCoo
	'Chatter'	wAit
	'Cheating Heart'	oMid
	'Cheers'	oSch
	'Cher'	wWal
	'Cherished'	wWal
	'Cherished Friendship'	oCoo
	'Cherokee Daybreak'	wAit
	'Cherokee Heritage'	oSch
	'Cherokee Sunrise'	wAit
	'Cherokee Tears'	wAit
	'Cherry'	wAit
	'Cherry Garden'	wWoS
	'Cherry Glen'	oSch oMid wAit
	'Cherry Lane'	last listed 99/00
	'Cherry Pop'	oSch
	'Cherry Smoke'	wWal
	'Cherry Sparkle'	wWal
	'Cherry Tart'	oSch
	'Chestnut Avenue'	oMid
	'Chevalier de Malte'	oMid
	'Chic Attire'	oCoo
	'Chickasaw Sue'	wWal oCoo
	'Chico Maid'	wWal
	'Chief Hematite'	wWal oCoo
	'Chiffon Ruffles'	oCoo
	'China Dragon'	wWal oGar
	'China Moon'	last listed 99/00
	'China Peach'	oMid wAit
	'China Walk'	oMid
	'Chinese Empress'	wWal
	'Chinese New Year'	oMid wAit
	'Chinese Treasure'	wWal

'Chip Shot'	wAit
'Chocolate Marmalade'	oMid wAit
'Chocolate Mint'	oMid
'Chocolate Swirl'	oMid
'Chocolate Vanilla'	oMid
'Christina'	wWal
'Christmas'	oSch wAit
'Christmas Time'	wWal
'Christopher Columbus'	oMid
chrysographes	oFor wCul wSwC wCol wHig wAit oBlo
chrysographes black form	oAmb oGar oSho
chrysographes 'Black Knight'	oNWe
chrysophylla	oFor wHer oBos
'Chubby Cheeks'	oSch oMid
'Chubby Cherub'	wAit
'Chuckatuck'	oMid
'Chuckles'	oCoo
'Cimarron Rose'	oSch wAit
'Cin Cin'	oMid
'Cinderella'	oMid wAit
'Cinderella Sunshine'	oSch
'Cinderella's Coach'	oCoo
'Cinema'	wWal
'Cinnamon'	wWal oCoo
'Cinnamon Apples'	wAit
'Cinnamon Flash'	wAit
'Cinnamon Glow'	oMid
'Cinnamon Sun'	oMid
'Circus Stripes'	wWal
'Circus World'	wWal oSch oMid
'Citoyen'	oMid
'Citron Frommage'	oMid
'Citrus Cooler'	oMid
'City Lights'	oSch oMid wAit
'City Slicker'	oCoo
'Cityscape'	oMid
'Clarence'	oSch oMid wAit oCoo
'Classic Bordeaux'	oCoo
'Classic Edition'	oMid
'Classic Hues'	oMid
'Classic Image'	oMid
'Classic Look'	oSch wAit
'Classic Navy'	wAit
'Classic Suede'	oMid
'Classical Music'	oMid
'Classy Babe'	oSch wAit
'Clear Creek'	oMid wAit
'Clear Morning Sky'	oCoo
'Clearwater River'	oCoo
'Clever Disguise'	oCoo
'Close Shave'	last listed 99/00
'Closed Circuit'	wWal
'Cloud Ballet'	oSch wAit
'Cloud Berry'	oMid
'Cloud Mistress'	oMid
'Clouds and Wine'	wAit
'Cloudy Skies'	oMid wAit
'Clue'	oMid
'Co-Ed Doll'	wWal
'Coalignition'	last listed 99/00
'Coastal Mist'	oSch oMid
'Cocoa Pink'	oMid wAit
'Codicil'	wWal
'Cold River'	oMid
'Colette'	oCoo
'Colette Thurillet'	wWal oMid wAit
'Collector's Art'	oMid wAit
'Color Brite'	wAit
'Color Carnival'	wWal oMid
'Color Glory'	oMid
'Color Me Blue'	oSch oMid wAit
'Color My World'	oMid
'Color Splash'	wWal
'Colorado Bonanza'	last listed 99/00
'Colorbration'	wWal
'Colorflick'	oMid
'Colortart'	oCoo
'Combustion'	wAit
'Comedian'	oSch
'Coming Attraction'	wWal
'Coming Up Roses'	oMid wAit
'Commandante'	last listed 99/00
'Commando'	last listed 99/00
'Compact Buddy'	oMid
'Compadre'	oMid

'Competitive Edge'	wWal wAit oCoo
'Complimentary'	oMid
'Concise'	oMid wAit
'Concoction'	wAit
'Condensed Version'	wAit
'Confederate Rose'	last listed 99/00
'Confederate Royalty'	oMid
'Confession'	oMid
confusa EDHCH 97241	wHer
confusa 'Martyn Rix'	oHed
'Congratulations'	oCoo
'Conjuration'	wWal oSch oMid wAit
'Conspiracy'	oMid wAit
'Constant Companion'	wAit
'Contemporary Art'	oMid
'Continuity'	wAit
'Cool Satin'	oSch
'Cool Treat'	oMid wAit
'Cooling Trend'	oCoo
'Copatonic'	oMid
'Copper and Snow'	oMid
'Copper Cymbal'	last listed 99/00
'Copper Mountain'	wWal
'Copyright'	oCoo
'Coral Chalice'	oCoo
'Coral Clouds'	wWal
'Coral Dreams'	oMid
'Coral Flush'	oMid
'Coral Magic'	wWal
'Coral Sunset'	wWal oSch
'Cordoba'	oMid wAit
'Corn Harvest'	last listed 99/00
'Corona Gold'	oMid
'Corps de Ballet'	oMid
'Cosmic Dawn'	wWal
'Cosmic Wave'	oMid
'Costa Rica'	oMid
'Costumed Clown'	last listed 99/00
'Cote d'Or'	wWal
'Cotton Blossom'	oSch
'Cotton Carnival'	wWal
'Cotton Charmer'	oMid wAit
'Cotton Club'	wWal
'Counting Sheep	wAit
'Country Charm'	oSch oMid
'Country Dance'	oMid wAit
'Country Diary'	oMid
'Country Manor'	wWal
'Country Moon'	oSch
'Countryman'	wWal
'Court Magician'	oSch
'Courtly Affair'	wAit
'Covert Action'	last listed 99/00
'Covet Me'	oMid
'Cozy Calico'	wWal
'Cracklin Burgundy'	wWal oCoo
'Crafty Lady'	last listed 99/00
'Cranapple'	oMid wAit
'Cranberry Cooler'	oMid
'Cranberry Crush'	wWal wAit oCoo
'Cranberry Delight'	oMid wAit
'Cranberry Ice'	wWal
'Crazy for You'	oMid
'Cream Taffeta'	wWal
'Creative Stitchery'	wWal oCoo
'Credible Justification'	wAit
'Credit Line'	oMid
'Crimson Fire'	wWal
'Crimson Snow'	wWal oSch
'Crimson Tiger'	wAit
'Crimson Twist'	oMid oCoo
cristata	oFor wCul wThG wCol oRus oSis oNWe wMag oTri
cristata 'Abbey's Violet'	last listed 99/00
cristata 'Alba'	oRus
'Critic's Choice'	last listed 99/00
'Crocus'	wAit
'Cross Current'	oSch oMid wAit
'Crown Sterling'	wWal
'Crowned Heads'	oMid wAit
'Crowning Touch'	wWal
'Cruise Control'	oMid wAit
'Cruzin'	wWal
'Crystal Blue'	wWal
'Crystal Morn'	oMid

	'Crystal Ruffles'	wAit
	'Crystalyn'	wWal
	'Cuddle Up'	wAit
	'Cup Race'	wWal
	'Cuss A'blue Streak'	wAit
	'Custom Made'	last listed 99/00
	'Cute or What'	wAit
	'Cutting Edge'	oMid
	'Cyber Net'	wAit
	'Cycles'	wWal
	'Daddy's Girl'	last listed 99/00
	'Dainty Morsel'	wAit
	'Dance for Joy'	wWal
	'Dance Hall Dandy'	oMid wAit
	'Dance Hall Dolly'	oMid
	'Dancing Beauty'	wWal
	'Dancing Fountain'	last listed 99/00
	'Daredevil'	wWal oSch
	'Dark Blizzard'	oSch
	'Dark Freeze'	oMid
	'Dark Passion'	oSch oMid
	'Dark Treasure'	oMid
	'Dark Triumph'	wWal
	'Dark Twilight'	oSch
	'Dark Vader'	wWoS oSch wAit
	'Dark Waters'	wAit oCoo
	'Darkness'	wAit
	'Dashing'	oMid wAit
	'Dauntless'	wAit
	'Davy Jones'	wWal oCoo
	'Dawn of Change'	oCoo
	'Dawn Princess'	wAit
	'Dawn Sky'	oMid
	'Dawning'	oMid wAit oCoo
	'Day Glow'	oMid wAit
	'Dazzling Gold'	wWal oSch
	'Dazzling Jewel'	oMid
	'Dear Jean'	oMid
	'Dear Marie'	wWal
un	'Dear Minnie'	wWal
	'Debrenee'	oMid wAit
	'Decipher'	oMid wAit
	'Decolletage'	wWal
	'Deep Dark Secret'	oMid
	'Deep Pacific'	wWal
	'Degas Dancer'	wWal oMid
	'Deity'	oMid
	delavayi	oTrP wCol
	delavayi DJHC 536	wHer
	'Delta Blues'	oSch oMid wAit
	'Deming Glacier'	wWal
	'Departure'	oMid
	'Depth of Field'	wWal oCoo
	'Descanso'	oMid
	'Desert Attire'	wAit
	'Desert Daybreak'	oMid
	'Desert Echo'	wWal
	'Desert Fury'	wAit
	'Desert Lullaby'	last listed 99/00
	'Desert Orange'	oSch oMid wAit
	'Desert Passion'	oMid
	'Desert Renegade'	oCoo
	'Deserts Rage'	last listed 99/00
	'Deserving Treasure'	oMid wAit
	'Designer Gown'	wWal
	'Designing Woman'	last listed 99/00
	'Details'	oMid
	'Devil May Care'	oMid
	'Devil's Lake'	oSch oMid
	'Devil's Riot'	oMid
	'Dewberry Dawn'	last listed 99/00
	'Diabolique'	oSch oMid wAit
	'Different Approach'	oMid wAit
	'Different World'	wWal
	'Dime a Dance'	oMid wAit
	'Dinky Circus'	oMid
	'Direct Flight'	oMid
	'Dirty Devil Canyon'	oMid
	'Discovered Gold'	oMid wAit
	'Discretion'	wWal
	'Distant Chimes'	wWal
	'Distant Fire'	wWal
	'Distant Roads'	oMid
	'Ditto'	oSch wAit
	'Diva Do'	last listed 99/00
	'Divine Duchess'	oMid
	'Divine Light'	oCoo
	'Dixie Desert'	wWal
	'Doctor Alan'	wAit
	'Dodge City'	wAit
	'Dolly Dancer'	oCoo
	'Donegal'	oMid
	'Don't Be Cruel'	oMid wAit
	'Don't Leave Me'	oMid
	'Doodle'	last listed 99/00
	'Doozey'	wAit
	'Dorothy Davis'	oMid
	'Dorothy Marquart'	oMid wAit
	'Dot Com'	oMid wAit
	'Dottie Joy'	wAit
	'Double Bubble'	oMid
	'Double Scoop'	last listed 99/00
	'Douce France'	oMid
	douglasiana	oFor oHan wCul oAls oRus wShR oGar oBos oSho wCCr oTri wBox wWld wSnq
un	***douglasiana*** 'Burnt Sugar'	oSis
	douglasiana 'Canyon Snow'	oNWe
un	***douglasiana*** 'Cape Ferrelo'	oSis
un	***douglasiana*** 'Cape Sebastian'	oSis
	douglasiana 'Pacific Moon'	wCSG
	'Dover Beach'	wWal
	'Dracula's Shadow'	oSch
	'Dragon's Fancy'	oMid
	'Dream Affair'	wWal
	'Dream Indigo'	oMid
	'Dream Lord'	oMid
	'Dream Lover'	wWal wAit
	'Dream Master'	oMid
	'Dream of Gold'	oMid
	'Dream of You'	last listed 99/00
	'Dreaming Lilac'	oMid
	'Dreamsicle'	oSch oMid wAit
	'Dress Blues'	wWal
	'Druid's Chant'	oMid
	'Drum Roll'	oSch oMid
	'Dublin'	oMid wAit
	'Dude Ranch'	oMid
	'Duke of Earl'	wAit
	'Duo Dandy'	oCoo
	'Dusky Challenger'	oSch oMid wAit oCoo
	'Dusky Dancer'	wWal
	'Dusky Evening'	wWal
	'Dutch Chocolate'	wWal
	'Dutch Treat'	wWal
	'Dynamite'	oSch oMid wAit
	'Eagle's Flight'	wWal oSch oCoo
	'Eagle's Wing'	oMid
	'Earl of Essex'	oSch oMid oCoo
	'Earliglo'	oMid
	'Early Wish'	wWal
	'Earth Song'	last listed 99/00
	'Earthborn'	wAit oCoo
	'Earth's Dark Angel'	oMid
	'East Indian Spice'	oMid oCoo
	'Easter'	oMid wAit
	'Easter Treasure'	last listed 99/00
	'Eastertime'	wWal oCoo
	'Easy to See'	oMid
	'Ebony Angel'	oCoo
	'Ebony Dream'	wAit
	'Echo de France'	wWal
	'Echoes'	last listed 99/00
	'Ecstatic Echo'	wWal oSch
	'Eden'	wWal
	'Edge of Winter'	wWal
	'Edith Wolford'	oSch wAit oCoo
	'Edna's Wish'	wWal oMid oCoo
	'Efreets'	oMid
	'Egret Snow'	oSch wAit
	'El Cerrito'	oMid
	'El Torito'	oMid wAit
	'Elainealope'	oMid wAit
	'Elaine's Angel'	last listed 99/00
	'Eleanor Kirkpatrick'	last listed 99/00
	'Eleanor's Pride'	wWal
	'Electrabrite'	wWal
	'Electric Shock'	oMid wAit
	'Electrique'	oMid wAit
	'Elegant Blue'	oMid
	'Elegant Charm'	wWal

	'Elegant Era'	wWal
	'Elegant Impressions'	wWal oSch oMid wAit
	'Elfin Magic'	wAit
	'Elisa Renee'	wWal
	'Elizabeth Poldark'	wWal oMid
	'Ellen Joy'	oMid
	'Elsiemae Nicholson'	last listed 99/00
	'Emanations'	oMid
	'Emblazoned'	oMid
	'Emperor's Concerto'	oMid
	'Emperor's Delight'	oMid
	'En Garde'	oMid wAit
	'En Pointe'	oMid
	'Enchanted April'	oMid wAit
	'Enchanted One'	oCoo
	'Enchanting'	wWal
	'Encre Bleue'	oMid
	'End Play'	wWal
	'Energizer'	wAit
	'Engaging'	oMid wAit
	'English Charm'	oMid wAit oCoo
	'Enhancement'	oMid wAit
	ensata	oHug oBlo wCCr
	ensata 'Abundant Display'	wAit
	ensata 'Admetus'	wAit
un	***ensata*** 'Agoga-kujyo'	oGar
	ensata 'Agrippinella'	wWal wAit
	ensata 'Akebono'	wWal wAit oGai
	ensata 'Alba'	wFai
qu	***ensata*** 'Alba Double White'	oGar
	ensata 'Amethyst Wings'	wWal
	ensata 'Anytus'	wWal wAit
	ensata 'Asato Biraki'	wWal
	ensata 'Asian Warrior'	wWal wAit
	ensata 'August Emperor'	wWal
	ensata 'Azure Ruffles'	wWal wAit
	ensata 'Banners on Parade'	wWal
	ensata 'Bellender Blue'	wWal
un	***ensata*** 'Beni Kohji'	wWal
un	***ensata*** 'Beni Toban'	wWal
	ensata 'Beni Tsubaki'	wWal wAit wBox
un	***ensata*** 'Big Flare'	wEde
	ensata 'Blue Lagoon'	oGar
	ensata 'Blue Marlin'	wAit
	ensata 'Bluetone'	wAit
	ensata 'Bridge of Dreams'	wAit
un	***ensata*** 'Bright Inspiration'	wWal
	ensata 'Brilliant Burgundy'	wWal
	ensata 'Butterflies in Flight'	wWal wAit oGai
	ensata 'Capaneus'	wWal
	ensata 'Caprician Butterfly'	wWoS wAit oGai
	ensata 'Cascade Crest'	wWal oGar wAit wBox
	ensata 'Cascade Spice'	wAit oGai
	ensata 'Center of Attention'	wWal wAit oGai
	ensata 'Chico Geisho'	wWal wAit wBox
	ensata 'Chidori'	wAit oGai
	ensata 'Chigokesho'	oGai
	ensata 'Chigosugata'	wAit
	ensata 'Confetti Shower'	wWal
	ensata 'Court Jester'	wWal
	ensata 'Crepe Paper'	wWal
	ensata 'Crystal Halo'	wWal
	ensata 'Dancing Waves'	wWal wAit oGai
	ensata 'Dark Drapery'	oFor wWal
	ensata 'Darling'	oAls
	ensata 'Dramatic Moment'	wWal
	ensata 'Ebb and Flow'	wWal
un	***ensata*** 'Edens Blue Pearl'	oGre
	ensata 'Edens Charm'	oGre
un	***ensata*** 'Edens Picasso'	oGre
	ensata 'Electric Glow'	wWal wAit
	ensata 'Electric Rays'	wWal oGar wAit oGai wBox
	ensata 'Enchanting Melody'	oGar
	ensata 'Enduring Pink Frost'	wWal wAit
	ensata 'Epimetheus'	wAit
	ensata 'Exuberant Chantey'	wWal wAit
	ensata 'Flying Tiger'	wWal wAit
	ensata 'Foreign Intrigue'	wWal
	ensata 'Freckled Geisha'	wWal oGar
	ensata 'Frilled Enchantment'	wWal wAit
	ensata 'Frosted Intrigue'	wWal
	ensata 'Frosted Pyramid'	wWal wBox
	ensata 'Garnet Royalty'	wWal oGar wAit
	ensata 'Geisha Gown'	wWal oGai
	ensata 'Geisha Obi'	wAit
	ensata 'Gekkei-kan'	see **I.** 'Laurel Crown'
	ensata 'Glitter and Glamour'	last listed 99/00
	ensata 'Good Omen'	wWal wAit oGai
	ensata 'Gracieuse'	oFor wWoS wSwC
	ensata 'Gusto'	wWal
	ensata 'Haku Botan'	oGar
	ensata 'Hatsu Kagami'	wWal wAit
	ensata 'Hegira'	wWal wAit
un	***ensata*** 'Henry's White'	oGar
	ensata 'Hidenishiki'	wAit
	ensata Higo hybrids	wSta
	ensata 'Hisakata'	wWal oGar
	ensata 'Hue and Cry'	wAit oGai
	ensata 'Iapetus'	oGai
	ensata 'Ike-No-Sazanami'	wWal wAit
	ensata 'Imperial Magic'	last listed 99/00
	ensata 'Imperial Presence'	wWal
	ensata 'Indigo Magic'	wWal
	ensata 'Ink on Ice'	wAit
	ensata 'Izu-No-Umi'	wWal
	ensata 'Jaciva'	wWal
	ensata 'Japanese Harmony'	wAit
	ensata 'Japanese Pinwheel'	wAit oGai
un	***ensata*** 'Jedo Jiman'	oAls oGar
	ensata 'Jewelled Sea'	wWal
	ensata 'Jocasta'	wWal
	ensata 'Joy Peters'	wWal wAit
	ensata 'Kalamazoo'	wWal
	ensata 'Kaleidoshow'	wWal
	ensata 'Kimboshi'	wWal
	ensata 'Kimiko'	wAit
	ensata 'King's Court'	wWal wAit oGai
	ensata 'Knight in Armor'	wWal wAit oGai
	ensata 'Kontaki-on'	wWal wAit
	ensata 'Lady in Waiting'	wWal wBox
un	***ensata*** 'Laurel Crown'	oSho wTGN
	ensata lavender	oHug
	ensata 'Le Cordon Bleu'	wWal
	ensata 'Light at Dawn'	wWal
	ensata 'Lilac Peaks'	wWal
	ensata 'Lion King'	wWal
	ensata 'Little Snowman'	wAit
	ensata 'Loyalty'	oFor
	ensata 'Magic Opal'	oGar
	ensata 'Magic Ruby'	oGar
	ensata 'Mai Ogi'	wWal
	ensata 'Maine Chance'	wAit
	ensata maroon	oGar
un	***ensata*** 'Matsuono Uki'	oNWe
	ensata 'Mauve Opera'	wWal
	ensata 'McKenzie Sunset'	wWal wAit
	ensata 'Michio'	wAit
	ensata 'Midnight Stars'	wWal wAit oGai
	ensata 'Midsummer Reverie'	wWal wAit oGai
	ensata 'Miss Coquette'	oGai
	ensata mixed	oSho wCoN
	ensata 'Moonlight Waves'	oHed
	ensata 'Murakumo'	oGar oSho
	ensata 'Night Angel'	wAit
	ensata 'Night Blizzard'	wWal
	ensata 'Nihonkai'	wWal
	ensata 'Ocean Mist'	wWal
	ensata 'Orchid Fawn'	wWal
	ensata 'Oriental Eyes'	wWal wAit oGai
	ensata 'Over the Waves'	wWal
	ensata 'Peacock Dance'	wWal oGai
	ensata 'Periwinkle Pinwheel'	wWal wAit oGai
	ensata 'Persian Rug'	wWal oNWe oGai
	ensata 'Picotee Princess'	oGar oGai
	ensata pink	oGar
	ensata 'Pink Dimity'	wWal wAit
un	***ensata*** 'Pink Lady'	wTGN
	ensata 'Pink Lips'	wWal
	ensata 'Pink Ringlets'	wWal
	ensata 'Pixie Won'	wWal
	ensata 'Pleasant Earlibird'	wAit oGai
	ensata 'Pleasant Sandman'	wAit
	ensata 'Pleasant Starburst'	wAit
	ensata 'Prairie Glory'	wWal
	ensata 'Prairie Noble'	wWal
	ensata 'Prairie Twilight'	wWal
	ensata 'Premier Danseur'	wWal
	ensata purple	oHug wHom
	ensata 'Raspberry Gem'	wWal
	ensata 'Reign of Glory'	oGar wAit

	ensata 'Rivulets of Wine'	wAit
	ensata 'Rose Adagio'	wWal
	ensata 'Rose Queen'	oFor oGar wAit oSha
	ensata 'Rose World'	wAit
	ensata 'Rosy Sunrise'	wWal
	ensata 'Royal Banner'	wSwC
	ensata 'Royal Crown'	wWal
	ensata 'Royal Game'	wWal wAit
	ensata 'Royal Lines'	oGre
	ensata 'Royal Ramparts'	wAit
	ensata 'Ruby Star'	wWal wAit oGai
	ensata 'Ruffled Dimity'	wWal oGar wAit
	ensata 'Sakurajishi'	wWal
	ensata 'Sapphire Crown'	wWal
	ensata 'Shihoden'	wWal
	ensata 'Silver Cascade'	wWal
	ensata 'Silverband'	wWal oAmb oGar wAit oNWe
	ensata 'Sky and Mist'	wWal
	ensata 'Skyrocket Burst'	wWal
un	***ensata*** 'Snow Queen'	wWal
	ensata 'Snowy Hills'	wAit
	ensata 'Sorceror's Triumph'	oGar oGai
	ensata 'Sparkling Sapphire'	wWal
	ensata var. ***spontanea***	wHer
	ensata 'Springtime Melody'	wAit
	ensata 'Springtime Prayer'	wAit
	ensata 'Springtime Snow'	wWal
	ensata 'Star at Midnight'	oGai
	ensata 'Stippled Ripples'	wWal
	ensata 'Stranger in Paradise'	wAit
	ensata 'Strut and Flourish'	wWal wAit oGai wBox
	ensata 'Summer Moon'	wWal
	ensata 'Summer Storm'	wWal wAit oGai
	ensata 'Swirling Waves'	wWal
	ensata 'Tender Trap'	wWal
	ensata 'The Great Mogul'	oGai
	ensata 'Thunder and Lightning'	oGai
	ensata 'Tuptim'	oGai
un	***ensata*** 'Uikaza'	oGai
	ensata 'Umi-botaro'	oGai
	ensata 'Valiant Prince'	wWal
	ensata 'Variegata'	oFor wWoS wGAc oHed oGar oSho oOut oGre wFai wBox oUps wSte
	ensata 'Whippoorwill'	wWal
	ensata white	oGar
	ensata 'Wilderness Snowball'	wAit
	ensata 'Wind Drift'	wWal
	ensata 'Windswept Beauty'	wWal
	ensata 'Wine Ruffles'	wAit oGai
	ensata 'Winged Sprite'	wWal
	ensata 'Wings Aflutter'	wWal
	ensata 'Woodland Brook'	wWal
	ensata 'Worley Pink'	wWal oGai
	ensata 'Wounded Dragon'	oGai
	ensata 'Yodo-No-Kawase'	wWal
	ensata 'Yuhi'	wWal
	'Ensemble'	oCoo
	'Envy'	oCoo
	'Envy of Dresden'	oCoo
	'Epicenter'	oMid wAit
	'Equestrian'	wWal
	'Erleen Richeson'	oCoo
	'Ermine Robe'	wWal
	'Erotic Touch'	oMid
	'Established Powers'	oMid
	'Esther, The Queen'	oSch
	'Eternal Bliss'	oSch wAit
qu	***evansiana***	oCis
qu	***evansiana*** white form	oCis
	'Evelyn Harris'	last listed 99/00
	'Evelyn Rose'	wAit
	'Evening Magic'	wWal
	'Evening Silk'	wAit
	'Ever After'	wAit oCoo
	'Everlasting Love'	wWal
	'Everything Plus'	wWal wAit oCoo
	'Everywhere'	oMid wAit
	'Exactitude'	oMid wAit
	'Excellency'	oSch oMid wAit
	'Exclusivity'	oMid
	'Excursion'	wWal
	'Exotic Isle'	wWal
	'Exotic Star'	wWal
	'Extol Peace'	wWal
	'Extravagant'	wWal
	'Eyebright'	wWoS oSch
	'Faberge'	oMid wAit
	'Fabio'	oMid
	'Fabrique'	wWal
	'Facsimilie'	wWal
	'Faded Love'	oMid
	'Faint Praise'	oMid
	'Fair Dinkum'	wWal oCoo
	'Fairmont'	wAit
	'Fairy Aire'	last listed 99/00
	'Fairy Favours'	oMid
	'Fairy Fun'	oMid wAit
	'Fairy Lore'	oMid wAit
	'Fairy Meadow'	oMid
	'Fairy Ring'	oMid
	'Fall Fiesta'	wWal oSch
	'Fallen Angel'	oMid wAit
	'Falling in Love'	oCoo
	'Fancy Dress'	oMid wAit
	'Fancy Lady'	oCoo
	'Fancy Tales'	wWal
	'Fancy Woman'	oSch oMid wAit
	'Fancy Wrappings'	oMid
	'Fanfaron'	oCoo
	'Fantasy Isle'	oSch
	'Far Corners'	oMid
	'Faraway Blue'	oCoo
	'Fashion Bug'	wAit
	'Fashion Designer'	oMid wAit
	'Fashion Statement'	oMid wAit
	'Fashionable Pink'	wWal
	'Fashionably Late'	oMid wAit
	'Fatal Attraction'	oMid
	'Fathom'	oMid wAit
	'Favorite Angel'	oSch wAit
	'Favorite Pastime'	oCoo
	'Feather Boa'	oMid
	'Feature Attraction'	oSch oMid wAit
	'Feed Back'	oSch oMid wAit
	'Feminine Charm'	wWal
	'Feminine Fire'	oCoo
	'Femme Fatale'	oCoo
	fernaldii	wHer
	'Festive Mood'	oMid wAit
	'Festive Skirt'	wWal
	'Feu du Ciel'	oMid
	'Fiction'	oCoo
	'Fiery Chariot'	wWal
	'Figure Head'	oMid
	'Filibuster'	oMid wAit
	'Film Festival'	oMid wAit
	'Final Decision'	oSch
	'Finalist'	oSch oMid wAit oCoo
	'Fine Blending'	oMid
	'Fine China'	wWal
	'Finsterwald'	wAit
	'Fire Bride'	oCoo
	'Fire Pit'	oCoo
	'Fire Power'	wWal
	'Firebreather'	oSch
	'Firebug'	wAit
	'Fireside Glow'	wWal
	'Firestorm'	oSch wAit
	'First Alert'	last listed 99/00
	'First Blush'	wWal
	'First Interstate'	wWal oSch
	'First Movement'	last listed 99/00
	'First Reunion'	oSch oCoo
	'First Romance'	wAit
	'First Waltz'	wWal
	'Five O'Clock World'	wAit
	'Five Star Admiral'	wWal
	'Fjord'	oMid wAit
	'Flambe'	wAit
	'Flamboyant Dance'	oMid
	'Flamenco'	oCoo
	'Flat Rate'	last listed 99/00
	'Flea Circus'	oMid wAit
un	'Flea Market Blue'	wWal
	'Flight of Fancy'	oSch oMid wAit
	'Flirty White Skirts'	wAit
	'Florentina'	oAls wFGN oGoo wFai oCoo iGSc oBar
	florentina	see **I.** 'Florentina'
	florentina alba	see **I.** 'Florentina'
	'Flow Blue'	oSch

'Flowerfield'	oMid
'Fly to Vegas'	wWal
'Fly with Me'	wAit
'Flyby'	oMid
'Flying Carpet'	oMid
foetidissima	oFor wHer oHan oJoy wCol oRus wAit oSho wCCr wMag wSta
foetidissima var. ***citrina***	wHer
foetidissima 'Variegata'	wHer wSwC wCol oGar oGre
'Fogbound'	oMid wAit
'Folksy Fun'	oMid
'Fondation Van Gogh'	wWal oMid
'Foolish Fancy'	wWal
'Footloose'	oSch oCoo
'Forbidden Fruit'	oMid
'Foreign Knight'	oMid
'Foreign Statesman'	oMid wAit
'Forever Yours'	oMid oCoo
'Forge Fire'	oCoo
'Forgiven'	oCoo
forrestii	wCCr
'Fort Apache'	oSch
'Fort Bragg'	oCoo
'Fortunata'	oCoo
'Fortune Teller'	oCoo
'Foundation'	last listed 99/00
'Fragrant Lilac'	oSch oCoo
'Freda Laura'	wWal
'Freedom's Bell'	wWal
'Freely Given'	last listed 99/00
'French Horn'	oMid
'French Melody'	last listed 99/00
'French Rose'	oMid
'Frenchi'	wAit
'Fresh Air'	oSch
'Fresh Start'	wAit
'Fresno Calypso'	wWal
'Friday Blues'	oMid wAit
'Friday Surprise'	wWal
'Fringe Benefits'	oSch wAit
'Fringe of Gold'	wWal oSch
'Frisco Follies'	wWal
'Frison-Roche'	oMid
'Frolicsome'	oMid
'From a Distance'	oCoo
'Frost Echo'	oMid wAit
'Frostbite'	oMid
'Frosted Sapphire'	oCoo
'Frosted Velvet'	oSch
'Frosting'	oMid wAit oCoo
'Frosty Elegance'	oMid
'Frothingslosh'	wAit
'Frozen Blue'	oMid oCoo
'Frugal'	oMid
'Fruit Cocktail'	oMid wAit
'Fuji Skies'	oSch
'Full Fashioned'	oMid
'Full Tide'	wWal
fulva	last listed 99/00
'Funny Girl'	oMid
'Furioso'	oMid
'Gala Gown'	wWal
'Gala Greetings'	oMid
'Gala Madrid'	wWal
'Gallant Moment'	wWal
'Gallant Rogue'	wWal oMid
'Galway'	oMid wAit
'Garden Garnet'	wAit
'Garnet Sport'	wWal
'Gawlee'	oMid
'Gay Motif'	oMid
'Gay Parasol'	wWal oCoo
'Gazoo'	oMid
'G'Day Mate'	wWal
'Geisha'	oMid wAit
'Gemstar'	oSch wAit
'Geniality'	wWal oMid wAit
'Gentle Grace'	oSch
'Gerald Darby'	see **I. *x robusta*** 'Gerald Darby'
germanica var. ***florentina***	see **I.** 'Florentina'
'Ghost Train'	oSch
'Giant Rose'	wWal
'Gift of Dreams'	wWal
'Gigolo'	wWal
'Ginger Swirl'	wWal
'Ginny's Cream'	oMid
'Giraffe Kneehiz'	oMid
'Girls' Favourite'	oMid
'Glacia Island'	wAit
'Glacier Point'	oMid
'Glad Choice'	last listed 99/00
'Glad Heart'	oMid oCoo
'Glass Slippers'	oMid
'Glazed Gold'	oCoo
'Glebe Brook'	oMid
'Glendale'	wWal
'Glitter Bit'	oMid
'Glittering Amber'	wWal
'Glitz 'n' Glitter'	wWal
'Glorious Morning'	last listed 99/00
'Glorious Sunshine'	wWal
'Glory Be'	last listed 99/00
'Glory Bound'	wWal
'Glowing Rubies'	oMid
'Gnu'	wAit
'Gnu Again'	wAit
'Gnu Blues'	oSch wAit
'Gnu Rayz'	wAit
'Gnus Flash'	oMid wAit
'Gnuz Spread'	wAit
'Goddess of Green'	wAit
'Goddess of Luck'	wAit
'Goddess of Pink'	oMid wAit
'Godsend'	oMid
'Going Bonkers'	last listed 99/00
'Going My Way'	wWal
'Gold Beach'	oMid
'Gold Canary'	oSch
'Gold Country'	oCoo
'Gold Frosting'	last listed 99/00
'Gold Galore'	wWal oCoo
'Gold Trimmings'	wWal
'Gold Velocity'	last listed 99/00
'Goldcinda'	oMid
'Golden Ecstasy'	wWal oSch
'Golden Eyelet'	oSch
'Golden Galaxy'	oSch
'Golden Immortal'	wAit
'Golden Muffin'	oCoo
'Golden Starlet'	last listed 99/00
'Golden Waves'	see **I.** Cal-Sibe hybrids 'Golden Waves'
'Goldfinger'	oMid wAit
'Goldie the Pirate'	oMid
'Goldkist'	oMid
'Gone Fission'	last listed 99/00
'Good and True'	oSch
'Good Day Oregon'	oCoo
'Good Day Sunshine'	wAit
'Good Humor'	oMid
'Good Looking'	oSch oMid
'Good Show'	wAit oCoo
'Good Vibrations'	oSch oMid
'Goodbye Girl'	wWal oSch wAit
'Goodbye Heart'	oMid
'Goodnight Moon'	oSch oMid wAit
'Goodwill Messenger'	wWal oCoo
'Got the Blues'	oMid
gracilipes	oJoy wCol
graminea	wCol oAls oRus wCSG wAit oSis
un ***graminea*** x 'Florentina'	oGar
'Grand Entrance'	wWal
'Grand Metallic'	oCoo
'Grand Old Opry'	wWal oCoo
'Grand Prix'	wWal oCoo
'Grand Waltz'	wWal
'Grape Jelly'	oMid
'Grape Orbit'	oSch
'Grapelet'	wAit
Graphic Arts'	oCoo
'Graphique'	wAit
'Grateful Citizen'	last listed 99/00
'Gratuity'	wAit
'Great America'	oMid
'Great Gatsby'	oSch oMid
'Great Swan River'	oMid
'Grecian Skies'	wWal oMid
'Green and Gifted'	wAit
'Green Prophecy'	wAit oCoo
'Grobswitcher'	last listed 99/00
'Groovy Grubworm'	oMid wAit

	'Guadalupe'	wAit
	'Guideword'	wAit
	'Guru'	wAit
	'Gypsy Caravan'	oCoo
	'Gypsy Queen'	wWal
	'Gypsy Romance'	wWal oSch oMid wAit
	'Gypsy Woman'	oCoo
	'Gyro'	wAit oCoo
	'Habit'	oMid
	'Hackmatack'	oMid
	'Halfpenny Green'	wAit
	'Halfway to Heaven'	oMid wAit
	'Halloween Halo'	wAit
	'Halo in Burgundy'	oMid
	'Halo in Peach'	oMid
	'Halo in Rosewood'	wAit
	'Halogram'	oMid
	'Hampton Gold'	oMid
	'Handiwork'	wWal
	'Handshake'	oMid
	'Handyman'	oMid
	'Happy New Year'	wAit
	'Harbor Master'	last listed 99/00
	'Harlem Nocturne'	wAit
	'Harlow Gold'	oSch oGar
	'Harmonics'	oMid
	'Harry Hite'	wAit
un	***hartwegii*** ssp. ***hartwegii***	wHer
	'Harvest Faire'	oMid wAit
	'Harvest Hues'	wAit
	'Harvest of Memories'	oSch oMid
	'Harvest Queen'	oCoo
	'Haute Couture'	oMid wAit
	'Hawaiian Halo'	wWal
	'Hazel's Pink'	oSch
	'He Man'	oMid
	'He-Man Blues'	last listed 99/00
	'Hearthstone'	oCoo
	'Heartland'	oMid
	'Heather Blush'	wWal
	'Heaven'	oMid
	'Heavenly Body'	oSch
	'Heavenly Rapture'	wWal
	'Heavenly Vision'	oSch oMid
	'Heaven's Edge'	oMid
	'Helen K. Armstrong'	last listed 99/00
	'Helen Leader'	oMid wAit
	'Helen Proctor'	last listed 99/00
	'Helen Ruth'	wWal
	'Helene C.'	wAit
	'Hellcat'	oSch wAit
	'Hello Darkness'	wWal oSch oMid wAit
	'Hell's Fire'	wWal
	'Her Kingdom'	oCoo
	'Her Royal Highness'	oMid
	'Here's Heaven'	wAit oCoo
	'Hey Dreamer'	oMid
	'Hey Looky'	wWal
	'Hey There'	wAit
	'Hi Calypso'	oMid
	'Hidden Glow'	oMid wAit
	'Hidden World'	oCoo
	'High Blue Sky'	oCoo
	'High Energy'	oMid wAit
	'High Impact'	oMid
	'High Life'	wWal
	'High Lonesome'	wAit
	'High Profile'	oSch
	'High Stakes'	oSch
	'High Wire'	oSch
un	'Hildegarde'(Dutch)	wRoo
	'Hilltop View'	oCoo
	'Hindenburg'	wWal
	'Hint'	wAit
	'Hippie'	oMid
	'Hippo'z Tutu'	oMid wAit
	'His and Hers'	oCoo
	'His Lordship'	wWal
	'Holden Clough'	wCul wAit
	'Holden's Child'	wAit
	'Holiday Lover'	oMid
	x hollandica 'Apollo'	see **I.** 'Apollo'
	'Holly Golightly'	oMid
	'Homecoming Queen'	wWal oCoo
	'Honey Cub'	oMid wAit

	'Honey Glazed'	wWal oSch
	'Honey Mustard'	oMid wAit
	'Honey Scoop'	wAit
	'Honeymoon Suite'	oSch wAit
	'Honky Tonk Blues'	wWal oSch oMid wAit oCoo
	'Honky Tonk Hussy'	wAit
	'Honorabile'	wAit
	'Hoodlum'	oMid wAit
	hookeri	oJoy oRus
	'Horatio'	oSch oMid
	'Hostess Royale'	oMid
	'Hot'	oSch wAit
	'Hot Buttons'	wAit
	'Hot Chocolate'	oMid
	'Hot Fudge'	oSch
	'Hot Gossip'	oSch oMid
	'Hot Jazz'	oMid
	'Hot Night'	last listed 99/00
	'Hot Spice'	oSch wAit
	'Hot Spiced Wine'	oSch
	'Hot Streak'	oCoo
	'Hot Summer Night'	oMid
	'Hot Wheels'	oSch
	'Hotdogs and Mustard'	wAit
	'Hotseat'	oMid
	'Hottentot'	oSch wAit
	'Houdini'	wWal
	'Howdy Folks'	wWal
	'Huckleberry Fudge'	oMid oCoo
	'Hula Hoop'	oMid wAit
	humilis	wMtT
	'Humohr'	oMid
	'Hurricane Season'	wAit
	'Hurrin' Hoosier'	last listed 99/00
	'Hustle'	oMid
	'Hyenasicle'	oMid
	'I Be Pink'	wAit
	'I Love You'	oMid
	'Ice and Indigo'	oSch oMid wAit
	'Ice Cream Treat'	oCoo
	'Ice Etching'	oMid
	'Iced Tea'	wAit
	'Iceland'	oMid
	'Ida Red'	wWal
	'Iditarod'	oMid oCoo
	'Idol's Dream'	wWal
	'I'll Fly Away'	last listed 99/00
	'I'm Pretty'	oCoo
	'I'm Yours'	oMid
	'Imagette'	oSch
	'Imagine Me'	last listed 99/00
	'Imbue'	oMid
	'Immortality'	oSch oMid oGar wAit oCoo
	'Impeccable Taste'	oCoo
	'Imperative'	oMid wAit
	'Impersonator'	oMid
	'Impressionist'	wWal oSch oMid wAit
	'Imprimis'	oMid
	'In Fashion'	wAit
	'In Flight'	last listed 99/00
	'In Limbo'	oMid wAit
	'In Person'	oCoo
	'In Reverse'	oSch wAit oCoo
	'In the Chips'	wAit
	'In the Mist'	oMid
	'In the Mood'	oMid
	'In the Red'	wAit
	'In Town'	wWal
	'Inca Doll'	wAit
	'Incantation'	wWal
	'Incendiary'	oMid wAit
	'Index'	oMid wAit
	'Indian Caper'	oSch
	'Indigo Doll'	oMid wAit
	'Indigo Princess'	wWal oSch oMid
	'Indiscreet'	wWal oSch
	'Infernal Fire'	oMid wAit
	'Inferno'	wWal oCoo
	'Infinite Grace'	wWal
	'Infinity Ring'	oMid
	'Inga Ivey'	wWal oCoo
	'Ingenious'	oMid
	'Inheritance'	wWal
	'Inky Dinky'	last listed 99/00
	'Inner Journey'	last listed 99/00

	'Inner Peace'	oSch
	'Innocent Blush'	oMid
	innominata	oHan oRus oSis oTri
	'Instant Hit'	oSch
	'Interesting Expression'	oMid wAit
	'Interpol'	oCoo
	'Into the Night'	wWal
	'Intuition'	wWal
	'Invitation'	last listed 99/00
	'Irene Frances'	wAit
	'Irene Nelson'	wWal
	'Iris Bohnsack'	wAit
	'Irish Moss'	wAit
	'Island Dancer'	oSch oMid wAit oCoo
	'Island Song'	oMid
	'Island Sunset'	wWal wAit
	'Island Surf'	wAit
	'Isn't This Something'	oMid wAit
	'Istanbul'	oMid
	'It's Delicious'	oMid
	'It's Magic'	oMid wAit
	'I've Got Rhythm'	oMid
	'Ivory Way'	oMid
	'Jacob's Well'	wAit
	'Jade Jewels'	wAit
	'Jade Maid'	wAit
	'Jaime Lynn'	wWal
	'Jam Session'	wWal
	'James Bond'	oMid wAit
	'James P.'	wAit
	'Jan Katz'	last listed 99/00
	'Jane Phillips'	wWal
	japonica	wAit oOEx
va	***japonica*** 'Aphrodite'	oGos oGre
	japonica 'Variegata'	wHer wAit wEde
	japonica white form	oGar
	'Jazz Festival'	wWal oSch oMid
	'Jazz Me Blue'	wWal oMid wAit
	'Jazzamatazz'	oSch
	'Jazzebel'	wWal
	'Jazzed Up'	wWal oSch oMid wAit
	'Je t'Aime'	oMid
	'Jean Hoffmeister'	wWal
	'Jelly Roll'	wWal
	'Jennifer Rebecca'	oSch oMid wAit oCoo
	'Jesse's Song'	oSch wAit oCoo
	'Jewel Baby'	wAit
	'Jeweled Ivory'	oMid
	'Jeweler's Art'	oSch oMid wAit
	'Jigsaw'	wWal
	'Jitterbug'	oSch oMid
	'Jo Pete'	oMid
	'Joe Cool'	oSch
	'John'	wAit
	'John Hoehner'	oMid
	'John Kearney'	oMid wAit
	'Johnny Reb'	wWal oSch wAit
	'Jolly Joey'	wAit
	'Joseph's Coat'	oMid wAit
	'Jovial Vagabond'	wWal
	'Joy Joy Joy'	oMid wAit
un	'Joy Terry'	wWal
	'Joyce McBride'	oSch
	'Joyful News'	wWal
	'Juan Valdez'	oMid
	'Juicy Fruit'	oMid
	'Jump for Joy'	oCoo
	'Jumping'	oSch
	'Jungle Princess'	wAit
	'Jungle Warrior'	oSch
	'Jurassic Park'	oSch oMid wAit
	'Juris Prudence'	wWal oCoo
	'Just A Flirt'	oMid
	'Just Dance'	last listed 99/00
	'Just Do It'	wAit
	'Just for Sophie'	oMid oCoo
	'Just Magic'	oMid
	'Just My Style'	oCoo
	kaempferi	see **I. *ensata***
	'Kah-nee-ta'	oCoo
	'Kaleidoscope'	wAit
	'Karen'	wWal
	'Kathleen Kay Nelson'	oMid wAit oCoo
	'Katie Pie'	oMid
	'Kati's Blush'	oCoo
	'Katy Petts'	oSch
	'Keeping Up Appearances'	oMid
	'Keiko's World'	oCoo
	'Keirith'	oMid wAit
	'Kelly Lynne'	oMid
	kemaonensis	oJoy
	'Kentucky Derby'	wWal oCoo
	'Kentucky Woman'	oSch oMid
	'Kermit'	wAit
	'Kevin's Theme'	oSch oMid oCoo
	'Kim's Melody'	oCoo
	'King's Castle'	wWal
	'Kiss Me Quick'	oMid
	'Kiss of Peace'	wWal oCoo
	'Kiss the Dawn'	oMid
	'Kitten'	oMid wAit
	'Kiwi Capers'	wAit
	'Kiwi Cheesecake'	oMid
	'Kiwi Wine'	oMid
	'Klondike Katie'	wWal
	'Klondike Lil'	wAit
	'Knick Knack'	wCul
	'Knight Templar'	last listed 99/00
	'Knock 'Em Dead'	oMid oCoo
	'Kona Nights'	wAit
	'Krugerand'	oCoo
	'Kuniko'	wWal
	'La Valse'	oSch oMid oCoo
	'Lace Artistry'	wWal oSch wAit
	'Lace Jabot'	wWal oCoo
	'Lace Legacy'	oMid wAit
	'Laced Cotton'	wWal oSch oCoo
	'Lacy Day'	oCoo
	'Lacy Primrose'	oMid
	'Lacy Snowflake'	wWal oCoo
	'Lady Bird Johnson'	oMid
	'Lady Celesta'	last listed 99/00
	'Lady Fire'	last listed 99/00
	'Lady Friend'	wWal oCoo
	'Lady in Red'	oSch oGar
	'Lady Juliet'	wAit
	'Lady Madonna'	wWal
	'Lady of Spain'	wWal
	'Lady of Waverly'	wWal
	laevigata 'Alba'	oGar oGre
	laevigata burgundy wine seedling	wAit
	laevigata 'Mountain Brook'	wAit
qu	***laevigata*** 'Queen of Violets'	oSsd
	laevigata 'Rose Queen'	see **I. *ensata*** 'Rose Queen'
	laevigata 'Snowdrift'	wGAc oGar
	laevigata 'Variegata'	oFor wCol oHed oGar wAit oNWe oOut oGre wFai wSte
	laevigata 'Wild Wine'	wWal oSsd
	'Lamb's Share'	oMid
	'Lanai'	oMid wAit
	'Lancer'	wAit
	'Land o' Lakes'	wWal
	'Land of Judah'	wWal
	'Laredo'	oMid
	'Lark Ascending'	oMid
	'Larry Gaulter'	oCoo
	'Larue Boswell'	oMid
	'Las Vegas'	oMid
	'Lascivious Dreams'	oMid
	'Laser Light'	wWal
	'Laser Print'	last listed 99/00
	'Lasso Lane'	last listed 99/00
	'Last Chance'	oSch oMid
	'Last Emperor'	wWal oCoo
	'Last Hurrah'	oSch
	'Lasting Memory'	wAit
	'Lasting Romance'	wAit
	'Late for School'	last listed 99/00
	'Late Liftoff'	wWal
	'Later On'	oMid
	'Latin Lady'	wWal wAit
	'Latin Lark'	oMid
	'Latin Lover'	wWal oCoo
	'Latin Rock'	wWal oSch
	'Lavalier'	oMid
	'Lavender Luck'	wWal
	'Lavish Lover'	oMid
	lazica	oJil
	'Leading Light'	oMid
	'Leaping Dolphin'	oSch

	'Leda's Lover'	wWal oSch
	'Legato'	oMid
	'Lemon and Spice'	wWal
	'Lemon Chess'	oMid
	'Lemon Dew'	oMid oCoo
	'Lemon Fever'	wWal
	'Lemon Lustre'	wWal
	'Lemon Mist'	wWal
	'Lemon Pop'	oSch wAit oCoo
	'Lemon Silence'	oMid
	'Lemon Up'	wAit
	'Lemon Whip'	wAit
	'Lemon Wine'	last listed 99/00
	'Lena Baker'	oMid
	'Lenora Pearl'	oSch wAit
	'Lenten Prayer'	wWal oMid
	'Letmeentertainu'	oMid
	'Let's Boogie'	oSch oMid
	'Letter from Paris'	oMid
	'Levity'	oSch wAit
	'Liaison'	wWal oCoo
	'Libation'	wAit
	'Liberty Light'	oMid
	'Life of Riley'	wAit
	'Light and Airy'	oMid
	'Lighted Window'	wWal
	'Lighted Within'	wWal
	'Lightning Bolt'	oMid
	'Lightning Streak'	oMid oCoo
	'Lightshine'	oSch
	'Lilac Flame'	wWal
	'Lilac Topper'	wWal
	'Lima'	oMid
	'Lime Fizz'	wWal
	'Lime Ruffles'	wAit
	'Lime Smoothy'	wAit
	'Lions Share'	wAit
	'Lipstick Lies'	oMid
	'Little Annie'	oSch
	'Little Clown'	wAit
	'Little Dream'	oSch
	'Little Episode'	oSch
	'Little Mermaid'	wAit
	'Little Much'	oCoo
	'Little Paul'	wAit
	'Little Showoff'	wAit
	'Little Snow Lemon'	oSch
	'Live Coals'	wAit
	'Live Jazz'	oSch
	'Live Music'	wWal
	'Living Legacy'	oCoo
	'Living Picture'	oCoo
	'Living Right'	oCoo
	'Liz'	wWal
	'Local Color'	oSch oMid wAit
	'Loganberry Squeeze'	oCoo
	'Lois Parrish'	oMid
	'Lois Ranier'	oSch
	'Londonderry'	oMid wAit
	'Lonely Hearts'	oMid
	'Lonely Street'	oMid wAit
	'Lonesome Dove'	oMid
	'Lonesome Stranger'	oCoo
	'Looking for Love'	oCoo
	'Lookout Point'	oMid
	'Loop the Loop'	wWal
	'Lord Baltimore'	wWal oMid
	'Lord Olivier'	oMid
	'Lorilee'	wWal
	'Los Coyotes'	oMid
	'Lotus Land'	oMid
	'Loudoun Lassie'	wWal
	'Louis d'Or'	oMid
	'Louise Blake'	wWal
	Louisiana hybrids	oHug
	Louisiana hybrids 'Acadian Miss'	oSch wAit
	Louisiana hybrids 'Alouette'	wAit
	Louisiana hybrids 'Ann Chowning'	oFor
	Louisiana hybrids 'Ann's Child'	wAit
	Louisiana hybrids 'Berenice'	wAit
	Louisiana hybrids 'Black Gamecock'	oSch wGAc oHug oGar oSsd oSho oGre wBox oBRG
	Louisiana hybrids 'Bonaparte'	wAit
	Louisiana hybrids "bout Midnight'	wAit
	Louisiana hybrids 'Byron Bay'	wAit

	Louisiana hybrids 'C'est Si Bon'	wAit
	Louisiana hybrids 'Cherry Cup'	oFor
	Louisiana hybrids 'Clara Goula'	oFor
	Louisiana hybrids 'Clyde Redmond'	wGAc wBox
	Louisiana hybrids 'Cotton Plantation'	wAit
	Louisiana hybrids 'Coup d'Etat'	wAit
	Louisiana hybrids 'Coupe de Ville'	wAit
	Louisiana hybrids 'Dorothea K. Williamson'	oGar
	Louisiana hybrids 'Dural Dreamtime'	wAit
	Louisiana hybrids 'Easter Tide'	oSch
	Louisiana hybrids 'Frank Chowning'	oSch
	Louisiana hybrids 'Full Eclipse'	oFor
	Louisiana hybrids 'Geisha Girl'	oFor
	Louisiana hybrids 'Glowlight'	oFor
	Louisiana hybrids 'Inner Beauty'	wAit
	Louisiana hybrids 'Koorawatha'	wAit
un	Louisiana hybrids 'Lola Comeaux'	oFor
	Louisiana hybrids 'Mac's Blue Heaven'	last listed 99/00
	Louisiana hybrids 'Malibu Magic'	wAit
	Louisiana hybrids 'Margaret Hunter'	oFor
	Louisiana hybrids mixed	oGar
	Louisiana hybrids 'Parade Music'	wAit
	Louisiana hybrids 'Rapport'	wAit
	Louisiana hybrids 'Sinfonietta'	oFor oSch wAit
	Louisiana hybrids 'Spanish Ballet'	wAit
	Louisiana hybrids 'Sun Fury'	oFor
	Louisiana hybrids 'Swiss Chalet'	wAit
	'Louisiana Lace'	wWal
	'Love Comes'	oMid
	'Love the Sun'	wWal oCoo
	'Love Theme'	wWal
	'Lovely Dawn'	oMid wAit
	'Lovely Jan'	wWal
	'Lover's Charm'	oCoo
	'Lover's Lane'	oMid
	'Low Ho Silver'	oSch wAit
	'Low Life'	wAit
	'Low Spirits'	oMid wAit
	'Lowell Storm'	last listed 99/00
	'Loyalist'	oSch oCoo
	'Lucille Richardson'	oMid
	'Lucky Draw'	oCoo
	'Lullaby of Spring'	wWal
	'Lumalite'	wAit
	'Luminosity'	wAit
	'Lunar Frost'	oMid wAit
	'Luxor Gold'	oSch oMid
	'Luxury Lace'	wWal
	'Lyme Tyme'	oMid wAit
	'Lyrique'	oMid
un	'Mad Mode'	wWal
	'Madame Froth'	oMid
	'Madeira'	wWal
	'Magharee'	wWal oSch oMid wAit oCoo
	'Magic Bubbles'	oMid wAit
	'Magic Child'	oMid
	'Magic Fountain'	oSch
	'Magic Man'	wWal
	'Magic Palette'	oMid
	'Magic Show'	oMid wAit
	'Magical Encounter'	oSch
	'Magician's Apprentice'	last listed 99/00
un	'Mahogany'	oGre
	'Maid of Orange'	oSch wAit
	'Make Believe Magic'	oMid wAit
	'Making Eyes'	oSch
	'Making Waves'	oCoo
	'Malaguena'	oCoo
	'Mallory Kay'	oMid oCoo
	'Mallow Dramatic'	oSch oMid wAit
	'Maltby Dream'	wWal
	'Man about Town'	oMid
	'Managua'	oMid wAit
	'Mango Entree'	oMid
	'Mango Tango'	oMid
	'Manhattan Blues'	oSch
	'Manor Born'	oMid
	'Many Happy Returns'	wAit
	'Many Thanks'	oCoo
	'Maple Treat'	oMid
	'Marbre Bleu'	oMid
	'Marcy Michele'	oMid wAit
	'Margaret Beaufort'	wAit
	'Margaret Inez'	oMid

'Margarita'	wWal
'Marginal Way'	wAit
'Maria Tormena'	wAit
'Mariah'	wWal oSch
'Mariposa Autumn'	oMid
'Mariposa Skies'	oMid
'Marksman'	oMid wAit
'Marmalade'	wWal
'Marriage Vows'	oCoo
'Marris'	wAit
'Martha's Gold'	wAit
'Marthella'	oMid
'Martini Mist'	oMid
'Mary Constance'	oMid
'Mary D'	oMid
'Mary Ellen Nichols'	oSch wAit
'Mary Frances'	wWal oSch wAit
'Mary's Lamb'	oSch
'Masked Bandit'	oMid wAit
'Master Plan'	oSch oMid wAit
'Master Sleuth'	oSch
'Master Touch'	wWal oCoo
'Matrix'	oMid
'Maui Gold'	wAit
'Maui Magic'	wAit
'Maui Moonlight'	oSch wAit
'Maui Surf'	wAit
'Mauna Kea'	last listed 99/00
'Mauvelous'	oMid
'Maverick's Game'	oMid wAit
'Mavis Waves'	oMid
'Maya Mint'	wAit
'Maybe an Angel'	oMid
'McKellar's Grove'	oCoo
'McKenzie Violet'	see I. Sino-Siberian hybrids 'McKenzie Violet'
'Meadow Court'	oHed oWnC
'Meadow Creek'	oMid
'Mean Streak'	last listed 99/00
'Mega Charm'	wWal
'Megabucks'	oSch oMid wAit
'Megglethrop'	oSch wAit
'Melancholy Man'	last listed 99/00
'Melba Hamblen'	oMid oCoo
'Melissa Sue'	wWal
mellita	see **I. *suaveolens***
'Mellow Magic'	oSch
'Mel's Honor'	oMid
'Melted Butter'	oMid
'Memory Song'	oMid
'Memphis Blues'	oCoo
'Merit'	wAit
'Merlot'	oSch oMid
'Merry Madrigal'	oCoo
'Merry Masque'	last listed 99/00
'Meshack'	oMid
'Mesmerizer'	oSch oMid
'Messenger'	oMid
'Metolius Blues'	oSch
'Michael Paul'	oSch
'Michele Taylor'	wWal
'Michelle Stadler'	oMid
'Michigan Pride'	wWal
'Midnight Dancer'	wWal
'Midnight Express'	wWal
'Midnight Lace'	wWal
'Midnight Madonna'	oMid
'Midnight Mist'	oSch
'Midnight Oil'	oMid wAit
'Midsummer Night's Dream'	oMid
'Milano'	oMid
milesii	last listed 99/00
'Mill Valley'	wWal oGar
'Millenium Sunrise'	oSch
'Million Miles'	wWal
'Mind Bend'	oMid
'Mind Reader'	oMid wAit
'Mind's Eye'	oMid
'Ming'	wAit
'Ming Rose'	oMid
'Mini Champagne'	oMid
'Minidragon'	oMid wAit
'Minnesota Glitters'	wWal
minutaurea	wCol oSis
'Miss Indiana'	wWal
'Miss Nellie'	oCoo
'Miss Sunshine'	wAit
'Missouri Smile'	oMid
missouriensis	oFor
'Mister Roberts'	oSch
'Misty Lady'	wAit
'Miz Mary'	oMid
'Mochaccino'	oMid
'Modern Classic'	wWal
'Modern Story'	oMid
'Mogul'	oMid
'Mohr Pretender'	oSch
'Monaco'	wWal
'Monet's Blue'	oSch
'Monet's Lady'	oMid
'Money'	oCoo
'Montevideo'	oCoo
'Mood Swing'	oMid
'Moomba'	wWal
'Moon Journey'	oSch oCoo
'Moon Love'	last listed 99/00
'Moonglade'	oMid wAit
'Moonlight Dance'	oMid
'Moonlit'	last listed 99/00
'Moon's Delight'	wWal
'More Refreshing'	wWal
'Morning Mood'	oMid wAit
'Morning Show'	wAit
'Morning's Blush'	oMid wAit
'Morocco'	last listed 99/00
'Morse Code'	oMid wAit
'Mostest'	oSch wAit
'Mother Earth'	oSch oMid wAit oCoo
'Mother's Little Helper'	oCoo
'Motto'	wAit
'Mountain Echo'	oMid
'Mountain Heather'	oSch
'Mountain Majesty'	oMid wAit
'Mountain Melody'	oCoo
'Mountain Shadows'	oMid
'Moustache'	last listed 99/00
'Move Over'	wAit
'Movie Magic'	oSch
'Much Obliged'	oMid
'Mulberry Echo'	oMid
'Mulberry Punch'	oSch
'Muse'	oMid wAit
'Music'	oMid
'Music Box Dancer'	oSch
'My Forte'	oCoo
'My Friend Jonathan'	oMid
'My Girl'	wWal oSch
'My Jodie'	oMid
'My Line'	oMid
'My Mauve'	last listed 99/00
'My Pretty Valentine'	oCoo
'My Souvenir'	oMid
'My Way'	oMid wAit
'Myst'	oMid
'Mysterious Balance'	oMid
'Mystic Glow'	wAit
'Mystic Lace'	oCoo
'Mystic Potion'	oMid
'Mystic Rites'	oMid wAit
'Mystic's Muse'	wWal oSch
'Mystique'	wWal wAit
'Nanny'	oSch
'Nate Rudolph'	oMid
'Native Warrior'	wHer
'Natural Reflexion'	oMid
'Nature's Own'	oMid
'Nautical Flag'	last listed 99/00
'Navajo Blanket'	wWal
'Navajo Jewel'	last listed 99/00
'Navy Blues'	last listed 99/00
'Navy Doll'	oSch
'Navy Strut'	wWal oCoo
'Near and Dear'	oMid
'Near Myth'	oMid wAit
'Neat Pleats'	last listed 99/00
'Nectar'	oSch wAit oCoo
'Neon Cowboy'	oMid
'Neon Rainbow'	wWal
'Neptune's Cloak'	oSch
'Neutron Dance'	wWal oSch

'New Centurion'	oSch wAit oCoo
'New Kid'	wAit
'New Leaf'	oMid
'New Lover'	last listed 99/00
'New Moon'	wWal wAit
'New Order'	last listed 99/00
'New Rochelle'	wWal
'New Tomorrow'	wWal
'Nigerian Raspberry'	oMid
'Night Attack'	wAit oCoo
'Night Edition'	wWal
'Night Fires'	wAit
'Night Flames	wAit
'Night Game'	oSch oMid wAit
'Night Owl'	oCoo
'Night Ruler'	oSch oMid oCoo
'Night Shift'	oSch
'Night to Remember'	oMid oCoo
'Nights of Gladness'	oCoo
'No Bikini Atoll'	oCoo
'No Contest'	oMid
'No Down Payment'	oMid wAit
'Noble House'	oCoo
'Noon Siesta'	wWal
'Nora Eileen'	oSch oMid
'Nordic Kiss'	wWal
'Nordica'	oSch oMid
'Noreen's Delight'	oMid
'North Pacific Seas'	oCoo
'Northern Jewel'	oSch
'Northwest Pride'	wWal
'Northwest Progress'	oSch oMid
'Notable'	oMid
'Nothing But Net'	oMid wAit
'Nothing to Lose'	oCoo
'Notorious'	oMid wAit
'Nut Ruffles'	wAit
'O. K. Corral'	last listed 99/00
'O'Brien's Choice'	oCoo
'O'So Pretty'	last listed 99/00
'Ocean Jewels'	oMid
'Ocean Pacific'	oCoo
'Ocelot'	oMid
'October Splendor'	oMid wAit
'Off Color Joke'	oMid
'Oh Jamaica'	oSch oMid wAit
'Oh James'	oMid
'Oktoberfest'	oCoo
'Ola Kala'	wWal wAit
'Old Black Magic'	oSch oMid
'Old Blue Eyes'	oSch
'Old Devil Moon'	oMid wAit
'Old Loyalties'	oMid
'Olive Garden'	last listed 99/00
'Olney Belle'	wAit
'Olympiad'	oCoo
'Olympic Challenge'	oCoo
'Olympic Pink'	wWal oCoo
'Ominous Stranger'	wAit
'On Edge'	wWal
'On Fire'	oSch
'One Desire'	wWal
'One Little Pinkie'	wAit
'Only You'	oMid
'Opal Brown'	oMid
'Opal's Legacy'	oMid
'Open Arms'	wWal
'Open Sky'	oSch
'Orageux'	oMid
'Orange Celebrity'	oCoo
'Orange Embers'	oMid wAit
'Orange Gumdrops'	wAit
'Orange Harvest'	wAit
'Orange Impact'	oMid
'Orange Jubilee'	oMid wAit
'Orange Outrage'	wAit
'Orange Petals'	oSch wAit
'Orange Plume'	oCoo
'Orange Pop'	oMid
'Orange Star'	wWal
'Orange Surprise'	oMid
'Orange Tiger'	oSch wAit
'Orbiter'	oCoo
'Orchard Brite'	see **I. *douglasiana*** 'Orchard Brite'
'Orchid Ensemble'	oMid

	'Orchidarium'	wWal
	'Oregold'	oSch
	'Oregon Skies'	oSch oCoo
	'Oriental Glory'	wWal
	orientalis	oFor
	'Orinoco Flow'	last listed 99/00
	'Orknies'	oMid
	'Ornament'	oSch
	'Osaka'	oMid
	'Ostentatious'	oMid
	'Ouragan'	oMid
	'Out of Control'	oMid
	'Out Yonder'	wWal
	'Outrageous Fortune'	wWal
	'Outreach'	wWal
	'Ovation'	wWal
	'Over the Blues'	last listed 99/00
	'Overdrawn'	oMid
	'Overjoyed'	wWal oSch oMid wAit
	'Overnight Sensation'	oSch oMid wAit
	'Overseas'	oMid wAit
	'Owyhee Desert'	oMid wAit
	'Oycz'	oMid
	'Ozark Dream'	wAit
	'Ozark Evening'	wAit
	'Ozark Jewel'	wAit
	'P. T. Barnum'	oCoo
	'Pacific Cloud'	oMid wAit
	Pacific Coast hybrids	oFor oNWe
un	Pacific Coast hybrids 'Ami Royale'	oSis
	Pacific Coast hybrids 'Big Money'	wAit oNWe
	Pacific Coast hybrids 'Big Wheel'	wAit
	Pacific Coast hybrids 'Billie Blue Jay'	wAit
	Pacific Coast hybrids 'Black Eye'	wAit
	Pacific Coast hybrids blue and white	oNWe
	Pacific Coast hybrids blue/lavender edge	oNWe
	Pacific Coast hybrids 'Blue Moment'	wAit
	Pacific Coast hybrids bronze/violet and buff	oNWe
un	Pacific Coast hybrids 'Buff Beauty'	oSis
	Pacific Coast hybrids 'Carrot Top'	wAit
	Pacific Coast hybrids 'Coastal Glow'	oGar oSis
	Pacific Coast hybrids 'Cozumel'	wAit
	Pacific Coast hybrids dark purple	oNWe
	Pacific Coast hybrids 'Drive You Wild'	last listed 99/00
	Pacific Coast hybrids 'Earthquake'	wAit
	Pacific Coast hybrids 'Endless'	last listed 99/00
	Pacific Coast hybrids 'Eye Patch'	wAit
	Pacific Coast hybrids 'Gamay'	wAit
	Pacific Coast hybrids 'Hot Number'	wAit
	Pacific Coast hybrids 'Idylwild'	wAit
	Pacific Coast hybrids 'Las Lomas'	wAit
	Pacific Coast hybrids light blue and white	oNWe
	Pacific Coast hybrids 'Los Californio'	wAit
	Pacific Coast hybrids maroon	oNWe
	Pacific Coast hybrids 'Mission Santa Cruz'	oSis
	Pacific Coast hybrids mixed	wCol oGar oNWe oGre
	Pacific Coast hybrids 'National Anthem'	wAit
	Pacific Coast hybrids 'Pacific Rim'	wAit
	Pacific Coast hybrids 'Pacific Snowball'	wAit
	Pacific Coast hybrids 'Pacific Snowflake'	wAit
	Pacific Coast hybrids purple and yellow	oNWe
	Pacific Coast hybrids 'Roaring Camp'	oNWe
	Pacific Coast hybrids 'Simply Wild'	wAit
	Pacific Coast hybrids 'Trancas'	wAit
	Pacific Coast hybrids 'Tulum'	wAit
	Pacific Coast hybrids 'Tunitas'	last listed 99/00
	Pacific Coast hybrids white	oNWe
	Pacific Coast hybrids #1	oNWe
	Pacific Coast hybrids #19	oNWe
	Pacific Coast hybrids #20	oNWe
	Pacific Coast hybrids #24	oNWe
	Pacific Coast hybrids #27	oNWe
	'Pacific Mist'	wWal oSch
	'Pacific Moon'	see **I. *douglasiana*** 'Pacific Moon'
	'Pacific Panorama'	wWal wAit
	'Pacific Shores'	wWal
	'Pacific Smoothie'	see **I.** Cal-Sibe hybrids 'Pacific Smoothie'
	'Pacific Starprint'	see **I.** Cal-Sibe hybrids 'Pacific Starprint'
	'Pagan Dance'	wWal oSch oMid wAit
	'Pagan Goddess'	oMid wAit
	'Pagan Mirth'	oMid
	'Pagan Pink'	last listed 99/00
	'Paint It Black'	oSch wAit
	'Paint the Scene'	oMid

'Painted Blue'	oMid
'Painted Clouds'	oCoo
'Painted Pictures'	wWal oCoo
'Painted Pink'	oMid
'Painted Softly'	last listed 99/00
'Paisano'	wWal
'Palace of Thoughts'	oCoo
pallida 'Argentea Variegata'	oAls oAmb oMid wAit oSis oGre
pallida 'Aurea Variegata'	see **I. *pallida*** 'Variegata'
pallida var. ***dalmatica***	see **I. *pallida*** ssp. ***pallida***
pallida ssp. ***pallida***	wFGN
pallida 'Variegata'	oFor wCul wWal oSch oHed oMid oGar wAit oSis oCoo wHom wBox
pallida 'Zebra'	see **I. *pallida*** 'Variegata'
'Panama Fling'	oCoo
'Pandora's Purple'	last listed 99/00
'Panic Button'	oMid
'Pansy Grace'	oMid
'Paradise'	wAit
'Paradise Found'	oCoo
'Pardner'	wAit
'Paris Lights'	wWal
'Parisian Flight'	wAit
'Partita'	last listed 99/00
'Parts Plus'	wAit
'Party Lights'	oMid
'Pass the Shades'	wWal
'Pass the Wine'	wWal oCoo
'Passion Flower'	oMid wAit
'Pastel Beau'	oMid
'Pastel Charm'	last listed 99/00
'Pastel Ribbons'	oMid wAit
'Patacake'	oSch
'Patent Leather'	wWal
'Patriot's Gem'	wAit
'Patriotic Banner'	oMid wAit
'Patriotic Colors'	oMid
'Pauline Cooley'	wWal oCoo
'Pauline Roderick'	oMid
'Pawnee Princess'	wAit
'Payoff'	wAit
'Peace and Harmony'	last listed 99/00
'Peaceful Moment'	oCoo
'Peach Bisque'	last listed 99/00
'Peach Cooler'	oMid wAit
'Peach Ice Cream'	oMid wAit
'Peach Jam'	wAit
'Peach Petals'	wAit
'Peach Picotee'	wWal
'Peachy Face'	last listed 99/00
'Pearls and Gold'	oMid
'Pearls of Autumn'	oCoo
'Pedigree'	oMid wAit
'Peignoir'	oMid
'Peking Summer'	wWal oSch
'Pele'	oSch oMid wAit
'Pemcaw'	last listed 99/00
'Penchant'	oMid
'Penny Lane'	oMid
'Perfect Gift'	oMid wAit
'Perfect Harmony'	oMid
'Perfect Pearl'	oMid
'Perfect Pitch'	oMid oCoo
'Perfume Shop'	oMid wAit
'Perhaps Love'	oMid
'Perilous Journey'	oCoo
'Persian Berry'	last listed 99/00
'Persian Gown'	oMid oCoo
'Persian Lantern'	wAit
'Personal Touch'	oMid
'Petite Ballet'	oMid wAit
'Petite Monet'	wAit
'Petite Polka'	oSch
'Petite Posy'	wWal
'Pewter Treasure'	oMid
'Phaeton'	oMid wAit
'Pharaoh's Spirit'	oMid
'Pheasant Feathers'	oCoo
'Phil's Pick'	oMid
'Photo Op'	oMid
'Physique'	last listed 99/00
'Piano Bar Melodies'	oMid
'Pibbling'	oMid
'Picante'	oMid wAit
'Picasso Moon'	oSch
'Picasso Print'	last listed 99/00
'Picture This'	oCoo
'Piggy Bank'	last listed 99/00
'Pillow Fight	oMid wAit
'Pink-All-Over'	wAit
'Pink Angel'	wWal
'Pink Attraction'	wAit
'Pink Belle'	oSch
'Pink Blink'	wAit
'Pink Bubbles'	oMid wAit oCoo
'Pink Buttons'	wAit
'Pink Caper'	oSch
'Pink Challenge'	oMid
'Pink Charming'	wAit
'Pink Confetti'	wWal
'Pink Flamingos'	oMid
'Pink Froth'	oCoo
'Pink Horizon'	wWal
'Pink Magnolia'	oMid
'Pink Pele'	oMid wAit
'Pink Pleasure'	wWal
'Pink Prevue'	oSch wAit
'Pink Reprise'	oCoo
'Pink Sapphire'	wAit
'Pink Sleigh'	wWal
'Pink Smash'	oMid
'Pink Starlet'	oMid wAit oCoo
'Pink Swan'	wWal oCoo
'Pink Taffeta'	wWal
'Pink Twilight'	oMid
'Piper's Flute'	wWal
'Piquant Fancy'	see **I.** Sino-Siberian hybrids 'Piquant Fancy'
'Pirate's Patch'	oSch wAit
'Pirate's Quest'	wWal
'Pitter Patter'	wAit
'Pixie Flirt'	wAit
'Pixie Pirate'	oMid
'Planned Treasure'	wWal
'Play Pretty'	wAit
'Play With Fire'	wWal oSch
'Playing Around'	oMid
'Pleasantly Warm'	oMid
'Pleasure Cruise'	wWal
'Pledge Allegiance'	oSch
'Plum Lucky'	wAit
'Plum Tart'	see **I. *graminea*** 'Plum Tart'
'Plume d'Or'	oMid
'Poco Taco'	wAit
'Poem of Ecstacy'	oMid wAit oCoo
'Poet Laureate'	oSch
'Point In Time'	oSch oCoo
'Point Made'	last listed 99/00
'Polar Queen'	last listed 99/00
'Polar Seas'	last listed 99/00
'Polynesian Flame'	wWal
'Pom Pom Girl'	oMid
'Pompeii Lady'	oMid
'Pond Lily'	oMid wAit
'Pontiff'	wWal
'Pooginook'	last listed 99/00
'Porcelain Frills'	wAit
'Portrait of Larrie'	wWal oCoo
'Post Time'	wWal
'Pour It On'	oMid
'Power of One'	oMid
'Power Surge'	oSch
'Prairie Thunder'	oMid
un 'Prancing Stallion'	wWal
'Precious Moments'	wWal
'Preferred Stock'	wWal
'Premier Edition'	last listed 99/00
'Prestige Item'	oMid
'Prettie Print'	wWal
'Pretty Beginning'	wAit
'Pretty In Pink'	oSch
'Pretty Is'	oMid
'Pretty Quirky'	wAit
'Priceless Pearl'	wWal
'Prince George'	oMid wAit
'Prince of Burgundy'	oSch wAit
'Prince of Tides'	oMid
'Princess Bluebeard'	oSch
'Princess Kiss'	oCoo
'Princess Pittypat'	oMid
'Princesse Caroline de Monaco'	oMid

	'Prism'	wAit
	prismatica	oFor wCul oRus wEde
	'Prissy Miss'	oCoo
	'Private Reserve'	wAit
	'Private Stock'	oMid
	'Private Treasure'	oMid wAit
	'Privileged Character'	oSch wAit
	'Progressive Attitude'	oMid wAit
	'Prom Night'	wWal
	'Promises Promises'	oMid
	'Prosperous Voyage'	oMid oCoo
	'Protocol'	oSch oMid wAit
	'Proton'	oMid
	'Proud Tradition'	oSch wAit oCoo
	'Proven Stock'	oMid wAit
	pseudacorus	oFor wSoo wGAc oRus oHug oGar oBos oBlo wFai oWnC oTri wBox oUps oSle
	pseudacorus 'Alba'	oCap
	pseudacorus cream	wRob
do	***pseudacorus*** 'Flore Pleno'	oFor oGar
	pseudacorus 'Holden Clough'	see **I.** 'Holden Clough'
	pseudacorus 'Holden's Child'	see **I.** 'Holden's Child'
	pseudacorus 'Linda West'	oGar
	pseudacorus 'Rising Sun'	oAls
	pseudacorus 'Roy Davidson'	wAit
	pseudacorus selected form	oRus
	pseudacorus 'Variegata'	oFor wCri wWal oSch wCol oHug oGar oSho wRob wFai oWnC wTGN oUps
	'Pulsar'	last listed 99/00
	pumila	last listed 99/00
un	***pumila alba***	oSho
	pumila atroviolacea	last listed 99/00
	pumila claret	oWnC
	pumila deep purple	oSho oWnC
	pumila 'Gelber Mantel'	wWoS
	pumila 'Lavendel Plicata'	wWoS
	'Pumpin' Iron'	oSch wAit
	'Pumpkin Center'	oSch
	'Pumpkin Cheesecake'	oMid wAit
	'Punch'	oSch oCoo
	'Punchline'	wWal
	'Puppet Baby'	wAit
	'Puppet Master'	last listed 99/00
	'Pure Allure'	wAit
	'Pure as Gold'	oMid wAit
	'Pure-as-the'	last listed 99/00
	'Purple Pepper'	wWal oSch
	'Purple Puma'	wAit
	'Purple Streaker'	wWal oMid
	'Quasar'	last listed 99/00
	'Queen Anne's Lace'	oMid
	'Queen Bee'	oMid
	'Queen Dorothy'	oSch
	'Queen in Calico'	wWal oCoo
	'Queen of Angels'	oSch oMid wAit
	'Queen of Hearts'	wWal oMid
	'Quicken'	oSch wAit
	'Quiet Elegance'	oMid wAit
	'Quiet Friendship'	wWal
	'Quiet Times'	wWal
	'Quietly'	wAit
	'Quito'	oMid
	'Radiant Apogee'	wWal
	'Radiant Burst'	oMid
	'Radiant Summer'	wWal
	'Ragtime'	wWal
	'Rain Dance'	wAit
	'Rain Man'	oSch oMid wAit
	'Rainbow Goddess'	oCoo
	'Rainbow Mountain'	wAit
	'Rainbow Sky'	oMid
	'Rainbow Tour'	oMid
	'Raindance Kid'	wAit
	'Rainy Pass'	wWal
	'Raku Blaze'	oMid wAit
	'Ramblin' Rose'	wWal
	'Rancho Grande'	oCoo
	'Rancho Rose'	wWal
	'Rapscallion'	wAit
	'Rapture in Blue'	oSch wAit oCoo
	'Rapture's Edge'	wWal
	'Rare Edition'	oSch oMid
	'Rare Occasion'	wAit
	'Rare Quality'	oSch
	'Rare Treat'	oSch oGar oCoo
	'Raspberry Blush'	oSch oGar
	'Raspberry Frills'	wWal
	'Raspberry Fudge'	last listed 99/00
	'Raspberry Rose'	oSch
	'Raspberry Splendor'	wAit
	'Raspberry Whip'	last listed 99/00
	'Rave On'	oSch
	'Rave Review'	wAit
	'Raven's Quote'	wWal oSch oCoo
	'Raven's Roost'	wWal
	'Ravishing'	oMid
	'Raziza'	wWal
	'Razzleberry'	wWal wAit
	'Reality'	oMid wAit
	'Really Mine'	oMid
	'Rebecca Perret'	oMid wAit oCoo
	'Rebus'	oMid wAit
	'Recurrent Event'	wAit
	'Recurring Dream'	oMid wAit
	'Recurring Ruffles'	wAit
	'Red at Night'	oCoo
	'Red Fringe'	oSch
	'Red Hawk'	oSch oMid wAit
	'Red Rider'	oMid
	'Red Rooster'	wAit
	'Red Tornado'	wWal
	'Red Zinger'	oSch
	'Refined'	wAit
	'Regal Affair'	wWal oCoo
	reichenbachii	wMtT
	'Reincarnation'	wWal oMid wAit
	'Rembrandt Magic'	oSch
	'Reminiscence'	oMid wAit
	reticulata	oSho
	reticulata 'Frank Elder'	oSis
	reticulata 'Harmony'	oGar oWoo
	reticulata 'J.S. Dijt'	oWoo
	reticulata 'Joyce'	oWoo
	reticulata 'Katharine Hodgkin'	oSis
	reticulata 'Natascha'	oWoo
	reticulata 'Pauline'	oGar
	reticulata 'Purple Gem'	oGar
	'Returning Rose'	wAit
	'Revival Meeting'	wWal oSch oMid
	'Revolution'	wWal
	'Rhapsody in Bloom'	oCoo
	'Rhonda Fleming'	oSch oMid wAit
	'Rich in Spirit'	oMid
	'Ride the Wind'	wWal oSch
	'Right Direction'	wWal
	'Right Honorable'	oSch
	'Ring Around Rosie'	oCoo
	'Ringer'	oSch oMid wAit
	'Ringo'	wWal wAit oCoo
	'Rinky-Dink'	last listed 99/00
	'Rip City'	oSch
	'Rippling River'	oSch oMid
	'Rippling Rose'	oCoo
	'Rising Moon'	wWal
	'Rising Sun'	oCap
	'Rite of Spring'	oMid oCoo
	'River Hawk'	wWal oSch
	'River Jordan'	last listed 99/00
	'River Runner'	oCoo
	'Riverboat Blues'	wAit oCoo
	'River's Edge'	oMid
	'Road Song'	oMid
	'Robin Goodfellow'	wAit
	'Robin's Egg'	oSch
	x robusta 'Gerald Darby'	wCul oGar oOut wFai
	'Rock Star'	oSch oMid wAit
	'Rockabye'	oMid
	'Rocket Master'	oMid
	'Rocky Mountain High'	wWal
	'Rocky Mountains'	wWal
	'Rococo'	wWal
	'Rodeo Star'	last listed 99/00
	'Rogue'	last listed 99/00
	'Role Model'	wWal oCoo
	'Roman Carnival'	oMid
	'Roman Lover'	wWal
	'Roman Song'	oMid
	'Romantic Evening'	oSch oMid wAit
	'Romantic Interlude'	wAit
	'Ron'	wWal

'Rondetta'	wWal
'Rondo'	wWal
'Rosalie Figge'	wAit
'Rose Princess'	wWal
'Rose Shiner'	wWal
'Rose Tattoo'	wWal
'Rosecraft'	oCoo
'Rosemary's Dream'	wAit
'Rosette Wine'	wWal oSch oMid
'Round Table'	oCoo
'Royal Crusader'	wWal oSch
'Royal Design'	oMid
'Royal Dolly'	wAit
'Royal Intrigue'	oCoo
'Royal Performance'	oMid
'Royal Regency'	wWal oCoo
'Royal Satin'	wWal
'Royal Touch'	wWal oGar
'Royal Viking'	oCoo
'Royal Yellow' (Dutch)	wRoo
'Roz'	oMid wAit
'Ruban Bleu'	oMid
'Ruby Eruption'	oMid
'Ruby Tuesday'	oSch oMid wAit
'Ruffled Ballet'	wWal oCoo
'Ruffled Copper Sunset'	oMid oCoo
'Ruffled Feathers'	wAit
'Ruffled Goddess'	oMid
'Ruffled Skirts'	last listed 99/00
'Ruffles and Lace'	wWal
'Ruffles Supreme'	oCoo
'Rumbleseat'	last listed 99/00
'Rumours'	last listed 99/00
'Rupaul'	oMid
'Rustic Dance'	wWal
'Rustic Royalty'	oSch
'Rustler'	wWal oSch
'Rusty Magnificence'	oMid wAit
'Ruth Black'	oMid
'Ruthie Girl'	oMid
'Sabrina's Kiss'	oMid
'Saharan Sun'	oMid
'Sailor'	oMid wAit
'Saint Helens' Wake'	wWal
'Saint Louis Blues'	wWal oCoo
'Saint Petersburg'	wAit
'Saint Teresa'	last listed 99/00
'Salsa Rio'	oSch
'Samson'	wAit
'Samurai Warrior'	oCoo
'San Francisco'	wAit
'San Juan Silver'	oMid
'San Miguel'	oMid
'Sandy Beach'	oMid
'Sandy Rose'	oMid oCoo
sanguinea	oFor
sanguinea 'Snow Queen'	oFor wCul oNat oJoy oGar oSho wBox oBRG oUps
'Santa'	oMid wAit
'Santafair'	oMid
'Santana'	wWal
'Santa's Helper'	oMid wAit
'Sapphire Gem'	oSch
'Sapphire Hills'	wWal
'Sapphire Jewel'	last listed 99/00
'Sarah Lauren'	oMid
'Sarah Taylor'	wWoS oSch
'Sara's Love'	wAit
'Sass with Class'	oSch
'Saturday Night Live'	oMid wAit
'Saucy'	oMid
'Saucy Sprite'	oMid
'Savannah'	last listed 99/00
'Savannah Sunset'	oSch
'Saxon'	oMid
'Say Hello'	oMid
'Scandia Delight'	wWal
'Scarlet O'Harlett'	oCoo
'Scene Stealer'	oMid
'Scented Bubbles'	oSch oMid
'Scented Nutmeg'	wWal
'Schortman's Garnet Ruffles'	oCoo
'Schubertiad'	oMid
'Scintillation'	wWal
'Scion'	oMid wAit

'Scorpio Star'	oMid
'Screen Play'	oMid wAit
'Sculptress'	wWal
'Sculptured Wild'	oMid
'Sea Cadet'	wAit
'Sea of Joy'	wWal oSch
'Sea Power'	oMid
'Sea Quest'	wWal
'Sea Splash'	wAit
'Sea Swells'	oMid
'Seakist'	oSch oMid
'Season Ticket'	oSch wAit
'Second Show'	oMid
'Second Time Around'	wWal
'Secret Weapon'	last listed 99/00
'Seize the Sizzle'	oMid wAit
'Select Circle'	oMid wAit
'Self Evident'	wAit
'Sentimental Mood'	wWal
'Sentimental Rose'	oSch
'Serendipity Elf'	wAit
'Serene Moment'	oSch
'Serengeti'	oMid
'Serenity Prayer'	oSch wAit
setosa	oFor oGar oGre wEde
setosa ssp. ***canadensis***	see **I. *hookeri***
setosa var. ***nana***	see **I. *hookeri***
'Shadow Box'	wAit
'Shaft of Gold'	wWal oSch
'Shakedown'	last listed 99/00
'Shaman'	wWal
'She Devil'	oMid wAit
'Sheba's Jewel'	wAit
'Sheer Bliss'	oCoo
'Sheer Class'	oSch
'Sheer Ecstasy'	wWal wAit
'Sheik'	wAit
'Shenanigan'	wAit
'Shipshape'	wWal
'Shirley M.'	oMid
'Shooting Stars'	wWal oCoo
'Shopper's Holiday'	last listed 99/00
'Shoreline'	wWal
'Shortbread Creme'	oMid
'Showcase'	wWal
'Shy Violet'	oSch
sibirica	wShR oBlo wFai wTGN oSle
un ***sibirica alba nana***	wCri wWal oGar wAit oSis oGre
sibirica 'Alice Mae Cox'	oGar
sibirica 'Angela'	last listed 99/00
sibirica 'Ann Dasch'	oBRG
sibirica 'Annick'	wWal wAit
sibirica 'Appaloosa Blue'	oMid
sibirica 'Aqua Whispers'	wWal oChe oAls oGar wAit
sibirica 'Augury'	wWal
sibirica 'Baby Sister'	wWoS wWal oChe oGar
sibirica 'Band of Angels'	wAit
qu ***sibirica*** 'Beautiful Blue'	oNat
sibirica 'Bellissima'	oChe oJoy oHed oMid wAit
sibirica 'Bennerup Blue'	oGar
sibirica 'Bernard McLaughlin'	last listed 99/00
sibirica 'Bickley Cape'	last listed 99/00
sibirica blue	oNWe wHom
sibirica 'Blue Burn'	wWal
sibirica 'Blue Butterfly'	see **I. *sibirica*** 'Butterfly'
sibirica 'Blue Kaleidoscope'	wHer wWoS oWnC
sibirica 'Blue King'	oFor oSho
sibirica 'Blue Pennant'	oHed
sibirica 'Blue Reverie'	oCap
sibirica 'Blue Snippit'	wWal
sibirica 'Blue Song'	oGar
sibirica blue tetraploid	oGre
sibirica 'Bountiful Violet'	oAls
sibirica 'Butter and Sugar'	wWal oSch wSwC oNat oJoy wCol oAls oGar oGre oBRG wSnq
sibirica 'Butterfly'	oBRG
sibirica 'Caesar's Brother'	oFor wCul wSwC oAls oRus oGar oSho oSqu oGre oUps wSnq
sibirica 'Cambridge'	wWal
sibirica 'Carmen Jeanne'	oMid wAit
sibirica 'Cascade Creme'	wWal
sibirica 'Chandler's Choice'	oChe
sibirica 'Charming Darlene'	oNat oGar
sibirica 'Chartreuse Bounty'	wWoS oWnC
sibirica 'Chatter Box Belle'	oChe

	sibirica 'Cheery Lyn'	oChe
	sibirica 'Chilled Wine'	wHer wWal oSch wSwC oNat oHed oGar oGre oBRG
	sibirica 'Circle Round'	oCap
	sibirica 'Cleve Dodge'	oFor oCap oGar
	sibirica 'Contrast in Styles'	wHer wAit
	sibirica 'Coronation Anthem'	wWal oSch oChe oMid wAit
	sibirica 'Countess Cathleen'	oMid
	sibirica 'Dance Ballerina Dance'	wHer wWal oJoy wAit
	sibirica 'Dancer's Fan'	wAit
	sibirica 'Dancing Nanou'	oAls
	sibirica dark blue	oRus oGar
	sibirica dark blue and white	wSta
	sibirica 'Dawn Waltz'	oMid
	sibirica 'Dear Dianne'	wWal wAit
	sibirica 'Devil's Dream'	wAit
	sibirica 'Dianne's Daughter'	wWal wAit
	sibirica 'Dreaming Orange'	wWal oJoy
	sibirica 'Dreaming Yellow'	wWal
	sibirica 'Drops of Brandy'	oMid
	sibirica 'Ego'	wHer
	sibirica 'Eric the Red'	wWal
	sibirica 'Ever Again'	oChe wAit
	sibirica 'Ewen'	wWoS
	sibirica 'Fairy Dawn'	wCul
	sibirica 'Fairy Fingers'	oChe
	sibirica 'Fine Tuned'	oChe oMid
	sibirica 'Fisherman's Morning'	oChe wAit
	sibirica 'Flight of Butterflies'	wWoS wSwC wCol oGar wAit oSis oWnC
	sibirica 'Foretell'	wWal
	sibirica 'Forrest McCord'	wHer wWal oJoy wAit
	sibirica 'Four Winds'	wAit
	sibirica 'Fourfold White'	oJoy
	sibirica 'Frosted Cranberry'	wWal oChe
	sibirica 'Frosty Rim'	wWal oChe
	sibirica 'Golden Edge'	oChe
	sibirica 'Gull's Wing'	oMid
	sibirica 'Harbor Mist'	oChe
	sibirica 'Harpswell Hallelujah'	wSwC
	sibirica 'Harpswell Happiness'	oChe oGar
	sibirica 'Harpswell Haze'	last listed 99/00
	sibirica 'Harpswell Snowburst'	oChe
	sibirica 'Harpswell Velvet'	wWal wAit
	sibirica 'Heliotrope Bouquet'	wAit
un	***sibirica*** 'Henry's White'	oWnC
	sibirica 'High Standards'	wWal oChe oJoy wAit
	sibirica 'Hubbard'	wAit
	sibirica 'Illini Charm'	wCul wWal oGar
	sibirica 'Indy'	oJoy
	sibirica 'Isabelle'	oChe oMid wAit
	sibirica 'Jamaican Velvet'	oChe oMid
	sibirica 'Jatinwane'	oMid
	sibirica 'Jaybird'	oJoy
	sibirica 'Jeweled Crown'	wWal oChe oMid wAit
	sibirica 'Just Because'	oChe oMid
	sibirica 'King of Kings'	wHer oSch oMid oGar wAit
	sibirica 'Lady of Quality'	wWal
	sibirica 'Lady Vanessa'	wWoS wCul wWal oChe oGar wAit
	sibirica 'Lang'	wWal oChe
	sibirica 'Laughing Brook'	wWal oJoy
	sibirica 'Lavender Bounty'	wWoS wWal oSch oSho
	sibirica 'Lavender Stipples'	oChe
	sibirica 'Lee's Blue'	oChe oMid
	sibirica 'Liberty Hills'	wAit
	sibirica light blue	oGar
	sibirica 'Lights of Paris'	oFor
	sibirica 'Lilting Laura'	wWal oChe
	sibirica 'Little White'	wCul wHig
	sibirica 'Lucky Lilac'	oJoy
	sibirica 'Mabel Coday'	wAit
	sibirica 'Mad Magenta'	oChe oMid wAit
	sibirica 'Marilyn Holmes'	wHer
	sibirica 'Marshmallow Frosting'	wAit
	sibirica 'Mary Louise Michie'	oChe wAit
	sibirica medium blue	oGar
	sibirica 'Mesa Pearl'	oMid
	sibirica mixed	oGar oSho wCoN
	sibirica 'Moon Silk'	oChe oGar wAit
	sibirica 'My Love'	wWal
	sibirica 'Navy Brass'	wWal
	sibirica 'Omar's Cup'	wWal
	sibirica 'Orville Fay'	oJoy oAls wSnq
	sibirica 'Outset'	last listed 99/00
	sibirica 'Over in Gloryland'	oChe oMid
un	***sibirica*** 'Painted Desert'	oBRG
	sibirica 'Pansy Purple'	oChe
	sibirica 'Papillon'	wWoS wCul
	sibirica 'Pas-de-Deux'	wAit
	sibirica 'Peg Edwards'	wWal
	sibirica 'Percheron'	wWal
	sibirica 'Perry's Pigmy'	oJoy
	sibirica 'Pink Haze'	wWal oGar wAit
	sibirica 'Pink Sparkle'	wWal oJoy
	sibirica 'Pirate Prince'	wHer wWal oCap oNWe
	sibirica 'Pleasures of May'	oChe oMid wAit
	sibirica 'Precious Doll'	last listed 99/00
	sibirica 'President Truman'	oChe
	sibirica purple	oGar oNWe
	sibirica 'Purple Prose'	oNat
	sibirica 'Purple Sand'	oChe wAit
	sibirica 'Reddy Maid'	wWal
	sibirica 'Regency Belle'	wWal oJoy wAit
	sibirica 'Regency Buck'	oJoy
	sibirica 'Reprise'	oJoy oMid wAit
	sibirica 'Rikugi Sakura'	wWal
	sibirica 'Rill'	wWal wAit
	sibirica 'Rimouski'	oJoy
	sibirica 'Roanoke's Choice'	wWal
	sibirica 'Roaring Jelly'	oChe oMid wAit
	sibirica 'Rosebud Melody'	oChe
	sibirica 'Ruffled Velvet'	wWoS oSch oJoy oGar oBRG
	sibirica 'Sailor's Fancy'	oMid
	sibirica 'Sally Kerlin'	oGar
	sibirica 'Sassy Kooma'	oChe
	sibirica 'Savoir Faire'	wHig
	sibirica 'Sea Shadows'	oSch oJoy
	sibirica 'Seneca Feather Dancer'	wAit
	sibirica 'Shadowed Eyes'	wAit
	sibirica 'Shaker's Prayer'	wWal oChe oCap wAit
	sibirica 'Shall We Dance'	oChe wAit
	sibirica 'Shirley Pope'	oJoy oMid wAit
	sibirica 'Shirley's Choice'	wWal oChe
	sibirica 'Showdown'	wWoS
	sibirica 'Silver Edge'	wWoS wWal oWnC
	sibirica 'Silver Rose'	wWal
	sibirica 'Sky Mirror'	wWal oChe
	sibirica 'Sky Wings'	oNat wEde
	sibirica 'Slightly Envious'	wAit
	sibirica 'Snow Prince'	oChe
	sibirica 'Snow Queen'	see **I. *sanguinea*** 'Snow Queen'
	sibirica 'Soft Blue'	wWal
	sibirica 'Sole Command'	wWal
	sibirica 'Sparkling Rose'	oFor wWoS
	sibirica 'Spirit of York'	oChe
	sibirica 'Springs Brook'	wWal oChe oMid oNWe oGre
	sibirica 'Sprinkles'	oChe wAit
	sibirica 'Star Glitter'	wAit
	sibirica 'Stars by Day'	wWal
	sibirica 'Starsteps'	wWal
	sibirica 'Strawberry Fair'	oMid
	sibirica 'Sultan's Ruby'	oJoy wAit
	sibirica 'Summer Sky'	wCul oChe wAit
	sibirica 'Super Ego'	wWal
	sibirica 'Sweet Surrender'	wWal oChe wAit
	sibirica 'Swirling Mist'	wWal
	sibirica 'Temper Tantrum'	wWal oMid
	sibirica tetraploid blue	oGre
	sibirica 'Tiffany Lass'	oChe
	sibirica 'Tropic Night'	wHer wCul
un	***sibirica*** 'Tufted Velvet'	oSho
	sibirica 'Variation in Blue'	wWal
	sibirica 'Vi Luihn'	last listed 99/00
	sibirica 'Vicki Ann'	wAit
	sibirica violet	oGar wBox
	sibirica violet blue	wCSG
	sibirica 'Violet Joy'	wWal wAit
	sibirica 'Violet Swirl'	oGar
	sibirica 'Visual Treat'	wAit
	sibirica 'Wall Street Blues'	wAit
	sibirica 'Waterloo'	wWal
un	***sibirica*** 'Weiser Zwerg'	oMid
	sibirica 'Welfenschatz'	oMid
	sibirica white	oRus wCSG oGar wHom wBox
	sibirica 'White Magnificence'	wWal
	sibirica 'White Prelude'	oChe
qu	***sibirica*** 'White Sails'	oGar
	sibirica 'White Swirl'	wRob oBRG wSnq
	sibirica 'White Triangles'	oMid wAit
	sibirica 'Windwood Serenade'	oChe oGar
	sibirica 'Windwood Spring'	wWal oChe oJoy oMid wAit

sibirica 'Wizardry'	oJoy
'Sidestitch'	last listed 99/00
'Sierra Grande'	wWal oSch oMid wAit
'Sierra Rim'	oMid
'Sigh of Colours'	oMid
'Sighs and Whispers'	oMid
'Significant Other'	oCoo
sikkimensis	wHer
'Silent One'	last listed 99/00
'Silent Screen'	oMid
'Silent Shadow'	oMid
'Siletz Bay'	oCoo
'Silk Brocade'	oMid
'Silken Shadows'	last listed 99/00
'Silver Fizz'	oCoo
'Silver Flow'	oCoo
'Silver Fox'	oCoo
'Silver Shower'	wWal
'Silver Years'	wWal
'Silverado'	wWal oSch oCoo
'Simmer'	oMid
'Sing Out'	last listed 99/00
'Sinister Desire'	oMid
Sino-Siberian hybrids 'Boop Eyes'	wAit
Sino-Siberian hybrids 'McKenzie Violet'	wAit
Sino-Siberian hybrids 'Piquant Fancy'	wAit
sintenisii	oFor wHer
'Sixtine C.'	oMid
'Skating Party'	wWal oSch oCoo
'Skiddle'	wAit
'Skiers' Delight'	wWal
'Skipalong'	oMid wAit
'Sky Blue Pink'	wAit
'Sky Dancing'	oMid
'Sky Search'	oCoo
'Skyblaze'	wWal
'Skyfire'	wWal
'Skylark's Song'	oMid wAit
'Skyray'	last listed 99/00
'Skywalker'	oSch oMid
'Slam Dunk'	wAit
'Slavegirl'	wAit
'Sleepy Time'	oSch wAit
'Slick Trick'	last listed 99/00
'Smart'	oSch
'Smart Move'	oMid wAit
'Smidget'	last listed 99/00
'Smiling Angel'	oMid
'Smiling Faces'	oMid
'Smiling Gold'	oCoo
'Smitten Kitten'	oSch wAit
'Smoke Rings'	wWal
'Smooth Move'	oMid
'Sneezy'	oSch oMid wAit
'Snickers'	oMid
'Snoopy'	oMid wAit
'Snow Blanket'	oCoo
'Snow Crown'	oMid
'Snow Job'	oMid
'Snow Plum'	oMid wAit
'Snow Season'	oMid wAit
'Snow Shoes'	wAit
'Snowbelt'	last listed 99/00
'Snowbrook'	wWal
'Snowcone'	wAit
'Snowmound'	wWal
'Snowy River'	oSch
'Snugglebug'	oSch wAit
'Snuggles'	wAit
'So Fine'	wWal oSch oMid wAit
'Soaring'	oMid
'Soaring Spirit'	oMid
'Social Event'	oSch oMid wAit
'Soft Cover'	oMid
'Solemn Promise'	oCoo
'Soloist'	oCoo
'Sombre Mood'	wWal
'Sombrero Way'	oMid
'Some Are Angels'	oMid
'Somersault'	oMid wAit
'Something Wonderful'	oMid
'Son of Star'	wWal
'Song of Norway'	wWal wAit
'Song of Spring'	wWal
'Sonja's Selah'	oSch wAit
'Sonoran Sands'	oMid
'Sooner Serenade'	wWal
'Sostenique'	wWal
'Sotto Voce'	oCoo
'Soul Sister'	oMid
'Sound of Gold'	wWal
'Southern Comfort'	wWal
sp.	oNWe
sp. DJHC 271	wHer
sp. DJHC 344	wHer
sp. DJHC 647	wHer
'Space Blazer'	wWal
'Space Viking'	oCoo
'Spanish Coins'	wAit
'Spanish Fireball'	oMid wAit
'Sparkle Berry'	oSch
'Sparkletts'	oMid
'Sparkling Dew'	wWal
'Sparkling Sunrise'	wWal
'Sparkplug'	oMid
'Sparky'	wAit
'Spartan'	oCoo
'Speakeasy'	wWal oCoo
'Special Feature'	oMid
'Special Friend'	last listed 99/00
'Spectral Challenge'	oCoo
'Speed Limit'	wWal oMid wAit oCoo
'Speeding Again'	oMid
'Spellbreaker'	wWal oMid
'Spiced Custard'	oSch
'Spiced Orange'	wWal
'Spiced Tiger'	oMid wAit
'Spin Again'	wAit
'Spin Doctor'	last listed 99/00
'Spin-off'	last listed 99/00
'Spinning Wheel'	oCoo
'Spirit of Memphis'	oSch
'Spirit World'	oSch oMid wAit
'Splash of Raspberry'	oMid wAit
'Splashacata'	oMid
'Spot It'	wAit
'Spot of Tea'	wAit
'Spreckles'	wWal
'Spree'	oMid
'Spring Fresh'	oMid wAit
'Spring Image'	oCoo
'Spring Kiss'	oMid
'Spring Parasol'	oCoo
'Spring Pleasure'	oCoo
'Spring Satin'	oMid
'Spring Serenade'	oMid
'Spring Shower'	oMid wAit
'Spring Tidings'	oCoo
'Spring Twilight'	oMid wAit
spuria	oFor wCSG
Spuria hybrids 'Alphaspu'	oChe
Spuria hybrids 'Baby Chick'	oChe
Spuria hybrids 'Bali Bali'	oChe
Spuria hybrids 'Belise'	oChe
Spuria hybrids 'Belissinado'	oChe
Spuria hybrids 'Betty Cooper'	oChe wAit
Spuria hybrids 'Blue Bunting'	oChe
Spuria hybrids 'Border Town'	wAit
Spuria hybrids 'Brass Beauty'	oChe
Spuria hybrids 'Bronzing'	oChe
Spuria hybrids 'Butterscotch Queen'	oChe wAit
Spuria hybrids 'Cafeine'	oChe
Spuria hybrids 'Cinnamon Stick'	oChe
Spuria hybrids 'Clara Ellen'	wAit
Spuria hybrids 'Cobalt Mesa'	oChe
Spuria hybrids 'Color Focus'	oChe wAit
Spuria hybrids 'Copper Trident'	oChe
Spuria hybrids 'Countess Zeppelin'	oChe
Spuria hybrids 'Cust'	oChe
Spuria hybrids 'Destination'	wAit
Spuria hybrids 'Easter Colors'	oChe
Spuria hybrids 'Eleanor Hill'	oChe
Spuria hybrids 'Elves Gold'	oChe
Spuria hybrids 'Finally Free'	oChe
Spuria hybrids 'Firemist'	oChe
Spuria hybrids 'Fixed Star'	oChe
Spuria hybrids 'Future Perfect'	oChe wAit
Spuria hybrids 'Goblin's Song'	oChe
Spuria hybrids 'Handsome Is'	oChe
Spuria hybrids 'Highline Coral'	oChe wAit

Spuria hybrids 'Highline Halo'	oSho
Spuria hybrids 'Highline Snowflake'	wAit
Spuria hybrids 'Ila Remembered'	oChe wAit
Spuria hybrids 'Imperial Seas'	oChe
Spuria hybrids 'Imperial Veil'	oChe
Spuria hybrids 'Infini'	oChe
Spuria hybrids 'Innovator'	oChe
Spuria hybrids 'Irene Benton'	oChe
Spuria hybrids 'Lavender Parade'	oChe
Spuria hybrids 'Lenkoran'	oChe
Spuria hybrids 'Lighted Signal'	oChe
Spuria hybrids 'Lucky Devil'	oChe wAit
Spuria hybrids 'Mahogany Lord'	oChe
Spuria hybrids 'Mary's Beau Brummel'	oChe
Spuria hybrids 'Midnight Rival'	oChe
Spuria hybrids 'Mighty Mauve'	oChe
Spuria hybrids 'Missouri Gal'	oRus
Spuria hybrids 'Missouri Lakes'	oChe
Spuria hybrids 'Moon Shrine'	oChe
Spuria hybrids 'Mystifier'	oChe
Spuria hybrids 'Offering'	oChe
Spuria hybrids 'Olinda'	oChe
Spuria hybrids 'Oro de Sonora'	oChe
Spuria hybrids 'Panacea'	oChe
Spuria hybrids 'Pixie Time'	oChe
Spuria hybrids 'Port of Call'	oRus
Spuria hybrids 'Prussian Magic'	oChe
Spuria hybrids 'Redwood Supreme'	wAit
Spuria hybrids 'Royal Cadet'	wAit
Spuria hybrids 'Sage'	oChe
Spuria hybrids 'Satinwood'	oChe
Spuria hybrids 'Shelford Giant'	oRus
Spuria hybrids 'Sonoran Caballero'	oChe
Spuria hybrids 'Sonoran Senorita'	oChe
Spuria hybrids 'Sonoran Skies'	oChe
Spuria hybrids 'Sultan's Sash'	oChe
Spuria hybrids 'Sun Singer'	oChe
Spuria hybrids 'Sunrise in Missouri'	oChe
Spuria hybrids 'Sunrise in Sonora'	wAit
Spuria hybrids 'Tassili'	oChe
Spuria hybrids 'Terra Nova'	oChe
Spuria hybrids 'Thrush Song'	oRus
Spuria hybrids 'Universal Peace'	oChe
Spuria hybrids white	oUps
Spuria hybrids 'White Olinda'	oChe
Spuria hybrids 'White Shimmer'	oChe
Spuria hybrids yellow	oRus oUps
Spuria hybrids 'Zamboanga'	oChe
Spuria hybrids 'Zulu Chief'	oChe wAit
spuria ssp. ***musulmanica***	oNWe
'Squiddler'	wAit
'Stair Stealer'	last listed 99/00
'Stairway to Heaven'	oSch oMid
'Stand and Salute'	wWal
'Stanza'	wAit
'Star Caper'	oSch
'Star Fleet'	oSch oMid
'Star Quality'	oMid wAit
'Star Sailor'	oSch oMid wAit
'Starbaby'	oSch wAit
'Starburst'	wWal
'Starlette Rose'	oMid
'Starlight Express'	wWal oMid wAit
'Starlit Velvet'	last listed 99/00
'Stars and Stripes'	oMid
'Starship'	oMid
'Starship Enterprise'	oSch oMid
'Startler'	wWal
'Starwoman'	oMid wAit
'Stately Art'	oMid wAit
'Stately Mansions'	wWal
'Static'	last listed 99/00
'Steffie'	wAit
'Stellar Lights'	wWal wAit
'Step Nicely'	wWal
'Stepping Out'	oSch wAit oCoo
'Sterling Blush'	wWal
'Stillness'	oMid
'Stinger'	oMid
'Stitch Witch'	oSch
'Stop the Music'	wWal
'Storm Center'	wWal
'Stormy Circle'	oMid
'Storybook'	wWal
'Storyland'	oMid

'Storyteller'	oMid
'Strawberry Cream'	wAit
'Strawberry Love'	oSch wAit
'Strawberry Swirl'	oMid
'Street Vendor'	last listed 99/00
'Streetwalker'	wWal
'Strictly Ballroom'	oSch oMid wAit
'Strike It Rich'	oCoo
'Striped Britches'	oMid
'Striped Pants'	wAit
'Struttin' High'	oSch
'Study in Black'	wWal
'Stylist'	oMid
suaveolens	wMtT
suaveolens 'Rubromarginata'	wMtT
'Subtle Hint'	oCoo
'Success Story'	wWal
'Sudden Impact'	oMid
'Sue Zee'	wAit
'Sugar Blues'	oSch
'Sugar Magnolia'	oSch
'Suky'	oSch oMid wAit
'Sultan's Daughter'	oMid
'Sultan's Palace'	oCoo
'Sultry Mood'	wWal
'Sumas'	wWal
'Summer Camp'	oMid
'Summer Luxury'	oMid
'Summer Olympics'	oSch
'Sun Doll'	oSch wAit
'Sun Fire'	wWal
'Sunbonnet Sue'	wAit
'Suncatcher'	oMid
'Sunday Chimes'	wWal
'Sunday Punch'	wWal
'Sunday Sunshine'	oCoo
'Sunkist Delight'	oMid
'Sunkist Frills'	wWal
'Sunkist Meadows'	oMid
'Sunmaster'	oMid wAit
'Sunny Bubbles'	oMid wAit
'Sunny Dawn'	oSch wAit
'Sunny Disposition'	wAit
'Sunny Glow'	oMid wAit
'Sunny Honey'	oSch
'Sunny Peach'	oMid
'Sunny Sacramento'	oMid
'Sunray Reflection'	oCoo
'Sunrise Seduction'	oMid
'Sunset Fires'	wWal
'Sunset Sonata'	wWal
'Sunsite'	wWal
'Sunso'	oMid
'Sunspinner'	wAit
'Superman'	wWal
'Supersimmon'	wWal
'Superstition'	wWal oCoo
'Supreme Sultan'	wWal oSch oMid wAit oCoo
'Surprising Wit'	oCoo
'Suspicion'	oMid
'Swazi Princess'	wWal
'Sweet Ballerina'	oCoo
'Sweet Bite'	oMid
'Sweet Delight'	oMid
'Sweet Lemonade'	oMid
'Sweet Lena'	wHol
'Sweet Musette'	wWal oSch
'Sweet Revenge'	oMid oCoo
'Sweet Thing'	last listed 99/00
'Sweeter than Wine'	wWal oSch oCoo
'Swing and Sway'	oMid oCoo
'Swingtown'	oSch oMid wAit
'Swirling Seas'	oCoo
'Swish'	oMid
'Symphony' (Dutch)	oGar
'Tacey'	oCoo
'Taco Belle'	wWal
'Taco Supreme'	oCoo
'Tact'	wAit
'Tahiti Sunrise'	wWal
'Talk to Me'	oCoo
'Tall Ships'	oMid wAit
'Tan Man'	oMid wAit
'Tangerine Sky'	wWal
'Tangfu'	oMid

	'Tangled Web'	oMid
	'Tantrum'	oMid wAit
	'Tanzanian Tangerine'	oMid wAit
	'Tarde'	wWal
	'Tarlatin'	wWal
	'Tattler'	oMid wAit
	'Taunt'	oMid
	'Taxi'	oMid
	'Tchin-Tchin'	oSch wAit
	'Tea Apron'	wWal
	'Teapot Tempest'	oMid
	tectorum	oFor wCol wAit
	tectorum 'Alba'	oFor wHer wWoS oJoy
	tectorum EDHCH 97249	wHer
	tectorum 'Variegata'	oAmb oUps
un	'Telstar' (Dutch)	wRoo
	'Temple Gold'	wWal
	'Tempting'	wAit
	'Tempting Fate'	oSch oMid oCoo
	tenax	oFor oHan wWoB wNot oJoy oRus oGar oBos wRai wWat oTri oAld wWld oSle
	'Tender Gender'	oMid
	'Tennessee Gentleman'	wAit
	'Tennessee Vol'	oSch
	'Tennessee Woman'	wAit
	'Tennison Ridge'	wWal wAit
	tenuissima	wCCr
	tenuissima NNS 95-302	wHer
	'Terre de Feu'	oMid
	'Test Pattern'	oCoo
	'Thai Orange'	oCoo
	'Theatre'	wWal
	'Theda Clark'	oSch
	'Therapy'	wAit
	'Theresa Lynn'	oMid
	'Thinking Out Loud'	oMid oCoo
	'Thirsty Oasis'	oMid
	'This and That'	oMid
	'Thornbird'	oSch wAit oCoo
	'Three Tokens'	wAit
	'Thrice Blessed'	last listed 99/00
	'Thriller'	wWal oSch
	'Thrillseeker'	oCoo
	'Throb'	oSch oMid
	'Thumkin'	oMid oCoo
	'Thunder Spirit'	oSch oMid
	'Thundercat'	wAit
	'Tickled Peach'	oSch wAit
	'Tide's In'	wWal oSch
	'Tiger Honey'	oSch oMid wAit
	'Tillie'	oSch
	'Time and Again'	wAit
	'Time Piece'	oSch
	'Time to Shine'	oMid
	'Time Will Tell'	oCoo
	'Timeless Moment'	wWal
	'Timeless Theme'	oMid
	'Timescape'	oCoo
	'Tingle'	oMid
	'Tink'	oMid wAit
	'Tintinnabulation'	oMid
	'Tiny Cherub'	oMid wAit
	'Titan's Glory'	wWal wAit
	'To the Point'	oMid
	'Toasted Watermelon'	oMid
	'Toastmaster'	wAit
	'Today's Fashion'	wWal
	'Todd'	wAit
	'Tokatee Falls'	oSch
	'Tokyo Blues'	oSch oMid
	'Tom Johnson'	oMid wAit
	'Tommyknocker'	wAit
	'Tomorrow's Child'	wWal
	'Too Sweet'	oCoo
	'Toon Town'	oMid
	'Tooth Fairy'	oMid wAit
	'Top Billing'	wWal
	'Top Gun'	wWal
	'Torchy'	oSch
	'Total Obsession'	oSch
	'Total Recall'	wAit oCoo
	'Touch of Class'	wWal
	'Touch of Mink'	oMid
	'Tracy Tyrene'	oCoo
	'Trading Places'	oMid
	'Traitor'	oMid wAit
	'Trajectory'	oMid
	'Tres Elegante'	wWal
	'Trifle'	oMid wAit
	'Triple Play'	wWal
	'Triple Whammy'	oSch wAit oCoo
	'Triplet'	oSch
	'Troll'	oSch
	'Tropical Encounter'	oMid oCoo
	'Tropical Magic'	oMid wAit
	'Tropical Night'	wWal
	'Tropical Punch'	oMid wAit
	'True Believer'	oMid
	'Truly'	oSch
	'Trusty Rusty'	oSch
	'Turkish Tangent'	oMid
	'Tut's Gold'	wWal
	'Tuxedo'	wWal
	'Tweety Bird'	oSch oMid wAit
	'Twice Blue'	last listed 99/00
	'Twice Thrilling'	oCoo
	'Twice Told'	oMid wAit
	'Twilight Blaze'	oMid wAit
	'Twilight Passage'	oMid wAit
	'Twilight Rain'	oMid
	'Twist of Fate'	wWal oCoo
	Two Rubies'	oSch
	'Two-Sided Coin'	oMid oCoo
	'Tyke'	wAit
	typhifolia	wHer wWal
	'Unforgettable Fire'	last listed 99/00
	unguicularis	oRus oAmb
	unguicularis ssp. ***cretensis***	last listed 99/00
	'Up Dancing'	oMid
	'Uptown Proper'	oCoo
	'Vague a l'Ame'	oMid
	'Vail'	oMid
	'Vain Frivolity'	oCoo
	'Valentine's Day'	oMid
	'Valley Sunset'	wWal
	'Vanda Song'	oMid
	'Vandal Spirit'	wWal
	'Vanilla Fudge'	oMid
	'Vanilla Kisses'	oMid
	'Vanity'	wWal wAit
	'Vanity's Child'	oSch
	'Vavoom'	oMid wAit
	'Velvet Toy'	wAit
	'Velvet Underground'	oMid wAit
	'Venetian Queen'	oMid
	'Venus Butterfly'	oMid
	verna	oFor wThG oRus oSis wCCr oTri
un	***verna alba***	wHer
	versicolor	oFor wSoo oJoy oRus oHug oSsd wFai oTri wBox
	versicolor 'Aquatic Alliance'	see **I.** 'Aquatic Alliance'
	versicolor 'Arctic Lavender'	see **I.** 'Arctic Lavender'
	versicolor 'Between the Lines'	wAit
	versicolor 'Candystriper'	wAit
un	***versicolor*** 'Jamie's Pink Delight'	oGar wBox
	versicolor 'Kermesina'	oSsd
	versicolor 'Party Line'	wAit
	versicolor 'Royal Dolly'	see **I.** 'Royal Dolly'
	versicolor x ***virginica*** 'Gerald Darby'	see **I. *x robusta*** 'Gerald Darby'
	versicolor 'Whodunit'	wAit
	versicolor white	oGre
	'Very Varied'	oMid wAit
	'Very Violet'	wAit
	'Vibrant'	oMid
	'Vibrant Rose'	wAit
	'Vibrations'	wWal wAit
	'Victoria Circle'	last listed 99/00
	'Victoria Falls'	wWal oSch wAit oCoo
	'Victorian Lace'	wWal
	'Vigilante'	wWal oSch
	'Viking Admiral'	wWal
	'Village Gossip'	oMid
	'Vintage Rose'	oMid wAit
	'Vintage Victorian'	oMid
	'Vintner'	oMid wAit
	'Violet Lass'	last listed 99/00
	'Violet Rings'	wWal oCoo
	'Violet Shimmer'	oMid
	'Virginia Lyle'	oSch
	'Virginia Rudkin'	oMid wAit

	virginica pink	last listed 99/00
	virginica var. ***shrevei***	oFor
	virginica white	last listed 99/00
	'Vision in Pink'	wWal oSch
	'Vista Dome'	wWal
	'Visual Arts'	wWal
	'Vitafire'	wWal
	'Vitality'	oSch
	'Viva Mexico'	oMid wAit
	'Vive La France'	last listed 99/00
	'Vivien'	wWal
	'Vizier'	oMid
	'Volatile'	oMid
	'Voltage'	oSch oMid wAit oCoo
	'Volts'	wAit
	'Volute'	oMid
	'Voodoo Spell'	oMid
	'Voyage'	last listed 99/00
	'Wabash'	wWal oSch oMid
	'Wacko'	wAit
	'Waffle Talk'	oCoo
	'Waiting for George'	oMid
	'Wake Up'	oSch
	'Walkara'	oMid
	'Walking Tall'	wWal
	'Walsterway'	wWal
	'Wando'	oMid
	'Wanted'	oMid
	'War Chief'	oSch
un	'War Lord'	wWal
	'War Sails'	oSch
	'Warm Breeze'	oMid
	'Warm Embrace'	wWal
	'Warm Memories'	oMid wAit
	'Warrior King'	oSch oCoo
	'Waterdragon'	oMid wAit
	'Waterworld'	oMid wAit
	'Webmaster'	oMid wAit
	'Wedding Candles'	last listed 99/00
	'Wedding Dance'	last listed 99/00
	'Wedding Vow'	wWal
	'Welch's Reward'	oMid wAit
	'Welder's Flame'	wAit
	'Well Endowed'	wWal oCoo
	'Well Suited'	oSch
	'Wench'	oMid wAit
	'West Coast'	wWal
	'Western Sage'	last listed 99/00
	'Westernaire'	oMid
	'Wet Silk'	oMid
	'What Again'	oSch wAit
	'Whatta Dream'	oMid
	'Whipped Honey'	oMid
	'Whispering'	wWal oMid
	'White China'	oMid
un	'White City Pass'	wWal
	'White Heat'	oMid
	'White Lightning'	wWal
	'White Wine'	oMid
	'Whole Cloth'	wAit
	'Who's Who'	oMid
	'Wicked Ways'	oMid
	'Wide Alert'	oMid
	'Widow's Veil'	wAit
	'Wiggle'	oMid
	'Wild Apache'	wWal
	'Wild Frontier'	oSch
	'Wild Ginger'	wWal
	'Wild Hair'	oMid
	'Wild Jasmine'	wWal oSch
	'Wild Lad'	oMid
	'Wild Thing'	wWal oSch oMid wAit
	'Wild Vision'	oMid
	'Wild Wings'	oMid
	'Wildest Dreams'	oMid
	wilsonii	wHer
	'Wind Spirit'	oMid
	'Winds of Change'	last listed 99/00
	'Windsor Rose'	wWal
	'Wine Time'	oMid
	'Wing Commander'	oMid
	'Wings of Gold'	wWal oMid
	'Wings of Peace'	oMid
	'Winifred Ross'	wWal
	'Winner Take All'	oCoo

	'Winning Debut'	oSch
	'Winning Edge'	oMid wAit
	'Winning Smile'	oMid
	'Winter Olympics'	wAit
	'Winter White'	wWal
	'Winterland'	oMid wAit
	'Winterscape'	last listed 99/00
	'Wish Waltz'	last listed 99/00
	'Wishful Thinking'	oSch oMid wAit
	'Witches' Sabbath'	oMid
	'Witching'	oMid wAit
	'Witch's Wand'	wWal oSch
	'With Castanets'	oSch
	'Wit's End'	last listed 99/00
	'Wizard's Return'	oMid
	'Woodwine'	wWal
	'Wooing'	last listed 99/00
	'Working Too Hard'	oCoo
	'World Class'	wWal oSch
	'World News'	wWal
	'World Premier'	oSch oMid wAit
	'World Tour'	last listed 99/00
	'World Without End'	oCoo
	'Wrangler'	oMid
	'Wrong Song'	oMid
	'Wych Way'	last listed 99/00
	xiphium	oTrP
	'Yak Attack'	wAit
	'Yankee Skipper'	oSch
	'Yaquina Blue'	wWal oSch oMid wAit oCoo
	'Yellow Brick Road'	last listed 99/00
	'Yellow Flirt'	oMid wAit
	'Yellow Tapestry'	wWal
	'Yes'	oSch oMid wAit
	'Yipee'	oSch
	'Yippy Skippy'	oMid
	'Yo'	wAit
	'Yo-yo'	oSch oWnC
	'You'll Love'	wWal
	'Young Blood'	oMid
	'Yours Free'	oMid
	'Zandria'	oMid
	'Zany'	wWal oCoo
	'Zebedee'	last listed 99/00
	'Zebra'	see **I. *pallida*** 'Variegata'
	'Zebra Night'	oMid
	'Zepherina'	oMid wAit
	'Zero'	oMid wAit
	'Zestful Miss'	oMid
	'Zillionaire'	oMid
	'Zinc Pink'	oSch wAit
	'Zing Me'	oSch oMid
	'Zipper'	oSch oMid wAit
	'Zowie'	oSch
	'Zula'	wAit
	'Zurich'	oMid
	ISATIS	
	tinctoria	oCrm
	ISMENE	see **HYMENOCALLIS**
	ISOLEPIS	
	cernua	oTrP wTGN
	ISOPLEXIS	
	canariensis	last listed 99/00
	ISOPOGON	
	formosus	wGra
	ISOPYRUM	
	biternatum	last listed 99/00
	ISOTOMA	
	axillaris	see **LAURENTIA *axillaris***
	ITEA	
un	***chinensis***	wHer oRiv oCis
	ilicifolia	oFor wHer oGos oGre wCCr oCis wCoN
	japonica 'Beppu'	oAls
un	***oldhamii***	wHer oCis
	virginica	oFor oBov
	virginica 'Henry's Garnet'	wHer wCul oGos oGar oRiv oOut oGre wCCr wWel
un	***virginica*** 'Little Henry'	oFor wWoS oGos oGre oBRG
un	***virginica*** 'Merlot'	oFor oDar oEdg
	virginica 'Sarah Eve'	oGos wWel
un	***virginica*** 'Saturnalia'	oFor wWel
	yunnanensis	oTrP wHer oCis
	IXIOLIRION	
	tataricum	wWoS

	JACARANDA	
	mimosifolia	oTrP
	JAMESIA	
	americana	oFor wHer
	JASIONE	
	amethystina	oSec
	crispa ssp. *amethystina*	see J. *amethystina*
	heldreichii	oFor
	humilis	see J. *amethystina*
	laevis	wTho oMis wCCr oEga
	laevis 'Blaulicht'	oJoy oUps
	laevis Blue Light	see J. *laevis* 'Blaulicht'
	perennis	see J. *laevis*
	JASMINUM	
	beesianum	oFor oJoy
	beesianum dark form	oHed
un	*beesianum* 'Marshall Olbricht'	oCis
	floridum	oGre wCCr oWnC
	fruticans	oCis
	grandiflorum	wCoN
	humile DJHC 98116	wHer
	humile 'Revolutum'	oFor oCis
	humile f. *wallichianum*	oFor
	mesnyi	oGar oSho oGre oWnC oCis oSle
	nudiflorum	oFor wHer oJoy wCSG oGar oGre iGSc oWnC wTGN oCis oBRG oWhS wCoN oUps wWel
un	*nudiflorum* 'Variegatum'	oCis
	officinale	oAls oGar oDan oSho wPir oUps
	officinale f. *affine*	oFor iGSc oCis
	officinale 'Argenteovariegatum'	oCis
	officinale 'Aureovariegatum'	see J. *officinale* 'Aureum'
	officinale 'Aureum'	oHed oCis
	officinale Fiona Sunrise / 'Frojas'	oHed oCis
	officinale 'Grandiflorum'	see J. *officinale* f. *affine*
	officinale 'Inverleith'	oHed oCis
	parkeri	oFor oHed oSis
	polyanthum	oAls wFGN oGar oSho oGre oWnC oWhS wCoN wSte wCoS wWel
	primulinum	see J. *mesnyi*
	x stephanense	oFor wCul wCri wCol oHed wCSG oGar oDan oSho wRai oGre wCCr oWnC wTGN oCis wSte wCoS wWel
	wallichianum	see J. *humile* f. *wallichianum*
	JATROPHA	
un	*capensis*	oRar
un	*cinerea*	oRar
	curcas	oRar
un	*dissecta*	oRar
un	*elba* BLM 0979	oRar
un	*mahafalensis*	oRar
	multifida	oRar
un	*platyphylla*	oRar
	podagrica	oRar
	JEFFERSONIA	
	diphylla	oRus oSis
	dubia	oSis
	JOVELLANA	
	sinclairii	oTrP wHer oHed
	violacea	last listed 99/00
	JOVIBARBA	
	allionii	oSto oSqu
	allionii from Aione	oSqu
	allionii from Esteng	oSqu
	allionii from Esteng x *hirta* from Biele Karpaty	oSqu
	allionii x *hirta*	oSto oSqu
	arenaria	oSto
	arenaria from Murtal	oSqu
un	*arenaria x hirta* ssp. *glabrescens* from Belansky Tatra	oSqu
	arenaria red and green	oSqu
	'Emerald Spring'	last listed 99/00
	heuffelii 'Aiolos'	oSqu
	heuffelii 'Alemene'	last listed 99/00
	heuffelii 'Apache'	last listed 99/00
	heuffelii 'Aquarius'	oSqu
un	*heuffelii* 'Avant Garde'	oSqu
	heuffelii 'Beacon Hill'	oSqu
	heuffelii 'Belcore'	oSqu
un	*heuffelii* 'Chocolate Sundae'	oSqu
un	*heuffelii* 'Chryseis'	oSqu
un	*heuffelii* 'Douce'	oSqu
	heuffelii 'Dunbar Road'	oSqu
un	*heuffelii* 'Faith'	oSqu
	heuffelii 'Fandango'	oSqu
	heuffelii from Kopaonikense	oSqu
	heuffelii from Urana E. Vogel	oSqu
	heuffelii 'Geronimo'	oSqu
un	*heuffelii* 'Giselle'	oSqu
	heuffelii var. *glabra* from Jakupica, Macedonia	oSqu
	heuffelii var. *glabra* from Osljak	oSqu
	heuffelii var. *glabra* from Pasina Glava	oSqu
	heuffelii var. *glabra* from Rhodope	oSqu
	heuffelii var. *glabra* from Treska Gorge, Macedonia	oSqu
	heuffelii var. *glabra* from Vitse	oSqu
un	*heuffelii* 'Gold Bug'	oSto
un	*heuffelii* 'Gommerina'	oSqu
	heuffelii 'Greenstone'	oSqu
	heuffelii 'Henry Correvon'	oSqu
un	*heuffelii* 'Hot Lips'	oSqu
un	*heuffelii* 'Hyacinth'	oSqu
	heuffelii 'Inferno'	oSto oSqu
	heuffelii 'Kapo'	oSqu
un	*heuffelii* 'Korstiana'	oSqu
un	*heuffelii* 'Lobates'	oSqu
un	*heuffelii* 'Lori Beth'	oSqu
un	*heuffelii* 'Maggie'	oSqu
	heuffelii 'Mary Ann'	oSqu
	heuffelii 'Mystique'	oSqu
	heuffelii 'Nannette'	oSqu
	heuffelii x *x nixonii* 'Jowan	oSqu
	heuffelii 'Nobel'	oSqu
	heuffelii 'Orion'	oSqu
un	*heuffelii* 'Randolph'	oSqu
	heuffelii 'Sundancer'	oSto oSqu
un	*heuffelii* 'Sunny Side Up'	oSqu
	heuffelii 'Sylvan Memory'	oSqu
	heuffelii 'Tan'	oSqu
	heuffelii 'Tancredi'	last listed 99/00
un	*heuffelii* 'Tjoklat'	oSqu
	heuffelii 'Torrid Zone'	oSqu
	heuffelii 'Vesta'	oSqu
	heuffelii 'Violet'	oSto oSqu
	heuffelii 'Vulcan'	oSqu
un	*heuffelii* 'Wotan'	oSqu
un	*hirta* 'Andreas Smits'	oSqu
	hirta ssp. *borealis*	oSqu
	hirta 'Csiki'	oSqu
un	*hirta* 'Farrah'	oSqu
	hirta from Biele Karpaty	oSqu
	hirta from Budai B	oSqu
	hirta from Bulgaria	oSqu
	hirta from Col d'Aubisque	oSqu
	hirta from Falkenstein	oSqu
	hirta from Hunsheimer Berg A	oSqu
	hirta from Mala Fatra	oSqu
	hirta from Mecsek B	oSqu
	hirta from Sierra Nevada	oSqu
	hirta from Wintergraben	oSqu
	hirta ssp. *glabrescens* from Belansky Tatra	oSqu
	hirta ssp. *glabrescens* from High Tatra	oSto
	hirta ssp. *glabrescens* from Smeryouka	oSto oSqu
un	*hirta* 'Histoni'	oSqu
un	*hirta* 'Moyouka'	oSqu
un	*hirta* 'Neilrechii'	oSqu
un	*hirta* 'Nicole'	oSqu
	hirta 'Preissiana'	oSqu
	hirta 'Prusssiana'	oSqu
un	*hirta* 'Sodollo'	oSqu
un	*hirta* 'White Knight'	oSto oSqu
	hirta yellow green form	oSqu
	sobolifera	oSqu
un	*sobolifera* 'Emerald Green'	oSqu
	sobolifera from Harmanschlag	oSqu
	sobolifera from Palausie	oSto oSqu
	sobolifera from St. Martin	oSqu
un	*sobolifera* from Taupize	oSqu
	sobolifera 'Green Globe'	oSqu
un	*sobolifera* x *hirta*	oSqu
un	*sobolifera* 'Madam Lorene'	oSqu
un	*sobolifera* 'Pale'	oSqu
	sobolifera russet	oSqu
	sp. from Huran	oSqu
	JUBAEA	
	chilensis	oCis

JUGLANS

	ailanthifolia	wBur oOEx
	ailanthifolia var. ***cordiformis***	wBCr
	Buartnut	wBCr wBur oOEx
	cinerea	wBCr wBur wRai oOEx
un	***cinerea*** 'Chamberlain' seedling	wBCr
	neotropica	oOEx
	nigra (nut)	oFor wBCr wClo wShR wRai
un	***nigra*** 'Cooksey'	oWhi wRai
un	***nigra*** 'Cranz'	wBur
un	***nigra*** 'Hare'	wBur
un	***nigra*** 'Rowher'	wBur
un	***nigra*** 'Sparrow'	wBur
un	***nigra*** 'Sparks 127'	wBur
un	***nigra*** 'Stambaugh'	wBur
un	***nigra*** 'Thomas'	oFor wBur oGar oOEx
un	***nigra*** 'Weschke'	wBur
	regia (nut)	wBur
	regia 'Ambassador'	last listed 99/00
	regia Carpathian Group	oAls oGar wKin wVan
	regia Carpathian Group 'Cascade'	wBur wClo wRai
un	***regia*** Carpathian Group 'Liddington's Late Leafing'	wBur
un	***regia*** 'Chambers'	wBur wRai
un	***regia*** 'Chandler'	wBur
un	***regia*** 'Chopaka'	wBur
un	***regia*** 'Chopaka' seedling	wBCr
	regia 'Cooke's Giant Sweet'	oGar oGre
	regia 'Franquette'	wBur
	regia 'Hansen'	last listed 99/00
	regia 'Hansen' seedling	wBCr
un	***regia*** 'Manregion'	wClo oGar oWhi
un	***regia*** 'Manregion' seedling	wVan
un	***regia*** 'Okanogan'	wBur
un	***regia*** 'Russian' seedling	wBCr
un	***regia*** 'Russian 3'	wBur
un	***regia*** 'Sejveno'	wBur
un	***regia*** 'Sommers'	wBur
un	***regia*** 'Sommers' seedling	wBCr
un	***regia*** 'Spurgeon'	wBur wRai

JUNCUS

un	***acuminatus***	wWat oTri oAld
	balticus	oHug
	'Carmen's Grey'	oTrP wHer oHug
	'Carman's Japanese'	oWnC
	effusus	wCli wSoo wNot wShR oHug oGar oSsd oBos wWin wWat oWnC oTri oAld wCoN
va	***effusus*** 'Gold Strike'	oHug
	effusus f. ***spiralis***	oTrP wCli wSwC wGAc oDar wCol oHug oGar oSsd oOut wWin oWnC oUps
un	***effusus*** 'Unicorn'	wSwC oGar
	ensifolius	wWoB wNot oHug oSsd oBos wWat oTri oAld
	filiformis	oTrP
un	'Occidental Blue'	wBox
	patens	oHug oSsd oBos oTri oAld
qu	***scirpus americanus***	oGar
	tenuis	oHug oGar oAld

JUNELLIA

	wilczekii F&W8447	wMtT

JUNIPERUS

	chinensis	oFor
qu	***chinensis*** 'Albovariegata'	oPor
un	***chinensis*** 'Aureovariegata'	oPor
un	***chinensis*** 'Bandatsuga Aurea'	wSta
	chinensis 'Blaauw'	wKin wSta
	chinensis 'Blue Alps'	oFor oPor oRiv oTPm
	chinensis 'Blue Point'	oGar oRiv wWel
va	***chinensis*** 'Expansa Variegata'	oPor
	chinensis Gold Coast®/ 'Aurea'	wKin oWnC
	chinensis 'Hetzii'	oRiv wKin
	chinensis 'Hetzii Glauca'	see **J. *chinensis*** 'Hetzii'
un	***chinensis*** 'Itowagawa'	oTin
	chinensis 'Kaizuka'	oFor oAls oGar oBlo oWnC wSta oBRG
	chinensis 'Kaizuka Variegata'	see **J. *chinensis*** 'Variegated Kaizuka'
	chinensis 'Keteleeri'	oGar wCCr
	chinensis 'Mac's Golden'	oFor
	chinensis 'Plumosa'	wSta
	chinensis var. ***procumbens***	see **J. *procumbens***
	chinensis 'Robust Green'	oFor oGar
	chinensis 'San Jose'	oAls oBlo wKin oWnC wSta
	chinensis 'Sea Green'	see **J. *x pfitzeriana*** 'Sea Green'
	chinensis 'Shimpaku'	oFor oPor wCol oSis oRiv oGre wSta oTin wWel
un	***chinensis*** 'Shoosmith'	oPor
	chinensis 'Spartan'	oGar wWel
un	***chinensis*** 'Spearmint'	oGar wWel
	chinensis 'Stricta'	oSho
	chinensis 'Torulosa'	see **J. *chinensis*** 'Kaizuka'
va	***chinensis*** 'Variegated Kaizuka'	oPor wSta
	chinensis 'Wintergreen'	oWnC
	communis 'Compressa'	oPor oGos wClo wCol oGar oDan oSis oGre oWnC oBRG wWel
	communis 'Depressa Aurea'	wHer
	communis 'Echiniformis'	last listed 99/00
	communis 'Effusa'	oPor oRiv oBlo oTPm
	communis 'Gold Cone'	oFor oPor wCol oAls oGar oDan oSis oRiv oSho oGre oWnC oBRG
un	***communis*** 'Golden Schnapps'	oPor oRiv
	communis 'Green Carpet'	oPor
	communis 'Hibernica'	oGar wAva
	communis var. ***jackii***	last listed 99/00
	communis var. ***montana***	wShR oGar oGoo
	communis 'Pencil Point'	oRiv
	communis 'Pendula'	oPor
	communis var. ***saxatilis***	oFor
	communis 'Sentinel'	oPor
	communis 'Silver Mist'	oRiv
un	***communis*** 'Suecica Nana'	oBRG
	communis 'Tage Lundell'	oPor
	conferta	wWoB oGar oSho oBlo oTPm wSta
un	***conferta*** 'Akebono'	oPor
	conferta 'Blue Pacific'	oAls oGar oTPm wWel
	conferta 'Emerald Sea'	oFor oPor oTPm wWel
	conferta 'Sunsplash'	last listed 99/00
	deppeana	oFor
un	***deppeana glauca***	oFor
	foetidissima	oFor
	formosana	oPor oGar
	horizontalis 'Bar Harbor'	oGar oSho oBlo oGre oTPm oWnC
	horizontalis 'Blue Chip'	wCSG oGar oSho oBlo oGre wKin oTPm oWnC wCoS
	horizontalis 'Blue Mat'	oTPm
un	***horizontalis*** 'Coast of Maine'	oPor
	horizontalis 'Glomerata'	last listed 99/00
	horizontalis 'Golden Spreader'	wKin
	horizontalis 'Green Acres'	oPor
un	***horizontalis*** 'Grey Forest'	oSis
	horizontalis 'Hughes'	wKin oWnC
	horizontalis Icee Blue™ / 'Monber'	oGar oSho wCoS
un	***horizontalis*** 'Lime Glow'	wHer
	horizontalis 'Mother Lode'	oPor oGar
un	***horizontalis*** 'Pancake'	oPor oSis
	horizontalis 'Prince of Wales'	oBlo wKin oWnC
	horizontalis 'Wiltonii'	oFor oAls oGar oGre wKin oTPm oWnC wBox wSta oGue
	horizontalis 'Youngstown'	oAls oGar wKin oTPm
	x media	see **J. *x pfitzeriana***
	occidentalis	oFor
	osteosperma	oFor
	oxycedrus	oFor oRiv wCCr
	x pfitzeriana	oAls oWnC
	x pfitzeriana 'Armstrongii'	oAls
qu	***x pfitzeriana*** 'Aureovariegata'	oRiv
	x pfitzeriana 'Daub's Frosted'	oFor wHer oPor oSis oRiv oSqu oGre oBRG
	x pfitzeriana 'Gold Star'	oGar
	x pfitzeriana golden	oTPm
	x pfitzeriana Mint Julep® / 'Monlep'	wKin oTPm oWnC
	x pfitzeriana 'Old Gold'	oAls oGar oSho oBlo oGre oTPm wSta oGue
	x pfitzeriana 'Pfitzeriana Aurea'	oAls oBlo wKin oWnC wSta
	x pfitzeriana 'Pfitzeriana Glauca'	oGar wKin
	x pfitzeriana 'Sea Green'	oGar oBlo wKin oWnC wSta oGue
	x pfitzeriana 'Sulphur Spray'	oPor
	pinchotii	oFor
	pingii	oRiv
	pingii 'Loderi'	see **J. *pingii*** var. ***wilsonii***
	pingii var. ***wilsonii***	oPor oRiv
qu	'Platte River'	oBlo
	procumbens	oGar
un	***procumbens*** 'Green Mound'	oAls oGar
	procumbens 'Nana'	oPor oAls oAmb oGar oRiv oSho oBlo wKin oTPm oWnC wSta oBRG wCoS wWel
	procumbens variegated	oTPm
	recurva	oFor
	recurva var. ***coxii***	oPor
	rigida	wHer wCCr
un	***rigida*** 'Blue Forest'	oPor

qu	***rigida*** 'Pendula'	oPor wTGN
	sabina	wKin
	sabina 'Blaue Donau'	wKin
	sabina Blue Danube	see **J. *sabina*** 'Blaue Donau'
	sabina 'Broadmoor'	wKin oTPm
	sabina 'Buffalo'	oGar oTPm
	sabina 'New Blue Tam'	see **J. *sabina*** 'Tamariscifolia New Blue'
un	***sabina*** 'Pepin'	wKin
	sabina 'Skandia'	wKin
	sabina 'Tamariscifolia'	oAls oGar oBlo oGre oTPm wSta
un	***sabina*** 'Tamariscifolia New Blue'	oTPm oWnC
va	***sabina*** 'Variegata'	oFor oPor oRiv
	scopulorum	oFor wShR wKin wPla
	scopulorum 'Cologreen'	oWnC
	scopulorum 'Gray Gleam'	oWnC
	scopulorum 'Moonglow'	oFor wCol oDan oRiv wKin oTPm
	scopulorum 'Pathfinder'	oWnC
	scopulorum 'Skyrocket'	oPor oGar oRiv oSho oBlo oGre wKin oTPm oWnC wTGN wSta
un	***scopulorum*** 'Sparkling Skyrocket'	oFor
	scopulorum 'Tabletop'	wKin
	scopulorum 'Tolleson's Blue Weeping'	oWnC wWel
	scopulorum 'Wichita Blue'	oGar oBlo wKin oTPm wWel
	silicicola	oFor
	sp. Daniel's dwarf	oPor
qu	'Spiney Creek'	oBlo
	squamata 'Blue Carpet'	oPor oJoy
	squamata 'Blue Star'	oFor wHer oPor oAls oGar oSis oSho oSqu oGre wKin oTPm oWnC wSta oBRG wAva wCoS wWel
	squamata 'Chinese Silver'	oPor
	squamata 'Holger'	oPor
	squamata 'Loderi'	see **J. *pingii*** var. ***wilsonii***
	squamata 'Meyeri'	wSta
	squamata 'Prostrata'	oSqu oWnC
	virginiana	oFor oRiv
	virginiana 'Blue Mountain'	oPor
	virginiana 'Burkii'	oPor oGar
	virginiana 'Grey Owl'	oBlo wCCr
	virginiana 'Hetzii'	oWnC
	virginiana silicicola	see **J. *silicicola***
	virginiana 'Skyrocket'	see **J. *scopulorum*** 'Skyrocket'
	JURINEA	
	mollis	oMis
	JUSSIAEA	see **LUDWIGIA**
	JUSTICIA	
un	***americana***	wGAc
	pectoralis	oOEx
	KADSURA	
	japonica	oFor oOEx
un	***japonica*** 'Cream'	oOEx
un	***japonica*** 'Delight'	oOEx
un	***japonica*** 'Dwarf'	oOEx
	japonica HC 970337	wHer
un	***japonica*** 'Peach'	oOEx
un	***japonica*** 'Sweet!'	oOEx
	japonica 'Variegata'	last listed 99/00
	KALANCHOE	
	blossfeldiana	oGar
	fedtschenkoi 'Marginata'	oSqu
	marmorata	last listed 99/00
	millottii	oSqu
	orgyalis	last listed 99/00
	pubescens	oSqu
	pumila	oSqu
	red leaf	oSqu
un	***synsepala lanceolata***	oSqu
	thyrsiflora	oSqu
	tomentosa	oSqu
	tomentosa dark	last listed 99/00
	KALIMERIS	
	mongolica	wWoS
	yomena 'Shogun'	wWoS wCol oCis
	yomena 'Variegata'	see **K. *yomena*** 'Shogun'
	KALMIA	
un	'Big Boy'	oAls
	latifolia	oAls wAva wCoN
	latifolia 'Alpine Pink'	wWhG
un	***latifolia*** 'Bridesmaid'	wWhG
	latifolia 'Bullseye'	oGre wWhG
	latifolia 'Carol'	wWhG
	latifolia 'Carousel'	oGar wWhG
un	***latifolia*** 'Comet'	wWhG
	latifolia 'Elf'	oFor oAls oGar oDan oSho oGre wWhG
	latifolia 'Freckles'	wCSG oGar wWhG
	latifolia 'Fresca'	oAls
un	***latifolia*** 'Galaxy'	oGre wWhG
	latifolia 'Goodrich'	oAls
un	***latifolia*** 'Goodshow'	oAls
	latifolia 'Heart of Fire'	oAls oGre wWhG
	latifolia 'Heart's Desire'	oGar
un	***latifolia*** 'Kaleidoscope'	oSho oGre wWhG
un	***latifolia*** 'Keepsake'	wTGN
	latifolia 'Little Linda'	oGos oGar oSho oGre wWhG
	latifolia 'Minuet'	oGar oSho oGre wWhG
un	***latifolia*** 'Nathan Hale'	oGre wWhG
	latifolia 'Nipmuck'	oAls
qu	***latifolia*** 'No Suchianum'	wWhG
	latifolia 'Olympic Fire'	oGar oSho oGre wWhG
un	***latifolia*** 'Olympic Wedding'	oAls
	latifolia 'Ostbo Red'	wCSG oDan wSta wWhG
un	***latifolia*** 'Paul Bosley'	wCSG oGar wWhG
	latifolia 'Peppermint'	oFor oGos oGar oSho oGre wWhG
	latifolia 'Pink Charm'	oDan wWhG
	latifolia 'Pink Frost'	oFor
	latifolia 'Pinwheel'	wCSG oGar oSho wWhG
un	***latifolia*** 'Pristine'	wWhG
	latifolia 'Raspberry Glow'	oGos oGar oSho oGre wWhG
	latifolia 'Richard Jaynes'	oGos oGre wWhG
	latifolia 'Sarah'	oGos oGar wWhG
un	***latifolia*** 'Sharon Rose'	oAls
	latifolia 'Snowdrift'	oGar oSho wWhG
	latifolia 'Tiddlywinks'	oGar oGre wWhG
un	***latifolia*** 'Willowleaf'	oBov
un	***latifolia*** 'Woodland Pink'	oAls
	microphylla	oSis oBos
qu	***microphylla*** var. ***microphylla***	oBov
	polifolia	oBos oSho wRho
	polifolia var. ***microphylla***	see **K. *microphylla***
	KALMIOPSIS	
	fragrans	see **K. *leachiana***
	leachiana	oFor wHer wThG oGar wRho oBov
	leachiana LePiniec form	see **K. *leachiana*** 'Marcel le Piniec'
	leachiana 'Marcel le Piniec'	oGos oBov oGre
	leachiana Umpqua Valley form	last listed 99/00
	KALOPANAX	
	pictus	see **K. *septemlobus***
	septemlobus	oFor oGar
	KEDROSTIS	
	africana	oRar
un	***capensis***	oRar
un	***punctulata***	oRar
un	***puniceus***	oRar
	KENNEDIA	
	coccinea	last listed 99/00
	eximia	last listed 99/00
	rubicunda	oFor
	KERRIA	
ch	***japonica*** (see Nomenclature Notes)	oRiv oBlo wCoN wCoS
	japonica 'Albiflora'	oFor wWoS oGos oGar
	japonica 'Golden Guinea'	wWoS oGos
va	***japonica*** 'Picta'	oFor wCri oAmb oGar oGre
	japonica 'Picta' prostrate form	oSis
do	***japonica*** 'Pleniflora'	oFor oAls oGar oOut oGre oWnC oCir wWel
un	***japonica*** 'Shannon'	oGos
	japonica 'Simplex'	oFor wCri oGar
un	***japonica*** 'Superba'	wCul
	japonica 'Variegata'	see **K. *japonica*** 'Picta'
	KETELEERIA	
	calcarea	see **K. *davidiana***
	davidiana	oFor wHer wCCr
	evelyniana	oFor wCol
un	***pubescens***	oCis
	KIGELIA	
	africana	wGra
	pinnata	see **K. *africana***
	KIRENGESHOMA	
	koreana	see **K. *palmata*** Koreana Group
	palmata	oFor wHer oHed oRus oAmb oMis oGar oNWe wNay wRob wMag wSte
	palmata Koreana Group	oFor wCol wRob
	KITAIBELA	
	vitifolia	oFor oUps
	KLEINIA	
	articulata	see **SENECIO *articulatus***
	KNAUTIA	
	arvensis	wCSG oMis oEga

	macedonica	oFor wHer wWoS wSwC wGAc oNat oAls oHed oMis oGar oDan oOut wRob oBlo oGre wFai oWnC wTGN oEga oUps
	macedonica Melton Pastels	wWoS
un	***macedonica*** 'Ruby Star'	oJoy
un	'Satchmo'	wMag
	KNIPHOFIA	
	'Alcazar'	oNWe wSnq
	'Apricot'	oGre wBox
	baurii	wHer
	'Border Ballet'	wWoS wMag wBWP
	'Bressingham Comet'	oFor wWoS
	caulescens	wHer oGoo oCis
	citrina	oGar oGoo oDan oGre wBWP
	'Cobra'	oFor
un	'Coral'	wCol
un	'Coral Glow'	oJoy oGar
	'Corallina'	wSnq
	'Dr. E. M. Mills'	last listed 99/00
	early hybrids	wSnq
	ensifolia	oDan
	'Flamenco'	oGar oUps
	foliosa	oDan oWnC
	galpinii (see Nomenclature Notes)	oDar oAmb
	hirsuta	oNWe
	hirsuta hybrid	oNWe
	hybrid, yellow	oNWe
	ichopensis	last listed 99/00
	laxiflora JA3.462.210	oNWe
	linearifolia	wHer oDan
	'Little Maid'	oFor wWoS wCol oSto oSis
	natalensis	last listed 99/00
	northiae	oNWe
	Pacific hybrids	oGar
	pauciflora	oHed
	'Percy's Pride'	oFor wHer wWoS oGos oHed
	'Pfitzeri'	oAls oGoo oGre oEga
	'Primrose Beauty'	wWoS oAls oGoo wSnq
un	'Red Castle Hybrid'	wMag
	'Royal Caste'	wMag
	sp. aff. ***rufa***	last listed 99/00
	sarmentosa	oTrP oDan
	'Shining Sceptre'	oFor oDar oGar oNWe oGre
	sp. OB3	oNWe
	splendida	last listed 99/00
	stricta	last listed 99/00
	'Sunningdale Yellow'	last listed 99/00
un	sp. cf. ***thodei***	oNWe
	triangularis ssp. ***triangularis***	oDan
	uvaria	oSto oGar oUps
	'Wayside Flame'	see **K.** 'Pfitzeri'
un	'Yellow Cheer'	oFor
	KOELERIA	
	cristata (see Nomenclature Notes)	oDar
	glauca	oSqu wWin wTGN oSec
	macrantha	oFor
	KOELREUTERIA	
	bipinnata	oFor oDan oGre wCCr
	bipinnata var. ***integrifoliola***	oGre oCis
	elegans	wCCr
	paniculata	oFor oAls oGar oRiv oSho oGre oWnC wCoN oSle wCoS
	paniculata 'Fastigiata'	last listed 99/00
un	***paniculata*** 'Rose Lantern'	oGar wWel
	KOHLERIA	
un	***glauca***	wBox
	KOLKWITZIA	
	amabilis	oFor oJoy oAls oGar oRiv oSho wTGN wCoN wCoS wWel
	amabilis deep pink form	oHed
	amabilis 'Pink Cloud'	oFor wWel
	KUNZEA	
	ambigua	last listed 99/00
	+ LABURNOCYTISUS	
	'Adamii'	oFor oGar oSho oGre
	LABURNUM	
	alpinum	oFor oDan
	alpinum 'Pendulum'	oFor oAls
	anagyroides	oFor wCoN
	anagyroides 'Quercifolium' seed grown	wHer
un	***anagyroides*** 'Sunspire'	wWel
	x watereri	oGar oWnC wCoN oSle
qu	***x watereri*** 'Pendula'	oGar
	x watereri 'Vossii'	oFor oAls oGar oSho oGre oWnC wTGN wCoS wWel
	LAGAROSTROBOS	
	franklinii	oFor wHer
	franklinii 'Pendulum'	wHer
	LAGERSTROEMIA (SEE NOMENCLATURE NOTES)	
	'Acoma'	oFor oAmb oGar oSho oGre oWnC
	'Catawba'	oSho
un	***checkiangense***	oCis
un	'Chickasaw'	oFor
	'Comanche'	oFor oDan
	dwarf pink	oAls
	dwarf purple	oAls
	'Hopi'	oFor oDan oWhi
un	***indica*** 'Centennial'	oFor
	indica Chica® Pink / 'Monink'	oGar oSho oGre
	indica Chica® Red / 'Moned'	oGar
	indica 'Glendora White'	oWnC
un	***indica*** 'Lipan'	oFor
un	***indica*** Little Chiefs mixed	oUps
un	***indica*** 'Nana'	oTrP
	indica 'Near East'	oGar
un	***indica*** 'New Orleans'	oWnC
	indica 'Peppermint Lace'	oSho oGre oWnC
	indica Petite Embers™ / 'Moners'	oGar oGre
	indica Petite Orchid™ / 'Monhid'	oGar
	indica Petite Pinkie™ / 'Monkie'	oGar
	indica Petite Plum™ / 'Monum'	oGar oWnC
	indica Petite Red Imp™ / 'Monimp'	oGar oGre
un	***indica*** 'Prairie Lace'	oFor
	indica seed grown	wCCr
un	***indica*** 'Tonto'	oFor
	indica 'Victor'	oFor
	indica 'Watermelon Red'	oGar oSho
un	***limii***	oGre oCis
	'Muskogee'	oFor
	'Natchez'	last listed 99/00
un	'Osage'	oFor
	'Pecos'	oAls oGar oSho oGre wWel
un	'Potomac'	oFor
	'Seminole'	oGar oSho
	'Sioux'	oFor
	'Tuscarora'	oWhi
	'Tuskegee'	oFor oWhi
un	'Yuma'	oFor
	'Zuni'	oFor oAls oGar oSho oGre oWnC wWel
	LAGUNARIA	
	patersonii	last listed 99/00
	LAMIASTRUM	see **LAMIUM**
	LAMIUM	
va	***album*** 'Friday'	oGar
	galeobdolon	oDar
un	***galeobdolon*** 'Compactum'	oFor
	galeobdolon 'Hermann's Pride'	oFor wHer oNat oJoy wRob wHom oLSG oUps
	galeobdolon ssp. ***montanum*** 'Florentinum'	oFor oInd oUps
un	***galeobdolon*** 'Petit Point'	oJoy
	galeobdolon 'Silver Spangled'	oFor
ch	***galeobdolon*** 'Variegatum'	see **L. *galeobdolon*** ssp. ***montanum*** 'Florentinum'
	maculatum	oJoy
	maculatum 'Album'	oFor oJoy
un	***maculatum*** 'Angel Wings'	oJoy
	maculatum 'Aureum'	wWoS oSis wHom oUps
	maculatum 'Beacon Silver'	oJoy oCir oLSG wWhG oUps
	maculatum 'Beedham's White'	wWoS
un	***maculatum*** 'Brocade'	oJoy
	maculatum 'Cannon's Gold'	last listed 99/00
	maculatum 'Chequers'	oFor
va	***maculatum*** 'Elisabeth de Haas'	wWoS
va	***maculatum*** 'Ickwell Beauty'	wHer
	maculatum 'Pink Nancy'	oJoy
un	***maculatum*** 'Pink Panther'	oWnC
	maculatum 'Pink Pewter'	oNat oSev wFai oLSG wWhG oUps
	maculatum 'Red Nancy'	wWoS oJoy
	maculatum 'Roseum'	oFor oJoy oLSG
	maculatum 'Shell Pink'	see **L. *maculatum*** 'Roseum'
	maculatum 'White Nancy'	oFor wCul oNat oJoy oAls oSev wHom oWnC oInd oCir oLSG wWhG oUps
	orvala	last listed 99/00
	orvala 'Album'	wHer
	LANTANA	
un	***camara*** 'American Red'	oWnC
	camara 'Irene'	oGar
	'Confetti'	wWel
	montevidensis	oGar

	Plant	Sources
	'Radiation'	oGar
	'Spreading Sunset'	oGar wWel
	Tangerine ™ / 'Mone'	oCir wWel
	LAPEIROUSIA	
	laxa	see ANOMATHECA laxa
	LARDIZABALA	
	biternata	oFor
	biternata HCM 98072	wHer
	LARIX	
	decidua	oFor oDar oGar oBlo oGre wKin oWnC wRav
un	decidua 'Girard's Dwarf'	oRiv
un	decidua 'Lanark'	oPor
	decidua 'Little Bogle'	oPor
	decidua 'Pendula'	oFor oPor wCol oAls oGar oRiv oBlo oGre oWnC wSta
	x eurolepis	see L. x marschlinsii
	gmelinii	oFor
	gmelinii var. olgensis	oDar oRiv oGre
	kaempferi	oFor oGar oRiv wCCr wKin oWnC wTGN wRav
	kaempferi 'Blue Dwarf'	oPor
	kaempferi 'Blue Rabbit'	oPor oRiv
	kaempferi 'Diane'	oFor oPor wCol oRiv wTGN wAva wWel
un	kaempferi 'Haverbeck'	oPor
	kaempferi 'Prostrata'	wCol
	kaempferi 'Wolterdingen'	oPor wCol
	laricina	oFor
un	laricina 'Craftsbury Flats'	oPor oRiv
	laricina 'Newport Beauty'	wCol oRiv
un	laricina 'Newport Beauty # 7'	oPor
un	laricina 'Newport Beauty # 9'	oPor
un	laricina 'Newport Beauty #17'	oPor
un	laricina 'Stubby'	wCol
	leptolepis	see L. kaempferi
un	x marschlinsii 'Morris Broom'	oPor
un	x marschlinsii 'Varied Directions'	oPor oRiv oGre
	mastersiana	oCis
	occidentalis	oFor wShR oGar oBos oWnC wPla
	olgensis	see L. gmelinii var. olgensis
	russica	see L. sibirica
	sibirica	oFor
un	sibirica 'Conica'	oPor
un	sibirica 'Rasputin'	wCol
	LATHYRUS	
	aureus	wHer
	cirrhosus	last listed 99/00
	davidii HC 970159	wHer
	latifolius	oFor oMis oLSG
	odoratus	wFGN oGar iArc wSte
	odoratus 'America'	oFra
un	odoratus 'Annie B. Gilroy'	oFra
un	odoratus 'Black Knight'	oFra
un	odoratus 'Blanche Ferry'	oFra
un	odoratus 'Butterfly'	oFra
	odoratus 'Captain of the Blues'	oFra
	odoratus 'Countess Cadogan'	last listed 99/00
	odoratus 'Cupani'	oFra
un	odoratus 'Donna Jones'	oFra
un	odoratus 'Dorothy Eckford'	oFra
un	odoratus 'Flora Norton'	oFra
un	odoratus 'Henry Eckford'	oFra
un	odoratus 'Indigo King'	oFra
un	odoratus 'Janet Scott'	oFra
un	odoratus 'King Edward VII'	oFra
un	odoratus 'Lady Grisel Hamilton'	oFra
un	odoratus 'Lord Nelson'	oFra
	odoratus 'Matucana'	oFra
un	odoratus 'Miss Willmott'	oFra
un	odoratus 'Mrs. Collier'	oFra
	odoratus 'Painted Lady'	oFra
un	odoratus 'Prince Edward of York'	oFra
un	odoratus 'Queen Alexandra'	oFra
un	odoratus 'Senator'	oFra
un	odoratus Spencer Group 'Alan Titmarsh'	oFra
un	odoratus Spencer Group 'Angela Ann'	oFra
un	odoratus Spencer Group 'Annabelle'	oFra
un	odoratus Spencer Group 'Annie Good'	oFra
un	odoratus Spencer Group 'Bandaid'	oFra
un	odoratus Spencer Group 'Beaujolais'	oFra
un	odoratus Spencer Group 'Black Diamond'	oFra
un	odoratus Spencer Group 'Blue Danube'	oFra
un	odoratus Spencer Group 'Blue Velvet'	oFra
un	odoratus Spencer Group 'Blushing Bride'	oFra
un	odoratus Spencer Group 'Brian Clough'	oFra
un	odoratus Spencer Group 'Buccaneer'	oFra
un	odoratus Spencer Group 'Burnished Bronze'	oFra
un	odoratus Spencer Group 'Cambridge Blue'	oFra
un	odoratus Spencer Group 'Candy Frills'	oFra
un	odoratus Spencer Group 'Claire Elizabeth'	oFra
un	odoratus Spencer Group 'Cream Southbournne'	oFra
un	odoratus Spencer Group 'Elizabeth Taylor'	oFra
un	odoratus Spencer Group 'Esther Rantzen'	oFra
un	odoratus Spencer Group 'Ethel Grace'	oFra
un	odoratus Spencer Group 'Fatima'	oFra
un	odoratus Spencer Group 'Felicity Kendall'	oFra
un	odoratus Spencer Group 'Firebrand'	oFra
un	odoratus Spencer Group 'Firecrest'	oFra
un	odoratus Spencer Group 'Frolic'	oFra
un	odoratus Spencer Group 'Gaiety'	oFra
un	odoratus Spencer Group 'Gypsy Queen'	oFra
un	odoratus Spencer Group 'Hampton Court'	oFra
un	odoratus Spencer Group 'Honeymoon'	oFra
un	odoratus Spencer Group 'Jilly'	oFra
un	odoratus Spencer Group 'Joker	oFra
un	odoratus Spencer Group 'Lady Diana'	oFra
	odoratus Spencer Group 'Lady Fairbairn'	oFra
un	odoratus Spencer Group 'Lilac Silk'	oFra
un	odoratus Spencer Group 'Lilac Time'	oFra
un	odoratus Spencer Group 'Midnight'	oFra
un	odoratus Spencer Group 'Mollie Rilstone'	oFra
	odoratus Spencer Group 'Mrs. Bernard Jones'	oFra
un	odoratus Spencer Group 'Myrtle Mann'	oFra
un	odoratus Spencer Group 'Nora Holman'	oFra
un	odoratus Spencer Group 'North Shore'	oFra
un	odoratus Spencer Group 'Oban Bay'	oFra
un	odoratus Spencer Group 'Our Harry'	oFra
un	odoratus Spencer Group 'Percy Thrower'	oFra
un	odoratus Spencer Group 'Pretty Polly'	oFra
un	odoratus Spencer Group 'Princess Elizabeth'	oFra
un	odoratus Spencer Group 'Restmorel'	oFra
un	odoratus Spencer Group 'Rosy Frills'	oFra
un	odoratus Spencer Group 'Sea Wolfe'	oFra
un	odoratus Spencer Group 'Southbourne'	oFra
un	odoratus Spencer Group 'Sue Pollard'	oFra
un	odoratus Spencer Group 'Terry Wogan'	oFra
un	odoratus Spencer Group 'The Doctor'	oFra
un	odoratus Spencer Group 'Vera Lynn'	oFra
un	odoratus Spencer Group 'Welcome'	oFra
	odoratus Spencer Group 'White Supreme'	oFra
un	odoratus Spencer Group 'Yashmin Khan'	oFra
	venetus	last listed 99/00
	vernus	oFor wHer wCol
	vernus 'Alboroseus'	oHed
	LAURENTIA	
	axillaris	wCul wTho oDar oAls oHed wFGN oGre wFai wHom oWnC wMag oBar wSta oCir wCoN oUps wCoS
	axillaris dark blue	oAls
	fluviatilis	see L. axillaris
	LAURUS	
	nobilis	oFor wCri wBur oDar oAls wFGN wCSG oGar oGoo oRiv oSho oWhi oOEx wFai wPir wCCr iGSc wHom oWnC oCrm oBar wTGN wNTP wCoN wSte wCoS wWel
	nobilis f. angustifolia	wHer
	nobilis 'Aurea'	last listed 99/00
	nobilis 'Saratoga'	oFor
	LAVANDULA	
	x allardii	wWoS wFGN oGoo oDut oBar
un	'Ana Luisa '	oDut oBar
	angustifolia	oAls wFGN wShR wCSG oSho oGre wFai iGSc wHom wMag oSec oCir wCoN oSle
	angustifolia 'Alba'	iGSc wHom
qu	angustifolia 'Alpine Alba'	wCul
un	angustifolia 'Baby Blue'	wCul wFGN oDut oBar
un	angustifolia 'Backhouse Nana'	oBar
un	angustifolia 'Betty's Blue'	oDut oBar
	angustifolia 'Bowles' Early'	oFor oAls wFGN oGar oDut wHom oBar wTGN
un	angustifolia 'Buena Vista'	wCul wFGN oGoo oDut iGSc oBar
un	angustifolia 'Carolyn Dille'	oDut
	angustifolia 'Compacta'	wFGN oGoo oBar
un	angustifolia 'Croxton's Wild'	wWoS oGoo oDut oBar
un	angustifolia 'Dark Supreme'	see L. angustifolia 'W. K. Doyle'
	angustifolia 'Dwarf Blue'	oGre

un	***angustifolia*** 'Empress Purple'	oDut
	angustifolia 'Fairie Garden Dwarf Pink'	wFai
	angustifolia 'Folgate'	oAls wFGN oDut
un	***angustifolia*** 'Gmerick'	wCCr
un	***angustifolia*** 'Graves'	wFGN oGoo oDut oBar
un	***angustifolia*** 'Grey Lady'	wCul wFGN oDut oBar
qu	***angustifolia*** 'H. B. Spica'	wCul oDut
	angustifolia 'Hidcote'	oFor wWoS wCul wSwC oDar oJoy oAls wFGN oGar oGoo oDut oSis oSho oOut oGre iGSc wHom oWnC wMag oCrm oBar wNTP oEga oUps oSle wSnq wCoS
	angustifolia 'Hidcote Blue'	see **L. *angustifolia*** 'Hidcote'
un	***angustifolia*** 'Hidcote Compact'	wWoS
	angustifolia 'Hidcote Pink'	wWoS wFGN oGoo oDut oSis wCCr oBar oCir wSnq
	angustifolia 'Imperial Gem'	oHed
un	***angustifolia*** 'Irene Doyle'	wCul wFGN oGoo oBar
	angustifolia 'Jean Davis'	wWoS oAls oHed wFGN oDut oSev oGre iGSc wHom wMag oBar wWhG oUps wSnq
	angustifolia 'Lady'	wFGN oSha wSnq
	angustifolia 'Lavender Lady'	oGar iGSc oWnC oBar wNTP oEga
	angustifolia 'Loddon Blue'	wWoS wFGN oGoo oDut oCrm oBar
qu	***angustifolia*** 'London Purple'	oDut
	angustifolia 'Maillette'	wFGN oDut oBar
un	***angustifolia*** 'Mary Medallion'	wMag
un	***angustifolia*** 'Mary Medelia'	wWoS
un	***angustifolia*** 'Melissa'	wCul oAls wFGN oDut oWnC oBar
un	***angustifolia*** 'Mitcham Grey'	wCul wFGN oDut oBar
	angustifolia 'Munstead'	oFor wWoS wCul wTho oAls wFGN oGar oGoo oSho oSha oBlo oGre iGSc wHom oWnC wMag oBar wNTP iArc oCir oLSG wWhG oUps oSle wSnq wCoS
	angustifolia 'Nana Alba'	wCul oGoo oDut oSis oBar
	angustifolia 'Nana Atropurpurea'	wFGN oSis oBar
un	***angustifolia*** 'Norfolk J2'	wFGN oBar
un	***angustifolia*** 'Pastor's Pride'	oBar
	angustifolia pink	wFGN
un	***angustifolia*** 'Premier'	wFGN oGoo oDut iGSc oBar
un	***angustifolia*** 'Purple Velvet'	oDut
	angustifolia 'Rosea'	oGoo iGSc wNTP
	angustifolia 'Royal Purple'	oAls wFGN oDut oBar
un	***angustifolia*** 'Royal Velvet'	wCul wSwC wFGN oGoo oDut oCrm oBar
un	***angustifolia*** 'Sachet'	wWoS wFGN oDut iGSc oCrm oBar wTGN
un	***angustifolia*** 'Sarah'	oBar wWhG
un	***angustifolia*** 'Seal's Seven Oaks'	wCul wFGN oGoo oDut oBar
un	***angustifolia*** 'Shakespeare's Garden'	wFGN
un	***angustifolia*** 'Sharon Roberts'	wSwC oGoo oDut oBar
un	***angustifolia*** 'Short 'n' Sweet'	wFGN oDut oBar
un	***angustifolia*** 'Silverleaf'	wCCr wMag
un	***angustifolia*** 'Sleeping Beauty'	wFGN
un	***angustifolia*** 'Summerland Supreme'	oBar
un	***angustifolia*** 'Susan Belsinger'	see **L. *angustifolia*** 'Short 'n' Sweet'
un	***angustifolia*** 'Tucker's Early Purple'	oDut oBar
un	***angustifolia*** 'Tuscan'	wSte
	angustifolia 'Twickel Purple' (see Nomenclature Notes)	oFor wWoS wFGN oGoo oDut iGSc oWnC wTGN oUps wSnq
un	***angustifolia*** 'Two Seasons'	see **L. *angustifolia*** 'Irene Doyle'
un	***angustifolia*** 'Victorian Amethyst'	wWoS
un	***angustifolia*** 'W. K. Doyle'	oAls wFGN oBar
	angustifolia 'White Dwarf'	see **L. *angustifolia*** 'Nana Alba'
un	***angustifolia*** 'Wycoff'	wFGN oBar
un	'Barbara Joan'	see **L. *x intermedia*** 'Dutch Mill'
	'Blue Cushion'	oFor oAls wFGN oDut oGre oBar wSnq
qu	***delphinensis***	wFGN
	dentata	wWoS oAls wFGN oGar oGoo oDut iGSc wHom oWnC wMag oBar wNTP oEga oUps
	dentata var. ***candicans***	oFor oAls wFGN oGoo oDut oOut iGSc oBar
	dentata var. ***candicans*** silver	see **L. *dentata*** var. ***candicans***
	dentata French Green	see **L. *dentata***
	dentata French Grey	see **L. *dentata*** var. ***candicans***
	dentata grey	see **L. *dentata*** var. ***candicans***
un	***dentata*** 'Lambkins'	wHer wWoS wFGN
	dentata 'Linda Ligon'	wHer wWoS wFGN oGoo oDut iGSc oBar oUps
	dentata variegated	see **L. *dentata*** 'Linda Ligon'
	'Devantville Cuche'	wHer
	'Dilly Dilly'	oDut
un	'Du Provence'	see **L. *x intermedia*** 'Provence'
un	'England'	wWoS oCrm oBar wSnq
	'Goodwin Creek'	see **L.** 'Goodwin Creek Grey'

	'Goodwin Creek Grey'	oFor wWoS wCul oJoy oAls oHed wFGN wCSG oAmb oGar oGoo oDut iGSc oCrm oBar oSec wNTP oUps oSle
un	***x heterophylla***	wCul oJoy oGoo iGSc oBar wNTP oUps
	x intermedia	wCul
un	***x intermedia*** 'Abbey'	oDut
	x intermedia 'Abrialii'	wWoS wFGN oGoo oDut oBar
	x intermedia 'Alba'	wWoS wCul oJoy oGoo oDut wMag oBar oUps wSnq
	x intermedia 'Arabian Night' (see Nomenclature Notes)	oDut
un	***x intermedia*** 'Barbara Joan'	oDut
	x intermedia 'Bogong'	oDut
un	***x intermedia*** 'Boisto'	oDut
un	***x intermedia*** 'Bridestow'	oDut
un	***x intermedia*** 'du Provence'	see **L. *x intermedia*** 'Provence'
	x intermedia Dutch Group	oAls wFGN oGar oGoo oDut oBar
un	***x intermedia*** 'Dutch Mill'	oJoy oAls wFGN oDut oGre oBar
un	***x intermedia*** 'Dutch Silver'	oDut
un	***x intermedia*** 'Elizabeth'	oDut
un	***x intermedia*** 'English Hedge'	oDut
	x intermedia 'Fred Boutin'	oFor wHer wWoS wCul wSwC oDar oJoy oAls wFGN oGar oGoo oDut oDan oOut oGre wCCr oWnC oBar oUps wSnq
un	***x intermedia*** 'Gertrude'	oDut
	x intermedia 'Grappenhall'	wWoS wCul oAls wFGN oGoo oDut wCCr iGSc wMag oBar
	x intermedia 'Grey Hedge'	oDut
	x intermedia 'Grosso'	oFor wWoS wCul wSwC oJoy oAls oHed wFGN oGar oGoo oDut oGre wCCr iGSc wHom oCrm oBar wTGN wNTP oUps wSnq
un	***x intermedia*** 'Grosso Alba'	wWoS
	x intermedia 'Hidcote Giant'	wCul wFGN oGoo oDut oBar
	x intermedia 'Impress Purple' (see Nomenclature Notes)	oDut
un	***x intermedia*** 'Jenniflera'	oDut
	x intermedia 'Lullingstone Castle'	oHed
un	***x intermedia*** 'Madonna	oDut
	x intermedia Old English Group	wWoS oDut
un	***x intermedia*** 'Pacific Blue'	oDut
	x intermedia 'Provence'	oFor wWoS wCul wSwC oDar oAls oHed wFGN oGar oGoo oDut oDan oOut wRai oGre iGSc wHom wMag oCrm oBar oCir oUps oSle wSnq
un	***x intermedia*** 'Scottish Cottage'	oDut
	x intermedia 'Seal'	oAls oGoo oDut oBar
un	***x intermedia*** 'Silver'	wWoS oBar
un	***x intermedia*** 'Spike'	wFGN oDut oBar
	x intermedia 'Super'	wWoS oJoy wFGN oGoo oDut oBar
un	***x intermedia*** 'Sussex'	oDut
	x intermedia 'Twickel Purple' (see Nomenclature Notes)	last listed 99/00
un	***x intermedia*** 'Waltham Giant'	oDut
un	***x intermedia*** 'Wilson Giant'	oDut
un	***x intermedia*** 'Yuulong'	oDut
	lanata	wFGN oGoo oDut wCCr iGSc oBar
qu	***lanata*** 'Richard Joy'	oDut
	latifolia	oGar oGoo wFai iGSc oWnC
un	'Lisa Marie'	oBar
un	'Martha Roderick'	oFor wWoS oJoy oAls oHed wFGN oAmb oGoo oDut oSis oSho oGre oWnC oCrm oBar wTGN
	minutolii	wWoS oAls iGSc
	multifida	wWoS oGoo wCCr wHom wMag oBar
	multifida W 126	oDut
un	'Pink Fairy'	oDut
	pinnata	oUps
un	'Purple Royale'	oDut
un	'Purple Twinkle'	oGre
	'Richard Gray'	oFor wWoS oGoo oBar
	'Sawyer's'	wWoS oAls oHed wFGN oDut oBar
un	'Silver Frost'	wFGN oGoo oDut iGSc oBar
	sp. tall white	oHed
	spica (see Nomenclature Notes)	oGar
	stoechas	wFGN wCSG oGoo oDut oSho iGSc wHom wMag oBar wNTP oEga
	stoechas 'Alba'	see **L. *stoechas*** f. ***leucantha***
	stoechas ssp. ***atlantica***	wCul wFGN oGoo oDut oBar
un	***stoechas*** 'Cy Hyde'	wWoS
	stoechas f. ***leucantha***	wCul wFGN oGoo oDut wCCr wMag
	stoechas ssp. ***lusitanica***	wCul wFGN oGoo oDut oCrm oBar
un	***stoechas*** 'Lutsko's Dwarf'	oJoy oSec
	stoechas 'Marshwood'	wWoS oBar
	stoechas 'Nana'	wFGN oDut oBar

	stoechas narrow leaved	wFGN oDut oBar
	stoechas 'Otto Quast'	wSwC oDar wFGN oGar oDut oSis oGre wCCr wHom oWnC oCir wWhG oUps wSnq wWel
	stoechas 'Papillon'	see **L. *stoechas*** ssp. ***pedunculata***
	stoechas ssp. ***pedunculata***	wWoS oJoy oAls wFGN oDut oBar
	stoechas ssp. ***pedunculata*** 'James Compton'	wHer
	stoechas 'Quasti'	see **L. *stoechas*** 'Otto Quast'
	stoechas 'Willow Vale'	wHer wWoS oHed wFGN oBar
un	***stoechas*** 'Wings of Night'	wHer wWoS wFGN oBar
	'Twickel Purple' (see Nomenclature Notes)	oJoy
	vera (see Nomenclature Notes)	oUps
un	'Victorian Amethyst'	oDut
	viridis	wHer wWoS oEdg oGoo wCCr oBar
	white	wFGN
	yellow	wFGN
	LAVATERA	
	arborea 'Variegata'	oFor
	'Barnsley'	oFor wWoS wCul wSwC oNat oDar oJoy oAls oHed wCSG oMis oGar oNWe oSho oBlo oGre wFai wPir wHom oWnC wMag oCir oEga oUps wSnq
	'Bicolor'	see **L. *maritima***
	'Bredon Springs'	oFor wHer wWoS oJoy oAls oHed oGar oSho oBlo wPir oEga oSle
	'Burgundy Wine'	oFor wWoS wSwC oNat oAls oGar oGre wMag oSle
	cachemiriana	wCSG
	'Candy Floss'	oFor wWoS oHed oGar wFai wMag oSle
	'Eye Catcher'	wFai
	'Kew Rose'	wHer wCul wCSG
	'Lavender Lady'	wHer
	maritima	oNat oGar wMag wCoS
	olbia	oDar oJoy
un	'Pink Beauty'	wHom
	'Pink Frills'	oFor wWoS oHed
un	'Rainbow'	oDar oGre
	'Rosea'	oNat oAls oMis oGar wFai oCir
	'Shorty'	wHer
	thuringiaca	wCul oNat oSho wMag
	thuringiaca 'Ice Cool'	oJoy
	'Wembdon Variegated'	last listed 99/00
	LAWSONIA	
	inermis	iGSc
	LEDEBOURIA	
	cooperi	oTrP oOut oGre
	socialis	oTrP oSqu
	X LEDODENDRON	
	'Arctic Tern'	oGre
	LEDUM	
	columbianum	see **L. *glandulosum*** var. ***columbianum***
	glandulosum	oGre wWat
	glandulosum var. ***columbianum***	oFor
	groenlandicum	oFor
	palustre	last listed 99/00
	palustre ssp. ***decumbens***	oSis
	LEEA	
	coccinea	see **L. *guineensis***
	guineensis	oGar
	LEIOPHYLLUM	
	buxifolium	oRiv oGre wSta
	buxifolium 'Nanum'	oSis
	LEITNERIA	
	floridana	oRiv
	LEMAIREOCEREUS	
	eruca	see **STENOCEREUS *eruca***
	LEMNA	
	sp.	last listed 99/00
un	**LENTINUS**	
	edodes	wRai
	LEONOTIS	
un	***latifolium***	oGoo
ch	***leonurus***	oAls iGSc oUps
	menthifolia	oCis
	nepetifolia	oCrm
un	***nepetifolia*** 'Naivasha Apricot'	wHer
	ocymifolia	oGoo oOEx oCrm oUps
	LEONTODON	
	rigens	oFor
	LEONTOPODIUM	
	alpinum	oRus oGoo oSis wFai
	LEONURUS	
un	***cardiaca***	wCul oAls wFGN oGoo oOEx wFai iGSc oCrm oBar oUps

	sibiricus	oCrm oBar
	LEPIDIUM	
un	***meyerii*** black medicinal	oOEx
un	***meyerii*** hesperada	oOEx
un	***meyerii*** purple	oOEx
un	***meyerii*** white	oOEx
un	***meyerii*** yellow	oOEx
		oOEx
	LEPTINELLA	
	minor	oTrP oSis
	pusilla	oAls wRob wFai
	pyrethrifolia	oTrP
	squalida	wCul oJoy oAls wRob wHom oWnC oCis
	LEPTODERMIS	
	oblonga	oEdg oGre
	LEPTOSPERMUM	
	grandiflorum	last listed 99/00
	humifusum	see **L. *rupestre***
	juniperinum	last listed 99/00
	laevigatum	wSte
	lanigerum	oHed oSis oSho oGre wCCr wBox oCis
	liversidgei	wCCr
	nitidum	oFor
	polygalifolium	oFor
	rupestre	oSis
	scoparium	oFor oGoo wCCr wSte
	scoparium (Nanum Group) 'Kiwi'	oSis
	scoparium (Nanum Group) 'Tui'	oSis
un	***scoparium*** 'Pictorgilli'	oSis
	scoparium red seedlings	wCCr
	sericeum	wHer oSis
	LESPEDEZA	
	bicolor	oFor
	thunbergii	oGre wCCr
	thunbergii 'Albiflora'	oFor
un	***thunbergii*** 'Avalanche'	oFor
un	***thunbergii*** 'Edo Shidori'	oFor wHer oAmb oGre wWel
un	***thunbergii*** 'Gempei'	wHer
	thunbergii 'Gibraltar'	oFor wHer oGos oAmb oNWe wWel
un	***thunbergii*** 'Pink Fountain'	oFor
un	***thunbergii*** 'White Fountain'	wHer
	LEUCAENA	
	pulverulenta	oFor
	LEUCANTHEMELLA	
	serotina	oRus
	LEUCANTHEMUM	
	x superbum	wCSG oSha iArc
do	***x superbum*** 'Aglaia'	oFor oAls oGar
	x superbum 'Alaska'	oFor wTho oAls oGar oSho oWnC oCir oSle
un	***x superbum*** 'Becky'	oFor oAls oWnC
un	***x superbum*** 'Brent Lewis'	wWoS
	x superbum 'Christine Hagemann'	wWoS
do	***x superbum*** 'Cobham Gold'	oEdg oAls
do	***x superbum*** 'Esther Read'	oFor oAls oRus wCSG oGar oWnC wMag wTGN oCir oSle
	x superbum 'Everest'	last listed 99/00
un	***x superbum*** 'Exhibition'	wTho
un	***x superbum*** 'Lady's Daisy'	wTho
	x superbum 'Marconi'	oAls wCSG oGar wMag
	x superbum 'Silberprinzesschen'	oFor wTho oAls oSho oBlo oWnC wMag
	x superbum 'Silver Princess'	see **L. *x superbum*** 'Silberprinzesschen'
	x superbum 'Snow Lady'	oAls oGar oGre oWnC wMag oSle
	x superbum 'Snowcap'	oFor oRus oGar oWnC
	x superbum 'Summer Snowball'	oFor wWoS
do	***x superbum*** 'T. E. Killin'	wWoS wTGN
	x superbum 'White Knight'	wMag
	x superbum 'Wirral Pride'	oFor wWoS
	vulgare	last listed 99/00
	LEUCOJUM	
	aestivum	wRoo oWoo
	aestivum 'Gravetye Giant'	oSis
	autumnale	oTrP oSis oNWe
	vernum var. ***carpathicum***	oSis
un	**LEUCOLEPIS**	
	acanthoneuron	wFFl
	L EUCOPHYTA	
	brownii	oFor oWnC
	LEUCOSCEPTRUM	
un	***japonicum*** 'Variegatum'	wHer
	LEUCOTHOE	
	axillaris	oFor oGar oBlo oGre oTPm wWel
un	***axillaris*** 'Nana'	wSta
	davisiae	oFor oGos oBos
	fontanesiana	see **L. *walteri***

	grayana	last listed 99/00
	keiskei	oFor
	populifolia	see **AGARISTA *populifolia***
	racemosa	wHer
	Scarletta® / 'Zeblid'	oGre
	walteri	oGar oRiv oBlo wCoN oSle
	walteri 'Nana'	oFor oGre
	walteri 'Rainbow'	oFor wHer oDar oAls oGar oDan oSho oBlo oGre oTPm oWnC oCir oUps wWel
	walteri 'Rollissonii'	wHer
	LEUZEA	
	centauroides	wHer
	LEVISTICUM	
	officinale	wCul oAls wFGN wCSG oGoo wFai iGSc wHom oWnC oCrm wNTP oUps
	LEWISIA	
un	'Blue Purple'	wTGN
	brachycalyx	last listed 99/00
	cantelovii	oJoy oSto
un	***columbiana*** ssp. ***columbiana***	oSto oSis oSqu
un	***columbiana*** ssp. ***columbiana*** BLM 0224	oRar
	columbiana 'Edithiae'	oSto oGar oSis
	columbiana ssp. ***rupicola***	oSto oGar oRar oGre
un	***columbiana*** ssp. ***rupicola*** Saddle Mountain form	wMtT
	columbiana ssp. ***wallowensis***	oSto oRar
	cotyledon	oRus oGar oSis oGre oWnC wMag oUps
	cotyledon f. ***alba***	oRar
	cotyledon Ashwood strain magenta	wHer
	cotyledon Ashwood strain orange/apricot	wHer
	cotyledon Ashwood strain pink	wHer
	cotyledon Ashwood strain white	wHer
	cotyledon Ashwood strain yellow	wHer
	cotyledon var. ***cotyledon***	oRar
	cotyledon flame orange hybrids	oRar
	cotyledon var. ***heckneri***	oSto oRar
	cotyledon var. ***howellii***	oSto oRar
	cotyledon mixed hybrids	oAls oRar wCCr
	cotyledon orange hybrids	oRar
	cotyledon var. ***purdyi***	see **L. *cotyledon*** var. ***cotyledon***
	cotyledon rainbow hybrids	wCol oSto
	cotyledon red hybrids	oRar
	cotyledon salmon pink hybrids	oRar
	cotyledon Sunset Group	wTGN oUps
	cotyledon yellow hybrids	oRar
	'George Henley'	oSis wMtT
	leana	oRar
	leana x ***columbiana*** ssp. ***rupicola***	wMtT
	longipetala	oSto
	longipetala BLM 0451	oRar
un	'Love Dream'	oLSG
	nevadensis	last listed 99/00
	nevadensis bernardina	see **L. *nevadensis***
un	'Norma Jean'	wMtT
un	'Orange Delight'	wMtT
	'Pinkie'	oSis
	pygmaea	oJoy oSto
	pygmaea BLM 0030	oRar
	redivia	wMtT wWhG
	redivia pink form	oSis
	redivia white form	oSis
	'Regensbergen'	oAls oGre
	tweedyi	oSto oSis oRar wMtT
un	***tweedyi*** 'Lovedream'	oSis
	LEYCESTERIA	
	crocothyrsos	oFor wHer wHig wCCr oCis
	formosa	oFor oTrP wHer wWoS oNat oJoy oEdg wCSG oAmb oGar oDan oOut oGre wFai wPir wTGN oCis oSle
	LEYMUS	
	arenarius	wCli oWnC
	arenarius 'Glaucus'	oFor oDar oGar
	condensatus 'Canyon Prince'	wWoS oSec
qu	***glaucus***	wShR
	LIATRIS	
	aspera	oJoy
	callilepis	see **L. *spicata***
	elegans	oJoy
	punctata	oGoo
	pycnostachya	oJoy oLSG
	scariosa 'Alba'	wFai
	scariosa 'September Glory'	wFai
	scariosa 'White Spire'	last listed 99/00
un	'Snow White'	oDar oSha
	spicata	oDar wShR oAmb oGar oSha wCCr iGSc wMag wCoN wWld oVBI wCoS
	spicata 'Alba'	oAmb oGar wRob oBlo
	spicata 'Floristan Violett'	oFor oBlo oWnC
	spicata 'Floristan Weiss'	oFor oGar oSho oGre wTGN oCir oEga
	spicata Floristan White	see **L. *spicata*** 'Floristan Weiss'
	spicata Goblin	see **L. *spicata*** 'Kobold'
	spicata 'Kobold'	oFor oJoy oAls oMis oGar oGoo oSis oSho oOut oBlo oGre oWnC wTGN oCir oEga oUps
	LIBERTIA	
	'Amazing Grace'	wHer
	formosa	oMis oDan
	grandiflora	oTrP oJoy
	ixioides	wCCr
	peregrinans	oTrP wHer
un	***peregrinans*** 'Bronze Sword'	oNWe
	sessiliflora (last listed as ***sessile***) HCM 98186	wHer
	LIBOCEDRUS	
	decurrens	see **CALOCEDRUS *decurrens***
qu	**LIEOPHYLLUM**	
	buxifolium	oSho
	LIGULARIA	
	dark leaf	wMag
	dentata	wCoN
	dentata 'Desdemona'	oGar wNay oGre wBox
	dentata 'Othello'	oJoy oAls oGar oDan wNay wFai wTGN oEga
	fischeri	last listed 99/00
	fischeri HC 970089	wHer
	'Gregynog Gold'	oAls wNay
	x hessei	oFor
	japonica	wRob
	x palmatiloba	oFor wNay
	przewalskii	oFor oAls oGar oSho oTwi wNay wRob wMag wTGN iArc oUps
	sibirica	last listed 99/00
	sp. DJHC 413	wHer
	sp. DJHC 479	wHer
	sp. DJHC 648	wHer
	sp. DJHC 98366	wHer
	sp. DW 45	wHer
	sp. EDHCH 97140	wHer
	'Sungold'	oFor
	'The Rocket'	oFor oAls oGar wNay oGre wFai wMag wTGN wBox oEga
	tussilaginea	see **FARFUGIUM *japonicum***
	veitchiana	last listed 99/00
	wilsoniana	wCSG wNay
	LIGUSTRUM	
	delavayanum	oCis
	ibota	wCCr
	japonicum	oWnC wSta wWel
un	***japonicum*** 'Korean Choice'	wWel
qu	***japonicum*** 'Nobilis'	wWel
	japonicum 'Texanum'	oAls oGar oSho oBlo oWnC wTGN wCoS wWel
	japonicum 'Variegatum'	oFor oGar
	lucidum	oFor oGre wCCr
	obtusifolium var. ***regelianum***	oGre
va	***ovalifolium*** 'Aureum'	oCis
	sinense 'Variegatum'	last listed 99/00
	sinense 'Wimbei'	wHer
	'Suwannee River'	oGar oSho
	texanum	see **L. *japonicum*** 'Texanum'
	'Vicaryi'	oFor wCul oAls oGar oGre oWnC wWel
	vulgare	oGoo
un	***vulgare*** 'Cheyenne'	oFor oGar wKin
	vulgare 'Chlorocarpum'	wCCr
un	***vulgare*** 'Pendulum'	oFor
un	***wallichii***	oOEx
	LILIUM	
	'Acapulco'	oGar wTLP
	African Queen Group	oGar
	'Allura'	wBDL
un	'Ambrosia' seed grown	wTLP
un	'America'	oGar
un	***amoenum*** DJHC 014	wHer
un	'Antarctica'	oVBI
	'Aphrodite'	wBDL wTLP
	Arctic Treasures tm Group	wBDL
un	'Arena'	oGar wTLP
	'Ariadne'	wBDL
un	'Aubade'	oVBI

LILIUM

un	'Barbaresco'	wBDL wTLP
	'Bessie'	wBDL
	'Black Beauty'	wBDL
un	'Black Bird'	oAls
	'Black Dragon' seed grown	wTLP
	'Bolero'	wBDL
	brownii	oOEx
un	'Brunello'	oAls
un	***bulbiferum*** var. ***bulbiferum***	wHer
	'Bums'	wHer
	'Butter Pixie'	oAls
	canadense var. ***coccineum***	oRus
	candidum	oRus
do	***candidum*** 'Plenum'	wHer
un	'Caress'	oAmb
un	'Caressa'	wTLP
	'Casa Blanca'	wSwC wBDL oAmb oGar wTLP oVBI
un	'Centennial Streamers'	wTLP
	Citronella Group	wBDL
	columbianum	wThG oBos wWat oAld wFFl
	concolor	wHer
	'Connecticut King'	oGar
	'Copper King' seed grown	wBDL wTLP
	'Coral Crest'	wBDL
un	'Coral Sunrise'	wTLP
	Coralbee Strain	wBDL
un	'Corrida'	oGar
	'Cote d'Azur'	wSwC oGar
un	'Crete'	oVBI
un	Crystal Palace Strain	wBDL
	davidii	wCol oRus
	davidii var. ***willmottiae***	wHer
un	'Dolce Vita'	wBDL wTLP
un	'Dream'	wTLP
	'Dubonette'	wBDL
	duchartrei DJHC 651	wHer
un	'Early Rose'	wTLP
	'Ed'	wHom
un	'Egypt'	wTLP
	'Emerald Angel'	wBDL
	'Fanfare'	wBDL
	'Fantango'	wBDL
	'Fata Morgana'	wTLP
	'Fellowship'	wBDL
un	'Firecracker'	oNWe
	'Firesong'	wBDL
	'Folklore Fable'	wBDL
	formosanum	oFor wBDL
	formosanum var. ***formosanum***	wHer
	formosanum var. ***pricei***	oRus oDan
	formosanum Southern US form	wHer
un	'Four Seasons'	oAls
un	Frangipani Strain	wTLP
	'Friendship'	wBDL
	'Genteel Lady'	wBDL
un	'Gipsy Eye'	oGar
	'Gold Eagle'	wBDL
	Golden Angel Group	wBDL
	'Golden Fiesta'	wBDL
un	Golden Goblets Strain	wTLP
	Golden Splendor Group	oGar
un	'Golden Stargazer'	oGar
	'Gran Paradiso'	oGar
	'Grand Cru'	oGar
	grayi	oRus
un	'Green-eyed Tigress'	wTLP
	Green Magic Group	wTLP
	hansonii	oRus oNWe
	'Harbor Star'	wBDL
	'Heart's Desire'	wBDL
un	Heart's Desire Group	wBDL
	Heavenly Trumpets™ Group	wBDL
	'Heavenly Word'	wBDL
	henryi	wHer oRus
	'Henry's Surprise'	wBDL
	humboldtii	wHer
	'Inspiration'	wBDL
un	'Italia'	oAls
	japonicum	oRus
un	'Jetset'	wBDL
un	'Jetstream'	wTLP
un	'Jolanda'	oVBI
	'Juan de Fuca'	wBDL
	kelleyanum	wHer
	'King Arthur'	wBDL

	'Kiss Proof'	oGar oVBI
un	'La Toya'	oGar
	'Lady Guinevere'	wBDL
do	***lancifolium*** 'Flore Pleno'	wHer wPle
	lankongense	oJil
	'Late Date'	wBDL
un	'Latvia'	wTLP
un	'Laura Lee'	wBDL
	leichtlinii var. ***maximowiczii*** DJH 238	wHer
	'Lemon Meringue'	wBDL
un	'Lemon Tree'	wBDL
	'Leslie Woodruff'	wBDL
	'Leslie's Regalia'	wBDL
	leucanthum	wHer wWoS
	'LeVern Friemann'	wBDL
	'Little Eve'	wBDL
un	'Little Yellow Kiss'	wBDL
	'Lollypop'	wTLP oVBI
	'Longidragon'	wBDL
	longiflorum DJHC 003	wHer
un	'Loreto'	oGar
	maculatum var. ***davuricum*** Japanese double	wHer
un	'Maharajah'	wBDL
	'Marco Polo'	wBDL oVBI
un	'Marseilles'	oGar
	martagon	wHer oNWe
	martagon var. ***pilosiusculum***	last listed 99/00
do	***martagon*** 'Plenum'	wHer
	martagon white parent	oNWe
	'Maureen'	wBDL
	michiganense	last listed 99/00
un	Midnight Strain	wTLP
	'Miss Burma'	oAls
	'Mona Lisa'	wTLP
un	'Monte Cristo'	wBDL
un	'Monte Negro'	oGar
un	'Montreal'	oGar
un	Moonlight Strain	wBDL
	Moonlight Strain 'Moonlight Belle'	wBDL
	Moonlight Strain 'Moonlight Starburst'	wBDL
	'Mr. Ed'	see **L.** 'Ed'
	'Muscadet'	wBDL
	nanum HWJCM 484	wHer
	nepalense	wHer
	nepalense 'Robusta'	last listed 99/00
un	'Nettuno'	wBDL
	'Nippon'	oAmb
	'Northwest Passage'	wBDL
	'Orange Delight'	oAls
un	'Oregon Mist'	wBDL
	'Pacific Sunset'	wBDL
un	'Paramount'	wTLP oVBI
	pardalinum	wHer
	pardalinum var. ***giganteum***	oRus oSec
	parryi	wHer
un	'Paula de Costa'	wBDL wTLP
	'Perfectly Pink'	wBDL
un	'Pink Paradise'	wTLP
	Pink Perfection Group	wSwC oGar
	'Pink Picotee'	wBDL
	'Pink Virtuoso'	wBDL
un	'Pixie'	wHom
	'Pollyana'	oVBI
	'Princess Anna'	wBDL
	'Princess Melissa'	wBDL
	pumilum	oFor
	'Queen's Tea'	wBDL
	'Raptura'	wBDL
	regale	oTrP oRus oGar wTLP
un	'Reinesse'	oAls
un	'Rodeo'	oVBI
un	'Romanesco'	wTLP
un	'Rosato'	oVBI
	'Rosine Nimbus'	wBDL
	Rosy Treasures™ Group	wBDL
un	'Royal Pink'	wTLP
un	'Royal Sunset'	oVBI
	'Ruby Empress'	wBDL
	'Ruby Jewels'	wBDL
un	'Salmon Classic'	oVBI
	'Salmon Jewels'	wBDL
un	San Gabrial Strain	wBDL
un	'San Sorbet'	oGar
	'Sans Souci'	wBDL

	Plant	Sources
	sargentiae	wRho
	sargentiae DJHC 708	wHer
	sargentiae SEH 142	wHer
	'Scarlett Delight'	wBDL
	Scarlet Treasures™ Group	wBDL
	'Scheherazade'	wBDL oAmb
	'Seabreeze'	wBDL
un	'Setpoint'	wBDL
un	'Seventy-six Trombones'	wBDL
un	'Siberia'	wBDL oAls
	'Showbiz'	wTLP
	'Snow Gem'	wBDL
	'Snowbells'	wBDL
un	'Sorbet'	oVBI
un	'Sorbonne'	wBDL wTLP
	speciosum var. ***gloriosoides***	wHer
	speciosum 'Uchida'	wBDL
un	***speciosum*** x 'Wing Dancer'	wBDL
	'Sphinx'	wTLP
un	'Spinoza'	wTLP
un	'Spirit'	wTLP
un	'Standing Ovation'	wBDL wTLP
	'Star Gazer'	oAls oGar wTLP oVBI
	'Star Step'	wBDL
un	'Starfighter'	wBDL
	Strawberry Vanilla Latte™	wBDL
	superbum	oTrP oJil oRus
	'Sweet Frost'	wBDL
	'Sweetie Pie'	wBDL
	'Sylvia'	wBDL
	taliense DJHC 306	wHer
un	'Tetra Impact'	wBDL
un	'Time Out'	wBDL
un	'Tom Pouce'	wTLP
un	'Trendsetter'	oGar
	tsingtauense HC 970158	wHer
	'Twinkles'	wBDL
un	'Venere'	oVBI
un	'Vermeer'	oGar
	'Victorian Petticoat'	wBDL
	'Victorian Ruffles'	wBDL
un	'Vino Red'	wHom
	vollmeri	wHer oNWe
	washingtonianum	oRus
	'Whistlin' Dixie'	wBDL
	'White Garden'	wBDL
un	'Woodriff's Memory'	wBDL oGar wTLP oVBI
un	'Yellow Grace'	wTLP
	LIMNANTHES	
	douglasii	oBos
	LIMNOBIUM	
	spongia	wGAc
	LIMNOCHARIS	
	flava	last listed 99/00
	LIMONIUM	
	bellidifolium	oSis
	binervosum	last listed 99/00
	'Blue Diamond'	see **GONIOLIMON *incanum*** 'Blue Diamond'
	gmelinii	oJoy
	latifolium	see **L. *platyphyllum***
	minutum	oSis
	perezii	oEga
	platyphyllum	oFor oJoy oAls oGoo wMag oSec oEga
	speciosum	see **GONIOLIMON *incanum***
	tataricum	see **GONIOLIMON *tataricum***
	LINARIA	
	aequitriloba	see **CYMBALARIA *aequitriloba***
	alpina	oLSG
	'Anstey'	last listed 99/00
	anticaria 'Antique Silver'	oSis
	cymbalaria	see **CYMBALARIA *muralis***
	dalmatica	last listed 99/00
	'Natalie'	wWoS
	purpurea	oFor wCul wCSG oAmb oOut wTGN oSec
	purpurea 'Canon Went'	wCul oNat oJoy oAls oHed wCSG
	purpurea 'Springside White'	last listed 99/00
	'Tony Aldis'	wHer wWoS oAls wTGN oSec
	triornithophora	last listed 99/00
	LINDELOFIA	
	longiflora	wHer
	LINDERA	
	aggregata	oCis
	benzoin	oFor oEdg oRiv oWhi oGre
un	***cheinii***	oCis

	Plant	Sources
	erythrocarpa	oGos oGar oGre oWnC wBox oCis wSte
	erythrocarpa HC 970452	wHer
	obtusiloba	oFor oGos oAmb oDan oRiv oWhi wCCr
	praecox HC 970646	wHer
un	***salicifolia***	wHer
	sericea DJH 470	last listed 99/00
	sp. EDHCH 97262	wHer
	umbellata	wHer
un	**LINDERNIA**	
	grandiflora	oHug
	LINNAEA	
	borealis	wThG wNot oRus oSis oBos wWat oTri wPla oAld
	borealis var. ***americana***	oBov
	borealis var. ***longiflora***	oSqu wFFl
	LINUM	
	campanulatum	oJoy oSis
	capitatum	oNWe
	flavum 'Compactum'	oJoy oBlo
	lewisii	see **L. *perenne*** ssp. ***lewisii***
	monogynum	oSis
	narbonense	oHed
	narbonense 'Heavenly Blue'	oFor oEga
	perenne	oGoo oSha oGre wFai iGSc
	perenne album	oFor
	perenne 'Blau Saphir'	oFor oSis oUps
	perenne ssp. ***lewisii***	oFor wCCr wWld
	perenne Sapphire	see **L. *perenne*** 'Blau Saphir'
	suffruticosum ssp. ***salsoloides***	oAls oWnC
	usitatissimum	oCrm
	LIPPIA	
	citriodora	see **ALOYSIA *triphylla***
	dulcis	oAls oGoo oOEx oCrm
	graveolens	oAls iGSc
	nodiflora	see **PHYLA *nodiflora***
	LIQUIDAMBAR	
	acalycina	oFor
un	***acalycina*** 'Burgundy Spray'	oCis
	formosana	oGar oRiv oGre
un	'Gold Dust'	wWel
	orientalis	wHer
	styraciflua	oFor oAls wShR wCSG oGar oSho oBlo oWnC wSta oWhS wCoN oSle wCoS
	styraciflua 'Burgundy'	oRiv oSho
un	***styraciflua*** 'Cherokee Ward'	oEdg
	styraciflua 'Festival'	oGar oRiv oSho
va	***styraciflua*** 'Golden Treasure'	oFor
	styraciflua 'Gumball'	oGar
	styraciflua 'Palo Alto'	oGar oRiv oSho
	styraciflua 'Rotundiloba'	oFor oGar oRiv oSho oWhi
	styraciflua 'Worplesdon'	oFor oGos oGar oRiv oBlo oGre wTGN wWel
	LIRIODENDRON	
	chinense	oFor oGre oWnC oCis
	tulipifera	oFor oAls wShR oGar oRiv oBlo oWnC wBox wCoN wRav oSle wCoS
	tulipifera 'Arnold'	oFor
	tulipifera Majestic Beauty™ / 'Aureomarginatum'	oFor
	LIRIOPE	
	exiliflora	oAls oGar wWel
	exiliflora Silvery Sunproof (see Nomenclature Notes)	oAls oGar oBRG wWel
	gigantea	wDav wWel
	japonica	see **OPHIOPOGON *japonicus***
	minor	wBox
	muscari	oMis oSho
	muscari 'Big Blue'	oGar oGre wWin oBRG oUps
	muscari 'Gold-banded'	oSec
va	***muscari*** 'John Burch'	oGar
	muscari 'Lilac Beauty'	oAls oGar
	muscari 'Majestic'	see **L. *exiliflora***
	muscari 'Monroe White'	wWin wWel
	muscari 'Royal Purple'	wWin
va	***muscari*** 'Silvery Midget'	oFor
	muscari 'Variegata'	oAls oMis oGar oSho oOut wWin oSec oBRG oUps
	muscari white	oAls
	spicata	wWin oBRG oEga
va	***spicata*** 'Gin-ryu'	oFor oAls oGar oGre oSec oBRG wWel
	spicata 'Silver Dragon'	see **L. *spicata*** 'Gin-ryu'
	LITCHI	
	chinensis	last listed 99/00
	LITHOCARPUS	
	densiflorus	oFor wBur oGar wCCr

un	***densiflorus*** f. ***attenuato-dentatus***	wHer
	densiflorus var. ***echinoides***	oFor wCol wSte
	edulis	oSho
	henryi	wHer oCis wSte
un	***verrulosus***	oCis
	LITHODORA	
	diffusa	wCoN
	diffusa 'Alba'	oBlo
un	***diffusa*** 'Blue Star'	oGre
	diffusa 'Grace Ward'	oFor oAls wCSG oGar oSis oGre oWnC wMag wTGN wSta oCir wWhG oUps wCoS
	diffusa 'Heavenly Blue'	oSis
	hispidula	oSis
	oleifolia	oSis
un	'White Swan'	oGre wTGN
	LITHOSPERMUM	
	diffusum	see **LITHODORA** ***diffusa***
	erythrorhizon	oCrm
	hispidulum	see **LITHODORA** ***hispidula***
	officinale	oGoo
	oleifolium	see **LITHODORA** ***oleifolia***
	LITSEA	
	aestivalis	oRiv
	LIVISTONA	
	chinensis	last listed 99/00
	LOBELIA	
	'Brightness'	oGar wFai oSec
	cardinalis	oFor oGar oGoo oBlo iGSc oWnC wMag oCrm oBar oSec oAld wCoN
	cardinalis 'Alba'	oGoo
	cardinalis pink	wHom
	cardinalis red	wHom
un	***chinensis***	oGoo
un	'Cinnabar Pink'	oLSG
un	Compliment Series	see **L.** Kompliment Series
	Compliment Blue	see **L.** 'Kompliment Blau'
	Compliment Deep Red	see **L.** 'Kompliment Tiefrot'
	Compliment Red	see **L.** 'Kompliment Scharlach'
	'Dark Crusader'	wSwC oNat oGar oNWe oGre wFai
	'Eulalia Berridge'	last listed 99/00
	Fan Cinnabar Rose	see **L.** 'Fan Zinnoberrosa'
	Fan Deep Red	see **L.** 'Fan Tiefrot'
	Fan Orchid Rose	oEga
	Fan Red	see **L.** 'Fan Scharlach'
	'Fan Scharlach'	oEdg oSec
	'Fan Tiefrot'	oEga
	'Fan Zinnoberrosa'	oGoo oEga
	fulgens	oSsd
	x gerardii	wSwC oSec
	x gerardii 'Vedrariensis'	oFor
un	'Gladys Lindley'	wWoS
	'Grape Knee-Hi'	oFor wHer wWoS wCol
	inflata	iGSc oCrm
	'Jack McMaster'	last listed 99/00
un	Kompliment Series	oBlo
	'Kompliment Blau'	oSis oEga oLSG
	'Kompliment Scharlach'	oJoy oAls oSha oLSG
	'Kompliment Tiefrot'	oJoy oGre oEga
	'La Fresco'	wHer wWoS wCol oSis wRob
	laxiflora	oSec
	laxiflora var. ***angustifolia***	wHer oJoy oHed oGar wBox wSte
	'Misty Morn'	wWoS
	'Monet Moment'	wHer
un	'Mourning Electra'	oGar
	'Pink Flamingo'	oFor oAls oMis oGar oGre wFai
	'Queen Victoria'	wHer oNat oJoy oAls oMis oGar oGoo oSis oOut oWnC wMag iArc wAva oEga oSle
	'Rose Beacon'	oFor wHer wWoS wSwC wCol wRob
	'Royal Fuchsia'	wHer
	'Ruby Slippers'	oFor wHer wWoS oJoy oGoo oDan
	'Russian Princess'	last listed 99/00
	sessilifolia	oJoy
	siphilitica	oFor wCul oNat oJoy oRus wCSG oMis oGoo oSha oBlo iGSc oSec oAld iArc wAva wCoN wWld oSle
	siphilitica 'Alba'	oMis oGoo wAva
	'Sparkle DeVine'	wHer wWoS oSis wRob
	splendens	see **L.** ***fulgens***
	'Summit Snow'	oFor wHer wWoS oJoy oDan oGre
	'Tania'	oJoy oAls oGar oOut wFai wBox wSte
	tupa	oJoy oHed
qu	'Violet Compliment'	oJoy
	'Wildwood Splendor'	oFor wCol oGar wMag
	LOMATIA	
	ferruginea	last listed 99/00
	hirsuta HCM 98076	wHer
	myricoides	wHer wCCr
	tinctoria	last listed 99/00
	LOMATIUM	
	columbianum	wWld
	dissectum	oWnC
	nudicaule	wWld
	utriculatum	wWld
	LONICERA	
	acuminata	oFor
	acuminata B&SWJ 3390	wHer
	alseuosmoides	oFor wHer
	x brownii 'Dropmore Scarlet'	oFor wCul oAls oGar oNWe oGre wFai oWnC oSle
	caerulea var. ***edulis***	wClo oWnC
un	***caerulea*** var. ***edulis*** 'Berry Blue'	oGre wBox
un	***caerulea*** var. ***edulis*** Blue Belle™	wCul wBur wRai
un	***caerulea*** var. ***edulis*** Bluebird™	wRai
un	***caerulea*** var. ***edulis*** 'Blue Velvet'	wCul wBur wBox
	caprifolium	oFor
	caprifolium 'Anna Fletcher'	last listed 99/00
	chrysantha	oFor wCCr
	ciliosa	oFor oHan wWoB wNot wShR oBos wWat wCCr oAld wFFl wWld
	conjugalis	last listed 99/00
un	***crassifolia*** SEH085	wMtT
	dioica	oGre
	edulis	see **L.** ***caerulea*** var. ***edulis***
un	'Emerald Mound'	wMag
	etrusca	oFor oGre oWnC
	etrusca 'Donald Waterer'	last listed 99/00
	etrusca 'Michael Rosse'	wHer
	etrusca 'Superba'	oJoy
	ferdinandii	oFor
	fragrantissima	oFor wClo oJoy oGar oGre oWnC oCis wSte
un	'Freedom'	wMag
	glabrata HWJCM 136	last listed 99/00
	gracilipes	oFor
	x heckrottii	oHan oGar oNWe wCoN
	x heckrottii 'Gold Flame'	oFor oDar oAls oGre wFai oTPm oWnC wMag wTGN oCir oUps oSle
	henryi	oFor wCul oGar oGre wTGN
	hildebrandiana	oGar
	hirsuta	oFor
	hispidula	oFor oHan oBos oRiv wWld oSle wSte
	interrupta	oFor
	involucrata	oFor wWoB wShR oGar oBos oRiv oGre wWat wCCr oTri oSec oAld wFFl oSle
un	***involucrata involucrata***	wNot
	x italica	oJoy
va	***x italica*** Harlequin / 'Sherlite'	wHer wWoS oHed oDan oNWe
	japonica	oJoy iGSc wSta wCoN
	japonica 'Aureoreticulata'	oFor oTrP wWoS wCul wCri oJoy wCSG oGar oGre wCCr oWnC wTGN oSec oBRG oUps
	japonica 'Halliana'	oFor wCul oJoy oAls oHed oGar oSho oSha oBlo oGre wHom wKin oTPm oWnC wMag wTGN oCir oUps wCoS wWel
	japonica 'Hall's Prolific'	oFor wEde
un	***japonica*** 'Hinlon'	oGar
un	***japonica*** 'Hinlon Honeydew'	oWnC
un	***japonica*** 'Purpurea'	oFor oDar oAls oGar oDan oSha oGre oWnC oUps oSle wCoS
un	'John Clayton'	wHer
	kamchatika	see **L.** ***caerulea*** var. ***edulis***
	korolkowii var. ***zabelii***	oJoy
	maackii	wCCr
	'Mandarin'	oFor oGar oGre wTGN
	maximowiczii var. ***sachalinensis***	oFor
un	***modesta*** var. ***lushanensis***	oDan oGre oWnC
	nigra	oFor
	nitida	oFor oGar oSho wCCr oBRG wCoN oSle wWel
	nitida 'Baggesen's Gold'	oFor oTrP wHer wWoS wCul oDar oJoy wCol oHed wCSG oGar oDan oNWe oOut wRob oSqu wFai wPir wCCr wBox oSec wSta oUps oSle wSte
	nitida 'Elegant'	oSec
	nitida 'Maigruen'	oFor wBWP
	nitida 'Red Tips'	last listed 99/00

va	***nitida*** 'Silver Beauty'	wHer wWoS wCul oGos wClo oAls oHed oSis oNWe oSho wRob oGre wPir wCCr oSec oCis oBRG wSte wWel
	nitida 'Silver Cloud'	wPir oCis
un	***nitida*** 'Silver Frost'	oJoy oAmb
	oblongifolia	oFor
	periclymenum	wCoN
	periclymenum 'Belgica'	oFor wHer oGar oGre wTGN
	periclymenum Berries Jubilee® / 'Monul'	wCSG oGar
	periclymenum 'Florida Serotina'	see **L. *periclymenum*** 'Serotina Florida'
	periclymenum 'Graham Thomas'	wHer oJoy oHed oBRG
	periclymenum 'Munster'	wHer
	periclymenum 'Serotina'	oFor wCul wCri oJoy oHed oGre oBRG
	periclymenum 'Serotina Florida'	wCul oAls oGar oGre wCCr wTGN wSte
	periclymenum 'Sweet Sue'	wHer
	pileata	oFor wClo oJoy oGar oOut oBlo wPir oWnC wSta
un	***pileata*** 'Hohenkrummer'	wHer oGar wWel
un	'Ponderings'	oJoy oGar oGre
	prolifera	oFor
	x purpusii	oGre
	rupicola var. ***syringantha***	oFor wHer wWoS
	ruprechtiana	oFor
	sempervirens	oFor oAlsoNWe wCoN
un	***sempervirens*** 'Blanche Sandman'	oFor wWoS oJoy oGre
un	***sempervirens*** 'Cedar Lane'	oFor
un	***sempervirens*** 'John Clayton'	oFor oGre
un	***sempervirens*** 'Magnifica'	oJoy oAls oGar oSho
	sempervirens f. ***sulphurea***	oFor
qu	***serotina***	wClo oBlo
	similis var. ***delavayi***	oHed
	sp. DJHC 639	wHer
	sp. HC 970396	wHer
	splendida	wHer
	standishii	oFor wCul oJoy oHed oRiv wCCr oCis
	syringantha	see **L. *rupicola*** var. ***syringantha***
	tatarica	wBCr oGoo oRiv oBlo
	tatarica 'Arnold's Red'	oFor oAls oGar oWnC
un	***tatarica*** 'Honeyrose'	oFor
	tatarica 'Rosea'	wMag
	x tellmanniana	oFor wHer wCul oGar oSho oGre wFai oBRG
	tomentella	wCCr
	tragophylla	wWoS
	vesicaria DJH 123	wHer
	x xylosteoides 'Clavey's Dwarf'	last listed 99/00
	x xylosteoides 'Miniglobe'	wKin
un	***xylosteum*** 'Emerald Mound'	oFor oGar oGre
	xylosteum 'Mollis'	oFor
un	***yunnanensis*** 'Pat's Variegated'	wCol
	LOPHOSPERMUM	
	erubescens	oTrP
	LOROPETALUM	
	chinense	oFor oTrP oBlo wCoN
	chinense 'Blush'	oFor wHer
	chinense 'Burgundy'	oFor
	chinense 'Fire Dance'	oGar oGre oWnC oCis
un	***chinense*** 'Pink Ice'	oBlo
un	***chinense*** 'Plum Delight'	oWnC oCis
	chinense f. ***rubrum***	wCoS
un	***chinense*** f. ***rubrum*** 'Daybreak's Flame'	oGar oGre oWnC wBox oCis
un	***chinense*** f. ***rubrum*** 'Hine's Purpleleaf'	oGar
un	***chinense*** f. ***rubrum*** 'Pipa's Red'	oGre wBox oCis
	chinense f. ***rubrum*** Razzleberri™ / 'Monraz'	oAls oGar oSho oBlo oWnC oCis
	chinense 'Ruby'	oFor oWhi
un	***chinense*** 'Sizzling Pink'	oFor oGar oRiv oSho oGre wWel
un	***chinense*** 'Snow Dance'	oGre oWnC wBox oCis wSte
	chinense 'Zhuzhou Fuchsia'	oCis
qu	'Hine's Plum'	oAls
	LOTUS	
un	'Amazon Sunset'	wTGN
	berthelotii	wTho wHom oUps
do	***corniculatus*** 'Plenus'	wMtT
un	***crassifolius*** var. ***subglaber***	oBos
	formosissimus	oBos
	'Gold Flash'	wTho wTGN
	hirsutus	oFor oDar oNWe wCCr
	LUDWIGIA	
	arcuata	oSsd
	palustris	last listed 99/00
	peploides	oWnC
	sp.	wSoo
	LUETKEA	
	pectinata	oRus
	LUMA	
	apiculata	oFor wSte
un	***gayana*** HCM 98201	wHer
	LUNARIA	
	annua	wFai
	annua variegata	oNWe
	rediviva	wHig
	LUPINUS	
	arboreus	wCul wBWP wAva wEde
	argenteus	wThG wPla
	bicolor	oAld wWld
un	Carnival strain	wSnq
un	Dwarf Russell	oFor
	Gallery Series	oAls oEga
	'Gallery Blue' (Gallery Series)	oAls oSho
	'Gallery Pink' (Gallery Series)	oAls oSho
	'Gallery Red' (Gallery Series)	oAls oSho
	'Gallery White' (Gallery Series)	oAls oSho wMag
	'Gallery Yellow' (Gallery Series)	oSho wMag
	latifolius	wNot oAld
	leucophyllus	last listed 99/00
	littoralis	oBos oAld
	Minarette Group	wTho oGar oUps
	mutabilis	oOEx
	'My Castle'	oWnC wMag
un	New Generation hybrids	oAls oNWe wMag
	'Pink Popsicle'	oFor
un	***polycarpus***	oBos
	polyphyllus	oBos oSha wWat
un	***polyphyllus superbus***	oFor
	Popsicle hybrids	wTho oLSG
	Red Flame	see **L.** 'Rote Flamme'
	'Red Popsicle'	last listed 99/00
	'Rote Flamme'	oFor
	rivularis	oBos oAld wWld
	Russell hybrids	wTho oGar wHom oCir oLSG oUps oSle
	sericeus	wPla
	'The Chatelaine'	oWnC wMag
	'The Governor'	oWnC wMag oUps
	'White Popsicle'	oFor
	'Yellow Popsicle'	oFor
	LUZULA	
	lactea	last listed 99/00
	luzuloides	wWin
	nivea	oFor wCli oDar oSis oOut oSqu wWin wPir wSte wSnq
un	***parviflora***	wNot
	pilosa	oHan
	sylvatica	wCli oDar oJoy oSqu wWin
	sylvatica 'Aurea'	wHer oSec
	sylvatica 'Auslese'	oFor
	sylvatica 'Marginata'	wHer wCli oNWe oOut wRob oGre wWin oSec wSte
	sylvatica 'Tauernpass'	oHan
	ulophylla	wWin
	LYCHNIS	
	alpina	wMag
	x arkwrightii 'Vesuvius'	oAls
	chalcedonica	oJoy oWnC wTGN oLSG
	chalcedonica var. ***albiflora***	oMis
do	***chalcedonica*** 'Flore Pleno'	wHer
	chalcedonica 'Morgenrot'	last listed 99/00
un	***chalcedonica*** Murky Puce Strain	wWoS
	chalcedonica pink	oMis
	chalcedonica red	oMis
	cognata DJH 97126	wHer
un	***cognata*** 'Filigree Red'	wHer wWoS
	coronaria	oTrP wTho oRus wFai oCrm oCir oLSG oUps oSle
	coronaria 'Alba'	wCul oWnC oSle
	coronaria 'Angel's Blush'	oGar wMag
	coronaria Atrosanguinea Group	wMag
	coronaria 'Dancing Ladies'	oAls
	coronaria Oculata Group	wCul oSec
	dioica	see **SILENE *dioica***
	flos-cuculi	oFor oHug
	flos-cuculi 'Nana'	oSec
	flos-jovis	oTrP
	flos-jovis 'Minor'	see **L. *flos-jovis*** 'Nana'
	flos-jovis 'Nana'	oSis oSle
	flos-jovis 'Peggy'	oSis oSec
un	***x haageana*** 'Orange Gnome'	wWoS
va	***miqueliana*** 'Variegated Lacy Red' seed grown	last listed 99/00
	'Molten Lava'	oAls

un	'Orange Zwerg'	oMis
	viscaria	oFor
do	viscaria 'Plena'	wTGN
	yunnanensis	oTrP wCol
	yunnanensis alba	see L. yunnanensis
	LYCIANTHES	
un	dejecta	oOEx
un	mociniana	oOEx
	rantonnetii	see SOLANUM rantonnetii
	LYCIUM	
	chinense	oGoo oOEx iGSc oCrm
	LYCOPODIUM	
	sp.	wFFl
	LYCOPUS	
	europaeus	oGoo wFai iGSc
	LYCORIS	
	africana	see L. aurea
	aurea	oTrP
	radiata	oTrP
	squamigera	oTrP
	LYGODIUM	
	japonicum	wFol wFan
	LYONIA	
	ligustrina	last listed 99/00
	lucida	oFor oGos oGre oWnC oCis
	mariana	oFor
	ovalifolia var. elliptica HC 970701	wHer
un	ovalifolia var. ovalifolia DJHC 079	wHer
	LYONOTHAMNUS	
	floribundus	oSho
	floribundus ssp. aspleniifolius	oFor wCCr
	LYSICHITON	
	americanus	wNot oSis oBos wFFl
	camtschatcensis	wHer wCol oRus
	LYSIMACHIA	
	atropurpurea	last listed 99/00
	atropurpurea 'Beaujolais'	iArc oCir
	barystachys	oFor
	ciliata	wCul
qu	ciliata atropurpurea	wWoS wCul wCol oRus oTwi
	ciliata 'Firecracker'	oFor oNat oHed wCSG oGar oSho wRob oWnC oSec oSec wAva oSle
	ciliata 'Purpurea'	see L. ciliata 'Firecracker'
	clethroides	oFor wCul wTho oNat oAls oRus wShR wCSG oMis oGar oSha oTwi wRob wMag oCir oEga oSle
	congestiflora 'Eco Dark Satin'	oFor
va	congestiflora Outback Sunset®	oJoy oOut wRob
	ephemerum	oFor wHer wCul oAls oHed wCSG oGar wCCr
un	'Golden Globe'	oGre
	henryi	wHer
	japonica var. minutissima	oAls oNWe
	lichiangensis	oFor
	minoricensis	wHer
	nummularia	wTho oAls oRus oHug oGar oSsd wHom oWnC wSta oCir oUps oSle
	nummularia 'Aurea'	oFor wHer wWoS wTho oDar oAls oHug oGar oSis oNWe wRob oGre wCCr wHom oWnC oSec oUps oSle
un	paridiformis var. stenophylla DJHC 704	wHer
	punctata	oFor oNat wShR oGar oDan oSho oCir oLSG oSle
va	punctata 'Alexander'	oFor wSwC oDar oJoy oEdg wCol oAls oRus oHug oDan oOut wRob oGre oWnC wTGN oEga oUps oSle
	LYTHRUM	
ch	virgatum 'Morden Gleam'	oWnC
	MAACKIA	
	amurensis	oFor oRiv oSho oGre
	chinensis	oFor
	hupehensis	see M. chinensis
	MACFADYENA	
	unguis-cati	last listed 99/00
	MACHAEROCEREUS	see STENOCEREUS
	MACHILUS	see PERSEA
	MACLEANIA	
	insignis BLM 0628	oRar
	MACLEAYA	
	cordata	oFor oAls oGar wFai wPir oSec wBWP wCoN oUps
un	cordata 'Plum Tassle' seed grown	wSte
	microcarpa	wCul oGoo
	microcarpa 'Kelway's Coral Plume'	oFor
un	microcarpa 'Summer Haze'	wFai wBox
un	'Plum Tassle'	oAmb
	MACLURA	
un	conchinensis	wRai oOEx
	pomifera	oFor oRiv
un	pomifera 'Whiteshield'	oFor
	MAGNOLIA	
	acuminata	oFor oRiv oWhi oGre oWnC wAva
	acuminata 'Koban Dori'	oGos oGre wWhG
ch	acuminata var. subcordata 'Miss Honeybee'	last listed 99/00
	'Albatross'	oGos
	amoena	oFor
	'Ann'	oFor oGos oGre wTGN oUps
	'Anne Rosse'	oGos oGre
	'Apollo'	oFor oGar oGre
ch	ashei	see M. macrophylla ssp. ashei
un	'Asian Artistry'	wClo wBox wWel
	'Athene'	oGos oGar oWhi oGre
	'Atlas'	oGos oWhi oGre wWel
	'Betty'	oFor oGre oUps wCoS
	'Big Dude'	oGos wBox
	biondii	oFor oGos oRiv oGre
	'Bovee'	see M. wilsonii
un	'Burncoose Tennis Court'	oGos
	'Butterflies'	oFor oGos wClo oWhi oGre wBox wWhG wWel
	'Caerhays Belle'	oGos oGar oGre
un	'Caerhays New Purple'	oGos
	campbellii	oCis
	campbellii ssp. mollicomata	oGos
un	campbellii ssp. mollicomata 'Hendricks Park'	oGos oGar oWhi
	campbellii ssp. mollicomata 'Lanarth'	last listed 99/00
	campbellii (Raffillii Group) 'Charles Raffill'	oGos oGre oWnC
un	campbellii 'Strybing Pink'	oGos
	campbellii 'Strybing White'	last listed 99/00
	'Charles Coates'	oGos oGre wBox wWel
	'Columbus'	oGos oGre
	cylindrica	oFor wHer wBCr oGar oDan oRiv oGre oWnC wBox wAva wWel
	'David Clulow'	oGos
	dawsoniana	oWhi
	dawsoniana 'Chyverton'	oGos
un	dawsoniana 'Chyverton Red'	oGre
	dawsoniana 'Clark's Variety'	last listed 99/00
un	'Daybreak'	wWel
	delavayi	wHer oCis
	denudata	oFor oGos wClo oEdg oGar oRiv oCis oGre oWnC wBox wWhG wWel
un	denudata 'Forrest's Pink'	oGos wWhG wWel
un	denudata 'Gere'	oGos oGre
	denudata hybrids, Korean origin	wHer
un	denudata 'Japonica'	oWhi
	'Elizabeth'	oGos wClo oGar oSho oGre oWnC wWhG wWel
un	'Emma Cook'	oGos
un	'Firefly'	oGos
	'Galaxy'	oFor oGos oGar oBlo oGre wBox wWhG
un	'Golden Glow'	oGos
	'Goldstar'	wClo oGre
	grandiflora	oFor oTrP wCCr oWnC wCoS
un	grandiflora 'D. D. Blanchard'	oGos wSte
	grandiflora 'Edith Bogue'	oFor oGos oGar oGre oWnC wWhG wSte
	grandiflora 'Goliath'	oGos oGre
	grandiflora 'Little Gem'	oAls oGar oSho oWnC wCoS wWel
	grandiflora Majestic Beauty®/ 'Monlia'	oAls oGar oSho oWnC wCoS
un	grandiflora 'Poconos'	oGos oGre
	grandiflora 'Saint Mary'	oGar oSho oWnC wSta wWel
	grandiflora 'Samuel Sommer'	oGos
	grandiflora 'Victoria'	oFor oGos oBlo oGre oWnC wSta wWhG wCoS
un	'Hattie Corthon'	wWel
	'Heaven Scent'	oGos wClo wBox
	heptapeta	see M. denudata
	hypoleuca	oWhi wCCr wAva
	hypoleuca DJHC 319	wHer
	'Iolanthe'	oGos oGre oWnC
un	'Ivory Chalice'	oGre wWhG
	'Jane'	oFor wBox
	'Judy'	oGos
	x kewensis 'Wada's Memory'	oFor wHer oGos oGre oWnC wBox wWhG wWel
	kobus	oFor oGar oRiv oWhi oGre wAva wWel
un	'Legacy'	oGos oGre

qu	***leuca***	oRiv
	liliiflora	oWnC
	liliiflora 'Gracilis'	last listed 99/00
	liliiflora 'Nigra'	oAls oGar oGre oWnC wWhG wWel
	liliiflora 'O'Neill'	oWnC wWhG
	x loebneri 'Ballerina'	oFor oGos oSho oGre wWhG
	x loebneri 'Leonard Messel'	oFor oGos oAmb oGar oBlo oGre oWnC wWhG wCoS wWel
	x loebneri 'Merrill'	oFor oGos oAls oGar oSho oBlo oGre oWnC wWhG wWel
	x loebneri 'Neil McEacharn'	oGos oGre
	x loebneri 'Spring Joy'	oGos oGre
	x loebneri 'Spring Snow'	oFor oGos oGre wWhG
qu	***macrocarpa***	wBox wWel
	macrophylla	oGar oSho oGre wCCr wTGN
ch	***macrophylla*** ssp. ***ashei***	oFor wHer oRed oGre
	'Manchu Fan'	oGos wWel
	'Mark Jury'	oGos oGre
	'Maryland'	oGos wWhG
	'Milky Way'	oGos oGre wBox
	mollicomata	see **M. *campbellii*** ssp. ***mollicomata***
	'Nimbus'	oGos
	officinalis	oFor oGos oGar oRiv oCis oGre oWnC wBox wAva wWel
	officinalis var. ***biloba***	last listed 99/00
un	'Orchid'	oFor oGos oGar oGre wWhG
un	'Paul Cook'	oGos oGre wBox
un	'Peachy'	oGos
	'Pegasus'	oGos
	'Peppermint Stick'	oGre
	'Pinkie'	oFor oGos oGre wWhG oUps
un	'Pristine'	wBox
	'Purple Prince'	last listed 99/00
	'Randy'	wWhG wCoS
	'Ricki'	oGre oUps
	'Royal Crown'	oFor oGos oGre
un	'Royal Flush'	oGos
	'Ruby'	oGar
	salicifolia	oFor
	salicifolia DJH 463	wHer
	salicifolia 'Else Frye'	oGos
un	***salicifolia*** 'Iufer'	oGos
un	***salicifolia*** 'Miss Jack'	oGos
	sargentiana var. ***robusta***	oGos oGre
un	***sargentiana*** var. ***robusta*** 'Bloodmoon'	oGar
	sargentiana var. ***robusta*** dark form	oGos
	'Sayonara'	oFor oGos oGre wBox
	'Serene'	oGos wClo oGre oWnC
	sieboldii	wBCr oGos wClo oEdg wCol oGar oDan oRiv oSho oGre wPir oWnC wTGN wBox wAva wRav wWhG wSte wWel
	sieboldii HC 970294	wHer
un	'Slavin's Snowy'	oGos
un	'Snow Queen'	oGos
	x soulangeana	oAls oGar oBlo oWnC wTGN oThr
	x soulangeana 'Alexandrina'	oAls oSho oBlo oGre oWnC wWhG wCoS
	x soulangeana 'Amabilis'	oGos
	x soulangeana 'Brozzonii'	oGre wBox
	x soulangeana 'Grace McDade'	oGre
ch	***x soulangeana*** 'Joe McDaniel'	oGos
	x soulangeana 'Lennei'	oAls
	x soulangeana 'Lennei Alba'	oGos oGre
	x soulangeana 'Lombardy Rose'	last listed 99/00
	x soulangeana 'Picture'	last listed 99/00
qu	***x soulangeana*** 'Purpleana'	oAls
	x soulangeana 'Rustica Rubra'	oGos oAls oGar oSho oBlo wWhG wWel
	x soulangeana 'San Jose'	oGos oAls
	x soulangeana 'Verbanica'	oGos oGre wWhG
	x soulangeana 'White Giant'	oGos
	'Spectrum'	oGos wBox
	sprengeri	last listed 99/00
	sprengeri var. ***diva***	oEdg oRiv oGre wWel
	sprengeri var. ***diva*** 'Burncoose'	oGos
	sprengeri 'Eric Savill'	oGos
	'Star Wars'	oGos
	stellata	oAls oWhi oGre wTGN
	stellata 'Centennial'	oFor oGos oGre wWel
	stellata 'Jane Platt'	oGos wClo wWhG wWel
	stellata 'King Rose'	oGre oWnC wWhG
	stellata pink	oAls oSho
	stellata 'Rosea'	oGos oGre wWhG
	stellata 'Royal Star'	oFor oGos oAls oGar oRiv oBlo oGre oWnC wBox wSta wWhG wCoS wWel
	stellata 'Rubra'	oGos oGar oGre wWhG
	stellata 'Waterlily'	oGos oGar oGre oWnC
	'Sundance'	oGos oGre wWhG
	'Sundew'	oGos oGre
	'Susan'	oGos wClo oGar oSho oGre oWnC wBox wWhG oUps wCoS
	Timeless Beauty®/ 'Monland'	oSho oGre
	'Todd Gresham'	oGos
	tripetala	oFor oRiv wCCr wBox
	x veitchii	oGos oWhi oGre
	x veitchii 'Rubra'	oGos
	virginiana	oGar oSho oBlo oGre wBox wAva wRav wSte
un	***virginiana*** var. ***australis***	oFor wCCr
	virginiana var. ***australis*** 'Henry Hicks'	oGos oGre wWhG
un	***virginiana*** 'Santa Rosa'	oGos
un	***virginiana*** 'Satellite'	oGos
	'Vulcan'	oFor oGos wClo oDan oGre wBox wWhG wWel
ch	'W. B. Clarke'	oGos
	x watsonii	see **M. *x wiesneri***
un	'White Stardust'	oFor
	x wiesneri	oGre
	wilsonii	oFor wHer oGre oWnC
un	'Woodsman'	oGre
	'Yellow Bird'	oFor wClo oGre wWhG wWel
	'Yellow Fever'	oGos wWhG
	'Yellow Lantern'	oGos oGre wWhG
	zenii	wHer oWnC
	X MAHOBERBERIS	
	miethkeana	oFor
	neubertii	wHer
	MAHONIA	
	aquifolium	oFor wBCr wThG wWoB wBur wNot oDar oAls wShR oGar oBos oRiv oSho oBlo oGre wFai wWat oTPm oWnC oTri wPla wCoN wFFl oSle wSte wSnq wWel
	aquifolium 'Compacta'	oFor oGar oBlo oTPm oWnC
	aquifolium 'Golden Abundance'	oFor
	aquifolium 'Mayhan Strain'	oTPm wSta
	bealei	see **M. *japonica*** Bealei Group
	dictyota	oFor
	fortunei	oFor oCis oGre
	gracilipes DJHC 755	last listed 99/00
	haematocarpa	wCCr
	japonica	wHer
	japonica Bealei Group	oFor wCul wClo oDar oEdg oGar oRiv oSho oGre oBRG wCoN wWel
	lomariifolia	wHer oSho wCoN
un	***mairei***	wHer
	x media 'Arthur Menzies'	oFor oTrP wHer wCol
un	***x media*** 'Cantab'	wHer
	x media 'Charity'	wHer wSte
	x media 'Hope'	wHer
	x media 'Lionel Fortescue'	wHer
	x media 'Underway'	wHer wSte
	x media 'Winter Sun'	wHer
un	***napaulensis*** 'Maharajah'	wHer
	nervosa	oFor wHer wWoB wBur wShR oGar oBos oRiv oSho oBlo oGre wWat oTPm oCrm oTri wSta wCoN wFFl wSnq wWel
	nevinii	wCul oCis wCCr
un	***pinnata*** 'Ken Howard'	wHer
	piperiana	wHer
	repens	oFor wCul wWoB oDar oAls wShR oGar oBos oRiv oBlo oGre wWat wKin oTPm wSta wPla
un	***vietchiorum***	wHer
	MAIANTHEMUM	
	bifolium	oRus
	bifolium ssp. ***kamtschaticum***	wThG wWoB wNot oBos wCoN wFFl wWld
	canadense	oFor
	dilatatum	see **M. *bifolium*** ssp. ***kamtschaticum***
	racemosum	see **SMILACINA *racemosa***
	MAIHUENIA	
	poeppigii	oSis
	poeppigii HCM 98030	wHer
	MALEOPHORA	
	crocea var. ***purpureocrocea***	oFor
	MALLOTUS	
	japonicus	oCis
	MALUS	
	'Adams'	oGar wBox
	'Adirondack'	oDan wWel
	baccata	oFor wPla
	baccata 'Dolgo'	wBCr wBur oEar wRai wCoS

MALUS

	Bechtel	see **M. *ioensis*** 'Plena'
	'Beverly'	last listed 99/00
	'Brandywine'	oDar oGar
	'Callaway'	oGre
	'Candied Apple'	oFor oGar wKin wTGN wAva
un	'Candy Mint'	oFor
un	'Candymint Sargent'	oGar
un	'Cardinal'	oFor
un	'Centennial'	wRai
	'Centurion'	oFor oGar oGre wBox
un	'Chestnut'	wBCr
	'Coralburst'	last listed 99/00
	coronaria	oFor
un	'Crimson Gold'	wC&O
	'David'	oFor wWel
	domestica 'Advance'	last listed 99/00
un	***domestica*** 'Airlie Red'	oRiv oGre
un	***domestica*** 'Akane'	wBCr wBur wClo oEar oAls oSho wRai
un	***domestica*** 'Akane Toyko Rose'	oRiv
	domestica 'Alexander'	wBCr
un	***domestica*** 'Alexis'	wBCr
	domestica 'Alkmene'	wRai
un	***domestica*** 'Almata'	wBCr
un	***domestica*** 'Amur Red'	wBCr
un	***domestica*** 'Apricot'	oWhi
	domestica 'Arkansas Black'	oFor wBCr oEar oGar oRiv
un	***domestica*** 'Arlie Red Flesh'	oRiv
un	***domestica*** 'Aroma'	wRai
	domestica 'Ashmead's Kernel'	wBCr wClo oEar oRiv wRai
un	***domestica*** 'Baker's Sweet'	wBCr
un	***domestica*** 'Baldwin'	wBCr
un	***domestica*** 'Beacon'	wBCr oWnC
un	***domestica*** 'Beautiful Arcade'	wBCr
	domestica 'Belle de Boskoop'	wBCr
un	***domestica*** 'Ben Davis'	wBCr
un	***domestica*** 'Benander's Hard Red'	wBCr
un	***domestica*** 'Benham'	wBCr
un	***domestica*** 'Bidy'	wBCr
un	***domestica*** 'Billy Red Flesh'	oRiv
un	***domestica*** 'Black Amish'	wBCr
un	***domestica*** 'Black Gilliflower'	wBCr oEar oRiv
un	***domestica*** 'Black Twig'	wBCr
	domestica 'Blenheim Orange'	wBCr
	domestica 'Blue Pearmain'	wBCr
un	***domestica*** 'Bottle Greening'	wBCr
un	***domestica*** 'Boughen's Delight'	wBCr
	domestica 'Braeburn'	wBCr oEar oAls oGar oSho oWnC wTGN wVan
un	***domestica*** 'Bramley'	wRai
	domestica 'Bramley's Seedling'	wBCr oEar
un	***domestica*** 'Breakey'	wBCr
un	***domestica*** 'Brewster's Twig'	wBCr
un	***domestica*** 'Brightness'	wBCr
un	***domestica*** 'Brock'	wBCr
un	***domestica*** 'Brook 27'	wBCr
un	***domestica*** 'Buckeye Gala'	wC&O
un	***domestica*** 'Burgundy'	oRiv
	domestica 'Calville Blanc d'Hiver'	wBCr oEar oRiv
un	***domestica*** 'Calville Rouge d'Automne'	wBCr
un	***domestica*** 'Cameo'	wVan
un	***domestica*** 'Campbell Red Delicious'	wVan
un	***domestica*** 'Canada Red'	wBCr
un	***domestica*** 'Canada Reinette'	wBCr
un	***domestica*** 'Century 21'	wBCr
un	***domestica*** 'Cestra Belter Kitaika'	wBCr
	domestica 'Chehalis'	wBur wClo oEar oAls oRiv wRai wCoS
un	***domestica*** 'Chenango Strawberry'	wBCr
un	***domestica*** 'Cherry Cox'	oRiv
un	***domestica*** 'Chinese Golden Early'	wBCr
un	***domestica*** 'Chipman'	wBCr
un	***domestica*** 'Cinnamon Spice'	wBCr
	domestica 'Claygate Pearmain'	wBCr
un	***domestica*** 'Cole's Quince'	wBCr
un	***domestica*** 'Compact Red McIntosh'	wVan
un	***domestica*** 'Compact Red Rome'	wVan
un	***domestica*** 'Compact Winter Banana'	wVan
un	***domestica*** 'Connell Red'	wBCr
un	***domestica*** 'Coos River Beauty'	oRiv
	domestica 'Cornish Gilliflower'	last listed 99/00
	domestica 'Cortland'	wBCr wVan wC&O
	domestica 'Court Pendu Plat'	wBCr
	domestica 'Cox's Orange Pippin'	wBCr oEar oWhi oWnC wTGN
un	***domestica*** 'Crimson Spire'	oAls oGar
	domestica Crispin	see **M. *domestica*** 'Mutsu'
un	***domestica*** 'Criterion'	wBCr oAls oWnC
un	***domestica*** 'Crow's Egg'	wBCr
un	***domestica*** 'Dayton'	wBur oRiv wRai
un	***domestica*** 'Dearborn's Unknown'	wBCr
un	***domestica*** Del™ Red Rome	wVan
	domestica 'Delicious'	oAls oGar wKin oWnC
	domestica 'Discovery'	wBCr oAls
	domestica 'Doctor Mathews'	wBCr
un	***domestica*** 'Downingland'	wBCr
	domestica 'Duchess of Oldenburg'	wBCr
un	***domestica*** 'Dudley'	wBCr
	domestica 'Duke of Devonshire'	wBCr
un	***domestica*** 'Dunning'	wBCr
	domestica 'Dutch Mignonne'	oRiv
un	***domestica*** 'Earliblaze'	wBCr
	domestica 'Earligold'	wC&O
un	***domestica*** 'Early Cortland'	wBCr
un	***domestica*** 'Early Geneva'	oRiv
un	***domestica*** 'Early Harvest'	wBCr
un	***domestica*** 'Early Joe'	wBCr
un	***domestica*** Early Red One™ Delicious	wVan
un	***domestica*** 'Early Sauce'	wBCr
un	***domestica*** 'Early Spur Rome'	wC&O
un	***domestica*** 'Early Thompson'	wBCr
	domestica 'Edward VII'	wBCr wRai
	domestica 'Egremont Russet'	wBCr wRai
	domestica 'Ellison's Orange'	wRai
	domestica 'Elstar'	wClo oEar wCoS
un	***domestica*** 'Emerald Spire'	oAls oGar
	domestica 'Empire'	wBCr oEar oWnC
	domestica 'Enterprise'	wBur wClo oSho wRai wCoS
un	***domestica*** 'Erwin Bauer'	oRiv
	domestica 'Esopus Spitzenberg'	see **M. *domestica*** 'Spitzenberg'
	domestica 'Fall Pippin'	wBCr
un	***domestica*** 'Fallawater Pippen'	wBCr
	domestica 'Fameuse'	wBCr oEar oWnC
un	***domestica*** 'Fameuse Snow'	oRiv
	domestica 'Fiesta'	wClo wRai
	domestica 'Fireside'	wBCr
	domestica 'Firmgold'	wVan
un	***domestica*** 'Florina'	wClo
	domestica 'Fortune'	see **M. *domestica*** 'Laxton's Fortune'
un	***domestica*** 'Fox'	wBCr
un	***domestica*** 'Freedom'	wBCr wBur wClo oEar
	domestica 'Freyberg'	last listed 99/00
	domestica 'Fuji'	wBur oEar oAls oGar oSho oWhi oGre wKin oWnC wTGN
	domestica 'Gala'	wBCr oEar oAls oGar oSho oGre wKin oWnC
	domestica 'Gala Royal'	see **M. *domestica*** 'Royal Gala'
un	***domestica*** Galasupreme®	wVan
un	***domestica*** Gale Gala™	wVan
un	***domestica*** 'Gano'	wBCr
un	***domestica*** 'Geneva Early'	oRiv
un	***domestica*** 'Gibson Gold Delicious'	wVan
un	***domestica*** 'Gilpin'	wBCr
	domestica Ginger Gold®	wVan
	domestica 'Gloria Mundi'	wBCr
un	***domestica*** 'Gloster'	wBCr
un	***domestica*** 'Gold Rush'	wBur
	domestica 'Golden Delicious'	wBCr oAls oGar oSho wKin oWnC wVan wC&O wCoS
	domestica 'Golden Noble'	wBCr
	domestica 'Golden Nugget'	wBCr
	domestica 'Golden Pearmain'	wBCr
	domestica 'Golden Russet'	wBCr wBur oRiv
un	***domestica*** 'Golden Sentinel'	wRai oWnC wBox
	domestica Golden Supreme®	wC&O
un	***domestica*** Goldspur® Delicious	wVan wC&O
un	***domestica*** 'Goodland'	wBCr
	domestica 'Granny Smith'	oFor wBCr wBur oAls oGar oSho wKin oWnC wVan wC&O wCoS
	domestica 'Gravenstein'	wBCr wBur wClo oEar oAls oGar oRiv oSho oWnC wTGN wC&O wCoS
	domestica 'Greensleeves'	wRai
un	***domestica*** 'Grimes Golden'	wBCr oRiv wVan
un	***domestica*** 'Grindstone'	wBCr
un	***domestica*** 'Haas'	wBCr
	domestica 'Haralson'	wBCr oRiv oWnC
un	***domestica*** 'Harcourt'	wBCr
	domestica 'Harry Master's Jersey'	last listed 99/00
un	***domestica*** 'Hawaii'	wBCr
un	***domestica*** 'Hawley'	wBCr
un	***domestica*** 'Hazen'	wBCr
un	***domestica*** 'Herefordshire Redstreak'	wBCr

un	***domestica*** 'Heyer'	wBCr
un	***domestica*** 'Heyer 2'	wBCr
un	***domestica*** 'Heyer 20'	wBCr
un	***domestica*** 'High Top Sweet'	wBCr
un	***domestica*** 'Hillwell Red Braeburn'	wC&O
un	***domestica*** 'Holiday'	wBCr oRiv
	domestica 'Holstein'	wBCr
un	***domestica*** Honey Crisp™	wBCr wClo wRai wVan wC&O
	domestica 'Honeygold'	wBCr
un	***domestica*** 'Hubbard's Nonesuch'	oRiv
un	***domestica*** 'Hubbardston Nonesuch'	wBCr
un	***domestica*** 'Hudson's Golden Gem'	wBCr oEar oRiv
un	***domestica*** 'Idamac'	wBCr
	domestica 'Idared'	wBCr wClo oAls wVan wC&O
un	***domestica*** 'Improved Golden'	wC&O
un	***domestica*** 'Improved McIntosh'	wC&O
	domestica 'Ingrid Marie'	oRiv
	domestica 'Irish Peach'	oRiv
	domestica 'James Grieve'	wBCr
un	***domestica*** 'Jefferis'	wBCr
	domestica 'Jerseymac'	wBCr wVan
un	***domestica*** 'Jesse Hall'	wBCr
un	***domestica*** 'Jewett Red'	wBCr
un	***domestica*** 'Jonafree'	wBCr wBur oRiv wVan
	domestica 'Jonagold'	wBCr wBur wClo oEar oAls oGar oWhi wRai wKin wTGN
un	***domestica*** Jonagold de Coster®	wVan
	domestica 'Jonagored'	wBCr
un	***domestica*** 'Jonamac'	wBCr wTGN wVan wC&O
	domestica 'Jonathan'	oFor wBCr wKin
	domestica Jonica™	wVan
	domestica 'Jonwin'	oRiv
un	***domestica*** Jubilee™ Fuji	wC&O
un	***domestica*** 'July Red'	wBCr
	domestica 'Kandil Sinap'	oRiv
	domestica 'Karmijn de Sonnaville'	wClo wRai
	domestica 'Katja'	wBCr
un	***domestica*** 'Keepsake'	wBCr oRiv
	domestica 'Kerry Pippin'	wBCr
	domestica 'Keswick Codling'	oRiv
	domestica 'Kidd's Orange Red'	wBCr oEar
un	***domestica*** 'King'	wBCr wBur oEar oAls oGar oRiv wTGN wCoS
un	***domestica*** 'King David'	wBCr
	domestica 'King Edward VII'	see **M. *domestica*** 'Edward VII'
	domestica 'King of the Pippins'	oRiv
	domestica 'Kingston Black'	wBCr wClo wRai
un	***domestica*** 'Lady'	wBCr oRiv
	domestica 'Lady Sudeley'	oRiv
un	***domestica*** 'Laketon'	wBCr
un	***domestica*** 'Laxton Superb'	wBCr
un	***domestica*** 'Laxton's Fortune'	wBCr oRiv
un	***domestica*** 'Law Red Rome'	wVan
	domestica 'Liberty'	wBCr wBur wClo oEar oAls oGar oSho oWhi wRai oWnC wCoS
	domestica 'Lobo'	wBCr
	domestica 'Lodi'	wBCr wBur oEar oAls oGar wVan
un	***domestica*** 'Lord's Seedling'	wBCr
un	***domestica*** 'Lowland Raspberry'	wBCr oEar
un	***domestica*** 'Lubsk Queen'	wBCr oGar
un	***domestica*** Lucky Rose™	wVan
un	***domestica*** 'Lurared'	wBCr
un	***domestica*** 'Lyman's Large Summer'	wBCr
un	***domestica*** 'Macfree'	wBCr
un	***domestica*** 'Machay'	oRiv
	domestica 'Macoun'	wBCr oRiv wC&O
	domestica 'Maiden's Blush'	wBCr oRiv
un	***domestica*** 'Malinda'	wBCr
un	***domestica*** 'Manalta'	wBCr
un	***domestica*** 'Mantet'	wBCr
	domestica 'Margil'	last listed 99/00
un	***domestica*** 'Mark's Sweet'	wBCr
un	***domestica*** 'Marlin Stephens'	wBCr
	domestica 'McIntosh'	wBCr oAls oGar oWnC wVan
	domestica 'Melrose'	wBCr wClo oEar oAls oGar oRiv wRai wVan
	domestica 'Merton Worcester'	oRiv
	domestica 'Michelin'	wRai
un	***domestica*** Midnight Spur™	wC&O
un	***domestica*** 'Milan'	wBCr
un	***domestica*** 'Minnesota 447'	wBCr
un	***domestica*** 'Minnesota 1743'	wBCr
	domestica 'Mollie's Delicious'	wBCr wVan
un	***domestica*** 'Mor-Spur McIntosh'	wC&O
un	***domestica*** 'Morden 359'	wBCr
un	***domestica*** 'Morden 363'	wBCr
	domestica 'Mother'	wBCr
un	***domestica*** 'Mountain Boomer'	wBCr
	domestica 'Mutsu'	wBCr oEar oAls oGar oRiv wVan wC&O
un	***domestica*** 'Myra Red Fuji'	wC&O
	domestica 'Newtown Pippin'	oFor wBCr oGar oRiv oWnC wVan wC&O
un	***domestica*** 'Noran'	wBCr
un	***domestica*** 'Noret'	wBCr
un	***domestica*** 'Norland'	wBCr
un	***domestica*** 'Norson'	wBCr
	domestica 'Northern Red Spy'	see **M. *domestica*** 'Northern Spy'
	domestica 'Northern Spy'	oFor wBCr oAls wVan wC&O
un	***domestica*** 'Northfield Beauty'	wBCr
un	***domestica*** 'Northpole'	wRai oGre oWnC wBox
un	***domestica*** 'Northwest Greening'	oRiv
un	***domestica*** 'Nospy'	oRiv
un	***domestica*** 'Nova Easygro'	wBCr
un	***domestica*** Nured®Jonathan Sport	wC&O
un	***domestica*** Nured® Rome	wC&O
un	***domestica*** Nured® Stayman	wC&O
un	***domestica*** Nured®Winesap	wC&O
un	***domestica*** 'Oley Olsen'	oRiv
	domestica 'Opalescent'	oRiv
un	***domestica*** Oregon Spur® II	wVan
	domestica 'Orin'	wC&O
un	***domestica*** 'Oriole'	wBCr
un	***domestica*** 'Ortley'	wBCr
un	***domestica*** Pacific Gala®	wC&O
un	***domestica*** 'Pacific Gold'	oRiv
un	***domestica*** 'Palouse'	wBCr oRiv
un	***domestica*** 'Parantene'	wBCr
un	***domestica*** 'Parkland'	wBCr
un	***domestica*** 'Park's Pleasant'	oRiv
un	***domestica*** 'Patton'	wBCr
	domestica 'Peck's Pleasant'	wBCr
un	***domestica*** 'PF 12'	wBCr
un	***domestica*** 'PF 51'	wBCr
un	***domestica*** 'Pink Excell Pink Flesh'	oRiv
un	***domestica*** 'Pink Pearl'	wBCr oEar oRiv oWhi wRai oGre
un	***domestica*** 'Pink Pearlman'	oRiv
un	***domestica*** 'Pink Sugar'	wBCr
un	***domestica*** Pinova™	wC&O
un	***domestica*** 'Porter'	wBCr
un	***domestica*** 'Potter Cox'	oRiv
un	***domestica*** 'Pound Sweet'	wBCr
un	***domestica*** 'Prairie Spy'	oEar
un	***domestica*** 'Prima'	wBur
un	***domestica*** 'Primate'	wBCr
un	***domestica*** Prime Gold®	wVan
un	***domestica*** 'Prime Gold Vurwell'	oRiv
un	***domestica*** 'Pristine'	wRai
	domestica 'Queen Cox'	wClo wRai
un	***domestica*** 'Quinte'	wBCr
un	***domestica*** Radiant Runkel®	wC&O
un	***domestica*** 'Ralls Janet'	wBCr
un	***domestica*** 'Ramey York'	wC&O
un	***domestica*** 'Ramsdell Sweet'	wBCr
un	***domestica*** 'Raritan'	wBCr
	domestica 'Red Alkmene'	wClo
	domestica 'Red Astrachan'	wBCr oRiv
un	***domestica*** 'Red Baron'	wBCr
un	***domestica*** 'Red Barton'	oRiv
un	***domestica*** 'Red Belle de Boskoop'	wRai
un	***domestica*** 'Red Butterscotch'	wBCr
	domestica 'Red Delicious'	see **M. *domestica*** 'Delicious'
un	***domestica*** 'Red Free'	wVan
	domestica 'Red Fuji'	wBCr oWnC wVan
un	***domestica*** 'Red Gold'	wBCr
un	***domestica*** 'Red Gravenstein'	wBCr wRai wVan
un	***domestica*** 'Red Hook'	wBCr
ch	***domestica*** 'Red Jonagold'	see **M. *domestica*** 'Jonagored'
un	***domestica*** 'Red June'	wBCr
un	***domestica*** 'Red Limber Twig'	wBCr
	domestica 'Red McIntosh'	see **M. *domestica*** 'McIntosh'
	domestica 'Red Melba'	wBCr
un	***domestica*** 'Red Ribston'	wBCr
	domestica 'Red Rome'	see **M. *domestica*** 'Rome Beauty'
un	***domestica*** 'Red Stayman'	wVan
un	***domestica*** 'Red Sumbo'	wBCr
un	***domestica*** 'Red Van Buren'	wBCr
	domestica 'Redfree'	wBur oRiv
un	***domestica*** 'Redwell'	wBCr
un	***domestica*** 'Reverend Morgan'	wBCr

un	***domestica*** 'Rhode Island Greening'	oRiv
	domestica 'Ribston Pippin'	wBCr oRiv
un	***domestica*** 'Richared Delicious'	wBCr
un	***domestica*** 'Roman Stem'	wBCr
	domestica 'Rome Beauty'	wBCr oAls oRiv
	domestica 'Ross Nonpareil'	wBCr
un	***domestica*** 'Rosthern 15'	wBCr
un	***domestica*** 'Rosthern 18'	wBCr
	domestica 'Roxbury Russet'	oEar oRiv
un	***domestica*** Royal Empire™	wVan wC&O
	domestica 'Royal Gala'	oRiv oWnC wVan
	domestica 'Rubergen Reinette'	oRiv
	domestica Rubinette®	wVan
un	***domestica*** Rubinstar®Jonagold	wC&O
un	***domestica*** 'Ruby Red Jonathan'	wVan
un	***domestica*** 'Rural Russet'	wBCr
un	***domestica*** 'Rusty Coat'	wBCr
un	***domestica*** 'Saint Lawrence'	wBCr
un	***domestica*** Sali™ Red Delicious	wVan
un	***domestica*** 'Sansa'	wVan
un	***domestica*** 'Sayaka'	wRai
un	***domestica*** 'Scarlet Gala'	wC&O
un	***domestica*** 'Scarlet Sentinel'	oWnC wBox
un	***domestica*** Scarlet Spur®	wVan
un	***domestica*** 'Schlect Spur Delicious'	wC&O
un	***domestica*** 'Sekai Ichi'	wBCr oRiv
un	***domestica*** 'Senshu'	wBCr wVan wC&O
un	***domestica*** 'Shafer'	wBCr
un	***domestica*** 'Shamrock'	wVan
un	***domestica*** 'Shay'	wRai
un	***domestica*** 'Shiawassee Beauty'	wBCr
	domestica 'Sierra Beauty'	oRiv
un	***domestica*** 'Signe Tillish'	wBCr
un	***domestica*** 'Simirenko Reinette'	wBCr
un	***domestica*** 'Sinta'	wBCr
un	***domestica*** 'Sir Prize'	oRiv
un	***domestica*** 'Skinner Seedling'	oRiv
un	***domestica*** 'Smokehouse'	wBCr oRiv
un	***domestica*** 'Smoother Apple'	oRiv
un	***domestica*** 'Snow Fameuse'	oFor
un	***domestica*** 'Sofstaholm'	wBCr
un	***domestica*** 'Somerset of Maine'	wBCr
	domestica 'Sops in Wine'	wBCr
	domestica 'Spartan'	wBCr wBur wClo oEar oGar oRiv oWhi wVan wC&O wCoS
un	***domestica*** 'Spartan x PL 255599'	wBCr
	domestica 'Spencer'	wBCr
un	***domestica*** 'Spigold'	wBCr oRiv wRai
un	***domestica*** 'Spilicious'	oRiv
	domestica 'Spitzenberg'	wBCr wBur oEar oAls oGar oRiv oWhi
	domestica 'Spitzenberg Red'	oRiv
un	***domestica*** 'Splendour'	wBCr oRiv
un	***domestica*** 'Spokane Beauty'	wBCr oRiv
un	***domestica*** Spur Goldblush™	wC&O
un	***domestica*** 'Spur Winter Banana'	wC&O
un	***domestica*** Spur York™	wVan
un	***domestica*** 'Stamps Red Delicious'	wC&O
un	***domestica*** 'Starr'	wBCr
un	***domestica*** 'State Fair'	wBCr
	domestica 'Stayman'	wBCr oEar oAls oGar oRiv
qu	***domestica*** 'Summer Apple'	oRiv
un	***domestica*** 'Summer Orange'	wBCr
un	***domestica*** 'Summer Rambo'	oRiv
un	***domestica*** 'Summer Rose'	wBCr
un	***domestica*** 'Summer Scarlet'	wBCr
	domestica 'Summerred'	wBCr oWnC
un	***domestica*** Sun Fuji®	wVan
	domestica 'Sunrise'	wRai
	domestica 'Suntan'	wBCr
un	***domestica*** Super Chief®	wVan
un	***domestica*** 'Super Jon'	oRiv
un	***domestica*** 'Sutton Beauty'	wBCr
un	***domestica*** 'Sweet Bough'	oRiv
	domestica 'Sweet Coppin'	wRai
un	***domestica*** 'Sweet Pippin'	wBCr
un	***domestica*** 'Sweet Sixteen'	wBCr oRiv wRai
un	***domestica*** Swiss Gourmet™	wVan
un	***domestica*** 'Sylvia'	wBCr
un	***domestica*** 'T. A. C. 114 Red Fuji'	wC&O
un	***domestica*** 'Tetofsky'	wBCr
un	***domestica*** Tex™ Red Winesap	wVan
un	***domestica*** 'Tolman Sweet'	wBCr
	domestica 'Tom Putt'	wBCr
un	***domestica*** 'Tompkins King'	wBCr oRiv wRai
un	***domestica*** TopExport™ Fuji	wC&O
un	***domestica*** 'Tsugaru'	wC&O
	domestica 'Twenty Ounce'	last listed 99/00
un	***domestica*** 'Twenty Ounce Pippin'	wBCr
un	***domestica*** 'Tydeman's Early'	wBCr
un	***domestica*** 'Tydeman's Red'	oRiv
un	***domestica*** 'Unity'	wBCr
un	***domestica*** 'Valentine'	wBCr
un	***domestica*** Valstar™	wVan
un	***domestica*** 'Veedum'	wBCr
un	***domestica*** 'Virginia Beauty'	wBCr
un	***domestica*** 'Virginia Winesap'	wBCr
	domestica 'Vista-bella'	wBCr
	domestica 'Wagener'	oRiv
un	***domestica*** 'Waltana'	wBCr
un	***domestica*** 'Washington Strawberry'	wBCr
	domestica 'Wealthy'	wBCr oAls oRiv wVan
un	***domestica*** 'Webster Pink'	oRiv
un	***domestica*** 'Westfield Seek No Further'	wBCr oRiv
un	***domestica*** 'Westland'	wBCr
un	***domestica*** 'White Pippin'	wBCr
	domestica 'White Winter Pearmain'	see **M. *domestica*** 'Winter Pearmain'
un	***domestica*** 'Williams'	wBCr
	domestica 'William's Pride'	wBur wClo wRai wCoS
	domestica 'Winesap'	see **M. *domestica*** 'Stayman'
	domestica 'Winter Banana'	wBCr wBur oGar oRiv oSho wCoS
	domestica 'Winter Pearmain'	oEar
un	***domestica*** 'Winter Queen'	wBCr
un	***domestica*** 'Winter Red Flesh'	wBCr oEar
un	***domestica*** 'Winterstein'	wRai
un	***domestica*** 'Wismer's Dessert'	wBCr
un	***domestica*** 'Wolf River'	wBCr oEar oRiv wRai
un	***domestica*** 'Wynooche'	wBur
un	***domestica*** 'Wynooche Early'	oWhi
	domestica 'Yarlington Mill'	wBCr
un	***domestica*** 'Yataka Fuji'	wBCr
un	***domestica*** 'Yates'	wBCr
un	***domestica*** 'Yeager Sweet'	wBCr
un	***domestica*** 'Yellow Bellflower'	wBCr oRiv
	domestica 'Yellow Delicious'	see **M. *domestica*** 'Golden Delicious'
un	***domestica*** 'Yellow June'	wBCr
	domestica 'Yellow Newtown'	see **M. *domestica*** 'Newtown Pippin'
un	***domestica*** 'Yellow Sheepnose'	wBCr
un	***domestica*** 'Yellow Transparent'	wBCr oRiv oSho wKin oWnC
un	***domestica*** 'Yellow Twig'	wBCr
un	***domestica*** 'York Imperial'	wBCr
un	***domestica*** #8906 Apple	wBCr
un	***domestica*** #8919 Apple	wBCr
un	***domestica*** #8921 Apple	wBCr
	'Donald Wyman'	oFor wBox
	'Doubloons'	oFor wWel
	'Evereste'	oGar wRai
	floribunda	oFor oGar oBlo wRai wBox
	fusca	oFor wWoB wBur wShR oBos wWat wCCr wWld oSle
un	'Golden Raindrops'	oFor oAls oDan
un	Guinevere™ / 'Guinzam'	wKin
	'Harvest Gold'	oFor wCul oGar wBox
	'Hopa'	oWnC
	hupehensis	oEdg oDan oRiv wAva
	'Indian Magic'	oFor oDan wBox wCoS
	'Indian Summer'	oFor wBox wC&O
	ioensis 'Plena'	oDar wCoS
	'Jewelberry'	oFor wBox
	'John Downie'	wRai
	kansuensis	oFor
	'Klehm's Improved Bechtel'	oFor oAls oWnC wWel
un	'Lancelot'	oGar
un	'Lollizam'	oGar
un	'Louisa'	oFor oAls oGar oGre
un	'Madonna'	oFor oWnC
un	'Manchurian'	wBCr oRiv wVan wC&O
	'Mary Potter'	oGar wBox
un	'Maypole'	oGar
	x micromalus	last listed 99/00
	x micromalus 'Kaido'	see **M. *x micromalus***
	x moerlandsii 'Liset'	oGar wCoS
	x moerlandsii 'Profusion'	oFor wBCr oDar oAls oWnC wCoS
	'Molten Lava'	oFor oGre
	'Narragansett'	last listed 99/00
	pear leaf	wC&O
un	'Pink Dawn'	oGar
	'Pink Perfection'	last listed 99/00
	'Pink Princess'	oFor oGar

	'Prairifire'	oAls oGar oDan oWnC wCoS wWel
	prunifolia	oRiv
	'Radiant'	wKin
	'Ralph Shay'	oFor
	'Red Baron'	oFor oBlo
	'Red Baron' bonsai	oGre
un	'Red Flesh'	wRai
	'Red Jade'	see **M. *x schiedeckeri*** 'Red Jade'
un	'Red Jewel'	oAls oWnC wBox wWel
	rivularis	see **M. *fusca***
un	'Robinson'	oFor oGar
	x robusta red flesh	oRiv
un	'Rose Glow'	oGre
un	'Royal Fountain'	oFor
	'Royal Gem'	oGar
	'Royalty'	oFor wKin oWnC
	'Royalty' bonsai	oGre
	sargentii	see **M. *toringo*** ssp. ***sargentii***
	'Sargent Tina'	see **M. *toringo*** ssp. ***sargentii*** 'Tina'
	x schiedeckeri 'Red Jade"	oFor oAls oWnC wTGN wSta wCoS
	'Sentinel'	oDan
	sieboldii	see **M. *toringo***
	sikkimensis	oTrP
un	'Simpson 10-35'	wC&O
un	'Sinai Fire'	oGar
	'Snowdrift'	oFor wBCr oAls oGar oRiv oBlo oWnC wBox wVan wC&O
	'Snowdrift' bonsai	oGre
	'Spring Snow'	wKin oWnC
	'Strawberry Parfait'	oGar
	'Sugar Tyme'	wCul oGar oWnC
	toringo	oFor wCCr
	toringo ssp. ***sargentii***	oFor oGar oRiv wKin wBox wWel
	toringo ssp. ***sargentii*** 'Tina'	oGar wKin wBox wCoS wWel
	toringoides	oRiv
un	'Transcendent'	oAls oGar oRiv
un	***transitoria*** 'Golden Raindrops'	oGar
	'Weeping Candied Apple'	see **M.** 'Candied Apple'
un	'Whitney'	oFor wBCr oGar wVan
un	'Wickson'	wBCr
	x zumi	oBlo oWnC wAva wCoS
	x zumi var. ***calocarpa***	oFor oAls oGar oDan wWel
	x zumi var. ***calocarpa*** bonsai	oGre
	MALVA	
	alcea	wCSG
	alcea var. ***fastigiata***	oFor oNat oAls oSho oWnC
	'Bibor Fehlo'	wSwC wCSG
	moschata	wCul wSwC wCSG oGoo wFai iGSc wMag wAva
	moschata f. ***alba***	oTrP wCul wSwC oAls oRus oSho oWnC wMag wTGN wSte
	moschata 'Romney Marsh'	see **ALTHAEA *officinalis*** 'Romney Marsh'
un	'Pink Perfection'	wHom
	sylvestris	oGoo wAva oLSG
	sylvestris 'Brave Heart'	wSwC wAva oUps
	sylvestris ssp. ***mauritanica***	oRus
un	***sylvestris*** ssp. ***mauritanica*** 'Elcanor'	wCSG
	sylvestris 'Primley Blue'	oGoo oGre
	sylvestris 'Zebrina'	wTho wSwC oAls wFGN oBlo oSec oCir
un	'Windsor Castle'	wTho wFGN
	MALVASTRUM	
	lateritium	wCul
	MALVAVISCUS	
	arboreus	oTrP
un	'Paquito Pink'	oTrP
	MAMMILLARIA	
	durispina	see **M. *kewensis***
	kewensis	oRar
un	***pectinifera*** f. ***pectinifera***	oRar
	MANDEVILLA	
	x amoena 'Alice du Pont'	oAls oGar oGre wWel
	boliviensis	oGar
	laxa	oTrP oGar wRai oGre oWnC wTGN wBox wWel
un	'Ruby Star'	oAls oGar
un	***sanderi*** 'Faire Lady'	oGar
	splendens	oGar wWel
	splendens 'Red Riding Hood'	see **M. *splendens***
	suaveolens	see **M. *laxa***
	Summer Snow™ / 'Monte'	wWel
	'White Delite'	wWel
	MANDRAGORA	
	officinarum	oCrm
	MANFREDA	
un	sp. aff. ***maculata*** BLM 1073	oRar

un	***nutida***	oRar
un	***scabra***	oRar
	virginica	wCCr
	MANGLIETIA	
un	***chingii***	wHer oGos oCis oGre wBox
	insignis	oFor oDan oCis oSho oGre oWnC wBox
	MANIHOT	
un	***angustiloba***	oRar
un	***caudata*** BLM 0696	oRar
	MARANTA	
	leuconeura var. ***erythroneura***	oGar
	leuconeura var. ***kerchoveana***	oGar
	MARGYRICARPUS	
	pinnatus HCM 98192	wHer
	setosus	see **M. *pinnatus***
	MARRUBIUM	
	cylleneum	wCul wFai iGSc
un	***rotundifolium***	oSis
	vulgare	oAls wFGN oGoo wFai iGSc wHom oWnC oCrm oBar oSec oUps
	vulgare Russian	wHom
	MARSHALLIA	
	trinerva	oFor
	MARSILEA	
	drummondii	oSsd
	mutica	wSoo wGAc oHug oGar oSsd oWnC wBox
	quadrifolia	oHug oGar
	schelpiana	oSsd oWnC wBox
	MATELEA	
	carolinensis	oFor
un	***cyclophylla***	oRar
un	***pavonii*** BLM 0601	oRar
	MATRICARIA	
	chamomilla	see **M. *recutita***
	recutita	oAls wFGN wFai iGSc wHom oWnC wBox oSec wNTP oUps
un	***recutita*** 'Santana Lemon'	oGar
	MATTEUCCIA	
	struthiopteris	oFor wFol oRus wHom wMag oTri wTGN wBox wCoN wSnq
un	***struthiopteris*** 'The King'	oRus
	MATTHIOLA	
un	***fruticulosa*** ssp. ***perennis*** 'Alba'	oNWe
	MAURANDYA	
	barclayana	oDar
	MAYTENUS	
	boaria	oFor
	magellanica HCM 98078	wHer
	MAZUS	
	radicans	oJoy wCol oAls oSle
	reptans	oFor oAls wTGN wCoN oUps
	reptans 'Albus'	oFor oJoy oAls
	MECONOPSIS	
	x beamishii	last listed 99/00
	betonicifolia	wHer wCri oHed oRus oNWe
	betonicifolia var. ***alba***	oFor wHer wMtT
	betonicifolia Harlow Carr strain	last listed 99/00
	cambrica	wThG oJoy wCol oAls wCSG oBov wMag oJil
	cambrica 'Frances Perry'	oJil
do	***cambrica*** 'Muriel Brown'	oNWe oJil
	cambrica orange	oRus
	cambrica yellow	oRus oNWe
	cambrica yellow/red	oNWe
	grandis	wHer oRus wMag
	horridula	wHer oJil
	integrifolia	wHer
	integrifolia x ***punicea***	oNWe
	napaulensis	wHer oGos oDan oNWe wAva
	napaulensis hybrids	oJil
	paniculata	wHer
	prattii	see **M. *horridula***
	x sheldonii	wHer oGos oRus oNWe wAva
	x sheldonii strain	oJil
	sp. ex ACE1474 M. Rudis	wMtT
	MEEHANIA	
	cordata	last listed 99/00
	MELALEUCA	
	alternifolia	oTrP oGoo iGSc oCrm oBar
	ericifolia	oTrP
	incana	oTrP
	lateritia	last listed 99/00
	leucadendra	last listed 99/00
	linariifolia	wCCr
	squamea	wCCr

	MELANOSELINUM	
	sp. aff. ***decipiens***	wHer
	MELIA	
	azedarach	oFor oRiv oOEx
	azedarach 'Umbraculiformis'	last listed 99/00
	MELIANTHUS	
	comosus	wHer oGre
	major	wHer wCCr oSec wCoN wSte
	minor	wHer
	pectinatus	wHer
	villosus	wHer
	MELICA	
	altissima 'Atropurpurea'	oGar oOut oWnC wSnq
	uniflora f. ***albida***	oHed
	MELICYTUS	
	alpinus	last listed 99/00
	crassifolius	oFor
	MELIOSMA	
	cuneifolia	see **M. *dilleniifolia*** ssp. ***cuneifolia***
	sp. aff. ***dilleniifolia*** ssp. ***cuneifolia*** EDHCH 97151	wHer
	dilleniifolia ssp. ***tenuis***	wHer
	myriantha	wHer
	oldhamii	see **M. *pinnata***
	pinnata HC 970298	wHer
	MELISSA	
	officinalis	wWoS oAls wFGN wCSG oGoo wFai iGSc wHom oWnC oBar oSec wNTP oCir wCoN oUps
	officinalis 'All Gold'	wBox
va	***officinalis*** 'Aurea'	wWoS wFai iGSc
	officinalis lime scented	iGSc
	officinalis 'Variegata'	see **M. *officinalis*** 'Aurea'
un	**MELLIODENDRON**	
	xylocarpum	oSho oGre
	MENISPERMUM	
	davuricum	oFor
	MENTHA	
	aquatica	oTrP wSoo oHug oGar oSsd oWnC wBox oHou
	arvensis	oHan oTDM oAls wFGN oHou
un	***arvensis*** 'Banana'	wWoS oAls oGoo iGSc wHom oBar wNTP oHou
	austriaca	see **M. *arvensis***
	'Blue Balsam'	oGoo iGSc oBar oHou
un	Butter Mint	oHou
	cervina	iGSc
un	Curly Lemon Mint	oHou
un	Double Mint	oHou
un	Dutch Apple Mint	oHou
un	Egyptian Mint	wWoS oHou
	emerald and gold	oAls wNTP
	x gentilis	see **M. *x gracilis***
	x gracilis	wWoS oTDM oAls wFGN oGoo iGSc oWnC oHou
	x gracilis 'Variegata'	iGSc wHom
un	Grapefruit	oTDM oAls wFGN oGoo iGSc oBar wNTP
	'Hillary's Sweet Lemon'	wHom
un	'Himalayan Silver'	wCul wFGN oGoo
un	Julip Mint	oAls oHou
	Lavender mint	wFGN iGSc oBar wNTP oHou
qu	Lemon Bergamot Mint	oAls
un	***longifolia*** 'Habak'	oAls
qu	***longifolia*** 'Himalayense'	wBox
un	***micromeria***	oAls
un	Orange Balsam	iGSc
qu	Orange Bergamot Mint	oHou
	x piperita	wTho oTDM oAls wFGN oGoo iGSc oWnC wBox wNTP oHou oCir oUps oSle
ch	***x piperita*** 'Black Mitchum'	wFGN oBar
	x piperita 'Candymint'	wFGN oHou
	x piperita f. ***citrata***	wFGN oWnC oBar wBox oHou oSle
	x piperita f. ***citrata*** 'Basil'	wFGN
	x piperita f. ***citrata*** 'Chocolate'	wWoS oTDM oAls wFGN oGoo iGSc wHom oCrm oBar wNTP oHou oCir
	x piperita f. ***citrata*** 'Lemon'	wFGN wNTP
	x piperita f. ***citrata*** 'Lime'	oAls wFGN oGoo iGSc wHom oBar wNTP oHou oCir
	x piperita f. ***citrata*** orange	oTDM oAls wFGN oGoo iGSc wHom oCrm oBar wNTP oHou oCir
	x piperita 'Todd Mitchum'	oAls
	x piperita 'Variegata'	oAls wFGN wHom oUps
	pulegium	oTDM oJoy oAls wFGN oGoo wFai iGSc oWnC oBar oHou oCir oUps
	pulegium 'Cunningham Mint'	wFGN wRai oCrm oBar
un	***pulegium*** 'Hart's'	wFai
	pulegium 'Nana'	see **M. *pulegium*** 'Cunningham Mint'
	requienii	wWoS oTDM oAls wFGN oGoo oSho iGSc wHom oWnC oBar wTGN wSta wNTP oHou oCir oSle
	rotundifolia	see **M. *suaveolens***
un	Scotch Mint	wFGN oGoo oCrm
un	Silver Mint	oGoo iGSc oHou
	spicata	wTho oAls wFGN wCSG oGoo iGSc wHom oWnC oBar wBox wNTP oUps oSle
	spicata 'Chewing Gum'	oAls oGoo oCrm oBar oHou
	spicata 'Crispa'	wCul oAls wFGN oGoo iGSc oHou
	spicata 'Kentucky Colonel'	oAls wFGN oHou
	spicata 'Moroccan'	oHou
	spicata narrow leaf	oHou
qu	***spicata*** 'Wrigley's Spearmint'	iGSc
	suaveolens	wTho oTDM wFGN wCSG iGSc wHom oBar wNTP oHou
	suaveolens Georgian apple mint	oAls
	suaveolens 'Variegata'	wTho oAls wFGN wCSG oGoo iGSc wHom oWnC wNTP oHou oCir oSle
	x villosa f. ***alopecuroides***	oGoo
	MENYANTHES	
	trifoliata	wSoo wGAc oSsd
	MENZIESIA	
	ciliicalyx	oBov
	ciliicalyx var. ***multiflora***	wRho
	ciliicalyx var. ***purpurea***	wRho
	ferruginea	oBos wFFl
	pilosa	last listed 99/00
	MERTENSIA	
	asiatica	see **M. *simplicissima***
un	***bella***	wCol oAls oGar oGre
	maritima ssp. ***asiatica***	see **M. *simplicissima***
	paniculata	oBos
	platyphylla	oBos
	pterocarpa	see **M. *sibirica***
	pulmonarioides	oRus oGar
	sibirica	oFor wHer oSho
un	***sibirica*** var. ***yezoensis***	oRus
	simplicissima	wHer oJoy oNWe oGre
un	***subcordata***	oRus
	virginica	see **M. *pulmonarioides***
	MESPILUS	
	germanica (fruit)	oFor wBur wRai
un	***germanica*** 'Breda Giant'	wBur wClo wRai oOEx
	germanica 'Bredase Reus'	wCul
	germanica 'Macrocarpa'	last listed 99/00
	germanica 'Nottingham'	wBur oOEx
	germanica 'Royal'	oOEx
un	***germanica*** 'Russian Giant'	oOEx
un	***germanica*** 'Westerfield'	oOEx
	MESTOKLEMA	
	arboriforme	oRar
	tuberosum	oRar
un	**METAPLEXIS**	
	japonicus HC 970462	wHer
	METASEQUOIA	
	glyptostroboides	oFor oTrP wBCr wWoB wBur wClo oDar oEdg wShR oAmb oGar oDan oRiv oSho oBlo wRai oGre wCCr wKin wSta wAva wCoN wRav wWhG wCoS wWel
un	***glyptostroboides*** 'Jack Frost'	oPor wWel
	glyptostroboides 'Sheridan Spire'	last listed 99/00
	METROSIDEROS	
	excelsus	wHer
	villosus	oTrP
	MEUM	
	athamanticum	wHer
	MICHELIA	
un	***chapensis***	wHer oGos oSho
un	***crassipes***	wHer
	figo	oFor oCis
	figo 'Port Wine'	oFor oCis
un	***floribunda***	oCis
un	***x foggii*** 'Jack Fogg'	oCis
un	***x foggii*** #2	oCis
un	***fulgens***	wBox
	fuscata	see **M. *figo***
un	***grandiflora***	oCis
un	***grandis***	oCis
un	***maudiae***	wHer oDan oCis oGre oWnC wBox
un	***megaphylla***	oCis
un	***platypetala***	wHer oDan oCis wBox wSte
	sinensis	wHer oGar oCis oSho oGre oWnC wBox

	wilsonii	see **M. *sinensis***
	yunnanensis	wHer oCis
	MICROBIOTA	
	decussata	oFor oPor wWoB wCol oAls oGar oSis oRiv oGre oTPm wTGN oBRG wCoN
un	***decussata*** 'UBC Clone'	wWel
	MICROCACHRYS	
	tetragona	oFor wHer oRiv wWel
	MICROCITRUS	
	australasica	oOEx
	MICRODERIS	see **LEONTODON**
	MICROMERIA	
	thymifolia	oGoo
	MIKANIA	
	dentata	oGar
	MILIUM	
	effusum 'Aureum'	oDar oAls oSis oOut oGre wWin wPir wCCr wBWP wCoN
	effusum 'Aureum' seed grown	wHer
	MILLETTIA	
	reticulata	oFor oCis oSho oGre oWnC wCoS
un	***taiwanensis***	wCoS
	MIMOSA	
	pudica	oTrP oAls
	MIMULUS	
	aurantiacus	oFor oAls oBos
	bifidus	oFor
	Calypso Series	oUps
	cardinalis	oFor oHan oNat oRus oSis oWnC oSle
	cardinalis yellow form	oSis
un	***dentatus***	oRus oBos
	guttatus	oFor wNot oHug oSsd oBos oTri
va	***guttatus*** 'Richard Bish'	oHug
	guttatus variegatus	see **M. *guttatus*** 'Richard Bish'
	lewisii	oHan oBos wCCr
	luteus 'Variegatus'	oFor wWoS wEde
	moschatus	oGoo
	'Puck'	wWoS
	puniceus	last listed 99/00
un	'Richard's Red'	oHed
	ringens	oRus oHug oGar oGoo oUps
	'Roter Kaiser'	oFor oSis
un	'Splash'	wWoS
	tilingii	wThG oWnC
qu	'Variegata'	oGre
	Verity hybrid	oFor
	MINA	see **IPOMOEA**
	MINUARTIA	
un	***kashmirica***	oSis
	obtusiloba	wMtT
un	***rubella*** 'Popcorn'	wMtT
un	***saxifraga*** ssp. ***tmolea***	wMtT
	stellata	wMtT
	verna	oUps
	MISCANTHUS	
qu	***floridulus***	oDar oGar oDan oOut wWin oTri wSnq
	'Giganteus'	wCli oOut oGre
	'Little Big Man'	oOut oGre wBox
	oligonensis 'Wetterfahne'	oGar oOut oGre wBox
	'Purpurascens'	oFor wCli oDar oJoy oHed oGar oDan wRob oGre wWin wHom oWnC oTri oBRG oUps wSnq
	sacchariflorus	oFor
	sinensis	oFor oSho wCoN wSnq
	sinensis 'Adagio'	wSwC oHed oGar oOut wWin oWnC oUps wSnq
	sinensis 'Arabesque'	oFor oDar oJoy oAls oGar wWin oWnC wSnq
	sinensis 'Autumn Light'	oGar wRob
un	***sinensis*** Blooming Wonder	see **M. *sinensis*** 'Bluttenwunder'
un	***sinensis*** 'Bluttenwunder'	oGar oGre wBox wSnq
un	***sinensis*** 'Burgunder'	oOut
un	***sinensis*** 'Central Park'	wWoS wWin
	sinensis var. ***condensatus***	oJoy oGar oOut wRob oTri
va	***sinensis*** var. ***condensatus*** 'Cabaret'	oGar wFai wWin oWnC oBRG wSnq
va	***sinensis*** var. ***condensatus*** 'Cosmopolitan'	wSwC oGar oDan wFai oWnC wBox oBRG wSnq
va	***sinensis*** 'Dixieland'	oOut wRob oGre wWin wSnq
un	***sinensis*** 'Far East'	wBox
	sinensis 'Flamingo'	wHer wWoS oGar oOut oGre wBox
un	***sinensis*** 'Gold and Silver'	oGre wBox
	sinensis Golden Feather	see **M. *sinensis*** 'Goldfeder'
va	***sinensis*** 'Goldfeder'	wBox
	sinensis 'Goliath'	wHer
	sinensis 'Gracillimus'	wHer wCli wCul wSwC oDar oAls oAmb oGar oDan oOut wRob wWin wHom oTPm oWnC oTri wBox oSec oBRG wWhG wSnq
	sinensis 'Gracillimus' dwarf	wBox oSec
	sinensis 'Graziella'	wWoS wSwC oDar oJoy oAls oAmb oGar oOut wWin wSnq
	sinensis 'Grosse Fontaene'	oGar
	sinensis 'Helga Reich'	wSnq
	sinensis 'Kirk Alexander'	wHer wWoS wSwC oDar oAmb oGar oOut wRob wBox wSte
	sinensis 'Kleine Fontaene'	oGar
va	***sinensis*** Little Nicky® / 'Hinjo'	oAls oOut wWin oWnC
	sinensis 'Malepartus'	oFor wHer wWoS oDar oAls oGar oOut oWnC oBRG wSnq
va	***sinensis*** 'Morning Light'	oFor wHer wCli wCul wSwC oJoy oAmb oGar oDan oOut wRob oGre wFai wWin wHom oWnC wTGN oSec oBRG wWhG wSnq
	sinensis 'Nippon'	wHer oOut wSnq
	sinensis 'November Sunset'	oGar
va	***sinensis*** 'Puenktchen'	oOut wWin wSnq
	sinensis var. ***purpurascens***	see **M.** 'Purpurascens'
un	***sinensis*** 'Red Feather'	oOut oGre
	sinensis Red Silver	see **M. *sinensis*** 'Rotsilber'
va	***sinensis*** 'Rotsilber'	wHer oAmb oGar oGre wSnq
	sinensis 'Sarabande'	oFor oDar oJoy oGar oSho oOut wFai wWin wBox oUps wSnq
ch	***sinensis*** 'Silberfeder'	oFor wHer wCul oJoy oGar oOut wRob oGre wWin oWnC oTri wBox wSnq
va	***sinensis*** 'Silberpfeil'	wFai wWin
	sinensis 'Silberspinne'	oOut oUps
va	***sinensis*** 'Strictus'	wCli oDar wHig oGar oOut wRob wFai wWin oTri oSec oBRG oCir oUps wSnq
	sinensis 'Undine'	wHer oOut oBRG
	sinensis 'Variegatus'	wCli wCul oDar oAls oRus wCSG oAmb oGar oDan oSho wRob oGre oWnC oTri wTGN wBox oSec oUps wSte wSnq
	sinensis Weathervane	see **M. *oligonensis*** 'Wetterfahne'
	sinensis 'Yaku Jima' (see Nomenclature Notes)	oFor wHer wWoS wCli oJoy wCol oAls oGar oOut wRob oTPm oWnC oTri wTGN wSnq
	sinensis 'Zebrinus'	wHer wCli oDar ccoAls oGar oDan wRob wFai wHom oWnC wTGN wBox wSta wSnq
	transmorrisonensis	wCli oGar wWin
	MITCHELLA	
	repens	wThG oRus oBov wCCr oCrm
	undulata HC 970254 small leaf form	wHer
	MITELLA	
	caulescens	wNot
un	***japonica*** 'Variegata'	wHer
	ovalis	oHan
	pentandra	wCol oTri
un	***pentanthera***	wWld
	trifida	wWld
	MITRARIA	
	coccinea	wHer oHed oBov
	MOLINIA	
	caerulea	oMis oGre wFai wHom oSec
	caerulea ssp. ***arundinacea*** 'Skyracer'	oFor wHer wCli wCul oDar oJoy oAls oGar oOut wWin oTri wBox oBRG wSnq
un	***caerulea*** ssp. ***arundinacea*** 'Tempest'	oGar oOut oGre
	caerulea ssp. ***arundinacea*** 'Transparent'	oOut
	caerulea ssp. ***caerulea***	oGar
	caerulea ssp. ***caerulea*** 'Moorflamme'	wHer wWoS
	caerulea ssp. ***caerulea*** 'Strahlenquelle'	wHer wWoS oSec
	caerulea ssp. ***caerulea*** 'Variegata'	wHer wWoS wCli oDar oJoy oAls oHed oGar oDan oNWe oOut wRob oGre wWin wCCr wHom oTPm oWnC oTri wTGN wBox oSec wSnq
	MOLUCCELLA	
	laevis	oUps
	MOMORDICA	
	charantia	oCrm
un	***dioica***	oRar
	rostrata	oRar
	MONADENIUM	
	coccineum	oRar
	ellenbeckii	oRar
un	***erubescens***	oRar
	magnificum	oRar
	rhizophorum	oRar
	schubei	oRar
	stapelioides	oRar

	Name	Sources
	MONARDA	
	'Adam'	last listed 99/00
	'Aquarius'	oFor wHer wFGN
	'Beauty of Cobham'	oFor
	'Blaustrumpf'	oFor oAls oRus oGar wFai oWnC wMag wTGN oLSG
	Blue Stocking	see **M.** 'Blaustrumpf'
	'Cambridge Scarlet'	oFor oGoo wFai oWnC wMag oLSG
	'Cherokee'	wHer wCul
	citriodora	oAls iArc oEga oUps
un	'Claire Grace'	oFor wWoS
	'Cobham Beauty'	see **M.** 'Beauty of Cobham'
	'Comanche'	oHed
	'Croftway Pink'	oFor oGar oGoo wMag oLSG
	'Dark Ponticum'	wCul
	didyma	wShR wFGN oGoo oSha
	didyma 'Alba'	last listed 99/00
un	***didyma*** 'Colraine Red'	oEga
un	***didyma*** 'Jacob Cline'	oFor wWoS oAls oRus oGoo oSho oWnC oCir oUps
	'Elsie's Lavender'	wHer
	'Fishes'	wHer wCul
	fistulosa	oFor oGoo wHom wWld oUps
	fistulosa 'Rose'	last listed 99/00
	'Gardenview'	oFor wFai
	'Gardenview Scarlet'	oDar oJoy oAls oHed oRus oGar oGoo oOut oWnC wMag wTGN
	'Lambada'	wMag
	'Loddon Crown'	last listed 99/00
	'Mahogany'	oFor oDar oAls oGoo oWnC wBox oLSG
	'Marshall's Delight'	oFor wHer wCul oAls oRus oGoo oSho wFai iGSc wHom oWnC wTGN oCir oEga oLSG oUps
	menthifolia	oGoo iGSc
	'Mildew Resistant'	oFor
	'Mohawk'	wHer
	'Oudolf's Charm'	oHed
	'Panorama'	wTho
	'Panorama' mixed	oBlo
	'Petite Delight'	oAls oSis oOut wMag wTGN oEga oUps
	'Pisces'	see **M.** 'Fishes'
	'Praerienacht'	oFor oJoy oRus oLSG
	Prairie Night	see **M.** 'Praerienacht'
	punctata	oFor oGoo oSis iGSc wHom
un	'Raspberry Wine'	oFor wWoS oAls oGar oWnC
	rose geranium scented	oGoo
	'Sagittarius'	wHer
	'Schneewittchen'	oFor wCul oGoo
	'Scorpio'	see **M.** 'Scorpion'
	'Scorpion'	oFor wHer wCul oHed oGar
	'Sioux'	wHer
	Snow White	see **M.** 'Schneewittchen'
	'Squaw'	wHer
un	'Stone's Throw Pink'	oFor wRob
	'Vintage Wine'	last listed 99/00
	'Violet Queen'	oFor oWnC oEga
	MONARDELLA	
	odoratissima	oHan oAls wFGN oGoo oNWe oBar
	MONESES	
	uniflora	wThG wFFl
	MONSTERA	
	deliciosa	oGar
	MONTIA	
	parvifolia	see **NAIOCRENE *parvifolia***
	perfoliata	see **CLAYTONIA *perfoliata***
	sibirica	see **CLAYTONIA *sibirica***
	MORAEA	
	alticola	wHer
	bicolor	see **DIETES *bicolor***
	irioides	see **DIETES *irioides***
un	***robusta***	wHer
	MORCHELLA	
	sp. morel	oOEx
	MORICANDIA	
	arvensis	oSec
	MORINA	
	longifolia	oFor wHer oHed oRus oJil
	persica	oFor wHer
	MORINGA	
un	***drouhardii***	oRar
	oleifera	oOEx
	pterygosperma	see **M. *oleifera***
	MORISIA	
	hypogaea	see **M. *monanthos***
	monanthos	wMtT
	MORUS	
	alba (fruit)	oFor wBur oAls oWhi oWnC oSle
	alba 'Chaparral'	wKin wCoS
	alba contorted	oWhi wRai oOEx
un	***alba*** 'Itoguwa'	oFor
	alba 'Kingan'	oGar
un	***alba*** 'Lavender'	wBur
un	***alba*** 'Oscar'	wBur wClo wRai oOEx
un	***alba*** 'Pakistan'	wBur oOEx
	alba 'Pendula'	oFor wBur oWhi wRai wTGN oSle
un	***alba*** 'Sugar Drop'	oOEx
un	***alba*** 'Sugarbaby'	oOEx
	alba var. ***tatarica***	wBCr
un	***alba*** 'Tehama'	oOEx
	australis	oFor
	australis 'Unryu'	oFor oGar
	cathayana	oFor oWhi
	'Illinois Everbearing'	oFor wBCr wBur oWhi wRai oOEx oSle
	nigra (nigra)	oFor wRai oOEx
un	***nigra*** 'Cox'	oWhi
un	***nigra*** 'Noir of Spain'	oWhi
un	***nigra*** 'Persian Fruiting'	wBur
	nigra 'Wellington'	oFor wClo oWhi wRai oOEx oSle
	rubra	oFor
un	'Shangri-la'	oOEx
	sp. contorted	oGre wTGN
	MUCUNA	
	pruriens	oCrm
	MUEHLENBECKIA	
	axillaris	oAls wPir
	axillaris 'Nana'	see **M. *axillaris*** Walpers
	axillaris Walpers	wHer oSho
	complexa	oFor
	complexa 'Nana'	see **M. *axillaris*** Walpers
	complexa var. ***trilobata***	wHer
	hastulata	wCCr
	MUHLENBERGIA	
	capillaris	last listed 99/00
	dumosa	wHer
	rigens	oFor oSec
	MUKDENIA	
	rossii	wCol oNWe wRob
	rossii DJH Sorak-san	wHer
	MUSA	
	acuminata 'Enano Gigante'	wCoS
	acuminata 'Zebrina'	oGar
	basjoo	oFor wHer oDan oCis oSho wRai oOEx wDav oCir
	lasiocarpa	oDan oCis oSho wRai wDav
un	'Super Dwarf'	oGar
	MUSCARI	
	armeniacum	wRoo oWoo
ch	***armeniacum*** 'Babies Breath'	see **M. *neglectum*** 'Baby's Breath'
do	***armeniacum*** 'Blue Spike'	wRoo oGar oWoo
	armeniacum 'Fantasy Creation'	last listed 99/00
	botryoides 'Album'	oWoo
	comosum 'Plumosum'	oDan oWoo wTGN
	latifolium	oGar
ch	***neglectum*** 'Baby's Breath'	oHed
	plumosum	see **M. *comosum*** 'Plumosum'
	MUSELLA	
ch	***lasiocarpa***	see **MUSA *lasiocarpa***
	MUTISIA	
	ilicifolia	oJoy
	spinosa	oSis
	MYOPORUM	
	parvifolium	oFor
	parvifolium 'Pink'	last listed 99/00
	parvifolium 'Putah Creek'	last listed 99/00
	MYOSOTIDIUM	
	hortensia	wHer oNWe oGre
	MYOSOTIS	
	alpestris	oUps
	'Gold 'n Sapphires'	wHer wWoS
un	***olympica***	wMtT
	palustris	see **M. *scorpioides***
	scorpioides	wGAc oHug oGar oSsd oUps
un	***scorpioides*** 'Bill Baker'	wHer
	scorpioides 'Mermaid'	wGAc
	scorpioides 'Pinkie'	oHed oGar
	scorpioides 'Snowflakes'	oHug
	scorpioides white	wGAc
	sylvatica alba	see **M. *sylvatica*** f. ***lactea***
	sylvatica 'Blue Ball'	oWnC
	sylvatica f. ***lactea***	oWnC

	sylvatica 'Rosea'	oWnC
	'Victoria Blue' (Victoria Series)	oAls wWhG
	'Victoria Rose' (Victoria Series)	wWhG
	white	wWhG
	MYRCEUGENIA	
un	***nannophylla*** HCM 98165	wHer
	MYRICA	
	californica	oFor wWoB wShR oGar oDan oBos oSho oOut oBlo oGre wWat wCCr oTri wCoN oSle wSte wWel
	cerifera	oFor oAls oGar wCCr
	gale	oFor wShR wWat
	heterophylla	oFor
un	***nagai***	oOEx
	pensylvanica	oFor oEdg oRiv oBlo oGre wFai iGSc oWnC wWel
un	***pumila***	oRiv
un	***pusilla***	oFor
	rubra	oSho
	MYRIOPHYLLUM	
	aquaticum	oHug oGar oSsd oWnC
	heterophyllum	wGAc
	MYRRHIS	
	odorata	wCri oAls wFGN oGoo wFai iGSc wMag oBar oUps
	MYRSINE	
	africana	oTrP
	semiserrata DJHC 450	wHer
	MYRTILLOCACTUS	
	sp.	oOEx
	MYRTUS	
	communis	oAls wFGN oRiv iGSc wSte
	communis 'Microphylla'	see **M. *communis*** ssp. ***tarentina***
	communis ssp. ***tarentina***	oAls oHed wFGN oGoo oBar
	communis ssp. ***tarentina*** 'Compacta'	oFor wNTP
	communis ssp. ***tarentina*** 'Microphylla Variegata'	oAls wFGN oGoo oBar
	communis 'Variegata'	oFor oHed oGoo iGSc oBar
un	**NABLONIUM**	
	calycoroides	wHer
	NAIOCRENE	
	parvifolia	oJoy oRus oBos oBov wFFl
	NANDINA	
	domestica	oGar oRiv oSho wPir wTGN wSta wCoN wWel
un	***domestica*** var. ***capillus***	oSis
un	***domestica*** var. ***capillus*** 'Senbazura'	oSis
un	***domestica*** var. ***capillus*** 'Tama Shishi'	oSis
	domestica 'Compacta'	oDar oGar oBlo wTGN wSta wCoS
qu	***domestica*** 'Compacta Nana'	oWnC
un	***domestica*** 'Filigree'	wSta
un	***domestica*** 'Filimentosa'	wTGN
	domestica 'Fire Power'	oFor wHer oDar oAls oGar oSho oGre oWnC
un	***domestica*** 'Gulf Stream'	oAls oGar oWnC wWel
	domestica 'Harbor Dwarf'	oFor oAls oGar oSho oGre oTPm oWnC nwWel
un	***domestica*** 'Lemon Hill'	oBlo
	domestica var. ***leucocarpa***	oSis wBox
un	***domestica*** 'Moon Bay'	oAls oGar oWnC wWel
	domestica 'Moyers Red'	oFor oDar oAls oGar oSho oBlo oGre oWnC wSta wCoS
	domestica 'Nana Purpurea'	oGar wSta
un	***domestica*** 'Ori-hime'	oGar
	domestica Plum Passion™ / 'Monum'	oAls oGar oWnC wWel
un	***domestica*** 'Red Select'	wSta
un	***domestica*** 'Royal Princess'	oFor oGre
un	***domestica*** 'San Gabriel'	oFor
un	***domestica*** 'Umpqua Chief'	oDar oRiv oGre oTPm
un	***domestica*** 'Umpqua Princess'	oFor oDar oGre
	domestica 'Umpqua Warrior'	oDar oRiv oTPm
	domestica 'Wood's Dwarf'	oGar oWnC wSta
	NARCISSUS	
	'Acapulco'	oOre
	'Acceleration'	oMit
	'Accord'	oOre
	'Ace'	oMit
	'Actaea'	oWoo
	'Acumen'	oMit
	'Afafura'	oMit
	'Affirmation'	oMit
	'Afterthot'	oMit
	'Akepa'	oMit
	'Alaskan Forest'	oMit
	'Albacore'	oOre

	'All American'	oMit
	'Allez'	oMit
	'Alumna'	oOre
	'Always'	oOre
	'Amadeus'	oMit
	'American Classic'	oMit
	'American Dream'	oMit
	'American Frontier'	oMit
	'American Goldfinch'	oMit
	'American Heritage'	oMit
	'American Shores'	oMit
	'American Songbird'	last listed 99/00
	'Amor'	wRoo oWoo
un	'Amy Linea'	oMit
	'Anatolia'	oMit
	'Ancestor'	oMit
	'Androcles'	oOre
	'Angel'	oBon
	'Angel Silk'	oMit
	'Angelic Choir'	oMit
	'Angkor'	last listed 99/00
	'Anvil Chorus'	oMit
	'Aplomb'	oOre
	'Apostle'	oOre
un	'Appalachian Star'	oOre
	'Apricot Sensation'	oWoo
un	'April Peach'	oMit
	'April Tears'	oBon oMit
	'Apropos'	oOre
	'Aquarius'	oOre
	'Arapaho'	oOre
	'Arawannah'	last listed 99/00
	'Arctic Char'	oOre
	'Array'	oOre
	'Arrowhead'	oMit
	'Artful'	oOre
un	'Ashland'	oOre
	'Ashmore'	oMit
	'Astrodome'	last listed 99/00
	'Astropink'	oMit
	'Autumn Gold'	oMit
	'Baby Moon'	oGar oWoo
	'Baby Star'	oBon
	'Bagatelle'	oBon
	'Bald Eagle'	oOre
	'Banker'	oMit
	'Bantam'	oBon oMit
	'Barbet'	oMit
	'Barbie Doll'	oOre
	'Barrett Browning'	oGar oWoo
	'Beau Monde'	oGar
	'Beautiful Dream'	oMit
	'Bebop'	oMit
	'Bella Coola'	oOre
	'Berceuse'	oMit
	'Big Gun'	last listed 99/00
	'Big John'	oOre
	'Big Sur'	oOre
	'Bionic'	oMit
	'Birthday Girl'	oBon
	'Bittern'	oMit
	'Blitz'	oMit
	'Bloemendaal'	oOre
	'Blue Mountains'	oMit
	'Blue Star'	oMit
	'Bobbysoxer'	oBon
	'Bon Bon'	oMit
	'Bon Voyage'	oOre
	'Bozely'	oMit
	'Bravoure'	oGar
	'Brer Fox'	oBon
	'Bridal Chorus'	oMit
	'Bridal Crown'	oWoo
	'Bright Candle'	oMit
	'Brindabella'	oMit
	'Brookdale'	oOre
	'Buchan'	oMit
	'Buckskin'	oOre
	bulbocodium ssp. ***bulbocodium*** var. ***conspicuus***	oWoo
	'Cabochon'	oOre
	'Calcite'	oOre
	canaliculatus	see **N. *tazetta*** ssp. ***lacticolor***
	'Candy Cane'	oOre
	'Canterbury'	oMit

	'Capistrano'	oOre
	'Carib'	oMit
	'Carlton'	oGar
	'Castanets'	oOre
	'Catalyst'	oMit
	'Cataract'	oOre
	'Cathedral Hill'	oOre
	'Cazique'	oMit
	'Cedar Hills'	oMit
	'Cedarbird'	oMit
	'Celilo'	last listed 99/00
	'Central Park'	oOre
	'Centre Ville'	oMit
	'Century'	oOre
	'Chapeau'	oOre
	'Charade'	last listed 99/00
	'Chatmoss'	oOre
	'Cheddar'	last listed 99/00
	'Cheerfulness'	wRoo oWoo
	'Chelan'	oOre
	'Chemeketa'	oOre
	'Cherry Bounce'	oMit
	'Chianti'	oOre
	'Chilito'	oOre
	'China Lake'	last listed 99/00
	'Chippewa'	oOre
	'Chiquita'	oOre
	'Chloe'	oOre
	'Chorale'	oOre
	'Chorine'	last listed 99/00
	'Chorus Line'	oOre
	'Christmas Valley'	oMit
	'Chromacolor'	oOre
	'Chukar'	oMit
	'Chutzpah'	oMit
	'Citron'	oMit
	'City Club'	oOre
	'Clare'	oBon
	'Class Act'	oMit
	'Class Ring'	oMit
	'Classic Delight'	last listed 99/00
	'Clavier'	oMit
	'Clearwater'	oMit
	'Close Encounter'	oOre
	'Clubman'	oMit
	'Cockatiel'	oMit
	'Coho'	oOre
	'Colonial Treasure'	oMit
	'Colonial White'	oMit
	'Colonnade'	oOre
	'Color Magic'	oMit
	'Concertina'	oMit
	'Conestoga'	oMit
	'Contravene'	last listed 99/00
	'Cool Evening'	oMit
	'Cool Pink'	oMit
	'Cool White'	oMit
	'Coral Crown'	oMit
	'Coral Springs'	oMit
	'Cordial'	oBon
	'Cornell'	oMit
	'Cotinga'	oMit
	'Cotton Candy'	oOre
	'Country Garden'	oMit
	'Cragford'	oWoo
	'Creation'	oMit
	'Crown Gold'	oMit
	'Crown Point'	oOre
	'Crystal Blanc'	oOre
	'Crystal Clear'	oOre
	'Crystal Star'	oMit
	'Culmination'	oMit
	cyclamineus RSBG	last listed 99/00
	'Dainty Miss'	oMit
	'Daiquiri'	oOre
	'Dawn Blush'	oMit
	'Dawn Light'	oOre
	'Dawncrest'	oMit
	'December Bride'	oMit
	'Decoy'	oMit
	'Deference'	oMit
	'Del Rey'	last listed 99/00
	'Deleena'	last listed 99/00
	'Delibes'	oGar oWoo
	'Delightful'	oMit
	'Delta Queen'	oOre
	'Demmo'	oMit
	'Denali'	oMit
	'Descanso'	oOre
	'Desert Bells'	oMit
	'Dewy Rose'	oOre
	'Di Hard'	oMit
	'Diablo'	oOre
	'Diamond Head'	oOre
	'Disquiet'	last listed 99/00
	'Distant Drums'	oMit
	'Dividend'	oOre
	'Doak's Stand'	oMit
	'Domingo'	oMit
	'Double Cream'	oOre
	'Dressy Bessie'	oOre
	'Drongo'	oMit
	'Drummer Boy'	last listed 99/00
	'Duke of Windsor'	oWoo
	'Durango'	oOre
	'Dutch Master'	wRoo oWoo
	'Early Arrival'	last listed 99/00
	'Easter Bonnet'	oGar
un	'Easter Surprise'	oMit
	'Edna Earl'	oWoo
	'Eggshell'	oOre
	'Elegant Lady'	oMit
	'Elixir'	oMit
	'Ellusive'	oMit
	'Elrond'	oBon
	'Emerald Empire'	oMit
	'Emerald Light'	oMit
	'Emerald Pink'	oMit
	'Emperor's Waltz'	oMit
	'Engagement Ring'	oMit
	'Entente'	oMit
	'Epona'	oMit
	'Equation'	oMit
	'Erlicheer'	oBon
	'Estuary'	oOre
	'Euphonic Grace'	oMit
	'Everpink'	oOre
	'Exalted'	oOre
	'Executive Pink'	oMit
	'Extrovert'	oMit
	'Fairy Chimes'	oMit
	'Falconet'	oMit
	'Falstaff'	oBon
	'February Gold'	wRoo
	'Ferral'	oMit
	'Fertile Crescent'	oMit
	'Fertile Plains'	oMit
	'Fidelity'	last listed 99/00
	'Finite'	oMit
	'Fiona Jean'	oMit
	'Fire Alarm'	oOre
	'Firestar'	oMit
	'First Formal'	oMit
	'First Impression'	oMit
un	'Flagship'	oOre
	'Flight'	oMit
	'Flower Carpet'	wRoo
	'Flower Record'	oWoo
	'Flower Waltz'	last listed 99/00
	'Flying Nun'	oOre
	'Folio'	oOre
	'Forest Park'	oOre
	'Fortissimo'	oGar oWoo
	'Fortune'	oGar
un	'Fortune Bowl'	wRoo
	'Foundation'	oOre
	'Foxfire'	oOre
	'Fragrant Rose'	oBon
	'Free Spirit'	oOre
	'Freedom Rings'	oMit
	'French Prairie'	oMit
	'Full Fashion'	oOre
	'Galactic'	oMit
	'Gallery'	oOre
un	'Gasparilla'	oOre
	'Genteel'	oOre
	'Geranium'	wRoo oWoo
	'Ghost'	last listed 99/00
	'Ghost Dancer'	oMit
	'Gigolo'	last listed 99/00

	'Gilead'	oMit
	'Ginger'	oOre
	'Girasol'	oOre
	'Glacier'	oWoo
	'Glen Echo'	oOre
	'Glissando'	last listed 99/00
	'Gloucester Point'	oOre
	'Gold Beach'	oMit
	'Gold Chain'	oMit
	'Gold Coin'	oMit
	'Gold Sails'	oMit
	'Gold Velvet'	oMit
	'Golden Chord'	oOre
	'Golden Falcon'	oOre
	'Golden Pond'	oMit
	'Golden Quince'	oBon oGar
	'Good Life'	oOre
	'Gracious Lady'	oBon
	'Graduation'	oMit
	'Grand Opening'	oOre
	'Great Gatsby'	oOre
	'Great Northern'	oMit
	'Grebe'	last listed 99/00
	'Greek Column'	oMit
	'Greenbrier'	oOre
	'Gull'	oMit
	'Habit'	oMit
	'Hacienda'	oOre
	'Happy Hour'	oBon
	'Harvard'	oMit
	'Hassle'	oMit
	'Hawera'	wRoo oHed oWoo
	'Hawk Eye'	last listed 99/00
	'Heart Throb'	oOre
	'Heartland'	oMit
	'Heiress'	oOre
	'High Cotton'	oOre
	'High Tea'	oOre
	'Highfield Beauty'	oBon
	'Highlite'	oOre
	'Homecoming'	last listed 99/00
	'Homestead'	oOre oBon
	'Honey Warbler'	last listed 99/00
	'Honeymoon'	oOre
	'Hoopoe'	oMit
	'Hummingbird'	oBon
	'Huon Chief'	oMit
	'Ice Age'	last listed 99/00
	'Ice Diamond'	oMit
	'Ice Follies'	wRoo oWoo
	'Ice King'	oWoo
	'Ice Wings'	oGar
	'Icelandic Pink'	oMit
	'Idealism'	oMit
	'Impeccable'	oMit
	'Impetuous'	oMit
	'Imprint'	oOre
	'Independence Day'	oMit
	'Indian Maid'	oOre oBon
	'Integer'	oMit
	'Intrigue'	oOre oBon
	'Invercauld'	oOre
	'Iroquois'	oMit
	'Irresistible'	oMit
	'Irvington'	oOre
	'Ivy League'	oOre
	'Jack Snipe'	wRoo oWoo
	'Jade'	oMit
	'Jamboree'	oOre
	'Janis Babson'	oOre
	'Javelin'	oOre
	'Jenny'	oGar
	'Jet Pink'	oMit
	'Jet Set'	oOre
	'Jetfire'	oGar oWoo
	'Jingle Bells'	oOre
	'Johnnie Walker'	oMit
	'Jolly Roger'	oOre
	jonquilla	oMit
	'Jovial'	oBon
	'Jumblie'	oBon oGar
	'June Bride'	oMit
	'Junior Prom'	oMit
	'Junne Johnsrud'	oOre
un	'Karelia'	oGar

	'Keepsake'	oOre
	'Ken's Favorite'	oOre
	'Key Largo'	oOre
	'Keystone'	oOre
	'Kissproof'	oWoo
	'Kokopelli'	oBon
	'Koomooloo'	oMit
	'Kurrewa'	oMit
	'La Mancha'	oOre
	'La Traviata'	oMit
	'Lalique'	oMit
	'Lara'	oOre
	'Lark'	oMit
un	'Leesburg'	oOre
	'Lemon Brook'	oMit
	'Lemon Honey'	oMit
	'Lemon Lyric'	oMit
	'Lemon Sails'	oMit
	'Lemon Sparks'	oMit
	'Lemon Sprite'	oMit
	'Lemon Supreme'	oMit
	'Lemon Tarts'	oMit
	'Lexington Green'	oMit
	'Liebeslied'	oMit
	'Life'	oMit
	'Limberlost'	oOre
	'Lime Chiffon'	oMit
	'Limequilla'	oMit
	'Limey Circle'	oMit
	'Lingerie'	oOre
	'Lintie'	oBon
	'Lissome'	oOre
	'Little Gem'	oBon
	'Lizzie Hop'	last listed 99/00
	'Lollipop'	oOre
	'Lone Star'	oOre
	'Lonesome Dove'	oOre
	'Lorikeet'	oMit
	'Lostine'	oOre
	'Love Boat'	oOre
	'Lovejoy'	oBon
	'Lyles'	oBon
	'Lynchburg'	oOre
	'Lyrebird'	oMit
	'Machan'	oMit
	'Macushla'	last listed 99/00
	'Magellan'	oMit
	'Magic Lantern'	oMit
	'Magic Step'	oMit
	'Magna Vista'	oOre
	'Manet'	oOre
	'Manna'	oOre
	'Marabou'	oOre
	'Marimba'	last listed 99/00
	'Marque'	oMit
	'Marshfire'	oOre
	'Mary Baldwin'	oOre
	'Mary Kate'	oBon
	'Mary's Pink'	oBon
	'Masada'	oMit
	'Maverick'	last listed 99/00
	'Maya Dynasty'	oMit
	'Meadow Lake'	oMit
	'Meditation'	last listed 99/00
	'Memoir'	oOre
un	'Millie Galyon'	oMit
	'Minikin'	oOre
	'Minnow'	oBon wRoo oWoo
	'Minute Waltz'	oMit
	'Minx'	oOre
	'Mirrabooka'	last listed 99/00
	'Misquote'	last listed 99/00
	'Mission Bells'	oMit
	'Mission Impossible'	oMit
	'Mistique'	last listed 99/00
	'Misty Glen'	oBon
	'Misty Morning'	oMit
	'Mobjack Bay'	oBon
	'Molten Lava'	oMit
	'Moneymaker'	oWoo
	'Monitor'	oMit
	'Monticello'	oOre
	'Monument'	last listed 99/00
	'Moomba'	last listed 99/00
	'Moonflight'	oMit

'Motmot'	oMit
'Mount Hood'	wRoo oGar oWoo
'Mountain Blue Bird'	oMit
'Mountain Dew'	oOre oBon
'Mrs. R. O. Backhouse'	oWoo
'Multnomah'	oOre
'Music'	last listed 99/00
un 'Music Hall'	oGar
'Muster'	oMit
'Mysterious'	oMit
'Nacre'	oOre
'Natural Beauty'	oMit
'Neahkahnie'	oOre
'Nehalem'	oOre
'New Penny'	oOre oBon
'Newcomer'	oOre
'Newport'	oOre
'Night Hawk'	oMit
'Night Life'	oMit
'Nile'	oMit
'Nonchalant'	oMit
'Nordic Rim'	oMit
'North River'	oBon
'Northwest'	oOre
'Noteworthy'	oOre
'Nowra'	oMit
'Nynja'	last listed 99/00
'Oakland'	last listed 99/00
obvallaris	last listed 99/00
'Odist'	oOre
x odorus	oBon
'Odyssey'	oOre
'Officiando'	oMit
'Old Spice'	oOre
'Omega'	oOre
'On Edge'	oMit
'Oneonta'	oOre
'Oomph'	oMit
'Oops'	oMit
'Orangery'	oWoo
'Orchard Place'	oMit
'Oregon Beauty'	oMit
'Oregon Bells'	oMit
'Oregon Cedar'	oMit
'Oregon Gold'	last listed 99/00
'Oregon Green'	oMit
'Oregon Lights'	oMit
'Oregon Pioneer'	oMit
'Oregon Rose'	oMit
'Oregon Snow'	last listed 99/00
'Oryx'	oMit
'Otago'	oMit
'Our Tempie'	oOre
'Outlook'	oOre
'Ouzel'	oMit
'Owyhee'	oMit
'Oxford'	oMit
'Pacific Rim'	oMit
'Paean'	oMit
'Painted Doll'	oOre
'Panache'	oBon
'Pantomime'	oOre
papyraceus	see **N.** 'Ziva'
'Parfait'	oOre
'Park Lane'	last listed 99/00
'Parkrose'	oOre
'Party Doll'	oOre
'Passionale'	oWoo
'Pasteline'	oMit
'Pay Day'	oMit
'Peace Pipe'	last listed 99/00
'Peach Garter'	oOre
'Peach Prince'	oOre
'Peacock'	oOre
'Peeping Tom'	oGar
'Perfect Spring'	oMit
'Peripheral Pink'	oMit
'Perpetuation'	oMit
'Personable'	oOre
'Petit Four'	oWoo
'Petrel'	oWoo
'Phalarope'	oMit
'Phoenician'	oMit
'Piano Concerto'	oMit
'Piedmont'	oOre
'Pinaroo'	last listed 99/00
'Pineapple Prince'	oMit
'Pink Angel'	oMit
'Pink Bomb'	oMit
'Pink China'	oMit
'Pink Declaration'	oMit
'Pink Evening'	oMit
'Pink Fire'	oMit
'Pink Flare'	last listed 99/00
'Pink Formal'	oMit
'Pink Frost'	last listed 99/00
'Pink Garden'	oOre
'Pink Glacier'	oMit
'Pink Holly'	oMit
'Pink Hummer'	last listed 99/00
'Pink Ice'	oMit
'Pink Migration'	last listed 99/00
'Pink Perfume'	last listed 99/00
'Pink Sails'	oMit
'Pink Satin'	oMit
'Pink Silk'	oMit
'Pink Sparkler'	oMit
'Pink Swan'	oMit
'Pink Tango'	oMit
'Pink Tea'	oOre
'Pink Valley'	oMit
'Pipestone'	oOre
'Pipit'	oWoo
'Piquant'	last listed 99/00
'Pittsburgh Someplace Special'	oMit
'Pixie's Sister'	oBon
'Pizarro'	oMit
'Plaza'	oOre
'Plover'	last listed 99/00
poeticus var. ***recurvus***	wRoo oGar
'Pongee'	oOre
'Pops Legacy'	oBon oMit
'Porcelain'	oOre
'Portfolio'	oOre oBon
'Portrait'	oOre
'Potential'	oMit
'Presidential Pink'	oMit
'Princeton'	oMit
'Prism'	last listed 99/00
'Professor Einstein'	oWoo
'Profile'	last listed 99/00
'Propriety'	oOre
'Prosperity'	oMit
'Protégé'	oOre
'Protocol'	oMit
'Puma'	oOre
'Punchline'	oMit
'Punter'	oMit
'Pyrite'	oOre
'Quark'	last listed 99/00
'Quasar'	oOre
'Queen City'	oMit
'Rain Dance'	oOre
'Ransom'	oMit
'Rapport'	oOre
'Rapture'	oMit
'Raspberry Creme'	oMit
'Razadas'	oMit
'Red Aria'	oMit
'Red Diamond'	oMit
'Red Fox'	last listed 99/00
'Red Sheen'	oMit
'Red Treasure'	oBon
'Redhill'	wRoo oGar oWoo
'Refrain'	oMit
'Regeneration'	oMit
'Relentless'	oMit
'Revelation'	oOre
'Rhine Wine'	oOre
'Rijnveld's Early Sensation'	oGar
'Rim Ride'	oOre
'Rimski'	oBon
'Ringing Bells'	oMit
'Riot'	last listed 99/00
'Rip van Winkle'	oBon oWoo
'Rippling Waters'	last listed 99/00
'Rising Star'	oOre oBon
'River Queen'	oOre
'Romance'	oWoo
'Rose City'	oOre

	'Rosebank'	oBon
	'Rosegarden'	oMit
un	'Rosy Cloud'	oGar
	'Roundelay'	oOre
	'Royal Coachman'	last listed 99/00
	'Royal Trophy'	oOre
	'Ruby Rim'	oMit
	'Ruby Romance'	oMit
	'Ruby Star'	oMit
	'Russian Chimes'	oMit
	'Sabine Hay'	oBon
un	'Sabot Hill'	oOre
	'Sailboat'	oBon
	'Salome'	wRoo oWoo
	'Sanction'	last listed 99/00
	'Satin Lustre'	oBon
	'Satsuma'	last listed 99/00
	'Saucy'	oOre
	'Scarlet Chord'	oMit
	'Scarlet Rim'	oMit
	'Scarlet Tanager'	oMit
	'Scarlett O'Hara'	wRoo
	'Sea Legend'	oMit
	'Seafoam'	oMit
	'Segovia'	oBon
	'Senior Ball'	oMit
	'Serape'	oOre
	'Serene Sea'	oMit
un	'Shadow'	oOre
	'Sherbet'	oOre
	'Shikellamy'	oBon
un	'Shiloh'	oOre
	'Shortcake'	oOre
	'Showbiz'	oMit
	'Showboat'	oOre
	'Shrike'	oMit
	'Shriner'	oOre
	'Siberian Pink'	oMit
	'Sidley'	oBon
	'Silent Pink'	oMit
	'Silk Purse'	oOre
	'Silken Wings'	last listed 99/00
	'Silver Falls'	oMit
	'Silver Snow'	oOre
	'Silver Thaw'	oOre
	'Skater's Waltz'	oMit
	'Skookum'	oOre
	'Sky Ray'	oOre
	'Skyfire'	oMit
	'Smooth Sails'	oMit
	'Smyrna'	oBon
	'Snow Frills'	last listed 99/00
	'Snow Pink'	oOre
	'Socialite'	oOre
	'Sonar'	oMit
	'Sophie Girl'	oBon
	'Soubrette'	oOre
	'Southern Hospitality'	oMit
	'Southwick'	oOre
	'Space Age'	last listed 99/00
	'Sparrow'	last listed 99/00
	'Spartan'	last listed 99/00
	'Spindletop'	oOre
	'Spinning Fire'	oMit
un	'Spring Break'	oOre
	'Spring Chimes'	oMit
	'Spring Morn'	oMit
un	'Spring Sensation'	oMit
	'Spring Tonic'	oMit
	'Springdale'	oBon
	'Stafford'	oBon
	'Standard Value'	wRoo
	'Star Wish'	oMit
	'Starbrook'	oMit
	'Starfall'	oMit
	'Starlet'	oOre
	'Starmount'	oOre
	'Starthroat'	oMit
	'Stinger'	last listed 99/00
	'Stint'	last listed 99/00
	'Straight Arrow'	oMit
	'Strawberry Ice'	oOre
	'Strawberry Soda'	oMit
	'Stunning'	oOre
	'Suave'	oMit
	'Suede'	oOre
	'Sugar Loaf'	oOre
	'Sun Disc'	oBon
	'Sunapee'	oOre
	'Sunday Chimes'	oMit
	'Sundial'	oBon
	'Sunny Thoughts'	last listed 99/00
	'Sunnyside'	oOre
	'Supreme Empire'	oMit
	'Surewin'	oMit
	'Surtsey'	oOre
	'Swain'	oOre
	'Swamp Fox'	oOre
	'Swedish Fjord'	oMit
	'Swedish Sea'	oMit
	'Sweet Orange'	oMit
	'Swift Arrow'	oMit
	'Swift Current'	oMit
	'Taffy'	oMit
	'Tahiti'	oWoo
	'Tahoe'	oOre
	'Tangent'	oMit
	'Tanglewood'	oOre
	'Tao'	oMit
	tazetta ssp. ***lacticolor***	oBon oWoo
	'Teal'	oMit
	'Temple Star'	oMit
	'Terminator'	oMit
	'Tete-a-tete'	wRoo oWoo
	'Texas'	wRoo
	'Thalia'	wRoo
	'The Benson'	oMit
	'Three of Diamonds'	oBon
	'Ticonderoga'	oMit
	'Tillicum'	last listed 99/00
un	'Timbuktu'	oOre
un	'Too Late'	oOre
	'Toto'	oBon
	'Toucan'	last listed 99/00
	'Treasure Valley'	last listed 99/00
	'Treasure Waltz'	oMit
	'Trevithian'	wRoo oWoo
	'Trigonometry'	oMit
	'Tripartite'	oMit
	'Trona'	oOre
	'Tropic Isle'	oMit
	'Truculent'	oMit
	'Truism'	oOre
	'Trumpet Warrior'	oMit
	'Tuckahoe'	oOre
	'Twerp'	oMit
	'Tyee'	oOre
	'Tynan'	oBon
	'Tyson's Corner'	oOre
	'Unique'	oWoo
	'Unity'	oOre
	'Unsurpassable'	last listed 99/00
	'Upshot'	oOre
	'Urbane'	oOre
	'Valley Forge'	oOre
	'Vantage'	oOre
	'Vapor Trail'	oOre
	'Velocity'	oMit
	'Velvet Spring'	oMit
	'Verdant Meadow'	oMit
	'Vermilion'	oOre
	'Vibrant'	oOre
	'Vice President'	last listed 99/00
	'Vienna Woods'	oMit
	'Viennese Waltz'	oMit
	'Virginia Walker'	oOre
	'Volare'	oOre
	'Wahkeena'	oOre
	'Wakefield'	oOre
	'Walden Pond'	oMit
	'Wampam'	oMit
	'Wannabe'	oMit
	'Warbler'	oMit
	'Wasco'	oOre
	'Watercolor'	oMit
	'Waterperry'	oGar
	'Well Worth'	oOre
	'Wendover'	oOre oBon
	'Whetstone'	oMit
	'Whetstone Tribute'	oMit

	'Whirlaway'	oMit
	'Whispering Winds'	oMit
	'White Hunter'	oOre
	'White Lion'	wRoo
	'White o' Morn'	oOre
	'White Satin'	oOre
	'White Tie'	oMit
	'Wilderness'	oMit
	'Williamsburg'	oOre
	'Wind Song'	last listed 99/00
	'Windsor Court'	oOre
	'Winged Flight'	oBon
	'Wings of Freedom'	oMit
	'Winter Evening'	oMit
	'Winter Waltz'	oMit
	'Wizard'	oOre
	'Wizbang'	oMit
	'Woods Pink'	oOre
	'Woodstar'	oMit
	'Woodthrush'	oMit
	'Woolaroo'	oMit
	'World Peace'	oMit
	'Yale'	oMit
	'Yazz'	oOre oBon
	'Yellow Cheerfulness'	wRoo
	'Yellowstone'	oOre
	'Yellowtail'	oOre
	'Young American'	oMit
	'Young Love'	oMit
	'Ziva'	oGar
	'Zombie'	last listed 99/00
	'Zulu'	oMit
	'Zumdish'	oMit
	NARDOSTACHYS	
	jatamansi	oOEx
un	**NASHIA**	
	inaguensis	iGSc
	NASSELLA	
un	***tenuissima***	wWin
	NASTURTIUM	
	officinale	wGAc oAls oHug oGar iGSc oWnC
	NECTAROSCORDUM	
	siculum	wHer wCul oNat wCol oRus
	siculum ssp. ***bulgaricum***	wIri
	NEILLIA	
	affinis	oFor wHer oGar oGre
	rubiflora HWJCM 017	last listed 99/00
	sinensis	oFor
	sp. EDHCH 97297	wHer
	thibetica	wCCr
	NELUMBO	
	lutea	oWnC
	nucifera 'Alba Grandiflora'	oWnC
	nucifera 'Alba Striata'	wGAc oWnC
	nucifera 'Charles Thomas'	oWnC
	nucifera 'Chawan Basu'	wGAc oWnC
un	***nucifera*** 'Hindu'	wGAc
	nucifera 'Momo Botan'	wGAc
	nucifera 'Mrs. Perry D. Slocum'	wGAc
	nucifera 'Pekinensis Rubra'	wGAc
	white	oSsd
	NEMATANTHUS	
	glaber	oGar
	glaber 'Tropicana'	oGar
	NEMESIA	
	Bluebird / 'Hubbird'	wTGN
	caerulea	oHed
un	'Compact Innocence'	wTGN
	fruticans (see Nomenclature Notes)	
un	***fruticans*** 'White Cloud'	oHed
	NEMOPHILA	
	menziesii 'Pennie Black'	oCir
	NEOALSOMITRA	
	sarcophylla	oRar
	NEODYPSIS	
	decaryi	see **DYPSIS** ***decaryi***
	NEOLITSEA	
	sericea	oCis oSho wSte
	sericea HC 970229	wHer
	NEOLLOYDIA	
	pseudopectinata	oRar
un	***pseudopectinata*** var. ***rubriflora***	oRar
	valdeziana	oRar
un	***valdeziana*** var. ***albiflora***	oRar
	NEOMARICA	
	caerulea	oTrP
	gracilis	oTrP
	NEPENTHES	
	bicalcarata	oGar
	sanguinea green	oGar
un	***thorelli***	oGar
	NEPETA	
un	'Blue Infinity'	oSha
	cataria	wTho oJoy oAls wFGN wCSG oGoo iGSc wHom oWnC oBar wNTP oCir oUps oSle
	cataria 'Citriodora'	oAls wFGN oGoo wFai iGSc wMag oBar wNTP wEde
	clarkei	oJoy oGoo
	x faassenii	wTho oNat oJoy wCSG oGar oDan wFai iGSc oSec wBWP wCoN oLSG
	x faassenii 'Alba'	wSnq
	x faassenii 'Blue Wonder'	wWoS wCul oJoy oGar oUps
	x faassenii 'Dropmore'	oFor wHer wWoS wCul wSwC oAls wFGN wHom wMag oLSG oUps wSnq
	govaniana	wHer
	grandiflora	wFGN oGoo
	grandiflora 'Bramdean'	wHer wWoS
	grandiflora 'Dawn to Dusk'	oFor wHer wWoS oJoy oAls oLSG wSnq
	grandiflora 'Pool Bank'	wHer wWoS oJoy oAls
un	'Mrs. Edgehill'	wWoS
ch	***mussinii***	see **N.** ***x faassenii***
	nervosa	oGoo
	phyllochlamys	last listed 99/00
	pink	wFGN
	'Porcelain'	see **N.** 'Porzellan'
	'Porzellan'	wHer wWoS
	racemosa	wCul oJoy
	racemosa 'Blue Ice'	wHer wWoS
	racemosa 'Little Titch'	wHer
un	***racemosa*** 'Planbessin'	wEde
	racemosa 'Snowflake'	oFor wCul wFGN oSis oLSG
	racemosa 'Superba'	wHer wWoS
	racemosa 'Walker's Low'	oFor wHer wWoS oNat oDar oJoy oAls wFGN oSis oBar oUps oSle wSnq
	reichenbachiana	see **N.** ***racemosa***
	sibirica	oFor wRob wFai wBWP oLSG
	sibirica 'Souvenir d'Andre Chaudron'	oFor wHer wWoS wSwC oNat oJoy
	'Six Hills Giant'	oFor wWoS wCri wSwC oNat oJoy wHig oAls oHed wFGN oSis wRob oGre wHom oWnC wMag oBar wTGN wBox oSec oInd oUps oSle wSnq
	sp. DJHC 148	wHer
	stewartiana DJHC 98025	wHer
	subsessilis	oFor wHer oSec wEde
	transcaucasica	oSha
	tuberosa	wHer oAls
	ucranica	wHer oSec
un	***yunnanensis***	oAmb oSec
	NEPHROLEPIS	
	exaltata	oWnC
	exaltata 'Bostoniensis'	oGar
un	***exaltata*** 'Dallasii'	oGar
un	***exaltata*** 'Rooseveltii Plumosa'	oGar
	obliterata 'Kimberly Queen'	oGar
	NEPTUNIA	
un	***aquatica***	oGar
	NERINE	
	crispa	see **N.** ***undulata***
	filifolia	last listed 99/00
	laticoma AHB 004	oRar
	undulata	last listed 99/00
	NERIUM	
	oleander 'Carneum Plenum'	oGar
	oleander deep pink	oCir
	oleander 'Hardy Pink'	oCis
	oleander 'Hardy Red'	oCis
un	***oleander*** 'Hardy White'	oCis
	oleander 'Mrs. Roeding'	see **N.** ***oleander*** 'Carneum Plenum'
	oleander 'Petite Pink'	oGar
	oleander 'Soeur Agnes'	oGar
	NEVIUSIA	
	alabamensis	oFor
	NICANDRA	
	physalodes	wSte
	NICOTIANA	
	alata	wSte
	alata Nicki Series	oUps
	glauca	wCCr
	knightiana	oSec

	langsdorffii	wBox wSte
va	***langsdorffii*** 'Cream Splash'	oGar oSec
un	***x sanderae*** 'Havana Appleblossom' (Havana Series)	oUps
	sylvestris	oGar oEga wSte
un	***sylvestris*** 'Only the Lonely'	oAls
un	***tabacum*** 'Burley'	oAls
	NIEREMBERGIA	
un	***caerulea*** 'Jessie'	oHed
	repens	oAls
	NIGELLA	
	damascena Persian Jewels Series	oUps
	NIPPONANTHEMUM	
	nipponicum	oFor wCSG wCCr oUps
	NOLINA	
	gracilis	oRar
	guatemalensis	oRar
	microcarpa	wCCr
	recurvata	oGar oRar
	stricta	oRar
	NOMOCHARIS	
	aperta	wCol
	sp. DJHC 409	wHer
un	**NOTHOCHELONE**	
	nemerosa	wShR
	NOTHOFAGUS	
	x alpina	wHer
	antarctica	oFor oAmb oRiv
	cunninghamii	oFor
	dombeyi	oFor oRiv
	fusca	last listed 99/00
	glauca	wCCr
	obliqua	oFor wHer oGar
ch	***procera*** (see Nomenclature Notes)	oDan wCCr
	NOTHOLIRION	
	bulbuliferum	wHer
	sp. aff. ***campanulatum*** DJHC 367	wHer
	NOTHOPANAX	see **POLYSCIAS**
un	**NOTHOPHOEBE**	
	cavalieri	oFor wHer oCis
	NYMPHAEA	
	'Albert Greenberg'	wGAc
	'Albida'	see **N.** 'Marliacea Albida'
un	'Almost Black'	wGAc
	'Arc en ciel'	wGAc oSsd oWnC
	'Atropurpurea'	wGAc
	'Attraction'	wGAc oSsd oWnC oTri
	'August Koch'	wGAc oSsd
	'Berit Strawn'	wGAc
	'Black Princess'	wGAc
	capensis	oWnC
	'Carolina Sunset'	wGAc
	'Charlene Strawn'	wGAc oSsd
	'Chromatella'	wGAc oGar oSsd oWnC wBox
	'Chrysantha'	oSsd
	'Colorado'	wGAc
	'Comanche'	wGAc oSsd
	'Emily Grant Hutchings'	wGAc
	'Evelyn Stetston'	oGar
	'Firecrest'	wGAc oTri wBox
	'Florida Sunset'	wGAc
	'Froebelii'	wGAc oSsd
	'Gloire du Temple-sur-Lot'	wGAc wBox
	'Gloriosa'	oWnC
	'Gonnere'	wBox
	'Green Smoke'	wGAc
	x helvola	see **N.** 'Pygmaea Helvola'
	'Hermine'	oSsd
un	'Hilary'	oSsd
	'Hollandia' (see Nomenclature Notes)	oGar
	hybrids	oGar oSsd
	'Indiana'	wGAc oWnC
	'James Brydon'	wGAc
	'Joey Tomocick'	wGAc
un	'Laydeckeri Alba'	wGAc
	'Laydeckeri Fulgens'	wGAc oSsd
	'Liou'	wGAc
un	'Little Champion'	wGAc
	'Louise'	wBox
	'Marliacea Albida'	oSsd oWnC
	'Masaniello'	oWnC wBox
	'Mayla'	wGAc oGar
un	'Midnight'	wGAc
un	'Missouri'	oSsd
	'Mrs. C. W. Thomas'	wBox
un	'Nolene'	wGAc
	odorata	oWnC oTri
	odorata alba	see **N. *odorata***
	'Odorata Sulphurea Grandiflora'	wGAc
	'Panama Pacific'	oSsd
	'Paul Hariot'	wGAc
	'Peaches and Cream'	wGAc
	'Perry's Baby Red'	wGAc oSsd
	'Perry's Black Opal'	wGAc
	'Perry's Crinkled Pink'	wGAc
	'Perry's Red Glow'	oGar
	'Perry's Wildfire'	wGAc
	'Peter Slocum'	wGAc
	'Pink Peony'	wGAc
	'Pink Sensation'	wGAc oSsd
	'Pink Sunrise'	wGAc
	'Pygmaea Helvola'	wSoo wGAc oSsd wBox
un	'Queen of Whites'	oSsd
	'Radiant Red'	wBox
	'Red Beauty'	wGAc
	'Red Flare'	wGAc
	'Rembrandt' (see Nomenclature Notes)	wGAc oGar oSsd
	'Rose Arey'	wGAc
	'Sioux'	wGAc oGar
	'Sirius'	wGAc
	sp.	wSoo
qu	'Sulphuria'	oWnC
	'Sunrise'	see **N.** 'Odorata Sulphurea Grandiflora'
un	***tetragona*** 'White Dwarf'	wBox
	'Texas Dawn'	wGAc oWnC oTri wBox
un	'Tina'	oSsd
	'Vesuve'	wGAc
	'Virginalis'	wGAc oGar
	'Virginia'	oWnC
	'Walter Pagels'	oSsd
	'William Falconer'	wGAc
	'Wow'	wGAc
	'Yul Ling'	wGAc
	NYMPHOIDES	
	crenata	oHug
	cristata	wGAc oSsd
un	***geminata***	wGAc oHug oSsd
	indica	last listed 99/00
	peltata	wSoo oGar oSsd oTri
	NYSSA	
	aquatica	oRiv oGre
	biflora	see **N. *sylvatica*** var. biflora
	ogeche	last listed 99/00
	sinensis	oFor oDan oRiv oCis oGre oWnC wWel
	sylvatica	oFor oEdg oAls wShR oAmb oGar oDan oRiv oGre wCCr wKin wAva wCoN oSle wSte wSnq wCoS wWel
	sylvatica var. ***biflora***	last listed 99/00
un	**OBETIA**	
	ficifolia	oRar
	OBREGONIA	
	denegrii	oRar
	OCIMUM	
	'African Blue'	oAls wFGN wFai iGSc oWnC oBar oUps
	americanum	last listed 99/00
un	'Aussie Sweetie'	oAls wFGN
	basilicum	wWoS wTho oAls wFGN wFai oWnC oCir
	basilicum 'Anise'	see **O. *basilicum*** 'Horapha'
	basilicum 'Cinnamon'	oAls wFai iGSc oWnC oUps
	basilicum 'Crispum'	oAls
un	***basilicum*** 'Cuban'	oAls
	basilicum 'Dark Opal'	oAls wFai iGSc oWnC
un	***basilicum*** 'Dwarf Bouquet'	oAls
un	***basilicum*** 'Fino Verde'	oUps
	basilicum 'Genovese'	oWnC oUps
un	***basilicum*** 'Green Goddess'	iGSc
	basilicum 'Horapha'	oUps
un	***basilicum*** Italian	iGSc
un	***basilicum*** Lettuce leaf	iGSc oUps
un	***basilicum*** Lime	iGSc oUps
un	***basilicum*** 'Mammoth'	oUps
un	***basilicum*** 'Mammoth Sweet'	oWnC
	basilicum 'Minette'	oAls
	basilicum 'Minimum'	see **O. *minimum***
un	***basilicum*** 'Mrs. Burns'	oAls
	basilicum 'Napolitano'	iGSc
	basilicum 'Purple Ruffles'	oAls
	basilicum 'Red Rubin'	wFGN iGSc oUps
	basilicum 'Rubin'	last listed 99/00
un	***basilicum*** 'Sweet Dani'	oAls oWnC oUps

un	***basilicum*** 'Sweet Italian'	oWnC
	basilicum 'Spicy Globe'	wFGN oWnC
	basilicum 'Thai'	see **O. *basilicum*** 'Horapha'
un	***basilicum*** 'Thai Lemon'	iGSc
	x citriodorum	oTDM wFai iGSc
	x citriodorum 'Siam Queen'	oAls iGSc oWnC oUps
	East Indian tree	oUps
	gratissimum	iGSc
un	'Green Bouquet'	oAls
	kilimandscharicum	oCrm
	minimum	oAls iGSc
	sanctum	see **O. *tenuiflorum***
	tenuiflorum	oAls iGSc oUps
	OEMLERIA	
	cerasiformis	oFor wWoB wNot wShR oGar oBos oRiv wWat oTri oAld wAva wFFl wRav oSle
	OENANTHE	
	javanica	wSoo wGAc oSsd oWnC wBox
	javanica 'Flamingo'	oFor wSoo wGAc oGoo oSsd wRob oSec
	sarmentosa	wWat
	OENOTHERA	
	'African Sun'	oAls
	berlandieri	see **O. *speciosa*** 'Rosea'
	biennis	oGoo oCrm oUps
	caespitosa	oBos wPla oAld
	fruticosa	oTrP wShR oGar oiSha oGre wCCr
	fruticosa Fireworks	see **O. *fruticosa*** 'Fyrverkeri'
	fruticosa 'Fyrverkeri'	oAls oGar wTGN oSle
	fruticosa ssp. ***glauca***	oGre wFai oWnC wSta oUps
va	***fruticosa*** ssp. ***glauca*** 'Erica Robin'	wHer wWoS oHed
	fruticosa ssp. ***glauca*** 'Sonnenwende'	oNWe
	fruticosa 'Youngii'	oAls oEga
	glaziouana	oUps
	kunthiana	oSis
	lamarckiana	see **O. *glaziouana***
	macrocarpa	oFor oAls oRus oGar oGre wFai oUps
un	***minima***	oHan
	missouriensis	see **O. *macrocarpa***
	odorata (see Nomenclature Notes)	wCul
	odorata 'Sulphurea'	see **O. *stricta*** 'Sulphurea'
	pilosella	wCSG
	speciosa	oFor oAls oSha oGre wFai wCCr
	speciosa 'Pink Petticoats'	wHom
	speciosa 'Rosea'	last listed 99/00
	speciosa 'Siskiyou'	oFor oAls oGar oDan oSis oWnC wMag wTGN wBWP oCir oEga oUps
	stricta 'Moonlight'	wMag
	stricta 'Sulphurea'	oJil
	'Sunburst'	last listed 99/00
	tetragona	see **O. *fruticosa*** ssp. ***glauca***
	versicolor 'Sunset Boulevard'	last listed 99/00
	'Woodside White'	oFor oSis
	OLEA	
	europaea Little Ollie™ / 'Montra'	oFor
un	***yunnanensis***	wHer
	OLEARIA	
	argophylla	oFor
	avicenniifolia	last listed 99/00
	x haastii	oHed
	ilicifolia	wHer oHed oCis
	macrodonta	oCis oGre wCCr
	x mollis (see Nomenclature Notes)	wHer
un	***monroei***	oCis
	nummulariifolia	wHer oCis
	phlogopappa	wCCr
	x scilloniensis (see Nomenclature Notes)	oHed
	x scilloniensis 'Master Michael'	oHed oCis
	solandri	oFor oCis
	OLSYNIUM	
	douglasii	wCol oRus oSis oBos oAld wWld
	filifolium	oTrP
	OMPHALODES	
	cappadocica	oFor wCri wCol oAls wCSG oSis oNWe oGre oHon
	cappadocica 'Anthea Bloom'	wHer
	cappadocica 'Cherry Ingram'	wHer
	cappadocica 'Lilac Mist'	wHer wWoS oDan oSis wRob wNay oGre wMag
	cappadocica 'Parisian Skies'	oFor wWoS oGar
	cappadocica 'Starry Eyes'	wHer wWoS wCol oHed oNWe oOut wRob wBox
	nitida	wWoS
	verna	wHer oNat wCol oAls wRob oHon oJil
	verna 'Alba'	wHer oHed wRob
	verna grandiflora	last listed 99/00
	ONCIDIUM	
	hybrids	oGar
	ONOCLEA	
	sensibilis	oFor oRus oNWe oTri wSnq
	ONONIS	
	spinosa	oFor oGoo
	ONOPORDUM	
	acanthium	wTGN
	ONOSMA	
	alborosea	oSis
	helvetica	oSis
	taurica	oSis
un	**OPERCULICARYA**	
	decaryi	oRar
	OPHIOPOGON	
	clarkei	wCol oSec
	intermedius 'Argenteomarginatus'	oHed
	jaburan	oAls
	japonicus	oTrP oGar oSho oOEx wWin wTGN wBox oSec oBRG
un	***japonicus*** 'Aritaki'	oFor
un	***japonicus*** 'Gyoku-ru'	wWin wTGN oBRG
	japonicus 'Kigimafukiduma'	wWin wTGN oBRG
	japonicus 'Kigimafukiduma' bonsai	oGre
	japonicus 'Kyoto Dwarf'	oFor oTrP
	japonicus 'Nanus'	oAls wCSG oGar oSqu wWin wPir oWnC wTGN wBox oSec wSta oBRG wWhG
	japonicus 'Silver Dragon'	oTrP
	planiscapus	wCCr
	planiscapus 'Arabicus'	see **O. *planiscapus*** 'Nigrescens'
	planiscapus 'Ebony Night'	see **O. *planiscapus*** 'Nigrescens'
	planiscapus 'Nigrescens'	oFor oTrP wHer oGos oDar oJoy wCol oAls oHed oGar oNWe oSho wFai wDav wPir wTGN wBox oSec wSta oCir wSte wWel
	OPLOPANAX	
	horridus	oBos wWat oAld wFFl
	OPUNTIA	
	compressa	oOEx
	ficus-indica	oOEx
un	***ficus-indica*** 'Honeydew'	oOEx
un	***ficus-indica*** 'Papaya'	oOEx
un	***ficus-indica*** ssp. ***supra noplaes***	oOEx
	ficus-indica white	oOEx
	fragilis	oSto
	humifusa	oGoo
	imbricata	oSto
	macrorhiza	oOEx
	polyacantha	oSto
	sp.	iGSc
	vulgaris	oSto
	vulgaris 'Variegata'	oSqu
	ORBEA	
	variegata	oRar
	ORBEOPSIS	
	melanantha	oRar
	ORCHIS	
	mascula	oRed
	ORIGANUM	
	'Barbara Tingey'	oAls wFGN oGoo
	calcaratum	oAls oGoo
	dictamnus	oAls wFGN oGoo oSqu iGSc wTGN
qu	***dictamnus*** x ***pulchellum***	oSis
	'Erntedank'	wFGN oGoo oBar oSec
	green flowering	wFGN
	heracleoticum (see Nomenclature Notes)	wNTP
un	'Hot and Spicy'	oTDM oAls
	x hybridinum	wFGN oUps
	'Ingolstadt'	last listed 99/00
un	'Jim's Best'	oAls wFGN oUps
un	'Kaliteri'	oAls iGSc oUps
	'Kent Beauty'	oFor oTrP oDar oAls wFGN oAmb oGoo oSis oNWe oWnC oBar oUps
	'Kent Beauty' seed grown	oNWe
un	'Khirgzstan'	wFGN
	laevigatum	wCul wFGN oGoo oUps
	laevigatum 'Herrenhausen'	oFor wCul oTDM oJoy wHig oAls wFGN oGar oGoo wFai oBar oCir oUps wSnq
	laevigatum 'Hopleys'	oFor oJoy wHig oAls wFGN oGoo wHom oUps
	laevigatum 'Hopley's Purple'	see **O. *laevigatum*** 'Hopleys'
	libanoticum	oAls oGoo
	majorana	oAls wFai iGSc oWnC oBar wNTP oCir
	x majoricum	wFGN oGoo wHom wNTP oUps
	microphyllum	oAls oGoo
	'Norton Gold'	oFor wHer wCul wFGN oAmb oGoo oSec

	'Nymphenburg'	wHer
	onites	iGSc
	onites 'Aureum'	oAls wFai oCir
	pulchellum	see **O. *x hybridinum***
un	'Richter's Finest'	wFGN
	'Rosenkuppel'	oFor wHer wWoS wCul oAls oHed wFGN oDan oBar oUps wSnq
	'Rotkugel'	wHer wWoS wHig wFGN oBar
	rotundifolium	wCul wFGN oGoo oBar
un	'Santa Cruz'	wHig oAls wFGN oGoo wFai
un	'Showy Dittany'	oGoo
qu	Sicilian	oUps
	sipyleum	last listed 99/00
un	***x suendermanii***	oSis
un	***tytanicum***	oAls
un	'UCSC'	wWoS oJoy
	vulgare	wCul wTho oTDM oAls wShR wCSG oGoo iGSc oWnC wMag wCoS
	vulgare var. ***album***	oAls
	vulgare 'Aureum'	oTDM oJoy oAls wFGN wHom oWnC wMag oSec oUps oSle
	vulgare 'Aureum Crispum'	wCul oTDM oJoy wFGN oGoo wFai oBar oUps
	vulgare 'Aureum Crispum' dwarf	oAls
	vulgare 'Compactum'	oTDM oJoy oAls wFGN oGar oGoo wFai wHom oBar oCir
	vulgare crinkled leaf	wFGN
	vulgare 'Gold Tip'	wCul oJoy wFGN wHom oBar wEde oUps
	vulgare ssp. ***hirtum***	oTDM oAls wFGN oGoo wFai iGSc wHom oWnC oBar oCir oUps
	vulgare ssp. ***hirtum*** dwarf	wFGN oUps
un	***vulgare*** 'Humile'	oAls oSis iGSc oBar
	vulgare 'Nanum'	wFGN oSle
	vulgare 'Thumble's Variety'	oAls
	vulgare variegated green and white	iGSc
	vulgare 'Variegatum'	see **O. *vulgare*** 'Gold Tip'
	vulgare ssp. ***viride***	oGoo
un	***vulgare*** 'White Anniversary'	oFor oTDM oDar oSis oUps
un	'Yellow Flicker'	oJoy
	ORIXA	
	japonica	oFor
	ORNITHOGALUM	
un	***nutans*** 'Silver Bells'	oDan
	orthophyllum ssp. ***kochii***	last listed 99/00
	pyrenaicum	oRar
	umbellatum	wThG oWoo
	ORONTIUM	
	aquaticum	wGAc oHug oSsd
	OROSTACHYS	
	aggregata	oSto
	furusei from Warenge	oSto
	malacophylla	see **O. *aggregata***
un	***minuta***	oSto
	OROXYLUM	
	indicum	oOEx
	ORTHILIA	
	secunda	wFFl
	ORTHROSANTHUS	
	laxus	oTrP
	multiflorus	oTrP
	OSMANTHUS	
	armatus	oCis wCCr
	x burkwoodii	oFor wHer wCul oHed oGar oDan oRiv oSho oBlo oGre wPir wCCr wSta wAva
	decorus	oFor oGre
	decorus 'Baki Kasapligil'	wHer
	delavayi	oFor wCul oDar oAls oGar oRiv oCis oSho oBlo oGre wPir wCCr oWnC wSta wAva wSte wWel
	delavayi var. ***latifolius***	oCis
un	***delavayi*** var. ***latifolius*** 'Nanjing's Beauty'	oGar oCis oWnC wBox
	x fortunei	oFor wHer wCCr
	x fortunei 'San Jose'	wCCr
	fragrans	oFor oGar wRai wPir
	fragrans f. ***aurantiacus***	wHer oCis wBox
un	***fragrans*** var. ***thunbergii***	oCis oGre oWnC wBox
	heterophyllus	oFor oGar oSho oBlo oGre wCCr oBRG wAva wCoS
va	***heterophyllus*** 'Goshiki'	oFor wHer oGos wClo oGar oSho wAva wWel
	heterophyllus 'Gulftide'	oFor oSho
	heterophyllus 'Kembu'	wHer
	heterophyllus 'Purpureus'	oFor wHer oCis wPir wAva
	heterophyllus 'Rotundifolius'	oGar oCis oBlo oGre
	heterophyllus 'Sasaba'	wHer
	heterophyllus 'Variegatus'	oFor wHer oGar oCis oBlo oGre wPir
	ilicifolius	see **O. *heterophyllus***
un	'Nenjing's Beauty'	oDan
	suavis HWJCM 129	wHer
un	'Thunbergii'	oDan
	yunnanensis	oCis
	X OSMAREA	
	burkwoodii	see **OSMANTHUS *x burkwoodii***
	OSMARONIA	see **OEMLERIA**
	OSMUNDA	
	cinnamomea	oFor oGar wNay oTri wTGN oBRG wSnq
	lancea	last listed 99/00
	regalis	oFor oGos wThG oRus oGar wNay wHom oTri wTGN oBRG wCoN oUps wSnq
	regalis 'Cristata'	wCul oRus
	regalis Cristata Group	see **O. *regalis*** 'Cristata'
	regalis 'Purpurascens'	wCul wFol wFan oRus
un	***regalis*** var. ***regalis***	wFan
	regalis 'Undulata'	wFol
	OSTEOMELES	
	schweriniae	oGar
	subrotunda	wCCr
	OSTEOSPERMUM	
	barberae	see **O. *jucundum***
un	'Brightside'	oGar
	'Cannington John'	wHer
	'Cannington Katrina'	last listed 99/00
	'Cannington Roy'	wHer
	'Cannington Vernon'	wHer
	deep purple	wHom
	fruticosum burgundy	oCir
un	'Highside'	oGar
	jucundum	oSis
	jucundum var. ***compactum***	oFor
un	***jucundum*** var. ***compactum*** 'Purple Mountain'	oSis
un	***juliana***	wCul
un	'Lavender Mist'	oSis
un	'Lusaka'	oCir
	'Mira'	oGar wTGN
un	'Namaqua'	wHom
un	'Nasinga Purple'	oCir
un	'Nasinga White'	oCir
	purple	oGar
un	'Seaside'	oGar
un	'Giant'	wHer
un	'Sonja'	wHom
	'Stardust'	oAmb
un	'Sunscape'	oGar
un	'Sunscape Volta'	oGar
	white	wHom
	white, variegated	wHom
un	'Wildside'	oGar
	OSTRYA	
	carpinifolia	oFor wCCr
	japonica	oFor oRiv
	japonica DJH 458	wHer
	knowltonii	oFor
	virginiana	oFor
	OTATEA	
	acuminata	oTra wBea oTBG
un	***acuminata acuminata*** seedlings	oTra
	acuminata ssp. ***aztecorum***	see **O. *acuminata***
	OTHONNA	
	cheirifolia	wHer
un	***ficicaulis***	oRar
	quercifolia	oRar
	OURISIA	
	coccinea	oBov
	'Loch Ewe'	wHer
	macrophylla	last listed 99/00
un	***poeppigii***	wMtT
	OXALIS	
	acetosella	wCSG
	acetosella var. ***subpurpurascens***	wCSG
	adenophylla	oOut
	brasiliensis	oAls oSis
un	***brasiliensis*** 'Alba'	oFor
	carnosa	see **O. *megalorrhiza***
	crassipes 'Alba'	oRus
	deppei	see **O. *tetraphylla***
	herrerae	oSqu
	hirta	oSis
	'Ione Hecker'	oNWe
un	***lasiandra chortii***	oSqu

	magellanica	oTrP oRus oNWe wPir
do	***magellanica*** 'Nelson'	wHer wWoS wCol wPir
	megalorrhiza	oTrP oRar
un	sp. aff. ***obliquifolia***	oNWe
	obtusa	oSis
	oregana	oFor oHan wCul wWoB wNot oJoy oRus oBos oSqu oGre wWat oWnC oTri wBox wSta oAld oWhS wRav oSle
	oregana evergreen form	wRob
	oregana pink	oRus oJil
un	***oregana*** 'Select Pink'	oNWe
	oregana selected form	oRus
	oregana f. ***smalliana*** red form	oAls
un	***oregana*** 'Smith River White'	oFor
	oregana white	oRus
	purpurea	oTrP
	purpurea 'Ken Aslet'	oSis
un	***pusillum***	oSqu
	regnellii	see **O. *triangularis*** ssp. ***papilionacea***
	siliquosa	see **O. *vulcanicola***
	succulenta	oTrP
	tetraphylla	wCSG oGar
	tetraphylla 'Iron Cross'	wCSG oNWe oSqu
	triangularis	oAmb
	triangularis ssp. ***papilionacea***	oTrP oAls
	triangularis ssp. ***papilionacea*** 'Atropurpurea'	oNWe oSqu
	tuberosa	oFor
	tuberosa blush	oOEx
	tuberosa red	oOEx
	tuberosa white	oOEx
	tuberosa yellow	oOEx
	valdiviensis	oEdg
	vulcanicola	wWoS
un	***vulcanicola*** 'Copper'	oSqu
un	'Wintergreen'	wCol wRob
	OXYCOCCUS	see **VACCINIUM**
	OXYDENDRUM	
	arboreum	wSwC oDar oAls oAmb oGar oRed oDan oRiv oSho oGre wSta wAva wCoN wSte wCoS wWel
	arboreum 'Chameleon'	oFor oGos wClo oGar oGre
	OXYLOBIUM	
	lanceolatum	last listed 99/00
	OXYPETALUM	
	caeruleum	see **TWEEDIA *caerulea***
	OXYTROPIS	
	chankaensis	last listed 99/00
un	***deflexa***	wMtT
	megalantha	wMtT
	pilosa	oSis
	shokanbetsuensis	oNWe wMtT
	OZOTHAMNUS	
	coralloides wild/UCSC	oCis
	'County Park Silver'	last listed 99/00
	hookeri	last listed 99/00
	ledifolius	last listed 99/00
	rosmarinifolius	oHed oCis
	selago	last listed 99/00
	PACHIRA	
	aquatica	oTrP oRar
un	**PACHYCORMUS**	
	discolor	oRar
	PACHYCYMBIUM	
	dummeri	oRar
	PACHYPHRAGMA	
	macrophyllum	wHer oJil
	PACHYPHYTUM	
	oviferum	oSqu
	PACHYPODIUM	
un	***baronii*** var. ***baronii***	oRar
	baronii var. ***windsori***	oRar
	brevicaule	oRar
	densiflorum	oRar
	densiflorum var. ***brevicalyx***	oRar
un	***griquense***	oRar
	lamerei	oRar
	rosulatum	oRar
	rosulatum var. ***gracilis***	oRar
	rosulatum var. ***horombense***	oRar
	rutenbergianum	oRar
	saundersii	oRar
	succulentum	oRar
	PACHYSANDRA	
	axillaris DJHC 782	wHer
	procumbens	wHer wCol wHig oRus
	stylosa	last listed 99/00
	terminalis	oAls oGar oSho wSta oCir wCoN oGue wCoS
	terminalis 'Green Carpet'	oFor oWnC
	terminalis 'Green Sheen'	wHer oGar
	terminalis 'Silver Edge'	wTGN oUps
	terminalis 'Variegata'	oAls oGar oSho oWnC
	PACHYSTIMA	see **PAXISTIMA**
	X PACHYVERIA	
	clavata	oSqu
	clavifolia	see **P. *clavata***
un	'Elaine Reinelt'	oSqu
un	***fittkaui***	oSqu
	glauca	last listed 99/00
un	'Ivory'	oSqu
	PACKERA	
	aurea	oFor
	PAEONIA	
	'Age of Gold' (T)	oBHP
	'Alhambra' (T)	oBHP
	'America'	last listed 99/00
un	'Anna Marie'	oBHP
	anomala	wHer
	'Athena' (H)	oPac
	'Audrey' (H)	oPac
	'Banquet' (T)	oBHP
	'Birthday' (H)	last listed 99/00
	'Black Panther' (T)	oBHP
	'Blaze' (H)	oPac
	'Blushing Princess' (H)	oCap
	'Boreas' (T)	oBHP
	'Border Charm'	oCap
	broteroi	last listed 99/00
	brownii	wHer
	'Buckeye Belle'	oGar
	californica NNS 98-423	wHer
un	'Callies Memory'	oCap
	cambessedesii	last listed 99/00
	'Campagna' (H)	oCap
un	'Canary Brilliants'	oCap
	'Carol'	oAde
	'Chinese Dragon'	oBHP
	'Claire de Lune'	last listed 99/00
	'Claudia'	last listed 99/00
	'Constance Spry' (H)	oCap
	'Cora Louise'	oCap
	'Coral Charm' (H)	oCap oPac oAde
	coriacea	last listed 99/00
	'Crusader'	oAde
	'Cytherea'	oAde
	daurica	see **P. *mascula*** ssp. ***triternata***
	'Dawn Glow' (H)	oCap
	delavayi	oFor wHer
	ex ***delavayi***	wCol
	delavayi hybrids	oJil
	delavayi var. ***ludlowii***	oFor wHer oBHP oGar oDan oGre wWel
	delavayi var. ***lutea***	wCCr
	delavayi Trollioides Group	last listed 99/00
	'Diana Parks' (H)	oPac
	'Early Windflower' (H)	oPac
	edulis	see **P. *lactiflora***
	'Eliza Lundy' (H)	oCap
	'Ellen Cowley'	oCap
	'Etched Salmon' (H)	last listed 99/00
	'Eventide' (H)	oCap
un	'Fedora' seed grown	wHer
	'First Arrival'	oCap
	'Flame'	oGre oPac oAde
	'Garden Treasure'	oAde
	'Gaugin' (T)	oBHP
	'Gold Sovereign' (T)	oBHP
	Grandma Kruse's	oMis
un	'Harkaway's Remembrance'	oGre
	'Helen Matthews' (H)	last listed 99/00
	'High Noon'	oBHP
	'Honor'	oCap
	'Horizon'	oAde
	'Huang Hua Kui' (T)	oBHP
	hybrid	oAls
	'Jin Yu Jiao Zhang' (T)	oBHP oGre
	'Joyce Ellen' (H)	wCSG
	'Julia Rose'	oCap
un	***kartilinika***	wHer
	kevachensis	see **P. *mascula*** ssp. ***mascula***

	lactiflora (H)	last listed 99/00
un	***lactiflora*** 'Ada Niva'	oPac
	lactiflora 'Adolphe Rousseau'	oAls iFid
un	***lactiflora*** 'Adrienne'	oPac
	lactiflora 'Albert Crousse'	iFid
un	***lactiflora*** 'Albert Niva'	oPac
	lactiflora 'Alecia Kunkel'	last listed 99/00
	lactiflora 'Alexander Fleming'	oFor oAls oGar wTGN oAde
	lactiflora 'Angel Cheeks'	oAde
	lactiflora 'Ann Cousins'	oAde
	lactiflora 'Asa Gray'	oPac
	lactiflora 'Augustin d'Hour'	oWnC
	lactiflora 'Avalanche'	oPac
	lactiflora 'Avis Varner'	oCap oPac
	lactiflora 'Baroness Schroeder'	last listed 99/00
	lactiflora 'Benjamin Franklin'	oPac
un	***lactiflora*** 'Betty Niva'	oPac
un	***lactiflora*** 'Biardi Hong'	oGre
	lactiflora 'Big Ben'	oAde
	lactiflora 'Bouchela'	oGar oWnC
	lactiflora 'Bouquet Perfect'	oCap
	lactiflora 'Bowl of Beauty'	oAls wCSG oAde
	lactiflora 'Bowl of Cream'	oGar oAde
un	***lactiflora*** 'Brand's Magnificent'	iFid
	lactiflora 'Camellia'	oCap
	lactiflora 'Cascade Gem'	last listed 99/00
	lactiflora 'Catherine Fontijn'	oGre
	lactiflora 'Charles McKellup	oPac
	lactiflora 'Charles' White'	oAde
	lactiflora 'Chestine Gowdy'	oPac
	lactiflora 'Chinook'	oCap
un	***lactiflora*** 'Claire'	oPac
	lactiflora 'Claire Dubois'	oPac
	lactiflora 'Couronne d'Or'	oAde
	lactiflora 'Cream Puff'	oCap
	lactiflora 'Dawn Pink'	oAde
	lactiflora 'Doctor Alexander Fleming'	see **P. *lactiflora*** 'Alexander Fleming'
	lactiflora 'Doctor F. G. Brethour'	iFid
	lactiflora 'Doreen'	oGar
	lactiflora double pink	wCSG oWnC wCoS
	lactiflora double red	oWnC wCoS
	lactiflora double white	wCSG
	lactiflora 'Douglas Brand'	oCap
	lactiflora 'Dresden Pink'	oAde
	lactiflora 'Duchesse de Nemours'	oCap oGar oBlo oGre oWnC oPac
	lactiflora 'Easy Lavender'	oCap
	lactiflora 'Eclatante'	see **P. *lactiflora*** 'L'Eclatante'
	lactiflora 'Edulis Superba'	oAls oGar oDan oPac oAde iFid
	lactiflora 'Elsa Sass'	oCap oPac oAde
un	***lactiflora*** 'Ethel White'	oPac
	lactiflora 'Fanny Crosby'	oPac
	lactiflora 'Feather Top'	oAde
	lactiflora 'Felix Crousse'	oFor oCap oAls oGar oSev oAde iFid
	lactiflora 'Felix Supreme'	oAde
un	***lactiflora*** 'Fen Yunu'	oGre
	lactiflora 'Festiva Maxima'	oFor oCap wCSG oGar oSev oGre oPac oAde oLSG
	lactiflora 'Frances Willard'	oPac
	lactiflora 'Francois Ortegat'	iFid
un	***lactiflora*** 'Francois Rousseau'	oPac
	lactiflora 'Gardenia'	oCap oAde
	lactiflora 'Gay Paree'	oCap oPac
	lactiflora 'General MacMahon'	see **P. *lactiflora*** 'Augustin d'Hour'
un	***lactiflora*** 'Geraldine'	oPac
	lactiflora 'Gold Mine'	oPac
	lactiflora 'Grover Cleveland'	oAde
	lactiflora 'Heidi'	oCap oPac
	lactiflora 'Henri Potin'	iFid
	lactiflora 'Imperial Divinity'	last listed 99/00
	lactiflora 'Imperial Parasol'	oCap
	lactiflora 'Inspecteur Lavergne'	oGar oGre
	lactiflora 'Irwin Altman'	last listed 99/00
un	***lactiflora*** 'Jacorma'	oGar
	lactiflora 'James R. Mann'	iFid
un	***lactiflora*** 'Jinzanciyu'	oGre
un	***lactiflora*** 'Joker'	oCap
un	***lactiflora*** 'Judge Berry'	oPac
	lactiflora 'Kansas'	oFor oCap oGar oAde
	lactiflora 'Karl Rosenfield'	oCap oAls oGar oWnC oPac wTLP oAde iFid oLSG
	lactiflora 'Krinkled White'	oGre oAde
un	***lactiflora*** 'La Boles'	oPac
	lactiflora 'La Lorraine'	oPac iFid
	lactiflora 'La Perle'	oPac
	lactiflora 'Lady Alexandra Duff'	oAls oGar oGre oAde

	lactiflora 'Laura Dessert'	oAls oPac
un	***lactiflora*** 'Lauren'	oPac
	lactiflora 'L'Eclatante'	wCSG oGar
un	***lactiflora*** 'Lemon Queen'	oPac
un	***lactiflora*** 'Leslie Peck'	oPac
	lactiflora 'Lillian Wild'	oAde
	lactiflora 'Longfellow'	oPac
	lactiflora 'Lora Dexheimer'	oPac
	lactiflora 'Lottie Dawson Rea'	oAde
	lactiflora 'Louise Marx'	oCap
	lactiflora 'Madame De Verneville'	oPac oAde iFid
	lactiflora 'Madame Ducel'	iFid
	lactiflora 'Mandarin's Coat'	oPac
	lactiflora 'Marie Lemoine'	oAls
	lactiflora 'Martha Bulloch'	oPac iFid
	lactiflora 'Mary Brand'	oPac iFid
	lactiflora 'Mary E. Nicholls'	oCap
	lactiflora 'Mikado'	oCap oPac
	lactiflora 'Minuet'	iFid
	lactiflora 'Miss America'	last listed 99/00
	lactiflora 'Miss Eckhart'	oAls
	lactiflora 'Mister Ed'	oCap oAde
un	***lactiflora*** 'Monsieur Dupont'	oPac
	lactiflora 'Monsieur Jules Elie'	oFor oCap oAls oGar oSev oGre wTGN oPac wTLP oAde iFid
	lactiflora 'Monsieur Martin Cahuzac'	oGar oPac oAde iFid
un	***lactiflora*** 'Monsieur Pailette'	oPac
	lactiflora 'Moonstone'	oCap
	lactiflora 'Mother's Choice'	last listed 99/00
	lactiflora 'Mount Saint Helens'	oCap
	lactiflora 'Mrs. Frank Beach'	oPac oAde iFid
	lactiflora 'Mrs. Franklin D. Roosevelt'	oCap
un	***lactiflora*** 'Mrs. Henry Bockstoce'	oAde
	lactiflora 'Myrtle Gentry'	oPac
	lactiflora 'Nick Shaylor'	oAde
	lactiflora 'Nippon Beauty'	oAde
	lactiflora 'Okinawa'	iFid
	lactiflora 'Paul M. Wild'	oAde
	lactiflora 'Peter Brand'	oGar
	lactiflora 'Petite Renee'	oCap
	lactiflora 'Philomele'	iFid
	lactiflora 'Pillow Talk'	oGar
	lactiflora 'Pink Parfait'	oCap oAde
	lactiflora 'Pink Pearl'	oGre
un	***lactiflora*** 'Polly Sharp'	oCap
un	***lactiflora*** 'Prairie Charm'	oAde
	lactiflora 'President Franklin D. Roosevelt'	oFor oAls oGar iFid oLSG
	lactiflora 'Primevere'	wCSG oDan
un	***lactiflora*** 'Princess Emerges'	oGre
	lactiflora 'Princess Margaret'	oCap
un	***lactiflora*** 'Qiao Ling'	oGre
un	***lactiflora*** 'Raspberry Ripple'	wTGN
	lactiflora 'Raspberry Sundae'	oCap oAls oGar oWnC oAde
un	***lactiflora*** 'Red Magic'	oGar
	lactiflora 'Reine Hortense'	oPac iFid
	lactiflora 'Renato'	oGar
	lactiflora 'Rivida'	oPac
un	***lactiflora*** 'Romance'	iFid
un	***lactiflora*** 'Rosette'	oPac
un	***lactiflora*** 'Roxy Peck'	oPac
un	***lactiflora*** 'Salmon Beauty'	oAde
	lactiflora 'Salmon Glory'	oCap
	lactiflora 'Sarah Bernhardt'	oFor oCap oNat oAls oGar oBlo oWnC oPac oAde iFid oLSG
	lactiflora 'Sea Shell'	oPac oAde
un	***lactiflora*** 'Setting Sun'	oPac
un	***lactiflora*** 'Shanhe Hong'	oGre
	lactiflora 'Shawnee Chief'	iFid
	lactiflora 'Shirley Temple'	oFor oAls wCSG oGar wTLP oAde
	lactiflora 'Solange'	oPac
	lactiflora 'Souvenir de Louis Bigot'	iFid
	lactiflora 'Tish'	oCap
	lactiflora 'Torchsong'	oAde
	lactiflora 'Tourangelle'	oAde iFid
un	***lactiflora*** 'Vivid'	oLSG
	lactiflora 'Vivid Rose'	last listed 99/00
	lactiflora 'Walter Faxon'	iFid
	lactiflora 'Walter Marx'	last listed 99/00
	lactiflora 'Westerner'	oPac
	lactiflora 'White Sands'	oCap oPac
	lactiflora 'Wilford Johnson'	oCap
	lactiflora 'Winnifred Domme'	oPac
un	***lactiflora*** 'Zi Ru Rong'	oGre
	'Laddie' (H)	last listed 99/00
	'Late Windflower'	oPac

	'Laura Magnuson' (H)	oCap oPac
	x lemoinei 'Alice Harding'	oBHP
	'Lovely Rose' (H)	oPac
	lutea	see **P. *delavayi*** var. ***lutea***
	lutea var. ***ludlowii***	see **P. *delavayi*** var. ***ludlowii***
	macrophylla	last listed 99/00
	'Many Happy Returns' (H)	oCap
	'Mary Jo Legare' (H)	oCap
	mascula	last listed 99/00
	mascula ssp. ***mascula***	wHer
	mascula ssp. ***russoi***	last listed 99/00
	mascula ssp. ***russoi*** 'Picotee'	oCap
	mascula ssp. ***triternata***	wHer wCol
	mlokosewitschii	wHer
	'Morning Lilac'	oCap
	'Nymphe'	oGar
	obovata	last listed 99/00
	obovata var. ***alba***	last listed 99/00
	officinalis	last listed 99/00
	officinalis 'Alba Plena'	oGre
un	***officinalis*** 'Anemoniflora'	oDan
	officinalis ssp. ***banatica***	last listed 99/00
	officinalis 'Rubra Plena'	wCSG oGar oGre
	officinalis single red	oGar
	officinalis single white	oGar
	officinalis ssp. ***villosa***	last listed 99/00
	'Old Faithful' (H)	last listed 99/00
	'Paula Fay'	oCap oAde
	peregrina	oJil
	'Pink Hawaiian Coral' (H)	oCap
	potaninii var. ***trollioides***	see **P. *delavayi*** Trollioides Group
	'Promenade' (H)	oCap
	'Red Charm'	oCap oAde
	'Red Comet' (H)	oPac
	'Roselette'	oPac
	'Royal Rose' (H)	last listed 99/00
un	***ruprechtiana***	wHer
	'Salmon Glow' (H)	oAde
	'Sanctus' (H)	last listed 99/00
	'Show Girl' (H)	oCap
	'Skylark' (H)	oPac
	'Sparkling Windflower' (H)	oCap
	steveniana	wHer
	suffruticosa (T)	oMac oDan oWnC wWel
un	***suffruticosa*** Aubergine Blue	wBox
un	***suffruticosa*** 'Bai He Wo Xue'	oBHP
	suffruticosa 'Bai Yu'	oFor
un	***suffruticosa*** 'Bei Guo Geng Guang'	oBHP
un	***suffruticosa*** Best Purple Orchid	wBox
	suffruticosa Big Deep Purple	see **P. *suffruticosa*** 'Da Zong Zi'
	suffruticosa Black Dragon Holds a Splendid Flower	see **P. *suffruticosa*** 'Wu Long Peng Sheng'
un	***suffruticosa*** Blue Flower Queen	wBox
	suffruticosa Blue Garden Jade	see **P. *suffruticosa*** 'Lan Tian Yu'
	suffruticosa Blue Sapphire	see **P. *suffruticosa*** 'Lan Bao Shi'
	suffruticosa Bright and Clear Sky after Rain	see **P. *suffruticosa*** 'Yu Guo Tian Qing'
	suffruticosa Bright Mountain Flowers in Full Bloom	oBHP
	suffruticosa 'Cang Zhi Hong'	oBHP
	suffruticosa Champion Black Jade	see **P. *suffruticosa*** 'Guan Shi Mo Yo'
un	***suffruticosa*** 'Chu E Huang'	oBHP
un	***suffruticosa*** 'Chui Tou Lan'	oBHP
	suffruticosa Cinnabar Ramparts	see **P. *suffruticosa*** 'Zu Sha Lei'
un	***suffruticosa*** Clear Up	wBox
un	***suffruticosa*** Color Printing	wBox
un	***suffruticosa*** 'Cong Zhong Xiao'	oBHP
un	***suffruticosa*** 'Da Jin Fen'	oBHP
	suffruticosa 'Da Zong Zi'	oBHP
un	***suffruticosa*** 'Dan Jin Fen'	oGre
un	***suffruticosa*** Dark Red Sand Fort	wBox
un	***suffruticosa*** Dew on Flower	wBox
	suffruticosa 'Dou Lu'	oBHP
un	***suffruticosa*** 'Eggplant Purple'	oBHP
	suffruticosa 'Er Qiao'	oBHP
	suffruticosa Family Wang's Red	see **P. *suffuticosa*** 'Wang Hong'
	suffruticosa Fancy Butterfly Flowers	see **P. *suffruticosa*** 'Xiao Ye Hue Die'
	suffruticosa 'Fen He Pian Jiang'	oBHP
	suffruticosa 'Fen Zhong Guan'	oBHP oGre
	suffruticosa 'Feng Dan'	oBHP
	suffruticosa Fire that Makes the Pills of Immortality	see **P. *suffruticosa*** 'Huo Lian Jin Dan'
un	***suffruticosa*** 'Fragrant Green Ball'	oBHP wBox
	suffruticosa Fragrant Green Ball	see **P. *suffruticosa*** 'Lu Xiang Qiu'
	suffruticosa Fragrant Jade	see **P. *suffruticosa*** 'Xiang Yu'
	suffruticosa 'Ge Jin Zi'	oBHP oAls
un	***suffruticosa*** Girl's Gown	wBox
	suffruticosa Gold Dusted Pink	see **P. *suffruticosa*** 'Da Jin Fen'
	suffruticosa Golden Jade	see **P. *suffruticosa*** 'Jin Yu Jiao Zhang'
	suffruticosa Gosling Yellow	see **P. *suffruticosa*** 'Chu E Huang'
	suffruticosa Grand Duke Dressed in Purple and Blue	see **P. *suffruticosa*** 'Zi Lang Kui'
	suffruticosa Green Dragon in Dark Pond	see **P. *suffruticosa*** 'Qing Long Wo Mo Chi'
	suffruticosa Green Hill Covered with Snow	see **P. *suffruticosa*** 'Qing Shan Guan Xue'
un	***suffruticosa*** Green on Jade House	wBox
	suffruticosa 'Guan Shi Mo Yu'	oBHP
	suffruticosa 'Hei Hue Kui'	oBHP
un	***suffruticosa*** Heroes	wBox
	suffruticosa Heze Red	see **P. *suffruticosa*** 'Lu He Hong'
un	***suffruticosa*** Honeydew from Heaven	oBHP
un	***suffruticosa*** 'Hong Bao Shi'	oBHP
	suffruticosa 'Hou Lian Jin Dan'	oBHP
	suffruticosa 'Hu Hong'	oAls
un	***suffruticosa*** 'Hua Hu Die'	oBHP
	suffruticosa hybrids	oFor
un	***suffruticosa*** Ice Covers Jade	wBox
	suffruticosa Iridescent Butterfly	see **P. *suffruticosa*** 'Hua Hu Die'
	suffruticosa Jade Seal of State	see **P. *suffruticosa*** 'Yu Xi Ying Yue'
	suffruticosa Jade Tower with Green Top	see **P. *suffruticosa*** 'Yu Lou Dian Cui'
	suffruticosa Jade White Plate	see **P. *suffruticosa*** 'Yu Ban Bai'
	suffruticosa 'Jin Pao Hong'	oBHP
un	***suffruticosa*** 'Jin Xiu Qiu'	oBHP
	suffruticosa King of the Black Flowers	see **P. *suffruticosa*** 'Hei Hua Kui'
	suffruticosa 'Kishu Caprice'	oBHP
un	***suffruticosa*** 'Kun Shan Ye Guang'	oBHP
un	***suffruticosa*** 'Lan Bao Shi'	oBHP
	suffruticosa 'Lan Tian Yu'	oBHP
un	***suffruticosa*** Lantian Jade	oAls wBox
un	***suffruticosa*** Lotus Red	wBox
	suffruticosa 'Lu He Hong'	oBHP
un	***suffruticosa*** 'Lu Xiang Qiu'	oBHP
	suffruticosa Lu's Red Lotus	see **P. *suffruticosa*** 'Lu He Hong'
	suffruticosa Maiden Dressed Fair	see **P. *suffruticosa*** 'Shu Nu Zhuang'
	suffruticosa 'Man Tang Hong'	oBHP
un	***suffruticosa*** 'Mu Dan'	wTGN
	suffruticosa Nodding Blue	see **P. *suffruticosa*** 'Chui Tou Lan'
	suffruticosa North Night Wind	see **P. *suffruticosa*** 'Bei Guo Feng Guang'
un	***suffruticosa*** Number Eighteen	wBox
	suffruticosa Number One Scholar's Red	see **P. *suffruticosa*** 'Zhuang Yuan Hong'
un	***suffruticosa*** One Flower Smiles	wBox
	suffruticosa Opium Purple	see **P. *suffruticosa*** 'Ya Pian Zi'
	suffruticosa Pea Green	see **P. *suffruticosa*** 'Dou Lu'
	suffruticosa Peach Blossom Flake	see **P. *suffruticosa*** 'Tao Hua Fei Xue'
un	***suffruticosa*** Phoenix White	oAls wBox
	suffruticosa pink	oGar
	suffruticosa Pink Lotus on River	see **P. *suffruticosa*** 'Fen He Pian Jiang'
un	***suffruticosa*** Precious Jewel Necklace	oAls
un	***suffruticosa*** Precious Offering of the Black Snake	wBox
	suffruticosa purple	oGar
un	***suffruticosa*** 'Purple Beauty'	oAls
	suffruticosa Purple in a Square	see **P. *suffruticosa*** 'Gi Jin Zi'
	suffruticosa Purple Pavilion of Fairies	see **P. *suffruticosa*** 'Zi Yao Tai'
	suffruticosa Purple Queen	see **P. *suffruticosa*** 'Zi Kui'
	suffruticosa Purple Silk Ball	see **P. *suffruticosa*** 'Jin Xiu Qiu'
	suffruticosa 'Qing Long Wo Mo Chi'	oBHP
	suffruticosa 'Qing Shan Guan Xue'	oBHP
un	***suffruticosa*** 'Qing Xiang Bai'	oBHP oGre
	suffruticosa red	oGar
un	***suffruticosa*** Red Cloud Faces Sun	wBox
un	***suffruticosa*** Red Gem	wBox
	suffruticosa Red Hall	see **P. *suffruticosa*** 'Man Tang Hong'
	suffruticosa Red Royal Robe	see **P. *suffruticosa*** 'Jin Pai Hong'
	suffruticosa Red Sapphire	see **P. *suffruticosa*** 'Hong Bao Shi'
	suffruticosa red/white	oGar wTGN
un	***suffruticosa*** Rich Red Fireplace	oAls
	suffruticosa ssp. ***rockii***	last listed 99/00
un	***suffruticosa*** Rolled Leaf Red	wBox
	suffruticosa 'Rou Fu Rong'	oGre
	suffruticosa 'Ruan-zhi-lan'	oBHP
	suffruticosa 'San Bian Sai Yu'	oBHP
	suffruticosa 'Saohime'	oAls
un	***suffruticosa*** Scholar's Red	see **P. *suffruticosa*** 'Zhuang Yuan Hong'
	suffruticosa seed red	wBox
un	***suffruticosa*** 'Shan Hu Tai'	oFor
un	***suffruticosa*** 'Sheng Dan Lu'	oFor
un	***suffruticosa*** 'Shi Yuan Bai'	oBHP oGre
	suffruticosa 'Shou An Hong'	oBHP oGre
un	***suffruticosa*** 'Shu Nu Zhuang'	oBHP
	suffruticosa Shy Girl	see **P. *suffruticosa*** 'Cang Zhi Hong'

un	***suffruticosa*** Silver Pink Glistening Gold	oAls
un	***suffruticosa*** Silver Powder on Golden Fish	wBox
	suffruticosa Silver Red	see **P. *suffruticosa*** 'Yin Hong Qiao Dui'
	suffruticosa Smiling in the Thickets	see **P. *suffruticosa*** 'Cong Zhong Xiao'
	suffruticosa Snow Kissed Peach	see **P. *suffruticosa*** 'Tao Hua Fei Xue'
	suffruticosa Snowy Flower from Icy Mountain	see **P. *suffruticosa*** 'Xue Hua Sha'
	suffruticosa Snowy Lotus	see **P. *suffruticosa*** 'Xue Lian'
	suffruticosa Soft Stem Blue	see **P. *suffruticosa*** 'Ruan Zhi Lan'
	suffruticosa Stone Garden White	see **P. *suffruticosa*** 'Shi Yuan Bai'
	suffruticosa Subtle White Fragrance	see **P. *suffruticosa*** 'Qing Xiang Bai'
	suffruticosa Supreme Pink	see **P. *suffruticosa*** 'Fen Zhong Guan'
	suffruticosa Supreme Red	see **P. *suffruticosa*** 'Shou An Hong'
	suffruticosa Swan Sleeps in Snow	see **P. *suffruticosa*** 'Bai He Wo Xue'
un	***suffruticosa*** 'Tao Hua Fei Xue'	oBHP
	suffruticosa Tricolor White Jade	see **P. *suffruticosa*** 'San Bain Sai Yu'
	suffruticosa Twin Beauty	see **P. *suffruticosa*** 'Er Qiao'
	suffruticosa Twin Beauty in Purple	see **P. *suffruticosa*** 'Zi Er Qiao'
un	***suffruticosa*** Uniform	wBox
	suffruticosa 'Wang Hong'	oBHP
	suffruticosa white	oGar wTGN
	suffruticosa White and Red Phoenix	see **P. *suffruticosa*** 'Feng Dan'
un	***suffruticosa*** White Crane in Snow	wBox
	suffruticosa White Flower with Purple Blotches	see **P. *suffruticosa*** 'Zi Ban Bai'
un	***suffruticosa*** White Jade	wBox
	suffruticosa White Light that Shines at Night	see **P. *suffruticosa*** 'Kun Shan Ye Guang'
	suffruticosa 'Wu Long Feng Sheng'	oFor oBHP oAls
un	***suffruticosa*** 'Xiang Yu'	oBHP wBox
un	***suffruticosa*** 'Xiao Ye Hue Die'	oBHP
un	***suffruticosa*** 'Xue Hua Sha'	oBHP
un	***suffruticosa*** 'Xue Lian'	oBHP
un	***suffruticosa*** 'Ya Pian Zi'	oBHP
	suffruticosa Yao's Yellow	see **P.** 'Yao Huang'
	suffruticosa yellow	oGar oGre
	suffruticosa Yellow Sunflowe	see **P. *suffruticosa*** 'Huang Hua Kui'
	suffruticosa 'Yin Fen Jin Lin'	oFor oGre
	suffruticosa 'Yin Hong Qiao Dui'	oBHP
	suffruticosa 'Ying Luo Bao Zhu'	oFor oBHP oGre
	suffruticosa 'Yoshinogawa'	oAls
un	***suffruticosa*** 'Yu Ban Bai'	oBHP
un	***suffruticosa*** 'Yu Guo Tian Qing'	oBHP
	suffruticosa 'Yu Lu Dian Cui'	oBHP
	suffruticosa 'Yu Xi Ying Xue'	oBHP
	suffruticosa 'Zhao Fen'	oFor oBHP oAls
	suffruticosa Zhao's Family Pink	see **P. *suffruticosa*** 'Zhao Fen'
	suffruticosa 'Zhi Hong	oGre
	suffruticosa 'Zhuang Yuan Hong'	oBHP oAls
	suffruticosa 'Zi Ban Bai'	oBHP
	suffruticosa 'Zi Er Qiao'	oBHP
un	***suffruticosa*** 'Zi Kui'	oBHP
	suffruticosa 'Zi Lan Kui'	oFor oBHP
un	***suffruticosa*** 'Zi Yao Tai'	oBHP
un	***suffruticosa*** 'Zu Kui'	oGre
un	***suffruticosa*** 'Zu Sha Lei'	oBHP oAls
	'Sunlight' (H)	last listed 99/00
	'Sweet May' (H)	oCap
	tenuifolia	wHer
	tenuifolia ssp. ***biebersteiniana***	wHer
	tenuifolia ssp. ***lithophila***	wHer
	tenuifolia 'Plena'	oSis
un	***tomentosa***	wHer
un	'Tria'	oBHP
	triternata	see **P. *mascula*** ssp. ***triternata***
	veitchii 'Alba' seed grown	wHer
	veitchii var. ***woodwardii***	wHer
un	'Vesuvian'	oBHP
	villosa	see **P. *officinalis*** ssp. ***villosa***
un	'Wine Angel'	oPac
	wittmanniana	last listed 99/00
	'Yao Huang'	oBHP
	'Your Majesty' (H)	oPac
un	'Zephyrus'	oBHP
	PALIURUS	
	ramosissimus	oOEx
	spina-christi	oFor
	PANAX	
	quinquefolius	oRus oGoo iGSc
	sp. DJHC 804	wHer
	PANDOREA	
	jasminoides	oWhS
va	***jasminoides*** 'Charisma'	oTrP
	jasminoides 'Variegata'	see **P. *jasminoides*** 'Charisma'
	pandorana	wTGN
	pandorana 'Alba'	last listed 99/00
	PANICUM	
	virgatum	oJoy oAls oGar oUps
	virgatum 'Cloud Nine'	wWin
un	***virgatum*** 'Dallas Blues'	oFor oAls oGar oDan wWin wSte wSnq
	virgatum 'Haense Herms'	oFor wHer wCli wBox
	virgatum 'Heavy Metal'	oFor wHer wWoS wCli oEdg oGar oSho oOut wRob wWin wBox oSec oBRG wSnq
	virgatum 'Pathfinder'	last listed 99/00
	virgatum 'Prairie Sky'	wHer oAmb oGar oOut oGre wWin wBox oSec oBRG wSnq
	virgatum 'Rehbraun'	oFor wHer oGar wRob oTPm oTri wBox wSnq
	virgatum 'Rotstrahlbusch'	wHer wWoS wCli oJoy oAmb oGar oSis oOut oGre wWin wHom wTGN wSte wSnq
	virgatum 'Rubrum'	oWnC
	virgatum 'Shenandoah'	oFor oSis oOut wSnq
	virgatum 'Strictum'	oJoy oUps
	virgatum 'Warrior'	oOut oSec
	PAPAVER	
	alpinum	wTho
un	***araitoensis***	oNWe
	atlanticum	wCul oNat wBWP
	burseri	last listed 99/00
	fauriei	oNWe
	lateritium	oDan
	miyabeanum 'Pacino'	wBWP
	nudicaule	oLSG
	nudicaule Champagne Bubbles Group	oAls wMag
un	***nudicaule*** 'Meadow Pastels'	oAls
un	***nudicaule*** 'Party Fun'	oAls
	nudicaule Wonderland hybrids	oWnC
	orientale	wCSG oSec iArc
	orientale 'Aglaja'	last listed 99/00
	orientale 'Allegro'	oDar oGar wHom oWnC oCir oLSG
un	***orientale*** 'Ballerina'	oNWe
	orientale 'Beauty of Livermere'	see **P. *orientale*** Goliath Group 'Beauty of Livermere'
	orientale 'Big Jim'	last listed 99/00
	orientale 'Black and White'	oGar
	orientale 'Bonfire Red'	wMag
	orientale 'Brilliant'	wTho oDar oAls oWnC oEga oLSG
	orientale 'Carneum'	oAls oWnC wMag
	orientale 'China Boy'	oAls
un	***orientale*** 'Degas'	oNat
un	***orientale*** 'Double Coral'	wHer
	orientale 'Doubloon'	oGre
	orientale 'Effendi'	oNWe
	orientale Goliath Group	wHer oAls wMag
	orientale Goliath Group 'Beauty of Livermere'	wTho oDar oGar wMag wBWP oEga oLSG
	orientale 'Graue Witwe'	wHer
	orientale 'Helen Elisabeth'	wWoS oGar
	orientale 'Juliane'	oAls
	orientale 'Lilac Girl'	last listed 99/00
	orientale 'Maiden's Blush'	wWoS
	orientale 'Mrs. Perry'	wCSG
	orientale 'Orange Glow'	oAls
	orientale 'Patty's Plum'	wHer oNWe
	orientale 'Perry's White'	wHer oGre
	orientale 'Pinnacle'	oAls oGre
	orientale 'Pizzicato'	wTho oAls wHom oWnC wMag wTGN
	orientale 'Prince of Orange'	oLSG
	orientale 'Prinzessin Victoria Louise'	oDar oAls oGar oBlo wMag wTGN oEga oLSG
	orientale 'Queen Alexandra'	wSwC wMag
	orientale 'Raspberry Queen'	wWoS oAls oGar wMag
	orientale 'Royal Wedding'	oAls oGar oDan wCCr oEga
	orientale 'Saffron'	wHer
	orientale 'Salmon Glow'	oCir
	orientale 'Springtime'	wWoS
	orientale 'Stormacet'	wHer
	orientale 'Tuerkenlouis	oAls oGar wMag
	orientale 'Watermelon'	oGar oNWe wMag
	orientale 'Wunderkind'	oAls
	rhoeas Angels' Choir	oUps
	rupifragum	oRus
	somniferum	oSec
	spicatum	oSec
	PAPHIOPEDILUM	
	hybrids	oGar
	PARADISEA	
	liliastrum	last listed 99/00
	lusitanica	wCol
	lusitanica large flowered form	wHer

	Name	Sources
	PARAHEBE	
	x bidwillii	last listed 99/00
	catarractae	oTrP oJoy oMis oNWe
	catarractae blue form	oHed oCis
	catarractae 'Delight'	oJoy
	formosa	last listed 99/00
	lyallii	oCis
	olsenii	oTrP oHed
	perfoliata	oDar oHed oCis
un	**PARAKMERIA**	
	lotungensis	oFor wHer oGos oGar oDan oCis oSho oGre oWnC wBox wSte
	yunnanensis	oCis
	PARAQUILEGIA	
	anemonoides	wMtT
	grandiflora	see **P. *anemonoides***
	PARATHELYPTERIS	
	novaeboracensis	oRus
	X PARDANCANDA	
	norrisii	oFor oTrP oAls oBlo
	norrisii 'Dazzler'	oUps
	PARIETARIA	
un	***diffusa***	wFai
	PARIS	
	sp. aff. ***bashanensis***	last listed 99/00
	lancifolia	last listed 99/00
	polyphylla	oTrP wHer oOEx
	sp. DJHC 221	wHer
	sp. aff. ***thibetica*** DJHC 504	wHer
	verticillata	wHer
un	***yunnanensis*** DJHC 103	wHer
	PARNASSIA	
	glauca	wThG
	palustris	wHer
	PAROCHETUS	
	communis	oTrP wWoS
	PARROTIA	
	persica	oFor wCul wFWB oGos wClo oAls oAmb oGar oDan oRiv oWhi oBov oGre oWnC wCoN oUps wWel
	persica 'Pendula'	wHer oGos wCol oGar oRiv oGre
un	***persica*** 'Select'	oGos oGar oRiv oGre wSte
	persica 'Vanessa'	oFor oGos wCol
	PARROTIOPSIS	
	jacquemontiana	oFor oRiv oGre
	PARRYA	
	nudicaulis	wMtT
	PARTHENIUM	
	integrifolium	wCCr
	PARTHENOCISSUS	
	henryana	oFor oTrP wCul oGos oDar oEdg oAmb oCis oGre wSte wWel
	himalayana var. ***rubrifolia***	oGar oGre
	inserta	oFor
	quinquefolia	oFor wCul oDar wCSG oGar oRiv oSho oWhi oGre oWnC wTGN oSle wSnq
	quinquefolia Star Showers™ / 'Monham'	oAls oGar oSho
	sp. DJHC 071	wHer
	sp. EDHCH 97050	wHer
	thomsonii	see **CAYRATIA *thomsonii***
	tricuspidata	oFor oAls oGre wSta oSle
	tricuspidata 'Fenway'	last listed 99/00
	tricuspidata 'Green Showers'	oGar
	tricuspidata 'Green Spring'	oGre
	tricuspidata 'Lowii'	last listed 99/00
	tricuspidata 'Veitchii'	oGar oGre oWnC wWel
	vitacea	see **P. *inserta***
	PASANIA	see **LITHOCARPUS**
	PASSIFLORA	
	x alatocaerulea	see **P. *x belotii***
	'Amethyst'	wGra
	ampullacea	wGra
	x belotii	wGra oUps
	bryonioides	last listed 99/00
	caerulea	oFor wWoS wClo oDar wCSG oMis oGar oOEx oGre wFai wPir wCCr oWnC wGra
un	***caerulea*** 'Alba'	wRai oWnC
un	***caerulea*** 'Blue Crown'	wCul wRai oOEx wTGN
	caerulea 'Constance Elliot'	wGra
	citrina	wGra
	coccinea	oDan oUps
	x decaisneana	wGra
	edulis	oOEx wGra
un	***edulis*** 'Black Knight'	wGra
	edulis f. ***flavicarpa***	wGra
	foetida	wGra
	gibertii	wGra
	incarnata	oTrP oGar oGoo oOEx oGre iGSc oCrm wGra
	'Incense'	wRai oWnC
un	'Indigo Dream'	wGra
	jamesonii	oAls wGra oUps
	'Lavender Lady'	see **P.** 'Amethyst'
	lutea	oOEx
un	***manicata*** 'Scarlet'	wGra
	mixta	wGra
	mollissima	oTrP oOEx wGra
	quadrangularis	oOEx wGra
	sanguinolenta	wGra
	sp. kuangxii	oOEx
	'Star of Bristol'	last listed 99/00
	suberosa	wGra
	subpeltata	wGra
	warmingii	last listed 99/00
	PATRINIA	
	saniculifolia HC 970137	last listed 99/00
	scabiosifolia	oUps
	scabiosifolia 'Nagoya'	last listed 99/00
	sp.	oSec
	sp. EDHCH 97255	wHer
	villosa	oFor
	villosa HC 970570	wHer
	PAULOWNIA	
	fortunei	oFor wCCr
	kawakamii	oFor wCCr
un	***taiwaniana***	wCCr
	tomentosa	wBur wCSG oGar oRiv oOEx wCCr oWnC wTGN wWhG wWel
	tomentosa 'Coreana'	last listed 99/00
	PAVONIA	
	hastata	oFor
	PAXISTIMA	
	canbyi	oFor wCul wThG oRiv wCCr wPla
	canbyi 'Compacta'	oSis
un	***canbyi*** 'Nana'	wHer
	myrsinites	see **P. *myrtifolia***
	myrtifolia	wHer wCul wNot wShR oBos oRiv wWat wCCr wPla wWel
	PEDICULARIS	
	canadensis	oAld
	PEDILANTHUS	
	sp. BLM 0810	oRar
	PEGANUM	
	harmala	oOEx
	PELARGONIUM	
	abrotanifolium	wKil
	'African Belle'	wKil
un	'African Queen'	wKil
	'Alcyone'	wKil
un	'Alliance'	wKil
un	'Almond'	wKil
un	'Alpha'	see **P.** 'Golden Harry Hieover'
	'Alpine Glow'	wKil
	'Altair'	wKil
	'Always'	wKil
un	Americana Series	oSev
un	'Angel'	oBar
un	'Applause'	wKil
un	'Apple Cider'	wKil
un	'Apple Fringed'	wKil
un	'Apple Mint'	wKil
un	'Appleblossom'	wKil
	'Apricot'	oTDM wKil iGSc oBar wNTP
	'Apricot Queen'	wKil
	'Arctic Star'	wKil
	aridum	oTDM oRar
	'Arizona'	wKil
	'Aroma'	oGoo wKil
	'Atomic Snowflake'	oGoo wKil wNTP
	'Attar of Roses'	oTDM oGoo wKil iGSc oBar wNTP
	'Attraction'	oGoo
	australe	oHed
un	'Azalea'	wKil
	'Aztec'	wKil
un	'Balcon Desrumeaux'	wKil
	'Balcon Royal'	see **P.** 'Roi des Balcons Imperial'
un	'Balsam'	wKil
	Barock / 'Fisrock'	wKil
	'Bashful'	wKil
un	'Batavia Star'	wKil

	'Bath Beauty'	wKil
	'Beauty'	wKil
	'Beauty of Eastbourne'	see **P.** 'Lachskoenigin'
un	'Belinda'	wKil
un	'Bella'	wKil
	'Ben Franklin'	wKil
un	'Benedict'	wKil
un	'Berlin'	wKil
un	'Big Lemon'	wNTP
un	'Bimbo'	wKil
	'Bird Dancer'	wKil
un	'Bird's Egg Double'	wKil
	'Bitter Lemon'	oGoo
un	'Black Dwarf'	wKil
un	'Black Jack'	wKil
un	'Black Lace'	wKil
	'Black Pearl'	wKil
un	'Black Sally'	wKil
	'Black Vesuvius'	see **P.** 'Red Black Vesuvius'
	'Blakesdorf'	wKil
un	'Blandes Musk'	wKil wNTP
un	'Blandfordianum Alba'	wKil
un	'Blandfordianum Roseum'	wKil oBar
	'Blazonry'	oGoo
un	'Blue Moon'	wKil
	'Bluebeard'	wKil
un	'Bode's Mint'	wKil
un	'Bode's Peppermint'	see **P.** 'Pungent Peppermint'
	'Both's Snowflake'	oGoo wKil
un	'Bougainvillea'	wKil
un	'Brandy'	wKil
un	'Brickshot'	wKil
	'Bridesmaid'	wKil
	'Brilliant'	wKil
un	'Broadleaf Peppermint'	wKil
un	'Bronze Beauty'	wKil
	'Bronze Queen'	wKil
	'Brown's Butterfly'	wKil
	'Brunswick'	oGoo wKil
	'Burnaby'	wKil
	'Butterfly Loreli'	wKil
un	'Butterfly Mrs. J. J. Knight'	wKil
	'Caligula'	wKil
	'Calypso'	wKil
	'Cameo'	wKil
	'Camphor Rose'	oGoo wKil wNTP
un	'Canary Island'	wKil
	'Candy'	wKil
un	'Candy Dancer'	oGoo wKil
	'Capella'	wKil
	capitatum	wKil
	'Capri'	wKil
	'Captain Starlight'	wKil
un	'Carl Gaffney'	oGoo
un	'Carlton's Pansy'	wKil
un	'Carmel'	wKil
	carnosum	oRar
	caylae	oTDM oRar
un	'Celery'	iGSc
	'Charity'	oGoo wKil
un	'Charmay Snow Flurry'	oGoo wKil
un	'Chattisham'	wKil
un	'Cherry Poinsettia'	wKil
	'Cherry Sundae'	oGoo wKil
un	'Chicago Rose'	wKil wNTP
un	'Chinese Cactus'	wKil
un	'Chocolate'	wHer oTDM
un	'Chocolate Joy'	wKil
	'Chocolate Peppermint'	oGoo wKil iGSc oBar
un	'Christchurch Beauty'	oGoo wKil
un	cinnamon	see **P.** 'Limoneum'
un	'Cinnamon Rose'	wKil wNTP
	'Citriodorum'	oBar
	'Citronella'	oTDM oGoo wKil
un	'Citrosa'	wKil iGSc
	x citrosum	oTDM oGoo wNTP
un	'Clackamas Star'	wKil
	'Cliff House'	see **P.** 'Double Lilac White'
	'Clorinda'	wKil oBar
	'Clorinda, Golden'	see **P.** 'Golden Clorinda'
	'Clown'	wKil
un	'Cocoa Mint Rose'	wKil
un	'Coconut' trailing	wKil
un	'Coconut' upright	wKil oBar
	'Coddenham'	wKil

	'Concolor Lace'	oGoo wKil oBar
un	'Congo'	wKil
	'Contrast'	oGoo wKil
	'Cook's Freckles'	wKil
	'Copthorne'	oGoo wKil iGSc wNTP
	cordifolium	oHed
	'Cornell'	wKil
	crispum	oGoo wKil oBar wNTP
	crispum 'Latifolium'	oGoo
	crispum 'Major'	oGoo
	crispum 'Minor'	oTDM wKil iGSc oBar
	crispum 'Variegatum'	oGoo oBar
	crithmifolium	wKil
	'Crocodile'	see **P.** 'The Crocodile'
un	'Crowfoot Rose'	oGoo wKil
	'Crystal Palace Gem'	oGoo wKil
	cucullatum	wKil iGSc wNTP
un	'Damon's Gold Leaf'	wKil
	'Dark Lady'	wKil
	dasycaule	wKil
	'Deacon Peacock'	wKil
un	'Dean's Delight'	oGoo wKil
	'Decora Lavender'	wKil
un	'Decora Pink'	wKil
un	'Decora Red'	wKil
un	'Deep Purple'	wKil
	denticulatum	oTDM wKil
	denticulatum 'Fernleaf'	see **P. *denticulatum*** 'Filicifolium'
	denticulatum 'Filicifolium'	oGoo wKil oBar
un	Designer Series	wTho
	dichondrifolium	oGoo wKil
	'Distinction'	wKil
	'Doctor Livingston'	oGoo wKil
un	***dolarnibon***	oCis
	'Dolly Read'	wKil
	'Dolly Varden'	oGoo wKil
	x domesticum	oGar
un	'Don's Seagold'	oGoo wKil
	'Dopey'	wKil
un	'Dorcas Brigham Lime'	wKil
	'Double Lilac White'	wKil
	'Dovedale'	wKil
	'Dubonnet'	wKil
	'Duke of Edinburgh'	see **P.** 'Hederinum Variegatum'
	'Dusty Rose'	wKil
	'Earliana'	wKil
	echinatum	wKil
un	'Edith North'	wKil
	'Edith Steane'	wKil
	'Elmsett'	wKil
	elongatum	oGoo
	'Els'	wKil
	'Elsie Hickman'	wKil
	'Emma Hoessle	see **P.** 'Frau Emma Hoessle'
un	'Empress of Russia'	wKil
	'Endsleigh'	wKil
	'Endsleigh Oak'	see **P.** 'Endsleigh'
	exhibens	oRar
un	'Exquisite'	wKil
	'Fair Ellen'	oGoo wKil wNTP
	'Fair Ellen Oak'	see **P.** 'Fair Ellen'
	'Falkland Hero'	wKil
	ferulaceum	oRar
	'Fifth Avenue'	wKil
	'Filicifolium'	see **P. *denticulatum*** 'Filicifolium'
un	'Firedancer'	wKil
un	'Fireworks'	wKil
un	'Fleur de Lis'	wKil
	'Fleurette'	wKil
	'Flower Basket'	wKil
	'Flower of Spring'	oGoo wKil
un	'Forest Maid'	wKil
	'Forever'	wKil
un	'Formosum'	wKil
un	'Forty Niner'	wKil
	Fragrans Group	oTDM oGoo wKil iGSc oBar
	Fragrans Group 'Fragrans Variegatum'	oGoo wKil oBar
	Fragrans Group 'Snowy Nutmeg'	see **P.** Fragrans Group 'Fragrans Variegatum'
	'Francis Parrett'	wKil
	'Frank Headley'	wKil
	'Frau Emma Hoessle'	wKil
un	'French Lace'	wKil wNTP
	'Frensham'	oGoo oBar
un	'Frensham Lemon'	wKil

	'Friary Wood'	wKil
	'Frills'	wKil
	'Fringed Apple'	oGoo
	'Fringed Aztec'	wKil
un	'Frolic'	wKil
un	'Fruit Angel'	wKil
un	'Fruity'	wKil
	fruticosum	oTDM oHed wKil
	fulgidum	wKil
	'Galilee'	wKil
	'Galway Star'	oGoo wKil
un	'Gardener's Joy'	wKil
	'Garnet'	wKil
	'Gay Baby'	wKil
	'Gemma'	wKil
	'Georgia Peach'	wKil
	'Geronimo'	wKil
	'Giant Oak'	wKil oBar
un	'Giant Oak Variegated'	wKil
un	'Giant Rose'	wKil
	gibbosum	wKil
un	'Gibson Girl'	wKil
	glaucifolium	wKil
	glutinosum	oGoo wKil oBar
	'Goblin'	oSev wKil
un	'Godfrey's Pride'	oGoo wKil
un	'Gold Dust'	oGoo wKil
	'Golden Brilliantissimum'	wKil
	'Golden Butterfly'	wKil
	'Golden Clorinda'	wKil
un	'Golden Deacon Lilac Mist'	wKil
	'Golden Ears'	oGoo
	'Golden Everaarts'	last listed 99/00
un	'Golden Formosum'	wKil
	'Golden Harry Hieover'	wKil
un	'Golden Lemon Crispum'	oGoo wKil
	'Golden Mr. Everaarts'	see **P.** 'Golden Everaarts'
un	'Golden Nutmeg'	oGoo
	'Gooseberry Leaf'	see **P.** 'Peach'
	'Grace Wells'	wKil
	'Grand Slam'	wKil
	'Grandma Fischer'	see **P.** 'Grossmutter Fischer'
	'Graveolens'	oGoo wKil iGSc
un	'Graves Staghorn Rose'	oGoo
un	'Green Gold'	wKil
	'Greetings'	wKil
un	'Grenchen'	wKil
	'Grey Lady Plymouth'	oTDM oGoo wKil oBar wNTP
	'Grey Sprite'	wKil
un	'Grossersorten'	wKil
	'Grossmutter Fischer'	wKil
	grossularioides	oTDM oGoo iGSc oBar
un	'Grumpy'	wKil
un	'Hanson's Wild Spice'	wKil
un	'Happy'	wKil
	'Happy Appleblossom'	wKil
	'Happy Thought'	oGoo wKil
	'Harvard'	wKil
	'Hederinum'	wKil
	'Hederinum Variegatum'	wKil
	'Heidi'	wKil
	hispidum	oGoo
un	'Horizon'	wKil
	x hortorum	oGar
	ionidiflorum	wKil
	'Ivory Snow'	wKil
	'Jack of Hearts'	wKil
	'Jack Read'	wKil
un	'Jackpot Lilac Mist'	wKil
un	'Jackpot Rose'	wKil
un	***jacobii***	oTDM oRar
un	'Jane Eyre'	see **P.** 'Jayne Eyre'
	'Janet Kerrigan'	wKil
un	'Janie'	wKil
	'Jaunty'	wKil
	'Jayne Eyre'	wKil
	'Jenifer Read'	wKil
un	'Jo Jo'	wKil
un	'Josephine'	wKil
	'Joy Lucille'	oTDM oGoo wKil iGSc oBar
un	'Joy Lucille Variegated'	see **P.** 'Variegated Joy Lucille'
un	'Jubilee'	wKil
un	'Juicy Fruit'	iGSc
un	'Jungle Treasure'	wKil
	'Juniper'	oGoo wKil
un	'Kaleidoscope'	wKil
	karrooense	wKil
	'Keepsake'	wKil
un	'Ken's Gold Stellar'	wKil
un	'Killdeer Rose'	wKil
	'King of Balcon'	see **P.** 'Hederinum'
	'Kleine Liebling'	wKil
	'Krista'	wKil
	'La France'	wKil
	'Lachskoenigin'	wKil
	'Lady Cullum'	wKil
	'Lady Mary'	wKil wNTP
	'Lady Plymouth'	oTDM oGoo wKil iGSc oBar
	'Lady Scarborough'	oTDM oGoo wKil iGSc oBar
un	'Large Flowered Crispum'	oGoo
	'Lass o'Gowrie'	wKil
un	'Laura Maid'	wKil
	'L'Elegante'	oMis wKil
un	'Lemon Balm'	oGoo wKil
un	'Lemon Meringue'	wKil
un	'Lemon Rose'	oBar
un	'Lemoneum'	wKil
	'Lilian Pottinger'	oGoo wKil oBar
	'Limoneum'	oTDM oGoo wKil iGSc
un	'Lita'	wKil
	'Little Darling'	see **P.** 'Kleine Liebling'
	'Little Gem'	oGoo wKil oBar
	'Little John'	wKil
un	'Little Rascal'	wKil
	'Longshot'	wKil
un	'Lotus Land'	wKil
un	'Luciflora'	wKil
un	'Lynnbrook Orchid'	wKil
	'Lyric'	wKil
	'Mabel Grey'	oTDM oGoo wKil iGSc oBar
	'Madame Auguste Nonin'	wKil
	'Madame Layal'	wKil
	'Madame Thibaut'	wKil
	'Magaluf'	wKil
un	'Magenta Rosebud'	wKil
	magenteum	oGar
	'Magic Lantern'	wHer
	'Mahogany'	wKil
un	'Mandee'	wKil
	'Mangles' Variegated'	wKil
	'Maple Leaf'	wKil
	'Marmalade'	wKil
un	'Marshall Macmahon'	wKil
	'Masterpiece'	oGoo wKil
	'Maureen'	wKil
	'Meditation'	wKil
	'Medley'	wKil
	'Melanie'	wKil
	'Memento'	wKil
	'Mere Cocktail'	wKil
	'Mexican Beauty'	wKil
un	'Mexican Sage'	oGoo wKil
un	'Mini Lilac Cascade'	wKil
un	'Mini Pink Cascade'	wKil
un	'Mini Red Cascade'	wKil
un	'Minipel Rosa'	wKil
un	'Mint Peacock'	wNTP
un	'Mint Rose Variegated'	see **P.** 'Variegated Mint Rose'
un	'Miss Australia Variegated'	wKil
	'Miss Burdett Coutts'	wKil
	'Miss Wackles'	wKil
	mollicomum	oCis
un	'Moon Beam'	wKil
	'Moor'	wKil
un	'Morning Glory'	wKil
	'Mr. Everaarts'	wKil
	'Mr. Henry Cox'	oGoo
un	'Mr. Robert'	wKil
	'Mr. Wren'	wKil
un	'Mrs. Banks'	wKil
un	'Mrs. Cox'	wKil
	'Mrs. J. C. Mappin'	wKil
	'Mrs. Kingsley'	wKil oBar oCir
	'Mrs. Parker'	wKil
	'Mrs. Pat'	oGoo wKil
	'Mrs. Pollock'	oGoo
un	'Mrs. Pollock' double	see **P.** 'Mrs. Strang'
	'Mrs. Strang'	wKil
un	'Mrs. Taylor'	wKil oBar wNTP
	'Needham Market'	wKil

	'Nervosum'	wKil iGSc oBar wNTP
un	'Nervosum Minor'	wKil
un	'New Calypso'	wKil
un	'New Gypsy'	wKil
un	'New Spice'	iGSc
	'Nicor Star'	wKil
un	'Nina'	wKil
	'Noel'	wKil
un	'Nutmeg Lavender'	wKil
	nutmeg, golden	wKil
	'Occold Embers'	wKil
	'Occold Shield'	oGoo wKil
un	'Ocean Wave'	wKil
	odoratissimum	oTDM oGoo wKil iGSc oBar wNTP
un	'Old Fashioned Rose'	wKil iGSc oBar
	'Old Rose'	wNTP
un	'Old Spice'	oHed wKil wNTP
	'Orange Parfait'	wKil
	orange scented	see **P.** 'Citriodorum'
	'Orion'	wKil
	PAC cultivars	see **P.** under cultivar names
	'Pagoda'	wKil
un	'Paint Box'	wKil
un	'Pandora'	wKil
	'Parisienne'	wKil
	'Paton's Unique'	oGoo wKil
	'Patricia Andrea'	wKil
un	'Paula's Lemon Scent'	wKil
	'Peace'	wKil
	'Peach'	oGoo wKil oBar
un	'Peacock'	oGoo iGSc
un	'Peacock Rose'	wKil
	peltatum	wTho oTDM oGar
un	***peltatum*** 'Mexicana'	wTGN
	'Peppermint Lace'	oGoo wKil
un	'Peppermint Rose'	wKil oBar
un	'Peppermint Spice'	wKil
	'Peppermint Star'	wKil
	'Persian Queen'	wHer wKil
	'Petals'	wKil
	'Petit Pierre'	see **P.** 'Kleine Liebling'
un	'Petite Peach'	wKil
un	'Pineapple'	wKil oBar
	'Pink Champagne'	wKil oBar wNTP
un	'Pink Poinsettia'	wKil
un	'Pink Queen'	wKil
	'Pink Rosebud'	wKil
	'Pixie'	wKil
	'Platinum'	oGoo wKil
	'Playmate'	wKil
	'Polka'	wKil
un	'Polka Dot'	wKil
	'Pompeii'	wKil
un	'Poquito'	oGoo wKil
	'Pretty Girl'	wKil
	'Pretty Polly'	oGoo wKil oBar
	'Prim'	wKil
	'Prince of Orange'	wKil iGSc
un	'Prince Rupert'	oTDM oGoo wKil iGSc oBar
un	'Prince Valiant'	wKil
	'Princess of Balcon'	see **P.** 'Roi des Balcons Lilas'
un	'Pungent Peppermint'	oGoo wKil oBar
un	'Purple Majesty'	wKil
	'Queen Esther'	wKil
	'Queen of Hearts'	wKil
	quercifolium	wKil iGSc
un	***quercifolium variegatum***	oGoo
	'Rads Star'	wKil
un	'Raspberry Sherbet'	wKil
qu	'Red'	wKil
	'Red Black Vesuvius'	wKil
	'Red Cascade'	wKil
un	'Red Flowered Rose'	oGoo wKil
un	'Red Heart'	wKil
un	'Red Rosebud'	wKil
	'Red Spider'	wKil
	'Red Witch'	wKil
	'Redondo'	wKil
	reniforme	oTDM wKil
	'Rigel'	wKil
	'Rita Scheen'	wKil
	'Rober's Lemon Rose'	oGoo wKil iGSc oBar
un	'Rococco'	wKil
	'Roger's Delight'	oGoo wKil
	'Rogue'	wKil
	'Roi des Balcons Imperial'	wKil
	'Roi des Balcons Lilas'	wKil
	'Roller's David'	wKil
	'Roller's Pioneer'	wKil
	'Rollisson's Unique'	wKil
un	'Rose Baby'	wKil
un	'Rose Lee'	wKil
un	'Rosie'	wKil
un	'Rosie O'Day'	wKil
	'Rosina Read'	wKil
	'Rosy Dawn'	wKil
un	'Round Leaf Orange'	iGSc oBar
un	'Round Leaf Rose'	oGoo wKil
un	'Royal Blood'	wKil
	'Royal Norfolk'	wKil
un	'Ruby Edged Oak'	wKil
un	'Ruth Lesley'	wKil
	'Salmon Black Vesuvius'	wKil
	'Salmon Queen'	see **P.** 'Lachskoenigin'
	'Samantha Stamp'	wKil
	'Samba'	wKil
	'Sancho Panza'	wKil
	'Saturn'	wKil
un	'Scarlet and White'	wKil
	'Scarlet Unique'	wKil iGSc wNTP
un	'Sharptooth Oak'	wKil oBar
	'Shelley'	wKil
	'Shottesham Pet'	wKil wNTP
	'Shrubland Rose'	oTDM oGoo wKil
	sidoides	wHer oHed oNWe wKil
un	'Silver Leaf Rose'	oGoo wKil
un	'Silver Ruby'	wKil
un	'Sirius'	wKil
	'Skeleton Leaf Rose'	see **P.** 'Doctor Livingston'
un	'Skeleton Rose'	iGSc oBar
	'Skelly's Pride'	wKil
un	'Skelton's Unique'	wKil
	'Skies of Italy'	oGoo wKil
	'Small Fortune'	wKil
un	'Smarty'	wKil
	'Snowbaby'	wKil
	'Snowflake'	oGoo wKil iGSc oBar
un	'Snowflake Rose'	wKil
	'Snowy Nutmeg'	see **P.** Fragrans Group 'Fragrans Variegatum'
un	'So Happy'	wKil
	'Somersham'	wKil
un	'Sonoma Lavender'	wKil
	'Sophie Dumaresque'	wKil
	sp.	oRar
un	'Spain'	wKil
un	'Spanish Lavender'	see **P. *cucullatum***
	'Spitfire'	wKil
	'Spring Park'	wKil iGSc oBar
un	'Staccato'	wKil
	'Stadt Bern'	wKil
un	'Staghorn Oak'	wKil
un	'Staghorn Peppermint'	wKil
un	'Staghorn Rose'	wKil
	'Star of Persia'	wKil
	'Stars and Stripes'	wKil
	'Stella Read'	wKil
	'Sugar Baby'	wKil
un	'Sugar Plum'	wKil
un	'Sundancer'	wKil
	'Sussex Delight'	wKil
	'Sweet Miriam'	wKil
	'Sybil Holmes'	wKil
	'Tammy'	wKil
	'Tangerine'	wKil
	'Tavira'	wKil
	'Ten of Hearts'	wKil
	tetragonum	wKil
	'The Crocodile'	wKil
	'Tiberias'	wKil
	'Tinkerbell'	wKil
un	'Tom Tit'	wKil
	tomentosum	oGar oGoo iGSc oBar wNTP
	'Torento'	oTDM oGoo wKil oBar wNTP
	'Tornado Lilac' (Tornado Series)	wHom
	'Tornado White' (Tornado Series)	wHom
	'Trudie'	wKil
	'Tunias Perfecta'	wKil
	'Turkish Delight'	wKil
un	'Tutone'	wKil

	'Vancouver Centennial'	wHer oGoo wKil
un	'Variegated Joy Lucille'	wKil oBar
	'Variegated Kleine Liebling'	wKil
	'Variegated Little Darling'	see **P.** 'Variegated Kleine Liebling'
un	'Variegated Mint Rose'	oBar
un	'Variegated Prince Rupert'	oBar
un	'Velvet Rose'	oGoo wKil wNTP
	'Venus'	wKil
	'Veronica'	wTho
un	'Veronica Contreras'	wKil
	'Village Hill Oak'	oGoo wKil
	'Violet Lambton'	wKil
	'Virginia'	wKil
	vitifolium	oGoo wKil iGSc
	'Voodoo'	wKil
	'Wantirna'	wKil
un	'Warrain'	wKil
	'Wedding Royale'	last listed 99/00
un	'White Champion'	wKil
	'White Chiffon'	wKil
un	'White Grossersorten'	wKil
	'White Mesh'	wKil
	'White Unique'	wKil
un	'Wildwood'	oTDM wKil
un	'Wilhelm Langguth'	oGoo wKil
	'Winnie Read'	wKil
un	'Yellow Geranium'	wKil
	PELECYPHORA	
	aselliformis	oRar
	pseudopectinata	see **NEOLLOYDIA *pseudopectinata***
	strobiliformis	oRar
	valdeziana	see **NEOLLOYDIA *valdeziana***
	PELLAEA	
	atropurpurea	oRus oSis
	PELTANDRA	
	virginica	wGAc oHug oSsd
	PELTIPHYLLUM	see **DARMERA**
	PELTOBOYKINIA	
	tellimoides	wGAc oGar wRob oGre wFai wBox
	watanabei	wCol oNWe
	PENIOCEREUS	
	greggii	oOEx
un	***rosea***	oRar
	striatus	oRar
	viperinus	oRar
	PENNISETUM	
	alopecuroides	wCli oJoy oAmb oGar oOut wFai wWin oWnC wTGN oBRG wSnq
	alopecuroides 'Cassian'	see **P.** 'Cassian's Choice'
	alopecuroides 'Caudatum'	oGar oWnC wSnq
	alopecuroides 'Hameln'	oFor oJoy oAmb oGar oDan oSho oOut wRob oBlo oGre wWin wPir wHom oWnC oTri wBox wWhG oUps wSnq
	alopecuroides 'Little Bunny'	oFor wCli wCul oJoy oGar oDan oSis oOut wRob oGre wWin wPir oTPm oWnC wBox oSec oBRG wEde oUps wSnq wCoS
	alopecuroides 'Little Honey'	wHer oOut oGre wWin wEde
un	***alopecuroides*** 'Morning Dew'	oGar
	alopecuroides 'Moudry'	wWoS wCli wSwC oJoy oGar oOut oGre wWin wPir wHom wBox oBRG wSnq
	alopecuroides 'Weserbergland'	wWin wBox
	'Burgundy Giant'	wWoS wWin
	'Cassian's Choice'	oFor oJoy oOut
	japonicum	see **P. *alopecuroides***
	orientale	oFor wCli oNat oJoy oGar oDan oSis oOut wRob wWin oTri wBox oSec oBRG wSnq
un	***orientale*** 'Tall Tales'	wSnq
	rubrum	see **P. *setaceum*** 'Rubrum'
	setaceum	oJoy wWin
	setaceum 'Eaton Canyon'	wWin wSnq
	setaceum 'Purpureum'	see **P. *setaceum*** 'Rubrum'
	setaceum 'Rubrum'	wWoS oNat oGar oSho wWin wHom oWnC oUps wSte wSnq wWel
	setaceum 'Rubrum Compactum'	see **P. *setaceum*** 'Eaton Canyon'
un	'Tall Tails'	oOut
un	'Tall Tales'	wWin
	villosum	wWin wSnq
	PENSTEMON	
	acuminatus	wWld
	'Alice Hindley'	oFor wHer wWoS wCri oJoy oGoo
	'Andenken an Friedrich Hahn'	oFor wHer wCul oNat oDar oJoy oAls oHed oGar oGoo oBos wFai oSle wSte
	'Apple Blossom'	wWoS wCri oDar oAls oHed oAmb oMis oSis wMag
un	***attenuatus*** var. ***attenuatus***	wCol

	barbatus	wCSG oSec
un	***barbatus*** 'Bashful Salmon'	oGar oUps
	barbatus 'Cambridge Mixed'	wTGN
	barbatus ssp. ***coccineus***	oGoo
	barbatus 'Elfin Pink'	oFor wCul oGar oSho oSha wSnq
un	***barbatus*** 'Petite Bouquet'	oEga
	barbatus var. ***praecox*** f. ***nanus*** 'Rondo'	wTGN oUps
	barrettiae	oBos oCis
	barrettiae x ***rupicola***	oNWe
un	'Bashful Salmon'	oFor
un	'Betty's Red'	wWoS
un	'Bev Jensen'	oDar oJoy
	'Blackbird'	wHer wWoS oNat oJoy oAls oHed oMis
un	'Blue Midnight'	oFor wCul oMis oGar
	broken top	oBos
	'Burgundy'	oAls
	caespitosus albus	wMtT
ch	***caespitosus*** 'Claude Barr'	wFGN
	caespitosus ssp. ***desertipictii*** NNS93-532	last listed 99/00
	campanulatus	wCul oJoy oAls oBos wFai
	canescens	oFor oSec
	cardinalis	oGoo
	cardwellii	wNot oBos oTri
	cardwellii 'Albus'	oRus oBos
	cardwellii purple	oBos
	cardwellii 'Roseus'	oBos
	'Cerise Kissed'	oJoy
	'Cherry'	last listed 99/00
un	'Cherry Glow'	oHed
un	'Cherry Red'	oSec
	clutei	oGoo oSec
	cobaea	oMis
qu	***coccineus***	wAva
	confertus	oBos wCCr wWld
	'Coral Kissed'	oJoy
	crandallii ssp. ***procumbens***	wMtT
	crandallii RM92-323	last listed 99/00
un	'Crystal'	oSis
	cyaneus	wSte
	davidsonii	wThG oBos wWld
	davidsonii 'Albus'	last listed 99/00
un	***davidsonii*** ssp. ***davidsonii*** 'Mount Adams Dwarf'	wMtT
	davidsonii var. ***menziesii***	oRus
	davidsonii var. ***menziesii*** 'Microphyllus'	oGoo oSis
	davidsonii var. ***menziesii*** pink	wMtT
	davidsonii var. ***menziesii*** 'Rampart White'	wMtT
	davidsonii var. ***prateritus***	oBos
	deep purple	oGoo
un	'Deepest Maroon'	wCul
	deustus	oBos wPla
	digitalis	oRus oGoo
	digitalis 'Husker Red'	oFor oEdg oAls oHed wFGN wCSG oMis oGoo oSho oSha oOut oBlo wFai wHom oWnC wMag wTGN oCir wAva oLSG oSle wSnq
un	'Dragontail'	wMtT
	eatonii	oGoo wPla
	eatonii var. ***undosus***	last listed 99/00
un	'Elizabeth Cozzens'	oJoy
	euglaucus	wNot oJoy wCol oBos
un	'Eureka White'	wCul oGoo
	'Evelyn'	wHer wCul oAls oHed oGoo oSis
	'Firebird'	see **P.** 'Schoenholzeri'
	fruticosus	wNot oRus oGoo oBos wWld
	fruticosus 'Purple Haze'	oSho wMag iArc
	fruticosus var. ***scouleri***	wThG
	'Garnet'	see **P.** 'Andenken an Friedrich Hahn'
	gentianoides	wTho wCol wCSG wHom
un	'Ghent'	oJoy
un	'Ghent Purple'	wWoS
	globosus	wCol
	x gloxinioides	see **P. *hartwegii***
	gormanii	oDan
	gracilis	wWld
	grandiflorus	oJoy oMis oGoo
un	'Grape Tart'	oSis
	hallii	last listed 99/00
	hartwegii	oMis
	hartwegii 'Albus'	oHed
	hartwegii hybrids	oSec
	harvardii	last listed 99/00
	heterodoxus NNS93-564	wMtT
	heterophyllus	wCCr iArc
	heterophyllus 'Blue of Zurich'	see **P.** 'Zueriblau'

	heterophyllus 'Catherine de la Mare'	wWoS oAls oWnC oSec wSnq
un	***heterophyllus*** 'Margarita Bop'	oHed
	heterophyllus 'True Blue'	see **P.** ***heterophyllus***
	heterophyllus 'Zueriblau'	oFor oMis
	'Hewell Pink Bedder'	oCis
	'Hidcote Pink'	wWoS wCul wCri oJoy oAls oHed oMis oGoo oSec
	hirsutus	oFor oJoy oRus oGoo oSec
	hirsutus var. ***pygmaeus***	oHed oSis oNWe
	'Holly White'	wWoS oAmb oGoo wTGN
	'Hopleys Variegated'	wHer wWoS oJoy wCol oHed oSis oGre oWnC wBox wSte
un	'Huntington Garden'	oGar oDan
	'Huntington Pink'	see **P.** 'Apple Blossom'
un	'Jerry's Purple'	oSis
	'Joy'	oJoy
	'King George V'	oAls oCis
	labrosus	oHed
	laetus var. ***roezlii***	wMtT
	laricifolius	last listed 99/00
un	'Lavender Lady'	wBox
un	'Lavender Ruffles'	oDar
un	'Lexington'	oSis
	linarioides ssp. ***coloradoensis***	oSis
	'Maurice Gibbs'	wAva
	menziesii	see **P.** ***davidsonii*** var. ***menziesii***
	Mexicali hybrids 'Pike's Peak Purple'	oFor wWoS
	Mexicali hybrids 'Red Rocks'	wWoS oGar oSis
	'Midnight'	wHer wWoS wCri oNat oDar oJoy oAls oGoo wMag oSec wAva wEde oEga
	'Mother of Pearl'	oFor wHer wWoS oNat oJoy wAva
	newberryi	oRus oBov
	newberryi var. ***sonomensis***	wMtT
	ovatus	wNot oAmb oSec wWld
	palmeri	oGoo
	'Papal Purple'	wHer wWoS oSec
	parvulus	wWld
	peckii	oHan oSis
	'Pershore Pink Necklace'	oAls oHed oMis
	pinifolius	oFor wHer wWoS oJoy wCol oRus wShR oGoo oDan oSis oBos oOut wCCr wAva
un	***pinifolius*** 'Magdalina Sunshine'	wCol
	pinifolius 'Mersea Yellow'	oFor wWoS oAls oRus oGoo oSis wEde
un	'Pink Holly'	oSis
	'Port Wine'	wHer
	'Prairie Dusk'	oFor oGar oSho wMag iArs oLSG wSnq
	'Prairie Fire'	oFor oGar
	procerus	wShR oBos wWld
un	***procerus*** 'Nisqually Cream'	oSis
	procerus var. ***tolmiei***	wMtT
un	***procerus*** var. ***tolmiei*** 'Hawkeye'	wMtT
	procumbens	see **P.** ***crandallii*** ssp. ***procumbens***
	pruinosus	oSis
	'Purple Passion'	oFor wWoS oDar oJoy oAls oGar oGoo oEga
un	'Purple Tiger'	oJoy
	'Raspberry Flair'	wWoS oJoy
	'Raven'	wWoS oJoy oAls
	'Razzle Dazzle'	oJoy
	'Red Emperor'	wWoS
un	'Red Trumpet'	wHer
	'Rich Ruby'	oHed oMis
	richardsonii	oJoy oHed oSis oBos wWld
un	***richardsonii*** var. ***richardsonii***	oRus
	roezlii	see **P.** ***laetus*** var. ***roezlii***
	rostriflorus	wCCr
	'Ruby'	see **P.** 'Schoenholzeri'
	rupicola	oBos wSta
	rupicola x ***cardwellii***	oBos
	rupicola 'Diamond Lake'	oSis
un	***rupicola*** 'Fiddler Mountain'	oSis
un	***rupicola*** 'Myrtle Hebert'	wMtT
	Scarlet Queen / 'Scharlachkoenigin'	oGoo oSis
	'Schoenholzeri'	oFor wHer wWos oNat oDar oJoy oAls oHed wFGN oGoo oCis oWnC oEga
	secundiflorus	oGoo
	serrulatus	oHan wNot oRus wShR oBos wWld wSte
	'Sissinghurst Pink'	see **P.** 'Evelyn'
un	'Sissinghurst White'	wHer
	smallii	oGoo wSnq
	'Snowflake'	see **P.** 'White Bedder'
	'Sour Grapes' (see Nomenclature notes)	oFor wCul oNat oDar oHed wFGN oGar oSho oGre oWnC wTGN oSec wAva oEga
	speciosus	wCCr
	'Stapleford Gem'	wHer oJoy oHed

	strictus	oFor wThG oGoo oSis oSha wPla wWld
	subserratus	oBos wWld
	'Sutton's Pink Bedder'	oJoy
un	***tenuis***	oJoy
	'Thorn'	wWoS oJoy oAls oHed oGoo oSis oNWe wCCr wSte
	tubaeflorus	wCCr
un	***utahensis albus***	oMis
	venustus	oMis wSte
	'Violet Kissed'	oJoy
	watsonii	oSec
un	'Wax Works'	oSis
	whippleanus	oHan oRus oSis wWld
	'White Bedder'	last listed 99/00
	'Windsor Red'	oAls
	'Wine Kissed'	oJoy
un	'Wisley Pink'	oJoy
un	**PENTHORUM**	
	sedioides	wGAc
un	**PENTOPETIA**	
	cotoneaster B63.042	oRar
	PEPEROMIA	
	asperula	last listed 99/00
	columnella	oSqu
	dolabriformis	oSqu
un	***ferreyrae***	oSqu
	nivalis	oSqu
	PERAPHYLLUM	
	ramosissimum	oFor
	PERESKIA	
un	***nicoyana***	oRar
	PERICALLIS	
	x hybridus 'Spring Glory'	oGar
	PERILLA	
	frutescens	wFai iGSc oUps wSte
	frutescens var. ***purpurascens***	oAls wFai wHom oCrm oUps
	PERIPLOCA	
	graeca	wHer
	PERNETTYA	see **GAULTHERIA**
	PEROVSKIA	
	atriplicifolia	wWoS wCul wSwC oNat oDar oJoy oAls wCSG oGar oGoo oSho oSha oBlo oGre wFai wPir iGSc wHom wMag wTGN wCoN oUps
	atriplicifolia 'Little Spire'	wWoS
	atriplicifolia 'Longin'	oFor oDar oJoy oDan oOut oWnC oEga oUps wSnq
	'Blue Spire'	oCir
	'Filigran'	oFor wHer wWoS wCul oDar oJoy oSis oOut oEga oUps wSnq
	PERSEA	
	borbonia	oFor wHer wCCr
un	'Carrizo Norte' (fruit)	oOEx
	'Fuerte' (fruit)	oOEx
	'Jim' (fruit)	oOEx
un	'Little Cado' (fruit)	oOEx
	'Mexicola' (fruit)	oOEx
	thunbergii	oFor oCis oSho
un	***yunnanensis***	oFor wHer oGos oRiv oCis wCCr wSte
	PERSICARIA	
	affinis 'Border Jewel'	oFor wHer oLSG
	affinis 'Darjeeling Red'	wHer wCul oGar oLSG oSle
	affinis 'Dimity'	see **P.** ***affinis*** 'Superba'
	affinis 'Donald Lowndes'	oHed
	sp. aff. ***affinis*** EDHCH 97098	wHer
	affinis 'Superba'	wHer oAls oLSG
	amplexicaulis	oGar wFai oSec
	amplexicaulis 'Atrosanguinea'	wHer wCul oHed
	amplexicaulis 'Firedance'	oHed
	amplexicaulis 'Firetail'	oFor wCul oGar
	amplexicaulis 'Inverleith'	wHer
	amplexicaulis 'Rosea'	wHer
	amplexicaulis Taurus / 'Blotau'	oFor
	bistorta	oRus
	bistorta 'Superba'	oFor wHer wCul wSwC oAls oGar wRob wNay wFai
	campanulata	oFor wCul wCSG
	campanulata var. ***lichiangense*** EDHCH 97108	last listed 99/00
	campanulata 'Rosenrot'	wHer oHed
	filiformis	see **P.** ***virginiana***
un	'Langthorn's Variety'	wWoS wRob
	macrophylla	last listed 99/00
	odorata	oAls oGoo iGSc wNTP
	polymorpha	oFor wNay

un	'Red Dragon'	wSwC
	tenuicaulis	wHer
	vacciniifolia	oFor wSta
	virginiana	oFor oTrP wHer wCri oJoy oGar wRob oSec
	virginiana Compton's form	wHer wWoS
	virginiana 'Lance Corporal'	wHer wSwC
un	***virginiana*** 'Langthornii'	oTrP
va	***virginiana*** 'Painter's Palette'	oFor oTrP wHer wWoS oNat oJoy oAls oDan oSho oSec
	virginiana Variegata Group	oFor wHer wRob wTGN
	wallichii	last listed 99/00
	weyrichii	last listed 99/00
	PETASITES	
	fragrans	last listed 99/00
	frigidus var. ***palmatus***	see **P. *palmatus frigidus***
	japonicus	wBlu oAls oOEx oGre oSle
un	***japonicus*** 'Akita-buki'	oOEx
	japonicus var. ***giganteus***	oFor wHer oGoo
	japonicus var. ***giganteus*** 'Nishiki-buki'	oFor wHer wWoS oGar oGoo wRob
	japonicus var. ***giganteus*** 'Variegatus'	see **P. *japonicus*** var. ***giganteus*** 'Nishiki-buki'
	japonicus f. ***purpureus***	wHer
	japonicus 'Variegatus'	see **P. *japonicus*** var. ***giganteus*** 'Nishiki-buki'
	palmatus	oAld wFFl
un	***palmatus*** 'Golden Palms'	oFor wWoS oEdg oDan oGre
un	***palmatus frigidus***	oFor oBos
un	***speciosa***	oBRG
un	**PETOPENTIA**	
	natalensis	oRar
	PETROMARULA	
	pinnata	wHer
	PETROPHYTUM	
	caespitosum	oSis
	cinerascens	oSis
	PETRORHAGIA	
	saxifraga	oDan
	saxifraga 'Pleniflora Rosea'	oWnC wTGN
	saxifraga 'Rosette'	oSis
	PETROSELINUM	
	crispum	wTho oAls wFGN wFai iGSc oWnC
	crispum 'Champion Moss Curled'	wCSG
	crispum var. ***crispum***	wTho wFai wNTP
un	***crispum*** var. ***crispum*** 'Triple Curled'	oCir
	crispum var. ***neapolitanum***	wTho oAls wFGN wFai iGSc oWnC wNTP oCir
	PETTERIA	
	ramentacea	oFor
	PETUNIA	
	integrifolia	oInd
	integrifolia white form	oInd
un	'Priscilla' (Supertunia Group)	wTGN
un	'Purple Sunspot' (Supertunia Group)	wTGN
un	Surfinia Blue Vein	wTGN
	Surfinia Pink Vein / 'Suntosol'	wTGN
	Surfinia White / 'Kesupite'	wTGN
	PEUCEDANUM	
	ostruthium	oCrm
	PHACELIA	
	bolanderi	oHan
un	***nemoralis***	wNot
	PHAIOPHELPS	
	nigricans	see **SISYRINCHIUM *striatum***
	PHALAENOPSIS	
	hybrids	oGar
	PHALARIS	
	arundinacea	wSoo wHom wBox
	arundinacea 'Dwarf's Garters'	see **P. *arundinacea*** 'Woods Dwarf'
	arundinacea var. ***picta***	oFor oAls wCSG oGre wSta oCir oUps
va	***arundinacea*** var. ***picta*** 'Feesey'	oFor wWoS wCli oDar oAls oSsd wWin oWnC wBox oUps
va	***arundinacea*** var. ***picta*** 'Luteopicta'	oDar oGre wWin
va	***arundinacea*** var. ***picta*** 'Picta'	wCli oGoo oWnC oTri
un	***arundinacea*** var. ***picta*** 'Rosea'	oGre oBRG
	arundinacea 'Strawberries and Cream'	see **P. *arundinacea*** var. ***picta*** 'Feesey'
	arundinacea 'Woods Dwarf'	oFor wTGN
	PHEGOPTERIS	
	connectilis	oRus oSis oGre
	decursive-pinnata	oFor wFan oGar oBRG
	PHELLODENDRON	
	amurense	oFor oRiv oGre
	amurense var. ***sachalinense***	oFor
	chinense	oFor oWnC
	japonicum	oFor
	lavalleei	oFor
un	***sinensis***	oCis
	PHILADELPHUS	
	argyrocalyx	oFor
	'Avalanche'	wHer oGre
	'Beauclerk'	wCCr
	'Belle Etoile'	oGos wCol oHed oGar oOut oGre wBox
	'Bouquet Blanc'	oFor wHer
do	'Buckley's Quill'	oFor wHer oGre
	'Burkwoodii'	last listed 99/00
	coronarius	oGar
	coronarius 'Aureus'	oFor wHer wCri oGos oJoy wCol oGar oDan oGre wWel
va	***coronarius*** 'Variegatus'	wHer
	delavayi	oTrP oCis
	delavayi var. ***calvescens***	see **P. *purpurascens***
	delavayi EDHCH 97170	wHer
	'Dwarf Minnesota Snowflake'	oFor wCul oGar oDan oBlo
	'Galahad'	oGos oGre
va	'Innocence'	oFor wHer oGre wWel
	laxiflorus	last listed 99/00
	lewisii	oFor oHan wWoB wBur wNot wShR oGar oBos oRiv oGre wWat wCCr iGSc oTri wPla oAld oSle
	lewisii 'Goose Creek'	oFor wHer
	lewisii var. ***gordonianus***	wFFl
	lewisii var. ***gordonianus*** 'Waterton'	wWel
do	'Manteau d'Hermine'	oHed
	mexicanus	oFor
	microphyllus	wHer wCCr
do	'Minnesota Snowflake'	oFor wHer oNat oJoy oGar oDan oRiv oSho oOut oBlo oGre wKin oWnC wTGN wWel
do	'Natchez'	oGar wWel
	pubescens	wCCr
	purpurascens	last listed 99/00
	schrenkii HC 970381	wHer
	schrenkii var. ***jackii***	oFor
	sp. aff. ***subcanus*** DJHC 805	wHer
	tomentosus	oFor
do	'Virginal'	oJoy oAls oHed oGar oDan
	x virginalis	wCoS
un	***x virginalis*** 'Snow Velvet'	oFor
	White Rock	oFor
	PHILLYREA	
	angustifolia	wHer
	latifolia	oFor wCCr
	media	see **P. *latifolia***
	PHILODENDRON	
	cordatum	oGar
	pertusum	see **MONSTERA *deliciosa***
un	'Xanadu'	oGar wCoS
	PHLOMIS	
	cashmeriana	oTrP oJoy
	ferruginea	wHer
	fruticosa	oFor wCul wSwC oGar oSho wFai wCCr wMag oSle
	fruticosa dwarf form	wHer wSwC oSis wCCr wBox
	italica	wHer oAls oHed wCCr
	lanata	oHed oCis oSho wEde
	longifolia	oFor
	macrophylla HWJCM 439	last listed 99/00
	purpurea	last listed 99/00
	russeliana	oFor oTrP wHer wCul wCri wSwC oJoy wHig oAls oRus wFGN wCSG oGar oCis wMag oSec
	samia	see **P. *russeliana***
	tuberosa	oTrP wHer oJoy wHig wBWP
	tuberosa 'Amazone'	oGar
	PHLOX	
un	***adsurgens*** 'Mary Ellen'	wMtT
un	***adsurgens*** 'Oregon Blush'	wMtT
	adsurgens 'Wagon Wheel'	wCol oSis
	x arendsii 'Anja'	oFor oSho
	x arendsii 'Suzanne'	last listed 99/00
	bifida	oSis
un	***bifida*** 'Betty Blake'	oSis
	borealis	oSis
	buckleyi	oFor oSis wHom oUps
	carolina 'Magnificence'	iArc
	carolina 'Miss Lingard'	wFai oWnC
	condensata	wMtT
	diffusa	oAls oBos
un	***diffusa*** 'Alba'	wMtT
un	***diffusa*** 'Goat Rocks Pink'	wMtT
	divaricata	wThG

	divaricata 'Blue Perfume'	wWoS
	divaricata 'Clouds of Perfume'	wWoS oNat oWnC
	divaricata 'Dirigo Ice'	wWoS
	divaricata 'Eco Texas Purple'	oSis
	divaricata ssp. ***laphamii*** 'Chattahoochee'	oFor
un	***divaricata*** 'London Grove'	oFor oWnC iArc
un	***divaricata*** 'London Grove Blue'	oSis
	divaricata 'Louisiana Purple'	last listed 99/00
un	***divaricata*** 'Montrose Tricolor'	wWoS
un	***divaricata*** 'Plum Perfect'	wWoS
un	***divaricata*** 'Sweet Lilac'	oFor wWoS oGar oSis
	divaricata 'White Perfume'	wWoS
un	***douglasii*** 'Appleblossom'	wMtT
	douglasii 'Crackerjack'	oFor oGar oSis
un	***douglasii*** 'May Snow'	oSis
	douglasii 'Rose Cushion'	oFor oSis
	douglasii 'Waterloo'	oSis
un	'Foxy Lady'	oSis
un	***glaberrima*** 'Anita Kistler'	oFor wWoS
un	***glaberrima*** 'Morris Berd'	oWnC wSnq
	glaberrima ssp. ***triflora***	last listed 99/00
un	***hendersonii***	wMtT
	maculata 'Alpha'	oFor oJoy wRob
	maculata 'Delta'	last listed 99/00
	maculata 'Natascha'	oFor wWoS
	maculata 'Omega'	oFor
	maculata 'Rosalinde'	oFor oGar
	'Millstream Jupiter'	last listed 99/00
un	'Miss Margie' (Springpearl hybrid)	wWoS
un	'Miss Mary' (Springpearl hybrid)	wWoS
	'Miss Wilma' (Springpearl hybrid)	wWoS
un	'Morris Berd'	wWoS
	muscoides	wMtT
	paniculata	oMis oSho
un	***paniculata*** 'Andre'	oSho
	paniculata 'Blue Boy'	oFor oNat oGar oWnC wMag wBox iArc oLSG
	paniculata 'Blue Moon'	oLSG
	paniculata 'Blue Paradise'	oWnC
	paniculata 'Brigadier'	oLSG
	paniculata 'Bright Eyes'	wHer oNat oWnC iArc oCir oLSG
un	***paniculata*** 'Charles Curtis'	oFor
	paniculata 'Darwin's Joyce'	see **P. *paniculata*** 'Nora Leigh'
	paniculata 'David'	oFor wWoS oAls oAmb oGar oSho wRob oWnC wMag oUps
	paniculata 'Dodo Hanbury Forbes'	oNat
	paniculata 'Eva Cullum'	oFor oGar oGre wMag
	paniculata 'Eventide'	oLSG
	paniculata 'Flamingo'	wWoS
	paniculata 'Franz Schubert'	oFor oJoy oAls oGar wFai oWnC iArc oLSG
	paniculata 'Fujiyama'	oFor wWoS wCul oJoy oDan oSha oGre wHom wMag wBox oLSG
un	***paniculata*** 'Garden Lavender'	oJoy
va	***paniculata*** 'Harlequin'	oFor wWoS
	paniculata 'Iris'	wHer
un	***paniculata*** 'Juliet'	oWnC oLSG
un	***paniculata*** 'Katherine'	wWoS
	paniculata 'Laura'	wWoS
	paniculata lavender	oSha
	paniculata 'Little Boy'	last listed 99/00
	paniculata 'Miss Elie'	last listed 99/00
	paniculata 'Miss Holland'	last listed 99/00
	paniculata 'Miss Kelly'	wWoS
	paniculata 'Miss Universe'	oGre
	paniculata mixed	wCSG wTGN wCoN
	paniculata 'Mother of Pearl'	wHer
	paniculata 'Mount Fuji'	see **P. *paniculata*** 'Fujiyama'
	paniculata 'Mount Fujiyama'	see **P. *paniculata*** 'Fujiyama'
	paniculata 'Nicky'	wWoS
va	***paniculata*** 'Norah Leigh'	oFor wHer wWoS oGos wCol wFGN oGar oNWe wRob oGre wFai wMag wBox oUps
	paniculata 'Orange Perfection'	oFor wWoS wMag iArc oLSG oUps
	paniculata 'Pax'	oNat
un	***paniculata*** 'Pink Gown'	oWnC wMag
	paniculata 'Prime Minister'	oFor wMag oLSG
un	***paniculata*** 'Red Super'	oSho
un	***paniculata*** 'Robert Poore'	wSnq
	paniculata 'Sandra'	oFor
un	***paniculata*** 'Shortwood'	wWoS
un	***paniculata*** 'Spring Delight'	oLSG wSnq
	paniculata 'Starfire'	oFor wHer oNat oJoy oGar oNWe oSha wMag oLSG oUps
un	***paniculata*** 'Starlight'	oFor wWoS
	paniculata 'The King'	oNat wFai oLSG oUps
un	***paniculata*** 'Thunderbolt'	oLSG
un	***pilosa*** 'Moody Blue'	oSis
	x procumbens 'Variegata'	oHed wTGN
un	'Sileneflora'	oSis wMtT
un	'Spring Delight'	oMis
	stolonifera	wCoN
	stolonifera 'Blue Ridge'	oNWe oGre oWnC oLSG
	stolonifera 'Bruce's White'	oLSG
un	***stolonifera*** 'Daybreak'	wWoS
	stolonifera 'Fran's Purple'	wCol
	stolonifera 'Home Fires'	oFor iArc
	stolonifera 'Pink Ridge'	oFor oSis oWnC oLSG
	stolonifera 'Sherwood Purple'	oFor oLSG
	stolonifera variegated	oGre
	subulata	wMag wCoN
	subulata 'Amazing Grace'	oSis
	subulata 'Atropurpurea'	wMag
	subulata blue	wTho wCSG oBlo iArc
un	***subulata*** 'Blue Hills'	oSis
	subulata 'Candy Stripe'	oFor oAls wCSG oGar wHom oWnC wMag wTGN oLSG wWhG
	subulata 'Coral Eye'	oSis wWhG
	subulata dark pink	wSta
un	***subulata*** 'Dirigo Arbutus'	oFor
	subulata 'Emerald Blue'	oAls oGar wHom oWnC wWhG
un	***subulata*** 'Emerald Green'	oAls
	subulata 'Emerald Pink'	oFor oAls wWhG
	subulata 'Fort Hill'	last listed 99/00
un	***subulata*** 'Late Red'	oAls
	subulata lavender	wSta
	subulata 'McDaniel's Cushion'	oSis
un	***subulata*** 'Millstream Daphne'	oFor
va	***subulata*** 'Nettleton Variation'	wWoS
	subulata 'Oakington Blue Eyes'	wMag
	subulata pink	oGar oBlo wBox wSta iArc oCir
	subulata red	oBlo iArc
	subulata 'Red Wings'	oSis oWnC wMag wTGN
un	***subulata*** 'Rosea'	oAls
	subulata 'Samson'	last listed 99/00
un	***subulata*** 'Santa Fe'	oWnC
	subulata 'Scarlet Flame'	oAls wHom wWhG
	subulata 'Tamaongalei'	oSis
	subulata 'Temiskaming'	oSis
	subulata white	oBlo iArc oCir
	subulata 'White Delight'	oWnC
un	'Sunrise'	wCol
un	'Tycoon'	oSis
un	**PHOEBE**	
	chekianjensis	wHer oCis
	formosana	wHer
	sheareri	see **P. *formosana***
	PHOENICAULIS	
	cheiranthoides	last listed 99/00
	PHOENIX	
	canariensis	oFor
	roebelonii	oGar
	PHORMIUM	
	'Amazing Red'	wHer oGos wSwC oCis
va	'Apricot Queen'	wHer wSwC
un	'Atropurpurea Nana'	oDan
	'Bronze Baby'	oOut
un	'Chocolate Baby'	wHer oGos
un	'Chocolate Fingers'	wHer oGos
	cookianum	wCli wBox
un	***cookianum*** 'Chocolate'	wHer oGos
	cookianum 'Flamingo'	wHer oGos wSwC
va	***cookianum*** ssp. ***hookeri*** 'Cream Delight'	wBox
	cookianum ssp. ***hookeri*** 'Tricolor'	wSwC oCis
	'Dark Delight'	wBox
un	'Dark Moon'	wHer
	'Dusky Chief'	wHer oCis
un	'Dwarf Burgundy'	wHer
va	'Jack Spratt'	wCli oGar oGoo oGre wBox oBRG wWel
	'Jester'	oGos
va	'Maori Chief'	last listed 99/00
va	'Maori Maiden'	wSwC oOut wBox oUps
va	'Maori Queen'	oCis
va	'Maori Sunrise'	wBox oUps
un	'Maori Sunset'	oUps
	'Pink Stripe'	wHer wCli oGos wSwC oCis wBox
	'Platt's Black'	wHer wCli oGos
	'Sea Jade'	wBox
va	'Sundowner'	oGos oUps
va	'Sunset'	oGar
	tenax	wCli oAls oGoo wCCr oCir

	tenax atropurpureum	see P. tenax Purpureum Group
	tenax bronze	wWoS wWin wCCr oUps wSnq
	tenax greens and reds	wPir
	tenax 'Nanum Purpureum'	wWel
	tenax Purpureum Group	oFor wCli oGar oSho oOut wTGN
un	tenax 'Rubrum'	see P. tenax Purpureum Group
	'Tom Thumb'	wCli oGos oGre
un	'Wings of Gold'	oGos
va	'Yellow Wave'	wHer wCli oGos wSwC oCis oGre wBox
	PHOTINIA	
	arbutifolia	oSho wCCr
un	arbutifolia 'Davis Gold'	oFor
	beauverdiana	wCCr
	beauverdiana var. notabilis	oJil
	davidiana	oFor oBlo oGre wPir wCCr wCoN
va	davidiana 'Palette'	last listed 99/00
	davidiana var. undulata	wPir wSta wWel
	davidiana var. undulata 'Fructu Luteo'	wHer
	davidsoniae	oCis
	x fraseri	oFor oAls oGar oSho oBlo oGre oTPm oWnC wSta oCir wCoN wWel
	x fraseri 'Red Robin'	oFor
	glabra	oBlo wCCr oWnC
va	glabra 'Parfait'	wHer
	glabra 'Variegata'	see P. glabra 'Parfait'
	serratifolia	wCSG oCis oGre wCCr
	serrulata	see P. serratifolia
	sp. EDHCH 97174	wHer
	villosa	oFor
	PHRAGMITES	
	australis 'Aureus'	see P. autralis 'Variegatus'
	australis 'Variegatus'	wSoo wGAc oSsd
	karka 'Candy Stripe'	wGAc oSsd oWnC
	PHUOPSIS	
	stylosa	wWoS
	PHYGELIUS	
	aequalis	last listed 99/00
un	aequalis 'Red Trumpet'	oBlo
	aequalis 'Yellow Trumpet'	oFor oTrP wWoS wCri oJoy oAls oAmb oMis oGoo oSis oBlo wCCr oUps wSte wSnq
	capensis	oTrP oJoy oMis oRiv wCCr oWnC
un	capensis 'Calvor's Variegated'	wHer
	capensis coccineus	see P. capensis
	capensis orange	wHer oHed
	x rectus	oAls wCSG wPir oWnC
	x rectus 'African Queen'	oFor wWoS wCri oNat oDar oJoy oAls oHed oGoo oSho oGre wAva wSnq
	x rectus 'Devil's Tears'	oDar oJoy oHed oAmb oGar oSho oOut oGre wBox oEga oUps wSte wSnq
un	x rectus 'Lemon Drop'	oJoy
	x rectus 'Moonraker'	oFor wHer wWoS oGos wSwC oNat oDar oJoy oAls oHed oGar oNWe oGre wHom wSte wSnq
	x rectus 'Pink Elf'	wHer wWoS oNat oJoy wHom oUps wSnq
	x rectus 'Salmon Leap'	wWoS oDar oJoy oHed wSnq
	x rectus 'Winchester Fanfare'	oFor wWoS wCri wSwC oNat oDar oJoy oHed wCSG oAmb oGar oGre wFai
	'Sensation'	oHed
	'Trewidden Pink'	wWoS oJoy oGar oGoo wAva oEga
	PHYLA	
	nodiflora	last listed 99/00
un	nodiflora var. rosea	wHer
	X PHYLLIOPSIS	
	hillieri 'Askival'	wRho
	hillieri 'Pinocchio'	oBov
	'Sugar Plum'	oFor wHer wClo oSis wMtT oGre
	PHYLLITIS	see **ASPLENIUM**
	PHYLLOCLADUS	
	aspleniifolius var. alpinus	wHer oPor wWel
	PHYLLODOCE	
	aleutica ssp. glanduliflora 'Flora Slack'	oOut
	empetriformis	oBos wMtT
	x intermedia 'Drummondii'	oOut
	PHYLLOSTACHYS	
	angusta	oTra wSus wCli oNor wBea wBox wBmG
	arcana	oTra wSus wBea wBox wBmG
	atrovaginata	wCli oNor wBea wBmG oTBG
	aurea	oFor oTra wCli oNor wBur wClo wBea oDar oAls oGar oSho oGre wSta wBam wBmG oTBG wCoS wWel
	aurea 'Albovariegata'	oTra wCli wBea
un	aurea 'Flavescens'	wBam
	aurea 'Flavescens Inversa'	oTra wCli oNor wBea oGre wBox wBmG oTBG
	aurea 'Holochrysa'	oTra wCli wBea wBox oTBG
	aurea 'Koi'	oTra wCli oNor wBea oAls wBox wBam wBmG oTBG
un	aurea 'Takemurai'	oTra
	aurea 'Variegata'	oNor
	aureosulcata	oTra wSus wCli oNor wBlu wClo oGar oOEx iGSc wTGN wBam wBmG oTBG
	aureosulcata f. alata	wSus wCli oNor wBlu wBea oAls wBox wBam wBmG oTBG
	aureosulcata 'Aureocaulis'	oTra wCli oNor wBea wRai wBmG oTBG
	aureosulcata 'Harbin'	oTra wCli wBea wBmG
	aureosulcata 'Spectabilis'	oTra wSus wCli oNor wBea oOEx wBox wBam wBmG oTBG
	bambusoides	oTra wSus wCli wBea oDar oGar oSho wRai oOEx wBox wBam oTBG
un	bambusoides 'Albovariegata'	wBea
	bambusoides 'Allgold'	oTra wCli wBea wBox wBmG oTBG
	bambusoides 'Castillonis'	oTra wCli oNor wBea wRai oOEx wBam wBmG oTBG
	bambusoides 'Castillonis Inversa'	oTra
un	bambusoides 'Golden Dwarf'	oTra
un	bambusoides 'Job's Spot'	oTra
	bambusoides 'Kawadana'	wBea oTBG
	bambusoides 'Marliacea'	wCli wBea oTBG
	bambusoides rib leaf	oTra
un	bambusoides 'Slender Crookstem'	oNor wBmG
	bambusoides 'Tanakae'	wBea
	bissetii	oTra wSus wCli oNor wBlu wClo wBea wTGN wBam wBmG oTBG
	bissetii dwarf	wBmG
	congesta	see P. atrovaginata
	decora	oTra wCli oNor wClo wBea wBmG oTBG
	dulcis	oTra wCli oNor wBea oTBG
	edulis	oTra wCli oNor wBea oOEx wBox wBam wBmG oTBG
	edulis gold stem Stan Terlitsky	wBea
un	edulis 'Goldstripe'	oTra
un	edulis 'Inodami'	wCli
un	edulis 'Savannah'	wBea
un	edulis 'Soft Gold'	wCli wBea
un	edulis 'Spring Beauty'	wCli
	edulis tall straight clone	wBea
	edulis white stripe clone	wBea
	flexuosa	oTra wCli oNor wBlu wBmG
	flexuosa seedlings	oTra
un	flexuosa 'Kimmei'	oTra
	flexuosa recovered flowering	wBea
	flexuosa 1992 seedling	wBea
	glauca	oTra oNor oGre wBox
	glauca 'Yunzhu'	oTBG
	heteroclada	oNor wBlu oAls wBam wBmG oTBG
	heteroclada 'Solid Stem'	see P. heteroclada 'Straight Stem'
	heteroclada 'Straight Stem'	oTra oNor wBea wBmG oTBG
	heterocycla f. pubescens	see P. edulis
	humilis	oTra wCli oNor wBlu wBea wBam wBmG
un	incarnata	wCli
	iridescens	oTra
	makinoi	oTra oNor wBea
	mannii	wBea
	meyeri	oTra oNor wBea wBmG oTBG
	nidularia	oTra wCli wBea wBmG
	nigra	oFor oTra wSus wCli oNor wBlu wBur wClo wBea oDar oGar wRai oGre wBox wBam wBmG oTBG wWel
	nigra 'Bory'	see P. nigra 'Boryana'
	nigra 'Boryana'	oTra wCli oNor wBlu wBea oOEx wBam wBmG oTBG
un	nigra 'Dikokuchiku'	wCli
	nigra 'Hale'	wSus wCli wBea
	nigra 'Henon'	see P. nigra var. henonis
	nigra var. henonis	oTra wSus wCli oNor wBlu wBea oAls wRai oOEx oGre wBox wBam wBmG oTBG
	nigra 'Megurochiku'	oTra wSus wCli oNor wBlu wBea wBox wSta wBam wBmG oTBG
	nigra f. punctata	oNor oOEx
un	nigra 'Shimadake'	oTra wCli
	nuda	oFor oTra wCli oNor wBlu wClo wBea oGar wRai wBam wBmG oTBG
	parvifolia	wCli
	platyglossa	oTra wCli oNor wBea oTBG
	praecox	oTBG
	pubescens	see P. edulis
	purpurata	see P. heteroclada

	Name	Sources
	rubromarginata	oTra wCli oNor wBlu wBea oOEx wBam wBmG
	sulphurea 'Houzeau'	wCli wBea wBox
	sulphurea 'Robert Young'	oTra wCli oNor wBlu wBea oAls oOEx wBox wBam oTBG
	sulphurea var. viridis	wCli
	violascens	oTra wCli oTBG
	viridiglaucescens	wCli oNor wBea
	viridis	see P. sulphurea var. viridis
	vivax	oTra wSus wCli oNor wCul wBur wClo wBea oGar wRai oOEx wBam wBmG oTBG
	vivax 'Aureocaulis'	oTra wCli wBea wBmG oTBG
	X PHYLLOTHAMNUS	
	erectus	oSis
	PHYSALIS	
	alkekengi	oGar oGoo oCir
	alkekengi var. franchetii	oAls oOEx oUps
	alkekengi var. franchetii 'Gigantea'	oFor
	PHYSOCARPUS	
	amurensis	oFor
	bracteatus	oFor
	capitatus	oFor oHan wWoB wNot oBos oRiv wWat oWnC oTri oAld wFFl oSle
	malvaceus	oFor wThG wPla
	opulifolius	wCCr wKin
	opulifolius 'Dart's Gold'	oFor wHer oAmb oGar oDan wRob oGre wFai wPir wKin oSec wWel
	opulifolius Diabolo™ / 'Monlo'	oFor wHer oGos oGar oNWe wRob wWel
	opulifolius 'Nanus'	oFor
un	opulifolius 'Nugget'	oFor
	PHYSOSTEGIA	
un	leptophylla	wGAc oSsd
	virginiana	oFor wCSG oWhS
	virginiana 'Alba'	oFor wCul wHom oEga
	virginiana 'Crown of Snow'	oFor oAls wCSG oWnC
	virginiana 'Pink Bouquet'	oAls oLSG
	virginiana 'Rosea'	oAls wHom oWnC iArc oCir
	virginiana ssp. speciosa 'Variegata'	wHer wWoS wSwC wCol oSho oOut wRob oWnC wMag oUps wSnq
	virginiana 'Summer Snow'	oFor oWnC wBox oSec oLSG oUps
	virginiana 'Variegata'	see P. virginiana ssp. speciosa 'Variegata'
	virginiana 'Vivid'	oFor wCul wMag oSec oLSG oUps
	PHYTEUMA	
	betonicifolium	oJoy
	scheuchzeri	wHer
	PHYTOLACCA	
	acinosa	oOEx
	americana	wHer oGoo oCrm
	esculenta DJH 298	wHer
	PICEA	
	abies	oGar oRiv oSho wKin oWnC wCoS
	abies 'Aarburg'	oPor
un	abies 'Abishenski Pendula'	oPor
	abies 'Acrocona'	oFor oPor oDan oRiv
un	abies 'Arnold Special'	oPor
	abies 'Aurea'	oPor wWel
un	abies 'Aurea Jacobsen'	oPor
un	abies 'Bergman's Flattop'	oPor
un	abies 'Bonitz'	oPor
	abies 'Cincinnata'	oPor
un	abies 'Clanbrassiliana Stricta'	oPor
	abies 'Compacta Asselyn'	oPor
	abies 'Conica'	oBRG
	abies 'Cupressina'	oFor oPor oGre wAva wWel
un	abies 'Dan's Dwarf'	oPor
	abies 'Diffusa'	oPor
	abies 'Echiniformis'	oPor
un	abies 'Elegantissima'	oPor oRiv
	abies 'Ellwangeriana'	oPor
	abies 'Emsland'	oPor
un	abies 'Eva'	oPor
	abies 'Farnsburg'	oPor
un	abies 'Finland'	oPor
	abies 'Formanek'	oPor wCol
	abies 'Frohburg'	oPor oGre
un	abies 'Glauca Pendula'	oGre oWnC
un	abies 'Gold Dust'	oPor oRiv
un	abies 'Gold Strike'	wCol
	abies 'Gregoryana Parsonsii'	oGar
un	abies 'Gull's Nest'	oPor
	abies 'Highlandia'	oPor
	abies 'Hillside Upright'	oPor
	abies 'Humilis'	oPor
un	abies 'Humphrey's Gem'	oPor
	abies 'Hystrix'	oPor
	abies 'Inversa'	oFor oPor wCol
un	abies 'Kellerman's Blue'	oGre
un	abies 'Kellerman's Blue Cameo'	oPor oRiv
	abies 'Kluis'	oPor
un	abies 'Kmak'	oPor
	abies 'Little Gem'	oFor oPor oGos oAls oGar oSis oRiv oSho oSqu oGre oTPm oWnC wSta oBRG wWel
	abies 'Lombarts'	oPor
un	abies 'Malena'	oPor
	abies 'Maxwellii'	oPor oGar wSta
	abies 'Merkii'	last listed 99/00
un	abies 'Mountain Dew'	oPor
un	abies 'Mrs. Cessanni'	oPor
	abies 'Mucronata'	oPor oRiv oGre oBRG
	abies 'Nana'	oPor
	abies 'Nidiformis'	oPor oAls oGar oRiv wKin oTPm oWnC wTGN wSta wAva wCoS wWel
	abies obovata	see P. obovata
	abies 'Ohlendorffii'	oPor
	abies 'Pachyphylla'	last listed 99/00
	abies 'Parsonii'	oPor wCol
	abies 'Pendula'	oPor oDar oGar oRiv oGre wKin oWnC wTGN wBox wSta wAva
un	abies 'Perry's Gold'	oPor wCol
	abies 'Procumbens'	oPor wCol
un	abies 'Pseudoprostrata'	oPor
	abies 'Pumila'	oFor oRiv oGre
	abies 'Pumila Nigra'	oPor oWnC
	abies 'Pusch'	oPor wCol
un	abies 'Pustertal'	oPor
	abies 'Pygmaea'	oPor oGar oSis wSta
un	abies 'Pyramidalis Nana'	oRiv
	abies 'Reflexa'	oPor wCol
	abies 'Remontii'	last listed 99/00
	abies 'Repens'	oFor oPor wWel
	abies 'Repens' gold sport	oPor
un	abies 'Rothenhaus'	oPor
	abies 'Rubra Spicata'	oPor
	abies 'Saint James'	oPor
un	abies 'Sherwood Compact'	oPor
un	abies 'Sherwood Multnomah'	oPor wSta
	abies 'Tabuliformis'	oPor
	abies 'Virgata'	oFor
	abies 'Wartburg'	oPor
un	abies 'Witches Brood'	oPor
	abies 'Wills Zwerg'	oPor
un	alcockiana 'Howell's Dwarf'	oPor oRiv oGre
	alcockiana 'Prostrata'	oPor oGre
	asperata	oFor
un	asperata var. aurantiaca 'Hunnewelliana'	oPor wCol
	aurantiaca	see P. asperata var. aurantiaca
	balfouriana	oPor oGre
	bicolor	see P. alcockiana
	brachytyla	oFor oPor wClo oGar
	breweriana	oFor oPor oRiv wTGN
	chihuahuana	oFor oPor oRiv
	engelmannii	oFor oRiv wKin
un	engelmannii 'Bravo'	oRiv
un	engelmannii 'Bush's Lace'	oPor wCol
un	engelmannii 'Compacta'	oPor
	engelmannii 'Snake'	wCol
un	engelmannii 'Stanley Mountain'	oPor
un	engelmannii 'Swan Creek'	oPor
un	engelmannii 'Vanderwolf Blue'	oPor oRiv
un	gemmata	oPor
	glauca	oFor oRiv
	glauca var. albertiana 'Alberta Globe'	oPor
	glauca var. albertiana 'Conica'	oFor oJoy oAls oGar oRiv oSho oBlo oSqu oGre wKin oTPm oWnC wSta oBRG oGue wCoS wWel
	glauca var. albertiana 'Gnome'	oPor
	glauca var. albertiana 'Laurin'	oPor oBRG
	glauca 'Arneson's Blue'	wCol oAls oGar
	glauca 'Arneson's Blue Variegated'	oPor
un	glauca 'Beehive'	oPor
	glauca 'Cecilia'	oPor wCol
	glauca var. densa	see P. glauca 'Densata'
	glauca 'Densata'	oFor oPor oGar oRiv wKin oWnC
un	glauca 'Ducharme'	oPor
	glauca 'Echiniformis'	oPor oSis oGre wSta oBRG
	glauca 'Elf'	last listed 99/00
un	glauca 'Hobbit'	oPor
	glauca 'Hudsonii'	oPor wTGN
	glauca 'Jean's Dilly'	oPor wBox

	glauca KBN variegated	oPor
un	***glauca*** 'Little Globe'	oPor wWel
un	***glauca*** 'Morton Arboretum Weeping'	oRiv
	glauca 'Pendula'	oPor wCol oRiv
	glauca 'Pixie'	wCol oSis
qu	***glauca*** 'Rainbow'	oFor
	glauca 'Rainbow's End'	oPor wCol oSis oRiv
	glauca 'Sander's Blue'	oPor oRiv oSho oGre
	glauca 'Sander's Fastigiate'	oPor oGre
	glauca 'Waconda'	oPor
	glauca 'Wild Acres'	oPor
	glauca witch's broom	oBRG
	glehnii	oFor oPor oRiv wSta
	x hurstii	last listed 99/00
	jezoensis	oFor wSta
	jezoensis 'Aurea'	last listed 99/00
	jezoensis ssp. ***hondoensis***	oPor
	jezoensis 'Yatsubusa'	oPor
ch	***koraiensis***	oFor
	koyamae	oFor oPor oRiv oGre
	likiangensis	oFor oPor oRiv
	likiangensis var. ***balfouriana***	see **P. *balfouriana***
	likiangensis var. ***purpurea***	see **P. *purpurea***
	mariana	oFor oRiv
	mariana 'Aureovariegata'	oPor wWel
	mariana 'Beissneri'	last listed 99/00
	mariana 'Doumetii'	oPor
	mariana 'Ericoides'	oPor oAls oGar oRiv
	mariana 'Nana'	oPor oGar oSis oRiv oTPm
	x mariorika 'Kobold'	oPor
	x mariorika 'Machala'	oPor
	maximowiczii	oPor
	meyeri	oFor oPor oRiv
	montigena	oFor oPor
	morrisonicola	oPor
	obovata	oFor oPor
	omorika	oFor oPor oDar oGar oRiv wCCr wKin wAva
	omorika 'Aurea'	oPor wCol
un	***omorika*** 'Berliner's Weeping'	oPor oGre
	omorika 'Frohnleiten'	oPor
un	***omorika*** 'Gotelli Weeping'	wCol
un	***omorika*** 'Hexenbesen'	oPor wCol
un	***omorika*** 'Kuck Weeping'	oPor
	omorika 'Nana'	oFor oPor oGar oDan oRiv oSho oGre oWnC
	omorika 'Pendula'	oFor oPor oRiv oGre oWnC
	omorika 'Pendula Bruns'	oPor wCol
	omorika 'Pimoko'	oPor wCol oGre oWnC
	omorika 'Riverside'	wHer wWel
	omorika 'Treblitsch'	oPor
un	***omorika*** 'Wodan'	oPor
	orientalis	oFor wHer oRiv wCCr wSte
	orientalis 'Atrovirens'	oPor
	orientalis 'Aurea'	oPor
	orientalis 'Aurea Compacta'	oAls oGar
	orientalis 'Aureospicata'	oFor oPor oGar oGre wWel
	orientalis 'Barnes'	oPor wCol oGre
	orientalis 'Bergman's Gem'	oPor wCol oWnC
	orientalis 'Compacta'	oPor
	orientalis 'Gowdy'	oPor
	orientalis 'Gracilis'	oPor
	orientalis 'Green Knight'	wWel
un	***orientalis*** 'Martin K'	oPor
	orientalis 'Mount Vernon'	oPor
	orientalis 'Nana'	oRiv
	orientalis 'Nutans'	oPor
	orientalis 'Pendula'	oFor
un	***orientalis*** 'Rariflora Monstrase'	oRiv
un	***orientalis*** 'Shadow Broom'	oPor
	orientalis 'Skylands'	oPor oGar oRiv oWnC
	orientalis 'Tom Thumb'	oPor
un	***orientalis*** van Spebrock	oPor
un	***orientalis*** 'Weeping Dwarf'	oPor
	polita	see **P. *torano***
	pungens	oGar oRiv
un	***pungens*** 'Baby Blue Eyes'	wSta
	pungens 'Bakeri'	oAls oGar oRiv oSho oTPm wSta
un	***pungens*** 'Blue Mist'	oPor
	pungens 'Blue Pearl'	oPor wCol
un	***pungens*** 'Blue Softie'	oPor
un	***pungens*** 'Buckwheat'	oPor
un	***pungens*** 'Colorado Select'	oGre
un	***pungens*** 'Conica'	oSle
	pungens 'Copeland'	oPor
un	***pungens*** 'Dani'	oPor
un	***pungens*** 'Dietz Prostrate'	oPor
	pungens 'Donna's Rainbow'	oPor
	pungens 'Edith'	oPor
un	***pungens*** 'Egyptian Pyramid'	oPor
	pungens 'Emerald Cushion'	oPor
	pungens 'Endtz'	oPor
	pungens 'Erich Frahm'	oPor
un	***pungens*** 'Fastigiata'	wCol oGar wSta
	pungens 'Fat Albert'	oFor oPor oGar oDan oGre
un	***pungens*** 'Foxtail' (see Nomenclature Notes)	oFor oPor
	pungens 'Fuerst Bismarck'	oPor
un	***pungens*** 'Gentry's Gem'	oPor
un	***pungens*** 'Girard Dwarf'	wCol
	pungens f. ***glauca***	oFor oGar oRiv oSho oWnC wPla wCoS
un	***pungens*** f. ***glauca*** 'Apache'	wKin
un	***pungens*** f. ***glauca*** 'Blue Wind'	oEdg
un	***pungens*** f. ***glauca*** 'Kaibab'	wKin
un	***pungens*** f. ***glauca*** 'Pagosa'	wKin
un	***pungens*** 'Glauca Compacta'	oPor
	pungens 'Glauca Globosa'	see **P. *pungens*** 'Globosa'
	pungens 'Glauca Pendula'	oFor oPor
	pungens 'Glauca Procumbens'	oPor oRiv
	pungens 'Glauca Prostrata'	oPor
	pungens 'Globosa'	oPor oAls oDan oRiv oSho oGre oTPm
	pungens green globe	oPor
	pungens 'Hoopsii'	oFor oPor oAls oGar oRiv oGre oTPm oWnC wSta
	pungens 'Hunnewelliana'	oPor oGar
	pungens 'Iseli Fastigiate'	oPor oTPm oWnC
	pungens 'Iseli Foxtail' (see Nomenclature Notes)	oWnC
	pungens 'Jean Iseli'	oPor wCol
	pungens 'Koster'	oGre wSta
un	***pungens*** 'Koster Pendula'	oTPm
	pungens 'Mesa Verde'	oPor
	pungens 'Mission Blue'	oPor
un	***pungens*** 'Misty Blue'	oTPm
	pungens 'Moerheimii'	oFor oPor oSho oWnC
	pungens 'Moll'	oPor
	pungens 'Montgomery'	oPor oAls oGar oRiv oGre oTPm wSta
	pungens 'Mrs. Cesarini'	wCol oGar oRiv oGre
	pungens 'Pendens'	oPor
	pungens 'Pendula'	wWel
	pungens 'Porcupine'	oPor
qu	***pungens*** 'Prostrata Kosteri Weeping'	oGre
	pungens 'R.H. Montgomery'	see **P. *pungens*** 'Montgomery'
	pungens 'Saint Mary'	see **P. *pungens*** 'Saint Mary's Broom'
	pungens 'Saint Mary's Broom'	oPor wCol oGre
	pungens 'Silver Falls'	oPor
	pungens 'Split Rock'	oPor
	pungens 'Spring Ghost'	oPor
	pungens Tallmadge Burley	oPor
	pungens 'Taponus'	last listed 99/00
	pungens 'Thomsen'	oPor
	pungens 'Thuem'	oPor wCol
	pungens 'Victor II'	oPor
	pungens 'Walnut Glen'	oPor
	pungens 'Yvette'	oPor
	pungens #3	wCol
	purpurea	oFor oPor oRiv
	retroflexa	oPor
	rubens	oFor
un	***rubens*** 'Pocono'	oPor
	schrenkiana	oFor
un	***schrenkiana*** 'Nana'	oPor
	sitchensis	oFor wWoB wNot oDar wShR oGar oBos oRiv oSho wWat oAld wAva wFFl wRav oSle
	sitchensis 'Papoose'	see **P. *sitchensis*** 'Tenas'
	sitchensis 'Sugar Loaf'	oFor oPor wCol
	sitchensis 'Tenas'	oFor oPor oAls oGar oDan oRiv oGre oWnC wWel
	sitchensis 'Virgata'	oPor
	smithiana	oFor oPor oRiv oGre wCCr
	torano	oPor
	wilsonii	oPor

PIERIS

	'Bert Chandler'	oBRG
	'Brouwer's Beauty'	oBov oGre
va	'Flaming Silver'	oGos oBRG
	'Flamingo'	oFor oGar oSho oBlo oGre oTPm
	'Forest Flame'	wSwC oAls oGar oRiv oSho oBlo oTPm oWnC oBRG

	Name	Sources
	formosa var. *forrestii*	last listed 99/00
	formosa HWJCM 134	wHer
	japonica	wCSG oGar oGre wBox
un	*japonica* 'Amamiana'	oGar oWnC
	japonica 'Bisbee Dwarf'	oGar oSis oBov wSta
	japonica 'Bonsai'	oSis
un	*japonica* 'Brookside'	oRiv oBRG
	japonica 'Cavatine'	oGos oSis oBRG wWel
	japonica 'Christmas Cheer'	oSis
	japonica 'Coleman'	oGre
	japonica 'Compacta'	last listed 99/00
	japonica 'Crispa'	oBov wSta
	japonica 'Daisen'	wSta
	japonica 'Debutante'	wHer oWnC
	japonica 'Dorothy Wyckoff'	oFor oGar
	japonica dwarf	wSta
un	*japonica* 'Esveld I'	oSis
va	*japonica* 'Little Heath'	oGos oSis oGre oBRG
	japonica 'Little Heath Green'	oSis
	japonica 'Little Heath Variegated'	see **P. *japonica*** 'Little Heath'
	japonica 'Mountain Fire'	oFor wSwC oAls oGar oRiv oSho oBlo oGre oTPm oWnC wSta oBRG oGue wCoS
	japonica 'Nana'	see **P. *nana***
un	*japonica* 'Nocturne'	wHer oGos oSis
	japonica 'Prelude'	oGos oAls oGar oSis oRiv oSho wSte
	japonica 'Purity'	oGar oWnC oThr
	japonica 'Pygmaea'	oFor wHer oSis oSho oBov wSta
un	*japonica* 'Red Head'	oBlo oGre
	japonica 'Red Mill'	oAls oGar
	japonica 'Sarabande'	oGre oBRG
	japonica 'Scarlett O'Hara'	oAls oGar oUps
un	*japonica* 'Shojo'	oGar
un	*japonica* 'Snowbells'	oFor
	japonica 'Snowdrift'	oGar oGre oWnC wSta
	japonica 'Temple Bells'	oBRG oThr
	japonica 'Valley Fire'	oFor oAls oGar oBlo oGre wSta
	japonica 'Valley Rose'	oFor oAls oGar oSis oSho oBlo oGre oTPm oWnC wSta oThr oGue
	japonica 'Valley Valentine'	wSwC oAls oGar oSis oRiv oSho oBlo oTPm oWnC wSta
	japonica 'Variegata' (see Nomenclature Notes)	oFor oAls oGar oRiv oSho oBlo oGre oWnC wSta oBRG wCoS
un	*japonica* 'Variegata Compacta'	oTPm
	japonica 'White Cascade'	oWnC
	japonica 'White Pearl'	last listed 99/00
	japonica var. *yakushimensis*	oFor oAls oGar
	'Karenoma'	oFor oGar oGre oTPm oWnC wWel
	nana	oBov oGre
	'Spring Snow'	last listed 99/00
	taiwanensis	see **P. *japonica***
	PILEA	
	depressa	oGar
	microphylla	oTrP
	PILEOSTEGIA	
	viburnoides	oFor wHer
	PILOSELLA	
	aurantiaca	oUps
	PILOSOCEREUS	
	leucocephalus	oRar
	PIMELEA	
	prostrata	oJoy oWnC
	PIMPINELLA	
	anisum	oAls oUps
	major 'Rosea'	oNWe
	PINCKNEYA	
	pubens	oFor
	PINELLIA	
	cordata	oFor oJoy wCol oRus oDan
	cordata green form	wHer
	cordata variegated leaf form	wHer wRob
	pedatisecta	wCol
	ternata	oCrm
	tripartita	wCol
un	*tripartita* 'Atropurpurea'	oRus
	PINGUICULA	
	ehlersiae x moranensis x *agnata #1*	wOud
un	*macroceras* ssp. *nortensis*	wOud
	moranensis	wOud
	'Sethos'	wOud
	PINUS	
	albicaulis 'Nana'	see **P. *albicaulis*** 'Noble's Dwarf'
	albicaulis 'Noble's Dwarf'	oRiv

	Name	Sources
	aristata	oFor oPor oEdg oAls oGar oDan oRiv oSho oGre wPir wCCr wKin oWnC wSta oSle wWel
un	*aristata* 'Grumpy'	oPor
un	*aristata* 'Jackson's Prostrate'	wCol
	aristata 'Sherwood Compact'	oGre
	armandii	oTrP wHer oGre wCCr
	armandii x *lambertiana*	oPor oRiv
	attenuata	oFor
	x attenuradiata	oFor
un	*ayacahuite brachycarpa*	oFor
	balfouriana	oRiv
	banksiana	oFor oRiv
un	*banksiana* 'Beehive'	oPor
	banksiana 'Chippewa'	oPor oRiv
	banksiana 'Manomet'	oPor
	banksiana 'Neponset'	oPor oRiv
un	*banksiana* 'Park 13 Broom'	oPor
	banksiana 'Schoodic'	oPor oRiv oGre wWel
	banksiana 'Uncle Fogy'	oPor oRiv
	brutia	see **P. *halepensis*** var. ***brutia***
	bungeana	oFor wHer oPor oEdg oGar oRiv oGre wCCr oWnC wAva wRav wSte
	bungeana 'Diamant'	oPor
un	*bungeana* 'Rowe'	oRiv
un	*bungeana* 'Rowe Arboretum'	oPor
un	*bungeana* 'Temple Gem'	oPor oGar
	canariensis	oFor
	cembra	oFor oRiv oGre oWnC wSta wAva
	cembra 'Aurea'	see **P. *cembra*** 'Aureovariegata'
	cembra 'Aureovariegata'	last listed 99/00
	cembra 'Blue Mound'	wHer oPor
	cembra 'Chalet'	oPor
un	*cembra* 'Chamolet'	oPor
	cembra 'Compacta Glauca'	oGre
un	*cembra* 'Ed Woods'	wCol
	cembra 'Pygmaea'	oPor oTin
	cembra ssp. *sibirica*	oFor
un	*cembra* ssp *sibirica* var. *altaica*	oRiv
	cembra 'Stricta'	last listed 99/00
	cembroides	oFor oOEx oGre
un	*clausa*	oFor
	contorta	oFor oAls oGar oBos oBlo oWnC oAld wAva oSle
un	*contorta* 'Chief Joseph'	oPor oRiv oWnC
un	*contorta* var. *contorta*	wWoB wNot wShR oRiv wWat wCCr wSte
	contorta var. *latifolia*	oFor oRiv wKin wSta
un	*contorta* var. *latifolia* 'Goose Pasture'	oPor
un	*contorta* var. *latifolia* 'Taylor's Sunburst'	oRiv
	contorta var. *murrayana*	oGar oGre wWel
	contorta 'Spaan's Dwarf'	oPor wClo oGar oRiv oGre wTGN oTin wWel
	coulteri	oFor wCol wCCr
	densata	wCol
	densiflora	oFor oGar oRiv wPir wCCr oWnC wSta
	densiflora 'Alice Verkade'	last listed 99/00
un	*densiflora* 'Arakawa' Robinson's	oTin
	densiflora 'Aurea'	oFor oPor wCol oGar oRiv oGre oWnC
un	*densiflora* 'Cesarini Variegated'	oPor
	densiflora 'Edsal Wood'	oPor
	densiflora 'Globosa'	oPor oRiv
	densiflora 'Jane Kluis'	oPor oRiv wWel
un	*densiflora* 'Little Christopher'	oPor
un	*densiflora* 'Low Glow'	oPor wWel
un	*densiflora* 'Low Glow' Waxman's	oTin
	densiflora 'Oculus Draconis'	oFor wHer oPor wCol oGar oDan oRiv
	densiflora 'Pendula'	oFor oPor wCol oGar oRiv wSta
un	*densiflora* 'Rata'	oPor oTin
un	*densiflora* 'Sunburst'	oPor
un	*densiflora* 'Tsukiyama'	oPor oTin
	densiflora 'Umbraculifera'	oPor wClo wCol oDan oRiv oSho oBlo oGre oWnC wSta oGue wCoS wWel
un	*densiflora* 'Umbraculifera Compacta'	oAls oGar oGre
	echinata	oFor
	edulis	oFor xwBur oEdg oGar oRiv oOEx wSta
	eldarica	see **P. *halepensis*** var. ***eldarica***
	elliottii	oFor wCCr
	engelmannii	oFor
	flexilis	oFor oRiv wCCr
un	*flexilis* 'Bergman's Dwarf'	oPor oTin
un	*flexilis* 'Cesarini Blue'	oPor oGar
un	*flexilis* 'Columnaris'	oRiv
	flexilis 'Extra Blue'	wCol oRiv
un	*flexilis* 'Glauca'	oPor oRiv
	flexilis 'Glauce Pendula'	wCol

un	***flexilis*** 'Glauca Reflexa'	oRiv
un	***flexilis*** 'Glenmore Silver'	oPor
un	***flexilis*** 'J. Michael'	oRiv
un	***flexilis*** 'Millcreek'	oPor
	flexilis 'Pendula'	oFor oPor wCol oRiv
	flexilis 'Vanderwolf's Pyramid'	oFor oPor wCol oAls oGar oRiv oTPm oWnC wSta wAva wWel
	gerardiana	oFor oPor oGar oRiv
	glabra	oFor
	greggii	oFor
	x hakkodensis	oPor
	halepensis	oFor
	halepensis var. ***eldarica***	oFor wCCr
	heldreichii	oFor oPor wClo oGar oRiv wPir wCCr wKin oWnC wAva
	heldreichii var. ***leucodermis***	see **P. *heldreichii***
	heldreichii var. ***leucodermis*** 'Aureospicata'	oPor oRiv
	heldreichii var. ***leucodermis*** 'Compact Gem'	oPor oGar oGre
un	***heldreichii*** var. ***leucodermis*** 'Compacta'	oRiv
	heldreichii var. ***leucodermis*** 'Groen'	oPor
	heldreichii var. ***leucodermis*** 'Malink'	oPor
	heldreichii var. ***leucodermis*** 'Satellit'	oPor
	heldreichii 'Smidtii'	oPor wCol
	x hunnewellii	oPor
	hwangshanensis	last listed 99/00
	jeffreyi	oFor wShR oRiv wCCr oSle
	koraiensis	oFor wBur oRiv oOEx oGre
	koraiensis 'Dragon Eye'	wHer oPor oGar oRiv oGre
	koraiensis 'Oculus Draconis'	see **P. *koraiensis*** 'Dragon Eye'
un	***koraiensis*** 'Rowe Arboretum'	oPor
	koraiensis 'Silveray'	oFor oPor oRiv oGre
un	***koraiensis*** 'Tabuliformis'	oPor
	koraiensis 'Winton'	oPor
	kwangtungensis	oPor oRiv oTin wWel
	lambertiana	last listed 99/00
	latifolia	see **P. *contorta*** var. ***latifolia***
	lawsonii	oFor
	leiophylla	oFor
ch	***leucodermis***	see **P. *heldreichii*** (H. Christ)
	leucodermis 'Zwerg Schneverdingen'	oPor
	longaeva	oRiv
	longaeva 'Sherwood Compact'	oPor wCol
	massoniana	wCCr
	maximartinezii	oFor
	monophylla	last listed 99/00
un	***monophylla*** 'Elegance'	oRiv
un	***monophylla*** 'Stanley Pyramid'	oRiv
	montezumae	wCCr
	monticola	oFor wWoB wNot wShR oGar oBos oRiv wCCr wKin wPla oAld oSle wCoS
un	***monticola*** 'Nana'	oPor
	monticola 'Pendula'	oRiv
un	***monticola*** 'Rigby's Weeping'	oPor wCol
	monticola 'Skyline'	oPor
	mugo	oGre wCoS
un	***mugo*** 'Allen'	oPor
	mugo 'Allgaeu'	oPor
un	***mugo*** 'Alpen Hexe'	oPor
	mugo 'Aurea'	oPor wCol oRiv oGre oWnC
un	***mugo*** 'Aurea Fastigiata'	oRiv
	mugo 'Big Tuna'	oPor
	mugo 'Corley's Mat'	oPor
un	***mugo*** 'Ed's #1 Frosty'	oTin
un	***mugo*** 'Elfingreen'	oPor oTin
	mugo 'Frisia'	oPor
	mugo 'Gnom'	oGar
un	***mugo*** 'Goldspire'	oPor
un	***mugo*** 'Gordon Bentham'	oPor
	mugo 'Green Candle'	oPor oGar oRiv oTin
	mugo 'Hesse'	oPor
	mugo 'Jacobsen'	oPor
	mugo 'Kissen'	oPor
	mugo 'Klosterkotter'	oPor
	mugo 'Kokarde'	oPor
	mugo 'Krauskopf'	oPor
un	***mugo*** 'Lew Hill'	oPor
	mugo 'Marand'	oPor
un	***mugo*** 'Mayfair Dwarf'	oRiv
	mugo 'Mitsch Mini'	oPor oRiv oGre oTin
	mugo 'Mops'	oPor oTin wWel
	mugo var. ***mughus***	see **P. *mugo*** var ***mugo***
	mugo var. ***mugo***	oFor wWoB oAls oGar oRiv oBlo wCCr wKin oTPm wBox oGue wCoS
un	***mugo*** 'Newport'	oPor
un	***mugo*** 'Northern Lights'	wCol
un	***mugo*** 'Oregon Jade'	oPor oGar oGre
	mugo 'Pal Maleter'	oPor
	mugo 'Paul's Dwarf'	oPor oRiv oTin wWel
un	***mugo*** 'Per Golden'	oPor
	mugo 'Prostrata'	oPor oRiv
	mugo Pumilio Group	oFor oPor wWoB oAls oGar oGre wCCr wKin oTPm oWnC wSta wAva wWel
un	***mugo*** 'Riesengeberg'	oTin
	mugo var. ***rostrata***	see **P. *mugo*** ssp. ***uncinata***
	mugo 'Sherwood Compact'	oPor oRiv oGre wWel
	mugo 'Sherwood Compact' seed grown	oPor
	mugo 'Slowmound'	wWel
	mugo 'Spaan's Pygmy'	oPor
un	***mugo*** 'Strang'	oPor
un	***mugo*** 'Svalklint'	oPor
un	***mugo*** 'Tatry'	oPor
	mugo 'Teeny'	oPor wCol wWel
un	***mugo*** 'Tyller'	oPor
un	***mugo*** 'Tyrol'	oPor oGar
un	***mugo*** 'Uelzen'	oPor
	mugo ssp. ***uncinata***	oFor
un	***mugo*** ssp. ***uncinata*** 'Fructata'	oPor
	mugo ssp. ***uncinata*** 'Gruene Welle'	oPor wCol
un	***mugo*** ssp. ***uncinata*** 'La Cabinase'	oPor
un	***mugo*** ssp. ***uncinata*** 'Offenpass'	oPor
	mugo ssp. ***uncinata*** 'Paradekissen'	oPor
un	***mugo*** ssp. ***uncinata*** var. ***rotunda*** 'Hrizdo'	oPor
un	***mugo*** ssp. ***uncinata*** var. ***rotunda*** 'Jesek'	oPor
un	***mugo*** ssp. ***uncinata*** var. ***rotunda*** 'Kosteinicek'	oPor
un	***mugo*** ssp. ***uncinata*** var. ***rotunda*** 'Loucky'	oPor
un	***mugo*** ssp. ***uncinata*** var. ***rotunda*** 'Novak'	oPor
un	***mugo*** ssp. ***uncinata*** 'Silver Candles'	oPor
un	***mugo*** 'Uplazy'	oPor
	mugo 'Valley Cushion'	oPor oGar oRiv wSta oTin wWel
un	***mugo*** 'Yellow Tip'	wCol
	mugo 'Zundert'	oPor
	muricata	oFor wShR
	murrayana	see **P. *contorta*** var. ***murrayana***
	nelsonii	oFor
	nigra	oAls oGar oRiv oBlo oGre wKin oWnC wSta wPla wAva oGue wCoS
	nigra 'Arnold Sentinel'	oPor wCol oRiv
	nigra 'Aurea'	oPor oRiv
	nigra 'Black Prince'	oPor
un	***nigra*** 'Compacta'	oGar
	nigra 'Frank'	oPor
	nigra 'Globosa'	oPor oRiv
	nigra 'Hornibrookiana'	oFor oPor wCol oGar oRiv oGre oTin wWel
un	***nigra*** 'Horstmann'	oPor
	nigra ssp. ***laricio***	oFor
	nigra ssp. ***laricio*** 'Pygmaea'	oPor oRiv
	nigra ssp. ***maritima***	see **P. *nigra*** ssp. ***laricio***
un	***nigra*** 'Nana Wurstie'	oPor
	nigra ssp. ***nigra*** 'Helga'	oPor oTin
	nigra 'Obelisk'	oPor oRiv
un	***nigra*** 'Pierick Bregeon'	wWel
	nigra 'Pygmaea'	see **P. *nigra*** ss. ***laricio*** 'Pygmaea'
	nigra 'Pyramidalis'	oPor
	oocarpa	last listed 99/00
	palustris	oFor wCCr
	parviflora	oFor oPor oRiv oGre wCCr oWnC
	parviflora 'Adcock's Dwarf'	oPor wClo wCol oRiv oTin
un	***parviflora*** 'Aizu-goyo'	oPor
	parviflora 'Al Fordham'	oPor oTin
	parviflora 'Aoba-jo'	oPor oTin
	parviflora 'Aoi'	oPor oTin
	parviflora 'Ara-kawa'	oPor oRiv oGre oTin
	parviflora 'Atko-goyo'	oPor oTin
un	***parviflora*** 'Azuma'	oRiv oTin
	parviflora 'Azuma-goyo'	oPor
	parviflora 'Baldwin'	oPor
	parviflora 'Bergman'	oPor oTin
un	***parviflora*** 'Bergman's US'	oPor
un	***parviflora*** 'Blue Wave'	wWel
	parviflora 'Bonnie Bergman'	oTin
	parviflora 'Brevifolia'	oTin
un	***parviflora*** 'Burk's Bonsai'	oPor oRiv oTin
un	***parviflora*** 'Burke's #3'	oTin
un	***parviflora*** 'Cleary'	oPor
un	***parviflora*** 'Ed Wood's Dwarf'	oTin
un	***parviflora*** 'Ed's Choice'	oTin
un	***parviflora*** 'Ei-ko-nishiki'	oPor oTin

	parviflora 'Fukai'	oPor
un	***parviflora*** 'Fukushima'	oTin
	parviflora 'Fukuzumi'	oPor oRiv oTin
un	***parviflora*** 'Gen Roku'	oTin
	parviflora 'Gimborn's Ideal'	oFor oPor wCol oRiv oTin
	parviflora 'Gimborn's Pyramid'	oPor oTin
un	***parviflora*** 'Gin Setsu'	oTin
	parviflora f. ***glauca***	oFor oPor wClo oRiv oTPm wSta wAva
	parviflora 'Glauca Nana'	oRiv oGre oTin wWel
un	***parviflora*** 'Glauca Nana Compacta'	oTin
un	***parviflora*** 'Go Gin'	oPor oRiv oTin
un	***parviflora*** 'Golden Candles'	oPor
un	***parviflora*** 'Goldilocks'	oPor
un	***parviflora*** 'Goyo Nishiki'	oTin
un	***parviflora*** 'Ha-tzumari-goyo'	oPor
	parviflora 'Hagoromo'	oPor wCol oTin
	parviflora 'Hagoromo Seedling'	wCol oRiv oGre
un	***parviflora*** 'Hime-goyo-matsu'	oPor
un	***parviflora*** 'Horai'	oTin
	parviflora 'Ibo-can'	oFor oPor oRiv oTin
	parviflora 'Iri-fune'	oPor
un	***parviflora*** 'Iseli Select'	oPor
un	***parviflora*** 'Ishizuchi'	oPor
un	***parviflora*** 'Kanrico'	oTin
	parviflora 'Kokonoe'	oPor oRiv oTin
	parviflora 'Kokuho'	oPor oRiv oTin
	parviflora 'Koraku'	oPor oTin
un	***parviflora*** 'Koru'	oPor
	parviflora 'Meiko'	oTin
un	***parviflora*** 'Miyajima'	oPor oGre oTin
	parviflora 'Myo-jo'	oPor
un	***parviflora*** 'Nana'	wWel
	parviflora 'Nasu-goyo'	oPor
un	***parviflora*** 'Ogon'	oRiv
un	***parviflora*** 'Ogon-goyo'	oPor
	parviflora 'Ogonjanome'	wHer oPor wCol oRiv oGre oTPm
un	***parviflora*** 'Pentaphyllum'	oTin
un	***parviflora*** 'Peterson'	oPor
un	***parviflora*** 'Pygmy Yatsubusa'	oPor
un	***parviflora*** 'Shirobana'	oPor oTin
	parviflora 'Shirobara'	oTin
un	***parviflora*** 'Tamayzima'	oTin
	parviflora 'Tani-mano-uki'	oPor wCol oGre oTin
	parviflora 'Tempelhof'	oGre wWel
un	***parviflora*** 'Tensyukazu'	oPor
	parviflora 'Tone'	last listed 99/00
un	***parviflora*** 'Unryu'	oTin
un	***parviflora*** 'Valavanis Yatsubusa'	oTin
	parviflora 'Venus'	oPor oTin
un	***parviflora*** 'Watnong'	oPor
qu	***parviflora*** 'Yatsubusa'	oPor oRiv oGre oTin
	parviflora 'Zui-sho'	oGre oTin
	patula	oFor wCCr
	peuce	oEdg
	peuce 'Arnold Dwarf'	oPor
	pinaster	oFor wCCr
	pinceana	oFor
	pinea	oFor wHer oOEx wCCr oWnC
	ponderosa	oFor wWoB wNot oDar wShR oGar oRiv oSho wWat wKin oWnC wPla oSle
un	***ponderosa*** 'Black Hill'	oRiv
un	***ponderosa*** 'Dixie'	oPor
un	***ponderosa*** 'Dusty Blue'	oPor
un	***ponderosa*** 'High Desert'	oPor
un	***ponderosa*** 'Jeff'	oPor
un	***ponderosa*** 'The Sphinx'	oRiv
	pseudostrobus	oFor
	pumila	oPor oRiv oGre
un	***pumila*** 'Dolina Dwarf'	oPor
	pumila 'Dwarf Blue'	see **P. *pumila*** 'Glauca'
	pumila 'Glauca'	oPor oTin
	pumila NA 20397	oPor
	pumila 'Nana'	oPor
	pungens	oFor
	radiata	wCCr
	resinosa	oFor
un	***resinosa*** 'Bennett's Aurea'	oPor
un	***resinosa*** 'Caspian # 1'	oPor
	resinosa 'Don Smith'	oPor oWnC
un	***resinosa*** 'Elkins'	oPor
	resinosa 'Quinobequin'	oPor
un	***resinosa*** 'Spaan's Fastigiate'	oPor
	resinosa 'University of Wisconsin'	oPor
	rigida	oFor oRiv
	rigida 'Sherman Eddy'	last listed 99/00
	roxburghii	oFor wCCr
	rudis	oFor
	sabineana	oFor wHer wCCr
	serotina	oFor oRiv
un	***shenkanensis***	oFor
	sibirica	see **P. *cembra*** ssp. ***sibirica***
	strobiformis	oFor oPor oRiv wCCr
un	***strobiformis*** 'Loma Linda'	oPor oTin
un	***strobiformis*** 'Pendula'	oPor
un	***strobiformis*** 'Undulata'	oPor
	strobus	oFor wCSG oGar oRiv oBlo wCCr oWnC wSta wCoS
	strobus 'Alba'	oPor
un	***strobus*** 'Albopicta'	oPor
	strobus 'Anna Fiele'	oPor
un	***strobus*** 'Baird Cutting Nana'	oPor
un	***strobus*** 'Baldwin'	oPor
un	***strobus*** 'Bennett Clumpleaf'	oPor oRiv
un	***strobus*** 'Bennett's Contorted'	oPor
	strobus 'Bennett's Dragon Eye'	oPor
un	***strobus*** 'Bergman's Variegated'	oPor
	strobus 'Billow'	oPor
	strobus 'Bloomer's Dark Globe'	oPor
	strobus 'Blue Shag'	oFor oPor oGar
	strobus 'Brevifolia'	oPor oRiv
un	***strobus*** 'Coney Island' Waxman	oTin
	strobus 'Contorta'	oGre
un	***strobus*** 'Curley's Dwarf'	oPor
un	***strobus*** 'David' Waxman	oTin
un	***strobus*** 'Diggy'	oPor
	strobus 'Dove's Dwarf'	oPor
un	***strobus*** 'Elkin's Dwarf'	oPor
	strobus 'Fastigiata'	oFor wClo oRiv oGre oTPm
un	***strobus*** 'Glauca Nana'	oGre wWel
un	***strobus*** 'Golden Candles'	oPor
	strobus 'Green Shadow'	oPor
	strobus 'Greg'	oPor wCol
un	***strobus*** 'Helen'	oPor
un	***strobus*** 'Hershey'	oPor
	strobus 'Hillside Gem'	oPor
	strobus 'Hillside Winter Gold'	oFor oPor oRiv oGre
	strobus 'Horsford'	oFor oPor wClo oRiv oGre oWnC oTin
un	***strobus*** 'Horsford's Sister'	oPor
un	***strobus*** 'Horsham'	oPor
	strobus 'Jericho'	oRiv
un	***strobus*** 'KBN Variegated'	wCol
un	***strobus*** 'Kelsey'	oPor
	strobus 'Kruegers Lilliput'	oPor
un	***strobus*** 'Laird's Broom'	oPor
un	***strobus*** 'Louie'	oPor wCol
	strobus 'Macopin'	oPor
un	***strobus*** 'Mary Butler'	oTin
	strobus 'Merrimack'	oPor oRiv
	strobus 'Minuta'	oPor oRiv oTin wWel
	strobus 'Nana'	see **P. *strobus*** 'Radiata'
un	***strobus*** 'National Life'	oPor
un	***strobus*** 'Northway'	oPor
	strobus 'Ontario'	oPor
un	***strobus*** 'Pacific Sunrise'	oPor wCol
un	***strobus*** 'Paul Waxman'	oPor
	strobus 'Pendula'	oFor oPor wClo oRiv oBlo oGre oTPm oWnC wSta wAva
un	***strobus*** 'Pine Acres'	oPor
	strobus 'Prostrata'	last listed 99/00
	strobus 'Pygmaea'	oPor wClo
	strobus 'Radiata'	oFor oPor wClo oGar oRiv oGre oTPm wSta
	strobus 'Sayville'	oPor
	strobus 'Sea Urchin'	oPor wCol oRiv oGre
	strobus 'Secrest'	oPor
un	***strobus*** 'Soft Touch'	oPor oRiv
un	***strobus*** 'Squirrel's Nest'	oPor
un	***strobus*** 'Stoneybrook'	oPor wCol
	strobus 'Torulosa'	oPor wClo oRiv wSta
	strobus 'U. Conn'	oPor oRiv
	strobus 'Uncatena'	oPor oTin
	strobus 'Verkade's Broom'	oPor
	strobus 'White Mountain'	oFor oPor wCol oRiv oGre oTin
un	***strobus*** 'Winter Gold'	last listed 99/00
	sylvestris	oAls wShR oGar oRiv oBlo oGre wCCr wKin oWnC wSta wPla wAva oGue
	sylvestris 'Albyn'	oPor wClo
un	***sylvestris*** 'Alderly Edge'	oPor
	sylvestris 'Argentea Compacta'	oPor
	sylvestris Aurea Group	oGar oRiv

un	*sylvestris* 'Auvergne'	oRiv wKin
	sylvestris 'Barrie Bergman'	oPor
un	*sylvestris* 'Beacon Hill'	oPor
	sylvestris 'Beuvronensis'	oFor oPor wClo oTin
	sylvestris 'Bonna'	oPor oRiv
	sylvestris 'Burghfield'	oPor
un	*sylvestris* 'Burgos'	wKin
un	*sylvestris* 'Byst's WB'	oPor
	sylvestris 'Chantry Blue'	oPor
un	*sylvestris* 'Clumber Hump'	oPor oRiv
	sylvestris 'Cutty Sark'	oPor oRiv
	sylvestris 'Doone Valley'	oPor
	sylvestris 'Fastigiata'	oPor oRiv oGre
un	*sylvestris* 'Fruhling's Gold'	oPor
	sylvestris 'Glauca Nana'	oGar oGre oTin wWel
	sylvestris 'Globosa Viridis'	oPor oWnC
	sylvestris 'Gold Coin'	oPor oRiv
qu	*sylvestris* 'Grensham'	oPor
un	*sylvestris* 'Guardarrama'	oTin
un	*sylvestris* 'Helms'	oTin
	sylvestris 'Hibernia'	oPor
un	*sylvestris* 'Hibernia Nana'	oRiv
	sylvestris 'Hillside Creeper'	oPor wCol oRiv wWel
va	*sylvestris* 'Inverleith'	oPor
	sylvestris 'Jeremy'	oPor
un	*sylvestris* 'Jukutsk'	oPor
	sylvestris KBN gold sport	oPor
un	*sylvestris* 'Kluis Pyramid'	oPor
	sylvestris Kristick witches' broom	oPor
un	*sylvestris* 'Little Ann'	oPor
	sylvestris 'Little Brolly'	oPor
	sylvestris 'Longmoor'	oPor
	sylvestris 'Mitsch Weeping'	oPor wCol oRiv
	sylvestris 'Moseri'	oPor
	sylvestris 'Nana'	see **P. *sylvestris*** 'Watereri'
	sylvestris 'Nana Compacta'	oPor
un	*sylvestris* 'Pendula	oRiv
	sylvestris 'Pixie'	oPor wCol oRiv oGre
	sylvestris 'Pygmaea'	oPor
	sylvestris 'Repens'	oPor oGar oRiv oGre
un	*sylvestris* 'Riverside Gem'	oPor oGre oWnC
	sylvestris 'Saxatilis'	oPor oTin
	sylvestris 'Sentinel'	oPor wCol oRiv
	sylvestris 'Spaan's Slow Column'	oPor oRiv
un	*sylvestris* 'Trooper'	oPor
	sylvestris 'Wartham'	oPor
	sylvestris 'Watereri'	oPor oGar oRiv oSho
un	*szemaoensis*	oFor
	tabuliformis	oFor oEdg
	tabuliformis var. *densata*	oPor
	tabuliformis var. *mukdensis*	oPor
	taeda	oRiv wCCr
	taiwanensis	oFor
	thunbergiana	see **P. *thunbergii***
	thunbergii	oFor oDar oAls oGar oRiv oSho oBlo oGre wKin oWnC wSta
	thunbergii 'Akame'	oTin
	thunbergii 'Aocha-matsu'	oPor oRiv
un	*thunbergii* 'Ara-kawa'	oRiv oGre
	thunbergii 'Banshosho'	oFor oPor wClo oGar oRiv oGre oWnC oTin wWel
	thunbergii 'Beni-kujaku'	oPor
un	*thunbergii* 'Eeohee'	oTin
un	*thunbergii* 'Fuji'	oPor oTin
un	*thunbergii* 'Gen-sekki-sho'	oPor
un	*thunbergii* 'Green Elf'	oPor
un	*thunbergii* 'Gyo Ku Ho'	oTin
un	*thunbergii* 'Ihara'	oTin
un	*thunbergii* 'Katsuga'	oTin
	thunbergii 'Ko-yo-sho'	oPor
	thunbergii 'Kotobuki'	oPor wCol oTin
un	*thunbergii* 'Kyokko'	oPor oTin
un	*thunbergii* 'Mikawa'	oGar oRiv oGre oTin
	thunbergii 'Mount Hood'	oPor
un	*thunbergii* 'Nishiki Eeehee'	oPor
	thunbergii 'Nishiki-ne	oTin
	thunbergii 'Nishiki Tsukasa'	oPor oRiv
un	*thunbergii* 'Nishiki Tsukasa' Yoshimura clone	oTin
	thunbergii var. *oculus draconis*	oGre
	thunbergii 'Ogon'	oPor oGar oWnC
un	*thunbergii* 'Porky'	oPor
	thunbergii 'Sayonara'	oPor oRiv wSta oTin
	thunbergii 'Shio-guro'	oPor
	thunbergii 'Shirome-janome'	oPor oRiv
un	*thunbergii* 'Suiken'	oPor
	thunbergii 'Taihei'	oTin
	thunbergii 'Thunderhead'	wHer oPor wCol oGar oRiv oGre wWel
un	*thunbergii* 'Torabu Matsu'	oFor
un	*thunbergii* 'Torafu-matsu'	oPor oRiv
un	*thunbergii* 'Yachio'	oPor oTin
un	*thunbergii* 'Yamaki'	oTin
	thunbergii 'Yatsubusa'	see **P. *thunbergii*** 'Sayonara'
un	*thunbergii* 'Yatsubusa Watnong'	oPor
	uncinata	see **P. *mugo*** ssp. ***uncinata***
	virginiana	oFor
un	*virginiana* 'Fanfare'	oPor
un	*virginiana* 'Sunset'	oPor
un	*virginiana* 'Topknot'	oPor
	virginiana 'Wate's Golden'	oFor oPor oGar oRiv oGre oWnC
	wallichiana	oFor wHer wCol oGar oRiv
	wallichiana 'Densa'	oPor
un	*wallichiana* 'Frosty'	oPor oGar
	wallichiana 'Nana'	oRiv
va	*wallichiana* 'Zebrina'	oFor wHer oPor wCol oGar oRiv
un	*wangii* var. *wilsonii*	oPor
	washoensis	oFor
un	'Wykoff'	oPor
	yunnanensis	oTrP wCCr
	PIPER	
	betle	oOEx
	methysticum	oOEx
	PIPTANTHUS	
	nepalensis DJHC 532	wHer
	PISTACIA	
	atlantica	last listed 99/00
	chinensis	oFor oTrP oGos oGar oRiv oCis oSho oGre wCCr oWnC
	vera	oFor
un	*vera* 'Far North' (nut)	oOEx
un	*vera* 'Kerman' (nut)	oOEx
un	*vera* 'Mountain Blue' (nut)	oOEx
un	*vera* 'Peters' (nut)	oOEx
	PISTIA	
	stratiotes	oTrP wSoo wGAc oHug oSsd oWnC
	stratiotes giant	oHug
	stratiotes 'Rosette'	wGAc oGar
	PITHECELLOBIUM	
	flexicaule	last listed 99/00
	PITTOSPORUM	
un	*argentea* 'Nana'	wHer
	bicolor	oFor
un	'Black Lace'	wHer
	eugenioides	oFor wHer wCCr wSte
va	'Garnettii'	oFor oJoy
	heterophyllum	oCis
un	*laterifolius*	oCis
	tenuifolium	oFor oSho wCCr
un	*tenuifolium* 'Rubra'	oCis
	tenuifolium 'Silver Magic'	oCis
un	*tenuifolium* 'Tough and Tall'	oCis
	tobira	oGar oCis oSho oBlo wCCr wCoS
	tobira Cream de Mint™ / 'Shima'	oGar
	tobira hardy	oCis
	tobira 'Variegatum'	oFor oGar oSho oBlo
	tobira 'Wheeler's Dwarf'	oGar oSho oBlo
	PLAGIANTHUS	
	regius	oFor
	PLAGIOMNIUM	
un	*insigne*	wFFl
	PLANTAGO	
	asiatica 'Variegata'	wCol
	cynops	last listed 99/00
	major	oGoo iGSc
	major 'Purpurea'	see **P. *major*** 'Rubrifolia'
	major 'Rosularis'	wCul wCol
	major 'Rubrifolia'	wCol oGoo oSec
	major 'Rubrifolia' seed grown	wHer
	maritima	last listed 99/00
	maxima	last listed 99/00
	nivalis	oHed
	psyllium	last listed 99/00
	sempervirens	last listed 99/00
un	*triandra*	last listed 99/00
	PLATANUS	
	x acerifolia	see **P. *x hispanica***
	x hispanica	wShR wKin wCoN
	x hispanica 'Bloodgood'	oFor oGar oSho oWnC
	mexicana	oFor
	occidentalis	oFor

	orientalis	wCCr wBox
	wrightii	oRiv wCCr
	PLATYCARYA	
	strobilacea	oFor wCCr
	strobilacea DJH 396	wHer
	PLATYCERIUM	
	bifurcatum	oGar
	PLATYCLADUS	
	orientalis	see **THUJA *orientalis***
	PLATYCODON	
	grandiflorus	oFor wCul wShR wCSG oGoo oSho iGSc oCrm oBar oSec oWhS wCoN
	grandiflorus albus	wCul
	grandiflorus 'Apoyama'	oSis oJil
	grandiflorus blue	oMis
	grandiflorus double blue	wSwC oGar
	grandiflorus 'Fuji Blue'	oFor
	grandiflorus Fuji mixed	oBlo
	grandiflorus 'Fuji Pink'	oGar wHom
	grandiflorus 'Hakone Blue'	oFor
	grandiflorus 'Hakone Double Blue'	oMis
un	***grandiflorus*** 'Hakone Double White'	oMis
	grandiflorus 'Hakone White'	oWnC
	grandiflorus 'Komachi'	oFor
	grandiflorus 'Mariesii'	oFor oAls oRus oGar oSho oWnC
	grandiflorus 'Misato Purple'	oSis
	grandiflorus Mother of Pearl	see **P. *grandiflorus*** 'Perlmutterschale'
	grandiflorus 'Perlmutterschale'	oFor oAls oSis oSho
	grandiflorus roseus	oMis
	grandiflorus 'Sentimental Blue'	oFor oDar oEdg oGar oSho oSha wHom wTGN
	grandiflorus 'Shell Pink'	see **P. *grandiflorus*** 'Perlmutterschale'
	grandiflorus white	oFor oRus oMis
	PLATYCRATER	
	arguta	wHer
	PLECOSTACHYS	
	serpyllifolia	oHed
	PLECTRANTHUS	
	amboinicus	oAls iGSc wHom oWnC oUps
	amboinicus 'Variegatus'	oAls iGSc wHom oUps
	argentatus	wWoS oHed oAmb
un	***argentatus*** 'Longwood Silver'	wHer
	australis	oGar
un	***barbatus***	oOEx
	ciliatus	oGar
un	***forsteri*** 'Green on Green'	wWoS
	madagascariensis 'Variegated Mintleaf'	last listed 99/00
	oertendahlii	last listed 99/00
	sp.	iGSc
	PLEIOBLASTUS	
	akebono	oTra wCli oNor oTBG
un	***amarus***	oTra
	argenteostriatus	oFor oTra wCli oNor wBlu wBea oDar
	auricomus	oFor oTra wSus wCli oNor wBlu wBea oDar oAls oGar oDan oOut wRob oGre wBox wBam wBmG oTBG wWel
	auricomus f. ***chrysophyllus***	oTra wCli oTBG
	chino	wSus wBmG
	chino 'Kimmei'	oTra wBea
	chino 'Murakamianus'	oTra wBmG oTBG
un	***chino vaginatus variegatus***	oTra wSus wCli oNor wBlu wBea wBam wBmG oTBG
	distichus	see **P. *pygmaeus*** var. ***distichus***
	fortunei	see **P. *variegatus***
	gramineus	oTra wCli wBea
	hindsii	wBea
	humilis	oTra wBlu
	kongosanensis	oNor
va	***kongosanensis*** 'Aureostriatus'	oTra
	linearis	oTra wCli wBea
	oleosus	wBea
	pygmaeus	oFor oTra wSus wCli oNor oDar oGar oGre oTBG wWel
	pygmaeus var. ***distichus***	wCli wBam oTBG
	pygmaeus var. ***distichus*** 'Mini'	oFor
qu	***pygmaeus pygmaeus***	oGar
va	***shibuyanus*** 'Tsuboi'	oTra wCli oNor wBea wBmG
	simonii	oTra wCli oNor wBmG oTBG
un	***simonii*** 'Medake' seed grown	wBea
	simonii f. ***variegatus***	oNor wBea
	sp.	oNor
	variegatus	oFor oTra wCli oNor wBea oGar oSho wRai oGre wBam wBmG oTBG wWel
	viridistriatus	see **P. *auricomus***

	PLEIONE	
	Alishan g	oRed
	aurita	oRed
	bulbocodioides	wThG oRed
un	***bulbocodioides*** 'Pogonoides'	oGos
	chunii	see **P. *aurita***
	formosana	oTrP oGos wThG wCol oRed
	formosana 'Achievement'	oRed
	formosana 'Blush of Dawn'	oGos oRed
un	***formosana*** 'Lilac Pearl'	oRed
	formosana 'Oriental Splendour'	oRed
	formosana 'Polar Sun'	oRed
	forrestii	oRed
	grandiflora	oRed
	grandiflora pink form	oRed
	hookeriana	last listed 99/00
	humilis	oRed
	Keith Rattray g	oRed
	limprichtii	last listed 99/00
	maculata	oRed
	praecox	oRed
	pricei	see **P. *formosana***
un	***saxicola***	oRed
	scopulorum	oRed
	Shantung g 'Ridgeway'	oRed
	sp. white w/pink	wSta
	speciosa 'Blakeway Phillips'	oRed
un	Tolima g	oGos
un	Vesuvius g No. 1	oRed
	yunnanensis (see Nomenclature Notes)	oRed
	PLEUROSPERMUM	
un	***yunnanense*** DJHC 98383	wHer
un	**PLEUROTUS**	
	ostreatus	wRai oOEx
	PLUMBAGO	
	auriculata	oGar
	capensis	see **P. *auriculata***
	POA	
	abbreviata	oTrP
	PODOCARPUS	
	acutifolius	wHer
	alpinus	last listed 99/00
	alpinus 'Blue Gem'	oPor oWnC
	alpinus female form	wHer
	chinensis	see **P. *macrophyllus***
	henkelii	oFor
	lawrencei	oFor wHer
un	***lawrencei*** 'Purple King'	oFor wHer oSho oBRG
un	'Lo Han Kuo'	oOEx
	macrophyllus	oFor oTrP wHer oGre wPir wBox wSta
	macrophyllus 'Maki'	oCis oSho wWel
	nivalis	oFor oPor oGar wSta
	nivalis female	wHer
	nivalis male	wHer
	salignus	wCCr
	salignus HCM 98057	wHer
	totara	wHer
	totara 'Aureus'	oFor
	totara dwarf	oOEx
	totara 'Pendulus'	wHer
	PODOPHYLLUM	
	emodi	see **P. *hexandrum***
	hexandrum	oTrP oJoy wCol oNWe
	hexandrum DJHC 257	wHer
	peltatum	oFor wHer wThG
	pleianthum	wHer
	pleianthum hybrid	wHer
	versipelle	wHer
	PODRANEA	
	ricasoliana	last listed 99/00
	POGOSTEMON	
	cablin	oTDM oAls iGSc oCrm
	heyneanus	oOEx
	patchouly	see **P. *cablin***
	POINCIANA	
	gilliesii	see **CAESALPINIA *gilliesii***
	POLEMONIUM	
un	'Blue Whirl'	iArc
	boreale	oSis wBWP
un	***boreale*** 'Heavenly Habit'	oMis
	caeruleum	oFor wCul wTho oNat oJoy oAls oRus wShR oGar oGoo oBlo wFai iGSc wHom oSec iArc oEga oUps oSle wSnq
	caeruleum var. ***album***	see **P. *caeruleum*** ssp. ***caeruleum*** f. ***album***

va	***caeruleum*** Brise d'Anjou / 'Blanjou'	oFor wWoS wSwC oDar oAls oGar oDan oNWe oBlo wFai wHom wTGN oCir oUps wSte
	caeruleum ssp. ***caeruleum*** f. album	oFor wCul oNat oJoy oRus oWnC oSec wBWP
	caeruleum ssp. ***himalayanum***	wTGN
	caeruleum 'Lace Towers'	last listed 99/00
	caeruleum ssp. ***nipponicum***	oDan
	carneum	oHan wCol oRus oMis oSis oNWe iArc oSle
	carneum 'Apricot Delight'	wWoS oHed oSec
	cashmerianum	wHer wCri wCCr
	foliosissimum (see Nomenclature Notes)	oHan
	'Hopleys'	oHed
	'Lambrook Mauve'	wHer
	nipponicum	see **P. *caeruleum*** ssp. ***nipponicum***
qu	***pacificum***	wHom
	pauciflorum	oHan oHed oMis oSis wBWP iArc
	pauciflorum silver leaved	last listed 99/00
	pulcherrimum	oHan iArc
	reptans	oRus
	reptans 'Blue Pearl'	oFor oAls oMis oSis oEga oLSG
	viscosum	oHan oSis
	yezoense 'Purple Rain'	last listed 99/00
	POLIOMINTHA	
	bustamanta	oHed
	POLLIA	
	japonica	oTrP
	POLYGALA	
	chamaebuxus	wHer oSis
	chamaebuxus var. ***grandiflora***	wHer oSis
	chamaebuxus 'Kamniski'	oSis
	chamaebuxus 'Rhodoptera'	see **P. *chamaebuxus*** var. ***grandiflora***
	POLYGONATUM	
	biflorum	oFor wHer wThG wCol oAls oRus wCSG oGar wRob
un	***biflorum*** 'Variegatum'	oGar
	canaliculatum	see **P. *biflorum***
	cirrhifolium DJHC 361	wHer
	cirrhifolium Himalayan form	wHer
	commutatum	see **P. *biflorum***
	curvistylum	last listed 99/00
	curvistylum DJHC 521	wHer
	falcatum HC 970662	wHer
	falcatum 'Variegatum'	oGar wNay
	geminiflorum	wHer
	hirtum	wHer
	hookeri	wHer wCol oRus
	humile	wHer wCol wHig oRus wRob oGre
	x hybridum	wHer
va	***x hybridum*** 'Striatum'	oGar
	japonicum	see **P. *odoratum***
	lasianthum	last listed 99/00
	lasianthum HC 970633	wHer
	multiflorum (see Nomenclature Notes)	wHer wMag
	multiflorum 'Variegatum'	see **P. *x hybridum*** 'Striatum'
	odoratum	oFor wHer
	odoratum var. ***pluriflorum***	wHer
un	***odoratum*** var. ***thunbergii*** 'Variegatum'	wCol
	odoratum 'Variegatum'	oFor wGAc oAls oRus oAmb oGar oDan oNWe
	oppositifolium HWJCM 067	wHer
	prattii	wHer wCol
	pubescens	wCol
	sp. aff. ***sibiricum*** DJHC 600	wHer
	sp. DJHC 256	wHer
	sp. DJHC 820	wHer
	verticillatum	wCol
	POLYGONUM	
	affine	see **PERSICARIA *affinis***
	amplexicaule	see **PERSICARIA *amplexicaulis***
	aubertii	see **FALLOPIA *baldschuanica***
	bistorta	see **PERSICARIA *bistorta***
	campanulatum	see **PERSICARIA *campanulata***
un	***coriaceum*** DJHC 98384	wHer
	cuspidatum	see **FALLOPIA *japonica***
	filiforme	see **PERSICARIA *virginiana***
	japonicum	see **FALLOPIA *japonica***
	multiflorum	see **FALLOPIA *multiflora***
	odoratum	see **PERSICARIA *odorata***
	'Painter's Palette'	see **PERSICARIA *virginiana*** 'Painter's Palette'
un	***paronychia***	oBos
un	***perfoliatum***	oOEx
	vacciniifolium	see **PERSICARIA *vacciniifolia***

	weyrichii	see **PERSICARIA *weyrichii***
	POLYMNIA	
un	'Purple Puma'	oOEx
un	***sonchifolia*** 'Pearl of Bolivia'	oOEx
	POLYPODIUM	
	australe	see **P. *cambricum***
	cambricum 'Wilharris'	oSis
	glycyrrhiza	wNot wFan oRus oSis oBos oTri wFFl
	glycyrrhiza 'Longicaudatum'	wFan
	interjectum	wFan
	vulgare	wFan oRus
	vulgare 'Cornubiense'	oSis
	vulgare Ramosum Group	oRus
	POLYSCIAS	
	'Elegans'	oGar
	POLYSTICHUM	
	acrostichoides	wFan oRus wNay oGre oWnC oTri wTGN oBRG
	andersonii	oRus
	braunii	oFor wFol wFan oRus oGar oGre oBRG
	falcinellum	wFol
	lachenense	wFol
	lentum	last listed 99/00
	makinoi	wFan
un	***microchlamys***	wFol
	munitum	oFor wThG wWoB wSwC wNot oJoy wFan oAls oRus wShR oGar oBos oBlo oSqu oGre wWat oTPm oWnC oTri oBRG oAld oCir oJil wCoN wFFl wRav oSle wSnq
	neolobatum	wFol wFan oRus oBRG
	polyblepharum	oFor oJoy wFan oRus oGar oSis oNWe wRob wNay oSev oSqu oGre oTPm oWnC wTGN oBRG wSnq
	proliferum	see **P. *setiferum*** Acutilobum Group
	richardii	last listed 99/00
	setiferum	oFor wFan oGar oGre oTPm wTGN oBRG wSnq wWel
	setiferum Acutilobum Group	wThG
un	***setiferum*** 'Barfod's Dwarf'	wFan
	setiferum Congestum Cristatum Group	wFan oSis wTGN oBRG wSnq
	setiferum Divisilobum Group	oFor wThG oRus oGar oSis
qu	***setiferum*** 'Divisilobum Multilobum'	oGos
	setiferum Plumosodivisilobum Group	wTGN oBRG
	setiferum 'Plumosomultilobum'	wSwC oRus oNWe
un	***setiferum*** 'Plumosomultiplumosum'	oFor
qu	***setiferum*** Plumosum Gracillium	oRus
	setiferum Plumosum Group	wThG oRus
	setiferum Rotundatum Group 'Cristatum'	wSwC oRus oGar
	sp. from China	wFan
	tsussimense	oFor wFan oRus oGar oSqu oGre oTPm wTGN oBRG
	xiphophyllum	wFol
un	**POLYTRICHUM**	
	juniperinum	wFFl
	PONCIRUS	
	trifoliata	oFor oRiv oSho oWhi wCCr
	trifoliata 'Flying Dragon'	oFor oOEx wWel
	PONTEDERIA	
	cordata	wGAc oHug oSsd oWnC oTri wTGN wBox
	cordata alba	oHug oSsd
un	***cordata*** 'Crown Point'	wGAc oGar
	cordata var. ***lancifolia***	wGAc
	cordata 'Pink Pons'	oHug
	cordata short form	wSoo
	cordata tall form	wSoo
	lanceolata	see **P. *cordata*** var. ***lancifolia***
	sp.	oSsd
	POPULUS	
	alba	oFor
	alba f. ***pyramidalis***	oFor
	alba 'Richardii'	oFor oGos
un	'Androscoggin'	oFor
	angustifolia	last listed 99/00
	balsamifera	oFor
	balsamifera trichocarpa	see **P. *trichocarpa***
	x canadensis 'Aurea'	wHer
	x canadensis 'Robusta' male	oFor wKin
va	***x candicans*** 'Aurora'	wCol
	x canescens 'Macrophylla'	oFor
	x canescens 'Tower'	oFor
	cathayana	oFor
	fremontii	oFor wCCr
	fremontii male	oFor
	x jackii 'Gileadensis'	wBCr
	maximowiczii	oFor

	nigra 'Italica' male	wBCr oAls oBlo wCoS
un	***nigra*** Red Caudina # 101	wBCr
	nigra 'Thevestina'	oGar wKin
	simonii 'Fastigiata'	oFor
	simonii 'Pendula'	wCCr
	szechuanica	oFor
	timber hybrid	wBCr
	tremula	oSle
	tremula 'Erecta'	oFor oGar oWnC
	tremuloides	oFor wBCr wWoB oDar oEdg oAls wShR oGar oGoo oBos oSho oWhi oBlo oGre wWat wKin oWnC wTGN wPla oAld wFFl oSle wCoS
	trichocarpa	oFor wWoB wBur wNot wShR wWat wKin wPla wFFl oSle
un	**POROPHYLLUM**	
	rudurale macrocephalum	iGSc
	PORTULACA	
	oleracea	oUps
	PORTULACARIA	
	afra green	oSqu
un	***africana*** 'Aureum'	oSqu
	POTENTILLA	
	alba	wCul oSqu
	alchemilloides	oHed
	argyrophylla	see **P. *atrosanguinea*** var. ***argyrophylla***
	atrosanguinea	wCul wCCr
	atrosanguinea var. ***argyrophylla***	wHig oHed
un	***atrosanguinea*** 'Rot'	oFor
un	***atrosanguinea*** 'Versicolor Plena'	oFor
	aurea	oLSG
	calabra	oHed
	cinerea	wCul oSis
	crantzii	oFor
	cuneata HWJCM 541	last listed 99/00
	egedei	oBos wWat oAld
	flabellifolia	oBos
	fragiformis	see **P. *megalantha***
	fruticosa	wShR oGoo wFai wSta oCir
	fruticosa 'Abbotswood'	oFor wCul oAls wCSG oAmb oGar oSho oGre wKin oTPm wSnq wWel
	fruticosa 'Annette'	oFor
un	***fruticosa*** 'Apricot Whispers"	oTPm
un	***fruticosa*** 'Bear Tooth Pass'	oSis
un	***fruticosa*** 'Cascade Cushion'	wMtT
un	***fruticosa*** 'Coronation'	wKin
	fruticosa 'Daydawn'	oFor oTPm oWnC
	fruticosa DJHC 98070	wHer
	fruticosa 'Farreri'	see **P. *fruticosa*** 'Gold Drop'
	fruticosa 'Floppy Disc'	oFor oGar wWel
	fruticosa Frosty™ / 'Monsidh'	oAls wWel
	fruticosa 'Gold Drop'	oGar oSho oBlo wKin oTPm oWnC
	fruticosa 'Goldfinger'	oAls oGar oBlo oGre wKin oTPm wTGN wWel
	fruticosa 'Goldstar'	oFor oAls oGar oTPm oWnC
un	***fruticosa*** 'Hollandia Gold'	oFor oTPm
	fruticosa 'Jackman's Variety'	oFor oWnC
	fruticosa 'Katherine Dykes'	wCul oGar oBlo oTPm wSnq
	fruticosa 'Klondike'	oTPm
	fruticosa 'Kobold'	oWnC
	fruticosa 'Longacre Variety'	oFor
	fruticosa Marian Red Robin / 'Marrob'	oFor
	fruticosa 'McKay's White'	oFor
	fruticosa 'Mount Everest'	oFor oAls oGar oBlo oTPm
un	***fruticosa*** 'Mount Townsend'	wHer
	fruticosa 'Pink Beauty'	oGar wTGN wSnq wWel
	fruticosa 'Pink Princess'	see **P. *fruticosa*** Princess / 'Blink'
un	***fruticosa*** 'Pink Whisper'	oFor wCSG oSho oTPm oWnC wTGN
	fruticosa 'Primrose Beauty'	oFor wCul wCCr oTPm wWel
	fruticosa Princess / 'Blink'	wCul
	fruticosa var. ***pumila***	wHer
	fruticosa Red Ace	oFor oAls oGar wWel
un	***fruticosa rubra***	wCul
	fruticosa 'Silver Schilling'	wHer
	fruticosa 'Snowbird'	wCul oGre oWnC
	fruticosa 'Snowflake'	oFor
	fruticosa 'Sunset'	oAls oTPm oWnC wWel
	fruticosa 'Sutter's Gold'	oFor oAls oBlo oTPm oWnC
	fruticosa 'Tangerine'	oFor wCul oAls oBlo oTPm oWnC
un	***fruticosa*** 'Triumph'	wKin
un	***fruticosa*** 'Variegata'	wHer
	fruticosa 'Yellow Bird'	oFor oGre
	fruticosa 'Yellow Gem'	oFor
	fulgens	see **P. *lineata***
	'Gibson's Scarlet'	oFor oLSG
	gracilis	wShR oBos oAld
	'Herzblut'	wFai
	lineata	oFor
	megalantha	oFor oJoy oAls wPir wCCr
	'Melton Fire'	wCul wHom
	'Monsieur Rouillard'	oGre
	nepalensis	oHed
	nepalensis 'Miss Willmott'	oFor wCul wTho oAls wCSG wFai wMag oLSG oUps
	nepalensis 'Roxana'	wBWP
	neumanniana	oAls wSta
	neumanniana 'Nana'	oAls wMag oLSG
un	***neumanniana*** 'Orange Flame'	oSis
	nitida 'Alannah'	oSis
	pacifica	see **P. *egedei***
	palustris	wCSG
	recta 'Warrenii'	oFor wFai oLSG
	rupestris	wCul
	sp. EDHCH 97105	wHer
	sp. (water plant)	wSoo
	speciosa	oMis
	tabernaemontani	see **P. *neumanniana***
un	***thurberi atrorubens***	oHan
	thurberi 'Monarch's Velvet'	oJoy oHed
	x tonguei	oFor wCul
	tridentata	see **SIBBALDIOPSIS *tridentata***
	verna	see **P. *neumanniana***
	villosa	see **P. *crantzii***
	'Volcan'	oHed
	'William Rollison'	oFor oAls wTGN
	'Yellow Queen'	oFor wWoS
	POTERIUM	see **SANGUISORBA**
	PRATIA	
	angulata	oFor oTrP oJoy oAls wFGN oGre
	pedunculata	oTrP
	pedunculata 'County Park'	oTrP wMtT
un	**PREMNA**	
	japonica	oFor
	PRIMULA	
	'Aire Mist'	wMtT
	'Aire Waves'	wMtT
	'Alan Robb'	oHed oRus
un	'Alejandra'	wHer wWoS oRus
	allionii	oSis
un	***allionii*** 'Agnes'	wMtT
	allionii 'Anna Griffith'	wMtT
	allionii 'Austen'	wMtT
	allionii 'Claude Flight'	wMtT
	allionii 'Crowsley Variety'	wMtT
	allionii 'Elizabeth Earle'	wMtT
un	***allionii*** Hartside 12 x 'Apple Blossom'	wMtT
	allionii 'Hemswell Blush'	see **P.** 'Hemswell Blush'
	allionii x ***hirsuta***	wMtT
	allionii JCA 4161/22	see **P. *allionii*** 'Jenny' JCA 4161/22
	allionii 'Jenny' JCA 4161/22	last listed 99/00
	allionii 'Ken's Seedling'	wMtT
	allionii x 'Lismore Treasure'	wMtT
	allionii 'Marjorie Wooster'	wMtT
	allionii 'Mrs. Dyas'	wMtT
un	***allionii*** 'Neon'	wMtT
	allionii 'Peggy Wilson'	last listed 99/00
	allionii 'Perkie' JCA4161/22	wMtT
	allionii 'Picton's Variety'	wMtT
un	***allionii*** 'S-MB94-7'	wMtT
	allionii 'Snowflake'	last listed 99/00
	allionii 'Tranquillity'	wMtT
	allionii 'Viscountess Byng'	wMtT
	allionii x 'White Linda Pope'	wMtT
	allionii 'William Earle'	wMtT
un	'Allure'	wMtT
	alpicola	oRus oNWe
	alpicola var. ***alba***	last listed 99/00
	alpicola var. ***alpicola***	last listed 99/00
	alpicola var. ***luna***	see **P. *alpicola*** var. ***alpicola***
	alpicola var. ***violacea***	oNWe oGre
un	'Amanda Gabrielle'	wMtT
	'April Rose'	oRus
	auricula	wCul oAls oRus wShR wTGN wAva
	auricula var. ***alpina***	oRus
	auricula 'Andrea Julie'	wMtT
	auricula 'Argus'	oSis wMtT
un	***auricula*** 'Arundel Stripe'	wMtT
	auricula autumn colors	wPle
	auricula blue	wPle
	auricula 'Brazil'	last listed 99/00

	auricula 'Brownie'	wMtT
	auricula ex 'Brownie'	wMtT
	auricula 'Camelot'	wMtT
	auricula 'Cortina'	wMtT
	auricula cream/white	wPle
	auricula dark pink	wPle
	auricula double cream	wWoS
	auricula 'Doublet'	wMtT
	auricula 'Emily'	wMtT
	auricula from Austria	wPle
	auricula 'Gleam'	wMtT
	auricula 'Gordon Douglas'	wMtT
	auricula 'Green Isle'	wMtT
	auricula 'Green Parrot'	wMtT
	auricula 'Green Shank'	wMtT
	auricula 'Greta'	wMtT
	auricula 'Hetty Woolf'	wMtT
	auricula mixed hybrids	oSis
	auricula 'Paradise Yellow'	wMtT
	auricula 'Parakeet'	wMtT
	auricula 'Petite Hybrid'	last listed 99/00
	auricula purple	wPle
	auricula red	wPle
	auricula 'Remus'	wMtT
un	***auricula*** 'RN-25'	wMtT
	auricula 'Rolts'	wMtT
	auricula 'Serenity'	wMtT
	auricula 'Shalford'	last listed 99/00
	auricula 'Sirius'	last listed 99/00
	auricula 'Sunflower'	last listed 99/00
	auricula 'Susannah'	last listed 99/00
	auricula 'Trouble'	wMtT
	auricula 'Trudy'	wMtT
	auricula 'Walton'	wPle
	auricula yellow	wPle
	beesiana	oFor wHer oNWe oGre oWnC wSnq
	x berninae 'Windrush'	wMtT
	bileckii	see **P. *x forsteri*** 'Bileckii'
	'Blue Sapphire'	oRus
	'Broadwell Pink'	wMtT
	x bulleesiana	oFor oEga
	bulleyana	oFor oTrP wSnq
	bulleyana DJHC 180	wHer
	Candelabra hybrids	oRus oNWe oGre
	capitata	wHer oMis wMag
	capitata ssp. ***mooreana***	oRus oDan
un	'Chinese Pagoda'	wMag
	'Clarence Elliott'	wMtT
	clarkei	oBov
	'Corporal Baxter'	oRus
	cortusoides	wPle
	Cowichan Amethyst Group	last listed 99/00
	Cowichan dark forms	oRus
	Cowichan Garnet Group	last listed 99/00
	Cowichan hybrid	wHig
	'Dawn Ansell'	wHer wWoS oHed oRus oNWe
	deflexa DJHC 98191	wHer
	denticulata	oFor oAls oRus wFai oHon wTGN
	denticulata var. ***alba***	oRus
un	***denticulata*** 'Blau Auslese'	oJoy oEga
	denticulata dark colors	oRus
un	***denticulata*** 'Grandiflora'	oRus
	denticulata mixed colors	oSis oSec
un	***denticulata*** 'Rubin Auslese'	oJoy
	'Dorothy'	wWoS oRus wPle oSis
	elatior	wPle
	'Ethel Barker'	last listed 99/00
	'Eugenie'	oRus
	firmipes	oJoy
	flaccida	oGre
	florindae	wHer wCul oJoy oRus wPle oGar oNWe wMag oSec wSnq
	florindae red seed grown	oNWe
un	***florindae*** 'Red Toy'	oRus
	forbesii	oTrP
	forrestii	oTrP
	x forsteri 'Bileckii'	wMtT
un	'Francesca'	wWoS
	'Fredies'	wHig
un	'Friday'	oRus
	frondosa	wPle
un	'Garryard'	oRus
	'Garryard Guinevere'	see **P.** 'Guinevere'
	'Gigha'	wCSG
un	'Gina'	oSis
	glaucescens	last listed 99/00
	'Glowing Embers'	oNWe
	Gold-laced Group	wHig
	'Granny Graham'	oFor wWoS oRus oSis wTGN
un	'Green Francesca'	oRus
	'Guinevere'	wWoS wHig oHed oRus oSis oNWe
un	'Hannah'	wMtT
un	HD-PH #1	wPle
un	HD-PH #2	wPle
	helodoxa	see **P. *prolifera***
	'Hemswell Blush'	wMtT
un	'High Point'	wMtT
	hirsuta	oSis wMtT
	japonica	oFor wHer wCul wCri oRus oOut wFai oEga wSnq
	japonica 'Carminea'	wTGN
	japonica 'Miller's Crimson'	wWoS
	japonica 'Miller's Crimson' seed grown	wHer
	japonica pink	wPle
	japonica 'Postford White'	wPle oNWe
un	'Jay I'	oSis
	'Jay-Jay'	wPle
	jesoana HC 970092	wHer
	'Johanna'	oSis
	juliae	wPle
	x juliana (see Nomenclature Notes)	
un	'Kelsey Ann'	wMtT
	'Ken Dearman'	oFor wHer wWoS oRus oSis wTGN
	'Kinlough Beauty'	oRus wPle
	kisoana	oSis oBov oGre
	kisoana alba	wHer oNWe
	latifolia	oSis
un	'Lilac Wanda'	wWoS
	lilacina plena	see **P. *vulgaris*** 'Lilacena Plena'
	'Lilian Harvey'	wHer wWoS oRus oSis
	'Linda Pope'	see **P. *marginata*** 'Linda Pope'
	'Linda Pope Alba'	see **P.** 'White Linda Pope'
un	'Lismore Jewel	wMtT
	'Lismore Yellow'	oSis wMtT
un	'Little Gem'	oRus
	luteola	wPle
un	'Mahogany'	wWoS
un	'Mahogany Sunrise'	oFor oRus oAmb oSis oOut wAva
	malacoides	oGar
	marginata alba	wMtT
un	***marginata*** 'Allan Jones'	wPle
	marginata 'Amethyst'	last listed 99/00
un	***marginata*** 'Bentham'	wMtT
un	***marginata*** 'Herb Dickson'	wPle
	marginata 'Holden Variety'	wMtT
	marginata hybrids	oSis
un	***marginata*** 'Jimmy Long'	wPle
	marginata 'Kesselring's Variety'	wMtT
	marginata 'Linda Pope'	wPle oSis wMtT
un	***marginata*** 'Lou Roberts'	wPle
	marginata 'Prichard's Variety'	wPle
	'Marie Crousse'	wHer
	'Mars'	oRus wMtT
	minima	wMtT
	minima var. ***alba***	wMtT
un	***minima*** 'Doug's Form'	wMtT
	'Miss Indigo'	oHed oRus oSis oNWe
	moupinensis	wRho
	moupinensis SEH 086	wHer
	moupinensis SEH086NA	wMtT
un	***obliqua complex*** DJHC 98329 seed grown	wHer
	'Our Pat'	last listed 99/00
un	'Pacific Giants'	oLSG
un	'Paragon'	wHer wWoS oRus wTGN
	parryi	last listed 99/00
	'Peardrop'	wMtT
	'Peter Klein'	oSis oBov
un	'Petite Honey'	wPle
	'Pink Ice'	wMtT
	poissonii	oTrP
	poissonii DJHC 339	wHer
	polyneura	oTrP
	prolifera	oDan oSho oEga
qu	***pruharuana***	wCul
	x pubescens 'Bewerley White'	oSis
	x pubescens 'Boothman's Variety'	wPle wMtT
	x pubescens 'Cream Viscosa'	wMtT
	x pubescens 'Freedom'	wPle
	x pubescens from Larry	wPle
	x pubescens 'Harlow Car'	last listed 99/00

	x pubescens 'Kath Dryden'	wMtT
	x pubescens 'Mrs. J. H. Wilson'	oSis
	x pubescens 'Pat Barwick'	last listed 99/00
	x pubescens 'Rufus'	wMtT
	pulverulenta	oFor wCul oJoy oGre
	'Purpurkissen'	oRus
	'Quaker's Bonnet'	see **P. *vulgaris*** 'Lilacena Plena'
	'Red Velvet'	wHer
	rosea	oRus oSis
	rosea 'Delight'	see **P. *rosea*** 'Micia Visser-de Geer'
	rosea 'Grandiflora'	wPle oSle
	rosea 'Micia Visser-de Geer'	last listed 99/00
un	'Rosetta Red'	wWoS oHed oRus oSis wTGN
	'Roy Cope'	oFor wWoS oRus oSis
	saxatilis	wPle
	secundiflora	wCul wCol oNWe oGre
	secundiflora DJHC 316	wHer
	sieboldii	wHer wCol oRus wPle oSis oNWe oGre
un	***sieboldii*** 'Akatonbo'	oSis
un	***sieboldii*** 'Benkeijo'	oSis
	sieboldii blue	wMtT
	sieboldii cerise pink	wPle
	sieboldii 'Cherubim'	wRob
un	***sieboldii*** 'Cotton Candy'	wMtT
un	***sieboldii*** 'Cover Girl'	wMtT
	sieboldii 'Fimbriated Red'	oSis
	sieboldii 'Geisha Girl'	wRob
	sieboldii 'Istaka'	oSis wMtT
	sieboldii lavender	wPle
un	***sieboldii*** 'Loie Benedict'	wMtT
	sieboldii 'Manakoora' seed grown	wPle
un	***sieboldii*** 'Mikunino-Homari'	oSis
	sieboldii 'Musashino'	oSis
un	***sieboldii*** var. ***purpurea***	oSis
	sieboldii 'Shi-un'	oSis
	sieboldii 'Sumina'	oSis
un	***sieboldii*** 'Thelma's Sieboldii'	wPle
	sieboldii white forms	wPle
un	***sieboldii*** 'White Lace'	wHer
	sieboldii 'Winter Dreams'	last listed 99/00
	sieboldii 'Yubisugata'	oHed oSis wMtT
	sikkimensis	last listed 99/00
	sikkimensis DJHC 300	wHer
	'Snow White'	oRus wPle
	Springtime Group	oRus
	'Sue Jervis'	wWoS oHed oRus
un	'Sunrise'	oRus wTGN
	'Sunshine Susie'	wHer oRus oSis
	'Tie Dye'	oFor wTGN
	'Val Horncastle'	wHer oNWe
un	'Velvet Moon'	wHer wWoS oRus
	veris	oFor wCul oAls oRus wPle oGoo wFai oWnC oCrm wTGN oSec oWhS wSnq
	veris 'Sunset Shades'	oNat oRus oNWe
	vialii	oFor oNat oDar oAls wCSG oAmb oNWe oSha wTGN oSec wAva oEga
	viscosa	see **P. *latifolia***
	vulgaris	oGoo wMag
	vulgaris double ruby red	wHer
	vulgaris 'Gigha White'	see **P.** 'Gigha'
qu	***vulgaris*** Grandiflora mixed	oUps
	vulgaris 'Lilacina Plena'	wWoS oRus oSis wRob
	vulgaris 'Quaker's Bonnet'	see **P. *vulgaris*** 'Lilacina Plena'
	vulgaris 'Red Giant'	oRus
	vulgaris ssp. ***sibthorpii***	wHer wPle
	'Wanda' (see Nomemclature Notes)	wWoS oJoy oRus wPle oNWe oHon wMag wTGN wSta
	Wanda hybrids	oRus
	'Wharfedale Butterfly'	wMtT
	'Wharfedale Crusader'	wMtT
	'Wharfedale Gem'	wMtT
	'Wharfedale Ling'	wMtT
	'Wharfedale Village'	last listed 99/00
	wilsonii	oFor
	wulfeniana	last listed 99/00
un	'Yvonne'	wHer wWoS oRus oAmb
un	'Yvonne Gold Lace'	wTGN
	PRINSEPIA	
	sinensis	oFor oOEx
	uniflora	oFor
	PRITZELAGO	
	alpina	wMtT
	PROBOSCIDEA	
un	***atheaefolia***	oRar

un	**PROPHOYLLUM**	
	ruderale	oOEx
	PROSOPIS	
	chilensis	see **P. *glandulosa***
	glandulosa	oFor
	juliflora	last listed 99/00
	PROSTANTHERA	
	cuneata	oFor
	incisa 'Rosea'	wHer
	lasianthos	last listed 99/00
un	***nivea*** 'Indica'	wWoS
	'Poorinda Ballerina'	wHer
	rotundifolia	wHer wWoS oAls oSho iGSc
	PROTEA	
	cynaroides	wGra
	lacticolor	wCCr
un	**PROTOASPARAGUS**	
	natalensis	oTrP
	PRUMNOPITYS	
	andina	wHer
	PRUNELLA	
	grandiflora	oFor oUps
un	***grandiflora*** 'Blueberry Ice'	oFor
	grandiflora 'Pink Loveliness'	oFor
	grandiflora rosea	oAls
	grandiflora 'White Loveliness'	oFor
	incisa	see **P. *vulgaris***
	laciniata	oHed wCSG
	vulgaris	oFor iGSc oCrm oBar oAld wFFl oUps
	PRUNUS	
	'Accolade'	oDar oGar oGre oWnC wSta
	'Amanogawa'	oFor oBlo oWnC wCoS wWel
	americana	oFor wPla
	amygdalus	see **P. *dulcis***
	andersonii	oFor
	angustifolia	oFor
	armeniaca 'Blenheim'	oAls oWnC wVan
	armeniaca 'Chinese'	wBCr oAls oWnC wVan
un	***armeniaca*** 'Goldbar'	wVan wC&O
	armeniaca 'Goldcot'	wVan
	armeniaca 'Goldrich'	wVan wC&O
un	***armeniaca*** 'Goldstrike'	wVan wC&O
un	***armeniaca*** 'Harcot'	wBCr
un	***armeniaca*** 'Harglow'	wClo wRai oWnC wCoS
un	***armeniaca*** 'Harogem'	wDCr
un	***armeniaca*** 'Lehrman'	wVan
un	***armeniaca*** 'Mandschurica'	wDCr
	armeniaca 'Moorpark'	wKin
	armeniaca 'Mormon'	see **P. *armeniaca*** 'Chinese'
	armeniaca 'Perfection'	wVan wC&O
	armeniaca 'Puget Gold'	wBur wClo oAls oGar wRai oOEx oGre wTGN wC&O wCoS
	armeniaca 'Rival'	wVan wC&O
un	***armeniaca*** 'Scout'	wBCr
	armeniaca 'Sun-Glow'	wC&O
	armeniaca 'Sungold'	wBCr wKin
un	***armeniaca*** 'Sweet Pit'	oOEx
	armeniaca 'Tilton'	oAls wKin wVan wC&O
un	***armeniaca*** 'Tomcot'	wVan wC&O
	armeniaca 'Wenatchee'	wVan wC&O
	armeniaca 'Wenatchee Moorpark'	see **P. *armeniaca*** 'Wenatchee'
un	'Assiniboine'	wBCr
un	***avium*** 'Angela'	wClo oGar wRai oWnC
un	***avium*** 'Attika'	wVan wC&O
un	***avium*** 'Bada'	oEar
	avium 'Bear Creek Early'	wBCr
	avium 'Bigarreau Napoleon''	wBCr oAls oGar oSho wKin oWnC wVan wC&O wCoS
	avium 'Bing'	wBCr oAls oGar oSho wRai wKin oWnC wVan wC&O wCoS
	avium 'Black Republican'	see **P. *avium*** 'Republican'
	avium 'Black Tartarian'	wKin oWnC wC&O wCoS
un	***avium*** 'Chelan'	wVan wC&O
	avium 'Corum'	last listed 99/00
un	***avium*** 'Early Burlat'	wClo oAls wRai
	avium 'Emperor Francis'	wRai wVan wC&O
	avium 'Glacier'	wTGN
un	***avium*** 'Gold'	wBCr wC&O
	avium 'Hardy Giant'	wClo oAls wRai wVan wC&O
un	***avium*** 'Hedelfingen'	wVan wC&O
un	***avium*** 'Index'	wC&O
	avium 'Kansas Sweet'	wRai
un	***avium*** 'Kristin'	wBCr wBur wClo oAls wRai
	avium 'Lambert'	oAls oGar wVan wC&O

	avium 'Lapins'	wBCr wBur wClo oGar oSho wRai oWnC wVan wC&O
	avium 'Napoleon'	see **P. *avium*** 'Bigarreau Napoleon'
un	***avium*** 'Queen Ann'	oEar
	avium 'Rainier'	wBCr wBur wClo oEar oAls oGar oSho wRai oWnC wTGN wVan wC&O wCoS
un	***avium*** 'Regina'	wVan wC&O
	avium 'Republican'	wVan wC&O
	avium 'Royal Ann'	see **P. *avium*** 'Bigarreau Napoleon'
	avium 'Sam'	wBur oAls oGar oSho wRai wCoS
un	***avium*** 'Schmidt'	wVan
un	***avium*** 'Schneider'	wVan wC&O
un	***avium*** 'Skeena'	wC&O
un	***avium*** 'Sonata'	wVan wC&O
	avium 'Stella'	oAls oGar oSho wRai oWnC wTGN wVan wCoS
un	***avium*** 'Sweet Ann'	wBur wRai
un	***avium*** 'Sweetheart'	wRai wVan wC&O
un	***avium*** 'Ulster'	wVan
	avium 'Utah Giant'	wVan
	avium 'Van'	wBCr wBur wClo oAls oWnC wVan wC&O
un	***avium*** '135-20-9'	wC&O
un	'Berry'	oWnC
	besseyi	oFor wBCr oOEx wPla oSle
	x blireana	oFor oAls oGar oSho oBlo oGre oWnC wTGN wSta wCoS
	campanulata	last listed 99/00
	caroliniana	oFor wHer oSho wCCr
	caroliniana 'Compacta'	oFor
	cerasifera 'Hollywood'	oEar oGar wRai oWnC wTGN
	cerasifera 'Krauter's Vesuvius'	see **P. *cerasifera*** 'Vesuvius'
	cerasifera 'Newport'	oGre wKin oWnC wCoS
un	***cerasifera*** 'Purple Pony'	oGre
	cerasifera 'Thundercloud'	oFor oDar oAls oGar oBlo wKin oWnC wTGN wSta oGue oSle wCoS wWel
	cerasifera 'Vesuvius'	oFor oGar oSho oGre oWnC oSle wCoS
un	***cerasus*** 'Balaton'	wC&O
	cerasus 'Meteor'	last listed 99/00
	cerasus 'Montmorency'	wBCr wBur wClo oAls oGar oSho wRai oWnC wC&O wCoS
	cerasus 'Morello'	wClo
	cerasus 'North Star'	wBCr oAls oGar oOEx wKin oWnC
un	***cerasus*** 'Suda Hardy Tart'	wBCr
un	***cerasus*** 'Surefire'	wRai
	x cistena	oFor wCul oRiv oWhi wRai wKin oWnC oGue oSle wCoS
un	'Comet of Kubansk'	wRai
	cyclamina	oFor oRiv
un	***x dasycarpa*** 'Mesch Mesch Amrah' (fruit)	wRai
un	***x dasycarpa*** 'Tlor-Tsiran' (fruit)	wRai
un	'Delight'	wRai
	dielsiana	oFor
	domestica x ***armeniaca*** (fruit)	oAls oGar
un	***domestica*** x ***armeniaca*** 'Dapple Dandy'	wRai oGre
un	***domestica*** x ***armeniaca*** 'Flavor King'	wRai
un	***domestica*** x ***armeniaca*** 'Flavor Queen'	wRai
un	***domestica*** x ***armeniaca*** 'Flavor Supreme'	oGar wRai oGre
	domestica 'Brooks'	oEar oAls oGar oSho oGre oWnC
	domestica 'Cambridge Gage'	last listed 99/00
un	***domestica*** 'Castleton'	wC&O
	domestica 'Coe's Golden Drop'	wRai
	domestica 'Damson'	see **P. *insititia***
un	***domestica*** 'Date Petite'	oWnC
un	***domestica*** 'Early Italian'	wBur oGar oSho wVan wC&O
	domestica 'Early Laxton'	wBCr wRai
un	***domestica*** 'Elma's Special'	wClo wRai
un	***domestica*** 'Empress'	wC&O
un	***domestica*** 'French Petite'	oAls
	domestica 'Golden Transparent'	last listed 99/00
ch	***domestica*** Green Gage Group	see **P. *domestica*** Reine-Claude Group
	domestica 'Imperial Epineuse'	wClo wRai
	domestica 'Italian'	see **P. *domestica*** 'Italian Prune'
	domestica 'Italian Prune'	wBCr oAls oGar wRai wKin wTGN wVan wC&O wCoS
	domestica 'Kirke's'	wRai
	domestica 'Mirabelle de Nancy'	see **P. *insititia*** 'Mirabelle de Nancy'
un	***domestica*** 'NY 6'	wC&O
un	***domestica*** 'NY 9'	wC&O
	domestica 'Opal'	last listed 99/00
	domestica 'Peach Plum'	wClo oAls oGar
	domestica 'President'	oAls wVan
	domestica Reine-Claude Group	wBCr wBur oEar oAls oGar oSho wTGN wVan

	domestica Reine-Claude Group 'Reine-Claude de Bavais'	wRai
un	***domestica*** 'Schoolhouse'	wRai
un	***domestica*** 'Seneca'	wBur wClo oEar wRai oWnC
	domestica 'Stanley'	oFor wBCr wClo oGar wKin wVan wC&O
un	***domestica*** 'Valor'	wRai
	domestica 'Victoria'	wRai
un	***domestica*** 'XX French Prune'	wBCr
un	***domestica*** 'Yellow Egg'	wClo oGar oSho
un	'Dream Catcher'	oFor
	dulcis 'All-in-One'	wBur oGar wRai
un	***dulcis*** 'Degn'	wVan
	dulcis 'Hall'	oFor oAls wRai oWnC
	dulcis 'Hall's Hardy'	see **P. *dulcis*** 'Hall'
	dulcis 'Ne Plus Ultra'	wVan
	dulcis 'Nonpareil'	wVan
un	***dulcis*** 'Titan'	wRai
	emarginata	wBur wWat oAld
un	***emarginata*** var. ***mollis***	wFFl
	fremontii	oFor
	fruticosa	oFor
	glandulosa	oSec
do	***glandulosa*** 'Alba Plena'	oFor oGre
	glandulosa 'Rosea'	wKin
	glandulosa 'Rosea Plena'	see **P. *glandulosa*** 'Sinensis'
do	***glandulosa*** 'Sinensis'	oFor wRav
	'Hally Jolivette'	oFor oWhi
	humilis	oFor
	ilicifolia	oFor wHer oOEx
	incisa	oFor
	insititia	wC&O
	insititia 'Mirabelle de Nancy'	wRai
un	***insititia*** 'Mirabelle 858'	wRai
	japonica	oFor
un	'Kaga'	wBCr
	'Kanzan'	oFor oDar oAls oGar oSho oBlo wRai oGre wKin oWnC wTGN wSta oSle wCoS wWel
	'Kiku-shidare-zakura'	oFor
	'Kwanzan'	see **P.** 'Kanzan'
	laurocerasus	oFor oAls oGar oSho oBlo oWnC wCoN wCoS wWel
	laurocerasus 'Camelliifolia'	wHer
va	***laurocerasus*** 'Castlewellan'	wCol
un	***laurocerasus*** 'Compacta'	oGar
qu	***laurocerasus*** 'Cream Marble'	oSec
un	***laurocerasus*** 'Kelpie'	wHer
un	***laurocerasus*** 'Marble Dragon'	oSho
un	***laurocerasus*** 'Marble Queen'	oCis
	laurocerasus 'Marbled White'	see **P. *laurocerasus*** 'Castlewellan'
	laurocerasus 'Mount Vernon'	wHer wCCr wSta
	laurocerasus 'Nana'	oAls oGar wSta wWel
	laurocerasus 'Otto Luyken'	oFor oAls oGar oSho oBlo oGre oTPm oWnC wSta oCir oGue wCoS
	laurocerasus 'Schipkaensis'	oAls oGar oBlo wSta wCoN
	laurocerasus 'Zabeliana'	oGar oBlo wSta wCoN
	lusitanica	wClo oAls oGar oSho oBlo wCCr oWnC wSta wCoN wCoS wWel
un	***lusitanica*** 'Center Stage'	wCol
	lusitanica 'Variegata'	wHer oGar oGre wSta wSte
	lyonii	oFor oSho oOEx wCCr
	maackii	oFor oGar oGre wKin wWel
	maritima	oFor
un	***maritima*** 'Jersey'	wRai
un	***maritima*** 'Premier'	wRai
un	'Methley'	wBCr wBur wClo oEar oGar wRai
un	***mongolica***	oFor
	'Mount Fuji'	see **P.** 'Shirotae'
	mume	oEdg oRiv
un	***mume*** 'Contorta'	oWhi
	mume 'Dawn'	oFor oGre oWnC
	mume 'Kobai'	oCis wRai
un	***mume*** 'Matsubara Red'	oFor
	mume 'Peggy Clarke'	oFor oGre oWnC wWel
	mume 'Rosemary Clarke'	oFor oGre
	mume 'W. B. Clarke'	oFor oGre
un	***mume*** 'White Christmas'	oCis
	nigra	last listed 99/00
	'Okame'	oFor oGar oBlo oWnC wWel
	padus	oFor wBCr wSta
un	***padus purpurea***	oFor
un	***padus*** 'Summer Glow'	oGar
un	'Pembina'	wBCr
	pendula	last listed 99/00
	pendula 'Pendula Rosea'	oFor oAls oGar oRiv oBlo oGre wKin oWnC wTGN wSta oGue wCoS

un	***pendula*** 'Pendula Rosea Plena'	oFor oAls wTGN
	pensylvanica	oFor
un	***persica*** 'Allstar'	wC&O
un	***persica*** 'Angelus'	wVan wC&O
un	***persica*** 'Arctic Gem'	wBCr wVan
un	***persica*** 'Arkansas 9'	wVan wC&O
un	***persica*** 'Babygold 5'	wVan wC&O
un	***persica*** 'Belle of Georgia'	wVan
un	***persica*** 'Blushingstar'	wC&O
un	***persica*** 'Camelliaeflora'	oFor
un	***persica*** 'Canadian Harmony'	wVan wC&O
un	***persica*** 'Coralstar'	wC&O
un	***persica*** 'Cresthaven'	wVan wC&O
un	***persica*** 'Delp Hale'	wVan
	persica 'Early Elberta'	oGar oSho wVan wC&O
	persica 'Early Redhaven'	wVan wC&O
	persica 'Elberta'	oAls wKin wVan wC&O
un	***persica*** 'Eldorado'	wRai
un	***persica*** 'Fairhaven'	wVan
	persica 'Flamecrest'	wC&O
	persica 'Flavorcrest'	wC&O
	persica Fortyniner'	oFor
	persica 'Frost'	oFor wBur wClo oAls oGar oSho wRai oWnC wTGN wCoS
un	***persica*** 'Giant Elberta'	oWnC
	persica 'Gleason Elberta'	see **P. *persica*** 'Early Elberta'
un	***persica*** 'Glohaven'	wVan wC&O
un	***persica*** 'Golden Glory'	oGar
	persica 'Golden Jubilee'	wVan
un	***persica*** 'Golden Monarch'	wVan
	persica 'Halehaven'	wVan
un	***persica*** 'Harken'	wBCr oGar wRai
	persica 'Improved Elberta'	see **P. *persica*** 'Early Elberta'
	persica 'Indian Free'	wRai
	persica 'J. H. Hale'	wVan wC&O
un	***persica*** 'Loring'	wVan wC&O
un	***persica*** 'M. A. Blake'	wVan wC&O
un	***persica*** 'Madison'	wVan
un	***persica*** 'Mary Jane'	wRai
un	***persica*** 'Monroe'	wC&O
un	***persica*** 'MP 1'	wRai
un	***persica*** 'Muir'	wRai
un	***persica*** var. ***nectarina*** 'Arctic Glow'	wBCr wVan
	persica var. ***nectarina*** 'Fantasia'	oAls wKin oWnC wVan wC&O wCoS
un	***persica*** var. ***nectarina*** 'Firebright'	wVan wC&O
	persica var. ***nectarina*** 'Flavortop'	wVan wC&O
un	***persica*** var. ***nectarina*** 'Harko'	wBCr wBur wClo oAls wCoS
	persica var. ***nectarina*** 'Independence'	oAls wVan
un	***persica*** var. ***nectarina*** 'Juneglo'	wRai
un	***persica*** var. ***nectarina*** 'Merricrest'	wBCr oAls
un	***persica*** var. ***nectarina*** 'Necta Zee'	wRai
un	***persica*** var. ***nectarina*** 'Nectar Babe'	wRai
un	***persica*** var. ***nectarina*** 'Red Gold'	oAls wVan wC&O
un	***persica*** var. ***nectarina*** 'Sunglo'	wC&O
un	***persica*** 'Newhaven'	wRai wVan
	persica 'O'Henry'	wVan wC&O
un	***persica*** 'P F #1'	wC&O
un	***persica*** 'P F # 5-B'	wC&O
un	***persica*** 'P F # 12-B'	wC&O
un	***persica*** 'P F # 15-A'	wC&O
un	***persica*** 'P F #17'	wC&O
un	***persica*** 'P F #23'	wC&O
un	***persica*** 'P F #27-A'	wC&O
un	***persica*** 'P F #20-007'	wC&O
un	***persica*** 'P F #24-007'	wC&O
un	***persica*** 'Pix Zee'	wRai
un	***persica*** 'Q-1-8'	wBur wRai
un	***persica*** 'Red Fremont'	wVan
un	***persica*** 'Red Lady'	wC&O
	persica 'Redglobe'	wVan wC&O
	persica 'Redhaven'	oFor wBCr oAls oGar wKin wVan wC&O
	persica 'Redskin'	wVan wC&O
un	***persica*** 'Redstar'	wC&O
un	***persica*** 'Regina'	wC&O
	persica 'Reliance'	wBCr wVan
	persica 'Rio Oso Gem'	wVan wC&O
un	***persica*** 'Rising Star'	wC&O
un	***persica*** 'Roza'	wVan wC&O
	persica 'Saturne'	wRai
un	***persica*** 'Siberian C' seed grown	wBCr
un	***persica*** 'Starfire'	wC&O
un	***persica*** 'Suncrest'	wVan wC&O
un	***persica*** 'Topaz'	wVan
	persica 'Veteran'	wBCr wBur oAls oSho wKin wVan
un	***persica*** 'Yakima Hale'	wVan

un	'Petite'	oAls
un	'Pipestone'	wBCr
un	'Plum Beauty'	oAls
	pumila	oFor
	pumila var. ***depressa***	wCol
	'Royal Burgundy'	oGar oGre wKin oWnC wTGN wWel
	salicifolia	oOEx
	salicina 'Beauty'	wBur wRai oWnC wCoS
	salicina 'Burbank'	oEar wCoS
un	***salicina*** 'Catalina'	wVan
un	***salicina*** 'Cocheco'	wRai
un	***salicina*** 'Duarte'	wVan
un	***salicina*** 'Early Golden'	wRai
	salicina 'Elephant Heart'	oAls oGar oSho wVan
un	***salicina*** 'Fortune'	wRai
	salicina 'Friar'	wVan wC&O
un	***salicina*** 'Improved Duarte'	wC&O
	salicina 'Santa Rosa'	oEar oAls oGar oGre oWnC wVan wC&O wCoS
un	***salicina*** 'Santa Rosa Weeping'	oAls wRai
	salicina 'Satsuma'	wBur oAls oGar oSho oWnC
un	***salicina*** 'Shiro'	wBur wClo oEar oAls oGar oSho wRai wTGN wVan wC&O
un	***salicina*** 'Simka'	wVan wC&O
	sargentii	oFor oGar
	sargentii 'Columnaris'	oWnC
	serotina	oFor wBCr wBur
	serrula	oFor oGar oDan oBlo oGre wSta wSte
	serrulata 'Beni Hoshi'	oGre
	serrulata 'Kwanzan'	see **P.** 'Kanzan'
qu	***serrulata*** weeping	oAls oBlo wTGN
	'Shirofugen'	oFor oAls oGar oSho oGre oWnC wTGN wSta wWel
	'Shirotae'	oFor oAls oGar oSho oBlo oGre oWnC wTGN wSta wCoS wWel
	'Shogetsu'	oFor oAls oSho oWnC wWel
un	Snow Fountains™ / 'Snowfozam'	oAls oGar oSho oBlo oGre wKin oWnC wSta wCoS
	'Snow Goose'	oGar oWnC
un	'South Dakota'	wBCr
	spinosa	oFor
	'Spire'	oFor
	'Sprite'	wRai
	subcordata	oFor wCCr
	x subhirtella	oGar
	x subhirtella 'Autumnalis'	oFor oDan oGar oSho oBlo oWnC wSta
	x subhirtella 'Pendula' hort.	see **P. *pendula*** 'Pendula Rosea'
	x subhirtella 'Pendula Rosca Plena'	see **P. *pendula*** 'Pendula Rosea Plena'
	x subhirtella 'Whitcombii'	wSta
	x subhirtella 'Yae-shidare-higan'	oGar oSho oSle
un	'Superior'	wBCr
	'Taihaku'	oFor wWel
	tenella	oFor wBCr
	tibetica	see **P. *serrula***
	tomentosa	oFor wBCr wRai
	triloba	oGre wKin
do	***triloba*** 'Multiplex'	wRav
	umbellata	oFor
un	'Underwood'	wBCr
	virginiana	wWoB wBur wShR wWat wKin wPla oSle
	virginiana 'Canada Red'	oGar oWnC
	virginiana var. ***demissa***	oFor
	virginiana 'Schubert'	oFor wBCr
un	***virginiana*** 'Schubert Select'	oWnC
un	'Waneta'	wBCr
un	'White Fountain'	oFor
	x yedoensis	oFor oWnC wSta
	x yedoensis 'Afterglow'	oGar oGre oWnC
	x yedoensis 'Akebono'	oFor oGar oSho oBlo oGre oWnC wSta oSle wWel
	x yedoensis 'Perpendens'	see **P. *x yedoensis*** 'Shidare-yoshino'
	x yedoensis 'Shidare-yoshino'	last listed 99/00
	x yedoensis 'Yoshino'	see **P. *x yedoensis***
	PSEUDOBOMBAX	
	ellipticum	oRar
	PSEUDOCYDONIA	
	sinensis	oFor
	PSEUDOLARIX	
	amabilis	oFor wHer oPor wClo oAmb oGar oRiv oWhi oGre wBox oBRG wCoN wSte wWel
	kaempferi	see **P. *amabilis***
	PSEUDOPANAX	
	arboreus	oFor
	ferox	oFor

	PSEUDOSASA	
	amabilis (see Nomenclature Notes)	oTra wCli wBea
un	***cantori***	oTra
	japonica	oFor oTra wSus wCli oNor wBea oSho oOEx oGre wBam wBmG oTBG
va	***japonica*** 'Akebonosuji'	wBea oTBG
	japonica seedling	wBea
	japonica 'Tsutsumiana'	oTra wSus wCli wBmG oTBG
	japonica 'Variegata'	see **P.** ***japonica*** 'Akebonosuji'
un	***longiligula***	oTra wBea
	owatarii	oTra wCli wBea wBam
	pleioblastoides	wBea
	usawai	oNor
un	***viridula***	oTra
	PSEUDOTAXUS	
	chienii	wHer
	PSEUDOTSUGA	
	japonica	wCol
	macrocarpa	last listed 99/00
	menziesii	oFor wWoB wNot oAls wShR oGar oRiv oSho oBlo oGre wWat wCCr wKin oWnC wTGN wSta wPla oAld wCoN wFFl wRav oSle wSnq wCoS
	menziesii Cecil Smith garden	oPor
	menziesii Dusek's witches' broom	oPor
	menziesii 'Elegans'	oPor
	menziesii 'Fletcheri'	oPor
	menziesii 'Fretsii'	oPor oRiv
	menziesii var. ***glauca***	oFor oPor wKin
	menziesii 'Glauca Pendula'	oPor oGar oRiv
	menziesii 'Graceful Grace'	oFor oPor oRiv wTGN
	menziesii 'Hess Select'	wCol
un	***menziesii*** 'Hillside Gold'	wHer oPor oRiv
	menziesii 'Hillside Pride'	oPor
un	***menziesii*** 'Hupp's Weeping'	oPor
	menziesii 'Idaho Gem'	oPor
un	***menziesii*** 'Idaho Weeper'	oPor
	menziesii 'Little Jon'	oPor oRiv
un	***menziesii*** 'Loggerhead'	oPor oRiv
	menziesii Pendula Group	wCol oGar oGre wSta
	menziesii 'Tempelhof Compact'	oPor
un	***menziesii*** 'Ziogus'	oPor
	PSIDIUM	
un	***guajava*** 'Bankok'	oOEx
	guajava 'Beaumont'	oOEx
un	***guajava*** 'Lemon'	oOEx
un	***guajava*** 'Mexican Cream'	oOEx
un	***guajava*** 'Pink'	oOEx
un	***guajava*** 'Red Indian'	oOEx
un	***guajava*** 'Thai'	oOEx
	littorale var. ***longipes***	oOEx
	PSORALEA	
	glandulosa HCM 98039	wHer
	PSYCHOTRIA	
	viridis	oOEx
	PTELEA	
	trifoliata	oFor
	PTERIDIUM	
	aquilinum	oBos oAld
	PTERIS	
un	***gallinopes***	wFol
	PTEROCARYA	
	fraxinifolia	last listed 99/00
	hupehensis	oFor
	macroptera	wCCr
	paliurus	**CYCLOCARYA** ***paliurus***
	x rehderiana	oFor
	sp. DJHC 98254	wHer
	stenoptera	oFor
	PTEROCELTIS	
	tatarinowii	oFor wCCr
	PTEROCEPHALUS	
	perennis	last listed 99/00
	pinardii	oSis
	PTEROSTYRAX	
	corymbosa	oFor oGar oWhi oGre wWel
	hispida	oFor oGos oDar oGar oDan oRiv oWhi oGre wPir oBRG wCoN wWel
	psilophylla	last listed 99/00
	PTILOTRICHUM	see **ALYSSUM**
	PULICARIA	
	dysenterica	wFai iGSc
	PULMONARIA	
	angustifolia	wHer wCSG oGar oSis oGre oHon wMag
	angustifolia 'Blaues Meer'	oFor wNay
un	'Apple Blossom'	oSle
	'Apple Frost'	wHer wWoS oJoy wNay
	'Barfield Regalia'	last listed 99/00
	'Benediction'	wHer wHig oHed oNWe oGre oJil
	'Berries and Cream'	wWoS wNay oGre
	'Blue Ensign'	oHed
	'British Sterling'	oGar wNay oUps
	'Coral Springs'	wHer wWoS wNay
	'Crawshay Chance'	oHed
	'De Vroomen's Pride'	wNay
	'Diana Chappell'	oHed
	'Excalibur'	wWoS wCol wHig oAmb oGar wNay wAva oUps
	'Fiona'	wNay
	'Golden Haze'	wHer wWoS wNWe wRob
	'Highdown'	see **P.** 'Lewis Palmer'
	hybrid	oNWe
un	'Ice Ballet'	wRob
un	'Janet Fisk'	wCul oGos oNWe wNay oHou oLSG
	'Lewis Palmer'	oFor wCul oGos wHig oHed oGar oSis oOut wNay oGre wFai wMag wBox
	'Little Star'	oFor wHer wWoS oSis wRob wNay
	longifolia	wEde
	longifolia 'Bertram Anderson'	wHer wCul wHig oGar oSis wNay oGre wFai wBox wSnq
	longifolia ssp. ***cevennensis***	oFor wHer wWoS wCol oAls oHed oGar oDan oSis oOut wRob wNay wFai oWnC oHou
un	***longifolia*** 'Little Blue'	wNay
	'Majeste'	wHer wWoS wNay
	'Margery Fish'	oFor wHig oHed oGar oNWe wNay oGre wFai oHou
	'Mary Mottram'	wWoS wNay
	'Merlin'	wHer oHed
	'Milky Way'	oFor wHer wWoS
	mollis	wHig
	mollis 'Samobor'	wNay
	'Mrs. Kittle'	wWoS wRob wNay
	'Nuernberg'	wHer
	officinalis	wCSG oCrm
	officinalis 'Blue Mist'	wWoS wNay
	officinalis Cambridge Blue Group	wNay
	officinalis 'Sissinghurst White'	wCul wHig oAls oGar wNay oGre wFai wBox oLSG
	officinalis 'White Wings'	wHer oJoy oDan wNay
	'Paul Aden'	wNay
un	'Pierre's Pure Pink'	oFor wCul oEdg oOut wNay
	'Polar Splash'	oFor wAva
	'Purple Haze'	wWoS wRob
	'Raspberry Splash'	oFor wHer oAmb oGar oDan wNay wAva
	'Regal Ruffles'	wNay
	'Roy Davidson'	oFor wHer wCul oJoy wCol wHig oHed wRob wNay oWnC wMag oHou oUps wSnq
	rubra	wCSG
	rubra var. ***albocorollata***	wHer
	rubra 'Barfield Pink'	wHer oGar wNay wFai
	rubra 'Bowles' Red'	wHer
va	***rubra*** 'David Ward'	oFor wHer wWoS oGos oHed oGar oDan oSis oNWe wRob wNay oGre wFai wMag wBox
	rubra 'Redstart'	wHer oJoy oGar oSis wNay wFai oHon oWnC wTGN oHou wAva wSnq
un	***saccharata*** 'Alan Leslie'	wNay
	saccharata Argentea Group	wHig wNay
	saccharata 'Bofar Red'	wNay
	saccharata 'Cotton Cool'	wWoS wRob wAva
	saccharata 'Dora Bielefeld'	oFor wHer oAls oGar oOut wRob wNay oGre wAva
	saccharata 'Leopard'	oFor wHer oHed oSis wNay
	saccharata 'Mrs. Moon'	oGar oSis oSho wNay wFai wMag oLSG
	saccharata 'Pink Dawn'	oFor wNay
	saccharata 'Reginald Kaye'	wHer wNay
	'Silver Lance'	oFor wNay
	'Silver Streamers'	wHer wWoS oGos oJoy wRob wNay
	'Smoky Blue'	oFor wWoS oNat oAls oGar wRob oGre oWnC oHou wAva
	sp. blue	wAva
	'Spilled Milk'	oFor wHer wNay oUps
	'Tim's Silver'	oGar oGre
	'Trevi Fountain'	wHer
	'Victorian Brooch'	wHer wWoS wGAc oJoy oHed oDan oNWe oOut wNay oGre
	PULSATILLA	
	albana	wMtT

	alpina 'Sulphurea'	see **P.** *alpina* ssp. *apiifolia*
	bungeana	last listed 99/00
	grandis	see **P.** *vulgaris* ssp. *grandis*
	halleri ssp. *grandis*	see **P.** *vulgaris* ssp. *grandis*
	halleri ssp. *slavica*	last listed 99/00
	montana	last listed 99/00
	patens	oFor wPla
	turczaninovii	oDan
	vernalis	wMtT
	vulgaris	wCul oNat oJoy oAls oRus wCSG oGar oNWe oSha oBlo oGre wFai wHom wCoN oUps
	vulgaris 'Alba'	oFor wCul oGar oSis wHom
	vulgaris dark shades	oHed
	vulgaris ssp. *grandis*	oFor oDan wMtT
	vulgaris Heiler hybrids	wCul oLSG
	vulgaris 'Papageno'	wFGN oSis
	vulgaris purple	oFor
	vulgaris var. *rubra*	oFor wCul oGar oSis wTGN
	PUNICA	
	granatum	wRai wCCr
un	*granatum* 'California Sunset'	oGar oCis oSho wWel
un	*granatum* 'Eightball'	oCis
	granatum 'Fleshman' (fruit)	oOEx
un	*granatum* 'Grenada' (fruit)	oOEx
un	*granatum* 'King' (fruit)	oOEx
	granatum var. *nana*	oFor oTrP oAls oGar oSho wRai oGre oWnC wCoS
un	*granatum* 'Phil's Best' (fruit)	oOEx
un	*granatum* Renans' (fruit)	oOEx
un	*granatum* 'Sweet' (fruit)	oOEx
	granatum 'Tayosho'	oGar oWnC
	granatum 'Tayosho' bonsai	oGre
	granatum twisted trunk, bonsai	oGre
	granatum white (fruit)	oOEx
	granatum white flower/fruit	oCis
	granatum 'Wonderful' (fruit)	oFor oGar oSho oOEx oWnC wWel
	granatum yellow flower/fruit	oCis
	PURSHIA	
	tridentata	wPla
	PUTORIA	
	calabrica	oSis
	PUYA	
un	*flavovirens*	oRar
	mirabilis	oRar
	PYCNANTHEMUM	
	muticum	oAls
	pilosum	oAls oGoo
	tenuifolium	iGSc
	virginianum	oFor oGoo oUps
	PYRACANTHA	
	angustifolia	wHer
	angustifolia 'Gnome'	wHer
	angustifolia Yukon Belle® / 'Monon'	last listed 99/00
	atalantioides 'Aurea'	wCCr
	coccinea 'Fiery Cascade'	oGar oWnC
	coccinea 'Government Red'	oBlo oTPm
	coccinea 'Kasan'	oGar
	coccinea 'Lalandei'	oBlo oTPm oWnC
	coccinea 'Lowboy'	oGar
	crenatoserrata	oFor
	crenatoserrata 'Cherri Berri'	oAls oGar oGre oWnC
	crenatoserrata 'Graberi'	oGar
	fortuneana	see **P.** *crenatoserrata*
	'Gold Rush'	oFor wHer
	'Golden Charmer'	last listed 99/00
va	'Harlequin'	oFor oGar oSho oGre
	koidzumii 'Victory'	oGar oSho oWnC
	'Mohave'	oGar oWnC
	'Navaho'	oBlo
	'Orange Glow'	oGar wTGN
	Red Elf™ / 'Monelf'	oGar oSho
	rogersiana	wCul
	sp. aff. *rogersiana* EDHCH 97017 (last listed as EDHCH 9701)	wHer
	'Ruby Mound'	oGar
	'Santa Cruz'	oGar oGre oTPm oWnC
	'Teton'	oGre oBRG
	'Watereri'	oAls
	X PYRACOMELES	
	vilmorinii	oFor oSho
	PYRETHROPSIS	see **RHODANTHEMUM**
	PYRETHRUM	
	roseum	see **TANACETUM** *coccineum*

	PYROLA	
	asarifolia	wThG oBos wFFI
	picta	oBos wFFI
	uniflora	see **MONESES** *uniflora*
	PYRROSIA	
un	*linearifolia*	oSis
	lingua	oSec
	PYRUS	
	calleryana	wCoN
	calleryana 'Aristocrat'	oGar oWnC wWel
	calleryana 'Autumn Blaze'	oFor oGar
	calleryana 'Bradford'	oAls oBlo oWnC
	calleryana 'Capital'	wBox wSta wWel
	calleryana 'Chanticleer'	oFor oDar oAls oGar oSho oBlo wKin oWnC wCoS wWel
	calleryana 'Cleveland Select'	see **P.** *calleryana* 'Chanticleer'
	calleryana 'Redspire'	oFor oAls oBlo oWnC wWel
	calleryana 'Trinity'	wSta
	communis 'Abbe Fetel'	wC&O
	communis 'Anjou'	oFor oEar oAls oGar oSho oWnC wVan wC&O
	communis 'Bartlett'	oFor oEar oAls oGar oSho wKin oWnC wVan wC&O wCoS
un	*communis* 'Basque'	oEar
un	*communis* 'Bennett'	wRai
	communis 'Beurre Bosc'	oFor wBur wClo oAls oGar oSho wRai oWnC wVan wC&O
un	*communis* 'Bierschmidt'	wBCr
un	*communis* 'Bojka'	wBCr
	communis 'Bosc'	see **P.** *communis* 'Beurre Bosc'
un	*communis* 'Bronze Beauty Bosc'	wC&O
un	*communis* 'Butirra'	oEar
	communis 'Butirra Precoce Morettini'	wRai
	communis 'Cascade'	wVan
un	*communis* 'Cebulka'	wBCr
	communis 'Clapp's Favorite'	wBCr
un	*communis* 'Columbia Red Anjou'	wVan
	communis 'Comice'	see **P.** *communis* 'Doyenne du Comice'
	communis 'Concorde'	wVan wC&O
	communis 'Conference'	wClo wRai
	communis 'D'Anjou'	see **P.** *communis* 'Anjou'
un	*communis* 'Doyenne de Juillet'	wBCr
	communis 'Doyenne du Comice'	wBur wClo oEar oAls oGar wRai wVan wC&O wCoS
	communis 'Du Comice'	see **P.** *communis* 'Doyenne du Comice'
un	*communis* 'Duchess'	wBCr
	communis 'Flemish Beauty'	wBCr oAls wVan
	communis 'Golden Russet Bosc'	see **P.** *communis* 'Beurre Bosc'
	communis 'Harrow Delight'	wRai
	communis 'Harvest Queen'	last listed 99/00
	communis 'Highland'	wBur oAls wRai
un	*communis* 'John'	wBCr
un	*communis* 'Kalle'	wVan
un	*communis* 'King Sobieski'	wBCr
un	*communis* 'Luscious'	wBCr
un	*communis* 'Magness'	wBCr
un	*communis* 'Maxine'	wBCr
	communis 'Moonglow'	wBCr
un	*communis* 'Noble Russet Bosc'	wC&O
un	*communis* 'Nova'	wBCr
un	*communis* 'Old Home'	wVan
un	*communis* 'Orcas'	wBur wClo wRai
	communis 'Packham's Triumph'	wC&O
un	*communis* 'Red Anjou'	wBCr oWnC wC&O
un	*communis* 'Red Bartlett'	oGar
un	*communis* 'Red Clapp's Favorite'	wBCr wClo
	communis 'Red Sensation'	see **P.** *communis* 'Sensation Red Bartlett'
un	*communis* 'Rescue'	wRai
un	*communis* 'Russian Early'	wBur
	communis 'Seckel'	wBCr wBur wTGN wVan wC&O
	communis 'Sensation Red Bartlett'	wBCr wC&O
un	*communis* 'Shannon'	oEar
un	*communis* 'Sherwood's Giant'	wBCr
un	*communis* 'Spalding'	wRai
un	*communis* 'Summer Blood Birne'	oEar wRai
un	*communis* 'Summer Crisp'	wBCr wKin
un	*communis* 'Tyson'	wBCr
un	*communis* 'Ubileen'	wBCr wBur wClo wRai oOEx
un	*communis* 'Warren'	wBur wRai
un	'Edgewood'	oGre
un	*fauriei* 'Korean Sun'	oWnC
	fusca	see **MALUS** *fusca*
	pyrifolia 'Chojura'	oFor wBCr wBur oEar oGar oSho oOEx oGre oWnC
un	*pyrifolia* 'Daisui Li'	oGre

un	***pyrifolia*** 'Hamese #1'	wRai
	pyrifolia 'Hosui'	wBCr oAls oOEx wBox wVan
un	***pyrifolia*** 'Ichaban Nashi'	wClo
un	***pyrifolia*** 'Ichiban'	wRai
	pyrifolia 'Kikusui'	oGre
un	***pyrifolia*** 'Korean Giant'	wBCr wBur wRai oOEx
un	***pyrifolia*** 'Kosui'	wBur wClo oEar oGar
un	***pyrifolia*** 'Meigetsu'	oEar
un	***pyrifolia*** 'Mishirasu'	wRai
	pyrifolia 'Niitaka'	oAls
	pyrifolia 'Nijisseiki'	wBur oEar oAls oGar oSho oOEx wKin oWnC wTGN wVan wCoS
un	***pyrifolia*** 'Seuri'	oEar oGar wRai oOEx
un	***pyrifolia*** 'Shin-li'	oGre
un	***pyrifolia*** 'Shinko'	wBur oEar oAls oGar oSho oOEx oGre
	pyrifolia 'Shinseiki'	wBCr wBur wClo oEar oAls oGar oSho wRai oOEx oGre oWnC wVan
	pyrifolia 'Shinsui'	wBCr
un	***pyrifolia*** 'Singo'	wBCr
	pyrifolia 'Tsu Li'	oOEx
	pyrifolia 'Ya Li'	oEar oOEx oGre
	pyrifolia 'Yakumo'	wBCr wBur
un	***pyrifolia*** 'Yoinashi'	wRai
un	***pyrifolia*** 'Yongi'	wRai
	pyrifolia '20th Century'	see **P. *pyrifolia*** 'Nijisseiki'
	pyrifolia '20th Century Nijisseiki'	see **P. *pyrifolia*** 'Nijisseiki'
	salicifolia 'Pendula'	oFor oAmb oGre
un	***salicifolia*** 'Pendula Silver Frost'	oGar
	'Shipova' (fruit)	wBur wRai oOEx wBox
	QUERCUS	
	acutissima	oFor wHer wCSG oDan oRiv wCCr
	agrifolia	oFor wShR oRiv
	alba	wBCr wCCr
	aliena	oRiv
	austrina	wCCr
	bicolor	oFor oRiv wBox oSle
	boissieri	see **Q. *infectoria*** ssp. ***veneris***
	borealis	see **Q. *rubra***
	calliprinos	see **Q. *coccifera*** ssp. ***calliprinos***
	canariensis	oFor
	castaneifolia	last listed 99/00
	cerris	oFor
un	***chenii***	oFor
	chrysolepis	oFor oBos oGre wSte
	chrysolepis Southern California	wCCr
	chrysolepis Southern Oregon	wCCr
	coccifera ssp. ***calliprinos***	wCCr
	coccinea	oFor oEdg wShR wCSG oGar oRiv oSho wCCr oWnC
un	***cornelius-mulleri***	wCCr
un	***cornelius-mulleri*** hybrids	wCCr
	dentata	oEdg
	douglasii	oFor wSte
	dumosa	oFor
	durata	oFor oRiv wCCr
	ellipsoidalis	oFor oRiv wCCr
	engelmannii	oFor
	facrist	see **Q. *robur*** 'Facrist'
	faginea	oFor
	falcata	oRiv
ch	***falcata*** var. ***pagodifolia***	see **Q. *pagoda***
	frainetto	oFor wCCr
un	***frainetto*** 'Forest Green'	oFor
	gambelii	oFor oRiv wCCr oSle
	garryana	oFor wWoB wBur wClo wNot wShR oGar oBos oRiv wWat wCCr oAld oSle wSte wSnq
	garryana var. ***breweri***	oFor wCol
	georgiana	oFor
	glandulifera	see **Q. *serrata***
	glauca	oFor wHer oGar oRiv oCis wCCr
	grisea	oFor
	hemisphaerica	oFor
	hypoleucoides	oCis wCCr wSte
	ilex	oFor oWhi wCCr
	imbricaria	oFor wCCr
	infectoria ssp. ***veneris***	oFor
	ithaburensis	oFor
	kelloggii	oFor wBCr oRiv
	laevis	last listed 99/00
	laurifolia	oFor oRiv wCCr
ch	***liaotungensis***	see **Q. *wutaishanica***
	lobata	oFor
	lyrata	oFor wCCr
	macrocarpa	oFor wBCr wBur oRiv wCCr oWnC
	marilandica	oFor
	mexicana	oFor oCis wCCr
	michauxii	oFor
	mongolica	oRiv
	muehlenbergii	oRiv wCCr
ch	***myrsinifolia***	see **Q. *glauca***
	myrtifolia	wCCr
	nigra	oFor oRiv wCCr
ch	***nuttallii***	see **Q. *texana***
ch	***obtusa***	see **Q. *laurifolia***
	pagoda	oFor wHer
	palustris	oFor wSwC wClo oAls wShR wCSG oGar wKin oWnC wBox wRav wCoS
un	***parvula shrevei***	wCCr
	phellos	oFor oDan oRiv wCCr
	phillyreoides	oFor oRiv wCCr
	phillyreoides HC 970596	wHer
un	***polymorpha***	oFor wHer
	prinus (see Nomenclature Notes)	wCCr
	pubescens	last listed 99/00
	pungens	oFor wCCr
	reticulata	oRiv wCCr
	robur	oFor wBCr wBur oRiv wCoS
	robur 'Atropurpurea'	last listed 99/00
	robur 'Concordia'	oDan
un	***robur*** 'Crimson Spire'	oDan
	robur 'Facrist'	oWhi
	robur f. ***fastigiata***	oAls
un	***robur*** 'Skyrocket'	oWnC
	rubra	oFor oEdg oAls oGar oSho wKin oSle
	rugosa	oCis
	rysophylla	last listed 99/00
	sadleriana	oFor wCol oBos oRiv wCCr wSte
	serrata	wCCr
	shumardii	oFor oRiv wCCr
	sp. NL, Mexico	oCis
	stellata	oFor
	suber	oFor wCCr
	Sweet Idaho Bur	wBCr
	texana	oFor wCCr
	turbinella x ***macrocarpa***	wCCr
	turbinella x ***robur***	wCCr
	vacciniifolia	oFor wCol oBos oRiv oGre wCCr
	variabilis	oFor oRiv
	velutina	oFor wCCr
	virginiana	oFor oRiv
	wislizeni	oFor oRiv oSho wCCr
	wutaishanica	oEdg oRiv
	QUILLAJA	
	saponaria	oFor oRiv
	QUIONGZHUEA	
	tumidinoda	see **CHIMONOBAMBUSA *tumidissinoda***
	RADERMACHERA	
	sinica	oGar
	RAMONDA	
	myconi	oBov
	RANUNCULUS	
do	***aconitifolius*** 'Flore Pleno'	wHer
do	***acris*** 'Flore Pleno'	oRus
	acris 'Stevenii'	wHer
	alpestris JJH921036	wMtT
	Bloomingdale Series	oGar
	bulbosus 'F. M. Burton'	wHer
do	***constantinopolitanus*** 'Plenus'	wHer wCul
	crenatus	last listed 99/00
	ficaria	last listed 99/00
	ficaria var. ***albus***	oGre
	ficaria var. ***aurantiacus***	wHer oSis oSec
	ficaria 'Brambling'	wHer
	ficaria 'Brazen Hussy'	oFor wHer wWoS oGos oAls oHed oSis oCis oNWe wMtT wRob oGre oSec
un	***ficaria*** 'Buttered Popcorn'	oCis
	ficaria ssp. ***chrysocephalus***	last listed 99/00
	ficaria 'Coffee Cream'	wHer
do	***ficaria*** 'Collarette'	wHer
	ficaria 'Coppernob'	oNWe
	ficaria 'Cupreus'	see **R. *ficaria*** var. ***aurantiacus***
	ficaria 'Damerham'	wHer
do	***ficaria*** 'Double Bronze'	wHer
do	***ficaria*** 'Double Mud'	wMtT
	ficaria 'Dusky Maiden'	last listed 99/00
do	***ficaria flore-pleno***	wHer oSis
	ficaria 'Fried Egg'	wHer
	ficaria 'Green Petal'	wHer
un	***ficaria*** 'Greencourt Gold'	wHer

qu	***ficaria*** 'Hoskin's Variegated'	wHer
	ficaria 'Limelight'	wHer
	ficaria 'Primrose'	wHer
	ficaria 'Randall's White'	wHer wCri
	ficaria 'Salmon's White'	oHed
	ficaria 'Yaffle'	wHer
	flammula	oHug
	gouanii	last listed 99/00
	gramineus	wHer wCol
un	***krasinovii*** JJH9204373	wMtT
	montanus 'Molten Gold'	oSis
	occidentalis	oTri oAld oSle
	repens	oHug oGar oAld
un	***repens*** 'Buttered Popcorn'	wRob oGre
do	***repens*** var. ***pleniflorus***	oFor
un	***yakushimanus***	oSis
un	***yakusimensis***	wHer
	RAOULIA	
	australis	wCoN oAls oHed oGre oWnC wTGN oUps
	hookeri	oHed
	hookeri var. ***alboscricea***	oSis
	hookeri var. ***apice-nigra***	oSis
	subsericea	wFGN
	RAPHIOLEPIS	see **RHAPHIOLEPIS**
	RAPHIONACME	
un	***flanganii***	oRar
	hirsuta	oRar
	sp.	oRar
	sp. Zimbabwe	oRar
	RATIBIDA	
	columnifera	oMis wWld
	REGNELLIDIUM	
	diphyllum	oGar
	REHDERODENDRON	
	macrocarpum	last listed 99/00
	REHMANNIA	
	angulata	see **R. *elata***
	elata	oFor oJoy oAmb oDan oNWe oUps
	elata 'Popstar'	last listed 99/00
	glutinosa	oSis oSec oInd oWhS
	sp. aff. ***glutinosa***	wHer
	REINECKEA	
	carnea	oTrP oHed
	carnea red form	oTrP
	carnea 'Variegata'	wHer wCol
	sp DR96-23	wCol
	RETAMA	
	raetam	last listed 99/00
un	**RHAMNELLA**	
	franguloides	oFor
	RHAMNUS	
	alaternus	oFor oSho wCCr
	alaternus 'John Edwards'	oSho
	californica	wShR oGre wCCr wSte
	californica 'Eve Case'	oFor
	californica ssp. ***crassifolia***	wCCr
	californica ssp. ***occidentalis***	oFor
	californica ssp. ***tomentella***	oFor wCCr
	cathartica	oGoo
	crocea ssp. ***ilicifolia***	oFor
	dahurica	oFor
	davurica	see **R. *dahurica***
	frangula 'Aspleniifolia'	oFor wCul oGar oDan wCCr oWnC wWel
	frangula 'Aspleniifolia' tree form	wCul
	frangula 'Columnaris'	oFor
	pallasii	oFor
	purshiana	oFor oHan wWoB wBur wNot wShR oGar oBos wRai wWat oTri oAld wFFl oSle wSte
	rubra	oFor
	tomentella	see **R. *californica*** ssp. ***tomentella***
	RHAPHIOLEPIS	
	x delacourii	wCCr
	x delacourii Majestic Beauty®/ 'Montic'	oGar oSho oGre oWnC
un	***indica*** 'Bay Breeze'	oWnC
	indica 'Clara'	oBlo
	indica Enchantress® / 'Moness'	oGar oSho oWnC
un	***indica*** 'Harbinger of 'Spring'	oFor
un	***indica*** 'Hines Darkleaf'	oGar
	indica Indian Princess® / 'Monto'	oSho oWnC
	indica Spring Rapture ® / 'Monrey'	oSho oWnC
	indica Springtime®/ 'Monme'	oSho
	indica White Enchantress™ / 'Monant'	oWnC
	ovata	see **R. *umbellata*** f. ***ovata***
un	'Pinkie'	wTGN
	umbellata	oGar
un	***umbellata*** 'Gulf Green'	wWel
	umbellata 'Minor'	oAls oGar wCoS
qu	***umbellata*** 'Minor Gulf Green'	oGar oWnC
	umbellata f. ***ovata***	oSho oBlo oGre wCCr
	RHAPHITHAMNUS	
	spinosus HCM 98190	last listed 99/00
	RHAPIDOPHYLLUM	
	hystrix	wHer
	RHAZYA	
	orientalis	see **AMSONIA *orientalis***
	RHEEDIA	see **GARCINIA**
	RHEUM	
	acuminatum	wHer oCis
	alexandrae	last listed 99/00
	australe	wHer oGar
	compactum	last listed 99/00
	emodi	see **R. *australe***
	x hybridum	oAls wCoS
	x hybridum 'Canada Red'	wBCr
un	***x hybridum*** 'Crimson'	wTGN
	x hybridum 'Crimson Cherry'	oGar
	x hybridum 'Strawberrry'	oSho
	x hybridum 'Victoria'	oFor wClo oGar oSho
	kialense	wHer
	palmatum 'Atrosanguineum' seed grown	wHer
	palmatum var. ***tanguticum***	oFor wWoS wRob wTGN
	rhabarbarum	see **R. *x hybridum***
	rhaponticum	last listed 99/00
	sp. DR97-10	wCol
	tanguticum	see **R. *palmatum*** var. ***tanguticum***
	tibeticum red form	oNWe
	RHEXIA	
	virginica	oGoo
	RHIPSALIDOPSIS	see **HATIORA**
	RHIPSALIS	
	baccifera	oSqu
	boliviana	oSqu
	elliptica	oSqu
un	***lorentziana***	oSqu
un	***madagascariensis***	oSqu
	micrantha	oSqu
	monacantha	oSqu
	paradoxa	oSqu
	pentaptera	oSqu
un	***platycarpa***	oSqu
	pulvinigera	oSqu
	puniceodiscus	oSqu
	rhombea	see **R. *elliptica***
	teres	see **R. *baccifera***
	virgata	oSqu
un	***wercklei***	oSqu
	RHODANTHEMUM	
	gayanum	oHed
	RHODIOLA	
un	***atuntensis***	oTrP
	fastigiata	oSto
	sp. aff. ***himalensis*** HWJCM 378	wHer
	integrifolia	see **R. *rosea*** ssp. ***integrifolia***
	ishidae	oSqu
	rosea	oSqu
	rosea ssp. ***atropurpurea***	see **R. *rosea*** ssp. ***integrifolia***
	rosea ssp. ***integrifolia***	oFor oSto
	sp. DJHC 330	wHer
	sp. DJHC 98144	wHer
	sp. DW 69	wHer
	yunnanensis EDHCH 97073	wHer
	RHODOCHITON	
	atrosanguineus	oHed oUps wSte
	RHODODENDRON (SEE NOMENCLATURE NOTES)	
	'A. Bedford'	see **R.** 'Arthur Bedford'
	'Abe Arnott'	oGre wHam wWhG
	'Abegail'	wHam
	'Abendsonne'	wHam wWhG
	aberconwayi	wWhG wWel
	aberconwayi compact form	oGre
	'Abraham Lincoln'	wHam
	'Accomplishment'	wWhG
	adenogynum Adenophorum Group	last listed 99/00
	adenophorum	see **R. *adenogynum*** Adenophorum Group
	adenopodum	oGre wWhG
	'Admiral Piet Hein'	wHam
un	'Adrastia'	wHam
un	***aequabile***	oBov
un	'Aesthetica'	wClo

RHODODENDRON

	aganniphum var. ***aganniphum***	wRho
	aganniphum var. ***flavorufum***	oGre
	x agastum	last listed 99/00
	'Aglo'	oGre wHam wWhG
un	'Aikoku' (EA)	oGre
un	'Aikoku Sport' (EA)	oGre
	'Airy Fairy'	oGos oGre wWhG wWel
	'Aksel Olsen'	oGre oWnC wHam
	alabamense (DA)	oGre
	Aladdin Group & cl.	wHam wWhG wWel
	'Alaska'	see **R.** 'Finlandia'
	Albatross Group & cl.	oGre
	'Albert'	wHam
	'Albert Close'	wHam
	albiflorum (DA)	wRho wHam
	albrechtii (DA)	wClo
un	***albrechtii*** 'Whitney'	oGos
	'Album Elegans'	oWnC wWhG
	'Album Novum'	oGre wWhG
un	'Alec G. Holmes'	wWhG
un	'Aleksandr Isayevich'	oBov
	'Alena'	wHam wWhG
	'Alexander' (EA)	oAls oGar oSis oGre oTPm wSta oBRG wWhG
	'Alice'	oBlo oGre wSta wHam oTal wWhG
	'Alice Franklin'	wHam
un	'Alice Springs'	wHam
	'Alice Street'	wHam
	'Alice Swift'	wWhG
	Alison Johnstone Group & cl.	oAls oGre
	'Allure'	wWhG
	'Aloha'	oAls oGre oWnC wHam oTal wWhG
un	'Alpine Glory'	oGre
	alutaceum	oGre
	alutaceum var. ***alutaceum*** Globigerum Group	oGos wWhG wWel
un	'Amaghasa' (EA)	wSta wWhG
un	'Ambassador'	wHam
un	'Amber Touch'	wWhG
	ambiguum	wRho wCCr
un	'Ambrosia'	wWhG
	'America'	oGre wSta wHam wWhG
	'Amity'	oGre wHam
do	'Amoenum' (EA)	oGre wSta oBRG wWhG
do	'Amoenum Coccineum' (EA)	oGre
	'Amy'	wWhG
	'Anah Kruschke'	oAls oGar oSho oBlo oGre oWnC wSta wHam wWhG wWel
qu	'Anah Rose'	oBlo
un	'Anastasia'	oBov
	'Anchorite' (EA)	wCCr
	'Andrew Patton'	wHam
	Angelo Group & cl.	oGre wHam
	'Angel's Dream'	last listed 99/00
un	'Anggi Lake'	oBov
	'Anica Bricogne'	wWhG
	'Anita Dunstan'	oGre
	'Ann Carey'	oTal
	'Anna'	oGre wHam
	'Anna' x 'Apricot Sherbet'	oGre
	'Anna H. Hall'	oGre wHam wWhG
un	'Anna Hybrid'	oGre
un	'Anna Kehr' (EA)	wWhG
	'Anna' x 'Loderi King George'	oGre
un	'Anna Pavlova' (DA)	wWhG
	'Anna Rose Whitney'	oAls oSho oGre oWnC wSta wHam wWhG wWel
	'Anna Vojtec'	wWhG
	'Anna' x ***yakushimanum*** #2	oGre wHam wWel
	'Anna' x ***yakushimanum*** #301	oGre wWhG
	'Annabella' (DA)	last listed 99/00
	'Anne's Delight'	wWhG
	'Annie Dring'	wHam
	'Anniversary Gold'	last listed 99/00
	anthopogon ssp. ***hypenanthum*** GLE	last listed 99/00
	anthosphaerum	wRho
	anthosphaerum DJHC 500	wHer
	'Antilope' (DA)	wCul oGre
	'Antoon van Welie'	oWnC wSta wHam oTal wWhG
	anwheiense (V)	last listed 99/00
	'Applause'	wHam
	'Apple Blossom'	wWhG
un	'Apple Blossom' (DA)	oGre
	'Apple Brandy'	wHam
	'Apricot Fantasy'	oGre wHam wWhG
	'Apricot Nectar'	oGar oGre wHam
	'Apricot Sherbet'	oGre wHam
	'Apricot Surprise' (DA)	oGar oGre wHam wWhG
	'April Chimes'	oGre
	'April Gem'	oGre wWhG
	'April Glow'	oAls
	'April Mist'	oSis oGre wHam
	'April Rose'	oGre wHam wWhG wWel
	'April White'	oGre wWhG
un	'Aravir'	oBov
	arborescens (DA)	wRho
	arboreum	wHam
	arboreum f. ***album***	see **R. *arboreum*** ssp. ***cinnamomeum*** var. ***allbum***
	arboreum ssp. ***arboreum***	last listed 99/00
	arboreum x ***barbatum*** HWJCM 145	wHer
	arboreum ssp. ***campbelliae***	see **R. *arboreum*** ssp. ***cinnamomeum*** var. ***cinnamomeum*** Campbelliae Group
	arboreum ssp. ***cinnamomeum***	wRho wWhG
	arboreum ssp. ***cinnamomeum*** var. ***album***	wWhG
	arboreum ssp. ***cinnamomeum*** var. ***cinnamomeum*** Campbelliae Group	oGre
	arboreum ssp. ***cinnamomeum*** var. ***roseum***	wRho wWhG
un	***arboreum*** 'Ihrig'	wWhG
un	***arboreum*** Lancaster's form	oGre
	arboreum ssp. ***nilagiricum***	last listed 99/00
	arboreum var. ***roseum***	see **R. *arboreum*** ssp. ***cinnamomeum*** var. ***roseum***
	'Arctic Pearl'	wHam
	'Arctic Tern'	see **X LEDODENDRON** 'Arctic Tern'
	argipeplum	wHam
	argyrophyllum	wCCr wHam wWhG
	argyrophyllum ssp. ***argyrophyllum***	last listed 99/00
	argyrophyllum ssp. ***hypoglaucum***	last listed 99/00
	argyrophyllum ssp. ***nankingense*** 'Chinese Silver'	oGre wWel
	armitii	last listed 99/00
un	'Arne Jensen'	oBov
	'Arneson Gem' (DA)	wHam wWhG
	'Arneson Ruby' (DA)	oGre
	'Arnold Piper'	wHam
un	'Aroma from Tacoma'	wWhG
	'Arthur Bedford'	oAls wSta wHam wWhG
	'Arthur J. Ivens'	wHam
	'Arthur Osborn'	wHam
	asterochnoum	wRho
	atlanticum (DA)	oRiv oGre wWhG
	'Atroflo'	wHam
un	'Atrosanguineum' (EA)	oGre
	'Attar'	last listed 99/00
	'August Lamken'	wHam wWhG
	'Auguste van Geerte'	wWhG
	augustinii	oFor oRiv oGre wSta wHam
	augustinii ssp. ***augustinii***	wRho
un	***augustinii*** ssp. ***augustinii*** 'Hobble'	wHam
	augustinii ssp. ***augustinii*** 'Smoke'	wWhG wWel
	augustinii 'Barto Blue'	oBlo wRho wWhG
	augustinii 'Blue Cloud'	oGre wWhG
	augustinii ssp. ***chasmanthum***	wRho wWhG wWel
	augustinii Electra Group & cl.	oGre wWhG wWel
	augustinii 'Green Eye'	see **R.** 'Green Eye'
	augustinii ssp. ***hardyi***	last listed 99/00
	augustinii 'Lackamas Blue'	wHam wWhG
	augustinii 'Marine'	wHam wWhG
un	***augustinii*** 'OSU'	wClo
un	***augustinii*** Species Foundation	wWhG
	augustinii 'Towercourt'	wHam wWhG wWel
	augustinii 'Whalley'	wWhG
	augustinii Windsor form	wWhG
	'Aunt Martha'	wHam
	aureum	last listed 99/00
	auriculatum	oGre wHam wWel
un	***aurigeranum***	wRho oBov
un	[(***aurigeranum*** x 'Dr. H. Sleumer') x ***konori***] x [(***macgreg.*** x ***aurig.***) x 'Dr. H. Sleumer']	oBov
un	***aurigeranum*** x 'Dr. H. Sleumer'	oBov
un	(***aurigeranum*** x 'Dr. H. Sleumer') x ***leucogigas***	oBov
un	***aurigeranum*** x ***laetum***	oBov
un	***aurigeranum*** x ***lochiae***	oBov
un	***aurigeranum*** x ***zoelleri*** 'Golden Gate'	oBov
	'Aurora'	oGre wHam
	austrinum (DA)	oGre

	'Autumn Gold'	wSta wHam oTal wWhG
	'Autumn Gold' x ***yakushimanum***	wWhG
	Avalanche Group & cl.	oGre
un	'Avocet' (DA)	wWhG
un	'Avocet's Friend' (DA)	oGre wWhG
	'Avondale'	wHam
	'Award'	oGre
	'Axel Olsen'	see **R.** 'Aksel Olsen'
	Azor Group & cl.	wSta oTal wWhG
	'Aztec'	wSta
	'Azurro'	oGre wHam wWhG
	'Azurwolke'	wWhG
	'Babylon'	wWhG
	'Bacher's Gold'	wHam oTal
	'Bad Eilsen'	wHam
	'Baden-Baden'	oAls oGar oGre oWnC wWhG wWel
	'Bagshot Ruby'	wHam
	bakeri	see **R. *cumberlandense***
un	'Balalaika'	oGre wHam wWhG
	balfourianum	wRho wHam
	'Ballerina' (DA)	wWhG
	'Ballet'	wWhG
	'Bali'	wHam
	'Balzac' (DA)	wWhG
	'Bambi'	oGos oGre wHam oBRG
	'Bambino'	oGre wHam wWhG wWel
	'Bananaflip'	wClo oGre wHam wWhG
un	'Baram Bay'	oBov
un	'Barbara' (EA)	wWhG
	'Barbara Behring'	wHam
	barbartum	wRho wHam
	barbartum HWJCM 485	wHer
	'Bariton'	oAls wHam wWhG wWel
	'Barmstedt'	oGre wHam wWhG
	'Baron Lionel'	wHam
	'Baron Lionel de Rothschild'	see **R.** 'Baron Lionel'
	'Barry Rodgers'	oGre wHam wWhG
	'Barto Alpine'	oGre wHam wWhG
	'Barto Blue'	see **R. *augustinii*** 'Barto Blue'
	'Barto Lavender'	oGre wHam
	'Bashful'	oAls wHam wWhG
	basilicum	oGos
	x bathyphyllum	oGre wHam
	beanianum	wHam
	Beau Brummel Group & cl	wSta wHam
	'Beaufort'	wHam
	'Beaulieu' (DA)	last listed 99/00
un	'Beautiful Bouquet'	wWhG
	'Beautiful Day'	wHam
	'Beauty of Littleworth'	wSta wHam wWhG
	beesianum	wRho
	'Belle Heller'	oGre wHam wWhG wWel
un	'Belona'	wHam wWhG
	'Belva's Joy'	oGre wWhG wWel
	'Ben Morrison' (EA)	oGar oSho oTPm oBRG wWhG
	'Bengal'	wHam
un	'Beni Kirishima' (EA)	oBlo oWnC
un	'Benji Yellow'	oBlo
	'Bergie Larson'	oAls oGar wHam wWel
un	'Berg's Queen Bee'	wWhG
	'Bernstein'	wHam wWhG
	'Bert Larson'	wHam
	'Besse Howells'	oGre wHam wWhG
	'Bessie Farmer'	last listed 99/00
	'Betsie Balcom'	wHam
	'Better Half'	oGre
	'Betty Anderson'	wHam wWhG
	'Betty Anne Voss' (EA)	oGre oTPm wWhG
	'Betty Hume'	wHam
	'Betty Sears'	wWhG
	'Betty Wormald'	wSta wWhG
	'Betty's Bells'	oGre
	beyerinckianum (V)	oBov
	Bibiani Group & cl.	oGre wWhG
	'Big Sam'	wHam
	'Big Sam' x 'Lem's Cameo'	wHam
	'Bikini'	wHam
un	'Bikini Island'	wWhG
un	'Bill'	wHam
	'Blaauw's Pink' (EA)	oTPm wWhG
	'Black Eye'	oGre wWel
	'Black Magic'	oAls oGar oGre wHam wWhG wWel
	'Black Satin'	oGre wHam wWhG wWel
	'Black Sport'	oGre wHam wWhG
	Blanc-mange Group & cl.	wHam wWhG

	'Blaney's Blue'	oGar oSho oGre wHam oBRG wWhG wWel
	'Blazen Sun'	last listed 99/00
	'Blewbury'	wHam wWhG
	'Blinklicht'	oGre oTal wWhG wWel
	'Blitz'	oAls oGar wWel
	'Blondie'	wWhG
	'Blue Bird'	wWhG
	'Blue Boy'	oAls oGre wHam wWhG wWel
	'Blue Danube'	oGar wHam
	'Blue Danube' (EA)	oAls oBlo oGre wSta oBRG wWhG
	'Blue Dawn'	wWel
	Blue Diamond Group & cl.	oGar oBlo oWnC wSta wHam oTal wWhG wWel
	'Blue Ensign'	wHam wWhG
	'Blue Frost'	wWhG
un	'Blue Jack'	wSta
	'Blue Jay'	oGar oSho wSta wHam wWhG wWel
	'Blue Lagoon'	last listed 99/00
	'Blue Pacific'	oGre oWnC wHam
	'Blue Pacific' x 'Lem's Cameo'	wHam
	'Blue Peter'	oGre wSta wHam oTal wWhG wWel
	'Blue Rhapsody'	oGre wWel
	'Blue Ridge'	wHam
	Blue Tit Group	oGar wCCr wHam oTal wWhG
	'Blue Wonder'	wHam
	Bluebird Group & cl.	oGre wSta wHam oBRG oTal
	'Bluenose'	wHam oTal wWhG
	'Bluette'	wSta wWhG
	'Blunique'	oGre wHam
	'Blutopia'	wHam wWhG wWel
	Bo-peep Group & cl.	oBlo wSta
	'Bob Bovee'	wHam wWhG
	'Bobolink'	wHam
	'Bob's Blue'	oGre oWnC wSta wHam wWel
	'Boddaertianum'	wHam
	'Bodega Crystal Pink'	wWel
	bodinieri	oGre
	'Bold Adventure'	wHam
	Bonito Group & cl.	wHam wWhG
	'Bonnie Babe'	oGre wHam
	bonsai, pink (A)	oGre
	bonsai, red (A)	oGre
	bonsai, salmon (A)	oGre
	bonsai, white (A)	oGre
un	'Border Gem' (A)	oWnC
un	'Boudoir' (EA)	wSta
	'Boule de Neige'	oGre oWnC wSta wHam wWhG wWel
	'Bouquet de Flore' (DA)	wHam
	'Bow Bells'	oGar oBlo oGre oWnC wSta wHam wWhG wWel
	brachyanthum	last listed 99/00
	brachyanthum ssp. ***hypolepidotum***	wRho
	brachycarpum	wRho oGre wHam
	brachycarpum HC 970147	wHer
	brachycarpum ssp. ***brachycarpum***	last listed 99/00
	'Brandt Red' x ***yakushimanum*** #2	wWhG
	'Brandt Red' x ***yakushimanum*** #18	oGre
	'Brandt's Tropicana' x 'Lem's Cameo #1'	wWel
	'Brasilia'	oGre wHam
	'Bravo!'	wHam
	'Brazil' (DA)	wWhG
	'Bremen'	oGre wHam
un	'Bretonne'	wWhG
	Bric-a-brac Group & cl.	oWnC wSta wWhG
	'Brickdust'	oGre wHam oBRG oTal wWhG wWel
	'Bridal Bouquet'	wWhG
	'Bridgeport'	wHam
	'Brigadoon'	wHam wWhG
un	'Brigg's Red Star'	oGre wWhG
	'Bright Forecast' (DA)	oGre wWhG
	'Britannia'	wHam wWhG
	'Brittany'	oGre wHam wWhG
	'Britton Hill'	oGre wHam wWhG
	'Britton Hill Bugle'	oGre wHam
	'Brocade'	wSta wHam
	'Bronze Wing'	oGre wWel
	brookeanum	wRho
	'Broughtonii Aureum'	wSta
	'Brown Eyes'	oGre wHam wWhG
	'Bruce Brechtbill'	wClo oGar oGre wHam oBRG oTal wWhG wWel
un	'Brunei Bay'	oBov
	'Bryce Canyon'	last listed 99/00
	bryophilum	oBov

	'Buccaneer' (EA)	oGar oSho oGre oTPm wSta
	'Buchanan Simpson'	wHam
	'Bud Flanagan'	wHam wWhG
un	'Bur Paw'	wWhG
	bureaui	wRho oGre wHam wWhG wWel
	bureaui EDHCH 97094	wHer
	bureaui x 'Ken Janeck'	wHam wWhG
	bureaui 'Lem's Variety'	oGre wWel
	bureaui x ***pronum***	oGre wWel
un	***bureaui*** 'Tigan Rugh'	wHam
	bureauoides	wHam
un	***bureauoides*** 'Reuthe'	wHam
	'Burgundy'	oGre
	'Burgundy Rose'	oGre
un	'Burnaby Centennial'	wHam
	'Butter Brickle'	wHam
	'Butter Yellow'	wHam
	'Buttered Popcorn'	wHam
	'Butterfly'	oWnC wSta wHam
	'Buttermint'	oGre wHam wWel
	'Buttons and Bows' (DA)	oFor oGre wWhG
	'C.I.S.'	wSta oTal
un	'C.O.D.'	wHam
	'Cadis'	oGre wHam
	caesium	last listed 99/00
un	'Calavar'	oBov
	calendulaceum (DA)	oFor oRed
	'California Gold'	last listed 99/00
	caliginis (V)	last listed 99/00
	callimorphum	wHam wWhG
	calophytum	wSta wHam wWel
	calophytum var. ***calophytum***	last listed 99/00
	calophytum pink	wWhG
	calophytum white	wWhG
un	***calophytum*** x 'Whitney Early Pink'	wWhG
	calostrotum	wHam wWhG
	calostrotum 'Gigha'	wRho
	calostrotum ssp. ***keleticum***	wClo oGar oSis oBlo oGre wSta wHam wWhG
	calostrotum ssp. ***keleticum*** Radicans Group	wSta oBRG wWhG
	calostrotum ssp. ***riparium*** Nitens Group	last listed 99/00
	calostrotum ssp. ***riparium*** Rock's Form R 178	wMtT
	calostrotum rose	wSta
	'Calsap'	oAls oGre wHam wWhG
un	'Camay'	wHam
	camelliiflorum	wRho
	campanulatum	wClo wHam wWhG wWel
	campanulatum ssp. ***aeruginosum***	wRho
	campanulatum ssp. ***aeruginosum*** RBGE	wHam
	campanulatum 'Knap Hill'	wWhG
	'Campfire' (EA)	wSta
	campylocarpum	wRho
	campylocarpum HWJCM 352	wHer
	campylogynum Celsum Group	wRho oGre wWhG
	campylogynum Charopoeum Group	wSta
	campylogynum Cremastum Group	wClo
	campylogynum Cremastum Group 'Bodnant Red'	wRho
	campylogynum Myrtilloides Group	wClo wWhG
	camtschaticum	wRho
un	'Can Can' (EA)	oGre
	'Canada'	oGre
	canadense (DA)	oFor oGre
	'Canadian Sunset'	oGar oGre wHam wWhG
	'Canary Islands'	wHam
	'Canby' (DA)	wWhG
	'Candystripe' (DA)	oGre
	canescens (DA)	last listed 99/00
do	'Cannon's Double' (DA)	oGar oGre wHam wWhG
un	'Cannon's Red' (DA)	oGre wWhG
un	'Cape Cod Cranberry'	oBov
un	'Capistrano'	wHam wWhG
	capitatum	last listed 99/00
	'Captain Jack'	wHam wWhG
	'Captain Jack' x ***yakushimanum***	wHam
	'Captain Kidd'	wSta
un	'Captivation' (EA)	wSta
	'Caractacus'	oGre wHam
un	'Carex Pink'	wHam
	Carita Group	wSta wHam
	'Carita Charm'	wHam
	'Carita Golden Dream'	wHam
	'Carl Phetteplace'	wHam
	'Carlene'	oGre wHam
	'Carmen'	wClo oAls oGar oGre wWhG wWel
	'Carmen' x 'Ken Janeck'	wHam
un	'Carol' (EA)	wHam wWhG
	'Carol Jean'	oAls
	'Caroline'	wHam
	'Caroline Gable' (EA)	oAls oGar oBlo wSta wWhG
	carolinianum	see **R. *minus*** var. ***minus***
	'Carolyn Grace'	wHam
un	'Carousel' (DA)	wWhG
	carringtoniae (V)	oBov
	'Carte Blanche'	wWhG
	'Cary Ann'	oAls oSho oWnC wHam
	'Casanova'	oGre wHam oTal wWhG wWel
un	'Cascade Pink' (DA)	oGre
un	'Cassie'	wWhG
	'Castanets'	wHam
	'Catalode'	wHam wWhG wWel
	catawbiense	oFor wRho
	'Catawbiense Album'	oGre oWnC wHam wWhG wWel
	'Catawbiense Boursault'	oSho oGre wWhG
	'Catawbiense Grandiflorum'	oGre oWnC wWhG
un	***catawbiense*** f. ***insularis***	wRho
	'Cathy Jo'	see **R. *'Kathie Jo'***
	caucasicum	wHam
	'Cavalcade'	wHam
	'Cecile' (DA)	oGar oGre oWnC wWhG
	'Celeste'	oGre wHam oTal
	'Centennial'	see **R.** 'Washington State Centennial'
	'Centennial Celebration'	wClo oGre wWhG
	cerasinum	wRho
	chamaethomsonii	wMtT wRho
un	'Chambri Lake'	oBov
	'Chanticleer' (EA)	wHam
	'Chapeau'	oGre wHam wWhG wWel
	chapmanii	see **R. *minus*** var. ***chapmanii***
	'Chapmanii Wonder'	oGre
	charitopes ssp. ***charitopes***	last listed 99/00
un	'Charmont'	wHam
	'Cheer'	oAls oGar oGre oWnC wSta wHam oTal wWhG
	'Cheerful Giant' (DA)	oGre wWhG
un	'Cherries 'n' Cream'	oGre wHam oBRG wWhG
un	'Cherry Drops' (EA)	oGre oBRG wWhG
	'Cherry Float'	oGre wHam wWel
	'Chetco' (DA)	oGre wHam wWhG
	'Chevalier Felix de Sauvage'	wSta wWhG
	'Chiffchaff'	last listed 99/00
	'Chikor'	oAls oGar oGre
	China Group & cl.	oGre wHam wWhG
	'China Doll' (see Nomenclature Notes)	wHam
un	'Chinzan' (EA)	oSis oGre oBRG
	'Chionoides'	oAls oGar oGre oWnC wSta wHam wWel
un	'Chiyo No Homane' (EA)	oGre
un	'Chojuho' (EA)	oGre oBRG wWhG
	'Choremia'	wWhG
un	'Chorus Line'	wWhG
	christi (V)	oBov
un	***christianae*** x ***aequabile***	oBov
un	***christianae*** RSF	oBov
un	***christianae*** 'Sunset'	oBov
	'Christmas Cheer'	oAls oGar wSta wHam wWhG wWel
	'Christmas Cheer' (EA)	see **R.** 'Ima-shojo'
	'Christy S.'	oGre
	chrysodoron	wRho wWhG
	ciliatum	oRiv wSta wHam wWhG
	Cilpinense Group	oGar oGre wSta wHam wWel
	cinnabarinum	wClo wHam wWel
	cinnabarinum ssp. ***cinnabarinum*** Blandfordiiflorum Group	wRho wWhG
	cinnabarinum ssp. ***cinnabarinum*** 'Conroy'	wHam
	cinnabarinum ssp. ***cinnabarinum*** Roylei Group	wRho wHam wWhG
	cinnabarinum hybrid, fragrant	oGre
	cinnabarinum ssp. ***xanthocodon***	wWhG
	cinnabarinum ssp. ***xanthocodon*** Concatenans Group	wRho wHam
	cinnabarinum ssp. ***xanthocodon*** Purpurellum Group	wRho
	'Cinnamon Bear'	wWhG wWel
	'Cinquero'	oGre wHam
	'Circus'	oGre wHam
	citriniflorum var. ***citriniflorum***	last listed 99/00
	citriniflorum var. ***horaeum***	wWel

un	'Clarke's Perfection'	wHam
	clementinae	oGre wWhG
	'Clementine Lamaire'	oGre wHam
un	'Cliva'	oGre wWhG
	'Clotted Cream'	wHam
un	'Cloud Cap' (EA)	wSta
	'Coccineum Speciosum' (DA)	wHam
	coelicum	wRho
	coeloneuron	wRho
do	'Colin Kendrick' (DA)	oGre
	'Colonel Coen'	oAls oSho oGre wHam wWhG wWel
un	***commonae***	oBov
	'Compacta'	oGre wHam
	'Comstock'	oGre
	concatenans	see **R. *cinnabarinum*** ssp. ***xanthocodon*** Concatenans Group
un	'Concho' (EA)	wSta
	concinnum	wRho oGre wHam
	concinnum 'Chief Paulina'	wRho
	concinnum Pseudoyanthinum Group	wCCr wSta wWhG
	'Conemaugh'	wSta oTal
	'Conewago'	wHam
	'Connecticut Yankee'	oGre wHam wWhG
un	'Conquistador'	wHam
	'Conroy'	see **R. *cinnabarinum*** ssp. ***cinnabarinum*** 'Conroy'
	'Consolini's Windmill'	oGre wHam wWhG
	'Constanze'	wWhG
un	'Content'	wCCr
	'Contina'	wWel
un	'Conversation Piece' (A)	oTPm
	'Cookie'	oGre wHam wWhG
un	'Copperman' (EA)	oGre
ch	'Coral Bells' (EA)	see **R.** 'Kirin'
	'Coral Glow'	wWhG
	'Coral Queen'	wHam
	'Cormid'	wWhG
do	'Corneille' (DA)	wWhG
	'Cornell Pink'	see **R. *mucronulatum*** 'Cornell Pink'
	Cornubia Group	wWhG
	'Corona'	wHam
	'Coronation Day'	last listed 99/00
	'Corringe' (DA)	wWhG
	'Corry Koster'	wHam
un	'Corsage' (EA)	wWhG
	coryanum	oGre
	'Cosmopolitan'	wHam
	'Cotton Candy'	wSta wHam oTal wWhG wWel
	'Cougar'	oGre
	'Countess of Derby'	wSta
	'County of York'	see **R.** 'Catalode'
	'Court Jester'	wHam
	'Cowbell'	last listed 99/00
	Cowslip Group	oGar
un	'Cranberry Lace'	wWhG wWel
	'Cranberry Swirl'	wWhG
un	***crassifolium***	wRho oBov
	'Crater Lake'	oAls oGar oGre wWhG wWel
	'Cream Crest'	oBlo oGre oWnC wSta wWhG wWel
	'Cream Glory'	oGre
	'Creamy Chiffon'	oGre wHam
	'Creeping Jenny'	oGre wHam
	'Creole Belle'	oGre wHam wWel
	'Crest'	wHam wWhG
un	'Crest Hawk'	wHam
	'Crete'	wClo oAls oGar oGre wHam wWhG wWel
	'Cricket'	wWhG
	'Crimson Pippin'	oAls oGre wHam wWel
un	'Crimson Tide' (DA)	wHam
	'Crossroads'	wWhG
un	'Crow Basin' (EA)	wSta
	cruttwellii (V)	oBov
un	'Cruttwell's Unknown Pink'	oBov
un	'Crystal Springs'	wWhG
	cumberlandense (DA)	wWhG
un	***cumberlandense*** 'Camp's Red' (DA)	wWhG
	cumberlandense 'Sunlight' (DA)	oFor oGos
	'Cunningham's Blush'	oAls oGre wHam
	'Cunningham's White'	oAls oGre oWnC wSta wHam oTal wWhG wWel
	'Cupcake'	oGos wClo oAls oGar oGre wHam oBRG oTal wWhG wWel
	'Curlew'	wClo oGre wHam wWhG wWel
	'Cutie'	wSta wHam
	'Cynthia'	oGre wSta wHam wWhG
	'Cyprus'	oGar
un	'Cyril'	oBov
	'Dad's Indian Summer'	oGre wWhG wWel
	'Dainty Jean'	last listed 99/00
	'Dairymaid'	wHam
	dalhousieae	wRho
	dalhousieae var. ***dalhousieae***	last listed 99/00
	dalhousieae var. ***rhabdotum***	last listed 99/00
un	'Damaris' (EA)	wHam wWhG
	'Dame Nellie Melba'	wWhG
	'Dan Laxdall'	last listed 99/00
	'Daniela'	oGre wHam oTal
	'Dan's Early Purple'	oGre wWel
	'Daphnoides'	oAls oBlo oGre oWnC wSta wHam oBRG oTal wWhG wWel
un	'Dappled Dawn'	oGre
un	'Dark Throat'	wWhG
	dauricum	oFor oGre wHam wWhG
	dauricum 'Album'	see **R. *dauricum*** 'Hokkaido'
	dauricum dwarf form	wRho
	dauricum 'Hokkaido'	last listed 99/00
	dauricum JJH9510229	wMtT
	'Dave Goheen'	wHam
	'David'	wHam wWhG
	davidsonianum	wHam
un	***davidsonianum*** 'Album'	wWhG
	davidsonianum 'Caerhays Pink'	last listed 99/00
	davidsonianum Exbury	oGre
	davidsonianum FCC form	wSta
	davidsonianum 'Ruth Lyons'	oGre wSta wHam wWhG
	davidsonianum 'Serenade'	wWhG
	'Daviesii' (DA)	wCul oWnC wHam wWhG
	Day Dream Group & cl.	wHam
	'Dazzler'	oGre wWhG
	'Debbie'	oGre wHam
un	'Debonaire' (EA)	wWhG
	'Debutante' (DA)	oGre wWhG
	decorum	wRho wCCr wHam wWhG
	decorum ssp. ***decorum***	last listed 99/00
	decorum ssp. ***decorum*** white form	wHam
	decorum ssp. ***diaprepes***	wHam
	decorum ssp. ***diaprepes*** 'Gargantua'	wWhG
	decorum DJHC 527	wHer
	decorum f. Emmet Lodge	oGre
	decorum pink	wClo oGre wHam
	decorum yellow form	oGre
	degronianum	wWhG
	degronianum 'Dalriada'	wRho
	degronianum ssp. ***heptamerum***	wClo wRho
un	***degronianum*** ssp. ***heptamerum*** var. ***kyomaruense***	wWhG
un	***degronianum*** ssp. ***heptamerum*** f. ***variegatum***	oGre
	degronianum ssp. ***yakushimanum***	see **R. *yakushimanum***
	'Del'	wWhG
	'Delaware Valley White' (EA)	oAls oTPm wSta wWhG
	'Deletissimo'	wHam
	'Denali'	wClo oGre wHam wWhG wWel
	denudatum	last listed 99/00
	'Desert Gold'	oGre oTal wWhG
	desquamatum	see **R. *rubiginosum*** Desquamatum Group
	'Devonshire Cream'	oGre
	'Dexter's Apricot'	wWhG
	'Dexter's Champagne'	wWhG
	'Dexter's Harlequin'	wHam wWhG
	'Dexter's Peppermint'	wHam wWhG
	'Dexter's Pink Satin'	wWhG
	'Dexter's Purple'	wWhG
	'Dexter's Spice'	oGre wHam wWel
un	'Dexter's Vanilla'	wHam wWhG
	'Dexter's Victoria'	wWhG
	'Diadem' (V)	wHam
	'Diana'	wWhG
	'Diane Titcomb'	wHam
un	***dianthosmum***	oBov
	diaprepes	see **R. *decorum*** ssp. ***diaprepes***
	dichroanthum ssp. ***apodectum***	last listed 99/00
	dichroanthum ssp. ***dichroanthum***	last listed 99/00
	dichroanthum ssp. ***scyphocalyx***	last listed 99/00
un	***dielsianum***	wRho
	'Diorama' (DA)	wCul
	'Disca'	wHam
	discolor	see **R. *fortunei*** ssp. ***discolor***
	Diva Group & cl.	wHam
	'Doc'	oGre wWhG

	'Doctor A. Blok'	wHam
	'Doctor H. C. Dresselhuys'	wHam
	'Doctor Herman Sleumer' (V)	oBov
	('Doctor H. Sleumer' x ***herzogii***) x (***laetum*** x ***zoelleri***)	oBov
	'Doctor M. Oosthoek' (DA)	last listed 99/00
	'Doctor Ross'	wHam
un	'Doctor Rudolph Henny' (DA)	oFor oGre wWhG
	'Doctor Stocker'	wHam
	'Doctor V. H. Rutgers'	oGre wSta wHam
un	'Dogwood' (EA)	wWhG
	'Dolly Madison'	oGre wHam wWhG
	'Donna Totten'	oGre
	'Dopey'	oAls oGre wWhG
un	'Dora'	wClo
	'Dora Amateis'	oAls oGar oBlo oGre oWnC wSta wHam oBRG oTal wWhG wWel
	'Dorinthia'	wWhG
un	'Doris Mossman'	oBov
	Dormouse Group	wHam
	'Dorothy Amateis'	oGre wHam
un	'Dorothy Clark' (EA)	wWhG
	'Dorothy Peste Anderson'	wWhG
un	'Dorothy Reese' (EA)	wSta
	'Dorothy Swift'	wHam wWhG
do	'Double Date'	oGre
do	'Double Delight' (DA)	oGre wWhG
	'Double Dip'	oGre
	'Double Winner'	wHam wWhG
	'Douglas McEwan'	wHam
	'Douglas R. Stephens'	wHam wWhG
un	'Dragon' (EA)	wWhG
un	'Dream' (EA)	wWhG
un	'Dream' (DA)	wWhG
	'Dreamland'	oAls oGre wSta wHam wWel
	'Dress Up'	wHam
	'Driven Snow' (EA)	wWhG
	drumonium	see **R.** ***telmateium***
	dryophyllum (see Nomenclature Notes)	oGre
	'Duchess of York'	oGre
un	'Duke' (DA)	wWhG
un	'Durango'	wWhG
	'Dusty Miller'	wWhG
un	'Dwarf Yellow Oliver'	wSta
	'Dynasty'	wHam
	'Earl Murray'	wWhG wWel
	'Early Splendor'	wWhG
	'Easter Bells'	wHam
	'Ebony Pearl'	wWhG
	edgeworthii	wRho oGre wWhG
	edgeworthii Bodnant selection	oGos
	'Edith Bosley'	oGar oGre wHam wWhG wWel
	'Edith Boulter'	oGre wHam wWel
	'Edmond Amateis'	wHam wWhG
	'Edward Dunn'	wHam wWhG
	'Edwin O. Weber'	oGre wHam
	'Egret'	oGre oBRG wWhG wWel
un	'Eikan' (EA)	oBlo oGre oWnC wSta
	'El Camino'	oGre wHam
	'Elby'	oGre wHam wWel
	'Eleanor Bee'	wHam
	Electra Group & cl.	see **R.** ***augustinii*** Electra Group & cl.
	'Electra's Son'	oGre
	elegantulum	wRho wWhG wWel
	elegantulum narrow leaf	oGre
	'Elisabeth Hobbie'	wClo oBlo oWnC wSta wHam
	Elizabeth Group	oGar oBlo oGre oWnC wHam wWhG wWel
un	'Elizabeth Ann Seton'	oBov
un	'Elizabeth Britt'	wSta
	'Elizabeth de Rothschild'	oGre
	'Elizabeth Lockhart'	wHam
	'Elizabeth Ostbo Cross'	see **R.** 'Elizabeth Red Foliage'
	'Elizabeth Red Foliage'	wClo oAls oGar
un	'Elizabeth Scott' (EA)	wSta
un	'Ellie Sather'	wWhG
	elliottii	wRho wWhG
un	'Elmer's Orphan'	wHam
	'Elrose Court'	wHam
un	'Elsbeth'	wHam
	'Else Frye'	last listed 99/00
	'Elsie Lee' (EA)	oGar oGre oTPm wWhG
	'Elsie Watson'	oGre wHam wWhG wWel
un	'Elvira'	wHam wWhG
	'Elya'	wWhG
	'Emanuela'	wHam wWhG

	'Emasculum'	wSta wWhG
un	'Emerald Ice'	oGre
	'Emily Allison'	wHam
un	'Emmanuel'	oBov
	'Enchanted Evening'	wHam wWhG
	'English Roseum'	oAls oGre oWnC wSta wWhG
un	'Equinox'	wHam
	'Erato'	oGre wHam
	'Ernie Dee'	oGre wHam wWel
un	'Eros' (EA)	oGar oSho oBlo oGre oWnC wSta
	erosum	last listed 99/00
	erubescens	see **R.** ***oreodoxa*** var. ***fargesii*** Erubescens Group
	'Esquire'	wHam
	'Estacada'	wWhG
	Ethel Group & cl.	wSta
	'Etta Burrows'	oGre wSta wHam
	'Eunie'	oGre wHam wWel
un	***euonymifolium***	wRho oBov
	'Eureka Maid'	see **R.** 'Countess of Derby'
	'Europa'	wHam wWhG
un	'Evelyn Maranville'	wWhG
	'Evening Glow'	oGre wSta wHam oTal wWhG
	'Everest' (EA)	oAls oGar oBlo oTPm oWnC wSta wWhG
	'Everything Nice'	oGre
	'Exbury Angelo'	last listed 99/00
	'Exbury Calstocker'	oGre
	'Exbury Cornish Cross'	wHam
un	'Exbury Gold'	oAls
un	Exbury hybrids (DA)	oFor oAls oSho oBlo wBox oGue
	'Exbury Naomi'	oGre wHam
un	'Exbury Orange'	oGar
	'Exbury Pink'	oGar
un	'Exbury Red'	oAls oGar
un	'Exbury' seed grown	wSta
	'Exbury White' (DA)	oGar
	Exbury white throat	oGar
un	'Exbury Yellow'	oGar
un	'Exhalted Ruler'	wHam
	'Exotic'	wHam
	'Extraordinaire'	oGre
	Fabia Group & cl.	oGre wSta wHam
	'Fabia' x ***bureaui***	wWel
	'Fabia' x ***bureaui*** x ***yakushimanum*** 'Koichiro Wada'	wWhG
	'Fabia Tangerine'	oGre wWhG
un	'Fabur'	oGre
	sp. aff. ***facetum***	wRho
	'Faggetter's Favourite'	wClo oGre wWhG wWel
	falconeri	wRho wHam
	falconeri ssp. ***falconeri*** USNA	oBov
	'Fancy'	wHam
	'Fanny'	see **R.** 'Pucella'
	'Fantastica'	oGre wHam wWhG wWel
	'Fantasy'	wHam
	fargesii	see **R.** ***oreodoxa*** var. ***fargesii***
	'Fascination' (EA)	oAls wWhG
	fastigiatum	wSta wWhG
do	'Fastuosum Flore Pleno'	oGre wSta wHam wWhG
	faucium	last listed 99/00
	'Fedora' (EA)	wSta
	'Felicitas'	oBov
	'Felix de Sauvage'	see **R.** 'Chevalier Felix de Sauvage'
	ferrugineum	wClo wRho wSta wHam wWhG
	'Festivo'	wHam wWhG
un	'Feuerschein'	wHam
	fictolacteum	see **R.** ***rex*** ssp. ***fictolacteum*** wWhG
	'Finlandia'	wHam oTal
	'Fire Cracker'	wWhG
un	'Fire Rim'	wWhG
	'Fireball' (DA)	oFor oGar oGre wHam wWhG
	'Fireman Jeff'	oAls oGar oGre wHam oTal
	'Firestorm'	wClo oGre oWnC wWhG
	'Firewine'	oGre
	'Flame Creeper' (EA)	oFor oAls oGar oSis oBlo oGre oTPm wSta oBRG
un	'Flaming Comet'	oGre
	'Flaming Star'	oGre
un	'Flamingo' (DA)	wWhG
un	'Flamingo Bay'	oBov
	flammeum (DA)	oGar
	Flava Group & cl.	wWhG wWel
	flavorufum	see **R.** ***aganniphum*** var. ***flavorufum***
	fletcherianum	last listed 99/00
	'Flora's Boy'	last listed 99/00

	'Fluff'	last listed 99/00
un	'Fluffy' (DA)	wHam
	'Forever Yours'	oGre wHam wWhG
	formosanum	last listed 99/00
	forrestii	wRho
	forrestii ssp. ***forrestii***	see **R. *forrestii***
	forrestii ssp. ***forrestii*** Repens Group	wClo wMtT
	fortunei	oFor oDan oGre wCCr wHam wWhG
	fortunei ssp. ***discolor***	oGre wHam wWhG
	fortunei ssp. ***discolor*** Houlstonii Group	wWhG
	fortunei ssp. ***discolor*** pink form	oGre
	fortunei 'Mrs. Charles Butler'	see **R. *fortunei*** 'Sir Charles Butler'
	fortunei 'Sir Charles Butler'	oGre
	'Fragrans Affinity'	oGre
	'Fragrantissimum'	oGos
	'Francesca'	oGre wHam
	'Frango'	wWhG
un	'Frank Abbott' (DA)	wHam wWhG
	'Frank Baum'	wHam
	'Frank Galsworthy'	oGre wHam wWhG wWel
	'Fred Hamilton'	wHam wWhG
	'Fred Peste'	oGos oGre wHam wWel
	'Fred Rose'	last listed 99/00
	'Freeman R. Stephens'	wWhG
	'Friday'	oGre
un	'Friday Surprise'	oGre
	'Frilled Yak'	wWhG
do	'Frills' (DA)	wWhG
do	'Frilly Lemon' (DA)	wHam wWhG
	'Frontier'	oGre wHam
	'Frosted Orange' (EA)	oAls oGar oSho oGre wSta oBRG
un	'Fruhlingsanfang'	wWhG
un	'Fuji Zakura' (EA)	oGre oBRG
	fulgens HWJCM 318 (last listed HWJCM 308)	wHer
	'Full Moon'	oGre wHam
	'Full Moon' x ***yakushimanum***	wWhG
	fulvum	oGre wHam
	fulvum ssp. ***fulvoides***	wHam
	'Furnivall's Daughter'	oAls oGar oGre wSta wHam wWel
	'Gable Rosebud' (EA)	see **R.** 'Rosebud'
un	'Gable's Bicolor'	wWhG
	'Gaiety' (EA)	oAls oGar oSho wSta
	'Gala'	oGre wHam
	galactinum	wRho
	'Gallipoli' (DA)	wWhG
un	'Gallipoli Red' (DA)	oGre
	'Gandy Dancer'	wHam
un	'Garland Pink'	oGar
	'Garnet'	wHam
	'Gartendirektor Glocker'	wHam wWel
	'Gartendirektor Rieger'	oGre wHam wWhG wWel
	'Geisha' (EA)	wSta wWhG
un	'Gekko' (A)	oBRG
do	'Gena Mae' (DA)	oGre
	'General Eisenhower'	wHam
un	'Genesis I'	wWhG
	'Genghis Khan'	wHam wWhG
un	'Gentle Giant'	wWhG
	'George Cunningham'	wWhG
	'George Reynolds' (DA)	oFor oGar oGre wHam wWhG
	'George Sweesy'	wWhG
	'George's Delight'	oGre wHam wWhG wWel
	x geraldii	oGre wSta
	'Germania'	wHam oTal wWhG
	Gertrud Schaele Group	wHam
	'Gertrude Bovee'	last listed 99/00
	'Getsutoku' (EA)	oGar
	'Gibraltar' (DA)	oFor oGar oGre oWnC wHam wWhG
	'Gigi'	oGre wHam
	'Gina'	oBov
	'Ginger' (DA)	wHam
	'Ginny Gee'	wClo oAls oSis oGre oWnC wHam oBRG oTal wWhG wWel
un	'Girard's Border Gem' (EA)	oGre wWhG
un	'Girard's Chiara' (EA)	oGar oGre wWhG
un	'Girard's Crimson' (EA)	oGre wWhG
un	'Girard's Dwarf Lavender' (EA)	wWhG
	'Girard's Fuchsia' (EA)	oAls oGar oGre wSta wWhG
	'Girard's Hot Shot' (EA)	oFor oGar oSho oGre wSta oBRG wWhG
un	'Girard's Leslie' (EA)	oGar
	'Girard's Lucky Stars' (***yakushimanum***)	wWhG
un	'Girard's National Beauty' (EA)	wWhG
un	'Girard's Pink' (EA)	wWhG
un	'Girard's Pleasant' (A)	oWnC
un	'Girard's Purple' (EA)	wWhG
un	'Girard's Red Pom Pom' (DA)	wWhG
un	'Girard's Renee Michelle' (A)	oAls
	'Girard's Roberta' (EA)	oAls
un	'Girard's Rose' (EA)	oAls oGar wWhG
un	'Girard's Scarlet' (EA)	oAls oWnC wWhG
un	'Girard's Variegated Hot Shot' (EA)	oFor oSis wWhG
	'Glacier' (EA)	oGar oBlo wCCr oWnC wSta
un	'Glamour' (EA)	oFor oAls oGar oBlo oGre oWnC wSta wWhG
	sp aff. ***glanduliferum***	last listed 99/00
	glandulosum	see **LEDUM *glandulosum***
	glaucophyllum	wHam
	glaucophyllum var. ***glaucophyllum***	wRho
	glaucophyllum ssp. ***tubiforme***	wRho
	'Glencora' (EA)	wSta
un	'Glenna'	wWhG
	'Glen's Orange'	oBlo
	'Gletschernacht'	oAls oGar wHam wWel
	globigerum	see **R. *alutaceum*** var. ***alutaceum*** Globigerum Group
	glomerulatum	see **R. *yungningense*** Glomeratum Group
	'Glory of Littleworth'	wWel
	'Glowing Embers' (DA)	wHam
	'Gold Mohur'	wSta
	'Goldbug'	wSta oTal
	'Goldbukett'	wHam wWhG
	'Golden Belle'	oBlo
un	'Golden Comet' (DA)	wWhG
	'Golden Dream'	wSta
	'Golden Eagle' (DA)	wWhG
	'Golden Flare' (DA)	oWnC
	'Golden Gala'	oGre
	'Golden Gate'	wClo oAls oGar oBlo oGre wHam wWhG wWel
	'Golden Genie'	wHam wWhG
un	'Golden Harvest'	wWhG
un	'Golden Hill'	oTal
	'Golden Jubilee'	wHam wWhG
	'Golden Lights' (DA)	oFor oGar wHam wWhG
	'Golden Princess'	last listed 99/00
	'Golden Ruby'	wHam
	'Golden Showers' (DA)	wHam
	'Golden Torch'	wClo oGre wHam wWhG wWel
	'Golden Wedding'	oGre wHam
	'Golden Wit'	oAls oGar wHam
	'Goldfee'	wHam wWhG
	'Goldfinch' (DA)	oGre
	'Goldfinger'	last listed 99/00
un	'Goldflamme'	oGar
va	'Goldflimmer'	wClo oSho oGre wWhG wWel
	'Goldfort'	oGre wHam
	'Goldilocks'	oGre
	'Goldkrone'	wClo oGre wHam wWhG
	'Goldstrike'	last listed 99/00
	'Goldsworth Crimson'	wHam
	'Goldsworth Orange'	wHam oTal wWhG
	'Goldsworth Pink'	wHam
	'Goldsworth Yellow'	wHam wWhG
	'Golfer'	oGos wClo oGre wHam wWhG wWel
	'Gomer Waterer'	oAls oGar oSho oBlo oGre wSta wHam wWhG wWel
	'Good News'	oGre wHam wWhG wWel
un	***goodenoughii***	oBov
	'Gordon Jones'	oGre wHam wWhG
	'Grace Seabrook'	oAls oGar oGre wHam wWhG wWel
	gracilentum (V)	oBov
	gracilentum x ***lochiae***	oBov
	'Graciosum' (DA)	wWhG
	'Graf Zeppelin'	oAls oGar oSho oGre wHam wWhG
	'Graf Zeppelin' x 'King of Shrubs'	wHam
	'Grand Slam'	oGre wHam wWel
	grande	wRho
	'Grandma's Hat'	oGre wWhG wWel
	'Graziella'	wWhG
	'Great Eastern'	wHam
un	'Great Expectations' (EA)	oAls oGar oTPm wWhG
	'Great Scott'	wHam
	'Green Eye'	wWhG
un	'Green Glow' (EA)	wWhG
	'Greenwood Orange' (EA)	oAls oGar oWnC wSta
	'Greer's Cream Delight'	oGre wHam
	'Greer's Starbright'	oGre
un	'Greeting'	wHam
	'Greta' (EA)	wSta

	'Gretchen'	wHam
	'Gretsel'	wWel
	'Grierosplendour'	wHam
	griersonianum	last listed 99/00
	griffithianum	wRho
	'Gristede'	wWhG
	groenlandicum	wThG
	Grosclaude Group & cl.	wHam
	'Grosclaude' x ***yakushimanum***	wHam
	'Grumpy'	oGre wHam wWhG
un	'Gumpo Fancy' (EA)	oAls oGre
	'Gumpo Pink' (EA)	oAls oGar oBlo oGre oTPm wSta
un	'Gumpo Splendor' (EA)	wSta
	'Gumpo White' (EA)	oAls oGar oSho oBlo oGre oTPm wSta
	'Gustav Luttge'	wWel
un	'Gustav Mehlquist'	oGre wWhG
	'Gwen Bell'	wHam
un	'Gyo Kudo' (EA)	oBRG
un	'Gypsy Queen'	oBlo oGre oTal
	'H. L. Larson'	see **R.** 'Hjalmar L. Larson'
un	'Haaga'	oGre wHam wWhG
	'Hachmann's Belona'	oGre
	'Hachmann's Brasilia'	oGre
	'Hachmann's Charmant'	oGre wWhG wWel
	'Hachmann's Feuerschein'	wWhG
	'Hachmann's Polaris'	oGre
	haematodes	wHam
	haematodes ssp. ***chaetomallum***	wRho
	haematodes ssp. ***haematodes***	last listed 99/00
un	'Hahn's Red' (EA)	oGar oBlo oGre oTPm oWnC
un	'Hakatashiro' (EA)	wSta
un	'Half and Half'	wHam
	'Halfdan Lem'	oAls oGre wHam wWhG wWel
	'Hallelujah'	oAls oBlo oGre wHam wWhG wWel
	'Hamlet' (DA)	last listed 99/00
	hanceanum	oGre
	hanceanum Nanum Group	oGos wWel
	hanceanum pink	wSta
	hanceanum white	wSta
un	'Hansa Bay'	oBov
	'Hansel'	oGre wHam
	'Hardijzer Beauty' (Ad)	oAls oGre oTPm oWnC wWhG wWel
do	'Hardy Gardenia' (EA)	oGar oGre wSta wWhG
	'Harnden's White'	oGre
	'Harold Amateis'	wWhG
	'Harry Carter'	oGre wHam
un	'Harry Wu'	oBov
	'Harvest Moon'	oTal
	'Hawaii'	oGre oTal wWhG
	'Hazel'	wHam wWel
	'Hazel Fisher'	wHam
	'Heart's Delight'	oAls oGre wHam wWhG
	'Heat Wave'	oGre
	'Heatherside Beauty'	last listed 99/00
un	'Heathwood'	wWhG
	'Heavenly Scent'	oGre wHam wWhG
un	'Helen' (EA)	oBlo oGre
	'Helen Child'	wWhG
	'Helen Close' (EA)	oAls oGar oBlo oGre oTPm
	'Helen Curtis' (EA)	wWhG
	'Helen Deehr'	oGre wHam wWhG
	'Helen Everitt'	wWhG
	'Helen Johnson'	oBlo
	'Helene Schiffner'	oGre wSta wHam wWhG wWel
	'Helene Schiffner' x ***yakushimanum***	wWhG
	heliolepis	wHam
	sp. aff. ***heliolepis*** DJHC 197	wHer
	heliolepis var. ***heliolepis***	wRho
	'Hellikki'	wHam wWhG wWel
	'Hello Dolly'	wSta wHam wWhG
un	'Helsinki University'	oGre wHam wWhG
	hemsleyanum	wHer
	'Hendrick's Park'	last listed 99/00
	'Henriette Sargent'	oGre wHam wWhG
	'Henry's Red'	oGre wHam wWhG wWel
	'Herbert' (EA)	oGre wSta
un	'Hershey's Red' (EA)	oGar oGre wSta wWhG
un	***herzogii***	oBov
	'Hexe' (EA)	oAls oGar oSho oBlo oGre oTPm wSta
	'Higasa' (EA)	oBlo oGre wSta wWhG
un	'High Fashion' (DA)	oGre wWhG
	'High Gold'	wHam
un	'High Sierras' (DA)	oGre
	'Hilda Niblett' (EA)	oGre oBRG wWhG
	'Hillcrest'	wWel

	'Hill's Bright Red'	oGre wHam wWel
	'Hindustan'	oGre wHam wWhG
	'Hino-crimson' (EA)	oAls oGar oSho oBlo oGre oTPm oWnC wSta oBRG wWhG
	'Hinode-giri' (EA)	oBlo oGre wSta
	hippophaeoides	oSis oRiv oWnC wHam wWhG wWel
	hippophaeoides DJHC 309	wHer
	hippophaeoides var. ***hippophaeoides***	oGre
	hirsutum	wWhG
do	***hirsutum*** 'Flore Pleno'	oGre oBRG
	'Hjalmar L. Larson'	wHam
	'Hockessin'	oGos
	hodgsonii	wRho wHam
	hodgsonii HWJCM 185	wHer
	'Holden'	oAls oSho oGre oWnC wHam wWhG wWel
un	'Holland' (EA)	wSta
	'Holy Moses'	oWnC wHam
do	'Homebush' (DA)	oGre wHam wWhG
	'Honey Bee'	last listed 99/00
	'Honeymoon'	last listed 99/00
	'Honeysuckle' (DA)	wWhG
	'Hong Kong'	oGre wHam
	'Honsu's Baby'	wWhG
	hookeri	wHam
	'Hoppy'	oGre wHam wWhG
un	'Horinouchi' x (***yakushimanum*** x 'President Roosevelt')	wWhG
	'Horizon Dawn'	oGre wHam wWhG
	'Horizon Lakeside'	oGre wHam wWhG wWel
	'Horizon Monarch'	oGre wHam wWhG wWel
	'Horizon Snowbird'	wHam wWhG
	'Hortulanus H. Witte' (DA)	last listed 99/00
	'Hot Dawn'	oGre wHam wWhG
	'Hot Shot'	see **R.** 'Girard's Hot Shot'
	'Hotei'	wClo oAls oGar oBlo oGre oWnC wSta wHam oBRG wWhG wWel
	'Hotei' x 'Crest'	wSta
	'Hotei' x 'Kubla Khan'	oGre
	Hotspur Group (DA)	wWhG
	houlstonii	see **R. *fortunei*** ssp. ***discolor*** Houlstonii Group
	huanum	wRho
	'Hudson Bay'	wWhG
	'Humboldt'	wWhG
qu	'Humboldt Surprise'	oGos
un	'Humbug'	wHam
	Humming Bird Group	oGre wWel
un	'Hummingbird Dwarf'	wHam
	hunnewellianum	oGre wWel
	hunnewellianum ssp. ***hunnewellianum***	see **R. *hunnewellianum***
	'Hurricane'	oGre wHam wWel
	'Hydon Dawn'	wHam wWhG
	'Hydon Harrier'	wHam
	'Hydon Hunter'	wHam
	'Hydon Mist'	wHam
	Hyperion Group	oAls oGar oGre wWhG
	hyperythrum	wRho wHam wWhG
	Ibex Group & cl.	wSta
	'Ice Cube'	wHam wWel
	Idealist Group & cl.	oGre wHam
	'Ightham Yellow'	wHam
	'Ignatius Sargent'	wSta wHam
do	'Il Tasso' (DA)	oGre wWhG
	'Ilam Alarm'	wHam
un	'Ilam Carmen' (DA)	oGre wWhG
	'Ilam Cream'	oGre
	'Ilam Orange' x ***macabeanum***	wHam
un	'Ilam Red Frills' (DA)	wWhG
	'Ilam Violet'	oBlo oGre wSta wHam wWhG
do	'Ima-shojo' (EA)	oBlo oGre oWnC wSta
	impeditum (see Nomenclature Notes)	oFor oAls oGar oSis oBlo oGre oTPm oWnC wSta wWhG wWel
	'Imperial'	oGre
	Impi Group & cl.	wHam wWhG
	'Independence Day'	wHam wWhG
	indicum (EA)	wCCr
do	***indicum*** 'Balsaminiflorum' (EA)	oBlo oGre wSta oBRG wWhG
	'Ingrid Mehlquist'	oGre wHam wWhG
	insigne	wHam wWhG
	intricatum	wClo oGre wWhG
	Intrifast Group	wWhG wWel
un	'Invitation'	wWhG
un	'Iora' (DA)	oGre
	'Irene Bain'	wHam wWhG
	'Irene Koster' (DA)	oFor oGar oGre wHam wWhG

	'Irene Stead'	wWhG
	'Irresistible Impulse'	oGre wHam wWel
	irroratum	wHam
	irroratum 'Polka Dot'	wWhG
	irroratum 'Spatter Paint'	oGre wWhG
	'Isabel Pierce'	oGre wHam
un	'Isso-no-haru' (A)	oAls oGar
	'Ivan D. Wood'	wHam
	'Ivory Coast'	oGre oTal
	'J. M. de Montague'	see **R.** 'The Hon. Jean Marie de Montague'
un	'J. T. Lovett'	oAls oGar
	'Jack A. Sand' (DA)	oGre wWhG
	'Jackie Ann'	last listed 99/00
	'Jacksonii'	wSta
	'Jackwill'	wHam
	'Jalisco Elect'	wHam
	'James Barto'	wHam
	'James Gable (EA)	wSta wWhG
un	'James Marschand'	wHam
un	'Jan' (EA)	oGre wSta wWhG
	'Jan Bee'	oGre wHam wWel
	'Jan Dekens'	wHam wWhG
	'Jane Abbott' (DA)	oGre wHam wWhG
	'Jane Martin'	wWhG
	'Janet'	last listed 99/00
	'Janet Blair'	oGre wHam wWhG
un	'Janet Rhea'	wWhG
	'Janet Scroggs'	wHam
	japonicum	see **R.** ***molle*** ssp. ***japonicum***
un	***jasminiflorum*** var. ***punctatum*** (V)	oBov
	javanicum	oBov
	Jean Group	wHam
	'Jean Marie'	see **R.** 'The Hon. Jean Marie de Montague'
	'Jean Marie de Montague'	see **R.** 'The Hon. Jean Marie de Montague'
	'Jean Rhodes'	wHam
	'Jeanne Church'	wWhG
un	'Jeanne Weeks' (EA)	oTPm
	'Jeannie's Black Heart'	wWhG
	'Jeff Hill' (EA)	last listed 99/00
un	'Jeremiah' (EA)	oGre
	'Jericho'	wHam
un	'Jessie's Song'	wWhG
un	'Jester' (DA)	wWhG
	'Jingle Bells'	wSta wHam wWhG
un	'Jo Ann Newsome'	oGre wHam
un	'Joan Garret' (EA)	wSta wWhG
un	'Joan Leslie'	wHam
	'Joan Scobie'	wHam
	Jock Group	oGar wSta wHam oTal
	'Jock Brydon' (DA)	wHam wWhG
un	'Jock's Cairn'	oBov
	'Jodi'	wWel
	'Jodie King'	wWhG
	'Joe Paterno'	wHam oTal
	'Johanna' (EA)	oBlo oGre wWhG
	'John Coutts'	oGre wWel
	'John Eichelser' (DA)	oGre wWhG
	'Johnny Bender'	wClo oAls oGar oGre wHam wWel
	johnstoneanum double	oGre
do	***johnstoneanum*** 'Double Diamond'	wRho wWel
	'Jolie Madame' (DA)	wCul
	'Jonathan Shaw'	oAls oGre wHam wWel
	'Joseph Hill' (EA)	wSta wWhG
	'Josephine'	wWhG
	'Juan De Fuca'	wWhG
	'Judy Spillane'	wHam wWhG
un	'Juko' (EA)	oAls oGar oBRG
	'Julischka'	wHam
	'June Bee'	wWhG
	'June Fire'	see **R.** 'Junifeuer'
	'Junifeuer'	oGre
un	'Junko' (EA)	oGre
un	'K. G. N. Orchid'	wWel
	kaempferi (EA)	wSta wWhG
un	'Kaempo' (EA)	oAls oGre
un	'Kagetsu Muji' (A)	oSho
	'Kalinka'	wHam oTal wWhG
un	'Kamrau Bay'	oBov
	'Karalee'	wWhG
	'Karen Triplett'	oGre wHam wWhG wWel
un	'Karens'	wWhG
	'Karin'	wClo wHam
	'Karin Seleger'	wHam wWel
qu	'Kasan (EA)	oGar
	'Kathie Jo'	wWhG wWel

	'Kathleen' (DA)	oGre wWhG
	'Kathryna'	wHam
un	'Kathy' (EA)	wSta
un	'Kathy Ann' (EA)	oGre wWhG
	'Kathy Van Veen'	oGre wHam wWhG
	'Katrina'	oGre wHam wWhG
	kawakamii (V)	last listed 99/00
qu	'Kazan' (EA)	oFor oGre oBRG wWhG
un	'Kazan Sport' (EA)	oSis
	keiskei	wRho oGre wHam wWhG
	keiskei x ***hanceanum***	wSta
un	***keiskei*** x ***MCP***	wSta
	keiskei var. ***ozawae*** 'Yaku Fairy'	wMtT wHam wWel
	keiskei short form	wHam
	keiskei tall form	wHam
	keleticum	see **R.** ***calostrotum*** ssp. ***keleticum***
	'Kelly'	wHam wWhG
	'Ken Janeck'	oGos oAls oGar oGre wHam oTal wWhG wWel
un	'Kennedy' (DA)	oGre wWhG
	'Kermesinum' (EA)	wWhG
	'Kermesinum Rose' (EA)	oGre oBRG
	'Kevin'	oGar
	keysii	wRho
	Kilimanjaro Group & cl.	oGre
	'Kilimanjaro' x ('The Hon. Jean Marie de Montague' x 'Leo')	oGre
	'Kimberly'	wClo oGre wHam wWhG wWel
	'Kimbeth'	wClo oAls oGar oBlo oGre wHam wWhG wWel
un	'Kimi-maru' (EA)	oAls oGar oGre oBRG
	'King Bee'	wHam
	'King of Shrubs'	wWhG
un	'King Red' (DA)	oGre wWhG
	'King Tut'	oGre wHam
un	'King's Crimson'	wWhG
	'King's Ride'	wWhG
	'Kingston'	wHam
un	'Kinpai' (EA)	oGre oBRG
do	'Kirin' (EA)	oAls oGar oBlo oTPm wSta wWhG
	kiusianum (EA)	wSta wWhG
	kiusianum 'Album' (EA)	oSis
	kiusianum 'Benichidori' (EA)	wSta wWhG
	kiusianum 'Betty Muir' (EA)	oGre wWhG
	kiusianum 'Harunokikari'	last listed 99/00
un	***kiusianum*** 'Hinode'	oSis oGre
un	***kiusianum*** 'Komo Kulshan'	oGre wWhG
	kiusianum lavender form	wClo wWhG
un	***kiusianum*** 'Mangetsu'	oBRG
	kiusianum 'Murasaki Shikibu'	oBRG
	kiusianum pink form	oGar oBlo wWhG
	kiusianum purple	oGar
	kiusianum red form	wWhG
un	***kiusianum*** 'Shirotae'	oWnC
un	***kiusianum*** 'Short Circuit'	oGre
	kiusianum white form	oBlo oGre wWhG
	'Kiwi Majic'	wWhG wWel
	'Klondyke' (DA)	oGre wWhG
	'Kluis Sensation'	wSta wHam oTal wWhG
	'Knap Hill Red' (DA)	oGre wWhG
un	'Ko kinsai' (EA)	oSis oGre oBRG
un	***kochii***	oBov
un	'Kohan-no-tsuki' (EA)	oGre oBRG
un	'Koho' (EA)	oGre
	'Kokardia'	oGre wHam wWhG
un	'Komane' (EA)	oGre
	'Komo Kutshan'	oTal
	'Koningin Emma' (DA)	last listed 99/00
	konori (V)	oBov
un	***konori*** var. ***konori***	wRho
un	***konori*** x ***laetum*** x ***aequabile***	oBov
	konori x ***lochiae***	oBov
	konori var. ***phaeopeplum*** (V)	last listed 99/00
	konori USNA	oBov
qu	'Korin' (EA)	oBRG
un	'Koromo Shikibu' (EA)	wWhG
	'Koster's Brilliant Red' (DA)	oGre
	'Kozan' (EA)	oGre
un	'Krakatoa' (DA)	oGre
	'Kristin'	oAls oGar oGre wWel
	'Kubla Khan'	oGre
	'Kubla Khan' x 'Cream Glory'	oGre
un	'Kurt Adler' x ***leucogigas***	oBov
un	'Kurt Herbert Adler'	oBov
qu	'Kusan' (A)	oGos

un	'Lace Valentine' (DA)	oGre wWhG
	'Lackamas Blue'	see **R. *augustinii*** 'Lackamas Blue'
	'Lackamas Firebrand'	wHam
	'Lackamas Ruby'	wHam
	'Lackamas Sovereign'	wHam
	lacteum	wRho
un	'Ladrillo'	wWhG
	'Lady Bligh'	wSta wHam wWhG wWel
	Lady Chamberlain Group & cl.	wHam
	'Lady Clementine Mitford'	oBlo oGre wSta wHam
	'Lady de Rothschild'	oGre wWhG
qu	'Lady Eleanor'	wHam
un	'Lady Jayne' (DA)	wHam
	'Lady Louise' (EA)	wSta
	'Lady Luck'	wHam
	'Lady of Spain'	oGre wHam
	'Lady Primrose'	wSta wHam
	'Lady Robin' (EA)	oSho wWhG
	'Lady Rosebery' (DA)	oGre wWhG
	Lady Rosebery Group & cl.	wSta
	'Lady Rosebery' *x* flavidum	wSta
	Ladybird Group & cl.	oBlo
	laetum (V)	wRho oBov
	laetum Black & Wood form	oBov
	laetum 'Golden Gate'	oBov
un	***laetum*** x ***phaeochitum***	oBov
	laetum x 'Pink Delight'	oBov
	laetum x ***zoelleri***	oBov
	'Lajka'	wWhG
un	'Lake Amaru'	oBov
un	'Lake Habbema'	oBov
un	'Lake Toba'	oBov
un	'Lake Wissel'	oBov
	'Lamplighter'	wSta wHam
	lanatum	wWhG
un	'Langley Park'	wHam
	'Langworth'	wWhG
	'Late Love' (EA)	wSta
	'Laura Morland' (EA)	last listed 99/00
	'Laurago'	wWhG
	'Laurie'	wWhG
	'Lavender Girl'	oGre wWel
	'Lavender Princess'	oGre oWnC wHam
	'Lavender Queen'	oGre oWnC wSta wWhG
	'Lavendula'	oGre wWel
un	'Lawrence'	oBov
un	'Lawrence Olsen'	oGre
	'Led Album' (EA)	see **R. *x mucronatum***
	ledebouri	see **R. *dauricum***
	ledifolium	see **R. *x mucronatum***
	ledifolium 'Album'	see **R. *x mucronatum***
	'Lee's Dark Purple'	oAls oGar oGre wHam wWhG
	'Lee's Scarlet'	oGre wHam wWhG wWel
	'Lemon Drop' (DA)	wHam wWhG
un	'Lemon Lights' (DA)	oGre wWhG
	'Lemonade'	oGre wHam wWhG
	'Lemonora' (DA)	oGre
un	'Lem's Bicolor'	oGre
	'Lem's Cameo'	oGre wSta wHam wWhG wWel
	'Lem's Cameo' x 'Gold Medal'	oGre
un	'Lem's Fluorescent Pink'	wHam
	'Lem's Monarch'	wClo oAls oGar oGre wHam wWhG wWel
un	'Lem's Rose'	wWel
	'Lem's Stormcloud'	oAls oGar oGre wHam wWhG wWel
	'Lem's Tangerine'	oGre wHam wWhG wWel
	'Lem's 121'	oGre wHam
un	'Len Living'	wHam
	'Leo'	oGre wHam wWhG wWel
	'Leona'	wHam
	'Leona Maud'	wWhG
	'Leonardslee Giles'	oGre wHam
	Leonore Group & cl.	wWel
un	'Leonore Frances'	oBov
	lepidostylum	oRiv wWhG
	lepidotum	wRho
	lepidotum Elaeagnoides Group	wSta
	lepidotum HWJCM 249	wHer
un	'Leprechaun' (EA)	oBRG
	leptanthum (V)	oBov
	leptothrium	wRho
un	'Leslie's Purple' (EA)	wWhG
	Letty Edwards Group & cl.	wHam
	leucaspis	wWhG
	'Leverett Richards'	wHam
	light pink	oAls
	light purple	wSta
	'Light Touch'	wHam
	'Lightly Lavender'	oAls oGre
un	'Lila' (DA)	oGre
	'Lila Pedigo'	oGre
	'Lillian Peste'	oGre wHam
	'Linda'	oGre wHam wWhG wWel
un	'Linda Jean' (EA)	wSta
	'Linda R' (EA)	wSta
ch	'Linearifolium'	see **R. *stenopetalum*** 'Linearifolium'
	Lionel's Triumph Group & cl.	oGre
	'Lissabon'	wHam
	litangense (see Nomenclature Notes)	wHam wWhG wWel
un	'Little Beaver'	wHam
un	'Little Beth'	wHam
	'Little Gem'	wHam
	'Little Red Riding Hood'	wHam
	'Little Sheba'	wHam
	'Little White Dove'	wHam
	'Liz Ann'	wHam wWhG
	lochiae (V)	wRho oBov
un	***lochiae*** x ***christianae***	oBov
un	***lochiae*** x ***commonae*** #2	oBov
un	***lochiae*** x ***culminicolum***	oBov
	lochiae x ***javanicum***	oBov
	lochiae x ***zoelleri***	oBov
un	***lochiae*** x ***zoelleri*** x ***commonae***	oBov
	x lochmium	wHam
	Lodauric Group	wHam
	'Lodauric Iceberg'	last listed 99/00
	Loderi Group	oGre wHam wWhG
	'Loderi Fairyland'	oGre
	'Loderi Game Chick'	oGre wHam wWhG wWel
un	'Loderi Horsham'	wHam
	'Loderi Ilam Cream'	see **R.** 'Ilam Cream'
	'Loderi Irene Stead'	see **R.** 'Irene Stead'
	'Loderi King George'	wClo wCSG oGre wHam wWhG wWel
un	'Loderi Olga'	wHam
	'Loderi Patience'	wHam
	'Loderi Pink Diamond'	oGre wHam wWel
	'Loderi Pretty Polly'	oGre wHam
un	'Loderi Queen May'	wHam
	'Loderi Sir Edmund'	oGre wHam wWhG
	'Loderi Sir Joseph Hooker'	wWhG
	'Loderi Superlative'	oGre wHam
	'Loderi Titan'	wHam
	'Loderi Venus'	oGre wHam wWhG wWel
	'Loderi White Diamond'	oGre
	'Loder's White'	oGre wSta wHam wWhG wWel
	'Lodestar'	wWhG
un	'Lofthouse Legacy'	wWhG
	'Lois'	oGre
un	'Lollipop' (DA)	oGre wHam wWhG
	longesquamatum	wRho
	longipes	wRho
	'Looking Glass'	oGre wHam wWhG
	loranthiflorum (V)	wRho oBov
	'Lord Roberts'	oSho oBlo oGre oWnC wSta wHam
	'Lori Eichelser'	wClo oGar oGre oWnC wWel
	'Lorna' (EA)	oBlo oGre oTPm wWhG
un	'Lorna Gable' (EA)	wSta
	'Louisa' (EA)	oGre
	'Louise Gable' (EA)	oGar oSho oBlo oGre oWnC
	'Love Story'	wWhG
un	'Lucie Sorensen'	oBov
	'Lucky Strike'	wSta wHam
	'Lucy Lou'	oWnC wWhG wWel
un	'Lugano'	wHam wWhG
qu	***lulodendron*** x ***azaleadendron***	wSta
	Luscombei Group	oGre
	lutescens	wRho wWhG
	lutescens FCC form	oGre wSta wWel
	luteum (DA)	oGre wHam wWhG wWel
un	***luteum*** 'Golden Comet' (DA)	wRho wWhG
	'Luxor'	wHam
	'Lydia'	last listed 99/00
	'Lynsey'	wWhG
	macabeanum	oGre wWhG wWel
	macabeanum Cecil Smith form selfed	oGre
	macgregoriae (V)	wRho
	'Macranthum' (EA)	oGar oWnC wSta wWhG
un	'Macranthum Double' (EA)	oGre oBRG
	'Macranthum Roseum' (EA)	oSho oTPm
	macrophyllum	oHan wWoB wBur wNot oBos wRho wWat oTri wHam oAld wRav

	macrophyllum 'Albion Ridge'	oGre wWel
	macrophyllum 'Seven Devils'	wWhG
	maculiferum ssp. ***anwheiense***	see **R.** ***anwheiense***
	'Madame Carvalho'	wHam wWhG
un	'Madame Cochet'	wHam
	'Madame Guillemot'	see **R.** 'Monsieur Guillemot'
	'Madame Jules Porges'	wHam
	'Madame Masson'	oAls oBlo oGre wSta wHam wWhG
	maddenii ssp. ***crassum***	wRho
	'Madras'	wHam
	'Madrid'	oGre wHam
un	'Madrigal' (EA)	oGre
	'Maggie Stoeffel'	wHam
	'Magic Moments'	wWhG
	'Mahmoud'	wWhG
un	'Maka Tsaru' (EA)	oBRG
	makinoi	last listed 99/00
	'Malahat'	oGre wHam
	'Malemute'	wHam
	mallotum	wWel
	'Malta'	oGre
	'Manda Sue'	oGre wHam wWel
un	'Mandarin Lights' (DA)	oFor oGre wHam wWhG
un	'Mandarin Spiced' (DA)	oGre wWhG
un	'Maneau Ra'	oBov
	'Manitou'	oGre wWhG wWel
	'Marchioness of Lansdowne'	wHam wWhG
	'Mardi Gras'	oAls oGar oGre wHam wWel
	Margaret Dunn Group & cl.	oGre wSta wHam wWel
	'Margaret Mack'	oGre wHam
un	'Maria Derby' (EA)	wWhG
do	'Maria Elena' (EA)	oGre
	'Maricee'	oSis wHam wWhG
	'Marie Forte'	wHam wWhG
	'Marie Starks'	wHam wWhG
	mariesii	wRho
	'Marilee' (EA)	oGre
un	'Marina' (DA)	oGre wWhG
	'Marka'	wWhG
	'Markeeta's Flame'	oGre wHam
	'Markeeta's Prize'	oAls oGar oGre wHam wWhG wWel
	'Marlene Peste'	wHam
	'Marley Hedges'	oGre wHam wWhG wWel
un	'Marlis'	wHam oTal
	'Mars'	wHam
	'Martha Hitchcock' (EA)	oBlo oGre wSta
	'Martha Isaacson' (Ad)	oGar wSta
	'Martha Robbins'	oAls oGar
	'Martian King'	wWel
	'Mary Belle'	oGre wHam
	'Mary Briggs'	oGre wHam
	'Mary Drennen'	wHam
	'Mary Fleming'	wClo oBlo oGre wSta wHam wWhG wWel
	'Mary Guthlein'	wHam
	'Mary Kittel'	wWhG
un	'Mary Margaret'	wWhG
	'Mary Poppins' (DA)	oFor oAls oGre wWhG
	'Maryke'	wHam wWhG
un	'Matsuyo' (EA)	oGre oBRG
un	'Maui Sunset'	wHam
	'Maureen'	oGar oGre wHam
	'Mavis Davis'	oGre wHam wWel
	maximum	last listed 99/00
un	***maximum*** 'Roseum'	oGre wHam wWhG
	'Maxine Childers'	oGos
un	'May Belle' (EA)	wWhG
	May Day Group & cl.	wSta wHam
	May Morn Group & cl.	wWel
	'Maya'	wHam wWel
	'Mazurka' (DA)	oGre wWhG
	Medusa Group	oGre wSta wHam wWhG wWel
	megeratum	wRho wWhG
	megeratum 'Bodnant'	wRho
	mekongense var. ***mekongense*** Viridescens Group	see **R.** ***viridescens***
	'Melidioso'	wHam wWhG
	'Melody'	wWhG
un	'Melon Pink'	wHam
	metternichii	see **R.** ***degronianum*** ssp. ***heptamerum***
un	'Michael' (EA)	wSta
	'Michael Hill' (EA)	wWhG
	'Michael Waterer'	wSta
	micranthum	last listed 99/00
	microgynum	last listed 99/00
	microgynum Gymnocarpum Group	last listed 99/00
	microphyton	wWhG
	'Midnight'	wSta wHam wWel
	'Midnight Mystique'	wHam wWhG
un	'Midnight Sky'	wWhG
	'Midsummer'	oGre wHam wWhG
	'Mike Davis'	oGre wHam
un	'Mikkeli'	oGre wHam wWhG wWel
	'Milestone'	oGre wWhG
un	'Milne Bay'	oBov
	mimetes	wHam
	'Minas Maid'	oGre
	'Minas Peace'	last listed 99/00
	'Minas Snow'	oGre wWhG
un	'Minato' (EA)	oGre
un	'Mindy's Love'	wWhG
	'Ming Toy'	last listed 99/00
	'Mini Bright'	last listed 99/00
un	'Minnetonka'	oAls oGar oGre oWnC wHam wWhG
	minus var. ***chapmanii***	wWhG
	minus var. ***minus***	wRho
	minus var. ***minus*** Carolinianum Group	wHam
un	'Miss Suzie' (EA)	wSta wWhG
	'Mission Bells'	wClo wHam wWhG
	'Mist Maiden'	see **R.** ***yakushimanum*** 'Mist Maiden'
	'Misty Moonlight'	oGre wHam wWel
	'Moerheim'	wWel
un	'Molalla Red' (DA)	oGre wHam wWhG
	molle ssp. ***japonicum*** (DA)	wRho wCCr
	'Mollie Coker'	oGre wHam wWhG
	Mollis orange (DA)	oGar
	Mollis yellow (DA)	oGar
	'Molly Ann'	oAls oGar oBlo oGre oWnC wSta
	'Molly Buckley'	wHam
	'Molly Forham'	wHam
	'Molly Smith'	oGre wWhG
un	'Mona Lisa'	wWhG
	'Monique Behring'	oGre wHam
	'Monsieur Guillemot'	wHam
un	'Montreal'	wHam
	montroseanum	wRho
	'Mood Indigo'	wWhG wWel
un	'Moonbeam' (EA)	wSta
un	'Moonlight Rose' (DA)	oGre wWhG
	Moonstone Group	wClo oGre wSta wHam oTal wWhG
	'Moonwax'	oGre
	'Morgenrot'	oAls oGar wHam
	morii	wWhG
	'Morning Cloud'	oGre
un	'Morning Mist' (EA)	wWhG
	'Morning Sunshine'	oGre
	'Moser's Maroon'	oGre wHam wWhG
un	'Mossman's Freckles'	wWhG
	'Mother Greer'	wClo wWhG
	'Mother of Pearl'	wHam wWhG
	'Mother's Day' (EA)	oAls oGar oGre oTPm wSta wWhG
	'Mount Clearview'	wHam wWhG
un	'Mount Cycloop'	oBov
	'Mount Everest'	wHam wWhG wWel
un	'Mount Kaindi'	oBov
un	'Mount Ophir'	oBov
un	'Mount Pirie'	oBov
	'Mount Rainier' (DA)	oGre wHam
	'Mount Saint Helens' (DA)	oFor oGar oGre wWhG
	'Mount Seven Star' (EA)	see **R.** ***nakaharae*** 'Mount Seven Star'
	'Mountain Flare'	wHam
	'Mountain Glow'	wHam
	moupinense	wHam wWhG
	'Mrs. A. C. Kenrick'	wWhG
	'Mrs. A. T. de la Mare'	wHam wWhG
un	'Mrs. Ann Pennington' (EA)	wSta
	'Mrs. Betty Robertson'	wClo wSta wHam wWhG
	'Mrs. C. B. van Nes'	wHam
	'Mrs. Calabash'	last listed 99/00
	'Mrs. Charles E. Pearson'	oGre wSta wWhG
	'Mrs. Charles S. Sargent'	oBlo
	'Mrs. Davies Evans'	wWhG
	'Mrs. E. C. Stirling'	wSta wHam wWhG
	'Mrs. Emil Hager' (EA)	wSta
	'Mrs. Furnivall'	oGar oGre wSta wHam wWhG wWel
	'Mrs. Furnivall' x ***yakushimanum***	wWhG
	'Mrs. G. W. Leak'	oBlo oGre wSta wHam oTal wWhG wWel
	'Mrs. Helen Jackson'	wWel
	'Mrs. Horace Fogg'	oGre
	'Mrs. Howard Phipps'	oGre
	'Mrs. J. C. Williams'	wHam

	'Mrs. J. G. Millais'	wHam
un	'Mrs. Jamie Fraser'	wHam
	'Mrs. Lammot Copeland'	wHam
	Mrs. Lionel de Rothschild Group & cl.	oTal wWhG
	'Mrs. R. S. Holford'	wHam
	'Mrs T. H. Lowinsky'	oAls oSho oGre wSta wHam oTal wWhG
un	'Mrs. Updike' (EA)	wSta
un	'Mrs. Villars' (EA)	wWhG
	x mucronatum (EA)	oGre wCCr wSta wWhG
	mucronatum var. ***mucronatum***	see **R.** ***x mucronatum***
	x mucronatum 'Roseum' (EA)	wWhG
	mucronulatum (DA)	wClo oRiv wSta wHam
un	***mucronulatum*** var. ***ciliatum***	wRho wCCr
	mucronulatum 'Cornell Pink' (DA)	oFor oGos oGar oGre wHam wWhG wWel
	mucronulatum 'Crater's Edge' (DA)	oGre wHam wWhG
	mucronulatum DJH 023 (DA)	wHer
	mucronulatum 'Mahogany Red' (DA)	wHam
	mucronulatum 'Winter Brightness' (DA)	wRho
	'Multimaculatum'	wHam
	'Muncaster Mist'	wHam wWel
un	'Murasaki Fuji'	oBRG
	'My Pretty One'	last listed 99/00
	myrtifolium	wClo wHam oBRG
	'Myrtifolium'	wClo oGar oGre oBRG wWel
un	'Nachi-no-tsuki' (EA)	oGre oBRG
	nakaharae (EA)	wSta wWhG
	nakaharae 'Mount Seven Star' (EA)	oSis oGre oBRG wWhG
	nakaharae pink (EA)	wSta
	nakaharae pink prostrate (EA)	wSta
	nakaharae salmon (EA)	wSta
	nakaharae sport	oBRG
un	'Nancy' (EA)	oGre
	'Nancy Evans'	wClo oAls oGar oGre wHam wWhG wWel
	'Nancy of Robin Hill' (EA)	oWnC wSta oBRG wWhG
un	'Nancy Pink' (EA)	oGre
	'Nantucket'	wHam
	Naomi Group & cl.	wHam
	'Naomi A. M.'	wHam
	'Naomi Astarte'	wHam
	'Naomi Carissima'	wHam wWhG
	'Naomi Early Dawn'	wHam wWhG
	'Naomi Exbury'	see **R.** 'Exbury Naomi'
	'Naomi Glow'	wHam wWhG
	'Naomi Hope'	wHam
	'Naomi Molly Buckley'	see **R.** ***'Molly Buckley'***
	'Naomi Nautilus'	oGre wHam wWhG
	'Naomi Nereid'	wHam
	'Naomi Pink Beauty'	wHam
	'Naomi Pixie'	wHam wWhG
	'Naomi' seedlings	wCCr
	'Naomi Stella Maris'	oGre wHam
do	'Narcissiflorum (DA)	wWhG
un	'Narnia'	oBov
	'Naselle'	oGre wHam wWhG wWel
	'Ne Plus Ultra'	last listed 99/00
	'Neat-O'	wHam wWhG
	'Nectarine'	wHam
	'Nelda Peach'	oGre wHam wWhG wWel
un	'Nelly'	wHam
un	'Nelson's Purple'	oGar
	neriiflorum 'Rosevallon'	see **R.** 'Rosevallon'
	'Nestucca'	oGre wWhG
	'Netty Koster'	wHam
	'New Hope'	oGre wWhG
	'Newcomb's Sweetheart'	wHam wWhG
un	'Next to Hawaiian Lei' (DA)	oGre
	'Nico' (EA)	wSta wWhG
un	'Nico Red' (EA)	oAls oBlo
	'Nicoletta'	wHam wWhG
un	'Nifty Fifty' (DA)	oGre wWhG
	'Night Editor'	see **R.** ***russatum*** 'Night Editor'
	'Nightwatch'	oGre wHam
un	'Nikaino Tsuaka' (EA)	oSis wSta
un	'Nike'	oGre
	'Nimbus'	wHam
un	'Niobe' (EA)	wSta
	'Nippon'	wHam
	nipponicum DJH 485	wHer
	nitens	see **R.** ***calostrotum*** ssp. ***riparium*** Nitens Group
	nivale ssp. ***boreale*** Stictophyllum Group	wSta
	nivale ssp. ***nivale***	oGre
	niveum	wWhG
	'Noble Mountain'	oGre oTal
	Nobleanum Group	wHam
	'Nobleanum Coccineum'	wHam
	'Nobleanum Venustum'	wHam wWhG
un	'Nocturne' (EA)	oBlo oGre
	'Nodding Bells'	wHam
	'Norman Behring'	oGre wHam wWhG
	'Normandy'	oGre wHam
un	'Norph'	wWhG
	'Northern Hi-Lights' (DA)	oGre wWhG
	Northern Lights Group (DA)	oFor
un	'Northern Starburst'	oFor oGre wHam wWhG wWel
	'Nosutchianum'	oGre wHam
	'Nova Zembla'	oAls oGre oWnC wSta oTal wWhG wWel
	'Noyo Brave'	oGos oAls oGar oGre wHam wWhG
	'Noyo Chief'	oBRG wWhG
un	'Noyo Don'	wWhG
un	'Noyo Dream'	oGre wWhG
	'Noyo Maiden'	wWhG
	'Nuance'	wHam
	nudiflorum	see **R.** ***periclymenoides***
	'Nugget'	wHam
un	'Nyohozan' (EA)	oGre oBRG
	oblongifolium	see **R.** ***viscosum***
	Obtusum Group (EA)	wWhG
	obtusum f. ***amoenum***	see **R.** 'Amoenum'
	occidentale (DA)	oFor oHan wWoB oGar oRed wRho oGre wWat wHam wWhG
	occidentale 'Leonard Frisbee' (DA)	wWhG
un	***occidentale*** 'Pistil Packin Mama' (DA)	wWhG
	'Oceanlake'	oFor oAls oGar oGre oWnC wSta wHam wWhG
	ochraceum	last listed 99/00
	'Odee Wright'	wClo oAls oGre wHam wWhG
	'Odoratum' (Ad)	oBlo wSta wHam
	'Oh Canada'	wHam
	'Oh! Kitty'	oGre wHam wWel
	'Old Copper'	oGre wSta wHam wWhG wWel
	'Old Gold' (DA)	oGre wWhG
	'Old Port'	oBlo wWhG
	oldhamii (EA)	wWhG
	'Olga'	oBlo wHam
	'Olga Mezitt'	wClo oAls oGar oGre oWnC wHam oBRG wWhG wWel
un	'Olga Niblett'	wWhG
	'Olin O. Dobbs'	oGre wWhG wWel
	'Olive'	oGre wSta wHam wWhG
	Olympic Lady Group	oGre wSta
un	'Olympic Snowball' (DA)	oGre
	'One Thousand Butterflies'	oAls oGre wHam wWhG
	'Ooh Gina'	wHam wWhG
	'Opal Fawcett'	wWhG
un	***openshawianum***	wWhG
un	'Orange' (EA)	wSta
	'Orange Marmalade'	oGre wHam
un	'Orange Splendour' (DA)	wWhG
	'Orangeade' (DA)	wWhG
	orbiculare	wHam wWhG
un	***orbiculatum***	oBov
	'Orchid Lights' (DA)	oFor oGar oGre wWhG
	oreodoxa var. ***fargesii***	wWhG
	oreodoxa var. ***fargesii*** Erubescens Group	oGos wWhG
	oreotrephes	wClo oRiv wRho wSta
	oreotrephes blue leaf	wWhG
	oreotrephes green leaf	wWhG
un	'Orient' (DA)	oGre wWhG
un	'Ormsby' (EA)	wSta
un	'Osakazuki' (EA)	see **R.** ***indicum***
	'Ostbo's Elizabeth'	see **R.** 'Elizabeth Red Foliage'
	'Ostbo's Low Yellow'	last listed 99/00
un	'Ostbo's Red'	oAls oWnC
un	'Otome' (EA)	oAls oGar oGre
un	'Otome-no-mai' (EA)	oBRG
	'Oudijk's Sensation'	oGar oGre wHam
	'Ovation'	wWhG
	ovatum (A)	wRho wWhG
	'Oxydol' (DA)	oGre wWhG
	pachypodum	wRho
	pachysanthum	oGos wRho oGre wWhG wWel
	pachysanthum Britt Smitt form	oGre
	'Pacific Gold'	oGre
	'Pacific Queen'	last listed 99/00
	'Pacific Sunset'	wHam wWhG wWel
	'Painted Lady'	oGre wWel
un	'Pamela Malland' (EA)	oGre wWhG
	'Panda' (EA)	oGos oGre oBRG
	'Papaya Punch'	oGre wHam wWhG wWel

	'Paprika Spiced'	oGre wHam wWhG wWel
	'Parade' (DA)	oFor oGre wHam
un	'Park Pink Triflorum'	wSta
	parmulatum 'Ocelot'	last listed 99/00
	'Parson's Gloriosum'	oWnC wSta wHam wWhG
	'Party Package'	wHam
	'Party Pink'	oGre wHam
un	'Partyglanz'	wWhG
	'Passionate Purple'	oGre wHam
	'Patricia'	wSta
	'Patty Bee'	oAls oGar oGre wHam wWhG wWel
	pauciflorum (V)	oBov
	'Paul Lincke'	oGre
	'Paul Molinari'	oGos
	'Paul R. Bosley'	oGre wHam wWhG
	'Pauline Bralit'	wHam wWhG
	'Pawhuska'	oGre wHam wWel
	'Peach Nugget'	wHam
	'Peach Satin'	wHam
	'Pearce's American Beauty'	wClo oGre wHam wWhG wWel
un	'Pearce's Apricot'	wHam
	'Pearce's Treasure'	wHam
un	'Pearl Bradford' (EA)	oAls oGar oBlo oGre oTPm wSta oBRG wWhG
un	'Pearl Bradford Sport' (EA)	oGre wSta
	'Peekaboo'	oAls oGar wWel
	'Peeping Tom'	oGre wHam wWhG wWel
un	'Peggy Ann' (A)	oAls
un	'Peggy Roberts'	wWhG
	'Peggy Zabel'	wHam
	'Peking'	last listed 99/00
	pemakoense	wSta wHam oBRG wWhG
	'Pematit Cambridge'	oGre
	pendulum	wWhG
	Penjerrick Group & cl.	wHam
un	'Pennsylvania' (DA)	wHam
	'Pepperpot'	wHam
un	'Peptalk'	wHam
	'Percy Wiseman'	oAls oGar oSho wWhG wWel
	'Perfectly Pink'	oGre
	'Perfume'	oGre wWhG
	periclymenoides (DA)	oFor oRiv wCCr
	'Persia'	last listed 99/00
	'Persil' (DA)	oGre
	'Peste's Fire Light'	oGre wWel
	'Peter Alan'	oGre oTal wWhG
	'Peter Behring'	wHam
	'Peter Faulk'	oGre wHam wWhG
	'Peter Koster'	wHam
	'Peter Tigerstedt'	oGre wWhG wWel
un	'Petersfehn'	oGre wWhG
un	'Petra'	oBov
un	***phaeochitum***	oBov
	phaeochrysum	wRho
	phaeochrysum var. ***levistratum***	wRho
	'Phalarope'	last listed 99/00
	'Phipps' Yellow'	wHam wWhG
	'Phyllis Ballard'	wHam
	'Phyllis Korn'	oGar oGre wHam oTal
	piercei	wWhG
	'Pierce's Apricot'	oGre
un	'Pietzold Pink' (EA)	wSta
	'Pikeland'	wHam
	Pilgrim Group & cl.	wWhG
	'Pillow Party'	wHam
un	'Pineapple Delight'	oGre wWhG
	'Pink and Sweet' (DA)	oFor oGre wHam wWhG
	'Pink Cameo'	oGre
	'Pink Cascade' (EA)	oFor wSta
	'Pink Cherub'	wHam
	'Pink Cloud'	wSta
un	'Pink Dawn' (EA)	oGre
un	'Pink Delight' (DA)	oGre wWhG
	'Pink Delight' x ***jasminiflorum***	oBov
	'Pink Drift'	oBlo oGre wHam
	'Pink Floss'	wHam
	'Pink Ice'	see **R.** 'Pink Sherbet'
	'Pink Jeans'	oGre wHam
	'Pink Pancake' (EA)	oFor oSis wSta wWhG
	'Pink Parasol'	oGre wWhG
	'Pink Pearl'	oGre wSta wHam wWhG
	'Pink Petticoats'	wHam
un	'Pink Plush' (DA)	oGre
un	'Pink Profusion' (EA)	wSta
un	'Pink Rosebud' (EA)	oSqu wWhG
	'Pink Ruffles' (DA)	oGre
	'Pink Sherbet'	oAls oGar oGre wHam wWhG wWel
	'Pink Snowflakes'	oGos oGar oGre wSta wHam wWel
un	'Pink Tenino' (EA)	oAls oGar wSta
	'Pink Tufett'	wWhG
	'Pink Twins'	oGre
un	'Pink Unique'	oAls
	'Pink Walloper'	see **R.** 'Lem's Monarch'
un	'Pink William' (DA)	wHam
un	'Pinky Pierce' (EA)	wSta
	'Pinnacle'	wHam
	'Pioneer'	oBlo
	'Pioneer Silvery Pink'	oGre
	'Pirouette'	wWhG
	PJM Group	oAls oGar oBlo oGre oWnC wSta wHam oBRG wWhG wWel
	'PJM Black Satin'	see **R.** 'Black Satin'
	'PJM Compact'	oAls wWhG wWel
	'PJM Elite'	oGar wHam
	'PJM - K'	wWel
	'PJM Regal'	oAls wHam
	'PJM Victor'	wSta
	x planecostatum (V)	oBov
	'Platinum Pearl'	oGre wWhG
un	'Pleasant Companion'	oBov
	'Pleasant Dream'	wWhG
qu	'Pleasant Girard White' (EA)	wSta
un	'Pleasant White' (EA)	oGre wWhG
	'Plum Beautiful'	oGre
un	'Plum High'	wWhG
	pocophorum var. ***hemidartum***	wRho
	pocophorum var. ***pocophorum***	wRho
un	'Pohjolas Daughter'	oGre wWhG
	'Point Defiance'	oAls oGar oGre wHam wWhG wWel
	'Point Fosdick'	wHam
	Polar Group & cl.	oGre wHam wWhG wWel
	'Polaris'	wClo wHam wWhG
un	***polyanthemum***	wRho
	polylepis	wRho
	'Polynesian Sunset'	wWhG
un	'Polypetalum' (EA)	oGre oTPm oBRG wWhG
	'Pom Pom'	oGre
	'Pontica' (DA)	see **R.** luteum
	ponticum (see Nomenclature Notes)	oBlo oGre wHam
	ponticum (A) 'Cheiranthifolium'	last listed 99/00
va	***ponticum*** (A) 'Silver Edge'	oGar wHam
un	***ponticum*** 'Silver Sword'	wHam
va	***ponticum*** (A) 'Variegatum'	wClo oAls oGar oSho oGre wHam oBRG wWhG wWel
	'Pontiyak'	wHam wWhG
	'Popeye'	wHam
un	'Popsicle' (DA)	oFor oGre wWhG
	'Porzellan'	oGre wHam wWhG
	'Potlatch'	oGre
	poukhanense	see **R.** ***yedoense*** var. ***poukhanense***
	'Powder Mill Run'	wWhG
	'Powder Puff'	wHam wWhG
un	'Powder Snow'	wWhG
	'Praecox'	wSta wHam wWhG
un	***praetervisum***	wRho oBov
	praevernum	wHam
	'Prairie Fire'	wWhG
un	'Prentice's Double Red'	wWhG
	'President Lincoln'	wSta
va	'President Roosevelt'	oAls oGar oGre oWnC wSta oTal wWhG wWel
	'Pretty Woman'	oGre wHam wWhG
	'Pridenjoy'	wWhG
	primuliflorum	last listed 99/00
	'Prince Camille de Rohan'	wHam
un	'Prince Charles de Raku'	wSta
	'Princess Anne'	oGre wHam oBRG wWel
un	'Princess Royal' (DA)	wWhG
	principis	oGre
	principis Vellereum Group	wWhG
	prinophyllum (DA)	oGre wWhG
un	'Prominent' (DA)	oGre
	'Promise of Spring'	wHam
	pronum	oGre wHam wWhG
	'Prostigiatum'	wWel
qu	'Prostrate Oly'	wSta
	prostrate pink (EA)	wSta
	prostrate white (EA)	wSta
	prostratum	see **R.** ***saluense*** ssp. ***chameunum*** Prostratum Group

	proteoides	wRho oGre wHam
	proteoides Caperei form	oGre
	proteoides Cecil Smith form	oGre
	proteoides large leaf form	wWhG
	proteoides small leaf form	wWhG
	protistum	last listed 99/00
	prunifolium (EA)	wHam
	pseudochrysanthum	wClo wRho oGre wHam
	pseudochrysanthum Ben Nelson's form	oGre oBRG
	pseudochrysanthum dwarf form	wWhG
	pseudochrysanthum Exbury form	oGos wWhG
	pseudochrysanthum 'Komo Kulshan'	oGre
	Psyche Group	see **R.** Wega Group
un	'PT 106'	wSta
	'Ptarmigan'	oGre wSta oBRG
	pubicostatum	wRho
	'Pucella' (DA)	wHam
	'Puget Sound'	oGre wHam wWhG wWel
	pumilum	wRho
	puralbum	see **R. *wardii*** var. ***puralbum***
	'Purple Gem	oFor oAls oGar oSis oGre oWnC wHam wWhG wWel
	'Purple Lace'	oAls oGar oSho oGre wSta wHam wWhG wWel
un	'Purple Majestic' (EA)	oGre
	'Purple Splendor' (EA)	oAls oGar oSho oBlo oGre oWnC wSta wWhG
	'Purple Splendor Compacta' (EA)	oAls oGar
	'Purple Splendour'	oAls oGar oGre wSta wHam wWhG wWel
	'Purpureum Elegans'	wHam oTal wWhG
	'Purpureum Grandiflorum'	wWhG
	'Quaker Girl'	wHam
un	'Quakeress'	wWhG
	Quaver Group	last listed 99/00
	'Queen Alice'	oGre wWhG wWel
	'Queen Anne's'	oGre wHam wWhG
	'Queen Anne's' x 'Plum Beautiful'	oGre
	'Queen Elizabeth II'	wHam
	'Queen Mary'	wWhG
	'Queen Nefertiti'	wHam
	Queen of Hearts Group & cl.	wHam
	'Queen of Sheba'	wHam
un	'Queensland'	oBov
	quinquefolium (DA)	wRho
	racemosum	oGos wClo oRiv wRho oGre wSta wHam wWhG
	racemosum x ***moupinense***	wSta
	racemosum 'Rock Rose'	wWel
	radicans	see **R. *calostrotum*** ssp. ***keleticum*** Radicans Group
	'Radium'	wSta
un	'Rain Fire' (EA)	oGre
	'Rainbow'	oGre
	'Ramapo'	oFor oAls oGar oBlo oGre oWnC wSta wHam oBRG wWhG wWel
	ramsdenianum	last listed 99/00
	'Rangoon'	oGre wWhG
	rarum (V)	last listed 99/00
do	'Raspberry Delight' (DA)	wHam wWhG
un	'Ray's Pink'	wHam
un	'Ray's Rose'	wHam
	'Razorbill'	wHam
	'Razzle Dazzle'	wHam
un	'Rebel'	wSta
	recurvoides	wRho oGre wHam
un	***recurvoides*** 'Branklyn'	wWhG
	recurvoides Exbury form	wWhG
un	***recurvoides*** 'Fisher's Dwarf'	oGre
	recurvoides narrow leaf form	oGre
un	***recurvoides*** 'Recurvum'	wWhG
	'Red Brave'	wHam wWhG
	'Red Cloud'	wHam
	'Red Delicious'	oGre wHam wWel
	'Red Eye'	oGre wHam wWel
un	'Red Feather' (EA)	wWhG
	'Red Fountain' (EA)	oGre oTPm wSta wWhG
un	'Red Gold'	oGre
un	'Red Letter' (DA)	oGre
	'Red Majesty'	oGre oBRG
	'Red Olympia'	oGre
	'Red Paint'	wHam
	'Red Petticoats'	wHam
un	'Red Prince'	oBov
	'Red Red' (EA)	oGre wWhG
	'Red River'	wHam wWhG
un	'Red Sport'	oGre
un	'Red Star'	wSta
	'Red Sunset' (DA)	oGre wHam wWhG
	'Red Walloper'	wClo wHam
	'Red Wood'	wClo
	'Redder Yet'	wHam
	'Redmond' (EA)	wSta
	'Redwing' (EA)	oAls oGar oBlo oTPm wSta
un	'Refrain'	wWhG
un	'Rejoice'	wHam
	'Relaxation'	wWel
	'Rendezvous'	oGre wHam wWhG
un	'Renee' (DA)	wHam
un	'Renee Michele' (EA)	wWhG
un	'Renne' (DA)	oGre oWnC
	'Renoir'	wHam wWhG
	Repose Group & cl.	last listed 99/00
	reticulatum (DA)	wWhG
	sp. aff. ***reticulatum*** HC 970602	wHer
	retusum (V)	oBov
un	'Reverend Paul'	wWhG
	rex	wHam wWhG
	rex ssp. ***fictolacteum***	wRho wWhG
	rex ssp. ***fictolacteum*** var. ***miniforme***	wRho
un	'Rhein's Luna'	wWhG
	'Rhonda Stiteler'	last listed 99/00
	'Ria Hardijzer'	oAls oGar oGre oWnC wSta
	'Rigging Slinger'	wHam
	rigidum	wHam wWel
	rigidum album	wWhG
un	'Rimfire'	wWel
	'Ring of Fire'	wClo oAls oGre wHam wWhG
un	'Rinpu' (EA)	oGar oSis oGre oBRG
	Riplet Group	wHam
	ririei	last listed 99/00
un	'Rising Star' (DA)	oGre wWhG
un	'Rising Sun'	oGre wWhG
	'Robert Allison'	wWhG
	'Robert Korn'	wHam
	'Robert Louis Stevenson'	wHam
un	'Roberta' (EA)	wWhG
	'Robin Hill Gillie' (EA)	wSta
	'Robinette'	wHam
	'Rocket'	oAls oSho oGre oWnC wHam wWhG
	'Rockhill Ivory Ruffles'	oGre wHam wWhG
	'Rockhill Parkay'	oGre wHam wWhG
	'Rockhill Sunday Sunrise'	oGre
un	'Rocky Point'	wWhG
	'Rocky White'	wHam
	'Rodeo'	wWhG
un	'Roehr's Peggy Ann' (EA)	oGar oGre oBRG wWhG
	'Roma Sun'	wHam
	'Roman Holiday'	wHam
	'Roman Pottery'	wHam
	'Romany Chal'	wWhG
	'Rosa Mundi'	oGar oGre oWnC wSta wHam wWhG wWel
	'Rosa Regen'	wWel
	'Rosalie Hall'	wHam
	'Rosata' (DA)	last listed 99/00
	'Rose Elf'	oAls oWnC wSta oTal
	'Rose Greeley' (EA)	wSta wWhG
	'Rose Lancaster'	wHam
	'Rose Ruffles' (DA)	wWhG
do	'Rosebud' (EA)	oAls oGar oOut oBlo oGre oTPm oWnC wSta
	'Rosemary Chipp'	last listed 99/00
un	'Rosepoint'	oGre
	roseum	see **R. *prinophyllum***
	'Roseum Elegans'	oSho oGre oWnC wHam oTal wWhG
	'Roseum Pink'	see **R.** 'English Roseum'
	'Roseum Superbum'	wWhG
	'Rosevallon'	wWhG
un	'Rosita'	wHam
	'Ross Maud'	wHam wWhG
	'Rosy Bell'	wSta
	'Rosy Dream'	oGre
	'Rosy Lights' (DA)	oFor oGre wHam wWhG
	'Rosy O'Grady'	see **R.** 'Whitney's Rosy O'Grady'
	'Rothenburg'	wHam
	roxieanum	oGre wHam
	roxieanum var. ***oreonastes***	oGre wWhG wWel
	roxieanum var. ***oreonastes*** rock 59589	oGre
	'Royal Command' (DA)	wWhG
	'Royal Decree'	wHam

	'Royal Lodge' (DA)	oGre wWhG
	'Royal Pink'	oAls oGar wWel
	'Royal Purple'	oBlo oGre wSta wHam
un	'Royal Robe' (EA)	oAls oGar wSta
	'Royal Ruby' (DA)	oWnC
un	'Royston Orange'	wHam
un	'Royston Peach'	wHam
un	'Royston Reverie'	wHam
	'Rubicon'	oGre wWhG
	rubiginosum	wHer wCCr
	rubiginosum Desquamatum Group	oGre wCCr wHam
	rubiginosum Desquamatum Group Whitney form	wWhG
un	***rubineiflorum***	oBov
	'Ruby F. Bowman'	oGre wHam wWel
	'Ruby Hart'	wSta wHam wWhG wWel
	'Rudy-Leona'	wHam
un	'Ruffles'	wWhG
	'Ruffles and Frills'	wHam
	rugosum (V)	last listed 99/00
	rupicola var. ***chryseum*** DJHC 276	wHer
	russatum	wSta wHam
un	***russatum*** 'Borchers'	oGre
	russatum 'Night Editor'	oGos wClo oAls oGre wHam oBRG wWhG wWel
	Russautinii Group	wHam oTal
	'Russell Harmon'	wHam
	'Ruth Davis'	wHam
	'Ruth Lyons'	see **R. *davidsonianum*** 'Ruth Lyons'
	'Ruth Mottley'	oGre wHam wWel
	'Sagamore Bayside'	wHam
un	'Sagittarius' (EA)	oGre
	'Saint Merryn'	oSis wWhG wWel
	'Saint Valentine'	oBov
un	'Sakuragata' (EA)	oGre
un	'Salmon Delight' (DA)	oGre wWhG
un	'Salmon Queen' (DA)	wWhG
	saluenense	wHam wWhG
	saluenense ssp. ***chameunum***	wRho
	saluenense ssp. ***chameunum*** Prostratum Group	wSta
ch	***saluenense*** ssp. ***riparioides*** Rock's form	see **R. *calostrotum*** ssp. ***riparium*** Rock's form R 178
	saluenense ssp. ***saluenense***	wRho
	sp. aff. ***saluenense*** ssp. ***saluenense***	wRho
	'Sammetglut'	wHam
un	'San Gabriel'	oBov
un	'Sandra Ann' (EA)	wWhG
	'Sandwich Appleblossom'	wHam wWhG
	'Sandy Petuso'	wWhG
	sanguineum	wHam
un	'Sanko No Tsuki' (EA)	oAls oGar oGre
un	'Santa Fe'	wWhG
	santapaui (V)	last listed 99/00
	'Sapphire'	wClo oGar oGre oWnC wSta wHam wWhG
	'Sappho'	oBlo oGre wSta wHam oTal wWhG wWel
	'Sapporo'	wHam wWhG
	sargentianum	wWhG
	Sarita Loder Group & cl.	oGre
	sataense	last listed 99/00
	'Satan's Fury'	wHam
	'Saturnus' (DA)	last listed 99/00
	'Sausalito'	last listed 99/00
	scabrifolium	wWhG
	scabrifolium var. ***spiciferum***	wWhG
	'Scandinavia'	wHam
	'Scarlet Pimpernel' (DA)	oGre
	'Scarlet Romance'	oGre wWhG
	'Scarlet Wonder'	oBlo oGre wSta wHam wWhG
un	'Schemson' (EA)	wSta
	schlippenbachii (DA)	oFor oGar oRiv wRho oGre wSta wHam wWhG wWel
	schlippenbachii DJH 042	wHer
	schlippenbachii 'Sid's Royal Pink' (DA)	oFor oGos
un	'Schneebukett'	oGre wHam
	'Schneekrone'	oGre wHam wWhG
	Schneespiegel®	wWhG
	'Schubert'	wHam
	'Scintillation'	oAls oGar oBlo oGre wSta wHam oTal wWhG wWel
	'Scintillation' Gem Garden	wHam
	'Scintillation' University of Rhode Island	wHam
	'Sea-Tac'	oGre
un	'Seabrook Glory'	wWhG
	searsiae	last listed 99/00
	'Seashell'	wHam
	'Seattle Gold'	wHam
un	'Seattle Pink'	wHam
	'Second Honeymoon'	wHam
	'Sefton'	wHam wWel
un	'Sefton Affinity'	wHam
	'Seikai' (EA)	oBRG
	semibarbatum	last listed 99/00
	'Senator Henry Jackson'	oGre wWhG wWel
	'Senora Meldon'	oGre wWhG wWel
	'Senorita Chere'	oAls oGar oGre wHam wWel
	'September Song'	oGar oGre wHam wWhG wWel
	'Serenata'	wHam
	'Serendipity'	wWhG
	serotinum	wHam
	serpyllifolium (DA)	oSis wSta wWhG
	serpyllifolium var. ***albiflorum*** (DA)	oGre
	serrulatum	see **R. *viscosum***
	Seta Group & cl.	oGre wSta oTal wWhG
un	'Settyu-no-matsu' (EA)	wWhG
	'Seven Stars'	wHam
	'Shaazam'	oGre
	'Shamrock'	oAls oGar oGre wSta oTal wWel
	'Sham's Juliet'	wHam
	'Sham's Pink'	wHam
un	'Shen-Nu-Tushki' (EA)	wSta
	sherriffii	last listed 99/00
un	'Sherwood Cerise' (EA)	wSta
	'Sherwood Orchid' (EA)	oAls oGar oBlo oTPm oWnC wSta wWhG
	'Sherwood Red' (EA)	oWnC wSta wWhG
un	'Shinkyo' (A)	oGre
	'Shinnyo-no-tsuki' (EA)	oSis oWnC
un	'Shira-fuji' (EA)	wWhG
	'Shirley'	wWhG
un	'Shiryu-no-homare' (EA)	oGre oBRG
un	'Shiryu-no-tsuki'	oBRG
un	'Shogun'	oGre
	'Shooting Star'	last listed 99/00
	'Show Boat'	oGre wHam wWhG wWel
	'Show-Off'	wHam
	'Shrimp Girl'	oGre wHam
	sichotense	see **R. *dauricum***
	'Sierra Sunrise'	wWhG
	'Siesta'	oGre
	sikangense var. ***exquisitum***	wRho
un	'Silberreif'	oGre wWhG
	'Silberwolke'	wWhG
un	'Silk Ribbon'	wWhG
un	'Silver Bear'	wWhG
ch	'Silver Edge'	see **R. *ponticum*** 'Silver Edge'
	'Silver Sixpence'	wHam wWel
	'Silver Skies'	wHam wWhG
	'Silver Slipper' (DA)	wHam
	'Silver Streak' (EA)	oAls oGar
un	'Silver Sword' (EA)	oGar oGre oBRG wWhG
un	'Silver Thimbles'	oBov
un	'Silvery Pink'	wClo
	simsii (EA)	oGar
	simsii var. ***eriocarpum*** (EA)	wWhG
	sinofalconeri	last listed 99/00
	sinogrande	wHam
	'Sir Charles Lemon'	wHam wWhG
	Sir Frederick Moore Group & cl.	wHam
un	'Sir Robert' (EA)	wSta
	'Sir Robert Peel'	wWhG
un	'Sirunki Lake'	oBov
	'Skipper'	wHam
	'Skookum'	oGre wHam wWel
un	'Skookum & Skookumchuck'	wWhG
	'Skookumchuck'	oGre wHam
	'Skyglow'	wWhG
	'Sleepy'	wHam wWhG
	'Small Gem'	oGre
	'Small Wonder'	wHam
	smirnowii	oGre wHam wWhG
	smirnowii x ***yakushimanum***	wWel
	smithii	see **R. *argipeplum***
	'Smokey # 9'	oGre wWel
	'Sneezy'	oGre wHam wWhG
	'Snipe'	oGar wHam wWel
	'Snow' (EA)	wSta
	'Snow Bird'	see **R.** 'White Bird'
un	'Snow Candle'	wWhG
	'Snow Crest'	last listed 99/00
un	'Snow Dwarf' (EA)	oGar

	'Snow Lady'	oAls oGar oBlo oGre oWnC wSta wWel
	Snow Queen Group & cl.	wWhG
	'Snow Sprite'	wHam
	'Snow White'	wHam
un	'Snowbird' (DA)	oGos oGar oGre wWhG wWel
un	'Snowcap' (EA)	oGar
	'Snowdrop'	wHam
	'Snowstorm'	wHam
un	'Soft Shimmer' (DA)	oGre
	'Soir de Paris' (DA)	wCul
	'Solidarity'	oAls oSho oGre oWnC wHam wWhG
un	***solitarium***	oBov
	'Sonata'	oGre wHam
	'Sonatine'	wHam wWhG
	'Songbird'	oGre wHam oBRG wWhG
	'Sonic Ray'	wHam
un	'Sonny's Love'	wWhG
	souliei	oFor wWhG
	'Souvenir of W. C. Slocock'	wSta wHam
	sp.	wSta
	sp. subsection ***argyrophylla***	wRho
	sp. DJH 485	wHer
	'Spanish Lady'	last listed 99/00
	'Sparkling Burgundy'	oGre wHam
	'Spellbinder'	wHam
	sphaeroblastum var. ***wumengense***	wRho
	spiciferum	see **R. *scabrifolium*** var. ***spiciferum***
	'Spicy Lights' (DA)	oFor
	spinuliferum	oRiv
	'Spitfire'	wHam
un	'Spring Dancer'	wWhG
	'Spring Dawn'	wClo wHam
	'Spring Frolic'	wHam
	'Spring Glory'	oAls
	'Spring Parade'	wHam wWhG
un	'Spring Salvo' (DA)	wWhG
un	'Spring Spirit'	wWhG
	'Spring Sun'	oGre
	'Springbok'	oBRG
	'Springfield'	oGre wHam
	'Springtime'	see **R.** 'Williams'
	'Spun Gold'	oGre wHam
	stamineum	wRho
	'Starburst'	oGre
	'Starcross'	wHam
	'Stardust'	wHam
un	'Starlight' (EA)	wWhG
	'Starry Night'	see **R.** 'Gletschernacht'
	'Startrek'	wHam
	stenopetalum 'Linearifolium ' (A)	oAls oGar wSta oBRG wWhG
	stenophyllum	see **R. *makinoi***
	'Stephanie'	wHam
	'Stephen Clarke'	wHam
	'Stewartstonian' (EA)	oFor oGar oSho oBlo oGre oTPm oWnC wSta oBRG wWhG
	stictophyllum	see **R. *nivale*** ssp. ***boreale*** Stictophyllum Group
	'Strawberry Ice' (DA)	oFor oGre oWnC wHam wWhG
	strigillosum	wRho wWhG
	strigillosum DJHC 806	wHer
un	***suaveolens***	oBov
	'Sue'	last listed 99/00
	'Sugar and Spice'	oAls wHam wWhG
	'Sugar Pink'	oGre wWhG
	'Sumatra'	oWnC wHam
	'Summer Cloud'	oGre
	'Summer Glow'	oGre wHam wWhG
un	'Summer Peach'	wWhG
	'Summer Rose'	wHam
	'Summer Snow'	wHam
	'Summer Summit'	wHam
	'Sun Chariot' (DA)	oGre oWnC
un	'Sun Star' (EA)	wSta wWhG
un	'Sundt Nectarine' (DA)	wHam
un	'Sunlight' (DA)	wWhG
	'Sunny Day'	oGre wWhG wWel
un	'Sunny's Brother'	oBov
un	'Sunset Bay'	wWhG
	'Sunset Pink' (DA)	oGar oGre
un	'Sunset Serenade'	wWhG
	'Sunsplash'	wHam
	'Sunspot'	oGre wHam
	'Sunspray'	oGre wHam wWel
un	'Sunstone'	wWhG
	'Sunstruck'	oGre wWhG

	'Sunup Sundown'	wHam
un	'Super Man'	wWhG
	superbum (V)	wRho oBov
	'Supergold'	wHam
	'Surrey Heath'	wHam
	'Susan'	oGre wHam wWhG wWel
	'Susannah Hill' (EA)	oTPm
	sutchuenense	oGre wHam wWhG
	sutchuenense var. ***geraldii***	see **R. *x geraldii***
	'Swamp Beauty'	oGre wHam
	'Swansdown'	oGre wWel
	'Sweet Mystery'	wWhG
	'Sweet Simplicity'	wSta wHam
	'Sweet Sixteen'	wHam
	'Sweet Sue'	oGre wHam wWel
	'Sweet Wendy'	oBov
un	'Sweetbriar' (A)	oWnC
	'Sweetie Pie'	wWhG
	'Swen'	wHam wWhG
un	'Sybil'	oBov
	'Sylphides' (DA)	oGar wWhG
un	'T. H. Williams'	oBlo
	'Tahitian Dawn'	oGre wHam wWhG wWel
	'Taku'	oGos oAls oGar oBRG
	taliense	last listed 99/00
	'Tally Ho'	wWhG
un	'Tamarindos'	oGre wHam wWhG wWel
	'Tanana'	wWhG
un	'Tangelo' (DA)	wWhG
	'Tanyosho'	oGre wHam wWhG wWel
	'Tapestry'	wHam
	tatsienense	last listed 99/00
	'Taurus'	wClo oAls oGar oGre wHam oTal wWhG wWel
un	'Taylori'	oBov
	'Teddy Bear'	oGos wClo wHam oBRG wWhG wWel
	'Ted's Orchid Sunset'	oGre wSta wWhG
un	'Telessa'	wWel
	telmateium	wRho
	temenium var. ***dealbatum*** Glaphyrum Group	oGre
	'Temple Belle'	wSta wHam oTal wWhG
	'Temptation'	wHam
un	'Tenino' (EA)	oBlo oGre wSta
	'Tennessee'	wHam
	'Tequila Sunrise'	wHam
un	'Terebinthia'	oBov
	Tessa Group & cl.	oBlo wHam
	'Tessa Bianca'	wHam wWhG wWel
	thayerianum	wClo oRiv wWhG
	'The Chief'	wHam
	'The Dream'	oGre
	'The General'	oGre wHam
	'The Hon. Jean Marie de Montague'	oAls oGar oSho oBlo oGre wSta wHam oTal wWhG wWel
un	'The Hon. Jean Marie de Montague' variegated	wWhG
	'The Rebel'	oGre
	thomsonii	wRho wHam wWhG
	thomsonii HWJCM 193	wHer
	thomsonii ssp. ***thomsonii***	last listed 99/00
	thomsonii white form	wHam
	Thomwilliams Group	wHam
	Thor Group & cl.	oGre wSta oBRG wWhG
	'Thousand Butterflies'	see **R.** 'One Thousand Butterflies'
	'Tiana'	oAls oGre wHam wWhG
	'Tickled Pink'	last listed 99/00
	'Tiddlywinks'	oGre wWel
	'Tiffany'	oGre wCCr
un	'Tina' (EA)	oGar oBlo oGre wSta
	'Titian Beauty'	oGre wHam
un	'Tochi-no-hikari' (EA)	oBRG
	'Today and Tomorrow'	wHam wWhG
	'Todmorden'	wWel
	'Tofino'	wHam wWhG
	'Tokatee'	wHam
	'Tom Everett'	wHam
	'Tom Williams'	wSta
un	***tomentosum*** 'Milky Way'	wRho
un	***tomentosum*** ssp. ***subarcticum***	wRho
un	'Tomo-O'	wWhG
un	'Tondelaya'	oGre
	'Tony'	wHam
	'Too Bee'	oGar wWhG
	'Top Banana'	oGre wHam wWhG wWel

	'Top Brass'	oGre
	'Top Hat'	wHam
	'Topsvoort Pearl'	oGre
	'Torero'	wHam wWhG
	'Tortoiseshell'	wHam
	'Tortoiseshell Wonder'	wHam oTal wWhG wWel
un	***tosaense*** 'Barbara' (EA)	wRho
	'Totally Awesome' (DA)	oFor oGre wWhG
	'Tow Head'	oGre wHam
	'Tracigo'	wHam
un	"Tradition' (A)	oTPm
	'Trail Blazer'	oGre wHam wWhG wWel
	traillianum	wRho
	traillianum var. ***traillianum***	see **R. *traillianum***
un	'Treasure' (EA)	oAls oGar oBlo oGre oTPm oWnC wSta
	'Tressa McMurry'	wHam
	trichanthum	wClo wRho
	triflorum var. ***triflorum***	wRho
	'Trilby'	oAls oGar oBlo oGre oWnC wSta wHam oTal wWhG
	'Trinidad'	oGre wHam
un	'Trooper' (EA)	wSta
un	'Troubador' (DA)	wWhG
	'Trude Webster'	oAls oGar oGre wHam wWhG wWel
un	'Trude Webster' x 'E. C. Stirling'	oGre
	'True Treasure'	wHam
un	'Trumphans'	oBov
	sp. aff. ***tsaii***	wRho
	tsariense	wWhG
	tschonoskii	oSis
	tsusiophyllum	last listed 99/00
un	***tuba***	oBov
	'Turkish Delight'	wHam
	'Tuscany'	oGre
	'Tweety Bird'	oGre wWhG
un	Twenty Grand Series	wHam
	'Twenty Grand' (EA)	oGar oSho oBlo oTPm
	'Twilight Pink'	oGre
un	'Twinkie' (DA)	wHam
	'Twinkles'	oTal
un	'Twisted Leaf' (DA)	oGre
va	'Uki Funei' (EA)	oGre
	'Umpqua Queen' (DA)	wWhG
	ungernii	wCCr wHam
	uniflorum	wRho
	'Unique'	oAls oGar oSho oBlo oGre oWnC wSta wHam oTal wWhG wWel
	'Unique Marmalade'	oGre wHam wWhG wWel
	'Unknown Warrior'	wSta wHam
	unnamed seedlings	wWel
un	'Unsurpassable' (EA)	oGre wWhG
	uvariifolium	wWhG
	uvariifolium var. ***uvariifolium***	last listed 99/00
	'Valley Forge'	wHam
	'Valley Sunrise'	oGre
	'Van'	oAls oGre wHam
	'Van Nes Sensation'	oGre wSta wHam wWhG wWel
	'Van Weerden Poelman'	wHam
	'Vancouver USA'	wWhG
	'Vanessa Pastel'	wHam
	vaseyi (A)	oFor
un	***vaseyi*** 'Alba'	oGre
	vellereum	see **R. *principis*** Vellereum Group
	venator	oGre
	'Venice'	wWhG
	vernicosum	wHam wWhG
	vernicosum var. ***rhantum***	last listed 99/00
	'Vernus'	oGre wWhG
	'Veronica Pfeiffer'	wHam
	'Very Berry'	oGar oGre wHam wWhG wWel
un	'Very Unique'	wWhG
	'Vibrant Violet'	oGre wHam wWhG wWel
	'Victor Frederick'	last listed 99/00
	'Viennese Waltz'	wHam wWhG
	villosum	see **R. *trichanthum***
	'Vincent van Gogh'	oGre wHam wWhG
	'Vinebelle'	wHam
un	'Vinecourt Dream' (DA)	wHam
do	'Vinecourt Duke' (DA)	wHam
	'Vinecrest'	wHam
	'Vinemount'	wHam
	'Vinewood'	wHam
	'Violetta' (EA)	wWhG
	Virginia Richards Group & cl.	oAls oGar oSho oBlo oGre oWnC wSta oTal wWhG
	'Virgo'	wHam
	viridescens	wRho
	viscosum (DA)	oFor oGar wRho oGre wWhG
un	***viscosum serrulatum***	wRho
	'Viscy'	oGre wHam wWhG wWel
	'Vivacious'	oGre wWhG
un	'Vladimir Bukowsky'	oBov
	'Voodoo'	wHam
	'Vulcan'	oFor oAls oGar oBlo oGre oWnC wSta wHam
	'Vulcan's Bells'	oBlo wHam
	'Vulcan's Flame'	oGar oGre wSta wHam wWhG wWel
un	'Vuyk's Joanna' (A)	oWnC
	'Vuyk's Rosyred' (EA)	oBRG
un	'Vuyk's Ruby' (A)	oWnC
	'Vuyk's Scarlet' (EA)	oTPm wSta wWhG
	'W. C. Slocock'	see **R.** 'Souvenir of W. C. Slocock'
un	'Wako' (EA)	oGre
	wallichii	wRho
	Walloper Group	wSta wHam
un	'Walloper # 10'	oAls
	'Wallowa Red' (DA)	oGre wHam wWhG
	'Wally Miller'	wHam wWhG
	'Walt Elliott'	wWhG
	'Walter Schmalscheidt'	wWhG
un	'Wanta Bee'	wWhG
	'War Dance'	oGre wHam wWhG wWel
un	'Warchant'	wHam
	wardii	wClo oGre wHam wWhG
	wardii var. ***puralbum***	wWhG
	wardii var. ***wardii***	wRho
	wardii var. ***wardii*** Litiense Group	wRho
	'Ward's Ruby' (EA)	oTPm
	'Warlock'	oGre wHam wWel
	'Warm Glow'	oGre wHam
	'Washington State Centennial' (DA)	oFor oGos oGre wHam wWhG
	wasonii	last listed 99/00
un	'Watchet' (EA)	wSta
	'Wee Bee'	oSis oGre wHam wWhG wWel
	'Wee Willie Winkie'	wWhG
	Wega Group	wHam
	'Weston's Innocence' (DA)	wHam
	'Weston's Pink Diamond'	oGre wWhG
	weyrichii (DA)	last listed 99/00
	'Wheatly'	wHam
	'Whidbey Island'	wHam wWhG
	'Whimsey'	last listed 99/00
un	'Whirlaway'	oGre wHam
	'Whisperingrose'	oGar oGre wHam oBRG wWhG
	'White Bird'	wSta
	'White Gold'	oGre wHam wWel
un	'White Heart' (EA)	wSta
un	'White Hino' (EA)	wSta wWhG
	'White Lights' (DA)	oFor oGos oGar oGre wHam wWhG
un	'White Lorna' (EA)	oAls oGar
un	'White Love' (EA)	oGre
un	'White Mimi' (EA)	wSta
un	'White Moon' (EA)	wSta
	'White Pearl'	wSta wHam
	'White Peter'	oGre wHam
	'White Pippin'	wHam wWhG
un	'White Rosebud' (EA)	oGar oSho oGre wSta oBRG wWhG
	'White Swan'	wSta wHam wWhG
un	'White Water'	wHam
do	'Whitethroat' (DA)	oGre
	'Whitney Appleblossom'	wWhG
	'Whitney Buff'	wWhG
	'Whitney Late Purple'	wWhG
	'Whitney Pink Mound'	wWhG
	'Whitney Pink Mound #2'	wWhG
	'Whitney Purple'	wWhG
	'Whitney Snow Queen Cross'	wWhG
	'Whitney Tiger Lily'	wWhG
	'Whitney White'	wWhG
	'Whitney 8305'	wWhG
	'Whitney's Dwarf Red'	oAls oGar wHam
	'Whitney's Georgeanne'	wHam wWhG
	'Whitney's Joyride'	wWhG
	'Whitney's Late Orange'	oAls oGre wWel
	'Whitney's Orange'	oGre wSta wHam wWhG wWel
	'Whitney's Peggy O'Neil'	wWhG
	'Whitney's Rosy O'Grady'	wHam wWhG
	'Wickiup'	oGre
	'Wickiup Sister'	oGre wHam
	'Wigeon'	wClo wHam

	wightii	wWhG
	'Wild Affair'	oGre wHam
un	'Wild Moon' (A)	oAls
un	'Wildwood Pixie Petticoat' (EA)	oSis
	'Wilgen's Ruby'	oAls oSho wHam
	'Willbrit'	wClo wHam
un	'Willgress' (EA)	wSta
	'Williams'	oAls oGar oGre
un	'Williamsburg' (EA)	wWhG
	williamsianum	oGos wRho oGre wHam wWhG
	williamsianum Caerhays form	wHam
	williamsianum x ***yakushimanum***	wClo
	Wilsonii Group	oGre wSta wWel
	wiltonii	oGre wWhG wWel
un	'Wind River'	wWhG
	'Windbeam'	wSta wHam oTal wWhG
	'Windjammer'	wHam
un	'Windsong'	wWhG
	'Windsor Lad'	oGre wHam wWhG
	'Wine Fuchsia'	oGre wWhG
un	'Wings of Gold'	oGre wWel
	Winsome Group & cl.	oAls oGar oGre wSta wHam oBRG wWhG wWel
	'Winter Snow'	oGre
	'Wintergreen' (EA)	wWhG
	'Wintonbury'	oGre wHam
un	'Wisely'	oAls
	'Wisp'	oGre wWhG
	'Wisp of Glory'	wHam
	'Wissahickon'	oGre
	'Witch Doctor'	oGre wHam
un	'Witches Butter'	oGre
	'Wizard'	oGre
	'Wojnar's Purple'	oGre wHam wWhG
	'Wombat' (EA)	oGar wWhG
un	***womersleyi***	oBov
	'Woody's Friggin Riggin'	wHam wWel
	'Wren'	oSis wMtT wHam wWhG wWel
un	***wrightianum***	wRho
un	***wrightianum*** x ***konori*** x ***laetum***	oBov
un	***wrightianum*** x ***lochiae***	oBov
	'Wyandanch Pink'	wWel
	'Wynterset White'	oGre wHam wWel
un	'Y B #2'	wHam
un	'Yachats' (DA)	oGre
	'Yachiyo Red' (EA)	oGre oBRG
	'Yaku Angel'	oGos wClo oAls oGar oWnC wWhG wWel
un	'Yaku Baron'	oGos oGre
un	'Yaku Corona'	wHam
	'Yaku Coronet'	wHam
	'Yaku Duchess'	wHam
	'Yaku Duke'	oAls wHam
	'Yaku Fantasia'	wHam
	'Yaku Frills'	wHam
un	'Yaku Kevin'	wHam
	'Yaku King'	wHam
	'Yaku Picotee'	wHam wWhG
	'Yaku Prince'	oGos oGre oWnC wHam oTal wWhG
	'Yaku Princess'	oAls oGre wHam oTal wWhG wWel
	'Yaku Queen'	oAls oGar oGre wHam
un	'Yaku Royale'	wHam
un	'Yaku Ruby'	wHam
	'Yaku Sensation'	wHam
	'Yaku Sunrise'	wClo oGar oGre wHam wWhG wWel
un	'Yaku White'	wHam
un	'Yaku #8402'	wWhG
	yakushimanum	wClo oGar wSta wHam
	yakushimanum 'Angel'	wHam
	yakushimanum Ben Nelson form	wWhG
	yakushimanum 'Berg'	wHam
	yakushimanum x ***bureaui***	oGos wWhG
	yakushimanum x ***campanulatum***	oGar oBRG
	yakushimanum Caperci form	wWhG
	yakushimanum x 'Corona'	wHam
	yakushimanum x 'Corona' x 'Hotei'	wHam
	yakushimanum x ***crinigerum***	wWhG
	yakushimanum x ***degronianum***	wWhG
	yakushimanum dwarf	wHam
un	***yakushimanum*** x ***eleganthum***	wWhG
un	***yakushimanum*** x 'Elsbeth'	wWhG
	yakushimanum Exbury form	wRho wHam wWhG
	yakushimanum x ***falconeri***	wWhG
un	***yakushimanum*** x 'Fancy'	oGre
	yakushimanum Fawcett form	wHam wWhG
	yakushimanum flat leaf form	oGre wWel

	yakushimanum Gossler narrow leaf form	oGos
un	***yakushimanum*** 'Heathcote'	wHam
un	***yakushimanum*** 'Incense'	wHam
	yakushimanum 'Ken Janeck'	see **R.** 'Ken Janeck'
	yakushimanum 'Koichiro Wada'	oGre wHam wWhG
un	***yakushimanum*** x 'Late Flowering'	oGre
	yakushimanum x ***makinoi***	wWhG
	yakushimanum x 'Medusa'	wWhG
	yakushimanum 'Mist Maiden'	oAls oGar oGre wHam wWhG
un	***yakushimanum*** x 'Norman Gill'	wWhG
	yakushimanum x ***pachysanthum***	wWhG wWel
un	***yakushimanum*** Phetteplace dwarf form	wHam
	yakushimanum Phetteplace tall form	wHam
	yakushimanum 'Pink Parasol'	wWhG
	yakushimanum x ***proteoides***	wWhG
	yakushimanum x ***recurvoides***	wWhG
	yakushimanum x 'Sir Charles Lemon'	oGre
	yakushimanum x ***smirnowii***	wWhG
	yakushimanum x ***strigillosum***	wWhG
un	***yakushimanum*** x ***sutchuenense*** var. ***geraldi***	wWhG
	yakushimanum x ***tsariense***	wWhG
un	***yakushimanum*** 'Van Ziele'	oGos
un	***yakushimanum*** Van Zile form	wWhG
un	***yakushimanum*** 'Van Zyle'	wHam
un	***yakushimanum*** 'Wellshimer'	oGre
	yakushimanum 'White Velvet'	wWhG
	yakushimanum Whitney form	wHam wWhG
	yakushimanum x ***williamsianum***	oGre oBRG
un	'Yamato No Hikari' (EA)	oGre
	'Years of Peace'	wHam
	yedoense var. ***poukhanense*** (EA)	oFor oGre wSta
	yedoense var. ***poukhanense*** HC 970195	wHer
	'Yellow Bells'	wHam
	'Yellow Cloud' (DA)	oGre wWhG
	'Yellow Hammer'	oGar oGre wSta wHam oTal
	'Yellow Petticoats'	wWhG
	'Yellow Pippin'	oGre wHam
	'Yellow Rolls Royce'	wHam
un	'Yellow Ruffles'	wHam
	'Yellow Saucer'	wHam
	'Yellow Spring'	wHam
un	'Yoshimi' (EA)	oTPm
un	'Yuka' (EA)	oAls oGar wWhG
	yungningense	oGre
un	***yungningense*** Glomeratum Group	wHam
	yunnanense	wCCr
un	***yunnanense*** 'Bodinieri'	wWhG
	yunnanense DJHC 195	wHer
	yunnanense white/gold-orange freckling	wWhG
	yunnanense white/red eye	wWhG
un	'Zeet'	wHam
	Zelia Plumecocq Group & cl.	oGre
	zoelleri (V)	wRho oBov
	zoelleri USNA (V)	oBov
	RHODOHYPOXIS	
	baurii	oTrP oAmb oGar oNWe oOut oSev oSec
un	***baurii*** 'Lily Jean'	oSis
	baurii pink	oFor oSis
un	***baurii*** 'Pinkei'	wHer
	baurii var. ***platypetala***	wHer oGre
	baurii red	last listed 99/00
	baurii rose	oFor
un	***baurii*** 'Rubella'	wHer
	baurii white	oFor oSis
	'Douglas'	oGre
	'Hebron Farm Red Eye'	see **X RHODOXIS *hybrida*** 'Hebron Farm Red Eye'
	'Helen'	wHer oSis oGre
	milloides	wHer
	milloides 'Claret'	wHer
	milloides 'Damask'	wHer
	'Picta'	wHer oGre
	'Stella'	oGre
	'Tetra Red'	oGre
	'Tetra White'	see **R.** 'Helen'
	thodiana	wHer
	RHODOMYRTUS	
	tomentosa	oOEx
	RHODOTYPOS	
	scandens	oFor oGos oRiv
	X RHODOXIS	
	hybrida 'Hebron Farm Red Eye'	wHer
	RHOICISSUS	
un	***tomentosa***	oTrP

	RHUS	
	aromatica	oFor wCul wBCr oGoo oRiv wCCr wKin wBox wPla
	aromatica 'Gro-low'	oFor wWel
	chinensis	oEdg
	copallina	oFor oGar oGre
	coriaria	oFor
	glabra	oFor wNot wShR oGre wCCr wSta wPla oAld oSle wCoS
ch	***glabra*** 'Laciniata' (see Nomenclature Notes)	oFor oGre wCoS
	integrifolia	oFor
	lancea	oFor wCCr
	ovata	oFor
	x pulvinata (Autumn Lace Group) 'Red Autumn Lace'	last listed 99/00
un	***punjabensis sinica***	oFor
	succedanea	oFor
	trichocarpa	wCCr
	trichocarpa HC 970361	wHer
ch	***trilobata***	see **R. *aromatica***
	typhina	oFor oAls oGar oGre wTGN wSta wCoS
	typhina 'Dissecta'	oFor oAls oGar oGre wKin wSta oSle wWel
	typhina 'Laciniata'	see **R. *typhina*** 'Dissecta'
	verniciflua	oFor
	RHYNCHOSPORA	
	colorata	wGAc oHug oGar oSsd
un	**RHYTIDIADELPHUS**	
	loreus	wFFl
	triquetrus	wFFl
	RIBES	
	alpinum	oRiv oWnC
	alpinum 'Aureum'	wHer
	aureum	see **R. *odoratum***
	bracteosum	wShR oBos oRiv wWat oAld wFFl
	cereum	oFor oBos wPla
ch	***cruentum***	see **R. *roezlii*** var. ***cruentum***
	x culverwellii Jostaberry (fruit0	wBCr wClo oAls oWhi wRai
un	***x culverwellii*** Red Jostaberry	oWhi
	divericatum	wBur wWat
	fasciculatum var. ***chinense***	oFor
	x gordonianum	wHer oHed
	grossularia	see **R. *uva-crispa*** var. ***reclinatum***
	henryi DJHC 777	wHer
	lacustre	oBos wWat wFFl
un	***lasiandra***	oFor
	laurifolium	oFor wHer oCis
	lobbii	last listed 99/00
	menziesii	wWat
	montigenum	oFor
	nevadense	oFor oRiv
	nigrum	wBCr oRiv
	nigrum 'Ben Lomond'	oWhi wRai
	nigrum 'Ben More'	oWhi
	nigrum 'Ben Nevis'	oOEx
	nigrum 'Ben Sarek'	oWhi wRai
	nigrum 'Black Reward'	oWhi
un	***nigrum*** 'Black September'	oWhi oOEx oWnC
	nigrum 'Blackdown'	wClo oWhi oOEx
	nigrum 'Boskoop Giant'	oWhi oOEx
un	***nigrum*** 'Brodtorp'	oWhi oOEx
un	***nigrum*** 'Champion'	oWhi
	nigrum 'Consort'	wBCr wBur wClo oGar oSho oWhi oOEx
	nigrum 'Crandall'	see **R. *odoratum*** 'Crandall'
un	***nigrum*** 'Crusader'	wBCr oWhi
un	***nigrum*** 'Green's Black'	oWhi
un	***nigrum*** 'Hilltop Baldwin'	oWhi wRai
un	***nigrum*** 'Invigo'	oWhi
	nigrum Jostaberry	see **R. *x culverwellii*** Jostaberry
	nigrum 'Laxton's Giant'	oWhi
un	***nigrum*** 'Magnus'	oWhi
	nigrum 'Mendip Cross'	oWhi
un	***nigrum*** 'Minaj Smyriou'	oWhi
un	***nigrum*** 'Mopsy'	oWhi
un	***nigrum*** 'Otelo'	oWhi
un	***nigrum*** 'Prince Consort'	wRai
un	***nigrum*** 'Purdy'	oWhi
un	***nigrum*** 'Risager'	oWhi
un	***nigrum*** 'Silver Geiter'	oWhi
un	***nigrum*** 'Strata'	oWhi oOEx
un	***nigrum*** 'Swedish Black'	oWhi
un	***nigrum*** 'Titania'	oWhi wRai
un	***nigrum*** 'Topsy'	oWhi
un	***nigrum*** 'Tsema'	oWhi
un	***nigrum*** 'Viola'	oWhi
	nigrum 'Wellington XXX'	wClo oWhi
un	***nigrum*** 'Willoughby'	oWhi oOEx
	odoratum	oFor wBCr wBur oJoy oHed wShR oGar oBos oRiv oWhi oGre wPla
	odoratum 'Crandall'	oFor wBCr wClo oAls oGar oWhi wRai oOEx
	roezlii	oFor
	roezlii var. ***cruentum***	last listed 99/00
un	***rubrum*** 'Amish Red' (R)	oWhi
	rubrum 'Blanka' (W)	oWhi
	rubrum 'Cascade' (R)	wBur oWhi wRai oOEx
	rubrum 'Cherry'	oFor wBCr wClo oAls oGar oSho oWhi oOEx oWnC
un	***rubrum*** 'Det Van'	oWhi
un	***rubrum*** 'European White'	oWhi oOEx
un	***rubrum*** 'Heros'	oWhi
	rubrum 'Jonkheer van Tets' (R)	wBur oWhi wRai oOEx
	rubrum 'Laxton's Number One' (R)	last listed 99/00
un	***rubrum*** 'Masons'	oWhi
un	***rubrum*** 'Minnesota 52'	oWhi
un	***rubrum*** 'Primus'	oWhi wRai
	rubrum 'Red Lake' (R)	wBCr wClo oWhi oOEx
	rubrum 'Redstart' (F)	oWhi
un	***rubrum*** 'Rolam'	oWhi
un	***rubrum*** 'Rosetta'	wRai
un	***rubrum*** 'Rotet'	oWhi
	rubrum 'Rovada' (R)	oWhi
un	***rubrum*** 'Tatran'	oWhi
un	***rubrum*** 'White Imperial'	wBur oSho oWhi oOEx
	rubrum 'White Pearl' (W)	wClo oGar
un	***rubrum*** 'White 1301'	oWhi
un	***rubrum*** 'Wilder'	wBCr oSho
	sanguineum	oFor oHan wWoB wThG wBur wClo oNat wNot wCoN oAls wShR oGar oGoo oBos oRiv oSho oBlo wWat wCCr oTri wPla oAld wFFl wRav oSle wSte wSnq wCoS wWel
	sanguineum 'Album'	wCri
un	***sanguineum*** 'Apple Blossom'	wHer
un	***sanguineum*** 'Aureum'	oGre
	sanguineum 'Claremont'	oGar oGre wSnq wWel
	sanguineum 'Elk River Red'	oFor oGos wBur oGar oDan oWhi oGre
un	***sanguineum*** 'Hannaman White'	oFor oGos oDar oWhi wRai oGre wSnq
un	***sanguineum*** 'Henry Henneman'	oHed
	sanguineum 'King Edward VII'	oFor wCul wBur oDar oAls oGar oDan oRiv oWhi oBlo oGre wRai wSta wAva oSle wSnq wWel
	sanguineum pink	wSta
	sanguineum 'Poky's Pink'	oFor wHer oGos oHed oRiv oWhi oBov oGre
ch	***sanguineum*** 'Porky Pink'	see **R. *sanguineum*** 'Poky's Pink'
	sanguineum 'Pulborough Scarlet'	oFor oDan oWhi wRai oGre wSnq wWel
	sanguineum red	wSta
	sanguineum 'Spring Showers'	oHed
un	***sanguineum*** 'Spring Snow'	oFor oGos oRiv oWhi wWel
un	***sanguineum*** 'Strybing Pink'	wHer
un	***sanguineum*** 'Variegatum'	wCol oSec
	sanguineum white	wCoN oBos
	sanguineum White Icicle / 'Ubric'	oFor wCul oGar oDan oRiv oGre oWnC wWel
	speciosum	wHer oRiv
	uva-crispa	oWnC
	uva-crispa var. ***reclinatum*** 'Achilles'	wClo oGar oWhi wRai oGre
un	***uva-crispa*** var. ***reclinatum*** 'Betty'	oWhi
un	***uva-crispa*** var. ***reclinatum*** 'Black Velvet'	oWhi wRai
	uva-crispa var. ***reclinatum*** 'Captivator'	wBCr wBur wClo oAls oGar oWhi wRai oOEx
	uva-crispa var. ***reclinatum*** 'Careless'	oWhi
	uva-crispa var. ***reclinatum*** 'Catherina'	oWhi
un	***uva-crispa*** var. ***reclinatum*** 'Clark'	oWhi
un	***uva-crispa*** var. ***reclinatum*** 'Colossal'	oWhi
	uva-crispa var. ***reclinatum*** 'Crown Bob'	oWhi
	uva-crispa var. ***reclinatum*** 'Early Sulphur'	oWhi
un	***uva-crispa*** var. ***reclinatum*** 'Fredonia'	oWhi oOEx
un	***uva-crispa*** var. ***reclinatum*** 'Friedl'	oWhi
	uva-crispa var. ***reclinatum*** 'Glenton Green'	last listed 99/00
	uva-crispa var. ***reclinatum*** Hinnonmaki Gold	see **R. *uva-crispa*** var. ***reclinatum*** 'Hinnonmaki Gul'
	uva-crispa var. ***reclinatum*** 'Hinnonmaki Gul'	wBur
	uva-crispa var. ***reclinatum*** Hinnonmaki Red	see **R. *uva-crispa*** var. ***reclinatum*** 'Hinnonmaki Rod'

	uva-crispa var. *reclinatum* 'Hinnonmaki Rod'	oWhi
un	*uva-crispa* var. *reclinatum* 'Hinnonmaki Yellow'	oWhi wRai oOEx
un	*uva-crispa* var. *reclinatum* 'Hoeing's Earliest'	oWhi
	uva-crispa var. *reclinatum* 'Howard's Lancer'	oWhi
	uva-crispa var. *reclinatum* 'Invicta'	wBur oWhi wRai
un	*uva-crispa* var. *reclinatum* 'Jahns Prairie'	oWhi
	uva-crispa var. *reclinatum* 'Keepsake'	last listed 99/00
un	*uva-crispa* var. *reclinatum* 'Leepared'	oWhi wRai oOEx
	uva-crispa var. *reclinatum* 'Leveller'	oWhi
	uva-crispa var. *reclinatum* 'Oregon Champion'	wBCr oGar oSho oWhi oOEx
un	*uva-crispa* var. *reclinatum* 'Orus 8'	oWhi wRai oOEx
	uva-crispa var. *reclinatum* 'Pixwell'	wBCroGar oOEx
	uva-crispa var. *reclinatum* 'Poorman'	oFor wBCrwClo oGar oWhi wRai oOEx
un	*uva-crispa* var. *reclinatum* 'Shulz'	oWhi
un	*uva-crispa* var. *reclinatum* 'Sylvia'	oWhi
	uva-crispa var. *reclinatum* 'Whinham's Industry'	oWhi
	uva-crispa var. *reclinatum* 'Whitesmith'	oWhi wRai
	viburnifolium	oFor oSho wCCr
	viscosissimum	oFor oTri
	RICINUS	
	communis 'Carmencita' seed grown	wSte
	communis 'Gibsonii'	wSte
	communis 'Sanguineus'	oNat
	RIGIDELLA	
	orthantha	wCol
	ROBINIA	
	x ambigua 'Decaisneana'	oFor
	x ambigua 'Idahoensis'	oFor wWhi
	hispida	oFor
	hispida fertilis	oFor
	hispida 'Macrophylla'	oGar
un	*hispida* 'Macrophylla Pendula'	wWel
	x holdtii 'Britzensis'	last listed 99/00
	x margaretta Casque Rouge	see **R.** ***x margaretta*** 'Pink Cascade'
	x margaretta 'Pink Cascade'	oFor
	neomexicana	oFor
	pseudoacacia	wBCr oWhi wRai wKin wPla
	pseudoacacia 'Bessoniana'	oFor
qu	*pseudoacacia* 'Contorta'	oRiv oWhi wWel
	pseudoacacia 'Fastigiata'	see **R.** ***pseudoacacia*** 'Pyramidalis'
	pseudoacacia 'Frisia'	oFor wHer wCul oGos oDar oGar oDan oRiv oWhi oGre wTGN wBox wAva wWel
	pseudoacacia 'Lace Lady'	see **R.** Twisty Baby™ / 'Lace Lady'
un	*pseudoacacia* 'Monophylla Pendula'	oFor
	pseudoacacia 'Pendula'	last listed 99/00
un	*pseudoacacia* 'Purple Robe'	oFor oDar oGar oDan oGre wWel
	pseudoacacia 'Pyramidalis'	oFor oRiv
	pseudoacacia 'Semperflorens'	oFor
	pseudoacacia 'Tortuosa'	oFor oWnC
qu	*pseudoacacia* 'Tortuosa Nana'	oGar
	pseudoacacia 'Umbraculifera'	oFor oGar
	pseudoacacia 'Unifoliola'	oFor
	x slavinii 'Hillieri'	oFor oWhi
	Twisty Baby™ / 'Lace Lady'	oFor oGar oRiv wTGN wWel
	RODGERSIA	
	aesculifolia	oFor wCul oAls oGar oNWe wNay wRob oGre wCCr
	aesculifolia bronze leaf	oNWe
	sp. aff. *aesculifolia* EDHCH 97112	wHer
	henrici	oAls
	henrici hybrids	oUps
	new hybrids	oGre
	pinnata	oFor wCol wFai wBox
	pinnata DJHC 126	wHer
	pinnata 'Elegans'	oTwi wNay wRob
	pinnata 'Superba'	oMis oGar oOut oGre wFai
	pinnata 'Superba' seed grown	wHer
	podophylla	oNWe
	podophylla HC 970494	wHer
	podophylla 'Rotlaub'	last listed 99/00
	sambucifolia	oFor oAmb oGar oGre
	sp. aff. *sambucifolia* DJHC 98246	wHer
	sp.	oNWe
	tabularis	see **ASTILBOIDES** ***tabularis***
	ROHDEA	
	japonica crested form	wCol
un	*japonica* 'Eco Olympic Torch'	wCol
un	*japonica* 'Godaishu'	wHer
	japonica 'Gunjaku'	wHer
	japonica 'Striata'	wCol
	japonica variegated form	wCol
	sp.	oCis
	ROMANZOFFIA	
	sitchensis	oSis
	ROMNEYA	
	coulteri	wGAc wCoN oDar oRus oGar oGre wFai wCCr
un	*coulteri* 'Butterfly'	wCoN oJoy
	coulteri var. *trichocalyx*	wHer wCCr
	ROMULEA	
	bulbocodium	oSis
	ramiflora	last listed 99/00
	RORIPPA	
	nasturtium-aquaticum	oSsd
	ROSA (SEE NOMENCLATURE NOTES)	
un	'A Shropshire Lass'	oHei wAnt
	Abbaye de Cluny™ / 'Meibrinpay'	wChr wCot oHei oGar oWnC oRNW wAnt
	'Abbotswood'	oHei
un	'Abracadabra'	wCot oGar oWnC oAls oRNW
	Abraham Darby®/ 'Auscot'	wChr wCot oHei oGar oSec oRNW wAnt
	Ace of Hearts / 'Korred'	wCot
un	'Acey Deucy'	oJus
	acicularis	oFor oAld
un	'Adam's Smile'	oJus
un	'Adelaide Hoodless'	oSle
	Admirable / 'Searodney'	oHei oJus
	'Admiral Rodney'	last listed 99/00
	Admired Miranda / 'Ausmira'	oRos
	Adolf Horstmann®	oEdm
	Agatha Christie / 'Kormeita'	oHei oRNW
	'Agnes'	wChr oRos wAnt
	'Agrippina'	see **R.** 'Cramoisi Superieur'
	'Aimee Vibert'	last listed 99/00
un	'Ain't Misbehavin'	oJus
un	'Ain't She Sweet'	oWnC
	'Alain Blanchard'	oHei
	x alba	oSha
	x alba 'Alba Maxima'	wChr oRos oHei
	x alba 'Alba Semiplena'	wChr oRos wCot oHei wCLG wAnt
	x alba 'Alba Suaveolens'	see **R.** ***x alba*** 'Alba Semiplena'
	Alba Meidiland®/ 'Meiflopan'	wChr oHei oRNW wAnt
	'Alberic Barbier'	oHei wAnt
	'Albertine'	wChr oRos oHei wAnt
	'Alchymist'	wChr oRos wCot oHei wAnt
	'Alec's Red'® / 'Cored'	oHei
	Alexander®/ 'Harlex'	oHei
	'Alexandre Girault'	oRos oHei wCLG
un	Alfie™ / 'Poulfi'	wCot
	'Alfred Colomb'	oRos
co	'Alfred de Dalmas'	see **R.** 'Mousseline'
un	'Alice Faye'	oJus
	'Alida Lovett'	oHei
un	'Alika'	oRos oHei
	'Alister Stella Gray'	wCLG
un	All Ablaze™ / 'Weksou'	wChr wCot oHei wAnt
un	'All That Jazz'	wChr wCot oGar
un	'Alleluia'	wCot
un	'Alliance Franco-Russe'	wCLG
un	'Allspice'	oGar oSho
un	'Almost Wild'	oSha
	'Aloha'	wChr wCot oHei oWnC oRNW
	Alpine Sunset®	wChr wCot oRNW
	altaica	see **R.** ***pimpinellifolia*** 'Grandiflora'
	Altissimo®/ 'Delmur'	oRos wCot oHei oEdm oGar oSha oRNW wAnt
un	'Always A Lady'	oJus
	'Amadis'	oRos
	'Amanda Kay'	oJus
	Amazing Grace™	see **R.** Myriam®/ 'Cocgrand'
	Amber Flash™ / 'Wildak'	oWnC
	Amber Queen®/ 'Harooyn'	wChr wCot oEdm oRNW
	Amber Queen®/ 'Harooyn' (tree)	oEdm
un	'Amber Waves'	wChr wCot oWnC oAls oRNW
un	'Ambiance'	oRNW
	Ambridge Rose / 'Auswonder'	wChr oHei oGar oRNW wAnt
	Ambridge Rose / 'Auswonder' (tree)	last listed 99/00
	'Amelia'	see **R.** 'Celsiana'
un	'Amelia Earhart'	wCot
	'America'	wCot oHei oGar oSho oWnC oRNW
	'American Beauty'	oRos oRNW
	'American Pillar'	wChr oRos wCot oHei oSha oRNW wAnt
un	'American Rose Centennial'	oJus
un	'America's Classic'	oRNW

	'Amethyste'	oHei
un	'Amiga Mia'	oHei
un	'Amy Rebecca'	oJus
	'Amy Robsart'	oHei oJoy
un	'Anastasia'	oRNW
	'Andersonii'	oHei
un	'Andrea'	oHei oJus
un	Andrea Stelzer / 'Korfachrit'	oEdm
	'Anemone'	oHei
un	'Angel Darling'	oJus
	'Angel Face'	wChr wCot oHei oGar oSho oWnC oAls oRNW oUps
	'Angel Face' (tree)	oGar
	Angela Rippon®/ 'Ocaru'	oHei
	'Angelita'	see **R.** 'Snowball'
	Anisley Dickson / 'Dickimono'	wChr oEdm oRNW
	Anisley Dickson / 'Dickimono' (tree)	oEdm
	Ann / 'Ausfete'	oHei
un	'Anna'	oRNW
	Anna Ford®/ 'Harpiccolo'	oHei oRNW
	Anna Livia / 'Kormetter'	oHei
	'Anna Pavlova'	wChr
	Anne Harkness®/ 'Harkaramel'	oHei oRNW
	Anne Moore / 'Morberg'	oJus
	'Anne of Geierstein'	oHei
	Anthony Meilland / 'Meitalbaz'	oWnC oRNW
	Anthony Meilland / 'Meitalbaz' (tree)	oWnC
	'Antigua'	last listed 99/00
	Antique	see **R.** Antique '89®/ 'Kordalen'
	Antique Artistry™ / 'Cleartful'	oHei
un	'Antique Lace'	oRNW
	Antique Tapestry / 'Cletape'	oHei
	Antique '89®/ 'Kordalen'	oRos wCot oHei oJoy oRNW
	'Antoine Rivoire'	wCLG
un	'Anytime'	oJus
un	'Apart'	oHei
un	Apple-a-Day™ Cherryberry™ (fruit)	oOEx
	'Apple Blossom'	oHei wAnt
un	Appleblossom Flower Carpet™ / 'Noamel'	wChr oAls oGar oSho oRNW wWel
	'Applejack'	oRos oHei
	'Apricot Nectar'	wChr oHei oRNW wCLG
un	'Apricot Perfection'	oHei
un	'April Moon'	oHei
	Arcadian / 'Macnewye'	oRNW
	'Archduke Charles'	oRos wCLG
co	'Archiduc Joseph' (see Nomenclature Notes)	wCLG
	Arctic Sunrise / 'Beararcsun'	oHei
	'Ardoisee de Lyon'	oRos oHei wCLG
	Arizona / 'Tocade'	oWnC
	'Arizona Sunset'	oJus
	arkansana	oFor oHei
un	***arkansana*** 'Alba'	wRob
un	'Arlene Francis'	wChr oWnC
	Armada®/ 'Haruseful'	wWoS oRos oHei
un	'Armide'	wCLG
	'Arthur Bell'	oRNW
	'Arthur de Sansal'	oRos oHei wAnt
	Artistry™ / 'Jacirst'	wCot oEdm oGar oSho oWnC oRNW
un	***arvensis plena***	oHei
	Ash Wednesday / 'Aschermittwoch'	oRos oHei
un	Aspen / 'Poulurt'	wChr
un	'Audrey Hepburn'	wChr oRNW
	Auguste Renoir®/ 'Meitoifar'	wChr wCot oHei oGar oWnC oRNW wAnt
un	'Aunt Honey'	oHei
un	'Aunt Ruth'	oHei
un	'Aurora'	oRos
	'Autumn Delight'	oRos oHei
	'Autumn Sunset'	wChr wCot oHei oEdm oGar oWnC oRNW wAnt
	Avon / 'Poulmulti'	oHei
	Awakening / 'Probuzini'	wChr oRos wCot oHei oRNW wCLG wAnt
	'Ayrshire Queen'	last listed 99/00
un	'Baby Betsy McCall'	oFor oJus
	Baby Blanket	see **R.** Oxfordshire / 'Korfullwind'
un	'Baby Cheryl'	oJus
un	'Baby Diana'	oJus
un	'Baby Eclipse'	oJus
	'Baby Faurax'	oRos oHei oRNW
un	'Baby Garnette'	oJus
un	'Baby Grande'	wChr oHei oJus
	'Baby Katie'	oJus
	Baby Love / 'Scrivluv'	wChr wCot oHei oRNW
	Baby Love / 'Scrivluv' (tree)	last listed 99/00
un	'Baby Mermaid'	see **R.** 'Happenstance'

	'Baby Michael'	oJus
	'Ballerina'	oFor wWoS wChr oRos wCot oHei wShR oGar oNWe oSha oRNW wAnt
	'Baltimore Belle'	oRos
	banksiae alba	see ***banksiae*** var. ***banksiae***
	banksiae alba plena	see ***banksiae*** var. ***banksiae***
	banksiae var. ***banksiae***	oFor wChr oHei oGar oGre wWel
	banksiae 'Lutea'	oFor wChr wCSG oGar oCis oSho oGre oRNW wWel
	banksiae var. ***normalis***	oRos
un	'Banshee'	oHei
	Bantry Bay®	oRos oSha
	Barbara Austin / 'Austop'	wChr wCot
un	'Barbara Bush'	wChr oGar oSho oWnC oAls oRNW
un	Barbara Streisand™ / 'Wekquaneze'	wCot oHei oEdm oRNW
un	'Barndance'	oHei
	'Baron de Bonstetten'	oHei
	'Baron Girod de l'Ain'	oRos oHei wAnt
	'Baroness Rothschild' (see Nomenclature Notes)	oHei
	'Baroness Rothschild' (HP)	see **R.** 'Baronne Adolph de Rothschild'
	'Baronne Adolph de Rothschild'	wChr wCLG
	Baronne Edmond de Rothschild®/ 'Meigriso'	wChr
	'Baronne Henriette de Snoy'	wCLG
	'Baronne Prevost'	wChr oRos wCot oHei oWnC wCLG wAnt
un	'Bayse's Myrhh Scented'	oHei
un	'Bayse's Purple'	oJoy
un	Beautiful Bride™	oHei
	'Beauty of Rosemawr'	oHei
un	'Beauty Secret'	wChr oJus
un	Bedazzled / 'Maccatsan'	oHei
un	Belami / 'Korhanbu'	wCot oEdm oRNW
	'Belinda'	last listed 99/00
un	'Belinda's Dream'	oHei
	bella	oFor
un	'Bella Donna'	oRos oHei
	'Belle Amour'	oHei
un	'Belle Blanca'	wChr
	'Belle de Crecy'	oRos wCot wAnt
	'Belle Isis'	oRos oHei
	'Belle of Portugal'	see **R.** 'Belle Portugaise'
	'Belle Poitevine'	wChr oRos wCot wAnt
	'Belle Portugaise'	wChr
un	'Belle Rouge'	oRNW
	Belle Story®/ 'Auselle'	oFor wChr wCot oHei oGar oRNW wAnt
	'Belvedere'	last listed 99/00
un	'Berkeley Beauty'	oJus
	Berkshire / 'Korpinka'	oHei
	'Berlin'	oAmb
un	'Bermuda Kathleen'	oHei
un	Berries 'n' Cream™ / 'Poulclimb'	wChr wCot oHei oEdm oGar oWnC oRNW wAnt
un	'Betty Bee'	oJus
un	Betty Boop™ / 'Wekplapic'	wChr wCot oHei oEdm oGar oSho oWnC oRNW wAnt
un	Betty Boop™ / 'Wekplapic' (tree)	wChr wCot oEdm oGar
	'Betty Prior'	wChr oRos wCot wCSG oSha oRNW
	'Betty Uprichard'	oRNW
un	'Bewitched'	wChr wCot oGar oWnC oRNW
un	***bhutan***	oHei
	Bibi Maizoon®/ 'Ausdimindo'	oHei oRNW
un	'Big John'	oHei oJus
	Big Purple®/ 'Stebigpu'	wChr wCot oHei oEdm oRNW wAnt
	Bill Warriner™ / 'Jacsur'	wChr wCot oGar oWnC oUps
	Bill Warriner™ / 'Jacsur' (tree)	oGar
	Billy Graham / 'Jacgray'	wChr wCot oGar oWnC oAls oRNW wAnt
un	'Bimboro'	wCot
un	'Birdie Blye'	oHei
	Birthday Girl / 'Meilasso'	oHei
	'Bishop Darlington'	oRos oHei wCLG
	'Bit o' Sunshine'	oJus
	Black Gold™ / 'Cleblack'	oHei
	'Black Ice'	oHei
	Black Jade™ / 'Benblack'	wChr oGar oJus
	Black Jade™ / 'Benblack' (tree)	oGar
	'Blairii Number Two'	oHei wCLG wAnt
	'Blanche de Vibert'	wCLG
	'Blanc Double de Coubert'	see **R.** 'Blanche Double de Coubert'
	'Blanche Double de Coubert'	wChr oRos wCot oHei oGar oWnC oRNW wAnt oUps
	'Blanche Moreau'	oRos
	'Blanchefleur'	wCLG

	blanda	oFor oHei
un	'Blast Off'	oWnC
	'Blaze'	wChr oRos oHei oGar oSha oWnC wAnt
	'Bleu Magenta'	oRos wCot
	'Bloomfield Abundance'	oHei
	'Bloomfield Courage'	oHei
	'Bloomfield Dainty'	wChr oRos oHei
	'Blossomtime'	wChr oHei
un	'Blue Girl'	wChr oGar oWnC
	Blue Moon®/ 'Tannacht'	oHei
un	'Blue Nile'	wChr
	Blue Parfum®/ 'Tanfifum'	oHei
	Blue Perfume	see **R.** Blue Parfum®/ 'Tanfifum'
	Blue Peter / 'Ruiblun'	oHei
un	'Blue Ribbon'	wChr oGar oSho
un	'Bluebell'	oRNW
un	Blueberry Hill / 'Wekcryplag'	wChr wCot oHei oEdm oGar oWnC oRNW wAnt
un	Blueberry Hill / 'Wekcryplag' (tree)	oEdm
un	'Blueblood'	oJus
un	'Bluesette'	oHei
un	'Blumenschmidt'	oHei
un	'Blush Hip'	wChr oRos
un	'Blush Moss'	oRos
	'Blush Noisette'	see **R.** 'Noisette Carnee'
	'Blush Rambler'	last listed 99/00
	'Blushing Lucy'	oRos oHei
	'Bobbie James'	wChr wCot oHei
un	'Bolero'	wCot
	'Bon Silene'	oHei oRNW
	Bonica®/ 'Meidomonac'	wChr wCot oHei oEdm oGar oWnC oRNW wAnt oUps oSle
	Bonica®/ 'Meidomonac' (tree)	last listed 99/00
	'Bonn'	oHei
	'Botzaris'	oRos
un	'Boudoir'	oHei
un	'Bougainville'	wCLG
	'Boule de Nanteuil'	wCot oHei
	'Boule de Neige'	wChr oRos wCot wCLG wAnt
	'Bouquet d'Or'	wCLG
un	'Bouquet Parfait'	oHei
	'Bouquet Tout Fait'	see **R.** 'Nastarana'
	'Bourbon Queen'	oRos oHei
	Bow Bells / 'Ausbells'	wChr wCot oHei oGar oRNW
un	'Boy Crazy'	oWnC
	bracteata 'Mermaid'	see **R.** 'Mermaid'
un	Brandenburg Gate™ / 'Jacgate'	oEdm
	Brandy™ / 'Arocad'	wChr wCot oGar oSho oAls oRNW
un	Brass Band™ / 'Jaccofl'	wChr oGar oWnC oRNW wAnt
un	Brass Band™ / 'Jaccofl' (tree)	oWnC
	'Brass Ring'	see **R.** Peek A Boo / 'Dicgrow'
	Braveheart™ / 'Clebravo'	oHei
	Breath of Life / 'Harquanne'	wCot oHei oSha oRNW
un	'Breathless'	oGar oWnC
	Bredon®/ 'Ausbred'	oHei
	'Breeze Hill'	oHei
un	'Breezy'	oJus
	Brenda Burg™ / 'Cleswan'	oHei
	'Brenda Colvin'	oHei
	Bridal Pink™ / 'Jacbri'	oRNW
un	'Bridal Pink Bridget'	oHei
un	'Bridal White'	oRNW
un	Bride's Dream / 'Koroyness'	wCot oEdm oWnC oRNW
un	Brigadoon / 'Jacpal'	wCot oEdm oGar oWnC oRNW
	Britannia / 'Frycalm'	oRos oHei
	'Broadway'	wChr wCot
un	'Bronze Star'	wChr
	Brother Cadfael / 'Ausglobe'	wChr oHei wAnt
	Brown Velvet / 'Macultra'	oRNW
	brunonii	wHer
	brunonii 'La Mortola'	wHer wCCr
un	'Bubble Bath'	oHei
	'Buff Beauty'	wChr oRos wCot oHei oSha oRNW wCLG
un	Buffalo Gal / 'Uhlater'	wChr wCot oGar
un	'Burbank Rambler'	oHei
	'Burgundiaca'	oRos oHei oRNW wAnt
un	'Burnet Irish Marbled'	oHei
	Bush Garden climber	oRos
un	'Butterflies'	wChr
un	'Buttons 'n' Bows'	oJus
	By Appointment / 'Harvolute'	oHei
un	'Cadillac deVille'	see **R.** 'Moonstone'
	caesia ssp. ***glauca***	oHei
un	Café Ole / 'Morole'	oHei
un	'Cajun Spice'	oRNW
un	'Cal Poly'	oJus
	californica	wShR
	californica 'Plena'	see **R. *nutkana*** 'Plena'
	'Camaieux'	oRos oHei
	Cambridgeshire / 'Korhaugen'	oHei
	'Camelot'	oRNW
	'Cameo'	oHei
	'Canary Bird'	see **R. *xanthina*** 'Canary Bird'
	Candelabra™ / 'Jaccinqo'	wChr wCot oEdm oGar oSho oWnC oRNW
	Candelabra™ / 'Jaccinqo' (tree)	oGar
	Candy Mountain / 'Jacchari'	oHei oGar oRNW
	Candy Rose®/ 'Meiranovi'	oHei
un	Candyman™ Cherryberry™ (fruit)	oOEx
	canina 'Abbotswood'	see **R.** 'Abbotswood'
	Canterbury / 'Ausbury'	oHei
un	Cape Cod / 'Poulfan'	wChr
	'Capitaine John Ingram'	oHei wCLG wAnt
un	'Captain Harry Stebbings'	oEdm
	'Captain Samuel Holland'	oHei oRNW
un	'Cara Mia'	oRNW
	'Cardinal de Richelieu'	oRos oHei oSha
	Cardinal Hume®/ 'Harregale'	oRos oHei
	Cardinal Song™ / 'Meimouslin'	oWnC oAls oRNW
	Carefree Beauty™ / 'Bucbi'	oHei oWnC oRNW oUps
	Carefree Delight / 'Meipotal'	wChr wCot oHei oEdm oRNW
	Carefree Wonder / 'Meipitac'	oGar
un	Caribbean™ / 'Korbirac'	wCot oEdm oGar oWnC oRNW
	'Carmenetta'	wChr oRos oHei
un	'Carnival Glass'	oHei
	'Carol Amling'	oRNW
	carolina	oFor oRos oHei
	Caroline de Monaco®/ 'Meipierar'	wChr wCot oWnC oAls
	Carpet of Color™	see **R.** Cambridgeshire / 'Korhaugen'
qu	'Carpet Pinky'	oSec
un	'Carrot Top'	oJus
un	'Carrot Top' (tree)	oWnC
	'Cary Grant'	oWnC oAls
	Casa Blanca / 'Meimainger'	oRos oHei
un	'Cascade'	oRNW
	Casino®/ 'Macca'	oHei
un	'Cassie'	oHei
	'Catherine Mermet'	oRNW wCLG
	'Cecile Brünner'	wChr oRos wCot oHei oJoy oGar oSho oSha oRNW wAnt
	'Celeste'	wChr oRos wCLG wAnt
	'Celestial'	see **R.** 'Celeste'
	'Celine Forestier'	oRNW wCLG
	'Celsiana'	oRos wCot oHei wCLG
un	'Centavo'	oHei
	Centenaire de Lourdes®/ 'Delge'	oRos oHei oRNW
	Centennial Star™ / 'Meinerau'	oAls oRNW
un	'Center Gold'	oJus
un	'Centerpiece'	oJus
	x centifolia	wChr oRos oHei
un	***x centifolia andrewsii***	oHei
	x centifolia 'Bullata'	oHei
	x centifolia 'Cristata'	wChr oRos wCot oHei wAnt
	x centifolia 'Muscosa'	wChr oRos oHei wAnt
	x centifolia 'Parvifolia'	see **R.** 'Burgundiaca'
	x centifolia var. ***pomponia***	see **R.** 'De Meaux'
	'Centifolia Variegata'	wChr oRos oHei wCLG
	'Cerise Bouquet'	oRos oHei
	Champagne Cocktail / 'Horflash'	oRos oHei oWnC oRNW
	'Champlain'	oHei
un	'Champlain's Pink'	oHei
	Chapeau de Napoleon	see **R. *x centifolia*** 'Cristata'
	Charisma / 'Peatrophy'	oWnC
	Charisma / 'Peatrophy' (tree)	oWnC
	'Charles Albanel'	oHei
	Charles Austin® /'Ausles'	wChr oRos wCot oHei oRNW wAnt
	Charles Aznavour®/ 'Meibeausai'	wChr oWnC
	Charles Aznavour®/ 'Meibeausai' (tree)	wChr
	'Charles de Mills'	wChr oRos wCot oHei wCLG
	Charles Kuralt / 'Cletraveler'	oHei oRNW
	'Charles Lawson'	oRos oHei wAnt
	Charles Rennie Mackintosh®/ 'Ausren'	oHei oWnC oRNW
	Charlotte / 'Auspoly'	wChr oHei oRNW
un	'Charm Bracelet'	oGar
un	'Charmglo'	oHei oJus
	Charmian®/ 'Ausmian'	oRos oHei
un	'Chain Rainbows'	oJus
un	'Chattem Centennial'	oJus

	Chaucer®/ 'Auscer'	wChr oRos oHei oRNW
un	'Cheerleader'	oJus
un	'Cheers'	oJus
	Chelsea Belle / 'Talchelsea'	oHei
un	'Chere Michelle'	oJus
	Cherish / 'Jacsal'	last listed 99/00
	Cherries Jubilee™ / 'Clecherry'	oHei
	Cherry Brandy '85®/ 'Tanryandy'	oHei
	Cherry Meidiland™ / 'Meirumour'	wChr wCot oHei oRNW wAnt
	'Chevy Chase'	wAnt
	'Chianti'	wChr oRos oHei
	Chicago Peace®/ 'Johnago'	wChr wCot oEdm oGar oWnC oAls
un	'Chick a Dee'	oJus
	Childhood Memories / 'Ferho'	oHei
	Child's Play™ / 'Savachild'	wCot oGar oWnC oJus oUps
	'China Doll'	wChr oHei oGar oWnC oRNW wAnt
	Chinatown®	oRNW
	chinensis 'Mutabilis'	see **R.** ***x odorata*** 'Mutabilis'
	chinensis 'Semperflorens'	see **R.** ***x odorata*** Old Crimson China
	chinensis 'Viridiflora'	see **R.** ***x odorata*** 'Viridiflora'
un	'Chiquita'	oJus
	'Chloris'	oRos oHei
un	'Choo Choo Centennial'	oJus
un	'Chorale'	oHei
un	'Chris Evert'	oGar wAnt
	Christian Dior / 'Meilie'	oGar oWnC
	Christine™ / 'Clealta'	oHei
	Christmas Snow™ / 'Clefrosty'	oHei
	Christopher Columbus®/ 'Meinronsse'	oHei oWnC oAls oRNW
	'Chrysler Imperial'	wChr oGar oWnC oRNW
	'Chrysler Imperial' (tree)	oWnC
	'Chuckles'	last listed 99/00
un	'Chula Vista'	oJus
	Cider Cup / 'Dicladida'	oHei
un	'Cimarosa'	oHei
	'Cinderella'	oJus
	City of London®/ 'Harukfore'	wChr wCot oHei oEdm oRNW
	City of York / 'Direktor Benschop'	wChr oRos oHei
	Clair Matin®/ 'Meimont'	oRos oHei oRNW
	'Claire Jacquier'	oRos
un	'Claire Renaissance'	wChr wCot
	Claire Rose®/ 'Auslight'	oRos oHei oRNW
	Clarissa®/ 'Harprocrustes'	last listed 99/00
	Classic Sunblaze®/ 'Meipinjid'	oWnC
un	'Claudia'	oRNW
un	'Claudia Cardinale'	wChr wCot
un	Cliffs of Dover / 'Poulemb'	oRNW
	Climbing Altissimo	see **R.** Altissimo®/ 'Delmur'
un	'Climbing Angel Face'	wChr oRos
	Climbing Autumn Sunset	see **R.** 'Autumn Sunset'
un	'Climbing Baronne Prevost'	oRos
un	Climbing Berries 'n' Cream	see **R.** Berries 'n' Cream™ / 'Poulclimb'
un	'Climbing Betty Prior'	last listed 99/00
un	'Climbing Blue Girl'	last listed 99/00
	'Climbing Cecile Brünner'	wChr oRos oHei oGar oWnC wAnt oUps
	'Climbing Cecile Brünner' (everblooming)	see **R.** Everblooming Climbing Cecile Brünner™
	Climbing City of York	see **R.** City of York / 'Direktor Benschop'
un	'Climbing Clotilde Soupert'	oHei
un	'Climbing Colette'	oGar
	'Climbing Crimson Glory'	oHei
un	'Climbing Dainty Bess'	oRos wCot oHei
	'Climbing Devoniensis'	oRos oHei
un	'Climbing Double Delight'	wChr
un	'Climbing Dream Weaver'	oGar
	Climbing Dublin Bay	see **R.** Dublin Bay®/ 'Macdub'
	'Climbing Ena Harkness'	oHei
	'Climbing Etoile de Hollande'	last listed 99/00
	'Climbing Fashion'	last listed 99/00
un	'Climbing First Prize'	wChr oHei oGar oWnC
	Climbing Fourth of July	see **R.** Fourth of July™ / 'Wekroalt'
	Climbing Gold Badge / 'Meigro-Nurisar'	see **R.** Climbing Gold Bunny / 'Meigro-Nurisar'
	Climbing Gold Bunny / 'Meigro-Nurisar'	oHei
	Climbing Handel	see **R.** Handel®/ 'Macha'
un	'Climbing Happy'	oRNW
un	'Climbing Heidelberg'	oRos
	'Climbing Iceberg'	wChr oRos oHei oEdm oGar oRNW
un	'Climbing King Tut'	oGar
	'Climbing la France'	oRos oHei
un	'Climbing Lace Cascade'	oGar
	'Climbing Lady Hillingdon'	last listed 99/00
	'Climbing Madame Butterfly'	oHei
	'Climbing Mademoiselle Cecile Brünner'	see **R.** 'Climbing Cecile Brünner'
	'Climbing Mrs. Herbert Stevens'	oRos oHei oRNW
	'Climbing Mrs. Sam McGredy'	last listed 99/00
	'Climbing x ***odorata*** Pallida'	see **R.** ***x odorata*** 'Pallida'
un	'Climbing Oklahoma'	last listed 99/00
	'Climbing Old Blush'	see **R.** ***x odorata*** 'Pallida'
	Climbing Orange Sunblaze / 'Meijikatarsar'	oHei
un	'Climbing Paprika'	oGar oWnC
un	'Climbing Peace'	wChr oGar
un	Climbing Pearly Gates™ / 'Wekmeyer'	oEdm
un	'Climbing Playgirl'	wChr oRNW
	'Climbing Red Fountain'	see **R.** 'Red Fountain'
un	'Climbing Renae'	oHei
qu	'Climbing Roulette'	oHei
un	'Climbing Royal America'	oRNW
un	'Climbing Royal Sunset'	see **R.** 'Royal Sunset'
un	'Climbing Sally Holmes'	oGar
un	'Climbing Shadow Dancer'	oWnC
	'Climbing Shot Silk'	wChr wCot oHei
un	'Climbing Snowflake'	oHei
	'Climbing Sombreuil'	see **R.** 'Sombreuil'
	'Climbing Souvenir de la Malmaison'	wChr wCot
	'Climbing Sutter's Gold'	wChr wCot
	'Climbing Talisman'	oRos oRNW
	'Climbing The Fairy'	see **R.** 'Lady Carolina'
	'Climbing The Queen Elizabeth'	wChr oRos oGar oWnC
un	'Climbing Tropicana'	last listed 99/00
	'Climbing Vicomtesse Pierre du Fou'	see **R.** 'Vicomtesse Pierre du Fou'
un	'Climbing Viking Queen'	oRNW
	Climbing Westerland	see **R.** Westerland™ / 'Korwest'
un	'Climbing Winifred Coulter'	oHei
	'Clotilde Soupert'	oRos oRNW
	'Clytemnestra'	oRos oHei wCLG
	Cocktail®/ 'Meimick'	oHei
un	'Cocorico'	oHei
un	'Coeur d'Alene'	wChr oGar
	'Coeur d'Amour'	see **R.** Red Devil®/ 'Dicam'
	Colette™ / 'Meiroupis'	wChr wCot oHei wRai oRNW wAnt
un	Cologne / 'Macsupbow'	oEdm oRNW
	'Colonel Fabvier'	oRos oRNW
ub	Color Magic / 'Jacmag'	wCot oEdm oRNW
un	'Colossus'	oRNW
un	'Colwyn Rose'	oHei
un	'Command Performance'	oWnC
	'Commandant Beaurepaire'	oRos oHei
un	'Commitment'	oHei
	'Communis'	see **R.** ***x centifolia*** 'Muscosa'
	Compassion®	wChr oRos wCot oHei oRNW
	'Complicata'	oFor wChr oRos oHei oNWe wRob wCCr
	'Comte Boule de Nanteuil'	see **R.** 'Boule de Nanteuil'
	'Comte de Chambord'	see **R.** 'Madame Knorr'
	'Comtesse Cecile de Chabrillant'	oHei
	'Comtesse de Murinais'	oRos oHei
un	'Comtesse de Rocquigny'	wCLG
un	'Concorde'	oRNW
	'Condesa de Sastago'	oRNW
	'Conditorum'	oHei
un	'Confetti'	oWnC oRNW
un	'Confidence'	oRNW
	'Conrad Ferdinand Meyer'	wChr oRos wCLG
	Conservation / 'Cocdimple'	oHei
	Constance Spry / 'Austance'	wChr oRos wCot oHei oGar wAnt
un	'Copper Sunset'	oJus
	'Coquette des Blanches'	wChr wCot
	Coral Dawn®	oHei oRNW
	Coral Fiesta®	see **R.** Maria Teresa de Esteban / 'Dotrames'
	Coral Meidiland™ / 'Meipopul'	wChr oHei oWnC oRNW
un	'Coral Sea'	oRNW
un	'Coral Treasure'	oHei
	'Cornelia'	wChr wCot oHei wCLG wAnt
	'Cornelia' (tree)	wChr
	'Corylus'	see **R.** 'Hazel Le Rougetel'
	corymbifera	oHei
un	'Cosette'	oHei
	Cotillion™ / 'Jacshok'	wChr wCot oWnC oRNW wAnt
	Cottage Garden / 'Haryamber'	oHei
	Cottage Rose™ / 'Ausglisten'	wChr wCot oHei oGar oRNW wAnt
un	'Cottontail'	oJus
un	Countess Celeste™	wChr oGar oRNW
un	Countess Celeste™ (tree)	oGar
	'Country Dancer'	wChr oRos oHei oRNW
	Country Living™ / 'Auscountry'	wChr oHei
un	'Country Music'	oHei

un	'Country Song'	oHei
un	'Countryman'	oRos oHei
	'Coupe d'Hebe'	oHei
un	'Crackling Fire'	oWnC
	'Cramoisi Picotee'	last listed 99/00
	'Cramoisi Superieur'	oRos oHei
un	'Crazy Dottie'	oJus
	Crazy for You	see **R.** Fourth of July™ / 'Wekroalt'
	'Crepuscule'	oRos oHei
	Cressida / 'Auscress'	wChr oRos wCot oHei wAnt
	'Crested Jewel'	oRos
un	'Crested Sweetheart'	oRNW
	Cri Cri / 'Meicri'	oHei
un	'Cricket'	oJus
un	Crimson Blush™ / 'Sieson'	oHei
un	Crimson Bouquet™ / 'Korbeteilich'	wChr wCot oEdm oWnC oAls oRNW
un	'Crimson Fuchsia'	oHei
	'Crimson Glory'	oHei oRNW
un	'Crimson Queen'	oHei
	'Crimson Rambler'	oRos oHei
	'Crimson Shower'	oHei wAnt
un	'Crimson Spire'	see **R.** 'Liebeszauber'
un	Crisp-n-Sweet™ (fruit)	oOEx
	Crystal Palace®/ 'Poulrek'	wChr
un	Crystalline™ / 'Arobipy'	oEdm
	Cupcake / 'Spicup'	wChr oUps
	Cupcake / 'Spicup' (tree)	oWnC
	'Cupid'	oHei
un	'Cuthbert Grant'	oHei oGar
	Cymbeline / 'Auslean'	wChr oRos oHei wCLG wAnt
	'Dainty Bess'	wChr oRos wCot oHei oEdm oSha oWnC oRNW
	'Dainty Maid'	last listed 99/00
	'Dairy Maid'	oHei
	'Daisy Hill'	oHei
un	'Daisy Mae Rogers'	wAnt
	x damascena var. ***bifera***	see **R.** ***x damascena*** var. ***semperflorens***
	x damascena var. ***semperflorens***	wChr oRos wCot oHei wAnt
co	***x damascena trigintipetala***	see **R.** 'Professeur Emile Perrot'
	x damascena var. ***versicolor***	wChr oRos
un	'Dame Bridget d'Olycart'	oHei
un	'Dame de Coeur'	oRNW
un	'Dame Prudence'	oRos oHei
	'Danae'	wChr oHei oJoy wCLG
un	'Dancing Doll'	oRNW
	Dancing in the Wind™ / 'Cledan'	oHei
	Danny Boy / 'Cleirish'	oHei
	Dapple Dawn / 'Ausapple'	oRos oHei oRNW
un	'Darby O'Gill'	oJus
un	'Dark Mirage'	oJus
un	'Darlow's Enigma'	wChr oRos oHei oNWe wAnt
un	'Dart's Dash'	oHei wRai
	'David Thompson'	wChr oHei oUps
	davidii	oFor oUps
	sp. aff. ***davidii*** DJHC 266	wHer
	davurica	oFor
	'De la Grifferaie'	oRos
	'De Meaux'	oHei
	'De Rescht'	wChr oRos wCot oHei oGar oSha wCLG wAnt
	'Dearest'	oHei
	Deb's Delight / 'Legsweet'	oHei
	Debut™ / 'Meibarke'	oWnC
	Dee Bennett™ / 'Savadee'	oJus
	'Deep Secret'	oHei
	'Delicata'	wChr wCot oHei
un	Della Balfour™	oHei
un	'Dentelle de Bruges'	oHei
un	'Denver's Dream'	oJus
	Desert Peace™ / 'Meinomad'	oWnC oAls oRNW
un	'Desiree Parmentier'	oRos
	'Desprez a Fleurs Jaunes'	oHei wCLG wAnt
	'Diamond Jubilee'	oRos oRNW wCLG
	Diana, Princess of Wales™ / 'Jacshaq'	wChr wCot oWnC oAls oRNW wAnt
	'Dick Koster'	oRos
	Dicky	see **R.** Anisley Dickson / 'Dickimono'
	Die Welt®/ 'Diekor'	wChr wCot
	Disco Dancer®/ 'Dicinfra'	oHei oRNW
un	'Distant Drums'	oRos oHei oRNW
un	'Doctor Brownell'	oRNW
un	'Doctor J. H. Nicholas'	oRNW
	Doctor Jackson™ / 'Ausdoctor'	wChr oRos oHei
un	Doctor Robert Korns / 'Letrob'	oHei
	'Doctor W. Van Fleet'	oSha oRNW
	'Dolly Parton'	oHei oGar oWnC
un	'Dominie Sampson'	oRos
	'Don Juan'	wChr wCot oHei oGar oSho oWnC oRNW
	'Doncasteri'	oHei
un	'Donna Faye'	oJus
	Donna Oehler™	oHei
un	'Dornroeschen'	see **R.** 'Sleeping Beauty'
	'Dorothy Perkins'	oHei
	Dortmund®	wChr oRos wCot oHei oEdm oRNW wAnt
	Double Delight®/ 'Andeli'	wChr wCot oHei oEdm oGar oSho oWnC oAls oRNW wAnt oUps
	Double Delight®/ 'Andeli' (tree)	wCot oEdm oGar
un	'Double Treat'	oJus
un	'Double Yellow'	oHei
	'Douglas Gandy'	oHei
	Dove®/ 'Ausdove'	wChr oRos oHei oRNW
	'Dream Girl'	oHei
un	Dream™ Orange / 'Twoaebi'	wChr
un	Dream™ Pink / 'Twojoan'	wChr
un	Dream™ Red / 'Twopaul'	wChr
un	Dream™ Yellow / 'Twoyel'	wChr
un	'Dream Sequence'	oHei
	Dream Weaver™ / 'Jacpicl'	wChr wCot oSho oWnC oRNW wAnt
un	'Dreamcatcher'	oJus
un	'Dreamcoat'	oJus
un	'Dreamer'	oJus
	'Dreamglo'	oJus
	'Dreaming Spires'	oRos oHei oRNW
	'Dresden Doll'	oHei
	Drummer Boy / 'Harvacity'	wChr oHei
un	'Dublin'	oEdm oWnC
	Dublin Bay®/ 'Macdub'	wChr oRos wCot oHei oEdm oWnC oRNW wAnt
	'Duc de Fitzjames'	oHei
	'Duc de Guiche'	wCLG
	'Duc de Rohan'	see **R.** 'Duchesse de Rohan'
un	'Duchess'	oRNW
	'Duchess of Portland'	see **R.** 'Portlandica'
	'Duchesse d'Angouleme'	wCLG
	'Duchesse de Brabant'	last listed 99/00
	'Duchesse de Montebello'	oHei
	'Duchesse de Rohan'	oRos oHei wCLG
	'Duet'	oRNW
	'Duftwolke'	see **R.** Fragrant Cloud / 'Tanellis'
	'Duke of Edinburgh'	oRos wAnt
	dumalis	oFor
	dunwichensis	see **R.** ***pimpinellifolia*** 'Dunwich Rose'
	'Dupontii'	oRos oHei
	'Dupuy Jamain'	oHei
un	'Dutch Provence'	oHei
un	'Dwarf Crimson Rambler'	see **R.** 'Madame Norbert Levasseur'
un	'Dwarf Pavement'	see **R.** 'Zwerg'
	Dynamite / 'Jacsat'	wChr wCot oWnC oRNW
un	'E. Veyrat Hermanos'	wCLG
un	'Earth Song'	oHei
	Easy Going / 'Harflow'	wChr wCot oEdm oRNW
	Easy Going / 'Harflow' (tree)	oEdm
un	'Echo'	oRos oHei
	'Eddie's Jewel'	wHer oHei
	Eden Climber™ / 'Meiviolin'	wChr wCot oHei wAnt
co	Eden Rose®(see Nomenclature Notes)	oGar oWnC oRNW
	Eden Rose 88	see **R.** Eden Climber™ / 'Meiviolin'
	Edith Holden / 'Chewlegacy'	oHei oRNW
	eglanteria	see **R.** ***rubiginosa***
	Eglantyne®/ 'Ausmak'	wChr oHei oGar oRNW wAnt
un	'Eiffel Tower'	oHei oRNW
	'Electron'	see **R.** 'Mullard Jubilee'
	'Elegance'	oRNW
un	'Elegante Gallica'	oHei
un	'Elie Beauvilain'	wCLG
	Elina®/ 'Dicjana'	wChr wCot oEdm oRNW
un	'Eliza'	oRNW
	Elizabeth of Glamis®/ 'Macel'	wCot
un	'Elizabeth Taylor'	wChr wCot
un	'Ellamae'	oJus
	Ellen®/ 'Auscup'	oRos oHei oGar oRNW wAnt
	'Ellen Poulsen'	oRos oHei
	'Ellen Willmott'	oRos
un	'Elmhurst'	oEdm oRNW
	'Elmshorn'	oHei
	'Else Poulsen'	oHei
un	'Elveshorn'	oHei

	Emanuel®/ 'Ausuel'	wAnt
un	'Ember'	oJus
	Emily / 'Ausburton'	oHei
	'Emily Gray'	oRos oHei
	'Empress Josephine'	see **R.** ***x francofurtana***
	'Ena Harkness'	oHei
	'Enfant de France'	wChr wCot oHei oSha wCLG wAnt
	Engineer's Rose	see **R.** 'Crimson Rambler'
	English Elegance®/ 'Ausleaf'	wChr wCot oHei oRNW
	English Garden®/ 'Ausbuff'	wChr wCot oHei oWnC oRNW wAnt
	'English Miss'	oHei oRNW
	English Sachet™ / 'Jacolfa'	wChr wCot oWnC oRNW
co	'English Violet'	oHei
	'Erfurt'	oHei
	'Erinnerung an Brod'	oRos oHei
	Escapade®/ 'Harpade'	oRos wCot oHei
	'Esmeralda'	see **R.** 'Keepsake'
	Essex / 'Poulnoz'	oHei oJoy
	'Etendard'	oHei oRNW
un	'Eternity'	oRNW
	'Ethel'	oHei
	'Etoile de Hollande'	oRos
un	'Eugene de Beauharnais'	wChr oRos wCot oHei wCLG
	'Eugene Fuerst'	oHei wCLG wAnt
	Euphrates / 'Harunique'	oHei
	Europeana®	wChr wCot oEdm oGar oWnC wAnt
	Europeana® (tree)	oWnC
un	'Eutin'	oRNW
	'Eva'	wChr wCot oHei
	'Evangeline'	wCul oRos oHei wAnt
	Evelyn®/ 'Aussaucer'	wChr wCot oHei oGar oRNW wAnt
	Evening Star®/ 'Jacven'	last listed 99/00
	Everblooming Climbing Cecile Brünner™	oHei oRNW
	'Everest Double Fragrance'	wChr wCot oHei oSha
un	'Excellenz von Schubert'	wChr oRos oHei
	'Excelsa'	oHei
un	'Exotica'	oRNW
	Eye Paint / 'Maceye'	oHei oEdm oRNW
	Eyeopener / 'Interop'	oHei
	'F. J. Grootendorst'	wChr oRos wCot oHei oGar oSho oRNW wAnt
	'Fabvier'	see **R.** 'Colonel Fabvier'
	Fair Bianca®/ 'Ausca'	wChr wCot oHei oGar oWnC oRNW wAnt
	Fair Bianca®/ 'Ausca' (tree)	last listed 99/00
un	'Fair Dinkum'	oJus
	Fairy Damsel / 'Harneaty'	oHei
un	'Fairy Frolic'	oHei
	'Fairy Lights'	see **R.** Avon / 'Poulmulti'
	'Fairy Moss'	oHei oJus
un	'Fairy Prince'	oHei
	Fame!™ / 'Jaczor'	wChr wCot oEdm oSho oWnC oAls oRNW
	Fancy Lady / 'Clclady'	oHei
	'Fantin-Latour'	wChr oRos oHei wCLG wAnt
un	'Favorite Dream'	wChr
	fedtschenkoana	oHei
	'Felicia'	wChr oRos oHei wCLG
	'Felicite Parmentier'	oRos oHei wAnt
	'Felicite Perpetue'	oRos oHei wAnt
	'Fellenberg'	oRos oHei
	'Ferdinand Pichard'	oRos wCot oHei oSha oRNW
	Ferdy®/ 'Keitoli'	oHei
	Festival / 'Kordialo'	last listed 99/00
un	'Festival Fanfare'	oHei
	Figurine™ / 'Benfig'	wChr oJus
	filipes	oFor
	filipes 'Kiftsgate'	wChr oRos oHei wAnt
	'Fimbriata'	wChr oRos wCot
	Financial Times Centenary / 'Ausfin'	oHei
un	'Fingerpaint'	oJus
	Fiona®/ 'Meibeluxen'	last listed 99/00
	Fire Meidiland™ / 'Meipsidue'	wChr oHei oRNW
	Fire Meidiland™ / 'Meipsidue' (tree)	wCot
un	'Fire 'n' Ice'	oRNW
un	'Fireworks'	oJus
un	'First Edition'	oWnC
un	'First Kiss'	wCot
un	First Light™ / 'Devrudi'	wChr oEdm oGar oWnC wAnt
	'First Prize'	wChr oGar oSho oWnC oAls
	Fisherman's Friend®/ 'Auschild'	oRos oSha
un	Five Colored Rose	see **R.** 'Fortune's Five Colored'
	Flamingo	see **R.** Margaret Thatcher / 'Korflueg'
un	'Flashfire'	oHei
un	'Flora Danica'	wCot
un	'Floradora'	oHei
un	'Floranne'	oJus
un	'Florence deLattre'	wChr wCot
un	'Flower Basket'	oHei oJus
	Flower Carpet	see **R.** Pink Flower Carpet™ / 'Noatraum'
un	Flower Carpet Appleblossom	see **R.** Appleblossom Flower Carpet™ / 'Noamel'
	Flower Carpet Pink	see **R.** Pink Flower Carpet™ / 'Noatraum'
un	Flower Carpet Red	see **R.** Red Flower Carpet
	Flower Carpet White	see **R.** White Flower Carpet®/ 'Noaschnee'
	'Flower Girl'	see **R.** 'Sea Pearl'
un	Flutterby / 'Wekplasol'	wChr wCot oHei oGar oCis wAnt
un	'Flying Colors'	oJus
un	'Focus'	see **R.** 'Sweet Bouquet'
	foetida 'Bicolor'	oFor wChr oRos
	foetida 'Persian Yellow'	see **R.** ***foetida*** 'Persiana'
	foetida 'Persiana'	wChr oRos oWnC
	foliolosa	oFor oRos
un	Folklore / 'Korlore'	wCot oEdm
un	'Folksinger'	wChr wCot oHei
un	'Fool's Gold'	oJus
un	Footloose / 'Tanotax'	oHei oGar oRNW
	'Forgotten Dreams'	oHei oEdm
un	'Fortune Cookie'	oJus
un	'Fortune Teller'	oWnC oRNW
	x fortuneana	oRos oHei
	Fortune's Double Yellow	see **R.** ***x odorata*** 'Pseudindica'
un	'Fortune's Five Colored'	oRos oHei
	'Fountain'	oHei
	Fountain Square / 'Jacmur'	wCot oGar oWnC oAls oRNW
	Fourth of July™ / 'Wekroalt'	wChr wCot oHei oEdm oGar oSho oWnC oRNW wAnt
un	'Foxi Pavement'	oHei
un	'Fragrant Apricot'	wChr wCot oGar oWnC oAls oRNW wAnt
un	'Fragrant Apricot' (tree)	wChr oGar
	Fragrant Cloud / 'Tanellis'	wChr wCot oHei oEdm oGar oSho oWnC oAls oRNW
	Fragrant Delight®	wChr oHei
un	'Fragrant Fantasy'	oRNW
	'Fragrant Hour'	wChr
un	'Fragrant Lady'	oRNW
	Fragrant Memory™	see **R.** Jadis / 'Jacdis'
un	'Fragrant Plum'	wChr wCot oHei oWnC oRNW wAnt
	'Francesca'	wChr oRos oHei wCLG wAnt
	Francine Austin®/ 'Ausram'	oHei
	'Francis Dubreuil'	oHei wCLG
	'Francis E. Lester'	oHei
	x francofurtana	wChr oHei
	'Francois Juranville'	oHei wCLG wAnt
un	'Francois Rabelais'	oRNW
	'Frau Dagmar Hartopp'	see **R.** 'Fru Dagmar Hastrup'
	'Frau Dagmar Hastrup'	see **R.** 'Fru Dagmar Hastrup'
	'Frau Eva Schubert'	oHei
	'Frau Karl Druschki'	oRos
	'Fred Loads'	oHei oRNW
	Frederic Mistral™ / 'Meitebros'	wChr wCot oHei oWnC oRNW
un	'Free Spirit'	oJus
un	'Freedom's Ring'	wChr
	'Freegold'	see **R.** Penelope Keith / 'Macfreego'
	'Freiherr von Marschall'	wCLG
	French Lace / 'Jaclace'	wChr wCot oGar oSho oAls oRNW wAnt
	French Lace / 'Jaclace' (tree)	last listed 99/00
un	French Perfume™ / 'Keibian'	wChr wCot oGar oSho oWnC oAls oRNW
un	French Perfume™ / 'Keibian' (tree)	oGar
un	Friendship®/ 'Linrick'	oAls oRNW
	'Fritz Nobis'	oHei
ub	Frohsinn / 'Tansinnroh'	wCot oEdm oRNW
	'Frontenac'	wChr
	'Fru Dagmar Hastrup'	wChr oRos wCot oHei wRai wAnt wWel
	'Fruehlingsanfang'	oHei
	'Fruehlingsgold'	wChr oRos oHei
	'Fruehlingsmorgen'	wChr wCot oHei
	'Fruehlingsschnee'	oHei
	'Fruehlingszauber'	oHei
	Fuchsia Meidiland™ / 'Meipelta'	wChr wCot oHei oRNW wAnt
	Fuchsia Meidiland™ / 'Meipelta' (tree)	oWnC
un	Full Sail / 'Maclanoflon'	oEdm oRNW
un	Full Sail / 'Maclanoflon' (tree)	oEdm
un	'Funny Girl'	wCot oGar oWnC
un	'Fuzzy Navel'	oHei
	'Gabriel Noyelle'	oRos wCLG
un	'Gabriel Privat'	oHei
un	***gallica grandiflora***	see **R.** 'Alika'
	gallica var. ***officinalis***	wChr oRos wCot oHei wCLG wAnt

	gallica 'Versicolor'	oRos wCot oHei wAnt
	Galway Bay®/ 'Macba'	oRos oHei oRNW
un	Garden Delicious™ (fruit)	oOEx
	Garden Party™ / 'Kormollis'	wChr wCot
	'Gardenia'	oRos oHei
un	'Garisenda'	oRos oHei
un	'Gartendirektor Otto Linne'	wChr oRos wCot oHei
un	'Gee Whiz'	oRNW
	Gemini™ / 'Jacnepal'	wChr wCot oEdm oWnC oAls oRNW
un	'Gene Boerner'	wChr wCot oWnC oRNW
	'General Fabvier'	see **R.** 'Colonel Fabvier'
	'General Jacqueminot'	oRos oHei
	'General Kleber'	oRos oHei
un	'General Washington'	oRos
co	***gentiliana***	oHei wCCr
un	Gentle Maid / 'Harvilac'	oHei oRNW
un	'Gentle Persuasion'	oHei
un	'George Burns'	wChr wCot oWnC oRNW
	'Georges Vibert'	oHei
	'Geranium' (***moyesii*** hybrid)	oRos oHei
	Gertrude Jekyll®/ 'Ausbord'	oFor wChr wCot oHei oGar oRNW wAnt
	'Geschwind's Most Beautiful'	see **R.** 'Geschwinds Schoenste'
	'Geschwinds Orden'	oHei
	'Geschwinds Schoenste'	oHei
un	'Ghirlande d'Amour'	oHei
	'Ghislaine de Feligonde'	oRos oHei oNWe wAnt
un	'Giant of Battles'	oHei
un	Gift of Life / 'Harelan'	see **R.** Poetry in Motion / 'Harelan'
	gigantea	oHei
	Gina Lollobrigida / 'Meilivar'	oGar oWnC oAls
un	'Ginger Hill'	oRNW
un	'Gingerbread Man'	oJus
	Gingernut / 'Coccrazy'	oHei
	Gingersnap / 'Arosnap'	wChr oWnC
	Gingersnap / 'Arosnap' (tree)	oWnC
	'Giselle'	oJus
un	Gitte / 'Korita'	oEdm
un	'Givenchy'	wChr
un	'Gizmo'	wChr wCot
	Glad Tidings / 'Tantide'	wChr wCot oHei oEdm oRNW
	Glad Tidings / 'Tantide' (tree)	oEdm
	Glamis Castle / 'Auslevel'	wChr wCot oHei oGar oSho oWnC oRNW wAnt
	Glamour Girl™ / 'Cleamour'	oHei
un	'Glastonbury'	oRos oHei
	glauca	oFor wHer oHan wWoS wCul wBCr oGos oRos wCot oHei wSwC oEdg wCSG oAmb oGar oDan oRiv oNWe wRob oOEx wFai wPir wCCr wAnt oUps oSle wSte wSnq
	'Gloire de Dijon'	wChr oHei oRNW wCLG wAnt
	'Gloire de Ducher'	wCLG
	'Gloire de France'	oHei wAnt
	'Gloire de Guilan'	oRos oHei
	'Gloire des Mousseuses	oRos oHei wCLG
	'Gloire des Rosomanes'	oHei
	'Gloire Lyonnaise'	wChr wCLG
	glutinosa	see **R. *pulverulenta***
	'Goethe'	oRos
un	'Gold Country'	oJus
	Gold Medal / 'Aroyqueli'	wChr wCot oEdm oGar oWnC oAls oRNW
un	'Gold Moss'	oHei
	Gold Star / 'Tantern'	oHei wAnt
	'Goldbusch'	oHei wCLG
un	'Golddust'	see **R.** 'Yellow Jacket'
	Golden Beauty / 'Clebeau'	oHei
	Golden Celebration™ / 'Ausgold'	wChr wCot oHei oGar oSho oWnC oRNW wAnt
un	'Golden Eagle'	oRNW
un	'Golden Fantasie'	oRNW
un	Golden Fire	see **R.** Perestroika / 'Korhitom'
	Golden Halo™ / 'Savahalo'	oJus
un	Golden Holstein / 'Kortikel'	wChr oEdm oRNW
un	'Golden Masterpiece'	oWnC
	'Golden Moss'	oRos oHei
un	'Golden Plover'	oHei
	Golden Showers®	wChr oRos wCot oGar oSho oSha oWnC wAnt
	Golden Showers® (tree)	oWnC
	Golden Star	see **R.** Gold Star / 'Tantern'
un	'Golden Touch'	oHei
	'Golden Wings'	wWoS wChr oRos oHei oNWe oRNW wCLG wAnt oSle
un	Goldencrisp Delicious™ (fruit)	oOEx
un	'Goldener Olymp'	oHei oRNW
	'Goldfinch'	oHei
	'Goldilocks'	oWnC
un	'Goldmoss'	oRNW
	Goldstern®	see **R.** Gold Star / 'Tantern'
un	'Gone Fishin'	oJus
	Good as Gold™ / 'Chewsunbeam'	oHei
	Good Ol' Summertime™ / 'Cleheat'	oHei
un	'Gourmet Pheasant'	oHei
	Gourmet Popcorn / 'Weopop'	wChr wCot oHei oGar oSha oJus
	Gourmet Popcorn / 'Weopop' (tree)	wChr wCot oGar
	Graceland / 'Kirscot'	last listed 99/00
un	'Gracie Allen'	wChr wCot oWnC oRNW
	Graham Thomas / 'Ausmas'	oFor wChr wCot oHei oGar oWnC oRNW wAnt
	Granada	wCot oHei oGar oWnC oAls oRNW
un	Grand Bouquet™	see **R.** 'Lyric'
	Grand Finale™ / 'Jacpihi'	wChr wCot oGar oSho oWnC oAls oRNW
	Grand Finale™ / 'Jacpihi' (tree)	oGar oSho
un	'Grand Gala'	oRNW
	'Grand Masterpiece'	last listed 99/00
un	'Grand Old Flag'	oHei
	Grandma's Lace™ / 'Clegran'	oHei
un	'Grandma's Pink'	oHei oRNW
un	'Granny Grimetts'	oHei
	'Great Maiden's Blush'	wChr oRos wCot oHei
	'Great News'	oRos
un	'Great Scott'	oEdm
	'Great Western'	oRos
	'Green Ice'	oHei oJus oRNW
	Green Snake®/ 'Lenwich'	oHei wAnt
	'Greenmantle'	oRos oHei
	Greensleeves®/ 'Harlenten'	oHei
	'Grootendorst Supreme'	oUps
co	'Gros Choux de Hollande'	oRos oHei
	Grouse / 'Korimro'	oHei
	'Gruss an Aachen'	wChr oRos wCot oHei oGar oRNW wCLG wAnt
	'Gruss an Aachen' (tree)	last listed 99/00
	'Gruss an Teplitz'	oHei wCLG
	'Guinee'	oRos oHei wAnt
qu	'Gurilande d'Amour'	oHei
	Guy de Maupassant™ / 'Meisocrat'	wChr wCot oHei oGar oWnC oRNW wAnt
	Gwent / 'Poulurt'	oHei
	gymnocarpa	oFor oHan wWoB oHei wNot wShR oBos wWat oTri oAld wFFl
	Gypsy Boy	see **R.** 'Zigeunerknabe'
un	Gypsy Carnival™ / 'Kiboh'	oWnC oAls
un	'Gypsy Dancer'	wChr oGar oWnC oRNW
	Hamburger Phoenix®	oHei
	Handel®/ 'Macha'	wChr wCot oHei oGar oWnC oRNW wAnt
un	'Hanini'	oJus
un	'Hans Christian Andersen'	oRNW
	'Hansa'	oFor wCul wChr oRos wCot oHei oGar oDan oSho oOEx oWnC oRNW wAnt oUps oSle
un	'Hanseat'	oHei
un	'Happenstance'	oHei
	'Happy'	oRNW
	Happy Child / 'Auscomp'	wChr wCot oHei
	Happy Hour' / 'Savanhour'	oJus
un	Happy Wanderer®	oHei
	x harisonii 'Harison's Yellow'	oFor wChr oHei
un	'Harlekin'	see **R.** Kiss of Desire / 'Korlupo'
	'Harry Wheatcroft'	wChr
un	'Hazel Le Rougetel'	oHei
un	'Heart o' Gold'	oWnC
	Heartbreaker / 'Weksibyl'	wChr oJus
	Heather Austin / 'Auscook'	wChr wCot oHei wAnt
un	'Heaven'	oRNW
un	'Heaven' (tree)	oWnC
un	'Heavenly Days'	oJus
	Heavenly Rosalind / 'Ausmash'	wChr oHei
	'Hebe's Lip'	oRos oHei oJoy
	'Heidelberg'	oHei
	'Heideroeslein'	see **R.** 'Nozomi'
un	'Heideschnee'	see **R.** 'Snow on the Heather'
un	'Heidi'	wChr oHei oWnC oJus
	'Heinrich Schultheis'	oHei
	'Heirloom'	wChr oGar oSho oWnC oRNW
un	Helen Naude / 'Kordiena'	wCot oEdm
	'Helen Traubel'	oRos oRNW
un	'Helena Renaissance'	wChr wCot
	helenae	oFor oHei wCCr
	Helmut Schmidt	see **R.** Simba / 'Korbelma'

	Helping Hands / 'Clehelp'	oHei
	hemsleyana	oHei
un	'Henri Barruet'	oRos oHei
	'Henri Fouquier'	oHei
	'Henri Martin'	oRos oHei oSha wAnt
un	'Henry Fonda'	wChr wCot oGar oSho oWnC oAls oRNW
	'Henry Hudson'	wCot oHei oDan oUps
	'Henry Kelsey'	oHei oRNW oSle
	Heritage®/ 'Ausblush'	wChr wCot oHei oRNW wAnt
	'Hermosa'	oRos oHei oRNW
	Hero®/ 'Aushero'	oRos oHei
	Hertfordshire / 'Kortenay'	oHei
	'Hi Neighbor'	oHei
	'Hiawatha'	oFor
	x hibernica	oHei
	'Hidcote Yellow'	see®'Lawrence Johnston'
un	'High Life'	oJus
	'High Noon'	wChr wCot oHei oWnC
un	'High Spirits'	oJus
	'Highdownensis'	oHei
	Hilda Murrell®/ 'Ausmurr'	oHei
	'Hippolyte'	oRos
un	Hoagy Carmichael / 'Mactitir'	oEdm
un	'Hokey Pokey'	oJus
un	'Hollywood'	oRNW
	holodonta	see **R. *moyesii*** f. ***rosea***
un	'Holy Toledo'	oGar
	'Hombre'	oJus
un	'Hondo'	oHei
un	'Honest Abe'	oHei oJus
un	'Honest Red'	oRNW
un	'Honey Bouquet'	wChr wCot oWnC oAls oRNW wAnt
un	'Honeysweet'	oHei
	Honor™ / 'Jacolite'	wChr wCot oGar oSho oWnC oAls oRNW
	Honor™ / 'Jacolite' (tree)	oWnC
	'Honorine de Brabant'	wChr oRos wCot oHei oRNW wAnt
un	'Hoot Owl'	oJus
un	'Hot Spot'	oRNW
	Hot Tamale / 'Jacpoy'	wChr wCot oGar oWnC oUps
	Hot Tamale / 'Jacpoy' (tree)	wCot oGar
	Hotline®/ 'Aromikeh'	wChr
	'Hugh Dickson'	wCot oHei
	hugonis	see **R. *xanthina*** f. ***hugonis***
un	'Humdinger'	oJus
	'Hunslet Moss'	oRos oHei
	'Hunter'	wChr wCot oHei oGar wAnt
un	'Huntington's Hero'	wChr wCot
un	'Hurdy Gurdy'	oHei oJus oRNW
	Ice Crystal™ / 'Cleice'	oHei
	Ice Meidiland™ /' Meivahyn'	wChr wCot
un	'Ice Queen'	oJus
	Iceberg / 'Korbin'	wChr oRos wCot oHei oEdm oGar oWnC oRNW wAnt oUps
	Iceberg / 'Korbin' (tree)	wChr wCot oEdm oGar
	'Illusion'	oHei
	'Immensee'	see **R.** Grouse
un	'Immortal Juno'	oHei
	Impatient / 'Jacdew'	wCot oWnC
un	'Imperatrice Eugenie'	oHei
un	'Impulse'	oJus
un	'In the Mood'	oJus
	Ingrid Bergman®/ 'Poulman'	wChr wCot oEdm oGar oSho oWnC oAls oRNW
un	'Ink Spots'	wChr
	Inner Wheel / 'Fryjasso'	oHei oRNW
un	'Innocence'	oJus
	'Interim'	see **R.** Red Trail / 'Interim'
	International Herald Tribune®/ 'Harquantum'	oRos oHei oRNW
un	'Intrepid'	oRNW
co	Intrigue (see Nomenclature Notes)	oGar oSho oWnC oAls
	Intrigue / 'Jacum'	wChr wCot oEdm
	Intrigue / 'Korlech'	wChr wCot oHei wAnt
	'Ipsilante'	oHei
	'Irene Watts'	oRos wCot oHei wAnt
	'Irish Elegance'	oRos oRNW
	'Irish Fireflame'	oRos oHei oRNW
un	'Irish Heartbreaker'	oJus
un	Irresistible / 'Tinresist'	oJus
un	'Isabel Renaissance'	wChr wCot
	'Isabella Sprunt'	oRos
	'Ispahan'	oRos wAnt
un	'Ivory Palace'	oJus
un	'Jackpot'	oJus
	x jacksonii 'Max Graf'	wChr oRos
	Jacqueline du Pre / 'Harwanna'	wCot oHei
	Jacquenetta / 'Ausjac'	oRos oHei oRNW
co	Jacques Cartier	see **R.** 'Marchesa Boccella'
	Jadis / 'Jacdis'	wChr oHei oSho oWnC oRNW
	'James Mason'	oHei wCLG
	'James Mitchell'	oHei
un	'Jane Pauley'	oEdm
	'Janet B. Wood'	last listed 99/00
un	'Janna'	oJus
un	Jan's Wedding	see **R.** Jan's Wedding Bouquet / 'Adajan'
un	Jan's Wedding Bouquet / 'Adajan'	oFor wWoS oHei
	Jardins de Bagatelle®/ 'Meimafris'	wChr oHei oGar oWnC oRNW
	'Jaune Desprez'	see **R.** 'Desprez a Fleurs Jaunes'
	Jayne Austin / 'Ausbreak'	wCot oHei
un	'Jazz Dancer'	oJus
	Jean Giono™ / 'Meirokoi'	wChr oHei oWnC oRNW wAnt
	Jean Kenneally™ / 'Tineally'	oJus
	'Jean Mermoz'	oRos oHei
	'Jeanne de Montfort'	oRos
	'Jeanne Lajoie'	wChr wCot oHei oGar oJus oRNW wAnt
un	'Jeff's Yellow Climber'	wAnt
un	'Jelly Bean'	oJus
	'Jema'	wChr
	Jennie Robinson / 'Trobette'	oHei
	Jennifer™ / 'Benjen'	oJus
co	'Jenny Duval'	see **R.** 'President de Seze'
	'Jens Munk'	oRos wCot oHei oDan oRNW wAnt
	'Jersey Beauty'	oHei
un	'Jet Trail'	oJus
un	'Jim Dandy'	oJus
un	'Jingle Bells'	wChr wCot oGar oSho
un	'Jingle Bells' (tree)	oGar
un	'Jitterbug'	wChr
un	'Joan Austin'	oJus
	Joan Fontaine™ / 'Clejoan'	oHei oRNW
	Johann Strauss™ / 'Meioffic'	wChr wCot oHei oGar oWnC oRNW wAnt
un	***johannensis***	oHei
	'John Cabot'	oHei wAnt
	John Clare / 'Auscent'	wChr wCot oHei wAnt
	'John Davis'	oHei oRNW wAnt
	'John Hopper'	last listed 99/00
	'John F. Kennedy'	wChr wCot oGar oSho oWnC oAls oRNW
	'John Franklin'	last listed 99/00
un	'John Phillip Sousa'	oHei
	'Joseph's Coat'	wChr wCot oHei oGar oSho oWnC oRNW wAnt
un	'Joycie'	oJus
ub	Joyfulness	see **R.** Frohsinn / 'Tansinnroh'
un	Jubilee™	wRai
	Jude the Obscure / 'Ausjo'	wChr oHei oRNW
	Judi Dench / 'Peahunder'	oRNW
un	'Judith Ann'	wCot oEdm
	'Judy Fischer'	oJus
	Judy Garland / 'Harking'	wChr wCot
un	'Julia Renaissance'	wCot
	Julia's Rose®	oHei oRNW
	'June Time'	oJus
	'Juno'	oHei
	Just Joey®	wChr wCot oHei oEdm oGar oWnC oAls oRNW wAnt
	Just Joey® (tree)	oEdm oGar
	'K of K'	oRos
	'Kaiserin Auguste Viktoria'	oRos
	Kaleidoscope / 'Jacbow'	wChr wCot oEdm oGar oSho oWnC
un	Kardinal™ / 'Korlingo'	wChr wCot oEdm oGar oWnC
un	Kardinal™ / 'Korlingo' (tree)	oEdm oWnC
un	Karen Blixen / 'Poulari'	oEdm
	'Karl Foerster'	oRos oHei
	Kateryna™ / 'Clekate'	oHei
	'Katharina Zeimet'	oRos wCLG wAnt
	'Kathleen'	wChr oRos wCot oHei wAnt
	'Kathleen Harrop'	oHei
	Kathryn Morley / 'Ausclub'	wChr wCot oHei wAnt
	'Kazanlik'	see **R.** 'Professeur Emile Perrot'
	Keepsake / 'Kormalda'	wChr wCot oEdm
	Kent®/ 'Poulcov'	wChr wCot
	Kent®/ 'Poulcov' (tree)	last listed 99/00
un	'Kentucky Derby'	oWnC
	'Kew Rambler'	oRos oHei
	'Kiftsgate'	see **R. *filipes*** 'Kiftsgate'
un	'King Tut'	wChr oGar oWnC
	King's Ransom®	wChr oSho oWnC
	Kiss 'n' Tell / 'Seakis'	oJus

un	Kiss of Desire / 'Korlupo'	oRos oHei
un	'Kiss the Bride'	oJus
	'Kitchener of Khartoum'	see **R.** 'K of K'
un	'Kiwi Sunrise'	oJus
un	Knock Out™ / 'Radrazz'	wChr wCot oEdm oWnC oAls wAnt
	x kochiana	oHei
	'Koenigin von Daenemark'	wChr oRos wCot oHei wCLG
	'Kordes Magenta'	see **R.** 'Magenta'
	'Kordes Perfecta'	see **R.** Perfecta / 'Koralu'
	'Korp'	last listed 99/00
	'Korresia'	wChr wCot oHei oEdm oGar oSho oWnC oAls oRNW wAnt
	'Korresia' (tree)	wChr wCot oEdm
un	'Ko's Yellow'	oJus
	Kristin™ / 'Benmagic'	oJus
	'Kronprinzessin Viktoria'	oRos oHei
	L. D. Braithwaite®/ 'Auscrim'	wChr wCot oHei oGar oWnC oRNW wAnt
	L. D. Braithwaite®/ 'Auscrim' (tree)	oFor
	'La Belle Sultane'	see **R.** 'Violacea'
un	'La Bonne Maison'	oHei
	'La France'	oRos wCot oRNW wAnt
un	'La Marne'	wChr oRos oHei oRNW
	'La Mortola'	see **R.** ***brunonii*** 'La Mortola'
un	'La Parisienne'	oRNW
	'La Reine'	oRos wCot oHei oRNW
	'La Reine des Violettes'	see **R.** 'Reine des Violettes'
	'La Reine Victoria'	see®'Reine Victoria'
	'La Ville de Bruxelles'	wChr oRos wCot oHei
	Lace Cascade / 'Jacarch'	wChr wCot oWnC oRNW
un	'Ladies View'	oJus
un	'Lady Bird Johnson'	oRNW
un	'Lady Carolina'	oHei oRNW
	'Lady Curzon'	wChr oHei
	'Lady Diana'	wChr oRNW
	'Lady Elgin'	see **R.** Thais / 'Memaj'
	'Lady Gay'	last listed 99/00
	'Lady Godiva'	oHei
	Lady Heirloom™ / 'Clejoy'	oHei
	'Lady Hillingdon'	oRos oHei oRNW
	Lady in Red / 'Sealady'	oJus
un	Lady Jane Grey™ / 'Harzazz'	oHei
	'Lady Mary Fitzwilliam'	wCLG
	'Lady Penzance'	oHei
un	'Lady X'	oWnC
	laevigata	oFor
	laevigata 'Anemone'	see **R.** 'Anemome'
un	***laevigata*** 'Romona'	oHei
	'Lafter'	oRos oHei oRNW
un	Lagerfeld™ / 'Arolaqueli'	wChr wCot oWnC wCLG
un	'Lal'	oRNW
	'Lamarque'	oRos oHei wCLG
	'Lanei'	oHei
	Lasting Peace™ / 'Meihurge'	oWnC oRNW
	'Latte'	oJus
	Laura Ashley / 'Chewharla'	oHei
	Laura Clements™ / 'Clespirit'	oHei
	Laura Ford®/ 'Chewarvel'	oRNW
un	Laurel Louise™	oHei
	'Lava Glow'	see **R.** Intrigue / 'Korlech'
	'Lavaglut'	see **R.** Intrigue / 'Korlech'
un	'Lavender Dream'	oGar oRNW
un	'Lavender Jade'	oJus
	'Lavender Jewel'	oJus
	'Lavender Lace'	oJus
	'Lavender Lassie'	wChr oRos wCot oHei oRNW wCLG wAnt
	'Lavender Pinocchio'	wChr wCot
	Lavinia / 'Tanklawi'	wCot
	'Lawrence Johnston'	oRos oHei
un	Lawrence of Arabia™ / 'Harverag'	oHei
un	'Le Perle'	oRos
	Leander®/ 'Auslea'	wChr oRos wCot oHei oRNW
un	LeAnn Rimes / 'Harzippee'	wChr wCot oEdm oRNW
	Leaping Salmon / 'Peamight'	oHei
	'Leda'	oRos wCot oHei oRNW wAnt
	Legend / 'Jactop'	oGar oWnC oRNW
	'Lemon Delight'	oHei
un	'Lemon Gems'	wCot oWnC
un	'Lemon Gems' (tree)	wChr wCot
	'Lemon Meringue'	oJus
un	'Lemon Spice'	wChr
	'L'Enfant de France'	see **R.** 'Enfant de France'
	Leonardo da Vinci / 'Meidauri'	wChr oHei oGar oWnC oRNW wAnt
	Leonidas / 'Meicofum'	wCot oRNW
	'Leonie Lamesch'	oRos

	'Leontine Gervais'	wChr wCot oHei wAnt
un	'Les Sjulin'	oHei
un	***leschenaultii***	oHei
	'Leverkusen'	wCot oHei
	'Leveson-Gower'	last listed 99/00
	Lexy™ / 'Clecharisma'	oHei
un	'Libby'	oJus
	Lichtkoenigin Lucia / 'Korlilub'	oRos oHei oRNW wAnt
un	Liebeszauber / 'Kormiach'	wChr wCot oEdm oRNW
un	'Lights of Broadway'	oJus
	'Lilac Charm'	oHei oRNW
un	'Lilac Pink Moss'	oHei
	Lilac Rose™ / 'Auslilac'	wChr wCot oHei oWnC oRNW wAnt
	Lilac Rose™ / 'Auslilac' (tree)	last listed 99/00
	Lilian Austin®/ 'Ausli'	wChr oRos wCot oHei oGar oRNW wAnt
	Lilian Austin®/ 'Ausli' (tree)	last listed 99/00
	'Lily the Pink'	wCot
un	Linda Campbell / 'Morten'	wChr wCot oGar oRNW wAnt
	Lipstick 'n' Lace / 'Clelips'	oHei
	Little Artist®/ 'Macmanley'	oHei oWnC
un	'Little Darling'	oEdm oRNW
un	'Little Flame'	wChr wCot oGar
un	'Little Girl'	oHei oJus oRNW
	'Little Green Snake'	see **R.** Green Snake®/ 'Lenwich'
	Little Jackie™ / 'Savor'	oJus
un	'Little Sir Echo'	oJus
	Little Sizzler / 'Jaciat'	wChr oGar oWnC
	Little Sizzler / 'Jaciat' (tree)	oGar
un	'Little Tiger'	oJus
un	'Live Wire'	oJus
	Liverpool Remembers / 'Frystar'	oEdm oRNW
ub	Livin' Easy / 'Harwelcome'	wChr wCot oEdm oWnC
un	Lloyd Center Supreme / 'Twoloy'	oEdm oRNW
	'Long John Silver'	wAnt
	sp. aff. ***longicuspis*** DJHC 080	wHer
un	'Looks Like Fun'	oJus
un	'Lord Don'	oHei
	Lord Mountbatten	see **R.** Mountbatten®/ 'Harmantelle'
	'Lord Penzance'	oRos
un	'Lordly Oberon'	oRos oHei
	'Lorraine Lee'	last listed 99/00
	'Los Angeles'	wCLG
un	'Louis Bugnet'	oHei
	'Louis Gimard'	oRos
	'Louis Philippe'	oRos
	Louise Clements™ / 'Clelou'	oHei
un	'Louise Estes'	wCot
	'Louise Odier'	wWoS wChr oRos oHei oRNW wAnt
	Love / 'Jactwin'	wChr wCot oGar oWnC oRNW
un	'Love Glow'	oJus
	Love Potion™ / 'Jacsedi'	wChr wCot oGar oWnC oAls
	Lovely Fairy®/ 'Spevu'	oRos wCot oHei oRNW wAnt
un	'Lovers Only'	oJus
	Love's Promise™ / 'Meisoyris'	oRNW
un	'Love's Song'	oHei
	Loving Touch™	wChr oJus
un	'Lowell Thomas'	oRNW
	Lucetta / 'Ausemi'	oRos wCot oHei oJoy oRNW
	luciae	oHei
un	***luciae fujisinensis***	oHei
	Luis Desamero / 'Tinluis'	oJus
	Lyda Rose™ / 'Letlyda'	oHei
	'Lykkefund'	oHei
un	'Lynette'	wChr
un	Lynn Anderson / 'Wekjoe'	wCot oGar oWnC
un	'Lyric'	oRos
	macounii	see **R.** ***woodsii***
	x macrantha	oRos oHei wAnt
un	***x macrantha*** 'Plum'	oHei
	macrophylla	last listed 99/00
	macrophylla doncasterii	see **R.** 'Doncasteri'
un	'Madame A. Labbey'	oHei
	'Madame Alfred Carriere'	wChr oRos wCot oHei wCLG wAnt
	'Madame Alice Garnier'	oHei
	'Madame Antoine Mari'	wCLG
	'Madame Butterfly'	oRos oHei
	'Madame Caroline Testout'	oRos oHei oRNW
	'Madame de Sombreuil'	see **R.** 'Sombreuil'
	'Madame Delaroche-Lambert'	wCLG
un	'Madame d'Enfert'	wCLG
	'Madame Ernest Calvat'	wChr wCot oHei oRNW
	'Madame Georges Bruant'	last listed 99/00
	'Madame Gregoire Staechelin'	oRos oHei oSha

	'Madame Hardy'	wChr oRos wCot oHei oWnC oRNW wCLG
	'Madame Isaac Pereire'	oFor wWoS wChr oRos wCot oHei oGar oWnC oRNW wCLG wAnt
un	'Madame Joseph Schwartz'	wChr wCLG
	'Madame Knorr'	wChr oRos oHei oGar wCLG wAnt
	'Madame Lauriol de Barny'	wCLG
	'Madame Legras de Saint Germain'	oRos wAnt
	'Madame Lombard'	oRos wCLG
	'Madame Louis Leveque'	wChr oRos wCot wAnt
un	'Madame Marie Curie'	oRos
	'Madame Norbert Levavasseur'	oRos
	'Madame Pierre Oger'	wChr oRos wCot oHei
	'Madame Plantier'	wChr oRos wCot oHei oGar oWnC wCLG oUps
un	'Madame Violet'	oRNW
	'Madame William Paul'	oHei
	'Madame Zoetmans'	oHei wAnt
	'Madeleine Selzer'	oHei
	'Mademoiselle Cecile Brünner'	see **R.** 'Cecile Brünner'
	'Maerchenland'	oHei
	Maestro®/ 'Mackinja'	oHei
	'Magenta'	oRos oHei oRNW wCLG
un	Maggie Barry / 'Macoborn'	oEdm oRNW
un	'Magic Blanket'	wChr wCot
	Magic Carpet™ / 'Jaclover'	wChr wCot oHei oGar
	Magic Carrousel®/ 'Moorcar'	oJus
un	'Magic Lantern'	oGar oWnC
	Magic Meidiland™ / 'Meibonrib'	wChr wCot oHei wAnt
	Magic Meidiland™ / 'Meibonrib' (tree)	wCot
	'Magna Charta'	oHei
	'Magnifica'	oFor wCul oHei wSwC oRNW
qu	'Maiden's Blush'	wChr
	'Maiden's Blush Great'	see **R.** 'Great Maiden's Blush'
	'Maiden's Blush Small'	see **R.** 'Small Maiden's Blush'
	'Maigold'	oRos oHei
un	'Make Believe'	oJus
	'Malaga'	oHei
	'Malmaison Rouge'	see **R.** 'Leveson-Gower'
	'Malton'	wAnt
	'Maman Cochet'	oRNW
	'Manettii'	oHei
	'Marbree'	oRos oHei wCLG
	'Marchesa Boccella'	oRos wCot oHei oWnC oRNW wCLG wAnt
	'Marchioness of Londonderry'	oRos oRNW wCLG wAnt
	Marco Polo / 'Meipalco'	oGar oWnC
	'Marechal Davoust'	oRos oHei
	'Marechal Niel'	oRNW wCLG
	Margaret Merrill / 'Harkuly'	wChr wCot oEdm oRNW wAnt
	Margaret Thatcher / 'Korflueg'	wChr
	'Margo Koster'	wChr oRos wCot oHei oRNW
	'Marguerite Hilling'	wChr oRos oHei
	Maria Callas	see **R.** Miss All-American Beauty / 'Meidaud'
un	'Maria Lisa'	oHei wAnt
un	'Maria Mathilda'	oRos oRNW
un	'Maria Stern'	oRNW
qu	'Maria Teresa'	oHei
	Maria Teresa de Esteban / 'Dotrames'	oEdm
	'Marie Bugnet'	oDan
un	'Marie Jean'	oHei
	'Marie Louise'	oHei wCLG
	'Marie Pavic'	wChr oRos wCot oHei oRNW wCLG wAnt
	'Marie van Houtte'	oHei
	'Marijke Koopman'	oEdm oRNW
	'Marina'	last listed 99/00
	Marinette / 'Auscam'	wChr wCot oHei oRNW
	Marjorie Fair®/ 'Harhero'	oRos oHei oSha oRNW wAnt
	'Martha'	oHei
un	Martha's Vineyard / 'Poulans'	wChr wCot oRNW
	'Martin Frobisher'	wChr wCot wAnt
un	'Martine Guillot'	wChr wCot
un	'Mary Kay'	oJus
un	'Mary Marshall'	oJus
	'Mary Queen of Scots'	oHei
	Mary Rose®/ 'Ausmary'	wChr wCot oHei oGar oRNW wAnt
	Mary Rose®/ 'Ausmary' (tree)	last listed 99/00
	'Mary Wallace'	oRos
	Mary Webb®/ 'Auswebb'	wChr wCot oHei wAnt
un	'Matador'	oAls
	'Max Graf'	see **R.** ***x jacksonii*** 'Max Graf'
	maximowicziana	oFor oHei
	'May Queen'	oHei
	Mayor of Casterbridge / 'Ausbrid'	wChr oHei oRNW wAnt
un	'Maytime'	oHei
	'Medallion'	wChr wCot oWnC oRNW
	'Meg'	wCot
	Melina®	see **R.** Sir Harry Pilkington / 'Tanema'
un	Melody Parfumee™ / 'Dorient'	wChr wCot oGar oSho oWnC oAls oRNW
un	Melody Parfumee™ / 'Dorient' (tree)	oGar oSho
	Memories™ / 'Clecats'	oHei
un	'Mendocino Delight'	oHei
	'Mermaid'	wChr oRos wCot oHei wCSG wCLG wAnt
un	'Merry England'	oHei
un	'Merryglo'	oJus
	'Merveille de Lyon'	oRos oHei
	'Mevrouw Nathalie Nypels'	oRos
un	'Michaelangelo'	wCot
	'Michele Meilland'	oRNW
un	***micranthosepium***	oHei
	x micrugosa 'Alba'	oHei
	Midas Touch™ / 'Jactou'	wChr wCot oEdm oGar oWnC oAls oRNW
	Midas Touch™ / 'Jactou' (tree)	last listed 99/00
	'Mignonette'	oRos oRNW
	Mikado / 'Kohsai'	oRNW
un	'Millie Walters'	oJus
	Minilights / 'Dicmoppet'	wWoS oHei
	Minnie Pearl®/ 'Savahowdy'	oWnC oJus
un	'Miranda'	wCLG
	'Mirandy'	oWnC oAls oRNW
	Miss All-American Beauty / 'Meidaud'	wChr oEdm oWnC oRNW
un	'Miss Daisy'	oSha
	Miss Harp / 'Tanolg'	wChr oWnC oRNW
	Mister Lincoln®	wChr wCot oEdm oGar oSho oWnC oAls oRNW wCLG
	Mister Lincoln® (tree)	oGar
	Mistress Quickly / 'Ausky'	oHei oRNW
un	'Mix 'n' Match'	wChr wCot wAnt
	'Mlle Cecile Brünner'	see **R.** 'Cecile Brünner'
un	'Modern Art'	oHei
	'Mojave'	oWnC
	Moje Hammarberg®	oRos oDan
	Molineux / 'Ausmol'	wChr wCot oHei wAnt
un	Mon Cheri™ / 'Arocher'	oWnC oRNW
	'Monsieur Tillier'	oHei
un	'Mont Blanc'	wCot
un	'Monte Cassino'	oHei
un	'Monte Rosa'	wChr wCot
un	'Montecito'	oHei
un	Moody Dream / 'Macmoodre'	wCot oEdm oRNW
un	'Moon River'	oGar
	Moon Shadow™ / 'Jaclaf'	wChr wCot oGar oWnC oAls oRNW
	Moonbeam / 'Ausbeam'	wChr oRos wCot oHei
	'Moonlight'	wChr oHei
un	'Moonsprite'	oRNW
un	Moonstone™ / 'Wekcryland'	wChr wCot oEdm oWnC oRNW
un	Moonstone™ / 'Wekcryland' (tree)	wCot
	'Morden Blush'	wChr wCot oGar
	'Morden Cardinette'	last listed 99/00
	'Morden Centennial'	wChr oGar wAnt oUps oSle
	'Morden Fireglow'	oWnC oRNW
	'Morey's Pink China'	oHei
un	'Morgenrot'	oHei
	'Morlettii'	oHei
un	Morning Blush™ / 'Siemorn'	oHei
	'Morning Colors'	see **R.** 'Radway Sunrise'
un	'Morning Greetings'	oHei
	Morning Has Broken™ / 'Clewedding'	oHei
	Morning Jewel®	oHei
	Morning Mist / 'Ausfire'	wAnt
un	'Morning Star'	wChr
	moschata	oHei
un	***moschata*** 'Plena'	oRos
un	'Most Unusual Day'	oRNW
	Mother's Love / 'Tinlove'	oJus
un	Mount Hood / 'Macmouhoo'	wCot oWnC oRNW
un	'Mountain Music'	oHei
	Mountbatten®/ 'Harmantelle'	oRos oHei
un	'Mountie'	oJus
	'Mousseline'	oHei wAnt
	Moye Hammarberg	see **R.** Moje Hammarberg®
	moyesii	oFor oHei oUps wSnq
	moyesii 'Geranium'	see **R.** 'Geranium'
	moyesii pink	oHei
	moyesii f. ***rosea***	oHei
	moyesii 'Sealing Wax'	see **R.** 'Sealing Wax'

	'Mozart'	oRos wCot oHei oRNW
	'Mr. Bluebird'	oHei
	'Mr. Lincoln'	see **R.** Mister Lincoln®
	'Mrs. B.R. Cant'	wCot wAnt
un	'Mrs. Bosanquet'	oRos wCLG
	Mrs. Doreen Pike / 'Austor'	oHei
	'Mrs. F. W. Flight'	last listed 99/00
un	'Mrs. J. H. Nicolas	wAnt
	'Mrs. John Laing'	wChr
	'Mrs. Oakley Fisher'	wChr oRos wCot oRNW
	'Mrs. Pierre S. duPont	last listed 99/00
un	'Mrs. W. H. Cutbush'	oHei
un	'Muenchen'	oHei
	'Mullard Jubilee'	wChr wCot oEdm oWnC oRNW
	mulliganii	oFor wHer oHei
	multiflora	oHei oSha
	multiflora var. ***cathayensis***	oHei
	multiflora 'Grevillei'	oRos oHei
	multiflora 'Platyphylla'	see **R.** ***multiflora*** 'Grevillei'
	mundi	see **R.** ***gallica*** 'Versicolor'
un	***muscifolia***	oHei
	'Mutabilis'	see **R.** ***x odorata*** 'Mutabilis'
	'My Honey'	oJus
un	'My Sunshine'	oJus
	Myriam®/ 'Cocgrand'	oHei
un	'Mystic'	wChr
	Mystic Meidiland™ / 'Meialate'	wChr oHei wAnt
un	Napa Valley / 'Poulino'	wChr
un	'Napoleon'	oRos oHei
un	'Narcisse de Salvandy'	oRos
	'Narrow Water'	oRos wCLG
	'Nastarana'	oRos
un	Natchez / 'Poullen'	wChr oRNW
un	'National Velvet'	oRNW
un	'Near You'	oJus
un	'Nearly Wild'	wWoS oRos oGar oDan
	'Nestor'	oHei
	'Nevada'	wChr wCot wAnt
un	'New Beginning'	oJus
	'New Dawn'	wChr oRos wCot oHei oGar oSha oInd oRNW wAnt oUps oSle
un	'New Day'	wChr wCot oWnC oRNW
un	'New Face'	oRos oHei oSha oRNW
un	'New Pristine'	see **R.** 'Helen Naude'
	New Year	see **R.** Arcadian / 'Macnewye'
	New Zealand / 'Macgenev'	wChr wCot oEdm oGar oWnC oAls oRNW
	New Zealand / 'Macgenev' (tree)	oEdm
un	Newport / 'Poudex'	wChr
un	'Newport Fairy'	see **R.** 'Non Plus Ultra'
	News®/ 'Legnews'	oHei
	Nice Day / 'Chewsea'	oHei
un	'Nickelodeon'	oJus
un	Nicole / 'Koricole'	wChr wCot oEdm oGar oRNW
un	Nicole / 'Koricole' (tree)	wCot
	Nigel Hawthorne / 'Harquibbler'	oRos
	Night Light®/ 'Poullight'	oHei
	nitida	oFor oHei
	Noble Antony / 'Ausway'	wChr oHei oRNW wAnt
	'Noisette Carnee'	wChr oRos wCot oHei oRNW wCLG wAnt
	'Non Plus Ultra'	wCot oHei
	Norfolk / 'Poulfolk'	oHei
	Northamptonshire / 'Mattdor'	oHei oRNW
un	'Northern Yellow'	oHei
un	'Northwest Sunset'	oRos oRNW
	'Norwich Castle'	wCot
	'Nova Zembla'	oRos
	'Nozomi'	oHei wAnt
	'Nuage Parfume'	see **R.** Fragrant Cloud / 'Tanellis'
	'Nuits de Young'	oRos oHei
	nutkana	oFor oHan wWoB wChr oHei wShR oBos oRiv oOEx wWat oTri wPla oAld wAva wFFl oSle wSnq
	nutkana 'Plena'	oHei wAnt
	'Nymphenburg'	wChr wCot oHei oRNW wCLG
	'Nypels' Perfection'	oHei
un	'Oceana'	wChr oRNW
	Octavia Hill / 'Harzeal'	wCot
un	Octoberfest™ / 'Maclanter'	wChr wCot oEdm oRNW wAnt
	x odorata 'Mutabilis'	oFor wChr oRos wCot oHei wPir wCCr oRNW wAnt
	x odorata 'Odorata'	wCLG
	x odorata Old Crimson China	oRos oHei
	x odorata 'Pallida'	oRos oHei wCLG wAnt
	x odorata 'Pseudindica'	oRos wAnt
	x odorata 'Viridiflora'	oRos oHei oRNW
	'Oeillet Flamand'	see **R.** 'Oeillet Parfait'
	'Oeillet Parfait'	oRos
	'Oklahoma'	wCot oHei oGar oSha oWnC oRNW
un	'Old Fashioned Lady'	oHei
	Old Glory™ / 'Benday'	oJus
	Old John / 'Dicwillynilly'	oHei
	Old Master / 'Macesp'	oHei oRNW
	Old Port / 'Mackati'	oRos oRNW
un	'Old Smoothie'	oRNW
un	'Olde Lace'	oWnC
	Olde Romeo / 'Hadromeo'	oWnC oRNW
un	'Olde Tango'	oWnC
un	'Ole'	oWnC oAls
	Olympiad®/ 'Macauck'	wChr wCot oEdm oGar oWnC oRNW oUps
	Olympiad®/ 'Macauck' (tree)	oGar oWnC oAls
	'Ombree Parfaite'	oRos wCLG
	omeiensis	see **R.** ***sericea*** ssp. ***omeiensis***
un	'Omni'	oHei
	'Opal Brunner'	see **R.** 'Perle d'Or'
	Opening Night™ / 'Jacolber'	wChr wCot oEdm oGar oWnC oAls wAnt
	Opening Night™ / 'Jacolber' (tree)	oSho
	'Ophelia'	last listed 99/00
un	'Opulence'	oRNW
	Ora Kingsley™ / 'Clehonor'	oHei
	Orange Altissimo™	see **R.** Christine™ / 'Clealta'
	Orange Sensation®	oHei
	Oranges and Lemons™ / 'Macoranlem'	wChr wCot oHei oEdm oGar
un	'Orchid Jubilee'	oJus
	Oregold	see **R.** Miss Harp / 'Tanolg'
un	'Oregon Rainbow'	oHei
un	'Oriental Charm'	oHei
un	'Oriental Simplex'	oJus
un	'Orlando'	oRNW
	'Orpheline de Juillet'	see **R.** 'Ombree Parfaite'
	Othello®/ 'Auslo'	oFor wChr wCot oHei oRNW wAnt
un	'Our Pearl'	oRNW
un	'Out of Yesteryear'	oRNW
	Oxfordshire / 'Korfullwind'	wChr wCot oHei oGar oRNW wAnt
	Oxfordshire / 'Korfullwind' (tree)	wChr
un	'Pacesetter'	oJus oUps
	'Painted Damask'	see **R.** 'Leda'
	Painted Moon / 'Dicpaint'	oRNW
	palustris	oFor oHei
un	'Pandemonium'	oJus
	'Papa Gontier'	oHei
	Papa Meilland®/ 'Meisar'	wChr oRos oGar oWnC oRNW
	'Papillon'	oRNW
	Paprika™ / 'Meiriental'	wChr oRNW
	'Parade'	oRos oHei oRNW wCLG
	Paradise®/ 'Weizeip'	wCot oGar oWnC oRNW
un	'Pareo'	oRNW
	Paris de Yves St. Laurent™ / 'Meivamo'	oGar oWnC oRNW
un	'Park Wilhelmshohe'	oHei
	Parkdirektor Riggers®	oHei
	'Parkjewel'	see **R.** 'Parkjuwel'
	'Parkjuwel'	oHei
un	'Parkzauber'	oHei
	Partridge / 'Korweirim'	oHei
	Party Girl®	oJus
	Pascali®/ 'Lenip'	wChr wCot oEdm oWnC oRNW
	Pascali®/ 'Lenip' (tree)	oEdm
	Pat Austin / 'Ausmum'	wChr wCot oGar oWnC oRNW wAnt
	Pat Austin / 'Ausmum' (tree)	wCot oGar
	Patriot Flame / 'Clescrub'	oHei
un	'Paul Bocuse'	wChr wCot
	'Paul Neyron'	wChr oRos wCot oHei oGar wCLG
un	'Paul Noel'	oRos
	'Paul Ricault'	oRos oHei
	Paul Shirville / 'Harqueterwife'	oHei oEdm
	Paul Shirville / 'Harqueterwife' (tree)	oEdm
	'Paul Transon'	oHei
	'Paul Verdier'	wCLG
	'Paulii Rosea'	last listed 99/00
	'Paul's Early Blush'	wChr wCLG
	'Paul's Himalayan Musk'	wHer wChr wCot oHei wAnt
	'Paul's Lemon Pillar'	oRos
	'Paul's Pride'	oHei
	'Paul's Scarlet Climber'	oRos oHei oRNW
	'Pax'	wChr oRos
	Peace / 'Madame A. Meilland'	wChr oRos wCot oHei oEdm oGar oSho oWnC oAls oRNW oUps
	Peace / 'Madame A. Meilland' (tree)	wCot oWnC

	Peach Blossom / 'Ausblossom'	wChr oRos wCot oHei wAnt
un	'Peaches 'n' Cream'	oJus
un	'Peachy Keen'	oJus
un	Pearl / 'Wekpearl'	wCot oEdm oRNW
	Pearl Drift®/ 'Leggab'	oHei
qu	'Pearl Groundcover'	oSec
	Pearl Meidiland / 'Meineble'	wChr oHei oRNW wAnt oUps
	Pearl Meidiland / 'Meineble' (tree)	last listed 99/00
	Pearl Sevillana™ / 'Meichonar'	wChr oRNW wAnt
un	'Pearlie Mae'	oHei
un	'Pearly Gates'	wChr wCot oRNW
	Peek A Boo / 'Dicgrow'	oWnC
	Pegasus / 'Ausmoon'	wChr
un	'Peggy Lee'	oRNW
	'Pelisson'	oHei
	pendulina	oRos oHei
	'Penelope'	wChr oRos wCot oHei oGar oRNW wCLG
	Penelope Keith / 'Macfreego'	oJus
un	'Penny Candy'	oJus
	Peppermint Ice / 'Bosgreen'	oHei
	Perdita®/ 'Ausperd'	wChr wCot oHei oGar oRNW wAnt
	Perdita®/ 'Ausperd' (tree)	wCot
	Perestroika / 'Korhitom'	oHei
	Perfect Moment / 'Korwilma'	wCot oEdm oAls oRNW
	Perfecta / 'Koralu'	oWnC
	Perfume Beauty™ / 'Meinaicin'	oWnC oAls
	'Perfume Delight'	wChr oGar oWnC oAls oRNW
un	'Perfumella'	oRNW
	'Perla de Alcanada'	oHei
	'Perle d'Or'	oRos oHei oRNW wAnt
	Peter Beales™ / 'Cleexpert'	oHei
un	'Peter Frankenfeld'	wCot oEdm oRNW
un	'Peter Mayle'	wChr
	Petit Four®/ 'Interfour'	oHei
un	'Petit Rat de l'Opera'	oHei
un	'Petite Carrousel'	oJus
	'Petite de Hollande'	wChr oRos oHei wCLG
	'Petite Lisette'	oRos
	'Petite Perfection'	wCot oWnC
un	'Petite Pink Scotch'	wAnt
	Pheasant / 'Kordapt'	oHei
	'Phyllis Bide'	oHei wAnt
	Piccolo / 'Tanolokip'	oHei
un	'Picture Perfect'	oRNW
	'Pierre de Ronsard'	see **R.** Eden Rose®
un	Pillow Fight ™ / 'Wekpipogop'	wChr oHei oWnC wAnt
	pimpinellifolia	oFor
	pimpinellifolia 'Dunwich Rose'	oHei
	pimpinellifolia 'Grandiflora'	oHei
	pimpinellifolia 'Marbled Pink"	oHei
	'Pinata'	wChr oHei oWnC wAnt
un	'Pink Bassino'	oHei
	Pink Bells®/ 'Poulbells'	oHei
un	'Pink Cascade'	oRNW
un	'Pink Champagne'	oHei
un	'Pink Cloud'	wAnt
un	'Pink Don Juan'	oRNW
	Pink Flower Carpet™ / 'Noatraum'	wChr oAls oGar oSho wWel
un	'Pink Freedom'	oRNW
un	'Pink Grootendorst'	wChr oRos wCot oHei oGar oSha oRNW wAnt
un	'Pink Gruss an Aachen'	oRos oHei
	Pink Meidiland®/ 'Meipoque'	wChr wCot oHei oRNW
	'Pink Parfait'	last listed 99/00
un	'Pink Pavement'	wCul wChr oHei wAnt
	Pink Peace / 'Meibil'	wChr oWnC oAls oRNW
	'Pink Perpetue'	last listed 99/00
	'Pink Petticoat'	oJus
	'Pink Prosperity'	last listed 99/00
un	'Pink Rambler'	oRos
un	'Pink Robin'	oHei
un	'Pink Robusta'	see **R.** The Seckford Rose / 'Korpinrob'
un	Pink Simplicity	see **R.** Simplicity®
	Pink Surprise / 'Lenbrac'	oHei
	'Pinkie Climber'	wAnt
	Pinocchio / 'Rosenmaerchen'	wCot
un	'Pinstripe'	wCot oWnC oUps
un	'Pipe Dreams'	oHei
un	'Pirette'	oHei
	pisocarpa	oFor oHan wWoB wNot wShR oBos wWat oTri oAld wSnq
un	Pixisticks™ Cherryberry™ (fruit)	oOEx
	'Playboy'	wChr oRos wCot oHei oEdm oWnC oRNW wAnt

	'Playboy' (tree)	wChr wCot oEdm
un	'Playfair'	wChr wCot
	Playgirl / 'Morplag'	wChr wCot oRNW
	Playtime (see Nomenclature Notes)	oRNW
	Pleasure / 'Jacpif'	oRNW
	Pleine de Grace / 'Lengra'	oHei
un	'Poema'	oRos oHei oRNW
	Poetry in Motion / 'Harelan'	wCot oEdm oRNW
un	'Poinsettia'	oRos oRNW
	Polar Star / 'Tanlarpost'	oEdm oWnC
	Polka™ / 'Meitosier'	wChr wCot oHei oGar wRai oWnC oRNW wAnt
	polyantha grandiflora	see **R. *gentiliana***
	pomifera	see **R. *villosa***
	'Pompon Blanc Parfait'	wChr oRos wCot oHei
	'Pompon de Bourgogne	see **R.** 'Burgundiaca'
	'Pompon des Princes'	see **R.** 'Ispahan'
	'Popcorn'	oSis oJus
un	Porcelain Bouquet™	oHei
	Portland Dawn / 'Seatip'	oJus
un	Portland Rose Festival®/ 'Dorjure'	oEdm
	'Portlandica'	oRos oHei
ub	Potter & Moore / 'Auspot'	oRos oHei oRNW
un	'Poulsen's Pearl'	oHei
un	'Prairie Fire'	oHei
un	'Prairie Harvest'	oHei
	'Prairie Princess'	wWoS
un	'Prairie Star'	oHei
	prattii	oFor
	'Precious Platinum'	oEdm
	'President de Seze'	oRos wCLG
un	'Presidential'	oHei
	Pretty Jessica / 'Ausjess'	oRos oHei oRNW wAnt
	Pretty Polly®/ 'Meitonje'	wCot
un	'Pretty Woman'	oJus
un	'Pride 'n' Joy'	wCot
	'Primevere'	last listed 99/00
qu	'Primrose	oHei
	primula	wChr
	Princess Alexandra / 'Pouldra'	wChr wCot
un	Princess Marianna / 'Poulusa'	wChr oHei oRNW
	Princesse de Monaco / 'Meimagarmic'	wChr wCot oWnC oRNW
	'Princesse Louise'	oRos oHei
	Priscilla Burton®/ 'Macrat'	wChr wCot oHei oRNW wAnt
	Pristine®/ 'Jacpico'	wChr wCot oEdm oGar oWnC oAls oRNW
co	'Professeur Emile Perrot'	oRos oHei
un	'Prom Date'	oJus
	'Prominent'	see **R.** 'Korp'
	'Prosperity'	wChr oRos oHei oRNW
	Prospero®/ 'Auspero'	wChr wCot oHei oRNW
	Prospero® (tree)	last listed 99/00
	'Proud Land'	oRNW
un	'Proud Titania'	oRos oHei
	'Pteracantha'	see **R. *sericea*** ssp. ***omeiensis*** f. ***pteracantha***
	pulverulenta	oFor
un	***pulverulenta dalmatica***	oHei
un	'Punkin'	oJus
un	'Puppy Love'	oJus
	Pure Poetry™ / 'Jacment'	oGar
un	Purple Dawn / 'Bridawn'	oHei
	'Purple Haze'	oJus
un	Purple Heart	wChr wCot oWnC oRNW
un	Purple Heart (tree)	wCot
	Purple Passion™ / 'Jacolpur'	wChr wCot oWnC oAls oRNW
un	'Purple Pavement'	wCul wChr oHei wSwC wAnt
un	'Purple Rain'	oRNW
	Purple Simplicity®/ 'Jacpursh'	wChr wCot oGar oWnC oRNW
	Purple Tiger / 'Jacpurr'	wChr oGar oSho oWnC oAls oRNW
	Purple Tiger / 'Jacpurr' (tree)	last listed 99/00
	Purple Velvet™ / 'Clewine'	oHei
	Quaker Star / 'Dicperhaps'	oEdm oRNW
	'Quatre Saisons Blanche Mousseuse'	wCot
un	'Queen Bee'	oHei
	Queen Elizabeth	see **R.** 'The Queen Elizabeth'
	'Queen Lucia'	see **R.** Lichtkoenigin Lucia / 'Korlilub'
un	Queen Margrethe™	wChr oHei oRNW
un	Queen Margrethe™ (tree)	wCot
	Queen Mother / 'Korquemu'	oHei
	Queen Nefertiti®/ 'Ausap'	oFor oRos oHei oRNW
	'Queen of Bourbons'	see **R.** 'Bourbon Queen'
	'Queen of Denmark'	see **R.** 'Koenigin von Daenemark'
un	'Queen of the Musks'	oHei oRNW
	'Queen of the Violets'	see **R.** 'Reine des Violettes'
	Radiant / 'Benrad'	oJus

	Radio Times / 'Aussal'	oHei
	Radox Bouquet / 'Harmusky'	wChr wCot
	'Radway Sunrise'	wHei
	'Rainbow'	last listed 99/00
	Rainbow's End™ / 'Savalife'	wCot oWnC oJus
un	'Raindrops'	oJus
un	Ralph's Creeper™	wChr oHei oGar oWnC oRNW wAnt
	'Rambling Rector'	oRos oHei
un	'Raphaela'	oRNW
un	'Raspberry Punch'	wChr wCot oGar oSho
un	'Raspberry Punch' (tree)	oGar
un	'Raspberry Ripple'	oHei
	Raspberry Sunblaze®	wCot
	'Raubritter'	oRos wCLG
un	Raven /'Frytrooper'	wChr oRNW
un	Raven /'Frytrooper' (tree)	wCot
un	Reba McEntire / 'Machahei'	wChr wCot oEdm oRNW
un	Reba McEntire / 'Machahei' (tree)	oEdm
	Rebecca Louise™ / 'Cleballet'	oHei
un	'Red Alert'	oJus
	'Red Ballerina'	see **R.** 'Marjorie Fair'
	Red Bells®/ 'Poulred'	oRos
	Red Blanket®/ 'Intercell'	oHei
un	'Red Blush'	oHei
	Red Cascade / 'Moorcap'	oHei oSha oRNW
	Red Coat / 'Auscoat'	wChr oRos oHei oRNW
	Red Devil®/ 'Dicam'	wCot
un	'Red Fairy'	wChr wCot oRNW
un	'Red Fairy' (tree)	wChr wCot
un	Red Flower Carpet™ / 'Noare'	wChr wWel
	'Red Fountain'	wChr wCot oHei oGar oWnC oRNW wAnt
un	'Red France'	oRNW
un	'Red Glory'	oHei
	'Red Grootendorst'	see **R.** 'F.J. Grootendorst'
un	'Red Hot'	oHei
	'Red Masterpiece'	oRNW
	'Red Max Graf'	see **R.** Rote Max Graf®/ 'Kormax'
	Red Meidiland®/ 'Meineble'	wChr wCot oHei oRNW
un	'Red Minimo'	oJus
	Red New Dawn	see **R.** 'Etendard'
un	'Red Perfume'	oHei oRNW
un	'Red Rambler'	oHei
	Red Rascal / 'Jacbed'	oGar oWnC
	Red Rascal / 'Jacbed' (tree)	oGar
	Red Ribbons / 'Kortemma'	wChr wCot oHei oGar oWnC oRNW wAnt
	Red Ribbons / 'Kortemma' (tree)	wChr wCot
	'Red Robin'	see **R.** 'Gloire des Rosomanes'
	Red Simplicity®/ 'Jacsimpl'	wChr oGar oWnC oRNW
	'Red Souvenir de la Malmaison'	see **R.** 'Leveson Gower'
	Red Trail / 'Interim'	oHei oSha
	'Red Velvet'	see **R.** Judi Dench / 'Peahunder'
un	'Redglo'	oJus
	Redoute / 'Auspale'	wChr wCot oHei oRNW wAnt
	'Regal'	oJus
	Regatta™ / 'Meinimo'	wChr oHei oGar oWnC oAls wAnt
	Regensberg®/ 'Macyou'	wChr oRos oHei oWnC wAnt
	Regensberg®/ 'Macyou' (tree)	last listed 99/00
	Regina Louise™ / 'Cleconcert'	oHei
un	'Regine'	oJus
	'Reine des Violettes'	wChr oRos wCot oHei oGar oRNW wAnt
	'Reine Victoria'	wChr oHei oGar oWnC oRNW
un	'Rembrandt'	oRos wCLG
	Remember Me®/ 'Cocdestin'	wChr wCot
un	'Renae'	oRNW
un	'Renne Danielle'	oHei
un	'Renny'	oJus
	'Reve d'Or'	oHei
	'Reveil Dijonnais'	oRos oHei
un	'Revelry'	oRNW
un	'Rhapsody'	oAls
un	'Rhode Island Red'	oRNW
	'Rhonda'	last listed 99/00
	x richardii	oRos wCLG
un	Rina Hugo / 'Pekvizo'	wCot oEdm oRNW
	Rio Samba™ / 'Jacrite'	wChr wCot oEdm oGar oSho oWnC oAls
	'Rise 'n' Shine'	oJus
un	'Rita Sammons'	oHei
	'Rival de Paestum'	oRos
un	'Robbie Burns'	wChr oHei
	'Robert le Diable'	oRos oHei
	'Robert Leopold'	oIlci
	'Robin Hood'	oFor oRos oHei oRNW wAnt
	Robin Redbreast®/ 'Interrob'	oHei oSha
	Robusta®/ 'Korgosa'	wChr oHei wAnt
un	'Rochester Cathedral'	oHei
	Rock 'n' Roll	see **R.** Tango / 'Macfirwal'
un	'Rocketeer'	wChr oRNW
un	'Rockin' Robin'	wChr oHei oWnC
	Romance®/ 'Tanezamor'	oHei
un	Romanze	see **R.** Romance®/ 'Tanezamor'
	Rosabell®/ 'Cocceleste'	last listed 99/00
	Rosarium Uetersen®/ 'Kortersen'	wChr wCot wAnt
	'Rose a Parfum de l'Hay'	wChr oHei oDan
	'Rose d'Amour'	oHei wCCr
	'Rose de Meaux White'	see **R.** 'White De Meaux'
	'Rose de Rescht'	see **R.** 'De Rescht'
co	'Rose des Maures'	see **R.** 'Sissinghurst Castle'
	'Rose du Roi'	oRos
	'Rose du Roi a Fleurs Pourpres'	oRos
	Rose Gaujard®/ 'Gaumo'	oHei
	'Rose-Marie Viaud'	oHei
un	'Rose Nuggets'	oJus
un	Rose-o-licious!™ (fruit)	oOEx
un	'Rose Odyssey 2000'	wCot
	Rose Rhapsody™ / 'Jacsash'	wChr wCot oWnC oRNW
	'Rosee du Matin'	'Chloris'
un	'Roselina'	wChr wCot oGar wAnt
	'Rosemary Rose'	oHei
un	'Rosenfest'	oHei
	'Roseraie de l'Hay'	wChr oRos wCot oHei oDan wAnt
	'Rosette Delizy'	oRNW
un	'Rosie'	oJus
un	Rosie O'Donnell / 'Wekwinwin'	wChr wCot oEdm oWnC oRNW
un	Rosie O'Donnell / 'Wekwinwin' (tree)	wCot
	Rosy Carpet / 'Intercarp'	oHei
	Rosy Cushion®/ 'Interall'	oHei oSha
un	'Rosy Dawn'	oJus
	Rosy Future / 'Harwaderox'	oHei
	Rote Max Graf®/ 'Kormax'	oHei
un	'Rouge Moss'	oHei
	'Roulettii'	oRos
	'Roundelay'	wCLG
	roxburghii	oRos wCot oHei wCCr
	roxburghii var. ***hirtula***	wHer oHei
	roxburghii f. ***normalis***	oHei wCLG
	roxburghii 'Plena'	see **R. *roxburghii*** f. ***roxburghii***
	roxburghii f. ***roxburghii***	oRos oHei
	Royal Amber™ / 'Cletivoli'	oHei
un	Royal Amethyst®/ 'Devmorada'	wChr wCot oEdm oRNW
un	Royal Bassino / 'Korfungo'	oHei
un	Royal Blush™ / 'Sieroyal'	oHei
	Royal Bonica™ / 'Meimodac'	wChr oHei oWnC oRNW
un	Royal Bouquet™ / 'Diadem'	oHei
	Royal Brompton Rose / 'Meivildo'	wChr wCot oGar oWnC oRNW wAnt
	Royal Brompton Rose / 'Meivildo' (tree)	last listed 99/00
	'Royal Gold'	wChr oWnC oRNW
	'Royal Highness'	wChr wCot wCLG
	'Royal Porcelain'	oHei
un	'Royal Sunset'	wChr wCot oEdm oRNW wAnt
	Royal William / 'Korzaun'	oEdm oRNW
	rubiginosa	oFor wCul wChr oRos wCot oHei oOEx oSec wAnt
	rubrifolia	see **R. *glauca***
	rubrifolia 'Carmenetta'	see **R.** 'Carmenetta'
un	'Ruby Pendant'	oJus
un	'Ruffles'	oRNW
	Ruffles 'n' Flourishes™ / 'Cleruff'	oHei
un	'Rugelda'	oHei oRNW wAnt
	rugosa	oFor oHan wBCr wWoB wChr oRos oHei wSwC wShR oGar wRai wFai wPir wAnt wAva oUps oSle wSte wSnq
	rugosa 'Alba'	wChr oRos wCot oHei wBur oGar oRiv wRai oOEx oRNW wAnt oSle wSnq wWel
	rugosa 'Alba' seed grown	wCul wBCr wSte
	rugosa var. ***alboplena***	wChr
un	***rugosa germanica***	oHei
	rugosa var. ***kamtschatica***	see **R. *rugosa*** var. ***ventenatiana***
un	***rugosa repens***	oHei
	rugosa 'Rubra'	oRos wCot oHei wBur oRiv oOEx oRNW oUps
	rugosa 'Scabrosa'	see **R.** 'Scabrosa'
un	***rugosa schalin***	oHei
	rugosa var. ***ventenatiana***	oHei
	Running Maid®/ 'Lenramp'	last listed 99/00
un	'Rural Rhythm'	oHei
	Rush®/ 'Lenmobri'	wChr

	Rushing Stream®/ 'Austream'	oHei
	'Russelliana'	oRos oHei wAnt
	Ruth Clements™ / 'Clemom'	oHei
	'Sabra'	oJus
un	'Sachet'	oJus
	'Sadler's Wells'	oHei
	'Safrano'	wCLG
	Saint Cecilia®/ 'Ausmit'	wCot oHei oRNW
un	Saint Patrick™ / 'Wekamanda'	wCot oEdm oGar oSho oWnC oAls oRNW
un	Saint Patrick™ / 'Wekamanda' (tree)	oEdm
	Saint Swithun / 'Auswith'	oHei oRNW wAnt
	'Salet'	oFor oRos oHei wAnt
	Sally Holmes®	wChr oRos wCot oHei oEdm oNWe oSha oRNW wAnt
	Samaritan / 'Harverag'	oRNW
un	'Samson'	oJus
	sancta	see **R. *x richardii***
	'Sander's White Rambler'	oHei
un	'Sangerhausen'	oHei
un	'Santa Claus'	wChr oJus
un	'Santa Claus' (tree)	oWnC
un	'Santa Fe'	oRNW
	Sarabande / 'Meihand'	wChr oRNW
	Sarah, Duchess of York	see **R.** Sunseeker / 'Dicracer'
	'Sarah van Fleet'	wChr oRos wCot oHei oRNW
un	'Sassy Lassie'	oJus
	Savoy Hotel / 'Harvintage'	oEdm oRNW
	'Scabrosa'	oRos oHei oNWe wRai wAnt
un	'Scamp'	oJus
	Scarlet Fire	see **R.** 'Scharlachglut'
	'Scarlet Grevillei'	see **R.** 'Russelliana'
	Scarlet Meidiland®/ 'Meikrotal'	wChr wCot oHei oWnC oRNW
	Scarlet Meidiland®/ 'Meikrotal' (tree)	oWnC
un	'Scented Dawn'	see **R.** Polka™ / 'Meitosier'
un	Scentimental / 'Wekplapep'	wChr wCot oHei oEdm oGar oSho oWnC oRNW wAnt
un	'Scentsational'	oJus
	Sceptre'd Isle™ / 'Ausland'	wChr oHei
	'Scharlachglut'	last listed 99/00
	'Schneezwerg'	last listed 99/00
	'Schoener's Nutkana'	oRos oHei
	'Schoolgirl'	last listed 99/00
	Scottish Special / 'Cocdapple'	last listed 99/00
	Scudbuster	see **R.** Patriot Flame / 'Clescrub'
	Sea Foam®	wChr oRos wCot oHei oGar oSha oRNW oSle
	Sea Foam® (tree)	last listed 99/00
	'Sea Pearl'	wChr wCot oHei oEdm
un	'Seabreeze'	oJus
	'Seagull'	oRos
	'Sealing Wax' (***moyesii*** hybrid)	oHei
	'Seashell'	oSho
un	'Seattle Scentsation'	oJus
	Secret / 'Hilaroma'	wCot oHei oEdm oGar oWnC oAls oRNW
un	'Seduction'	oRNW
un	'Selena'	oRNW
	sempervirens	oFor oHei
	'Senateur Amic'	oHei
un	'Senateur Lafollette'	oHei
	Send in the Clowns™ / 'Cleclown'	oHei
un	'Sensation'	oRNW
un	'September Song'	oHei
un	'Sequoia Gold'	oJus
	sericea	wCCr
	sericea HWJCM 322	wHer
	sericea ssp. ***omeiensis***	oHei
	sericea ssp. ***omeiensis*** EDHCH 97119	wHer
	sericea ssp. ***omeiensis*** f. ***pteracantha***	oFor wChr oRos oAmb oCis oSec
	'Serratipetala'	oHei
	sertata	oTrP
	setigera	oFor wCCr
un	***setigera*** 'Female Hips'	oHei
	setigera striped	oHei
	setipoda	oHei
	Sevillana™ / 'Meigekanu'	wChr wCot oHei oGar oRNW
	Sexy Rexy®/ 'Macrexy'	wChr wCot oEdm oWnC oRNW oUps
	Sexy Rexy®/ 'Macrexy' (tree)	wChr wCot oEdm
un	Shadow Dancer™ / 'Morstrort'	wChr oHei oRNW wAnt
	'Shailer's White Moss'	last listed 99/00
	Sharifa Asma®/ 'Ausreef'	wChr wCot oHei oGar oRNW wAnt
	Sheer Bliss / 'Jactro'	wChr wCot oEdm oGar oWnC oAls oRNW
	'Sheer Elegance'	oWnC oRNW
un	'Sheila MacQueen'	oRNW
	Sheila's Perfume / 'Harsherry'	wChr wCot oHei oEdm oWnC oRNW wAnt
	Sheila's Perfume / 'Harsherry' (tree)	wCot oEdm
un	'Shelly Renee'	oJus
un	'Shining Coral'	oWnC
un	'Shining Flare'	oWnC
un	'Shining Ruby'	oWnC
	Shocking Blue®/ 'Korblue'	wChr
un	'Show Garden'	oRNW
	Showbiz / 'Tanweieke'	wCot oWnC wAnt
	Showbiz / 'Tanweieke' (tree)	oWnC
un	'Showy Pavement'	oUps
un	'Shreveport'	oWnC
	'Shropshire Lass'	wChr oRos oHei
	'Sidonie'	see **R.** 'Sydonie'
	Signature®/ 'Jacnor'	wChr wCot oGar oWnC oAls oRNW
	Silver Jubilee®	oHei
	'Silver Moon'	oRos wAnt
	Simba / 'Korbelma'	wChr wCot oEdm oGar oWnC wAnt
	'Simon Frasier'	oHei
	Simon Robinson / 'Trobwich'	oHei
un	'Simplex'	oJus
	Simplicity®	wChr wCot oGar oWnC oRNW
un	'Simply Delightful'	oHei
	Singin' in the Rain / 'Macivy'	wCot oEdm oWnC oRNW
qu	'Single Sherry'	oHei
un	'Singles Better'	oJus
	Sir Clough / 'Ausclough'	oRos oHei oRNW
	Sir Edward Elgar / 'Ausprima'	oHei oRNW wAnt
	Sir Harry Pilkington / 'Tanema'	oRNW
	'Sir Thomas Lipton'	oFor wChr oRos wCot oHei oWnC oRNW wAnt oUps
	Sir Walter Raleigh®/ 'Ausspry'	last listed 99/00
	'Sissinghurst Castle'	oHei
	'Slater's Crimson China'	see **R. *x odorata*** Old Crimson China
un	'Sleeping Beauty'	oHei
	'Small Maiden's Blush'	oRos
un	'Small Miracle'	wChr wCot oGar oSho oWnC
un	'Small Miracle' (tree)	wChr oGar
un	'Small Virtue'	oJus
un	'Smoke Signals'	oJus
	Smooth Angel / 'Hadangel'	oWnC
	Smooth Lady / 'Hadlady'	oWnC
	Smooth Melody / 'Hadmelody'	oWnC
	Smooth Prince / 'Hadprince'	wChr
	Smooth Satin / 'Hadsatin'	wChr oWnC
un	'Snow Bride'	oJus
	Snow Carpet®/ 'Maccarpe'	wChr oHei wClo oOut oGre wBox oJus
	'Snow Dwarf'	see **R.** 'Schneezwerg'
	Snow Goose™ / 'Auspom'	oHei
un	'Snow Gosling'	oRos oHei oRNW
	'Snow Maiden'	oJus
un	'Snow on the Heather'	oHei
un	'Snow Owl'	oHei oGar
un	'Snow Pavement'	wCot oHei wAnt
un	'Snow Ruby'	oHei
un	'Snow Shower'	wChr wCot oHei oGar oWnC
	Snowball / 'Macangeli'	oJus
un	'Snowbride' (tree)	oWnC
	'Snowdon'	oHei
un	'Snowfall'	oJus
un	'Snowfire'	oGar oWnC oRNW
	'Snowflake'	wChr oHei
	Soaring Flight™	oHei
	'Soleil d'Or'	wChr
	Solitaire®/ 'Macyefre'	last listed 99/00
	'Sombreuil'	wChr oRos wCot oHei oGar oRNW wCLG wAnt
un	'Someday Soon'	oJus
un	'Something Else'	oJus
	'Sommermorgen'	see **R.** Oxfordshire / 'Korfullwind'
	Sonia	see **R.** Sweet Promise / 'Meihelvet'
un	'Sonia Rykiel'	wChr wCot
un	'Sorbet'	oRNW
un	'Sorcerer'	oJus
	soulieana	wHer oHei
un	'Southern Delight'	oJus
	'Souvenir de Brod'	see **R.** 'Erinnerung an Brod'
	'Souvenir de la Malmaison'	oFor wChr oHei oRNW wCLG wAnt
	'Souvenir de Madame Leonie Viennot'	oHei
	'Souvenir de Philemon Cochet'	wChr oRos oRNW
	'Souvenir de Pierre Vibert'	oHei
	'Souvenir de Saint Anne's	oRos oHei oRNW
	'Souvenir du Docteur Jamain'	oRos oHei oRNW wCLG wAnt
un	'Souvenir du President Lincoln'	oHei

	sp. DJHC 099	wHer
	sp. DJHC 359	wHer
un	'Space Walk'	oJus
	'Spanish Beauty'	see **R.** 'Madame Gregoire Staechelin'
un	Spanish Enchantress™	oHei oRNW
	Spanish Shawl	see **R.** Sue Lawley / 'Macsplash'
un	'Sparks'	oJus
	'Sparrieshoop'	oFor wChr oRos wCot oHei oRNW
un	'Spectra'	oRNW
un	'Spellcaster'	oRNW
co	'Spencer'	see **R.** 'Enfant de France'
un	'Spice Drop'	oJus
	Spice Twice™ / 'Jacable'	wChr wCot oGar oSho oWnC oRNW
	spinosissima	see **R.** ***pimpinellifolia***
	'Spong'	oRos
un	'Spring Hill's Freedom'	oRNW
	'Spring Morning'	see **R.** 'Fruehlingsmorgen'
un	'Spring Song'	oHei
un	Stainless Steel™ / 'Wekblusi'	wChr wCot oHei oEdm oWnC oRNW
	'Stanwell Perpetual'	wChr wCot oHei oDan oRNW wAnt
un	'Star Delight'	oHei
	Star of the Nile™ / 'Cleegypt'	oHei
	Starina®/ 'Megabi'	wChr wCot oJus oUps
	Starry Bouquet™	oHei
	'Stars 'n' Stripes'	oJus
	stellata var. ***mirifica***	oFor
	Sterling Silver™	wChr oGar oSho oWnC oRNW
un	'Stolen Moment'	oJus
	Strawberry Fayre / 'Arowillip'	oHei
	'Strawberry Ice'	oRNW
un	'Strawberry Swirl'	oJus
	Stephen's Big Purple	see **R.** Big Purple®/ 'Stebigpu'
un	'Street Wise'	oJus
	Stretch Johnson	see **R.** Tango / 'Macfirwal'
	'Striped Moss'	oHei
	Sue Lawley / 'Macsplash'	oHei
	Suffolk / 'Kormixal'	oHei
un	'Sugar Elf'	oHei
	Suma / 'Harsuma'	oHei oRNW
	Summer Dream / Jacshe'	last listed 99/00
	Summer Fashion / 'Jacale'	wChr wCot oWnC oAls
	Summer Snow / 'Weopop'	oRNW
	'Summer Sunset'	oHei
	'Summer Sunshine'	last listed 99/00
	Summer Wine / 'Korizont'	oHei wAnt
	Summer's Kiss™ / 'Meinivoz'	oWnC
	'Summerwind'	see **R.** Surrey / 'Korlanum'
	Sun Flare / 'Jacjem'	wChr oGar
	Sun Flare / 'Jacjem' (tree)	wChr
un	'Sun Goddess'	wChr oRNW
un	Sun Runner™ / 'Interdust'	wChr wCot oHei oGar oWnC oRNW wAnt
	Sunblest / 'Landora'	oWnC
	'Sunbright'	oWnC
un	'Sundancer'	oRNW
un	'Sundowner'	oRNW
un	'Sunny June'	wChr wCot
	Sunseeker / 'Dicracer'	oHei
	Sunset Celebration™	see **R.** Warm Wishes / 'Fryxotic'
un	'Sunshine Girl'	oJus
	Sunsplash / 'Jacyim'	last listed 99/00
	Sunsplash / 'Jacyim' (tree)	last listed 99/00
	'Sunsprite'	see **R.** 'Korresia'
un	'Suntan Beauty'	oJus
un	'Super Bowl'	see **R.** Cologne / 'Macsupbow'
un	Super Dorothy™	oRos oHei
	Super Excelsa®/ 'Helexa'	oHei
	Super Star®/ 'Tanorstar'	wChr oGar oWnC oAls
	'Superb Tuscan'	see **R.** 'Tuscany Superb'
un	'Superstar Supreme'	oRNW
	'Surpasse Tout'	oHei
	Surrey / 'Korlanum'	oHei
	Sussex / 'Poulave'	oHei
	'Sutter's Gold'	wChr wCot oRNW wAnt
un	'Suzy Q'	wChr
un	'S.W.A.L.K.'	oJus
	Swan®/ 'Auswhite'	wChr oHei
	'Swan Lake'	oHei oRNW
un	'Swansong'	oJus
	Swany®/ 'Meiburenac'	oRos oHei oRNW
un	'Sweet Afton'	wChr oHei oRNW
un	'Sweet Bouquet'	oHei oRNW
un	'Sweet Chariot'	wChr oWnC oJus
	Sweet Dream / 'Fryminicot'	oHei
un	Sweet Gesture / 'Maccarlto'	oEdm
un	Sweet Inspiration / 'Jacsim'	oRNW
un	'Sweet Inspiration' (tree)	oWnC
	Sweet Juliet®/ 'Ausleap'	wChr wCot oHei oRNW
un	Sweet Memories / 'Frymancot'	oHei
	Sweet Promise / 'Meihelvet'	wChr wCot oWnC oAls oRNW
	Sweet Shirley /Cleshir'	oHei
	Sweet Sunblaze®	see **R.** Pretty Polly®/ 'Meitonje'
	'Sweet Surrender'	oWnC oRNW
un	'Sweet Vivien'	wChr oRNW wAnt
	sweginzowii	oFor
	sweginzowii 'Macrocarpa'	oHei
	'Sydonie'	oRos
	Sympathie®	wChr wCot
	Symphony®/ 'Auslett'	wChr oRos wCot oHei oRNW wAnt
un	'Table Mountain'	wCot
un	Taboo™ / 'Tanelorak	wChr wCot oGar oSho oWnC oAls oRNW
un	'Tabris'	wChr oHei oRNW
un	'Taischa'	wCLG
	'Talisman'	oRos oRNW
	Tamora / 'Austamora'	wChr oRos wCot oHei oGar oRNW wAnt
	Tamora / 'Austamora' (tree)	wCot
	Tango / 'Macfirwal'	oHei oEdm oRNW
un	Tapis Rouge	see **R.** Eyeopener / 'Interop'
un	'Tara Allison'	oJus
un	'Tattooed Lady'	oJus
	'Tausendschoen'	wChr oRos oHei
	Tear Drop / 'Dicomo'	oHei
un	'Teddy Bear'	oJus
un	'Tempie Lee'	wCot oEdm
	'Temple Bells'	oHei
un	'Tempo'	oHei
un	'Temptress'	oRNW
un	'Tender Blush'	oHei
	Tequila Sunrise / 'Dicobey'	oRNW
	Terra Cotta / 'Meicobius	oRNW
un	'Texas'	oJus
un	'Texas' (tree)	oWnC
	'Texas Centennial'	last listed 99/00
	Thais / 'Memaj'	wCot
	The Alexandra Rose / 'Ausday'	wChr oHei
	The Alexandra Rose™ / 'Ausday' (tree)	wChr
	'The Bishop'	oRos oHei
	'The Bride'	wCLG
	The Countryman®/ 'Ausman'	oHei
	The Dark Lady / 'Ausbloom'	wChr wCot oRNW wAnt
	The Dragon's Eye™ / 'Cledrag'	oHei
	The Edwardian Lady	see **R.** Edith Holden / 'Chewlegacy'
	'The Fairy'	wChr oRos wCot oHei oGar oWnC oRNW wAnt oSle
	'The Fairy' (tree)	wChr wCot oGar
un	The Fawn™	oHei oRNW
un	The Friar	oHei
	'The Garland'	oRos oHei
un	'The Gift'	oHei
	The Herbalist™ / 'Aussemi'	wChr wCot oHei oRNW wAnt
	The Impressionist™ / 'Clepainter'	oHei
	The King's Rubies™ / ' Clegem'	oHei
	'The Knight'	oRos
	The Lady / 'Fryjingo'	last listed 99/00
	The McCartney Rose / 'Meizeli'	wChr wCot oGar oAls oRNW wAnt
	'The Miller'	oRos oHei oRNW
	The Nun / 'Ausnun'	wChr wCot oHei
	The Pilgrim / 'Auswalker'	wChr wCot oHei oWnC oRNW wAnt
	The Poet™ / 'Clepoetry'	oHei
un	'The Polar Star'	oHei
	The Prince®/ 'Ausvelvet'	wChr wCot oHei oGar oWnC oRNW wAnt
	The Prince®/ 'Ausvelvet' (tree)	last listed 99/00
	'The Prioress'	oRos oHei
	The Queen	see **R.** 'La Reine'
	'The Queen Elizabeth'	wChr wCot oHei oEdm oGar oWnC oAls oRNW
	The Reeve®/ 'Ausreeve'	wChr oHei oRNW wAnt
	The Seckford Rose / 'Korpinrob'	oHei wAnt
	The Squire®/ 'Ausquire'	wChr wCot oHei oGar oRNW
	The Wife of Bath / 'Ausbath'	oHei
	'The Yeoman'	oRos oHei
	'Thelma'	oRos oHei
	'Therese Bugnet'	oFor wCul wChr wCot oHei oGar oDan oRNW wAnt oUps oSle
	'Thisbe'	oHei
un	'Thor'	oHei
	Thora Hird / 'Tonybrac'	oHei
un	Thornbury Castle™ / 'Harmusky'	oHei

	Thousand Beauties	see **R.** 'Tausendschoen'
un	'Tickled Pink'	oRNW
	'Tiffany'	wChr oRos wCot oGar oWnC oRNW
	Tigris®/ 'Harprier'	oHei
	Timeless ™ / 'Jacecond'	wCot oEdm oSho oWnC oAls oRNW
	Timeless ™ / 'Jacecond' (tree)	oGar
un	Timeless Beauty™ / 'Korreahn'	oHei
un	Tineke	wCot oEdm oRNW
un	'Tiny Tears'	oJus
un	'Tipper'	oJus
	'Tipsy Imperial Concubine'	oHei
un	'Tivoli Gardens'	wCot
un	'Today'	oRNW
un	'Toffee'	oJus
un	'Tom Wood'	oHei
	tomentosa	oHei
	Topaz Jewel	see **R.** Yellow Dagmar Hastrup / 'Moryelrug'
un	'Torch of Liberty'	oJus
un	'Toro'	oWnC
	Touch of Class / 'Kricarlo'	wChr wCot oEdm oGar oWnC oAls oRNW
un	'Touch of Midas'	oHei oJus
	Toulouse-Lautrec®/ 'Meirevolt'	wChr wCot oHei oGar oWnC oRNW wAnt
	'Tour de Malakoff'	oRos
	Tournament of Roses / 'Jacient'	wChr wCot oEdm oGar oAls oRNW
un	Tower Bridge™	oHei
	Tradescant™ / 'Ausdir'	wChr wCot oHei oGar oSho oRNW
	Traviata™ / 'Meilavio'	wChr wCot oHei oGar oWnC oRNW wAnt
	'Treasure Trove'	oRos oHei
	Trevor Griffiths / 'Ausold'	wChr oHei
	'Tricolore de Flandre'	oHei
	Trier®	oRos oHei
	Troilus / 'Ausoil'	oRos oHei wCLG
un	'Tropical Paradise'	oRNW
un	'Tropical Passion'	oRNW
un	Tropical Sunset™ / 'Mactaurang	wChr wCot oWnC oAls oRNW wAnt
	Tropical Twist / 'Jacorca'	wChr wCot oGar oSho
	Tropicana	see **R.** Super Star®/ 'Tanorstar'
un	'True Vintage'	oJus
	Trumpeter®/ 'Mactru'	wChr wCot
	Trumpeter®/ 'Mactru' (tree)	wChr wCot oWnC
un	Turbo®/ 'Meirozrug'	oHei
un	Turbo Rugostar	see **R.** Turbo®/ 'Meirozrug'
	Turlock High / 'Clelock'	oHei
	'Turner's Crimson'	see **R.** 'Crimson Rambler'
	'Tuscany'	oHei
	'Tuscany Superb'	wCul oRos wCot oHei wCLG wAnt
un	Twilight	see **R.** 'Crepuscule'
un	'Tyler'	oRNW
un	'Typhoo Tea'	wCot
	'Uetersen'	wCLG
	Ultimate Pink™ / 'Jacval	wChr wCot oGar oSho oWnC oAls oRNW
un	'Unforgettable'	oWnC
	'Unique Blanche'	last listed 99/00
	Valencia®/ 'Koreklia'	wChr oGar
un	'Valerie Jeanne'	oJus
un	'Vanilla Perfume'	wChr wCot oWnC oAls oRNW wAnt
	'Vanity'	oRos oHei oRNW
	'Variegata di Bologna'	oRos oHei
	'Veilchenblau'	oRos wCot oHei wAnt
un	'Velvet Cloak'	oJus
	Velvet Fragrance / 'Fryperdee'	oHei
	'Venusta Pendula'	oHei
un	'Verdi'	oHei
un	'Versilia'	oRNW
un	'Very Cherry'	wCot oWnC
	'Vesuvius'	oRos oRNW
	Veterans' Honor™ / 'Jacopper'	wChr wCot oWnC oAls oRNW wAnt
	'Vick's Caprice'	wChr wCot
	'Vicomtesse Pierre du Fou'	oHei
un	'Victor Borge'	wChr wCot
	Victorian Charm™ / 'Clebliss'	oHei
un	'Victorian Lace'	wChr
un	Victorian Spice™ / 'Harzola'	wChr wCot oWnC oRNW
un	'Viking Queen'	last listed 99/00
	'Village Maid'	see **R.** 'Centifolia Variegata'
	villosa L.	oFor wHer oHei oGoo wCCr
un	***villosa engadinensis***	oHei
un	***villosa recondita***	oHei
un	'Vineyard Song'	oRNW
un	'Vintage Visalia'	oRNW
	'Violacea'	oRos
un	Violetta	see **R.** International Herald Tribune / 'Harquantum'
	'Violette'	oHei
un	'Virginia'	oRNW
un	Virginia Dare™	oHei
	virginiana	oFor oHei
un	***virginiana*** 'Harvest Song'	oHei
	virginiana 'Plena'	see **R.** 'Rose d'Amour'
un	'Vista'	oJus
	'Vivid'	oRos oHei
un	'Vogue'	oRNW
un	Voodoo / 'Aromiclea'	wChr wCot oGar oSho oWnC oRNW
	vosagiaca	see **R. *caesia*** ssp. ***glauca***
un	'Waikiki'	oRNW
	Warm Wishes / 'Fryxotic'	wChr wCot oEdm oGar oWnC oAls oRNW wAnt
	Warm Wishes / 'Fryxotic' (tree)	wCot
un	'Wartburg'	oHei
	Warwick Castle®/ 'Auslian'	wChr oRos oHei wAnt
un	'Watercolor'	oJus
	Watermelon Ice / 'Jacair'	oHei oGar
	webbiana	oFor
un	'Wedded Bliss'	oJus
	'Wedding Day'	oRos oHei
un	Wee Butterflies™	oHei
	'Weetwood'	oHei
	Weight Watchers Success™ / 'Jacbitou'	wChr wCot oWnC oAls oRNW
	'Weisse aus Sparrieshoop'	wCot
un	'Well's Climber'	wCot oRNW
un	Welsh Gold™	oHei
	Wenlock®/ 'Auswen'	wCot oHei oRNW
	'Werina'	see **R.** Arizona / 'Tocade'
	Westerland™ / 'Korwest'	wChr oRos wCot oHei oEdm oNWe oRNW wAnt
un	'Westfalenpark'	oHei
un	'Whimsical'	oJus
	'Whirlygig'	oJus
	Whisky Mac / 'Tanky'	last listed 99/00
	'White Bath'	see **R.** 'Shailer's White Moss'
	White Bells®/ 'Poulwhite'	oRos oHei
un	'White Cap'	oHei
	'White Cecile Brünner'	oHei
	White Cockade®	oHei
un	'White Dawn'	wChr oRos wCot oHei oSho oWnC oRNW wAnt
	'White de Meaux'	oRos oHei
un	'White Dorothy Perkins'	oHei
	'White Flight'	last listed 99/00
	White Flower Carpet®/ 'Noaschnee'	wChr oAls oGar oSho wWel
	'White Grootendorst'	oHei oRNW
un	'White Koster'	oRos oHei
un	'White Lightnin'	wChr oGar oRNW
	'White Maman Cochet'	last listed 99/00
	White Meidiland®/ 'Meicoublan'	wChr wCot oHei oRNW
	White Meidiland®/ 'Meicoublan' (tree)	wChr wCot
un	White New Dawn™	oHei
un	'White Pearl in Red Dragon's Mouth'	oRNW
	'White Pet'	wChr oRos wCot oHei
	'White Queen Elizabeth'	oHei
	White Simplicity®/ 'Jacsnow'	wChr wCot oGar oRNW
	'White Wings'	wChr oRos oHei
un	'Whiteout'	oJus
un	'Whoopi'	oJus
un	'Why Not'	oJus
	wichurana	oFor oRos oOEx
	wichurana 'Curiosity'	see **R. *wichurana*** 'Variegata'
	wichurana 'Hiawatha'	see **R.** 'Hiawatha'
un	***wichurana*** 'Poteriifolia'	oFor
	wichurana 'Variegata'	oFor wWoS oHei wRob oGre
	Wife of Bath®	see **R.** The Wife of Bath / 'Ausbath'
un	'Wild Dancer'	wChr oGar
	Wild Flower / 'Auswing'	oHei
	Wild Plum / 'Jacwiq'	wChr wCot oGar oUps
	Wild Plum / 'Jacwiq' (tree)	oGar
un	'Wild Spice'	wChr wCot wAnt
un	'Wildberry Breeze'	wChr wCot
	'Wilhelm'	oRos
	Will-o'-the-Wisp™ / 'Clemist'	oHei
	'Will Scarlet'	oHei
	'William Baffin'	wWoS wChr oRos wCot oHei oUps
co	'William Grant'	oHei
	'William Lobb'	wChr oRos wAnt
	William Shakespeare®/ 'Ausroyal'	wChr oRos oHei oRNW wAnt
	'William III'	oRos wAnt
	Wiltshire / 'Kormuse'	oHei

	Winchester Cathedral®/ 'Auscat'	wChr wCot oHei oWnC oRNW wAnt
	'Wind Chimes'	oHei
	Windflower / 'Auscross'	wChr oHei
	Windrush®/ 'Ausrush'	wChr oHei
un	'Windsong'	oRNW
	Wine and Roses™	oHei oRNW
un	'Winifred Coulter'	wCot
un	'Winnie Edmunds'	oRNW
un	'Winning Colors"	oRNW
	'Winnipeg Parks'	last listed 99/00
un	'Winsome'	wChr oJus
	Wise Portia / 'Ausport'	oRos oHei oRNW wAnt
un	'Wishful Thinking'	oJus
un	'Wit's End'	oJus
	'Woburn Abbey'	wChr
	Wonderstripe™ / 'Clewonder'	oHei
	woodsii	oFor oHan wChr oRos oHei wBur wShR wKin wPla oSle wSnq
	woodsii var. ***fendleri***	see **R. *woodsii***
un	'Work of Art'	oJus
	xanthina	oFor wBCr oHei oOEx
	xanthina 'Canary Bird'	wChr
	xanthina f. ***hugonis***	oFor wChr oRos oUps
	xanthina lindleyii	oHei
	xanthina f. ***spontanea***	oRos
un	'Yankee Lady'	wCot
un	'Yellow Blaze'	wChr
	Yellow Button®/ 'Auslow'	wChr oRos wAnt
	'Yellow Cecile Brünner'	see **R.** 'Perle d'Or'
	Yellow Charles Austin®/ 'Ausyel'	wChr oRos oHei
un	Yellow Chestnut™ (fruit)	oOEx
	'Yellow Cushion'	wCot
	Yellow Dagmar Hastrup / 'Moryelrug'	wCul wChr wCot oGar oSho wAnt oSle
un	'Yellow Fairy'	oHei wAnt
un	Yellow Jacket / 'Jacyepat'	oHei oGar oWnC
un	'Yellow Mozart'	oHei
	Yellow Simplicity®/ 'Jacyelsh'	wChr oGar oRNW
	Yesterday®	oHei
	'Yolande d'Aragon'	wChr oRos wCot oHei wAnt
	'York and Lancaster'	see **R. *x damascena*** var. ***versicolor***
un	'You 'n' Me'	oJus
	Yves Piaget®	see **R.** Royal Brompton Rose / 'Meivildo'
un	'Zebra	oRNW
un	'Zenaitta'	oHei
	'Zephirine Drouhin'	wChr oRos wCot oHei oGar oSho oWnC oRNW wAnt oUps
	'Zigeunerknabe'	oHei
un	'Zinger'	oJus
	Zitronenfalter®	oRos oHei
un	'Zwerg'	oHei oUps
	ROSCOEA	
	alpina	oTrP wHer wCri oJoy iArc
	auriculata	wCol oAls oSis
	'Beesiana'	oGos wCol
	cautleyoides	oTrP wCol oAls oRus oSis
	cautleyoides 'Kew Beauty' seed grown	wHer
	humeana	wHer oGos
	purpurea	oTrP wHer wCri oGos oRed oSha
	scillifolia	wHer oRus oSis
	sp.	oCis
	tibetica	wHer
	ROSMARINUS	
	angustissimus	see **R. *officinalis angustissimus***
	officinalis	oTDM wFGN oGoo oSho oSha oWnC wSta wNTP
	officinalis var. ***albiflorus***	wWoS wCul oJoy oAls wFGN oGar oGoo wCCr
	officinalis 'Albus'	oSis oGre oBar
	officinalis angustissimus	iGSc
	officinalis 'Arp'	wHer wWoS oJoy oAls wFGN oGoo wRai oGre wFai iGSc wHom oCrm oBar oUps oSle wSnq wWel
	officinalis 'Benenden Blue'	wHer oDar oJoy oAls wFGN oGar oGoo oGre wCCr oBar wWel
	officinalis 'Blue Boy'	wFGN oGoo iGSc oWnC oBar
un	***officinalis*** 'Blue Lady'	wWoS wFGN oBar
un	***officinalis*** 'Blue Spear'	oJoy wFGN oBar
un	***officinalis*** 'Blue Spire'	oFor wWoS oDar oAls oHed wFGN oGoo oSis oBar
	officinalis 'Collingwood Ingram'	see **R. *officinalis*** 'Benenden Blue'
un	***officinalis*** 'Constance deBaggio'	oAls wFGN oBar
un	***officinalis*** 'Dancing Waters'	wFGN oBar
un	***officinalis*** 'Dutch Mill'	wFGN oBar
un	***officinalis*** 'Flora Rosa'	wNTP
qu	***officinalis*** 'Forresteri'	oAls wFGN oGar iGSc oBar oUps
	officinalis 'Frimley Blue'	see **R. *officinalis*** 'Primley Blue'
	officinalis 'Golden Rain'	oFor wWoS wCul oJoy oAls wFGN oGoo oWnC oBar
un	***officinalis*** 'Goriza'	wHer oJoy oAls wFGN oGar oGoo oGre oBar oSec
un	***officinalis*** 'Green Rain'	oAls wFGN
un	***officinalis*** 'Herb Cottage'	wWoS oJoy oAls wFGN oGoo oBar
un	***officinalis*** 'Hill's Hardy'	wHer wWoS oJoy oAls wFGN oGoo oCrm oBar
un	***officinalis*** 'Howe'	oAls wFGN
un	***officinalis*** 'Hulka'	oGoo
un	***officinalis*** 'Huntington Blue'	oGoo
un	***officinalis*** 'Huntington Carpet'	wWoS oDar oGar oBar oUps wWel
un	***officinalis*** 'Ingramii'	oJoy
un	***officinalis*** 'Irene'	wWoS wSwC oAls oAmb oSis oUps
un	***officinalis*** 'Jen's Blush'	oHed
un	***officinalis*** 'Ken Taylor'	wWoS oJoy oAls oHed wFGN oGar wCCr oWnC oBar
	officinalis 'Lockwood de Forest'	oFor oAls wFGN oGar oGoo iGSc
	officinalis 'Logee's Blue'	wWoS oJoy wFGN oGre oBar oUps
un	***officinalis*** 'Madalene Hill'	oGar
	officinalis 'Majorca Pink'	wWoS oAls oGoo oWnC oBar
qu	***officinalis*** 'Miss Jessopp's	wFGN
	officinalis 'Miss Jessopp's Upright'	oAls oGar oGoo oBar wBWP
un	***officinalis*** 'Mount Vernon'	wFGN oCrm
un	***officinalis*** 'Mrs. Furneaux'	oAls wFGN
un	***officinalis*** 'Mrs. Howard's	wSwC oAls wFGN
un	***officinalis*** 'Nancy Howard'	oGoo
	officinalis pine scented	wHom
	officinalis pink	wFGN oGre iGSc wHom
	officinalis 'Primley Blue'	wWoS wCul wFGN oBar
	officinalis Prostratus Group	oFor wWoS wCul oTDM oJoy oAls wFGN wCSG oGar oGoo oSho oBlo oGre wFai iGSc wHom oWnC oCrm oBar wTGN wSta wNTP oCir oUps oSle
	officinalis red flowered	wWoS wCul wFGN oGoo oCrm oBar
	officinalis 'Roseus'	oJoy wHom wTGN oUps
un	***officinalis*** 'Salem'	wWoS wSwC oJoy oAls wFGN oAmb oGre oBar oUps oSle
un	***officinalis*** 'Salem Herb Cottage'	oCrm
	officinalis 'Santa Barbara'	wWoS oAls wFGN wCCr oBar wNTP oUps
un	***officinalis*** 'Sawyer's Blue'	wFGN oGoo oBar
	officinalis 'Severn Sea'	wWoS wFGN oGoo oCrm oBar
un	***officinalis*** 'Shimmering Stars'	wWoS oTDM oJoy wFGN oGoo oBar
	officinalis 'Sissinghurst Blue'	wWoS
un	***officinalis*** 'Spice Island'	oFor wWoS oAls iGSc oBar
un	***officinalis*** 'T. S.'	oAls wFGN oBar
un	***officinalis*** 'Taylor's Blue'	oJoy wFGN oGoo oBar
un	***officinalis*** 'Ticonderoga'	wFGN
	officinalis 'Tuscan Blue'	oFor wHer wWoS wCul oTDM oDar oAls wFGN wCSG oGar oGoo oBlo oGre iGSc oWnC oCrm oBar wTGN oInd wNTP oBRG wBWP oCir oUps wWel
un	***officinalis*** 'Very Oily'	oGoo
un	***officinalis*** 'Well Sweep Golden'	oGoo
un	***officinalis*** 'Wood's'	wFGN oBar
	ROSTRINUCULA	
	dependens GUIZ 18	wHer
	ROSULARIA	
	alpestris	oSto
	alpestris from Kashmir	oSto
	alpestris from Zozella	oSto
	chrysantha	oSto
	muratdaghensis	oSto
	platyphylla	see **R. *muratdaghensis***
	sedoides	last listed 99/00
	sedoides var. ***alba***	oSto
	serpentinica	oSto
	ROTHMANNIA	
	globosa	last listed 99/00
	RUBIA	
	cordifolia	oGoo oOEx
	tinctorum	oGoo wFai iGSc
	RUBUS	
	allegheniensis	oFor
	arcticus	oFor
	'Benenden'	wHer oGos wCol wCSG wCCr
	'Betty Ashburner'	oFor oTrP
	'Boysenberry' (fruit)	wBCr wCed oAls oGar oSho wRai wCoS
	'Boysenberry, Thornless' (fruit)	oFor oGar oOEx wTGN
	buergeri 'Variegatus'	oFor oCis
	calycinoides	see **R. *pentalobus***
un	***cockburnianus*** 'Aureus'	oSec
	cockburnianus Goldenvale™ / 'Wyego'	wWel

	'Coronarius'	see **R. *rosifolius*** 'Coronarius'
	deliciosus	oFor
	ellipticus	oOEx
un	***fruticosus*** 'Albovariegata'	wCul
un	***fruticosus*** 'Apache'	wCed wRai
un	***fruticosus*** 'Arapaho'	wCed wRai
un	***fruticosus*** 'Black Butte'	wCed oAls
un	***fruticosus*** 'Black Douglass'	wCed
	fruticosus 'Black Satin'	oFor oAls oOEx
	fruticosus 'Cascade'	wCed
un	***fruticosus*** 'Chester'	wCed
un	***fruticosus*** 'Chester, Thornless'	wClo
un	***fruticosus*** 'Chicksaw'	wCed
un	***fruticosus*** 'Hull'	wBCr wCed
un	***fruticosus*** 'Kiowa'	wCed wRai
un	***fruticosus*** 'Kotata'	wCed
	fruticosus 'Loch Ness'	wBur wRai
un	***fruticosus*** 'Navaho'	wCed
un	***fruticosus*** 'Navaho Thornless'	oOEx
un	***fruticosus*** 'Shawnee'	wCed
un	***fruticosus*** 'Siskiyou'	wCed
	fruticosus 'Sylvan'	wCed
un	***fruticosus*** 'Triple Crown'	wCed
un	***fruticosus*** 'Triple Crown Thornless'	wBur wRai
	fruticosus 'Variegatus'	last listed 99/00
un	***fruticosus*** 'Wild Cascade'	wCed
	fruticosus 'Waldo'	wCed
	glaucus	oOEx
un	Guanxi Superberry™	oOEx
	henryi var. ***bambusarum***	oCis
	ichangensis	oFor oCis wCCr
un	***idaeus*** 'Amity'	oFor wBCr wCed oAls oGar oSho wTGN
	idaeus 'Aureus'	wHer wWoS
	idaeus Autumn Bliss™	wBur wClo oGar
	idaeus 'Boyne'	oWnC
	idaeus 'Canby'	wCoS
un	***idaeus*** 'Caroline'	oSho
un	***idaeus*** 'Centennial'	wCed
un	***idaeus*** 'Chilcoten'	oGar
un	***idaeus*** 'Chilliwack'	wCed oAls wRai
un	***idaeus*** 'Dinkum'	wClo
	idaeus everbearing	wSta
	idaeus 'Fallgold'	oFor wBCr wClo oAls oGar
un	***idaeus*** 'Golden Summit'	oSho wRai
	idaeus 'Heritage'	wCed oAls oGar oSho wCoS
	idaeus 'Indian Summer'	oWnC
un	***idaeus*** 'Jewel'	wCed wRai
	idaeus 'Meeker'	wCed wClo oAls oGar wRai
	idaeus 'Newburgh'	wBCr oGar oWnC
un	***idaeus*** 'Royalty'	wRai
un	***idaeus*** 'Summit'	wCed oGar wRai oWnC wTGN
	idaeus 'Tulameen'	wCed wClo oSho wRai
	idaeus 'Willamette'	oAls oGar oSho wSta
un	***ikenoensis***	oFor
	irenaeus	oCis
un	***japonicus***	see **R. *ikenoensis***
un	***lambertianus hakonensis***	oFor
un	***lasiococcus***	oBos
	leucodermis	wPla wFFl
	lineatus	oFor wCol wCSG oAmb oCis wSte
	Loganberry Group (fruit)	oAls
	Loganberry Group Thornless (fruit)	wCed wBur wClo oGar wRai wTGN
	'Marionberry' (fruit)	wCed wClo oAls oGar oSho wRai wBox
	microphyllus 'Variegatus'	last listed 99/00
	nepalensis	oFor
	occidentalis (fruit)	wBCr
	occidentalis blackcap	wBur
	occidentalis 'Cumberland' (blackcap)	oWnC
	occidentalis 'Munger' (blackcap)	wCed oAls
	odoratus	oFor oGar
	'Olallieberry' (fruit)	wCed
	parviflorus	oFor oHan wWoB wBur wNot wShR oGar oBos oRiv wRai wWat wCCr oTri wPla oAld wFFl oSle
	parviflorus double form	oFor
	parvus	oFor
qu	***pectinatus*** var. ***tricolor***	oCis
	pectinellus var. ***trilobus***	oFor
	pedatus	wFFl
	pentalobus	oFor wCoN oDar oJoy oAls wShR wCSG oSho wRai wHom oWnC wTGN wSta oCir wRav oUps oSle
	pentalobus 'Emerald Carpet'	oFor oAls oGar oOEx
	phoenicolasius	oOEx
un	***roseus*** var. ***rocota***	oOEx

	rosifolius double cream	see **R. *rosifolius*** 'Coronarius'
	rosifolius 'Coronarius'	oFor wWoS wCol oHed oCis oNWe
un	***rosifolius*** 'Strawberries and Cream'	wTGN
	setchuenensis	oFor oSec
un	***sp.*** Burmese strazberry (fruit)	oOEx
	sp. evergreen, Columbia	oOEx
	sp. evergreen, Ecuador	oOEx
	spectabilis	wWoB wBur wNot wShR oBos wRai wWat wCCr oTri oAld wFFl oSle wSte wSnq
	spectabilis 'Flore Pleno'	see **R. *spectabilis*** 'Olympic Double'
do	***spectabilis*** 'Olympic Double'	oFor
	Tayberry Group (fruit)	wBCr wBur oAls oGar wRai
	tricolor	oFor oTrP oCis
	x tridel 'Benenden'	see **R.** 'Benenden'
	trifidus	oFor
	ursinus	wRai wWat wPla
un	***ursinus*** var. ***macropetalus***	wFFl
un	***ursinus*** 'Variegatus'	oGre
un	**RUCOLA**	
	silvatica	oGoo
	RUDBECKIA	
	Autumn Sun	see **R.** 'Herbstsonne'
	californica	oFor
	fulgida var. ***deamii***	oFor
	fulgida var. ***fulgida***	wSnq
	fulgida var. ***speciosa***	oMis
	fulgida var. ***sullivantii*** 'Goldsturm'	oFor wCul wSwC oNat oDar oAls oHed oAmb oMis oGar oGoo oSho oSha oOut oBlo oGre wHom oWnC wTGN oSec wBWP iArc oCir oEga oLSG oUps wSnq
	'Gold Drop'	see **R.** 'Goldquelle'
	'Goldilocks'	oAls wHom
	'Goldquelle'	oAls oAmb oSha wTGN oSle
	'Herbstsonne'	wWoS wCul oAls oGar oSho wRob oGre oSec oUps oSle
	hirta	wThG oNat wCoN oMis oAld oCir wWld
	hirta Becky mixed	oAls oGar
un	***hirta*** 'Indian Summer'	oAls oSho oEga oUps
	hirta 'Irish Eyes'	oAls
	hirta 'Sonora'	oGar
	laciniata	wHer wCul oGoo
	laciniata 'Golden Glow'	see **R. *laciniata*** 'Hortensia'
	laciniata 'Hortensia'	last listed 99/00
	maxima	oFor oUps
	nitida	oRus
	occidentalis	wWld
	occidentalis 'Green Wizard'	oSec oCir
	'Rustic Colors'	oAls
	speciosa	see **R. *fulgida*** var. ***speciosa***
	subtomentosa	oFor
	triloba	oNat oMis oGoo oWnC
	RUELLIA	
	brittoniana	oFor oTrP wGAc oHug oMis oSsd oOut oUps
un	***brittoniana*** 'Strawberries & Cream'	wGAc
	'Chi Chi'	oHug oSsd
	humilis	oNWe oGre oUps
	'Katie'	oHug
qu	***squarrosa alba***	oHug
	RUMEX	
	acetosa	oAls oGoo iGSc oBar oUps
un	***acetosa*** 'Variegatus'	wHer
	acetosella	oGoo iGSc
	hydrolapathum	wHer
	sanguineus	wWoS oNWe oGre oUps
	sanguineus var. ***sanguineus***	wBWP
	scutatus	wFGN wCSG oGoo wFai wHom oBar oUps
	scutatus French broadleaf	wFGN
	scutatus 'Silver Shield'	wWoS oGoo
	RUPICAPNOS	
	africana	last listed 99/00
	RUSCHIA	
un	***hamata***	oSis
	putterillii	wHer oSis
	RUSCUS	
	aculeatus	oFor
	hypoglossum	wHer
	RUTA	
	chalepensis	oOEx wCCr
	chalepensis 'Dimension Two'	oGoo
	graveolens	wCoN oAls wCSG oGoo iGSc wHom oWnC oBar oSec
un	***graveolens*** 'Blue Beauty'	wCul wFGN oBar
	graveolens 'Blue Curl'	iGSc
	graveolens 'Blue Mound'	wCul wFGN oBar

	Plant	Sources
	graveolens 'Curly Girl'	wWoS wFGN oSis
	graveolens 'Jackman's Blue'	wHer wCul wFGN oGoo wFai wCCr oCir oUps
	graveolens 'Variegata'	oSis
	RUYSCHIA	see **RUSCHIA**
	SABAL	
	bermudana	last listed 99/00
	mauritiiformis	last listed 99/00
	minor	oTrP oRiv oCis wDav wCCr
	palmetto	last listed 99/00
un	**SABIUM**	
	japonicum	oRiv
	sebiferum	oRiv
	SACCHARUM	
	ravennae	oFor wWoS oJoy oGar oSho oWnC wTGN wSnq
	SAGINA	
	subulata	oGre wFai wHom oWnC wCoN wWhG oSle
	subulata var. glabrata	oAls oCir
	subulata var. glabrata 'Aurea'	oAls oGre wHom oWnC wMag oCir oSle
	SAGITTARIA	
un	brevifolia	oSsd
un	chinensis	oOEx
	graminea	wGAc oHug oGar oSsd oWnC
un	graminea 'Crushed Ice'	oHug
	japonica	see **S. sagittifolia**
	latifolia	oFor oHug oGar oSsd oOEx wWat oTri oAld
un	latifolia 'Leopard Spot'	wGAc
	montevidensis	oHug
	natans	wGAc
	sagittifolia	oGar
	sagittifolia 'Bloomin Baby'	oGar
do	sagittifolia 'Flore Pleno'	wGAc oGar
	sinensis	see **S. graminea**
qu	subulata var. pusilla	wGAc oSsd
	SAINTPAULIA	
	ionantha	oGar
	SALIX	
un	acuminata var. microphylla	wCol
	acutifolia 'Pendulifolia '(m)	oFor
	alba f. argentea	see **S. alba** var. **sericea**
un	alba 'Belders'	wBCr
	alba 'Cardinalis' (f)	last listed 99/00
	alba 'Chermesina' hort.	see **S. alba** ssp. **vitellina** 'Britzensis'
	alba var. sericea	oFor wHer oGos oRiv wCCr
	alba 'Tristis'	oFor wBCr wBur oAls oGar wKin wTGN
	alba ssp. vitellina	oFor wBCr wBox
	alba ssp. vitellina 'Britzensis'	oFor oGre wCCr wTGN wAva
un	americana	wBox
	aquatica	see **S. cinerea**
	arctica	wBox
	arctica x reticulata	oGre
	arenaria	oFor oGar
	babylonica	oFor oGar wTGN wCoS
	babylonica 'Annularis'	see **S. babylonica** 'Crispa'
	babylonica 'Crispa'	oFor wCol oAls wSta
	babylonica var. pekinensis	oGar
	babylonica var. pekinensis 'Tortuosa'	wBCr oGoo wKin oWnC oCir wSte wCoS
	x bebbii	oFor
	bicolor	oFor
	bockii	oFor
un	boothii	oFor
	'Boyd's Pendulous' (m)	oFor oDar wCol oSqu
un	brachycarpa 'Blue Fox'	oFor
un	breweri	oFor
	caprea	oFor wBCr wCSG oGar oRiv wKin oWnC wTGN wBox wSta wWel
	caprea 'Curlilocks'	wHer
	caprea 'Kilmarnock' (m)	oRiv
	caprea var. pendula (see Nomenclature Notes)	oFor oAls oGar wKin oWnC
un	chaenomeloides	oFor wHer wBox
	chilensis	last listed 99/00
	cinerea	wBCr
	cinerea 'Variegata'	oFor wBox
	commutata	oFor
	x cottetii	oFor wBox
un	crenata	wCol wMtT
	daphnoides	oFor wHer
	discolor	oAls oBlo
	elaeagnos	oFor wWoS wCul oGos oHed wBox
	eriocephala 'American McKay'	oFor wBox
un	eriocephala mackenziana	oBos
	'Erythroflexuosa'	oFor wBur oRiv oWhi wBox wSta
	exigua	oFor wPla oAld
	fargesii	oFor oGos
	'Flame'	oFor oJoy oWnC wMag wBox
un	fluviatilis	oAld oSle
un	fluviatilis 'Multnomah'	oFor wBox
un	fluviatus	oAld
	fragilis	wBCr
un	fragilis 'Belgian Red'	wBox
	fragilis 'Bullata'	oFor wBox
	fruticulosa	wCol
un	geyeriana	oFor
un	gilgiana	oFor wBox
	'Golden Curls'	see **S.** 'Erythroflexuosa'
un	gooddingii	wCCr
qu	gracilis 'Nana'	oRiv
	gracilistyla 'Melanostachys' (m)	oFor wHer wCul wCri wClo oGar oRiv oWhi oGre wTGN wBox wAva
un	gracilistyla 'Variegata'	wRai oGre
un	gracilistyla 'Variegata Pendula'	oGar
	x grahamii 'Moorei' (f)	oFor wBox
	'Hakuro-nishiki'	see **S. integra** 'Hakuro-nishiki'
	helvetica	oNWe wMtT
un	hindsiana	oFor
	hookeriana	wWoB wFFl wWat wCCr oAld oSle
un	hookeriana 'Clatsop'	oFor
	humilis	oFor
	hylematica	see **S. fruticulosa**
	integra 'Albomaculata'	see **S. integra** 'Hakuro-nishiki'
va	integra 'Hakuro-nishiki'	oFor wHer wCul oInd wClo oDar oJoy wCol oAls oAmb oGar oDan oSis oRiv oNWe wRai oGre wFai oWnC wTGN wBox oSec oUps oSle wSte wSnq wWel
	irrorata	oFor oGar wCCr wBox
	japonica (see Nomenclature Notes)	oRiv
	koriyanagi	oGre
un	koriyanagi 'Rubykins'	wWoS
	lambertiana	see **S. purpurea** ssp. **lambertiana**
	lanata	wPir
	sp. aff. lanata	wHer
	lapponum	oFor oGos oJoy oGar oGre wBox
	lasiandra	oFor wWoB wBur wNot oGar oBos wWat wBox wPla oSle
un	lasiandra 'Nehalem'	oFor
	lasiolepis	oFor wWat
un	lasiolepis 'Rogue'	oFor
un	ligulifolia	wWat
un	ligulifolia 'Placer'	oFor
	lindleyana	oJoy oGar oSis oRiv oGre wBox oBRG
	lucida	oFor oAld
qu	lucida ssp. lasiandra	wFFl
	lutea	oBos
	magnifica	wHer oGos oJoy wCol oCis wSte
	matsudana	see **S. babylonica** var. **pekinensis**
	matsudana 'Umbraculifera'	oGar wBox
	melanostachys	see **S. gracilistyla** 'Melanostachys'
un	miyabeanum	oFor
	x moorei	see **S. x grahamii** 'Moorei'
	myrsinifolia	oFor wBox
un	myrsinifolia alpicola	oFor
	myrsinites	wSta
	myrtilloides 'Pink Tassels' (m)	oFor oGre
	nakamurana var. yezoalpina	oFor wCri oGos oDar wCol oGar oSis oRiv oOut oSqu oGre wPir wBox oSec oBRG wSte wWel
	nigra	oFor
	nigricans	see **S. myrsinifolia**
	x pendulina var. elegantissima	oFor
	pentandra	oFor wCCr wBox
	petiolaris	oFor wBox
	'Prairie Cascade'	oFor oGar
	procumbens	see **S. myrsinites**
	purpurea	wBCr oJoy wCCr
un	purpurea 'Blue Canyon'	oHed
un	purpurea 'Canyon Blue'	oAls oSis
	purpurea 'Dicky Meadows'	oFor
qu	purpurea 'Eugene'	oFor
	purpurea f. gracilis	see **S. purpurea** 'Nana'
	purpurea 'Green Dicks'	last listed 99/00
	purpurea ssp. lambertiana	wBox
	purpurea 'Nana'	oFor wHer wWoS wCul wClo oDar oGar oRiv oSqu oGre wPir wBox oCir wAva oUps oSle wWel
un	purpurea 'Nana Canyon Blue'	oJoy
un	purpurea 'Nana Pendula	oAls

	purpurea 'Pendula'	oFor wHer oAls wCCr
un	*purpurea* 'Streamco'	wBCr
	repens	oTrP wMtT
	repens var. *argentea*	oGos wCol oGar oSis oOut oSqu oGre wBox
	repens 'Iona' (m)	oSis
qu	*repens nitida*	oFor wBox
	reticulata	wMtT
	retusa	wMtT
	rigida	see S. *eriocephala*
un	'Rubykins'	oFor
	sachalinensis	see S. *udensis*
un	'Scarlet Curls'	oFor oGar oGre wTGN wBox wWel
	schraderiana	see S. *bicolor*
	scouleriana	oFor wWoB wNot oBos wFFl wWat wPla
	sitchensis	wWoB wShR wFFl wWat oAld oSle
un	'Snake'	oFor
	sp.	wBox
	triandra	oTrP wBCr
	triandra 'Black Maul'	oFor wBox
	udensis 'Sekka' (m)	oFor wHer oGar wBox oUps
	vitellina	see S. *alba* ssp. *vitellina*
	vitellina 'Pendula'	see S. *alba* 'Tristis'
	wolfii	last listed 99/00
	yezoalpina	see S. *nakamurana* var. *yezoalpina*
	SALVIA	
	aethiopis	last listed 99/00
	africana-caerulea wild collected	oCis
	africana-lutea	oGoo
	amplexicaulis	oSec
	apiana	oGoo iGSc oCrm oBar wNTP
	argentea	wCul oAls oAmb oGoo iGSc wHom wBWP oUps
	aurea	see S. *africana-lutea*
	austriaca	oSec
	azurea ssp. *pitcheri* var. *grandiflora*	last listed 99/00
un	*azurea* ssp. *pitcheri* var. *grandiflora* 'Nekan'	oSis oUps
	bertolonii	see S. *pratensis* Bertolonii Group
	blepharophylla	oAls
un	*blepharophylla* 'Diablo'	oGoo
	brandegei	wWoS oGoo
	buchananii	oAls oGoo
	bulleyana	oHed
	cacaliifolia	oAls oGoo
	caespitosa	oSis
	candelabrum	oAls wHom
	cardinalis	see S. *fulgens*
	chamaedryoides	wWoS oAls oHed oSis oNWe wCCr
	chiapensis	wWoS oAls oGoo
qu	*clethroides*	oSec
	clevelandii	oAls oGoo iGSc oWnC oCrm oUps
un	*clevelandii* 'Compacta'	oGoo
	clevelandii 'Winifred Gillman'	oGoo
	coahuilensis	oHed oGoo
	coccinea	oGoo oUps
	coccinea 'Brenthurst'	oAls oHed
	darcyi	oHed
	discolor	oHed wFGN oCrm oUps
	disjuncta	oGoo
	divinorum	oOEx oCrm
	dorisiana	iGSc
	elegans	wWoS wFai iGSc wHom oWnC oCrm oUps
	elegans creeping	oGoo
	elegans 'Frieda Dixon'	wWoS oAls wFGN iGSc
	elegans 'Honey Melon'	wWoS oTDM oAls wFGN oGoo oBar wNTP
	elegans 'Honeydew Melon'	see S. *elegans* 'Honey Melon'
un	*elegans* 'Peach Pineapple'	oAls
	elegans 'Scarlet Pineapple'	oTDM wFGN oGoo oBar wNTP oCir
	evansiana	see S. *mekongensis*
	farinacea	oBlo
	forsskaolii	oFor oJoy wCSG oGoo oDan oSec wBWP
	fruticosa	oGoo oCrm
	fulgens	oGoo
	gesneriiflora	oGoo
	glutinosa	oJoy oSec wBWP
	grahamii	see S. *microphylla* var. *microphylla*
	greggii	oGoo wPir oCir wAva
	greggii 'Alba'	oAls
un	*greggii* 'Brilliant Rose'	wWoS
un	*greggii* 'Chiffon'	oAls oSis
	greggii coral	wWoS oEga
	greggii 'Dark Dancer'	oAls
un	*greggii* 'Desert Blaze'	wWoS oAls
	greggii 'Furman's Red'	wWoS oGar oGoo oSis
un	*greggii* 'Hot Pink'	wWoS
un	*greggii* 'Lipstick'	wWoS oAls
	greggii mixed seedlings	wCCr
	greggii purple	wWoS oEga
un	*greggii* 'Raspberry Ripple'	oFor oJoy oSis
	greggii red	oAls oGar oEga
un	*greggii* 'San Takao'	wWoS
	greggii variegated	oAls
	greggii white	oEga
	guaranitica	wHer wSwC oAls oGar oGoo oSis oNWe oSho oGre wBox
	guaranitica 'Argentine Skies'	wHer oJoy oGoo
	guaranitica 'Black and Blue'	last listed 99/00
un	*guaranitica* 'Omaha'	wSwC oDan wBox
un	*guaranitica* 'Omaha Gold'	wWoS wCol oRus oSis oOut
	haematodes 'Indigo'	see S. *pratensis* 'Indigo'
	hians	oFor wBWP
	'Indigo Spires'	oAls oGoo
	involucrata	oAls oGoo
	iodantha	oGoo
	x jamensis	oSec
	x jamensis 'Cherry Queen'	oGoo
	x jamensis 'Cienega de Oro'	oGoo
un	*x jamensis* 'Cienega del Sol'	wWoS
	x jamensis 'Devantville'	oGoo
	x jamensis 'La Luna'	wWoS oHed
	x jamensis 'San Isidro Moon'	wWoS wPir
	japonica	oGoo
	judaica	wBWP
	jurisicii	oFor wHer oJoy oSis wCCr
	koyamae	last listed 99/00
	lavandulifolia	wWoS oSis iGSc
	lemmonii	see S. *microphylla* var. *wislizenii*
	leucantha	oFor oJoy oAls wFGN oGar oGoo oCrm oBar wTGN oUps wWel
	leucantha all purple	oGoo
	leucophylla	last listed 99/00
	leucophylla 'Point Sal'	oFor
	lyrata	oJoy oGoo iGSc oCrm oSec
	lyrata 'Burgundy Bliss'	oFor wWoS oRus
un	*lyrata* 'Purple Knockout'	oMis wBWP
	mekongensis DJHC 98126	wHer
	merjamie 'Mint-sauce'	last listed 99/00
	mexicana	oGoo
un	*mexicana* 'Gold Tip'	oGre wBox
	microphylla	oAls wFGN oGoo
un	*microphylla* 'Dennis Pink'	wFGN
	microphylla 'Maraschino'	wWoS oJoy oHed oMis oCis
	microphylla var. *microphylla*	oJoy
un	*microphylla* 'Wild Watermelon'	oGoo
	microphylla var. *wislizenii*	wCCr
qu	'Midnight'	oCir
	miltiorhiza	oGoo oCrm oBar
	miniata	oGoo
	moorcroftiana	oJoy oAls oSec
	muelleri	oGoo
un	*muhlerii*	wHer oAls
	nemerosa 'Amethyst'	oFor oJoy oAls oWnC oSle
	nemerosa Blue Hill	see S. *x sylvestris* 'Blauhuegel'
	nemerosa East Friesland	see S. *nemerosa* 'Ostfriesland'
	nemerosa 'Lubecca'	oFor oGar wMag oEga
	nemerosa May Night	see S. *x sylvestris* 'Mainacht'
	nemerosa 'Ostfriesland'	oFor wCul oAls oGar oSho wTGN oSec wEde oUps
	nemerosa 'Plumosa'	see S. *nemerosa* 'Pusztaflamme'
	nemerosa 'Pusztaflamme'	wWoS oAls oGar wTGN
	nemerosa Snow Hill	see S. *x sylvestris* 'Schneehuegel'
va	*nipponica* 'Fuji Snow'	oJoy wRob
	nubicola	wHer oSec
	officinalis	oTDM oAls wShR wFGN wCSG oGoo iGSc wHom oWnC wNTP oCir wCoN
	officinalis 'Albiflora'	oAls wFGN oGoo oSis iGSc
	officinalis 'Aurea' (see Nomenclature Notes)	oAls wFGN wHom oCrm wNTP wEde
qu	*officinalis* 'Aurea Variegata'	oJoy
	officinalis 'Berggarten'	wWoS wSwC oTDM oJoy oAls oHed wFGN oGoo wFai iGSc wHom oWnC oCrm oBar oSec wNTP wEde oUps
	officinalis broad-leaved	last listed 99/00
	officinalis x *clevelandii*	oCis
	officinalis 'Compacta'	wCul
	officinalis dwarf	wWoS wFGN oSle
qu	*officinalis fruticosa*	oAls
	officinalis grape scented	wFGN

	officinalis 'Holt's Mammoth'	oJoy wFGN oGoo wRai wFai oBar
va	*officinalis* 'Icterina'	wCul oTDM oAls wFGN wCSG oGoo wFai iGSc wMag oBar oCir oUps oSle
	officinalis latifolia	see S. *officinalis* broad-leaved
	officinalis lavender scented	wFGN
qu	*officinalis minimus*	oAls oGoo oBar wNTP
	officinalis 'Nana'	iGSc oUps
	officinalis Purpurascens Group	wWoS wCul wSwC oTDM oJoy oAls wFGN oGoo wFai iGSc wHom oWnC oCrm oBar oCir wEde oUps oSle
	officinalis 'Rosea'	wFGN oGoo
va	*officinalis* 'Tricolor'	wCul wTho oAls wFGN wCSG oGar oGoo wFai iGSc oWnC oBar oSec wNTP oUps oSle
	officinalis variegated black	wFGN
un	*officinalis* 'Woodcote Farm'	wWoS oJoy oAls wFGN oBar
un	*officinalis* 'Woodcote Farm Variegated'	oAls oBar
	patens	oHed oDan oNWe oUps wSte
	patens 'Guanajuato'	oHed
un	*phlomoides*	oSis
	pomifera	oGoo
	pratensis	iGSc
	pratensis 'Baumgartenii'	last listed 99/00
	pratensis Bertolonii Group	wBWP
	pratensis Haematodes Group	oNWe oSec
	pratensis 'Indigo'	oFor oHed oGar oNWe wBox
	przewalskii	wHer
	'Purple Majesty'	oAls oDan oSis
	purpurea	wNTP
	'Raspberry Royale'	wWoS oGoo
	repens	oCrm
	ringens	wCCr
	roemeriana	oHed
	rutilans	see S. *elegans* 'Scarlet Pineapple'
	scabiosifolia	oSec
	scabra	oSec
	sclarea	wHer wCul wFGN oGoo wFai wHom oCrm oBar wNTP oCir oUps
	sclarea silver	wFGN
	sclarea var. *turkestanica* (see Nomenclature Notes)	wHer oNWe wFai wBWP
un	'Sierra San Antonio'	oFor wWoS oGoo
	sinaloensis	oAls
	sonomensis	wCCr
un	*sonomensis* 'Dara's Choice"	wFGN wCCr
	sp. DJHC 259	wHer
	sp. DJHC 563 seed grown	wHer
	sp. DJHC 98315	wHer
	splendens	oUps
	staminea	wCCr oSec
	x superba	wCul wTho oUps
	x superba 'Amethyst'	see S. *nemerosa* 'Amethyst'
	x superba Blue Hill	see S. *x sylvestris* 'Blauhuegel'
	x superba Blue Queen	see S. *x sylvestris* 'Blaukoenigin
	x superba East Friesland	see S. *nemerosa* 'Ostfriesland'
	x superba 'Lubecca'	see S. *nemerosa* 'Lubecca'
	x superba May Night	see S. *x sylvestris* 'Mainacht'
ch	*x superba* Snow Hill	see S. *x sylvestris* 'Schneehuegel'
	x superba 'Viola Klose'	see S. *x sylvestris* 'Viola Klose'
	x sylvestris 'Blauhuegel'	oFor oAls oGar oWnC wAva oSle
	x sylvestris 'Blaukoenigin'	oFor oJoy oAls wFGN oMis oSha oSec iArc oEga
	x sylvestris 'Mainacht'	oFor wCul wSwC oNat oJoy oAls wFGN oGar oSis oOut oGre oWnC wMag wTGN wAva oEga
	x sylvestris May Night	see S. *x sylvestris* 'Mainacht'
	x sylvestris 'Rose Queen'	oFor oJoy wMag oSec wAva oEga
va	*x sylvestris* 'Schneehuegel'	wWoS oAls oGar oWnC wTGN wBox oEga
un	*x sylvestris* 'Stratford Blue'	oJoy
	x sylvestris 'Viola Klose'	wSwC
	transslyvanica	wSwC oSec
	uliginosa	oFor oJoy oAls oGoo oDan wFai oSec wSte
	'Van-Houttei'	oAls
	verbenacea	oGoo
	verticillata 'Alba'	oFor
	verticillata 'Purple Rain'	oFor wHer wWoS wCul oJoy oAls oHed oRus oNWe oGre oWnC oCir oEga oUps
	viscosa (see Nomenclature Notes)	oSec
un	'Waverly'	oCis
	yunnanensis DJHC 451 seed grown	wHer
	SALVINIA	
	auriculata	oTrP oWnC
	rotundifolia (see Nomenclature Notes)	wGAc oGar
	SAMBUCUS	
	adnata	last listed 99/00
	caerulea	oFor wThG wBur wNot oAls wShR oBos oRiv oWhi wRai wFFl wWat wCCr oWnC oCrm wBox wPla oAld oSle
	callicarpa	oFor oRiv
	canadensis	wBCr oGoo iGSc
	canadensis 'Adams' (F)	oFor wRai oOEx
	canadensis 'Aurea'	oFor wHer oJoy oGoo oGre wWel
	canadensis 'John's'	oFor oWhi wRai oOEx
	canadensis 'Maxima'	oFor
	canadensis 'Nova'	wBCr wBur wClo oAls oOEx
	canadensis 'York' (F)	oFor wBCr wBur wClo oAls oWhi oOEx
	melanocarpa	wPla
	mexicana	see S. *caerulea*
	nigra	oWnC wBox
	nigra 'Albopunctata'	oGre
	nigra 'Albovariegata'	wCul oJoy oGar
	nigra 'Atropurpurea'	see S. *nigra* 'Guincho Purple'
	nigra 'Aurea'	wCul
	nigra 'Aureomarginata'	oFor wHer wRai
	nigra 'Castledean'	wHer
	nigra 'Guincho Purple'	oFor wHer wWoS wCul oJoy oAls oAmb oGar oDan oNWe oGre wFai wPir wTGN oSec wAva oUps wWel
	nigra f. *laciniata*	oFor wHer wCul oGos oJoy oAls wRob oGre wWel
	nigra 'Linearis'	wHer oGre oSec wWel
va	*nigra* 'Madonna'	oFor wHer wWoS wCul oJoy wRob oGre oUps wWel
	nigra 'Marginata'	oFor wClo oNat oDar wCol wCSG oSho wRob oOEx wFai wPir wTGN oSec oUps
do	*nigra* 'Plena'	wHer
	nigra Porphyrifolia Group	wCCr
va	*nigra* 'Pulverulenta'	oFor wHer wCul oGos oSis oGre wTGN wWel
	nigra 'Purpurea'	see S. *nigra* 'Guincho Purple'
	nigra 'Pyramidalis'	wHer
	nigra 'Variegata'	see S. *nigra* 'Marginata'
	nigra 'Witches Broom'	wCol wAva
	pubens	wBur
	racemosa	wWoB oAls wShR oGar oBos oRiv wWat oWnC oTri wBox wPla oAld oSle wSte
un	*racemosa arborescens*	wNot
	racemosa 'Plumosa Aurea'	oHed oNWe oSec
qu	*racemosa* ssp. *pubens* var. *arborescens*	wFFl
	racemosa var. *sieboldiana*	oFor
	racemosa 'Sutherland Gold'	wCul oGos wSwC oAls oHed oGar oSis wWel
	sieboldiana	see S. *racemosa* var. *sieboldiana*
	SAMOLUS	
	parviflorus	oHug
	SANGUINARIA	
	canadensis	oFor wThG wCol oRus wFai iGSc oCrm wCoN oUps
do	*canadensis* f. *multiplex*	wHer oAmb oSis wCoN
	SANGUISORBA	
	canadensis	last listed 99/00
	hakusanensis	oJil wHig
	hakusanensis HC 970379	wHer
	menendezii	oJil
	minor	oFor oAls wFai iGSc
	obtusa	oFor wHig oGar oSho
un	*obtusa* 'Lemon Splash'	oFor wHer
	officinalis	wHig oAls oRus wFGN oGoo oDan oBar wBox
un	*officinalis* 'Shiro Fururin'	wHer
	officinalis 'Tanna'	wHer
	sp. DJHC 143	wHer
	sp. DJHC 535	wHer
	tenuifolia	oFor oDan
	tenuifolia 'Alba'	wHig
	tenuifolia 'Alba' seed grown	wHer
	tenuifolia 'Purpurea'	wHer
	SANSEVIERIA	
	cylindrica	oTrP
un	*trifasciata* 'Black Gold'	oGar
va	*trifasciata* 'Laurentii'	oGar
	SANTOLINA	
	chamaecyparissus	oFor oJoy oAls wShR wFGN oGoo oSho oBlo wFai iGSc wHom oWnC oBar wSta wNTP oCir oUps
	chamaecyparissus dwarf grey	see S. *chamaecyparissus* var. *nana*
un	*chamaecyparissus* 'Gey Saso Select'	oJoy
	chamaecyparissus 'Lemon Queen'	wWoS wCul oJoy oSis
	chamaecyparissus var. *nana*	oFor wWoS oDar wFGN oSis wFai oBar oSle wSnq

	chamaecyparissus 'Pretty Carol'	oJoy wFGN oGoo wFai oBar
un	chamaecyparissus 'Saso's Select'	oAls wFGN oBar
	chamaecyparissus Small-Ness'	wWoS oJoy oSis oSec
	incana	see S. chamaecyparissus
	neapolitana	see S. pinnata ssp. neapolitana
	pinnata ssp. neapolitana	oAls iGSc
	pinnata ssp. neapolitana 'Edward Bowles'	wCul oHed oGoo oSis wCCr
	rosmarinifolia	oAls oGoo wCCr iGSc
un	rosmarinifolia 'Morning Mist'	wWoS wCul oJoy oHed wFGN oGoo wFai oBar wSnq
	rosmarinifolia ssp. rosmarinifolia	oFor oJoy oAls wShR wFGN oGoo wFai iGSc wHom oBar wNTP oUps
un	rosmarinifolia ssp. rosmarinifolia 'Compact Green'	oGoo oBar oSle
un	serratifolia	wCSG
	virens	see S. rosmarinifolia ssp. rosmarinifolia
	SANVITALIA	
	procumbens	oUps
	SAPINDUS	
un	delavayi	oTrP
	drummondii	oFor
	mukorossi	oFor wCCr
	SAPIUM	
	japonicum	oFor wCCr
	japonicum HC 970268	wHer
	sebiferum	oFor oRiv oGre wCCr oWnC
	SAPONARIA	
	'Bressingham'	oSis wFai wTGN
	x lempergii 'Max Frei'	oFor wWoS oWnC
	lutea	oSis
	ocymoides	oFor wTho oAls oGar oSho wFai iGSc oWnC wTGN iArc oCir oEga oUps
	ocymoides 'Alba'	oWnC
	ocymoides 'Rubra Compacta'	last listed 99/00
	ocymoides 'Snow Tip'	oFor oWnC
	officinalis	wFGN iGSc wHom oSec wNTP oLSG oUps
do	officinalis 'Alba Plena'	last listed 99/00
va	officinalis 'Dazzler'	last listed 99/00
do	officinalis 'Rosea Plena'	oFor oAls wCSG oUps
	x olivana	wWoS oSis
un	x olivana 'Spring Cushion'	oAmb
	pumilio	oHed
un	**SAPOSHNIKOBA**	
	divaricata	oCrm
	SARCOCAULON	
	crassicaule	oRar
	SARCOCOCCA	
	confusa	oFor wHer oJoy oAls oGar wCCr wSta wSte wWel
	hookeriana	wAva
	hookeriana B&SWJ 2585	wHer
	hookeriana var. digyna	oFor
	hookeriana var. humilis	oFor wCul oNat oAls oGar oCis oSho oGre wPir wTGN wSta wSte wCoS wWel
	humilis	see S. hookeriana var. humilis
	orientalis	wHer
	ruscifolia	oFor wCul oHed oGar oSho oGre wPir wCCr oWnC wTGN wSta oCir wAva wCoS wWel
	ruscifolia var. chinensis	wHer
un	ruscifolia var. ruscifolia DJHC 717	wHer
	SARCOSTEMMA	
	socotranum	oRar
	viminale	oRar
	SARRACENIA	
	alata	wOud
	alata x flava	wOud
un	'Dixie Lace'	wOud
	flava	last listed 99/00
	flava x purpurea	wOud
	flava red blushed form	wOud
	flava veined form	wOud
un	'Judith Hindle'	wOud
un	'Ladies in Waiting'	wOud
	leucophylla	wGAc wOud oGar
	minor	wGAc oGar
	psittacina	wGAc oGar
	psittacina giant	wOud
	psittacina x minor	wOud
	purpurea	last listed 99/00
	purpurea ssp. venosa	wOud oGar
	rubra	oGar
	rubra ssp. rubra	wOud
	rubra ssp. wherryi	wOud

	SARUMA	
	henryi	wHer
	SASA	
	kurilensis	oFor wBmG oTBG
	kurilensis dwarf form	oTBG
va	kurilensis 'Shimofuri'	oTra wCli oNor wBea oTBG
	palmata	oTra wSus wCli oNor wBlu wBur wBea oOEx oGre wBox wBam wBmG oTBG
	pygmaea	see PLEIOBLASTUS pygmaeus
un	shimidzuama 'Asahinae'	wBea
	sp.	oNor
	tsuboiana	oTra wBox wBam oTBG
	veitchii	oFor oTra wSus wCli oNor oGos wBea oDar wCSG oGre wBox wBam oTBG
	veitchii 'Kumazasa'	see S. veitchii
un	yashadake	wBea
	SASAELLA	
	albostriata	see S. masamuneana f. albostriata
	glabra	see S. masamuneana
un	hidaensis muraii	oTra
	masamuneana	oNor oAls wBox wBam wBmG
va	masamuneana f. albostriata	oFor oTra wSus wCli wBlu wBea oGar oOut wRai oGre wBox wBmG oTBG wSte wWel
va	masamuneana f. aureostriata	wCli
un	masamuneana rhycantha	oTra
	ramosa	wCli wBea wBam wBmG oTBG
	SASSAFRAS	
	albidum	oFor oGar oRiv oGre wCCr
	tzumu	last listed 99/00
	SATUREJA	
	biflora	oAls wFGN iGSc oBar wNTP
un	byzantina	wHom
	douglasii	oGoo oBos wHom oCrm oBar oTri
	hortensis	oAls wFGN oSha iGSc wHom oWnC wNTP oUps
	montana	wWoS oAls wFGN oGoo oSha wFai iGSc wHom oBar wNTP oUps
	montana ssp. illyrica	oGoo oBar
	montana 'Nana'	wFGN oBar
	montana 'Procumbens'	oAls
	spicigera	wFGN wFai
	viminea	oAls oGoo
	SAUROMATUM	
	guttatum	see S. venosum
	venosum	wHer
	SAURURUS	
	cernuus	wSoo wGAc oHug oSsd wTGN
	chinensis	wGAc oHug
	SAXEGOTHAEA	
	conspicua	oFor wHer
	SAXIFRAGA	(see Nomenclature Notes)
	aizoon	see S. paniculata
	'Alba' (x apiculata)	oSis
	x andrewsii	oSis
	'Apple Blossom'	oAls wWhG
	x arendsii pink	oGar wWhG
	x arendsii red	wFai wWhG
	'Aretiastrum' (x boydii)	wWoS oSis wMtT
	'Ariel' (x hornibrookii)	oSis
	'Aureopunctata' (x urbium)	wHer wWoS oRus wFGN oNWe wRob oSqu wHom oSle
	'Becky Foster' (x borisii)	oSis
	bronchialis	wShR
	bronchialis ssp. funstonii	wMtT
	bronchialis var. vespertina	see S. vespertina
	x burnatii	oSis oNWe
	caespitosa	see S. cespitosa
	callosa	wShR wCSG oSis oNWe
	'Carmen' (x elisabethae)	oSis
	'Caterhamensis' (cotyledon)	oSis
	caucasica	oSis
	cespitosa	wTho oBos
	'Cherrytrees' (x boydii)	oSis
	'Clarence Elliott' (umbrosa)	oSis wRob oBov wMag
	cochlearis	oSis
	'Cockscomb' (paniculata)	oSis oSqu
un	confusa	oJoy
	cortusifolia var. fortunei	see S. fortunei
	cotyledon	oSis
	'Cranbourne' (x anglica)	oSis
	cuneifolia	oJoy
	cuscutiformis	wRob
	'Drakula' (ferdinandi-coburgi)	oSis
un	'Eco Butterfly'	oCis oNWe wRob

	'Elliott's Variety'	see **S.** 'Clarence Elliott'
	'Faldonside' ***(x boydii)***	oSis wMtT
	federici-augusti	see **S. *frederici-augusti***
	ferdinandi-coburgi var. ***rhodopea***	oSis
un	'Five Color' ***(fortunei)***	wHer oRus oNWe
un	'Flower Carpet' ***(cespitosa)***	oWnC
	fortunei	oRus wRob
	fortunei Clone 2	wHer
	'Foster's Gold' ***(x elisabethae)***	oSis
	'Franzii' ***(x paulinae)***	last listed 99/00
	frederici-augusti ssp. ***grisebachii***	oSis wMtT
	x fritschiana	wMtT
un	'G.M. Hopkins'	oSis
	x geum	wAva
un	'Go-nishiki' ***(fortunei)***	see **S.** 'Five Color'
	'Godiva' ***(x gloriana)***	oSis
	'Gratoides' ***(x grata)***	oSis
	'Gregor Mendel' ***(x apiculata)***	oSis
	grisebachii	see **S. *frederici-augusti*** ssp. ***grisebachii***
un	'Harry Marshall' ***(x irvingii)***	oSis
	'Harvest Moon' ***(stolonifera)***	oFor wHer oAls oSis oCis
	'Hindhead Seedling' ***(x boydii)***	oSis
un	'Hirsuta' ***(paniculata)***	oSis
	'Hocker Edge' ***(x arco-valleyi)***	oSis
	hostii ssp. ***hostii*** var. ***altissima***	oSis
un	***x hybrida***	oSis
	hypnoides	oNWe wTGN oCir
	'Icicle' ***(x elisabethae)***	oSis
	integrifolia	see **S. *oregana***
un	'Integrifolia' ***(androsacea)***	oSis
	iranica	wMtT
	'Jenkinsiae' ***(x irvingii)***	oSis
	'Johann Kellerer' ***(x kellereri)***	wMtT
	'Jupiter' ***(x megaseiflora)***	oSis
	'Kath Dryden' ***(x anglica)***	oSis wMtT
	'Kathleen Pinsent'	wMtT
	'Kestoniensis' ***(x salmonica)***	oSis
un	***kinlayi***	oSis oSqu
	x kochii	wMtT
	'Krasava' ***(x megaseiflora)***	oSis
un	'L. C. Godself' ***(x elisabethae)***	oSis
	'Labe' ***(arco-valleyi)***	oSis wMtT
	lingulata	see **S. *callosa***
	longifolia	wMtT
	'Luna' ***(millstreamiana)***	oSis
	'Lusanna' ***(x irvingii)***	oSis
	x macnabiana	oSis oSqu
	'Major' ***(cochlearis)***	oSis
	marginata var. ***boryi***	oSis
	marginata var. ***rocheliana***	oSis
	'Maria Luisa' ***(x salmonica)***	last listed 99/00
	'Marianna' ***(x borisii)***	oSis
	'Maroon Beauty' ***(stolonifera)***	wHer .
	'Mars' ***(x elisabethae)***	oSis
	mertensiana	wCol oBos oTri
	'Minor' ***(cochlearis)***	oSis wMtT
	'Minutifolia' ***(paniculata)***	oSis
	'Mona Lisa' ***(x borisii)***	wMtT
un	'Moon Beam' ***(boydilacina)***	oSis
	'Mother of Pearl' ***(x irvingii)***	oSis
	nepalensis	see **S. *cotyledon***
un	***occidentalis*** var. ***rufidula***	oBos
	'Opalescent'	oSis
	oregana	oBos
	paniculata	oSto oNWe
	paniculata var. ***brevifolia***	oSis
	paniculata ssp. ***cartilaginea*** pink	wMtT
	paniculata ssp. ***kolenatiana***	see **S. *paniculata*** ssp. ***cartilaginea***
	'Peach Blossom'	oSis
	'Penelope' ***(boydilacina)***	oSis wMtT
	'Peter Pan'	oAls oSis wWhG
	'Primrose Dame' ***(x elisabethae)***	oSis
	'Primulina' ***(x malbyana)***	wMtT
ch	***primuloides***	see **S.** 'Primuloides'
ch	'Primuloides' ***(umbrosa)***	wCul oHed oSis oSqu wFai
	'Princess' ***(burseriana)***	oSis
un	'Purple Robe'	oAls oSis wFai oWnC iArc
	'Purpurteppich'	wHom
	'Pyramidalis' ***(cotyledon)***	last listed 99/00
	'Rex' ***(paniculata)***	oSis
	rosacea	wCSG
	'Rosea' ***(paniculata)***	oSis wMtT oSqu
	'Rosemarie' ***(x anglica)***	wMtT
	'Rosina Suendermann' ***(x rosinae)***	oSis
	'Rubrifolia' ***(fortunei)***	wRob

	sancta	oSis
un	'Sandpiper'	wMtT
	scardica	oSis
	sempervivum	last listed 99/00
un	'Simplicity' ***(x ingwersenii)***	oSis wMtT
	spruneri	oSis
	spruneri var. ***deorum***	wMtT
	stolonifera	oFor wFGN oGar oNWe oGre
	stolonifera 'Cuscutiformis'	see **S. *cuscutiformis***
	stribrnyi	oSis
	'Suendermannii' ***(x kellereri)***	oSis
	'Suendermannii Major' ***(x kellereri)***	wMtT
	taygetea	last listed 99/00
	'Theoden' ***(oppositifolia)***	wMtT
un	'Theresia' ***(marie-theresiae)***	oSis
	x tiroliensis	oSis
	trifurcata	last listed 99/00
	'Triumph' ***(x arendsii)***	oGar wWhG
	umbrosa	oSto oAls wShR wBox oCir
	x urbium	oSto oRus wFai oWnC wWhG
un	***x urbium primuloides***	see **S.** 'Primuloides'
	x urbium primuloides 'Elliot's Variety'	see **S.** 'Clarence Elliott'
	valdensis	oSis
	'Valerie Finnis'	see **S.** 'Aretiastrum'
	'Variegata' ***(umbrosa)***	see **S.** 'Aureopunctata'
	veitchiana	wRob
un	'Velvet' ***(fortunei)***	wHer oNWe
	'Venetia' ***(paniculata)***	last listed 99/00
un	***versiculata***	oSis
	vespertina	wMtT
	'Vincent van Gogh' ***(x borisii)***	oSis
	'Walter Irving' ***(x irvingii)***	oSis wMtT
	wendelboi	oSis
	'Wendy' ***(x wendelacina)***	oSis
	'White Pixie'	oAls wWhG
	'Whitehill'	oSis oNWe
	'Winifred Bevington'	wWoS oSis oSqu
	x zimmeteri	wMtT
	SCABIOSA	
	alpina	see **CEPHALARIA *alpina***
un	'Alpine Pink'	wCSG
	atropurpurea 'Ace of Spades'	oNWe
	'Blue Butterfly'	see **S.** Butterfly Blue®
un	'Blue Mist'	wTGN
	Butterfly Blue®	oFor wWoS wCul wSwC oNat oAls oGar oSis oSho oGre oWnC wMag wTGN oCir wAva oEga oLSG oUps wSnq
	caucasica	wCul wCSG
	caucasica var. ***alba***	oFor wCul wMag oEga
	caucasica 'Compliment'	oGoo
	caucasica 'Fama'	oFor oMis oGar oSha oBlo wMag oEga
	caucasica House's hybrids	oWnC oEga
un	***caucasica*** 'Lavender Beauty'	oSec
	caucasica 'Perfecta'	oWnC
	caucasica 'Perfecta Alba'	oMis oGar oGre
un	***caucasica*** 'Perfecta Blue'	oGar oGre
	columbaria	oFor
	columbaria 'Nana'	oFor oJoy
	columbaria var. ***ochroleuca***	oFor wCul oCis wMag
	farinosa	oSis
	graminifolia	wHer wCul
	japonica var. ***alpina***	last listed 99/00
	lucida	oFor oSis oSec
	minoana	oHed
	ochroleuca	see **S. *columbaria*** var. ***ochroleuca***
	'Pink Mist'	oFor wWoS wCul oNat oJoy oAls oGar oSis oSho oWnC wMag wTGN wAva oEga oUps wSnq
	SCAEVOLA	
	aemula	wCoS
	SCHEFFLERA	
	arboricola	oGar
un	***arboricola*** 'Variegata'	oGar
	SCHIMA	
	argentea	see **S. *wallichii*** ssp. ***noronhae*** var. ***superba***
	wallichii	see **S. *wallichii*** ssp. ***noronhae*** var. ***superba***
	wallichii ssp. ***noronhae*** var. ***superba***	wHer oCis
	SCHINUS	
	molle	oFor
	SCHISANDRA	
	arisanensis B&SWJ 3050	last listed 99/00
	chinensis	oOEx oWnC oCrm wTGN
un	***chinensis*** 'Eastern Prince'	wCul wRai wBox wWel
	chinensis var. ***grandiflora***	see **S. *grandiflora***
	chinensis HC 970145	wHer

	chinensis male	oFor
	grandiflora EDHCH 97153	wHer
	grandiflora var. ***rubra***	see **S. *rubriflora***
un	***lancifolia*** DJHC 507	wHer
	propinqua	wHer oCis
	propinqua var. ***sinensis***	oFor
	rubriflora	oCis wRai oWnC wWel
	sp. DJHC 605	wHer
	SCHIZACHYRIUM	
	scoparium 'Blaze'	oFor
	scoparium 'The Blues'	oOut wSnq
un	**SCHIZONEPETA**	
	tenuifolia	oCrm
	SCHIZOPETALON	
	walkeri	wSte
	SCHIZOPHRAGMA	
	corylifolium	wHer
	hydrangeoides	oFor wHer oGos wCol wCSG oGre wTGN wCoN wWel
	hydrangeoides 'Brookside Littleleaf'	oFor oNWe wRob wWel
	hydrangeoides HC 970225	wHer
va	***hydrangeoides*** 'Moonlight'	oFor wHer wWoS wCul oGos wCol oAmb oGar oDan oRiv oCis oSho wPir wTGN oBRG wWel
un	***hydrangeoides*** 'Platt'	oGos
un	***hydrangeoides*** 'Platt's Dwarf'	oGre
	hydrangeoides 'Roseum'	wWoS
	integrifolium fauriei	last listed 99/00
	integrifolium fauriei B&SWJ 3690	wHer
	SCHIZOSTACHYUM	
	funghomii	wBea
	SCHIZOSTYLIS	
	alba	see **S. *coccinea*** f. ***alba***
	coccinea	oFor oTrP wCul oDar oAls oRus wCSG oDan oSis wCCr wCoN wSte wCoS
	coccinea f. ***alba***	oFor oTrP wWoS wCul wCSG oSis wAva wEde wSte
un	***coccinea*** 'Big Mama'	oGre
un	***coccinea*** 'Cherry Red'	wWoS oGre
	coccinea coral	wSta
	coccinea 'Fenland Daybreak'	last listed 99/00
	coccinea 'Major'	wHer wCSG
	coccinea 'Mrs. Hegarty'	oFor wHer oDar wCSG oSis wSte wSnq
	coccinea 'November Cheer'	wWoS wCul
qu	***coccinea*** 'Oregon Sunrise'	oNat
	coccinea 'Oregon Sunset'	oFor wWoS oGos oNat oDar oSis wHom wEde oEga oUps wSnq
	coccinea 'Pallida'	wWoS wSnq
	coccinea pink	wCol oAls wSta
	coccinea 'Snow Maiden'	oRus
	coccinea 'Sunrise'	wHer wCri wCSG oSis wAva wSnq
	coccinea 'Viscountess Byng'	wHer wWoS oAmb wAva wSnq
	coccinea white form	wSnq
	coccinea 'Zeal Salmon'	wHer
	SCHOENOPLECTUS	
	lacustris	wSoo
	lacustris ssp. ***tabernaemontani***	wCli
va	***lacustris*** ssp. ***tabernaemontani*** 'Albescens'	wCli oSsd
va	***lacustris*** ssp. ***tabernaemontani*** 'Zebrinus'	wCli wGAc oHug oGar oSsd oTri wTGN
	validus	oSsd wWat oAld
	SCHOTIA	
	afra	oTrP
	brachypetala	oTrP
	latifolia	oTrP
	SCHRANKIA	
	nuttallii	oFor
	SCIADOPITYS	
	verticillata	oFor wClo oMac oGar oRiv oBlo oGre wPir oWnC wSta oBRG oMac wAva wCoN wWhG wWel
un	***verticillata*** 'Joe Kozy'	oRiv
	verticillata 'Ossorio Gold'	oPor
	verticillata 'Variegata'	oPor
	SCILLA	
	adlamii	see **LEDEBOURIA *cooperi***
	autumnalis	wRob
	bifolia	last listed 99/00
	bithynica	last listed 99/00
	campanulata	see **HYACINTHOIDES *hispanica***
	litardierei	oNWe
	mischtschenkoana	wHer
	natalensis	last listed 99/00
	peruviana	oNWe
un	***peruviana*** var. ***gattefossei***	wHer
un	***peruviana*** var. ***hughii***	wHer
	scilloides DJH 277	wHer
	siberica	wRoo oGar oNWe oWoo
	siberica 'Spring Beauty'	wCul
	SCINDAPSUS	
	aureus	see **EPIPREMNUM *aureum***
va	***pictus*** 'Argyraeus'	oGar
	SCIRPUS	
un	***acutus***	wNot wWat oTri oAld
un	***albescens***	wGAc oHug
un	***americanus***	wShR wWat
un	***californicus***	wGAc
	cernuus	see **ISOLEPIS *cernua***
	fauriei var. ***vaginatus***	oSis
	lacustris	see **SCHOENOPLECTUS *lacustris***
	lacustris 'Albescens'	see **SCHOENOPLECTUS *lacustris*** ssp. ***tabernaemontani*** 'Albescens'
	lacustris 'Zebrinus'	see **SCHOENOPLECTUS *lacustris*** ssp. ***tabernaemontani*** 'Zebrinus'
	maritimus	see **BOLBOSCHOENUS *maritimus***
un	***microcarpus***	wWoB oHug oSsd oBos wWat oTri oAld
	sp.	oHug
	sylvaticus	wSoo
	tabernaemontani	see **SCHOENOPLECTUS *lacustris***
	validus	see **SCHOENOPLECTUS *validus***
	zebrinus	see **SCHOENOPLECTUS *lacustris*** ssp. ***tabernaemontani*** 'Zebrinus'
	SCLERANTHUS	
	biflorus	wFGN
	uniflorus	oNWe
	SCLEROCARYA	
	birrea	oRar
	SCOLIOPUS	
	hallii	oBos
	SCOPOLIA	
	carniolica brown-flowered form	wHer
	SCROPHULARIA	
	aquatica	see **S. *auriculata***
	auriculata 'Variegata'	oFor wHer oJoy oAls oHed wCSG oGar oCis oOut wBox oSec wAva
	californica	oBos
un	***californica*** var. ***oregana***	wShR
	chrysantha	oCis
	nodosa	oGoo iGSc
	nodosa 'Variegata'	see **S. *auriculata*** 'Variegata'
	SCUTELLARIA	
	alpina	oSis
	alpina 'Alba'	oJoy oSis
	alpina Jurasek list	wMtT
	altissima	oSec
un	***altissima*** 'Violet'	oEga
	angustifolia	oBos
un	***austinae***	oSec
	baicalensis	oJoy wHig wCSG oGoo oNWe wFai iGSc oCrm oBar
	formosana	oTrP
	indica	oNWe
	indica var. ***parvifolia*** 'Alba'	wMtT
	lateriflora	wCul oJoy oAls wFGN oGoo iGSc oCrm oBar
un	***oregana***	oHan
	orientalis	oSis
	prostrata	oSis wMtT
un	***romana***	iArc
	tournefortii	wMtT
un	**SECAMONE**	
	sp.	oRar
	SECURINEGA	
	suffruticosa	last listed 99/00
	X SEDADIA	
	amecamecana	oSqu
un	**SEDEVERIA**	
	'Abbey Brooke'	oSqu
	'Blue Giant'	oSqu
	'Golden Glow'	oSqu
	'Harry Butterfield'	oSqu
	x hummelii	oSqu
	'Jet Beads'	oSqu
	'Robert Grimm'	oSqu
	'White Stonecrop'	oSqu
	'Yellow Humbert'	last listed 99/00
	SEDUM	
	acre	oSqu wBWP oLSG
	acre 'Aureum'	oSto oSis oWnC iArc
un	***acre*** var. ***krajinae***	oSqu

SEDUM

	acre var. ***majus***	oSto
	acre 'Minus'	last listed 99/00
	acre var. ***sexangulare***	see **S. *sexangulare***
	adolphii	oSqu
	aizoon	oSqu wAva
	aizoon 'Euphorbioides'	oSqu
	alboroseum	wRob oSqu
	alboroseum cream	oSto
	alboroseum 'Frosty Morn'	oFor wHer wWoS oNat oJoy oAls oGar oDan oNWe wRob oGre wFai wTGN wBox wEde oLSG oUps wSnq
	alboroseum 'Mediovariegatum'	oFor wHer wWoS wCul oNat oJoy oSto oMis oGar oNWe wRob oSqu wEde oUps oSle wSnq
un	***alboroseum*** 'Mediovariegatum, Frilled Edge'	oJoy
un	***alboroseum*** 'Variegatum'	see **S. *alboroseum*** 'Mediovariegatum'
	album	oSto oSqu
	album athoum	see **S. *album***
un	***album*** var. ***balticum***	oSqu
un	***album*** 'Bella d'Inverno'	oSto oSqu
	album 'Chloroticum'	oSto oSqu
	album ssp. ***clusianum***	see **S. *gypsicola glanduliferum***
	album 'Coral Carpet'	oFor oSto oSis oWnC oUps
	album crested form	oJoy oSto
	album from France	oSqu
	album 'Laconicum'	see **S. *laconicum***
	album murale	see **S. *album*** ssp. ***teretifolium*** 'Murale'
un	***album serpentini***	oSqu
	album ssp. ***teretifolium*** 'Murale'	oSto oSqu wFai oCir
	allantoides	oSqu
	altissimum	see **S. *sediforme***
	x amecamecanum	see **X SEDADIA *amecamecana***
	amplexicaule	see **S. *tenuifolium***
	anacampseros	oSto oSqu
	anglicum	oSto oSqu
	anglicum 'Minus'	last listed 99/00
un	***anglicum*** ssp. ***pyrenaicum***	oSqu
	anopetalum	see **S. *ochroleucum***
	Autumn Joy	see S. 'Herbstfreude'
	batallae	last listed 99/00
	bellum	oSqu
	'Bertram Anderson'	wWoS oNWe wRob oSqu oGre wBox
	bithynicum	see **S. *hispanicum*** var. ***bithynicum***
	borissovae	oSto
un	***bourgaei***	oSqu
	brevifolium	oSto wHom
	brevifolium red form	oSis
	'Burrito'	oSqu
	caducum	oSqu
	'Carl'	last listed 99/00
qu	***carnea***	see **S. *spathulifolium***
	caucasicum	oSqu
	cauticola	oJoy oSto oAls oMis oNWe
	cauticola 'Lidakense'	oAls oSqu wHom oWnC wTGN oUps
	cauticola 'Robustum'	wHer oSec
	cepaea	oSqu
	chontalense	last listed 99/00
	clavatum	oSqu
	commixtum	oSqu
	compactum	oSqu
	confusum	oSto
	craigii	last listed 99/00
un	'Crocodile'	oSqu
	cyaneum Rudolph	oSto
	dasyphyllum	oSto wFGN oSqu wHom
	dasyphyllum 'Riffense'	oSto
	decumbens	oSqu
	dendroideum	oSqu
	divergens	oSto wShR oSis oSqu wWld
	ellacombeanum	see **S. *kamtschaticum*** var. ***ellacombeanum***
	erythrostichum	see **S. *alboroseum***
	erythrostichum f. ***variegatum***	see **S. *alboroseum*** 'Mediovariegatum'
	ewersii	oSto oWnC
	ewersii var. ***homophyllum***	oFor
	fastigiatum	see **RHODIOLA *fastigiata***
	floriferum	see **S. *kamtschaticum***
	forsterianum	oSto
	frutescens	oSqu
	furfuraceum	oSqu
	glaucophyllum	oSto
	gracile	oSto
un	'Green Rose'	oSqu
	greggii	oSqu
	grisebachii ssp. ***kostovii***	oSqu

	griseum	oSqu
	guadalajaranum	oSqu
	gypsicola	oSto
	gypsicola glanduliferum	oSqu
	'Harvest Moon'	oSto oSqu
	'Herbstfreude'	oFor wWoS wCul oNat oDar oJoy oSto wHig oAls wShR wCSG oMis oGar oDan oNWe oSho oOut wRob oGre wFai wHom oWnC wMag wTGN wBWP oCir wAva wEde oEga oLSG oUps oSle wSnq
un	***hernandezii***	oSqu
	hillebrandtii	see **S. *urvillei*** Hillebrandtii Group
	hintonii	last listed 99/00
	hirsutum ssp. ***baeticum***	oSqu
	hispanicum	oSto
	hispanicum var. ***bithynicum***	see **S. *pallidum*** var. ***bithynicum***
un	***hispanicum*** var. ***hispanicum***	oSqu
	hispanicum var. ***minus*** pink form	oSis
	hispanicum var. ***minus*** purple form	oAls
un	***hispanicum*** var. ***polypetalum***	oSqu
un	***hispanicum*** 'Purpureum'	oWnC
	hultenii	oSqu
	humifusum	oSqu
	hybridum 'Immergruenchen'	oSqu
	ishidae	see **RHODIOLA *ishidae***
	japonicum	oSto
	'Joyce Henderson'	wHer
	kamtschaticum	oFor wFGN oGar oSqu oLSG oUps
	kamtschaticum var. ***ellacombeanum***	oSto oSqu
	kamtschaticum var. ***floriferum***	oSto
	kamtschaticum var. ***floriferum*** 'Weihenstephaner Gold'	oSqu oWnC
	kamtschaticum from Korea	oSqu
un	***kamtschaticum*** 'Golden Carpet'	oSqu
	kamtschaticum var. ***kamtschaticum*** 'Variegatum'	oFor oSto wFGN oGar oGre oSec oUps
	kamtschaticum 'Takahira Dake'	oSto oSqu
	kamtschaticum 'Variegatum'	see **S. *kamtschaticum*** var. ***kamtschaticum*** 'Variegatum'
	kostovii	see **S. *grisebachii*** ssp. ***kostovii***
	laconicum	oSqu
	lanceolatum	wThG wFFl oSqu wWld
	laxum	oSto
un	***lineare*** 'Golden Teardrop'	oLSG
	lineare 'Variegatum'	oAls oSqu
un	'Little Gem'	oSqu
	Little Moor	see **S. *telephium*** 'Mohrchen'
	longipes	oSqu
	lucidum	oSqu
	lucidum crest	oSqu
un	***x luteolum***	oSqu
	luteoviride	oSqu
	lydium	oSto oSis oSqu oWnC oCir
un	***macdougallii***	oSqu
un	***makinoi*** 'Ogon'	wHer
	makinoi 'Variegatum'	oSqu
	maximum 'Atropurpureum'	last listed 99/00
	middendorffianum	oSto oLSG
	middendorffianum var. ***diffusum***	oSto oSqu
	middendorffianum from Manchuria	oSto
	middendorffianum 'Striatum'	oSqu
	'Mohrchen'	see **S. *telephium*** 'Mohrchen'
	monregalense	oSto oSqu
	montanum	see **S. *ochroleucum*** ssp. ***montanum***
	'Moonglow'	oSto
	morganianum	oGar oSqu
	multiceps	oTrP oSqu
un	***muscoideum***	oSqu
	nevii (see Nomenclature Notes)	oJoy wFGN
	nicaeense	see **S. *sediforme***
	nudum from Puerto Santo	oSqu
	nussbaumerianum	oSto oSqu
	oaxacanum	oSqu
	obtusatum	oSto
	ochroleucum	oSto oSqu
	ochroleucum ssp. ***montanum***	oSqu
	ochroleucum ssp. ***montanum*** blue form	oSqu
	ochroleucum ssp. ***montanum*** green form	oSqu
un	***ochroleucum*** ssp. ***montanum orientale***	oSqu
	ochroleucum Mt. Olympus form	oSis
	ochroleucum red leaf form	oSis
	oppositifolium	see **S. *spurium*** var. ***album***
	oreganum	oFor wNot oBos oSho oSqu oWnC oAld wWld oSle
	oreganum dwarf form	oSqu

	oreganum from Eagle Creek	oSqu
	oreganum from Otis-Salmon River	oSqu
	oreganum from Otter Crest	oSto
	oreganum from Willamina, OR	oSqu
	oreganum 'Procumbens'	see **S. *oreganum*** ssp. ***tenue***
	oreganum ssp. ***tenue***	oSto
	oreganum ssp. ***tenue*** from Santiam Pass	oSto
	oregonense	oSto
	oryzifolium tiny form	oSqu
un	***oxycoccoides***	oSqu
	oxypetalum	oRar oSqu
	pachyclados	oJoy oSto oSis oSqu
	pachyphyllum	wFGN oSqu wHom
	pallidum var. ***bithynicum***	oSto oSqu
	palmeri	wHer oSqu
un	***palmeri emarginatum***	oSqu
un	'Pearson's Puzzle'	oSqu
	pilosum	last listed 99/00
qu	***pinifolium*** 'Blue Spruce'	oFor
	pluricaule	oSto oAls oUps
	populifolium	oJoy oSto wRob oSqu
	potosinum	oSqu
un	***praealtum*** ssp. ***praealtum***	oSqu
un	***pruinatum*** 'Blue Spruce'	oLSG
	quevae	oSqu
un	'Red Carpet'	oLSG
	reflexum L.	see **S. *rupestre*** L.
	reptans	oSqu
	rosea	see **RHODIOLA *rosea***
un	'Rosy Glow'	wWoS oAls
	x rubrotinctum	oSqu
un	***x rubrotinctum*** 'Aurora'	oSqu
	'Ruby Glow'	oFor wCul oJoy oSto wShR oMis wMag wAva oLSG
	rupestre L.	oSto wFai wHom
	rupestre f. ***cristatum***	oSto
un	***rupestre erectum***	oSqu
	rupestre 'Monstrosum Cristatum'	last listed 99/00
un	***rupestre*** 'Silver Crest'	oSto
	ruprechtii	see ***telephium*** ssp. ***ruprechtii***
	sarmentosum	oSqu
	sarmentosum from Beijing	oSto
	sarmentosum from Kiansi	oSto
	sediforme	oTrP oSto
	selskianum	oSto oSqu
un	***selskianum*** 'Variegatum'	wFGN
	senanense	see **S. *japonicum***
un	***serpentini***	oSto
	sexangulare	oJoy oSto oSqu
un	***sexangulare elatum***	oSqu
un	***sexangulare*** 'Weisse Tatra'	oSqu
	sichotense	oSqu
	sieboldii	oFor oJoy oAls oHed wFGN oDan oSqu oWnC wTGN oLSG oUps oSle
	sieboldii 'Mediovariegatum'	oFor oSto oHed oGre oWnC oUps
un	***sieboldii*** var. ***nana***	oSto
un	***sieboldii*** 'Pink Jewel'	oMis wFai oCir
	sieboldii 'Variegatum'	oJoy oMis
	'Silver Moon'	oUps
	spathulifolium	oFor wCul wThG wNot oAls wShR oBos oTri oAld wWld
	spathulifolium 'Aureum'	oSto
	spathulifolium 'Cape Blanco'	oSto oAls oHed wFGN wCSG wHom oWnC wBWP wAva oUps
	spathulifolium 'Carnea'	oJoy oSto oSqu oGre
	spathulifolium ssp. ***pruinosum***	last listed 99/00
	spathulifolium 'Purpureum'	oFor oSto oAls oSis oWnC wBWP oUps
un	***spathulifolium*** ssp. ***spathulifolium***	oSto
un	***spathulifolium*** 'White Chalk'	wFGN
	spathulifolium ssp. ***yosemitense***	oSto
	spectabile	oSec
	spectabile 'Brilliant'	oFor oJoy oSto wHig oAls oGar oSqu oGre oWnC wMag oEga oLSG oUps wSnq
	spectabile 'Carmen'	oFor
	spectabile 'Iceberg'	wHer
	spectabile 'Indian Chief'	oFor oSto
	spectabile 'Meteor'	wCul oSto oAls oSho oWnC oLSG oUps oSle wSnq
un	***spectabile*** 'Neon'	oFor wSnq
	spectabile 'Rosenteller'	oSqu
	spectabile 'Stardust'	oJoy oSto oAls oGar wRob oSqu wSnq
	spectabile 'Steven Ward'	wHer
	spectabile 'Variegatum'	see **S. *alboroseum*** 'Mediovariegatum'
un	'Spiral Staircase'	oSqu
	spurium	oSho
	spurium var. ***album***	oSto oSqu
	spurium 'Atropurpureum'	oSqu
un	***spurium*** 'Bronze Beauty'	oSqu
	spurium 'Bronze Carpet'	oJoy oSto wFGN oGar oWnC oUps
	spurium 'Coccineum'	oSqu
	spurium 'Doctor John Creech'	see **S. *spurium*** 'John Creech'
	spurium Dragon's Blood	see **S. *spurium*** 'Schorbuser Blut'
un	***spurium*** 'Elizabeth'	oSis oSqu
un	***spurium*** 'Fool's Gold'	oSqu
	spurium from Turkey	oSqu
	spurium 'Fuldaglut'	oFor oWnC
	spurium 'Gold Carpet'	oSto
	spurium 'Green Mantle'	oSis
	spurium 'John Creech'	oFor oAls oSqu
un	***spurium*** 'Pearly Pink'	oSqu
un	***spurium*** 'Pink Jewel'	oJoy oSto oAls oSqu oWnC
	spurium Purple Carpet	see **S. *spurium*** 'Purpurteppich'
	spurium 'Purpurteppich'	oAls oSqu oWnC oUps
un	***spurium*** 'Raspberry Red'	oSqu
	spurium 'Ruby Mantle'	oSqu
	spurium 'Salmonium'	oSto
	spurium 'Schorbuser Blut'	oFor wCul oSto wFGN oSqu oGre wHom iArc wAva wEde oLSG
	spurium 'Tricolor'	see **S. *spurium*** 'Variegatum'
	spurium 'Variegatum'	oFor oSto oAls oHed wFGN wCSG oGar oSqu oGre wFai oWnC wAva wEde oUps
	stahlii	last listed 99/00
	stefco	oSto
	stenopetalum	wNot oJoy oSto wWld
	'Stewed Rhubarb Mountain'	oSqu
	stoloniferum	oSto
	stribrnyi	see **S. *urvillei*** Stribrnyi Group
	'Sunset Cloud'	wHer wWoS oGar oOut wRob oSec
	takasui	see **S. *cyaneum*** Rudolph
	takesimense	last listed 99/00
	takesimense HC 970204	wHer
un	***tectractinum***	oSto wTGN
	telephium 'Arthur Branch'	wHer oNWe wSnq
	telephium var. ***borderi***	wHer
	telephium 'Matrona'	oFor wHer oNat wHig oAls oGar wRob oWnC wTGN oLSG oUps oSle
	telephium ssp. ***maximum*** 'Atropurpureum'	oJoy wTGN
	telephium ssp. ***maximum*** 'Gooseberry Fool'	wHer oHed
	telephium 'Mohrchen'	oFor wCul oJoy oSto oAls oDan oNWe wRob oSec oLSG oUps
	telephium 'Munstead Red'	wHer wWoS
	telephium ssp. ***ruprechtii***	oJoy oSto oGar oNWe oOut wRob wBox
qu	***telephium*** 'Sunset Red'	oSec
	tenuifolium	oSto
	tenuifolium ssp. ***ibericum***	last listed 99/00
	ternatum	oFor oSto wFGN
un	***tetractinum***	wFGN
	treleasei	oSqu
un	***tschernokolevii***	oSto
	urvillei	last listed 99/00
	urvillei Hillebrandtii Group	oSto oSqu
	urvillei Stribrnyi Group	oSto
	'Vera Jameson'	oFor wWoS wCul oJoy oSto oAls wFGN oGar wHom wMag wTGN oSec wEde oEga oLSG oUps oSle wSnq
	winkleri	see **S. *hirsutum*** ssp. ***baeticum***
un	***zentaro-tashiroi***	oSto
	SELAGINELLA	
un	***bigelovii***	wFan
	braunii	wFan wTGN
	kraussiana	oTrP oAls wBox
	kraussiana 'Aurea'	oGre wBox oBRG
	kraussiana 'Brownii'	oTrP
un	***kraussiana*** 'Gold Tips'	wFan oSev
	martensii 'Variegata'	wBox
un	***mollendorfii***	wBox oBRG
un	***sanguinolenta*** var. ***compressa***	oSis
	sp.	wFFI
	uncinata	oGre wBox oBRG
	SELINUM	
	tenuifolium	see **S. *wallichianum*** EMAK 886
	wallichianum EMAK 886	wHer
	SELLIERA	
	radicans	oTrP
	SEMIAQUILEGIA	
	adoxoides	oFor
	ecalcarata	wCul wCri oHed oUps

	SEMIARUNDINARIA	
	fastuosa	oFor oTra oNor wBlu wBea oAls oOEx wBox wBam wBmG oTBG
	fastuosa var. ***viridis***	oTra wCli oNor wBam wBmG oTBG
un	***fortis***	oNor wBea
	okuboi	oTra wCli oNor wBlu wBea wBam
	villosa	see **S. *okuboi***
	yashadake	oTra oNor
	yashadake kimmei	oTra wCli wBea
	SEMPERVIVELLA	see **ROSULARIA**
	SEMPERVIVUM	
	'Abba'	oSqu
un	'Achalur'	oSqu
un	'Adelgonde'	oSqu
un	'Affine'	oSqu
	albidum	oSqu
	'Alcithoe'	last listed 99/00
	'Aldo Moro'	oSqu
	'Alluring'	last listed 99/00
	'Alpha'	oSqu
	altum	oSqu
	'Amanda'	oSqu
	'Ambergreen'	oSqu
un	'Amtmann Fischer'	oSqu
	andreanum	oSqu
un	'Andrenor'	oSqu
un	'Angustifolium'	oSqu
un	'Apetlon'	oSqu
un	'Aphrodite'	oSqu
	'Apple Blossom'	oSqu
	arachnoideum	oFor wEde
un	***arachnoideum*** Afterglow hybrid	oSto
un	***arachnoideum*** 'Albion'	oSqu
	arachnoideum var. ***bryoides***	oSto oSqu
un	***arachnoideum*** 'Cebanese'	oSto oSis
un	***arachnoideum*** 'Emily'	oSqu
	arachnoideum from Sion	oSqu
	arachnoideum var. ***glabrescens***	oSto oSis
	arachnoideum var. ***glabrescens*** 'Album'	oSqu
un	***arachnoideum*** 'Gladys'	oSqu
	arachnoideum hookeri	see **S. *x barbulatum*** 'Hookeri'
	arachnoideum x ***montanum*** ssp. ***stiriacum***	
	arachnoideum x ***nevadense***	oSto oSqu
	arachnoideum x ***pittonii***	oSqu
un	***arachnoideum*** 'Pygmalion'	oSqu
	arachnoideum 'Rubrum'	oSto oSis oSqu
qu	***arachnoideum*** 'Safari'	oSqu
un	***arachnoideum*** 'Sparkle'	oSqu
	arachnoideum 'Sultan'	oSqu
	arachnoideum ssp. ***tomentosum***	oSis oSqu wTGN
	arachnoideum ssp. ***tomentosum*** 'Minor'	oSqu
	arachnoideum ssp. ***tomentosum*** 'Stansfieldii'	oSto oSqu
un	***arachnoideum*** #3203	oSqu
un	'Arondina'	oSqu
	'Aross'	oSqu
	'Ashes of Roses'	oSqu
	atlanticum 'Edward Balls'	last listed 99/00
	atlanticum from Oukaimaden	oSqu
	'Atropurpureum'	oSqu
un	'Atrorubens'	oSqu
	'Atroviolaceum'	see **S. *tectorum*** 'Atroviolaceum'
un	'Aulis'	oSqu
	'Aureum'	see **GREENOVIA *aurea***
	'Aymon Correvon'	oSqu
	ballsii from Kambeecho	oSqu
	ballsii from Smolikas	oSqu
	ballsii from Tschumba Petzi	oSqu
	'Banderi'	last listed 99/00
un	'Banyon'	oSqu
	x barbulatum from Prague	oSqu
	x barbulatum from Valle Quarozzo	oSqu
	x barbulatum 'Hookeri'	oSto oSqu oUps
un	***x barbulatum rubrum*** Red Mountain	oSqu
un	'Barnattii'	oSqu
un	'Baronesse'	oSqu
	'Bascour Zilver'	oSqu
	'Bedivere'	oSqu
	'Bella Meade'	oSqu
	'Belladonna'	oSto oSqu
	'Bernstein'	oSqu
	'Beta'	oSqu
	'Bethany'	oSqu
	'Big Slipper'	oSqu
	'Birchmaier'	oSqu
	'Black Mini'	oSqu
	'Black Mountain'	oSqu
	'Blood Tip'	oSqu
	'Blue Moon'	oSto oSqu
	'Blush'	oSqu
	'Bold Chick'	oSqu
	'Booth's Red'	oSqu
	borissovae	oSqu
un	'Boule de Neige'	oSqu
un	'Britta'	oSqu
	'Brock'	oSqu
	'Bronco'	oSqu
	'Bronze Pastel'	oSqu
	'Brown Owl'	oSqu
	'Brownii'	oSto oSqu
un	'Bunny Girl'	oSqu
	'Butterbur'	oSto oSqu
un	'Butterfly'	oSqu
un	'C. William'	oSqu
	'Cafe'	oSqu
	calcareum	last listed 99/00
un	***calcareum*** ssp. ***calcareum***	oSto
	calcareum from Guillaumes, Mont Ventoux, France	last listed 99/00
	calcareum from Mont Ventoux, France	oSqu
	calcareum 'Greenii'	oSqu
	calcareum 'Grigg's Surprise'	oSto
	calcareum 'Monstrosum'	see **S. *calcareum*** 'Grigg's Surprise'
	calcareum 'Mrs. Giuseppi'	oSto oSqu
un	***calcareum*** 'Nigricans'	oSqu
	calcareum 'Pink Pearl'	oSqu
	calcareum 'Sir William Lawrence'	oSto oSis
	'Canada Kate'	oSqu
	'Candy Floss'	oSqu
un	***canni***	oSqu
	cantabricum from Lago de Enol #1	oSqu
	cantabricum from Leitariegos	oSqu
	cantabricum from Picos de Europa	oSqu
	cantabricum from San Glorio	oSqu
	cantabricum from Smoiedo #1	oSqu
	cantabricum from Valvernera	oSqu
	cantabricum ssp. ***guadarramense*** from Lobo No. 1	oSqu
	cantabricum x ***montanum*** ssp ***stiriacum***	oSqu
un	***cantabricum nigrum***	oSqu
	cantabricum ssp. ***urbionense***	oSqu
	'Carmen'	oSqu
	'Carneus'	oSqu
	'Carnival'	oSqu
un	'Casa'	oSqu
	caucasicum	oSqu
	'Cavo Doro'	oSqu
un	'Celon'	oSqu
un	'Centennial'	oSqu
un	***charolensis***	oSqu
	'Cherry Frost'	oSqu
un	'Cherry Vanilla'	oSqu
un	'Chivalry'	oSto oSqu
	'Chocolate'	oSqu
	ciliatum	see **AEONIUM *ciliatum***
	ciliosum	oSqu
	ciliosum var. ***borisii***	oSto
	ciliosum x ***erythraeum***	oSqu
	ciliosum from Ali Butus	oSto oSqu
	ciliosum var. ***galicicum*** 'Mali Hat'	oSqu
	'Circlet'	oSqu
un	'Circus'	oSqu
un	'Claey's Fluweel'	oSqu
	'Clara Noyes'	oSqu
	'Clare'	last listed 99/00
un	'Claret'	oSqu
	'Cleveland Morgan'	oSqu
	'Climax'	oSqu
un	'Clipper'	oSqu
un	'Cobweb'	wFGN
	'Cobweb Capers'	oSqu
un	'Cobweb Joy'	oSqu
un	'Cold Fire'	oSqu
	'Collage'	oSqu
	'Collecteur Anchisi'	oSqu
un	'Comillii'	oSqu
	'Compte de Congae'	oSqu
	'Congo'	oSqu
	'Cornstone'	oSqu
	'Coronet'	oSqu

	'Correvons'	see S. 'Aymon Correvon'
	'Corsair'	oSqu
un	'Crebben'	oSqu
un	'Crimson King'	oSqu
	'Crimson Velvet'	oSqu
un	'Crimson Web'	oSqu
un	'Crimsonette'	oSqu
	'Crispyn'	oSqu
	'Croton'	oSqu
	'Cupream'	oSqu
un	'Cynthian'	oSqu
	'Dallas'	oSqu
	'Damask'	oSqu
	'Dark Beauty'	oSqu
	'Dark Cloud'	oSqu
	'Dark Point'	oSqu
	davisii from Corah Gorge	oSqu
	'Deep Fire'	oSqu
	x degenianum	oSqu
	'Delta'	oSqu
	'Director Jacobs'	oSqu
un	'Dixie'	oSqu
un	'Dolo'	oSqu
	dolomiticum	oSqu
	dolomiticum x ***montanum***	oSqu
	'Donarrose'	oSqu
	'Duke of Windsor'	oSqu
	'Dusky'	oSqu
	'Dyke'	oSqu
un	'Dynamo'	oSqu
	'Edge of Night'	oSqu
un	'Educator Wollaert'	oSqu
	'El Greco'	oSqu
	'El Toro'	oSto oSqu
un	'Elene'	oSto oSqu
	'Elvis'	oSqu
un	'Emberly Pink'	oSqu
	'Emerald Giant'	oSto oSqu
un	'Emerald Heart'	oSqu
	'Emerson's Giant'	oSqu
un	'Eomer'	oSqu
	erythraeum	oSto oSqu
	erythraeum from Bulgaria	oSqu
	erythraeum from Pirin, Bulgaria	oSqu
	erythraeum from Rila	oSqu
	erythraeum 'Glasnevin'	last listed 99/00
	erythraeum SM 1031	oSqu
un	'Euphemia'	oSqu
	'Excalibur'	last listed 99/00
	'Exhibita'	oSqu
	'Exorna'	oSqu
un	'Fabiana Smits'	oSqu
	'Fair Lady'	oSqu
	'Fame'	oSqu
un	'Fantasia'	oSqu
un	'Fat Jack'	oSqu
	x fauconnettii	oSqu
	x fauconnetti thompsonii	oSqu
	'Finerpointe'	oSqu
	'Firebird'	oSto oSqu
	'Flaming Heart'	oSqu
	'Flander's Passion'	oSqu
	'Flasher'	oSqu
	'Ford's Spring'	oSqu
un	'Fredegar'	oSqu
	'Frost and Flame'	oSqu
	'Fuego'	oSqu
	x funckii	oSqu
un	'Fusiler'	oSqu
	'Fuzzy Wuzzy'	oSqu
	'Galahad'	oSqu
un	'Gamalea'	oSqu
	'Gambol'	oSqu
	'Gamma'	oSqu
	'Garnet'	oSqu
	'Gay Jester'	oSqu
	'Gazelle'	oSto
un	'Gildor'	oSqu
	'Ginnie's Delight'	oSqu
	giuseppii	oHed
	giuseppii from Pena Espiguete, Spain	oSqu
	giuseppii from Pena Prieta #1, Spain	oSqu
	giuseppii from Pena Prieta #2, Spain	oSqu
un	'Glaucum Minor'	oSis oSqu
un	'Gloconda'	oSqu
	'Gloriosum'	oSqu
	'Glowing Embers'	oSqu
un	'Godaert'	oSqu
un	'Goedele'	oSqu
	'Gollum'	oSqu
	'Granada'	oSqu
	'Granat'	oSqu
	grandiflorum	oSqu
	grandiflorum crested	oSqu
	grandiflorum from Kerguelen	oSqu
	grandiflorum from Pecceto di Sopra	oSqu
	'Grapetone'	oSqu
	'Greenwich Time'	last listed 99/00
un	'Grey Dawn'	oSqu
	'Grey Ghost'	oSqu
	'Grey Green'	oSqu
	'Grey Lady'	oSto
	'Greyfriars'	oSqu
un	***guillemottii***	oSqu
un	'Gypsy'	oSqu
	'Hall's Hybrid'	oSqu
	'Happy'	oSqu
	'Hart'	oSqu
	'Haullauer's Seedling'	oSqu
	x hausemanii (see Nomenclature Notes)	oSto oSqu
	'Heigham Red'	oSto oSqu
	'Heliotroop'	last listed 99/00
	helveticum	see S. ***montanum***
un	'Hepworth Hybrid'	oSqu
	'Hester'	oSqu
	'Hey-Hey'	oSqu
	'Hidde'	oSqu
	'Highland Mist'	oSqu
un	***hispidlum***	oSqu
	hookeri	see S. ***x barbulatum*** 'Hookeri'
	'Hot Shot'	last listed 99/00
	'Hullabaloo'	oSqu
un	'Hybrid Globe'	oSqu
	hybrid red and green crested	oSto
	'Icicle'	oSto oSqu oUps
	'Imperial'	oSqu
un	'Ineke'	oSqu
	ingwersenii	oSqu
	'Interlace'	oSqu
	'Irazu'	oSqu
	italicum from Mont Lepini	oSqu
	'IWO'	oSqu
	'Jack Frost'	last listed 99/00
	'Jane'	last listed 99/00
	'Jaspis'	oSqu
	'Jelly Bean'	oSqu
un	'Jeramia'	oSqu
	'Jet Stream'	oSqu
	'Jewel Case'	oSqu
un	'Jigger'	oSqu
	'Jolly Green Giant'	oSqu
un	'Joybelle'	oSqu
	'Jubilee'	oSto oSqu
	'Jungle Fires'	oSqu
un	'Jungle Shadows'	oSto wTGN oUps
un	'Jurrina'	oSqu
un	'Just Plain Crazy'	oSqu
	'Kalinda'	oSqu
	'Kappa'	oSqu
un	'Kautangel'	oSqu
	'Kelly Jo'	oSqu
	'Kimono'	oSqu
	kindingeri	oSqu
	'King George'	oSqu
	'Kip'	oSqu
	'Kismet'	oSqu
un	'Korresia'	oSqu
un	'Korump'	oSqu
	kosaninii from Koprivnik	oSqu
	'Kramers Purpur'	oSqu
	'Kramers Spinrad'	last listed 99/00
	'Lady Kelly'	oSto
un	'Laggeri'	oSqu
	lamottei	see S. ***tectorum***
	'Launcelot'	last listed 99/00
un	'Laura Lee'	oSto
	'Lavender and Old Lace'	oSto oSqu wTGN oUps
	Le Clair's hybrid No. 1	oSqu
	Le Clair's hybrid No. 2	oSqu
	'Leneca'	oSqu

	'Lentezon'	oSqu
un	'Leocadia'	oSqu
	'Leocadia's Nephew'	oSqu
	leucanthum	oSqu
un	***leucanthum rubrum***	oSqu
un	'Liliane'	oSqu
	'Lipari'	oSqu
	'Lipstick'	oSto
	'Lloyd Praeger	see **S. *montanum*** ssp. ***stiriacum*** 'Lloyd Praeger'
	'Lynne's Choice'	oSqu
	macedonicum	oSqu
	'Magic Spell'	oSqu
	'Magical'	oSqu
	'Magnificum'	oSqu
	'Maigret'	last listed 99/00
	'Majestic'	oSqu
	'Malby's Hybrid'	see **S.** 'Reginald Malby'
un	'Manuel'	oSqu
un	'Maria Laach'	oSqu
un	'Marietta'	oSqu
un	'Marjoleine'	oSqu
	'Marjorie Newton'	last listed 99/00
	marmoreum	oSqu
	marmoreum 'Brunneifolium'	oSto oSis oSqu
	marmoreum from Dinaricum	oSto
	marmoreum from Karawanken	oSto
	marmoreum from Monte Tirone	oSqu
	marmoreum from Okol	oSqu
	marmoreum ssp. ***marmoreum*** var. ***dinaricum*** 'Rubrifolium'	oSto
un	'Maroon Queen'	oSqu
	'Mary Ente'	last listed 99/00
	'Mate'	oSqu
	'Mauna Kea'	oSqu
un	'Mayfair'	oSto
un	'Mayfair Hybrid'	oSis
	'Meisse'	oSqu
	'Melanie'	last listed 99/00
	'Merkur'	oSqu
	'Merlin'	oSqu
	mettenianum	see **S. *tectorum***
un	'Michael'	oSqu
	'Midas'	oSqu
un	'Mike'	oSqu
	'Mila'	oSqu
un	'Minaret'	oSqu
	'Missouri Rose'	oSqu
	'Moerkerk's Merit'	oSqu
un	'Monasses'	oSto oSqu
	'Mondstein'	oSqu
un	'Money'	oSqu
	'Montague'	oSqu
	montanum	oSto oSqu
	montanum ssp. ***burnatii***	oSqu
	montanum carpaticum	oSqu
	montanum carpaticum 'Cmiral's Yellow'	last listed 99/00
	montanum from Andorra	oSqu
	montanum from Mass du Canigou	oSqu
	montanum 'Minimum'	last listed 99/00
	montanum ssp. ***montanum*** var. ***braunii***	oSqu
	montanum ssp. ***stiriacum***	oSto oSqu
	montanum ssp. ***stiriacum*** from Goldeck	oSqu
	montanum ssp. ***stiriacum*** 'Lloyd Praeger'	last listed 99/00
un	'Montfort'	oSqu
un	'Montgomery'	oSqu
	'More Honey'	oSqu
un	'Morellianum'	oSqu
	'Mulberry Wine'	oSqu
un	'Mustang'	oSqu
	'Mystic'	oSqu
	'Neptune'	oSqu
	nevadense	oSto oSqu
	nevadense var. ***hirtellum***	oSqu
	'Nico'	oSqu
un	'Nightwood'	oSqu
	'Nigrum'	see **S. *tectorum*** 'Nigrum'
	'Nixes 27'	oSqu
	'Norbert'	oSqu
	'Nortofts Beauty'	oSqu
	'Nouveau Pastel'	oSqu
un	'Oberon'	oSqu
	octopodes	oSqu
	octopodes var. ***apetalum***	oSto oSis oSqu
	'Oddity'	wFGN oOut oSqu oUps
	'Ohio Burgundy'	oSto oSis oSqu
un	'Ohioan'	oSqu
un	'Old Copper'	oSto oSqu
un	'Old Rose'	oSqu
	'Olivette'	oSqu
	'Omega'	oSqu
	'Ornatum'	last listed 99/00
	ossetiense	oSqu
	'Othello'	oSqu
un	'Pacific Blazing Star'	oSqu
un	'Pacific Charm'	oSqu
un	'Pacific Chuckles'	oSqu
un	'Pacific Clydsdale'	oSqu
un	'Pacific Daemon'	oSqu
un	'Pacific Dancer'	oSqu
un	'Pacific Dawn'	oSqu
un	'Pacific Daydream'	oSqu
un	'Pacific Deep'	oSqu
un	'Pacific Devils Food'	oSqu
un	'Pacific Drama'	oSqu
un	'Pacific Feather Power'	oSqu
un	'Pacific First Try'	oSqu
un	'Pacific Grace'	oSqu
un	'Pacific Green Rose'	oSqu
un	'Pacific Green Sleeves'	oSqu
un	'Pacific Hairy Hep'	oSqu
un	'Pacific Hazy Embers'	oSqu
un	'Pacific Hep'	oSqu
un	'Pacific Joyce'	oSqu
un	'Pacific Knight'	oSqu
un	'Pacific Majesty'	oSqu
un	'Pacific Mauve'	oSqu
un	'Pacific Mayfair Imp'	oSqu
un	'Pacific Mini Hep'	oSqu
un	'Pacific Opal'	oSqu
un	'Pacific Patches'	oSqu
un	'Pacific Phoenix Fire'	oSqu
un	'Pacific Plum Fuzzy'	oSqu
un	'Pacific Purple Shadows'	oSqu
un	'Pacific Red Hawk'	oSqu
un	'Pacific Red Rose'	oSqu
un	'Pacific Red Tide'	oSqu
un	'Pacific Rim'	oSqu
un	'Pacific Second Try'	oSqu
un	'Pacific Sexy'	oSqu
un	'Pacific Shadows'	oSqu
un	'Pacific Sonata'	oSqu
un	'Pacific Sparkler'	oSqu
un	'Pacific Spring Frost'	oSqu
un	'Pacific Taffy Pink'	oSqu
un	'Pacific Tart'	oSqu
un	'Pacific Teddy'	oSqu
un	'Pacific Thunder'	oSqu
un	'Pacific Tightwad'	oSqu
un	'Pacific Trails'	oSqu
un	'Pacific Velveteen'	oSqu
un	'Pacific Zoftic'	oSqu
	'Packardian'	oSqu
	'Painted Lady'	oSqu
	'Palissander'	oSqu
un	'Pallas'	oSqu
un	'Palouse'	oSqu
un	'Pandora'	oSqu
un	'Park Hay'	oSqu
	'Pastel'	oSqu
	'Patrician'	oSto oSqu
	'Pekinese'	oSto oSqu oUps
	'Peterson's Ornatum'	oSqu
un	'Picos de Uropa'	oSqu
	x piliferum	see **S. *x fauconnettii***
	'Pilosella'	oSqu
un	'Pink Charm'	oSqu
	'Pink Delight'	oSqu
	'Pink Puff'	oSqu
	'Pippin'	oSqu
	pittonii	oSto
	pittonii x ***montanum***	oSqu
	'Pixie'	oSto oSis oSqu
un	'Plastic'	oSqu
	'Plumb Rose'	last listed 99/00
	'Pluto'	oSqu
	'Poke Eat'	oSqu
	'Polaris'	oSqu
un	'Potsy'	oSqu
	'Powellii'	oSqu

	'Precious'	oSqu
	'Proud Zelda'	oSqu
	'Pruhonice'	oSqu
	pumilum from Adyl Su	oSto oSqu
	pumilum from Armchi	oSto
	pumilum from El'brus No. 1	oSqu
	pumilum x ***ingwersenii***	oSqu
un	'Purdy's Big Red'	oSqu
un	'Purdy's 60-1'	oSqu
un	'Purdy's 60-2'	oSqu
	'Purdy's 90-1'	oSqu
	'Purple Beauty'	wFGN oSqu
	pyrenaicum	see **S. *tectorum***
	'Quintessence'	oSqu
un	'Rachael'	oSqu
	'Racy'	oSqu
	'Raspberry Ice'	oSto oSis
un	'Rauhreif'	oSqu
	red	wFGN
	'Red Ace'	last listed 99/00
	'Red Beam'	last listed 99/00
un	'Red Cross'	oSqu
	'Red Delta'	oSqu
un	'Red Pluche'	oSqu
	'Red Rum'	last listed 99/00
	'Red Wings'	oSqu
	'Regal'	oSqu
	reginae-amaliae from Sarpun	oSqu
	reginae-amaliae from Vardusa	oSqu
	'Reginald Malby'	oSqu
	'Reinhard'	oSqu
	'Remus'	oSqu
	'Rhone'	oSqu
	'Risque'	oSqu
	'Rita Jane'	oSqu
	'Robin'	oSto
un	'Rocknoll Rosette'	oSqu
un	'Rogin'	oSto oSqu
	'Ronny'	oSqu
	'Roosemaryn'	oSqu
	'Rose Splendour'	oSqu
	x roseum 'Fimbriatum'	oSqu
	'Rosie'	oSto oSis oSqu
	'Rotkopf'	oSqu
	'Rotmantel'	oSqu
	'Rotund'	oSqu
	'Rouge'	oSqu
	'Royal Flush'	oSqu
	'Royal Opera'	oSqu
	'Royal Ruby'	oSqu
un	'Rubellum Mahogany'	oSqu
un	'Rubikon'	wTGN
	'Rubikon Improved'	oSto oSqu
	'Rubin'	oSqu
	'Rubrum Ash'	oSqu
un	'Rubrum Borsch'	oSqu
	'Ruby Heart'	last listed 99/00
	x rupicola	oSqu
	ruthenicum from Budapest	oSqu
	'Safara'	oSto
	'Saga'	oSqu
qu	'Sandfordii'	oSqu
	'Sanford's Hybrid'	oSqu
	'Sassy Frass'	oSqu
	'Saturn'	oSqu
	schlehanii	see **S. *marmoreum***
un	'Schnucki'	oSqu
	'Seminole'	oSqu
un	'Serena'	oSqu
un	'Serendipity'	oSqu
	'Shirley's Joy'	oSqu
	'Sideshow'	oSqu
un	'Siebenbergen'	oSqu
	'Silber Karneol'	see **S.** 'Silver Jubilee'
un	'Silver Cup'	oSqu
	'Silver Jubilee'	oSqu
un	'Silver Olympic'	oSqu
	'Silver Spring'	last listed 99/00
	'Silver Thaw'	oSto oSis oSqu
un	'Silver Web'	oSqu
un	'Silverina'	wFGN
un	'Silverine'	oSto
	'Sioux'	last listed 99/00
un	'Sirius'	oSqu
un	'Smits Seedling'	oSqu
	'Smokey Jet'	last listed 99/00
	'Snowberger'	oSqu
un	'Soft Line'	oSqu
	'Sopa'	oSqu
	sp.	iGSc
	'Spanish Dancer'	oSqu
un	***speciosum***	see **S. *tectorum***
	'Spherette'	oSqu
	'Spring Mist'	oSqu
	'Sprite'	oSqu
	'Starion'	oSqu
	'Starshine'	oSqu
un	'Strawberry Velvet'	oSqu
	'Stuffed Olive'	oSqu
un	'Sugary'	oSqu
un	'Sun Queen'	oSqu
	'Sun Waves'	oSqu
	'Sunkist'	oSqu
	'Super Dome'	oSqu
	'Superama'	oSqu
un	'Susan'	oSqu
	'Syston Flame'	oSqu
	'Tamberlane'	oSqu
	'Teck'	oSqu
	tectorum	oAls wFGN oSqu wWhG
	tectorum 'Atropurpureum'	oSto
	tectorum 'Atroviolaceum'	oSqu
	tectorum from Isella	oSqu
	tectorum green	oCir
un	***tectorum*** 'Marin'	oSqu
	tectorum 'Nigrum'	oSqu
	tectorum 'Purpureum'	see **S. *tectorum*** 'Atropurpureum'
	tectorum red	oCir
	tectorum 'Red-Purple'	oSqu
	tectorum 'Royanum'	oSto
	tectorum 'Sunset'	oSto
	tectorum ssp. ***tectorum*** 'Triste'	oSqu
un	'Tederheid'	oSqu
un	'Temby"	oSqu
un	'Tenbury'	oSqu
un	'Terracotta Baby'	oSqu
	'Thayne'	oSqu
un	'Theobaldi'	oSqu
	thompsonianum	oSqu
un	'Thunder'	oSto
un	'Thunder Cloud'	oSqu
un	'Tiara'	oSqu
	'Tiger Bay'	oSqu
un	'Tip Top'	oSqu
	'Titania'	oSqu
	'Topaz'	oSto oSqu
	'Tordeur's Memory'	oSqu
	'Traci Sue'	oSqu
	transcaucasicum	oSqu
	'Tristram'	oSqu
	'Truva'	oSqu
un	'Turmalin'	oSqu
	'Twilight Blues'	oSqu
un	'Utopian'	oSqu
	x vaccarii	oSqu
	'Vaughelen'	oSqu
un	'Velanovsky'	oSqu
	x versicolor	oSqu
	'Video'	oSqu
un	'Viking'	oSto
	villosum (see Nomenclature Notes)	oSqu
	'Violet Queen'	oSqu
	'Virgil'	oSqu
	'Virginus'	oSqu
un	'Viti'	oSqu
	'Vulcano'	last listed 99/00
	'Watermelon Rind'	oSqu
	'Webby Ola'	oSqu
	'Weirdo'	oSqu
	'Wendy'	last listed 99/00
	'Westerlin'	oSto oSqu
un	'Whirligig'	oSqu
	'Whitening'	oSqu
	'Wollcott's Variety'	oSqu
un	'Zarubianum'	oSqu
un	'Zebulon'	oSqu
	zeleborii	oSqu
	'Zenith'	oSqu
un	'Zenobia'	oSqu
un	'Zeppelin No. 3'	oSqu

un	'Zezette'	oSqu
un	'Zorba'	oSqu
	'Zulu'	oSqu
	SENECIO	
	acaulis	last listed 99/00
	articulatus	oTrP
	articulatus 'Variegatus'	last listed 99/00
	aureus	see **PACKERA *aurea***
un	***ballii***	oRar
	cineraria	oGoo
	cineraria 'Cirrus'	wAva
	cineraria 'Silver Dust'	wAva
un	***cristobalensis***	wHer
	x crustii	see **BRACHYGLOTTIS** Dunedin Group
	greyi (see Nomenclature Notes)	wCul wTGN wAva
	herreanus	oSqu
	kleiniiformis	last listed 99/00
	'Leonard Cockayne'	see **BRACHYGLOTTIS** 'Leonard Cockayne'
	rowleyanus	oGar oSqu
un	***rowleyanus*** 'Variegatus'	oSqu
	scandens DJHC 98155	wHer
	scaposus	last listed 99/00
	serpens	oSqu
	solandri var. ***rufiglandulosus***	last listed 99/00
	'Sunshine'	see **BRACHYGLOTTIS** (Dunedin Group) 'Sunshine'
	viravira	wWoS wSwC oSho oSec wAva
	SENNA	
	corymbosa	oCis oWnC wBox
	hebecarpa	oFor oUps
	lindheimeriana	wCCr
	multiglandulosa	wCCr
	SEQUOIA	
	giganteum	see **SEQUOIADENDRON *giganteum***
	sempervirens	oFor oDar wShR oGar wRai oGre wCCr wSta wCoN wRav wSte wCoS
	sempervirens 'Adpressa'	oFor wHer oPor oRiv oBov
un	***sempervirens*** 'Albo Spica'	oGos oGar oRiv oGre
	sempervirens 'Aptos Blue'	oFor wHer oDar oRiv oWhi oWnC
	sempervirens 'Cantab'	oFor oPor
un	***sempervirens*** 'Dawn'	oWnC
un	***sempervirens*** 'Emily Brown'	oRiv
un	***sempervirens*** 'Filoli'	oFor oRiv
	sempervirens 'Glauca'	oPor wWel
un	***sempervirens*** Kelly's prostrate	oPor
un	***sempervirens*** 'Loma Prieta Spike'	oRiv
un	***sempervirens*** 'Los Altos'	oFor
un	***sempervirens*** 'Los Altos Blue'	oDan
	sempervirens Majestic Beauty™ / 'Monty'	oGar oWnC
	sempervirens 'Nana Pendula'	see **S.** 'Prostrata'
	sempervirens 'Prostrata'	oGos oSis oRiv oGre oWnC oBRG
	sempervirens 'Santa Cruz'	oAls oGar oWnC
un	***sempervirens*** 'Simpson's Silver'	oFor oPor
un	***sempervirens*** 'Skyline'	oFor
	sempervirens 'Soquel'	oFor oAls oGar oSho
un	***sempervirens*** 'Winter Blue'	oFor
un	***sempervirens*** 'Woodside'	oOut wCCr
	SEQUOIADENDRON	
	giganteum	oFor oPor wWoB wBur oDar oEdg oAls wShR oGar oRiv oSho oGre wCCr oTPm wSta wAva wCoN wSte wCoS
un	***giganteum*** 'Aureovariegata'	oPor
	giganteum 'Glaucum'	oFor wHer oGre
	giganteum 'Hazel Smith'	oPor oRiv
un	***giganteum*** 'Moonie Minnie'	oFor oPor oRiv wWel
	giganteum 'Pendulum'	oFor oPor oAls oGar oRiv oBlo oGre oWnC wSta
un	***giganteum*** 'Phillip Curtis'	oPor
	SERENOA	
	repens	iGSc
	SERIPHIDIUM	
	canum	oFor
	novum	oFor
	nutans	wCul oGoo wFai
	tridentatum	oFor oGoo wPla
	SERISSA	
	foetida	see **S. *japonica***
	japonica 'Flore Pleno'	oFor oGre
	japonica 'Kyoto'	oGre
	japonica 'Mount Fuji'	oGre
un	***japonica*** 'Pink Mountain'	oGre
un	***japonica*** 'Pink Princess'	oGre
un	***japonica*** 'Sapporo'	oFor
un	***japonica*** 'Thousand Stars'	oGre
	japonica 'Variegata'	oFor
un	***japonica*** 'White Swan'	oGre
	SERRATULA	
	tinctoria	oCrm
	SESAMUM	
	indicum	iGSc
	SESELI	
	gummiferum	wHig oHed
	SESLERIA	
	autumnalis	oFor wHer wWin oSec wSnq
	caerulea	oFor oOut wWin oWnC oCir wSnq
	heufleriana	oHed wWin wBox oSec
	nitida	last listed 99/00
	SETARIA	
	palmifolia 'Rubra'	wHer
	SETCREASEA	see **TRADESCANTIA**
	SHEPHERDIA	
	argentea	oFor wBCr wShR oOEx wKin wPla wCoN
	canadensis	wThG wShR oOEx wPir wPla
	SHIBATAEA	
	chinensis	oTra oTBG
	kumasasa	oFor oTra wSus wCli oNor wBea oDar oAls oGar oSho oOut oGre wBam wBmG oTBG wWel
	lancifolia	wBea wBmG
	SHORTIA	
	galacifolia	oRus wMtT oBov
	uniflora var. ***orbiculatis*** 'Grandiflora'	oBov
	SIBBALDIA	
	procumbens	oFor oBos
	SIBBALDIOPSIS	
	tridentata	oFor oHed wCCr
	SIBIRAEA	
	altaiensis	see **S. *laevigata***
un	***angustata***	oTrP wCol
	laevigata	oFor
	SIDA	
un	***cordifolia***	oCrm
un	***fallax***	oTrP
	hermaphrodita	last listed 99/00
	SIDALCEA	
	'Brilliant'	oGar wMag
	campestris	oHan
	candida	oTrP
	candida 'Bianca'	oFor oAls oMis wMag
	cusickii	wCCr oAld
	'Elsie Heugh'	oFor oAls oGar oGre wMag wAva
	malviflora	wCSG oSec wAva
un	***malviflora*** 'Palustre'	oHed
	'Mr. Lindbergh'	oAls
	'My Love'	wHer
un	***nelsoniana***	oBos
	neomexicana	oTrP
	oregana	oHan oRus oBos
	'Party Girl'	wTho oAls oMis oSho oBlo oWnC wMag oUps
	purpetta	oWnC
	'Rosanna'	oWnC
	'Rosy Gem'	oEga
	Stark's hybrids	oAls
	'Sussex Beauty'	wHer oAls
	SIDERITIS	
	syriaca	oCrm
	SILENE	
	acaulis 'Frances'	wMtT
un	***acaulis*** 'Pink Pearl'	wMtT
	acaulis 'Tatoosh'	wMtT
	acaulis 'White Rabbit'	wMtT
	asterias	wHig
	caroliniana	wThG
do	***dioica*** 'Flore Pleno'	oSec
va	***dioica*** 'Graham's Delight'	oOut
	dioica 'Jade Valley'	oFor
do	***dioica*** 'Rosea Plena'	oWnC
	dioica 'Variegata'	see **S. *dioica*** 'Graham's Delight'
	fimbriata	oSec
	hookeri Ingramii Group	last listed 99/00
	maritima	see **S. *uniflora***
	mexicana	last listed 99/00
	petersonii	last listed 99/00
	regia	oFor oJoy oUps
	schafta	wTho oAls
un	***schafta*** 'Ralph Haywood'	wHer
	schafta 'Splendens'	oAls oSho

un	***scouleri***	wShR
	uniflora	oFor
	uniflora 'Druett's Variegated'	oFor wHer wWoS oHed
	uniflora 'Flore Pleno'	see **S. *uniflora*** 'Robin Whitebreast'
do	***uniflora*** 'Robin Whitebreast'	wCul oJoy oSec
un	***uniflora*** 'Swan Lake'	wCSG
	vallesia	oJoy
	vulgaris ssp. ***maritima***	see **S. *uniflora***
	vulgaris ssp. ***maritima*** 'Flore Pleno'	see **S. *uniflora*** 'Robin Whitebreast'
un	***yunnanensis*** DJHC 98331	wHer
	zawadskii	oJoy wHom
	SILPHIUM	
	integrifolium	oFor
	laciniatum	oCrm
	perfoliatum	wHer oJoy oGoo wWld
	SILYBUM	
	marianum	wFai iGSc oCrm oUps
	SIMMONDSIA	
	chinensis	oFor oGoo oOEx iGSc
	SINARUNDINARIA	
	nitida	see **FARGESIA *nitida***
	SINNINGIA	
	canescens	oRar
	cardinalis	oRar
	conspicua	oRar
	leucotricha	see **S. *canescens***
	speciosa	oGar
	SINOBAMBUSA	
	intermedia	oTra oNor wBea
	tootsik f. ***albostriata***	wBea
	SINOCALYCANTHUS	
	chinensis	oFor oTrP wHer oGos oEdg wCol oDan oRiv oCis oGre oWnC wBox wSte wWel
	SINOCRASSULA	
	yunnanensis	oSqu
	SINOFRANCHETIA	
	chinensis	wCCr
	SINOJACKIA	
	rehderiana	oFor wHer
	xylocarpa	oFor wHer wClo oAmb oGar oDan oRiv oCis oGre oWnC wSte
	SINOWILSONIA	
	henryi	last listed 99/00
	SISYMBRIUM	
	luteum	oFor
	SISYRINCHIUM	
un	***albiflos***	wCol
	angustifolium	oHan oRus wCCr wWld wSnq
	angustifolium album	oRus oDan
un	***angustifolium*** 'Lucerne'	wSnq
	bellum	see **S. *idahoense*** var. ***bellum***
	bermudianum	see **S. *angustifolium***
	californicum	oFor oTrP oHan oRus moGar oSsd oBos wWat wHom oSec oAld
	convolutum	oTrP wHer oJoy wSta
	douglasii	see **OLSYNIUM *douglasii***
	'E. K. Balls'	wWoS oSis oOut wAva
	elmeri	oRus
	filifolium	see **OLSYNIUM *filifolium***
	idahoense	wWat wTGN wWld oUps
	idahoense 'Album'	oSis oNWe
	idahoense var. ***bellum***	oTrP oHan wTho wNot oRus oBos wCCr wHom wSta oAld oCir
	idahoense var. ***bellum*** dwarf form	oSis
	macounii	see **S. *idahoense***
	macrocarpon	oSis
	'Mrs. Spivey'	wRob
	mucronatum 'Album'	oTrP oNWe
	patagonicum	oGar
	'Pole Star'	wSta
	'Quaint and Queer'	wCul oGar wTGN
	striatum	oFor oTrP wCul oJoy oRus wCSG oGar wCCr wBWP wAva
va	***striatum*** 'Aunt May'	oHed
	SIUM	
	sisarum	oGoo
	SKIMMIA	
	japonica	wTGN
	japonica 'Bowles' Dwarf Female' (f)	oGar oSho
	japonica 'Bowles' Dwarf Male' (m)	oFor oAls oGar oSho
	japonica female	wHer oGar oWnC wSta wCoN
	japonica male	oGar oGre oWnC wSta wCoN
	japonica ssp. ***reevesiana***	oGar oCis wCoS
	japonica 'Rubella' (m)	wHer
	japonica self fertile	wSta wCoN
	japonica white berried	oGre
	reevesiana	see **S. *japonica*** ssp. ***reevesiana***
	SLOANEA	
un	***sinensis***	oCis
	SMILACINA	
	atropurpurea	wHer
	fusca	wHer
	henryi	last listed 99/00
	oleracea	wHer
	racemosa	oFor oHan wThG wCol oRus oGar oBos wFFl oGre wFai wWat oTri oAld oSle wSte
	racemosa var. ***amplexicaulis***	wNot
	racemosa East Coast form	wHer
	racemosa yellow fruited form	wHer
	stellata	oFor oHan wNot wCol oRus oBos wFFl oTri wSte
	SMILAX	
	hispida	oFor
	sp. DJHC 196	wHer
	SMYRNIUM	
	olusatrum	oGoo
	perfoliatum	oNWe
	SOLANUM	
	atropurpureum	oTrP
	aviculare	wHer wCCr
	crispum	wCSG
	crispum 'Glasnevin'	oFor wHer wWoS wCul oGos oGar oCis oSho oGre wCCr wTGN oWhS
	crispum 'Variegatum'	oNWe
	dulcamara	oGoo iGSc
	dulcamara 'Variegatum'	wWoS oGar oGre oUps
	jasminoides	wTGN wWel
	jasminoides 'Album'	wCul oGre oWhS oUps
un	***jasminoides*** 'Grandiflorum'	oGar
qu	***jasminoides*** 'Variegatum'	oFor oGar
	laciniatum	wSte
	muricatum	oFor oOEx
un	***nelsoni***	oTrP
un	***paraguayensis***	oTrP
	pseudocapsicum	oGar
	quitoense (F)	oTrP
	rantonnetii	wWel
	'Royal Robe'	oGar
un	***tuberosum*** 'All Blue'	wIri
un	***tuberosum*** 'All Red'	wIri
un	***tuberosum*** 'Anoka'	wIri
un	***tuberosum*** 'Arran's Pilot'	wIri
un	***tuberosum*** 'Atlantic'	wIri
un	***tuberosum*** 'Augsburg Gold'	wIri
un	***tuberosum*** 'Austrian Crescent'	wIri
un	***tuberosum*** 'Bintje'	wIri
un	***tuberosum*** 'Bison'	wIri
un	***tuberosum*** 'Bliss Triumph'	wIri
un	***tuberosum*** 'Blue Mac'	wIri
un	***tuberosum*** 'Blue Pride'	wIri
un	***tuberosum*** 'Blue Victor'	wIri
un	***tuberosum*** 'Brigus'	wIri
un	***tuberosum*** 'Buffalo Red Ruby'	wIri
un	***tuberosum*** 'Butte'	wIri
un	***tuberosum*** 'Butterfinger'	wIri
un	***tuberosum*** 'Candy Stripe'	wIri
un	***tuberosum*** 'Caribe'	wIri
un	***tuberosum*** 'Caribe Sport'	wIri
un	***tuberosum*** 'Carola'	wIri
un	***tuberosum*** 'Catriona'	wIri
un	***tuberosum*** 'Cherokee'	wIri
un	***tuberosum*** 'Cherry Red'	wIri
un	***tuberosum*** 'Chieftain'	wIri
un	***tuberosum*** 'Cowhorn'	wIri
un	***tuberosum*** 'Daisy Gold'	wIri
un	***tuberosum*** 'Dark-red Norland'	wIri
un	***tuberosum*** 'Dazoc'	wIri
un	***tuberosum*** 'Desiree'	wIri
un	***tuberosum*** 'Early Ohio'	wIri
un	***tuberosum*** 'Elba'	wIri
un	***tuberosum*** 'Epicure'	wIri
un	***tuberosum*** 'French Fingerling'	wIri
un	***tuberosum*** 'Frontier'	wIri
un	***tuberosum*** 'German Butterball'	wIri
un	***tuberosum*** 'Gold Nugget'	wIri
un	***tuberosum*** 'Green Mountain'	wIri
un	***tuberosum*** 'Huckleberry'	wIri
un	***tuberosum*** 'Indian Pit'	wIri
un	***tuberosum*** 'Irish Cobbler'	wIri

un *tuberosum* 'Katahdin' wIri
un *tuberosum* 'Kennebec' wIri
un *tuberosum* 'Kerr's Pink' wIri
un *tuberosum* 'Krantz' wIri
un *tuberosum* 'Lemhi Russet' wIri
un *tuberosum* 'Levitt's Pink' wIri
un *tuberosum* 'Maris Piper' wIri
un *tuberosum* 'Morning Gold' wIri
un *tuberosum* 'Nooksack' wIri
un *tuberosum* 'Norgold M' wIri
un *tuberosum* 'Ozette' wIri
un *tuberosum* 'Pike' wIri
un *tuberosum* 'Pinto' wIri
un *tuberosum* 'Princesse™ La Ratte' wIri
un *tuberosum* 'Purple Chief' wIri
un *tuberosum* 'Purple Peruvian' wIri
un *tuberosum* 'Purple #5' wIri
un *tuberosum* 'Penta' wIri
un *tuberosum* 'Quaggy Joe' wIri
un *tuberosum* 'Ranger Russet' wIri
un *tuberosum* 'Red Cloud' wIri
un *tuberosum* 'Red Dale' wIri
un *tuberosum* 'Red Gold' wIri
un *tuberosum* 'Red Lasoda' wIri
un *tuberosum* 'Red Norland' wIri
un *tuberosum* 'Red Pontiac' wIri
un *tuberosum* 'Red Thumb' wIri
un *tuberosum* 'Rhine Gold' wIri
un *tuberosum* 'Rose Finn Apple' wIri
un *tuberosum* 'Rose Gold' wIri
un *tuberosum* 'Rote Erstling' wIri
un *tuberosum* 'Russet Burbank' wIri
un *tuberosum* 'Russet Norkotah' wIri
un *tuberosum* 'Russian Banana' wIri
un *tuberosum* 'Sangre' wIri
un *tuberosum* 'Sante' wIri
un *tuberosum* 'Shepody' wIri
un *tuberosum* 'Snowdon' wIri
un *tuberosum* 'Superior' wIri
un *tuberosum* 'Viking Purple' wIri
un *tuberosum* 'Viking Red' wIri
un *tuberosum* 'Warba Pink Eye' wIri
un *tuberosum* 'White Rose' wIri
un *tuberosum* 'Yellow Finn' wIri
un *xanthocarpum* oCrm

SOLDANELLA

alpina wWoS oRus wPle oSis oBov
carpatica oBov
cyanaster wPle
hungarica oSis oBov
un *hungarica* 'Major' wMtT oBov
x lungoviensis wMtT oBov
montana wHer oRus oSis oBov wPir
sp. oNWe
villosa oSis oBov

SOLEIROLIA

soleirolii oAls wHom oWnC wTGN wCoN
soleirolii 'Argentea' see **S. *soleirolii*** 'Variegata'
soleirolii 'Aurea' wHom
soleirolii 'Variegata' last listed 99/00

SOLENOPSIS

axillaris see **LAURENTIA *axillaris***

SOLENOSTEMON

un 'Big Red' wHer
un 'Black Magic' wHer
un 'Dipt in Wine' wHer
'El Brighto' wHer
'India Frills' wHer
va 'Inky Fingers' wHer
un 'Lime Queen' wHer
un 'Mardi Gras' wHer
un 'Texas Parking Lot' wHer
'The Line' wHer
un 'Touchelat' wHer
un 'Ulrich' wHer
unknown #1 wHer
unknown #4 wHer
un 'Victorian Ruffle' wHer
un 'Violet Ruffles' wHer
un 'Violet Tricolor' wHer
un 'Yellow in Sun' wHer

SOLIDAGO

un 'Angel Wings' wCul
bicolor oHan
caesia oGoo
canadensis oHan wNot wShR oGoo oSho wFFl iGSc oCrm oAld wWld
'Crown of Rays' oFor oAls oSis
flexicaulis 'Variegata' wHer
'Gold Spangles' oFor wHer wWoS wCol
Golden Baby see **S.** 'Goldkind'
'Golden Dwarf' oAls
'Goldkind' oFor oHan oJoy oSho oEga
un 'Judy's Giant' wCul
'Laurin' oAls
'Lemore' see **X SOLIDASTER *luteus*** 'Lemore'
un 'Lightning Rod' wWoS
multiradiata last listed 99/00
nemoralis oGoo
un *occidentalis* wNot
odora oHan oGoo
rigida last listed 99/00
rugosa 'Fireworks' oFor oGoo wSnq
sempervirens oFor oGoo
un *spathulata* var. *decumbens* wFFl
spathulata var. *nana* oFor wMtT
sphacelata 'Golden Fleece' oFor wHer oAls wSnq
Strahlenkrone see **S.** 'Crown of Rays'
un *tenuifolium* oCis

X SOLIDASTER

luteus oFor wWoS wCul
luteus 'Lemore' oFor oWnC oSle wSnq

SOLLYA

heterophylla oFor wGra
heterophylla 'Alba' oTrP

SOPHORA

davidii oRiv
flavescens last listed 99/00
japonica oFor oRiv wSte
japonica 'Regent' oGar
macrocarpa last listed 99/00
microphylla wHer
secundiflora last listed 99/00
un *secundiflora* 'Guadalupe Blue' seed grown wHer
sp. Easter Island wHer
tetraptera oFor

SORBARIA

aitchisonii see **S. *tomentosa*** var. ***angustifolia***
arborea see **S. *kirilowii***
kirilowii oFor oTrP
kirilowii DJHC 601 wHer
sorbifolia oFor oGar wKin
sorbifolia var. *stellipila* HC 970049 wHer
sp. aff. *tomentosa* var. *angustifolia* EDHCH 97257 wHer

SORBUS

alnifolia oFor wBCr wKin oSle
americana oFor
aria oFor wCCr
aria 'Lutescens' oFor
aria 'Magnifica' oGre
aria 'Majestica' wCSG
x arnoldiana oFor
aucuparia oFor wBCr oGar wKin wTGN wCoS
aucuparia 'Aspleniifolia' oFor
aucuparia 'Cardinal Royal' oGar
aucuparia 'Edulis' (F) oFor
aucuparia 'Pendula' oFor
aucuparia 'Rabina' wBur oOEx
austrica oFor
caloneura oFor
cashmiriana oFor oRiv
commixta oFor
cuspidata see **S. *vestita***
decora oFor wHer
un *degenii* wCCr
discolor (see Nomenclature Notes) oFor
domestica oFor wCCr
forrestii oFor wCCr
fruticosa last listed 99/00
glabrescens last listed 99/00
glabrescens EDHCH 97244 wHer
Harry Smith 12799 see **S.** sp. aff. ***koehneana*** Schneider Harry Smith 12799
helenae last listed 99/00
un *henryi* EDHCH 97191 wHer
hupehensis oFor oRiv wCCr
hupehensis 'Coral Fire' last listed 99/00
hupehensis DJHC 360 wHer
hupehensis DJHC 503 wHer

	hupehensis 'Pink Pagoda'	wCul oEdg oDan oWhi oWnC
	hybrida (see Nomenclature Notes)	oFor wKin
	intermedia 'Brouwers'	oGar
un	Ivan's Beauty™ (fruit)	wClo oOEx
un	Ivan's Belle™ (fruit)	wRai oOEx oGre
	'Joseph Rock'	wTGN
un	***khumbuensis***	wHer
	koehneana EDHCH 97071	wHer
	sp. aff. ***koehneana*** Schneider Harry Smith 12799	last listed 99/00
	latifolia	oFor wCCr
	meliosmifolia EDHCH 97346	wHer
	mougeotii	last listed 99/00
	pohuashanensis (see Nomenclature Notes)	oFor oOEx
	prattii	oFor wHer wCCr
	reducta	wHer oGar oSho
	reducta DJHC 98085	wHer
	rehderiana	oFor
un	***rufoferruginea*** 'Longwood Sunset'	wCul
	sargentiana EDHCH 97149	wHer
	scopulina (see Nomenclature Notes)	oFor oWoB wShR wPla
	setschwanensis DJHC 98407	wHer
	'Shipova'	see **PYRUS** 'Shipova'
	sitchensis (see Nomenclature Notes)	oWoB wBur wFFl wWat
	x thuringiaca 'Fastigiata'	oFor
	tianshanica	oFor
	torminalis	oFor wCSG
	vestita	wHer
	vilmorinii	oRiv wTGN
	wilsoniana	oFor
	sp. aff. ***wilsoniana*** EDHCH 97345	wHer
	SORGHASTRUM	
	avenaceum	oGoo wRob oTri
	avenaceum 'Sioux Blue'	oFor oOut wWin wSnq
	nutans	see **S. *avenaceum***
	SPARGANIUM	
	emersum	wWat oTri
	erectum	wSoo
	ramosum	see **S. *erectum***
	sp.	wGAc
	SPARTINA	
	pectinata 'Aureomarginata'	oFor
	SPARTIUM	
	junceum	oFor wCri oGar
	SPATHIPHYLLUM	
un	'Emerald Swirl'	oGar
	patinii	oGar
un	'Sensation'	oGar
	SPHAERALCEA	
	fendleri venusta	oGar
	munroana	oFor
	'Newleaze Coral'	oHed
	'Newleaze Pink'	oHed
	rivularis	oMis wBWP oAld wWld
	SPHAEROPTERIS	see **CYATHEA**
un	**SPIERANTHA**	
	convallarioides	wCol
	SPILANTHES	
	oleracea	wNTP
	SPIRAEA	
	alba	oFor
	albiflora	see **S. *japonica*** var. ***albiflora***
	'Arguta'	wHer oBlo
	x arguta 'Bridal Wreath'	see **S.** 'Arguta'
	x arguta 'Compacta'	see **S. *x cinerea***
	betulifolia var. ***lucida***	oFor wNot oBos
un	***betulifolia*** 'Tor'	oGos
	x billardii	oFor
qu	'Bridal Veil'	oBlo
	bullata	see **S. *japonica*** 'Bullata'
	x bumalda (see Nomenclature Notes)	
	cantoniensis	oFor
	x cinerea	oFor oBlo
	x cinerea 'Grefsheim'	oFor oHed
	densiflora	oFor oBos
	densiflora Summer Song® / 'Monvis'	oGre
un	***densiflora*** 'Trinity Rose'	oGos
	dolchica	see **S. *japonica*** 'Dolchica'
	douglasii	oHan wWoB wNot wShR oGar oBos wFFl wWat oTri wPla oAld oSle wSnq
un	***douglasii douglasii***	oFor
	douglasii ssp. ***menziesii***	oFor oGar
	fritschiana	last listed 99/00
	henryi	last listed 99/00
	japonica var. ***albiflora***	oFor oGar wKin
	japonica 'Alpina'	see **S. *japonica*** 'Nana'
va	***japonica*** 'Anthony Waterer'	oFor oDar oJoy oAls wCSG oGar oBlo oWnC wBox wSta wAva wWel
	japonica 'Bullata'	oFor wHer wCol oNWe
	japonica 'Bumalda'	oAls wCSG wCoS
	japonica 'Candle Light'	oFor wWoS
	japonica 'Coccinea'	oFor wKin
	japonica 'Crispa'	oFor oGar wKin
un	***japonica*** 'Dakota Gold Charm'	oGre
	japonica 'Dart's Red'	oJoy oAls oSev
qu	***japonica*** 'Dolchica'	wCSG oGar oGre oWnC wWel
	japonica 'Fire Light'	oFor wWoS oEdg
	japonica var. ***fortunei***	oJoy
	japonica 'Froebelii'	oFor wKin oWnC
	japonica 'Gold Mound'	oFor oJoy oAls oGar oSqu oGre wKin oBRG oCir oGue wWel
	japonica 'Goldflame'	oFor oDar oJoy oAls oMis oGar oDan oSho oBlo oGre wKin oWnC wMag wBox wSta wAva oUps wWel
un	***japonica*** 'Gumball'	oFor oGar
	japonica Limemound® / 'Monhub'	wCri oJoy oAls oGar oBlo oGre oWnC oSec wAva wWel
	japonica 'Little Princess'	oFor wWoS oJoy oAls wCSG oGar oBlo oGre wKin oWnC wMag wAva
	japonica Magic Carpet / 'Walbuma'	oFor wHer oNat oGar oSis oGre wFai wTGN oUps wSnq
un	***japonica*** 'Martyann'	oGre
	japonica 'Nana'	oFor oJoy oAls oGar oGre oWnC wWel
un	***japonica*** 'Norman'	oFor oGar
un	***japonica*** 'Pink Princess'	oGue
	japonica seedlings	wAva
	japonica 'Shiburi'	last listed 99/00
	japonica 'Shirobana'	oFor oJoy oAls wCSG oGar oBlo oGre oWnC wSta
un	'Jennifer Jean'	oBlo
	lucida	see **S. *betulifolia*** var. ***lucida***
un	'Neon Flash'	oFor wWoS oAls oGar oOut oGre oBRG oUps
	nipponica 'Halward's Silver'	oGre
	nipponica 'Snowmound'	oFor wHer oDar oJoy oAls oHed oGar oBlo oGre wKin wWel
	nipponica var. ***tosaensis*** hort.	see **S. *nipponica*** 'Snowmound'
do	***prunifolia***	oFor wHer wCSG oSho oGre wKin wTGN
	prunifolia 'Flore Plena'	see **S. *prunifolia***
	prunifolia 'Plena'	see **S. *prunifolia***
	x pyramidata	wSnq
	reevesiana	see **S. *cantoniensis***
	sp. DJHC 192	wHer
	thunbergii	oFor oGar
un	***thunbergii*** 'Fujino Pink'	oGar oGre
	thunbergii 'Mellow Yellow'	oFor
	thunbergii 'Mount Fuji'	wHer oCis
un	***thunbergii*** 'Ogon'	oAls oGar
	tomentosa	oFor
	trichocarpa 'Snow White'	oFor
	trilobata 'Fairy Queen'	oFor
	trilobata 'Swan Lake'	oGar
	x vanhouttei	oFor oJoy oGar oBlo oGre
va	***x vanhouttei*** Pink Ice	oFor oGre
	wilsonii	last listed 99/00
	SPIRODELA	
	polyrhiza	wGAc
	SPODIOPOGON	
	sibiricus	oJoy oGar oOut oGre wWin wSnq
	SPOROBOLUS	
	airoides	oFor
	heterolepis	oFor wBox
	SPRAGUEA	
	umbellata	wMtT
	STACHYS	
	affinis	oGoo oOEx
	albotomentosa	wHer wWoS oHed oMis oGoo oNWe wSte
	albotomentosa 'Hidalgo'	see **S. *albotomentosa***
	byzantina	oFor oNat oAls oRus wCSG oGoo iGSc wMag oSec wNTP oCir oUps oSle
	byzantina 'Big Ears'	oFor wWoS oJoy wCSG oGar oOut wFai oWnC wTGN wBox oSec oCir oLSG wSte wSnq
	byzantina 'Cotton Boll'	oFor wHer wWoS oAls oSec oSle
	byzantina 'Countess Helen von Stein'	see **S. *byzantina*** 'Big Ears'
un	***byzantina*** 'Countess von Zeppelin'	see **S. *byzantina*** 'Big Ears'
	byzantina 'Helene von Stein'	see **S. *byzantina*** 'Big Ears'
	byzantina 'Limelight'	oHed

	byzantina 'Primrose Heron'	oFor wHer wWoS wCul wCol oAls wFGN oOut wRob oGre wCCr oWnC wSte
qu	*byzantina* 'Purpurea'	oAls
	byzantina 'Silver Carpet'	oFor oRus wFGN oGar oGoo oGre oWnC oSec oLSG
va	*byzantina* 'Striped Phantom'	last listed 99/00
	candida	last listed 99/00
	chrysantha	oHed
	coccinea	oGoo
	coccinea 'El Salto'	last listed 99/00
ch	*coccinea* 'Hidalgo'	see **S. *albotomentosa***
un	*cooleyae*	wNot oBos wFFl
	cretica	oFor
	densiflora	see **S. *monieri***
	discolor	last listed 99/00
	grandiflora	see **S *macrantha***
	'Hidalgo'	see **S. *albotomentosa***
	lanata	see **S. *byzantina***
	macrantha	oFor wCul oGoo
	macrantha 'Hummelo'	wHer
un	*macrantha* 'Purpurea'	wWoS oAls
	macrantha 'Robusta'	last listed 99/00
	macrantha 'Rosea'	wHig oSec
	macrantha 'Superba'	wHig wAva
	monieri	wHer oJoy wHig oHed oSis
un	*monieri* 'Alba'	oSis oSle
	nivea	see **S. *discolor***
	officinalis	oGoo oOut oOEx wFai iGSc oCrm oBar oLSG
	palustris	last listed 99/00
	sieboldii	see **S. *affinis***
	spicata	see **S. *macrantha***
	sylvatica	oGoo iGSc
	STACHYURUS	
	sp. aff. ***chinensis*** EDHCH 97230	wHer
	himalaicus	oCis
va	'Magpie'	oCis
	praecox	oFor wCul oGos oDan oRiv oWhi oGre wPir wTGN wCoN wSte
un	*praecox* 'Aureomarginata'	wHer
	praecox 'Rubriflora'	wHer
un	*salicifolia*	wHer
un	*schetuanica*	wHer
	sp. EDHCH 97305	wHer
un	*szechuanensis*	wHer
	yunnanensis	wHer oCis
	STANLEYA	
	pinnata	oFor
	STAPELIA	
	desmetiana	oRar
	flavirostris	oRar
	gettliffei	oRar wGra
	gigantea	wGra
	hirsuta	wGra
	kwebensis	oRar wGra
	leendertziae	oRar
	longii	see **TRIDENTEA *longii***
	STAPELIANTHUS	
	decaryi	oRar
	STAPHYLEA	
	bumalda	oFor
	bumalda HC 970075	wHer
	colchica	oHed wCCr
	pinnata	oFor wHer
	trifolia	oFor
	STAUNTONIA	
	hexaphylla	oFor oCis oOEx
	hexaphylla HC 970600	wHer
	purpurea B&SWJ 3690	wHer
un	**STEGANOTAENIA**	
	araliaceae	oRar
	STELLARIA	
	media	wFai
	STENANDRIUM	
un	*dulce*	oNWe
	STENOCACTUS	
un	'San Felipe'	oGar
	STENOCARPUS	
	sinuatus	oTrP
	STENOCEREUS	
	eruca SB 1239	oRar
	STEPHANANDRA	
	incisa	wRob
	incisa 'Crispa'	oFor wHer wCul oRiv wCCr
un	*incisa* var. *incisa* DJH 301	wHer

	tanakae	wHer oHed oSho
	STEPHANOTIS	
	floribunda	oGar
	STERCULIA	
un	*africana*	oRar
un	*oldhamii*	oCis
un	*quinqueloba*	oRar
	STERNBERGIA	
	lutea	oSis
	STEVIA	
	rebaudiana	wWoS oAls wRai oOEx iGSc oWnC oCrm oBar oUps
	STEWARTIA	
un	'Ballerina'	wAva
	koreana	see **S. *pseudocamellia*** Koreana Group
	malacodendron	oRiv
	monadelpha	wHer oGar oRiv oBlo oBov oGre wCCr wSta wSte wWel
	ovata	wCCr
	pseudocamellia	wHer oGos wClo oDar oAls oAmb oGar oDan oRiv oSho oBlo oBov oGre wCCr oWnC wTGN wSta wAva wRav wSte wCoS wWel
	pseudocamellia hybrid	oGos
	pseudocamellia var. ***koreana***	see **S. *pseudocamellia*** Koreana Group
	pseudocamellia Koreana Group	oGar wSta wWel
	pseudocamellia Koreana Group HC 970398	wHer
	rostrata	oFor wHer
	serrata	oGre
	sinensis	oFor wHer oDan oRiv wAva wSte
	STIPA	
	arundinacea	see **CALAMAGROSTIS *arundinacea***
	barbata	wWin
	calamagrostis	wHig
	capillata	oFor wCli wWin
	elegantissima	last listed 99/00
	extremiorientalis	last listed 99/00
	gigantea	oFor wHer wWoS wCli oDar oJoy wHig oAls oHed oAmb oGar oDan oNWe oSho oOut oGre wWin oWnC oTri wTGN wSnq
	grandis	last listed 99/00
	pennata	wCCr
	robusta	last listed 99/00
	tenacissima	oFor
	tenuissima	wHer wWoS wCli wSwC oDar oJoy wCol wHig oAls oAmb oGar oNWe oSho oOut wRob wBWP oWhS wSnq
	tenuissima 'Pony Tails'	oAls
	tirsa	oCis
	STOKESIA	
	laevis	wCoS
	laevis 'Alba'	oFor wCSG
	laevis blue	wCSG oMis
	laevis 'Blue Danube'	oFor oAmb
un	*laevis* 'Klaus Jellito'	oFor oJoy oAls oGar oOut oBlo wTGN oEga
	laevis 'Mary Gregory'	oFor wSnq
	laevis 'Omega Skyrocket'	oFor wWoS oSis
	laevis 'Purple Parasols'	oFor wWoS oAls wCSG oGar oOut oEga
	laevis 'Silver Moon'	oFor
un	'Mischung'	oAls
	STRANVAESIA	see **PHOTINIA**
	STRELITZIA	
	nicolai	oGar
	reginae	wCoS
	STREPTOCARPUS	
	x hybridus (see Nomenclature Notes)	oGar
	STREPTOPUS	
	amplexifolius	wHer oBos
	sp.	wFFl
	STROBILANTHES	
	atropurpurea	wCul
	STYLOPHORUM	
	diphyllum	oFor wHer wCul wThG oRus oGar oNWe wFai
	STYRAX	
	americanus	oRiv
	dasyanthum	wHer
	grandifolius	oRiv
	hemsleyanus	wHer oGre
	japonicus	oFor wHer wCri oGos wClo oAls oGar oDan oRiv oSho oWhi oBlo oGre oWnC wTGN wSta wAva wCoN oGue wSte wCoS wWel
	japonicus Benibana Group	oGre wCoN

	japonicus (Benibana Group) 'Pink Chimes'	oGos wClo oDar oAls oGar oRiv oBlo oGre wSta wAva wWel
	japonicus 'Carillon'	oGos wTGN wSta wWel
un	***japonicus*** 'Emerald Pagoda'	wHer oGos wWel
	japonicus HC 970269	wHer
	japonicus 'Roseus'	see **S. *japonicus*** Benibana Group
	obassia	oFor wHer oGos oAls oAmb oGar oDan oRiv oWhi oBlo oGre wSta wAva wCoN wSte wWel
	odoratissimus	oGar oDan oGre wPir
	officinale	oAls
	officinale var. ***californicum***	oFor wHer
	shiraianum	last listed 99/00
	wilsonii	oDan
	SUCCISA	
	pratensis	oGoo wEde
	pratensis dwarf form	oSis
	SYCOPSIS	
	sinensis	oGos oCis oSho wCCr
un	***sinensis*** 'Variegata'	oCis
	SYMPHORICARPOS	
	albus	oFor wCul wWoB wBur oAls wShR oGar oBos oRiv oSho oGre wWat oTri wPla oAld wSte wSnq wWel
	albus 'Aureovariegatus'	see **S. *albus*** 'Taff's Variegated'
	albus var. ***laevigatus***	wNot wFFl
va	***albus*** 'Taff's Variegated'	wHer
va	***albus*** 'Taff's White'	oFor wHer wCol oGre wFai wWel
	albus 'Variegatus'	see **S. *albus*** 'Taff's White'
	x chenaultii	wCCr
	x chenaultii 'Hancock'	oFor oGar oGre wCCr
	x doorenbosii 'Magic Berry'	oFor wWoS oGos oDan wTGN
	x doorenbosii 'Mother of Pearl'	oFor
	x doorenbosii 'White Hedge'	oFor oGre
	mollis	oFor oHan oBos wFFl wWat oTri oAld
	occidentalis	oFor oSle
	orbiculatus	oFor wCul oGar oGre wCCr oWnC wWel
	orbiculatus 'Aureovariegatus'	see **S. *orbiculatus*** 'Foliis Variegatis'
	orbiculatus 'Foliis Variegatis'	oFor oGar oDan oNWe wRob wFai wPir wCCr wWel
	orbiculatus 'Variegatus'	see **S. *orbiculatus*** 'Foliis Variegatis'
	oreophilus	oAld
un	***vaccinoides***	oFor
	SYMPHYANDRA	
	hofmannii	last listed 99/00
	ossetica	oAls
	pendula	oFor
	wanneri	wTho
	SYMPHYTUM	
	azureum	oFor
	'Belsay'	wCol
	caucasicum	last listed 99/00
va	'Goldsmith'	wHer oGos wHig oAls oGar oOut wRob oGre wBox
	grandiflorum	see **S. *ibericum***
un	'Hidcote Beauty'	wWoS
	'Hidcote Blue'	oFor wHer oNat oAls oWnC wTGN
	ibericum	oFor wHer wCul oAls iGSc oLSG oSle
	ibericum 'Jubilee'	see **S.** 'Goldsmith'
	ibericum 'Variegatum'	see **S.** 'Goldsmith'
	ibericum 'Wisley Blue'	oFor
	officinale	oFor oAls wFGN oGoo iGSc oBar
	officinale dwarf	wHom
	'Rubrum'	oFor
	x uplandicum	wFai oBar
va	***x uplandicum*** 'Axminster Gold'	wHer
	x uplandicum 'Denford Variegated'	last listed 99/00
	x uplandicum 'Variegatum'	oNWe oOut wFai
	SYMPLOCOS	
un	***chinensis***	wCCr
un	***chinensis*** f. ***pilosa*** DJH 068	wHer
	sp. B&SWJ 3115	wHer
	SYNADENIUM	
	compactum var. ***rubrum***	oRar
	cupulare	oRar
	sp.	oRar
un	**X SYNCOPARROTIA**	
	semidecidua	oFor oGos wCCr
	semidecidua 'Variegata'	oCis
	SYNEILESIS	
	aconitifolia	wCol
	palmata HC 970627	wHer
	SYNTHYRIS	
	missurica	oSis
	pinnatifida laciniata	last listed 99/00
	reniformis	wNot oRus oSis oBos oTri oAld
	stellata	wCol oRus oSis
	stellata 'Alba'	oSis
	SYRINGA	
	afghanica	see **S. *protolaciniata***
	amurensis	see **S. *reticulata*** ssp. ***amurensis***
	x chinensis	oWnC wMag wWel
	x chinensis 'Alba'	oFor
	x chinensis 'Saugeana'	wMag
	dilitata	see **S. *oblata*** var. ***dilatata***
	x hyacinthiflora 'Blanche Sweet'	oFor oFra
	x hyacinthiflora 'Clarke's Giant'	oFor
	x hyacinthiflora 'Esther Staley'	last listed 99/00
	x hyacinthiflora 'Excel'	oFor
	x hyacinthiflora 'Maiden's Blush'	oFor oFra oWnC
	x hyacinthiflora 'Mount Baker'	wKin
	x hyacinthiflora 'Pocahontas'	oFor oFra oGar oGre wKin oWnC wMag
un	***x hyacinthiflora*** 'Royal Purple'	oFor oFra
	'Josee'	oFor
	x josiflexa 'Royalty'	oFor
	julianae	see **S. *pubescens*** ssp. ***julianae***
	komarovii EDHCH 97214	wHer
	komarovii ssp. ***reflexa***	wTGN
	x laciniata	oFor oFra wCul oGos oGar oDan oBlo oGre oGar oWhi oBlo
	meyeri	
	meyeri var. ***spontanea*** 'Palibin'	oFor wCul oAls oAmb oGar oSis oBov oSqu oGre wKin oWnC wSta oSle wWel
	microphylla	see **S. *pubescens*** ssp. ***microphylla***
	'Minuet'	oFra oAls oGar oGre
	'Miss Canada'	oFor wWhG wWel
	oblata	oFor oGre wCCr
	oblata var. ***dilatata*** DJH 103	wHer
	palibiniana	see **S. *meyeri*** var ***spontanea*** 'Palibin'
	patula hort.	see **S. *meyeri*** var ***spontanea*** 'Palibin'
	patula Nakai	see **S. *pubescens*** ssp. ***patula***
	pekinensis	see **S. *reticulata*** ssp. ***pekinensis***
	x persica	oWnC
	x prestoniae	wKin
un	***x prestoniae*** 'Alexander's Pink'	oWnC
	x prestoniae 'Donald Wyman'	oAls oGar oGre wKin oBRG wWel
	x prestoniae 'James MacFarlane'	oFor oAls oGar oBRG oUps
	protolaciniata	oTrP
	pubescens ssp. ***julianae***	last listed 99/00
	pubescens ssp. ***microphylla***	wCCr
	pubescens ssp. ***microphylla*** 'Superba'	oFor oGar oDan oWnC
	pubescens ssp. ***patula*** 'Miss Kim'	oFor oFra wCul oGos oAls oGar oSho oGre wKin oWnC wTGN oThr oGue wWhG oUps oSle wCoS wWel
	reflexa	see **S. *komarovii*** ssp. ***reflexa***
	reticulata	oRiv oWhi oBlo oGre wCCr wKin wTGN
	reticulata ssp. ***amurensis***	oFor wBCr
	sp. aff. ***reticulata*** ssp. ***amurensis*** DJH 331	wHer
	reticulata 'Ivory Silk'	oFor oGar oGre oWnC wMag oSle
ch	***reticulata*** var. ***mandschurica***	see **S. *reticulata*** ssp. ***amurensis***
	reticulata ssp. ***pekinensis***	oFor wBCr
un	***reticulata*** 'Snowcap'	wCul
un	***reticulata*** 'Summer Snow'	oFor
	rothomagenesis	see **S. *x chinensis***
	sweginzowii	last listed 99/00
	villosa	wBCr oGar wKin
	vulgaris	oFor wBCr oAls wShR wKin wSta wPla oGue
do	***vulgaris*** 'Adelaide Dunbar'	oFor oGar oWnC wWel
	vulgaris 'Agincourt Beauty'	oFra oWnC
	vulgaris var. ***alba***	oFor wSta
	vulgaris 'Albert F. Holden'	oFor oFra oGar
un	***vulgaris*** 'Alice Christenson'	oSho oGre oWnC wMag
	vulgaris 'Alphonse Lavallee'	oFor
un	***vulgaris*** 'Anabel'	oFra
	vulgaris 'Andenken an Ludwig Spaeth'	oFor oFra oAls oGar oGre oWnC oUps oSle wCoS
	vulgaris 'Angel White' (Descanso Hybrid)	oFor oAls oGar oWnC
un	***vulgaris*** 'Atheline Wilbur'	oFra oWnC
un	***vulgaris*** 'Aucubifolia'	oFor
	vulgaris 'Avalanche'	oFra oWnC
	vulgaris Beauty of Moscow	see **S. *vulgaris*** 'Krasavitsa Moskvy'
do	***vulgaris*** 'Belle de Nancy'	oFra oWnC wMag oUps
	vulgaris Blue Skies™ / 'Monore'	oWnC
un	***vulgaris*** 'Bridal Memories'	oFor oWnC wWhG
un	***vulgaris*** 'Burgundy Queen'	oFor
un	***vulgaris*** 'California Rose' (Descanso Hybrid)	oFor oGre wMag
do	***vulgaris*** 'Charles Joly'	oFor oFra wCSG oGar oSho wKin oWnC oGue oUps wCoS
	vulgaris 'Charles X'	oWnC wWel

	vulgaris 'Charm'	oGre oWnC
	vulgaris 'Congo'	oFra wMag wWhG
un	***vulgaris*** 'Crystale'	wMag
un	***vulgaris*** 'Dark Knight'	oFor oWnC wMag
un	***vulgaris*** 'Deepest Purple'	wCul
	vulgaris double lavender	oGre
	vulgaris 'Edith Cavell'	oUps
do	***vulgaris*** 'Edward J. Gardner'	oFor wCSG oGre wWhG
	vulgaris 'Ellen Willmott'	see **S. *vulgaris*** 'Miss Ellen Willmott'
un	***vulgaris*** 'F. K. Smith'	oWnC
un	***vulgaris*** 'Father John Fiala'	oFor oFra
un	***vulgaris*** 'Frank Klager'	oFra wMag
	vulgaris 'Glory'	oFor oFra oEdg oGre oWnC
do	***vulgaris*** 'Katherine Havemeyer'	oFra oGar oSho oGre oWnC wMag wWhG wWel
un	***vulgaris*** 'Klager's'	oWnC
un	***vulgaris*** 'Klager's Dark'	wTGN
un	***vulgaris*** 'Klager's Double Purple'	oGar
	vulgaris 'Krasavitsa Moskvy'	oFor oFra oGos wCSG oGar oGre oWnC oUps wWel
	vulgaris 'Lavender Lady' (Descanso Hybrid)	oFor oFra oGre oWnC
un	***vulgaris*** 'Leon Gambetta'	oWnC
un	***vulgaris*** 'Letha House'	oFra
un	***vulgaris*** 'Little Boy Blue'	oFor oFra
	vulgaris 'Ludwig Spaeth'	see **S. *vulgaris*** 'Andenken an Ludwig Spaeth'
do	***vulgaris*** 'Madame Lemoine'	oFor oFra wSwC oGar oGre oWnC wSta oGue wWhG wWel
un	***vulgaris*** 'Marie Finon'	oFor wMag
un	***vulgaris*** 'Marie Frances'	oFor oFra oWnC
do	***vulgaris*** 'Michel Buchner'	oGar oSho oWnC wMag
do	***vulgaris*** 'Miss Ellen Willmott'	oGar oSho oGre oWnC wMag
	vulgaris 'Monge'	oFra oGar oGre oWnC wWhG
	vulgaris 'Monique Lemoine'	oFra wMag
	vulgaris 'Mont Blanc'	oWnC
	vulgaris 'Montaigne'	oFra oWnC
do	***vulgaris*** 'Mrs. Edward Harding'	oFra oGar wKin
un	***vulgaris*** 'Mrs. Klager'	oFra oAls
un	***vulgaris*** 'My Favorite'	oGar oSho oWnC wMag
un	***vulgaris*** 'Nadezhda'	oFor oFra oWnC
	vulgaris 'Night'	oFor oFra
do	***vulgaris*** 'Paul Thirion'	oFra oSho oWnC
un	***vulgaris*** 'Peacock'	oSho
un	***vulgaris*** 'Pearl Martin'	oSho
un	***vulgaris*** 'Pink Elizabeth'	oFra oGre oWnC wMag
do	***vulgaris*** 'President Grevy'	oFor oFra oAls oGar oSho oGre wKin oWnC wMag wSta oGue wWhG oUps oSle wWel
	vulgaris 'President Lincoln'	oGar oSho wKin oWnC wWel
	vulgaris 'President Poincaire'	oFor oFra oGar oGre wWel
	vulgaris 'Primrose'	oFor oGos oGar wMag wWhG wWel
	vulgaris purple	oGar
un	***vulgaris*** 'R & B Mills'	wMag
	vulgaris 'Rochester'	oFor oFra oGar
un	***vulgaris*** 'Sarah Sands'	oWnC
	vulgaris 'Sensation'	oFor oFra oGos oGar oSho oGre wKin oWnC wMag wTGN wWhG oUps wCoS
	vulgaris true pink	oWnC
	vulgaris 'Vestale'	wMag
un	***vulgaris*** 'Wedgwood Blue'	oFor oFra oWnC
	yunnanensis	oFor wCCr
	TABERNAEMONTANA	
	divaricata	oOEx
	TAGETES	
	lemmonii	oGoo
	lucida	oAls oGoo oOEx oUps
un	***nelsonii***	oGoo
	signata	see **T. *tenuifolia***
	tenuifolia	iGSc
	tenuifolia 'Lemon Gem' (Gem hybrids)	wFGN
	tenuifolia 'Tangerine Gem' (Gem hybrids)	wFGN
	TAIWANIA	
	cryptomerioides	oFor wHer oPor oAmb oGar oCis
	cryptomerioides var. ***flousiana***	oCis
	flousiana	see **T. *cryptomerioides*** var. ***flousiana***
un	**TALINOPSIS**	
	fructescens DJF -1140	oRar
	sp. BLM 008	oRar
	TALINUM	
un	***napaforme***	oRar
	okanoganense	wMtT
	paniculatum	wMag
un	***sedoides***	see **T. *okanoganense***

	TAMARIX	
	parviflora	oFor oGar oOut
	ramosissima 'Pink Cascade'	oFor
	ramosissima 'Rubra'	oFor oGar wTGN
	ramosissima 'Summer Glow'	see **T. *ramosissima*** 'Rubra'
	tetrandra	oGar
	TANACETUM	
	argenteum	oSis
	balsamita	wFGN oGoo wFai iGSc oBar
	cinerariifolium	oGoo wFai iGSc
	coccineum	oGar oSho wFai oEga
	coccineum dark crimson	oGar
un	***coccineum*** Double Market hybrids	oWnC
	coccineum 'James Kelway'	oGar oLSG
un	***coccineum*** 'Rinjborg's Glory'	oWnC
	coccineum Robinson's hybrids	oGar wMag iArc oLSG
	coccineum 'Robinson's Red'	oDan oWnC oLSG
un	***coccineum*** 'Robinson's Rose'	oAls oDan oWnC wMag wTGN
	densum ssp. ***amani***	oSis oLSG
	macrophyllum	oFor wHer wWoS wCCr
	niveum	oFor oHed iGSc oSec
	parthenium	wCri oAls wCSG oGoo oSha wFai iGSc wHom oWnC oCrm oBar oSec wNTP
	parthenium 'Aureum'	oFor wWoS oAls wFGN oAmb oNWe wCCr wHom oSec wBWP
do	***parthenium*** double flowered	wFGN
un	***parthenium*** 'Dwarf Gold Button'	oUps
	parthenium 'Flore Pleno'	see **T. *parthenium*** 'Plenum'
	parthenium 'Golden Ball'	oLSG
do	***parthenium*** 'Plenum'	oGoo
un	***parthenium*** 'Tetra White Wonder'	oUps
un	***parthenium*** 'Ultra Double White'	oAls
	parthenium white	wFGN
	parthenium 'White Stars'	oLSG
	vulgare	oAls oGoo wFai iGSc
	vulgare var. ***crispum***	oFor oGar oGoo wFai iGSc
	TANAKAEA	
	radicans	last listed 99/00
	TAPISCIA	
	sinensis	oCis
un	***sinensis*** 'Eco-China Ruffles'	oCis
	TAVARESIA	
	barklyi	oRar
	TAXODIUM	
	ascendens	see **T. *distichum*** var. ***imbricatum***
	distichum	oFor oPor wBur oDar oEdg wShR oGar oRiv oWhi oGre wCCr wBox wCoN oSle wCoS
	distichum var. ***imbricatum***	oFor oPor oDar oGar oRiv oGre
	distichum var. ***imbricatum*** 'Nutans'	oPor wCCr wWel
	distichum 'Pendens'	oFor oRiv
	distichum 'Secrest'	oPor wCol oRiv
	distichum 'Shawnee Brave'	oFor
	TAXUS	
	baccata 'Amersfoort'	wHer oPor oAls oGar oSis oBRG
	baccata 'Dovastoniana' (f)	oPor
	baccata 'Dovastonii Aurea' (m)	oPor
	baccata 'Erecta' (f)	wBox
	baccata 'Fastigiata' (f)	oFor oAls oGar oSho oGre wSta wCoS wWel
	baccata 'Fastigiata Aurea'	oBlo oGre wSta oBRG
	baccata 'Fowle'	oPor
	baccata 'Nutans'	wCol
	baccata 'Repandens' (f)	oFor wCul oGar wBox wSta
qu	***baccata*** 'Repandens Aureomarginata'	wHer oAls oGar oNWe
	baccata 'Standishii' (f)	oPor oSis oRiv oWnC wWel
	baccata 'Stricta'	see **T. *baccata*** 'Fastigiata'
	baccata 'Stricta Aurea'	see **T. *baccata*** 'Fastigiata Aurea'
un	***baccata*** 'Watnong Gold	oPor
	brevifolia	oFor wHer wFFl wWat
	canadensis	last listed 99/00
	chinensis	oFor
	cuspidata	oGar wKin
qu	***cuspidata*** 'Aurea'	oFor
	cuspidata 'Aurea Low Boy'	see **T. *cuspidata*** 'Low Boy'
	cuspidata 'Aurescens'	oFor oAls oGar wWel
un	***cuspidata*** 'Bright Gold'	oFor wWel
	cuspidata Emerald Spreader™ / 'Monloo'	oAls wWel
	cuspidata 'Gold Queen'	oNWe
	cuspidata 'Low Boy'	oFor
	cuspidata 'Luteobaccata'	last listed 99/00
	cuspidata f. ***nana***	oPor oGar
	floridana	oFor wHer
	globosa	last listed 99/00

	globosa DJH 418	wHer
	mairei	wHer
	x media 'Beanpole'	wHer oBRG
un	*x media* 'Bobbink'	oFor
	x media 'Brownii'	oGar oGre wKin wWel
	x media dark green spreader	oGar
	x media 'Densiformis'	oAls oGar oBlo oGre oBRG
	x media 'Hicksii' (f)	oFor oAls oGar oBlo oGre wSta wWel
	x media 'Kelseyi'	oFor
	sumatrana	see **T.** ***mairei***
	TELEKIA	
	speciosa	oFor oGre
	TELLIMA	
	grandiflora	oFor oHan wNot wCol oAls wShR oGar oBos wFFl oSqu oGre wCCr oTri wBWP oAld wWld oSle
	grandiflora 'Forest Frost'	wHer wSwC oDan
	grandiflora Odorata Group	oRus wSte
	TELOPEA	
	speciosissima	oTrP wGra
	TERNSTROEMIA	
	gymnanthera	see **T.** ***japonica***
	japonica	oFor wCCr wCoN wWel
	TETRACENTRON	
	sinense	oFor oEdg oAmb oRiv oCis oGre
	TETRADENIA	
	riparia	oCri
	TETRADIUM	
	daniellii	last listed 99/00
	daniellii Hupehense Group	last listed 99/00
	hupehensis	see **T.** ***daniellii*** Hupehense Group
	TETRAPANAX	
	papyrifer	oCis
	TETRASTIGMA	
	sp.	oOEx
	voinierianum	oTrP oGar
	TEUCRIUM	
	ackermannii	oSis
	aroanium	oSis
	asiaticum	oFor wCul wCCr
	canadense	oFor iGSc oUps
	chamaedrys (see Nomenclature Notes)	oFor wCul oJoy oAls wFGN wCSG oGar oGoo oSho oOut wFai wCCr iGSc wHom wMag oCrm oBar wTGN oSec wNTP oEga oUps oSle wSnq
	chamaedrys 'Nanum'	oLSG
	chamaedrys 'Prostratum'	oFor oAls wHom
	cossonii majoricum	last listed 99/00
un	*cossonii* 'Mrs. Milsted'	oNWe
	flavum	wCCr
	fruticans	wCul wFGN wCSG wPir oBar oUps wSte
	fruticans 'Azureum'	oFor oGoo
	fruticans 'Compactum'	wCoS
	hircanicum	wWoS wHig oGoo wFai oUps
un	*longiflorum*	wHer
	x lucidrys	oFor
	majoricum	see **T.** ***polium*** f. ***pii-fontii***
	marum	wFGN oGoo oSis wCCr
	polium	wHer oGoo
un	*polium aureum* 'Mrs. Milsted'	oSis
	polium f. *pii-fontii*	oAls
	pulverulentum	see **T.** ***cossonii***
	pyrenaicum	oSis
	scorodonia	wWoS wFGN oGoo iGSc
	scorodonia 'Crispum'	oFor wCul wCri oAls wCSG wFai iGSc wBWP
	sp.	wCul
	subspinosum	last listed 99/00
un	**THADIANTHA**	
	dubia	oOEx
	THALIA	
	dealbata	wGAc wCol oHug oGar oSsd oWnC wBox
un	*geniculata* var. *ruminoides*	oSsd
	THALICTRUM	
	actaeifolium	last listed 99/00
	actaeifolium HC 970339	wHer
	adiantifolium	see **T.** ***minus adiantifolium***
	alpinum	wHer
	aquilegiifolium	oFor wCul wGAc oNat oJoy wShR oMis oGar oSis oNWe oOut oBlo wFai wMag oEga wWld
	aquilegiifolium var. *album*	oHan wHig
	aquilegiifolium 'Purpureum'	oGre
un	*brevisericeum* EDHCH 97007	wHer
	chelidonii	wHig
	coreanum	see **T.** ***ichangense***
	dasycarpum	oRus oAld
	dasycarpum purple	oMis
	delavayi	oFor oHan wCul oNat wHig oAls oRus oMis oGar oNWe oGre wFai wCCr oWnC wMag oEga oUps wSte
	delavayi 'Album'	oFor wHer oRus oNWe oGre
	delavayi DJHC 989081	wHer
do	*delavayi* 'Hewitt's Double'	oFor wHer oJoy oRus oAmb oNWe oGre wFai wMag wTGN
	delavayi 'Sternhimmel'	wHig
	diffusiflorum	wHer
	dioicum	oRus
	dipterocarpum	see **T.** ***delavayi***
un	'Elin'	wHer
	fendleri	oHan
	filamentosum GBG	oNWe
	filamentosum var. *tenerum* Heronswood form	wHer
	sp. aff. *finetii* DJHC 473	wHer
	flavum	oRus wFai
	flavum ssp. *glaucum*	oFor wHer wCul oJoy oMis oNWe wFai wCCr wBWP
	flavum 'Illuminator'	wHer
	flexuosum	see **T.** ***minus*** ssp. ***minus***
	foliolosum	oNWe
	ichangense	oSis
	isopyroides	wHer oRus oNWe
	kiusianum	oFor wCol oSis oNWe oGre wHom
	lucidum	wCul
	minus	oRus oSle
	minus adiantifolium	oFor wCul oJoy
	minus ssp. *minus*	wHer
	occidentale	oHan wNot oRus oBos oTri oSle
	polycarpum	oBos wCCr
	polygamum	oFor oRus
un	'Purple Mist'	oRus
	rochebruneanum	oFor wHer wWoS oNat oJoy oGar oDan oSis oNWe oGre wFai oWnC wTGN oUps oSle
	rochebruneanum 'Lavender Mist'	wCul oAls oGar
	rugosum	oHan
un	*sibericum*	wHer
	speciosissimum	see **T.** ***flavum*** ssp. ***glaucum***
un	*taquetii* HC 970255	wHer
	THAMNOCALAMUS	
	aristatus	oTra wBea wBmG oTBG
	crassinodus	oTra
	crassinodus 'Merlyn'	oTBG
	spathaceus	see **FARGESIA** ***murieliae***
	tessellatus	oTra wCli oNor wBea oTBG
	THEA	see **CAMELLIA**
	THELYPTERIS	
	decursive-pinnata	see **PHEGOPTERIS** ***decursive-pinnata***
un	*kunthii*	oFor oGre
	novaboracensis	see **PARATHELYPTERIS** ***novaeboracensis***
	palustris	oRus
	phegopteris	see **PHEGOPTERIS** ***connectilis***
	THERMOPSIS	
	caroliniana	see **T.** ***villosa***
	lanceolata	iArc
	lupinoides	oFor oGar
	villosa	oFor oLSG
	THLASPI	
	fendleri	wMtT
	montanum	wMtT
un	**THORNCROFTIA**	
	succulenta	oTrP
	THUJA	
	koraiensis	oGar wCCr
un	*koreana* 'Glauca Nana'	oPor
un	*koreana* 'Glauca Prostrata'	oFor
un	*koreana* x *standishii*	oPor
qu	'Lobbi'	oBlo
	occidentalis	wKin wCoN
un	*occidentalis* 'Abel Twa'	oPor
	occidentalis 'Aurea'	wHer
va	*occidentalis* 'Beaufort'	oPor
	occidentalis 'Boothii'	oPor
un	*occidentalis* 'Brandon'	oGue
	occidentalis 'Buchananii'	wCCr
	occidentalis 'Caespitosa'	oPor
un	*occidentalis* 'Columbia'	oPor
	occidentalis 'Danica'	oPor oGar

	occidentalis 'Degroot's Spire'	oPor wCol oBRG wAva wWel
un	***occidentalis*** 'Dirigo Dwarf'	oPor
	occidentalis 'Elegantissima'	oPor oRiv
	occidentalis 'Emerald'	see **T. *occidentalis*** 'Smaragd'
	occidentalis 'Emerald Green'	see **T. *occidentalis*** 'Smaragd'
	occidentalis 'Fastigiata'	wCoS
	occidentalis 'Filiformis'	oPor oGar
	occidentalis 'George Peabody'	oPor
	occidentalis 'George Washington'	oFor
	occidentalis 'Globosa'	oGar oWnC wCoS
	occidentalis 'Golden Globe'	oPor oGar oGue
	occidentalis 'Hetz Midget'	oPor oJoy oAls oGar oRiv oBlo wTGN
	occidentalis 'Holmstrup'	oPor
	occidentalis 'Hoveyi'	oPor
	occidentalis KBN select	oPor
un	***occidentalis*** 'Leptocladus'	oPor
	occidentalis 'Linesville'	oPor
	occidentalis 'Little Champion'	oPor
un	***occidentalis*** 'Little Elfie'	oPor
	occidentalis 'Little Giant'	oPor oRiv wKin
	occidentalis 'Lutea'	wCCr
	occidentalis 'Milleri'	oPor
	occidentalis 'Nigra'	wKin
	occidentalis 'Ohlendorffii'	oPor
	occidentalis 'Pendula'	oPor wSta
	occidentalis 'Pyramidalis'	wWoB oAls wCSG oGar oTPm oWnC wSta
	occidentalis 'Recurva Nana'	oPor oSis
	occidentalis 'Rheingold'	oFor oPor oAls oGar oDan oRiv oBlo oGre wKin oTPm oWnC oBRG wAva
	occidentalis 'Rosenthalii'	oPor
un	***occidentalis*** 'Sherwood Frost'	oFor oPor oJoy
	occidentalis 'Smaragd'	oFor oPor oAls oGar oRiv oGre wKin oTPm oWnC wTGN wSta oCir wAva oGue wCoS
	occidentalis 'Stolwijk'	oPor
un	***occidentalis*** 'Sudworth Gold'	wHer oPor
	occidentalis 'Sunkist'	oFor oPor oGre
	occidentalis 'Tiny Tim'	oAls oSis oBRG
	occidentalis 'Tom Thumb'	oPor
	occidentalis 'Umbraculifera'	oPor
va	***occidentalis*** 'Wansdyke Silver'	oFor oPor wTGN
	occidentalis 'Wareana Lutescens'	oPor
	occidentalis 'Woodwardii'	oPor wKin oTPm oWnC
	occidentalis 'Yellow Ribbon'	wKin oWnC
	orientalis	wCCr wKin wCoN
	orientalis 'Aurea Nana'	oAls oGar oRiv oWnC wSta
	orientalis 'Berckmannii'	see **T. *orientalis*** 'Aurea Nana'
	orientalis 'Bergmanii'	see **T. *orientalis*** 'Aurea Nana'
	orientalis 'Blue Cone'	last listed 99/00
	orientalis 'Raffles'	oPor oAls oGar
	orientalis 'Sanderi'	oPor
	orientalis 'Sunlight'	oWnC
	orientalis 'Westmont'	oGar
	plicata	oFor wBlu wWoB wBur wNot oDar oJoy wShR oGar oBos oRiv oBlo wFFl oGre wWat oWnC wSta wPla oAld wCoN wRav oSle wSnq wCoS
	plicata 'Atrovirens'	oDan oGre oBRG
	plicata 'Aurea'	oWnC
	plicata 'Brabant'	oPor wTGN
	plicata 'Can-can'	oPor
	plicata 'Collyer's Gold'	oFor
	plicata 'Cuprea'	oPor oAls oGar wCCr
qu	***plicata*** 'Emerald Cone'	wBWP
	plicata 'Excelsa'	wClo oGre
	plicata 'Fastigiata'	oGar
un	***plicata*** 'Filifera Nana'	wCol oGar
	plicata green sport	wAva
	plicata 'Grune Kugel'	oPor
	plicata 'Hillieri'	oFor wHer oGre
	plicata 'Hogan'	oBlo
un	***plicata*** 'Holly Turner'	wHer wCol
	plicata x ***koraiensis***	wCCr
	plicata 'Rogersii'	oPor
un	***plicata*** 'Rogersii' sport	oGre wAva
	plicata 'Stoneham Gold'	oFor oPor oJoy oGre
	plicata 'Sunshine'	oFor oDar oDan
un	***plicata*** 'Virescens'	oPor oGar oWhi oGre
un	***plicata*** 'Watnong'	wHer
un	***plicata*** 'Whipcord'	wHer
va	***plicata*** 'Zebrina'	oFor wHer oPor oJoy wCol oGar oRiv oWnC wAva wWel
	standishii	oFor oBRG

THUJOPSIS

	dolabrata	oFor oDar oJoy oRiv oSho oGre wCCr oWnC wAva
	dolabrata 'Nana'	oFor wHer oPor oJoy wCol oGre oBRG
	dolabrata 'Variegata'	oFor wHer oPor oDar oJoy wCol oGar oRiv oOut oGre wSte

THUNBERGIA

	alata	oGar oUps
	grandiflora 'Alba'	last listed 99/00

THYMUS

qu	***argaeus***	iGSc
	argenteus	see **T. *vulgaris*** 'Argenteus'
	broussonetii	last listed 99/00
	caespititius	oGoo iGSc oBar
	camphoratus	oAls oGoo iGSc oSle
	cherlerioides	oGoo
	cherlerioides Barn Owl seedling	wCul
	cilicicus	oSis
	x ***citriodorus***	wCul oTDM oJoy oAls wFGN oGoo iGSc oWnC oSec wNTP oCir wWhG oUps oSle
	x ***citriodorus*** 'Archer's Gold'	oFor wWoS wTho oDar oAls wFGN oGoo oBar
va	*x* ***citriodorus*** 'Argenteus'	oAls oGoo oGre oBar oSle
	x ***citriodorus*** 'Aureus'	oJoy oAls oGoo iGSc oWnC oBar
	x ***citriodorus*** 'Bertram Anderson'	oJoy
un	*x* ***citriodorus*** 'Fairie'	wFGN wFai
un	*x* ***citriodorus*** 'Gold Transparent'	wFGN oWnC
va	*x* ***citriodorus*** 'Golden King'	wWoS wFGN
va	*x* ***citriodorus*** 'Golden Lemon' (see Nomenclature Notes)	wFGN
un	*x* ***citriodorus*** 'Lemon Frost'	wCul wFGN oGoo
un	*x* ***citriodorus*** 'Lemon Mist'	wFGN oGoo oBar
un	*x* ***citriodorus*** 'Lime'	wWoS oAls wFGN iGSc oBar wNot
	x ***citriodorus*** silver edge	wFGN
un	*x* ***citriodorus*** 'Silver Lemon'	wFGN oGoo
qu	*x* ***citriodorus*** 'Variegatus'	oGar oGre wHom
un	'Coconut'	oAls wFGN
	doerfleri 'Bressingham'	oJoy oAls wFGN oBar
va	'Doone Valley'	wWoS wTho oDar oAls wFGN oGoo oSho oOut oGre iGSc oWnC oBar
un	'Doretta Klaber'	wWoS oSis
	'E. B. Anderson'	see **T. *x citriodorus*** 'Bertram Anderson'
un	'English Wedgwood'	wWoS wCul oAls wFGN oAmb oBar
	glabrescens ssp. ***decipiens***	oSis
	golden	iGSc
un	'Grey Hill'	wFGN oGoo
va	'Hartington Silver'	wFGN
	herba-barona	wCul oDar oAls wFGN oGoo oSis iGSc oBar wNTP oCir
	herba-barona 'Lemon-scented'	wFGN oGoo
	'Highland Cream'	see **T.** 'Hartington Silver'
	lanuginosus	see **T. *pseudolanuginosus***
	lavender	oBar wNTP
	leucotrichus	wCul wFGN oGoo oBar
un	'Linear Leaf Lilac'	oGoo iGSc
	longleaf grey	oGoo
un	'Longwood'	wFGN oGoo
	mastichina	iGSc
	mastichina hybrid	oGoo
un	'Mayfair'	wWoS oAls wFGN oGoo
qu	***microphyllum***	oOut
un	'Miniature'	oGoo
un	'Mongolian'	wFGN
	nummularius	oJoy
	nummularius 'Red Creeping'	see **T. *serpyllum coccineus***
un	'Pennsylvania Dutch Tea'	wFGN
	'Peter Davis'	oAls
un	'Pink Passion'	wFGN
	'Pink Ripple'	wWoS wCul wFGN oGoo oBar
	polytrichus ssp. ***britannicus***	wTho oTDM oJoy oAls wFGN oGoo oWnC oUps wCoS
	polytrichus ssp. ***britannicus*** 'Albus'	oGoo wMag oBar
	polytrichus ssp. ***britannicus*** 'Coccineus'	see **T. *serpyllum coccineus***
	polytrichus ssp. ***britannicus*** 'Minor'	oJoy oAls oGoo oBar
	polytrichus ssp. ***britannicus*** variegated	oGoo
	'Porlock'	oAls
	praecox	oUps
	praecox ssp. ***arcticus***	see **T. *polytrichus*** ssp. ***britannicus***
un	***praecox*** 'Evergold'	oSis
	pseudolanuginosus	wCul oAls oGoo oSis wRai oGre iGSc wHom oWnC wMag oBar wSta wNTP oCir wWhG oUps oSle wCoS
	pseudolanuginosus 'Hall's Variety'	oAls oGoo
	pulegioides	wWoS wTho oAls wFGN oGoo oSis iGSc oWnC oBar oSle

	pulegioides lemon	oBar
un	***quinquecostatus albiflorus***	oGoo
qu	***quinquecostatus ibukiensis alba***	iGSc
un	'Reiter's'	wFGN oSis
un	'Reiter's Red'	oSis
un	'Rose Williams'	oSis
	serpyllum	oAls wFGN oSho iGSc wNTP wWhG oUps
	serpyllum var. ***albus***	wCul oAls oSis oCir
	serpyllum 'Annie Hall'	wFGN oGoo
	serpyllum coccineus	wCul oTDM oJoy oAls wFGN wCSG oGar oGoo iGSc oWnC wWhG
	serpyllum 'Elfin'	oDar oAls wFGN oSis oGre iGSc wHom wMag oCir wWhG
un	***serpyllum*** 'Gold Transparent'	oJoy
va	***serpyllum*** 'Goldstream'	oJoy wFGN oGoo
	serpyllum 'Minor'	wFGN oSis oNWe wMag
	serpyllum 'Minus'	see **T. *serpyllum*** 'Minor'
	serpyllum 'Pink Chintz'	wWoS oAls wFGN oGar oSis oGre iGSc wHom oBar wWhG oUps
	serpyllum 'Roseus'	wCul
un	'Spicy Orange'	wWoS oDar wRai iGSc oBar
	thracicus	wWoS oJoy oAls wFGN iGSc
	vulgaris	oTDM oAls oSho iGSc oWnC wNTP oUps
	vulgaris 'Argenteus'	wWoS wTho oAls wFGN wCSG iGSc oWnC oBar wNTP oCir oUps
	vulgaris 'Aureus'	wMag
un	***vulgaris*** 'Dot Wells'	oAls
un	***vulgaris*** 'Dottie Jacobsen'	oBar
	vulgaris French	wFGN wNTP
un	***vulgaris*** 'Grey Hill'	wCul oAls oBar
	vulgaris narrowleaf French	oAls oGoo oBar
	vulgaris narrowleaf Provincial	oAls
	vulgaris 'Orange Balsam'	wWoS wCul oAls wFGN oGoo iGSc oWnC oBar
un	***vulgaris*** 'Passion Pink'	oGoo oBar
	vulgaris 'Pinewood'	oJoy oAls oGoo iGSc
	vulgaris 'Silver Posie' (see Nomenclature Notes)	wWoS oTDM
un	***vulgaris*** 'White Moss'	wFGN wNTP
un	***vulgaris*** 'Yellow Transparent'	oAls
un	'Wedgwood Blue'	wFGN oGoo wNTP
un	'Woolly Stemmed Sweet'	oGoo
qu	***zeranshanicus***	oAls

TIARELLA

	'Black Velvet'	wWoS wCol
	cordifolia	oFor wCSG oGoo oSqu wFai oTri
un	***cordifolia*** 'Eco Rambling Silhouette'	oSqu
	cordifolia 'Oakleaf'	wRob
	cordifolia 'Slick Rock'	last listed 99/00
un	***cordifolia*** 'Winter Glow'	oAls
	'Cygnet'	wHer wWoS oEdg oHed oDan wNay oSqu oGre wAva
	'Dark Eyes'	oAls oGar wRob wNay
	'Dark Star'	oJoy wNay oSqu
un	'Eco Eyed Glossy'	oSqu
	'Filigree Lace'	wNay
	'Freckles'	wNay
	'Inkblot'	oJoy oAls oDan wNay
	'Iron Butterfly'	oFor wWoS oDan wAva
	'Lacquer Leaf'	wWoS
	'Mint Chocolate'	wHer wNay oSev
	'Ninja'	wHer wWoS wSwC oJoy oEdg oHed oDan wNay wTGN
	'Pink Bouquet'	wWoS oAls oAmb wNay
	'Pinwheel'	wNay
	polyphylla	wCol
	polyphylla DJHC 574 seed grown	wHer
	polyphylla 'Filigran'	last listed 99/00
	polyphylla 'Moorgruen'	oRus oSqu
un	'Running Tapestry'	wWoS
	'Skeleton Key'	wHer wWoS oJoy wCol oDan wRob wNay oSqu oSle
	'Snowflake'	last listed 99/00
	'Spanish Cross'	oGar oSho wNay
	'Spring Symphony'	oEdg wCol oHed oNWe oOut
	'Tiger Stripe'	wWoS oGar wRob wNay oWnC
	trifoliata	oGar wFFl oAld oSle
un	***trifoliata*** ssp. ***trifoliata***	wNot
	unifoliata	oRus oBos
	wherryi	oHan oHed oRus oMis oSis oSha wFai oWnC wTGN oUps
	wherryi 'George Schenk'	oSqu
	wherryi 'Heronswood Mist'	oFor wHer wCol oOut

TIBOUCHINA

	urvilleana	oNat oGar oSho wMag oUps wWel

TIGRIDIA

	hybrids	oMis
	pavonia	wHer oVBI
	pink	wTGN
	red	wTGN
	white	wTGN

TILIA

	americana	oFor
	amurensis	oFor
	caucasica	oFor
	chinensis	oFor
	cordata	oFor oDar oRiv oBlo wKin oWnC
un	***cordata*** 'Corzam'	oGar
	cordata 'Greenspire'	oAls oGar oWnC
un	***cordata*** 'Pendula Nana'	oFor
	cordata 'Swedish Upright'	last listed 99/00
	cordata 'Winter Orange'	oFor
	x euchlora	last listed 99/00
	x europaea	oFor
	x europaea 'Wratislaviensis'	oFor
	henryana	oFor
	japonica	oFor oEdg
	kiusiana	oFor oWhi
	oliveri	oFor
	'Petiolaris'	oFor
	platyphyllos	oFor
	platyphyllos 'Laciniata'	oFor
	platyphyllos 'Rubra'	oFor
	platyphyllos 'Tortuosa'	oFor
	tomentosa	last listed 99/00
	tomentosa 'Sterling Silver'	oFor oGar oSho

TILLANDSIA

	aeranthos	oGar
un	***albertiana***	oGar
	araujuei	oGar
	argentea	oGar
	baileyi	oGar
	bergeri	oGar
	brachycaulos	oGar
	bulbosa	oGar
	butzii	oGar
	caput-medusae	oGar
un	***cauligera***	oGar
	concolor	oGar
	cyanea	oGar
un	***diaguitensis***	oGar
	duratii	oGar
un	***festucoides***	oGar
	filifolia	oGar
un	***gardneri rupicola***	oGar
un	***heteromorpha***	oGar
un	***incarnata***	oGar
	ionantha	oGar
	ixioides	oGar
un	***jucunda***	oGar
	juncea	oGar
	karwinskyana	oGar
	kolbii	oGar
	latifolia	oGar
un	***latifolia*** 'Divaricata'	oGar
un	***leonamiana***	oGar
un	***montana***	oGar
un	***purpurea***	oGar
un	***queroensis***	oGar
un	***schiedeana*** 'Major'	oGar
un	***streptocarpa***	oGar
	streptophylla	oGar
	stricta	oGar
	stricta hard leaf	oGar
	tectorum	oGar
	tenuifolia blue	oGar
un	***tenuifolia*** 'Silver Comb'	oGar
	tricolor	oGar
	xerographica	oGar

TITANOTRICHUM

	oldhamii	wHer

TOFIELDIA

	glutinosa	oBos

TOLMIEA

	menziesii	oFor oHan wNot oRus oBos wFFl wWat oTri oAld
va	***menziesii*** 'Taff's Gold'	oFor wWoS wCol oRus wCSG wRob

TOONA

	sinensis	oRiv oCis wRai oOEx wCCr wWel

	TORREYA	
	californica	oFor oWhi wCCr
	grandis	wHer oCis
	nucifera HC 970329	wHer
un	***nucifera*** var. ***sphaerica***	wHer
	TOVARA	see **PERSICARIA**
	TOWNSENDIA	
un	***alpingena***	wMtT
	exscapa	wMtT
	jonesii	last listed 99/00
	spathulata	wMtT
	TRACHELIUM	
	caeruleum	oFor
	TRACHELOSPERMUM	
	asiaticum	oFor oGar oCis oSho wPir wWel
	asiaticum dwarf	oCis
un	***asiaticum*** 'Red Top'	oFor
un	***asiaticum*** 'Theta'	oCis
	jasminoides	oFor oAls oGar oSho oGre wPir oWnC wTGN wCoN wSte wCoS wWel
	jasminoides 'Variegatum'	oFor
	jasminoides 'Wilsonii'	oCis
	TRACHYCARPUS	
	fortunei	oFor oTrP oDar oAls oGar oRiv oCis oSho oOEx wDav wCCr wCoN wCoS wWel
	latisectus	oCis
	wagnerianus	oCis
	TRACHYMENE	
	coerulea	wSte
	TRACHYSTEMON	
	orientalis	wHer wHig
	TRADESCANTIA	
qu	'Alba'	oFor
	Andersoniana Group	last listed 99/00
	Andersoniana Group 'Bilberry Ice'	wHer
	Andersoniana Group 'Blue and Gold'	wHer oHed
	Andersoniana Group "Blue Stone'	oFor oLSG
	Andersoniana Group 'Blushing Bride'	last listed 99/00
	Andersoniana Group 'Charlotte'	wGAc wFai wBox wSte
	Andersoniana Group 'Chedglow'	wHer
	Andersoniana Group 'Concord Grape'	oFor wHer wWoS oAls oSis oGre
	Andersoniana Group 'Double Trouble'	last listed 99/00
	Andersoniana Group 'Innocence'	oGar oSle
	Andersoniana Group 'Iris Prichard'	oFor oWnC
	Andersoniana Group 'J. C. Weguelin'	oGar
	Andersoniana Group 'Little Doll'	oSho wRob
	Andersoniana Group 'Osprey'	oFor oAls wHom oSle
	Andersoniana Group 'Pauline'	oFor
	Andersoniana Group 'Purple Dome'	oFor oAls wHom wTGN oCir
	Andersoniana Group 'Red Grape'	wWoS
	Andersoniana Group 'Snowcap'	wCul oGar wRob oLSG
	Andersoniana Group white	wHom
	Andersoniana Group 'Zwanenburg Blue'	oFor oGar wRob
un	'Angels Eyes'	wHer
	bracteata	oSis
un	'China Blue'	wRob
un	'Hawaiian Punch'	oAls wRob
un	'La Buffa'	oCis
un	'Navajo Princess'	wRob
	pallida 'Purpurea'	oUps
	'Purple Heart'	see **T. *pallida*** 'Purpurea'
un	'Purple Profusion'	oFor oHed oDan wSte
	subaspera	wThG
un	'True Blue'	wRob
	virginiana	oSec oEga
	virginiana blue	oGar
	virginiana 'Red Cloud'	oFor wCul oGar oGre oWnC wMag oLSG oSle
	TRAUTVETTERIA	
	carolinensis	oFor wCol
	TRICHODIADEMA	
	bulbosum	oSqu
	densum	oSqu
	TRICHOSANTHES	
	anguina	see **T. *cucumerina*** var. ***anguina***
	cucumerina var. ***anguina***	oOEx
	kirilowii	oCrm
	TRICYRTIS	
	affinis	wCol
un	***affinis*** 'Tricolor'	wHer oNWe oGre
	dilatata	see **T. *macropoda***
un	'Eco Gold Spangles'	wCol
un	'Empress'	wWoS wCol oAmb
	formosana	oSis oOut oGre wMag wSnq
	formosana 'Amethystina'	wWoS oUps

	formosana 'Dark Beauty'	last listed 99/00
	formosana 'Gates of Heaven'	last listed 99/00
	formosana 'Samurai'	wWoS wCol wRob
	formosana 'Seiryu'	last listed 99/00
	formosana Stolonifera Group	wRob
	formosana 'Variegata'	wHer wAva
	'Golden Gleam'	wWoS
	hirta	oDan oSho oOut wHom wMag
	hirta alba	wCol oGar oSis
un	***hirta*** 'Albescens'	wCol
un	***hirta*** 'Aurea'	last listed 99/00
	hirta gold leaf	wCol
un	***hirta*** 'Hakurakuten' seed grown	oNWe
	hirta HC 970733	wHer
	hirta 'Hotatagisa'	see **T.** Hototogisu
	hirta 'Hototogisu'	see **T.** Hototogisu
	hirta var. ***masamunei***	wHer
	hirta 'Miyazaki'	wWoS oJoy wCol oAls oGar oNWe oOut wRob wSte
un	***hirta*** 'Miyazaki Variegata'	oGre
un	***hirta*** 'Moonlight'	oFor wWoS
un	***hirta*** 'Ogon'	wRob
	hirta 'Shimona'	see **T.** 'Shimone'
qu	***hirta*** 'Shirohototagisa'	oFor
qu	***hirta*** 'Shirohotugisu'	wSnq
un	***hirta*** 'Taiwan Adbane'	oOut
	hirta 'Variegata'	wCol oSis oNWe wRob
	Hotatagisa	see **T.** Hototogisu
	Hototogisu (see Nomenclature Notes)	oFor oDan oOut oUps
	Japanese hybrids	oAls
	'Kohaku'	wWoS wCol oGar wAva
	latifolia	oFor oSis wRob
un	***latifolia*** 'Forbidden City'	wCol
	'Lemon Lime'	last listed 99/00
un	'Lightning Strike'	oFor wWoS wCol oGre
	'Lilac Towers'	oAmb
	macrantha	oOut
	macrantha ssp. ***macranthopsis***	wHer wCol oSis oGre
	macranthopsis	see **T. *macrantha*** ssp. ***macranthopsis***
	macropoda	oFor wWoS oJoy oRus oSis oGre
	macropoda HC 970450	wHer
	macropoda Yungi Temple Form	wHer wCol
	maculata HWJCM 431	wHer
un	***nana*** 'Chabo'	wHer
	ohsumiensis	wHer wCol
	'Outback's Blue Select'	oOut
un	***setochinensis***	wHer
	'Shimone'	oOut wRob
un	'Shiromotogisu'	oDan
	'Tojen'	wHer wWoS oJoy wCol oDan oSis wRob wAva wSte
va	'White Flame'	oFor
	'White Towers'	wHer wWoS oOut
	'White Towers' seed grown	oNWe
	TRIDENTEA	
	jucunda var. ***cincta***	oRar
	longii	wGra
	TRIENTALIS	
	latifolia	wThG oRus oBos wFFl
	TRIFOLIUM	
	pratense	iGSc
va	***pratense*** 'Susan Smith'	wHer wWoS oHed oGar oOut wBox
	repens 'Atropurpureum'	wHer oGre wBox
	repens 'Green Ice'	wHer wWoS oGre
	repens 'Purple Velvet'	wHom
	repens 'Purpurascens'	oFor wWoS wEde
	repens 'Purpurascens Quadrifolium'	oJoy oHed wCSG
va	***repens*** 'Wheatfen'	wGAc oHed oGar oGre wBox
	rubens	oHed
	TRIGONELLA	
	foenum-graecum	iGSc
	TRILLIDIUM	
	govanianum	see **TRILLIUM *govanianum***
	TRILLIUM	
	albidum	oRus oSis
	apetalon	wHer
	catesbyi	oTrP oGre wMag
	cernuum	oFor wThG
	chloropetalum	wThG oRus oBos oNWe oAld
un	***chloropetalum*** 'Volcano'	oRus
	cuneatum	oNWe
	erectum	oFor oTrP wThG oRus oSis oBos oNWe oGre wMag wTGN oAld
	erectum f. ***albiflorum***	oSis
	erectum 'Beige'	oGre

	erectum f. ***luteum***	oRus
	flexipes	wThG oRus
	govanianum	wHer
	grandiflorum	wThG oRus oSis oNWe wMag oTri wTGN oAld
	hibbersonii	see **T. *ovatum*** var. ***hibbersonii***
	kamtschaticum	wHer
	kurabayashii	wHer
	luteum	oFor wThG oRus oSis oBos oNWe oGre wMag
	nivale	wThG oGre
	ovatum	wHer wThG wWoB wNot oRus oSis oBos wFFl oTri oAld wAva oSle
	ovatum var. ***hibbersonii***	wHer
	parviflorum	wHer oSis
	recurvatum	wThG oGre
	rivale	wThG
	sessile	oFor oTrP wThG oGre wTGN
	smallii	wHer
	tschonoskii	wHer
	undulatum	wThG
	viride	oAld
	TRIOSTEUM	
	himalayanum	wHer
	pinnatifidum	wHer
	TRIPTEROSPERMUM	
	japonicum HC 970691	wHer
	TRIPTERYGIUM	
	regelii	oFor
	regelii DJH 327	wHer
	TRITELEIA	
	hyacinthina	wHer oBos
	laxa 'Koningin Fabiola'	wHer wCul oDan
	laxa Queen Fabiola	see **T. *laxa*** 'Koningin Fabiola'
	peduncularis	oFor
	TRITONIA	
	crocata	last listed 99/00
	disticha ssp. ***rubrolucens***	wHer
	lineata	wHer
	securigera	oTrP
	squalida	last listed 99/00
	TROCHETIOPSIS	
	ebenus	oHed
ch	***melanoxylon***	see **T. *ebenus***
	TROCHODENDRON	
	aralioides	wHer oGoo oCis oSho oDav oGre wCCr wWel
	TROLLIUS	
	asiaticus	oFor
	chinensis	oFor wHer wHig oRus oUps
	chinensis 'Golden Queen'	oFor oJoy oAls wCSG oGar oSho oBlo oWnC oEga
	x cultorum 'Alabaster'	last listed 99/00
	x cultorum 'Alabaster' seed grown	oNWe
	x cultorum 'Canary Bird'	wCSG
	x cultorum 'Cheddar'	oGar wNay
	x cultorum 'Commander-in-chief'	wNay oGre
	x cultorum 'Etna'	wWoS oGar wNay
	x cultorum 'Feuertroll'	wWoS
	x cultorum Fireglobe	see **T. *x cultorum*** 'Feuertroll'
	x cultorum 'Lemon Queen'	wWoS wHig oRus oAmb oGar oOut wNay oGre wBox
	x cultorum 'Orange Globe'	oSho
	x cultorum 'Orange Princess'	wWoS
	x cultorum 'Prichard's Giant'	wWoS
	x cultorum 'Superbus'	wWoS oGar wNay oGre
	x cultorum 'T. Smith'	wWoS
	europaeus	oFor oMis
	farreri	oNWe
	hondoensis	oNWe oGre
	ledebourii hort.	see **T. *chinensis***
un	'New Hybrids'	oRus
	pumilus	last listed 99/00
	yunnanensis	wNay
	sp. aff. ***yunnanensis*** DJHC 649	wHer
	yunnanensis DJHC 98072	wHer
	TROPAEOLUM	
un	'Blush Double'	wHer
	majus	oAls wFai
va	***majus*** Alaska Series	oAls wFGN oUps
do	***majus*** 'Darjeeling Double'	wHer
	majus 'Empress of India'	wFGN
un	***majus*** 'Glorious Gleam'	wFGN
do	***majus*** 'Hermine Grashoff'	wHer
	majus Jewel Series	wFGN oUps

un	***majus*** 'Mahogany'	oUps
	majus 'Peach Melba'	wFGN oUps
un	***majus*** 'Red Pygmy'	oSec
	majus Whirleybird hybrids	oUps
	peregrinum	wFGN wSte
	speciosum	wHer wCri oNWe
	tuberosum	oFor oGoo
	tuberosum var. ***lineamaculatum*** 'Ken Aslet'	wHer oRus
un	***tuberosum*** 'Muru-anu'	oOEx
un	***tuberosum*** 'Puca-anu'	oOEx
un	***tuberosum*** 'Sapu-anu'	oOEx
un	***tuberosum*** 'Yurac-anu'	oOEx
	TSUGA	
	canadensis	oDar oGar oRiv oWhi oGre oWnC wAva wCoN wCoS
	canadensis 'Abbott's Pygmy'	oPor
	canadensis 'Albospica'	oPor
un	***canadensis*** 'Aurora'	oPor
un	***canadensis*** 'Bacon's Cristate'	oPor oBRG
	canadensis 'Baldwin Dwarf Pyramid'	oPor
un	***canadensis*** 'Beaujean'	oPor
un	***canadensis*** 'Beehive'	oFor oPor oAmb
	canadensis 'Bennett'	oFor oPor wClo oAls oRiv wWel
	canadensis 'Betty Rose'	last listed 99/00
	canadensis 'Brandley'	oPor
un	***canadensis*** 'Burkitt's Dwarf'	oPor
un	***canadensis*** 'Burkitt's White Tip'	oPor
un	***canadensis*** 'Callicoon'	oPor
	canadensis 'Canby'	oPor oRiv
un	***canadensis*** 'Cappy's Choice'	oFor oPor oSis
	canadensis 'Cinnamonea'	last listed 99/00
	canadensis 'Cloud Prune'	oFor oSis
	canadensis 'Coffin'	oPor
	canadensis 'Cole's Prostrate'	oFor wHer oPor wWoB oSis oRiv wWel
	canadensis 'Compacta'	oPor
un	***canadensis*** 'Cotton Candy'	oPor
	canadensis 'Curley'	oFor oPor wClo oRiv wSta
	canadensis 'Devil's Fork'	oPor
un	***canadensis*** 'Elm City'	oGre
	canadensis Emerald Fountain™ / 'Monler'	oAls oGar
	canadensis 'Essex'	last listed 99/00
	canadensis 'Everitt's Dense Leaf'	oPor
	canadensis 'Everitt's Golden'	oFor oPor
	canadensis 'Fantana'	wHer
	canadensis 'Fastigiata'	wSta
un	***canadensis*** 'Freiburg'	oPor
	canadensis 'Fremdii'	oPor
	canadensis 'Frosty'	oPor
un	***canadensis*** 'Geneva'	oPor wClo wWel
un	***canadensis*** 'Gentsch Dwarf Globe'	oGre
va	***canadensis*** 'Gentsch White'	oFor wHer oPor wClo oGar oSis oSho oTPm wSta wWel
	canadensis 'Golden Splendor'	oFor wHer oPor oGar
	canadensis 'Gracilis'	oPor oAls oGre oTPm oBRG
un	***canadensis*** 'Green Cascade'	oPor
un	***canadensis*** 'Greenbrier'	oGar
un	***canadensis*** 'Greenlace'	oPor
	canadensis 'Greenspray'	last listed 99/00
	canadensis 'Greenwood Lake'	oPor
un	***canadensis*** 'Henry Hohman'	oPor
un	***canadensis*** 'Hicks'	oPor
	canadensis 'Horsford'	oPor
un	***canadensis*** 'Horsford Contorted'	oFor wClo oSis
un	***canadensis*** 'Humphrey Welch'	oPor oGre
	canadensis 'Hussii'	oFor wHer oPor oAmb oGar oTPm wSta oBRG
	canadensis 'Jacqueline Verkade'	oPor oSis oTPm
	canadensis 'Jeddeloh'	oFor oPor oAls oAmb oGar oSho oGre oTPm oWnC wWel
	canadensis 'Jervis'	oBRG
	canadensis 'Kelsey's Weeping'	oPor
	canadensis 'Lewis'	oGar oTPm
un	***canadensis*** 'Little Joe'	oPor
un	***canadensis*** 'Livingstone'	oPor
un	***canadensis*** 'Melville'	oPor
	canadensis 'Minima'	last listed 99/00
	canadensis 'Minuta'	oPor oTPm
un	***canadensis*** 'Molalla'	oPor
un	***canadensis*** 'Mount Bachelor'	oPor
un	***canadensis*** 'Mount Hood'	oPor
un	***canadensis*** 'Mount Jefferson'	oPor
un	***canadensis*** 'Mount Rainier'	oPor
un	***canadensis*** 'Mount Saint Helens'	oPor
un	***canadensis*** 'Mount Shasta'	oPor

	canadensis 'Nana'	last listed 99/00
	canadensis 'Nana Gracilis'	see **T. *canadensis*** 'Gracilis'
un	***canadensis*** 'New Gold'	oFor wWel
	canadensis 'Palomino'	oPor
	canadensis 'Pendula'	oFor oPor wClo oAls oGar oRiv oTPm wTGN wSta wAva wWhG wCoS wWel
	canadensis 'Perfecta Nana'	oPor
un	***canadensis*** 'Popeleski'	oPor
un	***canadensis*** 'Rankin'	oPor
un	***canadensis*** 'Rhapsody'	oPor
	canadensis 'Rugg's Washington Dwarf'	last listed 99/00
	canadensis 'Sargentii'	oSho oWnC wTGN
un	***canadensis*** 'Sherwood Compact'	oFor oPor
	canadensis 'Slenderella'	wClo oAmb oGar
	canadensis 'Snowflake'	oPor
un	***canadensis*** 'Spingard Littleleaf'	wHer
un	***canadensis*** 'Springarn'	oRiv
un	***canadensis*** 'Starker'	oPor
un	***canadensis*** 'Stewartii'	oSis
	canadensis 'Stewart's Gem'	oPor wClo oGar
un	***canadensis*** 'Stockman's Dwarf	oPor wWel
un	***canadensis*** 'Summer Snow'	oPor
	canadensis 'Taxifolia'	last listed 99/00
un	***canadensis*** 'Tualatin'	oPor
	canadensis 'Verkade Recurved'	oPor
	canadensis 'Von Helms'	last listed 99/00
	canadensis 'Watnong Star'	oPor oRiv
un	***canadensis*** 'Winds Way'	oPor
un	***canadensis*** 'Wodenethe'	oPor
	caroliniana	wHer
	caroliniana 'La Bar Weeping'	last listed 99/00
	chinensis	oPor oRiv
	diversifolia	oFor oPor wClo oJoy oAls oGar oRiv wSta
	heterophylla	oFor wWoB wNot oDar wShR oGar oBos oRiv oSho oWhi wFFl wWat wCCr wKin wSta oAld wCoN wRav oSle wCoS
	heterophylla 'Iron Springs'	wHer oPor wClo oDan oRiv oBRG wWel
un	***heterophylla*** 'Thorsen's Weeping'	wHer oPor wClo
un	***heterophylla*** 'Thoruson's	oRiv oBRG
	mertensiana	oFor wBlu oPor oGos wWoB wClo oGar oBos oRiv oSho oGre wCCr wSta oBRG oAld wAva wCoN oSle wSte wCoS wWel
	mertensiana 'Blue Star'	last listed 99/00
	mertensiana 'Elizabeth'	oPor oGos oGar
un	***mertensiana*** 'Mount Hood Blue'	oGre
	sieboldii	oFor wHer oPor
	yunnanensis	oFor wHer oPor
	TSUSIOPHYLLUM	
	tanakae	see **RHODODENDRON *tsusiophyllum***
	TULBAGHIA	
	cominsii	oHed
un	EBG 1974 4269	oHed
	fragrans	see **T. *simmleri***
	natalensis pink	oHed
un	***simmleri*** 'Alba'	oHed
	violacea	wCul oAls wFGN oMis oGoo wCCr iGSc oBar oSec oUps
va	***violacea*** 'Silver Lace'	oTrP oAls oHed oMis iGSc oBar
	violacea tricolor	wWoS oGoo
	violacea 'Variegata'	see **T. *violacea*** 'Silver Lace'
	TULIPA	
	'Abba'	oGar oWoo
un	'Abra'	wRoo
	'Abu Hassan'	wRoo
	'Ad Rem'	wRoo oWoo
	'Addis'	oWoo
un	'Adorno	oWoo
un	'Ajax'	oWoo
un	'Akela'	oWoo
	'Aladdin'	wRoo oWoo
	'Ali Baba'	oGar
	'Allegretto'	oWoo
	'Angelique'	wRoo oGar oWoo
un	'Anna Jose'	oWoo
	'Apeldoorn'	wRoo oWoo
	'Apeldoorn's Elite'	oWoo
	'Apricot Beauty'	wRoo oGar oWoo
	'Apricot Parrot'	oWoo
	'Aristocrat'	wRoo oWoo
	'Arma'	oWoo
un	'Asta Nielson'	oWoo
	'Attila'	wRoo oWoo
	'Ballade'	oWoo
	'Ballerina'	wRoo
	'Baronesse'	wRoo
un	'Bastogne'	oWoo
	batalinii	see **T. *linifolia*** Batalinii Group
	'Beauty of Apeldoorn'	oWoo
	Beauty Queen	oWoo
	'Bellflower'	oWoo
	'Bellona'	oWoo
	'Bing Crosby'	oWoo
	'Black Parrot'	wRoo oGar oWoo
	'Blenda'	wRoo oWoo
	'Bleu Aimable'	oWoo
	'Blue Heron'	oWoo
	'Blue Parrot'	wRoo oWoo
	'Burgundy Lace'	wRoo oGar oWoo
	'Candela'	oWoo
	'Cantor'	oWoo
	'Cape Cod'	oWoo
	'Capri'	oWoo
	'Carlton'	oWoo
	'Cassini'	oWoo
un	'Celebration'	oGar
	'Charles'	oWoo
	'China Pink'	wRoo oWoo
	'Christmas Dream'	oWoo
	'Christmas Marvel'	wRoo oGar oWoo
	clusiana	wRoo oWoo
un	'Compostella'	wRoo oWoo
	'Cordell Hull'	wRoo oWoo
un	'Coriolan'	oWoo
	'Corsage'	oWoo
	'Couleur Cardinal'	wRoo oGar oWoo
un	'Daydream'	oWoo
	'Don Quichotte'	wRoo oWoo
un	'Doorman's Record'	oWoo
	'Douglas Bader'	oWoo
	'Dreaming Maid'	wRoo
	'Dreamland'	wRoo
	'Dyanito'	wRoo
	'Elegant Lady'	oWoo
un	'Elite'	wRoo
	'Elizabeth Arden'	wRoo oWoo
	'Estella Rijnveld'	oWoo
un	'Esther'	oWoo
	'Fancy Frills'	oWoo
	'Fantasy'	oWoo
	'Fidelio'	wRoo oWoo
	'First Lady'	wRoo
	'Flaming Parrot'	oGar oWoo
un	'Francoise'	oWoo
	'Fringed Apeldoorn'	wRoo oWoo
	'Fringed Beauty'	oWoo
	'Gaiety'	oWoo
un	'Gander'	oWoo
un	'Gander's Rhapsody'	wRoo oWoo
	'Garden Party'	wRoo
	'Generaal de Wet'	oWoo
	'General Eisenhower'	oGar
	'Georgette'	oWoo
	'Golden Apeldoorn'	wRoo oWoo
	'Golden Artist'	oWoo
	'Golden Emperor'	oWoo
	'Golden Melody'	oWoo
un	'Golden West'	oGar
	'Gordon Cooper'	wRoo
	'Greenland'	see **T.** 'Groenland'
	'Groenland'	wRoo oGar oWoo
	'Gudoshnik'	oWoo
	'Halcro'	oWoo
	'Hamilton'	oWoo
	'Happy Family'	oWoo
	'Heart's Delight'	wRoo oWoo
un	'Heidrun Harden'	oWoo
un	'Hot Lips'	oGar
	humilis	wCul
	humilis var. ***pulchella*** Albocaerulea Oculata Group	wRoo
	humilis Violacea Group	wRoo
	humilis Violacea Group black base	oWoo
	'Ibis'	oWoo
	'Ile de France'	wRoo oWoo
	'Inzell'	oWoo
un	'Ivory Floradale'	oGar
	'Jacqueline'	oWoo
	'Jewel of Spring'	oWoo
un	'Judith Leyster'	oWoo
un	'Kaiserin Maria Theresia'	oWoo

	'Karel Doorman'	oWoo
	'Kees Nelis'	wRoo oWoo
	'Keizerskroon'	wRoo
	'Kingsblood'	wRoo oWoo
	'Leen van der Mark'	wRoo oWoo
	linifolia	oWoo
	linifolia (Batalinii Group) 'Bright Gem'	oWoo
	'London'	oWoo
un	'Los Angeles'	oWoo
	'Lucky Strike'	oGar oWoo
	'Lustige Witwe'	wRoo oWoo
	'Madame Lefeber'	wRoo oWoo
	'Magier'	wRoo oWoo
	'Maja'	wRoo oWoo
un	'Makassar'	wRoo
un	'Make-up'	oWoo
un	'Maria Theresia'	wRoo
	'Mariette'	oWoo
	'Marilyn'	oGar oWoo
	marjolletii	oWoo
	'Maureen'	wRoo oWoo
	'Maytime'	wRoo
	'Maywonder'	wRoo oWoo
un	'Meisner Porzellan'	oGar
	'Menton'	wRoo oWoo
	'Merry Christmas'	oWoo
	Merry Widow	see **T.** 'Lustige Witwe'
un	'Miranda'	wRoo oWoo
un	'Mirjoran'	oWoo
	'Mona Lisa'	oWoo
un	'Monsella'	oWoo
	'Monte Carlo'	wRoo oGar oWoo
	'Mount Tacoma'	oWoo
	'Mrs. John T. Scheepers'	wRoo oWoo
	'Murillo'	wRoo
	'Negrita'	wRoo oGar oWoo
	'New Design'	wRoo oGar oWoo
un	'Ollioules'	wRoo oWoo
	'Orange Bouquet'	oWoo
	'Orange Emperor'	wRoo
	'Orange Favorite'	wRoo oWoo
un	'Orange Goblet'	oWoo
	'Orange Monarch'	wRoo
un	'Orange Wonder'	oWoo
	'Oratorio'	oGar
	'Oriental Splendour'	wRoo
	'Oxford'	oWoo
un	'Pandion'	oWoo
	'Paul Richter'	oWoo
	'Pax'	oWoo
	'Peach Blossom'	wRoo oWoo
	'Peer Gynt'	oWoo
	'Peerless Pink'	oWoo
un	'Pink Attraction'	oWoo
	'Pink Diamond'	wRoo oGar
	'Pink Impression'	wRoo oGar oWoo
un	'Pink Supreme'	wRoo
	'Pinocchio'	oGar oWoo
	'Plaisir'	wRoo
	praestans 'Fusilier'	wRoo
	praestans 'Unicum'	oWoo
	'Preludium'	oWoo
un	'Primavera'	wRoo
un	'Princess Victoria'	oWoo
	'Prinses Irene'	wRoo oGar oWoo
	'Professor Roentgen'	oWoo
	'Prominence'	wRoo oWoo
	pulchella	see **T. *humilis*** var. ***pulchella*** Albocaerulea Oculata Group
	'Purissima'	wRoo oGar oWoo
	'Queen of Night'	wRoo oWoo
	'Queen of Sheba'	wRoo
	'Recreado'	oWoo
	'Red Emperor'	see **T.** 'Madame Lefeber'
un	'Red Gander'	oWoo
	'Red Georgette'	oWoo
	'Red Riding Hood'	wRoo oWoo
	'Red Shine'	oWoo
	'Renown'	oWoo
	'Rosalie'	wRoo
	'Rosario'	oWoo
	'Rosy Wings'	wRoo oWoo
un	'Salmon Parrot'	wRoo
un	'Scarlet Pimpernel'	wRoo
	'Scotch Lassie'	oWoo
	'Shakespeare'	oWoo
	'Shirley'	oWoo
	'Showwinner'	wRoo
	'Silentia'	wRoo oWoo
	'Snowflake'	oWoo
un	'Snowstar'	oWoo
	'Solva'	oWoo
	'Sorbet'	oWoo
	'Spring Green'	wRoo
	'Stresa'	wRoo oWoo
	'Striped Bellona'	oGar oWoo
	'Sweetheart'	oWoo
	sylvestris	oWoo
	tarda	wRoo oWoo
	'Temple of Beauty'	wRoo oWoo
un	'Tivoli'	oWoo
	'Toronto'	wRoo
	'Union Jack'	oGar
un	'Uptar'	oWoo
	'Valentine'	oWoo
	'West Point'	wRoo oWoo
	'White Dream'	wRoo
	'White Emperor'	see **T.** 'Purissima'
	'White Triumphator'	wRoo oWoo
	'Yellow Dawn'	wRoo oWoo
	'Yellow Emperor'	wRoo
	'Yokohama'	wRoo oWoo
	'Zombie'	oWoo
	TUPISTRA	
un	***chinensis*** 'Eco-China Ruffles'	oDan
	TUSSILAGO	
	farfara	oAls wFGN oGoo wFai oCrm oBar
	TWEEDIA	
	caerulea	oFor oHed wCoN wSte
	TYLECODON	
un	***bucholzianas***	oRar
un	***decipiens***	oRar
	TYPHA	
	angustifolia	oJoy oHug oSsd oTri
	latifolia	wSoo wGAc wNot oGar oSsd oBos wWat oTri oAld oSle
	latifolia 'Variegata'	oHug oGar oSsd wTGN
	laxmannii	oFor wSoo wGAc oHug oGar oSsd
	minima	oFor wHer wSoo oHug oGar oSsd oGre wTGN wBox wCoS
un	***minima*** 'Europa'	wGAc
	UEBELMANNIA	
un	***pectinifera*** var. ***pseudopectinifera***	oRar
	UGNI	
	molinae	wHer oOEx wGra
	ULLUCUS	
un	***tuberosus*** 'Como Verde de Monte'	oOEx
un	***tuberosus*** 'Pica de Pulga'	oOEx
un	***tuberosus*** 'Plata de Monte'	oOEx
un	***tuberosus*** 'Shrimp of the Earth'	oOEx
un	***tuberosus*** 'Yellow Jewels'	oOEx
	ULMUS	
	alata	oRiv
	crassifolia	oRiv
un	'Frontier'	oFor oWhi
	glabra	oFor
	glabra 'Camperdownii'	oFor wClo oDar oAls oGar oSho oBlo wTGN wSta wAva wCoS wWel
	x hollandica 'Jacqueline Hillier'	oFor wCol oAmb oGar oSho oGre
un	***minor gracilis***	oRiv
	parvifolia	oEdg oRiv oSle wSte
un	***parvifolia*** 'Corky'	oNWe
un	***parvifolia*** 'Corticosa'	oFor oGar oRiv oTin
un	***parvifolia*** 'Ed Wood'	oFor
	parvifolia 'Evergreen'	see **U. *parvifolia*** 'Sempervirens'
va	***parvifolia*** 'Frosty'	oFor oDar oSis
un	***parvifolia*** 'Golden Rey'	oFor
	parvifolia 'Hokkaido'	oFor oDar oSis oNWe oGre oBRG
	parvifolia 'Seiju'	oFor oDar oAmb oSis oGre oTin
	parvifolia 'Sempervirens'	oFor
	parvifolia 'Yatsubusa'	oFor oAmb oGar oRiv oGre oBRG oTin
	procera 'Louis van Houtte'	wHer
un	'Prospector'	oFor
	pumila	wShR
	rubra	oCrm
	UMBELLULARIA	
	californica	oFor wBur oEdg oAmb oGar oRiv oSho wRai oOEx oGre wWat wCCr wCoN wSte
	californica serpentine form	wCol

	Plant	Sources
un	**UNCARINA**	
	decaryi	oRar
	UNCINIA	
	uncinata	oHed wWin oSec
	UNGNADIA	
	speciosa	oFor wCCr
un	**UNICARIA**	
	tomentosa	oOEx
	UNIOLA	
	latifolia	see **CHASMANTHIUM *latifolium***
	paniculata	last listed 99/00
	URTICA	
	dioica	oGoo iGSc oUps oSle
	UTRICULARIA	
	dichotoma	wOud
	livida	wOud
	sandersonii	wOud
	UVULARIA	
	grandiflora	oFor wHer wCol oRus oSis wSte
	perfoliata	wHer oRus oSis
	sessilifolia	wCol oRus
	VACCINIUM	
un	***alaskaense***	oFor
	angustifolium	oFor wBCr oOEx
un	***angustifolium*** 'Brunswick'	oGre
un	***angustifolium*** 'Brunswick Dwarf'	wRai
	arboreum	last listed 99/00
	caespitosum	wThG wBur wCCr
	corymbosum (F)	oBlo wAva
	corymbosum 'Berkeley'	oAls oGar oSho wTGN
	corymbosum 'Bluecrop'	oFor wBur wClo oAls wCSG oGar oSho wRai oWnC oCir
	corymbosum 'Bluejay'	wBur wCSG oGre
	corymbosum 'Blueray'	oFor wBCr oAls oGar oSho oGre oWnC
	corymbosum 'Bluetta'	oGar oSho oGre oWnC
un	***corymbosum*** 'Brigitta'	oFor wSwC oGar oSho wRai
un	***corymbosum*** 'Brunswick'	wClo oWnC
un	***corymbosum*** 'Cape Fear'	oWnC
un	***corymbosum*** 'Chippewa'	wClo wRai oWnC
	corymbosum 'Collins'	oAls oGar
un	***corymbosum*** 'Colony'	oWnC
un	***corymbosum*** 'Darrow'	oFor wBur wClo oSho wRai oGre oWnC wTGN wBox oBRG
	corymbosum 'Duke'	wBur wClo oSho oWnC wTGN
	corymbosum 'Earliblue'	oFor wBur oAls oGar oSho
	corymbosum 'Elliott'	oFor wSwC wBur oGar oGre oWnC wBox
un	***corymbosum*** Georgia Gem'	oFor
un	***corymbosum*** 'Hardy Blue 1613-A'	oWnC
un	***corymbosum*** 'Hardyblue'	wBur
	corymbosum 'Herbert'	oGar oSho
	corymbosum 'Ivanhoe'	oGar oSho
	corymbosum 'Jersey'	oSho wRai oWnC
un	***corymbosum*** 'Legacy'	wBur wClo oGar wRai
un	***corymbosum*** 'Misty'	oGar wRai
	corymbosum 'Nelson'	oFor wBox
	corymbosum 'North Country'	oFor wClo oEdg oOEx oWnC
	corymbosum 'Northblue'	oFor wClo oGar oOEx
	corymbosum 'Northland'	oFor wBCr wBur wClo
	corymbosum 'Northsky'	oFor wCSG wRai oOEx oGre
un	***corymbosum*** 'Olympia'	oFor wClo oSho wRai oWnC wBox
un	***corymbosum*** 'O'Neal'	oFor oGre oWnC wTGN
	corymbosum 'Patriot'	wBCr wSwC wBur wClo oGar wRai oGre oWnC
un	***corymbosum*** 'Polaris'	wClo oGre
un	***corymbosum*** 'Reveille'	wRai
	corymbosum 'Rubel'	wSwC wBur oWnC oBRG
un	***corymbosum*** 'Sierra'	last listed 99/00
	corymbosum 'Spartan'	oFor wBur wCSG oGar oSho wRai oGre oWnC wBox
	corymbosum 'Toro'	wBur wClo oGar wRai oGre oWnC wBox
	crassifolium	oFor wSta
	crassifolium 'Bloodstone'	wWel
	crassifolium 'Well's Delight' (F)	oFor oGos oGar oGre oWnC wWel
	delavayi	oFor wHer oBov
	deliciosum	wRai
	floribundum	oOEx
	glaucoalbum	oGos oBov
	glaucoalbum HWJCM 099	wHer
	macrocarpon (F)	oFor wClo oAls oGoo oBRG wAva
	macrocarpon 'Hamilton'	oGos oGar oSis wMtT wRai oBov oGre oWnC wTGN oBRG
	macrocarpon 'Pilgrim'	last listed 99/00
un	***macrocarpon*** 'Stevens'	oFor wBur wRai oWnC
	membranaceum	oFor wThG wBur wShR wPla

	Plant	Sources
	moupinense	oFor oGos oGre wSta oBRG wWel
	myrtilloides	oFor
	nummularia	wHer oBov
	ovalifolium	oFor wRai wFFl oGre
	ovatum	oHan wThG wWoB wBur wClo wNot oAls wShR oGar oBos oRiv oSho oBlo wRai oOEx oGre wWat wCCr oWnC oTri wSta oAld wAva wRav oSle wWel
un	***ovatum*** 'Maureen's Select'	oGos
	ovatum x ***mortinia***	wCCr
un	***ovatum*** 'Thunderbird'	oGos wBur wTGN wBox
un	***ovatum*** 'Wunderlicht'	oGos
	oxycoccos	oGar oBos oWnC wSta
	parvifolium	oFor oHan wThG wWoB wBur wNot oAls wShR oBos oRiv wRai wFFl wWat wCCr oTri oAld wRav
	retusum	wMtT
	scoparium	oBos
	stramineum	oFor
un	'Sunshine Blue'	oGos wSwC wBur wClo oGar oSho wRai oGre oWnC wTGN oBRG
un	'Top Hat'	oFor wClo oSis oSho oOEx oBRG
	uliginosum	last listed 99/00
	vitis-idaea (fruit)	oFor wCul wBCr wClo oAls oGar oGoo oBlo oSqu wTGN wSta wWel
un	***vitis-idaea*** 'Erntdank'	oFor
un	***vitis-idaea*** 'Erntkrone'	oFor
	vitis-idaea Koralle Group	oFor wBur oOEx oGre oWnC
un	***vitis-idaea*** 'Masovia'	oFor oGos oBRG
	vitis-idaea var. ***minus***	wThG wShR oGoo oSis oBov oWnC wSta oBRG wWel
	vitis-idaea 'Red Pearl'	wBur wRai oGre oBRG wWel
un	***vitis-idaea*** 'Sanna'	wRai
un	***vitis-idaea*** 'Scarlet'	oOEx
un	***vitis-idaea*** 'Sussi'	wRai
	VALERIANA	
	arizonica	last listed 99/00
un	***occidentalis***	wWld
	officinalis	oFor wCul oAls wFGN oGoo wFai wCCr iGSc oWnC oCrm oBar oSec wCoN oEga
un	***officinalis*** 'Arterner Zuchtung'	oBar oUps
	phu 'Aurea'	wHer wCri oGar oOut wFai wBox oSec
un	***scouleri***	oBos
	supina	wMtT
	VALLISNERIA	
	americana	wSoo wGAc oHug oGar oSsd
ch	***gigantea***	see **V. *americana***
	VANCOUVERIA	
	chrysantha	oBov
	hexandra	wHer oHan wThG oNat wNot wCol oAls oRus oGar oSis oBos oBov wPir wWat oTri oAld wSte
	parviflora	see **V. *planipetala***
	planipetala	wHer wThG oJil wCol oRus oSis oNWe oBov wPir wSte
	VELTHEIMIA	
	bracteata	oTrP
	capensis	oTrP
	VERATRUM	
	californicum	oBos
	formosanum	wHer
	sp. DJHC 121	wHer
	viride	wHer oAld
	VERBASCUM	
	adzharicum	wBWP
	bakerianum	wBWP
	blattaria f. ***albiflorum***	wCul
	'Bold Queen'	wSnq
	bombyciferum	last listed 99/00
	bombyciferum Arctic Summer	see **V. *bombyciferum*** 'Polarsommer'wCul
	bombyciferum 'Polarsommer'	wFai wBWP
	bombyciferum 'Polarsommer' seed grown	oRus
	chaixii	last listed 99/00
	chaixii 'Album'	wHer wCul wSwC oNat wFai wBWP
	chaixii 'Cotswold Queen'	oAls
	chaixii 'Pink Domino'	wHer
	chaixii 'White Domino'	wGAc
	dumulosum	oSis
	'Golden Wings'	oHed oSis
	'Helen Johnson'	oFor wHer wWoS oJoy oGar oNWe oGre wMag oEga
	'Jackie'	oFor wHer wWoS wSwC oJoy oHed oRus oGar oNWe oGre
	'Letitia'	oSis
	nigrum	wCul

	olympicum	wCul wCSG oCrm wBWP
	phoeniceum	oFor wCul wTho wCSG oGar iGSc oUps
	phoeniceum 'Flush of White'	oSho oWnC wBWP
	'Silber Kandelaber' seed grown	oRus
	'Southern Charm'	oGar oNWe oCir
	'Spica'	wSwC
	'Spica' seed grown	oRus
	thapsus	iGSc oBar
	wiedemannianum	oSec
	VERBENA	
	'Abbeville'	wFGN
	'Apple Blossom'	wFGN
un	'Blue Princess'	wFGN
	bonariensis	oFor wCul wTho wSwC oDar wHig oAls oHed wCSG oGar oGoo oSis wRob oBlo oSqu oGre wPir wMag oSec wBWP oCir wAva oUps wSte
	canadensis	oGar
	canadensis pink	oAls
	'Carousel'	wFGN
	'Edith Eddleman'	oAls
un	'Greystone Daphne'	wFGN
	hastata	wCul oGoo oBos iGSc
	hastata 'Alba' seed grown	wHer
un	***hastata*** 'Simpler's Joy'	wCSG
	'Homestead'	see **V.** 'Homestead Purple'
	'Homestead Purple'	oFor wCul wTho oJoy oAls oHed wFGN wCSG oGar oGoo oSis oSha wRob oGre wHom wMag wTGN wCoS
	x hybrida pink	oBlo
	x hybrida purple	oBlo
	x hybrida red	oBlo
un	'Katie'	wFGN
	officinalis	wFGN oGoo wFai iGSc oCrm oBar oUps
un	'Old Royal'	see **V.** 'Old Royal Fragrance'
un	'Old Royal Fragrance'	wWoS wFGN oGoo
un	'Pastel Pink'	oHed
	patagonica	see **V. *bonariensis***
	peruviana red	oSis
	'Pink Parfait'	last listed 99/00
	rigida	wHig oGoo
	rigida 'Polaris'	last listed 99/00
	Romance Series scarlet	oCir
	'Silver Anne'	wFGN wCSG
	'Sissinghurst'	oJoy wFGN oSha
	'Snow Flurry'	wFGN
	stricta	oGoo
un	Tapien Blue-Violet	wTGN
	Tapien Pink / 'Sunver'	oJoy wTGN
un	'Taylortown Red'	oAls
	Temari Red Scarlet / 'Sunmarisu'	oGar
	Temari Violet / 'Sunmariba'	wTGN
	tenuisecta	wFGN
	tenuisecta 'Edith'	last listed 99/00
	VERBESINA	
	alternifolia	wCCr
	VERNONIA	
	arkansana	oFor
un	***chinensis***	oFor
	fasciculata	oFor
	noveboracensis	last listed 99/00
	VERONICA	
	allionii	oAmb oSis oSqu oWnC wSta
un	***alpina*** 'Alba'	oFor oWnC
	americana	oBos oAld
	armena	oSis
	austriaca 'Ionian Skies'	wEde
	austriaca ssp. ***teucrium***	wTho
	austriaca ssp. ***teucrium*** 'Crater Lake Blue'	wCul oJoy oSis oNWe oSho oOut wRob wFai oWnC wMag wEde oLSG oVBI
	austriaca ssp. ***teucrium*** 'Royal Blue'	oFor oWnC
	caespitosa	oSis
va	***chamaedrys*** 'Miffy Brute'	oFor wHer wWoS oSis wRob
	cinerea	oSis
	cusickii	oBos
	dwarf white	oSqu
un	***forrestii***	oFor
	gentianoides	oJoy wCol oHed wCSG wWld wSte
	gentianoides 'Alba'	oFor
	gentianoides 'Barbara Sherwood'	wWoS
	gentianoides 'Nana'	oSis
	gentianoides 'Tissington White'	oHed
	gentianoides 'Variegata'	wWoS oNWe wRob wSte
un	'Giles Van Hees'	oFor wWoS oSis

	'Goodness Grows'	oFor oJoy oAls wFGN oAmb oSis oWnC oSle
un	'Lavender Charm'	oFor wWoS
	liwanensis	oFor oAls oSis wFai oWnC oSle
	longifolia	wCul
	longifolia 'Blauriesen'	oJoy wEde
	longifolia 'Foerster's Blue	see **V. *longifolia*** 'Blauriesen'
va	***longifolia*** 'Joseph's Coat'	wHer
	longifolia 'Rosea'	oGoo
	macrostachya	oJoy
un	'Midnight'	oSha
va	***montana*** 'Corinne Tremaine'	oFor wWoS oDan wRob oGre
va	'Noah Williams'	oFor wHer oGar oGre wTGN
	officinalis	oGoo iGSc oCrm
	oltensis	oSis
	onoei	last listed 99/00
	orientalis ssp. ***orientalis***	last listed 99/00
	ornata	last listed 99/00
	pectinata	oSqu
	peduncularis	oSqu
	peduncularis 'Georgia Blue'	oFor wHer wWoS wSwC oJil oJoy wHig oAls oHed wFGN oSis wMag wEde oSle wSnq
	petraea	oWnC oSle
	petraea 'Madame Mercier'	oFor oJoy oSis
	'Pink Damask'	wHer wWoS
	prostrata	oSis
	prostrata 'Heavenly Blue'	oFor oLSG
	prostrata 'Mrs. Holt'	oFor oJoy
	prostrata 'Nana'	oSqu
un	***prostrata*** 'Nestor'	oWnC
	prostrata 'Trehane'	oFor wHer wWoS wCul wCSG oSis wEde oLSG
	repens	wMag oUps
	repens 'Alba'	oJoy
un	***repens*** 'Aurea'	oSec
un	***repens*** 'Sunshine'	oFor wWoS oAls
	saturejoides	oNWe
	schmidtiana	last listed 99/00
un	***schmidtiana*** 'Candida'	oSis
	scutellata	wFFl
un	'Snow White'	wFai oLSG
	spicata	oFor wTho wCSG oCir
	spicata 'Blue Carpet'	see **V. *spicata*** 'Nana Blauteppich'
	spicata 'Blue Charm'	oSle
	spicata 'Erika'	oAls wFai wTGN oLSG oSle
	spicata 'Icicle'	oFor wCul oAls oGar oSho oLSG
	spicata ssp. ***incana***	wTGN oSec oUps
	spicata ssp. ***incana*** 'Nana'	oSis
	spicata 'Minuet'	oFor wCul oSqu oLSG
	spicata 'Nana Blauteppich'	oWnC
	spicata var. nana 'Blue Carpet'	see **V. *spicata*** 'Nana Blauteppich'
	spicata Red Fox	see **V. *spicata*** 'Rotfuchs'
	spicata 'Romiley Purple'	last listed 99/00
	spicata rosea	see **V. *spicata*** 'Erika'
	spicata 'Rotfuchs'	oFor oGar oSho wFai oWnC oCir oLSG oUps
	spicata 'Sightseeing'	oUps
	'Sunny Border Blue'	oFor oNat oJoy oAls oSho oOut wFai oWnC wMag wTGN oEga
	surculosa	oAls oSis
	teucrium	see **V. *austriaca*** ssp. ***teucrium***
	turrilliana	wHer
	'Waterperry Blue'	wHer wWoS oHed wFGN oSis
	'White Icicle'	see **V. *spicata*** 'Icicle'
un	'Wood's Nymph'	wCol
	wormskjoldii	oFor oLSG
	VERONICASTRUM	
	virginicum	oFor oGoo wFai iGSc oCrm
	virginicum album	oFor wCSG oGar oOut oGre wBox
	virginicum 'Fascination'	wWoS
	virginicum var. ***incarnatum***	oGar oSho oWnC
	virginicum 'Lila Karina'	wFai
	virginicum 'Pink Glow'	oFor oUps
	virginicum roseum	see **V. *virginicum*** var. ***incarnatum***
	virginicum var. ***sibiricum***	wCol
	VESTIA	
	foetida	oCis wCCr
	lycioides	see **V. *foetida***
	VETIVERIA	
	zizanioides	oGoo iGSc oCrm
	VIBURNUM	
	acerifolium	oFor wCCr
un	***acuminatum***	oFor
	'Allegheny'	oFor wCCr

	atrocyaneum	last listed 99/00
	awabuki	oCis oGre oWnC
un	***awabuki*** 'Chindo'	wHer oCis oSho
	betulifolium	oFor wCCr
	betulifolium EDHCH 97266	wHer
	bitchiuense	oFor wHer wCCr
	x bodnantense	oDan wWel
	x bodnantense 'Charles Lamont'	oGos
	x bodnantense 'Dawn'	oFor wCul oGos oJoy oRiv oCis oWhi oGre wFai wAva wCoN wWel
	x bodnantense 'Deben'	wHer oGos
	x bodnantense 'Pink Dawn'	wClo oNat oDar oAls oGar oSho oBlo wSta wCoS
qu	'Bristol Snowflake'	wMag
	burejaeticum	oFor
	x burkwoodii	oAls oGar oDan oSho oWhi oBlo oGre wCCr oWnC wTGN wSta wAva
	x burkwoodii 'Anne Russell'	oFor oGos wClo
	x carlcephalum	oFor wCul oBlo oGre oSle wWel
	carlesii	wCSG oGar oOut oBlo wCCr oGue wCoS wWel
	carlesii 'Aurora'	oFor oGos
	carlesii 'Compactum'	wSta wWel
	carlesii DJH 288	wHer
	cassinoides	oFor wCCr
	'Cayuga'	oGos
	'Chesapeake'	oFor oGos oGre
un	'Chippewah'	oFor
	cinnamomifolium	oFor wHer wCSG oGar oCis oSho oBlo oGre wCCr wSta wWel
	'Conoy'	oFor wHer oGos
	cotinifolium	last listed 99/00
	cylindricum	oFor wCol oCis
	cylindricum HWJCM 434	wHer
	davidii	oFor oAls oGar oRiv oSho oBlo oGre wCCr oTPm oWnC wSta wCoS wWel
	dentatum	oFor oGar
un	***dentatum*** 'Autumn Jazz'	oGar oGre wWel
un	***dentatum*** 'Cardinal Candy'	wTGN
un	***dentatum*** 'Chicago Lustre'	oFor
un	***dentatum*** 'Northern Burgundy'	oFor
un	***dentatum*** 'Perle Bleu'	oGos
	dilatatum	oFor wCCr
un	***dilatatum*** 'Asian Beauty'	oFor
	dilatatum 'Catskill'	oFor
	dilatatum 'Erie'	last listed 99/00
	dilatatum 'Iroquois'	oGos
un	***dilatatum*** 'Michael Dodge'	oFor
	dilatatum 'Oneida'	oFor
	edule	wFFl wWat
	ellipticum	oFor
	erubescens DJHC 028	wHer
	'Eskimo'	oFor oGos oGar oGre wWel
un	***fargesii***	oCis
	farreri	oCis
	farreri 'Album'	see **V. *farreri*** 'Candidissimum'
un	***farreri*** 'Andangulari'	oSho
	farreri 'Candidissimum	oGos
	farreri 'Nanum'	oGos
	foetidum DJHC 489	wHer
un	***foetidum*** var. ***quadrangularum***	see **V. *foetidum*** var. ***rectangulatum***
	foetidum var. ***rectangulatum***	wCol oCis
	furcatum	oFor
	grandiflorum	oFor oGos
	harryanum	wHer
	henryi	wHer
	x hillieri 'Winton'	oFor wHer
	hupehense	oFor
un	'Huron'	oFor oGos
	japonicum	wHer oGar
	x juddii	wHer oGos oGue
	lantana	oFor wCCr
	lantana 'Mohican'	oFor oGos wClo oGar
	lantana 'Variegatum'	oFor oGar oSho oGre wWel
	lentago	oFor oGar oOEx oGre
	lobophyllum	wCCr
	macrocephalum	wHer
	macrocephalum f. ***keteleeri***	wHer
	macrocephalum 'Sterile'	see **V. *macrocephalum***
	'Mohawk'	oFor oGos wWel
	mullaha	oFor wCCr
	nervosum DJHC 635	wHer
	nudum	oGre
	nudum 'Winterthur'	oFor oGos oGre wTGN
	odoratissimum	oFor wCCr
	odoratissimum var. ***awabuki***	see **V. *awabuki***
	opulus	oOEx oGre oCrm wWel
	opulus 'Aureum'	oFor oGar oGre wFai wCCr
	opulus 'Compactum'	oFor wCul oGos oGar oRiv oGre wKin wTGN wAva
un	***opulus*** 'Leonard's Dwarf'	oFor
	opulus 'Nanum'	oFor oGos wCSG oGar oBov oGre oWnC wWel
	opulus 'Roseum'	oFor oAls wCSG oGar oSho oBlo oGre oTPm oWnC wTGN wSta wAva oSle wCoS wWel
	opulus 'Roseum' tree form	oAls
	opulus 'Sterile'	see **V. *opulus*** 'Roseum'
un	***opulus*** 'Ukraine'	oGre
	opulus 'Xanthocarpum'	oGre wCCr
	plicatum	wFai
un	***plicatum*** 'Kern's Pink'	oFor
	plicatum 'Lanarth'	oFor wCCr
	plicatum 'Mariesii'	oFor wCSG oAmb oGar oRiv oSho oGre oWnC wSta oBRG wAva wWel
	plicatum 'Nanum Semperflorens'	oFor wHer oGos oGue
un	***plicatum*** 'Newport'	wHer wCul oGos oGar wWel
	plicatum 'Pink Beauty'	wClo
	plicatum f. ***plicatum***	see **P. *plicatum*** 'Sterile'
	plicatum 'Popcorn'	oFor oGos
	plicatum 'Roseum'	wHer
	plicatum 'Rotundifolium'	oFor
	plicatum 'Shoshoni'	oFor wCul oAmb oGar oGre
	plicatum 'Sterile'	wHer
	plicatum 'Summer Snowflake'	oFor wCul oGos oAmb oGar oNWe oGre wTGN wSta wSnq
	plicatum f. ***tomentosum***	oAls oBlo wWel
un	***plicatum*** f. ***tomentosum*** 'Fugisanensis'	oFor
qu	***plicatum*** f. ***tomentosum*** 'Nanum'	oFor
	plicatum 'Watanabe'	see **V. *plicatum*** 'Nanum Semperflorens'
	'Pragense'	oFor wHer oGar oCis oBov wCCr
qu	***x pragense*** 'Decker'	oGos
	propinquum	oGar oRiv oCis oBlo wCCr wSta wSte
	prunifolium	oFor oGoo oOEx
	rafinesquianum	oFor
	x rhytidocarpum	oFor oGre
	rhytidophyllum	oGar oCis oBlo oGre wCCr wSta wCoN wWel
	rhytidophyllum 'Willowwood'	oFor oGos
	sargentii	wCSG oGre
	sargentii HC 970108	wHer
	sargentii 'Onondaga'	wCul oHed oGar oOut oGre wFai oWnC wTGN wWel
	sargentii 'Susquehanna'	oFor oGos
	schensianum	oFor
	setigerum	oFor oGar oOut
	'Shasta'	oFor oGos oGar oGre wWel
	sieboldii	oFor oGre
	sieboldii select form	wHer
	sieboldii 'Seneca'	last listed 99/00
un	***sp.*** Chinese Cherryberry (fruit)	oOEx
	sp. EDHCH 97070	wHer
	sp. from Madrona	oCis
	suspensum	oGar oSho
	tinus	wCoN
	tinus 'Bewley's Variegated'	oFor oGar oSho oGre wCCr
	tinus 'Compactum'	oFor oAls oGar oRiv oSho oBlo oGre oTPm oWnC wTGN wSta oCir wAva wCoS wWel
	tinus 'Lucidum'	wCCr
qu	***tinus*** 'Nanum'	oBRG
	tinus 'Robustum'	oGar oSho oGre wCCr oWnC wCoS wWel
	tinus 'Spring Bouquet'	see **V. *tinus*** 'Compactum'
	tinus 'Variegatum'	wHer oCis
	trilobum	oTrP wBCr wCSG oGoo oRiv wRai oWnC
	trilobum 'Alfredo'	oGar
	trilobum 'Bailey Compact'	oGar
	trilobum 'Compactum'	oGos oNat oRiv wTGN
	trilobum 'Wentworth'	oFor oGos wWel
	utile	wHer oCis
	wrightii	last listed 99/00
	VICIA	
	gigantea	wFFl
	unijuga HC 970165	last listed 99/00
	VINCA	
	major	oGar wCoN oUps wCoS
va	***major*** 'Maculata'	oAls
	major 'Variegata'	oUps
	minor	oAls oGar oSho wSta oCir wCoN oUps oSle
	minor f. ***alba***	oFor oAls oGar oWnC wSta oSle
	minor 'Argenteovariegata'	oFor oAls oGre oWnC wBox wCoS

	minor 'Atropurpurea'	oFor oAls oGre oWnC
	minor 'Aurea'	oAls
	minor blue	oGre wSta
	minor 'Bowles'	see **V. *minor*** 'La Grave'
un	***minor*** 'Double Bleu'	wTGN
	minor 'Gertrude Jekyll'	oFor oAls oGre
	minor 'La Grave'	oFor oAls oGar oWnC wCoS
do	***minor*** 'Multiplex'	oFor
	minor 'Purpurea'	see **V. *minor*** 'Atropurpurea'
un	***minor*** 'Ralph Shugert'	oAls oGar oGre
	minor 'Rubra'	see **V. *minor*** 'Atropurpurea'
un	***minor*** 'Sterling Silver'	oFor oAls
un	***minor*** 'Valley Glow'	oAls
	minor 'Variegata'	see **V. *minor*** 'Argenteovariegata'
qu	***minor*** 'Wine'	oGar
	VINCETOXICUM	
	nigrum	oFor
	VIOLA	
	adunca	oBos wMtT wFFl
	'Alice Witter'	oRus
un	'Alpine Summer'	oLSG
	'Ardross Gem'	oRus
	'Arkwright's Ruby'	oUps
	'Belmont Blue'	wWoS
un	'Better Times'	wWoS oRus
un	'Black Magic'	wWoS oHed
	'Blue Moon'	wHer wWoS
un	'Blue Remington'	oRus wTGN
	'Boughton Blue'	see **V.** 'Belmont Blue'
	calcarata	oSis
	canadensis	oRus
	'Cat's Whiskers'	wWoS
	'Chantreyland'	oUps
	'Columbine'	wWoS
	'Comte de Brazza'	wFGN
	cornuta	oUps
	cornuta 'Alba Minor'	wWoS
un	***cornuta*** 'Barford Blue'	wHer
un	***cornuta*** 'Helen Mount'	oLSG
	cornuta 'Minor Alba'	see **V.** 'Alba Minor'
un	***cornuta*** Toyland mixed	oUps
	corsica	oJoy
	cucullata	oRus
	cucullata 'Gloriole'	oRus
un	***cucullata*** 'Purple Showers'	oFor wWoS
	cucullata 'Red Giant'	wTGN
un	***cucullata*** 'Russian Blue'	oRus
	'Dancing Geisha'	wHer wWoS wCol oAmb oOut
	'Delicia'	wWoS oRus
	delphiniifolia	see **V. *pedatifida***
	'Desdemona'	wHer wWoS
	dissecta	oHed oOut wHom oUps
	'Duchesse de Parme'	oFor
un	'Eileen'	wWoS oRus
	'Elaine Cawthorne'	wWoS
un	'Elephant Ears'	wWoS
	'Etain'	wWoS oRus
	flettii	oSis
	'Gazelle'	wWoS oRus
	glabella	oHan wNot oRus oBos wFFl wWat oTri oAld
	grypoceras	wTGN
	grypoceras var. ***exilis***	oRus wEde
	grypoceras var. ***exilis*** 'Variegata'	wHom
	'Heaselands'	wHer wWoS
	hederacea	oTrP wFGN oUps
un	***hirsutula***	oSis
	'Irish Molly'	wHer wWoS
un	***japonica*** f. ***variegata***	oNWe
	'Jersey Gem'	oLSG
un	'Joel Strahm'	wWoS oRus
	jooi	last listed 99/00
	'Julian'	wWoS oRus
	koreana	see **V. *grypoceras*** var. ***exilis***
un	'La Violette'	wTGN
	labradorica (see Nomenclature Notes)	oFor wCul wCri oAls wFGN oGoo wPir wHom oWnC oBar oUps oSle wSnq
	labradorica 'Purpurea'	see **V. *riviniana*** Purpurea Group
	'Lianne'	oAls oRus oGoo wSnq
	'Lorna Cawthorne'	wWoS oRus
un	'Loveliana'	wCSG
	lutea 'Splendens'	oLSG
	'Maggie Mott'	oFor wWoS oRus
	missouriensis	oRus
	'Nellie Britton'	oRus
	nuttalii	oBos
	obliqua	see **V. *cucullata***
	odorata	wFGN wFai wHom wMag oBar wBox wSta oUps
	odorata 'Alba'	oCir
do	***odorata*** 'Alba Plena'	wHer
un	***odorata*** 'Alice de Rothschild'	wCul
un	***odorata*** 'Lamb's White'	oRus
	odorata pink	oRus wFGN
	odorata rosea	wCul
	odorata violet	oRus
	odorata white	oRus
	palmata	oRus
	palustris	wFFl
	papilionacea	see **V. *sororia***
un	'Partly Cloudy'	wCol
un	'Pebbles'	oRus
	pedata	oFor wCul oRus oUps wSnq
	pedata 'Bicolor'	oRus oSis
	pedata var. ***concolor***	oSis
	pedatifida	oFor oHed wFGN oBar
un	Perfection Series	oLSG
	pubescens	oRus
un	'Purple Showers'	oGar
	'Queen Charlotte'	oFor wCul
	'Raven'	wHer wWoS
	'Rebecca'	wWoS oRus oWnC oUps
	'Red Charm'	wTGN
	riviniana Purpurea Group	oHed wCCr
un	'Robin Dale'	wWoS
	'Rodney Davey'	wWoS wCol
	'Rosine'	oAls
	rotundifolia	oRus
	'Royal Robe'	oAls oRus oGoo oCir
	sagittata	oRus
	sempervirens	oHan wThG wNot oRus oBos wFFl wWat oTri
	sororia	oFor oRus
	sororia 'Freckles'	oFor wFGN oGar oGoo oOut wHom wMag wTGN oCir wEde
	sororia 'Priceana'	wFGN
	striata	oRus oUps
	'Swanley White'	see **V.** 'Comte de Brazza'
	'Sylvia Hart'	oSis oUps
	'Talitha'	wWoS oRus
un	'Tammy Dale'	wWoS
	tricolor	wCol oHed wFai iGSc
	variegata	see **V. *tricolor***
	verecunda var. ***yakusimana***	wCol oRus
	'Victoria Cawthorne'	wHer oHed
un	'Violin'	oLSG
	'Vita'	oHed
	'White Czar'	oFor oAls oRus
	x wittrockiana	oGar
	x wittrockiana Bambini Series	oLSG
un	***x wittrockiana*** Rococo mixed	oUps
un	***xanthopetala***	oNWe
	yakusimana	see **V. *verecunda*** var. ***yakusimana***
	yezoensis	last listed 99/00
	VITALIANA	
	primuliflora	oSis
un	***primuliflora*** ssp. ***cinerea***	wMtT
	VITEX	
	agnus-castus	oFor oGar oGoo oDan oRiv oSho oOEx oGre wFai wPir wCCr iGSc oBar oSec wCoN oUps wCoS wWel
un	***agnus-castus*** 'Abbeville Blue'	oFor
un	***agnus-castus*** 'Blushing Spires'	oUps
	agnus-castus var. ***latifolia***	oFor
un	***agnus-castus*** 'Shoal Creek'	oFor
	agnus-castus 'Silver Spire'	oGar wPir
	cannabifolia	see **V. *incisa***
	incisa	oFor oOEx wCCr iGSc
	incisa EDHCH 97038	wHer
	negundo var. ***heterophylla***	see **V. *incisa***
	rotundifolia	oFor oSho
	VITIS	
un	'Agawam'	oLon
un	'Alden'	wBCr oLon
un	'America'	oLon
	amurensis	oFor wCul wCCr
	amurensis DJH 196	wHer
	arizonica	oCis
un	'August Giant'	oLon
	'Aurore'	oLon

	'Baco Noir'	oLon
un	'Barry'	oLon
un	'Bath'	oLon
un	'Beaumont'	oLon
un	'Bell'	oLon
un	'Berckmans'	oLon
un	'Bokay'	oLon
	'Brilliant'	oLon
un	'Bronx Seedless'	oLon
un	'Buffalo'	oAls oLon
	californica	oFor wCCr
un	***californica*** 'Roger's Red'	wHer
	'Canadice'	wClo oAls oGar oWhi wTGN oLon
un	'Canandaigua'	oLon
	'Cascade'	oLon
	'Cayuga'	oLon
	'Chambourcin' JS 26.205	oLon
un	'Chancellor' S 7053	oLon
un	'Chelois' S 10.878	oLon
un	'Christmas'	oLon
	coignetiae	oGos oDar oAmb oGar oDan oCis oGre wCCr oWnC oBRG wAva wSte
	coignetiae Claret Cloak /'Frovit'	wHer
	coignetiae HC 970517	wHer
un	'Daitier Saint Vallier' SV 20.365	oLon
	davidii	last listed 99/00
un	'De Chaunac' Seibel 9549	oLon
un	'Delaware'	oLon
un	'Delicatessen'	oLon
un	'Diamond'	oLon
un	'Dobson'	oLon
un	'Dutchess'	oLon
un	'Edelweiss'	wBur oLon
un	'Einset'	wBur oAls oSho oWhi oWnC oLon
un	'Elizabeth'	oLon
un	'Eona'	oLon
un	'Erie'	oLon
un	'Esprit'	oLon
	ficifolia	see **V. *thunbergii***
	'Flame'	oAls oGar wRai oWnC oLon
qu	***flexons***	oRiv
	flexuosa HC 970731	wHer
	'Foch'	see **V.** 'Marechal Foch'
un	'Glenora'	oAls oGar oSho oOEx wTGN oLon
un	'Goff'	oLon
	'Golden Muscat'	oAls oGar oWnC oLon
un	'Greek Perfume'	oLon
	'Himrod'	wBur oAls oGar oSho wRai oWnC wCoN oLon
un	'Horizon'	oLon
	'Island Belle'	see **V. *labrusca*** 'Campbell Early'
un	'Ivan'	oLon
un	'John Viola'	oLon
un	JS 12.428	oLon
un	'Kay Gray'	oLon
un	'KeeWahDin'	oLon
un	'Kishwaukee'	oLon
un	***labrusca*** 'Alwood'	wBCr
	labrusca 'Campbell Early'	wBCr wBur oAls oLon
	labrusca 'Catawba'	oWnC
	labrusca 'Concord'	oAls oGar oWnC wCoS
	labrusca 'Concord Seedless'	wBur oGar wTGN
	labrusca 'Iona'	oLon
	labrusca 'Niagara'	oAls oWnC oLon
un	'LaCrosse'	oLon
un	'Lakemont'	wClo oWnC oLon
	'Leon Millot'	oLon
un	'Liberty'	oLon
un	'Lindley'	oLon
un	'Long John'	oLon
un	'Lucy Kuhlman'	oLon
un	'Lynden Blue'	wClo
	'Marechal Foch'	wBur oLon
un	'Mars'	oLon
un	'McCampbell'	wBCr oGar oLon
un	'Mitchell'	oLon
un	'Munson'	oLon
un	'Ontario'	oLon
un	'Osbu'	oLon
	piasezkii var. ***pagnuccii***	wHer
un	'President Van Buren'	wBCr
un	'Price'	wRai oLon
un	Ravat 262	oLon
un	'Rayon d'Or' Seibel 4986	oLon
un	'Reliance'	wClo oOEx oLon

un	'Remaily'	oAls oGar oOEx oLon
	riparia	oFor wTGN
un	'Rose Belle'	oLon
un	'Rougeon' S 5898	oLon
un	'Royal Blue'	oLon
un	'Ruby'	oLon
	'Ruby Seedless'	oGar oWnC oLon
un	'Saint Croix'	oLon
un	'Saint Pepin'	oLon
un	'Saturn'	oLon
un	Seibel 13666	oLon
	'Seneca'	oLon
un	Seyve Villard 5.247	oLon
un	Seyve Villard 23.512	oLon
un	'Shakoka'	oLon
un	'Sonoma'	oLon
un	'Steuben'	oLon
	'Suffolk Red'	wBCr oAls oGar oSho wRai oLon
un	'Sweet Seduction'	oWhi wRai
un	'Swenson Red'	wBur oLon
un	'Swenson White'	oLon
	'Thompson Seedless'	see **V. *vinifera*** 'Sultana'
	'Thornton'	oLon
	thunbergii	last listed 99/00
un	***thunbergii*** var. ***lobata***	wHer oRiv oCis
	thunbergii var. ***sinuata***	oFor
un	'Totmur' Baco 2-16	oLon
un	'Valiant'	oLon
un	'Van Buren'	oLon
un	'Vanessa'	wBur wRai oOEx oWnC oLon
un	'Variegata'	see **V. *vinifera*** 'Variegata'
un	'Ventura'	oLon
un	'Venus'	wBur oWhi wRai oOEx oLon
un	'Verdelet' S 9110	oLon
un	'Vidal Blanc'	oLon
	'Villard Blanc' SV 12.375	oLon
un	'Vineland'	oLon
un	***vinifera*** 'Beauty Seedless'	oLon
	vinifera 'Black Corinth'	oLon
	vinifera 'Black Monukka'	oGar wTGN
un	***vinifera*** 'Black Rose'	oLon
	vinifera 'Cabernet Sauvignon'	oGar wRai oWnC
	vinifera 'Chardonnay'	oGar
	vinifera 'Chasselas'	wClo
	vinifera 'Chenin Blanc'	oWnC
un	***vinifera*** 'Delight'	oLon
un	***vinifera*** 'Early Muscat'	oLon
un	***vinifera*** 'Fantasy Seedless'	oLon
	vinifera 'Gewuerztraminer'	oGar oLon
un	***vinifera*** 'Gold'	oLon
	vinifera 'Green Veltliner'	see **V. *vinifera*** 'Gruener Veltliner'
	vinifera 'Gruener Veltliner'	oLon
	vinifera 'Incana'	wHer
	vinifera 'Interlaken'	wBur wClo oAls oGar oSho wRai oWnC wTGN wCoN oLon wCoS
un	***vinifera*** 'Katta Kourgane'	oLon
	vinifera 'Madeleine Angevine'	wClo oWhi wRai
	vinifera 'Madeleine Silvaner'	oLon
un	***vinifera*** 'Merlot'	oGar oWnC
un	***vinifera*** 'Meunier'	oLon
un	***vinifera*** 'Monte Senario'	oLon
	vinifera 'Morio Muscat'	oLon
	vinifera 'Mueller-Thurgau'	wBur wClo wRai
un	***vinifera*** 'Muscadelle'	see **V. *vinifera*** 'Volga Dawn'
	vinifera 'Muscat'	see **V. *vinifera*** 'Muscat of Alexandria'
qu	***vinifera*** 'Muscat Blanc'	oLon
	vinifera 'Muscat Hamburg'	oLon
	vinifera 'Muscat of Alexandria'	oWnC
	vinifera 'Muscat Ottonel'	wBur
	vinifera 'New York Muscat'	wRai oLon
un	***vinifera*** 'Okanogan Riesling'	wBur oLon
un	***vinifera*** 'Orange Muscat'	oLon
	vinifera 'Pearle de Czaba'	oLon
un	***vinifera*** 'Pinot Chardonnay'	oWnC
	vinifera 'Pinot Gris'	wClo oWnC oLon
	vinifera 'Pinot Noir'	wClo oGar wRai oWnC
un	***vinifera*** 'Portuguese Blue'	oLon
	vinifera 'Purpurea'	oFor wHer wWoS wCul oGos oJoy oHed wCSG oAmb oGar oNWe wRai oGre wPir wTGN wBox oBRG wAva wWel
un	***vinifera*** 'Queen of the Vineyard'	oLon
un	***vinifera*** 'Schuyler'	wBur oLon
un	***vinifera*** 'Semillion'	oLon
un	***vinifera*** 'Siegerrebe'	wClo wRai
	vinifera 'Silvaner'	wBur

	vinifera 'Sultana'	oGar oWnC wCoS
un	***vinifera*** 'Utah Giant'	oLon
un	***vinifera*** 'Variegata'	wWoS oGos oDan oGre
un	***vinifera*** 'Volga Dawn'	oLon
un	'White Riesling'	oGar oWnC
un	'Wilder'	oLon
un	'Worden'	oLon
un	'Zinfandel'	oWnC
	WACHENDORFIA	
	thyrsiflora	oJoy
	WALDSTEINIA	
	fragarioides	oAls oRus wCSG oGre
	ternata	oFor
	WASHINGTONIA	
	filifera	last listed 99/00
	robusta	oFor wDav wSte
	WATSONIA	
	beatricis	see **W. *pillansii***
	distans	oTrP
	fulgens	oTrP
	pillansii	wHer wCCr
	WEDELIA	
	trilobata	oHug oGar
	WEIGELA	
	'Abel Carriere'	oGre
	'Boskoop Glory'	oGre
va	Briant Rubidor / 'Olympiade'	oFor oGos oJoy oHed oAmb oOut oSqu oGre wFai wBox
	'Bristol Ruby'	oAls wCSG oBlo oGre oWnC
	'Candida'	oFor oGos oGre
	Carnaval / 'Courtalor'	last listed 99/00
	'Centennial'	oGre
	coraeensis	oFor wWel
	'Eva Rathke'	last listed 99/00
	'Evita'	oFor oGre
	Feline / 'Courtamon'	oFor
	sp. aff. ***floribunda*** HC 970509	wHer
	florida	wCSG
un	***florida*** 'Alexandra'	oFor
	florida 'Bicolor'	wCul
un	***florida*** 'Cardinal'	wAva
	florida 'Foliis Purpureis'	oJoy
	florida HC 970053	wHer
	florida 'Java Red'	oFor wHer wWoS wCul oAls oGar oRiv oBlo oGre wKin oWnC wAva oGue oUps oSle wCoS
	florida 'Nana Variegata'	see W. 'Nana Variegata'
	florida pink	wAva
	florida 'Pink Princess'	oFor oJoy oGar oGre oTPm
	florida 'Sunny Princess'	oGos
	florida 'Tango'	oFor
	'Florida Variegata'	oFor wHer wWoS wCul oNat oJoy oHed wCSG oGar oRiv oOut wKin oWnC wTGN oSec wAva oGue oUps oSle
	florida 'Variegata Nana'	see W. 'Nana Variegata'
	florida 'Versicolor'	oFor wHer wAva
un	'French Lace'	wWoS oGos
	'Looymansii Aurea'	oFor wHer oGos oNWe oSqu
	middendorffiana	wHer oBov
	'Minuet'	oFor wHer wWoS oJoy oHed oAmb oGar wFai wKin oWnC wTGN wCoS
	Nain Rouge / 'Courtanin'	oGre
	'Nana Variegata'	oAls oGar oGre oTPm wWel
	'Olympiade'	see **W.** Briant Rubidor / 'Olympiade'
	'Polka'	oFor wWoS oJoy oGre
	'Praecox Variegata'	oFor
	'Red Prince'	oFor oGos oJoy oAls oGar oGre wKin
	Rubidor	see **W.** Briant Rubidor / 'Olympiade'
	'Rumba'	oFor wWoS oGoo oGre
	'Samba'	oFor wWoS
	'Snowflake'	oGre
qu	'Variegata'	oGre
	'Victoria'	oGar oOut wFai
un	'White Knight'	oFor wHer wCul oJoy oGar oGre
un	'Wine and Roses'	wHer oGos oGar oGre wTGN
	WELWITSCHIA	
	mirabilis	oRar
	WESTRINGIA	
	fruticosa 'Morning Light'	oRiv
	fruticosa 'Wynyabbie Gem'	last listed 99/00
	WHIPPLEA	
	modesta	oBos
	WIDDRINGTONIA	
	cedarbergensis	oFor oTrP
	cupressoides	see **W. *nodiflora***

	nodiflora	oFor wHer
	schwarzii	oTrP
	WIKSTROEMIA	
	sp. aff. ***gemmata***	last listed 99/00
	WILCOXIA	
	albiflora	see **ECHINOCEREUS *leucanthus***
	schmollii	see **ECHINOCEREUS *schmollii***
	striata	see **PENIOCEREUS *striatus***
qu	***tuberosa***	oRar
	WISTERIA	
un	'Anwen'	oBRG
	brachybotrys 'Alba'	oFor
	brachybotrys Murasaki-kapitan	oGre
	brachybotrys 'Shiro-kapitan'	oGar oCis oGre oCrm
	brachybotrys 'White Silk'	wHer oGre oBRG
	'Burford'	oDan
	'Caroline'	oFor wHer oGar oWhi oGre
	floribunda	oFor wCSG wTGN wCoN
	floribunda 'Alba'	oFor wHer oDar oAmb oGar oWhi oGre oWnC oBRG wAva
un	***floribunda*** 'Flore Pleno'	wHer
ch	***floribunda*** 'Longissima'	see **W. *floribunda*** 'Multijuga'
ch	***floribunda*** 'Longissima Alba'	see **W. *floribunda*** 'Alba'
	floribunda 'Macrobotrys'	see **W. *floribunda*** 'Multijuga'
	floribunda 'Multijuga'	oFor wHer oDar wCol oGar oWhi oGre oBRG wAva
	floribunda 'Pink Ice'	see **W. *floribunda*** 'Rosea'
	floribunda 'Purple Patches'	wHer
	floribunda 'Rosea'	oFor wHer wCol oGar oWnC wAva
	floribunda 'Royal Purple'	wHer oGre oBRG
	floribunda 'Shiro-noda'	see **W. *floribunda*** 'Alba'
ch	***floribunda*** 'Snow Showers'	see **W. *floribunda*** 'Alba'
un	***floribunda*** 'Texas Purple'	oAls oGar oWnC wCoS wWel
do	***floribunda*** 'Violacea Plena'	oFor oGar oGre oWnC
ch	***x formosa*** Black Dragon	see **W. *x formosa*** 'Yae-kokuryu'
un	***x formosa*** 'Issai Perfect'	oFor oGar oGre oWnC
do	***x formosa*** 'Yae-kokuryu'	wHer oAmb oBRG
	frutescens	oFor oGre
	frutescens 'Nivea'	oGos oGar oGre wWel
	frutescens white	see **W. *frutescens*** 'Nivea'
	'Lavender Lace'	oFor
	macrostachya	oFor
	macrostachya 'Clara Mack'	oFor wHer wCSG oGre
	sinensis	oFor wClo oDar oAls oGar oRiv oCis wPir oWnC wMag oCrm wTGN oBRG wAva wCoN wCoS wWel
	sinensis 'Alba'	oAls oGre oBRG
	sinensis 'Amethyst'	oFor oGre oBRG
un	***sinensis*** 'Anwen'	oGre
un	***sinensis*** 'Aunt Dee'	oDan wKin
	sinensis blue	oGar oSho oGre
	sinensis 'Blue Sapphire'	oFor wHer oGre
ch	***sinensis*** 'Cooke's Purple'	see **W. *sinensis*** 'Cooke's Special'
	sinensis 'Cooke's Special'	oFor oGar oGre oWnC
	sinensis pink	oGar oSho oGre
	sinensis 'Prolific'	oFor wCol oGar oGre oBRG wAva wWel
	sinensis purple	oSho oWnC
	sinensis 'Rosea'	oAls
un	***sinensis*** 'Texas White'	oFor oGre
qu	'Snow'	oBRG
ch	***venusta***	see **W. *brachybotrys*** 'Shiro-kapitan'
ch	***venusta*** var. ***violacea***	see **W. *brachybotrys*** Murasaki-kapitan
	WITHANIA	
	somnifera	oOEx iGSc wHom oCrm oBar
	WOODSIA	
	fragilis	last listed 99/00
	intermedia	last listed 99/00
	polystichoides	wFol wFan
	WOODWARDIA	
	fimbriata	oFor oRus oGar wNay wSnq wWel
	radicans	last listed 99/00
	WULFENIA	
	carinthiaca	oJoy
	XANTHOCERAS	
	sorbifolium	oFor wBur oEdg oDan oRiv oWhi oOEx oGre
	XANTHORHIZA	
	simplicissima	oFor wCul
	XANTHORRHOEA	
	australis	last listed 99/00
	XEROPHYLLUM	
	tenax	oFor wThG wWoB oEdg oRus oBos oGre oTri wPla oAld wSte wSnq
	XEROSICYOS	
	decaryi	oRar

	XYLOSMA	
un	***controversum***	oOEx
	YUCCA	
	aloifolia	oFor wCCr
un	***aloifolia*** 'Sunrise'	oFor
	baccata	oFor oGoo wCCr oWnC
	brevifolia	oRiv
	elata	oGoo wCCr
	filamentosa	oFor oSto wKin wBox oCir oLSG wCoS
va	***filamentosa*** 'Bright Edge'	oFor wHer oGos oEdg oOut wBox wAva
un	***filamentosa*** 'Gold Edge'	oFor wSwC oGre
	filamentosa 'Variegata'	last listed 99/00
va	***flaccida*** 'Golden Sword'	oAls oGar oNWe oSho wKin oWnC
un	***flaccida*** 'Ivory Tower'	wHer oAls oGar oWnC
va	'Garland's Gold'	oFor oBRG wWel
	glauca	oFor oEdg oGar iGSc oSle
	gloriosa 'Variegata'	wHer oGre
un	'Gold Sword'	see **Y. *flaccida*** 'Golden Sword'
	harrimaniae	wCCr
	recurvifolia	oDar oGar oSho oWnC oSec wWel
	rostrata	last listed 99/00
	schottii	wCCr
	YUSHANIA	
	anceps	wCli oNor
	anceps 'Pitt White'	oTra wCli
	maling	wBea oTBG
	ZALUZIANSKYA	
	capensis	wSte
un	***microsiphon***	oNWe
	ovata	oHed oNWe
	ZAMIOCULCAS	
	zamiifolia	oRar
	ZANTEDESCHIA	
	aethiopica	wCol oAmb oHug oGar oSsd oSho oCir wCoN oVBI wSnq
	aethiopica 'Childsiana'	oGar
	aethiopica 'Crowborough'	last listed 99/00
	aethiopica 'Green Goddess'	oTrP wCol oGre oVBI
	aethiopica 'Pershore Fantasia'	wHer
	albomaculata	wHer oVBI
un	'Deep Throat'	wHer
	elliottiana	oTrP oGar oVBI
un	'Flame'	oVBI
un	'Lavender Gem'	oVBI
un	***macrocarpa***	oRar
un	***odorata***	wHer
	rehmannii	wHer
un	***rehmannii*** 'Rubylite'	oVBI
	ZANTHOXYLUM	
	americanum	oTrP
	armatum	oOEx
	clava-herculis	oFor
un	sp. aff. ***diacanthoides*** EDHCH 97306	wHer
un	***nepalense*** HWJCM 103	wHer
	oxyphyllum	oOEx
	piperitum	oFor wPir wCCr
	piperitum HC 970013	wHer
	simulans	oFor wCCr
	ZAUSCHNERIA	
	arizonica	see **Z. *californica*** ssp. ***latifolia***
un	'Bowman'	oGoo
un	'Bowman Hybrid #1'	oSis
	californica	oJoy wCol oBos
	californica 'Albiflora'	oSis
un	***californica*** 'Calistoga'	oGoo
	californica ssp. ***cana***	wCCr
	californica compact form	oSis
	californica 'Dublin'	oHed oGoo oNWe
	californica etteri	oSis
	californica ssp. ***garrettii***	oJoy
	californica ssp. ***latifolia***	oFor oJoy oGar oGoo oBos
un	***californica*** 'Rogers Hybrid'	oGoo
un	***californica*** 'Silver Select'	oFor
	californica 'Solidarity Pink'	oFor oSis
un	***californica*** 'Wayne's Silver'	oSis
	californica 'Western Hills'	wHer
	latifolia	see **Z. *californica*** ssp. ***latifolia***
un	'Mattole Select'	oSis
	septentrionalis	oSis
	ZELKOVA	
	carpinifolia	oFor oRiv
un	***chinensis***	oRiv
	schneideriana	oRiv
	serrata	oFor oRiv oGre wCoN
	serrata 'Green Vase'	oFor oWnC wWel
	serrata 'Village Green'	oGar
	sinica	oFor oEdg wCCr
	ZENOBIA	
	pulverulenta	oFor wHer oGos oGar oBov oGre wBox wWel
un	***pulverulenta*** 'Woodlander's Blue'	wSte
	ZEPHYRANTHES	
	candida	oSsd oSec
un	**ZEUGITES**	
	americana var. ***mexicana***	oTrP
	ZIGADENUS	
	venenosus	oRus
	ZINGIBER	
	mioga	wHer
	mioga HC 970276	wHer
	officinale	iGSc oBar
	zerumbet	oTrP oOEx
un	***zerumbet*** 'Variegata'	oGar
	ZIZANIA	
	latifolia	oSsd
	ZIZIA	
	aurea	oFor
	ZIZIPHUS	
	jujuba 'Lang' (F)	wRai oOEx
	jujuba 'Li' (F)	wBur wRai oOEx
	mauritanica	oOEx
un	***recurva***	oOEx
un	***spinosa***	oOEx

NURSERY/CODE INDEX

This index contains all participating nurseries. Detailed information for those with codes is to be found in **Nurseries with Codes**, page 276. Information for those without codes will be found in **Additional Nurseries**, page 296. Nurseries without codes have not given us lists of the plants they carry and therefore do not appear in the **Plant Listing** section of the book.

A Lot of Flowers	
A Plethora of Primula	wPle
Adelman Peony Gardens	oAde
Aitken's Salmon Creek Garden	wAit
Alder View Natives	oAld
All Season Nursery, Gardens & Gifts	
Alpine Nursery Inc.	
Al's Fruit & Shrub Center	oAls
Amber Hill Nursery	oAmb
Antique Rose Farm	wAnt
Arcadia Greenhouses	iArc
Avalon Nursery	wAva
B & D Lilies	wBDL
Bainbridge Gardens	
Baker & Chantry Orchids	
Bamboo Gardener	wBmG
Bamboo Gardens of Washington	wBam
Barbara Ashmun's Jungle	
Barn Owl Nursery	oBar
Bear Creek Nursery	wBCr
Bear Creek Nursery	
Beauty and the Bamboo	wBea
Bell Family Nursery/Hydrangeas Plus	oBel
Bellevue Nursery, Inc.	
Bellwether Perennials	wBwP
Big Trees Today	
Bloom River Gardens	oBRG
Bloomer's Nursery	oBlo
Blue Heron Farm	wBlu
Bonsai Northwest	
Bosky Dell Natives, Inc	oBos
Bovees Nursery	oBov
Boxhill Farm	wBox
Brady's Nursery, Inc	
Briggs Hill Orchids, Inc.	
Brim's Farm & Garden	
Brothers Herbs and Tree Peonies	oBHP
Brown's Rose Lodge Nursery	
Burlingame Gardens	
Burnt Ridge Nursery	wBur
C & O Nursery	wC&O
Caprice Farm Nursery	oCap
Cascade Nursery	
Cedar Mill Nursery	
Cedar Valley Nursery, Inc.	wCed
Chehalem Gardens	oChe
Chets Garden and Pet Center	
Christianson's Nursery	wChr
Cindy's Plant Stand	
Circle B Nursery	oCir
Cistus Design Nursery	oCis
City People's Garden Store	
Classic Nursery	
Clinton Inc. Bamboo Growers	wCli
Cloud Mountain Farm and Nursery	wClo
Collector's Nursery	wCol
College Street Nursery	wCoS
Columbia Nursery & Design	
Colvos Creek Nursery	wCCr
Connell's Dahlias	wCon
Cooley's Gardens Inc	oCoo
Coos Grange Supply & Garden Center	
Cora's Nursery & Greenhouse	
Cornell Farm	
Cottage Creek Nursery	wCot
Country Gardens	wCou
Country Lane Gardens, Ltd.	wCLG
Country Nursery & Gardens	
Country Store and Gardens	wCSG
Courtyard Nursery	wCoN
Cricklewood	wCri
Crimson Sage Nursery	oCrm
Cultus Bay Nursery	wCul
D & B Fuchsia Nursery	wDBF
Dancing Oaks Nursery	oDan
Dan's Dahlias	
Daryll's Nursery	oDar
Dave's Taste of the Tropics	wDav
Decker Nursery	
Delta Farm & Nursery	oDel
Des Moines Way Nursery	
Dillards Nursery	
Don Smith's Tree Farm	
Down To Earth	
Downs By The Pond	
Duckworth's Nursery	
Dutch Mill Herbfarm	oDut
Eaden Gardens	
Earth's Rising	oEar
Eclectic Gardens	
Edelweiss Perennials	wEde
Edgewood Gardens	oEdg
Edmunds' Roses	oEdm
Egan Gardens	oEga
Elandan Gardens Ltd.	
Elk Meadows Nursery	
Elk Plain Nursery	
Exuberant Gardens	oExu
F W Byles, Co., Nursery	wFWB
Fairie Perennial and Herb Gardens	wFai
Fairlight Gardens Nursery	wFGN
Fancy Fronds	wFan
Ferguson's Fragrant Nursery	
Fessler Nursery	
Fiddler's Ridge Garden and Nature Store	iFid
Flat Creek Garden Center	
Foliage Gardens	wFol
Forest Flor Nursery	wFFl

NURSERY/CODE INDEX

NURSERY/CODE INDEX

Rhododendron Patch	
Rhododendron Species Botanical Garden	wRho
River Rock Nursery	oRiv
Roadhouse Nursery	
Robyn's Nest	wRob
Roozengaarde	wRoo
Roses Northwest	oRNW
Russell Graham, Purveyor of Plants	oRus
Sally Herman	
Sandy's Nursery	
Schreiner's Iris Gardens	oSch
Seasons Nursery	
Sea-Tac Gardens	wSea
Secret Garden Growers	oSec
Seven Mile Nursery	oSev
Shaman Garden	
Shannon's Perennials & Old Garden Roses	oSha
Shirley's Dahlias	
Shonnard's Nursery	oSho
Shore Road Nursery	wShR
Silver Lake Nursery	
Siskiyou Rare Plant Nursery	oSis
Skipper and Jordan Nursery	
Sky Nursery	
Sleepy Hollow Nursery	oSle
Snow Creek Daylily Gardens	wSno
Soos Creek Gardens	wSoo
Squak Mountain Greenhouses and Nursery	
Squaw Mountain Gardens	oSqu
Squirrel Heights Gardens	
Star Nursery & Landscaping, LLC	wSta
Steamboat Island Nursery	wSte
Stoller Farms	
StoneCrop Gardens	oSto
Summersun Greenhouse & Nursery	
Sundquist Nursery, Inc.	wSnq
Susandales Pond and Bog Plants	oSsd
Susan's Bamboo	wSus
Swan Creek Nursery	wSwC
Swan Island Dahlias	oSwa
Swansons Nursery	
Tallan Nursery	oTal
TDM Acres Herb Farm	oTDM
The Bamboo Garden	oTBG
The Berry Botanic Garden	
The Garden Shop at Lakewold Gardens	
The Gardens At Padden Creek	
The Greenery	wThG
The Greenhouse Nursery	wTGN
The Lily Pad	wTLP
The Pat Calvert Greenhouse	
The Plant Peddler	
The Plantsmen	oTPm
The Puget Sound Kiwi Co.	wPug
The Rose Guardians	oRos
Thompson's Greenhouse	wTho
Three Sisters Nursery	oThr
Tiny Tree Nursery	oTin
Tom's Garden Center	
Tower Perennial Gardens	wTow
Town and Country Gardens	
Tradewinds Bamboo Nursery	oTra
Trans-Pacific Nursery	oTrP
Trillium Gardens	oTri
Tsugawa Nursery	
Tualatin River Nursery	
Twinflower Nursery	oTwi
Upstarts! Growers	oUps
Valley Nursery Inc.	
Valley View Nursery	
Van Well Nursery	wVan
Vanveen Bulbs International	oVBI
Vittoria Nursery	
Walden-West	oWaW
Wallingford Garden Spot	
Walsterway Iris Gardens	wWal
Washco Lawn & Garden	
Watershed Garden Works	wWat
Wee Tree Farm	
Wells Medina Nursery	wWel
Wet Rock Gardens	
Whispering Springs	oWhS
Whitman Farms	oWhi
Whitney Gardens & Nursery	wWhG
Wight's Home & Garden	
Wild Bird Garden/Coastal Garden Center	
Wildside Growers	wWld
Wildwood Gardens	
Willamette Falls Nursery & Ponds	
Willow Lake Nursery	
Wind Poppy Farm and Nursery	wWin
Wine Country Nursery & Aquarium	oWnC
Wisteria Herbs & Flowers	
Woodbrook Nursery	wWoB
Wooden Shoe Bulb Company, Inc.	oWoo
Woodside Gardens	wWoS
Yamamoto's	
Z Oak Tree Hosta Farm	oZOT

NURSERIES WITH CODES

These nurseries have plants included in the Plant Section of this book, and are listed here in alphabetical order **by their codes**. A "w" preceding a code name indicates that the nursery is in Washington; an "o" indicates an Oregon nursery; an "i" indicates an Idaho nursery. The codes assigned are generally the first three letters of the nursery name; in cases where the names are similar, we have used a combination of letters from the beginning of additional words in the nursery name. Information about each nursery is as given to us by the proprietors. Since information may change, it is advisable to call the nursery before visiting.

oAde **Adelman Peony Gardens**
5690 Brooklake Rd NE
Brooks, OR 97305-9660
Proprietor(s): Carol Adelman
Phone: 503-393-6185
FAX: 503-393-3457
Email address: adelman-gardens@worldnet.att.net
Website: peonyparadise.com
Mail-order or walk-in: both
Credit cards: VISA, MC
Hours: May 1 - June 15: 9:00am-7:00pm
Specialties: Herbaceous peonies, over 100 varieties. Intersectional peonies (cross between tree peony and herbaceous peony).

wAit **Aitken's Salmon Creek Garden**
608 NW 119th St
Vancouver, WA 98685-3802
Proprietor(s): Barbara and Terry Aitken
Phone: 360-573-4472
FAX: 360-576-7012
Email address: aitken@e-z.net
Website: www.e-z.net/~aitken
Mail-order or walk-in: both
Catalogue: $2.00
Credit cards: VISA, MC
Hours: April-September: 8:00am-5:00pm
Specialties: Iris nursery; also orchids

oAld **Alder View Natives**
28315 SW Grahams Ferry Rd
Wilsonville, OR 97070
Proprietor(s): Keith Fitzgerald
Phone: 503-570-2894
FAX: 503-570-9904
Email address: natives1@gte.net
Mail-order or walk-in: both
Catalogue: SASE
Credit cards: no
Hours: 10:00am-5:00pm Saturday and Sunday; weekdays by appointment. By appointment only November through January.
Specialties: Native plants.

oAls **Al's Fruit & Shrub Center**
1220 N Pacific Hwy
Woodburn, OR 97071
Proprietor(s): Jack Bigej
Phone: 503-981-1245
FAX: 503-982-4608
Email address: alsfruit@alsfruit.com
Website: www.alsfruit.com
Mail-order or walk-in: walk-in only
Credit cards: MC, VISA, Discover
Hours: April-June: 9:00am-7:00pm 7 days/week; July-March: 9:00am-6:00pm 7 days/week
Specialties: Full service garden center. Wonderful selection of annuals, perennials, trees and shrubs. Delivery within 50 mile radius of store.

oAmb **Amber Hill Nursery**
11998 S Critcser Rd
Oregon City, OR 97045
Proprietor(s): Sherry Gardner
Phone: 503-657-9289
FAX: 503-657-1005
Mail-order or walk-in: walk-in only
Catalogue: $1.00
Credit cards: MC, VISA
Hours: Mid-March-October 31: 10:00am-5:00pm Wednesday-Saturday.
Specialties: Choice perennials, large selection of hydrangeas, ornamental grasses, a variety of unique small trees and shrubs, and 60 varieties of Japanese maples.

wAnt **Antique Rose Farm**
12220 Springhetti Rd
Snohomish, WA 98296
Proprietor(s): Don and Jackie McElhose
Phone: 360-568-1919
FAX: 360-568-1919
Mail-order or walk-in: walk-in only
Catalogue: free list
Credit cards: no
Hours: March 14-September: 10:00am-5:00pm Tuesday-Saturday
Specialties: Roses, including antique (albas, Bourbons, China, damasks, galllicas, hybrid musks, noisettes, moss, Portlands, teas), climbers, ramblers, David Austins, Romanticas, floribundas, species, hybrid teas, and much more.

iArc **Arcadia Greenhouses**
PO Box 1934
Sandpoint, ID 83864
Walk-in address: 143 Arcadia Lane
Sandpoint, ID 83864
Proprietor(s): A T O'Flynn
Phone: 208-263-8922
FAX: 208-263-8922
Email address: arcadia@nidlink.com
Mail-order or walk-in: both
Credit cards: yes
Hours: April 10-September 23: 9:00am-6:00pm 7 days/week
Specialties: Perennials and cold hardy trees. Beautiful park-like setting on 3 acres; pond, fountain, and display gardens.

wAva **Avalon Nursery**
16720 SR-9 SE
Snohomish, WA 98296
Walk-in address: (on corner of 168th SE and Highway 9)
Proprietor(s): Craig Tutt
Phone: 360-668-9696
Email address: avalonnursery@worldnet.att.net
Mail-order or walk-in: walk-in
Credit cards: VISA, MC, AmEx
Hours: Winter: 9:00am-8:00pm; Spring/Summer/Fall: 9:00am-6:00pm daily. Open all year.
Specialties: Retail and wholesale. Perennials, trees, shrubs, annuals, gifts, Christmas trees (live and cut).

wBam **Bamboo Gardens of Washington**
5016 192nd Pl NE
Redmond, WA 98053-4602
Walk-in address: 196th Ave NE & SR-202 (2 miles east of SR-520)
Redmond, WA 98053
Proprietor(s): Jeannine Florance

Phone: 425-868-5166
FAX: 425-868-5360
Website: bamboogardenswa.com
Mail-order or walk-in: walk-in; mail-order for poles only
Catalogue: list free
Credit cards: VISA, MC
Hours: Standard Time: 9:00am-4:00pm Monday-Saturday; Daylight Savings Time: 9:00am-5:00pm 7 days/week.
Specialties: Bamboo plants, bamboo poles, ornamental grasses, water garden plants, granite lanterns and basins; books and tools for building with bamboo.

oBar **Barn Owl Nursery**
22999 SW Newland Rd
Wilsonville, OR 97070
Proprietor(s): Christine Mulder
Phone: 503-638-0387
FAX: 503-638-0387
Website: barnowlnursery.citysearch.com
Mail-order or walk-in: walk-in only
Catalogue: none
Credit cards: VISA, MC
Hours: March - July and October - November: 4 days/week (Wednesday through Saturday), 10:00am-5:00pm, except holidays
Specialties: Herbs: many varieties of lavender and rosemary; around 400 varieties of herbs.

wBCr **Bear Creek Nursery**
PO Box 411
Northport, WA 99157
Proprietor(s): Hunter and Donna Carleton
Phone: 509-732-6219
FAX: 509-732-4417
Email address: bearcreek@plix.com
Website: bearcreeknursery.com
Mail-order or walk-in: mail-order only
Catalogue: $1.00, refundable
Credit cards: AmEx, VISA, MC, Discover
Specialties: 282 heirloom and new apple varieties, grafting and budding supplies - scionwood, budwood, and 24 rootstock choices. 84 page informative catalogue full of cold and drought hardy trees, shrubs and berries. 40 nut varieties including filazels, trazels, filberts, chestnuts, hickories, butternuts and buartnuts.

wBDL **B & D Lilies**
PO Box 2007
Port Townsend, WA 98368
Walk-in address: 330 P Street
Port Townsend, WA 98368
Proprietor(s): Bob and Dianna Gibson
Phone: 360-765-4341
FAX: 360-765-4074
Website: bdlilies.com
Mail-order or walk-in: both
Catalogue: $3.00
Credit cards: VISA, MC
Hours: 9:00am-5:00pm Monday-Friday
Specialties: Lilium bulbs and hemerocallis (daylilies).

wBea **Beauty and the Bamboo**
306 NW 84th St
Seattle, WA 98117-3117
Proprietor(s): Stan Andreasen
Phone: 206-781-9790
FAX: 206-297-2810
Email address: bambu501@aol.com
Mail-order or walk-in: walk-in only
Catalogue: none
Credit cards: VISA only
Hours: By appointment only
Specialties: Rare bamboos, installatlions, bamboo extraction. Discounts.

oBel **Bell Family Nursery/Hydrangeas Plus**
PO Box 389
Aurora, OR 97002
Proprietor(s): Kristin Van Hoose
Phone: 503-651-2887
FAX: 503-651-2648
Email address: bellfam@canby.com
Website: www.hydrangeasplus.com
Mail-order or walk-in: mail-order only
Catalogue: $4.50
Credit cards: VISA, MC
Hours: Not open to tours at this time. Seminars available through Bell Productions
Specialties: Slide show and garden club presentations also available.

oBHP **Brothers Herbs and Tree Peonies**
27015 SW Ladd Hill Rd
Sherwood, OR 97140
Proprietor(s): Richard W Rogers
Phone: 503-625-7548
FAX: 503-625-1667
Email address: rick@treony.com
Website: www.treony.com
Mail-order or walk-in: both
Catalogue: $2.00
Credit cards: VISA, MC, AmEx
Hours: 10:00am-4:00pm Monday-Saturday, spring and fall. In bloom 4/15-6/30. Please call for appointment.
Specialties: Largest collection of tree peonies this side of the Mississippi. All sold on their own roots, 3 years+. From America, China, Europe, Japan, New Zealand.

oBlo **Bloomer's Nursery**
89813 Sprague Rd
Eugene, OR 97408
Walk-in address: 89719 Armitage Rd
Eugene, OR 97408
Proprietor(s): Jim and Glenda Bloomer
Phone: 541-687-5919
FAX: 541-345-9021
Mail-order or walk-in: walk-in only
Catalogue: n/a
Credit cards: VISA, MC
Hours: Summer: 9:00am-6:00pm Monday-Saturday; 10:00am-5:00pm Sunday. Winter: 9:00am-5:00pm Monday-Saturday; 10:00am-5:00pm Sunday.
Specialties: Largest selection of landscape trees and shrubs in Lane County, OR.

wBlu **Blue Heron Farm**
12179 State Route 530
Rockport, WA 98283
Proprietor(s): Anne Schwartz and Mike Brondi
Phone: 360-853-8449
FAX: 360-853-8449
Mail-order or walk-in: walk-in only by appt. only
Catalogue: free
Hours: By appointment and during spring sales.
Specialties: Hardy bamboo - groundcovers to timber species; assorted native trees.

wBmG **Bamboo Gardener**
2609 NW 86th Street
Seattle, WA 98117
Proprietor(s): James Clever
Phone: 206-782-3490
Email address: bambuguru@earthlink.net
Website: home.earthlink.net/~bambuguru/

Mail-order or walk-in: both
Catalogue: on website
Credit cards: no
Hours: Daylight hours visits by appointment only.
Specialties: Bamboo plants, hardy temperate, running and clumping varieties. Groundcover, mid range and timber bamboos, bamboo poles, premade gate/fence/screen sections, rhizome barriers and books.

oBos **Bosky Dell Natives, Inc**
2020 SW 8th Ave #323
West Linn, OR 97068
Walk-in address: 23311 SW Bosky Dell Lane
West Linn, OR 97068
Proprietor(s): Lory Duralia
Phone: 503-638-5945
FAX: 503-638-8047
Email address: plants0000@aol.com
Mail-order or walk-in: both
Catalogue: free plant list
Credit cards: VISA, MC, AmEx, Discover
Hours: 10:00am-5:00pm 7 days/week
Specialties: Native plants exclusively; troughs and other yard ornaments.

oBov **Bovees Nursery**
1737 SW Coronado
Portland, OR 97219
Proprietor(s): Lucie Sorensen-Smith
Phone: 503-244-9341
Email address: bovees@teleport.com
Website: bovees.com
Mail-order or walk-in: both
Catalogue: $2.00
Credit cards: VISA, MC
Hours: 9:00am-5:00pm Monday-Saturday; noon-5:00pm Sunday
Specialties: Rhododendrons, rock garden plants, semi-tropical (vireya) rhododendrons.

wBox **Boxhill Farm**
14175 Carnation-Duvall Rd
Duvall, WA 98019
Phone: 425-788-6473
Email address: boxhillfarm@mindspring.com
Website: www.boxhillfarm.com
Mail-order or walk-in: both
Catalogue: on line only
Credit cards: VISA, MC, Discover
Specialties: Bamboo, boxwood, daylilies, ornamental grasses, perennials, tree peonies.

oBRG **Bloom River Gardens**
PO Box 177
Walterville, OR 97489
Walk-in address: 39744 Deerhorn Road
Springfield, OR 97478
Proprietor(s): Mark and Val Bloom
Phone: 541-726-8997
FAX: 541-726-4052
Email address: plants@bloomriver.com
Website: www.bloomriver.com
Mail-order or walk-in: mail-order
Catalogue: $2.00
Credit cards: VISA, MC
Hours: By appointment only.
Specialties: Hardy ferns, daylilies, dwarf conifers, hostas, specialty rhododendrons and azaleas, vacciniums, woody ornamentals.

wBur **Burnt Ridge Nursery**
432 Burnt Ridge Rd
Onalaska, WA 98570
Proprietor(s): Michael & Carolyn Dolan
Phone: 360-985-2873
FAX: 360-985-0882
Email address: burntridge@myhome.net
Website: landru.myhome.net/burntridge
Mail-order or walk-in: mail-order; walk-in by appointment
Catalogue: SASE
Credit cards: VISA, MC
Hours: Please call for appointment and directions.
Specialties: We specialize in unusual trees, shrubs, and vines that produce edible nuts or fruits, disease resistant stock, Northwest native plants.

wBwP **Bellwether Perennials**
4662 Center Road
Lopez Island, WA 98261
Proprietor(s): Jenny Harris
Phone: 360-468-3531
FAX: 360-468-3531
Email address: fruitandnut@rockisland.com
Mail-order or walk-in: both
Catalogue: list free
Credit cards: no
Hours: By appointment - or just call first to make sure I'm here. Also sell at the Farmers Market in the summer on Saturdays in Lopez Village.
Specialties: Drought tolerant, appropriate plants to our summer dry and sometimes extreme weather.

wC&O **C & O Nursery**
PO Box 116
Wenatchee, WA 98807
Walk-in address: 1700 N Wenatchee Ave
Wenatchee, WA 98801
Proprietor(s): Jack, Dick, Todd, Gary and Jim Snyder
Phone: 800-232-2636
FAX: 509-662-4519
Email address: tree@c-onursery.com
Website: www.c-onursery.com
Mail-order or walk-in: both
Catalogue: free
Credit cards: MC, VISA, Discover
Hours: 8:00am-5:00pm Monday-Friday
Specialties: Fruit trees.

oCap **Caprice Farm Nursery**
15425 SW Pleasant Hill Rd
Sherwood, OR 97140
Proprietor(s): Charlie and Cyndi Turnbow
Phone: 503-625-7241
FAX: 503-625-5588
Website: www.capricefarm.com
Mail-order or walk-in: both
Catalogue: $2.00
Credit cards: VISA, MC
Hours: April-November: 9:00am-4:00pm Monday-Saturday. May/June/July: 7 days/week.
Specialties: Peonies, daylilies, hosta, and interspecies peonies.

wCCr **Colvos Creek Nursery**
PO Box 1512
Vashon Island, WA 98070
Walk-in address: SW 240th and Pt. Robinson Rd
Vashon Island, WA 98070
Proprietor(s): Mike Lee
Phone: 206-749-9508
FAX: 206-463-3917
Email address: colvoscreek@juno.com
Mail-order or walk-in: both
Catalogue: $3.00
Credit cards: VISA, MC, Discover
Hours: Friday and Saturday: 10:00am-4:00pm

Specialties: NW natives, drought-hardy, rare trees and shrubs, maples, oaks, sorbus, eucalyptus, ilex, conifers.

wCed **Cedar Valley Nursery, Inc.**
3833 McElfresh Rd SW
Centralia, WA 98531
Proprietor(s): Charles Boyd
Phone: 360-736-7490
FAX: 360-736-6600
Email address: boyd@myhome.net
Mail-order or walk-in: mail-order only
Catalogue: free
Credit cards: no
Hours: Please call to schedule a visit.
Specialties: Tissue culture produced raspberries and blackberries.

oChe **Chehalem Gardens**
PO Box 74
Dundee, OR 97115
Walk-in address: 19105 NE Trunk Rd
Dundee, OR 97115
Proprietor(s): Ellen and Tom Abrego
Phone: 503-538-8920
Mail-order or walk-in: mail-order only
Catalogue: free price list
Credit cards: no
Hours: Open weekends during bloom season (May-early June) and by appointment.
Specialties: Siberian and spuria iris.

wChr **Christianson's Nursery**
15806 Best Rd
Mt Vernon, WA 98273
Proprietor(s): John and Toni Christianson
Phone: 360-466-3821
FAX: 360 466-2940
Email address: chrisnsy@fidalgo.net
Website: www.christiansons-nursery.com
Mail-order or walk-in: walk-in only
Credit cards: MC, VISA
Hours: Spring-fall: 9:00am-6:00pm 7 days/week; winter: 9:00am-5:00pm 7 days/week.
Specialties: Roses, perennials and uncommon trees and shrubs.

oCir **Circle B Nursery**
22745 NW Fisher Rd
Buxton, OR 97109
Proprietor(s): Barb and Bob Berini
Phone: 503-324-9274
FAX: 503-324-9274
Mail-order or walk-in: walk-in only
Catalogue: none
Credit cards: not yet
Hours: 10:00am-6:00pm Wednesday-Sunday. Will open mid-April; please call for date.
Specialties: Premium annuals, perennials, shrubs.

oCis **Cistus Design Nursery**
2827 NE 11th Ave
Portland, OR 97212
Walk-in address: 22711 NW Gillihan Rd
Portland (Sauvie Island), OR 97231
Proprietor(s): Sean Hogan and Parker Sanderson
Phone: 503-880-6011
FAX: 503-282-7766
Email address: info@cistus.com
Website: www.cistus.com
Mail-order or walk-in: both
Catalogue: on web
Credit cards: no
Hours: April-September: 10:00am-5:00pm Thursday-Sunday
Specialties: The home of zonal denial. Plants of all description that thrive in the Pacific Northwest. Tons of new introductions each season.

wCLG **Country Lane Gardens, Ltd.**
4407 NE 41st Street
Seattle, WA 98105
Proprietor(s): S Andrew Schulman
Phone: 206-525-0176
Email address: info@country-lane.com
Website: www.country-lane.com
Mail-order or walk-in: mail-order only
Catalogue: no charge
Credit cards: VISA, MC
Hours: No walk-ins. Mail order only.
Specialties: Old garden roses.

wCli **Clinton Inc. Bamboo Growers**
12260 1st Ave S
Seattle, WA 98168-2014
Proprietor(s): Vance Allen, Lee Gartner, Erika Harris
Phone: 206-242-8848
FAX: 206-444-9428
Email address: clintonbamboo@sprynet.com
Mail-order or walk-in: both
Catalogue: free
Credit cards: yes
Hours: May-September: 9:00am-5:00pm Friday, Saturday and Sunday; other times by appointment.
Specialties: Hardy temperate clumping and running bamboos, ornamental grasses and New Zealand flax (phormiums).

wClo **Cloud Mountain Farm and Nursery**
6906 Goodwin Rd
Everson, WA 98247
Proprietor(s): Tom & Cheryl Thornton
Phone: 360-966-5859
FAX: 360 966-0921
Email address: cloud-mt@pacificrim.net
Mail-order or walk-in: both
Catalogue: $1.00 donation
Credit cards: MC, VISA, Novus
Hours: February-early December: 10:00am-5:00pm Wednesday-Saturday; also, in February-June and September-October: 11:00am-4:00pm Sundays; March 1-May 27: open 7 days/week.
Specialties: Hardy fruits and choice ornamentals.

wCol **Collector's Nursery**
16804 NE 102nd Ave
Battle Ground, WA 98604
Proprietor(s): Bill Janssen and Diana Reeck
Phone: 360-574-3832
FAX: 360-571-8540
Email address: dianar@teleport.com
Website: www.collectorsnursery.com
Mail-order or walk-in: both
Hours: The following open houses for 2000 (or call for appointment): 3/17-19; 5/5-7; 6/3-4; 7/14-16; 8/25-27; 10/13-15; 11/3-5.
Specialties: Dwarf and rare conifers, uncommon perennials, especially shade plants; epimediums, tricyrtis, variegated plants, Asian plants; oddities.

wCon **Connell's Dahlias**
10616 Waller Rd E
Tacoma, WA 98446
Proprietor(s): Kim Connell, Kerry Connell, and Kirk Connell
Phone: 253-531-0292
FAX: 253-536-7725
Website: www.connells-dahlias.com
Mail-order or walk-in: both

Catalogue: $2.00
Credit cards: VISA, MC, Discover
Hours: October-March: 9:00am- 5:00pm Monday-Friday; March-September: 9:00am-5:00pm Monday-Saturday
Specialties: Dahlia tubers.

wCoN **Courtyard Nursery**
6400 Capitol Blvd SE
Tumwater, WA 98501
Proprietor(s): Robert Lee
Phone: 360-943-4360
FAX: 360-943-4360
Mail-order or walk-in: walk-in only
Catalogue: n/a
Credit cards: yes
Hours: Hours by appointment if you can't make it when I'm regularly open.
Specialties: Trees and shrubs

oCoo **Cooley's Gardens Inc**
PO Box 126 BE
Silverton, OR 97381-0126
Walk-in address: 11553 Silverton Rd NE
Silverton, OR 97381
Phone: 503-873-5463
FAX: 503-873-5812
Email address: cooleyiris@aol.com
Website: www.cooleysgardens.com
Mail-order or walk-in: mail-order only
Catalogue: $5.00
Credit cards: VISA, MC
Hours: Bloom festival mid-May to first week of June. Display beds, indoor cut flower show open 8:00am-7:00pm
Specialties: Tall bearded iris

wCoS **College Street Nursery**
3613 College Street SE
Lacey, WA 98503
Proprietor(s): Bill and Judy Pattison
Phone: 360-491-1688
FAX: 360-923-0274
Mail-order or walk-in: walk-in only
Credit cards: all
Hours: 9:00am-6:00pm Monday-Saturday; 11:00am-5:00pm Sunday. Closed Sundays in January.
Specialties: Vegetables, bulbs, annuals, groundcovers, houseplants, water plants, trees, shrubs, vines, perennials. Specialty - hanging baskets. Full service FTD and Teleflora florist.

wCot **Cottage Creek Nursery**
13300 Avondale Rd NE
Woodinville, WA 98072
Proprietor(s): Robert Nelson
Phone: 425-883-8252
FAX: 425-702-9243
Email address: cottagecreek@hotmail.com
Mail-order or walk-in: walk-in only
Credit cards: All major cards
Hours: Fall and winter: 10:00am-5:00pm everyday; Spring and summer: 10:00am-6:00pm
Specialties: Roses with over 500 varieties to ponder each year. Loads of perennials, too.

wCou **Country Gardens**
36735 SE David Powell Rd
Fall City, WA 98024
Proprietor(s): Keith Howe, Donald Howe, Willa Howe
Phone: 425-222-5616
FAX: 425-222-4827
Email address: daduck@nwlink.com
Website: www.nwlink.com/~dafox
Mail-order or walk-in: both
Hours: Call for appointment.
Specialties: Hydrangeas.

wCri **Cricklewood**
11907 Nevers Rd
Snohomish, WA 98290
Proprietor(s): Evie Douglas
Phone: 360-568-2829
Email address: cricklewod@aol.com
Mail-order or walk-in: mail-order; walk-in by appt.
Catalogue: $2.00 refunded with order
Credit cards: no
Hours: Walk-in by appointment.
Specialties: Hardy geraniums, hellebores, own-root old roses, Anemone nemerosa.

oCrm **Crimson Sage Nursery**
PO Box 337
Colton, OR 97017
Walk-in address: 30920 S Wall
Colton, OR 97017
Proprietor(s): Michelle DeFord
Phone: 503-824-4721
FAX: 503-824-7021
Email address: crimson@molalla.net
Website: www.crimson-sage.com
Mail-order or walk-in: mail-order only
Catalogue: $2.00
Credit cards: yes
Specialties: Medicinal herbs

wCSG **Country Store and Gardens**
20211 Vashon Hwy SW
Vashon Island, WA 98070
Proprietor(s): Vy Biel
Phone: 206-463-3655
FAX: 206-463-3679
Email address: csag@centurytel.net
Website: www.tcsag.com
Mail-order or walk-in: both
Credit cards: yes
Hours: 9:30am-5:30pm Monday-Saturday, 9:00am-5:00pm Sunday; self-guided tours anytime, guided tours every Saturday throughout the season.
Specialties: We have collections of hydrangea, spiraea, sorbus, viburnum, buddleia; display gardens, 10 acres of fields and old-fashioned flowers.

wCul **Cultus Bay Nursery**
7568 Cultus Bay Rd
Clinton, WA 98236
Proprietor(s): Mary Fisher
Phone: 360-579-2329
Mail-order or walk-in: walk-in only
Credit cards: VISA, MC
Hours: 9:00am-5:00pm daily; closed Tuesday
Specialties: Perennials, shrubs, vines, herbs, selected trees.

oDan **Dancing Oaks Nursery**
17900 Priem Rd
Monmouth, OR 97361
Proprietor(s): Fred Weisensee and Leonard Foltz
Phone: 503-838-6058
FAX: 503-838-6058, then 5 ** at message
Website: soon
Mail-order or walk-in: walk-in only
Catalogue: none
Credit cards: no
Hours: Hours are flexible; please call. Thursday, Friday, Saturday are best days for a visit.

Specialties: Unusual trees and shrubs, steel garden art, unusual perennials and some potted bulbs.

oDar **Daryll's Nursery**
15770 W Ellendale Rd
Dallas, OR 97338
Proprietor(s): Daryll Combs
Phone: 503-623-0251
FAX: 503-623-0251
Email address: coming soon
Website: coming soon
Mail-order or walk-in: walk-in only
Catalogue: price list $2.00
Credit cards: VISA, MC, Discover
Hours: Change by season; call for hours. Usually open March to November.
Specialties: Perennials, ornamental grasses, herbs, butterfly and hummingbird attracting plants, vines, shrubs, trees. Daryll is a propagator and grower of exciting plants.

wDav **Dave's Taste of the Tropics**
1618 NE 189th St
Shoreline, WA 98155
Proprietor(s): David Alvarez
Phone: 206-364-4428
FAX: 206-366-0604
Email address: tropics@wa.freei.net
Mail-order or walk-in: both
Catalogue: $2.00
Credit cards: no
Hours: By appointment only
Specialties: Hardy palms, bananas, tree ferns, agaves, and bamboo. Also other hardy tropicals.

wDBF **D & B Fuchsia Nursery**
PO Box 491
Brush Prairie, WA 98606
Proprietor(s): Betty Jean Wagoner
FAX: 520-223-8058
Email address: dbnsy@effectnet.com
Website: maxpages.com/dbnursery
Mail-order or walk-in: mail-order only
Catalogue: free on internet
Credit cards: no
Hours: By appointment only - make arrangements via email.
Specialties: Northwest-hardy perennial upright fuchsias, lax bush fuchsias, some annual varieties of fuchsias.

oDel **Delta Farm & Nursery**
3925 N Delta Hwy
Eugene, OR 97408-7100
Proprietor(s): Ron and Faye Spidell
Phone: 541-485-2992
FAX: 541-485-1985
Email address: deltafarm@nu-world.com
Website: www.deltafarm.com
Mail-order or walk-in: both
Catalogue: free
Credit cards: VISA, MC
Hours: March 1-July 15: 10:00am-6:00pm 7 days/week; other times, 10:00am-2:00pm or by appointment.
Specialties: Fuchsias, peppers by mail. Fuchsias, peppers (sweet and hot), herbs, hardy ferns, miscellaneous flowering plants.

oDut **Dutch Mill Herbfarm**
6640 NW Marsh Rd
Forest Grove, OR 97116
Proprietor(s): Barbara Remington
Phone: 503-357-0924
Mail-order or walk-in: both
Catalogue: SASE
Credit cards: no
Hours: Noon-6:00pm Wednesday through Saturday, or by appointment.
Specialties: Lavender, herbs - medicinal, fragrant, culinary, hops and interesting perennials. We have a Dutchmill lavender and Dutchmill rosemary, named for our farm in the trade today. Plants are organically grown since 1973, registered with Oregon State.

oEar **Earth's Rising**
PO Box 334
Monroe, OR 97456
Walk-in address: 25358 Cherry Creek Rd
Monroe, OR 97456
Proprietor(s): Delbert McCombs
Phone: 541-847-5950
FAX: 541-847-5950
Email address: earthsrising@juno.com
Website: www.wolfenet.com/~wind/delbert
Mail-order or walk-in: mail-order or by appointment
Catalogue: free
Credit cards: no
Hours: Visit possible, but not easy. 4 miles out of Monroe, 2 miles gravel. Call ahead to arrange.
Specialties: All stock organically grown, Tilth certified.

wEdc **Edelweiss Perennials**
13809 132nd Ave NE
Kirkland, WA 98034
Proprietor(s): Urs and Vickie Baltensperger
Email address: balts@msn.com
Website: foundaplant.com
Mail-order or walk-in: mail-order only
Catalogue: no
Credit cards: no
Hours: n/a
Specialties: Hardy geraniums, helleborus, hosta

oEdg **Edgewood Gardens**
5369 Donald St
Eugene, OR 97405
Proprietor(s): Jonathan Eeds and Glen Ray
Phone: 541-683-8307
Email address: jbeeds@pscnet.com
Mail-order or walk-in: mail-order
Catalogue: $1.00
Credit cards: no
Hours: Open by appointment.
Specialties: Rare and beautiful trees, shrubs and perennials.

oEdm **Edmunds' Roses**
6235 SW Kahle Rd
Wilsonville, OR 97070
Proprietor(s): Kathy and Phil Edmunds
Phone: 503-682-1476
FAX: 503-682-1275
Email address: edmunds@edmundsroses.com
Website: www.edmundsroses.com
Mail-order or walk-in: both
Catalogue: free
Credit cards: VISA, MC, Discover, AmEx
Hours: Rose field is open to the public in September. Visitors should call first to be sure the field is in bloom and nice for viewing.
Specialties: Exhibition and European rose varieties.

oEga **Egan Gardens**
9805 River Rd NE
Salem, OR 97303
Proprietor(s): Ellen Egan
Phone: 503-393-2131
FAX: 503-390-9020
Mail-order or walk-in: walk-in only

Credit cards: VISA, MC, AmEx, Discover
Hours: 1st weekend of March-mid July: 9:00am-6:00pm Monday-Saturday, 10:00am-5:00pm Sunday; Mid July-mid October: 9:00am-5:00pm Monday-Saturday, 10:00am-5:00pm Sunday. Closed mid October-start of March.
Specialties: Wide range of flowering annuals, perennials, baskets and planters. Top quality grower, retailing mostly our own product.

oExu **Exuberant Gardens**
22711 NW Gillihan Rd
Portland, OR 97231
Proprietor(s): David R and Dorothy S Rodal
Phone: 503-621-1164
FAX: 503-621-1164
Email address: drodal@pacifier.com
Mail-order or walk-in: walk-in only
Catalogue: free
Credit cards: no
Hours: April 21-July 1, 2001: Friday, Saturday and Sunday; or by appointment.
Specialties: Clematis.

wFai **Fairie Perennial and Herb Gardens**
6236 Elm St SE
Tumwater, WA 98501
Proprietor(s): David Baird and Steve Taylor
Phone: 360-754-9249
FAX: 360-943-7699
Email address: daveherbs@home.com
Website: www.hometown.aol.com/daveherbs
Mail-order or walk-in: walk-in only
Hours: Daily 10:00am-6:00pm
Specialties: Perennials, herbs, shrubs.

wFan **Fancy Fronds**
PO Box 1090
Gold Bar, WA 98251
Walk-in address: 40830 172nd St SE
Gold Bar, WA 98251
Proprietor(s): Judith I. Jones
Phone: 360-793-1472
FAX: 360-793-4243
Email address: judith@fancyfronds.com
Website: www.fancyfronds.com
Mail-order or walk-in: both
Catalogue: $3.00; list, SASE
Credit cards: no
Hours: By appointment only please.
Specialties: Temperate ferns from around the world, inclulding Victorian cultivars, tree ferns, xeric ferns.

wFFl **Forest Flor Nursery**
PO Box 89
Lummi Island, WA 98262
Proprietor(s): Wanda Cucinotta
Phone: 360-758-2778
FAX: 360-758-2778
Email address: forestflor@aol.com
Website: pending
Mail-order or walk-in: mail-order; retail in 2001
Catalogue: free wholesale only
Credit cards: no
Hours: Retail by appointment and sale on Memorial Day weekend only
Specialties: Specializing in Pacific Northwest native plants. Plants have been rescued from road construction and logging operations on DNR timber sales lands with a specical contract through the Department of Natural Resources and Forest Flor Nursery.

wFGN **Fairlight Gardens Nursery**
30904 164th Ave SE
Auburn, WA 98092
Proprietor(s): Judy and Don Jensen
Phone: 253-631-8932
FAX: 253-630-5630
Email address: bamboo1@foxinternet.net
Mail-order or walk-in: walk-in only
Credit cards: yes
Hours: March - October; 10:00am-6:00pm
Specialties: Herbs, bamboo, grasses, perennials, gazing globes, books, drieds, oils, soaps.

iFid **Fiddler's Ridge Garden and Nature Store**
1001 Fiddler's Ridge Loop
Potlatch, ID 83855
Proprietor(s): Theresa Greiner and John Madden
Phone: 208-875-1003
FAX: 208-875-0719
Email address: fiddler@turbonet.com
Mail-order or walk-in: both
Credit cards: MC, VISA, Discover
Hours: March 1-December 23: 9:00am-6:00pm Tuesday-Sunday; closed Mondays
Specialties: Hardy varieties for northern gardens. Hardy shrub roses, peonies, planting for the birds, perennials, herbs, bedding plants and vegetables, including heirlooms. We dig and sell over 30 varieties of peonies we grow here.

wFol **Foliage Gardens**
2003 128th Ave SE
Bellevue, WA 98005
Proprietor(s): Sue and Harry Olsen
Phone: 425-747-2998
Email address: foliageg@juno.com
Website: www.backyardgardener.com/Foliagegardens
Mail-order or walk-in: mail-order
Catalogue: $2.00
Credit cards: no
Hours: Open by appointment only; please call or write ahead.
Specialties: Hardy native and exotic spore grown ferns, dwarf and semi-dwarf Japanese maple cultivars.

oFor **Forestfarm**
990 Tetherow Rd
Williams, OR 97544
Proprietor(s): Ray and Peg Prag
Phone: 541-846-7269
FAX: 541-846-6963
Email address: forestfarm@rvi.net
Website: www.forestfarm.com
Mail-order or walk-in: mail-order; some walk-in
Catalogue: $4.00
Credit cards: yes
Hours: Please call - generally Wednesday, Thursday and Friday 9:00am-3:00pm, but we can adjust to some extent.
Specialties: Lots! Trees, shrubs, perennials, vines, rare and unusual plants from around the world, native plants, useful plants, plants for birds and wildlife, fragrant plants, etc, and etc.

oFra **Fragrant Garden Nursery**
PO Box 627
Canby, OR 97013
Proprietor(s): Pat Sherman
Phone: 503-263-6643
FAX: 503-266-2804
Email address: pats@canby.com
Website: fragrantgarden.com
Mail-order or walk-in: mail-order
Catalogue: $1.00
Credit cards: VISA, MC

Specialties: Sweet pea seed and starts - old fashioned and Spencer varieties.

oFre **Frey's Dahlias**
12054 Brick Rd
Turner, OR 97392
Proprietor(s): Sharon Frey
Phone: 503-743-3910
FAX: 503-743-3910 call first
Email address: freydahlia@juno.com
Website: landscapeusa.com
Mail-order or walk-in: both
Catalogue: free
Credit cards: VISA, MC
Hours: April-May: 9:00am-4:00pm Monday-Friday; August-October: 9:00am-6:00pm 7 days a week.
Specialties: Dahlias.

oFrs **Freshops**
36180 Kings Valley Hwy
Philomath, OR 97370
Phone: 541-929-2736
FAX: 541-929-2702
Email address: sales@freshops.com
Website: www.freshops.com
Mail-order or walk-in: mail-order only
Catalogue: free
Credit cards: yes
Hours: By appointment
Specialties: Hops - 10 varieties of ornamental and brewing hops.

wFWB **F W Byles, Co., Nursery**
PO Box 7705
Olympia, WA 98507
Walk-in address: The Barn Nursery, 9440 Old Hwy 99 SE
Olympia, WA 98501
Proprietor(s): Gudrun and Frank Byles
Phone: 360-352-4725
FAX: 360-352-1921
Email address: byles@juno.com
Website: www.theparagongroup.com/byles
Mail-order or walk-in: both
Catalogue: $2.00
Credit cards: soon
Hours: 10:00am-6:00pm Monday-Saturday; 10:00am-5:00pm Sunday; except holidays.
Specialties: Very large selection of rare and unusual Japanese and related maples for container, rock garden, patio and landscape. J. D. Vertrees collection plus many dwarfs and witches' brooms.

wGAc **Green Acres Gardens & Ponds**
15011 Vail Rd
Yelm, WA 98597
Proprietor(s): Jack and Sue Markham
Phone: 360-894-2940
Mail-order or walk-in: walk-in only
Catalogue: n/a
Credit cards: VISA, MC
Hours: April 1-August 31: 9:00am-6:00pm Friday-Saturday, 10:00am-5:00pm Sunday.
Specialties: Aquatic nursery. Everything for the pond. Specializing in aquatic plants, bamboo, rare and unusual perennials.

oGai **Gail Austin Garden Perennials**
8445 SW 80th Ave
Portland, OR 97223
Proprietor(s): Gail Austin
Phone: 503-246-5747
Email address: austing@internetcds.com
Mail-order or walk-in: mail-order only
Catalogue: $2.00 deductible from first order
Credit cards: no
Hours: No walk-ins, but we do have specific dates for open gardens and make appointments to view the garden during bloom season.
Specialties: Hemerocallis (daylilies)-over 900 varieties. Iris ensata (Japanese iris)-over 150 varieties. Official American Hemerocallis Society display garden.

oGar **Garland Nursery**
5470 NE Hwy 20
Corvallis, OR 97330
Proprietor(s): Don, Sandra, Brenda, Lee, Erica Powell
Phone: 541-753-6601
FAX: 541-753-3143
Email address: garlandnursery@proaxis.com
Website: www.proaxis.com/~garlandnursery
Mail-order or walk-in: walk-in only
Catalogue: none
Credit cards: VISA, MC, Discover
Hours: 9:00am-6:00pm Monday-Friday; 9:00am-5:00pm Saturday; 10:00am-5:00pm Sunday
Specialties: Huge selection of unusual and common plants: perennials, annuals, trees, conifers, shrubs, vines, tropicals, fruits, water plants, bonsai. Also arbors, decorative pots, fountains, bird baths, and gifts for gardeners.

oGoo **Goodwin Creek Gardens**
PO Box 83
Williams, OR 97544
Proprietor(s): Jim and Dotti Becker
Phone: 541-846-7357
FAX: 541-846-7357
Email address: info@goodwincreekgardens.com
Website: www.goodwincreekgardens.com
Mail-order or walk-in: mail-order
Catalogue: $1.00
Credit cards: yes
Hours: Appointment only.
Specialties: Herbs, plants for hummingbirds and butterflies, everlastings, pelargoniums.

oGos **Gossler Farms Nursery**
1200 Weaver Rd
Springfield, OR 97478
Proprietor(s): Roger, Marj, and Eric Gossler
Phone: 541-746-3922
FAX: 541-744-7924
Mail-order or walk-in: mail-order/walk-in by appt.
Catalogue: $2.00
Credit cards: VISA, MC
Hours: By appointment. Wholesale and retail.
Specialties: Magnolias, winter interest plants, hellebores, unusual trees and shrubs.

wGra **Grassy Knoll Exotic Plants**
2705 NW 6th Pl
Camas, WA 98607
Walk-in address: 1408 SE Coffey Rd
Washougal, WA 98607
Proprietor(s): Elizabeth Peters
Phone: 360-834-1170
Email address: grassyknollexoticplants@yahoo.com
Mail-order or walk-in: mail-order; walk-in by appt.
Catalogue: price list free
Credit cards: no
Hours: Walk-in by appointment only. I sell at the Vancouver Farmers' Market also.
Specialties: Over 20 kinds of passiflora, protea family, plants native to Australia and South Aftrica.

oGre **Greer Gardens, Inc.**
1280 Goodpasture Island Rd
Eugene, OR 97401
Proprietor(s): Harold Greer
Phone: 541-686-8266
FAX: 541-686-0910
Email address: orders@greergardens.com
Website: www.greergardens.com
Mail-order or walk-in: both
Catalogue: $3.00
Credit cards: yes, all major
Hours: 8:30am-5:30pm Monday-Saturday PST; 11:00am-5:00pm Sunday PST
Specialties: Wide selection including rhododendrons and azaleas, trees and shrubs, perennials.

IGSc **Good Scents**
1308 N Meridian Rd
Meridian, ID 83642
Proprietor(s): Lisa Doll
Phone: 208-887-1784
Email address: basil@micron.net
Website: netnow.micron.net/~basil
Mail-order or walk-in: both
Catalogue: $1.00
Credit cards: MC, VISA, Discover
Hours: March - September: 10:00am-6:00pm Monday-Saturday.
Specialties: Herbs - 400 varieties: culinary, medicinal, color dyeing, landscaping, tea.

oGue **Guerrero Nursery**
29040 SE Orient Dr
Gresham, OR 97030
Proprietor(s): Patricia and Enrique Guerrero
Phone: 503-663-2417
FAX: 503-663-2417
Mail-order or walk-in: walk-in only
Credit cards: no
Hours: 8:00am-4:00pm Tuesday-Sunday. Available Monday by appointment only.
Specialties: Hydrangeas, berberis; topiaries, animal and geometric, special order.

wHam **Hammond's Acres of Rhodies**
25911 70th Ave NE
Arlington, WA 98223
Proprietor(s): David G Hammond
Phone: 360-435-9206
FAX: 360-403-9177
Mail-order or walk-in: both
Catalogue: $2.00
Credit cards: yes
Hours: 10:00am-5:00pm each day; call before.
Specialties: Rhododendron hybrids and species, deciduous azalea hybrids and species.

oHan **Hansen Nursery**
PO Box 1228
North Bend, OR 97459
Walk-in address: Only available when making appointment
Proprietor(s): Robin L Hansen
Phone: 541-756-1156
Email address: rhansen@harborside.com
Mail-order or walk-in: mail-order; walk-in by appointment only
Catalogue: $1.00
Credit cards: no
Hours: Call for appointment.
Specialties: Cyclamen, native plants.

wHea **Heaths & Heathers**
502 E Haskell Hill Rd
Shelton, WA 98584
Proprietor(s): Karla Lortz
Phone: 360-427-5318
FAX: 360-432-9780
Email address: handh@heathsandheathers.com
Website: heathsandheathers.com
Mail-order or walk-in: mail-order only
Catalogue: free
Credit cards: VISA, MC, AmEx, Discover
Specialties: Heaths and heathers.

oHed **Hedgerows Nursery**
20165 SW Christensen Rd
McMinnville, OR 97128
Proprietor(s): David Mason and Susie Grimm
Phone: 503-843-7522
FAX: 503-843-7522
Email address: hedgerows@onlinemac.com
Mail-order or walk-in: walk-in only
Catalogue: no catalogue
Credit cards: no
Hours: Mid-March-September: 10:00am-5:00pm. March-June: Wednesday-Sunday; July-September: Thursday-Sunday; or by appointment.
Specialties: Unusual perennials, shrubs and climbing plants.

oHei **Heirloom Old Garden Roses**
24062 Riverside Dr NE
St Paul, OR 97137
Proprietor(s): John and Louise Clements
Phone: 503-538-1576
FAX: 503-538-5902
Website: www.heirloomroses.com
Mail-order or walk-in: both
Catalogue: $5.00
Credit cards: VISA, MC, Discover
Hours: Office hours: 8:00am-4:30pm Monday-Friday Pacific time. Nursery hours: 9:00am-4:00pm (May-September: 9:00am-5:00pm)
Specialties: Shrub roses, groundcover roses, fragrant roses, old garden roses, English roses, English Legend roses, climbing roses, rambling roses. We have our own breeding program with many wonderful new rose varieties of our own creation.

wHer **Heronswood Nursery Ltd**
7530 NE 288th St
Kingston, WA 98346
Proprietor(s): Daniel Hinkley and Robert Jones
Phone: 360-297-4172
FAX: 360-297-8321
Email address: heronswood@silverlink.net
Website: www.heronswood.com
Mail-order or walk-in: mail-order; walk-in by appointment only.
Catalogue: $5.00 1 year; $8.00 2 years
Credit cards: VISA, MC, AmEx
Hours: 9:00am-1:00pm by appointment only Monday-Friday; close at 4:00pm. Must make an appointment to come; 6 or more people is a paid tour.
Specialties: Rare and unusual plants: trees, conifers, shrubs, grasses, vines, perennials, and temperennials.

wHig **Highfield Garden**
4704 NE Cedar Creek Rd
Woodland, WA 98674
Proprietor(s): Irene and Gil Moss
Phone: 360-225-6525
Mail-order or walk-in: mail-order only
Catalogue: $1.00
Credit cards: no
Hours: n/a
Specialties: Hardy geraniums

wHol **Holland Gardens**
29106 Meridian E
Graham, WA 98338
Proprietor(s): Marty and Lorraine Holland
Phone: 253-847-5425
Email address: hollandgardens@compuserve.com
Website: www.geocities.com/hollandgardens
Mail-order or walk-in: both
Catalogue: Black and white - free; color brochure - $1.00
Credit cards: no
Hours: Any time, if not at home, honor system
Specialties: Specialize with 'Sweet Lena,' a very fragrant iris (it has a very pronounced sweet scented fragrance.).

wHom **Homestead Nursery**
18013 NE 25th Ave
Ridgefield, WA 98642
Proprietor(s): Lee and Pam Herrman
Phone: 360-576-6540
FAX: 360-576-3846
Mail-order or walk-in: walk-in only
Credit cards: yes
Hours: 9:00am-5:00pm; closed Wednesdays.
Specialties: Unusual to rare perennials and herbs.

oHon **Honeyhill Farms Nursery**
Proprietor(s): Jim and Audrey Metcalfe
Phone: 503-292-1817
Email address: honeyhill2@aol.com
Mail-order or walk-in: Appointments only
Catalogue: none
Credit cards: no
Hours: Open for tours and plant sales February and March when hellebores are in bloom. Please call for appointment.
Specialties: Hellebores and their companion plants.

oHou **House of Whispering Firs**
20080 SW Jaquith Rd
Newberg, OR 97132
Proprietor(s): Kathleen Thompson
Phone: 503-628-3695
FAX: 503-628-3553
Mail-order or walk-in: walk-in only
Credit cards: yes
Hours: 9:00am-6:00pm daily. Closed major holidays.
Specialties: Herbs, shade plants (hostas, fuchsias, ferns), and everlastings.

oHug **Hughes Water Gardens**
25289 SW Stafford Rd
Tualatin, OR 97062
Proprietor(s): Eamonn Hughes
Phone: 503-638-1709
FAX: 503-638-9035
Email address: eamonn@teleport.com
Website: www.watergardens.com
Mail-order or walk-in: both
Catalogue: no
Credit cards: VISA, MC
Hours: March-October: 9:00am-5:00pm Monday-Saturday, 10:00am-5:00pm Sunday; November-February: 9:00am-5:00pm Monday-Friday
Specialties: Waterlilies, lotus, bog and pond plants.

oInd **Independence Flower Farm**
4215 Independence Hwy
Independence, OR 97351
Proprietor(s): Diane Calabrese
Phone: 503-838-4414
Email address: c-brese@ncn.com
Mail-order or walk-in: by appt. only
Catalogue: no
Credit cards: no
Hours: By appointment only.
Specialties: Selected perennials.

wIri **Irish Eyes Inc**
PO Box 307
Thorp, WA 98946
Proprietor(s): Greg Anthony Lutovsky
Phone: 509-925-6025
FAX: 800-964-9210
Website: www.irish-eyes.com; www.hintofgarlic.com; www.inhotpursuit.com; www.beanandpea.com
Mail-order or walk-in: both
Catalogue: free
Credit cards: yes
Specialties: Potatoes, garlic, ornamental alliums, tomatoes, peppers, beans, and peas.

wJad **Jade Mountain Bamboo**
5020 116th St E
Tacoma, WA 98446
Proprietor(s): Dale Chesnut and Phil Davidson
Phone: 253-546-1129
FAX: 253-546-1129
Email address: phildavidson@worldnet.att.net or phildavidson@imajis.com
Website: business.fortunecity.com/ipo/180/index.html
Mail-order or walk-in: both
Catalogue: $1.00
Credit cards: yes
Hours: Please call first. Weekdays usually by appointment. Weekends 10:00am-4:00pm no appointment needed.
Specialties: Specialize in hardy bamboo - over 60 varieties. Installation barrier, bamboo removal, handmade granite garden lanterns, basins.

oJil **Jill Schatz Plants**
Proprietor(s): Jill Schatz
Phone: 503-297-8435
Email address: jschatz@xprt.net
Mail-order or walk-in: by appointment
Credit cards: no
Hours: Anytime, with an appointment; call for open garden dates.
Specialties: Choice and unusual small trees, shrubs, and perennials, including hydrangeas, meconopsis and arisaemas. Native plants also available. Limited species available in wholesale quantities.

oJoy **Joy Creek Nursery**
20300 NW Watson Rd
Scappoose, OR 97056
Proprietor(s): Maurice Horn, Mike Smith, Scott Christy
Phone: 503-543-7474
FAX: 503-543-6933
Website: www.joycreek.com
Mail-order or walk-in: both
Catalogue: $2.00
Credit cards: VISA, MC
Hours: March 1 - October 31: 9:00am-5:00pm daily; or by appointment
Specialties: Clematis, penstemon, hydrangeas, hostas, fuchsias, hebes, and many more perennials and small shrubs and trees.

oJus **Justice Miniature Roses**
5947 SW Kahle Rd
Wilsonville, OR 97070
Proprietor(s): Jerry G. Justice
Phone: 503-682-2370
Email address: justrose@gte.net
Mail-order or walk-in: both
Catalogue: free

Credit cards: no
Hours: 9:00am-4:00pm 7 days/week
Specialties: Miniature roses.

wKil **Killdeer Farms**
21606 NW 51st Ave
Ridgefield, WA 98642
Proprietor(s): Steve Barton
Phone: 360-887-1790
FAX: 360-887-3009
Mail-order or walk-in: both
Catalogue: free
Credit cards: VISA, MC
Hours: April-June: 10:00am-6:00pm Tuesday-Sunday; year round by appointment
Specialties: Geraniums/pelargoniums, over 500 varieties of geraniums, specialty tri-colors/bi-colors, rosebud flowered, tulip flowered, miniatures, dwarfs, ivy-leaf, Martha Washingon, scented leaf; also large selection of fuchsias/ferns.

wKin **Kinder Gardens Nursery**
1137 S Highway 17
Othello, WA 99344
Proprietor(s): Dennis and Claudia Kinder
Phone: 509-488-5017
FAX: 509-488-6513
Mail-order or walk-in: walk-in only
Catalogue: $3.95
Credit cards: yes
Hours: Seasonal hours are by the calendar year. Spring/Fall: 9:00am-6:00pm Monday-Saturday, Noon-5:00pm Sunday. Summer: 9:00am-6:00pm Tuesday-Saturday; closed Sunday and Monday.
Specialties: As one of the largest "grower nurseries" on the east side of the Cascades, our stock is acclimated to our climate. We offer a large variety of species and sizes.

oLon **Lon J. Rombough**
PO Box 365
Aurora, OR 97002
Proprietor(s): Lon Rombough
Phone: 503-678-1410
Email address: lonrom@hevanet.com
Website: www.hevanet.com/lonrom
Mail-order or walk-in: mail-order only
Catalogue: SASE
Credit cards: none
Specialties: Grape cuttings, 135+ varieties, all types.

oLSG **L & S Gardens**
50792 S Huntington Rd
La Pine, OR 97739
Proprietor(s): Linda Stephenson
Phone: 541-536-2049
FAX: 541-536-8634
Email address: lsgardens@hwy97.net
Credit cards: no
Hours: 9:00am-5:00pm Monday-Saturday, 10:00am-4:00pm Sunday, 7 days a week.
Specialties: Hardy perennials.

oMac **Macleay Perennial Gardens**
1420 Howell Prairie Rd SE
Salem, OR 97301
Proprietor(s): Robert W. Long
Phone: 503-581-3592
Mail-order or walk-in: walk-in by appt. only
Catalogue: none
Credit cards: no
Hours: Call for appointment.
Specialties: Giant lily (Cardiocrinum giganteum), Chinese paper bark maple (Acer griseum) seedlings and up to 8" diameter specimens (also seeds), Japanese umbrella pine (Sciadopitys verticillata), specimen tree peonies, unusual perennials and bog plants.

wMag **Magnolia Garden Center**
3213 W Smith St
Seattle, WA 98199
Second location: Greensleaves Nursery
16215 - 140th Place NE
Woodinville, WA 98072
Proprietor(s): Margaret and Chuck Flaherty
Phone: 206-284-1161
FAX: 206-284-0081
Email address: maggarcen@aol.com
Website: magnoliagarden.com
Mail-order or walk-in: both
Catalogue: no catalog
Credit cards: VISA, MC
Hours: 9:30am-6:00pm Monday-Friday; 9:00am-6:00pm Saturday; 10:00am-5:00pm Sunday
Specialties: Small trees, shrubs, perennials, specialty annuals.

oMid **Mid-America Garden**
PO Box 18278
Salem, OR 97305-8278
Walk-in address: 7185 Lakeside Dr NE
Salem, OR 97305
Proprietor(s): Paul Black and Tom Johnson
Phone: 503-390-6072
FAX: 503-390-6072
Mail-order or walk-in: both
Catalogue: $3.00
Hours: April - July: 10:00am-5:00pm Friday-Saturday
Specialties: All classes of bearded iris and Siberian iris, hostas and hemerocallis (daylilies). All plants are propagated at nursery site.

oMis **Miss Emily's Garden**
PO Box 1908
Wilsonville, OR 97070
Walk-in address: 6351 SW Advance Rd
Wilsonville, OR 97070
Proprietor(s): Jean C. Connolly
Phone: 503-682-7971
Email address: msemilys@gte.net
Mail-order or walk-in: walk-in only
Catalogue: no
Credit cards: no
Hours: 10:00am-5:00pm Wednesday-Saturday; closed Sunday-Tuesday. Can call for appointment at other times
Specialties: Unusual perennials.

oMit **Mitsch Daffodils**
PO Box 218
Hubbard, OR 97032
Walk-in address: 6247 S Sconce Rd
Hubbard, OR 97032
Proprietor(s): Richard and Elise Havens
Phone: 503-651-2742
FAX: 503-651-2792
Email address: havensr@web-ster.com
Website: www.web-ster.com/havensr/mitsch
Mail-order or walk-in: mail-order; walk-in only during flowering season
Catalogue: $3.00 deductible on order
Credit cards: VISA, MC, Discover
Hours: Open to visitors only during flower season, afternoons, mid-March to early April
Specialties: New varieties of daffodils - red pinks, lemon colors, species hybrids.

wMtT **Mt. Tahoma Nursery**
28111 112th Ave E
Graham, WA 98338
Proprietor(s): Rick Lupp
Phone: 253-847-9827
Email address: rlupp@aol.com
Website: www.backyardgardener.com/mttahoma
Mail-order or walk-in: both
Catalogue: $2.00
Credit cards: no
Hours: By appointment.
Specialties: Alpines - campanula, primula, fall gentiana, penstemon.

oNat **Natural Designs Nursery**
6112 SE Insley St
Portland, OR 97206
Proprietor(s): Marilyn Dubé
Phone: 503-774-0665
FAX: 503-774-0665
Email address: maridube@telcport.com
Mail-order or walk-in: walk-in only
Catalogue: none
Credit cards: no
Hours: Open one weekend per month; call for dates. Or by appointment.
Specialties: Many NW natives, shade garden plants and summer bulbs. Large selection of hydrangeas and other flowering shrubs.

wNay **Naylor Creek Nursery**
2610 West Valley Rd
Chimacum, WA 98325
Proprietor(s): Gary T Lindheimer and Jack E Hirsch
Phone: 360-732-4983
FAX: 360-732-7171
Email address: naylorck@olypen.com
Website: www.naylorcreek.com
Mail-order or walk-in: both
Catalogue: $1.00
Credit cards: yes
Hours: May 1-September 15: 10:00am-4:30pm, or by appointment
Specialties: Hostas, pulmonarias, epimediums, astilbes, and other shade tolerant perennials.

oNor **Northern Groves**
PO Box 1236
Philomath, OR 97370
Walk-in address: 23818 Henderson
Corvallis, OR 97333
Proprietor(s): Rick Valley
Phone: 541-929-7152
Email address: bamboogrove@cmug.com
Mail-order or walk-in: both
Catalogue: $3.00
Credit cards: no
Hours: By appointment.
Specialties: Bamboos.

wNot **Nothing But Northwest Natives**
14836 NE 249th St
Battle Ground, WA 98604
Proprietor(s): Dr. Kali Robson, PhD Botany
Phone: 360-666-3023
FAX: 360-666-3023
Mail-order or walk-in: walk-in only
Catalogue: SASE
Credit cards: no
Hours: March 2-July 30 and September 1-October 29: 10:00am-6:00pm; August and November-February: by appointment.
Specialties: Northwest native trees, shrubs, and perennials. Consulting for homeowners and reclamation professionals.

wNTP **No Thyme Productions**
8321 SE 61st St
Mercer Island, WA 98040
Walk-in address: By appointment only
Proprietor(s): Nancy Mencke
Phone: 206-236-8885
FAX: 206-230-8685
Email address: info@nothyme.com
Website: www.nothyme.com
Mail-order or walk-in: mail-order only
Catalogue: free
Credit cards: yes
Hours: Call; limited appointments
Specialties: 300 varieties of herbs and unusuals, pesticide free.

oNWe **Northwest Garden Nursery**
86813 Central Rd
Eugene, OR 97402-9284
Proprietor(s): Ernie and Marietta O'Byrne
Phone: 541-935-3915
FAX: 541-935-3915
Email address: nargsbs@efn.org
Mail-order or walk-in: walk-in only
Catalogue: no catalogue
Credit cards: no
Hours: March 1-October 15: 10:00am-6:00pm Thursday-Friday, 10:00am 5:00pm Saturday; or by arrangement at other times.
Specialties: Hellebores, meconopsis, unusual herbaceous shade plants and bulbs, perennials, vines

oOEx **Oregon Exotics Nursery**
1065 Messinger Rd
Grants Pass, OR 97527
Proprietor(s): Jerome R Black
Phone: 541-846-7578
FAX: 541-846-9488
Website: www.exoticfruit.com
Mail-order or walk-in: mail-order only
Catalogue: $4.00
Credit cards: yes - all
Hours: Please call in advance to make an appointment.
Specialties: Figs, pawpaws, persimmons, loquats, jujubes, hardy citrus, Himalayan fruit trees, Chinese medicinals, unusual garden seeds, Andean crops, rare perennial vegetables.

oOre **Oregon Trail Daffodils**
41905 SE Louden
Corbett, OR 97019
Proprietor(s): Bill and Diane Tribe
Phone: 503-695-5513
FAX: 503-695-5573
Email address: daffodil@europa.com
Mail-order or walk-in: mail-order only
Catalogue: free
Credit cards: no
Hours: Visitors welcome during flowering season (mid-March - early May). Please phone for conditions/directions.
Specialties: Exclusive distributor of many daffodil hybrids bred by Murray Evans and Bill Pannill.

wOud **Oudean's Willow Creek Nursery**
7421 137th Ave SE
Snohomish, WA 98290
Proprietor(s): Karen and Allan Oudean
Phone: 360-568-6024
FAX: 360-568-4904
Email address: cambrp@premier1.net
Website: www.firebirdz.net/willowcreek
Mail-order or walk-in: walk-in only
Catalogue: on the web
Hours: 10:00am-4:00pm Friday-Sunday

Specialties: Carnivorous plants, hardy and tropical.

oOut **Outback Redd's**
16595 SW 147th
Tigard, OR 97224
Proprietor(s): Trish and Jay Aldrich
Phone: 503-590-0734
Mail-order or walk-in: walk-in only
Hours: Please call first.
Specialties: Ornamental grasses and unusual perennials.

oPac **Pacific Peonies & Perennials**
11466 S Mulino Rd
Canby, OR 97013
Proprietor(s): Chris Baglien and Theresa Snelson
Phone: 503-263-6353
Email address: pacpeonies@cs.com
Mail-order or walk-in: both
Catalogue: free
Credit cards: VISA
Hours: April-June: 11:00am-5:00pm Thursday-Sunday; or by appointment
Specialties: Specialize in peonies - over 150 varieties - also perennial plants. Freeze-dried flowers, dried arrangements also. Special occasion freeze drying available.

wPir **Piriformis Nursery**
1051 N 35th St
Seattle, WA 98103
Proprietor(s): Tory Galloway
Phone: 206-632-1760
FAX: 206-632-5682
Website: www.piriformis.com
Mail-order or walk-in: walk-in only
Credit cards: yes
Hours: Wednesday-Sunday
Specialties: We sell only low water/drought tolerant plants and antique garden funk.

wPla **Plants of the Wild**
PO Box 866
Tekoa, WA 99033
Walk-in address: 123 Stateline Road
Tekoa, WA 99033
Proprietor(s): Kathy Hutton, Manager
Phone: 509-284-2848
FAX: 509-284-6464
Email address: kathy@plantsofthewild.com
Website: www.plantsofthewild.com
Mail-order or walk-in: both
Catalogue: free
Credit cards: MC, VISA, AmEx
Hours: 8:00am-4:30pm Monday-Friday
Specialties: Plants native to the Pacific Northwest

wPle **A Plethora of Primula**
244 Westside Hwy
Vader, WA 98593
Proprietor(s): April E. Boettger
Phone: 360-295-3114
FAX: 360-295-3900
Mail-order or walk-in: mail-order only and plant sales at Olympia Farmers Market, May-June
Catalogue: $2.00
Hours: Not open to public (on site) at all.
Specialties: Primula, especially garden auricula and assorted species.

oPor **Porterhowse Farms**
41370 SE Thomas Rd
Sandy, OR 97055
Proprietor(s): Don Howse and Lloyd Porter
Phone: 503-668-5834
FAX: 503-668-5834
Email address: phfarm@aol.com
Website: www.porterhowse.com
Mail-order or walk-in: both
Catalogue: free availability list
Credit cards: yes
Hours: Call ahead.
Specialties: Conifers - 1100 currently listed varieties and species. Conifer companions, broadleaf trees and shrubs, alpine and rock garden perennials.

wPug **The Puget Sound Kiwi Co.**
1220 NE 90th
Seattle, WA 98115-3131
Proprietor(s): Bob Glanzman
Phone: 206-523-6403
FAX: 206-523-6403
Mail-order or walk-in: both
Catalogue: SASE
Credit cards: no
Hours: By appointment only
Specialties: Kiwifruit (actinidia species), edible figs.

wRai **Raintree Nursery**
391 Butts Rd
Morton, WA 98356
Proprietor(s): Sam Benowitz
Phone: 360-496-6400
FAX: 888-770-8358
Email address: order@raintreenursery.com
Website: raintreenursery.com
Mail-order or walk-in: both
Catalogue: free
Credit cards: VISA, MC, Discover
Hours: Phone hours: 8:00am-4:30pm; at nursery 10:00am-4:00pm.
Specialties: Fruit trees, berries and unusual edibles for your backyard. Unique cultivars from around the world. Disease resistant varieties.

oRar **Rare Plant Research**
13245 SE Harold
Portland, OR 97236
Proprietor(s): Burl Mostul
FAX: 503-762-0289
Email address: rareplantr@aol.com
Website: www.rareplantresearch.com
Mail-order or walk-in: mail-order only
Catalogue: $2.00
Credit cards: VISA, MC
Hours: None
Specialties: Hardy bananas, lewisia, vary rare tropical xerophytes.

wRav **Raven Nursery**
22370 Indianola Rd
Poulsbo, WA 98370
Proprietor(s): Beatrice D. Idris
Phone: 360-598-3323
FAX: 360-598-6610
Email address: tworaven@ix.netcom.com
Mail-order or walk-in: walk-in only
Catalogue: n/a
Credit cards: no
Hours: By appointment only; call for directions.
Specialties: Trees: native conifers, maples, firs. Ferns, salal, huckleberries; daylilies.

oRed **Red's Rhodies**
15920 SW Oberst Ln
Sherwood, OR 97140

Proprietor(s): Dick and Karen Cavender
Phone: 503-625-6331
FAX: 503-625-6331
Email address: rhodies@pcez.com
Website: hardy-orchids.com
Mail-order or walk-in: both
Catalogue: $1.00
Credit cards: VISA, MC
Hours: Call first.
Specialties: Hardy terrestrial orchids: pleione, bletilla, dactylorhiza. Rhododendron occidentale.

wRho **Rhododendron Species Botanical Garden**
PO Box 3798
Federal Way, WA 98063
Walk-in address: 2525 S 336th St
Federal Way, WA 98003
Proprietor(s): Manager: Richie Steffen
Phone: 253-838-4646
FAX: 253-838-4686
Website: www..halcyon.com/rsf/
Mail-order or walk-in: both
Catalogue: $3.50
Credit cards: VISA, MC
Hours: March-May: 10:00am-4:00pm (closed Thursday), June-February: 11:00am-4:00pm (closed Thursday and Friday)
Specialties: Species rhododendrons in a 24 acre botanical garden, managed by the RSF, a non-profit organization. The garden contains one of the world's largest collections of rhododendron and azalea species. Garden admission: $3.50 adults, $2.50 seniors and students, under 12 free. Free admission November-February.

oRiv **River Rock Nursery**
19251 SE Hwy 224
Clackamas, OR 97015
Proprietor(s): Bob and Gretchen O'Brien
Phone: 503-658-4047
FAX: 503-658-6132
Email address: green@agora.rdrop.com
Mail-order or walk-in: both
Credit cards: yes
Hours: By appointment only please; call or email first.
Specialties: Unusual trees and shrubs of all sizes for all applications. Deciduous, dwarf conifers, heirloom apples, and more.

oRNW **Roses Northwest**
12155 SW Tualatin-Sherwood Rd
Tualatin, OR 97062-6828
Proprietor(s): Earl and Loris Itel; Barbara Itel, general manager
Phone: 503-692-3066
FAX: 503-692-0548
Mail-order or walk-in: both
Catalogue: Variety list $3.00
Credit cards: VISA, MC, Discover, AmEx, debit
Hours: April 1-October 31: 10:00am-6:00pm Tuesday-Sunday, closed Mondays except by appointment; November 1-March 31: 1:00pm-5:00pm Tuesday-Sunday, Mondays by appointment.
Specialties: Large selection of hard-to-find rose varieties. Plant locator service. Cut roses available daily or by special order. Knowledgeable staff. Retail/wholesale. Retail outlet for Carlton Roses.

wRob **Robyn's Nest**
7802 NE 63rd St
Vancouver, WA 98662
Proprietor(s): Robyn Duback
Phone: 360-256-7399
Email address: robyn@robynsnestnursery.com
Website: robynsnestnursery.com
Mail-order or walk-in: both
Catalogue: $2.00
Credit cards: VISA, MC
Hours: Mid-March-June and September-October: 10:30am-5:30pm Thursday-Friday, 10:30am-2:00pm Saturday; July-August: 10:30am-2:00pm Friday-Saturday.
Specialties: Hostas and other fine perennials.

wRoo **Roozengaarde**
15867 Beaver Marsh Rd
Mt Vernon, WA 98273
Phone: 800-732-3266, 360-424-8531
FAX: 360-424-3113
Email address: info@tulips.com
Website: www.tulips.com
Mail-order or walk-in: both
Catalogue: free
Credit cards: VISA, MC
Hours: 9:00am-5:00pm Monday-Saturday. Open 10:00am-5:00pm Sundays mid-March-April. Extended hours during Tulip Festival.
Specialties: Spring-flowering bulbs - tulips, daffodils, iris. Fresh-cut flowers year-round; we ship anywhere in U.S. Three acre show garden (spring-flowering).

oRos **The Rose Guardians**
PO Box 426
St Paul, OR 97137
Proprietor(s): Alice Stockfleth
Phone: 503-393-1051
FAX: 503-393-1051
Mail-order or walk-in: mail-order only
Catalogue: $5.00
Hours: No walk-ins
Specialties: Own-root roses. Senior discounts.

oRus **Russell Graham, Purveyor of Plants**
4030 Eagle Crest Rd NW
Salem, OR 97304
Proprietor(s): Russell and Yvonne Graham
Phone: 503-362-1135
Email address: grahams@open.org
Mail-order or walk-in: mail-order or by appointment
Catalogue: $2.00
Credit cards: no
Hours: By appointment or open weekends listed in catalog.
Specialties: Native perennials, cyclamen, hellebores, primula, trillium, ferns, less common companion plants.

oSch **Schreiner's Iris Gardens**
3625 Quinaby Rd NE
Salem, OR 97303
Proprietor(s): The Schreiner Family
Phone: 503-393-3232
FAX: 503-393-5590
Website: www.schreinersiris.com
Mail-order or walk-in: both
Catalogue: $5.00
Credit cards: VISA, MC only
Hours: May-first week of June: 8:00am-5:00pm (Iris blooming time)
Specialties: TB iris, Siberian iris and Louisiana iris; also dwarf bearded iris.

wSea **Sea-Tac Gardens**
20020 Des Moines Memorial Dr
Seattle, WA 98198
Proprietor(s): Louis and Patti Eckhoff
Phone: 206-824-3846
Email address: patheck@prodigy.net
Mail-order or walk-in: both
Catalogue: SASE
Credit cards: no

Hours: Daylight hours blooming season, August through October.
Specialties: Dahlias.

oSec **Secret Garden Growers**
29100 S Needy Rd
Canby, OR 97013
Proprietor(s): Patricia Thompson
Phone: 503-651-2006
Email address: secretgrwr@aol.com
Mail-order or walk-in: walk-in only
Catalogue: none
Credit cards: no
Hours: April-October: Saturday 10:00am-5:00pm; or by appointment.
Specialties: Perennials, perennials, perennials. Ornamental grasses, salvias. Demonstration gardens, trails, landscpae design and installation services. Country nursery.

oSev **Seven Mile Nursery**
34255 Seven Mile Ln SE
Albany, OR 97321-7238
Proprietor(s): Brian and Kelley Roth
Phone: 541-928-1145
FAX: 541-928-1145
Email address: sevenmi@dnc.net
Mail-order or walk-in: walk-in only
Catalogue: none
Credit cards: coming 2001
Hours: April-June: 10:00am-6:00pm Tuesday-Saturday
Specialties: Patio containers, mixed baskets, perennials, upscale annuals from Proven Winners™, Euroamerican ™ and The Flower Fields ™. Vegetable starts, bedding plants. We grow what we sell.

oSha **Shannon's Perennials & Old Garden Roses**
8061 Jordan St SE
Salem, OR 97301
Proprietor(s): Shannon Jorgenson
Phone: 503-370-8111
Email address: jorg@teleport.com
Mail-order or walk-in: both
Catalogue: free
Credit cards: no
Hours: Please call first.
Specialties: Cut flowers, perennials, antique and old roses, floral work.

oSho **Shonnard's Nursery**
6600 SW Philomath Blvd
Corvallis, OR 97333
Proprietor(s): Chris and Lynnette Shonnard
Phone: 541-929-3524
FAX: 541-929-6361
Mail-order or walk-in: walk-in only
Hours: 8:30am-6:00pm Monday-Friday, 9:00am-5:00pm Saturday, 10:00am-5:00pm Sunday
Specialties: Rare and unusual trees, shrubs, and perennials; houseplants; general nursery stock, full service landscape design, installation and maintenance; FTD florist.

wShR **Shore Road Nursery**
616 Shore Rd
Port Angeles, WA 98362
Proprietor(s): David and Julie Allen
Phone: 360-457-1536
FAX: 360-457-8482
Email address: plantman@olypen.com
Mail-order or walk-in: walk-in only
Catalogue: $1.00
Credit cards: no

Hours: March-December: 10:00am-5:00pm Tuesday-Saturday; advised to call ahead.
Specialties: Northwest native plants, contract growing, revegetation specialist.

oSis **Siskiyou Rare Plant Nursery**
2825 Cummings Rd
Medford, OR 97501
Proprietor(s): Baldassare Mineo
Phone: 541-772-6846
FAX: 541-772-4917
Email address: srpn@wave.net
Website: www.wave.net/upg/srpn
Mail-order or walk-in: both
Catalogue: $3.00
Credit cards: VISA, MC
Hours: July-August and November-December: 9:00am-5:00pm Monday-Friday. Also, by appointment, and the first and last Saturday of each month, March-November, 9:00am-2:00pm. Closed first Saturday in July and legal holidays.
Specialties: Alpines and rock garden plants, hardy perennials, shrubs and smaller conifers, Japanese maples (dwarf), hardy ferns.

oSle **Sleepy Hollow Nursery**
8350 Spring Valley Rd NW
Salem, OR 97304
Proprietor(s): N. Andrea Joy Riley
Phone: 503-363-1546
Mail-order or walk-in: walk-in only
Credit cards: no
Hours: March-November: 10:00am-4:00pm Thursday-Tuesday (closed Wednesdays) or by appointment
Specialties: An ever increasing selecton of native plants, perennials, unusual and edible plants, retailed in a "growing" show garden out in a country setting.

wSno **Snow Creek Daylily Gardens**
PO Box 2007, 330 P St
Port Townsend, WA 98368
Walk-in address: 330 P Street (business office)
Port Townsend, WA 98368
Proprietor(s): Bob and Dianna Gibson
Phone: 360-765-4342
FAX: 360-765-4074
Website: www.snowdaylily.com
Mail-order or walk-in: both
Catalogue: $3.00
Credit cards: VISA, MC
Hours: 9:00am-5:00pm Monday-Friday; at daylily field, mid-June through mid-August, Monday - Saturday
Specialties: Hemerocallis (daylily) plants

wSnq **Sundquist Nursery, Inc.**
P.O. Box 2451
Poulsbo, WA 98370
Proprietor(s): Nils Sundquist
Phone: 360-779-6343
FAX: 360-697-6971
Mail-order or walk-in: On Open Garden Days only
Credit cards: VISA, MC
Hours: Open only on the following 2000 days: April 8, July 15-16 (same as Bainbridge in Bloom), June 3, July 22, and September 9 (last 3 same as Heronswood): Hours on Open Garden Days: 9:00am-5:00pm. Call for address, directions, open days for 2001.
Specialties: The nursery is known for perennials, ferns, grasses, better trees and shrubs, and select natives, and is one of the region's best sources for hardy fuchsias. Display gardens include perennial beds, ornamental grasses, shrub roses (on trial without spraying), unusual trees, shrubs, natives, candelabra primroses, gunnera, and nearly 10,000 daffodils.

wSoo **Soos Creek Gardens**
12602 SE Petrovitsky Rd
Renton, WA 98058-6707
Proprietor(s): Helmut & Lourdes Brodka
Phone: 425-226-9308
FAX: 425-226-9308
Email address: sooscrk@w-link.net
Mail-order or walk-in: walk-in only
Catalogue: n/a
Credit cards: no
Hours: Summer only: Saturday and Sunday only 10:00am-6:00pm
Specialties: Water plants.

oSqu **Squaw Mountain Gardens**
PO Box 946
Estacada, OR 97023
Walk-in address: 36212 SE Squaw Mountain Rd
Estacada, OR 97023
Proprietor(s): Janis and Arthur Noyes, Joyce Hoekstra
Phone: 503-630-5458
FAX: 503-630-5849
Email address: HenNChicks@aol.com
Website: www.squawmountaingardens.com
Mail-order or walk-in: both
Catalogue: $2.00
Credit cards: VISA, MC
Hours: By appointment
Specialties: Sedums, sempervivums, succulents, ivies.

oSsd **Susandales Pond and Bog Plants**
23665 SE Borges Rd
Gresham, OR 97080
Proprietor(s): Sue T. Lane and Dale L. Hash
Phone: 503-661-4259
FAX: 503-661-1243
Email address: susandale@worldnet.att.net
Website: www.susandales.com
Mail-order or walk-in: both
Hours: Specialty grower - visitors welcome; call first.
Specialties: Pond and bog plants - native wetland plants, water lilies.

wSta **Star Nursery & Landscaping, LLC**
13916 42nd Ave S
Seattle, WA 98168
Proprietor(s): Mike Palmer
Phone: 206-241-2115
FAX: 206-241-2677
Mail-order or walk-in: walk-in only
Credit cards: no
Hours: 8:00am-4:30pm Monday-Friday
Specialties: Azaleas, rhodies, spec. trees, large selection of permanent landscape plants.

wSte **Steamboat Island Nursery**
8424 Steamboat Island Rd
Olympia, WA 98502
Proprietor(s): Laine McLaughlin
Phone: 360-866-2516
FAX: 360-866-2516 (call ahead)
Email address: steamboat@olywa.net
Website: www.olywa.net/steamboat
Mail-order or walk-in: walk-in only
Catalogue: $1.00
Credit cards: no
Hours: March-October: 10:00am-5:00pm Saturdays and Sundays; year-round by appointment
Specialties: Trees, shrubs, perennials, grasses, ferns, annuals, sweet pea starts, tender perennials and Pacific Northwest native plants. Plants with year-round interest, such as bark, fall color, and winter flowers. Drought tolerant and deer resistant plants.

oSto **StoneCrop Gardens**
2040 Crittenden St SW
Albany, OR 97321
Proprietor(s): Diane Hyde
Phone: 541-928-8652
FAX: 541-928-8652
Email address: sedums@dnc.net
Website: stonecropgardens.com
Mail-order or walk-in: both
Catalogue: free list
Credit cards: VISA, MC, AmEx
Hours: Noon-5:00pm Wednesday-Sunday; closed Monday and Tuesday
Specialties: Hardy succulents and rock garden plants.

wSus **Susan's Bamboo**
12608 Marine Dr
Marysville, WA 98271
Proprietor(s): Susan Pierson-Bonasera
Phone: 360-652-7765
Email address: jrbonasera@hotmail.com
Mail-order or walk-in: walk-in only
Catalogue: none
Credit cards: no
Hours: By appointment
Specialties: Bamboo, pygmy, medium and timber varieties.

oSwa **Swan Island Dahlias**
PO Box 700
Canby, OR 97013
Walk-in address: 995 NW 22nd Avenue
Canby, OR 97013
Proprietor(s): Nicholas and Ted Gitts
Phone: 503-266-7711
FAX: 503-266-8768
Email address: info@dahlias.com
Website: www.dahlias.com
Mail-order or walk-in: both
Catalogue: $3.00 refundable
Credit cards: VISA, MC, Discover, AmEx
Hours: Monday-Friday 9:00am-4:30 pm. August 1 through frost, fields are open daylight to dark.
Specialties: Dahlias.

wSwC **Swan Creek Nursery**
2709 64th St E
Tacoma, WA 98404
Proprietor(s): Barbara Menne
Phone: 253-536-6502
FAX: 253--536-6502
Mail-order or walk-in: walk-in only
Catalogue: price list free
Credit cards: no
Hours: Call first for appointment as it's our home.
Specialties: Less common perennials and shrubs, ornamental grasses, drought tolerant emphasis.

oTal **Tallan Nursery**
10770 S Barnards Rd
Canby, OR 97013
Proprietor(s): Otis and Joan Tallan
Phone: 503-651-2941
Mail-order or walk-in: walk-in only
Credit cards: no
Hours: Call for appointment and directions
Specialties: Container-grown rhododendrons, wide assortment of hybrids, some species, various sizes. Also Chamaecyparis pisifera 'Boulevard' and Abelia grandiflora, various sizes.

oTBG **The Bamboo Garden**
1507 SE Alder St
Portland, OR 97214

Walk-in address: 13822 SE Oatfield Rd
Milwaukie, OR 97222
Proprietor(s): Ned Jaquith
Phone: 503-654-0024
FAX: 503-231-9387
Email address: bamboo@bamboogarden.com
Website: www.bamboogarden.com
Mail-order or walk-in: mail-order and by appointment
Credit cards: no
Hours: Tuesday-Saturday: 8:00am-3:00pm. Please call a day in advance to set up an appointment.
Specialties: We specialize in hardy bamboos, timbers as well as groundcovers. We offer over 200 different varieties. *Plus we carry bamboo barrier.

oTDM **TDM Acres Herb Farm**
2889 Oak Knoll Rd NW
Salem, OR 97304
Proprietor(s): Marti Sohn
Phone: 503-399-1797
FAX: 503-399-1797
Email address: tdmacres@net.att.net
Mail-order or walk-in: walk-in only
Catalogue: none
Credit cards: no
Hours: May 1-September 15: Call for hours. Independence Saturday Market 9 - 1.
Specialties: Culinary herbs and pelargoniums.

wTGN **The Greenhouse Nursery**
81 S Bagley Creek Road
Port Angeles, WA 98362
Proprietor(s): Diana L. Politika
Phone: 360-417-2664
Email address: diana@olympus.net
Mail-order or walk-in: walk-in only
Credit cards: VISA, MC
Hours: 7 days a week. Closed late November through mid-January.
Specialties: "Hard-to-finds"; plants usually found only in mail-order catalogs.

wThG **The Greenery**
14450 NE 16th Pl
Bellevue, WA 98007
Proprietor(s): Lynn and Marilyn Watts
Phone: 425-641-1458
Email address: watts-greenery@msn.com
Mail-order or walk-in: walk-in only
Catalogue: n/a
Credit cards: no
Hours: By appointment only please.
Specialties: Species rhododendrons; natives.

wTho **Thompson's Greenhouse**
6412 State Route 9
Sedro Woolley, WA 98284
Proprietor(s): Stephen and Brenda Thompson
Phone: 360-856-2147
FAX: 360-856-4227
Hours: Seasonal: March-September
Specialties: Annuals, vegetables, perennials, hanging baskets; also trees, shrubs, roses, and berries.

oThr **Three Sisters Nursery**
21033 SE Yamhill
Gresham, OR 97030
Walk-in address: 37100 SE Highway 211
Sandy, OR 97055
Proprietor(s): Terry and Sue Powers
Phone: 503-665-8074
FAX: 503-665-9085
Email address: thrsis@aol.com
Mail-order or walk-in: walk-in only
Credit cards: no
Hours: Weekends only - need to call ahead of time
Specialties: Wholesale nursery, also sells retail. Many varieties of Japanese maples, dogwoods, smoke trees, weeping cherry, some shrubs.

oTin **Tiny Tree Nursery**
PO Box 1386
Sherwood, OR 97140
Proprietor(s): Ronald G. Yasenchak
Phone: 503-625-6002
Email address: cpq1699L@gte.net
Mail-order or walk-in: both
Catalogue: yes
Credit cards: no
Hours: Please call for appointment
Specialties: Miniature trees, dwarf conifers, maples, azaleas, rhododendrons and bonsai specimens.

wTLP **The Lily Pad**
PMB #374, 3403 Steamboat Island Rd NW
Olympia, WA 98502
Walk-in address: Olympia Farmers' Market, foot of Capitol Blvd (April-November)
Olympia, WA
Proprietor(s): Jan Detwiler
Phone: 360-866-0291
FAX: 360-866-7128
Email address: leepauling@earthlink.net
Website: www.lilypadbulbs.com
Mail-order or walk-in: mail-order only
Catalogue: $1.00
Credit cards: MC, VISA, AmEx, Discover
Hours: No walk-ins.
Specialties: Hybrid lilies (bulbs), tree peonies (bare root and potted), daylilies, and bush peonies.

wTow **Tower Perennial Gardens**
4010 East Jamieson Rd
Spokane, WA 99223
Proprietor(s): Alan and Susan Tower
Phone: 509-448-6778
FAX: 509-448-1661
Email address: tower@ior.com
Mail-order or walk-in: both
Catalogue: free
Credit cards: yes
Hours: Monday-Saturday 9:00am-5:00pm, Sunday 10:00am-5:00pm
Specialties: Hostas, shade plants, daylilies, dwarf conifers, extensive display gardens, landscape consultation.

oTPm **The Plantsmen**
19127 SE Highway 212
Clackamas, OR 97015
Proprietor(s): Roger and Elizabeth Hollingsworth
Phone: 503-658-3720
FAX: 503-658-3269
Mail-order or walk-in: walk-in only
Catalogue: free
Credit cards: no
Hours: Tuesday-Thursday: 8:00am-4:30pm. Phone first please!
Specialties: Conifers and broadleaf evergreens in containers

oTra **Tradewinds Bamboo Nursery**
28446 Hunter Creek Loop
Gold Beach, OR 97444
Proprietor(s): Gib and Diane Cooper
Phone: 541-247-0835
FAX: 541-247-0835

Email address: gib@bamboodirect.com
Website: bamboodirect.com
Mail-order or walk-in: mail-order
Catalogue: $3.00
Credit cards: MC, VISA, Discover, AmEx
Hours: Call for appointment.
Specialties: Bamboo plants, books, rhizome barrier, bamboo poles, tools, and fertilizer.

oTri **Trillium Gardens**
PO Box 803
Pleasant Hill, OR 97455
Walk-in address: Retail walk-in sales are by appointment. Call and we will give directions and address.
Proprietor(s): Shelia Klest
Phone: 541-937-3073
FAX: 541-937-2261
Email address: shelia@trilliumgardens.com
Website: www.trilliumgardens.com
Mail-order or walk-in: mail-order, walk-in by appt.
Catalogue: free
Credit cards: VISA, MC, Discover
Hours: 7:30am-4:00pm Monday-Friday
Specialties: Northwest native plants (wildflowers, groundcovers, shrubs, trees), wetland plants and ornamental grasses.

oTrP **Trans-Pacific Nursery**
16065 Oldsville Rd
McMinnville, OR 97128
Proprietor(s): Jackson Muldoon and Gerry Roe
Phone: 503-472-6215
FAX: 503-434-1505
Email address: groe@worldplants.com
Website: www.worldplants.com
Mail-order or walk-in: mail-order only
Catalogue: $2.00
Credit cards: VISA, MC
Specialties: Rare exotic, unusual plants, Japanese maples, plants of China, southern hemisphere.

oTwi **Twinflower Nursery**
40794 SE Kubitz Rd
Sandy, OR 97055
Proprietor(s): Bruce and Kathe Krohn
Phone: 503-668-4842
Mail-order or walk-in: walk-in only
Catalogue: free
Hours: Weekends all year; please phone ahead.
Specialties: Plants for shade; hostas, perennials, grasses, native plants.

oUps **Upstarts! Growers**
1177 Dairy Loop Rd
Roseburg, OR 97470
Proprietor(s): Debra Levings and Ron Breyne
Phone: 541-679-6530
FAX: 541-679-6530
Mail-order or walk-in: both
Catalogue: free list by mail or fax
Credit cards: yes
Hours: February 15-July 1: 9:00am-6:00pm Thursday, Friday, Saturday. Closed Sunday and Monday except by appointment. Tuesdays and Wednesdays and rest of the year is by chance or by appointment.
Specialties: We encourage retail customers and look forward to them! Specialties include unusual perennials and herbs, flowering shrubs and trees. Outstanding quality and advice. Chemical free. Artistic baskets and containers. We invite meetings at nursery: refreshments served, speaker available, 10% discount on that day

wVan **Van Well Nursery**
PO Box 1339
Wenatchee, WA 98807
Walk-in address: 2821 Grant Road
East Wenatchee, WA 98802
Proprietor(s): Peter, Joe, Tom and Dick Van Well
Phone: 509-886-8189, 800-572-1553
FAX: 509-886-0294
Email address: vanwell@vanwell.net
Website: www.vanwell.net
Mail-order or walk-in: both
Catalogue: $1.00
Credit cards: VISA, MC
Hours: 8:00am-5:00pm Monday-Saturday. Closed Saturdays June 1 - Ocotber 31.
Specialties: Fruit trees: apples, cherries, pears, peaches, nectarines, prunes, plums, Asian pears, and apricots.

oVBI **Vanveen Bulbs International**
PO Box 92052
Portland, OR 97292
Walk-in address: 889 NE Foothills Drive
Estacada, OR 97023
Proprietor(s): Yolanda Wilson
Phone: 503-970-2992
FAX: 503-630-7519
Email address: yolandavan@aol.com
Website: vanveenbulbs.com
Mail-order or walk-in: mail-order
Catalogue: free price list
Credit cards: yes
Hours: Bulbs available at garden shows, Beaverton Farmers Market and internet.
Specialties: Oriental and Asiatic lilies, calla lillies, liatris, unusual bulbs: gloriosa lily, pineapple lily, toad lily; veronica, Monte Cassino aster, gift baskets. Offer question and answer service.

wWal **Walsterway Iris Gardens**
19923 Broadway Ave
Snohomish, WA 98296
Walk-in address: Maltby area at Broadway and 200th SE
Maltby, WA
Proprietor(s): Ralph and Fran Walster
Phone: 360-668-4429
Mail-order or walk-in: both
Catalogue: $1.00
Credit cards: Discover, MC, VISA
Hours: May 20-September 1: 9:00am-5:00pm Wednesday through Sunday.
Specialties: 600 varieties of tall bearded iris, 120 varieties of Japanese iris, 70 varieties of Siberian iris.

wWat **Watershed Garden Works**
2039 44th Ave
Longview, WA 98632
Proprietor(s): Scott & Dixie Edwards
Phone: 360-423-6456
FAX: 360-423-6456
Email address: watershedgardenworks@compuserve.com
Mail-order or walk-in: mail-order only
Credit cards: VISA
Hours: Call first.
Specialties: NW natives.

oWaW **Walden-West**
5744 Crooked Finger Rd
Scotts Mills, OR 97375
Proprietor(s): Jay Hyslop and Charles Purtymun
Phone: 503-873-6875
Mail-order or walk-in: both
Catalogue: free

Credit cards: no
Hours: Closed Mondays
Specialties: Hostas.

wWel **Wells Medina Nursery**
8300 NE 24th St
Medina, WA 98039
Proprietor(s): Ned, Wendy, Alex, Lisa Wells
Phone: 425-454-1853
Mail-order or walk-in: walk-in only
Credit cards: VISA, MC
Hours: 9:00am-5:00pm Monday-Saturday, 10:00am-5:00pm Sunday; Daylight Savings Time: 9:00am-6:00pm Monday-Saturday, 10:00am-5:00pm Sunday.
Specialties: Japanese maples, rare and hard-to-find woody plants, perennials, and annuals.

wWhG **Whitney Gardens & Nursery**
PO Box 170, 306264 Highway 101
Brinnon, WA 98320-0080
Walk-in address: 306264 Highway 101
Brinnon, WA 98320-0080
Proprietor(s): Anne Sather
Phone: 360-796-4411
FAX: 360-796-3556
Email address: info@whitneygardens.com
Website: www.whitneygardens.com
Mail-order or walk-in: both
Catalogue: $4.00
Credit cards: yes
Hours: Garden: 9:00am-dusk daily; nursery: 10:00am-5:30pm daily. Both garden and nursery in November-January: 10:00am-4:30pm.
Specialties: Rhododendrons and azaleas, maples, magnolias, kalmias, camellias, conifers, ground covers and perennials. Seven acre garden established in 1955.

oWhi **Whitman Farms**
3995 Gibson Rd NW
Salem, OR 97304
Proprietor(s): Lucile Whitman
Phone: 503-585-8728
FAX: 503-363-5020
Website: www.whitmanfarms.com
Mail-order or walk-in: mail-order
Catalogue: free
Credit cards: no
Hours: Phone ahead.
Specialties: Ornamental trees in root control bags, gooseberries, currants, and other small fruits.

oWhS **Whispering Springs**
19425 Colby Ln
Hillsboro, OR 97123
Proprietor(s): Rebecca Snyder
Phone: 503-538-3942
Email address: flwrchld3@aol.com
Mail-order or walk-in: walk-in only
Catalogue: none
Credit cards: no
Hours: April-June: noon-7:00pm Wednesday-Friday; 9:00am-6:pm Saturday
Specialties: Perennials, unusual annuals, moss baskets and vines.

wWin **Wind Poppy Farm and Nursery**
3171 Unick Rd
Ferndale, WA 98248
Proprietor(s): Karen and Bruce Teper
Phone: 360-384-6804
FAX: 360-384-6804
Email address: windpoppy@aol.com
Mail-order or walk-in: walk-in only
Catalogue: $3.00 refunded with purchase
Credit cards: no
Hours: Call for appointment.
Specialties: Ornamental grasses, species poppies.

wWld **Wildside Growers**
6361 Hannegan Rd
Lynden, WA 98264
Proprietor(s): Susan Taylor and Veronica Wisniewski
Phone: 360-398-7158
FAX: 360-733-2581
Email address: staylor@telcomplus.net
Mail-order or walk-in: mail-order
Credit cards: no
Hours: By appointment only; call in advance. 2 - 4 retail sales/year
Specialties: Native plants with a focus on the Pacific Northwest.

oWnC **Wine Country Nursery & Aquarium**
4100 NE Portland Rd (Hwy 99W)
Newberg, OR 97132
Proprietor(s): Roger Yost
Phone: 503-538-1518
FAX: 503-537-9637
Email address: helloroger@aol.com
Website: winecountrynursery.com
Mail-order or walk-in: walk-in only
Credit cards: VISA, MC, Discover, AmEx
Hours: 9:00am-6:00pm daily; closed New Year's, Thanksgiving, and Christmas
Specialties: Bamboo, conifers, magnolias, Japanese maples, hostas, hydrangeas.

wWoB **Woodbrook Nursery**
1620 59th Ave NW
Gig Harbor, WA 98335
Walk-in address: By appt. only: 5919 78th Ave NW
Gig Harbor, WA 98335
Proprietor(s): Ingrid Wachtler
Phone: 253-265-6271
FAX: 253-265-6471
Email address: woodbrk@harbornet.com
Website: www.woodbrook.net
Mail-order or walk-in: by appointment only
Catalogue: free; call and leave address
Credit cards: no
Hours: Open by appointment only. Check website for special open hours for Christmas trees.
Specialties: Native plants (Pacific Northwest), u-cut and live Christmas trees.

oWoo **Wooden Shoe Bulb Company, Inc.**
PO Box 127
Mount Angel, OR 97362
Walk-in address: 33814 S Meridian Rd
Woodburn, OR 97071
Proprietor(s): Barb, Patti, Janet, Vicki and Denise Iverson and Karen Bever
Phone: 800-711-2006
FAX: 503-634-2710
Email address: iverson@molalla.net
Website: woodenshoe.com
Mail-order or walk-in: both
Catalogue: free
Credit cards: VISA, MC, AmEx
Hours: March and April during tulip bloom: 10:00am-4:30pm 7 days/week; September 15-October 31: 10:00am-4:30pm 7 days/week.
Specialties: Tulips and daffodils. Wooden Shoe Bulb Co is a mail order/retail nursery. We open our tulip fields (~40 acres) in the spring while they are in bloom. Display beds are also planted so visitors can see all the different varieties of tulips available and

can place their orders from their favorites. We also open in the fall from mid-September to October 31. Customers can either pick up their orders or buy bulbs from our extras that are available. We also grow a wide assortment of daffodils which can be viewed in March.

wWoS **Woodside Gardens**
1191 Egg & I Rd
Chimacum, WA 98325
Proprietor(s): Pamela West
Phone: 360-732-4754, 800-473-1152
FAX: 360-732-4754, 800-473-1152
Email address: woodside@olympus.net
Website: www.woodsidegardens.com
Mail-order or walk-in: both
Catalogue: $2.00
Credit cards: VISA, MC
Hours: April-October regular retail open 9:00am-5:00pm; other times by appointment only.
Specialties: Perennials: penstemons, phlox, primulas, violas, dianthus. Herbs: lavender, rosemary, oreganos. We also offer an unusual selection of ornamental grasses, shrubs, vines, and shrub roses.

oZOT **Z Oak Tree Hosta Farm**
1330 SW 345th
Hillsboro, OR 97123
Proprietor(s): Pat and David Zumwalt
Phone: 503-648-1960
FAX: 503-648-4327
Mail order or walk-in: walk-in only
Catalogue: n/a
Credit cards: no
Hours: May-September: call for hours.
Specialties: Hostas.

ADDITIONAL NURSERIES

These nurseries have not listed their specific plants in the plant section of this book, so they do not have codes. Please note that the nurseries are listed in alphabetical order by nursery name. Information about the nurseries is as given to us by the nursery proprietors. Since information may change, it is advisable to call the nursery before visiting.

A Lot of Flowers
1212 11th St
Bellingham, WA 98225
Proprietor(s): Penny Ferguson
Phone: 360-647-0728
Mail-order or walk-in: walk-in only
Catalogue: n/a
Credit cards: VISA, MC, AmEx
Hours: 10:00am-6:00pm Tuesday-Saturday, noon-5:00pm Sunday/Monday; closed for month of January
Specialties: Unique assortment of perennials, annuals, herbs, vines, groundcovers; beautiful setting, lots of statuary and gift items, great pot selection, cut flowers.

All Season Nursery, Gardens & Gifts
3829 Pleasant Hill Rd
Kelso, WA 98626-9781
Proprietor(s): Jim and Nancy Chennault
Phone: 360-577-7955
FAX: 360-577-1169
Email address: allseason@tdn.com
Website: www.tdn.com/allseason
Mail-order or walk-in: walk-in only
Catalogue: n/a
Credit cards: VISA, MC
Hours: 9:00am-6:00pm Monday-Saturday, noon-5:00pm Sunday, all year round
Specialties: Display gardens, teaching as well as hands-on workshops, year round; we grow much of our own annual bedding flowers and vegetable plants, baskets, moss baskets, and planters.

Alpine Nursery Inc.
16023 SE 144th St
Renton, WA 98059
Proprietor(s): William Spiry
Phone: 425-255-1598
FAX: 425-255-0709
Mail-order or walk-in: walk-in only
Credit cards: yes
Hours: 9:00am-5:30pm 7days/week
Specialties: Flowering trees, shade trees, shrubs, groundcovers, perennials, annuals, and flowering baskets.

Bainbridge Gardens
9415 Miller Rd NE
Bainbridge Island, WA 98110
Proprietor(s): Junkoh Harui
Phone: 206-842-5888
FAX: 206-842-7645
Mail-order or walk-in: walk-in only
Credit cards: yes
Hours: 9:00am-5:30pm Monday-Saturday, 10:00am-4:00pm Sunday; extended spring hours April-August
Specialties: Destination nursery with extensive perennial selection. Also unique garden art, aquatic department, large tree and shrub department, gourmet café, kids' playground and year-round classes.

Baker & Chantry Orchids
PO Box 554
Woodinville, WA 98072
Walk-in address: 18611 132nd Ave NE
Woodinville, WA 98072
Proprietor(s): Marlene Holl
Phone: 425-483-0345
Email address: bc_orchids@juno.com
Website: www.orchidmall.com or lennon.pub.csufresno.edu/~jms59/jcmain.html
Mail-order or walk-in: both
Credit cards: VISA, MC
Hours: 10:00am-5:00pm daily
Specialties: Orchids - paphiopedilums - phalaenopsis - phragmipedium - miltoniopsis -cattleyas and oncidiums - dendrobiums. Also many hybrids - from seedlings to flowering plants.

Barbara Ashmun's Jungle
8560 SW Fairway Dr
Portland, OR 97225
Proprietor(s): Barbara Ashmun
Phone: 503-297-1307
FAX: efax: 503-212-0500
Email address: barbarablossom@earthlink.net
Mail-order or walk-in: walk-in only
Catalogue: none
Credit cards: no
Hours: By appointment, or at scheduled open gardens; mail or email with address to get on mailing list.
Specialties: Own-root roses, cornus, datura, perennials, interesting new shrubs.

Bear Creek Nursery
40554 Cole School Rd
Scio, OR 97374
Proprietor(s): Dale and Thùy Chrestenson
Phone: 503-769-6974
FAX: 503-769-4879
Email address: bearcreekrhody@webtv.net
Mail-order or walk-in: walk-in only
Catalogue: no
Credit cards: no
Hours: Wholesale; retail customers welcome
Specialties: Rhododendrons, best represented by small leafed kinds, but have both small and large leafed. About 100 varieties.

Bellevue Nursery, Inc.
842 104th Avenue SE
Bellevue, WA 98004
Walk-in address: Bellevue Way and SE 10th Street
Bellevue, WA
Proprietor(s): Ken and Lori Smith
Phone: 425-454-5531
FAX: 425-453-9982
Email address: bellnurs@ix.netcom.com
Website: www.bellevuenursery.com
Mail-order or walk-in: both
Credit cards: VISA, MC
Hours: 9:00am-6:00pm Monday-Sunday (7days/week)
Specialties: Seasonal color and gifts.

Big Trees Today
PO Box 1402
Hillsboro, OR 97123
Walk-in address: 4820 SW Hillsboro Hwy
Hillsboro, OR 97123
Proprietor(s): Terry and Nancy Hickman
Phone: 503-640-3011
FAX: 503-640-2877
Email address: sales@bigtreestoday.com
Website: www.bigtreestoday.com
Mail-order or walk-in: both
Catalogue: none available
Credit cards: no
Hours: 8:00am-5:00pm Monday-Saturday
Specialties: We sell only large trees; transplanting services, have been in business (same owner) since 1979.

Bonsai Northwest
5021 S 144th St
Seattle, WA 98168
Proprietor(s): John Muth
Phone: 206-242-8244
FAX: 206-244-2301
Email address: johnbnw@aol.com
Website: www.bonsainw.com
Mail-order or walk-in: both
Catalogue: $2.00
Credit cards: VISA, MC
Hours: 10:00am-5:00pm Wednesday-Saturday; 11:00am-4:00pm Sunday; closed Monday and Tuesday
Specialties: Everything you need for bonsai.

Brady's Nursery, Inc
920 E Johns Prairie Rd
Shelton, WA 98584
Proprietor(s): Sharon and Keith Tibbits
Phone: 360-426-3747
FAX: 360-427-8306
Mail-order or walk-in: walk-in only
Credit cards: VISA, MC, Discover
Hours: Summer: 8:00am-6:00pm 7 days per week; Winter: 9:00am-5:00pm 7 days per week.

Briggs Hill Orchids, Inc.
27936 Briggs Hill Road
Eugene, OR 97405
Proprietor(s): Heléne and Barton Gendel
Phone: 541-431-3886
Email address: bgendel@aol.com
Mail-order or walk-in: walk-in only
Catalogue: n/a
Credit cards: soon
Hours: By appointment, Thursday-Sunday 10:00am-5:00pm
Specialties: Orchid plants for hobbyist or gift giver. Genera: paphiopedilums, phragmepidiums, cattleyas, phalaenopsis, miscellaneous oncidium, zygopetalums, miltoniopsis, angraecums, dendrobiums, species plant list.

Brim's Farm & Garden
34963 Highway 105
Astoria, OR 97103
Proprietor(s): Mike and Linda Brim
Phone: 503-325-1562
FAX: 503-325-9231
Email address: briml@pacifier.com
Mail-order or walk-in: walk-in only
Credit cards: VISA, MC
Hours: 8:30am-5:30pm Monday-Saturday; seasonally: 12:30pm-4:30pm Sunday
Specialties: Coastal selections and varieties that do well in maritime climates; roses, rhododendrons, apples, pears, berries, vegetables, well-rounded selection of broadleaf evergreens and deciduous flowering trees and shrubs.

Brown's Rose Lodge Nursery
5211 Salmon River Hwy
Otis, OR 97368
Proprietor(s): Wally and Karen Brown
Phone: 541-994-2953
Mail-order or walk-in: walk-in only
Catalogue: none
Credit cards: VISA, MC, Discover
Hours: Winter: 9:00am-5:00pm Monday-Saturday; spring and summer: 9:00am-6:00pm daily
Specialties: Shrubs, annuals and perennials, fuchsias, orchid cactus, plants for coastal gardens.

Burlingame Gardens
1389 Ocean Beach Rd
Hoquiam, WA 98550
Proprietor(s): Gail Johannes
Phone: 360-533-8463
FAX: 360-533-8463
Website: uswestdex.com/iyp/burlingamegardens
Mail-order or walk-in: walk-in only
Catalogue: none
Credit cards: soon
Hours: Mid-March - October: 10:00am-6:00pm Friday-Sunday
Specialties: Hamamelis x intermedia - many grafted cultivars; Acer palmatum - grafted cultivars; many other specialty grafted trees and vines (wisteria); new/old hard-to-find perennials, rhodies, choice trees (halesia, stewartia, cornus, etc.), cistus, hebe, euphorbia ssp., lavandula ssp. and cultivars.

Cascade Nursery
8921 55th Ave NE
Marysville, WA 98270
Proprietor(s): Fred and Elizabeth Huse
Phone: 360-659-2988
FAX: 360-653-6747
Mail-order or walk-in: walk-in only
Hours: February-June: 9:30am-6:00pm 7 days/week; July-December: 9:30am-5:00pm, closed Wednesdays.
Specialties: Full service nursery - specialty roses, perennials, rhododendrons and hanging baskets.

Cedar Mill Nursery
7520 B NE 219th St
Battle Ground, WA 98604
Proprietor(s): Johnny and Jerry Barnes
Phone: 360-687-5926
FAX: 360-687-0912
Mail-order or walk-in: walk-in only
Catalogue: none
Credit cards: VISA, M/C, AmEx, debit
Hours: Summer: 9:00am-6:00pm 7 days, Winter: 9:00am-5:00pm 5 days; closed on Wednesday and Thursday, open Friday-Tuesday.
Specialties: Japanese maples, statuary, gifts, bark, annuals, perennials, nursery stock, etc.

Chets Garden and Pet Center
229 SW H Street
Grants Pass, OR 97526
Proprietor(s): Cliff Bennett and Fred Rogers
Phone: 541-476-4424
FAX: 541-476-9421
Mail-order or walk-in: walk-in only
Credit cards: MC, VISA, AmEx, Discover
Hours: 9:00am-5:30pm Monday-Saturday, 10:00am-4:00pm Sunday
Specialties: Japanese maples, rhodies, conifers, perennials, houseplants, shade trees, and most all evergreen and deciduous shrubs for our zone.

Cindy's Plant Stand
1199 Monte-Elma Rd
Elma, WA 98541
Proprietor(s): Cindy L. Knight
Phone: 360-482-3258
Mail-order or walk-in: walk-in only
Credit cards: no
Hours: April-June: tomatoes, bedding plants and dahlia tubers available; August-frost: cut dahlias; Thanksgiving-Christmas: wreaths.
Specialties: Tomatoes, dahlias - both cut flowers and tubers, bedding plants, evergreen wreaths.

City People's Garden Store
2939 E Madison
Seattle, WA 98112
Proprietor(s): Steve Magley and Judith Gille
Phone: 206-324-0737
FAX: 206-328-6114
Mail-order or walk-in: walk-in only

Credit cards: yes
Hours: 9:00am-6:00pm Monday-Saturday; 10:00am-6:00pm Sunday
Specialties: Full service retail nursery with gifts, flower shop, landscape design, maintenance, and installation.

Classic Nursery
12526 Avondale Rd NE
Redmond, WA 98052
Proprietor(s): Leon and Linda Hussey
Phone: 425-885-5678
FAX: 425-869-6411
Email address: classicnursery@msn.com
Website: www.classicnursery.com
Mail-order or walk-in: walk-in only
Catalogue: n/a
Credit cards: VISA, MC, Discover
Hours: Spring-fall: 8:00am-6:00pm Monday-Saturday, 10:00am-5:00pm Sunday; Winter: 8:00am-5:00pm Monday-Saturday, 10:00am-5:00pm Sunday
Specialties: Unusual and usual annuals and perennials, deciduous and evergreen trees and shrubs, natives, bulbs, fruit trees, flowering trees, conifers, grasses (ornamental), gardening tools, books, gifts, chemicals, fertilizers, and soils.

Columbia Nursery & Design
14316 NW McCann Rd
Vancouver, WA 98685
Proprietor(s): Mary D Skinner
Phone: 360-573-4047
FAX: 360-573-7447
Mail-order or walk-in: walk-in only
Credit cards: no
Hours: April-September: 10:00am-5:00pm Saturdays, or by appointment
Specialties: Herbs and perennials; garden design.

Coos Grange Supply & Garden Center
1085 S Second St
Coos Bay, OR 97420
Proprietor(s): co-op
Phone: 541-267-7051
FAX: 541-267-6880
Mail-order or walk-in: walk-in only
Catalogue: n/a
Credit cards: VISA, MC
Hours: 9:00am-5:30 Monday-Saturday
Specialties: Complete nursery of shrubs, trees, fruit trees, annuals, perennials, and vegetables. The largest selection of blueberries.

Cora's Nursery & Greenhouse
902 24th St
Anacortes, WA 98221
Proprietor(s): Cora Zoberst
Specialties: Because I have taken up house sitting, I am dropping my nursery license. I still have some plants for sale.

Cornell Farm
8212 SW Barnes Rd
Portland, OR 97225
Proprietor(s): Deby Barnhart and Ed Blatter
Phone: 503-292-9895
FAX: 503-292-1051
Email address: cornell@hevanet.com
Mail-order or walk-in: walk-in only
Hours: Year round generally 9:00am-6:00pm; open every day.
Specialties: 5 acres of 1200 varieties of annuals and perennials grown on site. Eclectic container collection, designer potted gardens, connoisseur plant department (includes trees, shrubs, vines), garden décor. Roses: English, French, Romantica, ground cover, trees, miniatures, modern, antique.

Country Nursery & Gardens
2075 Seabeck Highway NW
Bremerton, WA 98312
Proprietor(s): Robin and Tracy Rodgers
Phone: 360-478-0288
FAX: 360-478-0263
Mail-order or walk-in: walk-in only
Credit cards: yes
Hours: April-September: 9:00am-6:00pm Monday-Saturday; October-March: 9:00am-5:00pm Monday-Saturday; Sundays 9:00am-5:00pm all year.
Specialties: Dwarf conifers - wide array of rhododendrons and azaleas.

Dan's Dahlias
994 S Bank Rd
Oakville, WA 98568
Proprietor(s): Dan Pearson
Phone: 360-482-2406
FAX: 360-482-2407
Email address: dansdals@centurytel.net
Website: www.dansdahlias.com
Mail-order or walk-in: mail-order; walk-in August 1-October 1
Catalogue: free
Credit cards: MC, VISA
Hours: Open seasonally August 1 to first frost
Specialties: Dahlia cut flowers and tubers. Retail sales only. Mail-order or seasonal walk-in.

Decker Nursery
PO Box 12
Alvadore, OR 97409
Walk-in address: 90808 B Street
Alvadore, OR 97409
Proprietor(s): Milton and Mary Decker
Phone: 541-688-8307
FAX: 541-688-8357
Mail-order or walk-in: walk-in only
Credit cards: yes
Hours: 9:00am-5:00pm Monday-Saturday, 10:30am-5:00pm Sunday; April, May, June: open till 6:00pm.
Specialties: Large landscaping projects - annuals and perennials for color. Fuchsia and other color baskets.

Des Moines Way Nursery
14634 Des Moines Memorial Dr S
Seattle, WA 98168
Proprietor(s): Fran Seike
Phone: 206-243-3011
Mail-order or walk-in: walk-in only
Credit cards: VISA, MC, AmEx, Discover
Hours: 9:00am-5:30pm Monday-Saturday, 10:00am-5:00pm Sunday
Specialties: Pruned pines, Japanese styling. Also full stock nursery, huge perennial selection, new and rare introductions, water plants/lilies, unusual conifers. Koi pond/Japanese tea garden open during business hours.

Dillards Nursery
13301 SE Bluff Rd
Sandy, OR 97055
Proprietor(s): David and Helen Dillard
Phone: 503-668-4063
FAX: 503-668-4063
Email address: dnursery@cnnw.com
Hours: Sunup to sundown 7 days a week.
Specialties: Conifers - pine, spruce; flowering and shade trees, groundcovers, broadleaf evergreens, perennials.

Don Smith's Tree Farm
22509 S Stormer Rd
Estacada, OR 97023
Proprietor(s): Don and Betty Smith
Phone: 503-631-2915
Hours: Come visit us anytime
Specialties: Alberta spruce and blueberries (berries, not plants).

Down To Earth

532 Olive St
Eugene, OR 97401
Proprietor(s): Jack Bates
Phone: 541-342-6820
FAX: 541-342-2261
Credit cards: yes
Hours: Open 10:00am-6:00pm Monday-Friday, 9:00am-6:00pm Saturday, 10:00am-5:00pm Sunday
Specialties: Hard to find, unusual perennials, tropicals for temperate gardens - bananas (Musa basjoo), gingers, etc. Organic vegetable starts.

Downs By The Pond

86667 Bailey Hill Rd
Eugene, OR 97405
Proprietor(s): Loretta and George Downs
Phone: 541-342-5887
FAX: 541-687-9286
Email address: h2olily@pond.net
Website: www.downsbythepond.com
Mail-order or walk-in: both
Catalogue: under development
Credit cards: yes, all major
Hours: March-October: 10:00am-6:00pm Monday-Saturday, 11:00am-4:00pm Sunday; Winter: by appointment.
Specialties: Water garden plants, supplies, fish, books; also consultation and installations.

Duckworth's Nursery

84846 S Willamette
Eugene, OR 97405
Proprietor(s): Peggy Duckworth
Phone: 541-345-5408
Mail-order or walk-in: walk-in only
Catalogue: none
Credit cards: all
Hours: 9:00am-6:00pm 7 days/week
Specialties: Large trees and shrubs.

Eaden Gardens

22551 S Eaden Rd
Oregon City, OR 97045
Proprietor(s): David and Gretchen Dewire
Phone: 503-631-4830
FAX: 503-631-4830
Email address: edenbonsai@aol.com
Website: www.bonsai.4mg.com
Mail-order or walk-in: mail-order
Catalogue: no
Credit cards: soon
Hours: By appointment
Specialties: We carry bonsai plus a line of larger specimen bonsai styled trees for the landscape. We could have any variety, but pines are our specialty.

Eclectic Gardens

35855 SW Orchaedia Dr
Hillsboro, OR 97123
Proprietor(s): Julie Holderith
Phone: 503-628-2139
Hours: Anytime by appointment
Specialties: Hemerocallis - daylilies.

Elandan Gardens Ltd.

3050 W State Hwy 16
Bremerton, WA 98312
Proprietor(s): Dan and Diane Robinson, Shanna Neimes, Will Robinson
Phone: 360-373-8260
FAX: 360-373-8260
Email address: elandan@elandangardens.com
Website: www.elandangardens.com
Mail-order or walk-in: both
Catalogue: website
Credit cards: yes, all
Hours: February-December: 10:00am-5:00pm Tuesday-Sunday; closed Mondays and January
Specialties: Bonsai, large Japanese maples. If it's ancient, crooked, big and beautiful and has character, we have it.

Elk Meadows Nursery

3485 Dosewallips Rd
Brinnon, WA 98320
Proprietor(s): Joe and Joy Baisch
Phone: 360-796-4886
Email address: elkmeadows@waypt.com
Website: www.northolympic.com/elkmeadows
Mail-order or walk-in: both
Catalogue: $3.00
Credit cards: MC, VISA
Hours: 8:00am-5:00pm Tuesday-Sunday; good idea to call first in case we're making a delivery.
Specialties: Hybrid 8" - 48" deciduous azaleas, rhodies, Kalmia latifolia, daphne and hydrangeas. We also source plantstock for those wishing to do their own landscape work.

Elk Plain Nursery

4020 224th Street E
Spanaway, WA 98387
Proprietor(s): Evelyn and Clayton Hoebelheinrich
Phone: 253-847-8071
FAX: 253-847-8071 call first
Email address: evelynkh@gte.net
Mail-order or walk-in: both
Credit cards: VISA, MC, Discover, AmEx
Hours: 9:30am-6:00pm Monday-Saturday, 11:00am-6:00pm Sunday
Specialties: Specializing in rhodies and azaleas, but have a good supply of trees, shrubs, perennials, annuals, and water gardening supplies.

Ferguson's Fragrant Nursery

21763 French Prairie Rd NE
St Paul, OR 97137
Proprietor(s): Danielle Ferguson
Phone: 503-633-4585
FAX: 503-633-4586
Mail-order or walk-in: walk-in only
Credit cards: yes
Hours: Spring/Summer: 9:00am-5:30pm; Fall: 9:00am-4:30pm; Winter: call ahead.
Specialties: Fragrant plants, premium annuals, unique mixed baskets and planters. Over 1000 roeses and many other wonderful fragrant shrubs.

Fessler Nursery

12666 Monitor McKee Rd NE
Woodburn, OR 97071
Proprietor(s): Ken, Marvin and Dale Fessler
Phone: 503-634-2448
FAX: 503-634-2866
Email address: mkfessler@peoplepc.com
Mail-order or walk-in: walk-in only
Catalogue: n/a
Credit cards: no
Hours: March-June: 6 days/week. November/December: 6 days/week for poinsettias.
Specialties: Fuchsias, geraniums (zonal, ivy, Martha Washington), verbenas, impatiens, heliotrope, petunias, gardenias, florist and outdoor/hardy azaleas, poinsettias. Other miscellaneous trailing annuals.

Flat Creek Garden Center

30039 SE Orient Dr
Gresham, OR 97080
Proprietor(s): Kathy Taggart
Phone: 503-663-4101
FAX: 503-663-4101
Website: flatcreekgarden.com

Specialties: Retail center that sells collector-type plants. Deciduous, woody shrubs, vines, wet (water, bog, etc.) plants, specialty trees, salix.

Foxglove Herb Farm
6617 Rosedale St
Gig Harbor, WA 98335
Proprietor(s): Michael Burkhart, Nancy Andrews, Jane Cooper, Stephen Burkhart
Phone: 253-853-4878
FAX: 253-853-3288
Email address: foxglovegifts@msn.com
Mail-order or walk-in: walk-in only
Credit cards: VISA, MC, Discover, AmEx
Hours: 10:00am-5:30pm Monday-Saturday, 11:00am-4:00pm Sunday; Closed: July 4th weekend, Labor Day weekend, and Tuesdays/Wednesdays from Labor Day through early November and January/February.
Specialties: Extensive line of herbs; unique perennials; gift shop - soaps candles, books, bath and skin care products, herbal oils and vinegars, garden art, tea and teapots, local artisans' creations.

Franz Witte Nursery
9770 W State St
Boise, ID 83703
Proprietor(s): Franz Witte
Phone: 208-853-0808
FAX: 208-853-4503
Mail-order or walk-in: walk-in only
Credit cards: VISA, MC
Hours: April-September: 9:00am-6:00pm Monday - Saturday, 10:00am-5:00pm Sunday. October: closed Sunday. November: 9:00am-5:00pm Monday-Friday
Specialties: Ornamental confiers, large caliper shade trees, perennials, shrubs, ornamental trees (flowering, Japanese maples), flamingos (plastic lawn ornament).

Fremont Gardens
4001 Leary Way NW
Seattle, WA 98107
Proprietor(s): Lorene Edwards and Karen M Souza
Phone: 206-781-8283
FAX: 206-781-7675
Email address: fremonted@aol.com
Mail-order or walk-in: limited mail-order; mostly walk-in
Credit cards: yes
Hours: 10:00am-6:00pm Tuesday-Saturday, Sunday 12; closed Mondays. Extended hours in spring and early summer, shorter hours January and February.
Specialties: Fremont Gardens specializes in unusual and uncommon varieties of perennials, annuals, vines and woody material. Also, a large selection of seeds, including many heirloom varieties. Unusual garden ornaments, books, and gardening supplies.

Friends & Neighbors Perennial Gardens
24708 NE 152nd Ave
Battle Ground, WA 98604-9771
Proprietor(s): Susan C. Henke
Phone: 360-687-2962
FAX: 360-687-0954
Mail-order or walk-in: walk-in only
Credit cards: VISA, MC
Hours: April 15-September 15: 9:00am-6:00pm 7 days/week
Specialties: Grow and sell only hardy perennials.

Garden Center West Inc
11500 Fairview Ave
Boise, ID 83713
Proprietor(s): Dieter and Celia Wiesemann
Phone: 208-376-3322
FAX: 208-376-7789
Mail-order or walk-in: walk-in only
Credit cards: VISA, MC, Discover, AmEx
Hours: 8:00am-6:00pm weekdays; 10:00am-6:00pm Sundays
Specialties: Garden statuaries, freshwater and marine fish, Christmas shop, gift and tools, bedding plants, nursery stock.

Gardens At Four Corners
321 Four Corners Rd
Port Townsend, WA 98368
Proprietor(s): Patti Kretzmeier and Gary Rohde
Phone: 360-379-0807
FAX: 360-379-8732
Mail-order or walk-in: walk-in only
Catalogue: none
Credit cards: all, including AmEx
Hours: Year round: 9:00am-6:00pm M-F; 9:00-5:00pm Sat-Sun
Specialties: Come wander our wonderful gardens. 3 acres of nursery stock and garden accessories. We're known for our quality and large selection of hardy, unusual perennials and other gardeners' delights.

George's Garden Center
16920 SE Sunnyside Rd
Clackamas, OR 97015
Proprietor(s): Richard E Barhoum
Phone: 503-658-5088
FAX: 503-658-8708
Email address: rbarhoum@netzero.net
Mail-order or walk-in: walk-in only
Credit cards: VISA, MC
Hours: 9:00am till dark
Specialties: Unusual plants, scupltured trees, grasses, large trees, Japanese maples.

Gibson's Nursery & Landscape Supply
South 1401 Pines Rd
Spokane, WA 99206
Proprietor(s): Gary Gibson
Phone: 509-928-0973
FAX: 509-926-4352
Mail-order or walk-in: walk-in only
Credit cards: MC, VISA
Hours: 8:00am-5:30pm Monday-Saturday
Specialties: Specializing in the highest quality nursery stock available. Special orders for unique or hard to find items are greatly welcomed.

Green Akres Garden Center, Inc.
545 E Chinden Blvd
Meridian, ID 83642
Phone: 208-895-8557
FAX: 208-895-8591
Email address: greenakres@msn.com
Mail-order or walk-in: both
Credit cards: yes
Hours: 9:00am-6:00pm Monday-Saturday; noon-5:00pm Sunday
Specialties: Full service garden center; we grow all our own stock. Specialties are perennials and hanging baskets. Also produce stand and wedding pavilion.

Green Thumb Garden Center
1475 S Main St
Lebanon, OR 97355
Proprietor(s): Pat and Deborah Gruebele
Phone: 541-451-5464
FAX: 541-451-5464
Mail-order or walk-in: walk-in only
Credit cards: MC, VISA, Discover
Hours: 9:00am-5:00pm Monday-Saturday; Winter: closed Sunday; Spring: 10:00am-4:00pm Sunday
Specialties: Fuchsia baskets and zonal geraniums.

Greenthumb Garden Center
256781 Highway 101
Port Angeles, WA 98362
Proprietor(s): Len and Chris Borchers
Phone: 360-452-2614

FAX: 360-417-5352
Mail-order or walk-in: walk-in only
Credit cards: VISA, MC, Discover
Hours: 9:00am-5:00pm Monday-Saturday, 11:00am-4:00pm Sunday
Specialties: Conifers, fruit trees, water gardening, display gardens with waterfalls and ponds.

Growers Garden & Florist
9155 SW Barbur Blvd
Portland, OR 97219
Proprietor(s): Jim and Marian Hantke
Phone: 503-245-6646
FAX: 503-246-6272
Mail-order or walk-in: walk-in only
Credit cards: VISA, MC
Hours: 8:30am-6:00pm Monday-Friday; 9:00am-6:00pm Saturday; 10:00am-5:00pm Sunday. Hours extended in spring and summer.
Specialties: Smaller growing ornamental trees and shrubs, collectible perennials, exceptional and unusual "others."

GrowScape Nurseries
PO Box 3939
Vancouver, WA 98662
Walk-in address: 10019 NE 72nd Avenue
Vancouver, WA 98662
Proprietor(s): Jerry Cates
Phone: 360-571-8082, 503-232-5404 (Portland)
FAX: 360-571-8087
Mail-order or walk-in: walk-in only
Credit cards: yes
Hours: March-October: 9:00am-5:00pm 7 days
Specialties: Retail and wholesale. We source plant lists for landscape contractors. Fax us your wish list, we'll give you a quote and availability. Orders over $500 free delivery.

Guentner Gardens
5780 Commercial St S
Salem, OR 97306
Proprietor(s): Michelle and Jerry Guentner
Phone: 503-585-7133
FAX: 503-585-4166
Mail-order or walk-in: walk-in only; mail-order soon
Credit cards: yes
Hours: 9:00am-6:00pm daily; closed December 26 - January 2.
Specialties: Largest garden center in Salem. Complete line, large selection, great line of garden gifts.

Harstine Heirloom Gardens
E 1190 Sunset Hill Rd
Shelton, WA 98584-8412
Proprietor(s): Sue and Bob Thompson
Phone: 360-427-2440
FAX: 360-427-7584
Email address: higardens@aol.com
Mail-order or walk-in: walk-in only
Catalogue: plant list
Credit cards: no
Hours: April-October, by appointment
Specialties: Ornamental grasses, natives and everlasting, lavender.

Heart of the Valley Greenhouse
7519 48th St East
Fife, WA 98424
Proprietor(s): Dale and Marla Rees
Phone: 253-922-7364
FAX: 253-922-1873
Mail-order or walk-in: walk-in only
Credit cards: VISA, MC
Hours: April 1-July 3: 9:00am-6:00pm 7 days/week
Specialties: Geraniums

Heartwood Garden & Gifts
923 Vernon Rd
Lake Stevens, WA 98258
Proprietor(s): Nell Steen and Martha Magee
Phone: 425-334-5351
FAX: 425-334-5361
Mail-order or walk-in: walk-in only
Credit cards: VISA, MC
Hours: 9:00am-6:00pm Monday-Saturday; 11:00am-5:00pm Sunday; some seasonal later hours
Specialties: General garden center: ornamentals, perennials, fruit trees, annuals, many roses.

Hidden Acres Nursery
19615 SW Cappoen Rd
Sherwood, OR 97140
Phone: 503-625-7390
Mail-order or walk-in: walk-in
Credit cards: no
Hours: Call please; hours change throughout the seasons.
Specialties: Specialize in rhododendrons - large sizes. Big, beautiful rhododendrons available 3 feet to 10 feet + and small sizes too. Large sales yard - 200 varieties - acres and acres. Also Pieris japonica (andromeda), ferns and blueberries. Mail order for large varieties, shipped by truck.

Hiland Trees & Nursery
2549 Highway 93 North
Arco, ID 83213
Proprietor(s): Will and Lori Beck
Phone: 208-527-3909
Mail-order or walk-in: walk-in only
Credit cards: no
Hours: 9:00am-6:00pm Monday-Saturday. Closed Sundays.
Specialties: Zone 3 cold hardy trees, shrubs and perennials. Bedding plants in season. Garden center featuring fertilome products.

Honeyman Nursery & Landscaping, Inc.
85089 Hwy 101 South
Florence, OR 97439
Proprietor(s): Lance and Barbara Rowland
Phone: 541-997-8522
FAX: 541-997-6211
Email address: honeymannsy@presys.com
Mail-order or walk-in: walk-in only
Credit cards: VISA, MC, Discover, AmEx
Hours: 9:00am-5:30pm Monday-Saturday, 10:00am-5:00pm Sunday
Specialties: Perennials, annuals, shrubs, trees; full retail nursery.

Irene's Plant Place
32914 S Kropf
Molalla, OR 97038
Proprietor(s): Irene Schriever
Phone: 503-651-2220
Mail-order or walk-in: walk-in only
Specialties: Baskets, bedding plants, perennials

Job's Nursery
4072 N Columbia River Rd
Pasco, WA 99301
Proprietor(s): Duane Job, Kathy Job
Phone: 509-547-4843
FAX: 509-547-4843
Mail-order or walk-in: walk-in only
Credit cards: VISA, MC
Hours: February 15-December 23: 9:00am-5:00pm
Specialties: Shade and flowering trees, evergreens, flowering shrubs, roses, and perennials.

Johnson Brothers Greenhouses
91444 Coburg Rd
Eugene, OR 97408
Proprietor(s): Vern Johnson and Robert Johnson
Phone: 541-484-1649

FAX: 541-343-6810
Email address: jbgh@cyber-dyne.com
Mail-order or walk-in: walk-in only
Catalogue: n/a
Credit cards: yes
Hours: Monday-Saturday.Winter:9:00am-5:00pm ; Rest of year: 9:00am-6:00pm
Specialties: Hanging baskets, huge assortment of perennials, poinsettias.

Julius Rosso Nursery Company
PO Box 80345
Seattle, WA 98108
Walk-in address: 6404 Ellis Ave S (north end of Boeing Field)
Seattle, WA 98108
Proprietor(s): Gene Rosso and Jerry Rosso
Phone: 800-832-1888
FAX: 206-762-2544
Email address: info@rossonursery.com
Website: www.rossonursery.com
Mail-order or walk-in: walk-in only
Hours: 8:00am-5:30pm Monday-Saturday; noon - 5:00pm Sunday
Specialties: Roses, groundcover, fruit trees, fruit bushes, perennials, ornamental grasses, Northwest natives, large caliper trees, Christmas trees and wreaths.

Kaija's
PO Box 630
Chehalis, WA 98532
Walk-in address: 623 NW State Avenue
Chehalis, WA 98532
Proprietor(s): Jerry and Matt Kaija
Phone: 360-748-4221, 800-717-4221(western WA only)
FAX: 360-740-9560
Email address: kaija@localaccess.com
Mail-order or walk-in: mostly walk-in
Credit cards: MC, VISA
Hours: 8:30am-6:00pm Monday-Saturday, 9:00am-5:30pm Sunday
Specialties: General nursery stock, fruit trees, roses, specialty annuals, perennials, vegetables starts, general garden supplies, organic and inorganic.

Kimberly Nurseries Inc.
2862 Addison Ave E
Twin Falls, ID 83301
Proprietor(s): Dave S Wright, President
Phone: 208-733-2717
FAX: 208-733-0043
Website: www.kimberlynurseries.com
Mail-order or walk-in: walk-in only
Credit cards: yes
Hours: 8:00am-6:00pm Monday-Saturday, 11:00am-5:00pm Sunday
Specialties: Retail and wholesale plant material, landscape design and installation, ponds/water features, hardscapes (fences, walls, pavers, mowstrips, sprinkler installation and repair), spraying and fertilization. Retail Christmas trees, gift shop.

Kingston Lumber & Garden
PO Box 393
Kingston, WA 98346
Walk-in address: 10900 NE State Hwy 104
Kingston, WA 98346
Phone: 360-297-3357
FAX: 360-297-3609
Website: www.kingstonlumber.com
Mail-order or walk-in: walk-in only
Credit cards: MC, VISA, AmEx
Hours: 9:00am-5:30pm Monday-Friday, 9:00am-4:30pm Saturday, 10:00am-2:00pm Sunday
Specialties: Roses, bamboo, clematis, annuals, perennials, herbs, groundcovers, bulbs. Gifts, tools, organics. Largest selection of wild bird supplies in Kitsap County.

K's Nursery
30891 S Oswalt Rd
Colton, OR 97017
Proprietor(s): Ken and Kathy Carroll
Phone: 503-824-3939
FAX: 503-824-3939
Mail-order or walk-in: walk-in only
Hours: Open March 1 - July 31 8:00am-6:00pm; September-October 9:00am-6:00pm everyday.
Specialties: Perennials, bedding plants, baskets, shrubs, trees, fruit.

Lake Grove Garden Center
PO Box 1409
Lake Oswego, OR 97035
Walk-in address: 15955 SW Boones Ferry Rd
Lake Oswego, OR 97035
Proprietor(s): Darryl and Chris Eddy
Phone: 503-636-2414
Mail-order or walk-in: walk-in only
Credit cards: VISA, MC, Discover, AmEx
Hours: Spring/Summer: 9:00am-6:00pm 7 days/week. Fall/Winter: 9:00am-6:00pm Monday-Saturday, 11:00am-6:00pm Sunday.
Specialties: Full line garden center. Bedding plants, perennials, trees, shrubs.

Leach Botanical Garden
6704 SE 122nd Ave
Portland, OR 97236
Proprietor(s): Leach Botanical Garden Friends, Scotty Fairchild, Staff Liaison
Phone: 503-761-9503
FAX: 503-762-4652
Email address: pkleach@ci.portland.or.us
Mail-order or walk-in: walk-in only
Catalogue: not available
Hours: Tuesday-Saturday: 9:00am-4:00pm; Sunday: 1:00pm-4:00pm. New stock every Tuesday at noon. Nursery supported plant sales April and September annually.
Specialties: Pacific Northwest natives, woodland perennials, alpines, special plants from Leach collection, unusual flowering shrubs. With annual seed exchange from over 30 botanical gardens and universities, we continually have new plant materials to offer.

Liskey Farms Inc.
4650 Lower Klamath Lake Rd
Klamath Falls, OR 97603
Walk-in address: 2525 Washburn Way - spring sales outlet, open April 15 - July 5 each year.
4000 Lower Klamath Lake Rd - growing greenhouse, open to the public April 10 - July 5.
Proprietor(s): Vickie Azcuenaga
Phone: 541-885-8517
FAX: 541-798-5355
Email address: liskeygh@aol.com
Mail-order or walk-in: walk-in only
Credit cards: yes, in season only
Hours: April 15-July 5: 9:00am-7:00pm at sales outlet; 8:00am-4:30pm at greenhouse.
Specialties: We grow a large selection of bedding plants, vegetable starts and perennials, all chosen especially for hardiness in our cold, high altitude climate.

Lorane Hills Farm & Nursery
PO Box 5464
Eugene, OR 97405
Walk-in address: 27634 Easy Acres Dr
Eugene, OR 97405
Proprietor(s): Evelyn and David Hess
Phone: 541-344-8943
Mail-order or walk-in: By appointment; mail-order soon
Hours: By appointment only at nursery; at Eugene Farmers' Market, Saturdays, April-October
Specialties: Beardless iris (especially Japanese and Sino-Sibe),a growing collection of natives and general herbaceous perennials.

M. D. Nursery & Landscaping
243 S Highway 33
Driggs, ID 83422
Proprietor(s): Mike Stears
Phone: 208-354-8816
FAX: 208-354-8733
Mail-order or walk-in: walk-in only
Credit cards: AmEx, VISA, MC
Hours: Daylight Savings Time: 9:00am-6:00pm Monday-Saturday; Standard Time: 10:00am-5:00pm Monday-Saturday
Specialties: Ball and burlap spruce, aspen; large sizes cold hardy perennials zones 2 and 3. Large ball and burlap cold hardy shrubs.

Malone's Landscape and Nursery
PO Box 1322
Maple Valley, WA 98038
Walk-in address: 24322 228th Avenue SE
Maple Valley, WA 98038
Proprietor(s): Jim and Debbie Malone
Phone: 425-413-0979
FAX: 425-413-0410
Mail-order or walk-in: both
Credit cards: MC, VISA
Specialties: Perennials, larger landscpae-size trees and general nursery stock; gift items.

Maritime Nursery
13119 SW 280th Street
Vashon Island, WA 98070
Proprietor(s): Kevin Gardener
Phone: 206-463-2971
FAX: 206-463-2930
Mail-order or walk-in: walk-in only
Catalogue: none
Credit cards: no
Hours: 8:00am-5:00pm Monday-Friday; mainly wholesale; retail by appointment.
Specialties: Heather, groundcovers; Pacific Northwest native trees, shrubs, groundcovers.

Mary's Country Garden
23628 172nd Ave SE
Kent, WA 98042
Proprietor(s): Mary Frey
Phone: 253-639-1243
Email address: mlfrey@aol.com
Mail-order or walk-in: walk-in only
Credit cards: no
Hours: April-June: 11:00am-5:00pm Tuesday-Saturday
Specialties: Perennials, vines, hardy fuchsia, hardy geranium, primula.

Max & Hildy's Garden Store
19350 NW Cornell Rd
Hillsboro, OR 97124
Proprietor(s): Robert Iwasaki
Phone: 503-645-5486
FAX: 503-645-5680
Mail-order or walk-in: walk-in only
Credit cards: VISA, MC, Discover
Hours: 9:00am-7:00pm Monday-Saturday; 10:00am-6:00pm Sunday. Tours available.
Specialties: No specialties. Great selection and mix. Natives, alpines, homegrown annuals and perennials. Garden art; gifts.

McComb Road Nursery
751 McComb Rd
Sequim, WA 98382
Proprietor(s): Neil Burkhardt and Jane Stewart
Phone: 360-681-2827
FAX: 360-681-7578
Email address: mccomb@olympus.net
Mail-order or walk-in: walk-in only
Credit cards: yes
Hours: Open all year - hours vary - please call.
Specialties: Display gardens, picnic area. Quality trees, shrubs, perennials. We grow much of our own stock.

Mitchell Bay Nursery
1071 Mitchell Bay Rd
Friday Harbor, WA 98250
Proprietor(s): Colleen Howe/Bruce Gregory
Phone: 360-378-2309
Hours: Wednesday-Saturday 11:00am-5:00pm (these will be new hours, not firmed up yet)
Specialties: Deer-resistant plants and cottage style plants seen in a demonstration garden.

Molbak's Inc.
13625 NE 175th St
Woodinville, WA 98072
Proprietor(s): Egon and Laina Molbak
Phone: 425-483-5000
FAX: 425-398-5190
Mail-order or walk-in: walk-in only
Credit cards: AmEx, Discover, MC, VISA
Hours: 9:00am-9:00pm Monday-Saturday; 9:00am-6:00pm Sunday; closed Thanksgiving, Christmas and New Year's Day
Specialties: Molbak's is a distinctive specialty retailer, providing an unusual retail shopping experience in gardening, fine gifts, floral, patio furniture and Chrismas décor. Family-owned and operated since 1956, Molbak's is one of the largest operations of its kind in the US.

Moorehaven Water Gardens
3006 York Rd
Everett, WA 98204
Proprietor(s): Val and Chris Moore
Phone: 425-743-6888
FAX: 425-514-5488
Email address: cmoore1023@aol.com
Mail-order or walk-in: both
Credit cards: VISA, MC
Hours: April-October: 10:00am-6:00pm everyday but Tuesday ; October-April: 10:00am-dusk Friday-Monday .
Specialties: Aquatic nursery - aquatic plants, imported koi and goldfish, pumps, liners, statuary, fountains, books, food, some bamboo gift items.

Moss Greenhouses
269 South 300 East
Jerome, ID 83338
Proprietor(s): Kevin and Dana Moss
Phone: 208-324-1000
Email address: becky.marshal@mossgreenhouses.com
Website: www.mossgreenhouses.com
Mail-order or walk-in: walk-in only
Catalogue: n/a
Credit cards: VISA, MC, Discover
Hours: March 1-October 31: 8:00am-6:00pm; closed November 1-February 28
Specialties: Hanging baskets, specialty containers, annuals and perennials.

Mountain Meadow Nursery, Inc.
7236 132nd Ave NE
Kirkland, WA 98033
Proprietor(s): George and Mary Anne LeDoux, Nancy Davidson Short
Phone: 425-885-0785
Mail-order or walk-in: By appointment only
Hours: Retail by appointment only; mostly wholesale
Specialties: Styrax obassia, Pacific Northwest natives, ferns (swords, deer, maidenhair, and many other NW natives), mostly shade plants.

Oregon Miniature Roses
8285 SW 185th Ave
Beaverton, OR 97007
Proprietor(s): Katherine Spooner

Phone: 503-649-4482
FAX: 503-649-3528
Mail-order or walk-in: both
Catalogue: $4.75 + shipping
Credit cards: yes
Hours: April 15-September 15: 9:00am-5:00pm
Specialties: All types of miniatures. Also potted large roses for walk-in customers.

Peninsula Gardens
5503 Wollochet Dr NW
Gig Harbor, WA 98335
Proprietor(s): Marlin and Bette Cram
Phone: 253-851-8115
FAX: 253-851-8104
Email address: pengard@ix.netcom.com
Website: www.peninsulagardens.com
Mail-order or walk-in: walk-in only
Catalogue: no
Credit cards: yes
Hours: Winter: 9:00am-5:30pm; Summer: 9:00am-6:00pm
Specialties: Perennials, trees, and shrubs, annuals, etc. Largest full service nursery, garden and gift center in Pierce County.

Peninsula Nurseries Inc
1060 Sequim-Dungeness Way
Sequim, WA 98382
Proprietor(s): Roger L. Fell
Phone: 360-683-6969
FAX: 360-681-2865
Email address: pennurseries@olylmpus.net
Website: www.peninsulanurseries.com
Mail-order or walk-in: walk-in only
Credit cards: VISA, MC, Discover
Hours: 9:00am-6:00pm Monday-Saturday; 9:00am-5:00pm Sunday
Specialties: Specialize in trees and perennials; extensive plant list. This is a large nursery.

Perennial Gardens
4221 South Pass Rd
Everson, WA 98247-9207
Proprietor(s): Donna Jensen
Phone: 360-966-2330
Mail-order or walk-in: walk-in only
Credit cards: no
Hours: Drop-ins any time I'm here. Call for appointment or take a chance. Open year round.
Specialties: Drought tolerant perennials, field tested for this area, that are beyond the common. Some natives, non bearded iris, cranesbill geraniums, penstemons, daylilies, potentillas, ground covers, white gentian, veronicas, sempervirens.

Pioneer West
710 N Tower Ave
Centralia, WA 98531
Proprietor(s): Rick and Deanna Schnatterly
Phone: 360-736-3872
FAX: 360-736-3930
Email address: PioneerWest@localaccess.com
Mail-order or walk-in: walk-in only
Credit cards: VISA, MC, Discover
Hours: Open 7 days a week. 8:30am-6:00pm Monday-Saturday; 10:00am-5:00pm Sunday
Specialties: Full line nursery; annuals and perennials, bulbs, gardening supplies, gardening art, water gardening, and gifts. Services of Washington Cerified Nursery Professionals. In business since 1927.

Plantasia Flower and Gardens
3938 88th Ave SW
Olympia, WA 98512
Proprietor(s): Evonne Peryea
Phone: 360-754-4321
FAX: 360-956-9228
Email address: plantasiax@aol.com
Mail-order or walk-in: mail-order soon
Hours: April-July: 11:00am-4:00pm Saturdays
Specialties: One acre display garden and gift shop. Unusual perennials and shrubs, new introductions, pond plants, garden gifts; pulmonaria, heuchera, hardy geranium, cryptomeria, Japanese maples.

Pleasant View Nursery
PO Box 467
McKenna, WA 98558
Walk-in address: 2410 336th St S
Roy, WA 98580
Proprietor(s): Marvin and Tina Potter
Phone: 253-843-2820
FAX: 253-843-1206
Mail-order or walk-in: walk-in only
Credit cards: VISA, MC, Discover
Hours: 9:00am-6:00pm Tuesday-Saturday; 11:00am-5:00pm Sunday, March-November
Specialties: Year-round color, unusual perennials, shrubs and trees.

Ponderings Plus
3360 N Pacific Hwy
Medford, OR 97502
Proprietor(s): GayLynn Dunagan and Cheryl Robertson
Phone: 541-773-3297
FAX: 541-772-2169
Mail-order or walk-in: walk-in only
Catalogue: none
Credit cards: VISA, MC, Discover, AmEx
Hours: March 1-September 30: 9:00am-5:30pm Monday-Saturday, noon-4:00pm Sunday; October 1-February 28: 10:00am-4:30pm Monday-Saturday, closed Sunday
Specialties: Perennials, pond plants and supplies.

Poole's Nursery
3518 6th Ave
Tacoma, WA 98406
Walk-in address: 6th and Union
Tacoma, WA 98406
Proprietor(s): Tony Palermo
Phone: 253-759-3519
FAX: 253-756-8153
Mail-order or walk-in: walk-in only
Credit cards: VISA, MC, Discover
Hours: Open 7 days a week year round
Specialties: Complete garden center located in Tacoma for over 110 years.

Portland Avenue Nursery
1409 E 59th
Tacoma, WA 98404
Phone: 253-473-0194
FAX: 253-473-4178
Mail-order or walk-in: walk-in only
Credit cards: VISA, MC, Discover, AmEx
Hours: Year-round: 9:00am-5:30pm Monday-Saturday, 10:00am-4:00pm Sunday . Call for seasonal hours.
Specialties: Three acre urban oasis specializing in unique and unusual Japanese maples and conifers. Excellent selection of roses, rhodies, trees, shrubs and perennials. Furniture, statuary, garden tools and accessories.

Portland Nursery
5050 SE Stark
Portland, OR 97215
Walk-in address: Second location: 9000 SE Division
Portland, OR 97266
Proprietor(s): Jon Denney
Phone: 503-231-5050 (Stark), 503-788-9000 (Division)
FAX: 503-231-7123 (Stark), 503-788-9002 (Division)
Website: www.portlandnursery.com
Mail-order or walk-in: walk-in only

Catalogue: n/a
Credit cards: VISA, MC, Discover
Hours: Open 7 days/week; spring/summer: hours vary; fall/winter 9:00am-6:00pm
Specialties: Wide selection of plants, tools, gifts for gardeners. Extensive conifer selection; Northwest native section (Stark Street); specialty perennials; houseplants and tropicals; pond supplies (Division Street)

Prune Hill Nursery
2531 NW 18th Ave
Camas, WA 98607
Proprietor(s): Gerald and Diana Christensen
Phone: 360-834-3765
FAX: 360-834-2081
Email address: prunehil@pacifier.com
Mail-order or walk-in: walk-in only
Credit cards: Discover, VISA, MC
Hours: 10:00am-6:00pm Tuesday-Saturday; 10:00am-4:00pm Sunday. Winter hours, please call.
Specialties: Bedding plants, hanging baskets, perennials, trees, shrubs, sod, cascading petunia baskets.

RainShadow Gardens
6298 Double Bluff Rd
Freeland, WA 98249
Proprietor(s): John C.Holbron, Jr. and Anne N. Davenport
Phone: 360-321-8003
Mail-order or walk-in: walk-in only
Credit cards: no
Hours: Early April-mid November: 11:00am-5:00pm Friday-Monday, or by appointment
Specialties: Unusual ornamental trees and shrubs, ornamental grasses and some uncommon perennials, bonsai. Display gardens of ornamental grasses, shade plants, dwarf conifers, secret garden, and bonsai collection.

Redmond Greenhouse
4101 S Highway 97
Redmond, OR 97756-9662
Walk-in address: between Bend and Redmond
Proprietor(s): Doug and Sherry Stott
Phone: 541-548-5418
FAX: 541-548-4634
Email address: info@redmondgreenhouse.com
Website: www.redmondgreenhouse.com
Mail-order or walk-in: both
Catalogue: website only
Credit cards: VISA, MC, Discover
Hours: Everyday except holidays, open 9:00am.
Specialties: Home and garden decor, statuary, fountains, florist, garden mercantile.

Rhododendron Patch
16428 88th Street E
Sumner, WA 98390
Proprietor(s): Evelyn and Clayton Hoebelheinrich
Phone: 253-863-0251
FAX: 253-863-0251
Email address: evelynkh@gte.net
Mail-order or walk-in: both
Credit cards: VISA, MC, Discover, AmEx
Hours: 9:30am-6:00pm Monday-Saturday, 11:00am-6:00pm Sunday
Specialties: Specializing in rhodies and azaleas, but have good supply of trees, shrubs, perennials, annuals,and water gardening supplies.

Roadhouse Nursery
12511 Central Valley Rd NW
Poulsbo, WA 98370-7016
Proprietor(s): Jan and George Bahr
Phone: 360-779-9589
Mail-order or walk-in: walk-in only
Credit cards: yes
Hours: February 15-November 15: 9:30am-5:30pm Tuesday-Saturday, 10:00am-5:00pm Sunday
Specialties: Aquatic plants, water gardening supplies, Japanese koi and goldfish.

Sally Herman
1190 74th Avenue SE
Salem, OR 97301
Proprietor(s): Sally Herman
Phone: 503-581-1750
Mail-order or walk-in: walk-in only
Hours: By appointment
Specialties: Clematis, grasses, miscellaneous unusual perennials and woody shrubs, and organically grown heirloom tomatoes and vegetable starts.

Sandy's Nursery
1830 Goodspeed
Tillamook, OR 97141
Proprietor(s): Bill and Sandy Howard
Phone: 503-842-7270
FAX: 503-842-7270
Mail-order or walk-in: both
Credit cards: VISA, MC, AmEx
Hours: Mail order for Christmas wreaths, swags and garlands.
Specialties: A changing variety of hard to find plants. A very wide selection of perennials, annuals, and landscape shrubs and trees.

Seasons Nursery
3990 SW Borland Rd
Tualatin, OR 97062
Proprietor(s): Holly Goble
Phone: 503-638-2701
Mail-order or walk-in: walk-in only
Credit cards: VISA, MC, Discover
Hours: 10:00am-6:00pm Wednesday-Friday, 10:00am-5:00pm Saturday-Sunday
Specialties: Ivy topiary, topiary frames, gift shop with home decor and garden treasures, creative classes.

Shaman Garden
31503 NE Lewisville Hwy
Battle Ground, WA 98604
Proprietor(s): Gerritt Oatfield and Larry Mullett
Phone: 360-666-3634
Mail-order or walk-in: walk-in only
Credit cards: no
Hours: February-October. 9:00am-6:00pm, except Wednesday and Thursday. Appointment only November-January.
Specialties: Groundcovers, vines and vegetable starts.

Shirley's Dahlias
3290 Brush College Rd NW
Salem, OR 97304-9522
Proprietor(s): Shirley and Bob Noteboom
Phone: 503-362-6302
Mail-order or walk-in: both
Hours: 10:00am-6:00pm Monday, Tuesday, Thursday, Friday, Sunday
Specialties: Dahlias; fresh cut flowers July-November; tubers to plant April-May; flower stand, special orders.

Silver Lake Nursery
11014 19th Avenue SE
Everett, WA 98208
Proprietor(s): Tom Sullivan
Phone: 425-337-6770
FAX: 425-337-5311
Mail-order or walk-in: walk-in only
Credit cards: VISA, MC, AmEx, Discover
Hours: 8:00am-7:00pm Monday-Saturday, 10:00am-6:00pm Sunday
Specialties: Roses, perennials, shrubs, seasonal color.

ADDITIONAL NURSERIES

Skipper and Jordan Nursery
7800 SE Short Rd
Gresham, OR 97080
Proprietor(s): Bob and Ilona Skipper and Brent and Teresa Jordan
Phone: 503-663-1125
FAX: 503-663-4245
Mail-order or walk-in: walk-in only
Credit cards: yes
Hours: 9:00am-5:00pm daily
Specialties: Wholesale and retail; shade and flowering trees and evergreen trees and shrubs: ash, birch, cherry, crabapple, pear, maple, oak, plum laurel, pine, spruce, arborvitae, cedar, and fir. We also locate and buy in for customers.

Sky Nursery
18528 Aurora Ave N
Shoreline, WA 98133
Proprietor(s): Susan Chaney
Phone: 206-546-4851
FAX: 206-546-8010
Website: www.skynursery.com
Mail-order or walk-in: walk-in only
Catalogue: n/a
Credit cards: MC, VISA, AmEx
Hours: 9:00am-6:00pm Monday-Saturday; 10:00am-5:00pm Sunday. Closed Thanksgiving, Christmas and New Year's Day.
Specialties: Open year round, featuring Northwest natives, annuals, perennials, water gardening; knowledgeable staff.

Squak Mountain Greenhouses and Nursery
7600 Renton-Issaquah Rd SE
Issaquah, WA 98027
Proprietor(s): Jim and Becky Pommer
Phone: 425-392-1025
Mail-order or walk-in: walk-in only
Credit cards: yes
Hours: 9:00am-6:00pm Monday-Saturday; Winter hours: 9:00am-5:00pm. Closed Sundays
Specialties: Greenhouse grows beautiful poinsettias for the holidays, a huge variety of color for spring and summer, including geraniums, fuchsias, impatiens, petunias and specialty annuals. We also grow seasonal color for winter and fall seasons. Customers select plants from our one-plus acre of greenhouses. Full service nursery features trees, shrubs, roses, grasses, groundcovers, vines and perennials (includes berries and fruit trees). We also sell some of these products bare root in early spring.

Squirrel Heights Gardens
6934 SE 45th Avenue
Portland, OR 97206
Proprietor(s): Betty Berdan and Dave Peterson
Phone: 503-771-6945
Email address: betsndave@cnnw.net
Mail-order or walk-in: walk-in only
Credit cards: no
Hours: Flexible hours - call anytime.
Specialties: Perennials, ornaamental grasses, specialty annuals, patio pots, cut flowers. We do most of our own propagating..

Stoller Farms
726 W Heintz St
Molalla, OR 97038
Walk-in address: 14682 S Herman Rd
Molalla, OR 97038
Proprietor(s): Marvin and Linda Stoller
Phone: 503-829-5385
Email address: stoller@molalla.net
Mail-order or walk-in: walk-in only
Catalogue: none
Credit cards: no
Hours: Monday-Saturday; best by appointment
Specialties: Fuchsia, begonia and moss-lined baskets. Annual bedding plants, vegetable starts.

Summersun Greenhouse & Nursery
4100 E College Way
Mt Vernon, WA 98273
Proprietor(s): Carl and Cheryl Loeb
Phone: 360-424-1663
FAX: 360-424-7934
Email address: summerr@cnw.com
Website: www.summersunnursery.com
Mail-order or walk-in: both
Credit cards: yes
Hours: Monday - Saturday 8:00am-6:00pm, Sunday 9:00am-5:30pm. Open all year.
Specialties: Perennials, bedding plants, nursery stock, indoor foliage, containers, and garden gifts.

Swansons Nursery
9701 15th Ave NW
Seattle, WA 98117
Proprietor(s): Wally Kerwin
Phone: 206-782-2543
FAX: 206-782-8942
Email address: garden@swansonsnursery.com
Website: www.swansonsnursery.com
Mail-order or walk-in: walk-in only
Catalogue: n/a
Credit cards: VISA, MC
Hours: Spring & summer: 9:00am-6:00pm; winter: 9:30am-5:00pm
Specialties: We have everything a gardener could desire: annuals, perennials, vines, groundcovers, trees, shrubs, seeds, tools, gifts for gardeners, pottery, statuary, arbors, etc. We specialize in ideas and helping gardeners be successful with gardens.

The Berry Botanic Garden
11505 SW Summerville Ave
Portland, OR 97219
Proprietor(s): Gael Varsi, Director of Horticulture
Phone: 503-636-4112
FAX: 503-636-7496
Email address: bbg@rdrop.com
Website: www.berrybot.org
Mail-order or walk-in: walk-in only, by appointment only for nonmembers
Catalogue: none
Credit cards: VISA, MC
Hours: Business hours by appointment 9:00am-4:30pm Monday-Friday
Specialties: Species rhododendrons, NW natives, flowering shrubs.

The Garden Shop at Lakewold Gardens
4524 N 18th
Tacoma, WA 98406
Walk-in address: 12317 Gravelly Lake Dr SW
Lakewood, WA 98499
Proprietor(s): Vickie Haushild
Phone: 253-584-3360
FAX: 2537-52-5998
Email address: vhaushild@aol.com
Website: www.lakewold.org
Mail-order or walk-in: walk-in only
Catalogue: none
Credit cards: VISA, MC
Hours: The Garden Shop has the same hours as Lakewold Gardens.
Specialties: Shade plants, unusual alpines, dwarf conifers, rare and unusual perennials.

The Gardens At Padden Creek
2014 Old Fairhaven Parkway
Bellingham, WA 98225
Proprietor(s): Mary Cragin
Phone: 360-671-0484
FAX: 360-671-0484
Mail-order or walk-in: walk-in only
Catalogue: n/a
Credit cards: yes

Hours: March 1 - mid-September: 9:00am-6:00pm everyday
Specialties: Perennials, shrubs, annuals, trees. Garden tools, artifacts, statuary, pots, seeds, etc. Display gardens, 2 ponds, and chickens.

The Pat Calvert Greenhouse
2300 Arboretum Drive E
Seattle, WA 98112-2300
Proprietor(s): Ann O'Mera, The Arboretum Foundation
Phone: 206-325-4510
FAX: 206-325-8893
Email address: gvc@arboretumfoundation.org
Website: nsccux.sccd.ctc.edu/~eaomera
Mail-order or walk-in: walk-in only
Catalogue: no
Credit cards: yes
Hours: Every Tuesday 10:00am-noon; April-October: second Saturday of the month 10:00am - 2:00pm; or by appointment
Specialties: We have a selection of trees, shrubs, vines, perennials and plants suitable for bonsai propagated from the Arboretum collections and members' gardens. All proceeds are used to support Arboretum programs and improve the collections.

The Plant Peddler
3022 E Burnside Street
Portland, OR 97214
Proprietor(s): Lisa Brending
Phone: 503-233-0384
FAX: 503-227-1245
Email address: plantped@aol.com
Website: www.theplantpeddler.citysearch.com
Mail-order or walk-in: walk-in only
Catalogue: n/a
Credit cards: yes
Hours: 10:00am-6:00pm Monday-Saturday, 11:00am-5:30pm Sunday
Specialties: Indoor/greenhouse plants. Exotic tropicals, cactus, succulents, flowering house plants, bromeliads, specimen plants; pottery from around the world.

Tom's Garden Center
410 SW Pacific Blvd
Albany, OR 97321
Proprietor(s): Tom and Annette Krupicka
Phone: 541-928-2521
Mail-order or walk-in: walk-in only
Credit cards: all
Hours: 8:30am-6:00pm Monday-Saturday, 10:00am-4:00pm Sunday
Specialties: Annuals, perennials, flowering and shade trees, fruit trees and vines, roses, huge hanging shade or sun baskets, and garden seeds.

Town and Country Gardens
5800 S Yellowstone Hwy
Idaho Falls, ID 83402
Proprietor(s): John Crook
Phone: 208-522-5247
FAX: 208-523-4508
Email address: tcgarden@townandcountrygardens.com
Website: www.townandcountrygardens.com
Mail-order or walk-in: walk-in only
Hours: 9:00am-6:00pm year round, extended hours during spring
Specialties: Perennials, zone 2 - 4 trees and shrubs

Tsugawa Nursery
410 Scott Ave
Woodland, WA 98674
Proprietor(s): Martin Tsugawa
Phone: 360-225-8750
FAX: 360-225-5086
Mail-order or walk-in: walk-in only
Hours: September-March: 9:00am-5:30pm Monday-Saturday; April-August: 9:00am-7:00pm Monday-Saturday. 10:00am- 5:30pm every Sunday.
Specialties: Bamboo plants and products. Japanese maples, bonsai and perennials. Home-grown annuals - geraniums. Six acres nursery stock and 10 acres growing area.

Tualatin River Nursery
65 Dollar St
West Linn, OR 97068
Proprietor(s): John and Lori Blair
Phone: 503-650-8511
FAX: 503-656-6646
Mail-order or walk-in: walk-in only
Credit cards: yes
Hours: 9:00am-6:00pm Monday-Saturday, 11:00am-4:00pm Sunday
Specialties: Heirloom vegetables, herbs, edible flowers, ornamental grasses, perennials, unique shrubs. Also, November brings specialty wreaths. Onsite cafe using nursery's organic greens for hungry gardeners!

Valley Nursery Inc.
20882 Bond Rd NE
Poulsbo, WA 98370
Proprietor(s): Brad Watts, President
Phone: 360-779-3806
FAX: 360-779-7426
Email address: valleynursery@silverlink.net
Mail-order or walk-in: walk-in; mail-order only for roses
Catalogue: For roses only, free
Credit cards: VISA, MC, Discover
Hours: Summer: 9:00am-6:00pm Monday-Friday; Winter: 9:00am-5:30pm Monday-Friday; Weekends: 9:00am-5:00pm
Specialties: Six acres of plant material - basic plant material, the rare and unusual. Dwarf conifers, perennials, water gardens, roses, complete garden shop and gift shop. Free quarterly newsletter; free mail-order bareroot rose catalogue.

Valley View Nursery
1675 N Valley View Rd (office)
Ashland, OR 97520
Walk in address: 1321 Center Drive, Medford, OR (garden center)
1675 N Valley View Rd, Ashland, OR (sales yard),
Phone: 541-488-2450 office; 541-488-1595 sales yard; 541-773-7972 garden center
FAX: 541-488-2454 office; 541-488-4281 sales yard; 541-772-8410 garden center
Email address: valleyviewnsy@mindspring.com
Mail-order or walk-in: walk-in only
Credit cards: yes
Specialties: The largest grower/broker/retailer between Salem, OR and Sacramento, CA - landscape trees and shrubs , 400 types of perennials. Special orders welcome.

Vittoria Nursery
23000 Fulquartz Rd
Dundee, OR 97115
Proprietor(s): Guy Vittoria
Phone: 503-538-3637
Mail-order or walk-in: walk-in only
Hours: 8:00am-4:00pm 7 days/week

Wallingford Garden Spot
1815 N 45th Street
Seattle, WA 98103
Proprietor(s): John and Marian Jarosz
Phone: 206-547-5137
FAX: 206-547-4409
Mail-order or walk-in: walk-in only
Credit cards: all major credit cards, including debit
Hours: 10:00am-8:00pm Monday-Friday; 10:00am-6:00pm Saturday; 10:00am-5:00pm Sunday.
Specialties: Small urban nursery with a passion for the rare and unusual; also the basics to meet the community's needs. Heirloom plants, organic gardening supplies, floral arrangements, specialty gardeners' gifts.

Washco Lawn & Garden
12725 SW Canyon Rd
Beaverton, OR 97005-2109
Proprietor(s): Dana Heimbecker
Phone: 503-646-1779
Mail-order or walk-in: walk-in only
Catalogue: none
Credit cards: VISA, MC
Hours: Hours vary by season
Specialties: Sedums, vegetables, groundcovers.

Wee Tree Farm
5340 NE Highway 20
Corvallis, OR 97330
Walk-in address: 5470 NE Highway 20
Corvallis, OR 97330
Proprietor(s): Diane Lund
Phone: 541-752-3430, 800-638-1098
FAX: 541-752-3431
Email address: diane@weetree.com
Website: www.weetree.com
Mail-order or walk-in: both
Credit cards: yes
Hours: At retail store (Garland Nursery): 9:00am-6:00pm Monday-Friday; 9:00am-5:00pm Saturday; 10:00am-5:00pm Sunday
Specialties: Bonsai plants, pots, tools, wire, classes

Wet Rock Gardens
1950 Yolanda Ave
Springfield, OR 97477
Proprietor(s): Kelly O'Neill
Phone: 541-746-4444
FAX: Call ahead to fax
Email address: gardens@wetrock.com
Website: www.wetrock.com
Mail-order or walk-in: walk-in only
Catalogue: n/a
Credit cards: not yet
Hours: By appointment and limited hours - call ahead please.
Specialties: U-pick cut flowers, cut flower and rock garden varieties.

Wight's Home & Garden
5026 196th St SW
Lynnwood, WA 98036
Proprietor(s): James A. Anderson
Phone: 425-775-3636
FAX: 425-672-1404
Mail-order or walk-in: walk-in only
Credit cards: VISA, MC, Discover, AmEx
Hours: 9:00am-9:00pm Monday-Friday, 9:00am-6:00pm Saturday and Sunday. Completely wheelchair accessible.
Specialties: Roses, perennials, Japanese maples, specialty conifers, bedding plants, rhododendrons and much more. Full-service nursery/garden center, including fountains and water gardening.

Wild Bird Garden/Coastal Garden Center
PO Box 2210
Westport, WA 98595-2210
Walk-in address: 4986 SR 105 S
Grayland, WA 98547
Proprietor(s): Cynthia Becker
Phone: 360-268-0804
Email address: wildbird@techline.com
Mail-order or walk-in: both
Credit cards: VISA, MC
Hours: May 1-September 30: 10:00am-5:00pm daily; October 1-April 30: Friday-Monday, 10:00am-5:00pm; Tuesday-Thursday, by chance.
Specialties: Landscape design, backyard songbird specialty shop, seasonal color, ocean gardening, herbs.

Wildwood Gardens
PO Box 250
Molalla, OR 97038-0250
Proprietor(s): Will Plotner
Phone: 503-829-3102
Email address: gardens@molalla.net
Mail-order or walk-in: mail-order; walk-in by appointment
Catalogue: $2.00
Credit cards: no
Hours: Call for appointment during bloom seasons
Specialties: Iris (Japanese, Siberian, spuria, species, PCI, all the bearded TB, BB, IB, SDB, MDB, MTB), daylilies, and hostas

Willamette Falls Nursery & Ponds
1720 Willamette Falls Dr
West Linn, OR 97068
Proprietor(s): John and Gloria Lightower
Phone: 503-656-7344
FAX: 503-650-9694
Email address: plantsman@home.com
Website: www.willamettefalls.com
Mail-order or walk-in: both
Catalogue: n/a
Credit cards: MC, VISA, AmEx, Discover, Diner's
Hours: 8:00am-7:00pm Monday-Friday, 9:00am-6:00pm Saturday/Sunday, year round
Specialties: New and unusual perennials, eucalyptus, magnoliaceae, aquatic plants, abutilons, passifloras, tropicals, variegated and color foliage, daturas and brugmansias.

Willow Lake Nursery
PMB 284, 4925 River Rd N
Keizer, OR 97303
Walk-in address: 5655 Windsor Island Rd N
Keizer, OR 97303
Proprietor(s): Bill, John and Lanora Blake
Phone: 503-390-3032
FAX: 503-393-0732
Email address: wlnursery@aol.com
Mail-order or walk-in: both
Credit cards: yes
Hours: year-round, seasonal hour changes
Specialties: Roses, perennials, water gardening, NW holiday gift packs.

Wisteria Herbs & Flowers
5273 S Coast Hwy
South Beach, OR 97366
Walk-in address: one mile south of the Newport Bridge
Proprietor(s): Linda Montgomery
Phone: 541-867-3846
Mail-order or walk-in: walk-in only
Catalogue: no
Credit cards: VISA, MC
Hours: March-October: 9:00am-5:00pm 7 days/week (March and October: 10:00am-4:00pm)
Specialties: Hosta (100+ varieties), astilbe, lavender (30+ varieties), and extensive selection of perennials and herbs (450+ varieties), some old favorites, some new introductions. About 200 varieties of annuals, various scented and fancy leaf geraniums, 30 fuchsias. Numerous display gardens that dash the myth that you cannot grow anything at the coast. All plants grown on site; 90% propagated on site.

Yamamoto's
5765 Sidney Rd SW
Port Orchard, WA 98367
Proprietor(s): Terry Yamamoto
Phone: 360-876-1889
FAX: 360-895-0315
Email address: landty@aol.com
Mail-order or walk-in: walk-in only
Credit cards: VISA, MC only
Hours: Summer and spring: 8:00am-4:30pm Monday-Saturday, 9:00am-4:30pm Sunday; or call for appointment.
Specialties: Japanese irises

COMMON NAME INDEX

The common name index is designed to help you find the correct botanical name for plants you are familiar with by their everyday names, such as Shasta Daisy. Common names are listed first, followed by their botanical counterpart(s). There may be more than one Latin name for a given common name, like Dusty Miller, which can refer to *Artemisia stellerana*, *Senecio cineraria*, *Senecio viravira*, or *Lychnis coronaria*. Or, on the other hand, there may be many common names for the same plant, like *Pulmonaria officinalis*, variously known as Blue Lungwort, Jerusalem Cowslip, Jerusalem Sage, Spotted Dog, and Soldiers and Sailors. This plethora of names makes it obvious why plantspeople use botanical rather than common names when referring to plants; since each plant has only one correct Latin name, it can immediately be identified through that name, rather than having to guess which Dusty Miller one might be talking about. However, many of us know certain plants by names we've heard since childhood, hence the need for a common name cross-reference. This is especially true for herbs, fruits, nuts, and vegetables.

Sometimes after a common name, only a genus name will be listed, with no species following. For example, Bleeding Heart refers to the whole genus *Dicentra*, and Bellflower is the general name for *Campanula*. There will be different names for different species, such as Western Bleeding Heart (*Dicentra formosa*), and Peach-leafed Bellflower (*Campanula persicifolia*), but all will be Bleeding Hearts or Bellflowers. In other genera, different species have very different common names; in the genus *Viburnum*, for example, *V. dentatum* is Arrowwood, *V. lantana* is the Wayfaring Tree, *V. prunifolium* is the Black Haw, and *V. opulus* is the Guelder Rose, so no one common name applies to the whole genus.

Occasionally a nursery will list a plant name with the common name used as a cultivar, for instance *Berberis verruculosa* 'Warty' (common name Warty Barberry), or *Oenothera fruticosa* ssp. *glauca* 'Sundrops' (common name for the genus Oenothera is Sundrops). Where we have been able to identify these instances, we have left off such common name cultivars since they are in fact not cultivar names.

Some common and botanical genus names are very similar; for instance, Rosemary is *Rosmarinus*, Rose is *Rosa*, Calamint is *Calamintha*, and Gentian is *Gentiana*. Sometimes common and botanical names, such as *Calla* and *Yucca*, are identical. Be sure also to look under the genus you are interested in for species names similar to common names, which are generally too alike to be included in this database. For example, the Laurel Oak is *Quercus laurifolia* (meaning laurel-leaved) and the Mongolian Oak is *Quercus mongolica*, Siberian Iris is *Iris sibirica*, etc.

It will be helpful to you if you can learn the meaning of some very common species names, to help you sort out, as you peruse the plant list, what a plant might look like or what conditions it might grow well under. These species names will appear again and again, and you will be able to tie them in to the common names. Examples are *pendula* (weeping), *nana* or *compacta* (a dwarf form), *laciniata* (cut-leafed), *japonica* (from Japan), *chinense* or *sinense* (from China), *occidentalis* (western), *orientalis* (eastern), *rubra* (red), *alba* (white), *nigra* (black), and *purpurea* (purple). We have generally not listed common names with species names that can easily be figured out, such as *Liquidambar orientalis* (the Oriental Sweet Gum), *Populus x canadensis* (the Canadian Poplar), or the Mongolian Oak mentioned above. We will list the genus names, however: *Quercus* = Oak, *Populus* = Poplar, *Liquidambar* = Sweet Gum; and variations of names that are not obvious, such as *Populus nigra* 'Italica', the Lombardy Poplar, or *Torreya grandis*, the Chinese Nutmeg-Yew.

Note that some commonly occuring names, such as Gum, Sage, or Bay will be alphabetized under these names, rather than Sweet Bay, Jerusalem Sage, or Blue Gum, as you may expect to find them. Be sure to look under several parts of a name if you don't find the name as you normally think of it.

Punctuation on the Latin name side of the list is as follows: when multiple species of the same genus have the same common name, they are listed with the word "or" between them, e.g., *Myrica cerifera* or *pensylvanica*. When multiple genera have the same common name, they are separated by a semicolon, e.g., *Senecio articulatus*; or *Plectranthus oertendahlii*.

Note: A few of the common names are phrased in ways that are now considered inappropriate. While we agree that they are objectionable, they are nevertheless the common names, and we have included them as such.

COMMON NAME INDEX

Aaron's Rod	Verbascum thapsus
Abele	Populus alba
Abelia, Glossy	Abelia x grandiflora
Acacia, Rose	Robinia hispida
Achocha	Cyclanthera
Acidanthera	Gladiolus callianthus
Action Plant	Mimosa pudica
Adam's Apple	Tabernaemontana divaricata
Adam's Needle	Yucca filamentosa
African Daisy	Osteospermum
Agapanthus, Pink	Tulbaghia simmleri
Ageratum, Hardy	Eupatorium coelestinum
Ajowan	Carum copticum
Alder, Black	Ilex verticillata; or Viburnum nudum
Alder, Mountain	Alnus incana ssp. incana
Alder, Red	Alnus rubra
Alder, Sitka	Alnus viridis ssp. sinuata
Alecost	Tanacetum balsamita
Alexanders	Smyrnium olusatrum
Algarrobo	Prosopis juliflora
Alkali Dropseed	Sporobolus airoides
Alkanet	Anchusa officinalis
All Heal	Prunella vulgaris
Alleluia	Oxalis acetosella
Allium, Turkish	Allium karataviense
Allium, Yellow	Allium moly
Allspice, Carolina	Calycanthus floridus
Allspice, Western	Calycanthus occidentalis
Almond	Prunus dulcis
Almond, Dwarf Russian	Prunus tenella
Almond, Flowering	Prunus glandulosa or japonica or triloba
Aloe, Cobweb	Haworthia arachnoidea
Aloe, Warty	Gasteria verrucosa
Alpine Calamint	Acinos alpinus
Alum Root	Heuchera
Alum Root, Small-flowered	Heuchera micrantha
Alyssum, Water	Samolus parviflorus
Amaryllis	Hippeastrum
Anacharis	Egeria densa
Andromeda	Pieris japonica
Anemone, European Wood	Anemone nemorosa
Anemone, Grape Leaf Japanese	Anemone tomentosa
Anemone, Japanese	Anemone x hybrida
Angel's Fishing Rod	Dierama pulcherrimum
Angel's Tears	Soleirolia soleirolii
Angel's Trumpet	Brugmansia
Angel's-Eye	Veronica chamaedrys
Angelica Tree	Aralia elata
Anise Hyssop	Agastache foeniculum
Anise, Common	Pimpinella anisum; or Foeniculum vulgare var. azoricum; or Myrrhis odorata
Anise, Purple	Illicium floridanum
Anise, Star	Illicium anisatum
Aniseed	Pimpinella anisum
Antelope Bitterbrush or Bush	Purshia tridentata
Antelope Ears	Platycerium
Apache Plume	Fallugia paradoxa
Apostle Plant	Neomarica northiana
Appalachian Tea	Viburnum cassinoides
Apple	Malus domestica
Apple of Peru	Nicandra physalodes
Apple, Chess	Sorbus aria
Apple, Chinese or Plum-Leaved	Malus prunifolia
Apple, Kangaroo	Solanum aviculare
Apple, Large Kangaroo	Solanum laciniatum
Apricot	Prunus armeniaca
Apricot Vine	Passiflora incarnata
Apricot, Black	Prunus x dasycarpa
Apricot, Desert	Prunus fremontii
Apricot, Flowering	Prunus mume
Apricot, Japanese	Prunus mume
Aralia, Castor or Tree	Kalopanax septemlobus
Aralia, Hardy	Eleutherococcus sieboldianus
Aralia, Japanese	Fatsia japonica
Aralia, Tree	Kalopanax septemlobus
Arborvitae, American	Thuja occidentalis
Arborvitae, Emerald	Thuja occidentalis 'Smaragd'
Arborvitae, False or Hiba	Thujopsis dolobrata
Arborvitae, Giant	Thuja plicata
Arborvitae, Japanese	Thuja standishii
Arrow Arum	Peltandra undulata
Arrow Arum, Green	Peltandra virginica
Arrowhead	Sagittaria
Arrowhead, Old World	Sagittaria sagittifolia
Arrowwood	Viburnum acerifolium
Arrowwood, Southern	Viburnum dentatum
Artemisia, Silver King	Artemisia ludoviciana ssp. mexicana var. albula
Artichoke, Chinese or Japanese	Stachys affinis
Artichoke, Globe	Cynara cardunculus Scolymus Group
Artichoke, Jerusalem	Helianthus tuberosus
Artillery Plant	Pilea microphylla
Arugula, Perennial	Rucola silvatica
Arum, Bog	Calla palustris
Arum, Dragon	Dracunculus vulgaris
Arum, Italian	Arum italicum
Arum, Narrow	Peltandra virginica
Ash, Flowering or Manna	Fraxinus ornus
Ash, Green	Fraxinus pennsylvanica
Ash, Mountain	see Mountain Ash
Ash, Northern Prickly	Zanthoxylum americanum
Ash, Oregon	Fraxinus latifolia
Ash, Prickly	Zanthoxylum
Ash, Red	Fraxinus pennsylvanica
Ash, Southern Prickly or Sea	Zanthoxylum clava-herculis
Ash, Utah	Fraxinus anomala
Ashwaganda	Withania somnifera
Asian Jacktree	Sinojackia xylocarpa
Asparagus, Cossack	Typha latifolia
Asparagus, Edible	Asparagus officinalis
Asparagus, Prussian or Bath	Ornithogalum pyrenaicum
Aspen, European	Populus tremula
Aspen, Quaking or Trembling	Populus tremuloides
Asphodel, False	Tofieldia
Aster, Beach	Erigeron glaucus
Aster, Bog	Aster nemoralis
Aster, Calico	Aster lateriflorus
Aster, California	Aster chilensis
Aster, Douglas	Aster subspicatus
Aster, Eastern Woodland	Aster divaricatus
Aster, Golden	Chrysopsis
Aster, Great Northern	Aster modestus
Aster, Japanese	Kalimeris yomena
Aster, New England	Aster novae-angliae
Aster, New York	Aster novi-belgii
Aster, Showy	Aster spectabilis
Aster, Stokes'	Stokesia
Aster, Tree	Olearia
Auricula	Primula auricula
Avens	Geum
Avens, Mountain	Dryas octopetala
Avens, Oregon	Geum macrophyllum
Avens, Purple	Geum triflorum
Avens, Water	Geum rivale
Avocado or Avocado Pear	Persea
Azalea, Chinese	Rhododendron molle
Azalea, Clammy	Rhododendron viscosum
Azalea, Coast or Dwarf	Rhododendron atlanticum
Azalea, Early	Rhododendron prinophyllum
Azalea, Flame	Rhododendron calendulaceum
Azalea, Florida or Florida Flame	Rhododendron austrinum
Azalea, Hiryu or Kirishima	Rhododendron Obtusum Group
Azalea, Hoary	Rhododendron canescens
Azalea, Indian	Rhododendron simsii
Azalea, Japanese	Rhododendron molle ssp. japonicum
Azalea, Korean	Rhododendron yedoense var. poukhanense
Azalea, Macranthum	Rhododendron indicum
Azalea, Mayflower	Rhododendron austrinum or prinophyllum
Azalea, Mock	Menziesia
Azalea, Oconee	Rnododendron flammeum
Azalea, Piedmont	Rhododendron austrinum or canescens or prinophyllum
Azalea, Pink-Shell	Rhododendron vaseyi
Azalea, Pinxterbloom	Rhododendron periclymenoides
Azalea, Plumleaf	Rhododendron prunifolium
Azalea, Pontic	Rhododendron luteum
Azalea, Rose-Shell	Rhododendron prinoplyllum
Azalea, Royal	Rhododendron schlippenbachii
Azalea, Smooth	Rhododendron arborescens
Azalea, Snow	Rhododendron x mucronatum
Azalea, Spider	Rhododendron stenopetalum 'Linearifolium'
Azalea, Summer	Pelargonium x domesticum
Azalea, Swamp or White Swamp	Rhododendron viscosum
Azalea, Sweet	Rhododendron arborescens or viscosum

COMMON NAME INDEX

COMMON NAME INDEX

Bilberry, Red	Vaccinium parvifolium
Bilberry, Tall or Oval-Leaved	Vaccinium ovalifolium
Bilberry, Thin-Leaved	Vaccinium membranaceum
Bindweed	Convolvulus
Birch-Bark Tree	Prunus serrula
Birch, Candle	Betula luminifera
Birch, Canoe	Betula papyrifera
Birch, Cut Leaf Weeping	Betula pendula 'Laciniata'
Birch, European White	Betula pendula
Birch, Gray	Betula populifolia
Birch, Himalayan	Betula utilis var. jacquemontii
Birch, Japanese White or Whitespire	Betula mandshurica var. japonica
Birch, Paper	Betula papyrifera
Birch, River	Betula nigra
Birch, Water	Betula occidentalis
Bird-bill	Dodecatheon
Bird of Paradise Flower	Strelitzia reginae
Bird of Paradise Tree	Strelitzia nicolai
Bird's-Eye	Veronica chamaedrys
Bird's Foot Trefoil	Lotus corniculatus
Birthroot	Trillium
Birthwort	Aristolochia clematitis
Bishop's-Cap	Mitella
Bishop's Hat	Epmedium
Bishop's Weed	Aegopodium podagraria
Bishop's Wort	Stachys officinalis
Bistort	Persicaria bistorta
Bitter Indian	Tropaeolum
Bitter Melon	Momordica charantia
Bitter Root	Lewisia rediviva
Bitter-Bark	Pinckneya pubens
Bittercress	Cardamine
Bittersweet	Celastrus; or Solanum dulcamara
Black Cohosh	Cimicifuga racemosa
Black-Eyed Susan	Rudbeckia hirta
Black-Eyed Susan Vine	Thumbergia alata
Black Sally	Eucalyptus stellulata
Black Swallowwort	Vincetoxicum nigrum
Black Twinberry	Lonicera involucrata
Black Widow	Geranium phaeum
Blackberry	Rubus fruticosus
Blackberry, Double Cream	Rubus rosifolius 'Coronarius'
Blackberry, Pacific	Rubus ursinus
Blackberry, Sow-Teat	Rubus allegheniensis
Blackcap	Rubus occidentalis
Blackroot	Veronicastrum virginicum
Blackthorn	Prunus spinosa
Bladdernut, European	Staphylea pinnata
Bladdernut, Japanese	Staphylea bumalda
Bladderwort	Utricularia
Blanket Flower	Gaillardia
Blazing Star	Liatris
Bleeding Heart	Dicentra
Bleeding Heart, Western	Dicentra formosa
Bloodflower	Asclepias curassavica
Blodroot	Sanguinaria
Bloody-Butcher	Trillium recurvatum
Blue Bean Shrub	Decaisnea fargesii
Blue Bells	Ruellia brittoniana
Blue Buttons	Knautia arvensis; or Succisa pratensis; or Vinca major
Blue-Chalksticks	Senecio serpens
Blue Cohosh	Caulophyllum thalictroides
Blue Fescue	Festuca glauca
Blue Lace Flower	Trachymene coerulea
Blue Mist Shrub	Caryopteris x clandonensis
Blue Potato Bush	Solanum rantonnetii
Blue Star Creeper	Laurentia axillaris; or Pratia
Blue Star Flower	Amsonia
Bluebeard	Caryopteris x clandonensis
Bluebell	Hyacinthoides
Bluebell Creeper or Australian Bluebell Creeper	Sollya heterophylla
Bluebell of Scotland	Campanula rotundifolia
Bluebells, Virginia	Mertensia pulmonarioides
Blueberry (fruit)	Vaccinium corymbosum
Blueberry, Andean	Vaccinium floribundum
Blueberry, Canadian	Vaccinium myrtilloides
Blueberry, Creeping	Vaccinium crassifolium
Blueberry, Highbush	Vaccinium corymbosum
Blueberry, Late Sweet	Vaccinium angustifolium
Blueberry, Lowbush or Low Sweet	Vaccinium angustifolium
Blueberry, Male	Lyonia ligustrina
Blueberry, Mountain	Vaccinium membranaceum
Blueberry, Red Alpine	Vaccinium scoparium
Blueberry, Sourtop	Vaccinium myrtilloides
Blueberry, Swamp	Vaccinium corymbosum
Blueberry, Velvet-Leaf	Vaccinium myrtilloides
Blueblossom	Ceanothus thyrsiflorus
Bluestar	Amsonia tabernaemontana
Bluestem or Little Bluestem	Schizachyrium scoparium
Bo Tree	Ficus religiosa
Boerboon, Karoo	Schotia afra
Boerboon, Weeping	Schotia brachypetala
Bog Bean	Menyanthes
Bog-Stars	Parnassia
Boneset	Eupatorium perfoliatum; or Symphytum officinale
Boojum Tree	Fouquieria columnaris
Borage	Borago officinalis
Borage, Country	Plectranthus amboinicus
Bottle Tree	Brachychiton
Bottlebrush	Callistemon; or Melaleuca
Bottlebrush, Crimson	Callistemon citrinus
Bouncing Bet	Saponaria officinalis
Bourtree	Sambucus nigra
Bow Wood	Maclura pomifera
Bower Plant	Pandorea jasminoides
Bowman's Root	Veronicastrum virginicum
Box Elder	Acer negundo
Box Thorn	Lycium
Box, Running	Mitchella repens
Box, Sweet	Sarcococca
Boxwood	Buxus
Boxwood, African	Myrsine africana
Boxwood, Dwarf	Buxus sempervirens 'Suffruticosa'
Boxwood, Oregon	Paxistima myrtifolia
Boysenberry	Rubus 'Boysenberry'
Brahmi	Bacopa monnieri
Bramble	Rubus
Bramble, Crimson or Arctic	Rubus arcticus
Brass Buttons	Cotula coronopifolia
Brass Buttons, New Zealand	Leptinella squalida
Breadfruit Vine	Monstera deliciosa
Bridal Wreath	Spiraea x vanhouttei; or Francoa; or Stephanotis floribunda
Brier, Sensitive	Schrankia
Brooklime	Veronica
Brooklime, American	Veronica americana
Brookweed	Samolus
Broom	Genista
Broom, Blue or Hedgehog	Erinacea anthyllis
Broom, Moonlight	Cytisus praecox 'Warminster'
Broom, Scotch	Cytisus
Broom, Spanish or Weavers'	Spartium junceum
Broom, White	Retama raetam
Brown Beth	Trillium erectum
Brown-Eyed Susan	Rudbeckia triloba
Bryony	Bryonia dioica
Buartnut	Juglans Buartnut
Buckbean	Menyanthes
Buckbrush	Ceanothus velutinus
Buckeye	Aesculus hippocastanum
Buckeye, Bottle Brush	Aesculus parviflora
Buckeye, California	Aesculus californica
Buckeye, False	Ungnadia speciosa
Buckeye, Mexican or Spanish or Texan	Ungnadia speciosa
Buckeye, Red	Aesculus pavia
Buckhorn	Osmunda cinnamomea
Buckthorn, Alder	Rhamnus frangula
Buckthorn, Common	Rhamnus cathartica
Buckthorn, Italian	Rhamnus alaternus
Buckwheat	Eriogonum
Buckwheat, Red	Eriogonum latifolium var. rubescens
Buffaloberry	Shepherdia canadensis
Buffaloberry, Silver	Shepherdia argentea
Bugbane	Cimicifuga
Bugbane, False	Trautvetteria carolinensis
Bugleweed	Ajuga
Bugloss	Anchusa officinalis
Bull Bay	Magnolia grandiflora
Bullace	Prunus insititia
Bullrush	Schoenoplectus; or Typha latifolia
Bullrush, Lesser	Typha angustifolia
Bunchberry	Cornus canadensis
Bunchgrass	Schizachyrium scoparium
Bunya Bunya Tree	Araucaria bidwillii
Burdock	Arctium lappa

COMMON NAME INDEX

COMMON NAME INDEX

COMMON NAME INDEX

COMMON NAME INDEX

COMMON NAME INDEX

COMMON NAME INDEX

COMMON NAME INDEX

COMMON NAME INDEX

COMMON NAME INDEX

COMMON NAME INDEX

Kumquat, Australian Dessert	Eremocitrus glauca
Kusamaki	Podocarpus macrophyllus
La-Kwa	Momordica charantia
Laburnum, Common	Laburnum anagyroides
Laburnum, Dalmatian	Petteria
Laburnum, Scotch	Laburnum alpinum
Lace Shrub	Stephanandra incisa
Lacebark	Hoheria
Lacquer Tree	Rhus verniciflua
Ladies' Delight	Viola x wittrockiana
Lady Bells	Adenophora confusa
Lady Fingers	Corydalis edulis
Lady's Bedstraw	Galium verum
Lady's Eardrops	Fuchsia magellanica
Lady's Mantle	Alchemilla mollis
Lady's Slipper	Paphiopedilum
Lamb's Ears or Tongue or Tails	Stachys byzantina
Lamb's Tail	Sedum morganianum
Larch, American	Larix laricina
Larch, Dahurian	Larix gmelinii
Larch, European	Larix decidua
Larch, Golden	Pseudolarix amabilis
Larch, Japanese	Larix kaempferi
Larch, Western	Larix occidentalis
Larkspur	Consolida ajacis
Lasiandra	Tibouchina urvilleana
Laurel Tree	Persea borbonia
Laurel, Australian	Pittosporum tobira
Laurel, Bay or True	Laurus nobilis
Laurel, Bog or Swamp	Kalmia polifolia
Laurel, California Bay	Umbellularia californica
Laurel, Cherry	Prunus laurocerasus or caroliniana
Laurel, Drooping	Leucothoe fontanesiana
Laurel, English	Prunus laurocerasus
Laurel, Great	Rhododendron maximum
Laurel, Indian	Ficus microcarpa
Laurel, Mountain	Kalmia latifolia
Laurel, Portugal	Prunus luscitanica
Laurel, Sierra	Leucothoe davisiae
Laurel, Swamp	Magnolia virginiana
Laurel, Texas Mountain	Sophora secundiflora
Laurustinus	Viburnum tinus
Lavatera, Tree	Lavatera olbia or thuringiaca
Lavender Cotton	Santolina chamaecyparissus
Lavender, English	Lavandula angustifolia
Lavender, Fern Leaf	Lavandula multifida
Lavender, French	Lavandula dentata
Lavender, Oregano Scented	Lavandula multifida
Lavender, Sea	Limonium platyphyllum
Lavender, Spanish	Lavandula stoechas
Lead Plant	Amorpha canescens
Leadwort	Plumbago
Leadwort, Cape	Plumbago auriculata
Leatherleaf	Chamaedaphne calyculata; or Viburnum wrightii
Leatherwood	Cyrilla racemiflora
Leek	Allium porrum
Lemon	Citrus x meyeri or limon
Lemon Balm	Melissa officinalis
Lemon Verbena	Aloysia triphylla
Lemon, Wild	Podophyllum peltatum
Lemon, Wild Water	Passiflora foetida
Lemonade Berry	Rhus integrifolia
Lemonwood	Pittosporum eugenioides
Leopard Flower	Belamcanda chinensis
Leopard Plant	Ligularia
Leopard Spot	Farfugium japonicum 'Aureomaculatum'
Leopard's Bane	Doronicum
Leverwood	Ostrya virginiana
Licorice	Glycyrrhiza glabra
Licorice Plant	Helichrysum petiolare
Licorice, Chinese	Glycyrrhiza uralensis
Licorice, Love Sick	Abrus precatorius
Lilac, Amur	Syringa reticulata ssp. amurensis
Lilac, California	Ceanothus
Lilac, Common	Syringa vulgaris
Lilac, Cut-Leaf	Syringa x laciniata
Lilac, Indian	Melia azedarach
Lilac, Japanese Tree	Syringa reticulata
Lilac, Korean	Syringa meyeri var. spontanea 'Palibin'
Lilac, Late	Syringa villosa
Lilac, Persian	Syringa x persica; or Melia azedarach
Lilac, Wild	Ceanothus sanguineus
Lily	Lilium
Lily of China or Sacred Lily of China	Rohdea japonica
Lily of the Field	Sternbergia lutea
Lily of the Nile	Agapanthus
Lily of the Valley	Convallaria majalis
Lily of the Valley Bush or Shrub	Pieris japonica
Lily of the Valley, False	Maianthemum bifolium
Lily of the Valley, Star Flowered	Smilacena stellata
Lily Tree	Magnolia denudata
Lily, American Bog	Crinum americanum
Lily, Arum	Zantedeschia aethiopica
Lily, Bamboo	Lilium japonicum
Lily, Bell	Lilium grayi
Lily, Blackberry	Belamcanda chinensis
Lily, Bugle	Watsonia
Lily, Calla	Zantedeschia
Lily, Candy	X Pardancanda norisii
Lily, Cape	Schizostylus coccinea
Lily, Cascade	Lilium washingtonianum
Lily, Checkered	Fritillaria meleagris
Lily, Chinese Edible	Lilium brownii
Lily, Chinese White	Lilium leucanthum
Lily, Cobra	Darlingtonia californica
Lily, Coral	Lilium pumilum
Lily, Corn	Veratrum viride
Lily, Crane	Strelitzia reginae
Lily, Cuban	Scilla peruviana
Lily, Devil	Lilium lancifolium
Lily, Easter	Lilium longiflorum
Lily, Fairy	Zephyranthes
Lily, Fawn	Erythronium
Lily, Fire	Cyrtanthus; or Xerophyllum tenax
Lily, Flax	Phormium
Lily, Fortnight	Dietes
Lily, Foxtail	Eremurus
Lily, Fragrant Water	Nymphaea odorata
Lily, Giant	Cardiocrinum giganteum
Lily, Ginger or Garland	Hedychium
Lily, Japanese Turk's Cap	Lilium hansonii
Lily, Kaffir	Schizostylus coccinea
Lily, Lemon	Hemerocallis lilioasphodelus; or Lilium parryi
Lily, Leopard	Lilium pardalinum
Lily, Madonna	Lilium candidum
Lily, Magic	Lycoris squamigera
Lily, Marble Martagon	Lilium duchartrei
Lily, Meadow or Wild Meadow	Lilium canadense
Lily, Morning Star	Lilium concolor
Lily, Natal	Moraea
Lily, One-Day	Tigridia
Lily, Orange	Lilium bulbiferum
Lily, Oregon	Lilium columbianum
Lily, Palm	Yucca gloriosa
Lily, Panther	Lilium pardalinum
Lily, Paradise	Paradisea liliastrum
Lily, Parrot	Alstroemeria psittacina
Lily, Peace	Spathiphyllum
Lily, Peruvian	Astroemeria
Lily, Pig	Zantedeschia aethiopica
Lily, Pineapple	Eucomis
Lily, Pink Fawn	Erythronium revolutum
Lily, Plantain	Hosta
Lily, Pond	Nymphaea odorata
Lily, Pygmy Water	Nymphaea tetragona
Lily, Queen	Curcuma petiolata
Lily, Rain	Zephyranthes
Lily, Red Spider	Lycoris radiata
Lily, Resurrection	Lycoris squamigera
Lily, Royal	Lilium regale
Lily, Saint Bernard's	Anthericum liliago
Lily, Saint Bruno's	Paradisea liliastrum
Lily, Showy or Showy Japanese	Lilium speciosum
Lily, Spider	Hymenocallis
Lily, Swamp	Saururus cernuus
Lily, Tiger	Lilium lancifolium
Lily, Toad-Cup	Neomarica
Lily, Torch	Kniphofia
Lily, Triplet	Triteleia laxa
Lily, Trout	Erythronium
Lily, Trumpet	Zantedeschia aethiopica; or Lilium longiflorum
Lily, Turk's Cap	Lilium martagon or superbum
Lily, Voodoo	Sauromatum venosum
Lily, Water	Nymphaea

COMMON NAME INDEX

Lily, White Water	Nymphaea odorata
Lily, Wood	Trillium
Lily, Yellow Flowered Cobra	Arisaema flavum
Lily, Yellow or Yellow-Bell	Lilium canadense
Lilyturf	Liriope
Lilyturf, Big Blue	Liriope muscari
Lilyturf, Dwarf	Ophiopogon japonicus
Lilyturf, White	Ophiopogon jaburan
Lime	Citrus latifolia or aurantiifolia
Lime Tree	Tilia
Lime, Australian Finger	Microcitrus australasica
Lime, Broad-Leafed	Tilia platyphyllos
Lime, Caucasian	Tilia x euchlora
Lime, Silver	Tilia tomentosa
Limequat	x Citrofortunella floridana
Linden	Tilia
Linden, Crimean	Tilia x euchlora
Linden, Large Leaf	Tilia platyphyllos
Linden, Little Leaf	Tilia cordata
Linden, Silver	Tilia tomentosa
Ling	Calluna
Lingberry	Vaccinium vitis-idaea var. minus
Lingen or Lingenberry	Vaccinium vitis-idaea var. minus
Lingonberry	Vaccinium vitis-idaea var. minus
Lion's Ear	Leonotis leonurus
Lion's-Heart	Physostegia
Lippia, Mexican	Lippia dulcis
Litchee	Litchi chinensis
Little Bluestem	Schizachyrium scoparium
Live-and-Die	Mimosa pudica
Live-Forever	Sedum telephium; or Sempervivum
Liver Berry	Streptopus amplexifolius
Liverleaf	Hepatica
Lizard Plant	Tetrastigma voinieranum
Lizard's Tail	Saururus
Lobelia, Blue or Great	Lobelia siphilitica
Lobelia, Scarlet	Lobelia cardinalis
Loblolly Bay	Gordonia lasianthus
Locust, Black	Robinia pseudoacacia
Locust, Bristly	Robinia hispida
Locust, Desert	Robinia neomexicana
Locust, Honey	Gleditsia triacanthos
Locust, Idaho	Robinia x ambigua 'Idahoensis'
Locust, Moss	Robinia hispida
Locust, Yellow	Robinia pseudoacacia
London Plane Tree	Platanus x hispanica
London Pride	Saxifraga umbrosa or x urbium
Loosestrife	Lysimachia; or Lythrum
Loosestrife, Gooseneck	Lysimachia clethroides
Loquat	Eriobotrya japonica
Loquat, Bronze	Eriobotrya deflexa
Lord's Candlestick	Yucca gloriosa
Lotus or Water Lotus	Nelumbo
Lotus, American	Nelumbo lutea
Lotus, Sacred or East Indian	Nelumbo nucifera
Lotus, Single Foot	Paris polyphylla
Lousewort	Pedicularis
Lovage	Levisticum officinale
Lovage, Black	Smyrnium olusatrum
Love-in-a-Mist	Nigella damascena; or Passiflora foetida
Love-In-Idleness	Viola tricolor
Love-Restorer	Sedum anacampseros
Lucky Bean Tree	Erythrina caffra
Lulo	Solanum quitoense
Lungwort	Pulmonaria; or Mertensia
Lungwort, Blue	Pulmonaria officinalis
Lupine, Andean	Lupinus mutabilis
Lupine, Beach	Lupinus littoralis
Lupine, Broad Leaf	Lupinus latifolius
Lupine, Carolina	Thermopsis villosa
Lupine, False	Thermopsis
Lupine, Silky	Lupinus sericeus
Lupine, Silvery	Lupinus argenteus
Lupine, Tree	Lupinus arboreus
Lupine, Western	Lupinus polyphyllus
Lychee	Litchi chinensis
Lyonia, Shiny	Lyonia lucida
Ma-Huang	Ephedra minima
Maca	Lepidium meyerii
Mace, Sweet	Tagetes lucida
Madake	Phyllostachys bambusoides
Madder	Rubia tinctorum
Madder, Indian	Rubia cordifolia
Madrone, Pacific	Arbutus menziesii
Magnolia Vine	Schisandra
Magnolia, Anhwei	Magnolia cylindrica
Magnolia, Anise	Magnolia salicifolia
Magnolia, Chinese	Magnolia x souleangeana
Magnolia, Cucumber	Magnolia acuminata
Magnolia, Great-Leaved	Magnolia macrophylla
Magnolia, Large-Flowered	Magnolia grandiflora
Magnolia, Oyama	Magnolia sieboldii
Magnolia, Saucer	Magnolia x soulangeana
Magnolia, Southern	Magnolia grandiflora
Magnolia, Star	Magnolia stellata
Magnolia, Sweet Bay	Magnolia virginiana
Magnolia, Umbrella	Magnolia tripetala
Magnolia, Willow-Leaf	Magnolia salicifolia
Magnolia, Yulan	Magnolia denudata
Mahogany, Florida	Persea borbonia
Mahonia, Holly	Mahonia aquifolium
Mahonia, Leatherleaf	Mahonia japonica Bealei Group
Maidenhair Tree	Ginkgo biloba
Maidenhair Vine	Muehlenbeckia complexa
Maidenhair, Golden	Polypodium vulgare
Male Berry	Lyonia ligustrina
Mallee, see also Eucalyptus or Gum	
Mallee, Black	Eucalyptus stellulata
Mallee, Round-leafed	Eucalyptus orbifolia
Mallow	Malva
Mallow, Checker	Sidalcea
Mallow, Confederate Rose	Hibiscus mutabilis
Mallow, Country	Sida cordifolia
Mallow, Globe or False	Sphaeralcea
Mallow, High or Tall	Malva sylvestris
Mallow, Musk	Malva moschata
Mallow, New Mexico Prairie	Sidalcea neomexicana
Mallow, Poppy	Callirhoe involucrata
Mallow, Prairie	Sidalcea
Mallow, Rose, Common or Marsh	Hibiscus moscheutos
Mallow, Sleepy	Malvaviscus
Mallow, Streambank Globe	Sphaeralcea rivularis
Mallow, Tree	Lavatera arborea
Mallow, Virginia	Sida hermaphrodita
Mallow, Wax	Malvaviscus arboreus
Mallow, White Prairie	Sidalcea candida
Mallow, Yugoslavian	Kitaibela vitifolia
Maltese Cross	Lychnis chalcedonica
Mandrake	Mandragora officinarum
Mandrake, Wild	Podophyllum peltatum
Mangosteen	Garcinia
Manio	Podocarpus salignus
Manuka	Leptospermum scoparium
Many-Fingers	Sedum pachyphyllum
Manzanita	Arctostaphylos
Manzanita, Hairy	Arctostaphylos columbiana
Maple, Amur	Acer tataricum ssp. ginnala
Maple, Bigleaf	Acer macrophyllum
Maple, Bigtooth	Acer grandidentatum
Maple, China	Acer grosseri
Maple, Colliseum	Acer cappodocicum
Maple, Douglas	Acer glabrum ssp. douglasii
Maple, Everred	Acer palmatum var. dissectum 'Dissectum Nigrum'
Maple, Fernleaf	Acer japonicum 'Aconitifolium'
Maple, Flowering	Abutilon
Maple, Full Moon	Acer shirasawanum
Maple, Green Laceleaf	Acer palmatum var. dissectum Viride Group
Maple, Hedge	Acer campestre
Maple, Japanese	Acer palmatum
Maple, Japanese Coral Bark	Acer palmatum 'Sango-kaku'
Maple, Korean	Acer tschonoskii ssp. koreanum
Maple, Laceleaf	Acer palmatum var. dissectum
Maple, Lion's Mane	Acer palmatum 'Shishigashira'
Maple, Manchurian Striped	Acer tegmentosum
Maple, Moosewood	Acer pensylvanicum
Maple, Norway	Acer platanoides
Maple, Painted	Acer mono
Maple, Paperbark	Acer griseum
Maple, Rocky Mountain	Acer glabrum
Maple, Rocky Mountain Sugar	Acer grandidentatum
Maple, Rough Bark	Acer palmatum 'Arakawa'
Maple, Scarlet or Red	Acer rubrum
Maple, Shantung	Acer truncatum
Maple, Silver	Acer saccharinum
Maple, Snakebark	Acer davidii

COMMON NAME INDEX

COMMON NAME INDEX

COMMON NAME INDEX

Oak, Spanish Red	Quercus falcata
Oak, Stone Mountain	Quercus georgiana
Oak, Swamp Chestnut	Quercus michauxii or prinus
Oak, Swamp Post	Quercus lyrata
Oak, Swamp Red	Quercus falcata
Oak, Swamp White	Quercus bicolor
Oak, Tanbark	Lithocarpus densiflorus
Oak, Truffle	Quercus robur
Oak, Turkey	Quercus cerris
Oak, Ubame	Quercus phillyreoides
Oak, Valley	Quercus lobata
Oak, Vallonea	Quercus ithaburensis
Oak, Water	Quercus nigra
Oak, Western	Quercus garryana
Oak, White	Quercus alba
Oak, White Leaf	Quercus hypoleucoides
Oak, Willow	Quercus phellos
Oak, Yellow Chestnut or Yellow	Quercus muehlenbergii
Oak, Yellow-Barked	Quercus velutina
Obedience	Physostegia virginiana
Obedient Plant	Physostegia
Oca	Oxalis tuberosa
Oceanspray	Holodiscus discolor
Oconee Bells	Shortia galacifolia
Ocotillo	Fouguieria splendens
October Daphne or October Plant	Sedum sieboldii
Oil of Ben Tree	Moringa oleifera
Okra, Wild	Viola palmata
Old Man	Artemisia absinthium
Old Man and Woman	Sempervivum tectorum
Old Man's Whiskers	Geum triflorum
Old Woman	Artemisia stelleriana
Oleander	Nerium oleander
Olive, Autumn	Elaeagnus umbellata
Olive, California	Umbellularia californica
Olive, Common or Edible	Olea europaea
Olive, Desert	Forestiera pubescens
Olive, Fragrant	Osmanthus fragrans
Olive, Holly	Osmanthus heterophyllus
Olive, Russian	Elaeagnus angustifolia
Olive, Sweet	Osmanthus fragrans
Olive, Tea	Osmanthus fragrans
Olive, Wild	Nyssa aquatica
Onion of Naples	Allium neapolitanum
Onion, Edible	Allium cepa
Onion, Egyptian Walking	Allium cepa Proliferum Group
Onion, Nodding	Allium cernuum
Onion, Siberian	Allium senescens ssp. montanum var. glaucum
Onion, Taper Tip	Allium acuminatum
Ookow	Dichelostemma congestum
Orange	Citrus sinensis
Orange, Bitter	Poncirus trifoliata
Orange, Hardy	Poncirus trifoliata
Orange, Mandarin	Citrus reticulata Mandarin Group
Orange, Osage	Maclura pomifera
Orange, Satsuma	Citrus unshiu
Orange, Trifoliate	Poncirus trifoliata
Orange, Wild	Prunus caroliniana
Orchid, Early Purple	Orchis mascula
Orchid, French Terrestrial	Orchis mascula
Orchid, Ground	Bletilla striata
Orchid, Lady's Slipper	Cypripedium calceolus; or Paphiopedilum
Orchid, Marsh	Dactylorhiza
Orchid, Moth	Phalaenopsis
Orchid, New Zealand Carpet	Pratia angulata
Orchid, Stream	Epipactus gigantea
Orchid, Terrestrial	Dactylorhiza
Orchid, White Fringed	Habenaria blephariglottis
Orchid, Woody	Magnolia liliiflora
Oregano	Origanum vulgare
Oregano, Cuban	Plectranthus amboinicus
Oregano, Golden	Origanum vulgare 'Aureum'
Oregano, Greek	Origanum vulgare ssp. hirtum
Oregano, Italian	Origanum x majoricum
Oregano, Mexican	Lippia graveolens
Oregano, Showy	Origanum laevigatum
Oregon Foetid Adder's-Tongue	Scoliopus hallii
Oregon Grape	Mahonia aquifolium
Oregon Grape, Cascade or Low	Mahonia nervosa
Oregon Grape, Creeping	Mahonia repens
Oregon Myrtle	Umbellularia californica
Oregon Sunshine	Eriophyllum lanatum
Oregon Tea	Ceanothus sanguineus
Organy	Origanum vulgare
Orpine	Sedum telephium
Orris Root	Iris florentina alba
Osier	Salix
Osier, Purple	Salix purpurea
Osoberry	Oemleria cerasiformis
Oswego Tea	Monarda didyma
Ougon-Kou Chiku or Robert Ougon-Chiku	Phyllostachys bambusoides 'Allgold'
Owl-Eyes	Huernia zebrina
Oxalis, Blue	Parochetus communis
Oxeye	Heliopsis
Oxlip	Primula elatior
Pachysandra, Alleghany	Pachysandra procumbens
Pachysandra, Japanese	Pachysandra terminalis
Pacific Dewberry	Rubus ursinus
Pagoda Tree, Japanese	Sophora japonica
Paint-Leaf	Euphorbia heterophylla
Painted Daisy	Tanacetum coccineum
Pale Trumpet	Sarracenia alata
Palm, Blue Fan	Brahea armata
Palm, Bottle	Nolina recurvata
Palm, Canary Island Date	Phoenix canariensis
Palm, Chinese Fan or Chinese Fountain	Livistona chinensis
Palm, Chusan	Trachycarpus
Palm, Cotton	Washingtonia filifera
Palm, Date	Phoenix
Palm, Desert Fan	Washingtonia filifera
Palm, Fan	Trachycarpus
Palm, Hemp	Trachycarpus fortunei
Palm, Honey	Jubaea
Palm, Jelly	Butia capitata
Palm, Kentia	Howea forsteriana
Palm, Mexican Blue	Brahea armata
Palm, Mexican Fan	Washingtonia robusta
Palm, Miniature Date	Phoenix roebelenii
Palm, Needle	Yucca filamentosa; or Rhapidophyllum hystrix
Palm, Northern or California Washingtonia	Washingtonia filifera
Palm, Orchid	Curculigo orchidoides
Palm, Parlor	Chamaedorea elegans
Palm, Petticoat	Washingtonia filifera
Palm, Pindo	Butia capitata
Palm, Porcupine	Rhapidophyllum hystrix
Palm, Pygmy Date	Phoenix roebelenii
Palm, Sentry	Howea forsteriana
Palm, Southern or Mexican Washingtonia	Washingtonia robusta
Palm, Syrup	Jubaea
Palm, Thread	Washingtonia robusta
Palm, Umbrella	Cyperus involucratus
Palm, Washington	Washingtonia
Palm, Windmill or Chinese Windmill	Trachycarpus fortunei
Palma Amarga or Palma de Vaca	Sabal mauritiiformis
Palma Christi	Ricinus communis
Palmella	Yucca elata
Palmetto	Sabal
Palmetto, Blue	Sabal palmetto; or Rhapidophyllum hystrix
Palmetto, Bush or Scrub	Sabal minor; or Serenoa repens
Palmetto, Cabbage	Sabal palmetto
Palmetto, Common	Sabal palmetto
Palmetto, Dwarf	Sabal minor
Palmetto, Saw	Serenoa repens
Palo Colorado	Luma apiculata
Pan	Piper betle
Panda Plant	Kalanchoe tomentosa
Pansy	Viola x wittrockiana
Pansy, Bedding	Viola cornuta
Pansy, Field	Viola tricolor
Pansy, Miniature	Viola tricolor
Pansy, Mountain	Viola lutea
Pansy, Wild or European Wild	Viola tricolor
Papaya	Carica papaya
Papaya, Mountain	Carica pubescens or x heilbornii
Papeda	Citrus hystrix or ichangensis
Paperbark	Melaleuca
Paperbark, Broad-Leaved	Melaleuca leucadendra
Paperbark, Swamp	Melaleuca ericifolia
Papoose Root	Caulophyllum thalictroides
Papyrus	Cyperus papyrus
Paraguay Tea	Ilex paraguariensis
Parrot Feather	Myriophyllum aquaticum
Parrot's Beak	Lotus berthelotii
Parsley, Black	Melanoselinum decipiens

COMMON NAME INDEX

COMMON NAME INDEX

COMMON NAME INDEX

Poplar, Gray	Populus x canescens
Poplar, Japanese	Populus maximowiczii
Poplar, Lombardy or Italian	Populus nigra 'Italica'
Poplar, Pickart's	Populus x canescens 'Macrophylla'
Poplar, Pyramidal	Populus nigra 'Italica'
Poplar, Silver-leafed	Populus alba
Poplar, Tulip	Liriodendron tulipifera
Poplar, Western Balsam	Populus trichocarpa
Poplar, White	Populus alba
Poplar, Willow-Leaved	Populus angustifolia
Poplar, Yellow	Liriodendron tulipifera
Poppy	Papaver
Poppy, Arctic	Papaver nudicaule
Poppy, Asiatic	Meconopsis
Poppy, Blue	Meconopsis betonicifolia
Poppy, California Tree	Romneya coulteri
Poppy, Celandine	Stylophorum diphyllum
Poppy, Corn or Field	Papaver rhoeas
Poppy, Flanders	Papaver rhoeas
Poppy, Himalayan Blue	Meconopsis
Poppy, Iceland	Papaver nudicaule
Poppy, Matilija	Romneya
Poppy, Opium	Papaver somniferum
Poppy, Plume	Macleaya
Poppy, Satin	Meconopsis napaulensis
Poppy, Snow	Eomecon chionantha
Poppy, Water	Hydrocleys nymphoides
Poppy, Welsh	Meconopsis cambrica
Poppy, Wood	Stylophorum diphyllum
Poppy, Yellow Chinese	Meconopsis integrifolia
Porcelain Berry or Vine	Ampelopsis brevipedunculata
Pork and Beans	Sedum x rubrotinctum
Potato	Solanum tuberosum
Potato Beans	Apios americana
Potato Bush, Blue	Solanum rantonnetii
Potato Vine	Solanum jasminoides
Potato, Air	Dioscorea bulbifera
Potato, Duck	Sagittaria latifolia
Potato, Swamp	Sagittaria
Potato, Swan	Sagittaria sagittifolia
Potato, Sweet	Ipomoea batatas
Potato, Wild	Chlorogalum pomeridianum
Potentilla, Shrubby	Potentilla fruticosa
Prairie Smoke	Geum triflorum
Prairie Tea	Potentilla rupestris
Prayer Plant	Maranta leuconeura
Prickly Pear	Opuntia vulgaris
Pride-of-China	Melia azederach
Pride-of-India	Melia azedarach; or Koelreuteria paniculata
Primrose	Primula vulgaris
Primrose Creeper	Ludwigia arcuata
Primrose, Baby	Primula forbesii or malacoides
Primrose, Candelabra	Primula japonica
Primrose, Cape	Streptocarpus
Primrose, Desert Evening	Oenothera caespitosa
Primrose, Drumstick	Primula denticulata
Primrose, English	Primula vulgaris
Primrose, Evening	Oenothera biennis
Primrose, Fairy	Primula malacoides
Primrose, Large Flowered Evening	Oenothera glaziouana
Primrose, Mealy Candelabra	Primula pulverulenta
Primrose, Mexican Evening	Oenothera speciosa 'Rosea'
Primrose, Texas	Calylophus drummondii
Primrose, White Evening	Oenothera speciosa
Prince's Plume	Stanleya
Princess Flower	Tibouchina urvilleana
Princess Tree	Paulownia tomentosa
Princess Vine	Cissus sicyoides
Privet, California	Ligustrum ovalifolium
Privet, Chinese	Ligustrum lucidum
Privet, Common	Ligustrum vulgare
Privet, Mock	Phillyrea
Privet, Wax Leaf	Ligustrum japonicum or lucidum
Proboscis Flower	Proboscidea
Protea, King or Giant	Protea cynaroides
Prune	Prunus domestica
Psyllium, Spanish	Plantago psyllium
Puccoon	Lithospermum
Purple Top	Verbena bonariensis
Purslane	Portulaca oleracea
Purslane, Winter	Claytonia perfoliata
Pusley	Portulaca oleracea
Pussy Paws	Spraguea umbellata
Pussy Toes	Antennaria
Pyrethrum	Tanacetum cinerariifolium or coccineum
Pyrethrum, Dalmatia	Tanacetum cinerariifolium
Queen Anne's Lace	Daucus carota
Queen of the Meadow	Filipendula ulmaria
Queen of the Prairie	Filipendula rubra
Queen's Bird of Paradise	Strelitzia reginae
Queen's Cup	Clintonia uniflora
Quercitron	Quercus velutina
Quickbeam	Sorbus aucuparia
Quicksilver Weed	Thalictrum dioicum
Quillay	Quillaja saponaria
Quince, Edible	Cydonia oblonga
Quince, Flowering	Chaenomeles
Quinine, Wild	Parthenium integrifolium
Quiverleaf	Populus tremuloides
Rabbit's Foot or Rabbit's Tracks	Maranta leuconeura var. kerchoveana
Rabbitbrush	Chrysothamnus
Raccoon Berry	Podophyllum peltatum
Radiator Plant	Peperomia
Ragged Robin	Lychnis flos-cuculi
Ragwort, Golden	Packera aurea
Rainhat-Trumpet	Sarracenia minor
Raisin Tree	Hovenia dulcis
Raisin, Wild	Viburnum cassinoides or lentago
Ramie	Boehmeria nivea
Rampion, German	Oenothera biennis
Rampion, Horned	Phyteuma
Raspberry, Black	Rubus occidentalis
Raspberry, Crinkle-leaf Creeping	Rubus pentalobus
Raspberry, European	Rubus idaeus
Raspberry, Mauritius	Rubus rosifolius
Raspberry, Purple Flowering	Rubus odoratus
Raspberry, Red or Wild	Rubus idaeus
Raspberry, Rocky Mountain	Rubus deliciosus
Raspberry, Trailing	Rubus pedatus
Raspberry, Whitebark	Rubus leucodermis
Rattlesnake Master	Eryngium yuccifolium
Rauram	Persicaria odorata
Razzlequat	Eremocitrus x Citrus
Red Bells	Fritillaria recurva
Red-Hot Poker	Kniphofia
Red Puccoon	Sanguinaria canadensis
Red Star	Rhodohypoxis
Red-Stemmed Wattle	Acacia rubida
Red-Veined Pie Plant	Rheum australe
Redbud, Eastern	Cercis canadensis
Redbud, Western	Cercis occidentalis
Redwood, Coast or California	Sequoia sempervirens
Redwood, Dawn	Metasequoia glyptostroboides
Redwood, Sierra or Giant	Sequoiadendron giganteum
Reed, Bur	Sparganium erectum
Reed, Common	Phragmites australis
Reed, Egyptian Paper	Cyperus papyrus
Reed, Giant	Arundo donax
Reed, Japanese Wind	Juncus filiformis
Reedmace	Typha
Reina-de-la-Noche	Peniocereus greggii
Restharrow	Ononis
Rhododendron, Bluet	Rhododendron intricatum
Rhododendron, Fujiyama	Rhododendron brachycarpum
Rhododendron, Honeybell	Rhododendron campylocarpum
Rhododendron, Leatherleaf	Rhododendron degronianum ssp. heptamerum
Rhododendron, Pacific	Rhododendron macrophyllum
Rhododendron, Piedmont	Rhododendron minus
Rhododendron, Silvery	Rhododendron grande
Rhododendron, Tree	Rhododendron arboreum
Rhododendron, West Coast	Rhododendron macrophyllum
Rhododendron, Willow-Leaved	Rhododendron lepidotum
Rhodora	Rhododendron canadense
Rhubarb, Edible or Garden	Rheum x hybridum
Rhubarb, Himalayan	Rheum australe
Rhubarb, Ornamental	Rheum palmatum
Rhubard, Sikkim	Rheum acuminatum
Rhus, Willow	Rhus lancea
Ribbonwood	Plagianthus regius
Ribwort	Plantago
Rice Flower	Pimelea
Rice-Paper Plant or Chinese Rice-Paper Plant	Tetrapanax papyrifer
Rice, Water or Manchurian Wild	Zizania latfolia
Roanoke Bells	Mertensia pulmonarioides
Roast Beef Plant	Iris foetidissima

COMMON NAME INDEX

Robin-Redbreast Bush	Melaleuca lateritia
Rock Brake, Purple	Pellaea atropurpurea
Rock Breaker	Saxifraga
Rock Cress	Arabis
Rocket	Eruca vesicaria ssp. sativa
Rocket, Dame's or Sweet	Hesperis matronalis
Rockfoil	Saxifraga
Rockfoil, Great Alpine	Saxifraga cotyledon
Rockrose	Cistus
Rockrose, Crimson Spot	Cistus ladanifer
Rockrose, Orchid	Cistus x purpureus
Rockrose, Sageleaf	Cistus salviifolius
Roman Candle	Yucca gloriosa
Roman Chamomile	Chamaemelum nobile 'Flore Pleno'
Ropebark	Dirca palustris
Roquette	Arugula
Rose Campion	Lychnis coronaria
Rose-of-China	Hibiscus rosa-sinensis
Rose Of Sharon	Hibiscus syriacus
Rose-Root	Rhodiola rosea
Rose, Alpine	Rosa pendulina; or Rhododendron ferrugineum
Rose, Apothecaries'	Rosa gallica var. officinalis
Rose, Austrian Briar or Austrian Yellow	Rosa foetida
Rose, Austrian Copper	Rosa foetida 'Bicolor'
Rose, Autumn Damask	Rosa x damascena var. semperflorens
Rose, Baldhip	Rosa gymnocarpa
Rose, Burgundian	Rosa 'Burgundiaca'
Rose, Burnet	Rosa pimpinellifolia
Rose, Cabbage	Rosa x centifolia
Rose, Chapeau de Napoleon (Napoleon's Hat)	Rosa x centifolia 'Cristata'
Rose, Cherokee	Rosa laevigata
Rose, Chestnut	Rosa roxburghii
Rose, Chinquapin	Rosa roxburghii
Rose, Christmas	Helleborus niger
Rose, Cliff	Cowania stanburyana
Rose, Climbing	Rosa setigera
Rose, Cluster	Rosa pisocarpa
Rose, Coffee	Tabernaemontana divaricata
Rose, Common Moss	Rosa x centifolia 'Muscosa'
Rose, Cotton	Hibiscus mutabilis
Rose, Crested Moss	Rosa x centifolia 'Cristata'
Rose, Cuisse de Nymphe	Rosa 'Great Maiden's Blush'
Rose, Desert	Adenium obesum; or Rosa stellata
Rose, Dunwich	Rosa pimpinellifolia 'Dunwich Rose'
Rose, Eglantine	Rosa rubiginosa
Rose, Egyptian	Scabiosa atropurpurea
Rose, Evergreen	Rosa sempervirens
Rose, Fairy	Rosa 'The Fairy'
Rose, Father Hugo's	Rosa xanthina f. hugonis
Rose, Five Colored	Rosa 'Fortune's Five Colored'
Rose, Fortune's Double Yellow	Rosa x odorata 'Pseudindica'
Rose, Four Seasons	Rosa x damascena var. semperflorens
Rose, French	Rosa gallica
Rose, Golden Rose of China	Rosa xanthina f. hugonis
Rose, Green	Rosa x odorata 'Viridiflora'
Rose, Guelder	Viburnum opulus
Rose, Hairy Alpine	Rhododendron hirsutum
Rose, Himalayan Musk	Rosa brunonii
Rose, Holland	Rosa x centifolia
Rose, Holy	Rosa x richardii
Rose, Hume's Blush	Rosa x odorata 'Odorata'
Rose, Incense	Rosa primula
Rose, Irish	Rosa x hibernica
Rose, Japanese	Kerria japonica; or Rosa rugosa
Rose, Kazanlik	Rosa 'Professeur Emile Perrot
Rose, Lady Banks' White	Rosa banksiae var. banksiae
Rose, Lady Banks' Yellow	Rosa banksiae 'Lutea'
Rose, Leafy	Rosa foliolosa
Rose, Lenten	Helleborus orientalis
Rose, Manchu	Rosa xanthina
Rose, Memorial	Rosa wichurana
Rose, Monthly	Rosa x damascena var. semperflorens
Rose, Moss	Portulaca
Rose, Musk	Rosa moschata
Rose, Napoleon's Hat	Rosa x centifolia 'Cristata'
Rose, Nootka	Rosa nutkana
Rose, Old Black	Rosa 'Nuits de Young'
Rose, Old Blush	Rosa x odorata 'Pallida'
Rose, Old Lilac	Rosa 'Lilac Rose'
Rose, Old Pink Moss	Rosa x centifolia 'Muscosa'
Rose, Old Spanish	Rosa 'Russelliana'
Rose, Old Velvet	Rosa 'Tuscany'
Rose, Old Velvet Moss	Rosa 'William Lobb'
Rose, Parson's Pink China	Rosa x odorata 'Pallida'
Rose, Pasture	Rosa carolina
Rose, Pea Hip or Clustered	Rosa pisocarpa
Rose, Perpetual White Moss	Rosa 'Quatre Saisons Blanche Mousseuse'
Rose, Persian Yellow	Rosa foetida 'Persiana'
Rose, Pompon	Rosa x centifolia var. pomponia or 'De Meaux'
Rose, Portland	Rosa 'Portlandica'
Rose, Prairie	Rosa arkansana or setigera
Rose, Prickly	Rosa acicularis
Rose, Provence	Rosa x centifolia
Rose, Quatre Saisons (Four Seasons)	Rosa x damascena var. semperflorens
Rose, Red Moss	Rosa 'Henri Martin'
Rose, Red Rose of Lancaster	Rosa gallica var. officinalis
Rose, Russell's Cottage or Russell's Old Cottage	Rosa 'Russelliana'
Rose, Saint John's	Rosa x richardii
Rose, Scotch	Rosa pimpinellifolia
Rose, Seven Sisters	Rosa multiflora 'Grevillei'
Rose, Snow	Serissa japonica
Rose, Summer Damask	Rosa x damascena
Rose, Sunshine	Rosa setigera
Rose, Swamp	Rosa palustris
Rose, Sweet Briar	Rosa rubiginosa
Rose, Sweetheart	Rosa 'Cecile Brunner'
Rose, Tomato	Rosa rugosa
Rose, Turkestan	Rosa rugosa
Rose, Western Wild	Rosa woodsii
Rose, White Prairie	Rosa foliolosa
Rose, White Rose of York	Roas x alba
Rose, Wood	Rosa gymnocarpa
Rose, York and Lancaster	Rosa x damascena var. versicolor
Rosebay	Rhododendron maximum; or Nerium oleander
Rosebay Willowherb	Epilobium angustifolium
Rosebay, California	Rhododendron macrophyllum
Rosebay, East Indian or Broad-Leaved	Tabernaemontana divaricata
Rosebay, Mountain	Rhododendron catawbiense
Rosemary	Rosmarinus officinalis
Rosemary, Australian or Victorian	Westringia fruticosa
Rosemary, Bog	Andromeda
Rosemary, Creeping	Rosmarinus officinalis Prostratus Group
Rosemary, Wild	Ledum palustre
Rosemyrtle, Downy	Rhodomyrtus tomentosa
Rosinweed	Silphium
Roundwood	Sorbus americana
Rowan	Sorbus aucuparia
Rubber Tree	Schefflera
Rubber Vine	Cryptostegia grandiflora
Rue	Ruta
Rue Anemone	Anemonella thalictroides
Rue Anemone, False	Isopyrum
Rue, Common	Ruta graveolens
Rue, Meadow	Thalictrum
Rue, Mid-Eastern	Ruta chalepensis
Rue, Syrian	Peganum harmala
Running Pop	Passiflora foetida
Rupturewort	Herniaria glabra
Rush	Juncus
Rush, Bull	Scirpus
Rush, California Gray	Juncus patens
Rush, Corkscrew	Juncus effusus 'Spiralis'
Rush, Creeping Spike	Eleocharis palustris
Rush, Dagger-leaf	Juncus ensifolius
Rush, Greater Wood	Luzula sylvatica
Rush, Hairy Wood	Luzula pilosa
Rush, Japanese	Acorus gramineus
Rush, Pink Flowering	Butomus umbellatus
Rush, Scouring	Equisetum
Rush, Slender	Juncus tenuis
Rush, Snowy Wood	Luzula nivea
Rush, Soft	Juncus effusus
Rush, Spreading	Juncus patens
Rush, Striped	Baumea rubiginosa
Rush, Sweet	Acorus calamus
Rush, White Bull	Schoenoplectus lacustris ssp. tabernaemontana 'Albescens'
Rush, Wood	Luzula
Rush, Wood Club	Scirpus sylvaticus
Rush, Zebra	Schoenoplectus lacustris ssp. tabernaemontani 'Zebrinus'
Rustyleaf	Menziesia ferruginea
Rye, Giant Wild	Leymus condensatus
Sacahuista	Nolina microcarpa

COMMON NAME INDEX

Safflower	Carthamus tinctorius
Sage Tree	Vitex agnus-castus
Sage, African	Salvia aethiopis
Sage, Autumn	Salvia gregii
Sage, Baby	Salvia microphylla
Sage, Bee	Salvia apiana
Sage, Bethelhem	Pulmonaria saccharata
Sage, Big	Seriphidium tridentatum
Sage, Black	Salvia discolor
Sage, Blue	Salvia clevelandii
Sage, California White	Salvia apiana
Sage, Chaparral	Salvia leucophylla
Sage, Clary	Salvia sclarea
Sage, Creeping	Salvia repens or sonomensis
Sage, Crimson	Salvia miltiorhiza
Sage, Culinary	Salvia officinalis
Sage, Divine	Salvia divinorum
Sage, Fringed	Artemisia frigida
Sage, Fruit	Salvia dorisiana
Sage, Garden or Common	Salvia officinalis
Sage, Gentian	Salvia patens
Sage, Golden	Salvia officinalis 'Icterina' or 'Aurea'
Sage, Gray	Salvia leucophylla
Sage, Greek	Salvia fruticosa
Sage, Jerusalem	Phlomis fruticosa; or Pulmonaria officinalis
Sage, Jim	Salvia clevelandii
Sage, Lavender Scented	Salvia lavandulifolia
Sage, Lilac	Salvia verticillata
Sage, Meadow	Salvia pratensis
Sage, Mealy or Mealy Cup	Salvia farinacea
Sage, Mexican Bush	Salvia leucantha
Sage, Peruvian	Salvia discolor
Sage, Pineapple or Pineapple-Scented	Salvia elegans
Sage, Prairie	Artemisia ludoviciana
Sage, Purple	Salvia officinalis Purpurascens Group
Sage, Red	Salvia miltiorhiza
Sage, Rosy-Leaf	Salvia involucrata
Sage, Russian	Perovskia atriplicifolia
Sage, San Luis Purple	Salvia leucophylla
Sage, Scarlet	Salvia coccinea or splendens
Sage, Silver	Salvia argentea
Sage, Spanish	Salvia lavandulifolia
Sage, Texas	Salvia coccinea
Sage, White	Salvia apiana
Sage, Wood	Teucrium scorodonia or canadense
Sagebrush	Seriphidium tridentatum
Sagebrush, Dwarf	Seriphidium novum
Sagebrush, Silver	Seriphidium canum
Saint John's Wort	Hypericum
Salal	Gaultheria shallon
Sallow	Salix caprea
Salmonberry	Rubus spectabilis or parviflorus
Saltbush, Four-Winged	Atriplex canescens
Samphire	Crithmum maritimum
Sandcherry, Purple Leaf	Prunus x cistena
Sandwort	Minuartia
Sandwort, Corsican	Arenaria balearica
Sandwort, Mountain	Arenaria montana
Sang	Panax quinquefolius
Sapote, White	Casimiroa edulis
Sarcococca, Fragrant	Sarcococca ruscifolia
Saskatoon	Amelanchier alnifolia
Sassafras	Sassafras albidum
Satin Flower	Clarkia amoena
Satin Pothos	Scindapsus pictus 'Argyraeus'
Sauer Klee	Oxalis adenophylla
Sausage Crassula	Senecio articulatus
Sausage Tree	Kigelia africana
Sausage Vine	Holboellia coriacea
Savory, Creeping	Satureja spicigera
Savory, Lemon	Satureja biflora
Savory, Summer	Satureja hortensis
Savory, Winter	Satureja montana
Saw Palmetto	Serenoa repans
Sawwort	Serratula tinctoria
Saxifrage, Dodder-Like	Saxifraga cuscutiformis
Saxifrage, Greater Burnet	Pimpinellla major
Saxifrage, Greater Evergreen	Saxifraga cotyledon
Saxifrage, Lifelong	Saxifraga paniculata
Saxifrage, Limestone	Saxifraga callosa
Saxifrage, Mossy	Saxifraga hypnoides
Saxifrage, Pyrenean	Saxifraga longifolia
Saxifrage, Shield-Leafed	Saxifraga cuneifolia
Scabious	Scabiosa; or Cephalaria
Scabious, Devil's Bit	Succisa pratensis
Scabious, Giant	Cephalaria gigantea
Scabious, Shepherd's	Jasione laevis
Scabious, Sweet	Scabiosa atropurpurea
Scarlet-Lightening	Lychnis chalcedonica
Scoke	Phytolacca americana
Scoot Berry	Streptopus amplexifolius
Scorpion Weed	Phacelia
Scottish Flame Flower	Tropaeolum speciosum
Sea Buckthorn	Hippophae rhamnoides
Sea Holly	Eryngium
Sea Holly, Moroccan	Eryngium variifolium
Sea Kale	Crambe maritima
Sea Lavender	Limonium platyphyllum
Sea Oats	Chasmanthium latifolium
Sea Oats, North American	Uniola paniculata
Seaberry	Hippophae rhamnoides
Seapink	Armeria
Seaside Oats	Uniola paniculata
Sedge, Black Flowering	Carex nigra
Sedge, Blue	Carex glauca
Sedge, Bowle's Golden	Carex elata 'Aurea'
Sedge, Bronze	Carex flagellifera
Sedge, Brown	Carex petriei
Sedge, Drooping	Carex pendula
Sedge, Giant Orange	Carex secta
Sedge, Japanese	Carex morrowiii
Sedge, Leather Leaf	Carex buchananii
Sedge, Orange	Carex testacea
Sedge, Palm	Carex muskingumensis
Sedge, Slender-Tufted	Carex acuta
Sedge, Slough	Carex obnupta
Sedge, Striped Umbrella	Carex phyllocephala 'Sparkler'
Sedge, Tufted Bronze	Carex secta var. tenuiculmis
Sedge, Weeping	Carex pendula
Sedge, White Top	Rhynchospora colorata
Sedum, Carpet	Sedum lineare
Sedum, Golden	Sedum adolphii
Sedum, Sakhalin Island	Sedum pluricaule
Sedum, Silver	Sedum treleasii
Self-Heal	Prunella vulgaris
Seneca	Polygala
Senna, Wild	Senna hebecarpa
Sensitive Plant	Mimosa pudica
September Bells	Rothmannia globosa
Sequoia, Creeping	Sequoia sempervirens 'Adpressa'
Sequoia, Giant	Sequoiadendron giganteum
Seri	Oenanthe javanica
Service Tree	Sorbus domestica
Service Tree, Wild	Sorbus torminalis
Serviceberry or Sarviceberry	Amelanchier
Serviceberry, Apple	Amelanchier x grandiflora
Serviceberry, Western	Amelanchier alnifolia
Sesame	Sesamum indicum
Shadbush	Amelanchier
Shallot	Allium cepa Aggregatum Group
Shame Plant	Mimosa pudica
Shamrock	Oxalis acetosella; or Trifolium repens
Shan Cha	Crataegus pinnatifida
Shang-Lu	Phytolacca acinosa
Shaving Brush Tree	Pseudobombax ellipticum
Sheep's Bit	Jasione
Sheepberry	Viburnum lentago
Shell Flower	Tigridia
Shellflower	Moluccella laevis; or Pistia stratiotes
Shingle Tree	Acrocarpus fraxinifolius
Shiso	Perilla frutescens
Shiya-Zasa	Sasaella masumuneana
Shoo-Fly Plant	Nicandra physalodes
Shooting Star	Dodecatheon
Shrub Yellowroot	Xanthorhiza simplicissima
Siberian Pea Shrub	Caragana arborescens
Siberian Pink Cups	Baldellia ranunculoides f. repens
Siberian Squill	Scilla siberica
Sidalcea, Meadow	Sidalcea campestris
Side-Saddle Flower	Sarracenia purpurea
Silk Tassel	Garrya
Silk Tree	Albizia julibrissin
Silk Vine	Periploca graeca
Silver-Dollar	Lunaria annua
Silver King	Artemisia ludoviciana ssp. mexicana var. albula
Silver Lace Vine	Fallopia baldschuanica; or Polygonum

COMMON NAME INDEX

COMMON NAME INDEX

Starfish Flower	Stapelia
Starfish Flower, Shaggy	Stapelia hirsuta
Starfish Plant	Cryptanthus acaulis; or Stapelia
Starflower	Trientalis borealis; or Smilacena stellata
Statice, German	Goniolimon tataricum
Statice, Perennial	Limonium
Steeplebush	Spiraea tomentosa
Stepmother's Flower	Viola x wittrockiana
Stewartia, Japanese	Stewartia pseudocamellia
Stewartia, Mountain	Stewartia ovata
Stinking Benjamin	Trillium erectum
Stitchwort	Stellaria
Stock	Matthiola
Stone Orpine	Sedum rupestre
Stonecrop	Sedum
Stonecrop, Wormleaf	Sedum stenopetalum
Stoneroot	Collinsonia canadensis
Storax	Styrax
Storksbill	Erodium
Strawbell	Uvularia perfoliata or sessilifolia
Strawberry Bush	Euonymus americanus
Strawberry Tree	Arbutus unedo
Strawberry, Alpine or Woodland	Fragaria vesca
Strawberry, Barren	Waldsteinia fragarioides
Strawberry, Coastal	Fragaria chiloensis
Strawberry, Edible	Fragaria x ananassa
Strawberry, Wild	Fragaria virginiana
String-of-Beads	Senecio rowleyanus
Sugar Leaf	Stevia rebaudiana
Sugar Scoop	Tiarella
Sugarberry	Celtis laevigata
Sugarbush	Rhus ovata
Sulfur Flower	Eriogonum umbellatum
Sumac, Dwarf	Rhus copallina
Sumac, Elm Leaved	Rhus coriaria
Sumac, Fernleaf	Rhus typhina 'Dissecta'
Sumac, Fragrant	Rhus aromatica
Sumac, Lemon	Rhus aromatica
Sumac, Lemonade	Rhus integrifolia
Sumac, Mountain	Rhus copallina
Sumac, Scarlet	Rhus glabra
Sumac, Shining	Rhus copallina
Sumac, Sicilian	Rhus coriaria
Sumac, Smooth	Rhus glabra
Sumac, Staghorn	Rhus typhina
Sumac, Sugar	Rhus ovata
Sumac, Sweet	Rhus aromatica
Sumac, Tanner's	Rhus coriaria
Sumac, Velvet	Rhus typhina
Sumac, Virginian	Rhus typhina
Sumac, Wing Rib	Rhus copallina
Summer Snowflake	Ornithogalum umbellatum
Summerberry	Viburnum trilobum
Summersweet	Clethra alnifolia
Summersweet, Japanese	Clethra barbinervis
Suncups	Oenothera
Sundew	Drosera
Sundrops	Oenothera fruitcosa
Sundrops, Ozark	Oenothera macrocarpa
Sunflower	Helianthus
Sunflower, Swamp	Helianthus angustifolius
Sunflower, Thin Leaf	Helianthus decapetalus
Sunrose	Helianthemum
Swamp Lantern	Lysichiton americanus
Swamp Pink	Helonias bullata
Sweet Amber	Hypericum androsaemum
Sweet Annie	Artemisia annua
Sweet Balm	Melissa officinalis
Sweet Bay	Laurus nobilis
Sweet Cicely	Myrrhis odorata
Sweet Flag, Common	Acorus calamus
Sweet Flag, Japanese	Acorus gramineus
Sweet Gale	Myrica gale
Sweet-Hurts	Vaccinium angustifolium
Sweet Leaf	Stevia rebaudiana
Sweet Mace	Tagetes lucida
Sweet Pepperbush	Clethra alnifolia
Sweet Shrub	Sinocalycanthus chinensis
Sweet Trumpet Plant	Sarracenia rubra
Sweet William	Dianthus barbatus
Sweet William, Wild	Phlox maculata or divaricata
Sweet Woodruff	Galium odoratum
Sweet Woodruff, Pink	Phuopsis stylosa
Sweetbells	Leucothoe racemosa
Sweetberry	Viburnum lentago
Sweetleaf	Symplocos
Sweetpea	Lathyrus odoratus
Sweetspire	Itea virginica
Swiss-Cheese Plant	Monstera deliciosa
Sword Plant	Echinodorus
Sycamore	Platanus
Sycamore, American or Eastern	Platanus occidentalis
Sycamore, Arizona	Platanus wrightii
Synthyris, Columbia	Synthyris stellata
Synthyris, Round-Leaved	Synthyris reniformis
Syringa	Philadelphus
Tabasco	Capsicum frutescens
Tacamahac	Populus balsamifera
Tallow Tree, Chinese or Vegetable	Sapium sebiferum
Tamarack	Larix laricina
Tamarisk	Tamarix
Tan Shen	Salvia miltiorhiza
Tangerine	Citrus reticulata
Tansy	Tanacetum vulgare
Tansy, Silver	Tanacetum niveum
Tapeworm Plant	Homocladium platycladum
Tapioca Plant	Manihot
Tarajo	Ilex latifolia
Tarata	Pittosporum eugenioides
Tare	Vicia
Taro, Cranberry	Colocasia rubra
Taro, Greenleaf	Colocasia esculenta
Taro, Imperial	Colocasia esculenta 'Illustris'
Tarragon, French	Artemesia dracunculus
Tarragon, Spanish	Tagetes lucida
Tartogo	Jatropha
Tarweed	Grindelia
Tarwi	Lupinus mutabilis
Tassel-White	Itea virginica
Tawhiwhi	Pittosporum tenuifolium
Tea	Camellis sinensis
Tea Oil Tree	Melaleuca alternifolia
Tea Plant	Viburnum lentago
Tea Tree, Australian	Leptospermum laevigatum
Tea Tree, Broad-Leaved	Melaleuca leucadendra
Tea Tree, New Zealand	Leptospermum scoparium
Tea Tree, River or Weeping	Melaleuca leucadendra
Tea Tree, Wooly	Leptospermum lanigerum
Tea, Crystal	Ledum palustre
Tea, Labrador	Ledum glandulosum
Tea, Mormon	Ephedra nevadensis
Tea, Sweet	Osmanthus fragrans
Tea, Trapper's	Ledum glandulosum
Teaberry	Gaultheria procumbens or pumila; or Viburnum cassinoides
Teasel	Dipsacus
Temu	Luma apiculata
Ten Commandments	Maranta leuconeura
Texas Ebony	Pithecellobium flexicaule
Texas Mud Baby	Echinodorus cordifolius
Thimbleberry	Rubus parviflorus or occidentalis or odoratus
Thistle, Argentine	Onopordum acanthium
Thistle, Blessed or Holy	Silybum marianum
Thistle, Cotton	Onopordum acanthium
Thistle, Giant	Onopordum acanthium
Thistle, Golden	Centaurea macrocephala
Thistle, Milk or Our Lady's Milk	Silybum marianum
Thistle, Oat	Onopordum acanthium
Thistle, Saint Mary's	Silybum marianum
Thistle, Scotch	Onopordum acanthium
Thistle, Silver	Onopordum acanthium
Thistle, Star	Centaurea
Thorn	Crataegus
Thoro-Wax	Bupleurum rotundifolium
Thousand-Mothers	Tolmiea menziesii
Three-Birds-Flying	Linaria triornithophora
Thrift	Armeria
Throatwort	Trachelium
Thuja 'Berckman Dwarf'	Thuja orientalis 'Aurea Nana'
Thyme	Thymus vulgaris
Thyme, Carraway	Thymus herba-barona
Thyme, Cat	Teucrium marum
Thyme, Cicilian	Thymus cilicicus
Thyme, Common or Garden	Thymus vulgaris
Thyme, Creeping	Thymus serpyllum or polytrichus ssp. britannicus
Thyme, English	Thymus polytrichus ssp. britannicus

COMMON NAME INDEX

COMMON NAME INDEX

COMMON NAME INDEX